A Guide to Choosing a Statistical Method

CATEGORICAL RESPONSE VARIABLE (ANALYZING PROPORTIONS)

1. If there is only one categorical response variable, use
   - descriptive methods of Chapter 2 (Sections 2.1 and 2.2)
   - inferential methods of Section 7.2 (confidence interval) and Section 8.2 (significance test) for proportions

2. To compare proportions of a categorical response variable for two or more groups of a categorical explanatory variable, use
   - descriptive methods of Chapter 3 (Sections 3.1 and 3.4)
   - inferential methods of Section 9.1 for comparing proportions between two groups
   - inferential methods of Chapter 10 for comparing two or more proportions or the independence of two categorical variables

3. If working with a binary response variable with quantitative and/or categorical explanatory variables (predictors), use logistic regression methods of Section 12.6

QUANTITATIVE RESPONSE VARIABLE (ANALYZING MEANS)

1. If there is only one quantitative response variable, use
   - descriptive methods of Chapter 2
   - inferential methods of Section 7.3 (confidence interval) and Section 8.3 (significance test) for a mean

2. To compare means of a quantitative response variable for two groups of a categorical explanatory variable, use
   - descriptive methods of Chapter 2
   - inferential methods of Sections 9.2 and 9.3 for independent samples
   - inferential methods of Section 9.4 for dependent samples
   - nonparametric tests in Section 14.1 for independent samples or Section 14.2 for dependent samples

3. To compare several means of a quantitative response variable for two or more groups of a categorical explanatory variable, use
   - ANOVA methods of Chapter 13 for independent samples, which are equivalent to regression methods with indicator variables for categorical predictors
   - nonparametric methods from Section 14.2

4. To analyze the association of a quantitative response variable and quantitative explanatory variable, use regression and correlation
   - descriptive methods of Chapters 3 and 11
   - inferential methods of Chapter 11 (Sections 11.3 and 11.4)

5. To analyze the association of a quantitative response variable and several explanatory variables (predictors), use
   - multiple regression methods of Chapter 12 (Sections 12.1- 12.4) for several quantitative predictors
   - multiple regression methods of Section 12.5 with indicator variables for categorical predictors

Statistics

The Art and Science
of Learning from Data

Statistics

The Art and Science of Learning from Data

Alan Agresti
University of Florida

Christine Franklin
University of Georgia

PEARSON

Prentice
Hall

Upper Saddle River, New Jersey 07458

Library of Congress Cataloging-in-Publication Data

Agresti, Alan.

Statistics: the art and science of learning from data/Alan Agresti, Christine A. Franklin.

p. cm.

ISBN 0-13-008369-0 — ISBN 0-13-045536-9 (IE)

    1.   Statistics — Textbooks. I. Franklin, Christine A. II. Title

QA276.12.A35 2007

519.5 — dc22   2005048929

Executive Acquisitions Editor: *Petra Recter*
Editor in Chief: *Sally Yagan*
Executive Project Manager: *Ann Heath*
Project Manager: *Michael Bell*
Production Editor: *Lynn Savino Wendel*
Senior Managing Editor: *Linda Mihatov Behrens*
Executive Managing Editor: *Kathleen Schiaparelli*
Manufacturing Manager: *Alexis Heydt-Long*
Manufacturing Buyer: *Alan Fischer*
Director of Marketing: *Patrice Jones*
Managing Editor, Digital Supplements: *Nicole M. Jackson*
Media Production Editor: *John Cassar*
Development Editor: *Donald Gecewicz*
Editorial Assistant / Supplements Editor: *Joanne Wendelken*
Art Director: *Maureen Eide*
Assistant to Art Director: *Dina Curro*
Interior Designer: *Susan Anderson*
Cover Designer: *Maureen Eide / Suzanne Behnke*
Art Editor: *Thomas Benfatti*
Creative Director: *Juan R. López*
Director of Creative Services: *Paul Belfanti*
Director, Image Resource Center: *Melinda Reo*
Manager, Rights and Permissions: *Zina Arabia*
Manager, Visual Research: *Beth Brenzel*
Manager, Cover Visual Research and Permissions: *Karen Sanatar*
Image Permission Coordinator: *Richard Rodrigues*
Photo Researcher: *Sheila Norman*
Manager, Composition: *Allyson Graesser*
Assistant Manager, Composition: *Clara Bartunek*
Electronic Production Specialist: *Julita Nazario*
Art Studio: Precision Graphics, Inc. / Laserwords
Cover Photo Credits: Jon Shireman / Stone / Getty Images, Inc.

 © 2007 by Pearson Education, Inc.
Pearson Prentice Hall
Pearson Education, Inc.,
Upper Saddle River, New Jersey 07458

Pearson Prentice Hall™ is a trademark of Pearson Education, Inc.

Printed in the United States of America

10 9 8 7 6 5 4 3 2 1

ISBN 0-13-008369-0

Pearson Education Ltd., London
Pearson Education Australia Pty. Limited, Sydney
Pearson Education Singapore Pte. Ltd.
Pearson Education North Asia, Ltd, Hong Kong
Pearson Education Canada, Ltd., Toronto
Pearson Educación de Mexico, S.A., de C.V.
Pearson Education, Japan, Tokyo
Pearson Education Malaysia, Pte. Ltd.

Dedication

To my wife Jacki for her extraordinary support, including making numerous suggestions and putting up with the evenings and weekends I was working on this book, and to my good friend Bruce Edwards for his continual encouragement over four years of weekly lunches.
Alan Agresti

To Corey and Cody, who have shown me the joys of motherhood and to my husband, Dale, for being a dear friend and a dedicated father to our boys.
Chris Franklin

Agresti/Franklin CD Contents

**Instructor to Instructor videos
(on the Instructor Edition CD only)**

Data Sets:

1. MINITAB
2. TI-83 Series, TI-84 Series, TI-89 Series
3. Excel
4. SPSS
5. JMP® INTRO
6. ASCII

> Look at the back endpapers of the book for a complete list of data files. Information is given on which examples, exercises, and activities require or reference the data files and applets.

Applets:

1. Sample from a population
2. Sampling distributions
3. Random numbers
4. Long run probability demonstrations:
   a. Simulating the probability of rolling a 6
   b. Simulating the probability of rolling a 3 or 4
   c. Simulating the probability of head with a fair coin
   d. Simulating the probability of head with an unfair coin (P(H) = 0.2)
   e. Simulating the probability of head with an unfair coin (P(H) = 0.8)
   f. Simulating the stock market
5. Mean versus median
6. Standard deviation
7. Confidence intervals for a proportion
8. Confidence intervals for a mean (the impact of confidence level)
9. Hypothesis tests for a proportion
10. Hypothesis tests for a mean
11. Correlation by eye
12. Regression by eye

Activities Manual — containing additional activities to those found within the text.

Contents

Preface

"Statistics ... the most important science in the whole world: for upon it depends the practical application of every other science and of every art; the one science essential to all political and social administration, all education, all organization based upon experience, for it only gives the results of our experience."

FLORENCE NIGHTINGALE (1820–1910)

One of Florence Nightingale's achievements as a nurse was to promote improved health care by using the art and science of statistics to help people learn from data. She was one of the individuals who helped to develop a new way of looking at the world that is now vital in this modern information age. Today, everyone from students and teachers to workers and employers has to be able to interpret data.

Why Did We Write This Book?

We have each taught introductory statistics for more than 25 years, and have witnessed the welcome evolution from the traditional formula-driven watered-down mathematical statistics course to a concept-driven approach. This concept-driven approach places more emphasis on why statistics is important in the real world and places less emphasis on probability. Our book's content and approach reflect our philosophy of teaching, our classroom experience, recent advances in technology, and student feedback about how they believe that they learn most easily. One of our main purposes in writing this book was to help make the conceptual approach more interesting and more readily accessible to the average college student. At the end of the course, we want more students to look back at their statistics course and think that they learned many worthwhile ideas that will serve them well for the rest of their lives.

What Approach Does This Book Take?

In 2005, the American Statistical Association (ASA) endorsed guidelines and recommendations for the introductory statistics course as described in the report, "Guidelines for Assessment and Instruction in Statistics Education (GAISE) for the College Introductory Course" (www.amstat.org/education/gaise). The reports states that the overreaching goal of all introductory statistics courses is to produce statistically educated students which means that students should develop statistical literacy and the ability to think statistically. The report gives six key recommendations for the college introductory course:

- Emphasize statistical literacy and develop statistical thinking.
- Use real data.
- Stress conceptual understanding rather than mere knowledge of procedures.
- Foster active learning in the classroom.
- Use technology for developing concepts and analyzing data.
- Use assessment to evaluate and improve student learning.

We wholeheartedly endorse these recommendations and have endeavored to write a textbook that supports these guidelines.

Here are the main ways that *Statistics: The Art and Science of Learning from Data* attempts to provide an improved context for teaching introductory statistics:

Learn to ask and answer interesting questions:

In presenting concepts and methods, we encourage students to think about the data and the appropriate analyses by continually posing questions. Our approach, learning by framing questions, is carried out in various ways, including (1) presenting a structured approach to examples that separates the question and the analysis from the scenario presented, (2) providing lots of homework problems that encourage students to think and write, (3) requesting that students perform simulations and hands-on activities to help them learn by doing, and (4) posing information in the form of questions in the section titles and in the figure captions that are answered within the text.

Keep students' needs in mind and reduce confusion:

In many class tests, our students have told us that this book is more "readable" than other introductory statistics texts. To achieve this readability, we have included a wide array of interesting real-data examples and exercises that intrigue students. We have also attempted to simplify our prose wherever possible, without sacrificing any of the accuracy that instructors expect in a textbook.

A serious source of confusion for students is the multiplicity of inference methods that derive from the many combinations of confidence intervals and tests, means and proportions, large sample and small sample, variance known and unknown, two-sided and one-sided inference, independent and dependent samples, and so on. We've placed the main emphasis on the most important cases for practical application of inference: large sample, variance unknown, two-sided inference, and independent samples. The many other cases are also covered (except for known variances), but more briefly, with the exercises focusing mainly on the way inference is commonly conducted in practice. We believe that this approach will help to reduce students' confusion.

Connect statistics to the real world through a more balanced presentation of means and proportions:

Every day in the media, we see and hear percentages and rates being used to summarize results of opinion polls, outcomes of medical studies, and economic reports. Less often do we hear about outcomes that are in any sense normally distributed or even continuous, yet that's the main emphasis of most statistics texts. In this text, we have increased the attention paid to the analysis of proportions. For example, we use contingency tables early in the text to illustrate the concept of association between two categorical variables and to show the potential influence of a lurking variable. When introducing sampling distributions and inference methods, we first explain using proportions for a binary outcome.

We don't neglect quantitative data, but we believe that students should become comfortable with data in the form that they will most often see in practice. This approach has pedagogical rewards. We have found that it is easier for a student to look at a sample of binary data or a distribution of sample proportions with a small sample size and understand what that distribution refers to than to do the same for a bell-shaped curve for a continuum of unobserved values.

How Have We Organized the Contents?

We've shaped the organization and content based on extensive feedback from reviewers and class testers and our beliefs about what works in the classroom and which topics merit inclusion in an introductory course. The main focus is on concepts and interpretation rather than on blind plugging of numbers into formulas. We also want students to come to appreciate that in practice, assumptions are not perfectly satisfied, models are not exactly correct, distributions are not exactly normal, and all sorts of factors should be considered in conducting a statistical analysis. The title of our book reflects the experience of data analysts, who soon realize that statistics is an art as well as a science.

Here's a brief chapter-by-chapter survey. We summarize the content of each chapter and also point out ways in which our presentation may differ somewhat from other texts:

Chapter 1: Statistics: The Art and Science of Learning from Data

Our first chapter gives an overview to students about how data and statistical analyses surround us in our everyday lives by presenting informative examples of real studies. The General Social Survey (GSS) is introduced, and instructions are given for downloading useful and interesting data that will be used throughout the textbook. We also introduce the use of technology through software and applets. An introductory overview of the field such as this chapter provides is missing from many statistics texts.

Chapter 2: Exploring Data with Graphs and Numerical Summaries

This first of two chapters on descriptive statistics introduces students to the univariate case—graphical and numerical methods for exploring and summarizing data on a single categorical or quantitative variable. The emphasis is on interpreting graphs and numerical summaries produced by computer or calculator output, rather than on constructing graphs and calculating numerical summaries by hand.

Chapter 3: Association: Contingency, Correlation, and Regression

This second of two chapters on descriptive statistics focuses on bivariate analyses. It introduces one of the most important topics of statistics—analyzing association between two variables. The chapter begins by exploring categorical variables through contingency tables. Students often see such summaries in their everyday lives, yet most books delay coverage of this material until near the end. Discussing contingency tables facilitates transition into Chapter 4 on producing data for topics such as survey sampling and comparisons in randomized experiments. It also provides background for the discussion of conditional probability in Chapter 5. We next introduce correlation and regression methods for quantitative variables. For many students, the first time that they see correlation is a "light-bulb" moment, the instant they understand the power and usefulness of statistics. The final section emphasizes important cautions, such as association not proving causation and the potential influence of lurking variables, two fundamental concepts that students must see early in the course.

Chapter 4: Gathering Data

This chapter introduces students to good ways and poor ways to sample and to experiment. Why an entire chapter on producing data? Valid analysis and interpretation of data depend on having an appropriate study design. When is it appropriate to make conclusions about causes and effects? How could other factors influence results? How important are issues such as nonresponse? What determines the extent to which conclusions can be generalized? Building on the presentation from Chapter 3 with proportions, we informally introduce the concept of margin of error—a concept and term seen daily by students reading about sample surveys in the media. Regardless of how sophisticated a statistical analysis is conducted, results may be nearly worthless if the study design is poor. However, even though randomized experiments are ideal, they are often not feasible or ethical. In practice, the types of studies we most often encounter are observational. Thus, we also devote detailed coverage to types of observational studies, with emphasis on sample surveys. The potential influence of lurking variables is further illustrated with examples of observational studies.

Chapter 5: Probability in Our Daily Lives

This chapter introduces students to the basic idea of probability and to how it can help us understand the forms that randomness takes in our lives. The unifying theme is how probability relates to our daily lives, using applications such as diagnostic disease testing and whether or not an apparent coincidence truly is coincidental. It's easy for students to get lost when probability is a major component of the course or is presented in a mathematical manner. So we attempt to offer "just enough" to help students understand probability distributions and sampling distributions in the following chapter and conditional probability in the analyses of contingency tables. The early introduction of contingency tables and conditional proportions in Chapter 3 lays the foundation for students to understand more easily the difficult concept of conditional probability. Students are also shown that simulation can estimate probabilities not easily calculated by simple rules.

Chapter 6: Probability Distributions

This chapter focuses on two of the most important probability distributions—the normal for continuous random variables and the binomial for discrete random variables—and introduces the fundamental concept of sampling distributions. The sampling distribution is a topic that students find hard to grasp, yet understanding it is crucial to their understanding of statistical inference. We believe sampling distributions are simplest to explain first for categorical data—using proportions in a small-sample example that identifies easily all the possible samples of binary data. The sampling distribution for a sample proportion is developed using a natural progression moving from the binomial distribution of counts. We believe that students who work through these sections and also the applet-related activities will have a stronger understanding of sampling distributions than students usually achieve in introductory-statistics courses. The end of the chapter links sampling distributions to the basic ideas of statistical inference, which is the focus of the following two chapters.

Chapter 7: Statistical Inference: Confidence Intervals

We believe that more emphasis should be placed on interval estimation than on significance testing in an introductory statistics course. We learn less from testing that a parameter takes some specific value than from finding a range of plausible values for it. Statisticians increasingly share the belief that significance tests have been overemphasized. For instance, many psychology journals now forbid reporting P-values without accompanying information about effect sizes, such as confidence intervals. Moreover, confidence intervals are simpler than tests for students to understand, especially with the use of an applet for simulation. Consequently, we devote this first inference chapter solely to estimation. An activity using simulation helps students visualize the role of the sampling distribution in forming confidence intervals. We begin the chapter with forming a confidence interval for a proportion. We believe that this is an easier transition for the students, because this method uses the normal distribution. We then move to the confidence interval for a mean, introducing the t distribution. We do not cover the case in which the population standard deviation is known, as this is very rarely true in practice and increases the profusion of methods that leads to student confusion.

Chapter 8: Statistical Inference: Significance Tests About Hypotheses

Although we believe it's important to emphasize interval estimation for inference, significance tests do play an important role in statistical inference. The main goal in many research studies is to check whether the data support certain statements, claims, or predictions. A significance test is a method of using data to summarize evidence about a claim or prediction. This chapter emphasizes the P-value approach, which is the common way to perform tests in practice and is more informative than simply making a reject-or-not-reject decision at an artificial alpha-level with a rejection region. Most journals prefer using P-values to report statistical results. As in the previous two chapters, we begin by introducing the significance test for a proportion, then move to the significance test for a mean. Both one-sided and two-sided alternatives are presented. We encourage more use of the two-sided alternative, however, because this is commonly seen in practice and because the results of using a two-sided alternative parallel (two-sided) confidence intervals. Also, in that case, t methods are robust to violations of the normality assumption. In this chapter, we also stress the limitations of significance tests. In doing so, we emphasize the distinction between *practical significance* and *statistical significance*, and we show the possible difficulties resulting when only "significant" results are reported. We also discuss Type I error, Type II error, and the power of a test.

Chapter 9: Comparing Two Groups

The chapter explains that a comparison of two groups is actually a bivariate analysis, in which the explanatory variable defines the groups to be compared. We avoid the case of known population standard deviation and we de-emphasize the analysis that assumes equal variances, showing how software can find the *df* value with the less restrictive analysis. We present methods for independent and for dependent samples (matched pairs), with our main emphasis on the independent samples. Because this text does not short-change categorical data, we also present methods for analyzing proportions using matched pairs. This topic is relatively simple to present but is rarely treated in introductory texts. This chapter also revisits the importance of taking into account potential lurking variables when comparing two groups, illustrating with an example showing Simpson's paradox.

Chapter 10: Analyzing the Association Between Categorical Variables

In our opinion, introductory statistics books place way too much emphasis on the chi-squared test. Like any significance test, it gives limited information. It should not be the sole or even the primary method taught in the inference chapter about bivariate categorical data. The chi-squared test addresses whether an association exists between categorical variables; however, it does not address the strength or the pattern of the association. We have included follow-up analyses such as measuring association using the difference and the ratio of proportions (relative risk), and inspecting residuals to detect which cells are responsible for a large chi-squared statistic. Doing so connects this material with regression methods presented in the following chapter for bivariate quantitative data.

Chapter 11: Analyzing Association Between Quantitative Variables: Regression Analysis

This chapter expands on the concepts of simple straight-line regression introduced in Chapter 3. After reviewing the introductory material from Chapter 3 on straight-line regression equations and correlation, this chapter explains what is meant by a model, connects this with the notion of residual unexplained variation around a regression line (and the related ANOVA table and *r*-squared measure and *F* statistic), discusses "regression toward the mean," and presents inference methods. The chapter also discusses ways of checking model assumptions.

Chapter 12–14: Advanced Topics

Chapters 12–14 contain more advanced topics for the first course or for a follow-up course. Chapter 12 covers multiple regression. It also shows how a regression model can include categorical explanatory variables (using indicator variables) or categorical response variables (logistic regression). Chapter 13 covers one-way and two-way ANOVA and shows their connection with multiple regression. Chapter 14 introduces nonparametric statistics.

How Do We Promote Student Learning?

In an effort to motivate students to think about the material, to ask appropriate questions, and to develop good problem-solving skills, we have created special features that distinguish this text.

A Vivid Chapter-Opening Example

Each chapter begins with a high-interest example that establishes some key questions and themes that are woven throughout the chapter. Illustrated with engaging photographs, this example is designed to 'grab students' attention and to draw them into the chapter. The issues discussed in the chapter's opening example are referred to and revisited in examples within the chapter.

Our Innovative Example Format

Recognizing that the worked examples are the key vehicle for engaging and teaching students, we have developed a unique structure to help students learn to model the question-posing and investigative thought process required to examine issues intelligently using statistics. The five components in this structure are as follow:

- **Picture the Scenario:** Background information is presented so students can visualize the situation. This step places the data to be investigated in context and often provides a link back to previous examples.

- **Questions to Explore:** Questions are posed that follow directly from information in the scenario. This step helps the student to focus on what is to be learned from the example and what types of questions are useful to ask.

- **Think It Through:** This step is the heart of each example as the questions posed are investigated and answered using appropriate statistical methods. Each solution is clearly matched to the questions so students can easily find the response to each posed Question to Explore.

- **Insight:** The overarching goal of the Insight step is to clarify the central ideas investigated in the example and to place them in a broader context. Many of the Insights also provide connections between seemingly disparate topics in the text by referring back to concepts learned previously and/or foreshadowing techniques and ideas to come later.

- **To Practice This Concept, Try Exercise XX:** Each example concludes by directing students to an end of section exercise that allows immediate practice of the concept or technique within the example, requiring the student to use similar reasoning as that employed in the example.

This example, from Chapter 2, demonstrates our example format:

EXAMPLE 9

Is There a Trend Toward Warming in New York City?

Picture the Scenario: Background information is presented so students can visualize the situation. This step places the data to be investigated in context and often provides a link back to previous examples.

Picture the Scenario

Each day, weather stations around the world record the daily average temperature, defined as the average of the maximum and minimum temperature over a midnight-to-midnight time frame. In a given year, the annual average temperature is the average of the daily average temperatures for that year. Let's analyze data on annual average temperature (in degrees Fahrenheit) in Central Park, New York City from 1901 to 2000. This is a continuous quantitative variable. The data are in the "Central Park Yearly Temps" data file on the text CD.

Questions to Explore: Questions are posed that follow directly from information in the scenario. Posing questions helps the student to focus on what is to be learned from the example and what types of questions are useful to ask about the data.

Question to Explore

What do we learn from a time plot of these annual average temperatures? Is there a trend toward warming in the United States?

Think It Through: This step is the heart of each example as the questions posed are investigated and answered using appropriate statistical methods. Each solution is clearly matched to the question so students can easily find the response to each posed Question to Explore.

Think It Through

Figure 2.8 shows a time plot, constructed using MINITAB software. The time plot constructed using the TI-83-84 calculator is in the margin. The observations fluctuate considerably, but the figure does suggest somewhat of an increasing trend in the annual average temperatures at this location.

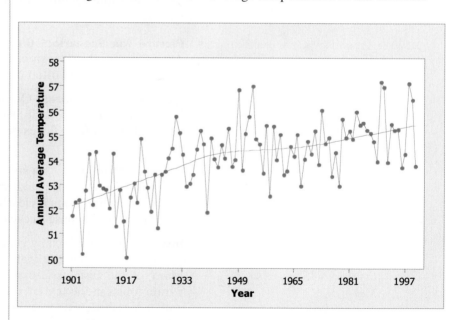

▲ **FIGURE 2.8** MINITAB **output for a time plot of Central Park, New York City average annual temperatures.** The annual average temperatures are plotted against the year from 1901 to 2000 A smoothing curve is superimposed. **Question:** Are the annual average temperatures tending to increase, decrease, or stay the same over time?

Insight: This step clarifies the central ideas investigated in the example and places them in a broader context that often states the conclusions in less technical terms. Many of the Insights also provide connections between seemingly disparate topics in the text by referring back to concepts learned previously and /or foreshadowing techniques and ideas to come.

To Practice This Concept: Each example concludes by directing students to an end-of-section exercise that allows immediate practice of the concept or technique within the example.

Insight

The short-term fluctuations in a time plot can mask an overall trend. It's possible to get a clearer picture by "smoothing" the data. This is beyond our scope here. MINITAB presents the option of smoothing the data to portray a general trend (Click on "data view" and choose the "lowess" option under "smoother"). This is the smooth curve passing through the data points in Figure 2.8. This curve goes from a level of about 52 degrees in 1901 to about 55 degrees in 2000.

The data reported in Figure 2.8 refer to one location in the United States. In a study of climate change, it would be important to explore other locations around the world to see if similar trends are evident there.

TO PRACTICE THIS CONCEPT, TRY EXERCISES 2.26 AND 2.27

A Broad Array of Useful and Engaging Exercises

The text contains a strong emphasis on real data both in the examples and exercises. Nearly all the chapters contain more than 100 exercises. These exercises are realistic and ask students to provide interpretations of the data or scenario rather than merely to find a numerical solution. We show how statistics addresses a wide array of applications, from opinion polls, to market research, to health and human behavior. Our goal is to have students come to believe the comment by the author H. G. Wells (1866–1946), "Statistical thinking will one day be as necessary for efficient citizenship as the ability to read and write."

Exercises are placed at the end of each section for immediate reinforcement, and a more comprehensive set of exercises is found at the end of the chapter. Each exercise has a descriptive label. Exercises for which technology is recommended are indicated with the icon, ■. Larger data sets used in examples and exercises are referenced in text, listed in the back endpapers and made available on the text CD. The exercises are divided into the following categories:

- **Practicing the Basics:** The section exercises and the first group of end-of-chapter exercises reinforce basic application of the methods.

- **Concepts and Investigations:** These exercises require that the student explore real data sets and carry out investigations for mini-projects, and explore concepts and related theory or extensions of the chapter's methods. This section contains some multiple-choice and true/false exercises to help students check their understanding of the basic concepts and to prepare for tests. It also contains a few more difficult, optional exercises (highlighted with the ◆◆ icon) that present some additional concepts and methods.

- **Class Explorations:** A brief third section provides exercises that are designed for group work based on investigations done by each of the students in a team. Additional activities may be found here as well.

Applying Statistics to Real Situations: In Practice

One of our goals is to help students understand some of the differences between proper "academic" statistics and what is actually done in practice. Data analysis in practice is an art as well as a science. Although statistical theory has foundations based on precise assumptions and conditions, in practice the real world is not so simple. This dicotomy makes it difficult to give general rules, always to be obeyed, for analyzing data. To that end, we have embedded numerous in-text references and separate *In Practice* boxes that alert students to the way statisticians actually analyze data in practice. These comments are based on our extensive consulting experience and by observing what well-trained statisticians do in practice.

Activities and Simulations for Hands-On Learning

Activities are an important pedagogical tool for promoting understanding by helping the student to learn statistical concepts through hands-on exploration. Although we believe the activities help to facilitate learning, we have made them optional (and placed most of them at the ends of the chapters) so that instructors may use them at their discretion. Each of Chapters 1–11 includes at least one activity. A distinction is made between those activities that are to be done in class and those activities that are to be done by students outside of class. The activity often involves simulation, commonly using an applet available on the text CD and within MyMathLab. Additional activities are available on the text CD and MyMathLab. We encourage the use of these activities and simulations to promote student learning.

Clearly Presented Definition and Summary Boxes

We highlight key definitions, guidelines, procedures, "In Practice" remarks, and other summaries in boxes throughout the text.

Immediate Help for the Student

Material presented in the margins or boxed off from the main text offers at-a-glance information to aid in comprehending material in the main text. During our class tests, students have given high ratings to these features in making the book user-friendly. The margin features include the following

- **In Words:** This box explains the definitions and symbolic notation found in the body of the text itself (which, for technical accuracy, must be more formal).

- **Recall:** As the student progresses through the book, concepts are presented that call back to information learned in previous chapters. The Recall box directs the reader back to a previous presentation in the text to review and reinforce concepts and methods already covered.

- **Activity/Applet:** The reader is directed to an activity or applet on the text CD or within MyMathLab. Applets have great value, because they vividly show students certain concepts visually. For example, creating a sampling distribution is accomplished more readily with applets than with a static text figure. As noted, the activities and applets are presented as optional explorations in the text.

Enhanced Visuals and a Graphical Approach to Learning Statistics

Increasingly, students are visual learners. Recognizing this fact, we have taken extra care to make the text figures informative. For instance, we've annotated many of the figures with labels that clearly identify the noteworthy aspects of the illustration. Further, most figure captions include a question (answered at the end of the chapter) designed to challenge the student to interpret and think about the information being communicated by the graphic. The graphics also feature a pedagogical use of color to help students recognize patterns and to distinguish between statistics and parameters. The use of color is explained in the inside front cover for easy access by students.

Up-to-Date Use of Technology

The ready availability of technology helps instructors to provide instruction that is less calculation based and more concepts oriented. Output from *computer software and calculators* is displayed throughout the textbook. The discussion focuses on interpretation of the output, rather than on the keystrokes needed to create the output. We use the generic-looking, popular, and reliable MINITAB statistical software for most of our output examples but also include selected screen captures from the TI-83+/84 graphing calculator, SPSS, and Microsoft EXCEL, as appropriate. Technology specific manuals are also available with this text (MINITAB, SPSS, Excel, and TI-83+/84). See the supplements listing within the front matter for more information.

An Invitation Rather Than a Conclusion

We hope that students using this textbook will gain a lasting appreciation for the vital role the art and science of statistics plays in analyzing data and helping us make decisions in our lives. As evidenced by the approach we've outlined above, our major goals for this textbook are that students should learn how:

- To produce data that can provide answers to properly posed questions

- To appreciate how probability helps us understand randomness in our lives, as well as to grasp the crucial concept of a sampling distribution and how it relates to inference methods

- To choose appropriate descriptive and inferential methods for examining and analyzing data and drawing conclusions

- To communicate the conclusions of statistical analyses clearly and effectively

- To understand the limitations of most research, whether because it was based on an observational study rather than a randomized experiment or survey, or because a certain lurking variable was not measured that could have explained the observed associations

We are excited about sharing through this text the insights that we have learned from our experience as teachers and from our students. Many students still enter statistics classes on the first day with dread because of its reputation as a dry, sometimes difficult, course. It is our goal to inspire a classroom environment that is filled with creativity, openness, realistic applications, and learning that students find inviting and rewarding. Many of our students have elected to take our second courses in statistics and even major or minor in statistics. We hope that this textbook will help the instructor and the students to experience a rewarding introductory course in statistics.

Alan Agresti,
University of Florida, Gainesville

Christine Franklin,
University of Georgia, Athens

Acknowledgments

We are indebted to the many reviewers, class testers, and students who gave us invaluable feedback and advice on how to improve the quality of the book. Many of our reviewers worked with us through several drafts of the manuscript and provided us with valuable insight into how to improve the manuscript at both a micro and macro level. They include the following individuals as well as others who prefer to remain anonymous:

ARIZONA Russel Carlson, The University of Arizona; Peter Flanagan-Hyde, Phoenix Country Day School • **CALIFORINIA** Robert Gould, University of

California Los Angeles; Dawn Holmes, University of California Santa Barbara; Rebecca Head, Bakersfield College; Susan Herring, Sonoma State University; Colleen Kelly, San Diego State University; Elaine McDonald, Sonoma State University; Helen Noble, San Diego State University • **CONNECTICUT** Anne Doyle, University of Connecticut; Dan Miller, Central Connecticut State University; Kathleen Mclaughlin, University of Connecticut; Stephen Sawin, Fairfield University • **DISTRICT OF COLUMBIA** Hans Engler, Georgetown University, Washington D.C. • **FLORIDA** James Lang, Valencia Community College; Megan Meece, University of Florida; Maria Ripol, University of Florida • **GEORGIA** Carrie Chmielarski, Student The University of Georgia; Charles LeMarsh, Lakeside High School; Broderick Oluyede, Georgia Southern University; Chandler Pike, The University of Georgia; Kim Robinson, Clayton State University; Jill Smith, University of Georgia; John Seppala, Valdosta State University; Joseph Walker, Georgia State University • **IOWA** John Cryer, The University of Iowa; William Duckworth, Iowa State University • **ILLINOIS** Jinadasa Gamage, Illinois State University; Richard Maher, Loyola University Chicago; Cathy Poliak, Northern Illinois University • **MASSACHUSETTS** Katherine Halvorsen, Smith College; Xiao-Li Meng, Harvard University; Daniel Weiner, Boston University • **MICHIGAN** Kirk Anderson, Grand Valley State University; Phyllis Curtiss, Grand Valley State University; Alla Sikorskii, Michigan State University • **MINNESOTA** Bob Dobrow, Carleton College • **MISSOURI** Larry Ries, University of Missouri–Columbia • **MONTANA** Jeff Banfield, Montana State University • **NORTH CAROLINA** Pamela Arroway, North Carolina State University; E. Jacquelin Dietz, North Carolina State University; Roger Woodard, North Carolina State University • **NEBRASKA** Linda Young, University of Nebraska • **NEW JERSEY** Harold Sackrowitz, Rutgers, The State University of New Jersey • **NEW YORK** Brooke Fridley, Mohawk Valley Community College ; Martin Lindquist, Columbia University; David Mathiason, Rochester Institute of Technology; Tian Zheng, Columbia University • **OHIO** John Holcomb, Cleveland State University; Jackie Miller, The Ohio State University; Stephan Pelikan, University of Cincinnati; Deborah Rumsey, The Ohio State University; Kevin Robinson, University of Akron • **OREGON** Henry Mesa, Portland Community College, Rock Creek; Qi-Man Shao, University of Oregon; Daming Xu, University of Oregon • **PENNSYLVANIA** Steven Gendler, Clarion University; Paul Lupinacci, Villanova University; Linda Myers, Harrisburg Area Community College • **SOUTH CAROLINA** Martin Jones, College of Charleston; Murray Siegel, The South Carolina Governor's School for Science and Mathematics; Webster West, University of South Carolina • **TENNESSEE** Robert Price, East Tennessee State University; Ginger Rowell, Middle Tennessee State University; Edith Seier, East Tennessee State University • **TEXAS** Larry Ammann, The University of Texas at Dallas; Tom Bratcher, Baylor University; Jianguo Liu, University of North Texas; Mary Parker, Austin Community College; Robert Paige, Texas Tech University; Diane Resnick, University of Houston-Downtown • **UTAH** Carolyn Cuff, Westminster College; Lajos Horvath, The University of Utah • **VIRGINIA** David Bauer, Virginia Commonwealth University; Ching-Yuan Chiang, James Madison University; Steven Garren, James Madison University; Debra Hydorn, Mary Washington College; D'Arcy Mays, Virginia Commonwealth University; Stephanie Pickle, Virginia Polytechnic Institute and State University • **WASHINGTON** J.R. Alldredge, Washington State University • **WYOMING** Burke Grandjean, University of Wyoming • **CANADA** David Loewen, University of Manitoba, Manitoba

Special thanks is due to the participants in our Chicago reviewer conference who spent two full days helping us think through the challenges associated with writing a first edition. They are Larry Amman, Phyllis Curtiss, Steve Garren, Martin Jones, James Lang, David Loewen, Megan Meece, Steve Pelikan, and Roger Woodard.

Class Testers:

We must also single out those individuals who braved an untested manuscript and asked their students to help work out the kinks in the manuscript through our many class tests:

ARIZONA
Peter Flanagan-Hyde, Phoenix Country Day School

CALIFORNIA
Amy McElroy, Corey Manchester, San Diego State University
Calvin Schmall, Solano Community College

COLORADO
David Most, Colorado State University

CONNECTICUT
Paul Bugl, University of Hartford
Dan Miller, Central Connecticut State University
Steve Sawin, Fairfield University
John Vangor, Fairfield University

FLORIDA
Nazanin Azarnia, Santa Fe Community College

GEORGIA
Ouida Dillon, Oconee County High School
Katherine Hawks, Meadowcreek High School
Charles LaMarsh, Lakeside High School
Steve Messig, Oconee County High School
Broderick Oluyede, Georgia Southern University
Chandler Pike, University of Georgia
John Seppala, Valdosta State University

MICHIGAN
Roy Erickson, Michigan State University
Mark Stevenson, Oakland Community College

NEVADA
Alison Davis, University of Nevado, Reno

NEW YORK
Steve Stehman, SUNY ESF

PENNSYLVANIA
Bonnie A. Green, East Stroudsburg University
Deborah Lurie, Saint Joseph's University

SOUTH CAROLINA
Martin Jones, Beverly Diamond, College of Charleston
Murray Siegel, The South Carolina Governor's School for Science and Mathematics

TENNESSEE
Bonney Daves, Christian Academy of Knoxville
Ginger Holmes Rowell, Middle Tennessee State University

VIRGINIA
Debra Hydorn, University of Mary Washington

CANADA
David Loewen, University of Manitoba

The following students also helped by way of a class test or focus group:

CONNECTICUT:
Central Connecticut State University—Chris Castaldo, Niki Chieka, Melissa Rocco, Angelita K. Tolly, Bryan Tracey, Jaime Valentine

FLORIDA:

Santa Fe Community College—Robin Adkins, Ifat Bilitzer, Staci Crawford, David Deal, Michelle Donovan, Whitney Edmunds, Tsion Gebru, Syed A. Hasan, Yuta Hasuko, Blake Hersch, Jessica Horne, David LaBorde, Mike Ligmanowski, Candace A. McCoy, Becky McDaniel, Allen Lee Pittman, Robert Rizzo, Tricia Robey, Kayla Smith, Eric Thomas, Latoya Warren, Jasma Williams

GEORGIA:

Georgia Southern University—Charles Allen, La Sheba Batiste, Justin A. Bitts, Camille Bolton, Jarred Crowder, Elizabeth L. Curry, Gregory Fernando, Jeremy Fordh, Sarah Hightower, James Jones, Ben Kellum, Cavarro Lewis, Sarah Moss, Charlie Moye, Ashley L. Scruggs, Shannon Stewart, Brandy Tyler

University of Georgia—Julie Ahern, Brad Alexander, William Barrow, Lindsay Blalock, Brittany Bloom, Kirk Braunius, A. Brennan, Ashley Brown, Stephen Austin Browne, Jeff Butt, Peter Caruana, Melissa Chaffin, Caroline Chapmna, Katie Chenowith, Rachael Clark, Benjamin Cobb, Maggie Cottingham, Peter Covara, Lauren Cramer, Paige Davies, Sara DiMauro, Anna Felz, Brian Ferslew, James Gable, Caitlin Gaddy, Scot Goins, Michelle Greene, Jeff Hall, Rachael Heller, Joseph Herbozo, Adam Hinman, Maggie Holahan, Lauren Howe, Betsy Hunter, Seela Jacobs, Kimberly Jakaitis, Patricia Johnston, Chase Jones, Jessica Kennedy, Lindsey Kenoe, Sharon Kim, David Kirby, Matthew Lampert, Hyeonmi Lee, Erin Lundberg, David Mahon, Bridget Mixson, Sheryl E. Moody, Chad Mumm, Shawn D. O'Connor, Ashley Owen, Gregrory Pak, Kristen Palmer, Sarah Pierce, Jessica Pratt, Luiza Raposo, Laura Rayne, Lindsay Ross, Michelle A. Santilli, Daniel Sattlemeyer, Kristen Schuelke, Mansi Sheth, Scott Smith, Dinean Stevens, Bryce Teale, Bjorn Thornton, Jacquelyn Tolliver, Anna Tyler, Andrew Vesper, Katie Walsh, Crystal Weaver, Liz Wells, Patrick West, Troy Wood, Christine Yuan

Valdosta State University—Christina Barraza, Ashley Creasy, Alicia D., Jessica Freeman, Tiffany L. Graham, Ashlee Kloss, Adrian Lancaster, Nicole L. List, James Dusty Martin, Jacqueline McGriff, Melanie Noel, Israel Saucedo, Christie Stephens, Logan Troutman

OHIO:

The Ohio State University—Sharina Harrington, Jackie Hauser, Donna L. Jenkins, Neneeva Prentiss, Mark Zito

University of Cincinnatti—Maria Ahmed, Jessica Aspra, Nathan Chiudiani, John FK Guy, Amy Ittikraicharoen, Hakimah Jabbar, Joy Kramer, Luqman Murtaza, Bryan Parkhurst, Kristina Schwartz, Rick Stockmeier, Paul M. Stone, Gia Tili, Elizabeth Trombly, Virginia Worthington

SOUTH CAROLINA:

College of Charleston—Meghan Alter, Ray Buckmaster, Emily Chaka, Caitrin Cunningham, Michelle Davis, Nakashia Dunner, Erendira Fabela, Lindsey Fairfax, Angelina Fletcher-Resendiz, Emily D. Franklin, Hazel Gibson, Anna Meacham, Kathryn Myers, Catherine Price, Morgan Rohrhofer, Takya Rogers, Shannon Sheehan, Melanie Shultis, Alicia Thompson, Emily Tyner, Sarah Voss, Sean Wilson

TENNESSEE:

Middle Tennessee State University—Chris Alexandera, Laura L. Andrews, Jacqueline Briones, Daniel Justin Cripps, Chris Du, Noel Edmonds, James A. Goffos, Ginny Harris, Sarah Hebert, Amanda M. Jenkins, Kalan, Kelch, Michael Larkins, Chun Park, Shelly N. Phillips, Rachel Sharp, Bryan Sizemore, Jerald M. Turner

CANADA:

University of Manitoba—Jazelle Barrian, Valene Bretand, N. Calanza, Wong Siew Cheong, Perry Cole, Kathryn Drepko, Ian Hackin, Michael Himbeault, Lissa Hinkelman, Jenna Howells, Megan Husak, Brenda Klassen, Darren Klassen, Jacqueline Macariola, Geraldine McMillan, Benjamin Meek, Brett Napier, Tara Nelson, Kaori Nonaka, Glen Pauch, Decris Providence, Raphael Santiago, Yu Song, Heather Spence, Rosana Tapia, Adam Virgilio, Sylvia Walder

Ancillaries:

Writing a first edition textbook today is made equally challenging by the myriad ancillaries and supplements that support the core text. Two critical ancillaries are the Instructor and Student Solution Manuals. Instructors and students rely upon these manuals for accuracy. Given the magnitude of exercises in our textbook, writing detailed and accurate solutions was a tremendous task. Susan Nolan (Seton Hall University) was meticulous and dedicated in developing solutions that foster better learning and understanding. We are also appreciative of Susan's willingness to serve as a CD Lecture Series Author for the text. We also want to thank Martin Jones (College of Charleston) for his contributions to the Instructor's Notes. Having both reviewed and class tested the manuscript, he has contributed many valuable insights on how to teach from our book. We are also grateful to Webster West (University of South Carolina) for his work in producing the applets which will provide a valuable resource for students and teachers alike.

We would also like to thank the following individuals for their excellent work on other supplements:

Kirk Anderson (Grand Valley State University), TI Technology Manual
Debra Hydorn (University of Mary Washington), SPSS Technology Manual
Kate McLaughlin (University of Connecticut), PowerPoint Slides
Megan Meece and Maria Ripol (University of Florida), Activities Manual
Linda Myers and Diane Benner (Harrisburg Area Community College), Excel and MINITAB Technology Manuals
Jill Smith (University of Georgia), Test Item File

Each of these individuals helped us to ensure that this first edition is as clean and accurate as possible.

A Word of Thanks

The detailed assessment of the text and solutions manual fell to our accuracy checker, Sarah Streett, who checked the manuscript in both the preliminary and final versions. Thank you for your fastidious attention to detail, Sarah.

We would like to thank our Prentice Hall team, individuals that have given countless hours in developing this text and without their guidance and assistance, the text would not have come to completion. We thank Ann Heath, Don Gecewicz, Petra Recter, Mike Bell, Lynn Savino Wendel, Joanne Wendelken, Tom Benfatti, Maureen Eide, and Patrice Jones, and from the University of Georgia, our research assistant, Kelly Kopf.

Alan Agresti would like to thank those who have helped us in some way, often by suggesting data sets or examples. These include Anna Gottard, Wolfgang Jank, Bernhard Klingenberg, Réne Lee-Pack, Jacalyn Levine, Megan Lewis, Megan Meece, Dan Nettleton and Yongyi Min. Many thanks also to Tom Piazza for his help with the General Social Survey. Finally, Alan Agresti would like to thank his wife Jacki Levine for her extraordinary support throughout the writing of this book. Besides putting up with the evenings and weekends he was working on this book, she offered numerous helpful suggestions for examples and for improving the writing.

Chris Franklin gives a special thank you to her husband and sons, Dale, Corey, and Cody Green. They have patiently sacrificed spending many hours with their spouse and mom over the past three years as she worked on this book. Corey made helpful suggestions reading the manuscript as a student in AP Statistics. She also appreciates the encouragement and support of her UGA colleagues, Chandler Pike and Jaxk Reeves and her many students who class tested the book, offering useful and practical suggestions for improvement. Many thanks to Michael Horvat for sharing the female athletic data sets in Chapters 11 and 12. Finally, Chris thanks her coauthor, Alan Agresti, for diligently working as a partner to complete the textbook they began planning many years ago.

Alan Agresti, *Gainesville, Florida*
Chris Franklin, *Athens, Georgia*

About the Authors

Alan Agresti is Distinguished Professor in the Department of Statistics at the University of Florida. He received his PhD at the University of Wisconsin and an Honorary Doctor of Science from De Montfort University in the UK. In 2003, he was named "Statistician of the Year" by the Chicago chapter of the American Statistical Association. He has written four other books, including *Categorical Data Analysis* (2nd Edition, Wiley, 2002) and *Statistical Methods for the Social Sciences* (3rd Edition, Prentice Hall, 1997), and more than 100 journal articles.

When not working at the University of Florida, Alan enjoys music, reading, and traveling around the world. He has taught courses or short courses for universities and companies in about 25 countries worldwide. His wife, Jacki Levine, is managing editor of *The Gainesville Sun* and editor of *Gainesville* Magazine.

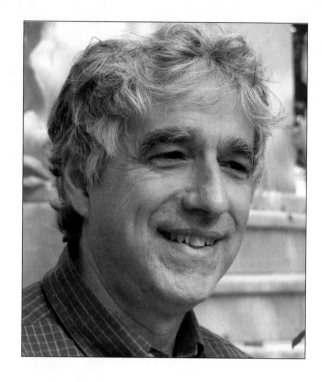

Chris Franklin is a Lecturer and Honors Professor in the Department of Statistics at the University of Georgia. She has been teaching statistics and actively involved with statistics education at the national level for over 25 years. Her work at the national level includes continued commitment to the AP Statistics program since its beginning in the 1990's, course development and writing national standards for Pre K-12 teachers, writing instructional materials (books and articles) for NCTM and other educational institutions and publications, and dedicated service to the American Statistical Association promoting strategic initiatives in statistics education. Chris is a Fellow of the American Statistical Association and has been honored with numerous teaching and advising awards at UGA.

Most important for Chris is spending time with her husband and two boys. She loves to run competitively, play the piano, read, and simply be outdoors. One of her most memorable experiences was backpacking in the backcountry for 11 days in Philmont, NM with her older son and the boy scouts.

STUDENT RESOURCES

STUDENT RESOURCES

Student Data CD

The text comes with a CD containing data sets (ASCII, SPSS, Excel, MINITAB, TI, and JMP), applets, which are useful for illustrating statistical concepts, and an Activity Manual in PDF format.

PRINTED RESOURCES
Student Study-Pack

Everything a student needs to succeed in one place.

- Text & Student Study Pack Bundle:
  ISBN:0-13-220914-4
- Student Study Pack Standalone
  ISBN: 0-13-168713-1

The Student Study-Pack contains:

- *Student Solutions Manual*

Fully worked solutions to odd-numbered exercises

- *CD Lecture Videos*

A comprehensive set of videos, tied to the textbook, in which examples from each chapter are worked out by a statistics instructor. The videos provide excellent support for students who require additional assistance, for distance learning and self-paced programs, or for students who missed class.

- *Technology Manual*

Contains detailed tutorial instructions and worked out examples & exercises for:

- TI-83+/84 Series Calculator
- Excel (including PHStat, an Excel plug-in)
- SPSS
- MINITAB

- *Pearson Tutor Center*

Tutors provide one-on-one tutoring for any problem with an answer at the back of the book. Students access the Tutor Center via toll-free phone, fax, or e-mail.

INTERNET RESOURCES

MyMathLab

- Text & MyMathLab bundle: ISBN: 0-13-220915-2
- MyMathLab Standalone ISBN: 0-13-147894-X

MyMathLab is a text-specific, online course that provides unlimited tutorial exercises closely correlated to the exercises in the text, a multimedia textbook with links to learning aids, such as animations and videos, student supplements in electronic format—the student solutions manual, PowerPoint Presentations and much more! Since 2001, over one million students have done better in math and statistics with MyMathLab.

MathXL®

- 12-month MathXL & Text bundle:
  ISBN: 0-13-238728-X
- 24-month MathXL & Text bundle:
  ISBN: 0-13-238727-1
- 12-month Standalone ISBN: 0-13-147892-3
- 24-month Standalone ISBN: 0-13-147897-4

MathXL offers unlimited practice exercises (algorithmically generated) that are closely correlated to the exercises in the text. MathXL provides students with the opportunity to practice concepts and techniques that will help them succeed.

Statistics

The Art and Science
of Learning from Data

1 | Statistics: The Art and Science of Learning from Data

moking ban OK'd
rohibition to take effect Jan. 31
in Columbus | **B1**

▶ Lawsuits fly on
Election Day | A9

▶ The wait to
vote in Gambier?
Six or seven
hours | B4

The Columbus Dispatch

WEDNESDAY, NOVEMBER 3, 2004

OME FINAL

134 | NO. 126
50 CENTS

CLIFFHANGER

Presidential race could hinge on Ohio's provisional ballots

Bush t
in vo
but
ha

By
ar

	Bush	Kerry	Pending
Electoral votes*	**254**	**225**	**59**

There are 538 total electoral
votes; 270 are needed to win

Popular vote* (89%)   Bush: 54,023,856   Kerry: 50,282,501

* As of 3:10 a.m.

TURNOUT AND TROUBLES

Waiting was the hardest part

THE COLUMBUS DISPATCH

They waited for hours, stead-
fast in their passion to vote and
undeterred by rain, sore feet or
Election Day problems com-
pounded by a record turnout.
More than 500,000 people
voted in Franklin County, and
thousands were still waiting in
line when the polls closed at
7:30 p.m., with the last person
voting at 11:20 p.m.
A three-hour wait was common across central Ohio. In the
Knox County
college village
of Gambier and
at least two Co-
lumbus pre-
cincts, hun-
dreds of people
waited as long
as six hours.
Democrats won a federal
court order last night to force
Franklin County Board of
...to give paper ballots
...forms to the
...who were

ld Rodrequs Turner waits with his mom, Shanne Heath, in the voting line outside Eas
ands in Franklin County waited in line to vote even after the 7:30 p.m. poll-clos

'date forges o

By Jonatha
and Jack K
THE COLUMB

EXAMPLE 1

How Can Statistics Help Us Learn About the World?

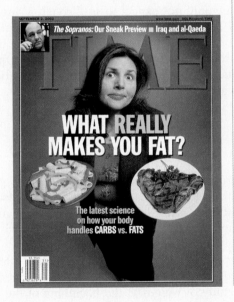

Picture the Scenario

In this book, we'll explore information from a wide variety of scenarios you'll encounter in your daily life. For example, you will need to evaluate reports in the media about opinion surveys, medical research studies, the state of the economy, and environmental issues. You'll face financial decisions such as having to choose between an investment with a sure return and one where you could make much more money but could possibly lose your entire investment. We'll learn how to analyze the available information to answer questions of interest in such scenarios. To read and question intelligently, an understanding of statistics is essential. A main purpose of this book is to show you why.

Questions to Explore

This book will show you how to collect appropriate information and how to apply statistical methods so you can better evaluate that information and answer the questions posed. Here's an example of questions we'll investigate in coming chapters:

- How can you evaluate evidence about global warming?
- Are cell phones dangerous to your health?
- What's the chance your tax return will be audited?
- How likely are you to win the lottery?
- Is there bias against women in appointing managers?
- What "hot streaks" should you expect in basketball?
- How can you analyze whether a diet really works?
- Are astrology predictions better than mere guessing?
- Do more people believe in heaven or in hell?
- What predicts whether or not you are happy?
- How can you predict the selling price of a house?
- How can you estimate a person's strength?
- Which tanning method gives you a better tan?

Thinking Ahead

Each chapter of this book begins by posing a question such as those above. Then the content of the chapter shows you how to answer that question. We'll see that **statistics** is the art and science of designing studies and analyzing the information that those studies produce. Statistics helps us to take information of various sorts and use it to better understand the world around us.

3

It's never been more the case than today: To *truly* make sense of the world around you, an understanding of statistics is essential.

In the business world, managers use statistics to analyze results of marketing studies about new products, to help predict sales, and to measure employee performance. Medical research studies use statistics to evaluate whether new ways to treat disease are better than existing ways. In fact, many, if not most, professional occupations today rely on statistical methods. In a competitive job market, an understanding of statistics provides an important advantage.

But it's important to understand statistics even if you will never use it in your job. Understanding statistics can help you navigate your world and make better choices. Why? Because every day you are bombarded with statistical information from news reports, advertisements, political campaigns, and surveys. How do you know what to heed and what to ignore? An understanding of the statistical reasoning underlying these pronouncements will help. For instance, this book will enable you to evaluate news stories of medical research studies more effectively and to know when you should be skeptical of their claims. Does regular, moderate drinking of alcohol truly lessen the chance of heart disease?

We realize that you are probably not reading this book in hopes of becoming a statistician. (That's too bad, because there's a severe shortage of statisticians—more jobs than trained people. And with the ever-increasing ways in which statistics is being applied, it's an exciting time to be a statistician.) You may even suffer from math phobia and fear what lies ahead. Please be assured that to learn the main concepts of statistics, logical thinking and perseverance are more important than high-powered math skills. Don't be frustrated if learning comes slowly and you need to read about a topic a few times before it starts to make sense. Just as you would not expect to sit through a single foreign language class session and be able to speak that language fluently, the same is true with the language of statistics. It takes time and practice. But we promise that work will be rewarded. Once you have completed even part of this text, you will understand much better how to make sense of statistical information, and hence, the world around you.

1.1 | How Can You Investigate Using Data?

Does a low-carbohydrate diet result in significant weight loss? Are people more likely to stop at a Starbucks if they've seen a recent TV advertisement for their coffee? Information gathering is at the heart of investigating answers to such questions. The information we gather with experiments and with surveys is collectively called **data**.

For instance, consider an experiment designed to evaluate the effectiveness of a low-carbohydrate diet. The data might consist of measurements for the people participating in the study of weight at the beginning of the study, weight at the end of the study, number of calories of food eaten per day, carbohydrate intake per day, body-mass index (BMI) at the start of the study, and gender. A marketing survey about the effectiveness of a TV ad for Starbucks could collect data on the percentage of people who went to a Starbucks since the ad aired and analyze how it compares for those who saw the ad and those who did not see it.

What Is Statistics?

You already have a sense of what the word "statistics" means. You hear statistics quoted about sports events (number of points scored by each player on a basketball team), statistics about the economy (median income, unemployment rate), and statistics about opinions and beliefs and behaviors (percentage of students who indulge in binge drinking). In this sense, a statistic is merely a number calculated from data. But we'll also use "statistics" in a broader sense, to refer to a field of study that deals with methods for producing and analyzing data.

Statistics

Statistics is the art and science of designing studies and analyzing the data that those studies produce. Its ultimate goal is translating data into knowledge and understanding of the world around us. In short, **statistics is the art and science of learning from data**.

Statistical methods help us investigate questions in an objective manner. The following examples ask questions that we'll learn how to answer using statistical investigations.

Scenario 1: Predicting an Election Using an Exit Poll   In elections,
television networks often declare the winner well before all the votes have been counted. They do this using "exit polling," interviewing voters after they leave the voting booth. Using an exit poll, a network can often predict the winner after learning how several thousand people voted, out of possibly millions of voters.

In California in October 2003, a special election was held to determine whether Governor Gray Davis should be recalled from office. In the same election, voters were asked to vote for the candidate who would replace Davis as governor if he were recalled. The four candidates included the Hollywood actor Arnold Schwarzenegger. If a majority voted yes for the recall, the candidate who received the most votes (which was Schwarzenegger) would be the new governor. The exit poll on which TV networks relied for their projections found after sampling 3160 voters that 54% of the sample voted to recall Gray Davis and 46% voted not to recall him.[1] Was this enough evidence to predict whether the recall effort succeeded, even though information was available about the votes of only 3160 of the 8 million voters? We'll learn how to answer that question in this book.

Scenario 2: Making Conclusions in Medical Research Studies   Statistical
reasoning is at the foundation of the analyses conducted in most medical research studies. Let's consider three examples of how statistics can be relevant.

Does regular aspirin intake reduce deaths from heart attacks? The Harvard Medical School conducted a landmark study to investigate. The people participating in the study regularly took either an aspirin tablet or a placebo tablet (a pill with no active ingredient). Of those who took aspirin, 0.9% suffered heart attacks during the study. Of those who took the placebo, 1.7% had heart attacks, nearly twice as many. Can you conclude that it's beneficial for people to take aspirin regularly? Or, could the observed difference be explained by how it was decided which people would receive aspirin and which would receive placebo? For instance, might those who took aspirin have had better results merely because they were healthier, on the average, than those who took placebo? Or, did those taking aspirin have a better diet or did they exercise more regularly, on the average?

---

[1]www.cnn.com/ELECTION/2003/recall

For years there has been controversy about whether regular intake of large doses of vitamin C is beneficial. Some studies have suggested that it is. But some scientists have criticized those studies' designs, claiming that the subsequent statistical analysis was meaningless. How do we know when we can trust the results of the statistics in a medical study reported in the media?

Suppose that you wanted to investigate whether smokers are more likely than nonsmokers to get lung cancer. You could pick half of the students from your class and tell them to smoke a pack of cigarettes each day for the next 50 years, and tell the other half of the class not to smoke. Fifty years from now you could see whether more smokers than nonsmokers got lung cancer. Obviously such a study is not ethical. And who wants to wait 50 years to get the answer? Well, 50 years ago a statistician in Britain figured out how to study the effect of smoking on lung cancer using data that were already available. How did he do this?

This book will show you how to answer questions such as these. You'll learn to be skeptical of results from many medical studies.

Scenario 3: Using a Survey to Investigate Peoples' Beliefs How similar are your opinions and lifestyle to those of others? It's easy to find out. Every other year, the National Opinion Research Center at the University of Chicago conducts the General Social Survey (GSS). This survey of about 2000 adult Americans provides data about the opinions and behaviors of the American public. You can use it to investigate how adult Americans answer a wide diversity of questions, such as, "Do you believe in life after death?," "Would you be willing to pay higher prices in order to protect the environment?," "Do you think a preschool child is likely to suffer if his or her mother works?," "How much TV do you watch per day?," and "How many sexual partners have you had in the past year?" Similar surveys occur in other countries, such as the Eurobarometer survey for countries in the European Union. We'll use data from such surveys to illustrate the proper application of statistical methods.

SDA 1.4: Tables

GSS 1972-2002 Cumulative Datafile

Aug 26, 2005 (Fri 02:58 PM PDT)

Variables					
Role	Name	Label	Range	MD	Dataset
Row	postlife	BELIEF IN LIFE AFTER DEATH	1-2	0,8,9	1

Frequency Distribution

Cells contain: -Column percent -N of cases		Distribution
	1: YES	79.3 20,982
postlife	2: NO	20.7 5,477
	COL TOTAL	100.0 26,459

Results on the Web from the General Social Survey (GSS)

Why Use Statistical Methods?

The scenarios just presented illustrate the three main aspects of statistics:

- **Design**: Planning how to obtain data to answer the questions of interest
- **Description**: Summarizing the data that are obtained
- **Inference**: Making decisions and predictions based on the data

Design refers to planning how to obtain the data. How could you conduct an experiment, so that the results are trustworthy, to investigate whether regular large doses of vitamin C are beneficial? In a marketing survey, how do you select the people to survey so you'll get data that provide good predictions about future sales?

Description means exploring and summarizing patterns in the data. Files of raw data are often huge. For example, over time the General Social Survey has collected data about hundreds of characteristics on many thousands of people. Such raw data are not easy to assess—we simply get bogged down in numbers. It is more informative to use graphs or numbers that summarize the data, such as an average amount of TV watching or the percentage of people who believe in life after death.

Inference means making decisions or predictions based on the data. Usually the decision or prediction refers to a larger group of people, not merely those in the study. For instance, in the exit poll described in Scenario 1, of 3160 voters sampled, 54% said they voted to recall Gray Davis of California from the governor's job, whereas 46% said they voted against the recall. Using these data, we'll see we can predict that the recall would be successful out of the 8 million people who voted (opening the door for Arnold Schwarzenegger to become governor). Stating the percentages for the sample of 3160 voters is *description*, while predicting the outcome for all 8 million voters is *inference*.

Statistical description and inference are the ways of analyzing the data. You can use them to investigate and answer questions that are important to society. For instance, "Has there been global warming over the past decade?" "Is having the death penalty available for punishment associated with a reduction in violent crime?" "Does student performance in school depend on the amount of money spent per student, the size of the classes, or the teachers' salaries?"

Long before we get to the stage of analyzing data, we need to give careful thought to posing the questions to be answered by that analysis. The nature of those questions has an impact on all stages—design, description, and inference. For example, in an exit poll, do we just want to predict which candidate won, or do we want to investigate *why* they won, for instance analyzing how voters' opinions about certain issues related to how they voted? We'll learn how questions such as these and the ones posed in the previous paragraph can be phrased in terms of statistical summaries (such as percentages and means) so we can use data to investigate their answers.

Finally, a topic that we have not mentioned yet but that is fundamental for developing statistical inference methods is **probability**. We'll study probability because it will help us to answer a question such as, "If the recall vote were actually supported by less than half of all voters, what's the chance it would get 54% support in an exit poll of 3160 voters?" If the chance were extremely small, we'd feel comfortable making the inference that the recall was supported by the majority of all 8 million voters.

IN WORDS

In English, the verb **infer** means to arrive at a decision or prediction by reasoning from known evidence.

IN WORDS

We'll see in Activity 1 that the term **variable** refers to the characteristic being measured, such as number of hours per day that you watch TV.

ACTIVITY 1: **Downloading Data from the Internet**

It is simple to get descriptive summaries of data from the General Social Survey (GSS). We'll demonstrate, using one question it asked in recent surveys, "On a typical day, about how many hours do you personally watch television?"

■ Go the Web site http://sda.berkeley.edu/GSS

■ Click on *Most recent SDA dataset for the GSS.**

■ In the *Select an action* menu, place the dot at the *Frequencies and crosstabulation* option. Click on *Start*.

■ The GSS name for the number of hours of TV watching is TVHOURS. Type TVHOURS as the *Row variable* name. Click on *Run the table*.

Now you'll see a table that shows the number of people and the percentage who made each of the possible responses. For all the years combined in which this question was asked, the most common response was 2 hours of TV a day (about 27% made this response).

What percentage of the people surveyed reported watching 0 hours of TV a day? How many people reported watching TV 24 hours a day?

Another question asked in the GSS is, "Taken all together, would you say that you are very happy, pretty happy, or not too happy?" The GSS name for this item is HAPPY. What percentage of people reported being very happy?

You might use the GSS to investigate what sorts of people are more likely to be very happy. Those who are happily married? Those who are in good health? Those who have lots of friends? Those who like their job? Those who attend religious services frequently? We'll see how to find out in this book.

If this doesn't work, your computer's firewall settings may be restricting access.

TO PRACTICE ACTIVITY 1, TRY EXERCISES 1.3 AND 1.4

1.1 **Aspirin and heart attacks:** The Harvard Medical School study mentioned in Scenario 2 used about 22,000 male physicians. The method of deciding whether a given person would receive aspirin or placebo was essentially made by flipping a coin. About 11,000 were assigned to take aspirin, and about 11,000 to take the placebo. The researchers described results by the percentage of each group that had a heart attack during the study. This was 0.9% for those taking aspirin and 1.7% for those taking placebo. They then used statistical methods to predict that if *all* male physicians could have participated in this study, the percentage having a heart attack would have been lower for those taking aspirin. Specify the aspect of this study that refers to (a) design, (b) description, and (c) inference.

1.2 **Poverty and race:** The **Current Population Survey (CPS)** is a monthly survey of households conducted by the U.S. Census Bureau for the Bureau of Labor Statistics. It provides a comprehensive body of data on the labor force, employment, and unemployment. A CPS of about 60,000 households indicated that of those households, 8.0% of the whites, 23.4% of the blacks, and 22.7% of the Hispanics had annual income below the poverty level (*Statistical Abstract of the United States*, 2000). Using these data, a statistical method makes the prediction that the percentage of *all* black households in the United States that had income below the poverty level was at least 22% but no greater than 25%. Specify the aspect of this study that refers to (a) description and (b) inference.

1.3 **GSS and heaven:** Go to the General Social Survey Web site, http://sda.berkeley.edu/GSS. By entering HEAVEN as the "row variable" name, find the percentages of people who said "yes, definitely," "yes, probably," "no, probably not," and "no, definitely not" when asked whether they believed in heaven.

1.4 **GSS and heaven and hell:** Refer to the previous exercise. You can obtain data for a particular survey year such as 1998 by entering YEAR(1998) in the "selection filter" option box before you click on *Run the Table*.

 a. Do this for HEAVEN in 1998, giving the percentages for the four possible outcomes. (This question was asked only in 1991 and 1998.)

 b. Summarize opinions in 1998 about belief in hell (row variable HELL). Was the percentage of "yes, definitely" responses higher for belief in heaven or in hell?

1.5 **GSS for subject you pick:** At the GSS Web site, click on *Open Extra Codebook Window*, click on *Standard Codebook* and then on *Sequential Variable List*. Find a subject that interests you and look up a relevant GSS code name to enter as the "row variable." Summarize the results that you obtain.

1.2 | We Learn about Populations Using Samples

We've seen that statistics consists of methods for **designing** investigative studies, **describing** (summarizing) data obtained for those studies, and making **inferences** (decisions and predictions) based on those data.

We Observe Samples but We Are Interested in Populations

The entities that we measure in a study are called the **subjects.** Usually the subjects are people, such as the individuals interviewed in a General Social Survey. But they need not be. For instance, the subjects could be schools, countries, or days. We might measure characteristics such as the following:

■ For each school: the per-student expenditure, the average class size, the average score of students on an achievement test

■ For each country: the percentage of residents living in poverty, the birth rate, the percentage unemployed, the percentage who are computer literate

■ For each day in an Internet café: the amount spent on coffee, the amount spent on food, the amount spent on Internet access

The **population** is the set of all the subjects of interest. In practice, we usually have data for only *some* of the subjects who belong to that population. These subjects are called a **sample**.

Population and Sample

The **population** is the total set of subjects in which we are interested. A **sample** is the subset of the population for whom we have (or plan to have) data.

In the 2002 General Social Survey (GSS), the sample was the 2765 people who participated in this survey. The population was the set of all adult Americans at that time—more than 200 million people.

EXAMPLE 2

The Sample and the Population for an Exit Poll

Picture the Scenario

Scenario 1 in the previous section discussed an exit poll. The purpose was to predict the outcome of the 2003 special election in California to consider whether the governor Gray Davis should be recalled from office. The exit poll sampled 3160 of the 8 million people who voted.

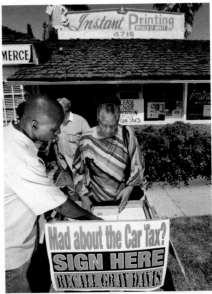

Question to Explore

For this exit poll, what was the population and what was the sample?

Think It Through

The population was the total set of subjects of interest, namely, the 8 million people who voted in this election. The sample was the 3160 voters who were interviewed in the exit poll. These are the people for whom the poll got data about their vote.

Insight

The ultimate goal of any study is to learn about the *population*. For example, the sponsors of this exit poll wanted to make an inference (prediction) about *all* 8 million voters, not just the 3160 voters sampled by the poll.

TO PRACTICE THIS CONCEPT, TRY EXERCISE 1.7 AND 1.8

Occasionally data are available from an entire population. For instance, every ten years the U.S. Census Bureau gathers data from the entire population of Americans (or nearly all of them). But the census is an exception. Usually, it is too costly and time consuming to obtain data from an entire population. It is more practical to get data for a sample. The General Social Survey and polling organizations such as the Gallup poll usually select samples of about 1000 to 2500 Americans to learn about opinions and beliefs of the population of *all* Americans. The same is true for surveys in other parts of the world, such as Eurobarometer in Europe.

Descriptive Statistics and Inferential Statistics

Using the distinction between samples and populations, we can now tell you more about the use of **description** and **inference** in statistical analyses.

> **Description in statistical analyses**
>
> **Descriptive statistics** refers to methods for summarizing the data. The summaries usually consist of graphs and numbers such as averages and percentages.

A descriptive statistical analysis usually combines graphical and numerical summaries. For instance, Figure 1.1 is a **bar graph** that shows the percentages of various types of households in the United States as of 2002. It summarizes a survey of 50,000 American households by the United States Census Bureau. The main purpose of descriptive statistics is to reduce the data to simple summaries without distorting or losing much information. Graphs and numbers such as percentages and averages are easier to comprehend than the entire set of data. It's much easier to get a sense of the data by looking at Figure 1.1 than by reading through the questionnaires filled out for the 50,000 sampled households. From this graph, it's readily apparent that the "traditional" household, defined as being a married man and woman with children in which only the husband is in the labor force, is no longer very common in the United States. In fact, "Other" households, which include female-headed households and households headed by young adults or older Americans who do not reside with spouses are by far the most common.

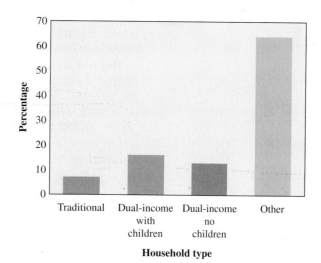

▲ **FIGURE 1.1: Types of U.S. households, based on sample of 50,000 households in the 2002 Current Population Survey.**

Although data are usually available only for a sample, descriptive statistics are also useful when data are available for the entire population, such as in a census. By contrast, inferential statistics are used when data are available only for a sample but we want to make a decision or prediction about the entire population.

Inference in statistical analyses

Inferential statistics refers to methods of making decisions or predictions about a population, based on data obtained from a sample of that population.

In most surveys, we have data for a sample, not for the entire population. We use descriptive statistics to summarize the sample data and inferential statistics to make predictions about the population.

EXAMPLE 3

Polling Opinion about Handgun Control

Picture the Scenario

Suppose we'd like to know what people think about controls over the sales of handguns. Let's consider how people feel who live in Florida, a state with a relatively high violent crime rate. The population of interest is the set of more than 10 million adult residents of Florida.

Since it is impossible to discuss the issue with all these people, we can study results from a recent poll of 834 residents of Florida conducted by the Institute for Public Opinion Research at Florida International University. In that poll, 54.0% of the sampled subjects said they favored controls over the sales of handguns. A newspaper article about the poll, using a method presented in this book, reports that the "margin of error" for how close this number falls to the population percentage is 3.4%. We'll see (in Chapter 7) that this means we can predict with high confidence (about 95% certainty) that the percentage of *all* adult Floridians favoring control over sales of handguns falls within 3.4% of the survey's value of 54.0%, that is, between 50.6% and 57.4%.

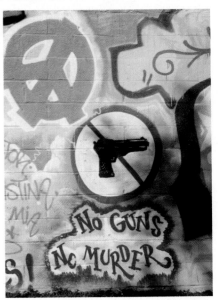

Question to Explore

In this analysis, what is the descriptive statistical analysis and what is the inferential statistical analysis?

Think It Through

The results for the sample of 834 Florida residents are summarized by the percentage, 54.0%, who favored handgun control. This is a descriptive statistical analysis. We're interested, however, not just in those 834 people but in the *population of all* adult Florida residents. The prediction that the percentage of *all* adult Floridians who favor handgun control falls between 50.6% and 57.4% is an inferential statistical analysis. In summary, we *describe* the *sample*, and we make *inferences* about the *population*.

Insight

The sample size of 834 was small compared to the population size of more than 10 million. However, because the values between 50.6% and 57.4% are all above 50%, the study concluded that a slim majority of Florida residents favored handgun control.

TO PRACTICE THIS CONCEPT, TRY EXERCISE 1.9 AND 1.10

An important aspect of statistical inference involves reporting the likely *precision* of a prediction. How close is the *sample* value of 54% likely to be to the true (unknown) percentage of the *population* favoring gun control? We'll see (in Chapters 4 and 6) why a well-designed sample of 834 people yields a sample percentage value that is very likely to fall within about 3 to 4% (the so-called *margin of error*) of the population value. In fact, we'll see that inferential statistical analyses can predict characteristics of entire populations quite well by selecting samples that are small relative to the population size. That's why most polls take samples of only about a thousand people, even if the population has millions of people. In this book, we'll see why this works.

Sample Statistics and Population Parameters

In Example 3, the percentage of the sample favoring handgun control is an example, of a **sample statistic**. We will regularly distinguish between sample statistics and the corresponding values for the population. The term **parameter** is used for a numerical summary of the population.

> **Recall**
>
> A **population** is the total group of individuals about whom you want to make conclusions. A **sample** is a subset of the population for whom you actually have data. ▪

Parameter and Statistic

A **parameter** is a numerical summary of the population.
A **statistic** is a numerical summary of a sample taken from the population.

For example, the percentage of the population of all adult Florida residents favoring handgun control is a parameter. Parameter values are usually unknown. The reason for taking a sample is to get sample statistics to estimate their values, because it's the population that we're interested in. Much of this text deals with ways of making inferences about unknown parameters using sample statistics.

Randomness and Variability

A sample tends to be a good reflection of a population when each subject in the population has the same chance of being included in that sample. That's the basis of **random sampling**. A simple example of random sampling is when a teacher puts each student's name on a slip of paper and places it in a hat and then draws names from the hat without looking.

- We'll see that random sampling allows us to make powerful inferences about populations.

- Randomness is also crucial to performing experiments well.

If, as in Scenario 2, we want to compare aspirin to placebo in terms of the percentage of people who later have a heart attack, it's best to randomly select who in the sample uses aspirin and who uses placebo. This tends to keep the groups balanced on the factors that could affect the results. Without randomizing, for example, the people who decided to use placebo might have tended to be healthier than those who picked aspirin, which could produce misleading results.

People are different from each other, so, not surprisingly, the measurements we make on them *vary* from person to person. For the GSS question about TV watching in Activity 1, different people reported different amounts of TV watching. In the exit poll of Example 1, not all people voted the same way. If subjects did not vary, we'd need to sample only one of them (and it would be a rather uninteresting world). We learn more about this variability by sampling more people. If we want to predict the outcome of an election, we're better off sampling 100 voters than one voter, and our prediction will be even more reliable if we sampled 1000 voters.

■ Just as people vary, so do samples vary.

Suppose you take an exit poll of 1000 voters to predict the outcome of an election. Suppose the Gallup organization also takes an exit poll of 1000 voters. Your sample will have different people than Gallup's. Consequently, the predictions will also differ. Perhaps your exit poll of 1000 voters has 480 voting for the Republican candidate, so you predict that 48% of all voters voted for that person. Perhaps their exit poll of 1000 voters has 440 voting for the Republican candidate, so they predict that 44% of all voters voted for that person. Activity 2 at the end of the chapter shows, however, that with random sampling the amount of variability from sample to sample is quite predictable. Both of your predictions are very likely to fall within 5% of the actual population percentage who voted Republican.

The Basic Ideas of Statistics

Now that you know the purpose of descriptive statistics and inferential statistics, we can tell you about the key concepts of statistics that you'll learn in this book:

Chapter 2: Exploring Data with Graphs and Numerical Summaries How can you present simple summaries of data? You replace lots of numbers with simple graphs and numerical summaries.

Chapter 3: Association: Contingency, Correlation, and Regression How does annual income ten years after graduation correlate with college GPA? You can find out by studying the **association** between those characteristics.

Chapter 4: Gathering Data How can you design an experiment or conduct a survey to get data to help you answer questions? You'll see why results may be misleading if you don't use **randomization**.

Chapter 5: Probability in Our Daily Lives How can you determine the chance of some outcome, such as your winning a lottery? Probability, the basic tool for evaluating chances, is also the key to how well inferential statistics work.

Chapter 6: Probability Distributions You've probably heard of the **normal distribution** or "bell-shaped curve" that describes peoples' heights or IQs or test scores. Why does it occur so often? You'll see why, and you'll learn its key role in statistical inference.

Chapter 7: Statistical Inference: Confidence Intervals How can an exit poll of 2232 voters possibly predict well the results for millions of voters? You'll find out by applying the probability concepts of Chapters 5 and 6 to make **statistical inferences** that show how closely you can predict summaries such as population percentages.

Chapter 8: Statistical Inference: Significance Tests About Hypotheses How can a medical study make a decision about whether a new drug is better than a placebo? You'll see how you can control the chance that a statistical inference makes a correct decision about what works best.

Chapters 9–14: Applying Descriptive and Inferential Statistics to Many Kinds of Data After Chapters 2–8 introduce you to the key concepts of statistics, the rest of the book shows you how to apply them in lots of situations. For instance, Chapter 9 shows how to compare two groups, such as using a sample of students from your university to make an inference about whether male and female students have different rates of binge drinking.

1.2 Practicing the Basics

1.6 Description and inference:

a. Distinguish between *description* and *inference* as reasons for using statistics. Illustrate the distinction using an example.

b. You have data for a population, such as obtained in a census. Explain why descriptive statistics are helpful but inferential statistics are not needed.

1.7 EPA: The Environmental Protection Agency (EPA) uses a few new automobiles of each model every year to collect data on pollution emission and gasoline mileage performance. For the Honda Accord model, identify what's meant by the (a) subject, (b) sample, and (c) population.

1.8 Aspirin inference: The medical study mentioned in Scenario 2 and in Exercise 1.1 conducted an experiment with 22,000 male physicians, to investigate whether regular intake of aspirin reduces the chance of heart attack compared to taking placebo. The study concluded that for *all* male physicians, aspirin would be more effective than placebo. Identify (a) the sample, (b) the population, and (c) the inference.

1.9 Graduating seniors' salaries: The job placement center at your school surveys *all* graduating seniors at the school. Their report about the survey provides numerical summaries such as the average starting salary and the percentage of students earning more than $30,000 a year.

a. Are these descriptive, or inferential, statistical analyses? Explain.

b. Are these numerical summaries better characterized as statistics, or as parameters?

1.10 At what age did women marry?: A historian wants to estimate the average age at marriage for women in New England in the early nineteenth century. Within her state archives she finds marriage records for the years 1800–1820. She takes a sample of those records, noting the age of the bride for each. The average age in the sample is 24.1 years. Using a statistical method,

she finds a margin of error and estimates that the average age of brides for the population was between 23.5 and 24.7.

a. What part of this example is descriptive, giving a summary of the data?

b. What part of this example is inferential, making a prediction about a population?

c. To what population does the inference refer?

d. Is the average age in the sample of 24.1 a statistic, or a parameter?

1.11 Age pyramids as descriptive statistics: The figure shown is a graph published by Statistics Sweden. It compares Swedish society in 1750 and in 2000 on the numbers of men and women of various ages, using "age pyramids." Explain how this indicates that

a. In 1750, few Swedish people were old.

b. In 2000, Sweden had many more people than in 1750.

c. In 2000, of those who were very old, more were female than male.

d. In 2000, the largest five-year group included "baby boomers" born right after World War II.

Graphs of number of men and women of various ages, in 1750 and in 2000.

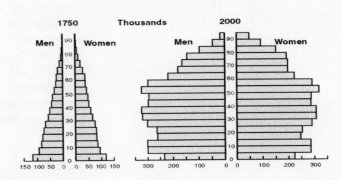

1.12 **Gallup polls:** Go to the Web site for the Gallup poll, www.gallup.com. From reports listed on their homepage, give an example of (a) a descriptive statistical analysis and (b) an inferential statistical analysis.

1.13 **National service:** Consider the population of all students at your school. A certain proportion support mandatory national service (MNS) following high school. Your friend randomly samples 20 students from the school, and uses the sample proportion who support MNS to predict the population proportion at the school. You take your own, separate random sample of 20 students, and find the sample proportion that supports MNS.

  a. For the two studies, are the populations the same?

  b. For the two studies, are the sample proportions necessarily the same? Explain.

1.14 **Samples vary less with more data:** We'll see that the amount by which results vary from sample to sample depends on the sample size. In an election with two candidates, Smith and Jones, suppose Smith receives close to 50% of all votes. An exit poll is taken on the day of the election to predict the winner.

  a. Which case would you find more surprising—taking an exit poll of 10 voters and finding that 0% or 100% of them voted for Smith, or taking an exit poll of 1000 voters and finding that 0% or 100% of them voted for Smith?

  b. In exit polls, samples of size 1000 voters each will tend to give more similar percentages voting for a particular candidate than samples of 10 voters each. Using part (a), explain why you would expect this to be true.

1.3 | What Role do Computers Play in Statistics?

Today's researchers (and students) are lucky: Unlike those in the previous generation, they don't have to do complex statistical calculations by hand. Powerful user-friendly computing software and calculators are now readily available for statistical analyses. New statistical methods continue to be developed all the time, and many modern methods are too complex to do by hand.

Using (and Misusing) Statistics Software and Calculators

MINITAB and **SPSS** are two popular statistical software packages on college campuses. The **TI-83+** and **TI-84**[2] graphing calculators, which have similar output, are useful as portable tools for generating simple statistics and graphs. The **Microsoft Excel** software can conduct some statistical methods, sorting and analyzing data with its spreadsheet program, but its capabilities are limited. Throughout this text and in exercises, we'll show examples of MINITAB and TI-83-84 output. Occasional exercises use SPSS or Excel. In your course, you'll probably use only one (or perhaps none) of these. But you won't have to know *how* to use the software to learn *what to look for* and *how to interpret* the output. You'll become accustomed to the kinds of output you'll see in the future if you use such products.

IN PRACTICE: You, not technology, must select valid analyses.

Given the current software capabilities, in practice why do we still have to learn about statistical methods? Can't computers do all this analysis for you? The problem is that a computer will perform the statistical analysis you request whether or not its use is valid for the given situation. Just knowing how to use software does not guarantee a proper analysis. You'll need a good background in statistics to understand which statistical method to use, which options to choose with that method, and how to interpret and make valid conclusions from the computer output. This text helps to give you this background.

---

[2]We will use the shorthand, TI-83+/84, to indicate output from a graphing calculator.

Data Files

To make statistical analysis easier, large sets of data are organized in a **data file**. This file usually has the form of a spreadsheet. It is the way statistical software expects to receive the data.

Figure 1.2 is an example of part of a data file. It shows how a data file looks in **MINITAB**. The file shows data for eight students on the following characteristics:

Gender (f = female, m = male)

Racial-ethnic group (b = black, h = Hispanic, w = white)

Age (in years)

College GPA (scale 0 to 4)

Average number of hours per week watching TV

Whether a vegetarian (yes, no)

Political party (dem = Democrat, rep = Republican, ind = independent)

Marital status (1 = married, 0 = unmarried).

Figure 1.2 shows the two basic rules for constructing a data file:

- Any one row contains measurements for a particular subject (for instance, person).

- Any one column contains measurements for a particular characteristic.

Some characteristics have numerical data, such as the values for hours of TV watching. Some characteristics have data that consist of categories or labels, such as the categories (yes, no) for whether a vegetarian or the labels (dem, rep, ind) for the categories of political affiliation.

Figure 1.2 resembles a larger data file from a questionnaire administered to a sample of students at the University of Florida. That data file, which includes other characteristics as well, is the one called "FL student survey" on the CD that comes with this text. (You may want to find this file on the CD now, to practice accessing data files used in some homework exercises.)

To construct a similar data file for your class, try Activity 3 in the Chapter Explorations at the end of the chapter.

▲ **FIGURE 1.2: Part of a MINITAB data file.**

EXAMPLE 4

Should Your Company Use Catalog Sales?

Picture the Scenario

You are the manager of a store that sells music CDs. Your sales have been shrinking because of competition from Internet companies (such as amazon.com) and the increasing practice of people downloading music from the Internet. You are considering trying to increase your sales by sending catalogs to potential customers. Because of the costs of sending the catalog, you would need to average at least $3 in orders per catalog mailed. To investigate whether you will be able to recover the costs and to find out which CDs are the most popular, you send a catalog to 500 potential customers. For the 10 CDs that you put in a special display on the cover of the catalog, the first person who was sent the catalog ordered nothing, and the second person ordered one copy of CD 3 and two copies of CD 5, at a cost of $15 each. After a month you create a data file summarizing the orders of the 500 potential customers.

Questions to Explore

a. How do you record the responses from the first two people on the data file for the 10 CDs on the cover display?

b. How many rows of data will your data file contain?

Think It Through

a. Each row refers to a particular person to whom you sent the catalog. Each column refers to a CD on the cover of the catalog. Each entry of data is the dollar amount spent on a given CD by a particular person. So, the first two rows of the data file, which give the results for the first two people, are

Person	cd1	cd2	cd3	cd4	cd5	cd6	cd7	cd8	cd9	cd10
1	0	0	0	0	0	0	0	0	0	0
2	0	0	15	0	30	0	0	0	0	0

b. The entire data file would consist of 500 lines of data, one line for each potential customer.

Insight

From the entire data file, you might determine that the average sales per person equaled $4. What could you then conclude about the average sales if you now send the catalog to tens of thousands of potential customers? Is it likely to be at least $3, so you'll break even? As we'll see (in Chapters 4 and 8), you would need other information from the data file and about the method of sampling to answer this question. If you decide to start using mail-order marketing, you'll want to see how statistics can help you use information that you gather about customers' demographics and past purchases to develop and maintain a promising list of potential customers.

TO PRACTICE THIS CONCEPT, TRY EXERCISE 1.16

Databases

Most studies design experiments or surveys to collect data to answer the questions of interest. Often, though, it is adequate to take advantage of existing archived collections of data files, called **databases**. Many databases are available on the Internet. By browsing various Internet addresses, you can obtain information on all sorts of things.

The General Social Survey, discussed in Scenario 3 and Activity 1, is one such database. The CD that comes with this text contains a database that has several data files that you'll use in some exercises. Here are some other databases that can be fun and informative to browse:

- Are you interested in pro baseball or basketball or football? Click on the statistics links at www.baseball.com or www.nba.com or www.nfl.com.

- Are you interested in what people believe around the world? Check out www.globalbarometer.org or www.europa.eu.int/comm/public_opinion or www.latinobarometro.org.

- Are you interested in results of Gallup polls about peoples' beliefs? See www.gallup.com.

Also very useful are search engines such as Google. Type in "data" in the search window at www.google.com, and it lists databases such as those maintained by the U.S. Census Bureau and Statistics Canada. A search engine also is helpful for finding data or descriptive statistics on a particular topic. For instance, if you want to find summaries of Canadians' opinions about the legalization of marijuana for medical treatment, try typing

poll legalize marijuana Canada medicine

in the search window.

IN PRACTICE: Always Check Sources

Not all databases or reported data summaries give reliable information. Before you give much credence to such data, verify that the data are from a trustworthy source and that the source provides information about the way the study was conducted.

For example, many news organizations and search Websites (such as cnn.com, abc.abcnews.go.com, netscape.com) ask you to participate in a "topic of the day" poll on their home page. Their summaries of how people responded are not reliable indications of how the entire population feels. They are biased, because it is unlikely that the people who respond are representative of the population. We'll explain why in Chapter 4.

Applets

Just like riding a bike, it's easier to learn statistics if you are actively involved, learning by doing. One way is to practice, by applying software or calculators to data files or data summaries that we'll provide. Another way is to perform activities that illustrate the ideas of statistics by using **applets**. Using an applet, you can take samples from artificial populations and analyze them to discover properties of statistical methods applied to those samples. This is a type of **simulation**—using a computer to mimic what would actually happen if you selected a sample and used statistics in real life. So, let's get started with your active involvement.

IN WORDS

Applet = *application* + et

An applet is a short *application* program for performing a specific task. We'll use them throughout the text.

ACTIVITY 2: Let's Use an Applet to Simulate Randomness and Variability

Let's get a feel for randomness and variability, the concepts introduced on page 13 of this chapter, that we'll see are key ones in the study of statistics. To do this, we will simulate taking an exit poll of voters. We will do this using the *sample from a population* applet on the text CD. This applet generates samples resembling those we'd get with random sampling. Select from the menu for the parent population, *binary*. Each observation from an individual in the population then has two possible outcomes, as in sampling a person who voted in an election with two candidates.

Let's see what would happen with exit polling when 50% of the entire population (a proportion of 0.50) voted for each candidate. We'll first use a small poll, only 10 voters. First, represent this by taking a coin and flipping it ten times. Let the number of heads represent the number who voted for the Democrat. What proportion in your sample of size 10 voted for the Democrat?

Now let's do this with the applet. We'll regard outcome 1 as voting for the Democratic candidate and outcome 0 as voting for the Republican candidate. Set the population proportion for outcome 1 at 0.50 by dragging the bars at 0 and 1. The applet automatically defaults to a population proportion = 0.5. To take a sample of size 10, you go to *the sample size* menu and select $n = 10$ and then click on *animate*. You will see that a certain number of outcomes occurred for each type. When we did this, we got outcome 1 four times and outcome 0 six times. This simulates sampling 10 voters in the exit poll, in which 4 said they voted for the Democratic candidate and 6

for the Republican candidate. It corresponds to a sample proportion of 0.40 voting for the Democrat.

To illustrate how samples vary, take your own sample of size 10 using this applet. When you do this, you will probably get a proportion different than 0.40 voting for the Democrat, because the process is random. What did you get? Now collect another sample of size 10 and find the sample proportion. What did you get? Repeat taking samples of size 10 at least 5 times. Note the sample proportion for each sample and how the sample proportions compare to the population proportion, 0.50. Are they always close?

Now repeat this for a larger exit poll, taking a sample of 1000 voters instead of 10 (set *sample n* = 1000 on the applet menu). What proportion voted for the Democrat? Repeat taking samples of size 1000 at least 5 times. Note the sample proportion for each sample and how these sample proportions compare to the population proportion, 0.50. Do the sample proportions tend to be close to 0.50? We predict that all of your sample proportions fell between 0.45 and 0.55. Are we correct?

We would expect that some sample proportions you generated using a sample size of 10 fell much farther from 0.50 than the sample proportions you generated using a sample size of 1000. This illustrates by simulation that sample proportions tend to be closer to the population proportion when the sample size is larger. In fact, we will discover as we move forward in the textbook that we do much better in making inferences about the population with larger sample sizes.

TO PRACTICE THIS ACTIVITY, TRY EXERCISES 1.20, 1.21, AND 1.33

1.3   Practicing the Basics

1.15 **Data file for friends:** Construct (by hand) a data file of the form of Figure 1.2, for two characteristics with a sample of four of your friends. One characteristic should take numerical values, and the other should take values that are categories.

1.16 **Shopping sales data file:** Construct a data file
TRY describing the purchasing behavior of five people, described below, who visit a shopping mall. Enter purchase amounts they spent on clothes, sporting goods, books, and music CDs as the data. Customer 1 spent $49 on clothes and $16 on music CDs, customer 4 spent $92 on books, and the other three customers did not buy anything.

1.17 **Internet poll:** The sample of people who register their opinion in a poll at an Internet site is not a random sample. Why not? (*Hint:* For a random sample, every person in the population has the same chance of being in the sample.)

1.18 **Create a data file with software:** Your instructor will show you how to create data files using the software for your course. Use it to create the data file you constructed by hand in Exercise 1.15 or 1.16. Print a copy of the data file.

1.19 **Use a data file with software:** You may need to learn how to open a data file from the text CD or download one from the Web for use with the software for your course. Do this for the "FL student survey" data file on the text CD, from the survey mentioned following Figure 1.2. Print a copy of the data file.

1.20 **Simulate with the *sample from a population* applet:**
TRY Refer to Activity 2 above.

   a. Repeat the activity using a population proportion 0.60: Take at least five samples of size 10 each, and observe how the sample proportions of the 1 outcome vary around 0.60, and then do the same thing with at least five samples of size 1000 each.

b. In (a), what seems to be the effect of the sample size on the amount by which sample proportions tend to vary around the population proportion, 0.60?

c. What is the practical implication of the effect of the sample size summarized in (b) with respect to making inferences about the population proportion when you collect data and observe only the sample proportion?

 1.21 **TRY** **Is a sample unusual?:** Suppose 70% of all voters voted for the Republican candidate. If an exit poll of 50 people were chosen randomly, would it be surprising if less than half that were sampled said they voted Republican? To reason an answer, use the applet described in Activity 2 to conduct at least ten simulations of taking samples of size 50 from a population with proportion 0.70. Note the sample proportions. Do you observe any sample proportion less than 0.50? What does this suggest?

CHAPTER REVIEW

CHAPTER SUMMARY

- Statistics consists of methods for conducting research studies and for analyzing and interpreting the data produced by those studies. **Statistics is the art and science of learning from data.**

- The first part of the statistical process involves **design**—planning an investigative study, such as how to obtain relevant data. The design often involves taking a **sample** from a **population** that contains *all* the **subjects** (usually, people) of interest. After we've collected the data, there are two types of statistical analyses:

  - **Descriptive statistics** summarize the sample data with numbers and graphs.

  - **Inferential statistics** make decisions and predictions about the entire population, based on the information in the sample data.

- With **random sampling**, each subject in the population has the same chance of being in the sample. This is desirable, because

then the sample tends to be a good reflection of the population. Randomization is also important to good experimental design.

- The measurements we make of a characteristic **vary** from individual to individual. Likewise, results of descriptive and inferential statistics **vary**, depending on the sample chosen. We'll see that the study of **variability** is a key part of statistics. **Simulation** investigations generate many samples randomly, often using an **applet**. They provide a way of learning about the impact of randomness and variability from sample to sample.

- The calculations for data analysis are simple using computer software. The data are organized in a **data file**. This has a separate row of data for each subject and a separate column for each characteristic. However, you'll need a good background in statistics to understand which statistical method to use and how to interpret and make valid conclusions from the computer output.

CHAPTER PROBLEMS

Practicing the Basics

1.22 **UW Student survey:** In a University of Wisconsin (UW) study about alcohol abuse among students, 100 of the 40,858 members of the student body in Madison were sampled and asked to complete a questionnaire. One question asked was, "On how many days in the past week did you consume at least one alcoholic drink?"

a. Identify the population and the sample.

b. For the 40,858 students at UW, one characteristic of interest was the percentage who would respond "zero" to this question. For the 100 students sampled, suppose 29% gave this response. Does this mean that 29% of the entire population of UW students would make this response? Explain.

c. Is the numerical summary of 29% a sample statistic, or a population parameter?

1.23 **ESP:** For several years, the General Social Survey asked subjects, "How often have you felt as though you were in touch with someone when they were far away from you?" Of 3887 sampled subjects who had an opinion, 1407 said never and 2480 said at least once. The proportion who had at least one such experience was 2480/3887 = 0.638.

a. Describe the population of interest.

b. Explain how the sample data are summarized using descriptive statistics.

c. For what population parameter might we want to make an inference?

1.24 **Presidential popularity:** Each month the Gallup organization conducts a poll for CNN and *USA Today* of the U.S. president's current popularity rating (see www.pollingreport.com). For a poll conducted March

21–23, 2005 of 1001 Americans, it was reported, "45% of people polled said they approve of how Bush is handling the presidency. The margin or error is plus or minus 3 percentage points." Explain how this provides an *inferential* statistical analysis.

1.25 Bush vs. Kerry in other countries: A BBC story (September 9, 2004) about a poll in 35 countries about whether people favored George W. Bush or John Kerry in the 2004 U.S. Presidential election stated that Kerry was clearly preferred in 30 countries and Bush was clearly preferred in 3 countries. For instance, the percentage breakdown for (Kerry, Bush) was reported as (74, 7) in Norway (with the other 19% undecided or not reporting), (74, 10) in Germany, (64, 5) in France, (58, 14) in Italy, (45, 7) in Spain, (47, 16) in the UK, (61, 16) in Canada, but (26, 31) in Poland and (32, 57) in the Philippines. The company that performed the polling (Globescan) reported, "The margin of error ranged from 2.3 to 5%."

a. Do these results summarize sample data, or population data?

b. Identify a descriptive aspect of this analysis.

c. Identify an inferential aspect of this analysis.

1.26 Reducing stress: Your school wants to make an inference about the percentage of students at the school who prefer having a several-day period between the end of classes and the start of final exams to help reduce the level of stress as students prepare for exams. A survey is taken of 100 students.

a. Identify the sample and the population.

b. For the study, explain the purpose of using (i) descriptive statistics, (ii) inferential statistics.

1.27 Marketing study: For the marketing study about CD sales in Example 4, identify the (a) sample and population, (b) descriptive and inferential aspects.

Select the correct response in the following two multiple-choice questions.

1.28 Believe in reincarnation?: In a survey of 750 Americans conducted by the Gallup organization, 24% indicated a belief in reincarnation. A method presented later in this book allows us to predict that for *all* adult Americans, the percentage believing in reincarnation falls between 21% and 27%. This prediction is an example of

a. descriptive statistics

b. inferential statistics

c. a data file

d. designing a study

1.29 Use of inferential statistics?: Inferential statistics are used

a. to describe whether a sample has more females or males.

b. to reduce a data file to easily understood summaries.

c. to make predictions about populations using sample data.

d. when we can't use statistical software to analyze data.

e. to predict the sample data we will get when we know the population.

1.30 True or false?: In a particular study, you could use descriptive statistics, or you could use inferential statistics, but you would rarely need to use both.

Concepts and Investigations

1.31 Statistics in the news: Pick up a recent issue of a national newspaper, such as *The New York Times* or *USA Today*. Identify an article that used statistical methods. Did it use descriptive statistics, or inferential statistics, or both? Explain.

1.32 What is statistics?: On a final exam that one of us recently gave, students were asked, "How would you define 'statistics' to someone who has never taken a statistics course?" One student wrote, "You want to know the answer to some question. There's no answer in the back of a book. You collect some data. Statistics is the body of procedures that helps you analyze the data to figure out the answer and how sure you can be about it." Pick a question that interests you, and explain how you might be able to use statistics to investigate the answer.

■ 1.33 TRY Surprising ESP data?: In Exercise 1.23, of 3887 sampled subjects, 63.8% said that at least once they felt as though they were in touch with someone when they were far away. Of 3887 sampled subjects who had an opinion, 1407 said never and 2480 said at least once. Suppose that only 20% of the entire population would report at least one such experience. If the sample of 3887 people were a random sample, would this sample proportion result of 0.638 be surprising? Investigate, using the applet described in Activity 2. Do this by simulating samples from the population with a proportion of 0.20. Use the sample size of 4000 as an approximation to the actual sample size of 3887.

■ 1.34 Create a data file: Using the statistical software that your instructor has assigned for the course, find out how to enter a data file. Create a data file using the data in Figure 1.2 and print a copy.

Class Explorations

■ 1.35 Activity 3: Your instructor will help the class create a data file consisting of the values for class members of characteristics based on responses to a questionnaire like the one on the following page. Alternatively, your instructor may ask you to use a data file of this type already prepared with a class of students at the

University of Florida, the "FL student survey" data file on the text CD. Using a spreadsheet program or the statistical software the instructor has chosen for your course, create a data file containing this information. Print the data. What are some questions you might ask about these data? Homework exercises in each chapter will use these data.

GETTING TO KNOW THE CLASS

Please answer the following questions. Do not put your name on this sheet. Skip any question that you feel uncomfortable answering. These data are being collected to learn more about you and your classmates and to form a database for the class to analyze.

1. What is your height (recorded in inches)? _____

2. What is your gender (M = Male, F = Female)? _____

3. How much did you spend on your last haircut? _____

4. Do you have a paying job during the school year at which you work on average at least 10 hours a week (y = yes, n = no)? _____

5. Aside from class time, how many hours a week, on average, do you expect to spend studying and completing assignments for this course? _____

6. Do you smoke cigarettes? (y = yes, n = no) _____

7. How many different people have you dated in the last 30 days? _____

8. What was (is) your high school GPA (based upon a 4.0 scale)? _____

9. What is your current college GPA? _____

10. What is the distance (in miles) between your current residence and this class? _____

11. How many minutes each day, on average, do you spend browsing the Internet? _____

12. How many minutes each day, on average, do you watch TV? _____

13. How many hours each week, on average, do you participate in sports or have other physical exercise? _____

14. How many times a week, on average, do you read a daily newspaper? _____

15. Do you consider yourself a vegetarian? (y = yes, n = no) _____

16. How would you rate yourself politically? (1 = very liberal, 2 = liberal, 3 = slightly liberal, 4 = moderate, 5 = slightly conservative, 6 = conservative, 7 = very conservative) _____

17. What is your political affiliation? (D = Democrat, R = Republican, I = Independent) _____

2 | Exploring Data with Graphs and Numerical Summaries

EXAMPLE 1

What Can Statistics Teach Us About Threats to Our Environment?

Picture the Scenario

As we've seen in Chapter 1, statistics plays a vital role in helping to inform us about issues in our daily lives.

One environmental issue of concern to many people is air pollution. The consumption of energy from fossil fuels affects air pollution. Most technologically advanced nations, such as the United States, use vast quantities of a diminishing supply of nonrenewable energy sources such as fossil fuels. Countries fast becoming technologically advanced, such as China and India, are greatly increasing their energy use. The result is increased emissions of carbon dioxide and other pollutants into the atmosphere. Emissions of carbon dioxide may also contribute to global warming, which is a potential danger to humankind in the future. Scientists use descriptive statistics to explore energy use, to learn whether or not air pollution has an effect on climate, to compare different countries in how they contribute to the worldwide problems of air pollution and climate change, and to measure the impact of global warming on the environment.

Questions to Explore

- How can we investigate which countries emit the highest amounts of carbon dioxide into the atmosphere or use the most non-renewable energy?
- Is climate change occurring, and how serious is it?

Thinking Ahead

In this chapter, we introduce **descriptive statistics** that help us to investigate questions like these. In Example 3 we'll use descriptive statistics to examine data on the dependence of the United States and Canada on fossil fuels for energy. Examples 11 and 18 explore carbon dioxide emissions for the world's largest nations and for nations in the European Union. Example 9 uses descriptive statistics to analyze some data about temperature change over time.

Chapter 1 distinguished between **descriptive statistics** (summarizing data) and **inferential statistics** (making a prediction or decision about a population, using sample data). Graphical and numerical summaries are the key elements of descriptive statistics.

Before we begin our journey into the art and science of analyzing data using descriptive statistics, we need to learn a few new terms. These pertain to the types of data we'll analyze. We'll then study ways of using graphs to describe data, followed by ways of summarizing the data numerically.

2.1 | What Are the Types of Data?

Statistical methods provide ways to measure and understand **variability**. For the characteristics observed in a study, variation occurs among the available subjects. For instance, there's variability among your schoolmates in weight, major, GPA, favorite sport, religious affiliation, and how long you spend studying today.

Variables

The characteristics observed to address the questions posed in a study are called **variables**.

> ### Variable
> A **variable** is any characteristic that is recorded for subjects in a study.

The terminology *variable* highlights that data values *vary*. For example, to investigate whether global warming has occurred where you live, you could gather and analyze data on the high temperature each day over the past century at the nearest weather station. The variable is the high temperature. Examples of other variables for each day are the low temperature, whether it rained that day, and the number of centimeters of precipitation.

Variables Can Be Quantitative (Numerical) or Categorical (in Categories)

The data values that we observe for a variable are referred to as the **observations**. Each observation can be numerical, such as number of centimeters of precipitation in a day. Or each observation can be a **category**, such as "yes" or "no" for whether it rained.

> ### Categorical and Quantitative Variables
> A variable is called **categorical** if each observation belongs to one of a set of categories.
> A variable is called **quantitative** if observations on it take numerical values that represent different magnitudes of the variable.

IN WORDS

Consider your classmates. From person to person, there is variability in age, GPA, major, and whether he or she is dating someone. So, these are **variables**. Age and grade point average (GPA) take numerical values, so are **quantitative**. Major and dating status are **categorical**, since their values are categories, such as (psychology, business, history) for major and (yes, no) for dating status.

For weather stations, for instance, the county in which a station is located and the daily observation of whether it rained (yes or no) are both categorical variables. The daily high temperature and the amount of precipitation are quantitative variables. For human subjects, examples of categorical variables include gender (with categories male and female), religious affiliation (with categories such as Catholic, Jewish, Muslim, Protestant, Other, None), place of residence (house, condominium, apartment, dormitory, other), and belief in life after death (yes, no). Other examples of quantitative variables are age, number of siblings, annual income, and number of years of education completed.

In the definition of a quantitative variable, why do we say that numerical values must *represent different magnitudes*? It's because we don't regard a variable such as telephone area code as quantitative, even though it is numerical. Quantitative variables measure "how much" there is of something (that is, *quantity* or *magnitude*). With them, you can find arithmetic summaries such as averages. Such summaries do not make sense for area codes, which are merely convenient numerical labels. Area codes do not differ in quantity, so a bank would not try to find the "average area code" of its customers.

Graphs and numerical summaries describe the main features of a variable:

- For **quantitative** variables, key features to describe are the **center** and the **spread** (variability) of the data. For instance, what's a typical annual amount of precipitation? Is there much variation from year to year?
- For **categorical** variables, a key feature to describe is the relative number of observations (for example, percentages) in the various categories.

Quantitative Variables Are Discrete or Continuous

For a quantitative variable, each value it can take is a number. Quantitative variables can be either **discrete** or **continuous**.

IN WORDS

A **discrete** variable is usually a count ("the number of ..."). A **continuous** variable has a continuum of infinitely many possible values (such as time, distance, or physical measurements such as weight and height).

Discrete and Continuous Variables

A quantitative variable is **discrete** if its possible values form a set of separate numbers, such as 0, 1, 2, 3, ...

A quantitative variable is **continuous** if its possible values form an interval.

Examples of discrete variables are the number of pets in a household, the number of children in a family, and the number of foreign languages in which a person is fluent. Any variable phrased as "the number of ..." is discrete. The possible values are separate numbers such as {0, 1, 2, 3, 4, ...}. The outcome of the variable is a count. *Any variable with a finite number of possible values is discrete.*

Examples of continuous variables are height, weight, age, and the amount of time it takes to complete an assignment. The collection of all the possible values of a continuous variable does not consist of a set of separate numbers, but rather an infinite region of values. The amount of time needed to complete an assignment, for example, could take the value 2.496631 ... hours. *Continuous variables have an infinite continuum of possible values.*

IN PRACTICE: Data Analysis Depends on Type of Variable

Why do we care whether a variable is *quantitative* or *categorical*, or whether a quantitative variable is *discrete* or *continuous*? We'll see that the method used to analyze a data set will depend on the type of variable the data represent.

Frequency Tables

The first step in numerically summarizing data about a variable is to look at the possible values and count how often each occurs. For a categorical variable, each observation falls in one of the categories. The category with the highest frequency is called the **mode**. We can use proportions or percentages to summarize the numbers of observations in the various categories.

EXAMPLE 2

Shark Attacks around the World

Picture the Scenario

A sensational news item of recent years has been shark attacks. In the United States, shark attacks most commonly occur in Florida. There were 289 reported shark attacks in Florida between 1990 and 2002. What regions of the world besides Florida have experienced shark attacks, and how many occur in each region? The International Shark Attack File (ISAF) provides data on shark attacks.

Table 2.1 classifies 735 shark attacks reported from 1990 through 2002, listing countries where attacks occurred as well as the three states in the United States where shark attacks were most common. The number of reported shark attacks for a particular region is a **frequency**, or count. The proportion is found by dividing the frequency by the total count of 735. The percentage equals the proportion multiplied by 100.

TABLE 2.1: Frequency of Shark Attacks in Various Regions for 1990–2002

Region	Frequency	Proportion	Percentage
Florida	289	0.393	39.3
Hawaii	44	0.060	6.0
California	34	0.046	4.6
Australia	44	0.060	6.0
Brazil	55	0.075	7.5
South Africa	64	0.087	8.7
Reunion Island	12	0.016	1.6
New Zealand	17	0.023	2.3
Japan	10	0.014	1.4
Hong Kong	6	0.008	0.8
Other	160	0.218	21.8
Total	**735**	**1.00**	**100**

Source: http://shark-gallery.netfirms.com/attack/statsw.htm

Questions to Explore

a. What is the variable? Is it categorical or quantitative?

b. Show how to find the proportion and percentage reported in Table 2.1 for Florida.

c. Identify the mode for these data.

Think It Through

a. Each observation (a shark attack) identifies a region for the attack. Region is the variable. It is categorical, with categories shown in the first column of Table 2.1.

b. Of the 735 reported shark attacks, 289 were in Florida. This is a proportion of 289/735 = 0.393, nearly 4 of every 10 attacks. The percentage is 100(0.393) = 39.3%.

c. For the regions listed, the greatest number of attacks was reported in Florida, with 39.3% of the reported worldwide attacks. Florida is the mode, because it has the greatest frequency of reported shark attacks.

Insight

Don't mistake the frequencies as values of a variable. They're merely a summary of how many times shark attacks occurred in each region. For each observation (shark attack), the variable summarized here is the region in which it took place.

In tables that summarize frequencies, *the total proportion is 1.0 and the total percentage is 100%*. In practice, the separate values may sum to a slightly different number (such as 99.9%) because of rounding.

TO PRACTICE THIS CONCEPT, TRY EXERCISE 2.8

Here's a summary of the descriptive statistics we found in Example 2:

Proportion and Percentage (Relative Frequencies)

The **proportion** of the observations that fall in a certain category is the frequency (count) of observations in that category divided by the total number of observations. The **percentage** is the proportion multiplied by 100. Proportions and percentages are also called **relative frequencies**.

A table like Table 2.1 that lists the possible values of a variable and their frequencies and/or relative frequencies is called a **frequency table**.

Frequency Table

A **frequency table** is a listing of possible values for a variable, together with the number of observations for each value.

For a quantitative variable, a frequency table usually divides the possible values into a set of intervals and displays the number of observations in each interval. For example, in the next section we'll analyze General Social Survey data on responses to the question, "On the average day, about how many hours do you personally watch TV?" This was measured as a discrete variable, with values 0, 1, 2, and so on. The table shown in the margin is one way of constructing a frequency table for the 905 subjects who responded. A larger table could list each distinct value for the variable.

Frequency Table: Daily TV watching

No. hours	Frequency	Percent
0–1	232	25.6
2–3	403	44.5
4–5	181	20.0
6–7	45	5.0
8 or more	44	4.9
Total	**905**	**100.0**

2.1 Practicing the Basics

2.1 Categorical/quantitative difference:

   a. Explain the difference between categorical and quantitative variables.

   b. Give an example of each.

2.2 U.S. married-couple households: According to the 2002 Current Population Survey of U.S. married-couple households, 7% are traditional (with children and with only the husband in the labor force), 16% are dual-income with children, 13% are dual-income with no children, and 64% are other (such as older married couples whose children no longer reside in the household). Is the variable "household type" categorical or quantitative? Explain.

2.3 Identify the variable type: Identify each of the following variables as categorical or quantitative.

   a. Number of pets in family

   b. County of residence

   c. Choice of auto to buy (domestic or import)

   d. Distance (in kilometers) of commute to work

2.4 Categorical or quantitative?: Identify each of the following variables as either categorical or quantitative.

   a. Choice of diet (vegetarian, nonvegetarian)

   b. Time spent in previous month attending place of religious worship

   c. Ownership of a personal computer (yes, no)

   d. Number of people you have known who have been elected to a political office

2.5 Discrete/continuous:

   a. Explain the difference between a discrete variable and a continuous variable.

   b. Give an example of each type.

2.6 Discrete or continuous?: Identify each of the following variables as continuous or discrete.

   a. The length of time to run a marathon

   b. The number of people in line at a box office to purchase theater tickets

   c. The weight of a dog

   d. The number of people you have dated in the past month

2.7 **Discrete or continuous 2:** Repeat the previous exercise for the following:

a. The total playing time of a CD

b. The number of courses for which a student has received credit

c. The amount of money in your pocket (*Hint:* You could regard a number such as $12.75 as 1275 in terms of "the number of cents.")

d. The distance between where you live and your statistics classroom, when you measure it precisely with values such as 0.5 miles, 2.4 miles, 5.38 miles.

2.8 **Number of children:** In the 2002 General Social Survey, respondents answered the question, "How many children have you ever had?" The results were

No. children	0	1	2	3	4	5	6	7	8+
Count		799	469	657	481	185	73	40	22	34

a. Is the variable, number of children, categorical or quantitative?

b. Is the variable, number of children, discrete or continuous?

c. Add proportions and percentages to this frequency table.

d. Which response is the mode?

2.2 | How Can We Describe Data Using Graphical Summaries?

Looking at a graph often gives you more of a feel for a data set than looking at the raw data or a frequency table. In this section, we'll learn about graphs for categorical variables and then graphs for quantitative variables. We'll find out what we should look for in a graph to help us understand the data better.

Graphs for Categorical Variables

The two primary graphical displays for summarizing a categorical variable are the **pie chart** and the **bar graph**.

- A **pie chart** is a circle having a "slice of the pie" for each category. The size of a slice corresponds to the percentage of observations in the category.

- A **bar graph** displays a vertical bar for each category. The height of the bar is the percentage of observations in the category.

EXAMPLE 3

How Much Electricity Comes from Renewable Energy Sources?

Picture the Scenario

On August 14, 2003, a major power outage hit the northeastern United States and parts of Canada. This major blackout brought attention to our dependence on energy and to the diminishing stockpile of natural resources. As the worldwide demand for energy increases, conservation and alternative sources of energy become more important. The so-called *renewable* energy sources produce electricity without depleting natural resources.

Table 2.2 shows sources of electricity and their use in the United States and Canada. Coal and natural gas are significant fossil fuel contributors to the emission of carbon dioxide, which is a form of air pollution and may lead to climate change.

Hydropower is a renewable source. The "other" sources refer to other renewable sources. These include solar power and wind power, which have received attention as energy sources that are less environmentally damaging.

TABLE 2.2: Sources of Electricity in the United States and Canada, and the Percentage of Electricity Use Generated by Each

Source	U.S. Percentage	Canada Percentage
Coal	51	16
Hydropower	6	65
Natural gas	16	1
Nuclear	21	16
Petroleum	3	1
Other	3	1
Total	**100**	**100**

Source: U.S. Department of Energy, Natural Resources Canada

Questions to Explore

a. Portray the information in Table 2.2 for the United States with a pie chart and a bar graph.

b. What percentage of electricity comes from renewable energy sources (hydropower or other) in the United States and in Canada?

Think It Through

a. Source of electricity is a categorical variable. Figure 2.1 is a pie chart of the U.S. data in Table 2.2. Sources of electricity with larger percentages have larger slices of the pie. The percentages are included in the labels for each slice of the pie.

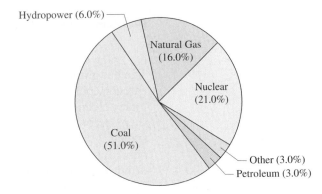

Percentage Use for Sources of Electricity

▲ **FIGURE 2.1: Pie chart of sources of electricity in the United States.** The label for each slice of the pie gives the category and the percentage of electricity generated by that source. The slice that represents the percentage generated by hydropower is 6% of the total area of the pie. **Question:** Why is it beneficial to label the pie wedges with the percent? (**Hint:** Is it always clear which of two slices is larger and what percent a slice represents?)

Figure 2.2 shows the bar graph. The categories with larger percentages have higher bars. The scale for the percentages is shown on a vertical axis. The width is the same for each bar.

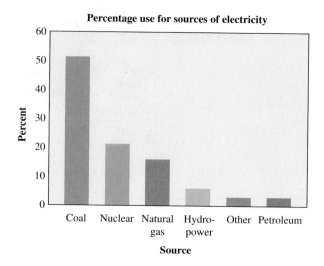

Percentage use for sources of electricity

▲ **FIGURE 2.2: Bar graph of sources of electricity in the United States.** The bars are ordered here from largest to smallest based on the percentage use.

b. In the United States, the renewable sources provide only about 9% (6 + 3 = 9) of the electricity. From either graph it is easy to see that the top source (the mode) is coal. By contrast, in Canada, Table 2.2 tells us that renewable sources provide about 66% of the electricity, and the mode is hydropower.

Insight

The pie chart and the bar graph both are simple to construct using software. The bar graph is more precise and more flexible. With a pie chart, when two slices are about the same size, it's often unclear which is larger. This is clearer from comparing heights of bars in a bar graph. We'll see that the bar graph can easily summarize how results compare for different groups (for instance, if we wanted to compare the United States and Canada for Table 2.2).

TO PRACTICE THIS CONCEPT, TRY EXERCISE 2.9.

The bar graph in Figure 2.2 displayed the categories in decreasing order of the category percentages. This order made it easy to separate the categories with high percentages visually. In some applications, it is more natural to display them according to their alphabetical order or some other criterion. For instance, if the categories have a natural order, such as if we're summarizing the percentages of grades (A, B, C, D, F) for students in a course, we'd use that order in listing the categories on the graph.

Pareto Charts

Figure 2.2 is a special type of bar graph called a **Pareto chart**. Named after an Italian economist (Vilfredo Pareto, 1848–1923) who advocated its use, it is a bar graph with categories ordered by their frequency, from the tallest bar to the shortest bar. The Pareto chart is often used in business applications to identify categories that warrant the most attention. The chart helps to portray the **Pareto principle**, which states that a small subset of categories often contains most of the observations. For example, Figure 2.2 shows that three categories (coal, nuclear, and natural gas) were responsible for about 88% of U.S. electricity generation.

Graphs for Quantitative Variables

Now let's explore how to summarize *quantitative* variables graphically. We'll look at three types of displays—the dot plot, stem-and-leaf plot, and histogram. We'll illustrate the graphs by analyzing data for a food product that many of us eat every day, as we continue our discussion of issues that influence our well-being.

Dot Plots

A **dot plot** shows a dot for each observation, placed just above the value on the number line for that observation. To construct a dot plot,

■ Draw a horizontal line. Label it with the name of the variable, and mark regular values of the variable on it.

■ For each observation, place a dot above its value on the number line.

Exploring the Health Value of Cereals: Dot Plot

Did You Know?

It is recommended that daily consumption should not exceed 2400 milligrams (mg) for sodium and (on a 2000-calorie-a-day diet) 50 grams for sugar. ■

Picture the Scenario

Obesity, high blood pressure, high cholesterol, and heart disease are partially caused by a poor diet. Nutritional labels on packaged foods give us information about the amount of fat, cholesterol, sodium, vitamins, and carbohydrates contained in a serving of the food. Let's investigate the amount of sugar and salt in breakfast cereals. Too much sugar contributes to weight gain. Too much sodium (the main ingredient of salt) contributes to high blood pressure.

How much sugar and how much sodium are in breakfast cereals? Table 2.3 lists 20 popular cereals and the amounts of sodium and sugar contained in a single serving (usually $\frac{3}{4}$ cup), according to what's listed on the cereal boxes. The sodium and the sugar amounts are both quantitative variables. The variables are continuous, because they measure amounts that can take any positive real number value. They are measured in this table by rounding (for instance, sugar measured to the nearest number of grams).

TABLE 2.3: Sodium and Sugar Amounts in 20 Breakfast Cereals.

The amounts refer to a single serving size. A third variable, Type, classifies the cereal as intended for adults (Type A) or children (Type C).

Cereal	Sodium (mg)	Sugar(g)	Type
Frosted Mini Wheats	0	7	A
Raisin Bran	210	12	A
All Bran	260	5	A
Apple Jacks	125	14	C
Capt Crunch	220	12	C
Cheerios	290	1	C
Cinnamon Toast	210	13	C
Crackling Oat Bran	140	10	A
Crispix	220	3	A
Frosted Flakes	200	11	C
Fruit Loops	125	13	C
Grape Nuts	170	3	A
Honey Nut Cheerios	250	10	C
Life	150	6	A
Oatmeal Raisin Crisp	170	10	A
Sugar Smacks	70	15	C
Special K	230	3	A
Wheaties	200	3	C
Corn Flakes	290	2	A
Honeycomb	180	11	C

Source: http://lib.stat.cmu.edu/DASL/Datafiles/Cereals.html

Questions to Explore

a. Construct a dot plot for the sodium values of the 20 breakfast cereals. (We'll consider sugar amounts in the exercises.)

b. What does the dot plot tell us about the data?

Think It Through

a. Figure 2.3 shows a dot plot. Each cereal sodium value is represented with a dot above the number line. For instance, the labeled dot above 0 represents the sodium value of 0 mg for Frosted Mini Wheats.

b. The dot plot gives us an overview of all the data. We see clearly that the sodium values fall between 0 and 290 mg, with most cereals falling between 125 and 250 mg.

▲ **FIGURE 2.3: Dot Plot for Sodium Content of 20 Breakfast Cereals.** The sodium value for each cereal is represented with a dot above the number line. **Question:** What does it mean when more than one dot appears above a value?

Insight

The dot plot portrays the individual observations. The number of dots above a value on the number line represents the frequency of occurrence of that value. From a dot plot, we'd be able to reconstruct (at least approximately) all the data in the sample.

TO PRACTICE THIS CONCEPT, TRY EXERCISE 2.14 AND 2.16 PART (B)

Stem-and-Leaf Plots

Another type of graph, called a **stem-and-leaf plot**, is similar to the dot plot in that it portrays the individual observations.

- Each observation is represented by a **stem** and a **leaf**. Usually the stem consists of all the digits except for the final one, which is the leaf.

- Place the stems in a column, starting with the smallest. Place a vertical line to their right. On the right side of the vertical line, indicate each leaf (final digit) that has a particular stem. List the leaves in increasing order.

Stems	Leaves
0	07
1	2245778
2	00112235699

Observation = 26, in
a sample of size 20

EXAMPLE 5

Exploring the Health Value of Cereals: Stem-and-Leaf Plot

Picture the Scenario

Let's re-examine the sodium values for the 20 breakfast cereals, shown again in the margin.

Cereal	Sodium
Frosted Mini Wheats	0
Raisin Bran	210
All Bran	260
Apple Jacks	125
Capt Crunch	220
Cheerios	290
Cinnamon Toast	210
Crackling Oat Bran	140
Crispix	220
Frosted Flakes	200
Fruit Loops	125
Grape Nuts	170
Honey Nut Cheerios	250
Life	150
Oatmeal Raisin Crisp	170
Sugar Smacks	70
Special K	230
Wheaties	200
Corn Flakes	290
Honeycomb	180

Questions to Explore

a. Construct a stem-and-leaf plot of the 20 sodium values.

b. How does the stem-and-leaf plot compare to the dot plot?

Think It Through

a. In the stem-and-leaf plot, we'll let the final digit of a sodium value form the leaf and the other digits form the stem. For instance, the sodium value for Sugar Smacks is 70. The stem is 7 and the leaf is 0. Each stem is placed to the left of a vertical bar. Each leaf is placed to the right of the bar. Figure 2.4 shows the plot. Notice that a leaf has only one digit, but a stem can have one or more digits.

   The Sugar Smacks observation of 70 is labeled on the graph. Two observations have a stem of 12 and a leaf of 5. These are the sodium values of 125 for Fruit Loops and Applejacks.

b. The stem-and-leaf plot looks like the dot plot turned on its side, with the leaves taking the place of the dots. In summary, we get the same information from a stem-and-leaf plot as from a dot plot.

Insight

A stem has no leaf if there is no observation at that value. These are the values at which the dot plot has no dots.

Stems	Leaves
0	0
1	
2	
3	
4	
5	
6	
7	0
8	
9	
10	
11	
12	55
13	
14	0
15	0
16	
17	00
18	0
19	
20	00
21	00
22	00
23	0
24	
25	0
26	0
27	
28	
29	00

> This data point is the Sugar Smacks sodium value, 70. The stem is 7 and the leaf is 0.

> These two leaf values are for Frosted Flakes and Wheaties. Each has 200 mg of sodium per serving. The stem is 20 and each leaf is 0.

▲ **FIGURE 2.4: Stem-and-leaf plot for cereal sodium values.** The final digit of a sodium value forms the leaf, and the other digits form the stem. **Question:** Why do some stems not have a leaf?

Truncated Data	
Frosted Mini Wheats	0
Raisin Bran	21
All Bran	26
Apple Jacks	12
Capt Crunch	22
Cheerios	29
Cinnamon Toast	21
Crackling Oat Bran	14
Crispix	22
Frosted Flakes	20
Fruit Loops	12
Grape Nuts	17
Honey Nut Cheerios	25
Life	15
Oatmeal Raisin Crisp	17
Sugar Smacks	7
Special K	23
Wheaties	20
Corn Flakes	29
Honeycomb	18

TO PRACTICE THIS CONCEPT, TRY EXERCISE 2.15

To make a stem-and-leaf plot more compact, we can **truncate** the data values: Remove the last digit, as shown in the margin, and plot the data as 0, 21, 26, 12, 22, and so on, instead of 0, 210, 260, 125, 220, ... Arranging the leaves in increasing order on each line, we then get the stem-and-leaf plot

$$0 \mid 07$$
$$1 \mid 2245778$$
$$2 \mid 00112235699$$

This is a bit *too* compact, as it does not portray where the data fall as clearly as Figure 2.4 or the dot plot. We could instead list each stem twice, putting leaves from 0 to 4 on the first stem and from 5 to 9 on the second stem. We then get

$$0 \mid 0$$
$$0 \mid 7$$
$$1 \mid 224$$
$$1 \mid 5778$$
$$2 \mid 0011223$$
$$2 \mid 5699$$

Like the dot plot, this gives us the sense that most sodium values are relatively high, with a couple of cereals being considerably lower than the others.

Histograms

With a dot plot or a stem-and-leaf plot, it's easy to reconstruct the original data set, because the plot shows the individual observations. This becomes unwieldy for large data sets. A more versatile way to graph the data, useful also for very large data sets, uses bars to summarize frequencies of outcomes.

Histogram

A **histogram** is a graph that uses bars to portray the frequencies or the relative frequencies of the possible outcomes for a quantitative variable.

EXAMPLE 6

Histogram of TV Watching

Picture the Scenario

The 2002 General Social Survey asked, "On the average day, about how many hours do you personally watch television?" Figure 2.5 shows the histogram of the 905 responses.

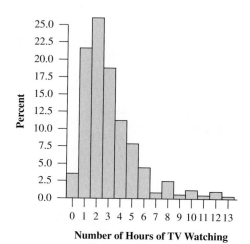

▲ **FIGURE 2.5: Histogram of GSS responses about number of hours of TV watching on the average day.**

Questions to Explore

a. What was the most common outcome?

b. What percentage of people reported watching no more than 2 hours per day?

Think It Through

a. The most common outcome (the mode) is the value with the highest bar. This is 2 hours per day of TV watching.

b. To find the percentage for "no more than 2 hours per day," we need to look at the percentages for 0, 1 and 2 hours per day. They seem to be about 3, 22, and 26. Adding these percentages together tells us that about 51% of the respondents reported watching no more than 2 hours of TV per day.

Insight

In theory, TV watching is a continuous variable. However, the possible responses subjects were able to make here were 0, 1, 2, ..., so it was measured as a discrete variable. Figure 2.5 is an example of a histogram of a discrete variable.

TO PRACTICE THIS CONCEPT, TRY EXERCISE 2.23, PARTS (A) AND (B)

For a discrete variable, a histogram usually has a separate bar for each possible value. An example is Figure 2.5 for the TV watching data. For a continuous variable, you need to divide the interval of possible values into smaller intervals formed with values grouped together. You can also do this when a discrete variable has a large number of possible values, such as a score on an exam. In such cases you can form a frequency table for the intervals and graph the frequencies or percentages for those intervals. Normally, the intervals should each have the same width.

Steps for Constructing a Histogram

- Divide the range of the data into intervals of equal width. For a discrete variable with few values, use the actual possible values.

- Count the number of observations (the frequency) in each interval, forming a frequency table.

- On the horizontal axis, label the values or the endpoints of the intervals. Draw a bar over each value or interval with height equal to its frequency (or percentage), values of which are marked on the vertical axis.

Cereal	Sodium
Frosted Mini Wheats	0
Raisin Bran	210
All Bran	260
Apple Jacks	125
Capt Crunch	220
Cheerios	290
Cinnamon Toast	210
Crackling Oat Bran	140
Crispix	220
Frosted Flakes	200
Fruit Loops	125
Grape Nuts	170
Honey Nut Cheerios	250
Life	150
Oatmeal Raisin Crisp	170
Sugar Smacks	70
Special K	230
Wheaties	200
Corn Flakes	290
Honeycomb	180

EXAMPLE 7

Exploring the Health Value of Cereals: Histogram

Picture the Scenario

Let's reexamine the sodium values of the 20 breakfast cereals. Those values are shown again in the margin.

Questions to Explore

a. Construct a frequency table.

b. Construct a histogram.

c. What information does the histogram not show that you can get from a dot plot or a stem-and-leaf plot?

Think It Through

a. To construct a frequency table, we divide the possible sodium values into separate intervals and count the number of cereals in each. The sodium values range from 0 to 290. We created Table 2.4 using eight intervals, each with a width of 40. With the interval labels shown in the table, for a continuous variable 0 to 39 actually represents 0 to 39.999999..., that is, 0 up to every number *below* 40. So, 0 to 39 is then shorthand for "0 to less than 40." Sometimes you will see the intervals written as 0 to 40, 40 to 80, 80 to 120, and so on. However, for an observation that falls at an interval endpoint, then it's not clear in which interval it goes.

TABLE 2.4: Frequency Table for Sodium in 20 Breakfast Cereals.

The table summarizes the sodium values using eight intervals and lists the number of observations in each, as well as the proportions and percentages.

Interval	Frequency	Proportion	Percentage
0 to 39	1	0.05	5%
40 to 79	1	0.05	5%
80 to 119	0	0.00	0%
120 to 159	4	0.20	20%
160 to 199	3	0.15	15%
200 to 239	7	0.35	35%
240 to 279	2	0.10	10%
280 to 319	2	0.10	10%

TI-83+/84 Histogram

b. Figure 2.6 shows the histogram for this frequency table. A bar is drawn over each interval of values, with height of the bar equal to the frequency. The histogram created using the TI-83+/84 calculator is in the margin.

▲ **FIGURE 2.6: Histogram of breakfast cereal sodium values.** The rectangular bar over an interval has height equal to the number of observations in the interval.

c. The histogram does not show the actual numerical values. For instance, we know that 1 observation falls below 40, but we do not know its actual value. In summary, with a histogram, we may lose the actual numerical values of individual observations, unlike a dot plot or a stem-and-leaf plot.

Insight

The histogram in Figure 2.6 labels the vertical axis with the frequencies. If it were labeled with proportions or percentages, the heights of the bars would be same. The histogram would be identical in appearance, and we would have the same graphical information about the sodium values. Either histogram would suggest that the cereal we choose to eat *does* matter if we wish to monitor our sodium intake.

TO PRACTICE THIS CONCEPT, TRY EXERCISE 2.19

The choice for the *number* of intervals in a histogram is up to you. If you use too few (fewer than about 5), the graph is usually too crude. If you use too many, you can lose information about the shape. There is no one right way to form a histogram, and different choices of intervals can reveal features of the data that just one histogram could miss.

Which Graph Should We Use?

We've now studied three graphs for quantitative variables—the dot plot, stem-and-leaf plot, and histogram. How do we decide which to use? Here are some guidelines:

- The dot plot and stem-and-leaf plot are more useful for small data sets, since they portray each individual observation. With large data sets, histograms usually work better, being more compact than the other displays.
- More flexibility is possible in defining the intervals with a histogram than in defining the stems with a stem-and-leaf plot.
- Data values are retained with the stem-and-leaf plot and dot plot but not with the histogram.

Unless the data set is small (say, about 50 or fewer observations), we usually prefer the histogram. When in doubt, create a histogram and the dot plot or stem-and-leaf plot, and then use whichever is clearer and more informative.

The Shape of a Distribution

For data on a quantitative variable, its graph is said to describe the "distribution of the data." Here are some things to look for in a graph showing a distribution:

- Look for the **overall pattern**. Do the data cluster together, or do one or more observations noticeably deviate from the rest, as in the histogram in the margin? We'll discuss such "outlier" observations later in the chapter.
- Do the data have a single mound? A distribution of such data is called **unimodal**. The highest point is at the **mode**. A distribution with *two* distinct mounds is called **bimodal**. A bimodal distribution results when a population is polarized on a controversial issue. Suppose each subject is presented with ten scenarios in which a person found guilty of murder may be given the death penalty. If we count the number of those scenarios in which subjects feel the death penalty would be just, many responses would be close to 0 (for subjects who oppose the death penalty generally) and many would be close to 10 (for subjects who think it's always or usually warranted for murder).

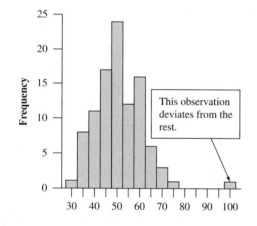

This observation deviates from the rest.

Unimodal

Bimodal

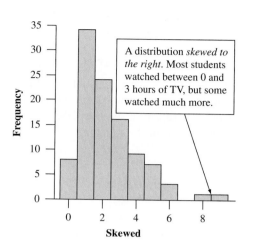

A distribution *skewed to the right.* Most students watched between 0 and 3 hours of TV, but some watched much more.

The **shape** of the distribution is often described as **symmetric** or **skewed**. A distribution is *symmetric* if the side of the distribution below a central value is a mirror image of the side above that central value. The distribution is *skewed* if one side of the distribution stretches out longer than the other side. For instance, the skewed distribution in the margin resulted from students in a class being asked how many hours they watched TV on the previous day.

In picturing features such as skew and symmetry, it's common to use smooth curves such as shown in Figure 2.7 to summarize the shape of a histogram. You can think of this as what can happen when you choose more and more intervals (making each interval narrower) and collect more data, so the histogram gets "smoother." The parts of the curve for the lowest values and for the highest values are called the **tails** of the distribution.

▲ **FIGURE 2.7:** Curves for data distributions illustrating symmetry and skew. **Question:** What does the longer tail indicate about the direction of skew?

IN WORDS

To **skew** means to pull in one direction.

Skewed Distribution

A distribution is **skewed to the left** if the left tail is longer than the right tail.

A distribution is **skewed to the right** if the right tail is longer than the left tail.

Identifying Skew

Let's consider some variables and think about what shape their distributions would have. How about IQ? Values cluster around 100, and tail off in a similar fashion in both directions. The appearance of the distribution on one side of 100 is roughly a mirror image of the other side, with tails of similar length. The distribution is approximately symmetric.

How about life span for humans? Most people in advanced societies live to be at least about 60 years old, but some die at a very young age, so life span would probably be skewed to the left.

What shape would we expect for the distribution of annual incomes of adults? Probably there would be a long right tail, with some people having incomes much higher than the overwhelming majority of people. The distribution would be skewed to the right.

IQ has a symmetric distribution

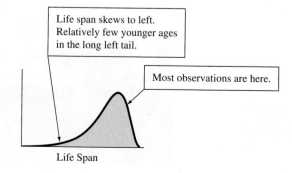

Life span skews to left. Relatively few younger ages in the long left tail.

Most observations are here.

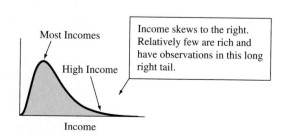

Most Incomes

High Income

Income skews to the right. Relatively few are rich and have observations in this long right tail.

EXAMPLE 8

What's the Shape of the Distribution of TV Watching?

Picture the Scenario

In Example 6, we constructed a histogram of the number of hours of TV watching reported in the GSS. It is shown again in the margin.

Question to Explore

How would you describe the shape of the distribution?

Think It Through

There appears to be a single mound of data clustering around the mode of 2. The distribution is unimodal. There also appears to be a long right tail, so the distribution is skewed to the right.

Insight

In surveys, the observation a subject reports is not necessarily the "true" value. In this distribution, the percentage is quite a bit higher for 8 hours than for 7 or 9, and for 10 than for 9 or 11. Perhaps subjects reporting high values tend to pick even numbers.

Number of Hours of TV Watching

TO PRACTICE THIS CONCEPT, TRY EXERCISE 2.21 AND 2.23, PART(C)

Time Plots: Displaying Data over Time

For some variables, observations occur over time. Examples include the daily closing price of a stock, and the population of a country measured every decade in a census. A data set collected over time is called a **time series**.

We can display time-series data graphically using a **time plot**. This charts each observation, on the vertical scale, against the time it was measured, on the horizontal scale. A common pattern to look for is a **trend** over time, indicating either a long tendency of the data to rise or a long tendency to fall. To see a trend more clearly, it is beneficial to connect the data points in their time sequence.

To illustrate, the figure in the margin is a time plot between 1995 and 2001 of the number of people worldwide who use the Internet[1] (in millions). In 1995, relatively few people were using the Internet, but by 2001 about 500 million people were doing so. There is a clear increasing trend over time. In practice, there's often not such a clear trend, as the next example illustrates.

---

[1] *Source:* www.netvalley.com/intvalstat.html

EXAMPLE 9

Is There a Trend Toward Warming in New York City?

Picture the Scenario

Each day, weather stations around the world record the daily average temperature, defined as the average of the maximum and minimum temperature over a midnight-to-midnight time frame. In a given year, the annual average temperature is the average of the daily average temperatures for that year. Let's analyze data on annual average temperature (in degrees Fahrenheit) in Central Park, New York City from 1901 to 2000. This is a continuous, quantitative variable. The data are in the "Central Park Yearly Temps" data file on the text CD.

Question to Explore

What do we learn from a time plot of these annual average temperatures? Is there a trend toward warming in the United States?

Think It Through

Figure 2.8 shows a time plot, constructed using MINITAB software. The time plot constructed using the TI-83+/84 calculator is in the margin. The observations fluctuate considerably, but the figure does suggest somewhat of an increasing trend in the annual average temperatures at this location.

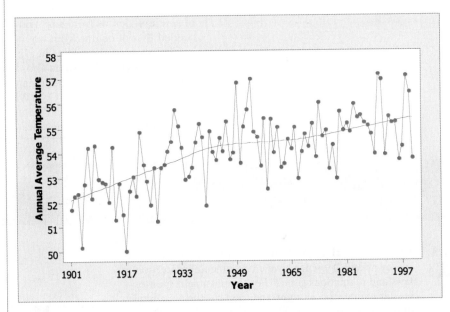

▲ **FIGURE 2.8: MINITAB output for a time plot of Central Park, New York City average annual temperatures.** The annual average temperatures are plotted against the year from 1901 to 2000. A smoothing curve is superimposed. **Question:** Are the annual average temperatures tending to increase, decrease, or stay the same over time?

Insight

The short-term fluctuations in a time plot can mask an overall trend. It's possible to get a clearer picture by "smoothing" the data. This is beyond our scope here. MINITAB presents the option of smoothing the data to portray a general trend (Click on "data view" and choose the "lowess" option under "smoother"). This is the smooth curve passing through the data points in Figure 2.8. This curve goes from a level of about 52 degrees in 1901 to about 55 degrees in 2000.

The data reported in Figure 2.8 refer to one location in the United States. In a study of climate change, it would be important to explore other locations around the world to see if similar trends are evident there.

TO PRACTICE THIS CONCEPT, TRY EXERCISES 2.26 AND 2.27

On the Shoulders of . . . Florence Nightingale

What relevance do graphs have to deaths from military combat?

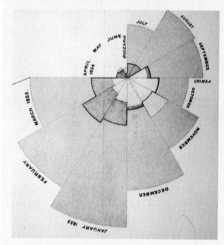

During the Crimean War in 1854, the British nurse Florence Nightingale (1820–1910) gathered data on the numbers of soldiers who died from various causes. She prepared graphical displays such as time plots and pie charts in an understandable format for policy makers. The graphs showed that more soldiers were dying from contagious diseases than from war-related wounds. The plots were revolutionary for her time. They helped her to promote her national cause of improving hospital conditions. After implementing sanitary methods, financed mostly with her own money, Nightingale showed with time plots that the relative frequency of soldiers' deaths from contagious disease decreased sharply and no longer exceeded that of deaths from wounds.

Throughout the rest of her life, Nightingale promoted the use of data for making informed decisions about public health policy. For example, she used statistical arguments to campaign for improved medical conditions in the United States during the Civil War in the 1860s (Franklin, 2002). Nightingale was once quoted as saying, "To understand God's thoughts we must study statistics, for these are the measure of His purpose" (Friendly, 2000).

2.2 Practicing the Basics

2.9 Environmental protection: A 2001 Roper organization survey asked: "How far have environmental protection laws and regulations gone?" For the possible responses—not far enough, about right, and too far—the percentages of responses were 44%, 35%, and 21%.

TRY

a. Sketch a bar chart to visually display the survey results.

b. Which is easier to sketch relatively accurately, a pie chart or a bar chart?

c. What is the advantage of using a graph to summarize the results instead of merely stating the percentages for each response?

2.10 What do alligators eat?: The bar chart (using MINITAB) is from a study[2] investigating the factors that influence alligators' choice of food. For 219 alligators captured in four Florida lakes, researchers classified the primary food choice (in volume) found in the alligator's stomach in one of the categories—fish, invertebrate (snails, insects, crayfish), reptile (turtles, baby alligators), bird, or other (amphibian, mammal, plants).

a. Is primary food choice categorical or quantitative?

b. What is the mode for primary food choice?

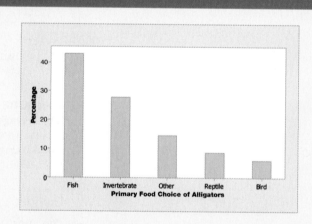

c. About what percentage of alligators had fish as the primary food choice?

d. This type of bar chart, with categories listed in order of frequency, has a special name. What is it?

2.11 Weather stations: The pie chart (constructed using EXCEL) shown portrays the regional distribution of weather stations in the United States.

a. Do the slices of the pie portray (i) variables or (ii) categories of a variable?

b. Identify what the two numbers mean that are shown for each slice of the pie.

---

[2]*Data courtesy of Clint Moore.*

c. From looking at the graph without inspecting the numbers, would it be easier to identify the mode using this graph or using the corresponding bar graph? Why?

Regional Distribution of Weather Stations

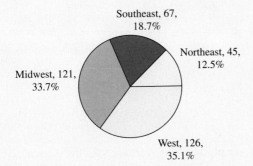

2.12   France is most popular holiday spot: Which countries are most visited by tourists from other countries? The table shows results according to *Travel and Leisure* magazine (2005).

a. Is country visited a categorical or a quantitative variable?

b. In creating a bar graph of these data, would it be most sensible to list the countries alphabetically, or in the form of a Pareto chart? Explain.

c. Does a dot plot or stem-and-leaf plot make sense for these data? Explain.

Most Visited Countries, 2005	
Country	**Number of Visits (millions)**
France	77.0
China	53.4
Spain	51.8
United States	41.9
Italy	39.8
United Kingdom	24.2
Canada	20.1
Mexico	19.7

2.13   Shark attacks worldwide: Table 2.1, part of which is shown again below, summarized shark attacks for different regions of the world. Using software or sketching, construct a bar graph, ordering the regions (i) alphabetically, (ii) as in a Pareto chart. Which do you prefer? Why?

Region	Frequency
Florida	289
Hawaii	44
California	34
Australia	44
Brazil	55
South Africa	64
Reunion Island	12
New Zealand	17
Japan	10
Hong Kong	6
Other	160

2.14   Sugar dot plot: For the breakfast cereal data given in Table 2.3, a dot plot (constructed using SPSS) for the sugar values (in grams) is shown:

a. Identify the minimum and maximum sugar values.

b. Identify the sugar outcome that occurs most frequently. What is this called?

SUGARg

2.15   eBay prices: eBay is a popular Internet company for personal auctioning. Let's consider some data from sales of the Palm M515 PDA, a popular handheld computer, during May 2003. The final prices (in dollars) at which the items sold were

235, 225, 225, 240, 250, 250, 210, 250, 249, 255, 200, 199,

240, 228, 255, 232, 246, 210, 178, 246, 240, 245, 225, 225

a. Construct a stem-and-leaf plot.

b. Summarize what this plot tells you about the data.

2.16   Graphing exam scores: A teacher shows her class the scores on the midterm exam in the stem-and-leaf plot shown:

```
6 | 588
7 | 01136779
8 | 1223334677789
9 | 011234458
```

a. Identify the number of students and their minimum and maximum scores.

b. Sketch how the data could be displayed in a dot plot.

c. Sketch how the data could be displayed in a histogram with four intervals.

2.17   Leaf unit: When the observations are large numbers, their final digits are not shown in a stem-and-leaf plot. The plot specifies a **leaf unit** by which to multiply each observation. For instance, for the cereal sugar data from Table 2.3 expressed in milligrams (an excerpt of which is shown on the next page), MINITAB software reports the figure shown here, indicating that "Leaf Unit = 1000." For instance, the observations of 14 and 15 in the final row of the plot represent observations of 14,000 and 15,000.

Cereal	Sugar (mg)
Frosted Mini Wheats	7,000
Raisin Bran	12,000
All Bran	5,000

```
0 | 1
0 | 23333
0 | 5
0 | 67
0 |
1 | 00011
1 | 2233
1 | 45
```

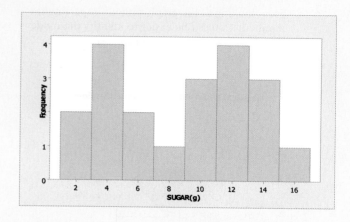

a. In milligrams, what is the observation in the first row of the plot?

b. Identify the sugar outcome (in mg) that occurs most frequently.

2.18 Truncated data and split stems: The figure below shows the stem-and-leaf plot for the cereal sodium values constructed after Example 5 using *split stems*, with leaves from 0 to 4 on the first split stem and leaves from 5 to 9 on the second.

Split stems for truncated cereal sodium values

```
0 | 0
0 | 7
1 | 224
1 | 5778
2 | 0011223
2 | 5699
```

a. Explain why the truncated data shown here go from 0 to 29.

b. Identify on the plot the observation of 230 for Special K.

c. What features of the data shown in a nontruncated plot are lost in this figure? (*Hint:* Do you have any detail of the large gap between the observations 0 and 70?)

2.19 Histogram for sugar: For the breakfast cereal data, the **TRY** figure shows a histogram (constructed using MINITAB) for the sugar values, in grams.

a. Identify the intervals of sugar values used for the plot.

b. Describe the shape of the distribution. What do you think might account for this unusual shape? (*Hint:* How else are the cereals classified in Table 2.3?)

c. What information can you get from the dot plot or stem-and-leaf plot of these data shown in Exercises 2.14 and 2.17 that you cannot get from this plot?

d. This histogram shows frequencies. If you were to construct a histogram using the *percentages* for each interval, how (if at all) would the shape of this histogram change?

2.20 Sugar plots: Using software with the "Cereal" data set on the text CD, construct a (a) dot plot, (b) stem-and-leaf plot, (c) histogram. Explain how to interpret each plot.

2.21 Shape of the histogram: For each of the following **TRY** variables, indicate whether you would expect its histogram to be symmetric, skewed to the right, or skewed to the left. Explain why.

a. Assessed value of houses in a large city

b. Number of times checking account overdrawn in the past year for the faculty in your school

c. IQ for the general population

d. The height of female college students

2.22 More shapes of histograms: Repeat the preceding exercise for

a. The scores of students (out of 100 points) on a very easy exam in which most score perfectly or nearly so, but a few score very poorly

b. The weekly church contribution for all members of a congregation, in which the three wealthiest members contribute generously each week

c. Time needed to complete a difficult exam (maximum time is 1 hour)

d. Number of music CDs (compact discs) owned, for each student in your school.

2.23 How often do students read the newspaper?: **TRY** Question 14 on the class survey (Activity 3 in Chapter 1) asked, "Estimate the number of times a week, on average, that you read a daily newspaper."

a. Is this variable continuous, or discrete? Explain.

b. The histogram shown gives results of this variable when this survey was administered to a class of 36 University of Georgia students. Report the (i) minimum response, (ii) maximum response, (iii) number of students who did not read the newspaper at all, (iv) mode.

c. Describe the shape of the distribution.

2.24 Shape of sodium values: Figure 2.6 showed a histogram of the 20 cereal sodium values, shown again below.

a. Would you call this skewed to the right, or skewed to the left?

b. Explain how the evidence of skew in (a) may be due to only two observations. (Hence, if we had data for *all* cereals, such skew might not exist.)

2.25 Central Park temperatures: The SPSS figure shows a histogram of the Central Park, New York annual average temperatures for the twentieth century.

Histogram of Central Park Annual Average Temperature

a. Describe the shape of the distribution.

b. What information can a time plot show that a histogram cannot provide?

c. What information does the histogram show that a time plot does not provide?

2.26 Is whooping cough close to being eradicated?: In the first half of the twentieth century, whooping cough was a frequently occurring bacterial infection that often resulted in death, especially among young children. A vaccination for whooping cough was developed in the 1940s. How effective has the vaccination been in eradicating whooping cough? One measure to consider is the **incidence rate** (number of infected individuals per 100,000 population) in the United States. The table shows incidence rates from 1925 to 1970.

Incidence Rates for Whooping Cough, 1925–1970[3]

Year	Rate per 100,000
1925	131.2
1930	135.6
1935	141.9
1940	139.6
1945	101.0
1950	80.1
1955	38.2
1960	8.3
1965	3.5
1970	2.1

a. Sketch a time plot. What type of trend do you observe? Based on the trend from 1945 to 1970, was the whooping cough vaccination proving effective in reducing the incidence of whooping cough?

b. The figure shown is a time plot of the data since 1980, reported in the Morbidity and Mortality Weekly Reports (MMWR).[4] What has the incidence rate been since about 1993? How does the incidence rate since 1993 compare with the incidence rate in 1970? Is the U.S. close to eradicating whooping cough?

c. Would a histogram of the incidence rates since 1935 address the question about the success of the vaccination for whooping cough? Why or why not?

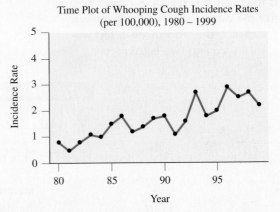

Time Plot of Whooping Cough Incidence Rates (per 100,000), 1980 – 1999

2.27 Warming in Newnan, Georgia?: Access the "Newnan, GA Temps" file on the text CD, which reports the average annual temperatures during the twentieth century for Newnan, Georgia. Construct a time plot to investigate a possible trend over time. Is there evidence of climate change?

[3]*Source: Historical Statistics of the United States, Colonial Times to 1970, U.S. Department of Commerce, p. 77*

[4]www.cdr.gov/mmwr/

2.3 | How Can We Describe the Center of Quantitative Data?

Section 2.2 introduced **graphical summaries** of data. For a quantitative variable, a graph shows the important feature of **shape**. It's always a good idea to look at the data first with a graph, to get a "feel" for the data. You can then consider **numerical summaries (statistics)**. For quantitative variables, we'll attempt to answer the questions, "What is a representative observation like?" and "Do the observations take similar values, or are they quite spread out?" Statistics that answer the first question describe the **center** of the distribution. Statistics that answer the second question describe the **spread**. We'll also study how the shape of the distribution influences the statistics and our choice of which are suitable for a given data set.

Describing the Center: The Mean and the Median

The best-known and most frequently used measure of the center of a distribution of a quantitative variable is the **mean**. It is found by averaging the observations.

Mean

The **mean** is the sum of the observations divided by the number of observations.

Another popular measure is the **median**. Half the observations are smaller than it, and half are larger.

IN WORDS

The **mean** refers to "averaging," adding up the data points and dividing by how many there are. The **median** is the point that splits the data in two, half the data below it and half above it (much like the median on a highway splits a road into two equal parts).

Median

The **median** is the midpoint of the observations when they are ordered from the smallest to the largest (or from the largest to the smallest).

EXAMPLE 10

What's the Center of the Cereal Sodium Data?

Picture the Scenario

In Examples 4, 5, and 7 in Section 2.2, we investigated the sodium level in 20 breakfast cereals and saw various ways to graph the data. Let's return to those data and learn how to describe their center. The observations (in mg) are

```
 0    70   125  125  140  150  170  170  180  200
200  210  210  220  220  230  250  260  290  290
```

Questions to Explore

a. Find the mean.

b. Find the median.

Think It Through

a. We find the mean by adding all the observations and then dividing this sum by the number of observations, which is 20:

Mean = (0 + 70 + 125 + ... + 290)/20 = 3710/20 = 185.5.

b. To find the median, we arrange the data from the smallest to the largest observation, as shown on the two data lines in the Picture the Scenario step. For the twenty observations, the smaller ten (on the first line) go from 0 to 200, and the larger ten (on the second line) go from 200 to 290. The midpoint is 200, which is the median. In summary, 200 is the midpoint of the ordered observations, and we get 185.5 by averaging all 20 of the observations.

Insight

The mean and median take different values. Why? The median measures the center by dividing the data into two equal parts, regardless of the actual numerical values above that point or below that point. The mean takes into account the actual numerical values of all the observations.

TO PRACTICE THIS CONCEPT, TRY EXERCISE 2.29

In this example, what if the smaller ten observations go from 0 to 200, and the larger ten go from 210 to 290? Then, the median is the average of the two middle observations, which is (200 + 210)/2 = 205.

How to Determine the Median

- Put the n observations in order of their size.
- When the number of observations n is odd, the median is the middle observation in the ordered sample.
- When the number of observations n is even, two observations from the ordered sample fall in the middle, and the median is their average.

A Closer Look at the Mean and the Median

Notation for the mean is used in formulas for the mean and for other statistics that describe spread around the mean.

IN WORDS

The formula

$$\bar{x} = \frac{\sum x}{n}$$

is short for "sum the values on the variable x and divide by the sample size."

NOTATION FOR A VARIABLE AND ITS MEAN

Variables are symbolized by letters near the end of the alphabet, most commonly x and y. The sample size is denoted by n. For a sample of n observations on a variable x, the mean is denoted by $\bar{x}$. Using the mathematical symbol $\sum$ for "sum," the mean has the formula

$$\bar{x} = \frac{\sum x}{n}.$$

For instance, the cereal data set has $n = 20$ observations. As we saw in Example 10,

$$\bar{x} = \left(\sum x\right)/n = (0 + 70 + 125 + \ldots + 290)/20 = 3710/20 = 185.5.$$

Here are some basic **properties of the mean**:

■ The mean is the *balance point* of the data: If we were to place identical weights on a line representing where the observations occur, then the line would balance by placing a fulcrum at the mean.

The Fulcrum Shows the Mean of the Cereal Sodium Data

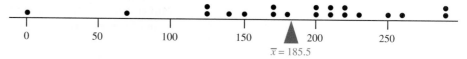

$\bar{x} = 185.5$

■ For a skewed distribution, the mean is pulled in the direction of the longer tail, relative to the median. The next example illustrates this.

■ The mean can be highly influenced by an **outlier**, which is an unusually small or unusually large observation.

Outlier

An **outlier** is an observation that falls well above or well below the overall bulk of the data.

EXAMPLE 11

CO_2 Pollution of the Eight Largest Nations

Picture the Scenario

The Pew Center on Global Climate Change[5] reports that possible global warming is largely a result of human activity that produces carbon dioxide (CO_2) emissions and other greenhouse gases. The CO_2 emissions from fossil fuel combustion are the result of the generation of electricity, heating, industrial processes, and gas consumption in automobiles. The *Human Development Report 2003*, published by the United Nations Development Programme,[6] reported the per capita CO_2 emissions in 1999 by country. For the eight largest countries in population size (which make up more than half the world's population) the values were, in metric tons per person:

China	2.3	Brazil	1.8
India	1.1	Russia	9.8
United States	19.7	Pakistan	0.7
Indonesia	1.2	Bangladesh	0.2

Questions to Explore

a. For these eight values, the mean is 4.6. What is the median?

---

[5] *Source:* http://www.pewclimate.org/global-warming-basics/facts_and_figures/

[6] hdr.undp.org

b. Is any observation a potential outlier? Discuss its impact on how the mean compares to the median.

c. Using this data set, explain the effect an outlier can have on the mean.

Think It Through

a. The CO_2 values have $n = 8$ observations. The ordered values are

$$0.2, \ 0.7, \ 1.1, \ 1.2, \ 1.8, \ 2.3, \ 9.8, \ 19.7.$$

Since n is even, two observations are in the middle, the fourth and fifth ones in the ordered sample. These are 1.2 and 1.8. The median is their average, 1.5.

Per Capita CO_2 Emissions

b. Let's consider a dot plot, as shown. The relatively high value of 19.7 falls well above the rest of the data. It is an outlier. This value as well as the value at 9.8 causes the mean, which is 4.6, to fall well above the median, which is 1.5.

c. The size of the outlier affects the calculation of the mean but not the median. If the observation of 19.7 for the United States had instead been the same as for China (2.3), the 8 observations would have had a mean of 2.4, instead of 4.6. The median would still have been 1.5. In summary, a single outlier can have a large impact on the value of the mean.

Insight

The mean may not be representative of where the bulk of the observations fall. This is fairly common with small samples when one observation is much larger or much smaller than the others. It's not surprising that the United States is an outlier, as the other nations are not nearly as economically advanced. Later in the chapter (Example 18) we'll compare carbon dioxide emissions for the United States to European nations.

> **TO PRACTICE THIS CONCEPT, TRY EXERCISE 2.30**

Comparing the Mean and Median

The shape of a distribution influences whether the mean is larger or smaller than the median. For instance, an extremely large value out in the right-hand tail pulls the mean to the right. It then usually falls above the median, as we just observed with the CO_2 data in Example 11. Generally, if the shape is

■ perfectly symmetric, the mean equals the median.

■ skewed to the right, the mean is larger than the median.

■ skewed to the left, the mean is smaller than the median.

As Figure 2.9 illustrates, the mean is drawn in the direction of the longer tail.

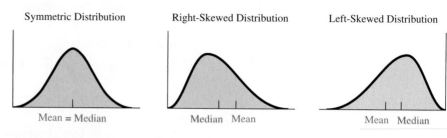

▲ **FIGURE 2.9: Relationship between the mean and median. Question:** For skewed distributions, what causes the mean and median to differ?

When a distribution is close to symmetric, with the tails of similar length, the median and mean are similar. For skewed distributions, the mean lies toward the direction of skew (the longer tail) relative to the median, as Figure 2.9 shows. Extreme observations in a tail affect the balance point for the distribution, which is the mean. The more highly skewed the distribution, the more the mean and median tend to differ.

Example 11 illustrated this property. The dot plot of CO_2 emissions shown there is skewed to the right. As expected, the mean of 4.6 falls in the direction of skew, above the median of 1.5. Another example is given by mean household income in the U.S. In 2001, the mean was $58,208 and the median was $42,228 (U.S. Bureau of the Census, *Current Population Reports*). This suggests that the distribution of household incomes in the United States is skewed to the right.

Why is the median not affected by an outlier?

■ How far an outlier falls from the middle of the distribution does not influence the median. The median is determined solely by having an equal number of observations above it and below it.

For the CO_2 data, for instance, if the value 19.7 for the United States were changed to 90, as shown in the margin, the median would still equal 1.5. The calculation of the mean uses *all* the numerical values, so unlike the median, it depends on how far observations fall from the middle. Because the mean is the balance point, an extreme value on the right side pulls the mean toward the right tail. Because the median is not affected, it is said to be **resistant** to the effect of extreme observations.

■ A numerical summary of the observations is called **resistant** if extreme observations have little, if any, influence on its value.

The median is resistant. The mean is not.

From these properties, you might think that it's always better to use the median than the mean. That's not true. The mean has other useful properties that we'll learn about and take advantage of in later chapters. Also, there are advantages to having a measure use the numerical values of *all* the data. For instance, *for discrete data that take only a few values, quite different patterns of data can give the same result for the median. It is then too resistant.*

Let's see an example of this situation. This example also shows how to find the mean and median from a frequency table.

THE MEDIAN IS RESISTANT TO OUTLIERS

1. Change 19.7 to 90 for U.S.

0.2, 0.7, 1.1, 1.2, 1.8, 2.3, 9.8, 90

median = (1.2 + 1.8)/2 = 1.5

2. Remove observation for U.S.

0.2, 0.7, 1.1, 1.2, 1.8, 2.3, 9.8

median = 1.2

EXAMPLE 12

How Many Times Have You Been Married

Picture the Scenario

A Census Bureau report[7] gave data on the number of times residents of the U.S. had been married, for subjects of various ages. Table 2.5 summarizes responses for subjects of age 20 to 24. The frequencies are actually *thousands* of people, for instance 7,074,000 men never married, but this does not affect calculations about the mean or median.

TABLE 2.5: Number of Times Married, for Subjects of Age 20 to 24

Number Times Married	Frequency	
	Women	Men
0	5861	7074
1	2773	1541
2	105	43
Total	**8739**	**8658**

Questions to Explore

a. Find the median and the mean for each gender.

b. Why is the median not particularly informative?

Think It Through

a. For the 8739 women, the ordered sample would be

$$0\ 0\ 0\ 0\ 0\ 0.....0 \quad 1\ 1\ 1\ 1\ 1\ 1...1 \quad 2\ 2\ 2\ 2\ 2\ 2...2$$
$$\text{(5861 0s)} \qquad \text{(2773 1s)} \qquad \text{(105 2s)}$$

Since n = 8739, the observations would be listed from 1st to 8739th, and the middle observation would be the $(1 + 8739)/2$ = 4370th. However, only three distinct responses occur, and more than 4370 of them are 0. So, the median is 0. Likewise, the median is 0 for men.

To calculate the mean for women, it is unnecessary to add the 8739 separate observations to obtain the numerator of $\bar{x}$, because each value occurred many times. We can find the sum of the 8739 observations by multiplying each possible value by its frequency and then adding:

$$\sum x = 5861(0) + 2773(1) + 105(2) = 2983.$$

Since the sample size is 8739, the mean is 2983/8739 = 0.34. Likewise, you can show that the mean for men is 0.19.

[7]*Source:* www.census.gov/prod/2002pubs/p70-80.pdf

b. The median is 0 for women and for men. This makes it seem as if there's no difference between men and women on this variable. By contrast, the mean uses the numerical values of all the observations, not just the ordering. In summary, in this age group we learn that, on the average, women have been married more often than men.

Insight

The median ignores too much information when the data are highly discrete—that is, when the data take only a few values. An extreme case occurs for **binary data**, which take only two values, 0 and 1. The median equals the more common outcome, but gives no information about the relative number of observations at the two levels. For instance, consider a sample of size 5 for the variable, number of times married. The observations $(1, 1, 1, 1, 1)$ and the observations $(0, 0, 1, 1, 1)$ both have a median of 1. The mean is 1 for $(1, 1, 1, 1, 1)$ and $3/5$ for $(0, 0, 1, 1, 1)$. When observations take values of only 0 or 1, the mean equals the *proportion* of observations that equal 1. It is much more informative than the median.

> **TO PRACTICE THIS CONCEPT, TRY EXERCISE 2.41**

IN PRACTICE: Effect of Shape on Choice of Mean or Median

- If a distribution is very highly skewed, the median is usually preferred over the mean because it better represents what is typical.

- If the distribution is close to symmetric or only mildly skewed or if it is discrete with few distinct values, the mean is usually preferred because it uses the numerical values of all the observations.

If you want to summarize the CO_2 emission data of Example 11, the median may be more relevant, because of the skew resulting from the extremely large value for the United States. But for the marriage data in Example 12, the median discards too much information and the mean is more informative. In practice, both the mean and median are useful, and frequently *both* values are reported.

The Mode

We've seen that the **mode** is the value that occurs most frequently. It describes a typical observation in terms of the most common outcome. The mode is most often used to describe the category of a categorical variable that has the highest frequency. With quantitative variables, it is most useful with discrete variables taking a small number of possible values. For the marriage data of Table 2.5, for example, the mode is 0 for each gender.

The mode need not be near the center of the distribution. It may be the largest or the smallest value, as in the marriage example. Thus, it is somewhat inaccurate to call the mode a measure of center, but often it is useful to report the most common outcome.

ACTIVITY 1: Let's Use an Applet to Simulate the Relationship between the Mean and Median

The *mean versus median* applet on the text CD allows you to add and delete data points from a sample. Different colored arrows indicate the mean and median after each new point is added. At the applet, set the number line to have a range of values from 0 to 20 and click on *Update*.

- In Example 11 on CO_2 use the per capita values (metric tons) for the 8 largest nations were $(0.2, 0.7, 1.1, 1.2, 1.8, 2.3, 9.8, 19.7)$. Create this sample using the applet. Now, update the line by changing the upper limit to 50 and

clicking "Update". Investigate what happens to the mean and median as you move the highest observation of 19.7 even higher, up to 50, or move it lower, down to 10. How does the outlier influence the relationship between the mean and median?

- Consider the effect of the sample size on the relationship between the mean and median, by adding to this distribution 20 values that are like the non-outliers. Does the outlier of 19.7 have as much effect on the mean when the sample size is larger?

> **TO PRACTICE THIS CONCEPT, TRY EXERCISE 2.36 AND EXERCISE 2.130**

2.28 Median versus mean: The mean and median describe the center.

   a. Why is the median sometimes preferred? Give an example.

   b. Why is the mean sometimes preferred? Give an example.

2.29 More on CO₂ emissions: The Pew Center on Global Climate Change reported the CO_2 emissions from fossil fuel combustion for the 7 countries in 1997 with the highest emissions. These values, reported as million metric tons of carbon equivalent, are 1490 (U.S.), 914 (China), 391 (Russia), 316 (Japan), 280 (India), 227 (Germany), and 142 (U.K.).

   a. Find the mean and median.

   b. The totals reported here do not take into account a nation's population size. Explain why it may be more sensible to analyze *per capita* values, as was done in Example 11.

2.30 Resistance to an outlier: Consider the following three sets of observations:

   Set 1: 8, 9, 10, 11, 12

   Set 2: 8, 9, 10, 11, 100

   Set 3: 8, 9, 10, 11, 1000

   a. Find the median for each data set.

   b. Find the mean for each data set.

   c. What do these data sets illustrate about the resistance of the median and mean?

2.31 Income and race: According to the U.S. Bureau of the Census, *Current Population Reports*, in 2001 the median household income was $44,517 for whites and $29,470 for blacks, whereas the mean was $60,512 for whites and $39,248 for blacks. Does this suggest that the distribution of household income for each race is symmetric, or skewed to the right, or skewed to the left? Explain.

2.32 Labor dispute: The workers and the management of a company are having a labor dispute. Explain why the workers might use the median income of all the employees to justify a raise but management might use the mean income to argue that a raise is not needed.

2.33 Cereal sodium: The dot plot shows the cereal sodium values from Example 4. What aspect of the distribution causes the mean to be less than the median?

Dot Plot of Sodium Values for 20 Breakfast Cereals

mean = 185.5 median = 200

Sodium (mg)

2.34 Center of plots: The figure shows dot plots for three sample data sets.

   a. For which, if any, data sets would you expect the mean and the median to be the same? Explain why.

   b. For which, if any, data sets would you expect the mean and the median to differ? Which would be larger, the mean or the median? Why?

2.35 Public transportation—center: The owner of a company in downtown Atlanta is concerned about the large use of gasoline by her employees due to urban sprawl, traffic congestion, and the use of energy inefficient vehicles such as SUVs. She'd like to promote the use of public transportation. She decides to investigate how many miles her employees travel on public transportation during a typical day. The values for her ten employees (recorded to the closest mile) are

     0   0   4   0   0   0   10   0   6   0

   a. Find and interpret the mean, median, and mode.

   b. She has just hired an additional employee. He lives in a different city and travels 90 miles a day on public transport. Re-compute the mean and median. Describe the effect of this outlier.

2.36 Public transportation—outlier: Refer to the previous exercise.

   a. Use the *mean versus median* applet to investigate what effect adding the outlier of 90 to the data set has on the mean and median.

   b. Now add 10 more data values that are near the mean of 2 for the original 10 observations. Does the outlier of 90 still have such a strong effect on the mean?

2.37 Student hospital costs: If you had data for all students in your school on the amount of money spent in the previous year on overnight stays in a hospital, probably the median and mode would be 0 but the mean would be positive.

   a. Explain why.

   b. Give an example of another variable that would have this property.

2.38 Net worth by degree: *Statistical Abstract of the United States 2000* reported that in 1998 for those with a college education, the median net worth was $146,400 and the mean net worth was $528,200. For those with a high school diploma only, the values were $53,800 and $157,800.

   a. Explain how the mean and median could be so different for each group.

   b. Which measure do you think gives a more realistic measure of a typical net worth, the mean or the median. Why?

2.39 Canadian income: According to Statistics Canada, in 2000 the median household income in Canada was $46,752 and the mean was $71,600. What would you predict about the shape of the distribution? Why?

2.40 Baseball salaries: According to sportingnews.com, in 2003 professional baseball players earned a mean salary of $2.5 million and a median salary of $800,000. What do you think causes these two values to be so different?

2.41 Knowing homicide victims: The table summarizes responses of 4390 subjects in recent General Social Surveys to the question, "Within the past month, how many people have you known personally that were victims of homicide?"

Number of People You Have Known Who Were Victims of Homicide

Number of Victims	Frequency
0	3944
1	279
2	97
3	40
4 or more	30
Total	**4390**

a. To find the mean, it is necessary to give a score to the "4 or more" category. Find it, using the score 4.5. (In practice, you might try a few different scores, such as 4, 4.5, 5, 6, to make sure the resulting mean is not highly sensitive to that choice.)

b. Find the median. Note that the "4 or more" category is not problematic for it.

c. If 1744 observations shift from 0 to 4 or more, how do the mean and median change?

d. Why is the median the same for (b) and (c), even though the data are so different?

2.42 Accidents: One variable in a study measures how many serious motor vehicle accidents a subject has had in the past year. Explain why the mean would likely be more useful than the median for summarizing the responses of the 60 subjects.

2.4 | How Can We Describe the Spread of Quantitative Data?

A measure of the center is not enough to describe a distribution well. It tells us nothing about the spread of the data. With the cereal sodium data, if we report the mean of 185.5 mg to describe the center, would the value of 250 mg for Honey Nut Cheerios be considered quite high, or are most of the data even farther from the median? To answer, we need numerical summaries for the spread of the distribution.

Measuring Spread: The Range

To see why a measure of the center is not enough, let's consider Figure 2.10.

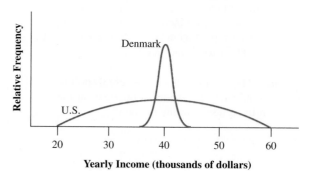

▲ **FIGURE 2.10: Income distributions for music teachers in Denmark and in the United States.** The distributions have the same mean, but the one for the United States is more spread out. **Question:** How would the range for Denmark change if one teacher earned $100,000?

This figure graphically compares hypothetical income distributions of music teachers in public schools in Denmark and in the United States. Both distributions are symmetric about $40,000 and have the same means. However, the annual incomes in Denmark (converted to U.S. dollars) go from $35,000 to $45,000, whereas those in the United States go from $20,000 to $60,000. Incomes are more uniform in Denmark and more spread out in the United States. A simple way to describe this is with the **range**.

Range

The **range** is the difference between the largest and the smallest observations.

In Denmark the range is $45,000–$35,000 = $10,000. In the United States the range is $60,000–$20,000 = $40,000. The range is a larger value when the data are more spread out.

The range is simple to compute and easy to understand, but it uses only the extreme values and ignores the other values. Also, it's affected severely by outliers. For example, if one teacher in Denmark made $100,000, the range would change from $45,000–$35,000 = $10,000 to $100,000–$35,000 = $65,000. The range is not a resistant statistic. It shares the worst property of the mean, not being resistant, and the worst property of the median, ignoring the numerical values of nearly all the data.

Measuring Spread: The Standard Deviation

The range uses only the largest and smallest observations. The most popular summary of spread uses *all* the data. It describes a typical value of how far the data fall from the mean. It does this by summarizing **deviations** from the mean.

■ The **deviation** of an observation x from the mean $\bar{x}$ is $(x - \bar{x})$, the difference between the observation and the sample mean.

For the cereal sodium values, the mean is $\bar{x} = 185.5$. The observation of 250 for Honey Nut Cheerios has a deviation of $250 - 185.5 = 64.5$. The observation of 70 for Sugar Smacks has a deviation of $70 - 185.5 = -115.5$. Figure 2.11 shows these deviations. Each observation has a deviation from the mean.

▲ **FIGURE 2.11: Dot plot for cereal sodium data, showing deviations for two observations. Question:** When is a deviation positive, and when is it negative?

■ A deviation $x - \bar{x}$ is *positive* when the observation falls *above* the mean. A deviation is *negative* when the observation falls *below* the mean.

■ The interpretation of the mean as the balance point implies that the positive deviations counterbalance the negative deviations. Because of this, *the sum of the deviations always equals zero.* Summary measures of spread use either the squared deviations or their absolute values.

■ The average of the squared deviations is called the **variance**. Because the variance uses the *square* of the units of measurement for the original data, its square root is easier to interpret. This is called the **standard deviation**.

IN WORDS

The symbol $\sum (x - \bar{x})^2$ is called a **sum of squares**. It represents finding the deviation for each observation, squaring each deviation, and then adding them up. Roughly, the standard deviation s represents a *typical distance* of an observation from the mean.

The Standard Deviation s

The **standard deviation s** of n observations is

$$s = \sqrt{\frac{\sum (x - \bar{x})^2}{n - 1}} = \sqrt{\frac{\text{sum of squared deviations}}{\text{sample size} - 1}}.$$

This is the square root of the **variance s^2**, which is an average of the squares of the deviations from their mean,

$$s^2 = \frac{\sum (x - \bar{x})^2}{n - 1}.$$

Although its formula looks complicated, a calculator can compute the standard deviation s easily. Its interpretation is quite simple: *We'll regard s as a typical distance of an observation from the mean.* The most basic property of the standard deviation is this:

■ The larger the standard deviation s, the greater the spread of the data.

A small technical point: You may wonder why the denominator of the variance and the standard deviation use $n - 1$ instead of n. We said that the variance was an *average* of the n squared deviations, so should we not divide by n? This book is not intended to present the mathematical theory of statistics, so we won't go into details. But basically it means that the deviations have only $n - 1$ pieces of information about variability: That is, $n - 1$ of the deviations determine the last one, because the deviations sum to 0. With $n = 2$ observations, for example, suppose the first observation has deviation $(x - \bar{x}) = 5$. Then the second observation must have deviation $(x - \bar{x}) = -5$, because the deviations must add to 0. With $n = 2$, there's only $n - 1 = 1$ nonredundant piece of information about variability.

EXAMPLE 13

Comparing Women's and Men's Ideal Number of Children

Picture the Scenario

Students in a discussion class led by a teaching assistant for one of the authors were asked on a questionnaire at the beginning of the course, "How many children do you think is ideal for a family?" The observations, classified by student's gender, were

Men: 0, 0, 0, 2, 4, 4, 4

Women: 0, 2, 2, 2, 2, 2, 4

Question to Explore

Both men and women have a mean of 2 and a range of 4. Are the distributions equally spread out? If not, which is more spread out?

Think It Through

Let's check dot plots for the data.

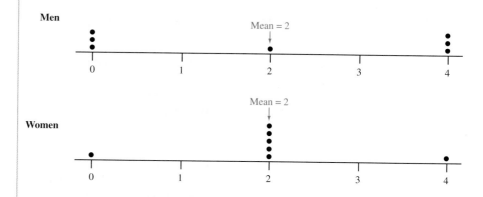

The typical deviation from the mean for the male observations appears to be about 2. The observations for females mostly fall right at the mean, so their typical deviation is smaller.

Let's calculate the standard deviation for men. Their observations are 0, 0, 0, 2, 4, 4, 4. The deviations and squared deviations about their mean of 2 are

Value	Deviation	Squared deviation
0	$(0 - 2) = -2$	4
0	$(0 - 2) = -2$	4
0	$(0 - 2) = -2$	4
2	$(2 - 2) = 0$	0
4	$(4 - 2) = 2$	4
4	$(4 - 2) = 2$	4
4	$(4 - 2) = 2$	4

The sum of squared deviations equals

$$\sum (x - \bar{x})^2 = 4 + 4 + 4 + 0 + 4 + 4 + 4 = 24.$$

The standard deviation of these $n = 7$ observations equals

$$s = \sqrt{\frac{\sum (x - \bar{x})^2}{n - 1}} = \sqrt{\frac{24}{6}} = \sqrt{4} = 2.0.$$

This means that for men, a typical distance of an observation from the mean is 2.0. By contrast, you can check that the standard deviation for women is $s = 1.2$. Since $2.0 > 1.2$, the observations for males tended to be farther from the mean than those for females. In summary, the men's observations were more spread out.

Insight

The standard deviation is more informative than the range. For these data, it detects that the women were more consistent than the men in their viewpoints about the ideal number of children. The range does not detect this, as it equals 4 for each gender.

IN PRACTICE: Rounding

Statistical software and calculators can find the standard deviation s for you. Try calculating s for a couple of small data sets to help you understand what it represents. After that, rely on software or a calculator. To ensure accurate results, don't round off while doing a calculation. (For example, use a calculator's memory to store intermediate results.) When presenting the solution, however, round off to two or three significant digits. In calculating s for women, you get $s = \sqrt{1.33333\ldots} = 1.1547005\ldots$. Present the value as $s = 1.2$ or $s = 1.15$ to make it easier for a reader to comprehend.

TO PRACTICE THIS CONCEPT, TRY EXERCISE 2.43

EXAMPLE 14

What's a Plausible Standard Deviation for Exam Scores?

Picture the Scenario

The first exam in your Statistics course is graded on a scale of 0 to 100. Suppose that the mean score in your class is 80.

Question to Explore

Which value is most plausible for the standard deviation s: 0, 10, or 50?

Think It Through

The standard deviation s is a *typical distance* of an observation from the mean. A value of s = 0 seems unlikely. For that to happen, every deviation would have to be 0. This means that every student must then score 80, the mean. A value of s = 50 is implausibly large, since 50 or −50 would not be a typical distance of a score from the mean of 80. We would instead expect to see a value of s such as 10. With s = 10, a typical distance is 10, as occurs with the scores of 70 and 90.

Insight

In summary, we've learned that s is a typical distance from the mean, larger values of s represent greater spread, and s = 0 means that all observations take the same value.

TO PRACTICE THIS CONCEPT, TRY EXERCISES 2.44 AND 2.45

PROPERTIES OF THE STANDARD DEVIATION, s

- The greater the spread of the data, the larger is the value of s.
- $s = 0$ only when all observations take the same value. For instance, if the reported ideal number of children for seven people is 2, 2, 2, 2, 2, 2, 2, then the mean equals 2, each of the seven deviations equals 0, and $s = 0$. This is the minimum possible spread for a sample.
- s can be influenced by outliers. It uses the mean, which we know can be influenced by outliers. Also, outliers have large deviations, and so they tend to have *extremely* large squared deviations. These can blow up the value of s and make it sensitive to outliers.

Interpreting the Magnitude of s: The Empirical Rule

Suppose that a distribution is unimodal and approximately symmetric with a **bell shape**, as in the margin figure. The value of s then has a more precise interpretation. Using the mean and standard deviation, we can form intervals that contain certain percentages (approximately) of the data.

Bell-Shaped Distribution

EMPIRICAL RULE

If a distribution of data is bell-shaped, then approximately

- 68% of the observations fall within 1 standard deviation of the mean, that is, between $\bar{x} - s$ and $\bar{x} + s$ (denoted $\bar{x} \pm s$).
- 95% of the observations fall within 2 standard deviations of the mean $(\bar{x} \pm 2s)$.
- All or nearly all observations fall within 3 standard deviations of the mean $(\bar{x} \pm 3s)$.

IN WORDS

$\bar{x} - s$ denotes the value one standard deviation below the mean, $\bar{x} + s$ denotes the value one standard deviation above the mean, and $\bar{x} \pm s$ denotes all values falling *within* one standard deviation of the mean.

Figure 2.12 is a graphical portrayal of the Empirical Rule.

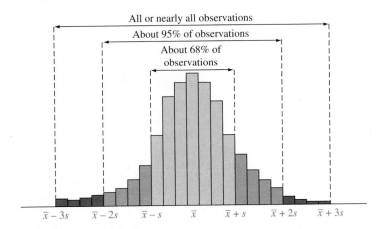

Did You Know?

The **Empirical Rule** has this name because many distributions of data observed in practice (*empirically*) are approximately bell-shaped.   ■

▲ **FIGURE 2.12: The Empirical Rule.** For bell-shaped distributions, this tells us approximately how much of the data fall within one, two, and three standard deviations of the mean. **Question:** About what percentage would fall *more than* two standard deviations from the mean?

EXAMPLE 15

Describing Female College Student Heights

Picture the Scenario

Many human physical characteristics have approximately a bell-shaped distribution. Let's explore height. Question 1 on the student survey in Activity 3 of Chapter 1 asked for the student's height. Figure 2.13 shows a histogram of the heights from responses to this survey by 261 female students at the University of Georgia. (The data are in the "Heights" data file on the text CD.) Table 2.6 presents some descriptive statistics, using MINITAB.

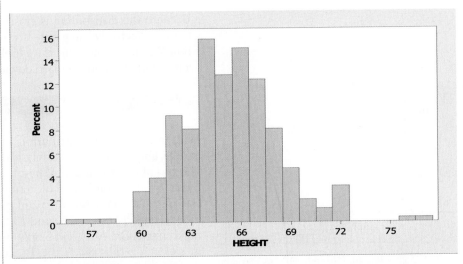

▲ **FIGURE 2.13: Histogram of female student height data.** This summarizes heights, in inches, of 261 female college students. **Question:** How would you describe the shape, center, and spread of the distribution?

Question to Explore

Can we use the Empirical Rule to describe the spread of these data? If so, how?

TABLE 2.6: MINITAB Output for Descriptive Statistics of Student Height Data

Variable	N	Mean	Median	StDev	Minimum	Maximum
HEIGHT	261	65.284	65.000	2.953	56.000	77.000

Think It Through

Figure 2.13 has approximately a bell shape. The figure in the margin shows a bell-shaped curve that approximates the histogram. From Table 2.6, the mean and median are close, about 65 inches, which reflects an approximately symmetric distribution. The Empirical Rule is applicable.

From Table 2.6 the mean is 65.3 inches and the standard deviation (labeled StDev) is 3.0 inches (rounded). By the Empirical Rule, approximately

■ 68% of the observations fall between

$$\bar{x} - s = 65.3 - 3.0 = 62.3 \text{ and } \bar{x} + s = 65.3 + 3.0 = 68.3,$$

that is, within the interval (62.3, 68.3), about 62 to 68 inches.

■ 95% of the observations fall within

$$\bar{x} \pm 2s, \text{ which is } 65.3 \pm 2(3.0), \text{ or } (59.3, 71.3), \text{ about 59 to 71 inches.}$$

■ All or nearly all observations fall within $\bar{x} \pm 3s$, or (56.3, 74.3).

Of the 261 observations, by actually counting we find that

■ 187 observations, 72%, fall within (62.3, 68.3).
■ 248 observations, 95%, fall within (59.3, 71.3).
■ 258 observations, 99%, fall within (56.3, 74.3).

In summary, the percentages predicted by the Empirical Rule are near the actual ones. Here, height is a continuous variable, but it is measured as discrete by rounding to the nearest inch. This somewhat adversely affects how close the actual sample percentages are to the values predicted by the Empirical Rule.

Insight

Because the distribution is close to bell-shaped, we can predict simple summaries effectively using only two numbers—the mean and the standard deviation. We can do the same if we look at the data only for the 117 males, who had a mean of 70.9 and standard deviation of 2.9.

TO PRACTICE THIS CONCEPT, TRY EXERCISE 2.46

With a bell-shaped distribution for a large data set, the observations usually extend about 3 standard deviations below the mean and about 3 standard deviations above the mean. When the distribution is highly skewed, the most extreme observation in one direction may not be nearly that far from the mean. For instance, on a recent exam given to introductory statistics students, the scores ranged between 30 and 100, with median = 88, $\bar{x}$ = 84 and s = 16. The maximum score of 100 was only one standard deviation above the mean (that is, $100 = \bar{x} + s = 84 + 16$). By contrast the minimum score of 30 was more than three standard deviations below the mean. This happened because the distribution of scores was highly skewed to the left. (See the margin figure.)

Recall

From Section 1.2, a **population** is the total group about whom you want to make conclusions. A **sample** is a subset of the population for whom you actually have data. A **parameter** is a numerical summary of the population, and a **statistic** is a numerical summary of a sample.■

In Chapter 6, we'll see why the Empirical Rule "works." For a special family of smooth, bell-shaped curves (called the **normal distribution**), we'll see how to find the percentage of the distribution in *any* particular region by knowing only the mean and the standard deviation.

Sample Statistics and Population Parameters

Of the numerical summaries introduced so far, the mean $\bar{x}$ and the standard deviation s are the most commonly used in practice. We'll use them frequently in the rest of the text. The formulas that define $\bar{x}$ and s refer to *sample* data. They are *sample statistics*.

We will distinguish between sample statistics and the corresponding *parameter* values for the population. The population mean is the average of all observations in the population. The population standard deviation describes the spread of the population observations about the population mean. These are usually unknown. Inferential statistical methods help us to make decisions and predictions about the population parameters based on the sample statistics.

In later chapters, to help distinguish between sample statistics and population parameters, we'll use different notation for the parameter values. Often, Greek letters are used to denote the parameters. For instance, we'll use μ (the Greek letter, mu) to denote the population mean and σ (the Greek letter lowercase sigma) to denote the population standard deviation.

2.4 Practicing the Basics

2.43 Sick leave: [TRY] A company decides to investigate the amount of sick leave taken by its employees. A sample of eight employees yields the following numbers of days of sick leave taken in the past year:

$$0 \quad 0 \quad 4 \quad 0 \quad 0 \quad 0 \quad 6 \quad 0$$

a. Find and interpret the range.

b. Find and interpret the standard deviation s.

c. Suppose the 6 was incorrectly recorded and is supposed to be 60. Redo (a)-(b) with the correct data and describe the effect of this outlier.

2.44 Shape of home prices?: [TRY] According to the National Association of Home Builders, the median selling price of new homes in the United States in 2002 was $187,100. Which of the following is the most plausible value for the standard deviation: −$15,000, $1000, $60,000, or $1,000,000? Why?

2.45 Exam standard deviation: [TRY] For an exam given to a class, the students' scores ranged from 35 to 98, with a mean of 74. Which of the following is the most realistic value for the standard deviation: −10, 0, 3, 12, 63? Clearly explain what's unrealistic about each of the other values.

2.46 Heights: [TRY] For the sample heights of Georgia college students in Example 15, the males had $\bar{x} = 71$ and $s = 3$ and the females had $\bar{x} = 65$ and $s = 3$.

a. Use the Empirical Rule to describe the distribution of heights for males.

b. The standard deviation for the overall distribution (combining females and males) was 4. Why would you expect it to be larger than the standard deviations for the separate male and female height distributions?

2.47 Histograms and standard deviation: The figure shows histograms for three different samples, each with sample size $n = 100$.

a. Which sample has the (i) largest, (ii) smallest standard deviation?

b. To which sample(s) is the Empirical Rule relevant? Why?

Histograms and Relative Sizes of Standard Deviations

2.48 Female strength: The "high school female athletes" data file on the text CD has data for 57 female high school athletes on the maximum number of pounds they were able to bench press. The data are roughly bell shaped, with $\bar{x} = 79.9$ and $s = 13.3$. Use the Empirical Rule to describe the distribution.

2.49 Female body weight: The "college athletes" data file on the text CD has data for 64 female college athletes. The data on weight (in pounds) are roughly bell shaped with $\bar{x} = 133$ and $s = 17$.

  a. Give an interval within which about 95% of the weights fall.

  b. Identify the weight of an athlete who is three standard deviations above the mean in this sample. Would this be a rather unusual observation? Why?

2.50 Shape of cigarettes taxes: A recent summary for the distribution of cigarette taxes (in cents) among the 50 states in the United States reported $\bar{x} = 73$ and $s = 48$. Based on these values, do you think that this distribution is bell shaped? If so, why? If not, why not, and what shape would you expect?

2.51 Empirical Rule and skewed, highly discrete distribution: Example 12 gave data on the number of times married. For the observations for men, shown again below, $\bar{x} = 0.19$ and $s = 0.40$.

No. Times	Count
0	7074
1	1541
2	43
Total	**8658**

  a. Find the actual percentages of observations within one, two, and three standard deviations of the mean. How do these compare to the percentages predicted by the Empirical Rule?

  b. How do you explain the results in (a)?

2.52 How much TV?: The 2002 General Social Survey asked, "On the average day, about how many hours do you personally watch television?" Of 905 responses, the mode was 2, the median was 2, the mean was 2.98, and the standard deviation was 2.36. Based on these statistics, what would you surmise about the shape of the distribution? Why?

2.53 How many friends?: A recent General Social Survey asked respondents how many close friends they had. For a sample of 1467 people, the mean was 7.4 and the standard deviation was 11.0. The distribution had a median of 5 and a mode of 4.

  a. Based on these statistics, what would you surmise about the shape of the distribution? Why?

  b. Does the Empirical Rule apply to these data? Why or why not?

2.54 Judging skew using $\bar{x}$ and s: If the largest observation is less than one standard deviation above the mean, then the distribution tends to be skewed to the left. If the smallest observation is less than one standard deviation below the mean, then the distribution tends to be skewed to the right. A professor examined the results of the first exam given in her statistics class. The scores were

$$35 \quad 59 \quad 70 \quad 73 \quad 75 \quad 81 \quad 84 \quad 86$$

The mean and standard deviation are 70.4 and 16.7. Using these, determine if the distribution is either left or right skewed. Construct a dot plot to check.

2.55 EU data file: The "European Union unemployment" data file on the text CD contains unemployment rates in December 2003 for the 25 countries that were in the European Union in 2004. Using software,

  a. Construct a graph to describe these values.

  b. Find the standard deviation. Interpret.

2.56 Create data with a standard deviation: Use the *standard deviation* applet on the text CD to investigate how the standard deviation changes as the data change.

  a. Create ten observations that have a mean of 5 and a standard deviation of about 2.

  b. Create ten observations that have a mean of 5 and a standard deviation of about 4.

  c. Placing ten values between 0 and 10, what is the largest standard deviation you can get? What are the values that have that standard deviation value?

2.5 | How Can Measures of Position Describe Spread?

The mean and median describe the center of a distribution. The range and the standard deviation describe the spread. We'll now learn about some other ways of describing a distribution using measures of **position**.

One type of measure of position tells us the point where a certain percentage of the data fall above or fall below that point. The median is an example. It specifies a location such that half the data fall below it and half fall above it. The range uses two other measures of position, the maximum value and the minimum value. Another type of measure of position tells us *how far* an observation falls from a particular point, such as the number of standard deviations an observation falls from the mean.

*p*th Percentile

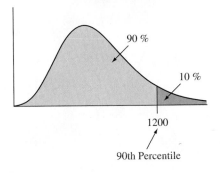

90th Percentile

Measures of Position: The Quartiles and Other Percentiles

The median is a special case of a more general set of measures of position called **percentiles**.

Percentile

The **pth percentile** is a value such that p percent of the observations fall below or at that value.

Suppose you're informed that your score of 1200 on the SAT college entrance exam falls at the 90th percentile. Set $p = 90$ in this definition. Then, 90% of those who took the exam scored between the minimum score and 1200. Only 10% of the scores were higher than yours.

Substituting $p = 50$ in this definition gives the 50th percentile. For it, 50% of the observations fall below or at it and 50% above it. But this is simply the median. *The 50th percentile is usually referred to as the median.*

Three useful percentiles are the **quartiles**. The **first quartile** has $p = 25$, the 25th percentile. The lowest 25% of the data fall below it. The **second quartile** has $p = 50$, the 50th percentile, which is the median. The **third quartile** has $p = 75$, the 75th percentile. The highest 25% of the data fall above it. *The quartiles split the distribution into four parts, each containing one quarter (25%) of the observations.* See Figure 2.14.

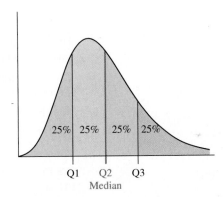

▲ **FIGURE 2.14: The quartiles split the distribution into four parts.** 25% is below the first quartile (Q1), 25% is between the first quartile and the second quartile (the median, Q2), 25% is between the second quartile and the third quartile (Q3), and 25% is above the third quartile. **Question:** Why is the second quartile also the median?

The quartiles are denoted by Q1 for the first quartile, Q2 for the second quartile, and Q3 for the third quartile. Notice that one quarter of the observations fall below Q1, two quarters (one half) fall below Q2 (the median), and three quarters fall below Q3.

FINDING QUARTILES

- Arrange the data in order.
- Find the median (the midpoint). This is the **second quartile, Q2**.
- Consider the lower half of the observations. The median of these observations is the **first quartile, Q1**.
- Consider the upper half of the observations. Their median is the **third quartile, Q3**.

EXAMPLE 16

What Are the Quartiles for the Cereal Sodium Data?

Picture the Scenario

Let's again consider the sodium values for the 20 breakfast cereals.

Questions to Explore

a. What are the quartiles for the 20 cereal sodium values?

b. Interpret the quartiles in the context of the cereal data.

Think It Through

a. From Table 2.3 the sodium values, arranged in ascending order, are

$$\text{Q1} = 145$$

0	70	125	125	**140**	**150**	170	170	180	**200**
200	210	210	220	**220**	**230**	250	260	290	290

$$\text{Q3} = 225$$

- The median of the 20 values is the average of the 10th and 11th observations, 200 and 200, which is **Q2 = 200 mg**.
- The first quartile Q1 is the median of the 10 smallest observations (in the top row), which is the average of 140 and 150, **Q1 = 145 mg**.
- The third quartile Q3 is the median of the 10 largest observations (in the bottom row), which is the average of 220 and 230, **Q3 = 225 mg**.

b. The quartiles tell you how the data split into four parts. The sodium values range from 0 mg to 145 mg for the first quarter, 145 mg to 200 mg for the second quarter, 200 mg to 225 mg for the third quarter, and 225 mg to 290 mg for the fourth quarter.

Insight

The quartiles also give information about shape. The distance of 55 from the first quartile to the median exceeds the distance of 25 from the median to the third quartile. This commonly happens when the distribution is skewed to the left, as shown in the margin figure. Although each quarter of a distribution may span different lengths, each quarter contains the same number (25%) of the observations.

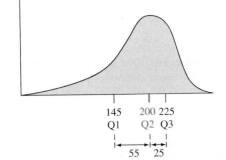

145   200 225
Q1   Q2 Q3

55   25

TO PRACTICE THIS CONCEPT, TRY EXERCISE 2.57

IN PRACTICE: Finding percentiles using technology

Percentiles other than the quartiles are reported usually only for large data sets. Software can do the computational work for you, and we won't go through the details. Precise algorithms for the calculations use interpolations, and different software often uses slightly different rules. This is true even for the quartiles Q1 and Q3: Most software, but not all, does not use the median observation itself in the calculation when the number of observations n is odd.

Measuring Spread: The Interquartile Range

The quartiles are also used to define a measure of spread that is more r
than the range and the standard deviation. This measure summarizes the range
for the *middle half* of the data. The middle 50% of the observations fall between the first quartile and the third quartile—25% from Q1 to Q2 and 25%
from Q2 to Q3. The distance from Q1 to Q3 is called the **interquartile range**,
denoted by **IQR**.

Interquartile Range (IQR)

The **interquartile range** is the distance between the third and first quartiles,

$$IQR = Q3 - Q1.$$

For instance, for the breakfast cereal sodium data, we just saw in Example 16 that

- Minimum value = 0
- First quartile Q1 = 145
- Median = 200
- Third quartile Q3 = 225
- Maximum value = 290

The range is $290 - 0 = 290$. The interquartile range is $Q3 - Q1 = 225 - 145 = 80$.

As with the range and standard deviation s, the more spread out the data, the
larger the IQR tends to be. But unlike those measures, the IQR is not affected by
any observations below the first quartile or above the third quartile. In contrast,
the range depends solely on the minimum and the maximum values, the most extreme values, so the range changes as either extreme value changes. For example,
if the highest sodium value were 1000 instead of 290, the range would change dramatically from 290 to 1000, but the IQR would not change. So, it's often better to
use the IQR instead of the range or standard deviation to compare spreads for distributions that are very highly skewed or that have severe outliers.

Detecting Potential Outliers

In any statistical analysis, you should examine the data for unusual observations, such as outliers. Can you think of a reason they differ so much from the
other observations? We've talked about outliers in an informal sense, but is
there a formula for flagging an observation as potentially being an outlier?
One way uses the interquartile range.

THE 1.5 × IQR CRITERION FOR IDENTIFYING POTENTIAL OUTLIERS

An observation is a potential outlier if it falls more than 1.5 × IQR below
the first quartile or more than 1.5 × IQR above the third quartile.

Recall

Ordered Sodium Values

0 70 125 ... 260 290 290 ■

From Example 16, the breakfast cereal sodium data has Q1 = 145 and Q3 = 225. So, IQR = Q3 − Q1 = 225 − 145 = 80. For those data

- $1.5 \times IQR = 1.5 \times 80 = 120$.

- $Q1 - 1.5 \times IQR = 145 - 120 = 25$ (lower boundary, potential outliers below),

and

$Q3 + 1.5 \times IQR = 225 + 120 = 345$ (upper boundary, potential outliers above).

By the 1.5 × IQR criterion, observations below 25 or above 345 are potential outliers. The only observation below 25 or above 345 is the sodium value of 0 mg for Frosted Mini Wheats. This is the only potential outlier.

Why do we identify an observation as a *potential* outlier rather than calling it a *definite* outlier? When a distribution has a long tail, some observations may be more than 1.5 × IQR below the first quartile or above the third quartile even if they are not outliers, in the sense that they are not separated far from the bulk of the data. For instance, in a long right tail there need not be a long gap between the largest observation and the rest of the data.

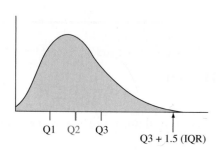

The Box Plot: Graphing a Five-Number Summary of Positions

The quartiles and the maximum value and the minimum value are five numbers often used as a set to summarize positions that describe center and spread.

THE FIVE-NUMBER SUMMARY OF POSITIONS

The **five-number summary** of a dataset is the minimum value, first quartile Q1, median, third quartile Q3, and the maximum value.

The five-number summary is the basis of a graphical display called the **box plot**. The **box** of a box plot contains the central 50% of the distribution, from the first quartile to the third quartile (see the margin figure). A line inside the box marks the median. The lines extending from the box are called **whiskers**. These extend to encompass the rest of the data, except for potential outliers, which are shown separately.

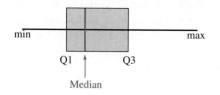

CONSTRUCTING A BOX PLOT

- A **box** goes from the lower quartile Q1 to the upper quartile Q3.

- A line is drawn inside the box at the median.

- A line goes from the lower end of the box to the smallest observation that is not a potential outlier. A separate line goes from the upper end of the box to the largest observation that is not a potential outlier. These lines are called **whiskers**. The potential outliers (more than 1.5 IQR below the first quartile or above the third quartile) are shown separately.

EXAMPLE 17

Box plot for the Breakfast Cereal Sodium Data

Picture the Scenario

Example 7 constructed a histogram for the breakfast cereal sodium values. That figure is shown again in the margin. The two leftmost bars resulted from sodium values of 0 and 70, which were considerably smaller than the others. Figure 2.15 shows a box plot for the sodium values. Labels are also given for the five-number summary of positions.

Questions to Explore

a. Explain how this box plot was constructed and how to interpret it.

b. Are the sodium values of 0 and 70 outliers?

Think It Through

a. The five-number summary of sodium values shown on the box plot is

   minimum = 0, Q1 = 145, median = 200, Q3 = 225, and maximum = 290.

   The middle 50% of data range from Q1 = 145 mg to Q3 = 225 mg, which are the two outer lines of the box. The median of 200 mg is shown by the center line in the box.

   As we saw above, based on the lower and upper boundaries of 25 and 345 for identifying potential outliers using the $1.5 \times$ IQR criterion, the sodium value of 0 mg for Frosted Mini Wheats is the only potential outlier. On the box plot, an asterisk identifies it as a potential outlier. The whisker from Q1 is drawn down to 70, which is the smallest observation that is not below the lower boundary of 25.

b. We've identified the observation of 0 as a potential outlier. In fact, there is a large gap between 0 and 70, the nearest observation. This indicates that 0 is an outlier. The observation 70 is not highlighted as a potential outlier.

Insight

Most software identifies observations that are more than 1.5 IQR from the quartiles by a separate symbol, such as *. (Some software uses a separate symbol for observations more than 3 IQR from the quartiles.) The TI-83+/84 output for the box plot is shown in the margin. The left whisker is drawn to 70 with the outlier at 0 on the y-axis.

Why show potential outliers separately? One reason is to identify them for further study. Was the observation perhaps incorrectly recorded? Was that subject fundamentally different from the others in some way? Often it makes sense to repeat a statistical analysis without an outlier, to make sure the results are not overly sensitive to a single observation.

▲ **FIGURE 2.15: Box plot and five-number summary for 20 breakfast cereal sodium values.** The central box contains the middle 50% of the data. The line in the box marks the median. Whiskers extend from the box to the smallest and largest observations, except to highlight potential outliers. **Question:** Why is the left whisker drawn down only to 70 rather than to 0?

TI − 83+/84 Box Plot

TO PRACTICE THIS CONCEPT, TRY EXERCISE 2.66

Another reason for keeping outliers separate in a box plot is that they do not provide much information about the shape of the distribution, especially for large data sets. Some software also can provide a box plot that extends the whiskers to the minimum and maximum, even if outliers exist. However, an extreme value can then cause the box plot to give the impression of severe skew when actually the remaining observations are not at all skewed.

The Box Plot Compared with the Histogram

A box plot does not portray certain features of a distribution, such as shape, as clearly as does a histogram. For example, the histogram and box plot in the margin refer to the same data. The histogram suggests that the distribution is bimodal (two distinct mounds), but we could not learn this from the box plot. This comparison shows that more than one type of graph may be needed to summarize a data set well.

A box plot does indicate skew from the relative lengths of the whiskers and the two parts of the box. The side with the larger part of the box and the longer whisker usually has skew in that direction. And as we've seen, box plots are useful for identifying potential outliers. We'll see next that they're also useful for graphical comparisons of distributions.

Side-by-Side Box Plots Help to Compare Groups

In Example 15 we looked at female college student heights, from the "Heights" data file on the text CD. To compare heights for females and males, we could look at side-by-side box plots, as shown in Figure 2.16.

The box plots suggest that both distributions are approximately symmetric. The median (the center line in a box) is approximately 71 inches for the males and 65 inches for the females. Although the centers differ, the spread is similar, as indicated by the width of the boxes (which is the IQR) being similar. Both samples have heights that are unusually short or tall, flagged as potential outliers. The upper 75% of the male heights are higher than the lower 75% of female heights. That is, 75% of the female heights fall below their third quartile, about 67 inches, whereas 75% of the male heights fall above their first quartile, about 69 inches.

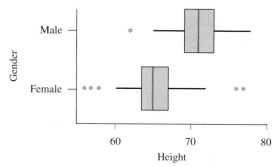

▲ **FIGURE 2.16: Box plots of male and female college student heights.** The box plots use the same scale for height. **Question:** What are approximate values of the quartiles for the two groups?

The z-Score Also Identifies Position and Potential Outliers

The Empirical Rule tells us that for a bell-shaped distribution it is unusual for an observation to fall more than three standard deviations from the mean. An alternative criterion for identifying potential outliers uses the standard deviation.

■ An observation in a bell-shaped distribution is regarded as a potential outlier if it falls more than three standard deviations from the mean.

How do we know the number of standard deviations that an observation falls from the mean? When $\bar{x} = 84$ and $s = 16$, it is clear that a value of 100 is one standard deviation above the mean, since $(100 - 84) = 16$. Alternatively,

$$(100 - 84)/16 = 1.$$

Taking the difference between an observation and the mean and dividing by the standard deviation specifies the number of standard deviations that the observation falls from the mean. This number is called the **z-score**.

---

z-score

The **z-score** for an observation is the number of standard deviations that it falls from the mean. For sample data, the z-score is calculated as

$$z = \frac{\text{observation} - \text{mean}}{\text{standard deviation}}.$$

EXAMPLE 18

Detecting Countries That Are Pollution Outliers

Picture the Scenario

In Example 11, we examined CO_2 emissions for the world's eight most populous nations. We will now consider air pollution data for the European Union (EU). The "energy-eu" data file[8] on the text CD contains data on per capita carbon-dioxide (CO_2) emissions, in metric tons, for the 25 nations in the EU. The mean was 8.3 and the standard deviation was 3.3. Figure 2.17 shows a box plot of the data. The maximum of 19.4, representing Luxembourg, is highlighted as a potential outlier.

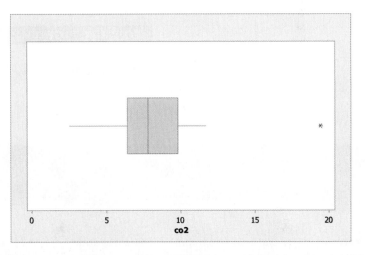

▲ **FIGURE 2.17: MINITAB box plot of carbon dioxide emissions for European Union nations. Question:** Can you use this plot to approximate the five-number summary?

Questions to Explore

a. How many standard deviations from the mean was the CO_2 value of 19.4 for Luxembourg?

b. The CO_2 value for the United States was 19.8. According to the three standard deviation criterion, is the United States an outlier on carbon dioxide emissions relative to the EU?

Think It Through

a. Since $\bar{x} = 8.3$ and $s = 3.3$ inches, the z-score for the observation of 19.4 is

$$z = \frac{\text{observation} - \text{mean}}{\text{standard deviation}} = \frac{19.4 - 8.3}{3.3} = \frac{11.1}{3.3} = 3.4.$$

---

[8]*Source: Human Development Report 2004, see* hdr.undp.org/statistics/data

The carbon dioxide emission (per capita) for Luxembourg is 3.4 standard deviations above the mean. By the three standard deviation criterion, this is a potential outlier. Since it's well removed from the rest of the data, we'd regard it as an actual outlier. However, Luxembourg has only 350,000 people, so in terms of the *amount* of pollution it is not a major polluter in the EU.

b. The CO_2 value for the United States is even higher, so its z-score is higher. We find that $z = (19.8 - 8.3)/3.3 = 3.5$. The United States is also a CO_2 outlier, relative to EU nations.

Insight

The z-scores of 3.4 and 3.5 are positive. This indicates that the observations are *above* the mean, because an observation above the mean has a positive z-score. The z-score is negative when the observation is *below* the mean. For instance, France has a CO_2 value of 6.2, which is below the mean of 8.3 and has a z-score of -0.6.

The United States has faced a challenge for some time to reduce its CO_2 emissions. Ten years ago, the President's Council on Sustainable Development[9] reported to then President Clinton that the United States, with 5% of the world's population, used about 25% of global energy and was responsible for about 25% of the global emissions of CO_2.

TO PRACTICE THIS CONCEPT, TRY EXERCISES 2.68 AND 2.69

_____

[9]*In Sustainable America,* released Feb. 1996. *See* clinton2.nara.gov/PCSD/Publications.

2.5 Practicing the Basics

2.57 **Vacation days:** *National Geographic Traveler* magazine
TRY recently presented data on the annual number of vacation days averaged by residents of eight different countries. They reported 42 days for Italy, 37 for France, 35 for Germany, 34 for Brazil, 28 for Britain, 26 for Canada, 25 for Japan, and 13 for the United States.

a. Report the median.

b. By finding the median of the four values below the median, report the first quartile.

c. Find the third quartile.

d. Interpret the values found in parts (a)–(c) in the context of these data.

2.58 **European unemployment:** In recent years many European nations have suffered from relatively high unemployment. For the 15 nations that made up the European Union in 2003, the table shows the unemployment rates reported by Eurostat as of December 2003.

a. Find and interpret the median.

b. Find the first quartile (Q1) and the third quartile (Q3).

c. Find and interpret the mean.

European Union 2003 unemployment rates (www.europa.eu.int/comm/eurostat)		
Belgium 8.3	France 9.5	Italy 8.5
Denmark 6.0	Portugal 6.7	Finland 8.9
Germany 9.2	Netherlands 4.4	Austria 4.5
Greece 9.3	Luxembourg 3.9	Sweden 6.0
Spain 11.2	Ireland 4.6	U.K. 4.8

2.59 **Female strength:** The "high school female athletes" data file on the text CD has data for 57 high school female athletes on the maximum number of pounds they were able to bench press, which is a measure of strength. For these data, $\bar{x} = 79.9$, Q1 $= 70$, median $= 80$, Q3 $= 90$.

a. Interpret the quartiles.

b. Would you guess that the distribution is skewed, or roughly symmetric? Why?

2.60   Female body weight: The "college athletes" data file on the text CD has data for 64 college female athletes. The data on weight (in pounds) has $\bar{x} = 133$, Q1 = 119, median = 131.5, Q3 = 144.

   a.   Interpret the quartiles.

   b.   Would you guess that the distribution is skewed, or roughly symmetric? Why?

2.61   Ways to measure spread: The standard deviation, the range, and the interquartile range (IQR) summarize spread of the data.

   a.   Why is the standard deviation s usually preferred over the range?

   b.   Why is the IQR sometimes preferred to s?

   c.   What is an advantage of s over the IQR?

2.62   Variability of cigarette taxes: Here's the five-number summary for the distribution of cigarette taxes (in cents) among the 50 states in the United States.

   Minimum = 2.5, Q1 = 36, Median = 60,

   Q3 = 100, Maximum = 205

   a.   About what proportion of the states have cigarette taxes (i) greater than 36 cents, (ii) greater than $1?

   b.   Between what two values are the middle 50% of the observations found?

   c.   Find the interquartile range. Interpret it.

   d.   Based on the summary, do you think that this distribution was bell shaped? If so, why? If not, why not, and what shape would you expect?

2.63   Sick leave: Exercise 2.43 showed data for a company that investigated the annual number of days of sick leave taken by its employees. The data are

   0   0   4   0   0   0   6   0

   a.   The standard deviation is 2.4. Find and interpret the range.

   b.   The quartiles are Q1 = 0, median = 0, Q3 = 2. Find the interquartile range.

   c.   Suppose the 6 was incorrectly recorded and is supposed to be 60. The standard deviation is then 21.1 but the quartiles do not change. Redo (a)–(b) with the correct data and describe the effect of this outlier. Which measure of spread, the range, IQR, or standard deviation, is least affected by the outlier? Why?

2.64   Computer use: During a recent semester at the University of Florida, students having accounts on a mainframe computer had drive use (in kilobytes) described by the five-number summary, minimum = 4, Q1 = 256, median = 530, Q3 = 1105, and maximum = 320,000.

   a.   Would you expect this distribution to be symmetric, skewed to the right, or skewed to the left? Explain.

   b.   Use the 1.5 × IQR criterion to determine if any potential outliers are present.

2.65   Central Park temperature distribution revisited: Exercise 2.25 showed a histogram for the distribution of Central Park annual average temperatures for the twentieth century. The box plot generated in SPSS for these data is shown here.

   a.   If this distribution is skewed, would you expect it to be skewed to the right or to the left? Explain.

   b.   Approximate each component of the five-number summary, and interpret.

2.66   Box plot for easy exam: The scores on an easy exam have mean = 88, standard deviation = 10, minimum = 65, Q1 = 77, median = 85, Q3 = 91, maximum = 100. Sketch a box plot, labeling which of these values are used in the plot.

2.67   Public transportation: Exercise 2.35 described a survey about how many miles per day employees of a company use public transportation. The sample values were

   0   0   4   0   0   0   10   0   6   0

   a.   Identify the five number summary, and sketch a box plot.

   b.   Explain why Q1 and the median share the same line in the box.

   c.   Why does the box plot not have a left whisker?

2.68   Energy statistics: The United Nations publication, *Energy Statistics Yearbook*, lists per capita consumption of energy. The 2001 data for the 25 nations that now make up the European Union are in the "energy-EU" data file on the text CD. The energy values (in kilograms) have a mean of 4998 and a standard deviation of 1786 and are roughly bell shaped, except for the value of 9814 for Luxembourg.

   a.   Using the MINITAB box plot shown, give approximate values for the five-number summary and indicate whether any countries were judged to be potential outliers according to that plot.

   b.   Italy had a value of 4222. How many standard deviations from the mean was it?

c. The United States is not in this data file, but its value was 11,067. Relative to the distribution for the EU nations, how many standard deviations from the mean was it?

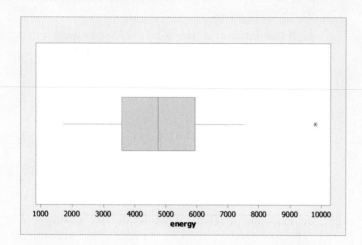

2.69 European Union unemployment rates: The 2003 [TRY] unemployment rates of countries in the European Union shown in Exercise 2.58 ranged from 3.9 to 11.2, with Q1 = 4.6, median = 6.7, Q3 = 9.2, a mean of 7.1, and standard deviation of 2.3.

a. In a box plot, what would be the values at the outer edges of the box, and what would be the values to which the whiskers extend?

b. Spain had the highest unemployment rate of 11.2. Is it an outlier according to the three standard deviation criterion? Explain.

c. What unemployment value for a country would have a z-score equal to 0?

2.70 Air pollution: Example 18 discussed EU carbon dioxide emissions, which had a mean of 8.3 and standard deviation of 3.3.

a. Canada's observation was 14.4. Find its z-score relative to the distribution of values for the EU nations, and interpret.

b. Sweden's observation was 5.3. Find its z-score, and interpret.

2.71 Female heights: For the 261 female heights shown in the box plot in Figure 2.16, the mean was 65.3 inches and the standard deviation was 3.0 inches. The shortest person in this sample had a height of 56 inches.

a. Find the z-score for the height of 56 inches.

b. What does the negative sign for the z-score represent?

c. Is this observation a potential outlier according to the three standard deviation distance criterion? Explain.

2.72 Hamburger sales: The manager of a fast-food restaurant records each day in a year the amount of money received from sales of food that day. Using software, he finds a bell-shaped histogram with a mean of $1165 and a standard deviation of $220. Today the sales equaled $2000. Is this an unusually good day? Answer by providing statistical justification.

2.73 Florida students again: Refer to the "FL student survey" data set on the text CD and the data on weekly hours of TV watching.

a. Use software to construct a box plot. Interpret the information on the plot, and use it to describe the shape of the distribution.

b. Using a criterion for outliers, investigate whether there are any potential outliers.

2.6 | How Are Descriptive Summaries Misused?

In this chapter, we've learned how to portray and describe data. We've learned that the nature of the data affects how we can effectively describe them. For example, if a distribution is very highly skewed or has extreme outliers, we've seen that some numerical summaries of the data such as the mean can be misleading.

To illustrate, during a recent semester at the University of Florida, students having accounts on a mainframe computer had drive usage (in kilobytes) described by the five-number summary:

minimum = 4, Q1 = 256, median = 530, Q3 = 1105, maximum = 320,000

and by the mean of 1921 and standard deviation of 11,495. The maximum value was an extreme outlier. You can see what a strong effect that outlier had on the mean and standard deviation. For these data, it's misleading to use these two values to summarize the distribution.

To finish this chapter, let's look at how we also need to be on the lookout for misleading graphical summaries.

Beware of Poor Graphs

With modern computer-graphic capabilities, newspapers and other periodicals use graphs in an increasing variety of ways to portray information. The graphs are not always well designed, however, and you must look at them skeptically.

EXAMPLE 19

College Enrollments "Heading in Different Directions"

Picture the Scenario

Look at Figure 2.18. According to the title of the graph and the two-sentence caption, the graph is intended to display how total enrollment has risen at the University of Georgia in recent years while the number of African American students has "dipped just about every year." The graph is a time plot showing the enrollment between 1996 and 2004, using white human figures to portray total enrollment and black human figures to portray African American enrollment.

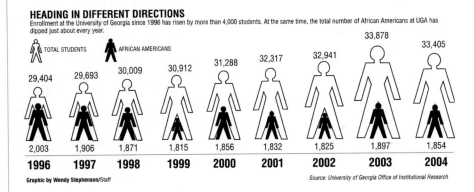

HEADING IN DIFFERENT DIRECTIONS
Enrollment at the University of Georgia since 1996 has risen by more than 4,000 students. At the same time, the total number of African Americans at UGA has dipped just about every year.

TOTAL STUDENTS   AFRICAN AMERICANS

29,404	29,693	30,009	30,912	31,288	32,317	32,941	33,878	33,405
2,003	1,906	1,871	1,815	1,856	1,832	1,825	1,897	1,854
1996	**1997**	**1998**	**1999**	**2000**	**2001**	**2002**	**2003**	**2004**

Graphic by Wendy Stephenson/Staff   Source: University of Georgia Office of Institutional Research

▲ **FIGURE 2.18: An example of a poor graph. Question:** What's misleading about the way the data are presented?

Questions to Explore

a. Do the heights of the human figures accurately represent the counts? Are the areas of the figures in accurate proportions to each other? Answer by comparing the two observations for 1996 and by comparing those for 1996 to those to for 2004.

b. Is an axis shown and labeled as a reference line for the enrollments displayed?

c. What other design problems do you see?

Think It Through

a. In the year 1996, the African American count (2003) is about 7% of the total of 29,404. However, the height of the African American figure is about 2/3 the height of the total enrollment figure. The graphs also mislead by using different widths for the figures. Compare, for instance, the figures for total enrollment in 1996 and 2004. The total enrollments are not nearly as different as the areas taken up by the figures make them appear.

b. Figure 2.18 does not provide a vertical axis, much less a labeling of that axis. It is not clear whether the counts are supposed to be represented by the heights or by the areas of the human figures.

c. It is better to put the figures next to each other, rather than to overlay the two sets of figures as was done here. The use of solid black figures and outlined figures can easily distort your perception of the graph. While it may not have been deliberate, the design obscures the data.

Insight

The intent of Figure 2.18 was to provide an intriguing visual display that quickly tells the story of the newspaper article. However, deviating from standard graphs gave a misleading representation of the data.

TO PRACTICE THIS CONCEPT, TRY EXERCISE 2.77

Here are some things to keep in mind when you construct a graph:

GUIDELINES FOR CONSTRUCTING EFFECTIVE GRAPHS

- Label both axes and provide a heading to make clear what the graph is intended to portray. (Figure 2.18 does not even provide a vertical axis.)
- To help our eyes visually compare relative sizes accurately, the vertical axis should usually start at 0.
- Be cautious in using figures, such as people, in place of the usual bars or points. It can make a graph more attractive, but it is easy to get the relative percentages that the figures represent incorrect.
- It can be difficult to portray more than one group on a single graph when the variable values differ greatly. (This is the case with Figure 2.18, in which one frequency is very small compared to the other.) Consider instead using separate graphs, or plotting relative sizes such as ratios or percentages.

EXAMPLE 20

What Are More Effective Graphs of the "Heading in Different Directions" Data?

Picture the Scenario

Let's look at Figure 2.18 again—because it's such a good example of how things can go wrong. The graph is intended to show how enrollment has risen at the University of Georgia in recent years while the number of African American students has decreased. Suppose the newspaper asks you to repair the graph, providing an accurate representation of the data.

Questions to Explore

a. Show a better way to portray graphically the same information about the enrollment counts that is in Figure 2.18.

b. Describe two alternative ways to portray the data that more clearly show the African American counts relative to the size of the student body over time.

Think It Through

a. Since the caption of Figure 2.18 indicates that counts are the desired data display, we can construct a time plot showing total and African American enrollments on the same scale. Figure 2.19 shows this revised graph. This time plot clearly labels the horizontal and vertical axes. There is the natural reference starting point of 0 for the vertical axis, allowing for better relative comparison of the two groups. We can see that the African American enrollment is a small part of the total enrollment. Between 1996 and 2004 there's been some tendency upward in the total enrollment, with relatively little change in the African American enrollment.

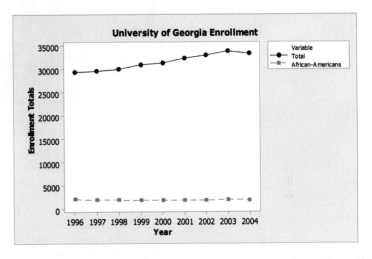

▲ **FIGURE 2.19: A better graph for the data in Figure 2.18. Question:** What trends do you see in the enrollments from 1996 to 2004?

b. Figure 2.19 is better than Figure 2.18 but is not ideal. Any trend that may be there for African Americans is not clear, because their counts are small compared to the total enrollment counts. Given that both the total and the African American counts vary from year to year, it's more meaningful to look at the African American *percentage* of the student body and plot that over time. See Exercise 2.76, which shows that this percentage went down from 1996 to 2004. Even if we plot the counts, as in Figure 2.19, it may be better to show a separate plot on its own scale for the African American counts. Or, in a single graph, rather than compare the African American counts to the *total* enrollment (which contains the African American group), it may be more meaningful to compare them to the *rest* of the enrollment, that is, the total enrollment minus the number of African Americans.

Insight

In constructing graphs, strive for clarity and simplicity. Many books have been written in recent years showing innovative ways to portray information clearly in graphs. Examples are *The Visual Display of Quantitative Information* and *Envisioning Information*, by Edward Tufte (Graphics Press).

TO PRACTICE THIS CONCEPT, TRY EXERCISE 2.76

When you plan to summarize data by graphical and numerical descriptive statistics, think about questions such as the following: What story are you attempting to convey with the data? Which graphical display and numerical summaries will most clearly and accurately convey this story ?

On the Shoulders of: . . . John Tukey (1915–2000)

"The best thing about being a statistician is that you get to play in everyone's backyard."
—John Tukey

John Tukey developed methods of exploratory data analysis.

In the 1960s, John Tukey of Princeton University was concerned that statisticians were putting too much emphasis on complex data analyses and ignoring simpler ways to examine and learn from the data. Tukey developed new descriptive methods, under the title of **exploratory data analysis (EDA)**. These methods make few assumptions about the structure of the data and emphasize data display and ways of searching for patterns and deviations from those patterns. Two graphical tools that Tukey invented were the stem-and-leaf plot and the box plot.

Initially, few statisticians promoted EDA. However, in recent years, some EDA methods have become common tools for data analysis. Part of this acceptance has been inspired by the availability of computer software and calculators that can implement some of Tukey's methods, although Tukey emphasized that his methods could be done with paper and pencil, especially for smaller data sets. Over his career, Tukey made many other important contributions to statistics and to science, including coining terms such as *software* and *bit*.

Tukey's work illustrates that statistics is an evolving discipline. Almost all of the statistical methods used today were developed in the past century, and new methods continue to be created, largely because of increasing computer power. John Tukey died recently, but he will be remembered for his unparalleled contributions to statistics and science and for his unusual creativity.

2.6 Practicing the Basics

2.74 Great pay (on the average): The six full-time employees of Linda's Tanning Salon near campus had annual incomes last year of $8900, $9200, $9200, $9300, $9500, $9800. Linda herself made $250,000.

   a. For the seven annual incomes at Linda's Salon, report the mean and median.

   b. Why is it misleading for Linda to boast to her friends that the average salary at her business is more than $50,000?

2.75 Market share for food sales: The pie chart shown was displayed in an article in *The Scotsman* newspaper (Jan. 15, 2005) to show the market share of different supermarkets in Scotland.

   a. Pie charts can be tricky to draw correctly. Identify two problems with this chart.

   b. From looking at the graph without inspecting the percentages, would it be easier to identify the mode using this graph or using a bar graph? Why?

Market Share of Scottish Supermarket Chain

Market share of Scottish supermarket chains

2.76 Enrollment trends: Examples 19 and 20 presented graphs showing the total student enrollment at the University of Georgia between 1996 and 2004 and the data for African American enrollment during that same time period. The data are repeated here.

Enrollment at University of Georgia

Year	Total Students	African American
1996	29,404	2,003
1997	29,693	1,906
1998	30,009	1,871
1999	30,912	1,815
2000	31,288	1,856
2001	32,317	1,832
2002	32,941	1,825
2003	33,878	1,897
2004	33,405	1,854

Source: University of Georgia Office of Institutional Research

a. Construct a graph for only African American enroll-ments over this period. Describe the trend in these enrollment counts.

b. Find the percentage of enrolled students each year who are African Americans, and construct a time plot of the percentages.

c. Summarize what the graphs constructed in (a) and (b) tell you that you could not learn from Figures 2.18 and 2.19 in Examples 19 and 20.

2.77 **Terrorism and war in Iraq:** In 2004, a college newspaper reported results of a survey of students taken on campus. One question asked was, "Do you think going to war with Iraq has made Americans safer from terrorism, or not?" The figure shows the way the magazine reported results.

a. Explain what's wrong with the way this bar chart was constructed.

b. Explain why you would not see this error made with a pie chart.

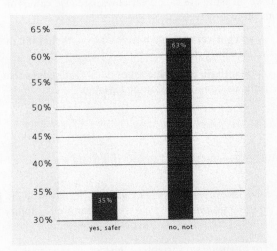

2.78 **Bad graph:** Search some publications and find an exam-ple of a graph that violates at least one of the principles for constructing good graphs. Summarize what's wrong with the graph and explain how it could be improved.

CHAPTER REVIEW

Answers to the Chapter Figure Questions

Figure 2.1: It is not always clear exactly what portion or percentage of a circle the slice represents, especially when a pie has many slices. Also, some slices may be representing close to the same percentage of the circle; thus, it is not clear which is larger or smaller.

Figure 2.3: The number of dots above a value on the number line represents the frequency of occurrence of that value.

Figure 2.4: There's no leaf if there's no observation having that stem.

Figure 2.7: The direction of the longer tail indicates the direction of the skew.

Figure 2.8: The annual average temperatures are tending to increase over time.

Figure 2.9: The mean is pulled in the direction of the longer tail since extreme observations affect the balance point of the distribution. The median is not affected by the size of the observation.

Figure 2.10: The range would change from $10,000 to $65,000.

Figure 2.11: A deviation is positive when it falls above the mean and negative when it falls below the mean.

Figure 2.12: Since about 95% of the data fall within 2 standard devi-ations of the mean, about 100% − 95% = 5% fall more than 2 stan-dard deviations from the mean.

Figure 2.13: The shape is approximately bell shaped. The center as measured by the mean and median is about 65 inches. The heights range from 56 to 77 inches.

Figure 2.14: The second quartile has two quarters (25% and 25%) of the data below it. Therefore, Q2 is the median since 50% of the data falls below the median.

Figure 2.15: The left whisker is drawn only to 70 since the next low-est sodium value, which is 0, is identified as a potential outlier.

Figure 2.16: The quartiles are about Q1 = 69, Q2 = 71, and Q3 = 73 for males and Q1 = 63, Q2 = 65, and Q3 = 67 for females.

Figure 2.17: Yes. The left whisker extends to the minimum (about 2), the left side of the box is at Q1 (about 7), the line inside the box is at the median (about 8), the right side of the box is at Q3 (about 10), and the outlier identified with a star is at the maxi-mum (about 19).

Figure 2.18: Neither the height nor the area of the human figures accurately represents the frequencies reported.

Figure 2.19: The total enrollment goes up until 2003. The African-American enrollment seems fairly stable.

CHAPTER SUMMARY

This chapter introduced **descriptive statistics**—graphical and numerical ways of **describing** data. The characteristics of interest that we measure are called **variables**, because values on them vary from subject to subject.

■ A **categorical variable** has observations that fall into one of a set of categories.

■ A **quantitative variable** takes numerical values that represent different magnitudes of the variable.

■ A quantitative variable is **discrete** if it has separate possible values, such as the integers 0, 1, 2, ... for a variable expressed as "the number of ...." It is **continuous** if its possible values form an interval.

When we explore data, key features to describe are the **shape**, using graphical displays, and the **center** and **spread**, using numerical summaries.

Overview of Graphical Methods

■ For categorical variables, data are displayed using **pie charts** and **bar graphs**. Bar graphs provide more flexibility and make it easier to compare categories having similar percentages.

■ For quantitative variables, a **histogram** is a graph of a frequency table. It displays bars that specify frequencies or relative frequencies (percentages) for possible values or intervals of possible values. The **stem-and-leaf plot** (a vertical line dividing the final digit, the leaf, from the stem) and **dot plot** (dots above the number line) show the individual observations. They are useful for small data sets. These three graphs all show shape, such as whether the distribution is approximately bell shaped, skewed to the right (longer tail pointing to the right), or skewed to the left.

■ The **box plot** has a box drawn between the first quartile and third quartile, with a line drawn in the box at the median. It has whiskers that extend to the minimum and maximum values, except for potential **outliers**. An **outlier** is an extreme value falling far below or above the bulk of the data.

■ A **time plot** graphically displays observations for a variable measured over time. This plot can visually show **trends** over time.

Overview of Measures of the Center

Measures of the center attempt to describe a "typical" or "representative" observation.

■ The **mean** is the sum of the observations divided by the number of observations. It is the balance point of the data.

■ The **median** divides the ordered data set into two parts of equal numbers of observations, half below and half above that point. The median is the 50th percentile (second quartile). It is a more representative summary than the mean when the data are highly skewed.

■ The lower quarter of the observations fall below the **first quartile (Q1)**, and the upper quarter fall above the **third quartile (Q3)**. These are the 25th percentile and 75th percentile. These quartiles and median split the data into four equal parts.

Overview of Measures of Spread

Measures of spread describe the variability of the data.

■ The **range** is the difference between the largest and smallest observations.

■ The **deviation** of an observation x from the mean is $x - \bar{x}$. The **variance** is an average of the squared deviations. Its square root, the **standard deviation**, is more useful, describing a typical distance from the mean.

■ The **Empirical Rule** states that for a bell-shaped distribution: About 68% of the data fall within one standard deviation of the mean. About 95% of the data fall within two standard deviations, $\bar{x} \pm 2s$.

Nearly all, if not all, the data fall within three standard deviations, $\bar{x} \pm 3s$.

■ The **interquartile range (IQR)** is the difference between the third and first quartiles, which span the middle half of the data. It is more **resistant** than the range and standard deviation, being unaffected by extreme observations.

■ An observation is a potential outlier if it falls (a) more than $1.5 \times \text{IQR}$ below Q1 or above Q3, or (b) more than 3 standard deviations from the mean. The **z-score** is the number of standard deviations that an observation falls from the mean.

SUMMARY OF NOTATION

Mean $\bar{x} = \dfrac{\sum x}{n}$, where x denotes the variable, n is the sample size, and Σ indicates to sum the observations

Standard deviation $s = \sqrt{\dfrac{\sum (x - \bar{x})^2}{n - 1}}$

$$z\text{-score} = \dfrac{\text{observed value} - \text{mean}}{\text{standard deviation}}$$

CHAPTER PROBLEMS

Practicing the Basics

2.79 **Categorical or quantitative?:** Identify each of the following variables as categorical or quantitative.

   a. Number of children in family

   b. Amount of time in football game before first points scored

   c. College major (English, history, chemistry, ...)

   d. Type of music (rock, jazz, classical, folk, other)

2.80 **Continuous or discrete?:** Which of the following variables are continuous, when the measurements are as fine as possible?

   a. Age of mother

   b. Number of children in a family

   c. Cooking time for preparing dinner

   d. Latitude and longitude of a city

   e. Population size of a city

2.81 **Immigration into United States:** The table shows the number (in thousands) of the foreign-born population of the United States in 2002, by place of birth.

Foreign-Born Population in the United States	
Place of Birth	**Number (in Thousands)**
Europe	4548
Caribbean	3102
Central America	11819
South America	2022
Asia	8281
Other	2680
Total	**32452**

Source: U.S. Statistical Abstract, 2003

   a. Is "Place of birth" quantitative or categorical? Show how to summarize results by adding a column of percentages to the table.

   b. Which of the following is a sensible numerical summary for these data: Mode, mean, median? Explain, and report whichever is/are sensible.

   c. How would you order the place of birth categories for a Pareto chart? What's its advantage over the ordinary bar graph?

2.82 **Cool in China:** A recent survey[10] asked 1200 university students in China to pick the personality trait that most defines a person as "cool." The possible responses allowed, and the percentage making each, were individualistic and innovative (47%), stylish (13.5%), dynamic and capable (9.5%), easygoing and relaxed (7.5%), other (22.5%).

   a. Identify the variable being measured.

   b. Classify the variable as categorical or quantitative.

   c. Which of the following methods could you use to describe these data?: (i) bar chart, (ii) dot plot, (iii) box plot, (iv) median, (v) mean, (vi) mode, (vi) IQR, (vii) standard deviation.

2.83 **Chad voting problems:** The 2000 U.S. Presidential election had various problems in Florida. One was "overvotes"—people mistakenly voting for more than one Presidential candidate. (There were multiple minor-party candidates.) There were 110,000 overvote ballots, with A1 Gore marked on 84,197 and George W. Bush on 37,731. These ballots were disqualified. Was overvoting related to the design of the ballot? The figure shows MINITAB dot plots of the overvote percentages for 65 Florida counties organized by the type of voting machine—optical scanning, Votomatic (voters manually punch out chads), and Datavote (voter presses a lever that punches out the chad mechanically)—and the number of columns on the ballot (1 or 2).

   a. The overvote was highest (11.6%) in Gadsden County. Identify the number of columns and the method of registering the vote for that county.

   b. Of the six ballot type and method combinations, which two seemed to perform best in terms of having relatively low percentages of overvotes?

   c. How might these data be summarized further by a bar graph with six bars?

Overvote Percentages by Number of Columns on Ballot and Method of Voting

Source: A. Agresti and B. Presnell, Statistical Science, vol. 17, pp. 1–5, 2002.

[10]*Source:* Public relations firm Hill & Knowlton, as reported by *Time* magazine.

2.84 Number of children: For the question "How many children have you ever had?" in the 2002 General Social Survey, the results were

No. children	0	1	2	3	4	5	6	7	8+
Count	799	469	657	481	185	73	40	22	34

a. Which is the most appropriate graph to display the data—dot plot, stem-and-leaf plot, or histogram? Why?

b. Based on sketching or using software to construct the graph, characterize this distribution as skewed to the left, skewed to the right, or symmetric. Explain.

2.85 Newspaper reading: Exercise 2.23 gave results for the number of times a week a person reads a daily newspaper for a sample of 36 students at the University of Georgia. The frequency table is shown below.

a. Construct a dot plot of the data.

b. Construct a stem-and-leaf plot of the data. Identify the stems and the leaves.

c. The mean is 3.94. Find the median.

d. Is the distribution skewed to the left, skewed to the right, or symmetric? Explain.

No. Times	Frequency
0	2
1	4
2	4
3	8
4	4
5	5
6	2
7	4
8	2
9	1

2.86 Match the histogram: Match each histogram with one of the following descriptions: Skewed to the left, symmetric and bimodal, symmetric and unimodal, skewed to the right.

2.87 Enchiladas and sodium: The March 2000 *Consumer Reports* compared various brands of supermarket enchiladas in cost and sodium content, as shown in the table.

Brand and Type	Cost (per serving)	Sodium Content (mg)
Amy's Black Bean	$3.03	780
Patio Cheese	$1.07	1570
Banquet Cheese	$1.28	1500
El Charrito Beef	$1.53	1370
Patio Beef	$1.05	1700
Banquet Beef	$1.27	1330
Healthy Choice Chicken	$2.34	440
Lean Cuisine Chicken	$2.47	520
Weight Watchers Chicken	$2.09	660

a. Construct a dot plot of the sodium amounts in the various enchilada brands.

b. What is the advantage(s) of using the dot plot instead of a histogram?

c. Construct a stem-and-leaf plot of the sodium values. What does it show?

d. Summarize your findings from these graphs.

2.88 Enchiladas and cost: Refer to the previous exercise. Repeat parts (a), (c), and (d) for the cost per serving. Summarize your findings.

2.89 What shape do you expect?: For the following variables, indicate whether you would expect its histogram to be bell shaped, skewed to the right, or skewed to the left. Explain why.

a. Number of times arrested in past year

b. Time needed to complete difficult exam (maximum time is 1 hour)

c. Assessed value of home

d. Age at death

2.90 Sketch plots: For each of the following, sketch roughly what you expect a histogram to look like, and explain whether the mean or the median would be greater.

a. The selling price of new homes in 2005

b. The number of children ever born per woman age 40 or over

c. The score on an easy exam (mean = 88, standard deviation = 10, maximum = 100, minimum = 50)

d. Number of months in which subject drove a car last year

2.91 Median vs. mean income: Based on the 2000 census, the U.S. Census Bureau reported that the median U.S. household income in 1999 was $41,994. Would you expect the mean household income to have been larger, or smaller? Explain.

2.92 Rich get richer?: A U.S. Federal Reserve study in 2000 indicated that for those families with annual incomes above $100,000, their median net worth was about $500,000 both in 1995 and in 1998, but their mean net worth rose from $1.4 million in 1995 to $1.7 million in 1998.

a. Is the distribution of net worth for these families likely to be symmetric, skewed to the right, or skewed to the left? Explain.

b. A newspaper story about this said that the mean uses "a calculation that captures the huge gains made by the wealthiest Americans." Why would the median not necessarily capture those gains?

c. The same report indicated that families earning less than $10,000 annually saw their median net worth fall by 25% over these three years to $3600. Show that their median net worth in 1995 was $4800.

2.93 Baseball salaries: During the strike of professional baseball players in 1994, two quite different numbers were reported for the central tendency of players' annual salaries. One was $1.2 million and the other was $500,000. One of these was the median and one was the mean. Which value do you think was the mean? Why?

2.94 Weekly profits: You've started your own small business. Each week, after paying your employees and taking a nominal salary for yourself, you record the remaining profit. A negative value represents a loss that week. The ordered sample (in dollars) after eight weeks is

$$-5 \quad 19 \quad 30 \quad 33 \quad 35 \quad 41 \quad 44 \quad 46$$

Find the mean and the standard deviation. Interpret.

2.95 What does _s_ equal?:

a. For an exam given to a class, the students' scores ranged from 35 to 98, with a mean of 74. Which of the following is the most realistic value for the standard deviation: $-10, 1, 12, 60$? Why?

b. The sample mean for a data set equals 80. Which of the following is an impossible value for the standard deviation? $200, 0, -20$? Why?

2.96 Female heights: According to a recent report from the U.S. National Center for Health Statistics, females with age between 25 and 34 years have a bell shaped distribution for height, with mean of 65 inches and standard deviation of 3.5 inches.

a. Give an interval within which about 95% of the heights fall.

b. What is the height for a female who is three standard deviations below the mean? Would this be a rather unusual height? Why?

2.97 Energy and water consumption: In (a) and (b), what shape do you expect for the distributions of electricity use and water use in a recent month in Gainesville, Florida? Why? (Data supplied by N. T. Kamhoot, Gainesville Regional Utilities.)

a. Residential electricity used had mean = 780 and standard deviation = 506 kilowatt hours (Kwh). The minimum usage was 3 Kwh and the maximum was 9390 Kwh.

b. Water consumption had mean = 7100 and standard deviation = 6200 (gallons).

2.98 Stock fund volatility: An article in the March 2005 issue of _T. Rowe Price Investor_ notes that its Capital Appreciation Fund is less volatile than most multi-cap value funds. For the five years 2000–2004, the standard deviation of the fund's annual return was 9.6, compared to 16.1 for the Lipper Multi-Cap Value Funds Average. The article mentions that "Standard deviation is a statistical measure of volatility and indicates how much an investment's actual performance during a specific period varies from its average performance over a longer period." How would you explain what this means to someone who has never studied statistics?

2.99 Student heights: The "Heights" data file on the text CD has heights for female and male students (in inches). For males, the mean is 70.9 and the standard deviation is 2.9. For females, the mean is 65.3 and the standard deviation is 3.0.

a. Interpret the male heights using the Empirical Rule.

b. Compare the center and spread of the height distributions for females and males.

c. The lowest observation for males was 62. How many standard deviations below the mean is it?

2.100 Cigarette tax: How do cigarette taxes per pack vary from one state to the next? The data set of 2003 cigarette taxes for all 50 states is in the "Cigarette tax" data file on the text CD.

a. Use software to construct a histogram. Write a short description of the distribution, noting shape and possible outliers.

b. Find the mean and median for the cigarette taxes. Which is larger, and why would you have expected that from the histogram?

c. Find the standard deviation, and interpret it.

2.101 Cereal sugar values: Revisit the sugar data for breakfast cereals that are given in the table below.

a. Interpret the box plot in the figure (EXCEL output) by giving approximate values for the five-number summary.

Boxplot of SUGAR(g)

b. What does the box of the box plot suggest about possible skew?

c. The mean is 8.20 and the standard deviation is 4.56. Find the *z*-score associated with the minimum sugar value of 1. Interpret.

Cereal	Sugar (g)
Frosted Mini Wheats	7
Raisin Bran	12
All Bran	5
Apple Jacks	14
Capt Crunch	12
Cheerios	1
Cinnamon Toast	13
Crackling Oat Bran	10
Crispix	3
Frosted Flakes	11
Fruit Loops	13
Grape Nuts	3
Honey Nut Cheerios	10
Life	6
Oatmeal Raisin Crisp	10
Sugar Smacks	15
Special K	3
Wheaties	3
Corn Flakes	2
Honeycomb	11

2.102 Stock prices positions: The data values below represent the prices per share of the 10 most actively traded stocks from the New York Stock Exchange (rounded to the nearest dollar) on Oct. 1, 2003.

2 4 11 12 13 15 31 31 37 47

a. Sketch a dot plot or construct a stem-and-leaf plot.

b. Find the median, the first quartile, and the third quartile.

c. Sketch a box plot. What feature of the distribution displayed in the plot in (a) is not obvious in the box plot? (*Hint*: Are there any gaps in the data?)

■ 2.103 Temperatures in Central Park: Access the "Central Park yearly temps" on the text CD. Using software,

a. Construct a histogram and interpret, noting shape, center, and spread.

b. Construct a box plot. Identify any potential outliers.

c. Find and interpret the five-number summary of positions.

d. Find and interpret the mean and standard deviation.

2.104 Teachers' salaries: According to *Statistical Abstract of the United States, 2003*, average salary (in dollars) of secondary school classroom teachers in 2002 in the United States varied among states with a five-number summary of: minimum = 31,200, Q1 = 37,400, Median = 40,000, Q3 = 48,400, Maximum = 57,200.

a. Find and interpret the range and interquartile range.

b. Sketch a box plot, marking the five-number summary on it.

c. Predict the direction of skew for this distribution. Explain.

d. If the distribution, although skewed, is approximately bell shaped, which of the following would be the most realistic value for the standard deviation: (i) 100, (ii) 1000, (iii) 6000, or (iv) 25,000? Explain your reasoning.

2.105 Health insurance: In 2002, the five-number summary of positions for the statewide percentage of people without health insurance had a minimum of 7.5% (Iowa), Q1 = 9.6, Median = 12.4, Q3 = 15.9, and maximum of 23.5% (Texas) (*Statistical Abstract of the United States, 2003*).

a. Do you think the distribution is symmetric, skewed right, or skewed left? Why?

b. Which is most plausible for the standard deviation: −16, 0, 4, 15, or 25? Why?

2.106 What box plot do you expect?: For each of the following variables, sketch a box plot that would be plausible.

a. Exam score (min = 0, max = 100, mean = 87, standard deviation = 10)

b. IQ (mean = 100 and standard deviation = 16)

c. Weekly religious contribution (median = $10 and mean = $17)

2.107 High school graduation rates: The distribution of high school graduation rates in the United States in 2002 had a minimum value of 78.1 (Texas), first quartile of 81.6, median of 86.5, third quartile of 88.1, and maximum value of 92.2 (Alaska) (*Statistical Abstract of the United States, 2003*).

a. Report the range and the interquartile range.

b. Would a box plot show any potential outliers? Explain.

2.108 SAT scores revisited: The U.S. statewide average total SAT scores (math plus verbal) for 2003 are summarized in the box plot shown.

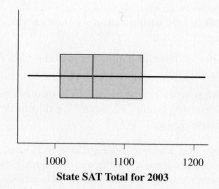

State SAT Total for 2003

a. Explain how the box plot gives you information about the distribution shape.

b. Using the box plot, give the approximate value for each component of the five-number summary. Interpret each.

c. A histogram for these data is also shown. What feature of the distribution would be missed by using only the box plot?

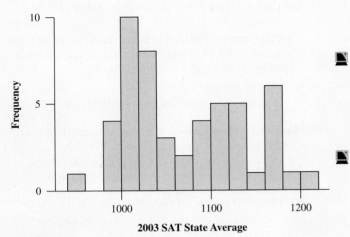

2003 SAT State Average

2.109 Blood pressure: A World Health Organization study (the MONICA project) of health in various countries reported that in Canada, systolic blood pressure readings have a mean of 121 and a standard deviation of 16. A reading above 140 is considered to be high blood pressure.

a. What is the z-score for a blood pressure reading of 140? How is this z-score interpreted?

b. The systolic blood pressure values have a bell-shaped distribution. Report intervals within which about 95% of the systolic blood pressure values fall.

2.110 No cereal sodium: The cereal sodium values have a mean of 185.5 and a standard deviation of 71.2. Find the z-score for the cereal that has a sodium value of 0. Interpret.

2.111 Who was Roger Maris?: Roger Maris, who spent most of his professional baseball career with the New York Yankees, held the record for the most home runs in one season (61) from 1961 until 1998, when the record was broken by Mark McGwire. Maris played in the major leagues from 1957 to 1968. The number of home runs he hit in each year that he played is summarized in MINITAB output as shown.

Variable	N	Mean	Median	StDev
RMHR	12	22.92	19.50	15.98

Variable	Minimum	Maximum	Q1	Q3
RMHR	5.00	61.00	10.00	31.75

a. Use the three standard deviation criterion to determine if any potential outliers are present.

b. The criterion in (a) requires the distribution to be approximately bell-shaped. Is there any evidence here to contradict this? Explain.

c. A sports writer commented that Roger Maris hit *only* 13 home runs in 1966. Was this unusual for Maris? Comment, using statistical justification.

Concepts and Investigations

2.112 Baseball's great home run hitters: The file "Baseball's HR Hitters" on the text CD contains data on the number of home runs hit each season by some of baseball's great home run hitters. Analyze these data using techniques introduced in this chapter to help judge statistically which player might be considered the best. Specify the criterion you use to compare the players.

2.113 How much spent on haircuts?: Is there a difference in how much males and females spend on haircuts? Access the "Georgia student survey" data file on the text CD or use your class data to explore this question using appropriate graphical and numerical methods. Write a brief report describing your analyses and conclusions.

2.114 Controlling asthma: A study of 13 children suffering from asthma (*Clinical and Experimental Allergy*, vol. 20, pp. 429–432, 1990) compared single inhaled doses of formoterol (F) and salbutamol (S). Each child was evaluated using both medications. The outcome measured was the child's peak expiratory flow (PEF) eight hours following treament. Is there a difference in the PEF level for the two medications? The data on PEF follow:

Child	F	S	Child	F	S	Child	F	S	Child	F	S
1	310	270	5	410	380	9	330	365	13	220	90
2	385	370	6	370	300	10	250	210			
3	400	310	7	410	390	11	380	350			
4	310	260	8	320	290	12	340	260			

a. Construct plots to compare formoterol and salbutamol. Write a short summary comparing the two distributions of the peak expiratory flow.

b. Consider the distribution of differences between the PEF levels of the two medications. Find the 13 differences and construct and interpret a plot of the differences. If on the average there is no difference between the PEF level for the two brands, where would you expect the differences to be centered?

2.115 Back-to-back stem-and-leaf plot: To compare two groups graphically, one can use the same stems for each and put leaves for one group on one side and for the other group on the other side. This is called a **back-to-back stem-and-leaf plot**. The figure shown compares sugar amounts for cereals listed in Table 2.3 according to whether they were intended either for adults or children.

Back-to-Back Stem-and-Leaf Plot of Sugar for Adult and Children Cereals	
Adult	Children
|100|	0
0 |200|	
0 0 0 |300|	0
|400|	
0 |500|	
0 |600|	
0 |700|	
|800|	
|900|	
0 0 |1000|	0
|1100|	0 0
0 |1200|	0
|1300|	0 0
|1400|	0
|1500|	0

Cereal	Type	Sodium
Frosted Mini Wheats	A	0
Raisin Bran	A	210
All Bran	A	260
Apple Jacks	C	125
Capt Crunch	C	220
Cheerios	C	290
Cinnamon Toast	C	210
Crackling Oat Bran	A	140
Crispix	A	220
Frosted Flakes	C	200
Fruit Loops	C	125
Grape Nuts	A	170
Honey Nut Cheerios	C	250
Life	A	150
Oatmeal Raisin Crisp	A	170
Sugar Smacks	C	70
Special K	A	230
Wheaties	C	200
Corn Flakes	A	290
Honeycomb	C	180

a. Is the distribution of sugar values similar, or different, for the two cereal types? If they are different, describe the difference.

b. Construct a back-to-back stem-and-leaf plot for the sodium values, shown again above. Is the distribution of sodium values similar, or different, for the two cereal types?

2.116 You give examples: Give an example of a variable that you'd expect to have a distribution that is

a. Approximately symmetric
b. Skewed to the right
c. Skewed to the left
d. Bimodal
e. Skewed to the right, with a mode and median of 0 but a positive mean

2.117 Political conservatism and liberalism: Where do Americans tend to fall on the conservative-liberal political spectrum? The General Social Survey asks, "I'm going to show you a seven-point scale on which the political views that people might hold are arranged from extremely liberal, point 1, to extremely conservative, point 7. Where would you place yourself on this scale?" The table shows the seven-point scale and the distribution of 1331 responses for a recent survey (2002).

Score	Category	Frequency
1.	Extremely liberal	47
2.	Liberal	143
3.	Slightly liberal	159
4.	Moderate, middle of road	522
5.	Slightly conservative	209
6.	Conservative	210
7.	Extremely conservative	41

This is a categorical scale with ordered categories, called an **ordinal scale**. Ordinal scales are often treated in a quantitative manner by assigning scores to the categories

and then using numerical summaries such as the mean and standard deviation.

a. Using the scores shown in the table, the mean for these data equals 4.12. Using the reasoning from Example 12, set up the way this would be calculated.

b. Identify the mode.

c. In which category does the median fall? Why?

2.118 Mode but not median and mean: The previous exercise showed how to find the mean and median when a categorical variable has ordered categories. A categorical scale that does *not* have ordered categories (such as choice of religious affiliation or choice of major in college) is called a **nominal scale**. For such a variable, the mode applies, but not the mean or median. Explain why.

For the multiple choice questions 2.119–2.121, select the best answer.

2.119 GRE scores: In a study of graduate students who took the Graduate Record Exam (GRE), the Educational Testing Service reported that for the quantitative exam, U.S. citizens had a mean of 529 and standard deviation of 127, whereas the non-U.S. citizens had a mean of 649 and standard deviation of 129.

a. Both groups had about the same amount of variability in their scores, but non-U.S. citizens performed better, on the average, than U.S. citizens.

b. If the distribution of scores was approximately bell shaped, then almost no U.S. citizens scored below 400.

c. If the scores range between 200 and 800, then probably the scores for non-U.S. citizens were symmetric and bell shaped.

d. A non-U.S. citizen who scored three standard deviations below the mean had a score of 200.

2.120 Facts about s: Which statement about the standard deviation s is false?

a. s can never be negative.

b. s can never be zero.

c. For bell-shaped distributions, about 95% of the data fall within $\bar{x} \pm 2s$.

d. s is a nonresistant (sensitive to outliers) measure of spread, as is the range.

2.121 Relative GPA: The mean GPA for all students at a community college in the fall 2005 semester was 2.77. A student with a GPA of 2.0 wants to know her relative standing in relation to the mean GPA. A numerical summary that would be useful for this purpose is the

a. standard deviation

b. median

c. interquartile range

d. number of students at the community college

2.122 Soccer true-false: According to a story in the *Guardian* newspaper (football.guardian.co.uk) in the U.K., the mean wage for a Premiership player in 2001–2002 in the U.K. was 600,000 pounds. True or false: If the income distribution is skewed to the right, then the median salary was even larger than 600,000 pounds.

◆◆ **2.123** **Mean for grouped data**: Refer to the calculation of the mean in Example 12 or in Exercise 2.117. Explain why the mean for grouped data can be expressed as a sum, taking each possible outcome times the *proportion* of times it occurred.

◆◆ **2.124** **Male heights:** According to a recent report from the U.S. National Center for Health Statistics, for males with age 25–34 years, 2% of their heights are 64 inches or less, 8% are 66 inches or less, 27% are 68 inches or less, 39% are 69 inches or less, 54% are 70 inches or less, 68% are 71 inches or less, 80% are 72 inches or less, 93% are 74 inches or less, and 98% are 76 inches or less. These are called **cumulative percentages**.

   a. Which category has the median height? Explain why.

   b. Nearly all the heights fall between 60 and 80 inches, with fewer than 1% falling outside that range. If the heights are approximately bell-shaped, give a rough approximation for the standard deviation of the heights. Explain your reasoning.

◆◆ **2.125** **Range and standard deviation approximation:** Use the Empirical Rule to explain why the standard deviation of a bell-shaped distribution for a large data set is often roughly related to the range by Range ≈ 6*s*. (For small data sets, one may not get any extremely large or small observations, and the range may be smaller, for instance about 4 standard deviations.)

◆◆ **2.126** **Range the least resistant:** We've seen that measures such as the mean, the range, and the standard deviation can be highly influenced by outliers. Explain why the range is worst in this sense. (*Hint*: As the sample size increases, explain how a single extreme outlier has less effect on the mean and standard deviation, but can still have a large effect on the range.)

◆◆ **2.127** **Using MAD to measure variability:** The standard deviation is the most popular measure of spread. It uses squared deviations, since the ordinary deviations sum to zero. An alternative measure is the **mean absolute deviation**, $\sum |x - \bar{x}|/n$.

   a. Explain why greater spread tends to result in larger values of this measure.

   b. Would this be more, or less, resistant than the standard deviation? Explain.

◆◆ **2.128** **Rescale the data:** The mean and standard deviation of a sample may change if data are rescaled (for instance, temperature changed from Fahrenheit to Celsius). For a sample with mean $\bar{x}$, adding a constant c to each observation changes the mean to $\bar{x} + c$, and the standard deviation s is unchanged. Multiplying each observation by $c > 0$ changes the mean to $c\,\bar{x}$ and the standard deviation to cs.

   a. Scores on a difficult exam have a mean of 57 and a standard deviation of 20. The teacher boosts all the scores by 20 points before awarding grades. Report the mean and standard deviation of the boosted scores. Explain which rule above you used, and identify c.

   b. Suppose that annual income for some group has a mean of $39,000 and a standard deviation of $15,000. Values are converted to British pounds for presentation to a British audience. If one British pound equals $2.00, report the mean and standard deviation in British currency. Explain which rule above you used, and identify c.

   c. Adding a constant and/or multiplying by a constant is called a **linear transformation** of the data. Do linear transformations change the *shape* of the distribution? Explain your reasoning.

Class Explorations

  2.129 **The average student:** Refer to the data file you created in Activity 3 in Chapter 1. For variables chosen by your instructor, describe the "average student" in your class. Prepare a one-page summary report. In class, students will compare their analyses of what makes up an average student.

  2.130 **Create own data:** For the *mean versus median* applet, your instructor will give you a data set to illustrate the effect of extreme observations on the mean and median. Write a short summary of your observations.

  2.131 **GSS:** Access the General Social Survey at http://sda.berkeley.edu/GSS.

   a. Find the frequency table and histogram for Example 6 on TV watching. (*Hint*: Place a dot at "frequencies and crosstabulation," click on "start," enter TVHOURS as the row variable, YEAR(2002) as the selection filter, choose "bar chart" for Type of Chart, and click on "run the table.")

   b. Your instructor will have you obtain graphical and numerical summaries for another variable from the GSS. Students will compare results in class.

BIBLIOGRAPHY

Franklin, Christine A. (2002). "The Other Life of Florence Nightingale," *Mathematics Teaching in the Middle School* 7(6): 337–339.

Friendly, Michael. (2000) "Florence Nightingale-Statistical Links." www.math.yorku.ca/SCS/Gallery/flo.html.

Tukey, J. W. (1977). *Exploratory Data Analysis*. Reading, MA: Addison-Wesley.

3 | Association: Contingency, Correlation, and Regression

EXAMPLE 1

Is Smoking Actually Beneficial to Your Health?

Picture the Scenario

Numerous studies have investigated the potential harm of cigarette smoking. It is now generally believed that smoking is harmful to your health in many ways. However, one study found conflicting evidence. This study conducted a survey, of 1314 women in the United Kingdom during 1972–1974, asking each woman whether she was a smoker. Twenty years later, a follow-up survey observed whether each woman was deceased or still alive. The researchers studied the possible link between whether a woman smoked and whether she survived the 20-year study period. During that period, 24% of the smokers died and 31% of the nonsmokers died. The higher death rate for the nonsmokers is surprising.

Questions to Explore

- Is smoking actually beneficial to your health, since a smaller percentage of smokers died?

- What descriptive statistical methods can we use in exploring the data?

- If we observe a link between smoking status and survival status, is there something that could explain how this happened?

Thinking Ahead

In Chapter 2 we distinguished between categorical and quantitative variables. In this study, we can identify two categorical variables. One is smoking status—whether a woman was a smoker (yes or no). The other is survival status—whether a woman survived the 20-year study period (yes or no). In practice, research investigations almost always need to analyze more than one variable. The link between two variables is often the primary focus.

This chapter presents descriptive statistics for examining data on two variables. We'll analyze what these data suggest about the link between smoking and cancer. We'll learn that a third variable can influence the results. In this case, we'll see that the age of the woman is important. We'll revisit this example in Example 16 and analyze the data while taking age into account as well.

---

[1]Described by D. R. Appleton et al., *American Statistician*, vol. 50, pp. 340–341 (1996).

Example 1 has categorical variables, but we'll learn also how to examine links between pairs of quantitative variables. For instance, we might want to answer questions such as "What's the relationship between the daily amount of gasoline use by automobiles and the amount of air pollution?" or "Do high schools that have higher per-student funding tend to have higher mean SAT scores for their students?"

Response Variables and Explanatory Variables

When we analyze data on two variables, our first step is to distinguish between the **response variable** and the **explanatory variable**.

IN WORDS The data analysis examines how the outcome on the **response** variable *depends on* or is *explained by* the value of the **explanatory** variable.	**Response Variable and Explanatory Variable** The **response variable** is the outcome variable on which comparisons are made. The **explanatory variable** defines the groups to be compared with respect to values on the response variable.

In Example 1, survival status (whether a woman is alive after 20 years) is the response variable. Smoking status (whether the woman was a smoker) is the explanatory variable. In a study of air pollution in several countries, the carbon dioxide (CO_2) level in a country's atmosphere might be a response variable, and the explanatory variable could be the country's amount of gasoline use for automobiles, or its gross domestic product, or its energy use.

The main purpose of a data analysis with two variables is to investigate whether there is an **association** and to describe the nature of that association.

IN WORDS When there's an **association**, the likelihood of a particular value for one variable depends on the value of the other variable. The chance of a college GPA above 3.5 is greater for those with high school GPA = 4.0 than for those with high school GPA = 3.0. So high school GPA and college GPA have an association.	**Association between Two Variables** An **association** exists between two variables if a particular value for one variable is more likely to occur with certain values of the other variable.

In Example 1, surviving the study period was more likely for smokers than for nonsmokers. So, there is an association between survival status and smoking status. For higher levels of energy use, does the CO_2 level in the atmosphere tend to be higher? If so, then there is an association between energy use and CO_2 level.

This chapter presents methods for studying whether associations exist and for describing how strong they are. We explore associations between categorical variables in Section 3.1 and between quantitative variables in Sections 3.2 and 3.3.

3.1 | How Can We Explore the Association between Two Categorical Variables?

How would you respond to the question, "Taken all together, would you say that you are very happy, pretty happy, or not too happy?" We could summarize the percentage of people who have each of the three possible outcomes for this categorical variable using a table, a bar graph, or a pie chart. However, we'd probably want to know how the percentages depend on the values for other variables, such as a person's marital status (married, unmarried). Is the percentage who report being very happy higher for married people than for unmarried people?

We'll now look at ways to summarize the association between two categorical variables. You can practice the concepts with some data on personal happiness in Exercises 3.3, 3.8, and 3.56.

EXAMPLE 2

Are Pesticides Present Less Often in Organic Foods?

Picture the Scenario

Do you ever worry about whether pesticide residues are on your foods?

One appeal of eating organic foods is the belief that they are pesticide-free and thus healthier. However, relatively little fruit and vegetable acreage is cultivated as organic (only 2% in the United States), and consumers pay a premium for organic food.

How can we investigate how the percentage of foods carrying pesticide residues compares for organic foods and conventionally grown foods? The Consumers Union led a study based on sampling carried out by the United States Department of Agriculture (USDA) and the state of California.[2] The sampling was conducted as part of regulatory monitoring of foods for pesticide residues before being distributed for sale. For this study, Table 3.1 displays the frequencies of foods for all possible category combinations of the two variables, "food type" and "pesticide status."

> **TABLE 3.1: Frequencies for Food Type and Pesticide Status.**
>
> The row totals and the column totals are the frequencies for the categories of each variable. The counts inside the table give information about the association.
>
> | Food Type | Pesticide Status | | Total |
> |---|---|---|---|
> | | Present | Not Present | |
> | Organic | 29 | 98 | 127 |
> | Conventional | 19485 | 7086 | 26571 |
> | **Total** | **19514** | **7184** | **26698** |

Questions to Explore

a. What is the response variable, and what is the explanatory variable?

b. How does the proportion of foods that had pesticides present compare for the two food types?

c. What proportion of all sampled produce items contained pesticide residues?

Think It Through

a. Pesticide status, namely whether pesticide residues are present, is the outcome of interest. The food type, organic or conventionally grown, is the variable that defines two groups to be compared on their pesticide status. So, pesticide status is the response variable and food type is the explanatory variable.

---

[2]*Source: Food Additives and Contaminants* 200, Vol.19 No., 427–446.

b. From Table 3.1, 29 out of 127 organic foods contained pesticide residues. The proportion with pesticides is 29/127 = 0.228. Likewise, 19,485 out of 26,571 conventionally grown foods contained pesticide residues. The proportion is 19,485/26,571 = 0.733, much higher than for organic foods. Since 0.733/0.228 = 3.2, the relative occurrence of pesticide residues for conventionally grown produce is approximately 3 times that for organically grown produce.

c. The overall proportion of sampled produce that contained pesticide residues is the total number with pesticide residues out of the total number of food items. This is

$$(29 + 19,485)/(127 + 26,571) = 19,514/26,698 = 0.731.$$

We can find this result using the column totals of Table 3.1.

Insight

The value of 0.731 for the overall proportion containing pesticide residues is close to the proportion for conventionally grown foods alone, which we found to be 0.733. This is because conventionally grown foods make up a very high percentage of the sample. (From the row totals, the proportion of the sampled items that were conventionally grown was 26,571/26,698 = 0.995.) In summary, pesticide residues occurred in more than 73% of the sampled items, and they were much more common (about 3 times as common) for conventionally grown than organic foods.

TO PRACTICE THIS CONCEPT, TRY EXERCISE 3.3

Contingency Tables

Table 3.1 has two categorical variables: Food type and pesticide status. We can analyze the categorical variables separately, through the column totals for pesticide status and the row totals for food type. These totals are the category counts for the separate variables, for instance (19514, 7184) for the (present, not present) categories of pesticide status. We can also study the association between them, for instance by using the counts for the category combinations to find proportions, as in Example 2, part (b). Table 3.1 is an example of a **contingency table**.

Contingency Table

A **contingency table** is a display for two categorical variables. Its rows list the categories of one variable and its columns list the categories of the other variable. Each entry in the table is the frequency of cases in the sample with certain outcomes on the two variables.

	Pesticides	
Food Type	Yes	No
Organic	29	98
Conventional	19485	7086
	cell ↑	

Each row and column combination in a contingency table is called a **cell**. For instance, the first cell in the second row of Table 3.1 (shown again in the margin) has the frequency 19,485, the number of observations in the conventional category of food type that had pesticides present. The process of taking a data file and finding the frequencies for the cells of a contingency table is referred to as **cross-tabulation** of the data. Table 3.1 is formed by cross-tabulation of food type and pesticide status for the 26,698 sampled food items.

Conditional Proportions

Consider the question, "Do organic and conventionally grown foods differ in the proportion of food items that contain pesticide residues?" To answer, we find the proportions on pesticide status within each category of food type, and then compare them. From Example 2, the proportions are 29/127 = 0.23 for organic foods and 19,485/26,571 = 0.73 for conventionally grown foods.

These proportions are called **conditional proportions**. The proportions are formed **conditional** upon (that is, given) food type. Given that the food type is organic, the proportion of food items with pesticides present equals 0.23. Table 3.2 shows the conditional proportions.

Recall

Using Table 3.1, shown again here, we obtain 0.23 from 29/127 and 0.77 from 98/127, the cell counts divided by the first row total. ■

	Pesticides		
Food Type	Yes	No	**Total**
Organic	29	98	**127**
Conven.	19485	7086	**26571**

TABLE 3.2: Conditional Proportions on Pesticide Status, for Two Food Types.

These treat pesticide status as the response variable. The sample size *n* in a row shows the total on which the conditional proportions in that row were based.

Food Type	Pesticide Status		**Total**	*n*
	Present	Not Present		
Organic	0.23	0.77	**1.000**	127
Conventional	0.73	0.27	**1.000**	26571

The conditional proportions in each row sum to 1.0. The sample size *n* for each set of conditional proportions is listed in the margin, so the reader can determine the frequencies on which the conditional proportions were based. *Whenever we distinguish between a response variable and an explanatory variable, it is natural to form conditional proportions for categories of the response variable.*

EXAMPLE 3

How Can We Compare Pesticide Residues for the Food Types Graphically?

Picture the Scenario

For the data on food type and pesticide status, we've now seen two ways to display the data. Table 3.1 showed cell frequencies in a contingency table, and Table 3.2 showed conditional proportions.

Questions to Explore

a. How can we use a single graph to show the relationship?

b. What does the graph tell us?

Think It Through

a. We've seen that to compare the two food types on pesticide status, the response variable, we can find conditional proportions for the pesticide status categories (shown in Table 3.2). Figure 3.1 shows bar graphs of the

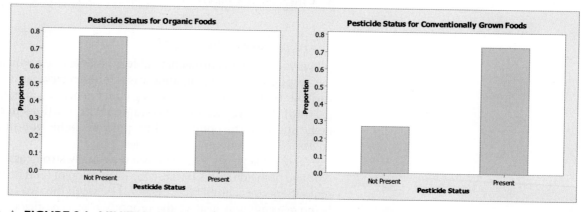

▲ **FIGURE 3.1: MINITAB output of conditional proportions on pesticide status.** In this graph, the conditional proportions are shown separately for each food type. **Question:** Can you think of a way to display the results that makes it easier to compare the food types on the pesticide status?

conditional proportions, one for organic foods and another for conventionally grown foods. More efficiently, we can construct a single bar graph that shows **side-by-side bars** to compare the conditional proportion of pesticide residues in conventionally grown and organic foods. Figure 3.2 shows this graph, which conveys the same information as the two separate bar graphs in Figure 3.1. The advantage of the side-by-side bar graph is that it allows for easy comparison of the two food types.

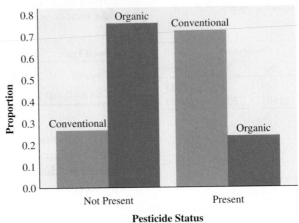

Pesticide Status for Organic vs. Conventional Foods

▲ **FIGURE 3.2: Conditional proportions on pesticide status, given the food type.**
For a particular pesticide status category, the side-by-side bars compare the two food types.
Question: Comparing the bars, how would you describe the difference between organic and conventionally grown foods in the conditional proportion with pesticide residues present?

b. Figure 3.2 clearly shows that that the proportion of foods having pesticides present is much higher for conventionally grown food than for organically grown food.

Insight

Chapter 2 used bar graphs to display proportions for a single variable. With two variables, bar graphs usually display conditional proportions, as in Figure 3.2. This is useful for making comparisons, such as the way Figure 3.2 compares organic and conventionally grown foods in terms of the proportion of pesticide residues.

TO PRACTICE THIS CONCEPT, TRY EXERCISE 3.4

Is There an Association?

When you form a contingency table, first decide on which variable to treat as the response variable. In some cases, either variable could be the response variable, such as in cross-tabulating belief in heaven (yes, no) with belief in hell (yes, no). Then you can form conditional proportions in either or both directions. Studying the conditional proportions helps you judge whether there is an *association* between the variables.

Table 3.2 suggests that there is a reasonably strong association between food type and pesticide status, because the proportion of food items with pesticides present differs considerably (0.23 vs. 0.73) between the two food types. There would be *no association* if, instead, the proportion with pesticides present had been the same for each food type. For instance, suppose that for each food type, 60% had pesticides present and 40% did not have pesticides present, as shown in Table 3.3. Then, the food types have the same pesticide status distribution.

TABLE 3.3: Conditional Proportions on Pesticide Status for Each Food Type, Showing No Association.

The conditional proportions for the response variable (pesticide status) categories are the same for each food type.

| | Pesticide Status | | |
Food Type	Present	Not Present	Total
Organic	0.40	0.60	127
Conventional	0.40	0.60	26571

Figure 3.3 is a side-by-side display of the conditional proportions from Table 3.3. Compare Figure 3.3, no association, with Figure 3.2, which shows an association. In Figure 3.3, for any particular response category, the bars are the same height, indicating no association: Pesticide status does not depend on food type. By contrast, in Figure 3.2 the bars have quite different heights.

IN PRACTICE: Comparing "population" and "sample" conditional proportions

When sampling from a population, even if there is *no* association between the two variables in the *population* you can't expect the *sample* conditional proportions to be *exactly* the same because of ordinary random variation from sample to sample. Later on in the text, we'll present inferential methods to determine if observed sample differences between conditional proportions are large enough to indicate that the variables are associated in the population.

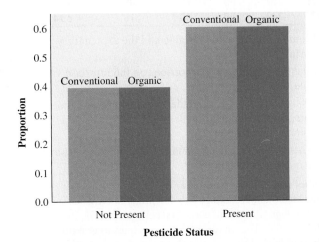

▲ **FIGURE 3.3:** Conditional proportions on pesticide status, given food type, showing no association. **Question:** What's the difference between Figures 3.2 and 3.3 in the pattern shown by the bars in the graph?

3.1 Practicing the Basics

3.1 Which is the response/explanatory variable?: For the following pairs of variables, which more naturally is the response variable and which is the explanatory variable?

  a. College grade point average (GPA) and high school GPA

  b. Number of children and mother's religion

  c. Happiness (not too happy, pretty happy, very happy) and whether married (yes, no)

3.2 Sales and advertising: Each month, the owner of Fay's Tanning Salon records in a data file the monthly total sales receipts and the amount spent that month on advertising.

  a. Identify the two variables.

  b. For each variable, indicate whether it is quantitative or categorical.

  c. Identify the response variable and the explanatory variable.

3.3 Does higher income make you happy?: Every General Social Survey includes the question "Taken all together, would you say that you are very happy, pretty happy, or not too happy?" The table uses the 2002 survey to cross-tabulate happiness with family income, measured as the response to the question, "Compared with American families in general, would you say that your family income is below average, average, or above average?"

Happiness and Family Income, from 2002 General Social Survey

| | Happiness | | | |
Income	Not Too Happy	Pretty Happy	Very Happy	Total
Above average	17	90	51	158
Average	45	265	143	453
Below average	31	139	71	241
Total	**93**	**494**	**265**	**852**

a. Identify the response variable and the explanatory variable.

b. Construct the conditional proportions on happiness at each level of income. Interpret and summarize the association between these variables.

c. Overall, what proportion of people reported being very happy?

3.4 **Religious activities:** In a recent General Social Survey, respondents answered the question, "In the past month, about how many hours have you spent praying, meditating, reading religious books, listening to religious broadcasts, etc.?" The responses on this variable were cross-tabulated with the respondent's gender. The table shows the results.

Religious Activity by Gender

	Number of Hours of Home Religious Activity					
Gender	0	1–9	10–19	20–39	40 or more	**Total**
Female	229	297	88	103	49	**766**
Male	276	243	59	40	16	**634**

a. Identify the response variable and the explanatory variable.

b. Find the conditional proportions on the response variable, and interpret them.

c. Using software, create a side-by-side bar graph that compares males and females on the response variable. Summarize results.

d. What would be a disadvantage of using two pie charts to make this comparison?

3.5 **Alcohol and college students:** The Harvard School of Public Health, in its College Alcohol Study Survey, has surveyed college students in about 200 colleges in 1993, 1997, 1999, and 2001. The survey asks students questions about their drinking habits. Binge drinking is defined as 5 drinks in a row for males and 4 drinks in a row for females. The table shows results from the 2001 study, cross-tabulating subjects' gender by whether they have participated in binge drinking.

Binge Drinking by Gender

	Binge Drinking Status		
Gender	Binge Drinker	Non-Binge Drinker	**Total**
Male	1908	2017	**3925**
Female	2854	4125	**6979**
Total	**4762**	**6142**	**10904**

a. Identify the response variable and the explanatory variable.

b. Report the cell counts of subjects who were (i) male and a binge drinker, (ii) female and a binge drinker.

c. Can you compare the counts in (b) to answer the question, "Is there a difference between male and female students in the proportion who binge drink?" Explain.

d. Construct a contingency table that shows the conditional proportions of sampled students who do or do not binge drink, given gender. Interpret.

e. Based on (d), does it seem that there is an association between binge drinking and gender? Explain.

3.6 **How to fight terrorism?:** A survey of 1000 adult Americans (*Rasmussen Reports*, April 15, 2004) asked each whether the best way to fight terrorism is to let the terrorists know we will fight back aggressively or to work with other nations to find an international solution. The first option was picked by 53% of the men but by only 36% of the women in the sample.

a. Identify the response variable and the explanatory variable, and their categories.

b. Use a contingency table to display the results.

c. Explain why the percentages reported here are *conditional* percentages.

d. Give an example of how results would have to differ from these for you to conclude that there's *no* evidence of association between these variables.

3.7 **Gender gap in party ID:** In recent election years, political scientists have analyzed whether a "gender gap" exists in political beliefs and party identification. The table shows data collected from the 2002 General Social Survey on gender and party identification (ID).

Party ID by Gender

	Party Identification			
Gender	Democrat	Independent	Republican	**Total**
Male	356	460	369	**1185**
Female	567	534	395	**1496**
Total	**923**	**994**	**764**	**2681**

a. Identify the response and explanatory variables.

b. What proportion of sampled individuals is (i) male and Republican, (ii) female and Republican?

c. What proportion of the overall sample is (i) male, (ii) Republican?

d. The figure (MINITAB output) displays the proportion of individuals identifying with each political party, given gender. What are these proportions called? Is there a difference between males and females in the proportions that identify with a particular party? Summarize whatever gender gap you observe.

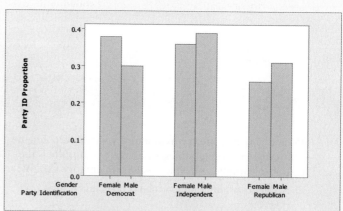

3.8 **Use the GSS:** Go to the GSS Web site http://sda.berkeley.edu/GSS, click on "frequencies and crosstabulation," then "start," type "sex" for the row variable and "happy" for the column variable (without the quotation marks), put a check in the "row" box only for percentaging in the table options, and click on "Run the table."

a. Report the contingency table of counts.

b. Report the conditional proportions to compare the genders on reported happiness.

c. Are females and males similar, or quite different, in their reported happiness?

3.2 | How Can We Explore the Association between Two Quantitative Variables?

In practice, when we investigate the association between two variables, there are three types of cases:

■ The variables could both be *categorical*, such as food type and pesticide status. The data can then be displayed in a contingency table. We've seen we can then explore the association by comparing conditional proportions.

■ One variable could be *quantitative* and one could be *categorical*, such as income and gender. We can then compare the categories (such as females and males) using summaries of center and spread for the quantitative variable (such as the mean and standard deviation of income) and graphics such as side-by-side box plots.

■ Both variables could be *quantitative*, such as in Example 4 below. We then analyze how the outcome on the response variable tends to change as the value of the explanatory variable changes. The rest of the chapter considers this case.

In exploring the relationship between two quantitative variables, we'll use the principles introduced in Chapter 2 for a single variable. We first use graphics to look for an overall pattern. We follow up with numerical summaries. We check also for unusual observations that deviate from the overall pattern and that may affect results.

Recall

Figure 2.16 in Chapter 2 used **side-by-side box plots** to compare heights for females and males. ■

EXAMPLE 4

Worldwide Use of the Internet

Picture the Scenario

In the past ten years the number of people who use the Internet has mushroomed. Although its use is now common in Westernized countries, in some countries few people use it. In 2001, for instance, about half of North Americans used the Internet, but only about 1% used it in Saudi Arabia, and only about 0.1% in Nigeria.

Table 3.4 shows recent data for 39 nations on Internet use,[3] measured as the percentage of adult residents who use the Internet. What other variables are likely to be associated with Internet use? Table 3.4 shows four possibilities:

GDP: Gross domestic product, per capita, in thousands of U.S. dollars

CO₂: Carbon-dioxide emissions, per capita (in metric tons)

Cellular: Percentage of adults who are cellular-phone subscribers

Fertility: Mean number of children per adult woman

Recall

Carbon dioxide emissions, which depend strongly on energy use, is an index of a nation's air pollution. ■

---

[3]*Source: Human Development Report 2003*, published by the United Nations Development Programme, http://hdr.undp.org

TABLE 3.4: Data for Several Nations from Human Development Report, 2003

Nation	Internet	GDP	CO2	Cellular	Fertility
Algeria	0.6	6.1	3.0	0.3	2.8
Argentina	10.1	11.3	3.8	19.3	2.4
Australia	37.1	25.4	18.2	57.4	1.7
Austria	38.7	26.7	7.6	81.7	1.3
Belgium	31.0	25.5	10.2	74.7	1.7
Brazil	4.7	7.4	1.8	16.7	2.2
Canada	46.7	27.1	14.4	36.2	1.5
Chile	20.1	9.2	4.2	34.2	2.4
China	2.6	4.0	2.3	11.0	1.8
Denmark	43.0	29.0	9.3	74.0	1.8
Egypt	0.9	3.5	2.0	4.3	3.3
Finland	43.0	24.4	11.3	80.4	1.7
France	26.4	24.0	6.1	60.5	1.9
Germany	37.4	25.4	9.7	68.2	1.4
Greece	13.2	17.4	8.2	75.1	1.3
India	0.7	2.8	1.1	0.6	3.0
Iran	1.6	6.0	4.8	3.2	2.3
Ireland	23.3	32.4	10.8	77.4	1.9
Israel	27.7	19.8	10.0	90.7	2.7
Japan	38.4	25.1	9.1	58.8	1.3
Malaysia	27.3	8.8	5.4	31.4	2.9
Mexico	3.6	8.4	3.9	21.7	2.5
Netherlands	49.0	27.2	8.5	76.7	1.7
New Zealand	46.1	19.2	8.1	59.9	2.0
Nigeria	0.1	0.8	0.3	0.3	5.4
Norway	46.4	29.6	8.7	81.5	1.8
Pakistan	0.3	1.9	0.7	0.6	5.1
Philippines	2.6	3.8	1.0	15.0	3.2
Russia	2.9	7.1	9.8	5.3	1.1
Saudi Arabia	1.3	13.3	11.7	11.3	4.5
South Africa	6.5	11.3	7.9	24.2	2.6
Spain	18.3	20.2	6.8	73.4	1.2
Sweden	51.6	24.2	5.3	79.0	1.6
Switzerland	30.7	28.1	5.7	72.8	1.4
Turkey	6.0	5.9	3.1	29.5	2.4
United Kingdom	33.0	24.2	9.2	77.0	1.6
United States	50.2	34.3	19.7	45.1	2.1
Vietnam	1.2	2.1	0.6	1.5	2.3
Yemen	0.1	0.8	1.1	0.8	7.0

The data are in the "Human development" data file on the text CD. This example uses only Internet use and the gross domestic product (GDP), which is a summary description of a nation's wealth. We'll explore the other variables in exercises.

Question to Explore

Use numerical summaries and graphics to describe the center and spread of the distributions of Internet use and GDP.

Think It Through

Using software, we find that for these 39 nations,

> Internet (Percent use): Mean = 21.1, standard deviation = 18.5

> GDP (1000$): Mean = 16.0, standard deviation = 10.6.

The standard deviation for Internet use is almost as large as the mean. This suggests skew to the right, since the smallest value (0.1) is only slightly more than a standard deviation below the mean. Figure 3.4 portrays the distributions using box plots. The box plot for Internet use also suggests some skew to the right, since the whisker is much longer in the direction of the larger values.

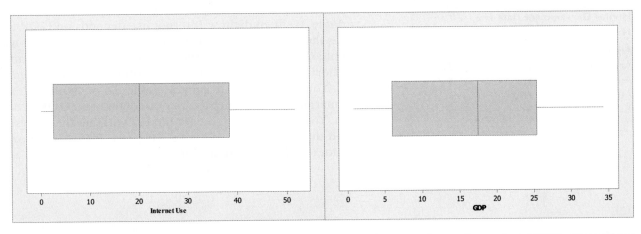

▲ **FIGURE 3.4: MINITAB box plots for Internet use and per capita gross domestic product (GDP). Question:** Approximate the five-number summary for Internet use, and indicate what the box plot suggests about the shape of its distribution.

Insight

The box plots portray each variable *separately*. How can we portray the *association* between Internet use and GDP, showing both variables at the same time? We'll study how to do that next.

TO PRACTICE THIS CONCEPT, TRY EXERCISE 3.11, PART (A)

Looking for a Trend: The Scatterplot

With two quantitative variables it is common to denote the response variable by *y* and the explanatory variable by *x*. This is because graphical plots for examining the association use the *y*-axis (the vertical one) for values of the response variable and the *x*-axis (the horizontal one) for values of the explanatory variable. This graphical plot is called a **scatterplot**.

Scatterplot

A **scatterplot** is a graphical display for two quantitative variables. It uses the horizontal axis for the explanatory variable *x* and the vertical axis for the response variable *y*. The values of *x* and *y* for a subject are represented by a point relative to the two axes. The observations for the *n* subjects are *n* points on the scatterplot.

EXAMPLE 5

Constructing a Scatter-plot for Internet Use and GDP

Picture the Scenario

We return to the data from Example 4 for 39 nations on Internet use and the GDP (gross domestic product).

Questions to Explore

a. Portray the relationship with a scatterplot.

b. What do we learn about the association by inspecting the scatterplot?

Think It Through

a. The first step is to identify the response variable and the explanatory variable. We'll study how Internet use depends on GDP. So Internet use is the response variable. We use y to denote Internet use and x to denote GDP. We plot Internet use on the vertical axis and GDP on the horizontal axis. Any statistical software can construct a scatterplot. You use a data file such as Table 3.4 or the table excerpted in the margin with Internet use in one column and GDP in another column. You select from the data file the variable that plays the role of x and the variable that plays the role of y. Figure 3.5 shows the scatterplot that MINITAB produces.

Human Development Data File

Nation	Internet	GDP
Algeria	0.6	6.1
Argentina	10.1	11.3
Australia	37.1	25.4
Austria	38.7	26.7
Belgium	31.0	25.5
. . .		

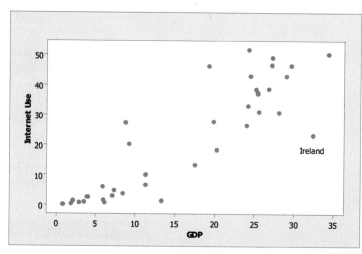

▲ **FIGURE 3.5: MINITAB scatterplot for y = Internet use and x = gross domestic product (GDP) for 39 nations.** The point for Ireland is given by the ordered pair $x = 32.4$ and $y = 23.3$ and is labeled. **Question:** Is there any point you'd identify as standing out in some way? Which country does it represent, and how is it unusual in the context of these variables?

For example, consider the observation for the United States. It has GDP $x = 34.3$ thousand dollars and Internet use $y = 50.2\%$. Its point has co-ordinate 34.3 for the x-axis and 50.2 for the y-axis. This is the top right point in Figure 3.5. The scatterplot is also easily generated by the TI-83+/84 calculator (see the margin, where the point for Ireland has been highlighted with the TRACE key.)

b. Here are some things we learn by inspecting the scatterplot:

- There is a clear trend. Nations with larger values of GDP tend to have larger values of Internet use.

- For countries with very low GDP (below about 8), there is little variability in Internet use. The Internet use is close to 0 for each such country.

- The point for Ireland seems a bit unusual. Its Internet use is relatively low (23.3%), considering its high GDP (32.4). Based on values for other countries with similarly high GDP, we might expect Internet use between about 40% and 50% rather than 23.3%. Likewise, Internet use for Malaysia (27.3%) is a bit higher than we'd expect for a country with such a low GDP (8.8). Can you identify the point for Malaysia on the scatterplot?

Insight

There is a clear association. The countries with lower GDP tend to have lower Internet use, and the countries with higher GDP tend to have higher Internet use.

TO PRACTICE THIS CONCEPT, TRY EXERCISE 3.10, PARTS (A) AND (B)

How to Examine a Scatterplot

We examine a scatterplot to study **association**. How do values on the response variable tend to change as values of the explanatory variable change? As GDP gets higher, for instance, we've seen that Internet use tends to get higher. When there's a trend in a scatterplot, what's the direction? Is the association **positive** or **negative**?

IN WORDS

Positive association: As x goes up, y tends to go up.

Negative association: As x goes up, y tends to go down.

Positive Association and Negative Association

Two quantitative variables x and y are said to have a **positive association** when high values of x tend to occur with high values of y, and when low values of x tend to occur with low values of y. They are said to have a **negative association** when high values of one variable tend to pair with low values of the other variable, and low values of one pair with high values of the other.

Figure 3.5 displays a positive association, because high (low) values of GDP tend to occur with high (low) values of Internet use. If we were to study the association between x = weight of car and y = gas mileage (MPG = miles per gallon) we'd expect a negative association: Heavier cars would tend to get poorer gas mileage. The figure in the margin illustrates, using data from the "car weight and mileage" data file on the text CD.

Here are some questions to explore when you examine a scatterplot:

- Does there seem to be a positive association, a negative association, or no clear evidence of any association?

- Can the trend in the data points be approximated reasonably well by a straight line? In that case, do the data points fall close to the line, or do they tend to scatter quite a bit?

- Are some observations unusual, falling well apart from the overall trend of the data points? What do the unusual points tell us?

EXAMPLE 6

Did the Butterfly Ballot Cost Al Gore the 2000 Presidential Election?

Picture the Scenario

In the 2000 Presidential election in the United States, the Democratic candidate was Al Gore and the Republican candidate was George W.

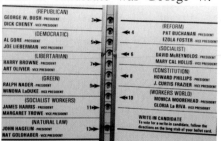

Bush. In Palm Beach County, Florida, initial election returns reported 3407 votes for the Reform party candidate, Pat Buchanan. Some political analysts thought that this total seemed surprisingly large. They felt that most of these votes may have actually been intended for Gore (whose name was next to Buchanan's on the ballot) but wrongly cast for Buchanan because of the design of the "butterfly ballot" used in that county, which some voters found confusing. On the butterfly ballot, Bush appeared first in the left column, followed by Buchanan in the right column, and Gore in the left column (see the figure above).

Question to Explore

For each of the 67 counties in Florida, the "Buchanan and the butterfly ballot" data file on the text CD includes the Buchanan vote and the vote for the Reform party candidate in 1996 (Ross Perot). How can we explore graphically whether the Buchanan vote in Palm Beach County in 2000 was in fact surprisingly high, given the Reform party voting totals in 1996?

Think It Through

Figure 3.6 is a scatterplot of the county-wide vote for the Reform party candidates in 2000 (Buchanan) and in 1996 (Perot). Each point represents a county. This figure shows a strong positive association statewide: Counties with a high

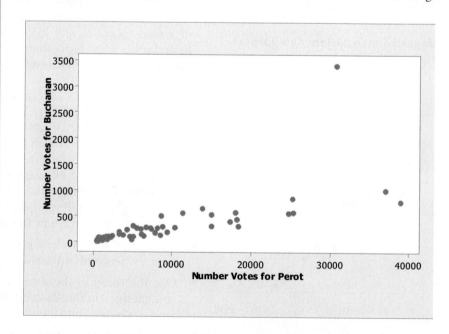

▲ **FIGURE 3.6: MINITAB scatterplot of Florida countywide vote for Reform Party candidates Pat Buchanan in 2000 and Ross Perot in 1996. Question:** Why is the top point, but not each of the two rightmost points, considered an outlier relative to the overall trend of the data points?

Perot vote in 1996 tended to have a high Buchanan vote in 2000, and counties with a low Perot vote in 1996 tended to have a low Buchanan vote in 2000. The Buchanan vote in 2000 was roughly only 3% of the Perot vote in 1996.

In Figure 3.6, one point falls well above the others. This severe outlier is the observation for Palm Beach County, the county that had the butterfly ballot. It is far removed from the overall trend for the other 66 data points, which follow approximately a straight line.

Insight

Alternatively you could plot the Buchanan vote against the Gore vote or against the Bush vote (Exercise 3.21). These and other analyses conducted by statisticians[4] predicted that fewer than 900 votes were truly intended for Buchanan in Palm Beach County, compared to the 3407 votes he actually received. Bush won the state by 537 votes and, with it, the Electoral College and the election. So, this may have been a pivotal factor in determining the outcome of this election. Other factors that played a role were 110,000 disqualified "overvote" ballots in which people mistakenly voted for more than one Presidential candidate (with Gore marked on 84,197 ballots and Bush on 37,731) often because of confusion from names being listed on more than one page of the ballot, and 61,000 "undervotes" caused by factors such as "hanging chads" from manual punch-card machines in some counties.

TO PRACTICE THIS CONCEPT, TRY EXERCISE 3.21

In practice, data points in a scatterplot sometimes fall close to a straight line trend, as we saw for all data except Palm Beach County in Figure 3.6. This represents a strong association, in the sense that we can predict the y value quite well from knowing the x value. The next step in analyzing a relationship is summarizing the strength of the association.

How Can We Summarize Strength of Association? The Correlation

When the data points follow roughly a straight line trend, the variables are said to have an approximately **linear** relationship. In some cases the data points fall close to a straight line, but more often there is quite a bit of variability of the points around the straight-line trend. A summary measure called the **correlation** describes the strength of the linear association.

Correlation

The **correlation** summarizes the direction of the association between two quantitative variables and the strength of its straight-line trend. Denoted by r, it takes values between −1 and +1.

- A positive value for r indicates a positive association and a negative value for r indicates a negative association.

- The closer r is to ±1, the closer the data points fall to a straight line, and the stronger is the linear association. The closer r is to 0, the weaker is the linear association.

---

[4]For further discussion of these and related data, see Exercise 2.83 and the article by A. Agresti and B. Presnell, *Statistical Science*, vol. 17, pp. 1–5, 2002.

Let's get a feel for the correlation r by looking at its values for the scatterplots shown in Figure 3.7:

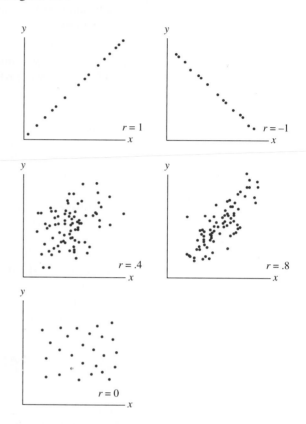

▲ **FIGURE 3.7: Some scatterplots and their correlations.** The correlation gets closer to ±1 when the data points fall closer to a straight line. **Question:** Why are the cases in which the data points are closer to a straight line considered to represent stronger association?

The correlation r takes its extreme values of $+1$ and -1 only when the data points follow a straight line pattern *perfectly*. These are the first two cases shown in Figure 3.7. Then $r = +1$ occurs when the line slopes upward. The association is then positive, since higher values of x tend to occur with higher values of y. The value $r = -1$ occurs when the line slopes downward, corresponding to a negative association.

In practice, don't expect the data points to fall perfectly on a straight line. However, the closer they come to that ideal, the closer the correlation is to 1 or -1. For instance, the scatterplot in Figure 3.7 with correlation $r = 0.8$ shows a stronger association than the one with correlation $r = 0.4$, for which the data points fall farther from a straight line.

Recall

The **absolute value** of a number gives the distance the number falls from zero on the number line. The correlation values of -0.9 and 0.9 both have an absolute value of 0.9. They both represent a stronger association than correlation values of -0.6 and 0.6, for example. ■

Properties of the Correlation

■ The correlation r always **falls between -1 and $+1$.** The closer the value to 1 in absolute value (see the margin comments), the stronger the **linear (straight-line) association**, as the data points fall nearer to a straight line.

■ A **positive correlation** indicates a **positive association**, and a **negative correlation** indicates a **negative association**.

■ The value of the correlation **does not depend on the variables' units**. For example, suppose one variable is the income of a subject, in dollars. If we change the observations to units of euros or to units of thousands of dollars, we'll get the same correlation.

■ Two variables have **the same correlation no matter which is treated as the response variable**.

The correlation can be calculated by statistics software and by many calculators.

EXAMPLE 7

What's the Correlation Between Internet Use and GDP?

Picture the Scenario

Example 5 displayed a scatterplot for Internet use and GDP (gross domestic product) for 39 nations, shown again in the margin. We observed a positive association.

Questions to Explore

a. What value does software give for the correlation?

b. How can we interpret the correlation value?

Think It Through

a. Since the association is positive, we expect to find $r > 0$. If we input the columns of Internet use and GDP values into MINITAB and request the correlation from the Basic Statistics menu, we get

```
Correlations: gdp, internet

Pearson correlation of gdp and internet = 0.888.
```

b. The correlation of $r = 0.888$ is not far below 1.0, the strongest possible positive correlation. This confirms the strong positive association that we observed in the scatterplot. In summary, a nation's extent of Internet use seems to be strongly associated with its GDP, with higher GDPs tending to correspond to higher Internet use.

Insight

The identifier *Pearson* for the correlation in the MINITAB output refers to the British statistician, Karl Pearson. In 1896 he provided the formula that is used to compute the correlation from sample data. This formula is shown next.

TO PRACTICE THIS CONCEPT, TRY EXERCISES 3.12 AND 3.13

Recall

From Section 2.5 the *z*-**score** for an observation is the number of standard deviations that the observation falls from the mean. It equals the difference between the observation and the mean, divided by the standard deviation. ■

What Formula Gives the Correlation?

Although software can compute the correlation for us, it helps to understand it if you see its formula. For an observation x on the explanatory variable, let z_x denote the z-score that represents the number of standard deviations that x falls from the mean of x. That is,

$$z_x = \frac{\text{observed value} - \text{mean}}{\text{standard deviation}} = \frac{(x - \bar{x})}{s_x},$$

where s_x denotes the standard deviation of the x-values. Similarly, let z_y denote the number of standard deviations that an observation y on the response variable falls from the mean of y. To obtain r, you calculate the product $z_x z_y$ for each observation, then find a typical value (a type of average) of those products.

CALCULATING THE CORRELATION r

$$r = \frac{1}{n-1} \sum z_x z_y = \frac{1}{n-1} \sum \left(\frac{x - \bar{x}}{s_x} \right) \left(\frac{y - \bar{y}}{s_y} \right),$$

where n is the number of points, $\bar{x}$ and $\bar{y}$ are means, and s_x and s_y are standard deviations for x and y. The sum is taken over all n observations.

For $x =$ GDP and $y =$ Internet use, Example 7 found the correlation $r = 0.888$, using statistical software. To visualize how the formula works, let's revisit the scatterplot. This is reproduced in Figure 3.8, with a vertical line at the mean of x and a horizontal line at the mean of y. These lines divide the scatterplot into four **quadrants**. The summary statistics are

$$\bar{x} = 16.0 \qquad \bar{y} = 21.1$$

$$s_x = 10.6 \qquad s_y = 18.5.$$

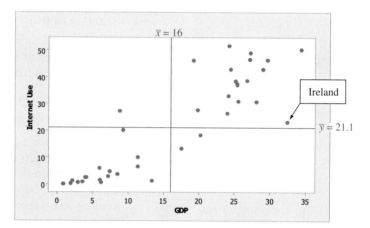

▶ **FIGURE 3.8: MINITAB scatterplot of Internet use and GDP divided into quadrants at $(\bar{x}, \bar{y})$.** Of the 39 data points, 36 lie in the upper-right quadrant (above the mean on each variable) or the lower-left quadrant (below the mean on each variable). **Question:** Do the points in these two quadrants make a positive, or a negative, contribution to the correlation value? (*Hint:* Is the product of z-scores for these points positive or negative?)

The point for Ireland, ($x = 32.4$, $y = 23.3$), has as its z-scores ($z_x = 1.55$, $z_y = 0.12$). This point is labeled in Figure 3.8. Since $x = 32.4$ is to the right of the mean for x and $y = 23.3$ is above the mean of y, it falls in the upper-right quadrant. Both its z-scores are positive. All except 3 of the 39 nations have points that fall in the upper-right and lower-left quadrants.

■ The product of the z-scores for the points in the upper-right quadrant (such as Ireland) is positive. The product is also positive for points in the lower-left quadrant. Such points contribute to a positive correlation.

■ The product of the z-scores for the points in the upper-left and lower-right quadrants is negative. Such points contribute to a negative correlation.

The overall correlation reflects the number of points in the various quadrants and how far they fall from the means.

Graph the Data to Check Whether the Correlation Is Appropriate

The correlation is an efficient way to summarize the association shown by lots of data points with a single number. But be careful to use it only when it is appropriate. Figure 3.9 illustrates why. It shows a scatterplot in which the data points follow a U-shaped curve. There is an association, because as x increases, y first tends to decrease, and then it tends to increase. For example, this might happen if x = age of person and y = annual medical expenses. Medical expenses tend to be high for newly born and young children, then they tend to be low until the person gets old, when they tend to become high again. However, $r = 0$ for the data in Figure 3.9.

The correlation is designed for straight-line relationships. For Figure 3.9, $r = 0$ and it fails to detect the association. The correlation is not valid for describing association when the points cluster around a curve rather than around a straight line.

This figure highlights an important point to remember about *any* data analysis:

■ **Always plot the data.**

If we merely used software to calculate the correlation for the data in Figure 3.9 but did not plot the data, we might mistakenly conclude that the variables have no association. They *do* have one, but it is not a straight-line association.

▲ **FIGURE 3.9: The correlation poorly describes the association when the relationship is curved.** For this U-shaped relationship, the correlation is 0 (or close to 0), even though the variables are strongly associated. **Question:** Can you use the formula for r, in terms of how points fall in the quadrants, to reason why the correlation would be close to 0?

IN PRACTICE: Always construct a scatterplot

Always **construct a scatterplot** to display a relationship between two quantitative variables. The correlation indicates the direction and strength only of an approximate *straight-line* relationship.

3.2 Practicing the Basics

3.9 **Used cars and direction of association:** For the 100 cars on the lot of a used-car dealership, would you expect a positive association, negative association, or no association between each of the following pairs of variables? Explain why.

a. The age of the car and the number of miles on the odometer

b. The age of the car and the resale value

c. The age of the car and the total amount that has been spent on repairs

d. The weight of the car and the number of miles it travels on a gallon of gas

3.10 **Cell phones and GDP:** Table 3.4 showed data on the percent of the population in a country that uses cell phones, the variable called "cellular." Let y = cell-phone use and x = GDP.

a. The SPSS output shows a scatterplot. Describe this plot in terms of (i) the variability of cell-phone use values for nations close to 0 on GDP, (ii) identifying two nations that have less cell-phone use than you'd expect, given their GDP.

b. Give the approximate x- and y-coordinates for the nation that has highest (i) cell-phone use, (ii) GDP.

c. Would the correlation be positive, or negative? Explain what it means for the correlation to have this sign.

d. Suppose you considered the correlation only for those nations having GDP above 15. Would the correlation be stronger, or weaker, than for all 39 nations?

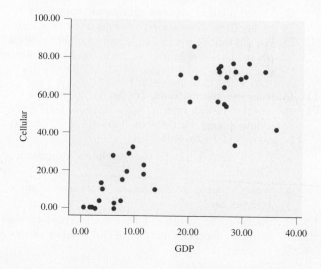

3.11 **Economic development and air pollution:** Table 3.4
TRY also has data on the carbon dioxide emissions in several countries, denoted by CO_2. This is a measure of the extent to which a country contributes to air pollution and to problems (such as global warming) that may be associated with air pollution. In this sample, its mean is 6.8 and its standard deviation is 4.7 (metric tons per capita).

   a. The five-number summary of positions for CO_2 is minimum = 0.3, Q1 = 3.0, median = 6.8, Q3 = 9.7, maximum = 19.7. Sketch a box plot. Based on this and the mean and standard deviation, describe the shape of the distribution of CO_2 values.

   b. The MINITAB output below shows a scatterplot relating CO_2 to a nation's per capita gross domestic product (GDP). Identify the approximate coordinates of the nation that is highest on each of these variables.

   c. The correlation between these variables equals 0.79. Would you call this association strong or weak? Explain.

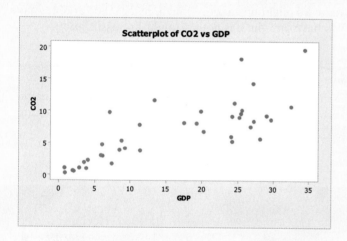

3.12 **Politics and newspaper reading:** For the "FL student
TRY survey" data file on the text CD, the correlation between y = political ideology (scored 1 = very liberal to 7 = very conservative) and x = number of times a week reading a newspaper is -0.07.

   a. Would you call this association strong or weak? Explain.

   b. The correlation between political ideology and religiosity (how often attend religious services) is 0.58. For this sample, which explanatory variable, newspaper reading or religiosity, seems to have a stronger association with y? Explain.

3.13 **Internet use correlations:** For the nations in Table 3.4,
TRY the correlation with Internet use is 0.888 for GDP, 0.818 for cellular-phone use, 0.669 for literacy, -0.551 for fertility, and 0.680 for CO_2 emissions.

   a. Which variable has the *strongest* linear association with Internet use?

   b. Which variable has the *weakest* linear association with Internet use?

   c. Interpret the correlation between Internet use and fertility (the mean number of children per adult woman).

3.14 **Match the scatterplot with *r*:** Match the scatterplots below with the correlation values.

   1. $r = -0.9$

   2. $r = -0.5$

   3. $r = 0$

   4. $r = 0.6$

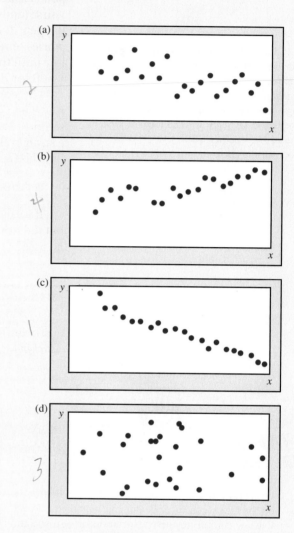

3.15 **What makes *r* = 1?:** Consider the data:

x	3	4	5	6	7
y	8	13	12	14	16

   a. Sketch a scatterplot.

   b. If one pair of (x, y) values is removed, the correlation for the remaining four pairs equals 1. Which pair is it?

   c. If one y value is changed, the correlation for the five pairs equals 1. Identify the y value and how it must be changed for this to happen.

3.16 **z-scores for *r* = 1:** Use the points from the previous exercise with $x = 3, 5, 6, 7$.

   a. Find the z-scores on x and on y for each point. Comment on how z_x and z_y relate to each other for each point.

   b. Compute r using the z-scores from part (a) for the four observations. Is this the value you expected to get for r? Why?

3.17 *r* = 0: Sketch a scatterplot for which *r* > 0, but *r* = 0 after one of the points is deleted.

3.18 Correlation inappropriate: Describe a situation in which it is inappropriate to use the correlation to measure the association between two quantitative variables.

3.19 Which mountain bike to buy?: Is there a relationship between the weight of a mountain bike and its price? A lighter bike is often preferred, but do lighter bikes tend to be more expensive? The following table, from the "mountain bike" data file on the text CD, gives data on price, weight, and type of suspension (FU = full, FE = front end) for 12 brands.

Mountain Bikes

BRAND AND MODEL	PRICE($)	WEIGHT(LB)	TYPE
Trek VRX 200	1000	32	FU
Cannondale Super V400	1100	31	FU
GT XCR-4000	940	34	FU
Specialized FSR	1100	30	FU
Trek 6500	700	29	FE
Specialized Rockhop	600	28	FE
Haro Escape A7.1	440	29	FE
Giant Yukon SE	450	29	FE
Mongoose SX 6.5	550	30	FE
Diamondback Sorrento	340	33	FE
Motiv Rockridge	180	34	FE
Huffy Anorak 36789	140	37	FE

Source: Consumer Reports, June 1999.

a. You are shopping for a new bike. You are interested in whether and how weight affects the price. Which variable is the logical choice for the (i) explanatory variable, (ii) response variable?

b. Sketch a scatterplot of price and weight. Does the relationship seem to be approximately linear? In what way does it deviate from linearity?

c. The correlation equals −0.32. Interpret it in context. Does weight appear to affect the price strongly in a linear manner?

3.20 Enchiladas and sodium-revisited: Is there a relationship between the sodium content and the cost of enchiladas? Use software to analyze the data from *Consumer Reports* (March 2000) in the following table, contained in the "Enchiladas" file on the text CD.

BRAND AND TYPE	COST (PER SERVING)	SODIUM CONTENT(MG)
Amy's Black Bean	$3.03	780
Patio Cheese	$1.07	1570
Banquet Cheese	$1.28	1500
El Charrito Beef	$1.53	1370
Patio Beef	$1.05	1700
Banquet Beef	$1.27	1330
Healthy Choice Chicken	$2.34	440
Lean Cuisine Chicken	$2.47	520
Weight Watchers Chicken	$2.09	660

a. Construct a scatterplot to show how sodium depends on cost. Is the association positive, or negative? Do you notice any unusual observations?

b. What might explain the gap observed in the scatterplot? (*Hint:* How do the health-oriented products compare in sodium content to the others?)

c. Obtain the correlation, *r*. Interpret this value in context.

 3.21 Buchanan vote: Refer to Example 6 and the "Buchanan and the butterfly ballot" data file on the text CD. Let *y* = Buchanan vote and *x* = Gore vote.

a. Construct a box plot for each variable. Summarize what you learn.

b. Construct a scatterplot. Identify any unusual points. What can you learn from a scatterplot that you cannot learn from box plots?

c. For the county represented by the most outlying observation, about how many votes would you have expected Buchanan to get if the point followed the same pattern as the rest of the data?

d. Repeat (a) and (b) using *y* = Buchanan vote and *x* = Bush vote.

3.3 | How Can We Predict the Outcome of a Variable?

We've seen how to explore the relationship between two quantitative variables graphically with a scatterplot. When the relationship has a straight-line pattern, the correlation describes it numerically. We can analyze the data further by finding an equation for the straight line that best describes that pattern. This equation predicts the value of the response variable from the value of the explanatory variable.

IN WORDS

The **symbol ŷ,** which denotes the predicted value of *y*, is pronounced *y-hat.*

Regression Line: An Equation for Predicting the Response Outcome

The **regression line** predicts the value for the response variable *y* as a straight-line function of the value *x* of the explanatory variable. Let $\hat{y}$ denote the **predicted value** of *y*. The equation for the regression line has the form

$$\hat{y} = a + bx.$$

In this formula, *a* denotes the **y-intercept** and *b* denotes the **slope**.

EXAMPLE 8

How Can Anthropologists Predict Height Using Human Remains?

Picture the Scenario

Anthropologists often try to reconstruct information using partial human remains at burial sites. For instance, after finding a femur (thighbone), they may want to predict how tall an individual was. An equation they use to do this is the regression line, $\hat{y} = 61.4 + 2.4x$, where $\hat{y}$ is the predicted height and x is the length of the femur, both in centimeters.

Questions to Explore

How can we graph the line that depicts how the predicted height depends on the femur length? A femur found at a particular site has length of 50 cm. What is the predicted height of the person who had that femur?

Think It Through

The formula $\hat{y} = 61.4 + 2.4x$ has y-intercept 61.4 and slope 2.4. It has the straight-line form $\hat{y} = a + bx$ with $a = 61.4$ and $b = 2.4$.

Each number x, when substituted into the formula $\hat{y} = 61.4 + 2.4x$, yields a value for $\hat{y}$. For simplicity in plotting the line, we start with $x = 0$, although in practice this would not be an observed femur length. The value $x = 0$ has $\hat{y} = 61.4 + 2.4(0) = 61.4$. Now, points on the y-axis have $x = 0$, so the line has height 61.4 at the point of its intersection with the y-axis. Because of this, the constant 61.4 in the equation is called the **y-intercept**. The line intersects the y-axis at the point with (x, y) coordinates (0, 61.4), which is 61.4 units up the y-axis.

The value $x = 50$ has $\hat{y} = 61.4 + 2.4(50) = 181.4$. When the femur length is 50 cm, the predicted height of the person is 181.4 cm. The coordinates are (50, 181.4). We can plot the line by connecting the points (0, 61.4) and (50, 181.4). Figure 3.10 plots the straight line for x between 0 and 50. In

▲ **FIGURE 3.10: Graph of the regression line for x = femur length and y = height of person. Questions:** At what point does the line cross the y-axis? How can you interpret the slope of 2.4?

summary, the predicted height $\hat{y}$ increases from 61.4 to 181.4 as x increases from 0 to 50.

Insight

A regression equation is often called a **prediction equation**, since it predicts the value of the response variable y at any value of x. Sadly, this particular prediction equation had to be applied recently with bones found in mass graves in Kosovo, to help identify Albanians who had been executed by Serbians in 1998.[5]

> **TO PRACTICE THIS CONCEPT, TRY EXERCISE 3.23, PART (A), AND 3.24, PART (A)**

Recall

Math facts:

y-intercept is the value of the line in the y direction when $x = 0$.

Slope = rise/run = change in y / change in x ▪

Interpreting the y-Intercept and Slope

The **y-intercept** is the predicted value of y when $x = 0$. This fact helps us plot the line, but it may not have any interpretative value if no observations had x values near 0. It does not make sense for femur length to be 0 cm, so the y-intercept for the equation $\hat{y} = 61.4 + 2.4x$ is not a relevant predicted height.

The **slope** b in the equation $\hat{y} = a + bx$ equals the amount that $\hat{y}$ changes when x increases by one unit. For two x values that differ by 1.0, the $\hat{y}$ values differ by b. For the line $\hat{y} = 61.4 + 2.4x$, we've seen that $\hat{y} = 181.4$ at $x = 50$. If x increases by 1.0 to $x = 51$, we get $\hat{y} = 61.4 + 2.4(51) = 183.8$. The increase in $\hat{y}$ is from 181.4 to 183.8, which is 2.4, the slope value. For each 1-cm increase in femur length, height is predicted to increase by 2.4 cm. Figure 3.11 portrays this interpretation.

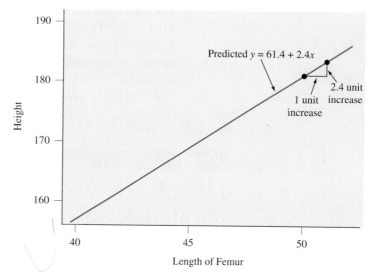

▲ **FIGURE 3.11: The slope of a straight line.** The slope is the change in the predicted value $\hat{y}$ of the response variable for a 1-unit increase in the explanatory variable x. For an increase in femur length from 50 cm to 51 cm, the predicted height increases by 2.4 cm. **Question:** What does it signify if the slope equals 0?

Recall

Table 3.5 had entries like the following: ▪

	GDP	Fertility
China	4.0	1.8
U.S.	34.3	2.1
Yemen	0.8	7.0

When the slope is negative, the predicted value $\hat{y}$ *decreases* as x increases. The straight line then goes downward, and the association is *negative*. For instance, for the data from Table 3.4 on y = fertility (mean number of children per woman) and x = GDP (gross domestic product), the regression equation is $\hat{y} = 3.5 - 0.07x$. This equation shows a negative association, with slope -0.07. As GDP increases, the predicted fertility decreases, as shown in the margin figure.

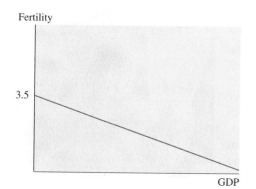

[5]"The Forensics of War," by Sebastian Junger in *Vanity Fair*, October 1999.

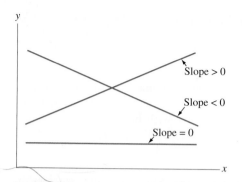

▲ FIGURE 3.12: Three regression lines showing positive association (slope > 0), negative association (slope < 0), and no association (slope = 0). Question: Would you expect a positive, or negative, slope when *y* = annual income and *x* = number of years of education?

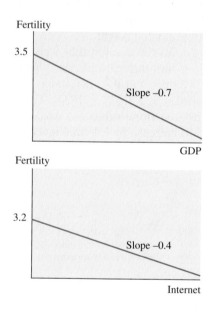

When the slope = 0, the regression line is horizontal (parallel to the *x*-axis). The predicted value $\hat{y}$ of *y* stays constant at the *y*-intercept for any value of *x*. Then the predicted value $\hat{y}$ does not change as *x* changes, and the variables do not exhibit an association. Figure 3.12 illustrates the three possibilities for the sign of the slope.

The absolute value of the slope describes the *magnitude* of the change in $\hat{y}$ for a 1-unit change in *x*. The larger the absolute value, the steeper the regression line. A line with *b* = 4.2, such as $\hat{y} = 61.4 + 4.2x$, is steeper than one with *b* = 2.4. A line with *b* = −0.07 is steeper than one with *b* = −0.04.

For instance, when *y* = fertility and *x* = GDP, the regression line is $\hat{y} = 3.5 − 0.07x$. When instead *x* = Internet use, the regression line is $\hat{y} = 3.2 − 0.04x$. This slope of −0.04 is smaller in absolute value than the slope of −0.07 with GDP as the predictor. There is a slightly greater decrease in predicted fertility with an increase of $1000 in per capita GDP than with an increase of 1% in Internet use. This does *not* mean that GDP has a *stronger* effect than Internet use on fertility, because it does not make sense to equate an increase of $1000 in GDP with an increase of 1% in Internet use.

Depending on the units of measurement, a 1-unit increase in a predictor *x* could be a trivial amount, or it could be huge. However, the slope projects to any size of increase in *x*. For example, in predicting *y* = fertility from *x* = Internet use by $\hat{y} = 3.2 − 0.04x$, a 1-unit increase in *x* is pretty small. If the percentage using the Internet is 50% higher in one country than in another (e.g., 50% compared to 0%, as we see in comparing the United States to Yemen), we'll predict that its fertility changes by 50 times the amount for a 1-unit change. The predicted change in fertility is then 50(−0.04) = −2.0; that is, two children fewer per woman, a considerable decrease.

Using Data to Find the Slope and *y*-Intercept of a Regression Line

How can we use the data to find the regression equation? We should first construct a scatterplot to make sure that the relationship has roughly a straight line trend. If so, then software or calculators can easily find the straight line that best fits the data.

EXAMPLE 9

How Can We Predict Baseball Scoring Using Batting Average?

Picture the Scenario

In baseball, two summaries of a team's offensive ability are the team batting average (the proportion of times the team's players get a hit, out of the times they get a hit or instead are "out") and team scoring (the team's mean number of runs scored per game). Table 3.5 shows these statistics for the 2003 season for American League teams, from the "AL team statistics" data file on the text CD.

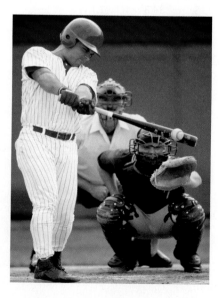

TABLE 3.5: Team Batting Average and Team Scoring (Mean Number of Runs Per Game) for American League Teams in 2003[6]

Team	Batting Average	Team Scoring
Boston	.289	5.9
Toronto	.279	5.5
Minnesota	.277	4.9
Kansas City	.274	5.2
Seattle	.271	4.9
New York	.271	5.4
Anaheim	.268	4.5
Baltimore	.268	4.6
Texas	.266	5.1
Tampa Bay	.265	4.4
Chicago	.263	4.9
Oakland	.254	4.7
Cleveland	.254	4.3
Detroit	.240	3.6

Scoring runs is a result of hitting, so team scoring is the response variable y and team batting average is the explanatory variable x. Figure 3.13 is the scatterplot. There is a straight-line trend summarized by a strong positive correlation, $r = 0.875$.

▲ **FIGURE 3.13: MINITAB** output for scatterplot of team batting average and team scoring, with regression line superimposed. **Question:** How can you find the prediction error that results when you use the regression line to predict team scoring for a team?

---

[6]*Source:* www.usatoday.com/sports/baseball

TI-83+/84 output

Questions to Explore

a. According to software, what is the regression equation?

b. If a team has a batting average of 0.275 next year, what is their predicted mean number of runs per game?

c. How do you interpret the slope in this context?

Think It Through

a. With the software package MINITAB, when we choose the regression option in the Regression part of the Statistics menu, part of the output tells us:

The regression equation is TEAM_SCORING = −6.1 + 41.2 BAT_AVG

The y-intercept is $a = -6.1$. The slope is the coefficient of the explanatory variable, denoted in the data file by BAT_AVG. It equals $b = 41.2$ The regression equation is

$$\hat{y} = a + bx = -6.1 + 41.2x.$$

The TI-83+/84 calculator provides the output shown in the margin.

b. We predict that an American League team with a team batting average of 0.275 will score an average of $\hat{y} = -6.1 + 41.2(0.275) = 5.2$ runs per game.

c. Since the slope $b = 41.2$ is positive, the association is positive: The predicted team scoring increases as team batting average increases. The slope refers to the change in $\hat{y}$ for a 1-unit change in x. However, $x = $ team batting average is a proportion. In Table 3.5, the team batting averages fall between about 0.24 and 0.29, a range of 0.05. An increase of 0.05 in x corresponds to an increase of $(0.05)41.2 = 2.1$ in predicted team scoring. The mean number of runs scored per game is predicted to be about 2 higher for the best hitting teams than for the worst hitting teams.

Insight

Figure 3.13 shows the regression line superimposed over the scatterplot. It applies only over the range of observed batting averages. For instance, it's not sensible to predict that a team with batting average of 0.0 will average −6.1 runs per game.

> **TO PRACTICE THIS CONCEPT, TRY EXERCISES 3.23 AND 3.28**
> **TO GET A FEEL FOR FITTING A REGRESSION LINE, USE THE**
> **"REGRESSION BY EYE" APPLET DISCUSSED IN EXERCISE 3.105**

Residuals Measure the Size of Prediction Errors

The regression equation $\hat{y} = -6.1 + 41.2x$ predicts team scoring for a given level of $x = $ team batting average. We can compare the predicted values to the actual team scoring to check the accuracy of those predictions.

For example, New York had $y=5.4$ and $x=0.271$. The prediction for $y=$mean number of runs per game at $x=0.271$ is $\hat{y}=-6.1+41.2x=-6.1+41.2(0.271)=5.1$. The prediction error is the difference between the actual y value of 5.4 and the predicted value of 5.1, which is $y - \hat{y} = 5.4 - 5.1 = 0.3$. For New York, the regression equation under predicts y by 0.3 runs per game. For Tampa Bay, $x = 0.265$ and $\hat{y} = -6.1 + 41.2(0.265) = 4.8$. The actual value is $y = 4.4$, so the prediction is too high. The prediction error is $4.4 - 4.8 = -0.4$. The prediction errors are called **residuals**.

Residual

The **prediction error** for an observation, which is the difference $y - \hat{y}$ between the actual value and the predicted value of the response variable, is called a **residual**.

Each observation has a residual. Some are positive and some are negative. A positive residual occurs when the actual y is larger than the predicted value $\hat{y}$, so that $y - \hat{y} > 0$. A negative residual results when the actual y is smaller than the predicted value $\hat{y}$. The smaller the absolute value of a residual, the closer the predicted value is to the actual value, so the better is the prediction.

Graphically in the scatterplot, *the residual for an observation is the vertical distance between the point and the regression line.* Figure 3.14 illustrates this for the positive residual for New York. The residuals are vertical distances because the regression equation predicts y, the variable on the vertical axis, at a given value of x. Notice the parallel with analyzing contingency tables by studying values of the response variable, *conditional* on (given) values of the explanatory variable.

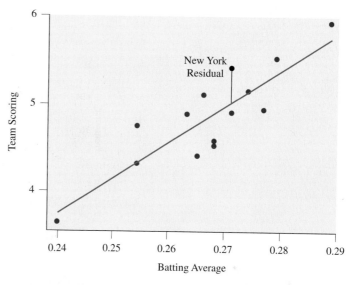

▲ **FIGURE 3.14: Scatterplot of team batting average and team scoring, with the residual for New York at the point ($x = 0.271$, $y = 5.41$).** The residual is the prediction error, which is the vertical distance of the point from the regression line. **Question:** Why is a residual a *vertical* distance from the regression line?

How Can We Detect an Unusual Vote Total?

Picture the Scenario

Example 6 investigated whether the vote total in Palm Beach County, Florida in the 2000 Presidential election was unusually high for Pat Buchanan, the Reform party candidate. We did this by plotting, for all 67 counties in Florida, Buchanan's vote against the Reform party candidate (Ross Perot) vote in the 1996 election.

Question to Explore

If we fit a regression line to y = Buchanan vote and x = Perot vote for the 67 counties, would the residuals help us detect any unusual vote totals for Buchanan?

Think It Through

A county with a large residual has its predicted vote far from the actual vote. This would indicate an unusual vote total. We can easily have software find the regression line and the residuals for the 67 counties. Then we can quickly see whether any of the residuals are particularly large by constructing their histogram. Figure 3.15 shows this histogram. The residuals cluster around 0, but one is very large, greater than 2000. Inspection of the results shows that this residual applies to Palm Beach County, for which the actual Buchanan vote was $y = 3407$ and the predicted vote was $\hat{y} = 1100$. Its residual is $y - \hat{y} = 3407 - 1100 = 2307$. In summary, in Palm Beach County Buchanan's vote was much higher than predicted.[7]

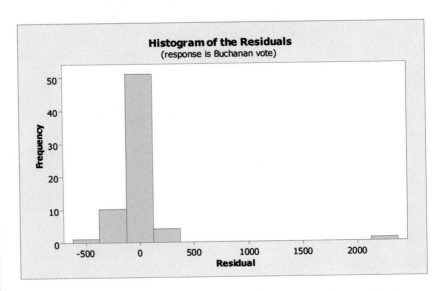

▲ FIGURE 3.15: MINITAB output of residuals for predicting 2000 Buchanan presidential vote in Florida counties using 1996 Perot vote. Question: What does the right-most bar represent?

Insight

As we'll discuss in the next section, an extreme outlier can pull the regression line toward it. Because of this, it's a good idea to fit the data *without* the Palm Beach County observation and see how well that line predicts results for Palm Beach County. We'd then get the regression equation $\hat{y} = 45.7 + 0.02414x$. Since the Perot vote in Palm Beach County was 30,739, this line would predict a Buchanan vote there of $\hat{y} = 45.7 + 0.02414(30,739) = 788$. This compares to the actual Buchanan vote of 3407 in Palm Beach County, for a residual of $3407 - 788 = 2619$. Again, the actual Buchanan vote seems surprisingly high.

TO PRACTICE THIS CONCEPT, TRY EXERCISE 3.29

[7]A more complex analysis accounts for larger counts tending to vary more. However, our analysis is adequate for highlighting data points that fall far from the linear trend.

The Method of Least Squares Yields the Regression Line

We've seen that software finds the regression line for you. It chooses the optimal line to fit through the data points by making the residuals as small as possible. This involves compromise, because a line can perfectly predict one point but do poorly for many other points. The summary measure used to evaluate predictions is

$$\text{Residual sum of squares} = \sum (\text{residual})^2 = \sum (y - \hat{y})^2.$$

This squares each vertical distance between a point and the line and then adds them up. The better the line, the smaller the residuals tend to be, and the smaller the residual sum of squares tends to be. Each line has a set of predicted values, a set of residuals, and a residual sum of squares. The line that software reports is the one having the *minimum* residual sum of squares. This way of selecting a line is called the **least squares method**.

Least Squares Method

Among the possible lines that can go through data points in a scatterplot, the regression line results from the **least squares method**. This gives the line that has the smallest value for the residual sum of squares in using $\hat{y} = a + bx$ to predict y.

It's simple for software to use the least squares method to find the regression line for us. Besides making the errors as small as possible in this summary sense, this line

- Has some positive residuals and some negative residuals, but the sum (and mean) of the residuals equals 0.
- Passes through the point, $(\bar{x}, \bar{y})$.

The first property tells us that the too-low predictions are balanced by the too-high predictions. The second property tells us that the line passes through the center of the data.

The method of least squares provides formulas for the y-intercept and slope, based on summary statistics for the sample data. Let $\bar{x}$ denote the mean of x, $\bar{y}$ the mean of y, s_x the standard deviation of the x values and s_y the standard deviation of the y values. The slope b is directly related to the correlation r.

Regression Formulas for y-Intercept and Slope

The **slope** equals $b = r\left(\dfrac{s_y}{s_x}\right)$.

The **y-intercept** equals $a = \bar{y} - b(\bar{x})$.

For example, the baseball data in Example 9 have $\bar{x} = 0.267$ for batting average, $\bar{y} = 4.865$ for team scoring, $s_x = 0.0121$, and $s_y = 0.568$. The correlation is $r = 0.875$, so

$$b = r\left(\frac{s_y}{s_x}\right) = 0.875(0.568/0.0121) = 41.2,$$

$$\text{and } a = \bar{y} - b(\bar{x}) = 4.865 - 41.2(0.267) = -6.1.$$

The regression line to predict team scoring from batting average is $\hat{y} = -6.1 + 41.2x$. These formulas for a and b help us to interpret the regression line $\hat{y} = a + bx$ (for example, see Exercises 3.100 and 3.101), but you should rely on software for calculations.

The Slope, the Correlation, and the Units of the Variables

We've used the correlation to describe the strength of the association. Why can't we use the *slope* to do this, with bigger slopes representing stronger associations? The reason is that the numerical value of the slope depends on the units for the variables.

For example, the regression line between y = Internet use and x = GDP is $\hat{y} = -3.6 + 1.5x$. GDP was measured in *thousands* of U.S. dollars (per capita). Suppose we instead measure GDP in dollars, such as $x = 19,200$ for New Zealand instead of $x = 19.2$. A one-unit increase in GDP then refers to a single dollar (per capita). This is only 1/1000 as much as a thousand-dollar increase, so the change in the predicted value of y would be 1/1000 as much, or $(1/1000)1.5 = 0.0015$. Thus, if x = GDP in dollars, the slope of the regression equation is 0.0015 instead of 1.5. The strength of the association is the same in each case, since the variables and data base are identical. Only the units of measurement for one variable changed.

The slope b doesn't tell us whether the association is strong or weak, since we can make b as large or as small as we want by changing the units. By contrast, *the correlation does not change when the units of measurement change.* It is 0.888 between Internet use and GDP, whether we measure GDP in dollars or in thousands of dollars.

In summary, we've learned how the correlation describes the strength of the association. We've also seen how the regression line predicts the response variable y using the explanatory variable x. Although correlation and regression methods serve different purposes, there are strong connections between them:

■ They are both appropriate when the relationship between two quantitative variables can be approximated by a straight line.

■ The correlation and the slope of the regression line have the same sign. If one is positive, so is the other one. If one is negative, so is the other one. If one is zero, the other is also zero.

There are some differences between correlation and regression methods. With regression we must identify response and explanatory variables. We get a different line if we use x to predict y than if we use y to predict x. By contrast, correlation does not make this distinction. We get the same correlation either way. Also, the values for the y-intercept and slope of the regression line depend on the units, whereas the correlation does not. Finally, the correlation falls between -1 and $+1$, whereas the regression slope can equal any real number.

r-Squared (r^2)

When we predict a value of y, why should we use the regression line? We could instead predict y using the center of its distribution, such as the sample mean, $\bar{y}$. The reason for using the regression line is that if x and y have an association, then we can predict most y values more accurately by substituting x values into the regression equation, $\hat{y} = a + bx$, than by using the sample mean $\bar{y}$ for prediction.

Another way to describe the strength of association refers to how well you can predict y using the regression equation. The square of the correlation summarizes how much less prediction error there is when you use the regression line to predict y, compared to using $\bar{y}$ to predict y. The prediction error is summarized by the sum of squared distances between the actual y values and the predicted values. We use the notation r^2 for this measure because, in fact, it equals the square of the correlation r.

Specifically, r^2 is interpreted as the **proportional reduction in error**. For Internet use and GDP, $r = 0.888$, so $r^2 = (0.888)^2 = 0.79$. This means that the prediction error using the regression line to predict y is 79% smaller than the prediction error using $\bar{y}$ to predict y. Because the calculation of prediction error in r^2 refers to *squared* distances, it relates to comparing the *variance* of the y values to the variance around the regression line. It is more difficult to interpret the *variance* than the *standard deviation*, so we wait until Chapter 11 to introduce r^2 fully and explain it further. We mention it here merely because you'll see it listed on regression software output, and we wanted you to have a rough idea of what it represents.

Associations with Quantitative and Categorical Variables

In this chapter we've learned how to explore an association between categorical variables and between quantitative variables. It's also possible to mix the variable types or add other variables. For example, with two quantitative variables, we can identify points in the scatterplot according to their values on a relevant categorical variable. This is done by using different symbols or colors on the scatterplot to portray the different categories.

EXAMPLE 11

What's the Gender Difference in Winning Olympic High Jumps?

Picture the Scenario

The summer Olympic Games occurs every four years. One of the track and field events is the high jump. Men have competed in the high jump since 1896 and women since 1928. The "high jump" data file on the text CD contains the winning heights (in meters) for each year.[8]

Questions to Explore

a. How can we display the data on these two quantitative variables (winning height, year) and the categorical variable (gender) graphically?

b. How have the winning heights changed over time? How different are the winning heights for men and women in a typical year?

Think It Through

a. Figure 3.16 shows a scatterplot with x = year and y = winning height. The data points are displayed with a circle for men and a square for women. There were no Olympic games during World War II, so no observations appear for 1940 or 1944.

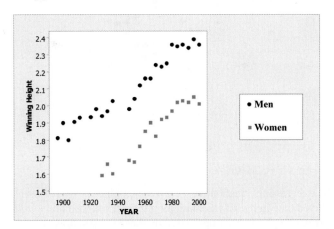

▲ **FIGURE 3.16: MINITAB scatterplot for the winning high jumps (in meters) in the Olympics.** The dots represent men and the squares represent women. **Question:** In a typical year, what is the approximate difference between the winning heights for men and for women?

---

[8]From *New York Times Almanac, 2004*, pp. 895–897.

P1:L2.L1

X=1896 _____ Y=1.81 _____

TI-83+/84 Scatterplot

b. The scatterplot shows that for each gender the winning heights have an increasing trend over time. Men have consistently jumped higher than women, between about 0.3 and 0.4 meters in a given year. The women's winning heights are similar to those for the men about 60 years earlier — for instance, about 2.0 meters in 1990–2000 for women and in 1930–1940 for men. A TI-83+/84 output of this scatterplot appears in the margin.

Insight

We could describe these trends by fitting a regression line to the points for men and a separate regression line to the points for women. However, note that in recent Olympics, the winning distances have leveled off somewhat. We should be cautious in using regression lines to predict future winning heights.

TO PRACTICE THIS CONCEPT, TRY EXERCISE 3.36

3.3 Practicing the Basics

3.22 **Sketch plots of lines:** Identify the values of the y-intercept a and the slope b, and sketch the following regression lines, for values of x between 0 and 10.
  a. $\hat{y} = 7 + 0.5x$
  b. $\hat{y} = 7 + x$
  c. $\hat{y} = 7 - x$
  d. $\hat{y} = 7$

3.23 **Sit-ups and the 40-yard dash:** Is there a relationship between how many sit-ups you can do and how fast you can run 40 yards? The EXCEL output shows the relationship between these variables for a study of female athletes to be discussed in Chapter 11.

Excel Scatterplot of Time to Run 40-Yard Dash by number of Sit-Ups

  a. The regression equation is $6.71 - 0.024x$. Find the predicted time in the 40-yard dash for a subject who can do (i) 10 sit-ups, (ii) 40 sit-ups. Based on these, explain how to sketch the regression line over this scatterplot.
  b. Interpret the y-intercept and slope of the equation in (a), in context.
  c. Based on the slope in (a), is the correlation positive, or negative? Explain.

3.24 **Home selling prices:** The "house selling prices" data file on the text CD lists selling prices of homes in Gainesville, Florida in 2003 and some predictors for the selling price. For y = selling price in thousands of dollars and x = size of house in thousands of square feet, $\hat{y} = 9.2 + 77.0x$.

  a. How much do you predict a house would sell for if it has (i) two thousand square feet, (ii) three thousand square feet?
  b. Using results in (a), explain how to interpret the slope.
  c. Is the correlation between these variables positive or negative? Why?
  d. One home that is three thousand square feet sold for $300,000. Find the residual, and interpret.

3.25 **Internet in Ireland:** For 39 nations, Example 7 found a correlation of 0.888 between y = Internet use (%) and x = gross domestic product (GDP, in thousands of dollars per capita). The regression equation is $\hat{y} = -3.61 + 1.55x$.

  a. Based on the correlation value, the slope had to be positive. Why?
  b. Ireland had a GDP of 32.4 thousand dollars and Internet use of 23.3%. Find its predicted Internet use based on the regression equation.
  c. Find the residual for Ireland. Interpret.

3.26 **Cell phones and GDP:** Table 3.4 showed data on the percent of the population in a country that uses cell phones, the variable called "cellular" in the table. Let y = cell-phone use and x = GDP. The regression equation is $\hat{y} = -0.13 + 2.62x$.

  a. Predict cell-phone use at the (i) minimum x value of 0.8, (ii) maximum x = 34.3.
  b. Interpret the slope of the prediction equation. Is the association positive or negative? Explain what this means.
  c. For the United States, x = 34.3 and y = 45.1. Find the predicted cell-phone use and the residual. Interpret the large negative residual.

3.27 Air pollution and GDP: Exercise 3.11 analyzed a scatterplot relating per capita carbon dioxide emissions in several countries, denoted by CO_2, and per capita gross domestic product (GDP).

    a. Let $y = CO_2$ use and x = GDP. The regression equation is $\hat{y} = 1.26 + 0.346x$. Predict CO_2 at the (i) minimum x value of 0.8, (ii) maximum $x = 34.3$.

    b. For the United States, $x = 34.3$ and $y = 19.7$. Find the predicted CO_2 value and the residual. Interpret the large positive residual.

    c. Switzerland is nearly as economically advanced as the United States, having $x = 28.1$. For it, $y = 5.7$. Find the predicted CO_2 value and the residual. Interpret.

3.28 How much do seat belts help?: A study in 2000 by the
[TRY] National Highway Traffic Safety Administration estimated that failure to wear seat belts led to 9200 deaths in the previous year, and that that value would decrease by 270 for every 1 percentage point gain in seat belt usage. Let $\hat{y}$ = predicted number of deaths in a year and x = percentage of people who wear seat belts.

    a. Report the slope b for the equation $\hat{y} = a + bx$.

    b. If the y-intercept equals 28,910, then predict the number of deaths in a year if (i) no one wears seat belts, (ii) 73% of people wear seat belts (the value in 2000), (iii) 100% of people wear seat belts.

3.29 Regression between cereal sodium and sugar: The
[TRY] figure shows the result of using MINITAB to conduct a regression analysis of y = sodium and x = sugar for the breakfast cereal data considered in Chapter 2 (the "Cereal" data file on the text CD).

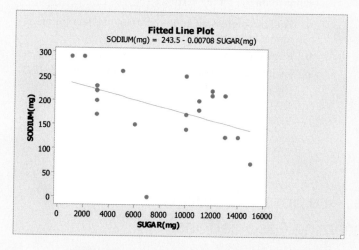

Fitted Line Plot
SODIUM(mg) = 243.5 - 0.00708 SUGAR(mg)

    a. Suppose you had just fitted a line to the scatterplot by eyeballing. In what sense would the line chosen by software be better than your line?

    b. Find the predicted value and residual for *Frosted Mini Wheats* ($x = 7000$, $y = 0$).

    c. Identify the observation and the residual from (b) on the scatterplot that shows the regression line. Does this residual appear to be large relative to the residuals for the other points on the scatterplot?

    d. Now let's look at a histogram of the residuals. Tell what you learn from the MINITAB histogram shown, and explain how its leftmost bar relates to a point on the scatterplot.

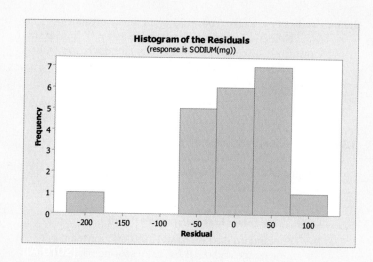

Histogram of the Residuals
(response is SODIUM(mg))

3.30 Regression and correlation between cereal sodium and sugar: Refer to the previous exercise. Show the algebraic relationship between the correlation of -0.453 and the slope of the regression equation, using the fact that the standard deviations are 4561 mg for sugar and 71.2 mg for sodium.

3.31 Advertising and sales: Each month, the owner of Fay's Tanning Salon records in a data file y = monthly total sales receipts and x = amount spent that month on advertising, both in thousands of dollars. For the first three months of operation, the observations are as shown in the table.

ADVERTISING	SALES
0	4
1	6
2	8

    a. Sketch a scatterplot.

    b. From inspection of the scatterplot, state the correlation and the regression line. (*Note:* You should be able to figure them out without using software or formulas.)

    c. Find the mean and standard deviation for each variable.

    d. Using (c), find the regression line, using the formulas for the slope and the y-intercept. Interpret the y-intercept and the slope

3.32 Midterm-final correlation: For students who take Statistics 101 at Lake Wobegon College in Minnesota, both the midterm and final exams have mean = 75 and standard deviation = 10. The regression equation relating y = final exam score to x = midterm exam score is $\hat{y} = 30 + 0.60x$.

    a. Find the predicted final exam score for a student who has (i) midterm score = 100, (ii) midterm score = 50. Note that in each case the predicted final exam score "regresses toward the mean" of 75. (This is a property of the regression equation that is the origin of its name, as Chapter 11 will explain.)

    b. Show that the correlation equals 0.60, and interpret it. (*Hint:* Use the relation between the slope and correlation.)

3.33 Predict final exam from midterm: In an introductory statistics course, x = midterm exam score and y = final exam score. Both have mean = 80 and standard deviation = 10. The correlation between the exam scores is 0.70.

a. Find the regression equation

b. Find the predicted final exam score for a student with midterm exam score = 80.

3.34 NL baseball: Example 9 related y = team scoring (per game) and x = team batting average for American League teams. For National League teams in 2004, $\hat{y} = -6.54 + 42.7x$.

a. The team batting averages fell between 0.247 and 0.277. Explain how to interpret the slope in context.

b. The standard deviations were 0.01002 for team batting average and 0.501 for team scoring. The correlation between these variables was 0.854. Show how the correlation and slope of 42.7 relate, in terms of these standard deviations.

c. Software reports $r^2 = 0.73$. Explain how to interpret this measure.

3.35 Mountain bikes revisited: Is there a relationship between the weight and price of a mountain bike? This question was considered in Exercise 3.19. Analyze the "mountain bike" data file on the text CD.

a. Construct a scatterplot. Interpret.

b. Find the regression equation. Interpret the slope in context. Does the y-intercept have contextual meaning?

c. You decide to purchase a mountain bike that weighs 30 pounds. What is the predicted price for the bike?

3.36 Mountain bike and suspension type: Refer to the previous exercise. The data file contains price, weight, and type of suspension system (FU = full, FE = frontend in the scatterplot shown.)

a. Do you observe a linear relationship? Is the single regression line, which is $\hat{y} = 1896 - 40.45x$, the best way to fit the data? How would you suggest fitting the data?

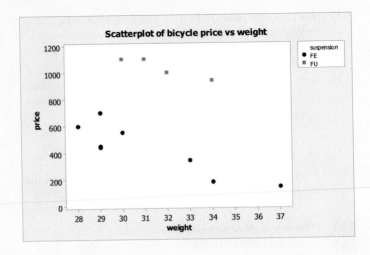

Scatterplot of bicycle price vs weight

b. Find separate regression equations for the two suspension types. Summarize your findings.

c. The correlation for all 12 data points is $r = -0.32$. If the correlations for the full and front end suspension bikes are found separately, how do you believe the correlations will compare to $r = -0.32$? Find them, and interpret.

d. You see a mountain bike advertised for $700 that weighs 28.5 lb. The type of suspension is not given. Would you predict that this bike has a full or a front end suspension? Statistically justify your answer.

3.37 Olympic high jump: The data discussed in Example 11 on winning high jumps in the Olympics are in the "high jump" data file on the text CD.

a. Find the regression equation using the data for men alone. Interpret the slope.

b. Repeat (a) using the data for women alone.

c. Using the regression equations from (a) and (b), find the predicted winning high jump for men and for women in the 2004 Olympic games.

3.4 | What Are Some Cautions in Analyzing Associations?

This chapter has introduced ways to explore **associations** between variables. In using these methods, you need to be cautious about certain potential pitfalls.

Extrapolation Is Dangerous

Extrapolation refers to using a regression line to predict y values for x values outside the observed range of data. This is riskier as we move farther from that range. If the trend changes, extrapolation gives poor predictions. For example, regression analysis is often applied to observations of a quantitative response variable over time. The regression line describes the time trend. But it is dangerous to use the line to make predictions far into the future.

EXAMPLE 12

How Can We Forecast Future Global Warming?

Picture the Scenario

In Chapter 2, we explored trends in temperatures over time using time

plots. Let's use regression with the "Central Park yearly temps" data file on the text CD to describe the trend over time of the mean annual temperatures from 1901–2000 for Central Park, New York City.[9]

Questions to Explore

a. What does a regression line tell us about the trend over the twentieth century?

b. What does it predict about the mean annual temperature in the year (i) 2005, (ii) 3000? Are these extrapolations sensible?

Think It Through

a. Figure 3.17 shows a time plot of the Central Park mean annual temperatures. For a regression analysis, the mean annual temperature is the response variable. Time (the year) is the explanatory variable. Suppose we set $x = (\text{year} - 1900)$, so $x = 0$ for 1900, $x = 1$ for 1901, $x = 2$ for 1902, and so forth. Then, x represents the number of years after 1900. Software tells us that the regression line is $\hat{y} = 52.5 + 0.031x$. Figure 3.17 superimposes the regression trend line over the time plot.

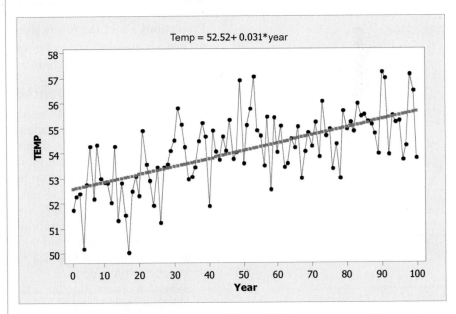

▲ **FIGURE 3.17: MINITAB time plot of Central Park mean annual temperature versus time, showing fitted regression line.** The years 1900–2000 are coded as times 0 through 100. **Question:** If the present trend continues, what would you predict for the mean annual temperature for 2005 (time x = 105)?

---

[9]*Source:* http://www.ncdc.noaa.gov/oa/ncdc.html

- The y-intercept, $a = 52.5$, indicates that for the year 1900 (coded as $x = 0$), the predicted mean annual temperature was 52.5 degrees Fahrenheit.

- The slope, $b = 0.031$, indicates that for each 1-year increase, the predicted mean annual temperature increases by 0.031 degrees Fahrenheit.

The positive slope reflects an increasing trend: The mean annual temperature tended upward over the century. The slope value of 0.031 seems close to 0, indicating little warming. However, 0.031 is the predicted change *per year*. Over the century, the predicted change is $100(0.031) = 3.1$ degrees, which is quite significant.

b. Using the trend line, for the year 2005 (coded as $x = 2005 - 1900 = 105$) the predicted mean annual temperature is

$$\hat{y} = 52.5 + 0.031(105) = 55.8 \text{ degrees Fahrenheit.}$$

Farther into the future we see more dramatic increases. At the next millennium, the year 3000 has $x = 3000 - 1900 = 1100$. The forecast is $\hat{y} = 52.5 + 0.031(1100) = 86.6$. If this is accurate, it would be exceedingly uncomfortable to live in New York!

Insight

It is dangerous to extrapolate far outside the range of observed x values. There's no guarantee that the relationship will have the same trend outside that range. It seems reasonable to predict for 2005. That's not looking too far into the future from the last observation in 2000. However, it is foolhardy to predict for 3000. It's not sensible to assume that the same straight line trend will continue for the next 1000 years. As time moves forward, the mean annual temperatures may level off or even decrease.

TO PRACTICE THIS CONCEPT, TRY EXERCISE 3.39

Predictions about the future using time series data are called **forecasts**. When we use a regression line to forecast a trend for future years, we must make the assumption that the past trend will remain the same in the future. This is risky.

Be Cautious of Influential Outliers

One reason to plot the data before you do a correlation or regression analysis is to check for unusual observations. Such an observation can tell you something interesting, as in Examples 6 and 10 about the Buchanan vote in the 2000 U.S. Presidential election. Furthermore, a data point that is an outlier on a scatterplot can have a substantial effect on the results, especially with small data sets.

What's relevant here is not whether an observation is an outlier in its x value, relative to the other x values, or in its y value, relative to the other y values. Instead, we search for observations that are **regression outliers**, being well removed from the trend that the rest of the data follow. The margin figure shows an observation that is a regression outlier, although it is not an outlier on x alone or on y alone.

When an observation has a large effect on results of a regression analysis, it is said to be **influential**. For an observation to be influential, two conditions must hold:

- Its x value is relatively low or high compared to the rest of the data.

- The observation is a regression outlier, falling quite far from the trend that the rest of the data follow.

When both of these happen, the line tends to be pulled toward that data point and away from the trend of the rest of the points.

Regression Outlier

Figure 3.18 shows two regression outliers. The correlation without these two points equals 0.00. The first regression outlier is near the middle of the range of *x*. It does not have much potential for tilting the line up or down. It has little influence on the slope or the correlation. The correlation changes only to 0.03 when we add it to the data set. The second regression outlier is at the high end of the range of *x* values. It is influential. The correlation changes to 0.47 when we add it to the data set.

▲ **FIGURE 3.18: An observation is a regression outlier if it is far removed from the trend that the rest of the data follow.** The top two points are regression outliers. Not all regression outliers are influential in affecting the correlation or slope. **Question:** Which regression outlier in this figure is influential?

Is Higher Education Associated with Higher Murder Rates?

Picture the Scenario

Table 3.6 shows data[10] for the 50 states and the District of Columbia on

Violent crime rate: The annual number of murders, forcible rapes, robberies, and aggravated assaults per 100,000 people in the population

Murder rate: The annual number of murders per 100,000 people in the population

Poverty: Percentage of the residents with income below the poverty level

High school: Percentage of the adult residents who have at least a high school education

College: Percentage of the adult residents who have a college education

Single parent: Percentage of families headed by a single parent

The data are in the "U.S. statewide crime" data file on the text CD. Let's look at the relationship between *y* = murder rate and *x* = college. We'll look at other variables in the exercises.

---

[10]From *Statistical Abstract of the United States 2003*

TABLE 3.6: Statewide Data on Several Variables

State	Violent Crime	Murder Rate	Poverty	High School	College	Single Parent
Alabama	486	7.4	14.7	77.5	20.4	26
Alaska	567	4.3	8.4	90.4	28.1	23.2
Arizona	532	7	13.5	85.1	24.6	23.5
Arkansas	445	6.3	15.8	81.7	18.4	24.7
California	622	6.1	14	81.2	27.5	21.8
Colorado	334	3.1	8.5	89.7	34.6	20.8
Connecticut	325	2.9	7.7	88.2	31.6	22.9
Delaware	684	3.2	9.9	86.1	24	25.6
District of Columbia	1508	41.8	17.4	83.2	38.3	44.7
Florida	812	5.6	12	84	22.8	26.5
Georgia	505	8	12.5	82.6	23.1	25.5
Hawaii	244	2.9	10.6	87.4	26.3	19.1
Idaho	253	1.2	13.3	86.2	20	17.7
Illinois	657	7.2	10.5	85.5	27.1	21.9
Indiana	349	5.8	8.3	84.6	17.1	22.8
Iowa	266	1.6	7.9	89.7	25.5	19.8
Kansas	389	6.3	10.5	88.1	27.3	20.2
Kentucky	295	4.8	12.5	78.7	20.5	23.2
Louisiana	681	12.5	18.5	80.8	22.5	29.3
Maine	110	1.2	9.8	89.3	24.1	23.7
Maryland	787	8.1	7.3	85.7	32.3	24.5
Massachusetts	476	2	10.2	85.1	32.7	22.8
Michigan	555	6.7	10.2	86.2	23	24.5
Minnesota	281	3.1	7.9	90.8	31.2	19.6
Mississippi	361	9	15.5	80.3	18.7	30
Missouri	490	6.2	9.8	86.6	26.2	24.3
Montana	241	1.8	16	89.6	23.8	21.4
Nebraska	328	3.7	10.7	90.4	24.6	19.6
Nevada	524	6.5	10.1	82.8	19.3	24.2
New Hampshire	175	1.8	7.6	88.1	30.1	20
New Jersey	384	3.4	8.1	87.3	30.1	20.2
New Mexico	758	7.4	19.3	82.2	23.6	26.6
New York	554	5	14.7	82.5	28.7	26
North Carolina	498	7	13.2	79.2	23.2	24.3
North Dakota	81	0.6	12.8	85.5	22.6	19.1
Ohio	334	3.7	11.1	87	24.6	24.6
Oklahoma	496	5.3	14.1	86.1	22.5	23.5
Oregon	351	2	12.9	88.1	27.2	22.5
Pennsylvania	420	4.9	9.8	85.7	24.3	22.8
Rhode Island	298	4.3	10.2	81.3	26.4	27.4
South Carolina	805	5.8	12	83	19	27.1
South Dakota	167	0.9	9.4	91.8	25.7	20.7
Tennessee	707	7.2	13.4	79.9	22	27.9
Texas	545	5.9	14.9	79.2	23.9	21.5
Utah	256	1.9	8.1	90.7	26.4	13.6
Vermont	114	1.5	10.3	90	28.8	22.5
Virginia	282	5.7	8.1	86.6	31.9	22.2
Washington	370	3.3	9.5	91.8	28.6	22.1
West Virginia	317	2.5	15.8	77.1	15.3	22.3
Wisconsin	237	3.2	9	86.7	23.8	21.7
Wyoming	267	2.4	11.1	90	20.6	20.8

Questions to Explore

a. Construct the scatterplot between y = murder rate and x = college. Does any observation look like it could be influential in its effect on the regression line?

b. Use software to find the regression line. Check whether the observation identified in (a) actually is influential by finding the line again without that observation.

Think It Through

a. Figure 3.19 shows the scatterplot. The observation out by itself is D.C. It has $x = 38.3$ and $y = 41.8$. This is the largest observation on both these variables. It satisfies both conditions for an observations to be influential: It has a relatively extreme value on the explanatory variable (college), and it is a regression outlier, falling well away from the linear trend of the other points.

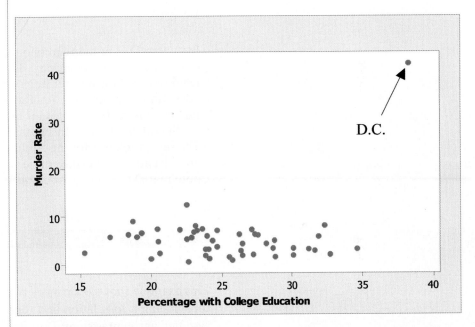

▲ **FIGURE 3.19: MINITAB scatterplot relating murder rate to percentage with college education. Question:** How would you expect the slope to change if D.C. is excluded from the regression analysis?

b. Using software, the regression line fitted to all 51 observations, including D.C., equals $\hat{y} = -3.1 + 0.33x$. The slope is *positive*, as shown in the first plot in Figure 3.20. You can check that the predicted murder rates *increase* from 1.9 to 10.1 as the percentage with a college education increases from $x = 15\%$ to $x = 40\%$, roughly the range of observed x values. By contrast, when we fit the regression line only to the 50 states, excluding the observation for D.C., $\hat{y} = 8.0 - 0.14x$. The slope of -0.14 reflects a *negative* trend, as shown in the second plot in Figure 3.20. Now, the predicted murder rate *decreases* from 5.9 to 2.4 as the percentage with a college education increases from 15% to 40%.

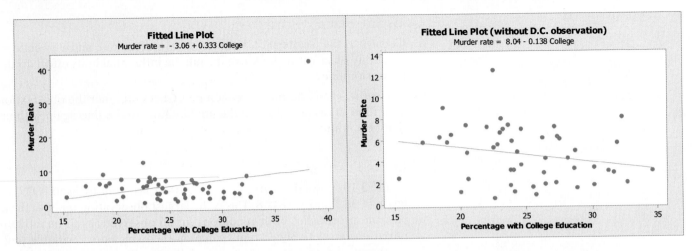

▲ **FIGURE 3.20: MINITAB scatterplots relating murder rate to percentage with college education, with and without observation for D.C. Question:** Which line better describes the trend for the 50 states?

Insight

Including D.C. in the regression analysis has the effect of pulling the slope of the regression line upward. The regression line then makes it seem, misleadingly, as if the murder rate *increases* when the percentage with a college education increases. In fact, for the rest of the data the predicted murder rate *decreases*. The regression line including D.C. distorts the relationship for the other 50 states. In summary, the D.C. data point is highly influential. The regression line for the 50 states alone better represents the overall negative trend. In reporting these results, we should show this line and then note that D.C. is a regression outlier that falls well outside this trend.

TO PRACTICE THIS CONCEPT, TRY EXERCISE 3.41

This example shows once more that you should *construct a scatterplot* first before finding a correlation or a regression line. The correlation and the regression line are **nonresistant**: They are prone to distortion by outliers. *Investigate any regression outlier.* Was the observation possibly recorded incorrectly, or is it merely different from the rest of the data in some way? It is often a good idea to refit the regression line without it to see if it has a large effect, as we did in this last example.

Correlation Does Not Imply Causation

In a regression analysis, suppose that as x goes up, y also tends to go up (or go down). Can we conclude that there's a *causal* connection, with changes in x causing changes in y?

The concept of causality is central to science. We are all familiar with it, at least in an informal sense. We know, for example, that exposure to a virus can cause the flu. But just observing an association between two variables is not enough to imply a causal connection. There may be some alternative explanation for the association.

Does More Education Cause More Crime?

Picture the Scenario

Figure 3.21 shows recent data on y = crime rate and x = education for Florida counties, from the "FL crime" data file on the text CD. Education was measured as the percentage of residents aged at least 25 in the county who had at least a high school degree. Crime rate was measured as the number of crimes in that county in the past year per 1000 residents.

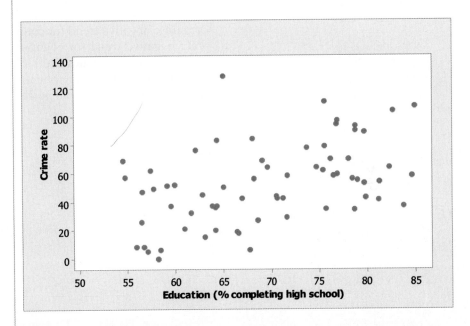

▲ **FIGURE 3.21: MINITAB scatterplot of crime rate and percentage with at least a high school education.** There is a moderate positive association (r = 0.47). **Question:** Does more education cause more crime, or does more crime cause more education, or possibly neither?

As the figure shows, these variables have a positive association. The correlation is r = 0.47. Unlike the previous example, there is no obviously influential observation causing the positive correlation. Another variable measured for these counties is urbanization, measured as the percentage of the residents who live in metropolitan areas. It has a correlation of 0.68 with crime rate and 0.79 with education.

Question to Explore

From the positive correlation between crime rate and education, can we conclude that having a more highly educated populace causes the crime rate to go up?

Think It Through

The strong correlation of 0.79 between urbanization and education tells us that counties that are highly urban tend to have higher education levels. The moderately strong correlation of 0.68 between urbanization and crime rate tells us that the highly urban counties also tend to have higher crime. So, perhaps the reason for the positive correlation between education and crime rate is that education tends to be greater in more highly urbanized counties, but crime rates also tend to be higher in such counties. In summary, a correlation could occur without any causal connection.

Insight

For counties with similar levels of urbanization, the association between crime rate and education may look quite different. You may then see a *negative* correlation. Figure 3.22 portrays how this could happen. It shows a negative trend between crime rate and education for counties having urbanization = 0 (none of the residents living in a metropolitan area), a separate negative trend for counties having urbanization = 50, and a separate negative trend for counties having urbanization = 100. If we ignore the urbanization values and look at all the points, however, we see a positive trend—higher crime rate tending to occur with higher education levels, as reflected by the overall positive correlation.

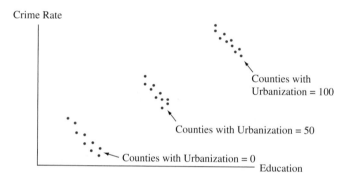

▲ **FIGURE 3.22: Hypothetical scatter diagram relating crime rate and education.** The points are also labeled by whether urbanization = 0, 50, or 100. **Question:** Sketch lines that represent (a) the overall positive relationship between crime rate and education, (b) the negative relationship between crime rate and education for counties having urbanization = 0.

TO PRACTICE THIS CONCEPT, TRY EXERCISES 3.47 AND 3.48

Whenever two variables are associated, other variables may have influenced that association. In Example 14, urbanization influenced the association between crime rate and education. This illustrates an important point:

Correlation does not imply causation.

In Example 14, crime rate and education were positively correlated, but that does not mean that having a high level of education causes a county's crime rate to be high. Whenever we observe a correlation between variables x and y, there may be a third variable correlated with both x and y that is responsible for their association. Let's look at another example to illustrate this point.

EXAMPLE 15

Ice Cream Positively Correlated with Drowning

Picture the Scenario

The "Gold Coast" of Australia, south of Brisbane, is famous for its magnificent beaches. Because of strong rip tides, however, each year many people drown. Data collected each month show a positive correlation between y = number of people who drowned in that month and x = number of gallons of ice cream sold in refreshment stands along the beach in that month.

Question to Explore

Identify another variable that could be responsible for this association.

Think it Through

In the summer in Australia (especially January and February), the weather is hot. People tend to buy more ice cream in those months. They also tend to go to the beach and swim more in those months, and more people drown. In the winter, it is cooler. People buy less ice cream, fewer people go to the beach, and fewer people drown. So, the mean temperature in the month is a variable that could be responsible for the correlation. As mean temperature goes up, so does ice cream sales and so does the number of people who drown.

Insight

If we looked only at months having similar mean temperatures, probably we would not observe any association between ice cream sales and the number of people who drown.

TO PRACTICE THIS CONCEPT, TRY EXERCISE 3.47 AND 3.48

A third variable that is not measured in a study (or perhaps even known about to the researchers) but that influences the association between the response variable and the explanatory variable is referred to as a **lurking variable**.

Lurking Variable

A **lurking variable** is a variable, usually unobserved, that influences the association between the variables of primary interest.

In interpreting the positive correlation between crime rate and education for Florida counties, we'd be remiss if we failed to recognize that the correlation could be due to a lurking variable. This could happen if we observed those two variables but not urbanization, which would then be a lurking variable. Likewise, if we got excited by the positive correlation between ice cream sales and the number drowned in a month, we'd fail to recognize that the monthly mean temperature is a lurking variable.

Simpson's Paradox

The statement that *correlation does not imply causation* can be expressed more generally as **association does not imply causation**. This warning holds whether we analyze associations between quantitative variables or between categorical variables. Let's illustrate by revisiting Example 1, which presented a study indicating that smoking could apparently be beneficial to your health. Could a lurking variable be responsible for this association?

EXAMPLE 16

Is Smoking Actually Beneficial to Your Health?

Picture the Scenario

Example 1 mentioned a survey[11] of 1314 women in the United Kingdom

during 1972–1974 that asked each woman whether she was a smoker. Twenty years later, a follow-up survey observed whether each woman was dead or still alive. Table 3.7 is a contingency table of the results. The response variable is survival status after 20 years. We find that 139/582, which is 24%, of the smokers died, and 230/732, or 31%, of the nonsmokers died. There was a greater survival rate for the smokers.

TABLE 3.7: Smoking Status and 20-Year Survival in Women

Smoker	Survival Status		Total
	Dead	Alive	
Yes	139	443	582
No	230	502	732
Total	**369**	**945**	**1314**

Could the age of the woman at the beginning of the study explain the association? Presumably it could if the older women were less likely than the younger women to be smokers. Table 3.8 shows separate contingency tables relating smoking status and survival status for these 1314 women separated into four age groups.

TABLE 3.8: Smoking Status and 20-Year Survival, for Four Age Groups

	Age Group						
	18–34 Survival?		35–54 Survival?		55–64 Survival?		65+ Survival?	
Smoker	Dead	Alive	Dead	Alive	Dead	Alive	Dead	Alive
Yes	5	174	41	198	51	64	42	7
No	6	213	19	180	40	81	165	28

[11]Described in article by D.R. Appleton et al., *American Statistician*, 50, 340–341 (1996).

Questions to Explore

a. Show that the counts in Table 3.8 are consistent with those in Table 3.7.

b. Use conditional percentages to describe the association in Table 3.8 between smoking status and survival status for each age group.

c. How can you explain the association in Table 3.7, whereby smoking seems to help women live a longer life? How can this association be so different from the one shown in Table 3.8?

Think It Though

a. If you add the counts in the four separate parts of Table 3.8, you'll get the counts in Table 3.7. For instance, from Table 3.7, we see that 139 of the women who smoked died. From Table 3.8, we get this from $5 + 41 + 51 + 42 = 139$, taking the number of smokers who died from each age group. Doing this for each cell, we see that the counts in the two tables are consistent with each other.

b. Table 3.9 shows the conditional percentages who died, for smokers and nonsmokers in each age group. Figure 3.23 plots them. For each age group, a higher percentage of smokers than non-smokers died.

TABLE 3.9: Conditional Percentages of Deaths for Smokers and Nonsmokers, by Age.

For instance, for smokers of age 18–34, from Table 3.8 the proportion who died was $5/(5 + 174) = 0.028$, or 2.8%.

Smoker	Age Group			
	18–34	35–54	55–64	65+
Yes	2.8%	17.2%	44.3%	85.7%
No	2.7%	9.5%	33.1%	85.5%

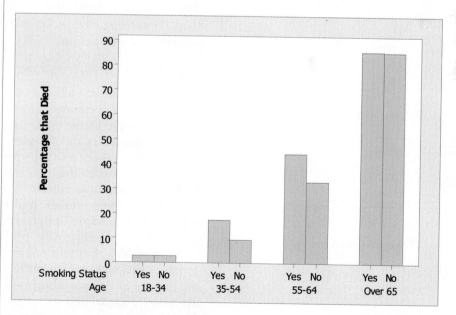

▲ **FIGURE 3.23: MINITAB bar graph comparing percentage of deaths for smokers and nonsmokers, by age.** This side-by-side bar graph shows the conditional percentages from Table 3.9.

c. Could age explain the association? From Table 3.8, the proportion of women who smoked in 1972–1974 tended to be higher for the younger women. The percentage of smokers was 45% in the 18–34 age group (that is, $(5 + 174)/(5 + 174 + 6 + 213) = 0.45$), but only 20% in the

65+ age group. At the same time, younger women were less likely to die during the 20-year study period. For instance, the proportion who died was $(5 + 6)/(5 + 174 + 6 + 213) = 0.03$ in the 18–34 age group but 0.86 in the 65+ age group. In summary, the overall association in Table 3.7 could merely reflect younger women being more likely to be smokers, while the younger women were also less likely to die during this time frame.

Table 3.7 indicated that smokers had *higher* survival rates than non-smokers. When we looked at the data separately for each age group in Table 3.8, we saw the reverse: Smokers had *lower* survival rates than nonsmokers. The analysis using Table 3.7 did not account for age, which strongly influences the association.

Insight

Because of the reversal in the association after taking age into account, the researchers did *not* conclude that smoking is beneficial to your health. This example illustrates the dramatic influence that a lurking variable can have, unknown to researchers if they fail to include a certain variable in their study. An association can look quite different after adjusting for the effect of a third variable by grouping the data according to its values.

TO PRACTICE THIS CONCEPT, TRY EXERCISE 3.51

The fact that the direction of an association between two variables can change after we include a third variable and analyze the data at separate levels of that variable is known as **Simpson's paradox**.[12] We observed Simpson's paradox in Figure 3.22 in Example 14, in which a positive correlation between crime rate and education changed to a negative correlation when data were considered at separate levels of urbanization. This example serves as a warning: *Be cautious about interpreting an association.* Always be wary of lurking variables that may influence the association.

How Can Lurking Variables Affect Associations?

Lurking variables can affect associations in many ways. For instance, a lurking variable may be a **common cause** of both the explanatory and response variable. In Example 15, the mean temperature in the month is a common cause of both ice cream sales and the number of people who drown.

In practice, there's usually not a single variable that causally explains a response variable or the association between two variables. More commonly, there are **multiple causes**. When there are multiple causes, the association among them makes it difficult to study the effect of any single variable. For example, suppose someone claims that growing up in poverty causes crime. Realistically, probably lots of things contribute to crime. Many variables that you might think of as possible causes, such as a person's educational level, whether the person grew up in a stable family, and the quality of the neighborhood in which the person lives, are themselves likely to be associated with whether a person grew up in poverty. Perhaps people growing up in poverty tend to have poorly educated parents and grow up in high-crime neighborhoods and achieve low levels of education, and perhaps all these factors make a person more likely to become a criminal. Growing up in poverty may have a direct effect on crime but also an indirect effect through these other variables.

---

[12]The paradox is named after a British statistician who in 1951 investigated conditions under which this flip-flopping of association can happen.

It's especially tricky to study cause and effect when two variables are measured over time. The variables may be associated merely because they both have a *time trend*. Suppose that both the divorce rate and the crime rate have an increasing trend over a 10-year period. They will then have a positive correlation: Higher crime rates occur in years that have higher divorce rates. Does this imply that an increasing divorce rate *causes* the crime rate to increase? Absolutely not. They would also be positively correlated with all other variables that have a positive time trend, such as annual average house price and the annual use of cell phones. There are likely to be other variables yet that are themselves changing over time and that have causal influences on the divorce rate and the crime rate.

Confounding

When two explanatory variables are both associated with a response variable but are also associated with each other, there is said to be **confounding**. It is difficult to determine whether either of them truly causes the response, because a variable's effect could be at least partly due to its association with the other variable. Example 16 illustrates a study with confounding. Over the 20-year study period, smokers had a greater survival rate than nonsmokers. However, age was a confounding variable. Older subjects were less likely to be smokers, and older subjects were more likely to die. Within each age group, smokers had a *lower* survival rate than nonsmokers. Age had a dramatic influence on the association between smoking and survival status.

What's the difference between a confounding variable and a lurking variable? A lurking variable is not measured in the study. It has the *potential* for confounding. If it were included in the study and if it were associated both with the response variable and the explanatory variable, it would become a confounding variable. It would affect the relationship between the response and explanatory variables.

The potential for lurking variables to affect associations is the main reason it is difficult to study many issues of importance, whether it be medical issues such as the effect of smoking on cancer, or social issues such as what causes crime or what causes the economy to improve or what causes students to succeed in school. It's not impossible—statistical methods can analyze the effect of an explanatory variable after adjusting for confounding variables (as we adjusted for age in Example 16)—but there's always the chance that an important variable was not included in the study.

3.4 Practicing the Basics

3.38 Extrapolating murder: The SPSS figure shows the data and regression line for the 50 states in Table 3.6 relating
  x = percentage of single-parent families to
  y = annual murder rate (number of murders per 100,000 people in the population).

  a. The lowest x value was for Utah and the highest was for Mississippi. Using the figure, approximate those x-values.

  b. Using the regression equation stated on the figure, find the predicted murder rate at $x = 0$. Why is it not sensible to make a prediction at $x = 0$ based on these data?

Fitted Line Plot
Murder = –8.25 + 0.56 Single Parent

R sq Linear = 0.459

3.39 Men's Olympic long jumps: The Olympic winning men's long jump distances (in meters) from 1896 to 2000 and the fitted regression line for predicting them using x = year are displayed in the MINITAB output below. (The data, in the "Long Jump" data file on the text CD, are from *2004 New York Times Almanac, pp. 895–897.*)

a. Identify an observation that may influence the fit of the regression line. Why did you identify this observation?

b. Which do you think is a more reasonable prediction for the year 2008—the sample mean of the y values in this plot, or $\hat{y} = -24.7 + 0.0168(2008)$? Why?

c. Would you feel comfortable using the regression line shown to predict the winning long jump for men in the year 3000? Why or why not?

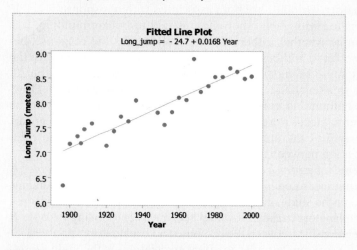

Fitted Line Plot
Long_jump = - 24.7 + 0.0168 Year

3.40 U.S. average annual temperatures: Use the "U.S. Temperatures" data file on the text CD.

a. Fit a trend line, and interpret the slope.

b. Predict the mean annual U.S. temperature for the year (i) 2010, (ii) 3000.

c. In which prediction in (b) do you have more faith? Why?

3.41 Murder and education: Example 13 found the regression line $\hat{y} = -3.1 + 0.33x$ for all 51 observations on y = murder rate and x = percent with a college education.

a. Show that the predicted murder rates increase from 1.85 to 10.1 as percent with a college education increases from $x = 15\%$ to $x = 40\%$, roughly the range of observed x values.

b. When the regression line is fitted only to the 50 states, $\hat{y} = 8.0 - 0.14x$. Show that the predicted murder rate decreases from 5.9 to 2.4 as percent with a college education increases from 15% to 40%.

c. D.C. has the highest value for x (38.3) and is an extreme outlier on y (41.8). Is it a regression outlier? Why?

d. What causes results to differ so according to whether D.C. is in the data set? Which line is more appropriate as a summary of the relationship? Why?

3.42 Murder and poverty: For Table 3.6, the regression equation for the 50 states and D.C. relating y = murder rate and x = percent of people who live below the poverty level is $\hat{y} = -4.1 + 0.81x$. For D.C., $x = 17.4$ and $y = 41.8$.

a. When the observation for D.C. is removed from the data set, $\hat{y} = 0.4 + 0.36x$. Does D.C. have much influence on this regression analysis? Explain.

b. If you were to look at a scatterplot, based on the information above, do you think that the poverty value for D.C. would be relatively large, or relatively small? Explain.

3.43 TV watching and the birth rate: The figure shows recent data on y = the birth rate per 1000 people and x = the number of televisions per 100 people, for six African and Asian nations. The regression line, $\hat{y} = 29.8 - 0.024x$ applies to the data for these six countries. For illustration, another point is added at $(81, 15.2)$ which is the observation for the United States. The regression line for all seven points is $\hat{y} = 31.2 - 0.195x$. The figure shows this line and the one without the U.S. observation.

a. Does the U.S. observation appear to be (i) an outlier on x, (ii) an outlier on y, (iii) a regression outlier relative to the regression line for the other six observations?

b. State the two conditions under which a single point can have a dramatic effect on the slope, and show they apply here.

c. This one point also drastically affects the correlation, which is $r = -0.051$ without the United States but $r = -0.935$ with the United States. Explain why you would conclude that the association between birth rate and number of televisions is (i) very weak without the U.S. point, (ii) very strong with the U.S. point.

d. Explain why the U.S. residual for the line fitted using that point is very small. This shows that a point can be influential even if it's residual is not large.

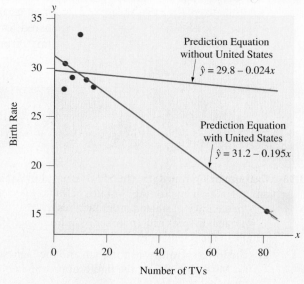

Regression Equations for Birth Rate and Number of TVs per 100 People

3.44 Looking for outliers: Using software, analyze the relationship between x = college education and y = percentage single-parent families, for the data in Table 3.6, which are in the "U.S. statewide crime" data file on the text CD.

a. Construct a scatterplot. Based on your plot, identify two observations that seem quite different from the others, having a y value relatively very large in one case and somewhat small in the other case.

b. Find the regression equation (i) for the entire data set, (ii) deleting only the first of the two outlying observations, (iii) deleting only the second of the two outlying observations.

c. Is either observation influential on the slope? Summarize the influence.

d. Including D.C., $\hat{y} = 21.2 + 0.089x$, whereas deleting that observation, $\hat{y} = 28.1 - 0.206x$. Find $\hat{y}$ for D.C. in the two cases. Does the predicted value for D.C. depend much on which regression equation is used?

3.45 Regression between cereal sodium and sugar: Let $x =$ sodium and $y =$ sugar for the breakfast cereal data in the "Cereal" data file on the text CD.

a. Construct a scatterplot. Do any points satisfy the two criteria for a point to be potentially influential on the regression? Explain.

b. Find the regression line and correlation using all the data points, and then using all except a potentially influential observation. Summarize the influence of that point.

3.46 TV in Europe: An article (by M. Dupagne and D. Waterman, *Journal of Broadcasting and Electronic Media*, vol. 42, pp. 208–220, 1998) studied variables relating to the percentage of TV programs in 17 Western European countries that consisted of fiction programs imported from the United States. One explanatory variable considered was the percentage of stations in the country that are private. The data are in the "TV Europe" data file on the text CD.

a. Construct a scatterplot. Describe the direction of the association.

b. Find the correlation and the regression line. Draw the line on the scatterplot.

c. Do you observe an outlier? If so, what makes the observation unusual? Would you expect it to be an influential observation?

d. Delete the observation identified in part (c) from the data set. Find the regression line and the correlation. Compare your results to those in part (b). Was the observation influential? Explain.

3.47 Height and vocabulary: Do tall students tend to have better vocabulary skills than short students? We might think so looking at a sample of students from grades 1, 6, and 12 of Lake Wobegon school district. The correlation was 0.81 between their height and their vocabulary test score: Taller students tended to have higher vocabulary test scores.

a. Is there a causal relationship, whereby being taller gives you a better vocabulary?

b. Explain how a student's age could be a lurking variable that might be responsible for this association, being a common cause of both height and vocabulary test score.

c. Sketch a hypothetical scatterplot (as we did in Figure 3.22 for the example on crime and education), labeling points by the child's grade (1, 6, and 12), such that overall there is a positive trend, but the slope would be about 0 when we consider only children in a given grade. How does the child's age play a role in the association between height and test score?

3.48 More firefighters cause worse fires?: Data are available for all fires in Chicago last year on $x =$ number of firefighters at the fire and $y =$ cost of damages due to the fire.

a. Would you expect the correlation to be negative, zero, or positive? Why?

b. If the correlation is positive, does this mean that having more firefighters at a fire causes the damage to be worse? Explain.

c. Identify a third variable that could be a common cause of x and y. Construct a hypothetical scatterplot (like Figure 3.22 for crime and education), identifying points according to their value on the third variable, to illustrate your argument.

3.49 Antidrug campaigns: An Associated Press story (6/13/2002) reported that "A survey of teens conducted for the Partnership for a Drug Free America found kids who see or hear anti-drug ads at least once a day are less likely to do drugs than youngsters who don't see or hear ads frequently? When asked about marijuana, kids who said they saw the ads regularly were nearly 15 percent less likely to smoke pot."

a. Discuss at least one lurking variable that could affect these results.

b. Explain how multiple causes could affect whether a teenager smokes pot.

3.50 Death penalty and race: The table shows results of whether the death penalty was imposed in murder trials in Florida between 1976 and 1987. For instance, the death penalty was given in 53 out of 467 cases in which a white defendant had a white victim.

Death Penalty, by Defendant's Race and Victim's Race				
Victim's Race	Defendant's Race	Death Penalty		
		Yes	No	Total
White	White	53	414	467
	Black	11	37	48
Black	White	0	16	16
	Black	4	139	143

Source: M. Radelet and G. Pierce, *Florida Law Review,* Vol. 43, pp. 1–34, 1991

a. First, consider only those cases in which the victim was white. Find the conditional proportions that got the death penalty when the defendant was white and when the defendant was black. Describe the association.

b. Repeat (a) for cases in which the victim was black. Interpret.

c. Now add these two tables together to get a summary contingency table that describes the association between the death penalty verdict and defendant's race, ignoring the information about the victim's race. Find the conditional proportions. Describe the association, and compare to (a) and (b).

d. Explain how these data satisfy Simpson's paradox. How would you explain what is responsible for this, to someone who has not taken a statistics course?

e. In studying the effect of defendant's race on the death penalty verdict, would you call victim's race a confounding variable? What does this mean?

3.51 NAEP scores: Eighth-grade math scores on the National Assessment of Educational Progress had means of 277 in Nebraska and 271 in New Jersey (H. Wainer and L. Brown, *American Statistician*, Vol. 58, p. 119, 2004).

a. Identify the response variable and the explanatory variable.

b. For white students, the means were 281 in Nebraska and 283 in New Jersey. For black students, the means were 236 in Nebraska and 242 in New Jersey. For other nonwhite students, the means were 259 in Nebraska and 260 in

New Jersey. Identify the third variable given here. Explain how it is possible for New Jersey to have the higher mean for each race, yet for Nebraska to have the higher mean when the data are combined. (This is a case of Simpson's paradox for a quantitative response.)

3.52 Age a confounder?: A study observes that the subjects in the study who say they exercise regularly reported only half as many serious illnesses per year, on the average, as those who say they do not exercise regularly. One paragraph in the results section of an article about the study starts out, "We next analyzed whether age was a confounding variable in studying this association."

a. Explain what this sentence means and how age could potentially explain the association between exercising and illnesses.

b. If age was not actually measured in the study and the researchers failed to consider its effects, could it be a confounding variable or a lurking variable? Explain the difference between a lurking variable and a confounding variable.

CHAPTER REVIEW

Answers to the Chapter Figure Questions

Figure 3.1: Construct a single graph with side-by-side bars (as in Figure 3.2) to compare the conditional proportions of pesticide residues in the two types of food.

Figure 3.2: The proportion of foods having pesticides present is much higher for conventionally grown food than for organically grown food.

Figure 3.3: For a particular response, the bars in Figure 3.2 have different heights, in contrast to Figure 3.3 for which the bars have the same height.

Figure 3.4: The approximate five-number summary for the distribution of Internet use is *minimum* = 0, $Q1$ = 2, *Median* = 20, $Q3$ = 39, *and maximum* = 51. The box plot suggests some skew to the right.

Figure 3.5: The point labeled "Ireland" appears unusual. Its Internet use (about 23%) is lower than what's expected based on its high GDP of about 32.

Figure 3.6: The top point is far above the overall straight-line trend of the other 66 points on the graph. The two rightmost points fall in the overall increasing trend.

Figure 3.7: If data points are closer to a straight line, there is a smaller amount of variability around the line; thus, we can more accurately predict one variable knowing the other variable, indicating a strong association.

Figure 3.8: These 36 points make a positive contribution to the correlation, since the product of their z-scores will be positive.

Figure 3.9: Yes, the points fall in a balanced way in the quadrants. The positive cross products in the upper-right quadrant are roughly counterbalanced by the negative cross products in the upper-left quadrant. The positive cross products in the lower-left quadrant are roughly counterbalanced by the negative cross products in the lower-right quadrant.

Figure 3.10: The line crosses the y-axis at 61.4. For each femur length increase of 1 cm, the height is predicted to increase 2.4 cm.

Figure 3.11: The height is predicted to be the same for each femur length.

Figure 3.12: We would expect a positive slope.

Figure 3.13: The prediction error is represented by a vertical line from the prediction line to the observation. The distance between the point and line is the prediction error.

Figure 3.14: The residuals are vertical distances because the regression equation predicts y, the variable on the vertical axis.

Figure 3.15: The rightmost bar represents a county with a large positive residual. For it, the actual vote for Buchanan was much higher than what was predicted.

Figure 3.16: The difference in winning heights for men and women is typically about 0.3 to 0.4 meters.

Figure 3.17: $\hat{y} = a + bx = 52.52 + 0.031(105) = 55.8$.

Figure 3.18: The observation in the upper right corner will influence the tilt of the straight-line fit.

Figure 3.19: The slope will decrease if D.C. is excluded from the regression analysis.

Figure 3.20: The fitted line to the data that does not include D.C.

Figure 3.21: Possibly neither. There may be another variable influencing the positive association between crime rate and amount of education.

Figure 3.22: Line (a) will pass through all the points, with a positive slope. In case (b), the lines passes only through the left set of points with *urbanization* = 0. It has the same negative slope that those points show.

CHAPTER SUMMARY

This chapter introduced descriptive statistics for studying the **association** between two variables. We explored how the value of a **response variable** (the outcome of interest) depends on the value of an **explanatory variable**.

■ For two *categorical variables*, **contingency tables** summarize the counts of observations at the various combinations of categories of the two variables. Bar graphs can plot **conditional proportions** or percentages for the response variable, at different given values of the explanatory variable.

■ For two *quantitative variables*, **scatterplots** display the relationship and show whether there is a **positive association** (upward trend) or a **negative association** (downward trend). The **correlation**, r, describes this direction and the strength of linear (straight-line) association. It satisfies $-1 \le r \le 1$. The closer r is to -1 or 1, the stronger the linear association.

■ When a relationship between two quantitative variables approximately follows a straight line, it can be described by a **regression line**, $\hat{y} = a + bx$. The slope b describes the direction of the association (positive or negative, like the correlation) and gives the effect on $\hat{y}$ of a 1-unit increase in x.

■ The correlation and the regression equation can be strongly affected by an **influential observation**. This takes a relatively small or large value on x and is a **regression outlier**, falling away from the straight-line trend of the rest of the data points. Be cautious of **extrapolating** a regression line to predict y at values of x that are far above or below the observed x values.

■ **Association does not imply causation.** A **lurking variable** may influence the association. It is even possible for the association to reverse in direction after we adjust for a third variable. This phenomenon is called **Simpson's paradox.**

SUMMARY OF NOTATION

Correlation $r = \dfrac{1}{n-1}\sum z_x z_y = \dfrac{1}{n-1}\sum \left(\dfrac{x-\overline{x}}{s_x}\right)\left(\dfrac{y-\overline{y}}{s_y}\right)$

where n is the number of points, $\overline{x}$ and $\overline{y}$ are means, s_x and s_y are standard deviations for x and y.

Regression equation: $\hat{y} = a + bx$, for predicted response $\hat{y}$, y-intercept a, slope b.

Formulas for the slope and y-intercept are

$$b = r\left(\dfrac{s_y}{s_x}\right) \text{ and } a = \overline{y} - b(\overline{x}).$$

CHAPTER PROBLEMS

Practicing the Basics

3.53 Choose explanatory and response: For the following pairs of variables, identify the response variable and the explanatory variable.

   a. Number of square feet in house and assessed value of house

   b. Political party preference (Democrat, Independent, Republican) and gender

   c. Annual income and number of years of education

   d. Number of pounds lost on a diet and type of diet (low-fat, low-carbohydrate)

3.54 Graphing data: For each case in the previous exercise,

   a. Indicate whether each variable is quantitative or categorical.

   b. Describe the type of graph that could be used to portray the results.

3.55 Life after death for males and females: In the 1998 General Social Survey, respondents answered the question, "Do you believe in a life after death?" The table shows the responses cross-tabulated with gender.

Opinion about Life after Death by Gender

	Opinion about life after death			
Gender	Yes, Definitely	Yes, Probably	No, Probably Not	No, Definitely Not
Male	280	118	53	51
Female	380	129	57	59

a. Construct a table of conditional proportions.

b. Summarize results. Is there much difference between responses of males and females?

■ **3.56 God and happiness:** Go to the GSS Web site http://sda.berkeley.edu/GSS click on "frequencies and crosstabulation," then "start," type "god" for the row variable and "happy" for the column variable (without the quotation marks), and click on "Run the table."

a. Report the contingency table of counts.

b. Treating reported happiness as the response variable, find the conditional proportions. For which opinion about God are subjects most likely to be very happy?

c. To analyze the association, is it more informative to view the proportions in (b) or the frequencies in (a)? Why?

3.57 Degrees and income: The mean annual salaries earned in 1997–1999 by year-round workers with various educational degrees are given in the table:

Degree	Mean Salary
No diploma	$23,400
High school diploma	$30,400
Bachelor's degree	$52,200
Master's degree	$62,300
Doctoral degree	$89,400
Professional degree	$109,600

Source: U.S. Census Bureau and the U.S. Bureau of Labor Statistics.

a. Identify the response variable? Is it quantitative or categorical?

b. Identify the explanatory variable? Is it quantitative or categorical?

c. Explain how a bar graph could summarize the data.

■ **3.58 Whooping cough:** Consider the "whooping cough" data file on the text CD.

a. Identify the response variable and the explanatory variable.

b. Construct a bar graph to display the incidence rate of whooping cough contingent upon year. Interpret.

3.59 Women managers in the work force: The following side-by-side bar graph appeared in a 2003 issue of the *Monthly Labor Review* about women as managers in the work force. The graph summarized the percentage of managers in different occupations who were women, for the years 1972 and 2002.

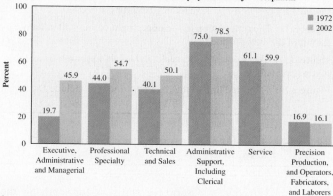

Source: Monthly Labor Review (2003) Vol. 126, No. 10, p. 48.

a. Consider the first two bars in this graph. Identify the response variable and explanatory variable.

b. Express the information from the first two bars in the form of conditional proportions in a contingency table for two categorical variables.

c. Based on (b), does it seem as if there's an association between these variables? Explain.

d. The entire graph shows two explanatory variables. What are they?

3.60 African droughts and dust: Is there a relationship between the amount of dust carried over large areas of the Atlantic and the Caribbean and the amount of rainfall in African regions? In an article (by J. Prospero and P. Lamb, *Science*, vol. 302, p. 1024, Nov. 7, 2003) the following scatterplots were given along with corresponding regression equations and correlations. The precipitation index is a measure of rainfall.

Source: J. M. Prospero and P.J. Lamb, *Science*, 302: 1024–1027 (2003).

a. Match the following regression equations and correlations with the appropriate graph.

   (i) $\hat{y} = 14.05 - 7.18x$; $r = -0.75$

   (ii) $\hat{y} = 16.00 - 2.36x$; $r = -0.44$

   (iii) $\hat{y} = 12.80 - 9.77x$; $r = -0.87$

b. Based on the scatterplots and information in part (a), what would you conclude about the relationship between dust amount and rainfall amounts?

3.61 Crime rate and urbanization: For the data in Example 14 on crime in Florida, the regression line between y = crime rate (number of crimes per 1000 people) and x = percentage living in an urban environment is $\hat{y} = 24.5 + 0.56x$.

a. Using the slope, find the difference in predicted crime rates between counties that are 100% urban and counties that are 0% urban. Interpret.

b. Interpret the correlation of 0.67 between these variables.

c. Show the connection between the correlation and the slope, using the standard deviations of 28.3 for crime rate and 34.0 for percentage urban.

3.62 Predict crime using poverty: A recent analysis of data for the 50 U.S. states on y = violent crime rate (measured as number of violent crimes per 100,000 people in the state) and x = poverty rate (percent of people in the state living at or below the poverty level) yielded the regression equation, $\hat{y} = 209.9 + 25.5x$.

a. Interpret the slope.

b. The state poverty rates ranged from 8.0 (for Hawaii) to 24.7 (for Mississippi). Over this range, find the range of predicted values for the violent crime rate.

c. Would the correlation between these variables be positive or negative? Why?

3.63 Height and paycheck: The headline of an article in the *Gainesville Sun* (October 17, 2003) stated, "Height can yield a taller paycheck." It described an analysis of four large studies in the U.S. and Britain by a University of Florida professor on subjects' height and salaries. The article reported that for each gender, "an inch is worth about $789 a year in salary. So, a person who is 6 feet tall will earn about $5,523 more a year than a person who is 5 foot 5."

a. For the interpretation in quotes, identify the response variable and explanatory variable.

b. State the slope of the regression equation, when height is measured in inches and income in dollars.

c. Explain how the value $5,523 relates to the slope.

3.64 Predicting college GPA: An admissions officer claims that at his college the regression equation $\hat{y} = 0.5 + 7x$ approximates the relationship between y = college GPA and x = high school GPA, both measured on a four-point scale.

a. Sketch this equation between $x = 0$ and 4, labeling the x- and y-axes. Is this equation realistic? Why or why not?

b. Suppose that actually $\hat{y} = 0.5 + 0.7x$. Predict the GPA for two students having GPAs of 3.0 and 4.0. Interpret, and explain how the difference between these two predictions relates to the slope.

3.65 College GPA = high school GPA: Refer to the previous exercise. Suppose the regression equation is $\hat{y} = x$. Identify the y-intercept and slope. Interpret the line in context.

3.66 What's a college degree worth?: In 2002, a Census Bureau survey reported that the mean total earnings that a full-time worker in the United States can expect to earn between ages 25 and 64 is $1.2 million for those with only a high-school education and $2.1 million for those with a college degree but no advanced degree.

a. Assuming four years for a college degree and a straight-line regression of y = total earnings on x = number years of education, what is the slope? Interpret it.

b. If y instead measures earnings per year (rather than for 40 years), then what is the slope? Interpret.

3.67 Car weight and gas hogs: The table shows a short excerpt from the "car weight and mileage" data file on

the text CD. That file lists several 2004 model cars with automatic transmission and their x = weight (in pounds) and y = mileage (miles per gallon of gas). The prediction equation is $\hat{y} = 45.6 - 0.0052x$.

Automobile Brand	Weight	Mileage
Honda Accord Sedan LX	3164	34
Toyota Corolla	2590	38
Dodge Dakota Club Cab	3838	22
Jeep Grand Cherokee Laredo	3970	21
Hummer H2	6400	17

Source: auto.consumerguide.com, honda.com, toyota.com, landrover.com, ford.com

a. Interpret the slope in terms of a 1000 pound increase in the vehicle weight.

b. Find the predicted mileage and residual for a Hummer H2. Interpret.

3.68 Predicting Internet use from cell phone use: We now use data from the "human development" data file on cell-phone use and Internet use for 39 countries.

a. The MINITAB output below shows a scatterplot. Describe it in terms of (i) identifying the response variable and the explanatory variable, (ii) indicating whether it shows a positive or a negative association, (iii) describing the variability of Internet use values for nations that are close to 0 on cell phone use.

b. Identify the approximate x- and y-coordinates for a nation that has less Internet use than you would expect, given its level of cell-phone use.

c. The prediction equation is $\hat{y} = 1.27 + 0.475x$. Describe the relationship by noting how $\hat{y}$ changes as x increases from 0 to 90, which is roughly its minimum and maximum.

d. For the United States, $x = 45.1$ and $y = 50.15$. Find its predicted Internet use and residual. Interpret the large positive residual.

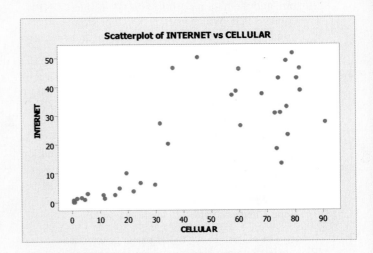

3.69 Income depends on education?: For a study of counties in Florida, the table shows part of a printout for the regression analysis relating y = median income (thousand of dollars) to x = percent of residents with at least a high school education.

a. County A has 10% more of its residents than County B with at least a high school education. Find their difference in predicted median incomes.

b. Find the correlation. (*Hint*: use the relation between the correlation and the slope of the regression line.) Interpret the (i) sign, (ii) strength of association.

Variable	Mean	Std Dev
INCOME	24.51	4.69
EDUCATION	69.49	8.86

The regression equation is income = −4.63 + 0.42 education

3.70 Fertility and GDP: Refer to the "Human development" data file on the text CD. Use x = GDP and y = fertility (mean number of children per adult woman).

a. Construct a scatterplot, and indicate whether regression seems appropriate.

b. Find the correlation and the regression equation.

c. With x = percent using contraception, $\hat{y} = 6.7 - 0.065x$. Can you compare the slope of this regression equation with the slope of the equation with GDP as a predictor to determine which has the stronger association with y? Explain.

d. Contraception has a correlation of −0.887 with fertility. Which variable has a stronger association with fertility: GDP or contraception?

3.71 Women working and birth rate: Using data from several nations, a regression analysis of y = crude birth rate (number of births per 1000 population size) on x = women's economic activity (female labor force as a percentage of the male labor force) yielded the equation $\hat{y} = 36.3 - 0.30x$ and a correlation of −0.55.

a. Describe the effect by comparing the predicted birth rate for countries with $x = 0$ and countries with $x = 100$.

b. Suppose that the correlation between the crude birth rate and the nation's gross national product (GNP) equals −0.35. Which variable, GNP or women's economic activity, seems to have the stronger association with birth rate? Explain.

3.72 Education and income: The regression equation for a sample of 100 people relating x = years of education and y = annual income (in dollars) is $\hat{y} = -20,000 + 4000x$, and the correlation equals 0.50. The standard deviations were 2.0 for education and 16,000 for annual income.

a. Show how to find the slope in the regression equation from the correlation.

b. Suppose that now we let x = annual income and y = years of education. Will the correlation or the slope change in value? If so, show how.

3.73 Income in euros: Refer to the previous exercise. Results in the regression equation $\hat{y} = -20,000 + 4000x$ for y = annual income were translated to units of euros, at a time when the exchange rate was $1.25 per euro.

a. Find the intercept of the regression equation. (*Hint*: What does 20,000 dollars equal in euros?)

b. Find the slope of the regression equation.

c. What is the correlation when annual income is measured in euros? Why?

3.74 Changing units for cereal data: Refer to the "Cereal" data file on the text CD, with x = sugar and y = sodium (both in mg), for which $\hat{y} = 243.5 - 0.0071x$.

a. Convert the amounts measured in milligrams to grams. Compare your results with the previous analysis. What changes, and what does not change? Explain why the results are what you would expect intuitively.

b. Now suppose only y is changed to grams. What changes compared to measuring both in milligrams? Explain why you expect such a change.

3.75 Murder and single-parent families: For Table 3.6 on the 50 states and D.C., the EXCEL figure below shows the relationship between the murder rate and the percentage of single-parent families.

a. For D.C., percentage of single-parent families = 44.7 and murder rate = 41.8. Identify D.C. on the scatterplot, and explain the effect you would expect it to have on a regression analysis.

b. The regression line fitted to all 51 observations is $\hat{y} = -21.4 + 1.14x$. The regression line fitted only to the 50 states is $\hat{y} = -8.2 + 0.56x$. Summarize the effect of including D.C. in the analysis.

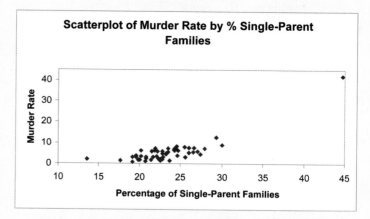

3.76 Violent crime and college education: For the "U.S. statewide crime" data file from the text CD, let y = violent crime rate and x = percent with a college education.

a. Construct a scatterplot. Identify any points that you think may be influential in a regression analysis.

b. Fit the regression line using all 51 observations. Interpret the slope.

c. Fit the regression line after deleting the observation identified in (a). Interpret the slope, and compare results to (b).

3.77 Violent crime and high school education: Repeat the previous exercise using x = percent with at least a high school education. This shows that an outlier is not especially influential if its x value is not relatively large or small.

3.78 Crime and urbanization: For the "U.S. statewide crime" data file on the text CD, using MINITAB to analyze y = violent crime rate and x = urbanization (percentage of the residents living in metropolitan areas) gives the results shown:

Variable	N	Mean	StDev	Minimum	Q1	Median	Q3	Maximum
violent	51	441.6	241.4	81.0	281.0	384.0	554.0	1508.0
urban	51	68.36	20.85	27.90	49.00	70.30	84.50	100.00

The regression equation is violent = 36.0 + 5.93 urban

a. Using the five-number summary of positions, sketch a box plot for y. What do this and the reported mean and standard deviation of y tell you about the shape of the distribution of violent crime rate?

b. Construct a scatterplot. Does this show any potentially influential observations? Would you predict that the slope would be larger, or smaller, if you delete this observation? Why?

c. Fit the regression without the observation highlighted in (b). Describe the effect on the slope.

3.79 High school graduation rates and health insurance: Access the "HS Graduation Rates" file on the text CD, which contains statewide data on x = high school graduation rate and y = percentage of individuals without health insurance.

a. Construct a scatterplot. Describe the relationship.

b. Find the correlation. Interpret.

c. Find the regression equation for the data. Interpret the slope, and summarize the relationship.

3.80 Women's Olympic high jumps: Example 11 discussed how the winning height in the Olympic high jump changed over time. Using the "High Jump" data file on the text CD, MINITAB reports

Women_Meters = −11.9 + 0.007 Year_Women

for predicting the women's winning height (in meters) using the year number.

a. Predict the winning Olympic high jump distance for women in (i) 2008, (ii) 3000.

b. Do you feel comfortable making either prediction in (a)? Explain.

3.81 Income and height: A survey of adults revealed a positive correlation between the height of the subject and their income in the previous year.

a. Explain how gender could be a potential lurking variable that could be responsible for this association.

b. If gender had actually been one of the variables measured in the study, would it be a lurking variable, or a confounding variable? Explain.

3.82 More TV watching goes with fewer babies?: For United Nations data from several countries, there is a strong negative correlation between the birth rate and the per capita television ownership.

a. Does this imply that having higher television ownership causes a country to have a lower birth rate?

b. Identify a lurking variable that could provide an explanation for this association.

3.83 More sleep causes death?: An Associated Press story (2/15/2002) quoted a study at the University of California at San Diego that reported, based on a nationwide survey, that those who averaged at least 8 hours sleep a night were 12 percent more likely to die within six years than those who averaged 6.5 to 7.5 hours of sleep a night.

a. Explain how the subject's age could be positively associated both with time spent sleeping and with an increased death rate, and hence could be a lurking variable responsible for the observed association between sleeping and the death rate.

b. Explain how the subject's age could be a common cause of both variables.

3.84 Ask Marilyn: Marilyn vos Savant writes a column for *Parade* magazine to which readers send questions, often puzzlers or questions with a twist. In one column, a reader posed the following question (*Parade Magazine*, 28 April 1996, p 6): "A company decided to expand, so it opened a factory generating 455 jobs. For the 70 white-collar positions, 200 males and 200 females applied. Of the females who applied, 20% of the females and only 15% of the males were hired. Of the 400 males applying for the blue-collar positions, 75% were hired. Of the 100 females applying, 85% of were hired. A federal Equal Employment enforcement official noted that many more males were hired than females, and decided to investigate. Responding to charges of irregularities in hiring, the company president denied any discrimination, pointing out that in both the white collar and blue collar fields, the percentage of female applicants hired was greater than it was for males. But the government official produced his own statistics, which showed that a female applying for a job had a 58% chance of being denied employment while male applicants had only a 45% denial rate. As the current law is written, this constituted a violation.... Can you explain how two opposing statistical outcomes are reached from the same raw data?"

a. Construct two contingency tables giving counts relating gender to whether hired (yes or no), one table for white-collar jobs and one table for blue-collar jobs.

b. Construct a single contingency table for gender and whether hired, combining all 900 applicants into one table. Verify that the percentages not hired are as quoted above by the government official.

c. Comparing the tables in (a) and (b), explain why this is an example of Simpson's paradox.

Concepts and Investigations

3.85 NL baseball team ERA and number of wins: Is a baseball team's earned run average (ERA = the average number of earned runs they give up per game) a good predictor of the number of wins that a team has for a season? The data for the National League teams in 2003 are available in the "NL team statistics" file on the text CD. Conduct a correlation and regression analysis, including graphical and numerical descriptive statistics. Summarize results in a two-page report.

3.86 Time studying and GPA: Is there a relationship between the amount of time a student studies and a student's GPA? Access the "Georgia Student Survey" file on the text CD or use your class data to explore this question using appropriate graphical and numerical methods. Write a brief report summarizing results.

3.87 Warming in Newnan, Georgia: Access the "Newnan GA Temps" file on the text CD, which contains data on average annual temperatures for Newnan, Georgia during the twentieth century. Fit a regression line to these temperatures and interpret the trend. Compare the trend to the trend found in Example 12 for Central Park, New York temperatures.

3.88 Regression for dummies: You have done a regression analysis for the catalog sales company you work for, using monthly data for the last year on y = total sales in the month and x = number of catalogs mailed in preceding month. You are asked to prepare a 200-word summary of what regression does under the heading "Regression for Dummies," to give to fellow employees who have never taken a statistics course. Do this, being careful not to use any technical jargon with which the employees may not be familiar.

3.89 Fluoride and AIDS: An Associated Press story (8/25/1998) about the lack of fluoride in most of the water supply in Utah quoted antifluoride activist Norma Sommer as claiming that fluoride may be responsible for AIDS, since the water supply in San Francisco is heavily fluoridated and since that city has an unusually high incidence of AIDS. Describe how you could use this story to explain to someone who has never studied statistics that association need not imply causation.

3.90 Fish fights Alzheimer's: An AP story (July 22, 2003) described a study conducted over four years by Dr. Martha Morris and others from Chicago's Rush-Presbyterian-St. Luke's Medical Center involving 815 Chicago residents aged 65 and older (*Archives of Neurology*, July 21, 2003). Those who reported eating fish at least once a week had a 60% lower risk of Alzheimer's than those who never or rarely ate fish. However, the story also quoted Dr. Rachelle Doody of Baylor College of Medicine as warning, "Articles like this raise expectations and confuse people. Researchers can show an association, but they can't show cause and effect." She said it is not known whether those people who had a reduced risk had eaten fish most of their lives, and whether other dietary habits had an influence. Using this example, describe how you would explain to someone who has never taken statistics (a) what a lurking variable is, (b) how there can be multiple causes for any particular response variable, and (c) why they need to be skeptical when they read new research results.

3.91 What's wrong with regression?: Explain what's wrong with the way regression is used in each of the following examples:

a. Winning times in the Boston marathon (at www.bostonmarathon.org) have followed a straight line decreasing trend from 160 minutes in 1927 (when the race was first run at the Olympic distance of about 26 miles) to 130 minutes in 2004. After fitting a regression line to the winning distances, you use the equation to predict that the winning time in the year 2300 will be about 13 minutes.

b. Using data for several cities on x = % of residents with a college education and y = median price of home, you get a strong positive correlation. You conclude that having a college education causes you to be more likely to buy an expensive house.

c. A regression between x = number of years of education and y = annual income for 100 people shows a modest positive trend, except for one person who dropped out after 10th grade but is now a multimillionaire. It's wrong to ignore any of the data, so we should report all results including this point, such as the correlation $r = -0.28$.

Choose the best response in multiple-choice questions 3.92–3.97:

3.92 Correlate GPA and GRE: In a study of graduate students who took the Graduate Record Exam (GRE), the Educational Testing Service reported a correlation of 0.37 between undergraduate grade point average (GPA) and the graduate first year GPA.[13] This means that

a. As undergraduate GPA increases by one unit, graduate first-year GPA increases by 0.37 units.

b. Since the correlation is not 0, we can predict a person's graduate first-year GPA perfectly if we know their undergraduate GPA.

c. The relationship between undergraduate GPA and graduate first-year GPA follows a curve rather than a straight line.

d. As one of these variables increases, there is a weak tendency for the other variable to increase also.

3.93 Properties of r: Which of the following is *not* a property of r?

a. r is always between -1 and 1.

b. r depends on which of the two variables is designated as the response variable.

c. r measures the strength of the linear relationship between x and y.

d. r does not depend on the units of y or x.

e. r has the same sign as the slope of the regression equation.

3.94 Interpreting r: One can interpret $r = 0.30$ as

a. a weak, positive association

b. 30% of the time $\hat{y} = y$

c. $\hat{y}$ changes 0.30 units for every one-unit increase in x

d. a stronger association than two variables with $r = -0.70$

3.95 Correct statement about r: Which one of the following statements is correct?

a. The correlation is always the same as the slope of the regression line.

b. The mean of the residuals from the least-squares regression line is 0 only when $r = 0$.

c. The correlation is the percentage of points that lie in the quadrants where x and y are both above the mean or both below the mean.

[13]*Source: GRE Guide to the Use of Scores*, 1998–1999, www.gre.org

d. The correlation is inappropriate if a U-shaped relationship exists between x and y.

3.96 Describing association between categorical variables: You can summarize the data for two categorical variables x and y by

   a. drawing a scatterplot of the x and y values.

   b. constructing a contingency table for the x and y values.

   c. calculating the correlation between x and y.

   d. constructing a box plot for each variable.

3.97 Slope and correlation: The slope of the regression equation and the correlation are similar in the sense that

   a. they do not depend on the units of measurement.

   b. they both must fall between -1 and $+1$.

   c. they both have the same sign.

   d. neither can be affected by severe regression outliers.

3.98 True-false: The variables y = annual income (thousands of dollars), x_1 = number of years of education, and x_2 = number of years experience in job are measured for all the employees having city-funded jobs in Knoxville, Tennessee. Suppose that the following regression equations and correlations apply:

   (i) $\hat{y} = 10 + 1.0x_1, \quad r = 0.30.$

   (ii) $\hat{y} = 14 + 0.4x_2, \quad r = 0.60.$

The correlation is -0.40 between x_1 and x_2. Which of the following statements are true?

   a. The weakest association is between x_1 and x_2.

   b. The regression equation using x_2 to predict x_1 has negative slope.

   c. Each additional year on the job corresponds to a $400 increase in predicted income.

   d. The predicted mean income for employees having 20 years of experience is $4000 higher than the predicted mean income for employees having 10 years of experience.

◆◆ 3.99 Correlation doesn't depend on units: Suppose you convert y = income from British pounds to dollars, and suppose a pound equals 2.00 dollars.

a. Explain why the y values double, the mean of y doubles, the deviations $(y - \bar{y})$ double, and the standard deviation s_y doubles.

b. Using the formula for calculating the correlation, explain why the correlation would not change value.

◆◆ 3.100 When correlation = slope: Consider the formula $b = r(s_y/s_x)$ that expresses the slope in terms of the correlation. Suppose the data are equally spread out for each variable. That is, suppose the data satisfy $s_x = s_y$. Show that the correlation and the slope are the same. (In practice, the standard deviations are not usually identical. However, this provides an interpretation for the correlation as representing what we would get for the slope of the regression line if the two variables were equally spread out.)

◆◆ 3.101 Center of the data: Consider the formula $a = \bar{y} - b\bar{x}$ for the y-intercept.

a. Show that $\bar{y} = a + b\bar{x}$. Explain why this means that the predicted value of the response variable is $\hat{y} = \bar{y}$ when $x = \bar{x}$.

b. Show that an alternative way of expressing the regression model is as $(\hat{y} - \bar{y}) = b(x - \bar{x})$. Interpret this formula.

◆◆ 3.102 Final exam "regresses toward mean" of midterm: Let y = final exam score and x = midterm exam score. Suppose that the correlation is 0.70 and that the standard deviation is the same for each set of scores.

a. Using part (b) of the previous exercise and the relation between the slope and correlation, show that $(\hat{y} - \bar{y}) = 0.70(x - \bar{x})$.

b. Explain why this means that the predicted difference between your final exam grade and the class mean is 70% of the difference between your midterm exam score and the class mean. Your score is predicted to "regress toward the mean." (The concept of *regression toward the mean*, which is responsible for the name of regression analysis, will be discussed in Section 11.2.)

Class Explorations

■ 3.103 Analyze your data: Refer to the data file the class created in Activity 3 in Chapter 1. For variables chosen by your instructor, conduct a regression and correlation analysis. Prepare a one-page report summarizing your analyses, interpreting your findings.

■ 3.104 Activity: Effect of moving a point: The *regression by eye* applet on the text CD lets you add and delete points on a scatterplot. The regression line is automatically calculated for the points you provide.

   a. Your instructor will give the students five data points that have an approximate linear relation between x and y but a slope near 0. Plot these in a scatterplot.

   b. Add a sixth observation that is influential. What did you have to do to get the slope to change noticeably from 0?

   c. Now consider the effect of sample size on the existence of influential points. Start by creating 20 points with a

similar pattern as in (a) that have a slope near 0. Is it now harder to add a single point that makes the slope change noticeably from 0? Why? Compare your results in part (c) to the results from part (b). Your class will discuss the results.

   d. Try this activity, with the *correlation by eye* applet, looking at the effect of an influential point on the correlation coefficient, r. Repeat parts (a)-(c) in terms of the correlation instead of the slope.

■ 3.105 Activity: Guess the correlation and regression: The *regression by eye* applet and the *correlation by eye* applet on the text CD allow you to guess the correlation r and the regression line for a scatterplot with randomly generated data points. Your instructor will give you instructions on how to use these applets to practice matching the correct value of r to a scatterplot and to estimate the correct regression line.

4 | Gathering Data

EXAMPLE 1

Are Cell Phones Dangerous to Your Health?

Picture the Scenario

Cell phones have become the must-have communication gadget of the new millennium. There's no doubt about it: The use of cell phones has become a routine part of our lives. But how safe are they? Cell phones emit electromagnetic radiation, produced in the form of nonionizing radio-frequency energy. A cellular phone's antenna is the main source of this energy. The closer the antenna is to the user's head, the greater the exposure to cell phone radiation.

With the increase in the popularity of cell phones has come a growing concern that heavy use of cell phones may increase a person's risk of getting cancer. Several studies have explored whether there's an association between cell phone use and the occurrence of cancer. For instance,

Study 1: A German study (Stang et al., 2001) compared 118 patients with a rare form of eye cancer called uveal melanoma to 475 healthy patients who did not have the eye cancer. The patients' cell phone use was measured using a questionnaire. The eye cancer patients used cell phones more often, on the average.

Study 2: A U.S. study (Muscat et al., 2000) compared 469 patients with brain cancer to 422 patients who did not have brain cancer. The patients' cell-phone use was measured using a questionnaire. The two groups' use of cell phones was similar.

Study 3: An Australian study (Repacholi, 1997) conducted an experiment with 200 transgenic mice, specially bred to be susceptible to cancers of the immune system. One hundred mice were exposed for two half-hour periods a day to the same kind of microwaves with roughly the same power as the kind transmitted from a cell phone. The other 100 mice were not exposed. After 18 months, the brain tumor rate for the mice exposed to cell phone radiation was twice as high as the brain tumor rate for the unexposed mice.

Studies 1 and 3 found an association between cell phone use and cancer. Study 2 did not.

Questions to Explore

- Why is it that results of different medical studies sometimes disagree?
- What is the best study design for gathering data to explore whether there is an association between cell phone use and the occurrence of cancer? Can such a study design establish a direct causal link between cell phone use and cancer? Is this study design ethically feasible to conduct?

Chapters 2 and 3 introduced graphical and numerical ways of describing data. We described shape, center, and spread, looked for patterns and unusual observations, and explored associations between variables. For these and other statistical analyses to be useful, we must have "good data." But what's the best way to gather data to ensure they are "good"? The studies described in Example 1 used two different methods. The first two studies merely *observed* subjects, whereas the third study *conducted an experiment* with them. We'll now learn about such methods, we'll study their pros and cons, and we'll see both good ways and bad ways of using them. This will help us understand when we can trust the conclusions of a study and when we need to be skeptical.

4.1 | Should We Experiment or Should We Merely Observe?

Recall

From Section 1.2, the **population** consists of *all* the subjects of interest. The **sample** is the subset of the population for whom the study collects data.

From the beginning of Chapter 3, the **response variable** measures the outcome of interest. Studies investigate how that outcome depends on the **explanatory variable**. For instance, does whether or not subjects get brain cancer depend on their cell phone use? ■

We use statistics to learn about a **population**. The German and the U.S. studies mentioned in Example 1 examined subjects' cell phone use and whether or not the subjects had brain or eye cancer. Since it would be too costly and time consuming to seek out *all* individuals in the populations of Germany or the United States, the researchers used **samples** to gather data. These studies, like many, have two variables of primary interest—a **response variable** and an **explanatory variable**. The response variable is whether or not a subject has cancer (eye or brain). The explanatory variable is the amount of cell phone use.

Types of Studies: Experimental and Observational

Some studies, such as the Australian study in Example 1, perform an **experiment**. Subjects are assigned to experimental conditions, such as receiving radiation or not receiving radiation, that we want to compare on the response outcome. These experimental conditions are called **treatments**. They correspond to different values of the explanatory variable, such as whether the subject receives radiation.

Experiment

A researcher conducts an **experiment** by assigning subjects to certain experimental conditions and then observing outcomes on the response variable. The experimental conditions, which correspond to assigned values of the explanatory variable, are called **treatments**.

For example, the Australian study used mice as the subjects. Each mouse was assigned to one of two treatments, receiving radiation or not receiving radiation. These are the categories of the explanatory variable—whether or not the subject is exposed to cell phone radiation. The purpose of the experiment was to examine the association between this variable and the response variable—whether or not the mouse developed a brain tumor.

IN WORDS

An **observational study** merely *observes* rather than *experiments* with the study subjects. An **experimental study** assigns to each subject a treatment (such as exposure or nonexposure to cell phone radiation) and then observes the outcome on the response variable.

An experiment assigns each subject to a treatment and then observes the response. By contrast, many studies merely *observe* the values on the response variable and the explanatory variable for the sampled subjects without doing anything to them. Such studies are called **observational studies**.

Observational Study

In an **observational study**, the researcher observes values of the response variable and explanatory variables for the sampled subjects, without anything being done to the subjects (such as imposing a treatment).

EXAMPLE 2

Different Types of Cell Phone Studies

Picture the Scenario

Example 1 described three studies about whether there's an association between cell phone use and cancer. We've seen that the Australian study was an experiment, using mice. The German and the U.S. studies both observed the amount of cell phone use for cancer patients and for non-cancer patients using a questionnaire.

Questions to Explore

a. Were the German and U.S. studies experiments or observational studies?

b. How were these studies fundamentally different from the Australian study?

Think It Through

a. In both the German and the U.S. studies, information on the amount of cell phone use was gathered by giving a questionnaire to the sampled subjects. The subjects (people) decided how much they would use a cell phone and, thus, their amount of radiation exposure. The studies merely observed this exposure. No experiment was performed. The studies were therefore *observational* studies.

b. In the Australian study, the researcher rather than the subject (the mouse) decided how much radiation each subject would receive. The researchers did not merely observe the subjects but determined which treatment each would receive. The study was therefore an *experimental* study rather than an observational study.

Insight

One reason that results of different medical studies sometimes disagree is that they are not the same *type* of study. An experimental study with mice is not directly comparable to an observational study with humans, and as we'll see, there are different types of observational studies, some more trustworthy than others.

TO PRACTICE THIS CONCEPT, TRY EXERCISE 4.2, PART (A), AND EXERCISE 4.3

Let's consider another study. From its description, we'll see if we can figure out whether it was an experiment or an observational study.

EXAMPLE 3

Does Drug Testing Reduce Students' Drug Use?

Picture the Scenario

"Student Drug Testing Not Effective in Reducing Drug Use," proclaimed the headline in a May 19, 2003 news release from the University of Michigan. It reported results from a study of 76,000 students in 497 high schools and 225 middle schools nationwide.[1] Each student in the study filled out a questionnaire. One question asked whether the student used drugs. The study found that drug use was similar in schools that tested for drugs and in schools that did not test for drugs. For instance, the table in the margin shows the conditional proportions on drug use, given whether drug testing is used, for the sampled twelfth graders from the two types of schools.

Conditional Proportions on Drug Use

Drug Tests?	Drug Use Yes	No	n
Yes	0.37	0.63	5653
No	0.36	0.64	17,437

Questions to Explore

a. What were the response and explanatory variables?

b. Was this an observational study, or was it an experiment?

Think It Through

a. The purpose of the study was to compare the percentage of students who used drugs in schools that tested for drugs and in schools that did not test for drugs. Whether the student used drugs was the response variable. The explanatory variable was whether the student's school tested for drugs. Both variables were categorical, with categories "yes" and "no." For each grade, the data were summarized in a contingency table, as summarized above in the margin for twelfth graders.

b. For each student, the study merely observed whether his or her school tested for drugs and whether he or she used drugs. So this was an observational study.

Insight

An experiment would have assigned schools to use or not to use drug testing, rather than leaving the decision to the schools. For instance, the study could have "randomly" selected half the schools to do drug testing and half not to do it, and then a year later measured the student drug use in each school. As we will soon discuss, an experimental study gives the researcher more control over outside influences. This control allows more accuracy in studying the association.

TO PRACTICE THIS CONCEPT, TRY EXERCISE 4.7

Recall

From Section 3.4, a **lurking variable** is a variable, not observed in the study, that influences the association between the response and explanatory variables due to its own association with each of those variables. ■

Advantage of Experiments over Observational Studies

In an observational study, lurking variables can affect the results. The German study in Example 1 found an association between cell phone use and eye cancer. However, there could be lifestyle, genetic, or health differences between the subjects with eye cancer and those without it, and between those who use cell phones a lot and those who do not. This could affect the association. For example, a possible lifestyle lurking variable is computer use. Perhaps those who use cell phones a lot also use computers a lot. Perhaps high exposure to computer screens increases the chance of eye cancer. In that case, the higher

---

[1] Study by R. Yamaguchi et al., reported in *Journal of School Health*, vol. 73, pp. 159–164, 2003.

prevalence of eye cancer for heavier users of cell phones could be due to their higher use of computers, not their higher use of cell phones.

By contrast, an experiment reduces the potential for lurking variables to affect the results. Why? We'll see that with a type of "random" selection to determine which subjects receive each treatment, the groups of subjects receiving the different treatments are "balanced"; that is, the groups have similar distributions on other variables, such as lifestyle, genetic, or health characteristics. For instance, if we randomly determine which mice receive radiation and which do not, the two groups of mice have similar distributions on health. One group will not be much healthier, on the average, than the other group. When the groups are balanced on a lurking variable, there is no association between the lurking variable and the explanatory variable (for instance, between health and whether the mouse receives radiation). Then, the lurking variable will not affect how that explanatory variable is associated with the response variable. One treatment group will not tend to have more cancer cases simply because its mice have poorer health, on the average.

Establishing **cause and effect** is central to science. But it's not possible to establish cause and effect definitively with observational studies. There's always the possibility that some lurking variable could be responsible for the association. If people who make greater use of cell phones have a higher rate of eye cancer, it may be because of some variable that we failed to measure in our study, such as computer use. As we learned in Section 3.4, **association does not imply causation**.

Because it's easier to adjust for lurking variables in an experiment than in an observational study, *we can study the effect of an explanatory variable on a response variable more accurately with an experiment than with an observational study.* With an experiment, the researcher has control over which treatment (e.g., radiation or no radiation) each subject receives. If we find a higher rate of brain cancer for those exposed to radiation, we can be more sure that we've discovered a causal relationship than if we merely find an association in an observational study. Because of this, the best method for determining causality is to conduct an experiment.

What Type of Study Is Possible?

If experiments are preferable, why ever conduct an observational study? Why bother to measure whether cell phone usage was greater for those with cancer than for those not having it? Why not instead conduct an experiment, such as the following: Pick half the students from your school "at random" and tell them to use a cell phone each day for the next 50 years. Tell the other half of the student body not to ever use cell phones. Fifty years from now, analyze whether cancer was more common for those who used cell phones.

You can see obvious difficulties with such an experiment. First, it's not ethical for a study to expose part of the sample to something (such as cell phone radiation) that may potentially be harmful. Second, it's difficult in practice to make sure that the subjects do as told. How can you monitor them to ensure that they adhere to their treatment assignment over the 50-year experimental period? Third, who wants to wait 50 years to get an answer?

For these reasons, medical experiments are often performed over a short time period, and often with animals rather than humans. In the Australian study, mice were exposed to radiation similar to that emitted from a cell phone. Because inferences about human populations are more trustworthy when we use samples of humans than when we use samples of animals, however, scientists often resort to observational studies. We'll study methods for designing a good observational study. This can yield useful information when an experiment is not practical.

Finally, another reason observational studies are common is that many questions of interest do not involve trying to assess causality. For instance, if we want to gauge the public's opinion on some issue, or if we want to conduct a marketing study to see how people rate a new product, it's completely adequate to use an observational study that samples (appropriately) the population of interest.

Using Data Already Available

Of course, you yourself will not conduct a study every time you want to answer some question, such as whether cell phone use is dangerous. It's human nature to rely instead on data that are already available. The most readily available data come from your personal observations. Perhaps a friend recently diagnosed with brain cancer was a frequent user of cell phones. Is this strong evidence that frequent cell phone use increases the likelihood of getting brain cancer?

Informal observations of this type are called **anecdotal evidence**. Unfortunately, there is no way to tell if they are representative of what happens for an entire population. Sometimes you hear people give anecdotal evidence to attempt to disprove causal relationships. "My Uncle Geoffrey is 85 years old and his entire adult life he's smoked a pack of cigarettes a day, yet he's as healthy as a horse." An association does not need to be perfect, however, to be causal. Not all people who smoke a pack of cigarettes each day will get lung cancer, but a much higher proportion of them will do so than people who are non-smokers. Perhaps Uncle Geoffrey is lucky to be in good health, but that should not encourage you to tempt the fates by smoking regularly.

Instead of using anecdotal evidence to draw conclusions, you should rely on data from reputable research studies. You yourself can find research results on topics of interest by entering keywords in search engines on the Internet. This directs you to published results, such as in medical journals for medical studies.[2] Results from well-designed studies are more trustworthy than anecdotal evidence.

The previous chapters have already mentioned many sources of available data. For instance, Activity 1 in Chapter 1 introduced the General Social Survey (GSS), a popular observational database for tracking opinions and behaviors of the American public. Likewise, federal governments collect an enormous amount of demographic and economic data. Some central sites for such data are www.fedstats.gov for the United States, www.statcan.ca for Canada (Statistics Canada), www.inegi.gob.mx for Mexico (Instituto Nacional de Estadistica Geografia e Informatica), www.statistics.gov.uk for the United Kingdom, and www.abs.gov.au for Australia.

The Census and Other Sample Surveys

The General Social Survey (GSS) is an example of a **sample survey**. It gathers information by interviewing a sample of subjects from the U.S. adult population to provide a snapshot of that population. The study discussed in Example 3 on student drug use in schools that tested or did not test for drugs also used a sample survey.

Sample Survey

A **sample survey** selects a sample of people from a population and interviews them to collect data.

A sample survey is a type of observational study. The subjects provide data on the variables measured. There is no assignment of subjects to different treatments.

---

[2]Surveys of research about cell phones suggest that there is no convincing evidence yet of adverse radiation effects (e.g., D. R. Cox, *J. Roy. Statist. Soc., Ser. A*, vol. 166, pp. 241–246, 2003). The primary danger appears to be people using cell phones while driving!

Most countries conduct a regular **census**. This is a survey that attempts to count the number of people in the population and to measure certain characteristics about them. It is different from most surveys, which sample only a small part of the entire population.

In Article 1, Section 2, the U.S. Constitution states that a complete counting of the U.S. population is to be taken every 10 years. The first census was taken in 1790. It counted 3.9 million people, while today the U.S. population is estimated to be nearly 300 million. Other than counting the population size, here are three key reasons for conducting the U.S. census:

■ The Constitution mandates that seats in the House of Representatives be apportioned to states based on their portion of the population measured by the census. When the 1910 census was completed, Congress fixed the number of seats at 435. With each new census, states may gain or lose seats depending on how their population size compares with other states.

■ Census data are used in the drawing of boundaries for electoral districts.

■ Census data are used to determine the distribution of federal dollars to states and local communities.

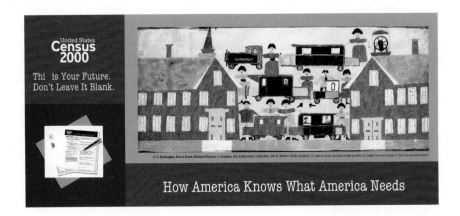

Methods for taking the U.S. census have varied since horseback riders served as enumerators (the people who make the counts) in the first censuses. From the 1930s to the 1960s, women who were homemakers served as the enumerators. But as women increasingly took jobs outside the home, too few enumerators were available, so the Census Bureau began to mail forms to households in 1970. The mail-back rate for the forms was 90%. However, this decreased to 65% for the 1990 and 2000 censuses. The 35% who did not respond required follow-ups. The Census Bureau sends a reminder requesting that the form be returned. If this is unsuccessful, enumerators make personal visits to the household address.

Although it's the intention of a census to sample *everyone* in a population, in practice this is not possible. Some people do not want to be counted. Some do not have known addresses for the Census Bureau to send a census form: Some people are homeless; others are transient. Fortunately, the U.S. founding fathers realized that it was impractical to attempt a census every year. Yet data are needed regularly on economic variables, such as the unemployment rate. Thus, the U.S. Census Bureau and other government agencies continually take smaller samples of the population, rather than relying solely on the complete census. An example is the monthly Current Population Survey of about 50,000 households.

It is usually much more practical to take a sample rather than to try to measure everyone in a population. In fact, in the late 1990s statisticians at the Census Bureau proposed using sampling methods to estimate numbers of people missed by the ordinary census. Their proposal became a political battleground between Republicans (who did not want sampling used) and Democrats (who supported sampling). In 1999, the Supreme Court ruled 5 to 4 that sampling could not be used to adjust the counts for apportioning seats in Congress but could be used to adjust counts for federal funding.

For a sample survey to be informative, it is important that the sample reflect the population well. As we'll discuss next, "random" selection—letting chance determine which subjects are in the sample—is the key to getting a good sample.

4.1   Practicing the Basics

4.1   Cell phones: Consider the U.S. cell phone study described in Example 1.

a. Identify the response variable and the explanatory variable.

b. Was this an observational study or an experiment? Explain why.

4.2   Acoustic neuroma: A study identified nearly all cases in Sweden in 1999–2000 of acoustic neuroma, a benign tumor on the auditory nerve (S. Lonn et al., *Epidemiology*, vol. 15, p. 653, 2004). It matched these subjects with others of similar age, gender, and residential area who did not have acoustic neuroma and compared the two groups on their mobile phone use. When considering tumors on the same side of the head as the phone was normally used, those with tumors were more likely to have used mobile phones for at least 10 years than those without tumors.

a. Was this an experimental study or an observational study? Explain.

b. Explain why this does not prove that greater use of mobile phones causes subjects to be more likely to develop acoustic neuroma.

4.3   GSS: Activity 1 in Chapter 1 described the General Social Survey. Is the GSS an experiment or an observational study? Explain.

4.4   Experiments vs. observational studies: When either type of study is feasible, an experiment is usually preferred over an observational study. Explain why, using an example to illustrate.

4.5   School testing for drugs: Example 3 discussed a study comparing high schools that tested for drugs with high schools that did not test for drugs, finding similar levels of student drug use in each. State a potential lurking variable that could affect the results of such a study. Describe what the effect could be.

4.6   Hormone therapy and heart disease: Since 1976 the Nurses' Health Study has followed more than 100,000 nurses. Every two years, the nurses fill out a questionnaire about their habits and their health. Results from this study indicated that post-menopausal women have a reduced risk of heart disease if they take a hormone replacement drug.

a. Suppose the hormone-replacement drug actually has no effect. Identify a potential lurking variable that could explain the results of the observational study.

(*Hint*: Suppose that the women who took the drug tended to be more conscientious about their personal health than those who did not take it.)

b. Recently a randomized experiment called the Women's Health Initiative was conducted by the National Institutes of Health to see if hormone therapy is truly helpful. The study, planned to last for eight years, was stopped after five years when analyses showed that women who took hormones had 30% more heart attacks. This study suggested that rather than reducing the risk of heart attacks, hormone replacement drugs actually increase the risk.[3] How is it that two studies could reach such different conclusions? (For attempts to reconcile the studies, see a story by Gina Kolata in *The New York Times*, April 21, 2003.)

c. Explain why randomized experiments, when feasible, are preferable to observational studies.

4.7   Hairdressers at risk: In a study by Swedish researchers (*Occupational and Environmental Medicine* 2002;59: 517–522), 2410 women who had worked as hairdressers and given birth to children were compared to 3462 women from the general population who had given birth. The hairdressers had a slightly higher percentage of infants with a birth defect.

a. Identify the response variable and the explanatory variable.

b. Is this study an observational study or an experiment? Explain.

c. Can we conclude that there's something connected with being a hairdresser that causes higher rates of birth defects? Explain.

4.8   Helping low-income smokers quit: A study (*Journal of Family Practice* 2000; 50:138–144) with 233 low-income adult smokers evaluated the effectiveness of usual care (physician advice and follow-up) for smokers wishing to quit to the usual care enhanced by computer-assisted telephone counseling sessions. Each subject was assigned randomly either to the usual care or to the usual care plus counseling, and their smoking status (still smoking or quit smoking) was observed after 3 months. The percentage who had quit smoking was higher for the group receiving counseling.

---

[3]See article by H. N. Hodis et al., *New England Journal of Medicine*, August 7, 2003.

a. Identify the response variable and the explanatory variable.

b. Was this study an observational study, or an experiment? Explain.

c. How does the design of this study take into account potential lurking variables?

4.9 Experiment or observe?: Explain whether an experiment or an observational study would be more appropriate to investigate the following:

a. Whether or not smoking has an effect on coronary heart disease

b. Whether or not higher SAT scores tend to be positively associated with higher college GPAs

c. Whether or not a special coupon attached to the outside of a catalog makes recipients more likely to order products from a mail-order company

4.10 Physician likes "cold turkey": A physician who has practiced for more than 20 years believes that going "cold turkey" ' is more effective than nicotine patches and other smoking cessation methods in helping smokers to quit. The physician gives as evidence his observation of patients in his practice.

a. Explain the limitations of such anecdotal evidence.

b. Mention a lurking variable that could affect the association the physician observed.

4.11 Seat belt anecdote: Andy once heard about a car crash victim who died because he was pinned in the wreckage by a seat belt he could not undo. Because of this, Andy refuses to wear a seat belt when he rides in a car. How would you explain to Andy the fallacy behind relying on this anecdotal evidence?

4.12 Census every 10 years?: Give at least two reasons why the U.S. takes a census only every 10 years. What are reasons for taking the census at all?

4.13 Canadian census: Using the Internet, research the date that the Canadian census was first conducted. Find the Canadian population size then and now according to the most recent Canadian census.

4.2 | What Are Good Ways and Poor Ways to Sample?

The sample survey is a very common type of observational study. The first step of a sample survey is to define the population targeted by the study. Usually, this is straightforward. For instance, the Gallup organization (www.gallup.com) conducts a monthly survey of about 1000 adult Americans to report the percentage of those sampled who respond "approve" when asked, "Do you approve or disapprove of the way [the current president of the United States] is handing his job as president?" The population consists of all adults living in the United States.

Sampling Frame and Sampling Design

Once you've identified the population, the second step is to compile a list of subjects in it, so you can sample from it. This list is called the **sampling frame**.

Sampling frame

The **sampling frame** is the list of subjects in the population from which the sample is taken.

Ideally, the sampling frame lists the entire population of interest. In practice, as in a census, it's usually hard to identify every subject in the population.

Suppose you plan to sample students at your school, asking students to complete a questionnaire about various issues. The population is all students at the school. One possible sampling frame is the student directory. Another one is a list that the registrar keeps of registered students.

Once you have a sampling frame, you need to specify a method for selecting subjects from it. The method you use is called the **sampling design**. Here's one possible sampling design for sampling students at your school: You sample all students in your statistics class. It's easy to conduct such a sample survey, because the class is a readily available sample. But do you think your class is necessarily reflective of the entire student body? Do you have a representative mixture of freshmen through seniors, males and females, athletes and nonathletes, working and nonworking students, political party affiliations, and so on? With this sampling design, it's doubtful.

When you pick a sample merely by convenience, the results may not be representative of the population. Some response outcomes may occur much more than in the population, and some may occur much less. Consider a survey question about the number of hours a week that you work at an outside job. If your class is primarily juniors and seniors, they may be more likely to work than freshmen or sophomores. The mean response in the sample may be much larger than in the overall student population. Information from the sample may be misleading.

Simple Random Sampling

You're much more likely to obtain a representative sample if you let *chance*, rather than *convenience*, determine the sample. The sampling design should give each student an equal chance to be in the sample. It should also enable the data analyst to figure out how likely it is that descriptive statistics (such as sample means) fall close to corresponding values we'd like to make inferences about for the entire population. These are reasons for using **random sampling**.

You have probably many times been part of a selection process in which your name was put in a box or hat along with many other names, and someone blindly picked a name for a prize. If the names were thoroughly mixed before the selection, this emulates a *random* type of sampling, called a **simple random sample**.

Recall

As in Chapters 2 and 3, n denotes the *number* of observations in the sample, called the **sample size**. ■

> **Simple Random Sample**
>
> A **simple random sample** of n subjects from a population is one in which each possible sample of that size has the same chance of being selected.

A *simple random sample* is often just called a **random sample**. The "simple" adjective distinguishes this type of sampling from more complex random sampling designs presented in Section 4.4.

EXAMPLE 4

Sampling Club Officers for a New Orleans Trip

Picture the Scenario

A campus club can select two members to attend the club's annual conference in New Orleans, with all expenses paid by the national organization. The club members decide that two of the five club officers should attend. The five officers are President (P), Vice-President (V), Secretary (S), Treasurer (T), and Activity Coordinator (A). They make up the population and the sampling frame. The club members want the selection process to be fair, so they decide to select a simple random sample of size $n = 2$. The five names are written on identical slips of paper, placed in a hat, mixed, and then a neutral person blindly selects two slips from the hat.

Questions to Explore

a. What are the possible samples of two officers?

b. What is the chance that a particular sample of size 2 will be drawn?

c. What is the chance that the Activity Coordinator will be chosen?

Think It Through

a. Using the letters that denote each officer, the possible samples of two officers are

(P,V)	(P,S)	(P,T)	(P,A)	(V,S)
(V,T)	(V,A)	(S,T)	(S,A)	(T,A)

b. There are 10 possible samples. This process of randomly selecting two slips from a hat ensures that each sample has an equal chance of occurring. Since there are 10 possible samples, the chance of any one sample being selected is 1 out of 10.

c. The Activity Coordinator (A) occurs in 4 of the samples. So the chance that the Activity Coordinator is selected for a free trip is 4 out of 10, or 2/5.

Insight

The chance of 2/5 for the Activity Coordinator is the same as for the President or any other officer. In practice, especially with larger populations, it is difficult to mix the slips so that each sample truly has an equal chance of selection. There are better ways of selecting simple random samples, as we'll learn next.

TO PRACTICE THIS CONCEPT, TRY EXERCISE 4.14

Did You Know?

The *random numbers* applet on the text CD can also be used to generate random numbers. ▪

How Can We Select a Simple Random Sample?

What's a better way to take a simple random sample than blindly drawing slips of paper out of a hat? You first number the subjects in the sampling frame. You then generate a set of these numbers randomly. Finally, you sample the subjects whose numbers were generated.

You can generate numbers randomly with a **random number table** or with software or a statistical calculator. A random number table contains a sequence of digits such that any particular digit is equally likely to be any of the numbers 0, 1, 2, ..., 9 and does not depend on the other numbers generated. If a particular digit is a 6, for instance, the next digit is just as likely to be a 6 (or a 0) as any other number. The numbers fluctuate according to no set pattern. Table 4.1 shows part of a random number table.

TABLE 4.1: A Portion of a Table of Random Numbers

Line/Col.	(1)	(2)	(3)	(4)	(5)	(6)	(7)	(8)
1	10480	15011	01536	02011	81647	91646	69179	14194
2	22368	46573	25595	85393	30995	89198	27982	53402
3	24130	48360	22527	97265	76393	64809	15179	24830
4	42167	93093	06243	61680	07856	16376	39440	53537
5	37570	39975	81837	16656	06121	91782	60468	81305

USING RANDOM NUMBERS TO SELECT A SIMPLE RANDOM SAMPLE

To select a simple random sample,

■ Number the subjects in the sampling frame, using numbers of the same length (number of digits).

■ Select numbers of that length from a table of random numbers or using software or a calculator with a random number generator.

■ Include in the sample those subjects having numbers equal to the random numbers selected.

EXAMPLE 5

Auditing the Accounts of a School District

Picture the Scenario

Local school districts must be prepared for annual visits from state auditors whose job is to verify the actual dollar amount of the accounts within the school district and verify that the money is being spent appropriately. A school district may have many accounts, sometimes 100 or more. It is too time-consuming and expensive for auditors to review all accounts, so they typically review only some of them. So that a school district cannot anticipate which accounts will be reviewed, the auditors often take a simple random sample of the accounts.

Questions to Explore

a. How can the auditors use random numbers to select 10 accounts to audit in a school district that has 60 accounts?

b. Why is it important for the auditors not to use personal judgment in selecting the accounts to audit?

Think It Through

a. The sampling frame consists of the 60 accounts. We first number the accounts 01 through 60. In a random number table, we select two digits at a time until we have 10 different two-digit numbers between 01 and 60. (We number the accounts 01 through 60 rather than 1 through 60 because we select two-digit numbers from the table.) Any pair of random digits has the same chance of selection.

Choose a random starting place in the table. We will illustrate by starting with row 1, column 1 of Table 4.1. The random numbers in this row are

10480   15011   01536   02011   81647   91646   69179   14194

The first 10 two-digit numbers are

10   48   01   50   11   01   53   60   20   11

The first account chosen is the one numbered 10. The second chosen is the one numbered 48, and so forth. We observe that 01 and 11 occur twice. We discard the two repeats, and we need two more numbers between 01 and 60. The next 10 two-digit numbers are

81   64   79   16   46   69   17   91   41   94

We skip the numbers 81, 64, 79, since no account in the sampling frame of 60 accounts has an assigned number that large. The last two accounts sampled are those numbered 16 and 46. In summary, the auditors should audit the accounts numbered 01, 10, 11, 16, 20, 46, 48, 50, 53, and 60.

b. By using simple random sampling, the auditors hold the school district responsible for all accounts. If the auditors personally chose the accounts to audit, they may tend to select certain accounts each year. A school district would soon learn which accounts they need to have in order and which accounts can be given less attention.

Insight

Likewise, the Internal Revenue Service (IRS) randomly selects some tax returns for audit, so taxpayers cannot predict ahead of time whether they will be audited.

TO PRACTICE THIS CONCEPT, TRY EXERCISE 4.15

▲ **FIGURE 4.1:** This **MINITAB** screen is used to randomly generate 10 integers between 01 and 60, inclusively. The 10 random numbers will be scored in column 1 of the **MINITAB** spreadsheet.

Methods of Collecting Data in Sample Surveys

In Example 5, the subjects sampled were accounts. More commonly, they are people. For any survey, it is difficult to get a good sampling frame, since a list of all adults in a population typically does not exist. So, it's necessary to pick a place where almost all people can be found. This is not a shopping mall or a rock concert or a college campus but instead a person's place of residence.

Once we identify the desired sample for a survey, how do we contact the people to collect the data? The three most common methods are

- Personal interview
- Telephone interview
- Self-administered questionnaire

Personal interview In a personal (face-to-face) interview, an interviewer asks prepared questions and records the subject's responses. An advantage is that subjects are more likely to agree to participate. A disadvantage is the cost. Also, some subjects may not answer sensitive questions pertaining to opinions and to lifestyle, such as alcohol and drug use, that they might answer on a printed questionnaire.

Telephone interview A telephone interview is like a personal interview but conducted over the phone. A main advantage is lower cost, since no travel is involved. A disadvantage is that the interview might have to be short. Subjects aren't as patient on the phone and may hang up before starting or completing the interview.

Self-administered questionnaire Subjects are requested to fill out a questionnaire mailed to them by post or e-mail. An advantage is that it is cheaper than a personal interview. A disadvantage is that more subjects may fail to participate.

So which method is most used in practice? Most major national polls that survey regularly to measure public opinion use the telephone interview. An example is the polls conducted by the Gallup Organization. The General Social Survey is an exception, using personal interviews for their questionnaire, which is quite long.

For telephone interviews, since the mid-1980s, many sample surveys have used **random digit dialing** to select households. The survey can then obtain a sample without having a sampling frame. In the United States, typically the area code and the 3-digit exchange are randomly selected from the list of all such codes and exchanges. Then, the last four digits are dialed randomly, and an adult is selected randomly within the household. Although this sampling design incorporates randomness, it is not a simple random sample, because each sample is not equally likely to be chosen. (Do you see why not? Does everyone have a telephone, or exactly one telephone?)

How Accurate Are Results from Surveys with Random Sampling?

The most common use of data from sample surveys is to estimate population percentages. For instance, a Gallup poll recently reported that 30% of Americans worried that they might not be able to pay health-care costs during the next 12 months. How good is such a sample estimate of the population percentage? When you read results of surveys, you'll often see a statement such as, "The **margin of error** is plus or minus 3 percentage points." This means that it's very likely that the population percentage is no more than 3% lower or 3% higher than the reported sample percentage. So if Gallup reports that 30% worry about health-care costs, it's very likely that in the entire population the percentage that worry about health-care costs is between about 27% and 33% (that is, within 3% of 30%). "Very likely" means that 95 times out of 100 such statements are correct.

Chapters 6 and 7 will show details about the margin of error and how to calculate it. For now, we'll use a rough approximation. When using a simple random sample of n subjects, the margin of error is approximately

$$\frac{1}{\sqrt{n}} \times 100\%.$$

EXAMPLE 6

Finding the Margin of Error for a Gallup Poll

Picture the Scenario

In April 2003, there was worldwide concern about the spread of the viral respiratory illness called *severe acute respiratory syndrome* (SARS). It is spread by close person-to-person contact and has symptoms of high fever and difficulty in breathing. The majority of reported cases were in China, but many cases were reported in other countries, including the United States and Canada. The World Health Organization (WHO) implemented measures to control its spread, so that an epidemic would

Did You Know?

In 2005, there was much publicity about a new influenza virus, an Avian flu strain, and concern of a potential pandemic. SARS and this influenza strain are both examples of how new diseases arise because of microorganisms undergoing new mutations in the midst of a population with a large number of possible hosts. ■

not occur. The outbreak was contained, with WHO reporting 8098 cases resulting in 774 deaths.

A Gallup poll conducted for CNN and *USA Today* in three successive weeks monitored Americans' concern about SARS. The poll questioned 1001 adults using telephone interviews. Figure 4.2 shows bar graphs that the Gallup organization used to report results.

▲ **FIGURE 4.2:** Results[4] of Gallup polls about SARS.

Questions to Explore

Find an approximate margin of error for these results. How is it interpreted?

Think It Through

The sample size was $n = 1001$. The Gallup poll was a random sample, but not a simple random sample. It used random digit dialing, which can give results nearly as accurate as with a simple random sample. So the margin of error was approximately

$$\frac{1}{\sqrt{n}} \times 100\% = \frac{1}{\sqrt{1001}} \times 100\% = 0.03 \times 100\% = 3\%.$$

In the final poll, Gallup reported that 43% were somewhat or very worried about exposure to SARS. This refers to the *sample*. The margin of error of 3% suggests that in the *population* of adult Americans, it's likely that between about 40% and 46% were worried about SARS exposure.

Insight

You may be surprised or skeptical that a sample of only 1001 people out of a huge population (such as 200 million adult Americans) can provide such precise inference. That's the power of random sampling. We'll see *why* this works in Chapter 6. The more precise formula shown there will give a somewhat smaller margin of error value when the sample percentage is far from 50%.

**TO PRACTICE THIS CONCEPT TRY EXERCISES 4.20 AND 4.21
AND TRY USING THE "SAMPLE FROM A POPULATION" APPLET
AS DESCRIBED IN EXERCISE 4.104**

Be Wary of Sources of Potential Bias in Sample Surveys.

A variety of problems can cause responses from a sample to tend to favor some parts of the population over others. Then, results from the sample are not representative of the population and are said to exhibit **bias**.

---

[4]*Source:* http://www.gallup.com/poll/releases/pr030425.asp.

IN WORDS

There is **bias** if, because of the way the study was designed, certain outcomes will occur more often in the sample than they do in the population. For example, for the population of adults in your home town, results of an opinion survey may be biased in a liberal direction if you sample only educators or biased in a conservative direction if you sample only business owners.

IN PRACTICE: Be alert for potential bias

A sound sampling design can prevent sampling bias. It cannot prevent response bias and nonresponse bias. With any sample survey, carefully scrutinize the results. Look for information about how the sample was selected, how large the sample size was, the nonresponse rate, how the questions were worded, and who sponsored the study. The less you know about these details, the less you should trust the results.

Bias may result from the sampling method. The main way this occurs is if the sample is not random. We'll see examples later in the section. Another way it can occur is due to **undercoverage**—having a sampling frame that lacks representation from parts of the population. A telephone survey will not reach homeless people or prison inmates or other people not having a telephone. If its sampling frame consists of the names in a telephone directory, it will not reach those having an unlisted number. Responses by those who are not in the sampling frame might tend to be quite different from those who are in the frame. Bias resulting from the sampling method, such as nonrandom sampling or undercoverage, is called **sampling bias**.

Sampling bias is merely one type of possible bias in a sample survey. When some sampled subjects cannot be reached or refuse to participate, there is **nonresponse bias**. The subjects who are willing to participate may be different from the overall sample in some way, perhaps having strong, emotional convictions about the issues being surveyed. Even those who do participate may not respond to some questions, resulting in nonresponse bias due to **missing data**. All major surveys suffer from some nonresponse bias. The General Social Survey has a nonresponse rate of about 20–30%. The nonresponse rate is much higher for many telephone surveys. By contrast, government-conducted surveys often have lower nonresponse rates. The Current Population Survey, which measures factors related to employment, has a nonresponse rate of only about 7%. To reduce nonresponse bias, investigators try to make follow-up contact with the subjects who do not return questionnaires.

Results have dubious worth when there is substantial nonresponse. For instance, in her book *Women and Love* (1987), author Shere Hite presented results of a survey of adult women in the United States. One of her conclusions was that 70% of women who had been married at least five years have extramarital affairs. She based this conclusion on responses to questionnaires returned from a sample of 4500 women. This sounds impressively large. However, the questionnaire was mailed to about 100,000 women. We cannot know whether this sample of 4.5% of the women who responded is representative of the 100,000 who received the questionnaire, much less the entire population of adult American women.

Besides sampling bias and nonresponse bias, a third type of potential bias is in the actual responses made. This is called **response bias**. An interviewer might ask the questions in a leading way, such that subjects are more likely to respond a certain way. Or, subjects may lie because they think their response is socially unacceptable, or they may give the response that they think the interviewer prefers.

If you design an interview or questionnaire, you should strive to construct questions that are clear and understandable. *Avoid questions that are confusing, long, or leading.* The wording of a question can greatly affect the responses. A Roper Poll was designed to determine the percentage of Americans who express some doubt that the Holocaust occurred in World War II. In response to the question, "Does it seem possible or does it seem impossible to you that the Nazi extermination of the Jews never happened?" twenty-two percent said it was possible the Holocaust never happened. The Roper organization later admitted that the question was worded in a confusing manner. When they asked, "Does it seem possible to you that the Nazi extermination of the Jews never happened, or do you feel certain that it happened?" only 1% said it was possible it never happened.[5]

Even the order in which questions are asked can dramatically influence results. One study[6] asked, during the Cold War, "Do you think the U.S. should let Russian

[5]*Newsweek*, July 25, 1994.
[6]Described in Crossen (1994).

newspaper reporters come here and send back whatever they want?" and "Do you think Russia should let American newspaper reporters come in and send back whatever they want?" For the first question, the percentage of yes responses was 36% when it was asked first and 73% when it was asked second.

TYPES OF BIAS IN SAMPLE SURVEYS

Sampling bias occurs from using nonrandom samples or having undercoverage.

Nonresponse bias occurs when some sampled subjects cannot be reached or refuse to participate or fail to answer some questions.

Response bias occurs when the subject gives an incorrect response (perhaps lying), or the question wording or the way the interviewer asks the questions is confusing or misleading.

Volunteer and Other Convenience Samples: Poor Ways to Sample

Have you even been stopped on the street or at a shopping mall to participate in a survey? Such a survey is not a random sample but rather a **convenience** sample. It is easy for the interviewer to obtain data relatively cheaply. But the sample may poorly represent the population. Biases may result because of the time and location of the interview and the judgment of the interviewer about whom to interview. For example, working people might be underrepresented if the interviews are conducted on workdays between 9 A.M. and 5 P.M. Poor people may be underrepresented if the interviewer conducts interviews at an upscale shopping mall.

Have you ever answered a survey you've seen posted on the Internet, such as at the home page for a news organization? A sample of this type, called a **volunteer sample**, is the most common type of convenience sample. As the name implies, subjects *volunteer* for the sample. One segment of the population may be more likely to volunteer than other segments because they have stronger opinions about the issue or are more likely to visit that Internet site. This results in sampling bias. For instance, a survey by the Pew Research Center (1/6/2003) estimated that 46% of Republicans said they like to register their opinions in online surveys, compared with only 28% of Democrats. (Why? One possible lurking variable is personal income: On the average, Republicans are wealthier than Democrats, and wealthier people are more likely to have access to and use the Internet.) Thus, results of online surveys may be more weighted in the direction of Republicans' beliefs (for instance, more conservative) than the general population.

Convenience samples are not ideal. Sometimes, however, they are necessary, both in observational studies and in experiments. This is often true in medical studies. Suppose we want to investigate how well a new drug performs compared to a standard drug, for subjects who suffer from high blood pressure. We're not going to be able to find a sampling frame of all who suffer from high blood pressure and take a simple random sample of them. We may, however, be able to sample such subjects at certain medical centers or using volunteers. Even then, randomization should be used wherever possible. For the study patients, we can randomly select who is given the new drug and who is given the standard one.

Some samples are poor not only because of their convenience but also because they use an inappropriate sampling frame. An example of this was a poll in 1936 to predict a Presidential election.

EXAMPLE 7

Taking a Poor Sample— The *Literary Digest* Poll

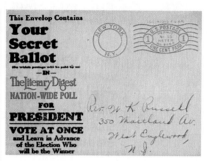

Picture the Scenario

The magazine called *Literary Digest* conducted a poll to predict the result of the 1936 Presidential election between Franklin Roosevelt (Democrat and incumbent) and Alf Landon (Republican). At the time, their poll was famous, because they had correctly predicted three successive elections. In 1936 they mailed questionnaires to 10 million people and asked how they planned to vote. The sampling frame was constructed from telephone directories, country club memberships, and automobile registrations. Approximately 2.3 million people returned the questionnaire. The *Digest* predicted that Landon would win, getting 57% of the vote. Instead, Landon actually got only 36%, and Roosevelt won in a landslide.

Questions to Explore

a. What was the population?

b. How could the *Literary Digest* poll make such a large error, especially with such a huge sample size? What type of bias could have occurred in this poll?

Think It Through

a. The population was all registered voters in the United States in 1936.

b. This survey had two severe problems:

■ *Sampling bias due to undercoverage of the sampling frame and a nonrandom sample*: In 1936, the United States was in the Great Depression. Those who had cars and country club memberships and thus received questionnaires tended to be relatively wealthy. The wealthy tended to be primarily Republican, the political party of Landon. Many potential voters were not on the lists used for the sampling frame. There was also no guarantee that a subject in the sampling frame was a registered voter.

■ *Nonresponse bias*: Of the 10 million people who received questionnaires, 7.7 million did not respond. As might be expected, those individuals who were unhappy with the incumbent (Roosevelt) were more likely to respond.

Insight

For this same election, a pollster who was getting his new polling agency off the ground surveyed 50,000 people and predicted that Roosevelt would win. Who was this pollster? George Gallup. The Gallup Organization is still with us today. However, the *Literary Digest* went out of business soon after the 1936 election.[7]

TO PRACTICE THIS CONCEPT, TRY EXERCISE 4.27

---

[7]Bryson, M. C. (1976), *American Statistician*, vol. 30, pp. 184–185.

A Large Sample Size Does Not Guarantee an Unbiased Sample

Many people think that as long as the sample size is large, it doesn't matter how the sample was selected. This is incorrect, as illustrated by the *Literary Digest* poll. A sample size of 2.3 million did not prevent poor results. Many Internet surveys have thousands of respondents, but a volunteer sample of thousands is not as good as a random sample, even if that random sample is much smaller. *We're almost always better off with a simple random sample of 100 people than with a volunteer sample of thousands of people.*

KEY PARTS OF A SAMPLE SURVEY

- Identify the **population** of all the subjects of interest.
- Construct a **sampling frame**, which attempts to list all the subjects in the population.
- Use a **random sampling design**, implemented using random numbers, to select *n* subjects from the sampling frame.
- Be cautious about **sampling bias** due to nonrandom samples (such as volunteer samples) and sample undercoverage, **response bias** from subjects not giving their true response or from poorly worded questions, and **nonresponse bias** from refusal of subjects to participate.

4.2 Practicing the Basics

4.14 Choosing officers: In Example 4, "Sampling Club Officers for a New Orleans Trip," two of the officers denoted by (P, V, S, T, A) were selected to attend a conference in New Orleans. Suppose instead that three officers can get fully paid trips. For a simple random sample of three of the five officers:

   a. Show all the possible samples.

   b. What is the chance that a particular sample of size three will be drawn?

   c. What is the chance that the Activity Coordinator will be chosen?

4.15 Sample students: A class has 50 students. Use the second row of digits in the random number table (Table 4.1) to select a simple random sample of three students. If the students are numbered 01 to 50, what are the numbers of the three students selected?

4.16 Random numbers applet: Use the *Random Numbers* applet on the text CD or use software to (a) answer the previous exercise, (b) select a simple random sample of three students out of a class of 500 students.

4.17 Auditing accounts—applet: Use the *Random Numbers* applet on the text CD to select 10 of the 60 school district accounts described in Example 5. Explain how you did this, and identify the accounts to be sampled.

4.18 Auditing accounts—software: Use either software or a statistical calculator to select 10 of the 60 school district accounts described in Example 5. Explain how you did this, and identify the accounts to be sampled.

4.19 Sampling from a directory: A local telephone directory has 50,000 names, 100 per page for 500 pages. Explaining how you found and used random numbers, select 10 numbers to identify subjects for a simple random sample of 10 names.

4.20 Comparing polls: The table shows the result of the 2000 Presidential election along with the vote predicted by several organizations in the days before the election. The sample sizes were typically about 1000 to 2000 people. The percentages for each poll do not sum to 100 because of voters who indicated they were undecided or preferred another candidate.

   a. Treating the sample sizes as 1000 each, find the approximate margin of error.

   b. Do most of the predictions fall within the margin of error of the actual vote percentages? Considering the relative sizes of the sample and the population and the undecided factor, would you say that these polls had good accuracy?

Predicted Vote

Poll	Gore	Bush	Nader
Gallup/CNN/USA	46	48	4
Harris	47	47	5
ABC/Wash Post	45	48	3
CBS	45	44	4
NBC/WSJ	44	47	3
Pew Research	47	49	4
Actual vote	48.4	47.9	2.7

Source: www.ncpp.org/1936-2000.htm

4.21 **Margin of error and n:** The Gallup poll in Example 6
[TRY] reported that 43% of Americans worried that they would
be exposed to SARS. Find the approximate margin of
error if (a) $n = 100$, (b) $n = 400$, (c) $n = 1600$. Explain
how the margin of error changes as n increases.

4.22 **Bias due to perceived race:** A political scientist at the
University of Chicago studied the effect of the race of the
interviewer.[8] Following a phone interview, respondents
were asked whether they thought the interviewer was
black or white (all were actually black). Perceiving a
white interviewer resulted in more conservative opinions.
For example, 14% agreed that "American society is fair to
everyone" when they thought the interviewer was black,
but 31% agreed to this statement when posed by an interviewer that the respondent thought was white. Which
type of bias does this illustrate: Sampling bias, nonresponse bias, or response bias? Explain.

4.23 **Confederates:** Some southern states in the United States
have wrestled with the issue of a state flag that is sensitive to
African Americans and not divisive. Suppose a survey asks,
"Do you oppose the present state flag that contains the
Confederate symbol, a symbol of past slavery in the South
and a flag supported by extremist groups?"

    a. Explain why this is an example of a leading question.

    b. Explain why a better way to ask this question would
be, "Do you favor or oppose the current state flag
containing the Confederate symbol?"

4.24 **Call in your opinion:** Many TV news and entertainment
programs ask viewers to offer their opinions on an issue
of the moment by calling an 800 or 900 phone number or
voting on the Internet. For instance, one night the ABC
program *Nightline* asked viewers whether the United Nations should continue to be located in the United States.
Of more than 186,000 callers, 67% wanted the United
Nations out of the United States. At the same time, a scientific poll using a random sample of about 500 respondents estimated the true percentage wanting the United
Nations out of the United States to be about 28%
(D. Horvitz et al., *Chance*, vol. 8, pp. 16–25, 1995).

    a. The random sample is much smaller. Do you think it
is more trustworthy, or less trustworthy, than the results of the call-in poll for *Nightline*? Explain your
reasoning.

    b. Is the call-in poll a random sample? If not, what type
of sample is it?

4.25 **Job market for MBA students:** An article from the
February 2, 2003 *Atlanta Journal Constitution* about the
bleak job market for MBA students graduating in 2003
described an opinion survey conducted by a graduate student at a major state university. The student polled 1500
executive recruiters, asking their opinions on the industries most likely to hire. He received back questionnaires
from 97 recruiters, of whom 54 indicated that health care
was the industry most likely to see job growth.

    a. What is the population for this survey?

    b. What was the intended sample size? What was the
sample size actually observed? What was the percentage of nonresponse?

    c. Describe two potential sources of bias with this survey.

4.26 **Gun control:** More than 75% of Americans answer *yes*
when asked, "Do you favor cracking down against illegal
gun sales?" but more than 75% say *no* when asked,
"Would you favor a law giving police the power to decide
who may own a firearm?"

    a. Which statistic would someone who opposes gun con-
trol prefer to quote?

    b. Explain what is wrong with the wording of each of
these statements.

4.27 **Stock market associated with poor mental health:**
[TRY] An Internet survey of 545 Hong Kong residents suggested that close daily monitoring of volatile financial
affairs may not be good for your mental health (*J. Social and Clinical Psychology* 2002;21:116–28). Subjects
who felt that their financial future was out of control
had the poorest overall mental health, whereas those
who felt control over their financial future had the best
mental health.

    a. What is the population of interest for this survey?

    b. Describe why this is an observational study.

    c. Briefly discuss the potential problems with the sam-
pling method used and how these problems could
affect the survey results.

4.28 **"What rots beneath":** This was the headline in a *New
York Times* article (May 19, 2003). The Gowanus Canal in
Brooklyn has become famous for its contamination from
sewage and industrial waste. Scientists and Army Corps
of Engineers technicians have used augurs, drills, and a
split-spoon, which sucks up the muck of the canal bottom,
to analyze what is 'living' in the canal.

    a. Describe the population under study.

    b. Explain why a census is not practical for this study.
What advantages does sampling offer?

4.29 **Identify the bias:** A newspaper in Los Angeles designs a
survey to estimate the proportion of the city's adult residents who favor a proposal to legalize casinos in the city.
It takes a list of the 1000 people who have subscribed to
the paper the longest, and sends each of them a questionnaire that asks, "Do you think it is a good idea to legalize
casinos in Los Angeles, which will broaden the tax base
and contribute money to education?" After analyzing results from the 50 people who reply, they report that 90%
of the local citizens are in favor of the proposal. Identify
the bias that results from

    a. Sampling bias due to undercoverage

    b. Sampling bias due to the sampling design

    c. Nonresponse bias

    d. Response bias due to the way the question was asked

4.30 **Types of bias:** Give an example of a survey that would
suffer from

    a. Sampling bias due to the sampling design

    b. Sampling bias due to undercoverage

    c. Response bias

    d. Nonresponse bias

---

[8]Study by Lynn Sanders, as reported by the *Washington Post*, June 26, 1995.

4.3 | What Are Good Ways and Poor Ways to Experiment?

We've just learned about good ways and poor ways to gather a sample in an observational study. Likewise, there are good ways and poor ways to conduct an experiment. First, let's recall from Section 4.1 what's meant by an experiment: We assign each subject to an experimental condition, called a **treatment**. We then observe the outcome on the response variable. The goal of the experiment is to investigate the association—how the treatment affects the response. An advantage of an experiment over an observational study is that it provides stronger evidence for causation.

In an experiment, the subjects are often referred to as **experimental units**. This is merely to emphasize that the objects measured need not be human beings. They could, for example, be schools, stores, mice, or computer chips. The Australian study described in Example 1 conducted an experiment on 200 mice. The mice were the experimental units.

The Elements of a Good Experiment

Let's consider another example to help us learn what makes a good experiment. It is common knowledge that smoking is a difficult habit to break. Studies have reported that regardless of what smokers do to quit, most relapse within a year. Some scientists have suggested that smokers are less likely to relapse if they take an antidepressant regularly after they quit. How can you design an experiment to study whether antidepressants help smokers to quit?

For this type of study, as in most medical experiments, it is not feasible to randomly sample the population (all smokers who would like to quit). It is necessary to use a convenience sample. For instance, a medical center might advertise to attract volunteers from the smokers in a community who would like to try to quit.

Control comparison group   Suppose you have 400 volunteers who would like to quit smoking. You could ask them to quit, starting today. You could have each start taking an antidepressant, and then a year from now check how many have relapsed. Perhaps 42% of them would relapse. But this is not enough information.[9] You need to be able to compare this to the percentage who would relapse if they were *not* taking the antidepressant.

An experiment normally has a primary treatment of interest, such as receiving an antidepressant. But it should also have a second treatment for comparison to help you analyze the effectiveness of the primary treatment. So the volunteers should be split into two groups: One group receives the antidepressant, and the other group does not. You could give the second group a pill that looks like the antidepressant but that does not have any active ingredient—a placebo. This second group, using the placebo treatment, is called the **control group**. After subjects are assigned to the treatments and observed for a certain period of time, the relapse rates are compared.

Why bother to give the placebo to the control group, if the pill doesn't contain any active ingredient? This is partly so that the two treatments appear identical to the subjects. (As we'll discuss below, subjects should not know which treatment they are receiving.) This is also because people who take a placebo tend to respond better than those who receive nothing, perhaps for psychological reasons. This is called the **placebo effect**. For instance, of the subjects not receiving the antidepressant, perhaps 75% would relapse within a

---

[9]Chapter 11 explains another reason a control group is needed. The "regression effect" implies that, over time, poor subjects tend to improve and good subjects tend to get worse, in relative terms.

year, but if they received a placebo pill perhaps 55% of them would relapse. If the relapse rate was 42% for subjects who received antidepressants, then in comparing relapse rates for the antidepressant and control groups it makes a big difference how you define the control group.

In some experiments, a control group may receive an existing treatment rather than a placebo. For instance, a smoking cessation study might analyze whether an antidepressant works better than a nicotine patch in helping smokers to quit. It may not be necessary to include a placebo group if the nicotine patch has already been shown in previous studies to be more effective than a placebo. Or the experiment could compare all three treatments: antidepressants, nicotine patch, and a placebo.

Randomizing in an experiment In a smoking cessation experiment, how should the 400 study subjects be assigned to the treatment groups? Should you personally decide which treatment each subject receives? This could result in bias. If you are conducting the study to show that the antidepressant is effective, you might consciously or subconsciously place smokers who you believe will be more likely to succeed into the group that receives the antidepressant.

It is better to use **randomization** to assign the subjects: Pick a simple random sample of 200 of the 400 subjects to receive the antidepressant. The other 200 subjects will form the control group. Randomization helps to prevent bias from one treatment group tending to be different from the other in some way, such as having better health or being younger. In using randomization, we attempt to *balance the treatment groups* by making them similar with respect to their distribution on potential lurking variables. This enables us to attribute any difference in their relapse rates to the treatments they are using, not to lurking variables or to researcher bias.

You might think you can do better than randomization by using your own judgment to assign subjects to the treatment groups. For instance, when you identify a potential lurking variable, you could try to balance the groups according to its values. If age is that variable, every time you put someone of a particular age in one treatment group, you could put someone of the same age in the other treatment group. There are two problems with this. First, *many* variables are likely to be important, and it is difficult to balance groups on *all* of them at once. Second, you may not have thought of other relevant lurking variables. Even if you can balance the groups on the variables you identified, the groups could be unbalanced on these other variables, causing the overall results to be biased in favor of one treatment.

When you hear about new research findings, you can feel more confident about their worthiness if they come from a randomized experiment with a control group than if they come from an experiment without a control group or from an observational study. An analysis[10] of published medical studies about treatments for heart attacks indicated that the new therapy provided improved treatment 58% of the time in studies without randomization and control groups but only 9% of the time in studies having randomization and control groups. Although this does not prove anything (as this analysis itself was an observational study), it does suggest that studies conducted without randomization or other ways to reduce bias may produce results that tend to be overly optimistic.

In summary, use randomization to assign subjects to the treatments

■ To eliminate bias that may result if you (the researchers) assign the subjects

■ To balance the groups on variables that you know affect the response

■ To balance the groups on lurking variables that may be unknown to you

---

[10]See Crossen (1994), p. 168.

Blinding the study   It is important that the treatment groups be treated as equally as possible. Ideally, the subjects are **blind** to the treatment to which they are assigned. In the smoking cessation study, the subjects should not know whether they are taking the antidepressant or a placebo. Whoever has contact with the subjects during the experiment, including the data collectors who record the subjects' response outcomes, should also be blind to the treatment information. Otherwise they could intentionally or unintentionally provide extra support to one group. When neither the subject nor those having contact with the subject know the treatment assignment, the study is called **double-blind**. That's ideal.

EXAMPLE 8

Design the Study to Assess Antidepressants for Quitting Smoking

Picture the Scenario
To investigate whether it helps smokers to quit smoking if they take antide-

pressants, one study[11] used 429 men and women who were 18 or older and had smoked 15 cigarettes or more per day for the previous year. The subjects were highly motivated to quit and in good health. They were assigned to one of two groups: One group took 300 mg daily of bupropion, an antidepressant that has the brand name Zyban. The other group did not take an antidepressant. At the end of a year, the study observed whether each subject had successfully abstained from smoking or had relapsed.

Questions to Explore
a. Identify the response and explanatory variables, the treatments, and the experimental units.
b. How should the researchers assign the subjects to the two treatment groups?
c. Without knowing more about this study, what would you identify as a potential problem with the study design?

Think It Through
a. This experiment has

  Response variable: Whether the subject abstains from smoking for 1 year

  Explanatory variable: Whether the subject received Zyban (yes or no)

  Treatments: Zyban, no Zyban

  Experimental units: The 429 volunteers who are the study subjects

b. The researchers should randomize to assign subjects to the two treatments. They could use random numbers to select a simple random sample of half (215) of the subjects to form the group that uses Zyban. The procedure would be as follows:

---

[11]*Annals of Internal Medicine* 2001; 135:423–433.

■ Number the study subjects from 001 to 429.

■ Pick a three-digit random number between 001 and 429. If the number is 392, then the subject numbered 392 is put in the Zyban group.

■ Continue to pick three-digit numbers until you've picked 215 distinct values between 001 and 429. This determines the 215 subjects who will receive Zyban.

c. The description of the experiment in "Picture the Scenario" did not say whether the subjects who did *not* receive Zyban were given a placebo or whether the study was blinded. If not, these would be potential sources of bias.

Insight

In the actual reported study, the subjects were randomized to receive Zyban or a placebo for 45 weeks. The study *was* double-blinded. At the end of 1 year, 55.1% of the subjects receiving Zyban were not smoking, compared with 42.3% in the placebo group. After 18 months, 47.7% of the Zyban subjects were not smoking compared to 37.7% for the placebo subjects. However, after 2 years, the percentage of nonsmokers was similar for the two groups, 41.6% versus 40%.

TO PRACTICE THIS CONCEPT, TRY EXERCISE 4.32

Did You Know?

A randomized experiment comparing medical treatments is often referred to as a **clinical trial**. ■

IN PRACTICE: The randomized experiment in medicine

In medicine, the randomized experiment (clinical trial) has become the "gold standard" for evaluating new medical treatments. The Cochrane Collaboration (www.cochrane.org) is an organization devoted to synthesizing evidence from medical studies all over the world. According to this organization, there have now been hundreds of thousands of randomized experiments comparing medical treatments. In most countries, a company cannot get a new drug approved for sale unless it has been tested in a randomized experiment.

Source: Senn, S. (2003). *Dicing with Death.* Cambridge University Press, p. 68.

Sample Size and Statistical Significance

Now suppose you've conducted an experiment, and it's a year later. You find that 44% of the subjects taking Zyban have relapsed, whereas 55% of the control group relapsed. Can you conclude that Zyban was effective in helping smokers quit?

Not quite yet. You must convince yourself that this difference between 44% and 55% cannot be explained by the variation that occurs naturally just by chance. Even if the effect of Zyban is no different from the effect of placebo, the sample relapse rates would not be *exactly* the same for the two groups. Just by ordinary chance, in the random assignment of subjects to treatment groups, on the average one group may be slightly more committed to quitting smoking than the other group.

In a randomized experiment, the variation that could be expected to occur just by chance alone is roughly like the margin of error with simple random sampling. So if a treatment has $n = 215$ observations, this is about $\left(1/\sqrt{215}\right) \times 100\% = 7\%$ for a percentage. If the population percentage of people who relapse is

Recall

For simple random sampling, the margin of error is approximately $\dfrac{1}{\sqrt{n}} \times 100\%$, which is 10% for $n = 100$, 7% for $n = 200$, and 3% for $n = 1000$. ■

50%, the sample percentage of 215 subjects who relapse is very likely to fall between 43% and 57% (that is, within 7% of 50%). The difference in a study between 44% relapsing and 55% relapsing could be explained by the ordinary variation expected for this size of sample.

By contrast, if each treatment had $n = 1000$ observations, the ordinary variation we'd expect due to chance is only about 3% for each percentage. Then, the difference between 44% and 55% could not be explained by ordinary variation: There is no plausible common population percentage for the two treatments such that 44% and 55% are both within 3% of its value. Zyban would truly seem to be better than placebo. The difference expected due to ordinary variation is smaller with larger samples. You can be more confident that sample results reflect a true effect when the sample size is large than when it is small. Obviously, we cannot learn much by using only $n = 1$ subject for each treatment. The process of assigning *several* experimental units to each treatment is called **replication**.

When the difference between the results for the two treatments is so large that it would be rare to see such a difference by ordinary random variation, we say that the results are **statistically significant**. For example, the difference between 55% relapse and 44% relapse when $n = 1000$ for each group is statistically significant. We can then conclude that the observed difference is truly due to the effect of the treatments, rather than merely due to ordinary random variation. Thus, in the population, we infer that the response truly depends on the treatment. How to determine this and how it depends on the sample size are topics we'll study in later chapters.[12] Suffice it to say for now, the larger the sample size, the better.

When a study has statistically significant results, it's still possible that the observed effect was merely due to chance. Even if the treatments are identical, one may seem better just because, by sheer luck, random assignment gave it many subjects who tend to respond better, perhaps being healthier on the average. For this reason, another type of replication is also important. You can feel more confident if other researchers perform similar experiments and get similar results.

Generalizing Results to Broader Populations

We've seen that random samples are preferable to convenience samples, yet convenience samples are more feasible in many experiments. When an experiment uses a convenience sample, be cautious about the extent to which results generalize to a larger population. Look carefully at the characteristics of the sample. Do they seem representative of the overall population? If not, the results have dubious worth.

Many medical studies use volunteers at *several* medical centers to try to obtain a broad cross section of subjects. But some studies mistakenly try to generalize to a broader population than the one from which the sample was taken. A psychologist may conduct an experiment using a sample of students from an introductory psychology course. For the results to be of wider interest, however, the psychologist might claim that the conclusions generalize to *all* college students, to all young adults, or even to all adults. Such generalizations may well be wrong, since the sample may differ from those populations in fundamental ways, such as in average socioeconomic status or the distribution of race or gender.

---

[12] It's a bit more complicated than the reasoning shown above. The margin of error for comparing two percentages differs from the one for estimating a single percentage.

SUMMARY: KEY PARTS OF A GOOD EXPERIMENT

- A good experiment has a **control comparison group**, **randomization** in assigning experimental units to treatments, **blinding**, and **replication**.
- The **experimental units** are the subjects—the people, animals, or other objects to which the treatments are applied.
- The **treatments** are the experimental conditions imposed on the experimental units. One of these may be a **control** (for instance, either a placebo or an existing treatment) that provides a basis for determining if a particular treatment is effective. The treatments correspond to values of an explanatory variable.
- **Randomize** in assigning the experimental units to the treatments. This tends to balance the comparison groups with respect to lurking variables.
- **Replicating** the treatments on many experimental units helps, so that observed effects are not due to ordinary variability but instead are due to the treatment. Repeat studies to increase confidence in the conclusions.

Carefully assess the scope of conclusions in research articles, the mass media, and advertisements. Evaluate critically the basis for the conclusions by noting the experimental design or the sampling design upon which the conclusions are based.

4.3 Practicing the Basics

4.31 **Smoking affects lung cancer?:** You would like to investigate whether smokers are more likely than nonsmokers to get lung cancer. For the students in your class, you pick half at random and tell them to smoke a pack of cigarettes each day, and you tell the other half not to ever smoke. Fifty years from now, you will analyze whether more smokers than nonsmokers got lung cancer.

    a. Is this an experiment or an observational study? Why?

    b. Summarize at least three practical difficulties with this planned study.

4.32 **Never leave home without duct tape:** There have [TRY] been anecdotal reports of the ability of duct tape to remove warts. In an experiment conducted at the Madigan Army Medical Center in the state of Washington (*Archives of Pediatric and Adolescent Medicine* 2002; 156: 971–974), 51 patients between the ages of 3 and 22 were randomly assigned to receive either duct-tape therapy (covering the wart with a piece of duct tape) or cryotherapy (freezing a wart by applying a quick, narrow blast of liquid nitrogen). After two months, the percentage successfully treated was 85% in the duct tape group and 60% in the cryotherapy group.

    a. Identify the response variable, the explanatory variable, the experimental units, and the treatments.

    b. Describe the steps of how you could randomize in assigning the 51 patients to the treatment groups.

4.33 **Rats and cell phones:** The Science Daily Web site (www.sciencedaily.com) described a study at Washington University that exposed rats to two common types of cell phone radiation for four hours a day, five days a week, for two years. One third of the rats were exposed to analog

cell phone frequency, one third to digital cell phone frequency, and one third served as controls and received no radiation. At the end of two years, the brain, spinal cord, and other organs were examined for cancerous tumors. No statistically significant difference was found among groups in the percentage of tumors.

    a. For this study, identify the experimental units, explanatory and response variables, and treatments.

    b. What is a difference between this study and the Australian cell phone study discussed in Example 1 that could explain the different conclusions?

4.34 **No statistical significance:** A randomized experiment investigates whether an herbal treatment is better than a placebo in treating subjects suffering from depression. Unknown to the researchers, the herbal treatment has no effect: Subjects have the same score on a rating scale for depression (for which higher scores represent worse depression) no matter which treatment they take.

    a. The study will use eight subjects, numbered 1 to 8. Using random numbers, pick the four subjects who will take the herbal treatment. Identify the four who will take the placebo.

    b. After taking the assigned treatment for three months, subjects' results on the depression scale are as follows:

Subject	Response	Subject	Response	Subject	Response	Subject	Response
1	85	2	60	3	44	4	95
5	69	6	78	7	50	8	75

Based on the treatment assignment in (a), find the sample mean response for those who took the herbal supplement and for those who took the placebo.

c. Using the means in (b), explain how (i) sample means can be different even when there is "no effect" in a population of interest, and (ii) a difference between two sample means may not be "statistically significant" even though those sample means are not equal.

4.35 **Blind study:** Why do you think that, ideally, subjects should be blind to the treatment to which they are assigned? Why should it matter whether they know?

4.36 **Colds and vitamin C:** For some time there has been debate about whether regular large doses of vitamin C reduce the chance of getting a common cold.

   a. Explain how you could design an experiment to test this. Describe all parts of the experiment, including (i) what the treatments are, (ii) how you assign subjects to the treatments, (iii) how you could make the study double-blind.

   b. An observational study indicates that people who take Vitamin C regularly get fewer colds, on the average. Explain why these results could be misleading.

4.37 **Reducing high blood pressure:** A pharmaceutical company has developed a new drug for treating high blood pressure. They would like to compare its effects to those of the most popular drug currently on the market. Two hundred volunteers with a history of high blood pressure and who are currently not on medication are recruited to participate in a study.

   a. Explain how the researchers could conduct a randomized experiment. Indicate the experimental units, the response and explanatory variables, and the treatments.

   b. Explain what would have to be done to make this study double-blind.

4.4 | What Are Other Ways to Conduct Experimental and Observational Studies?

In this chapter, we've learned the basics of good ways to conduct observational and experimental studies. This final section shows ways these methods are often extended in practice so they are even more powerful. We'll first learn about how to incorporate more than one explanatory variable in an experiment. Then we'll study alternatives to simple random sampling in observational studies.

Multifactor Experiments

Categorical explanatory variables in an experiment are often referred to as **factors**. For example, the experiment described in Example 8 had a factor measuring whether a subject used the antidepressant called Zyban.

Many experiments have more than one factor. Suppose the researchers also wanted to study a second factor, using a nicotine patch versus not using one. The experiment could then have four treatment groups that result from cross-classifying the two factors. See Figure 4.3. The four treatments are Zyban alone, nicotine patch alone, Zyban and nicotine patch, neither Zyban nor nicotine patch.

Why use both factors at once in an experiment? Why not do one experiment about Zyban and a separate experiment about the nicotine patch? The reason is that we can learn more from a two-factor experiment. For instance, the combination of using both a nicotine patch and Zyban may be more effective than using either method alone.

> **IN WORDS**
>
> A **factor** is a categorical explanatory variable (such as whether the subject takes an antidepressant) having as categories the experimental conditions (the **treatments**, such as Zyban or no Zyban).

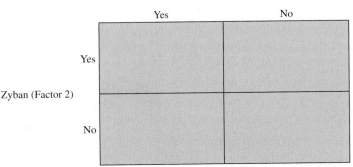

▲ **FIGURE 4.3: An experiment can use two (or more) factors at once.** The treatments are the combinations of categories of the factors.

Do Antidepressants and/ or Nicotine Patches Help Smokers Quit?

Picture the Scenario

Example 8 analyzed a study about whether Zyban helps a smoker quit cigarettes. Let's now also consider nicotine patches as another possible cessation aide.

Question to Explore

How can you design a single study to investigate the effects of nicotine patches and Zyban on whether a subject relapses into smoking?

Think It Through

You could use the two factors with four treatment groups shown in Figure 4.3. Figure 4.4 portrays the design of a randomized experiment, with placebo alternatives to Zyban and to the nicotine patch.

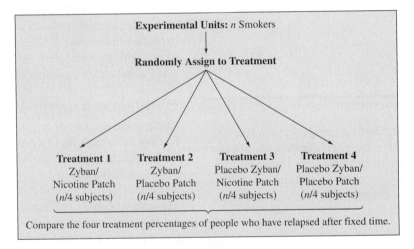

Experimental Units: *n* Smokers

↓

Randomly Assign to Treatment

Treatment 1 Zyban/ Nicotine Patch (*n*/4 subjects)	**Treatment 2** Zyban/ Placebo Patch (*n*/4 subjects)	**Treatment 3** Placebo Zyban/ Nicotine Patch (*n*/4 subjects)	**Treatment 4** Placebo Zyban/ Placebo Patch (*n*/4 subjects)

Compare the four treatment percentages of people who have relapsed after fixed time.

▲ **FIGURE 4.4: Diagram of a randomized experiment with two factors (whether use Zyban and whether use a nicotine patch). Question:** A 3-factor design could also incorporate whether a subject receives counseling to discourage smoking. How many treatments would such a design have?

The study should be double-blind: Subjects should not know whether or not the pill is active or whether or not the patch is active, and neither should the individuals who evaluate them during the experiment. After a fixed amount of time, you compare the four treatments as to the percentages who have relapsed.

Insight

In fact, a two-factor experiment *has* been conducted to study both these factors.[13] The most effective treatments were Zyban alone or in combination

---

[13]Jorenby, D. et al. (1999), *New England Journal of Medicine*, 340(9): 685–691.

with a nicotine patch. We'll explore the results in Exercise 4.67. If we included a third factor of whether or not the subject receives counseling, we would have three factors with $2 \times 2 \times 2 = 8$ treatments to compare.

TO PRACTICE THIS CONCEPT, TRY EXERCISE 4.38

Matched Pairs Designs

The experiments described in Examples 8 and 9 used **completely randomized designs**: The subjects were randomly assigned to one of the treatments. Sometimes it is feasible to use an alternative experimental design in which *each treatment is observed for each subject*. Medical experiments often do this to compare treatments for chronic conditions that do not have a permanent cure.

To illustrate, suppose a study plans to compare an oral drug with a placebo for treating migraine headaches. The subjects could take one treatment the first time they get a migraine headache and the other treatment the second time they get a migraine headache. The response is whether the subject's pain is relieved. The first three subjects might contribute the following results to the data file:

Subject	Drug	Placebo	
1	Relief	No relief	← first *matched pair*
2	Relief	Relief	
3	No relief	No relief	

For the entire sample, we would compare the percentage of subjects who have pain relief with the drug to the percentage who have pain relief with the placebo. The two observations for a particular subject are called a **matched pair**, because they both come from the same person.

A matched-pairs design in which subjects cross over during the experiment from using one treatment to using another treatment is called a **crossover design**. It provides the benefit that it helps to remove certain sources of potential bias. Using the same subjects for each treatment keeps potential lurking variables from affecting the results, because those variables take the same values for each treatment. For example, any difference observed between the drug and the placebo responses is not because subjects taking the drug had better overall health, because each subject received both treatments.

Another application of a matched pairs design occurs in agricultural experiments that compare the crop yield for two fertilizer types. In each field, two plots are marked, one for each fertilizer. Two plots in the same field are likely to be much more similar in soil quality and moisture than two plots from different fields. If we instead used a completely randomized design to assign entire fields to each fertilizer treatment, there's the possibility that by chance one fertilizer tends to get applied to fields that have better soil. This is especially true when the experiment can't use many fields.

Blocking in an Experiment

The matched pairs design extends to the comparison of more than two treatments. The migraine headache crossover study could investigate relief for subjects who take a low dose of an oral drug, a high dose of that drug, and a placebo. The crop yield study could have several fertilizer types.

In experiments with matching, a set of matched experimental units is referred to as a **block**. In the migraine headache study, each person is a block. In the crop yield study, each field is a block. A purpose of using blocks is to provide similar experimental units for the treatments, so they're on an equal footing in any subsequent comparisons.

To reduce possible bias, treatments are usually randomly assigned within a block. In the migraine headache study, the order in which the treatments are taken would be randomized. Bias could occur if all subjects received one treatment before another. An example of potential bias in a crossover study is a positive "carry-over" effect if the first drug taken has lingering effects that can help improve results for the second drug taken. Likewise, in a crop yield study, the fertilizers are randomly assigned to the plots in a field. A block design with random assignment of treatments to units within blocks is called a **randomized block** design.

Sample Surveys: Other Random Sampling Designs Are Useful in Practice

We've seen that an experiment can better investigate cause and effect than an observational study. With an observational study, it's possible that an association is due to some lurking variable. However, observational studies do have advantages. It's often not practical to conduct an experiment. For instance, with human subjects it's not ethical to design an experiment to study the effect of cell phone use on getting cancer, and we don't know whether experimental results from an artificial laboratory environment for mice can be extrapolated to humans.

It is possible to have a well-designed and informative observational study. Sample surveys that select subjects randomly are an example. Simple random sampling gives every possible sample the same chance of selection. In practice, more complex random sampling designs are often easier to implement. Sometimes they are even preferable to simple random sampling.

Cluster sampling   To use simple random sampling, we need a sampling frame—the list of all, or nearly all, subjects in the population. Unfortunately, this is often not available. It is often easier to identify **clusters** of subjects. A study of residents of a country can identify counties or census tracts. A study of students can identify schools. A study of the elderly living in institutions can identify nursing homes. We can obtain a sample by randomly selecting the clusters and observing each subject in the clusters chosen.

For instance, suppose you would like to sample about 1% of the families in your city. You could use city blocks as clusters. Using a map to label and number city blocks, you could select a simple random sample of 1% of the blocks and then select every family on each of those blocks for your observations.

> **Cluster Random Sample**
>
> Divide the population into a large number of **clusters**, such as city blocks. Select a simple random sample of the clusters. Use the subjects in those clusters as the sample. This is a **cluster random sample**.

For personal interviews, when the subjects within a cluster are close geographically, cluster sampling is less expensive per observation than simple random sampling. By interviewing every family in a particular city block, you can obtain many observations quickly and with little travel.

In summary, cluster random sampling is a preferable sampling design if

■   A reliable sampling frame is not available, or

■   The cost of selecting a simple random sample is excessive.

A disadvantage is that we usually need a larger sample size with a cluster sample than with a simple random sample in order to achieve a particular margin of error. (It's beyond the scope of this book to learn how to find this margin of error.)

Stratified sampling You're conducting a small survey to estimate the mean number of hours a week that students in your school work on outside employment. Using the student directory, you plan to take a simple random sample of $n = 40$ students. You plan to analyze how the mean compares for freshmen, sophomores, juniors, and seniors. Merely by chance, you may get only a few observations from one of the classes, making it hard to estimate the mean well for that class. If the student directory also identifies students by their class, you could amend the sampling scheme to take a simple random sample of 10 students from each class, still having an overall $n = 40$. The four classes are called **strata** of the population. This type of sample is called a **stratified random sample**.

Stratified Random Sample

A **stratified random sample** divides the population into separate groups, called **strata**, and then selects a simple random sample from each stratum.

A limitation of stratification is that you must have a sampling frame and know the stratum into which each subject in the sampling frame belongs. You might want to stratify students in your school by whether the student has a job, to make sure you get enough students who work and who don't work. However, this information may not be available for each student.

In summary, stratified random sampling has the

▪ Advantage that you can include in your sample enough subjects in each group (stratum) you want to evaluate.

▪ Disadvantage that you must have a sampling frame and know the stratum into which each subject belongs.

What's the difference between a stratified sample and a cluster sample? A stratified sample *uses every stratum*. By contrast, a cluster sample *uses a sample of the clusters*, rather than all of them. Figure 4.5 illustrates the distinction among sampling subjects (simple random sample), sampling clusters of subjects (cluster random sample), and sampling subjects from within strata (stratified random sample).

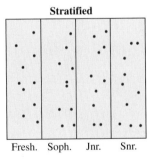

▲ **FIGURE 4.5: Ways of randomly sampling 40 students.** The figure is a schematic for a simple random sample, a cluster random sample of 7 clusters of students who live together, and a stratified random sample of 10 students from each class (Fresh., Soph., Jnr., Snr.). **Question:** What's the difference between clustering and stratifying?

Comparison of different random sampling methods

A good sampling design ensures that each subject in a population has an opportunity to be selected. The design should incorporate randomness. Table 4.2 summarizes the random sampling methods we've presented.

TABLE 4.2: Summary of Random Sampling Methods

Method	Description	Advantages
Simple random sample	Each possible sample is equally likely	Sample tends to be a good reflection of the population
Cluster random sample	Identify clusters of subjects, take simple random sample of the clusters	Do not need a sampling frame of subjects, less expensive to implement
Stratified random sample	Divide population into groups (strata), take simple random sample from each stratum	Ensures enough subjects in each group that you want to compare

In practice, sampling designs often have two or more stages. When carrying out a large survey for predicting national elections, the Gallup Organization often (1) identifies election districts as clusters and takes a simple random sample of them, and (2) takes a simple random sample of households within each selected election district, rather than sampling every household in a district, which might be infeasible. Marketing companies also commonly use *two-stage cluster sampling*, first randomly selecting test market cities (the clusters) and then randomly selecting consumers within each test market city. Many major surveys, such as the General Social Survey, Gallup polls, and the Current Population Survey, incorporate both stratification and clustering.

In future chapters, when we use the term "random sampling," we'll mean *simple random sampling*. The formulas for most statistical methods assume simple random sampling. Similar formulas exist for other types of random sampling, but they are complex and beyond the scope of this text.

Types of Observational Studies

Sample surveys attempt to take a cross section of a population at the current time. Such observational studies are sometimes called **cross-sectional** studies. Rather than taking a cross section at some time, some observational studies are *backward looking* (**retrospective**) or *forward looking* (**prospective**). Observational studies in medicine are often retrospective. How can we study whether there is an association between cell phone use and brain cancer if we cannot perform an experiment? We can form a sample of subjects who have brain cancer and a similar sample of subjects who do not and then compare the past use of cell phones for the two groups. This approach was first applied to study smoking and lung cancer.

EXAMPLE 10

Retrospective Linking of Lung Cancer with Smoking

Picture the Scenario
When was the link first shown between smoking and lung cancer? In 1950 in London, England, the medical statisticians Austin Bradford Hill and Richard Doll conducted one of the first studies. In 20 hospitals, they matched 709 patients admitted with lung cancer in the preceding year with 709 noncancer patients at the same hospital of the same gender and within the same 5-year grouping on age. All patients were queried about their smoking behavior. A smoker was defined as a person who had smoked at

least one cigarette a day for at least a year. The study used a *retrospective* design to "look into the past" in measuring the patients' smoking behavior.

Table 4.3 shows the results. The 709 *cases* in the first column of the table were the patients with lung cancer. The 709 *controls* in the second column were the matched patients without lung cancer.

TABLE 4.3: Results of Retrospective Study of Smoking and Lung Cancer

The cases had lung cancer. The controls did not. The *retrospective* aspect refers to studying whether subjects had been smokers in the past.

| | **Lung Cancer** | |
Smoker	Cases	Controls
Yes	688	650
No	21	59
Total	**709**	**709**

Question to Explore

Compare the proportions of smokers for the lung cancer cases and the controls. Interpret.

Think It Through

For the lung cancer cases, the proportion who were smokers was $688/709 = 0.970$, or 97%. For the controls (not having lung cancer), the proportion who were smokers was $650/709 = 0.917$, or about 92%. The lung cancer cases were more likely than the controls to have been smokers.

Insight

An inferential analysis showed that these results were statistically significant. This suggested that an association exists between smoking and lung cancer.

TO PRACTICE THIS CONCEPT, TRY EXERCISE 4.46

Case-control studies The type of retrospective study used in Example 10 to study smoking and lung cancer is what's called a **case-control study**.

IN WORDS

In Example 10, the response outcome of interest was having lung cancer. The cases had lung cancer, the controls did not have lung cancer, and they were compared on the explanatory variable—whether the subject had been a smoker (yes or no).

Case-Control Study

A **case-control study** is a retrospective observational study in which subjects who have a response outcome of interest (the cases) and subjects who have the other response outcome (the controls) are compared on an explanatory variable.

This is a popular design for medical studies in which it is not practical or ethical to perform an experiment. We can't randomly assign subjects into a smoking group and a nonsmoking group—this would involve asking some subjects to start smoking. Since we can't use randomization to balance effects of potential lurking variables, usually the cases and controls are matched on such variables. In Example 10, cases and controls were matched on their age, gender, and hospital.

In a case-control study, the number of cases and the number of controls is fixed. The random part is observing the outcome for the explanatory variable. For instance, in Example 10, for the 709 patients of each type we looked back in time to see whether they smoked. We found percentages for the categories of the explanatory variable (smoker, nonsmoker). It is not meaningful to form a percentage for a category of the response variable. For example, we cannot estimate the population percentage of subjects who have lung cancer, for smokers or non-smokers. By the study design, *half* the subjects had lung cancer. This does not mean that about half the population had lung cancer.

EXAMPLE 11

Case-Control Studies about Cell Phone Use

Picture the Scenario

The U.S. and German studies about cell phone use mentioned in Example 1 were both case-control studies. In the U.S. study, the cases were brain cancer patients. The controls were hospitalized for benign conditions but were matched with the cases on age, gender, race, and month of admission. In the German study, the cases had a type of eye cancer. In forming a sample of controls, the German study did not attempt to match subjects with the cases.

Question to Explore

Why might researchers decide to match cases and controls on characteristics such as age?

Think It Through

We've seen that one way to balance groups on potential lurking variables such as age is to randomize in assigning subjects to groups. However, these were observational studies, not experiments. So it was not possible for the researchers to use randomization to balance treatments on potential lurking variables. Matching is an attempt to achieve the balance that randomization provides.

When researchers fail to use relevant variables to match cases with controls, those variables could influence the results. They could mask the true relationship. The results could suggest an association when actually there is not one, or the reverse. Without matching, results are more susceptible to effects of lurking variables.

Insight

The lack of matching in the German study may be one reason that the results from the two studies differed, the U.S. study not finding an association and the German study finding one. For example, in the German study suppose the eye cancer patients tended to be older than the controls, and suppose older people tend to be heavier users of cell phones. Then, age could be responsible for the observed association between cell-phone use and eye cancer.

TO PRACTICE THIS CONCEPT, TRY EXERCISE 4.47

Prospective studies A *retrospective* study, such as a case-control study, looks into the past. By contrast, a *prospective* study follows its subjects into the future.

EXAMPLE 12

The Prospective Nurses' Health Study

Picture the Scenario

The Nurses' Health Study, conducted by researchers at Harvard University,[14] began in 1976 with 121,700 female nurses age 30 to 55. The purpose of the study was to explore relationships among diet, hormonal factors, smoking habits, and exercise habits and the risk of coronary heart disease, pulmonary disease, and stroke. Since the initial survey in 1976, the nurses have filled out a questionnaire every two years.

Question to Explore

What does it mean for this observational study to be called *prospective*?

Think It Through

The retrospective smoking study in Example 10 already knew whether subjects had lung cancer, and it looked into the past to see whether they had been smokers. By contrast, when the Nurses' Health Study began, it was not known whether a particular nurse would eventually have an outcome such as lung cancer. The study followed her into the future to see whether a nurse developed that outcome and to analyze whether certain explanatory variables (such as smoking) were associated with it.

Insight

Over the years, several analyses have been conducted using data from the Nurses' Health Study. A recent finding (reported in *The New York Times*, Feb. 11, 2003) was that highly overweight 18-year-olds were five times as likely as young women of normal weight to need hip replacement later in life.

TO PRACTICE THIS CONCEPT, TRY EXERCISE 4.48

Observational Studies and Causation

Can we *ever* definitively establish causation with an observational study? No, we can't. For example, because the smoking and lung cancer study in Example 10 was observational, cigarette companies argued vociferously that a lurking variable could have caused this association. So why are doctors so confident in declaring that smoking causes lung cancer? For a combination of reasons: Experiments conducted using animals have shown an association. In many countries, over time female smoking has increased relative to male smoking (it was once rare for women to smoke), and the incidence of lung cancer has increased in women compared to men. Other studies, both retrospective and prospective, have added more and more evidence. For example, a prospective study began in 1951 with 35,000 doctors, and at its completion in 2001 the researchers[15] estimated that cigarettes took an average of 10 years off the lives of smokers who never quit. This study estimated that at least half the people who smoke from youth are eventually killed by their habit. (In fact, Richard Doll and a colleague recently estimated that 30% of *all* cancer cases can be attributed to smoking.)

---

[14] More information on this study can be found at clinicaltrials.gov/show/NCT00005152 and at www.channing.harvard.edu/nhs.

[15] See article by R. Doll et al., *British Med. J.*, June 26, 2004, p. 1519.

Most importantly, studies carried out on different populations of people have *consistently* concluded that smoking is associated with lung cancer, even after adjusting for all potentially confounding variables that researchers have suggested. As more studies are done that adjust for confounding variables, this reduces the chance that a lurking variable remains that can explain the association. As a consequence, although we cannot definitively conclude that smoking causes lung cancer, physicians will not hesitate to tell you that they believe it does.

On the Shoulders of . . .
Austin Bradford Hill and Richard Doll

"How did statisticians become pioneers in examining the effects of smoking on lung cancer?"

In the mid-twentieth century, Austin Bradford Hill was the leading medical statistician in Britain. In 1946, to assess the effect of streptomycin in treating pulmonary tuberculosis, he designed what is now regarded as the first controlled randomized medical experiment. About that time, doctors noticed the alarming increase in the number of lung cancer cases. The cause was unknown but suspected to be atmospheric pollution, particularly coal smoke. In 1950, Bradford Hill and the medical doctor and epidemiologist Richard Doll published results of the case-control study (Example 10) that identified smoking as a potentially important culprit. They were also pioneers in developing a sound theoretical basis for the case-control method. They followed their initial study with other case-control studies and with a long-term prospective study that showed that smoking is not only the predominant risk factor for lung cancer but also correlated with other diseases.

At first, Bradford Hill was cautious about claiming that smoking causes lung cancer. As the evidence mounted from various studies, he accepted the causal link as being overwhelmingly the most likely explanation. An influential article written by Bradford Hill in 1965 proposed criteria that should be satisfied before you conclude that an association reflects a causal link. Over the past 50 years, the statistical work of Doll and Hill has resulted in saving millions of lives, as more and more people become aware of the dangers of smoking. For his work, in 1961 Austin Bradford Hill was knighted by Queen Elizabeth. Richard Doll was knighted in 1969. The world owes grateful thanks to Sir Austin (1897–1991) and Sir Richard (1912-2005) for their statistical contributions to determining answers to important medical issues.

4.4 Practicing the Basics

4.38
TRY
Two factors helpful?: A two-factor experiment designed to compare two diets and to analyze whether results depend on gender randomly assigns 20 men and 20 women to the two diets, 10 of each to each diet. After three months the sample mean weight losses are as shown in the table.

Sample Mean Weight Loss by Diet Type and Gender

Diet	Female	Male	Overall
Low carb	12	0	6
Low fat	0	12	6
Overall	6	6	6

The header "Gender" spans the Female and Male columns.

a. Identify the two factors and the response variable.

b. What would the study have concluded if it did only a one-factor analysis on diet, looking at the overall results and not having the information about gender?

c. What could the study learn from the two-factor study that it would have missed by doing a one-factor study on diet alone or a one-factor study on gender alone?

4.39 **Caffeine jolt:** A recent study (*Psychosomatic Medicine* 2002; 64: 593–603) claimed that people who consume caffeine regularly may experience higher stress and higher blood pressure. In the experiment, 47 regular coffee drinkers consumed 500 milligrams of caffeine in a pill form (equivalent to four 8-oz cups) during one workday, and a placebo pill during another workday. The researchers monitored the subjects' blood pressure and heart rate, and the subjects recorded how stressed they felt.

a. Identify the response variable(s), explanatory variable, experimental units, and the treatments.

b. Is this an example of a completely randomized design, or a crossover design? Explain.

4.40 Allergy relief: An experiment is being designed to compare relief from hay fever symptoms given by a low dose of a drug, a high dose of the drug, and a placebo. Each subject who suffers from hay fever and volunteers for the study is observed on three separate days, with a different treatment used each day. There are two days between treatments, so that a treatment does not have a carry-over effect for the next treatment assigned.

a. What are the blocks in this block design? What is this type of block design called?

b. Suppose the study is conducted as a double-blind study. Explain what this means.

c. Explain how randomization within blocks could be incorporated into the study.

4.41 Effect of partner smoking in smoking cessation study: Smokers may have a more difficult time quitting smoking if they live with another smoker. How can an experiment explore this possibility in a study to compare Zyban with placebo? Suppose the researchers split the subjects into two groups: those who live with another smoker, and those who do not live with smokers. Within each group, the subjects are randomly assigned to take Zyban or a placebo. The figure shows a flow chart of this design, when 250 of the 429 study subjects live with nonsmokers and 179 live with another smoker.

a. Is this design a *completely* randomized design? Why or why not? (*Hint:* Is the smoking status of the person a subject lives with randomly determined?)

b. Does this experiment have blocks? If so, identify them.

c. Is this design a randomized block design? Explain why or why not.

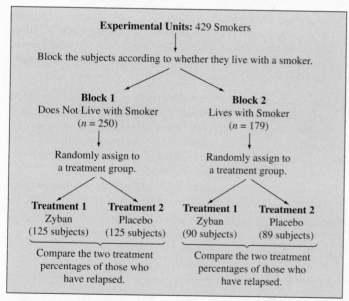

Diagram of an experiment to account for smoking status of partner

4.42 Health insurance: A researcher is planning a survey to estimate the proportion of employers in the United States that do not offer health insurance. She wants to

study how the results depend upon the type of company for which a person works.

a. Suppose she considers three types of companies (individual-owned with less than 50 employees, individual-owned with more than 50 employees, public company) and can take a simple random sample of 100 companies from each type. Is this an example of a stratified sample, or a cluster sample? Explain.

b. Suppose instead she decides to randomly sample 300 companies and then cross-classify them on the type of company and on whether they offer health insurance. Would this be a stratified sample, a cluster sample, or neither? Explain.

4.43 Club officers again: In Example 4, "Sampling Club Officers for a New Orleans Trip," two officers were selected to attend a conference in New Orleans. Three of the officers are female and two are male. It is decided to send one female and one male to the convention.

a. Labeling the officers as 1, 2, 3, 4, 5, where 4 and 5 are male, draw a stratified random sample using random numbers. Explain how you did this.

b. Explain why this sampling design is not a simple random sample.

4.44 Clusters vs. strata:

a. With a cluster random sample, do you take a sample of (i) the clusters? (ii) the subjects within every cluster?

b. With a stratified random sample, do you take a sample of (i) the strata? (ii) the subjects within every stratum?

c. Summarize the main differences between cluster sampling and stratified sampling in terms of whether one samples the groups or samples from within the groups that form the clusters or strata.

4.45 More school accounts to audit: Example 5 discussed how to sample 10 accounts to audit in a school district that has 60 accounts.

a. Accounts 1–30 are large and accounts 31–60 are small. Explain why simple random sampling cannot guarantee sampling both types of accounts. (*Hint:* Is it possible that *all* the accounts sampled could be small?)

b. To ensure sampling large and small accounts, select a stratified random sample of 5 accounts of each type. Show the steps of your procedure.

4.46 German cell phone study: The contingency table shows results from the German study about whether there was an association between mobile phone use and eye cancer (Stang et al., 2001).

a. The study was "retrospective." Explain what this means.

b. Explain what is meant by "cases" and "controls" in the heading of the table.

c. What proportion had used mobile phones, of those in the study who (i) had eye cancer, (ii) did not have eye cancer?

Eye Cancer and Use of Mobile Phones

Mobile Phones	Cases	Controls
Yes	16	46
No	102	429
Total	**118**	**475**

4.47 Smoking and lung cancer: Refer to the smoking case-control study in Example 10. Since subjects were not matched according to *all* possible lurking variables, a cigarette company can argue that this study does not prove a causal link between smoking and lung cancer. Explain this logic, using diet as the lurking variable.

4.48 Smoking and death: Example 1 in Chapter 3 described a survey of 1314 women during 1972–1974, in which each woman was asked whether she was a smoker. Twenty years later, a follow-up survey observed whether each woman was deceased or still alive. Was this study a cross-sectional study, or a retrospective study, or a prospective study? Explain.

CHAPTER REVIEW

Answers to the Chapter Figure Questions

Figure 4.4: The design would have 2 x 2 x 2 = 8 treatments.

Figure 4.5: In a stratified sample, every stratum is used. A simple random sample is taken within each stratum. A cluster sample takes a simple random sample of the clusters. All the clusters are *not* used.

CHAPTER SUMMARY

This chapter introduced methods for gathering data.

- An **experiment** assigns subjects to experimental conditions (such as drug or placebo) called **treatments**. These are categories of the explanatory variable. The outcome on the response variable is then observed for each subject.

- An **observational study** observes subjects on the response and explanatory variables. The study samples the population of interest, but merely observes rather than applies treatments to those subjects.

Since association does not imply causation, with observational studies we must be aware of potential **lurking variables** that influence the association. In an experiment, a researcher uses **randomization** in assigning experimental units (the subjects) to the treatments. This helps to balance the treatments on lurking variables. A randomized experiment can provide support for causation. To reduce bias, experiments should be **double-blind**, with neither the subject nor the data collector knowing to which treatment a subject was assigned.

A **sample survey** is an observational study that samples a population. Good methods for selecting samples incorporate random sampling.

- A **simple random sample** of *n* subjects from a population is one in which each possible sample of size *n* has the same chance of being selected.

- A **cluster random sample** takes a simple random sample of clusters (such as city blocks) and uses subjects in those clusters as the sample.

- A **stratified random sample** divides the population into separate groups, called **strata**, and then selects a simple random sample from each stratum.

Be cautious of results from studies that use a convenience sample, such as a **volunteer sample**, which Internet polls use. Even with random sampling, **biases** can occur due to sample undercoverage, nonresponse by many subjects in the sample, and responses that are biased because of question wording or subjects' lying.

Most sample surveys take a cross section of subjects at a particular time. A **census** is a complete enumeration of an entire population. **Prospective** studies follow subjects into the future, as is true with many experiments. Medical studies often use **retrospective** observational studies, which look at subjects' behavior in the past.

- A **case-control study** is an example of a retrospective study. Subjects who have a response outcome of interest, such as cancer, serve as cases. Other subjects not having that outcome serve as controls. The cases and controls are compared on an explanatory variable, such as whether they had been smokers.

A categorical explanatory variable in an experiment is also called a **factor**. **Multifactor experiments** have at least two explanatory variables. With a **completely randomized design**, subjects are assigned randomly to categories of each explanatory variable. Some designs instead use **blocks**—sets of experimental units that are matched. **Matched pairs designs** have two observations in each block. Often this is the same subject observed for each of two treatments, such as in a crossover study.

Data are only as good as the method used to obtain them. Unless we use a good study design, conclusions may be faulty.

CHAPTER PROBLEMS

Practicing the Basics

4.49 Observational vs. experimental study: Without using technical language, explain the difference between observational and experimental studies to someone who has not studied statistics. Illustrate with an example, using it also to explain the possible weaknesses of an observational study.

4.50 Unethical experimentation: Give an example of a scientific question of interest for which it would be unethical to conduct an experiment. Explain how you could instead conduct an observational study.

Explain source of bias in study

4.51 Breaking up is hard on your health: Married people seem to have better health than their single peers. If a marriage ends, the healthy edge tends to disappear, with divorced or separated people reporting higher rates of illness. This conclusion was based on a nationwide survey conducted by Statistics Canada of 9775 people, aged 20 to 64, about physical and mental health and relationship status, at 2-year intervals beginning in 1990.

a. Identify the explanatory variable and the response variable.

b. Explain why this is an example of an observational study.

c. Would it be practically possible to design this study as an experiment? Explain why or why not.

Source: Journal of Marriage and Family 2002; 64: 420–432

4.52 Dogs vs. cats: What is the preferred American pet? According to *Statistical Abstract of the United States* 2002 and the American Veterinary Medical Association, of pet-owning households, 36% have a dog and 32% have a cat. In 2002, vet care was sought by 85% of households owning a dog and by 67% of households owning a cat.

a. What is the population being considered?

b. Do you think these statistics were based on an (i) observational study? (ii) experiment? (iii) sample survey? (iv) census?

4.53 Fear of asbestos: Your friend reads about a study in an epidemiology journal that estimates that the chance is 15 out of 1 million that a teacher who works for 30 years in a school with typical asbestos levels gets cancer from asbestos. However, she also knows about a teacher who died recently who may have had asbestos exposure. In deciding whether to leave teaching, should she give more weight to the study or to the story she heard about the teacher who died? Why?

4.54 Dress in bathing suit?: In a recent Miss America beauty pageant, television viewers could cast their vote on whether to cancel the swimwear parade by phoning a number the network provided. Of the 1 million viewers who called and registered their opinion, 79% said they wanted to see the contestants dressed as bathing beauties. Since everyone had a chance to call, was this a simple random sample of all the viewers of this program? Why or why not?

4.55 Sampling your fellow students: You are assigned to direct a study on your campus to discover factors that are associated with strong academic performance. You decide to identify 20 students who have perfect GPAs of 4.0, and then measure explanatory variables for them that you think may be important, such as high school GPA and average amount of time spent studying per day.

a. Explain what is wrong with this study design.

b. Describe a study design that would provide more useful information.

4.56 Beware of Internet polling: An Internet survey at a newspaper Web site reports that only 14% of respondents believe in gun control. Mention a lurking variable that could bias the results of such an online survey, and explain how it could affect the results.

4.57 Comparing female and male students: You plan to sample from the 3500 undergraduate students who are enrolled at the University of Rochester in order to compare the proportions of female and male students who would like to see the United States have a female President.

a. Describe the steps for how you would proceed, if you plan a simple random sample of 80 students. Illustrate, by picking the first three students for the sample.

b. Suppose that you use random numbers to select students, but stop selecting females as soon as you have 40, and you stop selecting males as soon as you have 40. Is the resulting sample a simple random sample? Why or why not?

c. What type of sample is the sample in (b)? What advantage might it have over a simple random sample?

4.58 Football discipline: A large southern university had problems with 17 football players being disciplined for team rule violations, arrest charges, and possible NCAA violations. The online *Atlanta Journal Constitution* ran a poll with the question, "Has the football coach lost control over his players?" having possible responses, "Yes, he's been too lenient," and "No, he can't control everything teenagers do."

a. Was there pontential for bias in this study? If so, what types of bias?

b. The poll results after two days were

YES	6012	93%
NO	487	7%

Does this large sample size guarantee that the results are unbiased? Explain.

4.59 Smallpox fears: A Gallup poll indicated that concerns over the possibility of terrorists using smallpox were greater in 2003 than in the weeks following the terrorist attacks on September 11, 2001. The poll, conducted January 23–25, 2003, found that 17% were very worried. A similar poll conducted November 9–13, 2001 found that 11% were very worried. (*Source:* www.usatoday.com/news/nation/2003–01–27-smallpox-usat_x.htm)

a. The 2003 poll used telephone interviews with 1000 adults. What is the population of interest? What are possible sources of bias for this survey?

b. The article states that the margin of error was ± 3 percentage points. Explain what this means and show how an approximate margin of error can be calculated.

4.60 What's your favorite poem?: In fall 1995, the BBC in Britain requested viewers to call the network and indicate their favorite poem. Of 7500 callers, more than twice as many voted for Rudyard Kipling's *If* than for any other poem. The BBC then reported that this was the clear favorite.

a. Since any person could call, was this sample a simple random sample? Explain.

b. Was this a volunteer sample? Explain.

c. If the BBC truly wanted to determine Brits' favorite poem, how could they more reliably do so?

4.61 Video games mindless?: "Playing video games not so mindless." This was the headline in a *CNN* news report[16] about a study that concluded that young adults who play video games regularly demonstrated better visual skills than young adults who do not play regularly. Sixteen young men took a series of tests that measured their visual skills. The men who had played video games in the previous six months performed better on the tests than those who hadn't played.

a. What are the explanatory and response variables?

b. Was this an observational study or an experiment? Explain.

c. Specify a potential lurking variable. Explain your reasoning.

4.62 Physicians' health study: Read about the first Physicians' Health Study at phs.bwh.harvard.edu.

a. Explain whether it was (i) an experiment or an observational study, (ii) a retrospective or prospective or cross-sectional study.

b. Identify the response variable and the explanatory variable(s), and summarize results.

4.63 Aspirin prevents heart attacks?: During the 1980s approximately 22,000 physicians over the age of 40 agreed to participate in a long-term study called the Physicians' Health Study. One question investigated was whether aspirin helps to lower the rate of heart attacks. The physicians were randomly assigned to take aspirin or take placebo.

a. Identify the response variable and the explanatory variable.

b. Explain why this is an experiment, and identify the treatments.

c. There are other explanatory variables, such as the amount of exercise a physician got, that we would expect to be associated with the response variable. Explain how such a variable is dealt with by the randomized nature of the experiment.

4.64 Exercise and heart attacks: Refer to the previous exercise. One potential confounding variable was the amount of exercise the physicians got. The randomization should have balanced the treatment groups on exercise. The contingency table shows the relationship between whether the physician exercised vigorously and the treatments.

	Exercise Vigorously?		
Treatment	Yes	No	**Total**
Aspirin	7910	2997	10907
Placebo	7861	3060	10921

a. Find the conditional proportions (recall Section 3.1) in the categories of this potential confounder for each treatment group. Are they similar?

b. Do you think that the randomization process did a good job of achieving balanced treatment groups in terms of this potential confounder? Explain.

4.65 Smoking and heart attacks: Repeat the previous exercise, considering another potential confounding variable — whether the physicians smoked. The contingency table shown cross-classifies treatment group by smoking status.

	Smoking Status			
Treatment	Never	Past	Current	**Total**
Aspirin	5431	4373	1213	11017
Placebo	5488	4301	1225	11014

4.66 Aspirin, beta-carotene, and heart attacks: Refer to the previous three exercises. This completely randomized study actually used two factors: whether received aspirin and whether received beta-carotene. Draw a table to portray the four treatments for this study.

4.67 Zyban and nicotine patch study results: The subjects for the study described in Example 9 were evaluated for abstinence from cigarette smoking at the end of 12 months. The table shows the percentage in each group that were abstaining.

Group	Abstinence Percentage	Sample Size
Nicotine Patch Only	16.4	244
Zyban Only	30.3	244
Nicotine Patch with Zyban	35.5	245
Placebo Only	15.6	160

a. Find the approximate margin of error for the abstinence percentage in each group. Explain what a margin of error means.

b. Based on the results in (a), does it seem as if the difference between the Zyban only and placebo only treatments is "statistically significant"? Explain.

c. Based on the results in (a), does it seem as if the difference between the Zyban only and nicotine patch with Zyban treatments is "statistically significant"? Explain.

d. Based on the results in parts (a), (b), and (c), how would you summarize the results of this experiment?

4.68 Prefer Coke or Pepsi?: You would like to conduct an experiment with your class to see if students tend to prefer Coke or Pepsi.

a. Explain how you could do this, incorporating ideas of blinding and randomization, (i) with a completely randomized design, (ii) with a matched pairs design.

b. Which design would you prefer? Why?

4.69 Comparing gas brands: The marketing department of a major oil company wants to investigate whether cars get better mileage using their gas (Brand A) than from an independent one (Brand B) that has cheaper prices. The department has 20 cars available for the study.

a. Identify the response variable, the explanatory variable, and the treatments.

b. Explain how to use a completely randomized design to conduct the study.

c. Explain how to use a matched pairs design to conduct the study. What are the blocks for the study?

d. Give an advantage of using a matched pairs design.

4.70 Samples not equally likely in a cluster sample?: In a cluster random sample with equal-sized clusters, every subject has the same chance of selection. However, the sample is not a simple random sample. Explain why not.

4.71 Nursing homes: You plan to sample residents of registered nursing homes in your county. You obtain a list of all 97 nursing homes in the county, which you number from 01 to 97. Using random numbers, you choose 5 of the nursing homes. You obtain lists of residents from those 5 homes and interview all the residents in each home.

a. Are the nursing homes clusters, or strata?

b. Explain why the sample chosen is not a simple random sample of the population of interest to you.

4.72 Multistage health survey: A researcher wants to study regional differences in dental care. He takes a multistage sample by dividing the United States into four regions, taking a simple random sample of ten schools in each region, randomly sampling three classrooms in each school, and interviewing all students in those classrooms about whether they've been to a dentist in the previous year. Identify each stage of this sampling design, indicating whether it involves stratification or clustering.

4.73 Hazing: Hazing within college fraternities is a continuing concern. Before a national meeting of college presidents, a polling organization is retained to conduct a survey among fraternities on college campuses, gathering information on hazing for the meeting. The investigators from the polling organization realize that it is not possible to find a reliable sampling frame of all fraternities. Using a list of all college institutions, they randomly sample 30 of them, and then interview the officers of each fraternity at each of these institutions that has fraternities. Would you describe this as a simple random sample, cluster random sample, or stratified random sample of the fraternities? Explain.

4.74 TV linked to obesity in young children: A study taking a sample survey of 2761 white, black, and Hispanic low-income adults having pre-school children reported that the prevalence of overweight children was significantly related to the number of hours per day reported that the children spend watching TV or videos. (*Source: Pediatrics* 2002; 109(6): 1028–1035)

a. Identify the response and explanatory variables.

b. Was this observational study cross-sectional, prospective, or retrospective?

c. State a possible lurking variable, and describe its potential effect.

4.75 Cell phone study revisited: The American study reported in Example 1 about cell phones used 891 subjects with age ranging from 18 to 80 years. Of them, 469 people had brain cancer, and 422 people did not have brain cancer.

a. Explain why this was a retrospective study. Identify the cases and the controls.

b. How could researchers compare the groups, once data were collected?

c. Describe a potential source of bias for this study.

4.76 Twins and breast cancer: Excessive cumulative exposure to ovarian hormones is believed to cause breast cancer. A study (*New England J. Medic.* 2003; 348: 2313–2322) used information from 1811 pairs of female twins, one or both of whom had breast cancer. Paired twins were compared with respect to age at puberty, when breast cancer was first diagnosed, and other factors. Their survey did not show an association between hereditary breast cancer and hormone exposure.

a. What type of observational study was used by the researchers?

b. Why do you think the researchers used this design instead of a randomized experiment?

Concepts and Investigations

4.77 Cell phone use: Using the Internet, find a study about cell phone use and its potential risk when used by drivers of automobiles.

a. Was the study an experiment, or was it an observational study?

b. Identify the response and explanatory variables.

c. Describe any randomization or control conducted in the study as well as attempts to take into account lurking variables.

d. Summarize conclusions of the study. Do you see any limitations of the study?

4.78 Read a medical journal: Go to a Web site for an online medical journal, such as *British Medical Journal* (www.bmj.com). Pick an article in a recent issue.

a. Was the study an experiment, or was it an observational study?

b. Identify the response variable and the primary explanatory variable(s).

c. Describe any randomization or control conducted in the study as well as attempts to take into account lurking variables.

d. Summarize conclusions of the study. Do you see any limitations of the study?

4.79 Internet poll: Find an example of results of an Internet poll. Do you trust the results of the poll? If not, explain why not.

4.80 Search for an observational study: Find an example of an observational study from a newspaper, journal, the Internet, or some other medium.

a. Identify the explanatory and response variables and the population of interest.

b. What type of observational study was conducted? Describe how the data were gathered.

c. Were lurking variables considered? If so, discuss how. If not, can you think of potential lurking variables? Explain how they could affect the association.

d. Can you identify any potential sources of bias? Explain.

4.81 Search for an experimental study: Find an example of a randomized experiment from a newspaper, journal, the Internet, or some other media.

a. Identify the explanatory and response variables.

b. What were the treatments? What were the experimental units?

c. How were the experimental units assigned to the treatments?

d. Can you identify any potential sources of bias? Explain.

4.82 Judging sampling design: In each of the following situations, summarize negative aspects of the sample design.

a. A newspaper wants to determine whether its readers believe that government expenditures should be reduced by cutting social programs. They ask for readers to vote at their Internet site. Based on 1434 votes, they report that 93% of the city's residents believe that social programs should be reduced.

b. A congresswoman reports that letters to her office are running 3 to 1 in opposition to the passage of stricter gun control laws. She concludes that approximately 75% of her constituents oppose stricter gun control laws.

c. An anthropology professor wants to compare attitudes toward premarital sex of physical science majors and social science majors. She administers a questionnaire to her class of Anthropology 437, Comparative Human Sexuality. She finds no appreciable difference in attitudes between the two majors, so she concludes that the two student groups are about the same in their views about premarital sex.

d. A questionnaire is mailed to a simple random sample of 500 household addresses in a city. Ten are returned as bad addresses, 63 are returned completed, and the rest are not returned. The researcher analyzes the 63 cases and reports that they represent a "simple random sample of city households."

4.83 More poor sampling designs: Repeat the previous exercise for the following scenarios:

a. A principal in a large high school wants to sample student attitudes toward a proposal that seniors must pass a general achievement test in order to graduate. She lists all of the first-period classes. Then, using a random number table, she chooses a class at random and interviews every student in that class about the proposed test.

b. A new restaurant opened in January. In June, after six months of operation, the owner applied for a loan to improve the building. The loan application asked for the annual gross income of the business. The owner's record book contains receipts for each day of operation since opening. She decides to calculate the average daily receipt based on a sample of the daily records and multiply that by the number of days of operation in a year. She samples every Friday's record. The average daily receipt for this sample was then used to estimate the yearly receipts.

4.84 Age for legal alcohol: You want to investigate the opinions students at your school have about whether the age for legal drinking of alcohol should be 18.

a. Write a sentence to ask about this in a sample survey in such a way that results would be biased. Explain why it would be biased.

b. Now write an alternative sentence that should result in unbiased responses.

4.85 Quota sampling: An interviewer stands at a street corner and conducts interviews until obtaining a quota in various groups representing the relative sizes of the groups in the population. For instance, the quota might be 50 factory workers, 100 housewives, 60 elderly people, 30 Hispanics, and so forth. This is called **quota sampling**. Is this a random sampling method? Explain, and discuss potential advantages or disadvantages of this method. (The Gallup Organization used quota sampling until it predicted, incorrectly, that Dewey would easily defeat Truman in the 1948 Presidential election.)

4.86 Smoking and heart attacks: A Reuters story (4/2/2003) reported that "The number of heart attack victims fell by almost 60 percent at one hospital six months after a smoke-free ordinance went into effect in the area (Helena,

Montana), a study showed, reinforcing concerns about second-hand smoke." The number of hospital admissions for heart attack dropped from just under seven per month to four a month during the six months after the smoking ban.

a. Did this story describe an experiment or an observational study?

b. In the context of this study, describe how you could explain to someone who has never studied statistics that association does not imply causation. For instance, give a potential reason that could explain this association.

4.87 Issues in clinical trials: A randomized clinical trial is planned for AIDS patients to investigate whether a new treatment provides improved survival over the current standard treatment. It is not known whether it will be better, or worse.

a. Why do researchers use randomization in such experiments, rather than letting the subjects choose which treatment they will receive?

b. When patients enrolling in the study are told the purpose of the study, explain why they may be reluctant to be randomly assigned to one of the treatments.

c. If a researcher planning the study thinks the new treatment is likely to be better, explain why he or she may have an ethical dilemma in proceeding with the study.

4.88 Compare smokers with nonsmokers?: Example 10 and Table 4.3 described a case-control study on smoking and lung cancer. Explain carefully why it is not sensible to use the study's proportion of smokers who had lung cancer (that is, $688/(688 + 650)$) and proportion of non-smokers who had lung cancer $(21/(21 + 59))$ to estimate corresponding proportions of smokers and nonsmokers who have lung cancer in the overall population.

4.89 Is a vaccine effective?: A vaccine is claimed to be effective in preventing a rare disease that occurs in about one of every 100,000 people. Explain why a randomized clinical trial comparing 100 people who get the vaccine to 100 people who do not get it is unlikely to be worth doing. Explain how you could use a case-control study to investigate the efficacy of the vaccine.

For multiple-choice questions 4.90–4.97, select the best response.

4.90 What's a simple random sample?: A simple random sample of size *n* is one in which

a. Every *n*th member is selected from the population.

b. Each possible sample of size *n* has the same chance of being selected.

c. There is *exactly* the same proportion of women in the sample as is in the population.

d. You keep sampling until you have a fixed number of people having various characteristics (e.g., males, females).

4.91 Getting a random sample: When we use random numbers to take a simple random sample of 50 students from the 20,000 students at a university,

a. It is impossible to get the random number 00000 or 99999, since they are not random sequences.

b. If we get 20001 for the first random number, for the second random number that number is less likely to occur than the other possible five-digit random numbers.

c. The draw 12345 is no more or less likely than the draw 11111.

d. Since the sample is random, it is impossible that it will be nonrepresentative, such as having only females in the sample.

4.92 **Be skeptical of medical studies?:** An analysis of published medical studies about heart attacks (Crossen, 1994, p. 168) noted that in the studies having randomization and strong controls for bias, the new therapy provided improved treatment 9% of the time. In studies without randomization or other controls for bias, the new therapy provided improved treatment 58% of the time.

a. This result suggests it is better not to use randomization in medical studies, because it is harder to show that new ideas are beneficial.

b. Some newspaper articles that suggest a particular food, drug, or environmental agent is harmful or beneficial should be viewed skeptically, unless we learn more about the statistical design and analysis for the study.

c. This result shows the value of case-control studies over randomized studies.

d. The randomized studies were poorly conducted, or they would have found the new treatment to be better much more than 9% of the time.

4.93 **Opinion and question wording:** A recent General Social Survey asked subjects if they supported legalizing abortion in each of seven different circumstances. The percentage who supported legalization varied between 45% (if the woman wants it for any reason) to 92% (if the woman's health is seriously endangered by the pregnancy). This indicates that

a. Responses can depend greatly on the question wording.

b. Observational studies can never be trusted.

c. The sample must not have been randomly selected.

d. The sample must have had problems with response bias.

4.94 **Campaign funding:** When the Yankelovich polling organization asked, "Should laws be passed to eliminate all possibilities of special interests giving huge sums of money to candidates?" 80% of the sample answered *yes*. When they posed the question, "Should laws be passed to prohibit interest groups from contributing to campaigns, or do groups have a right to contribute to the candidate they support?" only 40% said *yes* (Source: *A Mathematician Reads the Newspaper,* by J. A. Paulos, New York: Basic Books, 1995, p. 15). This example illustrates problems that can be caused by

a. Nonresponse

b. Bias in the way a question is worded

c. Sampling bias

d. Undercoverage

4.95 **Emotional health survey:** An Internet poll conducted in the U.K. by Netdoctor.co.uk asked individuals to respond to an "Emotional health survey" (see www.hfienberg.com/clips/pollspiked.htm). There were 400 volunteer respondents. Based on the results, the British Broadcasting Corporation (BBC) reported that "Britons are miserable—it's official." This conclusion reflected the poll responses, of which one-quarter feared a "hopeless future," one in three felt "downright miserable," and nearly one in ten thought "their death

would make things better for others." Which of the following is *not* correct about why these results may be misleading?

a. Many people who access a medical website and are willing to take the time to answer this questionnaire may be having emotional health problems.

b. Some respondents may not have been truthful or may have been Internet surfers who take pleasure in filling out a questionnaire multiple times with extreme answers.

c. The sample is a volunteer sample rather than a random sample.

d. It's impossible to learn useful results about a population from a sample of only 400 people.

4.96 **Sexual harassment:** In 1995 in the United Kingdom, the Equality Code used by the legal profession added a section to make members more aware of the dangers of sexual harassment. It states that "research for the Bar found that over 40 percent of female junior tenants said they had encountered sexual harassment during their time at the Bar." This was based on a study conducted at the University of Sheffield that sent a questionnaire to 334 junior tenants at the Bar, of whom 159 responded. Of the 159, 67 were female. Of those females, 3 said they had experienced sexual harassment as a major problem, and 24 had experienced it as a slight problem.

a. The quoted statement might be misleading because the nonresponse was large.

b. No one was forced to respond, so everyone had a chance to be in the sample, which implies it was a simple random sample.

c. This was an example of a completely randomized experiment, with whether a female junior tenant experienced sexual harassment as the response variable.

d. This was a retrospective case-control study, with those who received sexual harassment as the cases.

4.97 **Effect of response categories:** A study (N. Schwarz et al., *Public Opinion Quarterly*, vol. 49, p. 388, 1985) asked German adults how many hours a day they spend watching TV on a typical day. When the possible responses were the six categories (up to $\frac{1}{2}$ hour, $\frac{1}{2}$ to 1 hour, 1 to $1\frac{1}{2}$ hours, ..., more than $2\frac{1}{2}$ hours), 16% of respondents said they watched more than $2\frac{1}{2}$ hours per day. When the six categories were (up to $2\frac{1}{2}$ hours, $2\frac{1}{2}$ to 3 hours, ..., more than 4 hours), 38% said they watched more than $2\frac{1}{2}$ hours per day.

a. The samples could not have been random, or this would not have happened.

b. This shows the importance of question design, especially when people may be uncertain what the answer to the question really is.

c. This study was an experiment, not an observational study.

♦♦ 4.98 **Sample size and margin of error:**

a. Find the approximate margin of error when $n = 1$.

b. Show the two possible percentage outcomes you can get with a single observation. Explain why the result in (a) means that with only a single observation, you have essentially no information about the population percentage.

c. How large a sample size is needed to have a margin of error of about 5% in estimating a population percentage? (*Hint*: Take the formula for the approximate margin of error, and solve for the sample size.)

♦♦ 4.99 **Systematic sampling:** A researcher wants to select 1% of the 10,000 subjects from the sampling frame. She selects subjects by picking one of the first 100 on the list at random, and then skipping 100 names to get the next subject, skipping another 100 names to get the next subject, and so on. This is called a **systematic random sample**.

a. With simple random sampling, (i) every subject is equally likely to be chosen, and (ii) every possible sample of size *n* is equally likely. Indicate which, if any, of (i) and (ii) are true for systematic random samples. Explain.

b. An assembly-line process in a manufacturing company is checked by using systematic random sampling to inspect 2% of the items. Explain how this sampling process would be implemented.

♦♦ 4.100 **Complex multistage GSS sample** Go to a Web site for the GSS, http://www.icpsr.umich.edu/GSS, click on *Appendix*, and then click on *Sampling Design and Weighting*. There you will see described the complex multistage design of the GSS. Explain how the GSS uses (a) clustering, (b) stratification, (c) simple random sampling.

♦♦ 4.101 **Mean family size:** You'd like to estimate the mean size of families in your community. Explain why you'll tend to get a smaller sample mean if you sample *n* families than if you sample *n* individuals (asking them to report their family size). (*Hint*: When you sample individuals, explain why you are more likely to sample a large family than a small family. To think of this, it may help to consider the case *n* = 1 with a population of two families, one with 10 people and one with only 2 people.)

♦♦ 4.102 **Capture–recapture:** Biologists and naturalists often use sampling to estimate sizes of populations, such as deer or fish, for which a census is impossible. Capture-recapture is one method for doing this. A biologist wants to count the deer population in a certain region. She captures 50 deer, tags each, and then releases them. Several weeks later, she captures 125 deer and finds that 12 of them were tagged. Let *N* = population size, *M* = size of first sample, *n* = size of second sample, *R* = number tagged in second sample. The table shows how results can be summarized.

		In First Sample?		
		Yes (tagged)	No (not tagged)	**Total**
In Second	Yes	R		*n*
Sample?	No			
	Total	**M**		**N**

a. Identify the values of *M*, *n*, and *R* for the biologist's experiment.

b. One way to estimate *N* lets the sample proportion of tagged deer equal the population proportion of tagged deer. Explain why this means that

$$\frac{R}{n} = \frac{M}{N},$$

and hence that the estimated population size is $N = (M \times n)/R$.

c. Estimate the number of deer in the deer population using the numbers above.

d. The U.S. Census Bureau uses capture-recapture to make adjustments to the census by estimating the undercount. The capture phase is the census itself (persons are 'tagged' by having returned their census form and being recorded as counted) and the recapture phase (the second sample) is the Post-Enumerative Survey (PES) conducted after the census. Label the table in terms of the census application.

Class Explorations

4.103 **Munchie capture–recapture:** Your class can use the capture-recapture method described in the previous exercise to estimate the number of goldfish in a bag of Cheddar Goldfish. Pour the Cheddar Goldfish into a paper bag, which represents the pond. Sample 10 of them. For this initial sample, use Pretzel Goldfish to replace them, to represent the tagged fish. Then select a second sample and derive an estimate for *N*, the number of Cheddar Goldfish in the original bag. See how close your estimate comes to the actual number of fish in the bag. (Your teacher will count the population of cheddar goldfish in the bag before beginning the sampling.) If the estimate is not close, what could be responsible, and what would this reflect as difficulties in a real-life application such as sampling a wildlife population?

4.104 **Margin of error:** Activity 2 in Chapter 1 used the "Sample ⟨TRY⟩ from a population" applet to simulate randomness and variability. We'll use that applet again here, but with a much larger sample size, 1000.

a. For a population proportion of 0.50 for outcome 1, simulate a random sample of size 1000. What is the sample proportion of outcome 1? Do this 10 separate times, keeping track of the 10 sample proportions.

b. Find the approximate margin of error for a sample proportion based on 1000 observations.

c. Using the margin of error found in part (b) and the 10 sample proportions found in part (a), form ten intervals of believable values for the true proportion. How many of these intervals captured the actual population proportion, 0.50?

d. Collect the 10 intervals from each member of the class. (If there are 20 students, 200 intervals will be collected.) What percentage of these intervals captured the actual population proportion, 0.50?

4.105 **Activity: Sampling the states:** This activity illustrates how sampling bias can result when you use a non-random sample, even if you attempt to make it representative: You are in a geography class, discussing center and spread for various characteristics of the states in the contiguous United States. A particular value of center is the mean area of the states. A map and a list of the states with their areas (in square miles) are shown in the figure and table below. Area for a state includes dry land and permanent inland water surface.

Although we could use these data to calculate the actual mean area, let's explore *how well sampling performs in estimating the mean area* by sampling 5 states and finding the sample mean.

a. The most convenient sampling design is to use our eyes to pick 5 states from the map that we think have areas representative of all the states. Do this, picking five states that you believe have areas representative of the actual mean area of the states. Compute their sample mean area.

b. Collect the sample means for all class members. Construct a dot plot of these means. Describe the distribution of sample means. Note the shape, center, and spread.

c. Another possible sampling design is simple random sampling. Select five random numbers between 01 and 48. Go to the table below and take a simple random sample of five states. Compute the sample mean area.

d. Collect the sample means from (c) of all class members. Construct a dot plot of the sample means using the same horizontal scale as in (b). Describe this distribution of sample means. Note the shape, center, and spread.

e. The true mean total land area for the 48 states can be calculated from the table below, by dividing the total at the bottom of the table by 48. Which sampling method, using your eyes or using random selection, tended to be better at estimating the true population mean? Which method seems to be less biased? Explain.

f. Write a short summary comparing the two distributions of sample means.

Table: Areas of the 48 States in the Continental U.S.	
State	**Area**
Alabama	52,419
Arizona	113,998
Arkansas	53,179
California	163,696
Colorado	104,094
Connecticut	5,543
Delaware	2,489
Florida	65,755
Georgia	59,425
Idaho	83,570
Illinois	57,914
Indiana	36,418
Iowa	56,272
Kansas	82,277
Kentucky	40,409
Louisiana	51,840
Maine	35,385
Maryland	12,407
Massachusetts	10,555
Michigan	96,716
Minnesota	86,939
Mississippi	48,430
Missouri	69,704
Montana	147,042
Nebraska	77,354
Nevada	110,561
New Hampshire	9,350
New Jersey	8,721
New Mexico	121,589
New York	54,556
North Carolina	53,819
North Carolina	70,700
Ohio	44,825
Oklahoma	69,898
Oregon	98,381
Pennsylvania	46,055
Rhode Island	1,545
South Carolina	32,020
South Dakota	77,116
Tennessee	42,143
Texas	268,581
Utah	84,899
Vermont	9,614
Virginia	42,774
Washington	71,300
West Virginia	24,230
Wisconsin	65,498
Wyoming	97,814
U.S. TOTAL	**3,119,819**

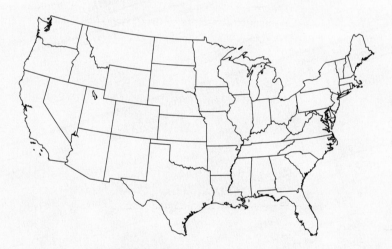

Map of the Continental United States

BIBLIOGRAPHY

Burrill, G., C. Franklin, L. Godbold, and L. Young (2003). *Navigating through Data Analysis in Grades 9–12,* NCTM

Crossen, C. (1994). *Tainted Truth: The Manipulation of Fact in America.* New York: Simon & Schuster.

Leake, J. (2001) "Scientist Link Eye Cancer to Mobile Phones," *The London Times,* Jan. 14. www.rfsafe.com/pages/articles/the_times_011401.htm

Muscat, J., et al. (2000). "Handheld Cellular Telephone Use and Risk of Brain Cancer." *Journal of the American Medical Association* 284: 3001–3007.

Repacholi, M.H. (1997). Radiofrequency field exposure and cancer. *Environ. Health Prospect* 105: 1565–1568.

Stang A, et al. (2001). "The possible role of radio frequency radiation in the development of uveal melanoma." *Epidemiology* 12(1): 7–12.

5 | Probability in Our Daily Lives

EXAMPLE I

What's the Chance Your Tax Return Is Audited?

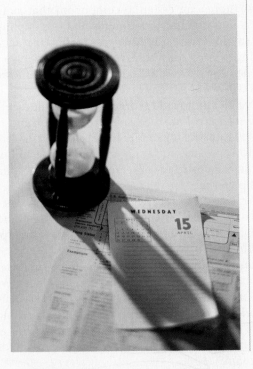

Picture the Scenario

For many Americans, April 15 — "Tax Day" — is a day filled with dread as income taxes from the previous year are due. Tax preparation, collection, and defense represent a huge business in the United States as CPAs, tax attorneys, and software programs crunch through earnings and deductions to determine how much is owed. It is assumed that taxpayers will honestly report all income and take only legitimate deductions. Each year, a small percentage of Americans are subjected to an audit of their tax return to determine whether they did, indeed, pay their fair share of taxes. In fact, in 2004, audits performed by the Internal Revenue Service (IRS) raised an additional $43 billion dollars. According to IRS Commissioner Mark Everson, audits increased 40% in 2004 for taxpayers with adjusted gross incomes in excess of $100,000 and the total number of audits of taxpayers exceeded a million for the first time since 1999.

Questions to Explore

- What is the likelihood of an American taxpayer being audited?
- What is the likelihood of being audited if your annual income is (i) $100,000 or higher, (ii) below $25,000?

Thinking Ahead

In Example 5 in Section 5.2 we'll look at some recent data on numbers of people who are audited. We'll see that the chance of an audit is quite small. But does that chance depend on your income level? At the beginning of Section 5.3 we'll see how to use data from the IRS to answer this.

The answers to these questions are found using **probability** methods. In this chapter, we'll learn how to find probabilities and how to interpret them. This knowledge will help you, as a consumer of information, understand how to assess probabilities in many aspects of your life that involve uncertainty.

In everyday life, you often must make decisions when you are uncertain about the outcome. Should you invest money in the stock market? Should you get dental insurance, or should you get extra collision insurance on your car? Should you start a new business, such as opening a pizza place across from campus? Should you take an advanced course that may be hard to pass? Daily, you face even mundane decisions, such as whether or not to buy a lottery ticket, to place a bet on a sporting event, or to carry an umbrella with you in case it rains.

This chapter introduces **probability**—the way we quantify uncertainty. You'll learn how to measure the chances of the possible outcomes for **random phenomena**—the many things in your life for which the outcome is uncertain. Using probability, for instance, you can find the chance you'll win a lottery. You can find the likelihood that an employer's drug test correctly detects whether or not you've used drugs. You can measure the uncertainty that comes with randomized experiments and with random sampling in surveys. The ideas in this chapter set the foundation for how we'll use probability in the rest of the book to make inferences based on data.

5.1 | How Can Probability Quantify Randomness?

As we discovered in Chapter 4, there's an essential component that statisticians rely on to try to avoid bias in designing experiments and in sampling in observational studies. This is **randomness**—randomly assigning subjects to treatments or randomly selecting people for a sample. Randomness also applies to the outcomes of a response variable. The possible outcomes are known, but it's uncertain which will occur for any given observation.

Since we were children, we've all employed randomization in games we've played. Some popular randomizers are rolling dice, spinning a wheel, and flipping a coin. Randomization helps to make a game fair. Rolls of dice and flips of coins are also simple ways to represent the randomness that occurs with randomized experiments and with sample surveys. For instance, the "head" and "tail" outcomes of a coin flip can represent "drug" and "placebo" when a medical study randomly assigns a subject to receive one of two treatments.

With a *small* number of observations, outcomes of random phenomena may look quite different from what you expect. For instance, you may expect to see a random pattern with different outcomes; but instead, exactly the same outcome may happen a few times in a row. That's not necessarily a surprise, as unpredictability for any given observation is the essence of randomness. We'll discover, however, that with a *large* number of observations, summary statistics become surprisingly *nonrandom*. For instance, as we make more and more observations, the proportion of times that a particular outcome occurs gets closer and closer to a certain number. This long-run proportion provides the basis for the definition of *probability*.

EXAMPLE 2

Is a Die Fair?

Picture the Scenario

The board game you've been playing has a die that determines the number of spaces moved on the board. After you've rolled the die 100 times, the number 6 has appeared 23 times, more frequently than each of the other 5 numbers, 1 through 5. At one stage, it turns up three times in a row, resulting in your winning a game. Your opponent then complains that the die favors the number 6 and is not a fair die.

Did you Know?

The singular of *dice* is *die*. For a proper die, numbers on opposite sides add to 7, and when 4 faces up the die can be turned so that 2 faces the player, 1 is at the player's right, and 6 is at the player's left.

(*Ainslie's Complete Hoyle,* New York: Simon & Schuster, 2003) ■

APPLET

You can try this yourself on the text CD using the *simulating the probability of rolling a 6* applet. This is designed to generate "binary" data, which means that each trial has only *two* possible outcomes, such as "6" or "not 6."

Questions to Explore

a. If a fair die is rolled 100 times, how many 6s do you expect?

b. Would it be unusual to get 23 6s out of 100 rolls? Would it be surprising to roll three 6s in a row at some point?

Think it Through

a. With many rolls of a fair die, you would expect the six numbers to appear about equally often. A 6 should occur about one-sixth of the time. In 100 rolls, we expect a 6 to come up about $(1/6)100 = 16.7 \approx 17$ times.

b. How can we determine whether or not it is unusual for 6 to come up 23 times out of 100 rolls, or three times in a row at some point? One way uses simulation. We could roll the die 100 times and see what happens, roll it another 100 times and see what happens that time, and so on. Does a 6 appear 23 (or more) times in many of the simulations? Does a 6 occur three times in a row in many simulations?

This simulation using a die would be tedious. Fortunately, we can use an applet or other software to simulate rolling a fair die. Each simulated roll of a die is called a **trial**. After each trial we record whether a 6 occurred. We will also keep a running record of the proportion of times that a 6 has occurred. At each value for the number of trials, this is called a **cumulative proportion**. Table 5.1 shows partial results of one simulation of 100 rolls. To find the cumulative proportion after a certain number of trials, divide the number of 6s at that stage by the number of trials. For example, by the eighth roll (trial), there had been three 6s in eight trials, so the cumulative proportion is $3/8 = 0.375$. Figure 5.1 plots the cumulative proportions against the trial number.

TABLE 5.1: Simulation Results of Rolling a Fair Die 100 Times

Each trial is a simulated roll of the die, with chance 1/6 of a 6. At each trial, we record whether a 6 occurred as well as the cumulative proportion of 6s by that trial.

Trial	6 Occurs?	Cumulative Proportion of 6s	
1	yes	1/1	= 1.0
2	no	1/2	= 0.500
3	yes	2/3	= 0.667
4	no	2/4	= 0.500
5	yes	3/5	= 0.600
6	no	3/6	= 0.500
7	no	3/7	= 0.429
8	no	3/8	= 0.375
...	...	...	...
30	no	9/30	= 0.300
31	no	9/31	= 0.290
32	yes	10/32	= 0.313
33	yes	11/33	= 0.333
34	yes	12/34	= 0.353
35	no	12/35	= 0.343
...	...	...	...
99	no	22/99	= 0.222
100	yes	23/100	= 0.230

▲ **FIGURE 5.1: The Cumulative Proportion of Times a 6 Occurs, for a Simulation of 100 Rolls of a Fair Die.** The horizontal axis of this MINITAB figure reports the number of the trial, and the vertical axis reports the cumulative proportion of 6s observed by then. **Question:** The first four rolls of the die were 6, 2, 6, and 5. How can you find the cumulative proportion of 6s after each of the first four trials?

In this simulation of 100 rolls of a die, a 6 occurred 23 times, different from the expected value of about 17. From Table 5.1, a 6 appeared three times in a row for trials 32 through 34.

Insight

One simulation does not prove anything. It suggests, however, that rolling three 6s in a row out of 100 rolls may not be highly unusual. It also shows that 23 6s out of 100 trials can occur. To find out whether 23 6s is unusual, we need to repeat this simulation many times. In Chapter 6, we will learn about the binomial distribution, which will allow us to compute the likelihood for observing 23 (or more) 6s out of 100 trials.

TO PRACTICE THIS CONCEPT, TRY EXERCISE 5.10

ACTIVITY 1: **Simulate Many Rolls of a Die**

Since one simulation is insufficient to tell us what is typical, go to the *simulating the probability of rolling a 6* applet on the text CD, which can conduct simulations of this type in which each trial has two possible outcomes. Set the sample size equal to 100. How many 6s did you observe out of the 100 simulated rolls? After simulating 100 rolls, how close was the cumulative proportion of 6s to the expected value of 1/6?

Do the same simulation 25 times to get a feeling for how the sample cumulative proportion at 100 simulated rolls compares to the expected value of 1/6 (that is, 16.7%). You've probably seen that it's possible for 23 or more (that is, at least 23%) of the 100 rolls to result in 6s. If you keep doing this simulation over and over again, about 6.3% of the time you will get 23 or more 6s out of the 100 rolls. Also, about 30% of the time you will see at least three 6s in a row somewhere out of the 100 rolls.

Now, change the sample size for each simulation to 1000, simulating rolling the die 1000 times. Here's a prediction: The cumulative proportion of 6s at 1000 simulated rolls will tend to fall *closer* to its expected value of 1/6. We'll now be very surprised if at least 23% of the 1000 rolls result in 6s.

Long-Run Behavior of Random Outcomes

We've seen that the number of 6s in 100 rolls of a die can vary. One time we might get 19 6s, another time we might get 22, another time 13, and so on. So what do we mean when we say that there's a one-in-six chance of a 6 on any given roll?

Let's continue the simulation from Example 2 and Table 5.1. In that example we stopped after only 100 rolls, but now let's continue simulating for a very large number of rolls. Figure 5.2 shows the cumulative proportion of 6s plotted against the trial number from trial 100, up to a total of 10,000 rolls. As the trial number increases, the cumulative proportion of 6s gradually settles down. After 10,000 simulated rolls, Figure 5.2 shows that the cumulative proportion is *very* close to the value of 1/6.

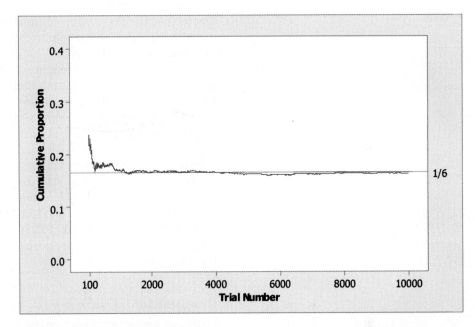

▲ **FIGURE 5.2: The Cumulative Proportion of Times that a 6 Occurs for a Simulation of 10,000 Rolls of a Fair Die.** Figure 5.1 showed the first 100 trials, and this MINITAB figure shows results of trials 100 through 10,000. As the trial number gets larger, the cumulative proportion gets closer to 1/6. The probability of a 6 on any single roll is defined to be this *long-run* value. **Question:** What would you expect for the cumulative proportion of heads after you flipped a balanced coin 10,000 times?

With a relatively short run, such as 10 rolls of a die, the cumulative proportion of 6s can fluctuate a lot. It need not be close to 1/6 after the final trial. However, as the number of trials keeps increasing, the proportion of times the number 6 occurs becomes more predictable and less random. It gets closer and closer to 1/6. *With random phenomena, the proportion of times that something happens is highly random and variable in the short run but very predictable in the long run.*

Did You Know?

The **law of large numbers** is what helps keep casinos in business. For a short time, a gambler may be lucky and make money, but in the long run the casino knows that it's going to come out ahead. ■

Probability Quantifies Long-Run Randomness

In 1689, the Swiss mathematician Jacob Bernoulli proved that as the number of trials increases, the proportion of occurrences of any given outcome approaches a particular number (such as 1/6) "in the long run." To show this, he assumed that the outcome of any one trial does not depend on the outcome of any other trial. Bernoulli's result is known as the **law of large numbers**.

We will interpret the **probability** of an outcome to represent "long-run" results. Imagine a randomized experiment or a random sampling of subjects that provides a very long sequence of observations. Each observation does or does not have that outcome. The probability of the outcome is the proportion of times that it occurs, in the long run.

Probability

With a randomized experiment or a random sample or other random phenomenon (such as a simulation), the **probability** of a particular outcome is the proportion of times that the outcome would occur in a long run of observations.

When we say that a roll of a die has outcome 6 with probability 1/6, this means that the proportion of times that a 6 would occur in a long run of observations is 1/6. The probability would also be 1/6 for each of the other possible outcomes: 1, 2, 3, 4, or 5.

A weather forecaster might say that the probability of rain today is 0.70. This means that in a large number of days with atmospheric conditions like those today, the proportion of days in which rain occurs is 0.70. Since a probability is *a proportion*, it takes a value between 0 and 1. Sometimes probabilities are expressed as percentages, such as when the weather forecaster reports the probability of rain as 70%. Probabilities then fall between 0 and 100, but we'll mainly use the proportion scale.

Why does probability refer to the *long run*? Because we can't accurately assess a probability with a small number of trials. If you sample ten people and they are all right-handed, you can't conclude that the probability of being right-handed equals 1.0. It takes a much larger sample of people to predict accurately the proportion of people in the population that are right-handed.

Independent Trials

With random phenomena, many believe that when some outcome has not happened in a while, it is *due* to happen: Its probability goes up until it happens. In many rolls of a fair die, if a particular value (say, 5) has not occurred in a long time, some think it's due and that the chance of a 5 on the next roll is greater than 1/6. If a family has four girls in a row and is expecting another child, are they due to get a boy? Does the next child have more than a 1/2 chance of being a boy?

Example 2 showed that over a short run, observations may well deviate from what is expected (remember three 6s in a row?). But with many random phenomena, such as outcomes of rolling a die or having children, what happens on previous trials does not affect the trial that's about to occur. The trials are **independent** of each other.

Independent Trials

Different trials of a random phenomenon are **independent** if the outcome of any one trial is not affected by the outcome of any other trial.

With independent trials, whether you get a 5 on one roll of a fair die does not affect whether you get a 5 on the following roll. It doesn't matter if you had 20 rolls in a row that are not 5s, the next roll still has probability 1/6 of being a 5. The die has no memory. If you have lost many bets in a row, don't assume that you are due to win if you continue to gamble. Many gamblers have gone broke that way. The *law of large numbers*, which gamblers invoke as the *law of averages*, only guarantees *long-run* performance. Over the short amount of time in which a gambler's money can disappear, the variability may well exceed what you expect.

How Can We Find Probabilities?

In practice, how can we find probabilities? Sometimes we can find them by making assumptions about the nature of the random phenomenon. For instance, by symmetry, it may be reasonable to assume that the possible outcomes are *equally likely*. In rolling a die, we might assume that based on the physical makeup of the die, each of the six sides has an equal chance. Thus, we take the probability of rolling any particular number to equal 1/6. If we flip a coin and assume that the coin is balanced, we take the probability of flipping a tail (or a head) to equal 1/2. Notice that, like proportions, *the total of the probabilities for all the possible outcomes equals 1*.

IN PRACTICE: The sample proportion estimates the actual probability

In theory, we could observe several trials of a random phenomenon and use the sample proportion of times an outcome occurs as its probability. In practice, this is imperfect. The sample proportion merely *estimates* the actual probability, and only for a *very large* number of trials is it necessarily close. In Chapters 6 and 7, we'll see how the sample size determines just how good that estimate is.

Types of Probability: Relative Frequency and Subjective

We've defined the probability of an outcome as a long-run proportion (relative frequency) of times that the outcome occurs in a very large number of trials. However, this definition is not always helpful. Before the launch of the first space shuttle, how could NASA scientists assess the probability of success? No data were available about long-run observations of past flights. If you decide to start a new type of business when you graduate, you won't have a long run of trials with which to estimate the probability that the business is successful. Likewise, no long run of trials is available if you want to assess the probability of life after death, the probability that a defendant being judged in a jury trial is truly guilty, or the probability that a marriage you are contemplating will be a success and will not end in divorce.

In such situations, you must rely on **subjective** information rather than solely on **objective** information such as data. You assess the probability of an outcome by taking into account all the information you have. Such probabilities are not based on a long run of trials. In this **subjective definition of probability**, the probability of an outcome is defined to be your degree of belief that the outcome will occur, based on the available information. A branch of statistics uses subjective probability as its foundation. It is called **Bayesian statistics**, in honor of Thomas Bayes, a British clergyman who discovered a probability rule on which it is based. The subjective approach is less common than the approach we discuss in this text and is beyond our scope. We'll merely warn you to be wary of anyone who gives a subjective probability of 1 (certain occurrence) or of 0 (certain nonoccurrence) for some outcome. As Benjamin Franklin said, nothing is certain but death and taxes!

5.1 Practicing the Basics

5.1 **Probability:** Explain what is meant by the long-run relative frequency definition of probability.

5.2 **Testing a coin:** Your friend decides to flip a coin repeatedly to analyze whether the probability of a head on each flip is 1/2. He flips the coin 10 times and observes a head 7 times. He concludes that the probability of a head for this coin is 7/10 = 0.70.

    a. Your friend claims that the coin is not balanced, since the probability is not 0.50. What's wrong with your friend's claim?

    b. If the probability of flipping a head is actually 1/2, what would you have to do to ensure that the cumulative proportion of heads falls very close to 1/2?

5.3 **Vegetarianism:** You randomly sample 10 people in your school, and none of them is a vegetarian. Does this mean that the probability of being a vegetarian for students at your school equals 0? Explain.

5.4 **Fear of flying?:** As of the year 2000, the Boeing 737 airplane had had 10 fatal accidents in 31 million flights.

    a. Can you consider this a long run or short run of trials? Explain.

    b. Estimate the probability of a fatal accident on a particular flight. (By contrast, for a trip by *auto* in a Western country, the probability of death in a 1000-mile trip is on the order of 50 times this probability.)

5.5 **Due for a basket:** True or false? If a pro basketball player misses 10 shots in a row, his chance of making a shot keeps going up until he makes one. Explain your answer.

5.6 **Random digits:** Consider a random number generator designed for equally likely outcomes. Which of the following is *not* correct, and why?

    a. For each random digit generated, each integer between 0 and 9 has probability 0.10 of being selected.

    b. If you generate 10 random digits, each integer between 0 and 9 must occur exactly once.

    c. If you generated a very large number of random digits, then each integer between 0 and 9 would occur close to 10% of the time.

    d. The cumulative proportion of times that a 0 is generated tends to get closer to 0.10 as the number of random digits generated gets larger and larger.

5.7 **Polls and sample size:** A pollster wants to estimate the proportion of Canadian adults who support the prime minister's performance on the job. He comments that by the law of large numbers, to ensure a sample survey's accuracy, he does not need to worry about the method for selecting the sample, only that the sample has a very large sample size. Do you agree with the pollster's comment? Explain.

5.8 **Heart transplant:** Before the first human heart transplant, Dr. Christiaan Barnard of South Africa was asked to assess the probability that the operation would be successful. Did he need to rely on the relative frequency definition or the subjective definition of probability? Explain.

5.9 **Life on other planets?:** Is there intelligent life on other planets in the universe? If you are asked to state the probability that there is, would you need to rely on the relative frequency or the subjective definition of probability? Explain.

5.10 **Applet for coin flipping:** Use the *simulating the probability of head with a fair coin* applet on the text CD or other software to illustrate the long-run definition of probability by simulating short-term and long-term results of flipping a balanced coin.

    a. Set the sample size to $n = 10$. Run the applet ten times, and record the cumulative proportion of heads for each of the 10 simulations.

    b. Now set the sample size $n = 100$. Run the applet 10 times, and record the ten cumulative proportions of heads for the separate sets. Do they vary much?

    c. Now set $n = 1000$. Run the applet 10 times, and record the ten cumulative proportions of heads. Do they vary more, or less, than the proportions in (b) based on $n = 100$?

    d. Summarize the effect of the number of trials n on the variability of the proportion. How does this reflect what's implied by the law of large numbers?

5.11 **Unannounced pop quiz:** A student's worst fear is going to class and having the teacher announce a pop quiz for which the student is completely unprepared. The quiz consists of 100 true-false questions. The student has no choice but to guess the answer randomly for all 100 questions.

    a. Simulate taking this quiz by random guessing. Number a sheet of paper 1 to 100 to represent the 100 questions. Write either a T (true) or F (false) for each question, by predicting what you think would happen if you repeatedly flipped a coin and let a tail represent a T guess and a head represent an F guess. (Don't actually flip a coin, but merely write down what you think a random series of guesses would look like.)

    b. How many questions would you expect to answer correctly simply by guessing?

    c. The table shows the 100 correct answers. The answers should be read across rows. How many questions did you answer correctly?

Pop Quiz Correct Answers

T	F	T	T	F	F	T	T	T	T	T	F	T	F	F	T	T	F	T	F
F	F	F	F	F	F	F	T	F	F	T	F	T	F	T	F	F	T	F	T
T	F	F	F	F	F	T	F	T	T	F	T	T	T	F	F	F	F	F	T
T	F	F	T	F	F	T	T	T	T	F	F	F	F	F	F	F	F	T	F
F	F	T	F	F	T	T	F	F	T	F	T	F	T	T	T	T	F	F	F

d. The above "answers" were actually randomly generated by the *simulating the probability of head with a fair coin* applet on the text CD. What percentage were true, and what percentage would you expect? Why are they not necessarily identical?

e. Are there groups of numbers within the sequence of 100 answers that appear nonrandom? For instance, what is the longest run of Ts or Fs? By comparison, which is the longest run of Ts or Fs within your sequence of 100 answers? (There is a tendency in guessing what randomness looks like to identify too few long runs in which the same outcome occurs several times in a row.)

5.12   Stock market randomness: An interview in an investment magazine (*In the Vanguard*, Autumn 2003) asked mathematician John Allen Paulos, "What common errors do investors make?" He answered, "People tend not to believe that markets move in random ways. Randomness is difficult to recognize. If you have people write down 100 Hs and Ts to simulate 100 flips of a coin, you will always be able to tell a sequence generated by a human from one generated by real coin flips. When humans make up the sequence, they don't put in enough consecutive Hs and Ts, and they don't make the lengths of those runs long enough or frequent enough. And that is one of the reasons people look at patterns in the stock market and ascribe significance to them."

a. Suppose that on each of the next 100 business days the stock market has a 1/2 chance of going up and a 1/2 chance of going down, and its behavior one day is independent of its behavior on another day. Use the *simulating the stock market* applet on the text CD or other software to simulate whether the market goes up or goes down for each of the next 100 days. What is the longest sequence of consecutive Hs (represented in the applet with a 1) or Ts (represented in the applet with a 0) that you observe?

b. Run the applet 9 more times, with 100 observations for each run, and each time record the longest sequence of consecutive Hs or Ts that you observe. For the 10 runs, summarize the proportion of times that the longest sequence was 1, 2, 3, 4, 5, 6, 7, 8, or more. (Your class may combine results to estimate this more precisely.)

c. Based on your findings, explain why if you are a serious investor you should not get too excited if sometime in the next few months you see the stock market go up for five days in a row or go down for five days in a row.

5.2 | How Can We Find Probabilities?

We've learned that probability enables us to quantify uncertainty and randomness. Now, let's explore some basic rules that can help us find probabilities.

Sample Spaces

The first step is to list all the possible outcomes. The set of possible outcomes for a random phenomenon is called the **sample space**.

> **Sample Space**
>
> For a random phenomenon, the **sample space** is the set of all possible outcomes.

For example, when you roll a die once, the sample space consists of the six possible outcomes, {1, 2, 3, 4, 5, 6}. When you flip a coin twice, the sample space consists of the four possible outcomes, {HH, HT, TH, TT}, where, for instance, TH represents a tail on the first flip and a head on the second flip.

Sample Space for a Multiple-Choice Pop Quiz

Picture the Scenario

The instructor in your statistics course decides to give a short unannounced pop quiz (not a favorite class activity for students) with three multiple-choice questions. Each question has five options, and the student's answer is either correct (C) or incorrect (I). If a student answered the first two questions correctly and the last question incorrectly, the student's outcome on the quiz can be symbolized by CCI.

Question to Explore

What is the sample space for the correctness of a student's answers on this pop quiz?

Think It Through

One technique for listing the outcomes in a sample space is to draw a **tree diagram**, with branches showing what can happen on different trials. For student performance on three questions, the tree has three sets of branches, as Figure 5.3 shows.

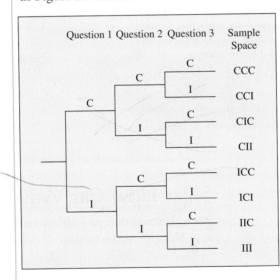

◀ **FIGURE 5.3: Tree Diagram for Student Performance on a Three-Question Pop Quiz.** Each path from the first set of 2 branches to the third set of 8 branches determines an outcome in the sample space. **Question:** How many possible outcomes would there be if the quiz had four questions?

From the tree diagram, a student's performance has 8 possible outcomes. They are

$$\{CCC, CCI, CIC, CII, ICC, ICI, IIC, III\}.$$

Insight

The number of branches doubles at each stage. There are 2 branches for question 1, $2 \times 2 = 4$ branches at question 2, and $2 \times 2 \times 2 = 8$ branches at question 3.

How many possible outcomes are in a sample space when there are repeated trials? To determine this, multiply together the number of possible outcomes for each trial. A pop quiz with 3 questions has $2 \times 2 \times 2 = 8$ possible outcomes denoting whether each answer is correct or incorrect. With four questions there are $2 \times 2 \times 2 \times 2 = 16$ possible outcomes.

What if we want to consider the actual responses made? With four multiple-choice questions and five possible responses on each, there are $5 \times 5 \times 5 \times 5 = 625$ possible response sequences. The tree diagram is ideal for visualizing a small number of outcomes. As the number of trials or the number of possible outcomes on each trial increases, it becomes impractical to construct a tree diagram. We'll learn about methods for finding probabilities without having to list entire sample spaces.

Events

We'll sometimes need to specify a particular group of the outcomes in a sample space, that is, a *subset* of the outcomes. An example is the subset of outcomes for which a student passes the pop quiz, by answering at least two of the three questions correctly. A subset of a sample space is called an **event**.

IN WORDS

An **event** corresponds to a particular outcome or a group of possible outcomes. An example is the event of passing the pop quiz.

Event

An **event** is a subset of the sample space.

Events are usually denoted by letters from the beginning of the alphabet, such as A and B, or by a letter or string of letters that describes the event. For a student taking the three-question pop quiz, some possible events are

A = student answers all 3 questions correctly = {CCC}

B = student passes (at least 2 correct) = {CCI, CIC, ICC, CCC}.

Finding Probabilities of Events

Each outcome in a sample space has a probability. So does each event. To find such probabilities, we list the sample space and specify assumptions that are plausible about its outcomes. Sometimes we can assume that the outcomes are equally likely. The probabilities for the outcomes in a sample space must follow two basic rules:

Probabilities for a Sample Space

■ The probability of each individual outcome is between 0 and 1.
■ The total of all the individual probabilities equals 1.

EXAMPLE 4

Assigning Subjects to Echinacea or Placebo for Treating Colds

Picture the Scenario

The University of Wisconsin is conducting a multicenter

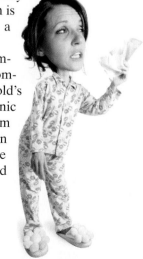

randomized experiment[1] to compare an herbal remedy (Echinacea) to a placebo for treating the common cold. The response variables are the cold's severity and its duration. Suppose a particular clinic in Madison, Wisconsin has four volunteers, of whom two are men (Jamal and Ken) and two are women (Linda and Mary). Two of these volunteers will be randomly chosen to receive the herbal remedy, and the other two will receive the placebo.

---

[1]See www.fammed.wisc.edu/research/news0503d.html

Questions to Explore

a. Identify the possible samples to receive the herbal remedy. For each possible sample, what is the probability that it is the one chosen?

b. What's the probability of the event that the sample chosen to receive the herbal remedy consists of one man and one woman?

Think It Through

a. The six possible samples to assign to the herbal remedy are {(Jamal, Ken), (Jamal, Linda), (Jamal, Mary), (Ken, Linda), (Ken, Mary), (Linda, Mary)}. This is the sample space for randomly choosing two of the four people. For a simple random sample, every sample is equally likely. Since there are 6 possible samples, each one has probability 1/6. These probabilities fall between 0 and 1, and their total equals 1, as is necessary for probabilities for a sample space.

b. The event in which the sample chosen has one man and one woman consists of the outcomes {(Jamal, Linda), (Jamal, Mary), (Ken, Linda), (Ken, Mary)}. These are the possible pairings of one man with one woman. Each outcome has probability 1/6, so the probability of this event is 4(1/6) = 4/6 = 2/3.

Insight

When each outcome is equally likely, the probability of a single outcome is simply 1/(number of possible outcomes), such as 1/6 in (a). The probability of an event is then (number of outcomes in the event)/(number of possible outcomes), such as 4/6 in (b).

TO PRACTICE THIS CONCEPT, TRY EXERCISE 5.19

This example shows that to find the probability for an event, we can (1) find the probability for each outcome in the sample space, and (2) add the probabilities of each outcome that the event contains.

Probability of an Event

- The probability of an event A, denoted by P(A), is obtained by adding the probabilities of the individual outcomes in the event.

- *When all the possible outcomes are equally likely,*

$$P(A) = \frac{\text{number of outcomes in event A}}{\text{number of outcomes in the sample space}}.$$

In Example 4, to find the probability of choosing one man and one woman, we first determined the probability of each of the six possible outcomes. Because the probability is the same for each, 1/6, and because the event contains four of those outcomes, the probability is 1/6 added 4 times. The answer equals 4/6, the number of outcomes in the event divided by the number of outcomes in the sample space.

IN PRACTICE: Equally likely outcomes are unusual

Except for simplistic situations such as simple random sampling or flipping balanced coins or rolling fair dice, different outcomes are not usually equally likely. Then, probabilities are often estimated, using sample proportions from simulations or from large samples of data.

EXAMPLE 5

What Are the Chances of a Taxpayer Being Audited?

Picture the Scenario

April 15 is tax day in the United States—the deadline for filing federal income tax forms. The main factor in the amount of tax owed is a taxpayer's income level. Each year, the IRS audits a sample of tax forms to verify their accuracy. Table 5.2 is a contingency table that cross-tabulates the 80.2 million long-form federal returns received in 2002 by the taxpayer's income level and whether the tax form was audited.

Recall

Section 3.1 introduced **contingency tables** to summarize the relationship between two categorical variables. ■

TABLE 5.2: Contingency Table Cross-Tabulating Tax Forms by Income Level and Whether Audited

There were 80.2 million forms, and for simplicity the frequencies are reported in thousands and rounded. For example, 90 represents 90,000 tax forms that reported income under $25,000 and were audited.

Income Level	Whether Audited		Total
	Yes	No	
Under $25,000	90	14010	**14100**
$25,000–$49,999	71	30629	**30700**
$50,000–$99,999	69	24631	**24700**
$100,000 or more	80	10620	**10700**
Total	310	79890	**80200**

Source: Statistical Abstract of the United States: 2003.

Questions to Explore

a. What is the sample space?

b. For a randomly selected taxpayer in 2002, what is the probability of (i) an audit, (ii) an income of $100,000 or more?

Think It Through

a. The sample space is the set of possible outcomes. These are the 8 cells for the income and whether audited combinations in this table, such as (Under $25,000, Yes), (Under $25,000, No), ($25,000–$49,999, Yes), and so forth.

b. If a taxpayer was randomly selected, from Table 5.2,

(i) The probability of an audit was 310/80200 = 0.004.

(ii) The probability of an income of $100,000 or more was 10700/80200 = 0.133.

Insight

Even though audits are not welcome news to a taxpayer, in 2002 only about 0.4% of taxpayers were audited. Is this typical? Inspecting similar data for earlier years shows that this percentage decreased considerably in recent years. For instance, in 1996 the percentage audited was 1.2%, three times the value in 2002.

TO PRACTICE THIS CONCEPT, TRY EXERCISE 5.23 PARTS (A) AND (B)

Basic Rules for Finding Probabilities about a Pair of Events

Some events are expressed as the outcomes that (a) are *not* in some other event, or (b) are in one event *and* in another event, or (c) are in one event *or* in another event. We'll next learn how to calculate probabilities for these three cases.

The complement of an event   For an event A, the rest of the sample space that is *not* in that event is called the **complement** of A.

Complement of an Event

The **complement** of an event A consists of all outcomes in the sample space that are *not* in A. It is denoted by A^c. The probabilities of A and of A^c add to 1, so

$$P(A^c) = 1 - P(A).$$

In Example 5, for instance, the event of having an income *less than* $100,000 is the complement of the event of having an income of $100,000 or more. Because the probability that a randomly selected taxpayer had an income of $100,000 or more is 0.13, the probability of income less than $100,000 is $1 - 0.13 = 0.87$.

Figure 5.4 illustrates the complement of an event. The box represents the entire sample space. The event A is the oval in the box. The complement of A, which is shaded, is everything else in the box that is not in A. Together, A and A^c cover the sample space. Because an event and its complement contain all possible outcomes, their total probability is 1, and the probability of one of them is 1 minus the probability of the other. A diagram like Figure 5.4 that uses areas inside a box to represent events is called a **Venn diagram**.

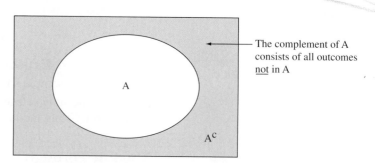

▲ **FIGURE 5.4: Venn Diagram Illustrating an Event A and its Complement A^c.**
Question: Can you sketch a Venn diagram of two events A and B such that they share some common outcomes, but some outcomes are not in A or in B?

To find the probability of an event, it's sometimes easier to find the probability of its complement and then subtract that probability from 1. An example is when we need to find the probability that *at least one* of several events will occur. It's usually easier to find the probability of its complement, that *none* of these events will occur.

EXAMPLE 6

Should We Be Surprised by Finding No Women on a Jury?

Picture the Scenario

A jury of 12 people is chosen for a trial. The defense attorney claims it must have been chosen in a biased manner, because 50% of the city's adult residents are female yet the jury contains no women.

Questions to Explore

If the jury were randomly chosen from the population, what is the probability the jury would have (a) no females, (b) at least one female?

Think It Through

Let's use a symbol with 12 letters, with F for female and M for male, to represent a possible jury selection. For instance, MFMMMMMMMMMM denotes the jury in which only the second person selected is female. The number of possible outcomes is $2 \times 2 \times 2 \times \cdots \times 2$, that is, 2 multiplied 12 times, which is $2^{12} = 4096$. This is a case in which listing the entire sample space is not practical. Since the population is 50% male and 50% female, these 4096 possible outcomes are equally likely.

a. Only 1 of the 4096 possible outcomes corresponds to a no-female jury, namely, MMMMMMMMMMMM. So the probability of this outcome is 1/4096, or 0.00024. This is extremely unlikely, if a jury is truly chosen by random sampling.

b. As noted previously, it would be tedious to list all possible outcomes in which at least one female is on the jury. But this is not necessary. The event that the jury contains *at least one* female is the complement of the event that it contains *no* females. Thus,

$$P(\text{at least one female}) = 1 - P(\text{no females}) = 1 - 0.00024 = 0.99976.$$

Insight

You might instead let the sample space be the possible values for the *number* of females on the jury, namely 0, 1, 2, ..., 12. But these outcomes are not equally likely. For instance, only one of the 4096 possible samples has 0 females, but 12 of them have 1 female: The female could be the first person chosen, or the second (as in MFMMMMMMMMMM), or the third, and so on. Chapter 6 will show a formula (the "binomial") that gives probabilities for this alternative sample space.

TO PRACTICE THIS CONCEPT, TRY EXERCISE 5.16

Disjoint events Events that do not share any outcomes in common are said to be **disjoint**.

Did You Know?

Disjoint events are also referred to as **mutually exclusive** events. We will use only the term "disjoint." ■

Disjoint Events

Two events, A and B, are **disjoint** if they do not have any common outcomes.

Example 3 discussed a pop quiz with three questions. The event that the student answers exactly one question correctly is {CII, ICI, IIC}. The event that the student answers exactly two questions correctly is {CCI, CIC, ICC}. These two events have no outcomes in common, so they are disjoint. In a Venn diagram, they have no overlap. See Figure 5.5. By contrast, neither is disjoint from the event that the student answers the first question correctly, which is {CCC, CCI, CIC, CII}, because this event has outcomes in common with each of the other two events.

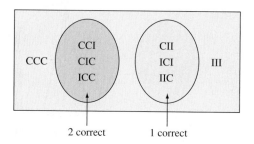

▲ **FIGURE 5.5: Venn Diagram Illustrating Disjoint Events.** The event of a student answering one question correctly is disjoint from the event of answering two questions correctly. **Question:** Identify on this figure the event that the student answers the first question correctly. Is this event disjoint from either of the two events identified in the Venn diagram?

Consider an event A and its complement, A^c. They share no common outcomes, so they are disjoint events.

Intersection and union of events Some events are composed from other events. For instance, for two events A and B, the event that *both* occur is an event. Called the **intersection** of A and B, it consists of the outcomes that are in both A and B. By contrast, the event that the outcome is in A *or* B or both is the **union** of A and B. It is a larger set, containing the intersection as well as outcomes that are in A but not in B and outcomes that are in B but not in A. Figure 5.6 shows Venn diagrams illustrating the intersection and union of two events.

IN WORDS

Intersection means A and B (the "overlap" of the events).
Union means A or B or both.

Intersection and Union of Two Events

The **intersection** of A and B consists of outcomes that are in both A *and* B

The **union** of A and B consists of outcomes that are in A *or* B. In probability, "A *or* B" denotes that A occurs or B occurs or both occur.

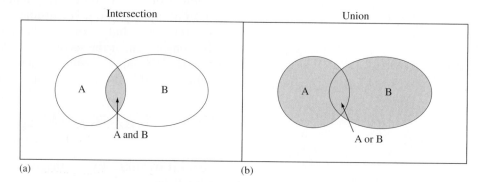

▲ **FIGURE 5.6: The Intersection and the Union of Two Events.** Intersection means A occurs *and* B occurs, denoted "A and B." The intersection consists of the shaded "overlap" part in Figure 5.6(a). Union means A occurs *or* B occurs *or* both occur, denoted "A or B." It consists of all the shaded parts in Figure 5.6(b). **Question:** How could you find P(A or B) if you know P(A), P(B), and P(A and B)?

For instance, for the three-question pop quiz, consider the events:

A = student answers first question correctly = {CCC, CCI, CIC, CII}
B = student answers two questions correctly = {CCI, CIC, ICC}.

Then the intersection, A *and* B, is {CCI, CIC}, the two outcomes common to A and B. The union, A *or* B, is {CCC, CCI, CIC, CII, ICC}, the outcomes that are in A or in B or in both A and B.

How do we find probabilities of intersections and unions of events? Once we identify the possible outcomes, we can use their probabilities. For instance, for Table 5.2 (reproduced in the margin) for 80.2 million tax forms, let

A denote {audited = yes}
B denote {income ≥ $100,000}

The intersection A and B is the event that a taxpayer is audited *and* has income ≥$100,000. This probability is simply the proportion for the cell in which these two events occurred, namely P(A and B) = 80/80200 = 0.001. Another way to find probabilities of intersections and unions uses the probabilities of the events themselves, as we'll see in the next two subsections.

| | Audited | |
Income	Yes	No
Under $25,000	90	14010
$25,000–$49,999	71	30629
$50,000–$99,999	69	24631
≥ $100,000	80	10620

How Can We Find the Probability that Event A or Event B Occurs?

Since the union A or B contains outcomes from A and from B, we can add P(A) to P(B). However, this sum counts the outcomes that are in *both* A and B (their intersection) twice. See Figure 5.7. We need to subtract the probability of the intersection from P(A) + P(B) so that it is only counted once. If there is no overlap, that is, if the events are disjoint, no outcomes are common to A and B. Then we can simply add the probabilities.

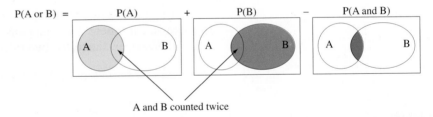

P(A or B) =   P(A)   +   P(B)   −   P(A and B)

A and B counted twice

▲ **FIGURE 5.7: The Probability of the Union, Outcomes in A or B or Both.**
Add P(A) to P(B) and subtract P(A and B) to adjust for outcomes counted twice.
Question: When does P(A or B) = P(A) + P(B)?

Addition Rule: Probability of the Union of Two Events

For the **union** of two events, P(A or B) = P(A) + P(B) − P(A and B).

If the events are **disjoint**, then P(A and B) = 0, so P(A or B) = P(A) + P(B).

For example, consider a family with two children. The sample space possibilities for the genders of the two children are {FF, FM, MF, MM}, where the first letter in a symbol is the first child's gender and the second letter is the second child's gender (F = female, M = male). Let A = {first child a girl} and B = {second child a girl}. Then, assuming the four possible outcomes in the sample space are equally likely, P(A) = P({FF, FM}) = 0.50, P(B) = P({FF, MF}) = 0.50, and P(A and B) = P({FF}) = 0.25. The event A or B is the event that the first child is a girl, or the second child is a girl, or both, that is, *at least* one child is a girl. Its probability is

P(A or B) = P(A) + P(B) − P(A and B) = 0.50 + 0.50 − 0.25 = 0.75.

How Can We Find the Probability that Events A and B Both Occur?

The probability of the intersection of events A and B has a formula to be introduced in Section 5.3. In the special case discussed next, it equals $P(A) \times P(B)$.

Consider a basketball player who shoots two free throws. Let M1 denote making free throw 1, and let M2 denote making free throw 2. For any given free throw, suppose he has an 80% chance of making it, so $P(M1) = P(M2) = 0.80$. What is the probability of M1 and M2, making free throw 1 *and* free throw 2? In the long run of many pairs of free throws, suppose that for 80% of the cases in which he made the first three throw, he also made the second. Then the percentage of times he made both is the 80% of the 80% of times he made the first one, for a probability of $0.80 \times 0.80 = 0.64$ (that is, 64%).

This multiplication calculation is valid only under an assumption, **independent trials**: Whether a player makes the second free throw is independent of whether he makes the first. The chance of making the second is 80%, regardless of whether or not he made the first. (For pro basketball players, independence is approximately true: Whether a player makes his first shot has almost no influence on whether he makes the second one. See Exercise 5.32 for some data.)

To find the probability of the intersection of two events, we can multiply probabilities whenever the events are independent. We'll see a formal definition of independent events in the next section, but it essentially means that whether one event occurs does not affect the probability that the other event occurs.

Multiplication Rule: Probability of the Intersection of Independent Events

For the **intersection** of two **independent** events, A and B,

$$P(A \text{ and } B) = P(A) \times P(B).$$

The paradigm for independent events is repeatedly flipping a coin or rolling a die, where what happens on one trial does not affect what happens on another. For instance, for two rolls of a die,

$$P(6 \text{ on roll 1 } and \text{ 6 on roll 2}) = P(6 \text{ on roll 1}) \times P(6 \text{ on roll 2}) = \frac{1}{6} \times \frac{1}{6} = \frac{1}{36}.$$

This multiplication rule extends to more than two independent events.

EXAMPLE 7

Guessing yet Passing a Pop Quiz

Picture the Scenario

For a three-question multiple-choice pop quiz, a student is totally unprepared and randomly guesses the answer to each question. If each question has five options, then the probability of selecting the correct answer for any given question is 1/5, or 0.20. With guessing, the response on one question is not influenced by the response on another question. Thus, whether one question is answered correctly is independent of whether or not another question is answered correctly.

Questions to Explore

a. Find the probabilities of the possible student outcomes for the quiz, in terms of whether each response is correct (C) or incorrect (I).

b. Find the probability the student passes, answering *at least two* questions correctly.

Think It Through

a. For each question P(C) = 0.20 and P(I) = 1 − 0.20 = 0.80. The probability that the student answers all three questions correctly is

P(CCC) = P(C) × P(C) × P(C) = 0.20 × 0.20 × 0.20 = 0.008.

This would be unusual. Similarly, the probability of answering the first two questions correctly and the third question incorrectly is

P(CCI) = P(C) × P(C) × P(I) = 0.20 × 0.20 × 0.80 = 0.032.

This is the same as P(CIC) and P(ICC), the other possible ways of getting two correct. Figure 5.8 is a tree diagram showing how to multiply probabilities to find the probabilities for all eight possible outcomes.

▲ **FIGURE 5.8: Tree Diagram for a Student Guessing on a Three-Question Pop Quiz.** Each path from the first set of branches to the third set determines one sample space outcome. Multiplication of the probabilities along that path gives its probability, when trials are independent. **Question:** Would you expect trials to be independent if a student is *not* merely guessing on every question? Why or why not?

b. The probability of *at least* two correct responses is

P(CCC) + P(CCI) + P(CIC) + P(ICC) = 0.008 + 3(.032) = 0.104.

In summary, there is only about a 10% chance of passing when a student randomly guesses the answers.

Insight

As a check, you can see that the probabilities of the eight possible outcomes sum to 1.0. The probabilities indicate that it is in a student's best interests not to have to rely on random guessing.

TO PRACTICE THIS CONCEPT, TRY EXERCISE 5.15

Events Often Are Not Independent

In practice, events need not be independent. For instance, on a quiz given earlier in the course that had only two questions, the instructor found the following proportions for the actual responses of her students (again, I = incorrect, C = correct):

Outcome:	II	IC	CI	CC
Probability:	0.26	0.11	0.05	0.58

Recall

From Section 3.1, in contingency table form the proportions can be expressed as

1st Question	2nd Question	
	C	I
C	0.58	0.05
I	0.11	0.26

A and B

Let A denote {first question correct} and let B denote {second question correct}. Based on these probabilities,

$$P(A) = P(\{CI, CC\}) = 0.05 + 0.58 = 0.63$$
$$P(B) = P(\{IC, CC\}) = 0.11 + 0.58 = 0.69$$

and

$$P(A \text{ and } B) = P(\{CC\}) = 0.58.$$

If A and B were independent, then

$$P(A \text{ and } B) = P(A) \times P(B) = 0.63 \times 0.69 = 0.43.$$

Since P(A and B) actually equaled 0.58, A and B were not independent.

Responses to different questions on a quiz are typically not independent. Most students do not guess randomly. Students who get the first question correct may have studied more than students who do not get the first question correct, and thus they may also be more likely to get the second question correct.

IN PRACTICE: Make sure that assuming independence is realistic

Don't assume that events are independent unless you have given this assumption careful thought and it seems plausible. In Section 5.3, you will learn more about how to find probabilities when events are not independent.

Probability Rules

In this section we have developed several rules for finding probabilities. Let's summarize them.

Rules for Finding Probabilities

- The probability of each individual outcome is between 0 and 1, and the total of all the individual probabilities equals 1. The **probability of an event** is the sum of the probabilities of the individual outcomes in that event.

- For an event A and its **complement** A^c (not in A), $P(A^c) = 1 - P(A)$.

- The **union** of two events (that is, A occurs or B occurs or both) has

$$P(A \text{ or } B) = P(A) + P(B) - P(A \text{ and } B).$$

- When A and B are **independent**, the **intersection** of two events has
$$P(A \text{ and } B) = P(A) \times P(B).$$

- Two events A and B are **disjoint** when they have no common elements. Then

$$P(A \text{ and } B) = 0, \text{ and thus } P(A \text{ or } B) = P(A) + P(B).$$

5.2   Practicing the Basics

5.13 Election study: An election study records for each sampled subject either D = voted for Democratic candidate, R = voted for Republican candidate, I = voted for an independent candidate, or N = did not vote.
   a. Show how to summarize possible outcomes for two elections with a tree diagram. For example, one possible outcome is DR, voting for the Democrat in the first election and for the Republican in the second election.

   b. If the study considers three elections, how many possible outcomes are in the sample space?

5.14 Random digit: A single random digit is selected using software or a random number table.

a. State the sample space for the possible outcomes.

b. State the probability for each possible outcome, based on what you know about the way random numbers are generated.

c. Each outcome in a sample space must have probability between 0 and 1, and the total of the probabilities must equal 1. Show that your assignment of probabilities in (b) satisfies this.

5.15 Pop quiz: A teacher gives a four-question unannounced true-false pop quiz, with two possible answers to each
TRY question.

a. Use a tree diagram to show the possible response patterns, in terms of whether any given response is correct or incorrect. How many outcomes are in the sample space?

b. An unprepared student guesses all the answers randomly. Find the probabilities of the possible outcomes on the tree diagram.

c. Refer to (b). Using the tree diagram, evaluate the probability of passing the quiz, which the teacher defines as answering *at least* three questions correctly.

5.16 More true-false questions: Your teacher gives a true-
TRY false pop quiz with 10 questions.

a. Show that the number of possible outcomes for the sample space of possible sequences of 10 answers is 1024.

b. With random guessing, show that the probability of getting *at least* one question wrong is 0.999.

5.17 Rain tomorrow?: Tomorrow, it might rain sometime, or it might not rain at all. Since the sample space has two possible outcomes, each must have probability 1/2. True or false? Explain.

5.18 Two girls: A couple plans to have two children. Each child is equally likely to be a girl or boy, with gender independent of that of the other child.

a. Construct a sample space for the genders of the two children.

b. Find the probability that both children are girls.

c. Answer (b) if in reality for a given child the chance of a girl is 0.49.

5.19 All girls in a family: A family has four children, all girls.
TRY Is this unusual?

a. Construct a sample space for the possible genders of four children. (*Hint:* The outcome FMFF represents the case where only the second child born is a male.)

b. If each child is equally likely to be a girl or a boy, find the probability that all four children are female.

c. What other assumption is necessary for the calculation in (b) to be valid?

5.20 Wrong sample space: Refer to the previous exercise. What's wrong with the following logic for part (b)? A family with 4 children can have 0, 1, 2, 3, or 4 girls. Since there are five possibilities, the probability of 4 girls is 1/5.

5.21 Insurance: Every year the insurance industry spends considerable resources assessing risk probabilities.

To accumulate a risk of about one in a million of death, you can drive 100 miles, take a cross country plane flight, work as a police officer for 10 hours, work in a coal mine for 12 hours, smoke two cigarettes, be a nonsmoker but live with a smoker for two weeks, or drink 70 pints of beer in a year (Wilson and Crouch, 2001, pp. 208–209). Show that a risk of about one in a million of death is also approximately the probability of flipping 20 heads in a row with a balanced coin.

5.22 True-blooded Republicans: Of those who voted in the 2000 and 2004 U.S. Presidential elections, about 48% voted Republican in 2000 and about 51% voted Republican in 2004. True or false: Since $0.48 \times 0.51 = 0.24$, about 24% voted for the Republican (George W. Bush) in both elections. If true, explain why. If false, explain the flaw in the logic.

5.23 Seat belt use and auto accidents: Based on records of
TRY automobile accidents in a recent year, the Department of Highway Safety and Motor Vehicles in Florida reported the counts who survived (S) and died (D), according to whether they wore a seat belt (Y = yes, N = no). The data are presented in the contingency table shown.

Outcome of auto accident by whether subject wore seat belt

Wore Seat Belt	Survived (S)	Died (D)	Total
Yes (Y)	412,368	510	412,878
No (N)	162,527	1601	164,128
Total	574,895	2111	577,006

a. What is the sample space of possible outcomes for a randomly selected individual involved in an auto accident? Use a tree diagram to illustrate the possible outcomes. (*Hint:* One possible outcome is YS).

b. Using these data, estimate (i) P(D), (ii) P(N).

c. Estimate the probability that an individual did not wear a seat belt and died.

d. Based on (b), what would the answer to (c) have been if the events N and D were independent? So, are N and D independent, and if not, what does that mean in the context of these data?

5.24 Catalog sales: You are the marketing director for a museum that raises money by selling gift items from a mail-order catalog. For each catalog sent to a potential customer, the customer's entry in the data file is Y if they ordered something and N if they did not (Y = yes, N = no). After you have mailed the fall and the winter catalogs, you estimate the probabilities of the buying patterns based on those who received the catalog as follows:

Outcome (fall, winter):	YY	YN	NY	NN
Probability:	0.30	0.10	0.05	0.55

a. Let F denote buying from the fall catalog and W denote buying from the winter catalog. Find P(F) and P(W).

b. Explain what the event "F and W" means, and find P(F and W).

c. Are F and W independent events? Explain why you would not normally expect customer choices to be independent.

5.25 Arts and crafts sales: A local downtown arts and crafts shop found from past observation that 20% of the people who enter the shop actually buy something. Three potential customers enter the shop.

a. How many outcomes are possible for whether the clerk makes a sale to each customer? Construct a tree diagram to show the possible outcomes. (Let Y = sale, N = no sale.)

b. Find the probability of at least one sale to the three customers.

c. What did your calculations assume in (b)? Describe a situation in which that assumption would be unrealistic.

5.3 | Conditional Probability: What's the Probability of A, Given B?

Many employers require potential employees to take a diagnostic test for drug use. The diagnostic test has two categorical variables of interest: (1) whether the person has recently used drugs (yes or no) and (2) whether the diagnostic test says the person has used them (yes or no). Suppose the diagnostic test says the person has recently used drugs. What's the probability that the person truly did use drugs?

This section introduces **conditional probability**. This deals with finding the probability of an event when you know that the outcome was in some particular part of the sample space. Most commonly, it is used to find a probability about a category for one variable (for instance, a person being a drug user), when we know the outcome on another variable (for instance, a test result that indicates drug use).

Finding the Conditional Probability of an Event

Example 5 showed a contingency table on income and whether a taxpayer is audited by the Internal Revenue Service. The table is shown again in the margin. We found that the probability that a randomly selected taxpayer was audited equaled $310/80200 = 0.004$. Were the chances higher if a taxpayer was at the highest income level? From the margin table, the number having income ≥$100,000 was 10700. Of them, 80 were audited, for a probability of $80/10700 = 0.007$. This is higher than 0.004, but still less than 1%.

In practice, tables often provide probabilities rather than counts for the outcomes in the sample space. Table 5.3 shows probabilities for the cells, based on the cell frequencies in the contingency table, for the eight possible outcomes. For example, the probability of income ≥$100,000 *and* being audited was $80/80200 = 0.0010$. The probability of income ≥$100,000 was $10700/80200 = 0.1334$. So, of those at the highest income category, the proportion $0.0010/0.1334 = 0.007$ were audited. This is the same answer we obtained above using the cell frequencies.

	Audited		
Income	Yes	No	Total
<$25,000	90	14010	14100
$25,000–49,000	71	30629	30700
$50,000–99,000	69	24631	24700
≥$100,000	80	10620	10700
Total	**310**	**79890**	**80200**

TABLE 5.3: Probabilities of Taxpayers at the Eight Possible Combinations of Income and Whether Audited

Each frequency in Table 5.2 was divided by 80200 to obtain the cell probabilities shown here, such as $90/80200 = 0.0011$.

Income Level	Whether Audited		Total
	Yes	No	
	These 8 probabilities sum to 1.0		
Under $25,000	0.0011	0.1747	0.1758
$25,000–$49,999	0.0009	0.3819	0.3828
$50,000–$99,999	0.0009	0.3071	0.3080
≥$100,000	0.0010	0.1324	0.1334
Total	**0.0039**	**0.9961**	**1.0000**

Let event A denote {audited = yes} and let event B denote {income ≥ $100,000}. Of the highest income group, also being audited is the intersection event A and B. So, given B, the probability of A is the proportion of A and B cases out of the B cases. This is P(A and B)/P(B). This ratio is the **conditional probability** of the event A, given the event B.

IN WORDS

P(A | B) is read as "the probability of event A, given event B." The vertical slash represents the word "given". Of the times that B occurs, P(A | B) is the proportion of times that A also occurs.

Conditional Probability

For events A and B, the **conditional probability** of event A, given that event B has occurred, is

$$P(A \mid B) = \frac{P(A \text{ and } B)}{P(B)}.$$

From Table 5.3 with events A (audited = yes) and B (income ≥ $100,000),

$$P(A \mid B) = \frac{P(A \text{ and } B)}{P(B)} = \frac{0.0010}{0.1334} = 0.007.$$

Given that a taxpayer has income ≥ $100,000, the chances of being audited are 0.007.

In Section 3.1, in learning about contingency tables we saw that we could find **conditional proportions** for a categorical variable, at any particular category of a second categorical variable. These enable us to study how the outcome on a response variable depends on the outcome on an explanatory variable. The conditional probabilities just found are merely conditional proportions. They refer to the population of taxpayers, treating audit status as a response variable and income level as an explanatory variable.

We could find similar conditional probabilities on audit status at each given level of income. We'd then get the results shown in Table 5.4. Using the cell probabilities in Table 5.3 (which refer to intersections of income events and audit events, as shown in the margin table), we get each conditional probability by dividing a cell probability for a particular audit status by the row total that is the probability of income at that level. In each row of Table 5.4, the conditional probabilities sum to 1.0.

	Audited	
Income	Yes	No
Under $25,000	.0011	.1747
$25,000–$49,999	.0009	.3819
$50,000–$99,999	.0009	.3071
≥$100,000	.0010	.1324

The total of these cell probabilities over the 8 cells is 1.0.

TABLE 5.4: Conditional Probabilities on Whether Audited, Given the Income Level

Each cell probability in Table 5.3 was divided by the row marginal total probability to obtain the conditional probabilities shown here, such as 0.0010/(0.0010 + 0.1324) = 0.0010/0.1334 = 0.007.

	Whether Audited		
Income Level	Yes	No	Total
Under $25,000	0.006	0.994	**1.000**
$25,000–$49,999	0.002	0.998	**1.000**
$50,000–$99,999	0.003	0.997	**1.000**
≥$100,000	0.007 ← 0.007 = 0.0010/0.1334	0.993	**1.000**
	from Table 5.3		

Figure 5.9 is a graphical illustration of the definition of conditional probability. "Given event B" means that we restrict our attention to the outcomes in that event. This is the set of outcomes in the denominator. The proportion of those cases in which A occurred are those outcomes that are in event A as well as B. So the intersection of A and B is in the numerator.

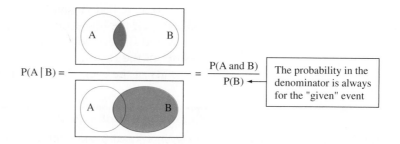

▲ **FIGURE 5.9: Venn Diagram of Conditional Probability of Event A, Given Event B.** Of the cases in which B occurred, P(A|B) is the proportion in which A also occurred. **Questions:** Sketch a representation of P(B|A). Is P(A|B) necessarily equal to P(B|A)?

$$P(A \mid B) = \frac{}{} = \frac{P(A \text{ and } B)}{P(B)}$$

The probability in the denominator is always for the "given" event

EXAMPLE 8

Evaluating the Triple Blood Test for Down Syndrome

Picture the Scenario

Diagnostic testing is used to detect many medical conditions. Some of these tests use blood samples, such as the Prostate-Specific Antigen (PSA) test for prostate cancer, the Enzyme-linked Immunosorbent Assay (ELISA) screening test for HIV, and selected tests for pregnancy. Some tests use X-rays or other imaging devices such as the mammogram for diagnosing breast cancer and the Magnetic Resonance Imaging (MRI) body scan. Diagnostic testing is also used for drug detection of employees in a workplace or athletes in a competition, often with urine samples.

A diagnostic test for a condition is said to be **positive** if it states that the condition is present and **negative** if it states that the condition is absent. How accurate are diagnostic tests? One way to assess accuracy is to measure the probabilities of the two types of possible error:

False positive: Test states the condition is present, but it is actually absent.

False negative: Test states the condition is absent, but it is actually present.

The Triple Blood Test screens pregnant women for the genetic disorder, Down syndrome. This syndrome, which occurs in about 1 in 800 live births, arises from an error in cell division that results in a fetus having an extra copy of chromosome 21. It is the most common genetic cause of mental impairment. The chance of having a baby with Down syndrome increases after a woman is 35 years old.

A study[2] of 5282 women aged 35 or over analyzed the Triple Blood Test to test its accuracy. It was reported that of the 5282 women, "48 of the 54 cases of Down syndrome would have been identified using the test and 25 percent of the unaffected pregnancies would have been identified as being at high risk for Down syndrome (these are false positives)."

Questions to Explore

a. Construct the contingency table that shows the counts for the possible outcomes of the blood test and whether the fetus has Down syndrome.

b. Assuming the sample is representative of the population, estimate the probability of a positive test for a randomly chosen pregnant woman 35 years or older.

[2]J. Haddow et al., *New England Journal of Medicine*, vol. 330, pp. 1114–1118, 1994.

c. Given that the diagnostic test result is positive, estimate the probability that Down syndrome truly is present.

Think It Through

a. We'll use the following notation for the possible outcomes of the two variables:

Down syndrome status: D = Down syndrome present, D^c = unaffected

Blood test result: POS = positive, NEG = negative.

Table 5.5 shows the four possible combinations of outcomes. From the article quote, there were 54 cases of Down syndrome. This is the first row total. Of them, 48 tested positive, so $54 - 48 = 6$ tested negative. These are the counts in the first row. There were 54 Down cases out of $n = 5282$, so $5282 - 54 = 5228$ cases were unaffected, event D^c. That's the second row total. Now, 25% of those 5228, or $0.25 \times 5228 = 1307$, would have a positive test. The remaining $5228 - 1307 = 3921$ would have a negative test. These are the counts for the two cells in the second row.

TABLE 5.5: Contingency Table for Triple Blood Test of Down Syndrome

	Blood Test Result		
Down Syndrome Status	POS	NEG	Total
D (Down)	48	6	**54**
D^c (unaffected)	1307	3921	**5228**
Total	**1355**	**3927**	**5282**

b. From Table 5.5, the estimated probability of a positive test is $P(\text{POS}) = 1355/5282 = 0.257$.

c. The probability of Down syndrome, given that the test is positive, is the conditional probability, $P(D|\text{POS})$. Conditioning on a positive test means we consider only the cases in the first column of Table 5.5. Of the 1355 who tested positive, 48 cases actually had Down syndrome, so $P(D|\text{POS}) = 48/1355 = 0.035$. Let's see how to get this from the definition of conditional probability,

$$P(D|\text{POS}) = \frac{P(D \text{ and } \text{POS})}{P(\text{POS})}.$$

Since $P(\text{POS}) = 0.257$ from part (b) and $P(D \text{ and } \text{POS}) = 48/5282 = 0.0091$, we estimate $P(D|\text{POS}) = 0.0091/0.257 = 0.035$. In summary, of the women who tested positive, fewer than 4% actually had fetuses with Down syndrome. This is somewhat comforting news to a woman who has a positive test result.

Insight

So why should a woman undergo this test, since most positives are false positives? From Table 5.5, $P(D) = 54/5282 = 0.0102$, so we estimate about a 1% chance of Down syndrome for women aged 35 or over. Also from Table 5.5, $P(D|\text{NEG}) = 6/3927 = 0.0015$, a bit more than 1 in a thousand. A woman can have much less worry about Down syndrome if she has a negative test result, because the chance of Down is then a bit more than 1 in a thousand, compared to 1 in a hundred overall.

TO PRACTICE THIS CONCEPT, TRY EXERCISES 5.32 AND 5.33

IN PRACTICE: Conditional probabilities in the media

When you read or hear a news report that uses a probability statement, be careful to distinguish whether it is reporting a conditional probability. Most statements are conditional on some event, and must be interpreted in that context. For instance, probabilities reported by opinion polls are often conditional on a person's gender, or race, or age group.

Multiplication Rule for Finding P(A and B)

From Section 5.2, when A and B are independent events, $P(A \text{ and } B) = P(A) \times P(B)$. The definition of conditional probability provides a more general formula for P(A and B) that holds regardless of whether A and B are independent. We can rewrite the definition $P(A \mid B) = P(A \text{ and } B)/P(B)$, multiplying both sides of the formula by P(B), as

$$P(A \text{ and } B) = P(A \mid B) \times P(B).$$

Multiplication Rule for Evaluating P(A and B)

For events A and B, the probability that A and B both occur equals

$$P(A \text{ and } B) = P(A \mid B) \times P(B).$$

Applying the conditional probability formula to $P(B \mid A)$, we also see that

$$P(A \text{ and } B) = P(B \mid A) \times P(A).$$

EXAMPLE 9

How Likely Is a Double Fault in Tennis?

Picture the Scenario:

In a tennis match, on a given point, the player who is serving has two chances to hit the ball in play. The ball must fall in the correct marked box area on the opposite side of the net. A serve that misses that box is called a *fault*. Most players hit the first serve very hard, resulting in a fair chance of making a fault. If they do make a fault, they hit the second serve less hard and with some spin, making it more likely to be successful. Otherwise, with two misses—a *"double fault"*—they lose the point.

Question to Explore

The 2004 men's champion in the Wimbledon tennis tournament was Roger Federer of Switzerland. During that tournament he made 64% of his first serves. So, he faulted on the first serve 36% of the time $(100 - 64 = 36)$. Given that he made a fault with his first serve, he made a fault on his second serve only 6% of the time. Assuming these are typical of his serving performance, when he serves, what is the probability that he makes a double fault?

F2(.06)

F1(.36)

(.94)

Make Second Serve

(.64)

Make First Serve

Think It Through

Let F1 be the event that Federer makes a fault with the first serve and let F2 be the event that he makes a fault with the second serve. From the information above, we know $P(F1) = 0.36$ and $P(F2 \mid F1) = 0.06$, as shown in the margin figure. The event that Federer makes a double fault is "F1 and F2". From the multiplication rule, its probability is

$$P(F1 \text{ and } F2) = P(F2 \mid F1) \times P(F1) = 0.06 \times 0.36 = 0.02.$$

Insight

Federer makes a fault on his first serve 36% of the time, and in 6% of those cases he makes a fault on his second serve also. So, he makes a double fault in 6% of the 36% of points in which he faults on the first serve, or in the proportion $0.06 \times 0.36 = 0.02$, 2% of the points.

TO PRACTICE THIS CONCEPT, TRY EXERCISE 5.36

Sampling with or without Replacement

In many sampling processes, once subjects are selected from a population, they are not eligible to be selected again. This is called *sampling without replacement*. At any stage of such a sampling process, probabilities of potential outcomes depend on the previous outcomes. Conditional probabilities are then used in finding probabilities of the possible samples.

EXAMPLE 10

How Likely Are You to Win Lotto?

Picture the Scenario

Many states have lotteries. The biggest jackpot, typically millions of dollars, usually comes from the Lotto game. In Lotto South, available in Georgia, Kentucky, and Virginia, six numbers are randomly sampled without replacement from the integers 1 to 49. For example, a possible sample is $(4, 9, 23, 26, 40, 46)$. Their order of selection is not important.

Question to Explore

You buy a Lotto South ticket. What is the probability that it is a winning ticket, having the six numbers chosen?

Think It Through

The probability of winning is the probability that the six numbers chosen are the six that you have on your ticket. For the first number chosen, 6 of the 49 numbers that can be selected are ones you hold, so $P(\text{you have 1st number}) = 6/49$. Given that you have the first number, for the second trial there are 5 numbers left that you hold out of 48 possible, so $P(\text{have 2nd number} \mid \text{have 1st number}) = 5/48$. Given that you have the first two, for the third trial there are 4 numbers left that you hold out of 47 possible, so $P(\text{have 3rd number} \mid \text{have 1st and 2nd numbers}) = 4/47$. Continuing with this logic, using an extension of the multiplication rule with conditional probabilities,

Did You Know?

Comparable to the probability of 0.00000007 of winning Lotto South are the annual probabilities of about 0.0000004 of being hit by a meteorite, 0.0000002 of dying in a tornado, and 0.00000016 of dying by a lightning strike (Wilson and Crouch 2001, p. 200). It's also roughly the probability that a person of average mortality will die in the next 3 minutes. ■

$P(\text{have all 6 numbers}) = P(\text{have 1st and 2nd and 3rd and 4th and 5th and 6th})$

$= P(\text{have 1st})P(\text{have 2nd} \mid \text{have 1st})P(\text{have 3rd} \mid \text{have 1st and 2nd}) \dots$

$\dots P(\text{have 6th} \mid \text{have 1st and 2nd and 3rd and 4th and 5th})$

$= (6/49) \times (5/48) \times (4/47) \times (3/46) \times (2/45) \times (1/44)$

$= 720/10{,}068{,}347{,}520 = 0.00000007.$

This is about 1 chance in 14 million.

Insight

Let's give this small number some perspective. The chance of winning the jackpot in Lotto South is less than your chance of being hit by a meteorite in the next year. The probability that a given person will die during the next year in a car crash is about 0.00015, 2100 times the chance of winning Lotto South. If you have money to spare, go ahead and play the lottery, but understand why many call it "sport for the mathematically challenged." The lottery is a regressive way for states to collect money. Many players cannot afford it. Mark Twain said,[3] "There are two times in a man's life when he should not speculate: when he can't afford it, and when he can."

TO PRACTICE THIS CONCEPT, TRY EXERCISE 5.40

Lotto South uses *sampling without replacement*. Once a number is chosen, it cannot be chosen again. If, by contrast, Lotto South allowed numbers to be picked more than once, the sampling scheme would be *sampling with replacement*.

After each observation without replacement, the population remaining is reduced by one and the conditional probability of a particular outcome changes. It can change considerably when the population size is small, as we saw in the Lotto example. With large population sizes, reducing the population by one does not much affect the probability from one trial to the next. In practice, when selecting random samples, we normally sample without replacement. With populations that are large compared to the sample size, the probability at any given observation depends little on the previous observations. Probabilities of possible samples are then quite similar for sampling without replacement and sampling with replacement.

Independent Events Defined Using Conditional Probability

Two events A and B are **independent** if the probability that one occurs is not affected by whether or not the other event occurs. This is expressed more formally using conditional probabilities.

Independent Events, in Terms of Conditional Probabilities

Events A and B are **independent** if $P(A \mid B) = P(A)$, or equivalently, if $P(B \mid A) = P(B)$. If either holds, then the other does too, and also

$$P(A \text{ and } B) = P(A) \times P(B).$$

For instance, let's consider the genders of two children in a family (F = female, M = male). The sample space is {FF, FM, MF, MM}. Suppose these four outcomes are equally likely, which is approximately true in practice.

[3]In *Following the Equator*.

Let A denote {first child is female} and let B denote {second child is female}. Then $P(A) = 1/2$, since two of the four possible outcomes have a female for the first child. Likewise, $P(B) = 1/2$. Also, $P(A \text{ and } B) = 1/4$, since this corresponds to the single outcome, FF. So, from the definition of conditional probability,

$$P(B|A) = P(A \text{ and } B)/P(A) = (1/4)/(1/2) = 1/2.$$

Thus, $P(B|A) = 1/2 = P(B)$, so A and B are independent events. Given A (that is, that the first child is female), the probability that the second child was female is 1/2, since one outcome (FF) has this out of the two possibilities (FF, FM). Intuitively, the gender of the second child does not depend on the gender of the first child.

In sampling without replacement, outcomes of different trials are dependent. For instance, in the Lotto game described in Example 10, let A denote {your first number is chosen} and B denote {your second number is chosen}. Then $P(A) = 6/49$, but $P(A|B) = 5/48$ (since there are 5 possibilities out of 48 numbers), which differs slightly from $P(A)$.

EXAMPLE 11

How to Check Whether Two Events Are Independent

Picture the Scenario

Table 5.5 showed a contingency table relating the result of a diagnostic blood test (POS = positive, NEG = negative) to whether a pregnant woman has Down syndrome (D = Down syndrome, D^c = unaffected). The estimated cell probabilities based on the frequencies in that table are shown in the table in the margin.

Status	Blood Test POS	NEG	Total
D	0.009	0.001	**0.010**
D^c	0.247	0.742	**0.990**
Total	**0.257**	**0.743**	**1.00**

Questions to Explore

a. Are the events POS and D independent, or dependent?

b. Are the events POS and D^c independent, or dependent?

Think It Through

a. The probability of a positive test result is 0.257. However, the probability of a positive result, given Down syndrome, is

$$P(POS|D) = P(POS \text{ and } D)/P(D) = 0.009/0.010 = 0.90.$$

Since $P(POS|D) = 0.90$ differs from $P(POS) = 0.257$, the events POS and D are dependent.

b. Likewise,

$$P(POS|D^c) = P(POS \text{ and } D^c)/P(D^c) = 0.247/0.990 = 0.250.$$

This differs slightly from $P(POS) = 0.257$, so POS and D^c are also dependent events.

Insight

As we'd expect, the probability of a positive result depends on whether the fetus has Down syndrome, and it's much higher if the fetus does. The diagnostic test would be worthless if the disease status and the test result were independent.

Generally, if A and B are dependent events, then so are A and B^c and so are A^c and B and so are A^c and B^c. For instance, if A depends on whether B occurs, then A also depends on whether B does not occur. So, once we find that POS and D are dependent events, we know that POS and D^c are dependent events also.

TO PRACTICE THIS CONCEPT, TRY EXERCISE 5.41

We can now justify the formula given in Section 5.2 for the probability of the intersection of two independent events, namely P(A and B) = P(A) × P(B). This is a special case of the multiplication rule for finding P(A and B),

$$P(A \text{ and } B) = P(A \mid B) \times P(B).$$

If A and B are independent, then P(A | B) = P(A), so the multiplication rule simplifies to

$$P(A \text{ and } B) = P(A) \times P(B).$$

Checking for Independence

Here are three ways to check whether events A and B are independent:

- Is P(A | B) = P(A)?
- Is P(B | A) = P(B)?
- Is P(A and B) = P(A) × P(B)?

If any of these is true, the others are also true and the events A and B are independent.

EXAMPLE 12

Are Disjoint Events Independent?

Question to Explore

Are they independent events also?

Think It Through

Since A and B are disjoint, they have no outcomes in common. So, P(A and B) = 0. Thus P(A) × P(B) cannot equal P(A and B), so they are dependent events. In fact, P(B | A) = P(A and B)/P(A) = 0, so given that A occurs, B *cannot* occur.

Insight

Consider events POS = positive diagnostic test and NEG = negative diagnostic test. They are disjoint, since they have no common outcomes. Knowing that a test result was negative, it cannot have been positive, so P(POS | NEG) = 0. Thus, these events, like all pairs of disjoint events, are very strongly *dependent*.

Picture the Scenario

Suppose A and B are disjoint events, each with positive probability.

TO PRACTICE THIS CONCEPT, TRY EXERCISE 5.41

5.26   Alcohol and college students: For the results of the 2001 Harvard School of Public Health College Alcohol Study Survey about binge drinking, the estimated probability of being a binge drinker was 0.49 for males and 0.41 for females. Introducing notation, express each of these as a conditional probability.

5.27   Spam: Because of the increasing nuisance of spam e-mail messages, many start-up companies have emerged to develop e-mail filters. One such filter was recently advertised as being 95% accurate. This could mean that (a) 95% of spam is blocked, (b) 95% of valid e-mail is allowed through, (c) 95% of the e-mail allowed through is valid, or (d) 95% of the blocked e-mail is spam. Let S denote {message is spam}, and let B denote {filter blocks message}. Using these events and their complements, identify each of these four possibilities as a conditional probability.

5.28   Audit and low income: Table 5.3 on audit status and income is shown below. Show how to find the probability of:

a. Being audited, given that the taxpayer is in the lowest income category.

b. Being in the lowest income category, given that the taxpayer is audited.

Income	Audited	
	Yes	No
<$25,000	0.0011	0.1747
$25,000–49,999	0.0009	0.3819
$50,000–99,999	0.0009	0.3071
≥$100,000	0.0010	0.1324

5.29   Commuting to work: The 2000 U.S. census reported the frequencies in the following table for how workers 16 years and over commute to work.

Method Used to Commute to Work	Frequency
Car, truck, or van—drove alone	97,102,050
Car, truck, or van—carpooled	15,634,051
Public transportation (including taxicab)	6,067,703
Walked	3,758,982
Other means	1,532,219
Worked at home	4,184,223
Total	128,279,228

a. What is the probability that a randomly selected worker commuted to work in a car, truck, or van?

b. Given that a person commuted to work in a car, truck, or van, what is the probability of car pooling?

5.30   Cancer deaths: Current estimates are that about 25% of all deaths are due to cancer, and of the deaths that are due to cancer, 30% are attributed to tobacco, 40% to diet, and 30% to other causes.

a. Define events, and identify which of these four probabilities refer to conditional probabilities.

b. Find the probability that a death is due to cancer and tobacco.

5.31   Revisiting seat belts and auto accidents: The table below is from Exercise 5.23 classifying auto accidents by survival status (S = survived, D = died) and seat belt status of the individual involved in the accident.

Belt	Outcome		
	S	D	Total
Yes	412,368	510	412,878
No	162,527	1601	164,128
Total	574,895	2111	577,006

a. Estimate the probability that the individual died (D) in the auto accident.

b. Estimate the probability that the individual died, given that they (i) wore, (ii) did not wear a seat belt. Interpret results.

c. Are the events of dying and wearing a seat belt independent? Justify your answer.

5.32   Go Celtics!: Larry Bird, who played pro basketball for the Boston Celtics, was known for being a good shooter. In games during 1980–1982, when he missed his first free throw, 48 out of 53 times he made the second one, and when he made his first free throw, 251 out of 285 times he made the second one.

a. Form a contingency table that cross tabulates the outcome of the first free throw (made or missed) in the rows and the outcome of the second free throw (made or missed) in the columns.

b. For a given pair of free throws, estimate the probability that Bird (i) made the first free throw, (ii) made the second free throw.

c. Estimate the probability that Bird made the second free throw, given that he made the first one. Does it seem as if his success on the second shot depends strongly, or hardly at all, on whether he made the first?

5.33   Down syndrome again: Example 8 discussed the triple blood test for Down syndrome, using data summarized in a table shown again below.

Down	Blood Test		
	POS	NEG	Total
D	48	6	54
D^c	1307	3921	5228
Total	1355	3927	5282

a. Given that a test result is negative, show that the probability the fetus actually has Down syndrome is $P(D \mid NEG) = 0.0015$.

b. Is $P(D \mid NEG)$ equal to $P(NEG \mid D)$? If so, explain why. If not, find $P(NEG \mid D)$.

5.34 America, the fast food nation: A Gallup Organization poll conducted in July 2003 reported the results of a survey asking approximately 1000 randomly chosen Americans (18 years or older) questions about fast food consumption. It reported that 59% of the male respondents and 46% of the female respondents eat fast food at least once weekly.

a. What type of percentages (probabilities) were reported by Gallup: ordinary or conditional? Explain, by specifying events to which the probabilities refer.

b. Suppose that half those polled were female and half were male. Construct a contingency table showing counts for gender and fast food status.

5.35 Happiness in marriage: Are people happy in their marriages? The table shows results from the 2002 General Social Survey for married adults classified by gender and level of happiness.

| | Level of Happiness | | | |
| | Very Happy | Pretty Happy | Not too Happy | Total |
Gender				
Male	221	95	9	325
Female	149	120	9	278
Total	370	215	18	603

a. Estimate the probability that a married adult is very happy.

b. Estimate the probability that a married adult is very happy, (i) given that their gender is male, (ii) given that their gender is female.

c. For these subjects, are the events being very happy and being a male independent? (Your answer will be merely for this sample. Chapter 10 will show how to answer this for the population.)

5.36 Serena Williams serves: In the 2004 Wimbledon tennis championship, Serena Williams made 63% of her first serves. When she faulted on her first serve, she made 93% of her second serves. Assuming these are typical of her serving performance, when she serves, what is the probability that she makes a double fault?

5.37 Shooting free throws: Pro basketball player Shaquille O'Neal is a poor free-throw shooter. Consider situations in which he shoots a pair of free throws. The probability that he makes the first free throw is 0.50. Given that he makes the first, suppose the probability that he makes the second is 0.60. Given that he misses the first, suppose the probability that he makes the second one is 0.40.

a. What is the probability that he makes both free throws?

b. Find the probability that he makes *one* of the two free throws (i) using the multiplicative rule with the two possible ways he can do this, (ii) by defining this as the complement of making neither or both of the free throws.

c. Are the results of the free throws independent? Explain.

5.38 Drawing cards: A standard card deck has 52 cards consisting of 26 black and 26 red cards. Three cards are dealt from a shuffled deck, *without replacement*.

a. True or false: The probability of being dealt 3 black cards is $(1/2) \times (1/2) \times (1/2) = 1/8$. If true, explain why. If false, show how to get the correct probability.

b. Let A = first card red and B = second card red. Are A and B independent? Explain why or why not.

c. Answer (a) and (b) if each card is replaced in the deck after being dealt.

5.39 Moral values: A correction in the February 14–21, 2005 issue of *The New Yorker* magazine stated that the January 3 issue that year "contained the incorrect statistic that four-fifths of Bush voters identified moral values as the most important factor in their decision. In fact, four-fifths of those identifying moral values as the most important factor of their decision were Bush voters."

a. Identify two events A and B such that *The New Yorker* reported the result as $P(A \mid B)$ when it should have been $P(B \mid A)$.

b. If $P(A) = 0.20$ and $P(B) = 0.51$ with $P(B \mid A) = 4/5$, find $P(A \text{ and } B)$.

c. Using (b), find $P(A \mid B)$. This shows that $P(A \mid B)$ is quite different from $P(B \mid A)$ when $P(A)$ is quite different from $P(B)$.

5.40 Big loser in Lotto: Example 10 showed that the probability of having the winning ticket in Lotto South was 0.00000007. Find the probability of holding a ticket that has zero winning numbers out of the 6 numbers selected (without replacement) for the winning ticket out of the 49 possible numbers.

5.41 Family with two children: For a family with two children, let A denote {first child is female} and let C denote {both children are female}.

a. Show that $P(C \mid A) = 1/2$.

b. Are A and C independent events? Why or why not?

5.42 Checking independence: In three independent flips of a balanced coin, let A denote {first flip is a head}, B denote {second flip is a head}, C denote {first two flips are heads}, and D denote {three heads on the three flips}.

a. Find the probabilities of A, B, C, and D.

b. Which, if any, pairs of these events are independent? Explain.

5.4 | Applying the Probability Rules

You may not have thought about it, but probability relates to many aspects of your daily life—for instance, when you make decisions that affect your financial well being, and when you evaluate risks due to lifestyle decisions (wearing a seat belt while in a vehicle, smoking, dieting, exercising, traveling somewhere). Objectively or subjectively you need to consider questions such as: What's the chance that the business you're thinking of starting will succeed? What's the chance that you will get cancer if you continue to smoke? What's the chance that the extra collision insurance you're thinking of getting for your car will be needed?

We'll now apply the basics of probability to a few more examples to practice key concepts from this chapter. We'll start by seeing what probability tells us about coincidence in our lives.

Is a "Coincidence" Truly an Unusual Event?

Some events in our lives seem to be highly coincidental. One of the authors, who lives in Georgia, once spent a summer vacation in Newfoundland, Canada. One day during that trip, she made an unplanned stop at a rest area and observed a camper with a Georgia license tag from a neighboring county to her hometown. She discovered that the camper's owner was a patient of her physician husband. Was this meeting as highly coincidental as it seemed?

Events that seem highly coincidental are often not so unusual when viewed in the context of *all* the possible random occurrences at all times. Lots of topics can trigger an apparent coincidence—the person you meet having the same last name as yours, the same birth place, the same high school or college alma mater, the same profession, the same birthday, the same make, year and color of car, a common friend, and so on. Think of the huge number of possible coincidences that can potentially occur. It's really not so surprising that in your travels you will sometime have such a coincidence, such as seeing a friend or meeting someone who knows somebody you know.

With a large enough sample of people or times or topics, seemingly surprising things are actually quite sure to happen. Events that are rare per person occur rather commonly with large numbers of people. If a particular event happens to one person in a million each day, then in the United States we expect about 300 such events a day and more than 100,000 every year. The one in a million chance regularly occurs, however surprised we may be if it should happen to us.

If you take a coin now and flip it 10 times, probably you would be surprised to get 10 heads. But if you flipped the coin for a long time, would you be surprised to get 10 heads in a row at some stage? Perhaps you would, but you should not be. For instance, if you flip a balanced coin 2000 times, then you can expect the longest run of heads during those flips to be about 10. When a seemingly unusual event happens to you, think about whether it is like seeing 10 heads on the next ten flips of a coin, or more like seeing 10 heads in a row sometime in a long series of flips. If the latter, it's really not such an unusual occurrence.

Coincidence and Seemingly Unusual Patterns

Once we have data, it's easy to find patterns: ten heads in a row, ten tails in a row, five tails followed by five heads, and so forth for lots of other possibilities. Our minds are programmed to look for patterns. Out of the huge number of things that can happen to us and to others over time, it's not surprising to see patterns that seem unusual occasionally.

To illustrate that an event you may perceive as a coincidence is actually not that surprising, let's answer the question, "What is the chance that at least two people in your class have the same birthday?"

Did You Know?

If something has a very large number of opportunities to happen, occasionally it will happen, even if it's highly unlikely at any one observation. This is a consequence of the **law of large numbers**. ▪

ACTIVITY 2: **Matching Birthdays**

If your class is small to moderate in size (say, fewer than about 60 students), the instructor may have all your classmates state their birth dates. Does a pair of students in the class share the same birthday?

EXAMPLE 13

Is a Matching Birthday Surprising?

Picture the Scenario

Suppose a class has 25 students. Since there are 365 possible birth dates (without counting February 29), our intuition tells us that the probability is small that there will be any birthday matches. Assume that the birth date of a student is equally likely to be any one of the 365 days in a year and that students' birth dates are independent (for example, there are no twins in the class).

Question to Explore

What is the probability that *at least* two of the 25 students have the same birthday? Is our intuition correct that the probability of a match is small?

Think It Through

The event of *at least one* birthday match includes 1 match, 2 matches, or more. To find the probability of *at least one* match, it is simpler to find the complement probability of *no* matches. Then

$$P(\text{at least one match}) = 1 - P(\text{no matches}).$$

To begin, suppose a class has only 2 students. The first student's birthday could be any of 365 days. Given that student's birthday, the chance that the second student's birthday is different is 364/365, since 364 of the possible 365 birthdays are different. The probability is $1 - 364/365 = 1/365$ that the two students share the same birthday.

Now, suppose a class has 3 students. The probability that all three have different birthdays is

P(no matches) = P(students 1 and 2 and 3 have different birthdays) =

P(students 1 and 2 different) × P(student 3 different|students 1 and 2 different)

= (364/365) × (363/365).

The second probability in this product equals 363/365 because there are 363 days left for the third student that differ from the different birthdays of the first two students.

For 25 students, similar logic applies. By the time we get to the 25th student, for that student's birthday to differ from the other 24, there are 341 choices left out of 365 possible birthdays. So

P(no matches) = P(students 1 and 2 and 3 . . . and 25 have different birthdays)

= (364/365) × (363/365) × (362/365) × . . . . × (341/365).

   ↑   ↑   ↑   ↑   ↑

student 2,   student 3,   student 4,   next 20 students   student 25,

given 1   given 1, 2   given 1, 2, 3   given 1, . . . , 24

This product equals 0.43. Using the probability for the complement of an event,

$$P(\text{at least one match}) = 1 - P(\text{no matches}) = 1 - 0.43 = 0.57.$$

The probability exceeds 1/2 of at least one birthday match in a class of 25 students.

Insight

Is this higher than you expected? It should not be once you realize that with 25 students, there are 300 *pairs* of students who can share the same birthday (see Exercise 5.45.) Remember that with lots of opportunities for something to happen, coincidences are really not so surprising.

Did you have a birthday match in your class? Was it coincidence? If the number of students in your class is at least 23, the probability of at least one match is greater than 1/2. For a class of 50 students, the probability of at least one match is 0.97. For 100 students, it is 0.9999997 (there are then 4950 different *pairs* of students). Here are a couple of other facts about matching birthdays that may surprise you:

■ With 88 people, there's a 1/2 chance that at least 3 people have the same birthday.

■ With 14 people, there's a 1/2 chance that at least two people have a birthday within a day of each other in the calendar year.

TO PRACTICE THIS CONCEPT, TRY EXERCISE 5.43

Sometimes a cluster of occurrences of some disease in a neighborhood will cause worry in residents that there is some environmental cause. But *some* disease clusters will appear around a nation just by chance. If we look at a large number of places and times, we should expect some disease clusters. By themselves, they seem unusual, but viewed in a broader context they may not be. Epidemiologists are statistically trained scientists who face the difficult task of determining which such events can be explained by ordinary random variation and which cannot.

In Practice, Probability Models Approximate Reality

We've now found probabilities in many rather idealized situations. In practice, it's often not obvious when different outcomes are equally likely or different events are independent. Then, when calculating probabilities, it is advisable to specify a **probability model** that spells out all the assumptions made.

Probability Model

A **probability model** specifies the possible outcomes for a sample space and provides assumptions on which the probability calculations for events composed of those outcomes are based.

The next example illustrates a probability model used together with the rules for finding probabilities.

EXAMPLE 14

Assessing Safety of the Space Shuttle

Picture the Scenario
Out of the first 113 space shuttle missions, there were two failures, one of which was the Columbia disaster on flight 113. Since then, much attention has focused on estimating probabilities of success or failure (disaster) for a mission. But before the first flight, there were no trials to provide data for estimating probabilities.

Question to Explore
Based on all the information available, a scientist is willing to predict the probability of success for any particular mission. How can you use this to find the probability of *at least one* failure in a total of 100 missions?

Think It Through

Let S1 denote the event that the first mission is successful, S2 the event that the second mission is successful, and so on up to S100, the event that mission 100 is successful. If all these events occur, there is no failure in the 100 flights. The event of *at least one* failure in 100 flights is the complement of the event that they are all successful (0 failures). By the probability for complementary events,

$$P(\text{at least 1 failure}) = 1 - P(0 \text{ failures})$$
$$= 1 - P(S1 \text{ and } S2 \text{ and } S3 \ldots \text{ and } S100).$$

The intersection of events S1 through S100 is the event that *all* 100 missions are successful.

We now need a probability model in order to evaluate P(S1 and S2 and S3...and S100). If our probability model assumes that the 100 shuttle flights are *independent*, then the multiplication rule for independent events implies that

$$P(S1 \text{ and } S2 \text{ and } S3 \ldots \text{ and } S100) =$$
$$P(S1) \times P(S2) \times P(S3) \times \cdots \times P(S100).$$

If our probability model assumes that each flight has the *same probability* of success, say P(S), then this product equals $[P(S)]^{100}$. To proceed further, we need a value for P(S). According to a discussion of this issue in the excellent PBS video series *Against All Odds: Inside Statistics*, a risk assessment study by the Air Force in 1982 used the estimate P(S) = 0.971. With it,

$$P(\text{at least 1 failure}) = 1 - [P(S)]^{100} = 1 - [0.971]^{100} = 1 - 0.053 = 0.947.$$

Different estimates of P(S) can result in very different answers. For instance, the *Against All Odds* episode mentioned that a 1985 NASA study estimated that P(S) = 0.9999833. Using this,

$$P(\text{at least 1 failure}) = 1 - [P(S)]^{100} = 1 - [0.9999833]^{100}$$
$$= 1 - 0.998 = 0.002.$$

This was undoubtedly overly optimistic. We see that different probability models can result in drastically different probability assessments.

Insight

The answer above depended strongly on the assumed probability of success for each mission. Since there were 2 failures in the first 113 flights, after that flight an estimate of P(S) was 111/113 = 0.982. But the assumptions of independence and of the same probability for each flight may also be suspect. For instance, other variables (e.g., temperature at launch, experience of crew, age of craft used, quality of O-ring seals) could affect that probability.

TO PRACTICE THIS CONCEPT, TRY EXERCISE 5.50

In practice, probability models merely *approximate* reality. They are rarely *exactly* satisfied. For instance, in the matching birthday example our probability model ignored February 29, assumed each of the other 365 birthdays are equally likely, and assumed that students' birth dates are independent. So, the answer found is only approximate. Whether probability calculations using a particular probability model are accurate depends on whether assumptions in that model are close to the truth or quite unrealistic.

Probabilities and Diagnostic Testing

We've seen the important role that probability plays in diagnostic testing, illustrated in Section 5.3 (Example 8) with the pregnancy test for Down syndrome. Table 5.6 summarizes conditional probabilities about the result of a diagnostic test, given whether some condition or state (such as Down syndrome) is present. We let S denote that the state is present and S^c denote that it is not present.

Recall

The vertical slash means "given." **P(POS | S)** denotes the conditional probability of a positive test result, given that the state is present.▪

State present?	Diagnostic Test Result		Total Probability
	Positive (POS)	**Negative (NEG)**	
Yes (S)	Sensitivity P(POS|S)	False negative rate P(NEG|S)	1.0
No (Sᶜ)	False positive rate P(POS|Sᶜ)	Specificity P(NEG|Sᶜ)	1.0

TABLE 5.6: Probabilities of Correct and Incorrect Results in Diagnostic Testing

The probabilities in the body of the table refer to the test result, conditional on whether the state (S) is truly present. The sensitivity and specificity are the probabilities of the two types of correct diagnoses.

In Table 5.6, **sensitivity** and **specificity** refer to correct test results, given the actual state. For instance, given that the state tested for is present, sensitivity is the probability the test detects it by giving a positive result, that is, $P(POS|S)$.

Medical journal articles that discuss diagnostic tests commonly report the sensitivity and specificity. However, what's more relevant to you once you take a diagnostic test are the conditional probabilities that condition on the test result. If a diagnostic test for Down syndrome is positive, you want to know the probability that Down syndrome is truly present. If you know the sensitivity and specificity and how often the state occurs, can you find $P(S|POS)$ using the rules of probability?

The easiest way to do this is with a tree diagram, as shown in the margin. The first branches show the probabilities of the two possible states, $P(S)$ and $P(S^c)$. The next set of branches show the known conditional probabilities, such as $P(POS|S)$, for which we are given the true state. Then the products $P(POS|S)P(S)$ and $P(POS|S^c)P(S^c)$ give intersection probabilities $P(S$ and $POS)$ and $P(S^c$ and $POS)$, which can be used to get probabilities such as

$$P(POS) = P(S \text{ and } POS) + P(S^c \text{ and } POS).$$

Then you can find $P(S|POS) = P(S \text{ and } POS)/P(POS)$.

EXAMPLE 15

Random Drug Testing of Air Traffic Controllers

Picture the Scenario

Air traffic controllers have a job that affects the lives of millions of people each day—monitoring the flights of aircraft and helping to ensure safe takeoffs and landings. In the United States, because the Federal Aviation Administration (FAA) considers it important to promote a drug-free environment, air traffic controllers are required to undergo periodic random drug testing. A urine test is used as an initial screening test due to its low cost and ease of implementation. One such urine test is the *Triage Panel for Drugs of Abuse plus TCA*,[4] which detects the presence of drugs such as

---

[4]Screening assay from Biosite Diagnostics, San Diego, California.

amphetamines, opiates, and barbiturates. Its sensitivity and specificity have been reported[5] as 0.96 and 0.93. Based on past drug testing of air traffic controllers, the FAA reports that the probability of drug use at a given time is approximately 0.007 (less than 1%). This is called the **prevalence** of drug use.

Questions to Explore

a. A positive test result puts the air traffic controller's job in jeopardy. What is the probability of a positive test result?

b. Find the probability an air traffic controller truly used drugs, given that the test is positive.

Think It Through

a. We've been given the probability of drug use, $P(S) = 0.007$, the sensitivity $P(POS|S) = 0.96$, and the specificity $P(NEG|S^c) = 0.93$. Figure 5.10 shows a tree diagram that is useful for visualizing these probabilities and for finding $P(POS)$.

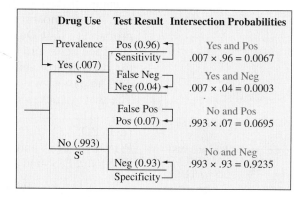

▲ **FIGURE 5.10: Tree Diagram for Random Drug Testing of Air Traffic Controllers.** The first set of branches show the probabilities for drug use. The second set of branches shows the conditional probabilities for the test result, given whether used drugs. Multiplication of the probabilities along each path gives the probabilities of intersections of the events.

From the tree diagram and the multiplicative rule for intersection probabilities,

$$P(S \text{ and } POS) = P(POS|S)P(S) = 0.96 \times 0.007 = 0.0067.$$

The other route with a positive test result has probability

$$P(S^c \text{ and } POS) = P(POS|S^c)P(S^c) = 0.07 \times 0.993 = 0.0695.$$

To find $P(POS)$, we add the probabilities of these two possible positive test routes. Thus,

$$P(POS) =$$

$$P(S \text{ and } POS) + P(S^c \text{ and } POS) = 0.0067 + 0.0695 = 0.0762.$$

There's nearly an 8% chance of the test suggesting that the person used drugs.

b. The probability of drug use, given a positive test, is $P(S|POS)$. From the definition of conditional probability,

$$P(S|POS) = P(S \text{ and } POS)/P(POS) = 0.0067/0.0762 = 0.09.$$

When the test is positive, only 9% of the time had the person actually used drugs.

[5]M. Peace et al., *Journal of Analytical Toxicology*, vol. 24 (2000).

If you're uncertain about how to answer this and part (a) with a tree diagram, you can construct a contingency table, as shown in Table 5.7. The table shows the summary value of P(S) = 0.007 in the right margin, the other right-margin value P(S^c) = 1 − 0.007 = 0.993 determined by it, and the intersection probabilities P(S and POS) = 0.0067 and P(S^c and POS) = 0.0695 found in (a). From that table, of the proportion 0.0762 of positive cases, 0.0067 truly had used drugs, so the conditional probability is P(S|POS) =P(S and POS)/P(POS) = 0.0067/0.0762 = 0.09.

TABLE 5.7: Contingency Table Cell Probabilities for Air Controller Drug Test

State Present	Drug Test Result		Total
	POS	NEG	
Yes (S)	0.0067	0.0003	0.007
No (S^c)	0.0695	0.9235	0.993
Total	0.0762	0.9238	1.000

Insight

If the prevalence rate is truly near 0.007, the chances are low that an individual who tests positive is actually a drug user. Does a positive test mean the individual will automatically lose their job? No, if the urine test comes back positive, the individual is given a second test that is more accurate but more expensive than the urine test.

TO PRACTICE THIS CONCEPT, TRY EXERCISES 5.51

Are you surprised that P(S|POS) is so small (only 0.09) for the air controllers? To help understand the logic behind how this could be, it's a good idea to show on a tree diagram what you'd expect to happen with a typical group of air controllers. Figure 5.11 shows a tree diagram for what we'd expect to happen for 1000 of them.

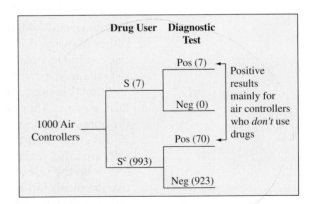

▲ **FIGURE 5.11: Expected Results of Drug Tests for 1000 Air Controllers.** This shows typical results when the proportion 0.007 (7 in 1000) are drug users, a positive result has probability 0.96 for those who use drugs, and a negative result has probability 0.93 for those who do not use drugs. Most positive results occur with individuals who are *not* using drugs, since there are so many such individuals (more than 99% of the population). **Question:** How would results change if a higher percentage of the population were using drugs?

Since the proportion 0.007 (which is 7 in 1000) of the target population uses drugs, the tree diagram shows 7 drug users and 993 non–drug users out of the 1000 air controllers. For the 7 drug users, there's a 0.96 chance the test detects the drug use. So we'd expect all 7 or perhaps 6 of the 7 to be detected with the test (Figure 5.11 shows 7). For the 993 non–drug users, there's a 0.93 chance the test is negative. So we'd expect about $0.93 \times 993 = 923$ individuals to have a negative result and the other $993 - 923 = 70$ non–drug users to have a positive result, as shown on Figure 5.11.

In summary, Figure 5.11 shows $7 + 70 = 77$ individuals having a positive test result, but only 7 actually were drug users. Of those with a positive test result, the proportion who truly were drug users is $7/77 = 0.09$. What's happening is that the 7% of errors for the large majority of individuals who *do not* use drugs is much larger than the 96% of correct decisions for the small number of individuals who *do* use drugs.

If the prevalence rate were $P(S) = 0.15$ (let's hope it's not really this high!) instead of 0.007, you can verify that $P(S \mid POS) = 0.71$, rather than 0.09. The probability that the person truly used drugs is then much higher, 0.71. In fact, the chance that a positive test result is truly correct depends very much on the prevalence rate. The lower the prevalence rate, the lower the chance. The more drug-free the population, the less likely an individual who tests positive for drugs truly used them.

In Example 15, we started with probabilities of the form $P(POS \mid S)$ and used them to find a conditional probability $P(S \mid POS)$ that reverses what is given and what is to be found. The method used to find this reverse conditional probability can be summarized in a single formula (see Exercise 5.106). It is known as **Bayes rule**, named after the person who originally discovered it. We have not shown that formula here because it is easier to understand the logic behind evaluating this conditional probability using tree diagrams or contingency tables.

Probability Answers to Questions Can Be Surprising

The results in Example 15 on drug testing of air controllers may seem surprising, but actually they are not uncommon. For instance, consider the mammogram for detecting breast cancer in women. One recent study[6] estimated sensitivity = 0.86 and specificity = 0.88. Of the women who receive a positive mammogram result, what proportion actually have breast cancer? In Exercise 5.51 you can work out that it may be only about 0.07.

These diagnostic testing examples point out that when probability is involved, answers to questions about our daily lives are sometimes not as obvious as you may think. Consider the question, "Should a woman have an annual mammogram?" Medical choices are rarely between certainty and risk, but rather between different risks. If a woman fails to have a mammogram, this is risky—she may truly have breast cancer. If she has a mammogram, there's the risk of a false positive or false negative, such as the risk of needless worry and additional invasive procedures (often biopsy and other treatments) due to a false positive. There's also the risk from the X-ray itself and from traveling to the appointment, small as these may be.

Likewise, with the Triple Blood Test for Down syndrome, most positive results are false, yet the recommendation after receiving a positive test may be to follow up the test with an amniocentesis, a procedure that gives a more definitive diagnosis but has the risk of causing a miscarriage. With some diagnostic tests (such as the PSA test for prostate cancer), testing can detect the disease, but the evidence is unclear about whether or not early detection has much, if any, effect on life expectancy.

Simulation to Estimate a Probability

Some probabilities are very difficult to find with ordinary reasoning. In such cases, one way to approximate an answer is to simulate. We have carried out simulations in previous chapters, in chapter exercises, and in Activity 1 in this chapter. The steps for a simulation are as follows:

---

[6]W. Barlow et al., *J. Natl. Cancer Inst.*, 2002, vol. 94, p. 1151.

Carrying Out a Simulation

■ Identify the random phenomenon to be simulated.

■ Describe how to simulate observations of the random phenomenon.

■ Carry out the simulation many times (ideally, at least 1000 times).

■ Summarize results and state the conclusion.

EXAMPLE 16

Be Bold or Be Cautious to Save Your Business?

Picture the Scenario

The business you started last year has only $5000 left in capital. A week from now you need to repay a $10,000 loan, or you will go bankrupt, so you are desperate. You see two possible ways to raise the money you need. You can approach a large company and see if they will invest $10,000 in your company. You would spend the $5000 you have to woo them, and your guess is that the probability of success is 0.20. Alternatively, you can approach in sequence several small companies and see if each of them will invest $2000. You would spend $1000 each to woo them, and again you guess that the probability of success is 0.20 for each appeal.

Questions to Explore

What's the probability you will raise enough money to repay the loan, with each strategy? Which is the better strategy, to be bold with one large bet or to be cautious and use several smaller bets?

Think It Through

With the bold strategy, the probability of success is 0.20, the chance that the large company you approach agrees to back you. With the cautious strategy, it's not easy to find the probability, so you can use simulation. For a randomizer, you can use a random number table or generator on a computer or calculator, or an applet such as the *random numbers* applet on the text CD.

Carrying Out the Simulation

■ *Identify the random phenomenon to be simulated*: Observing the success or failure of each appeal until you have $10,000 or have gone broke.

■ *Describe how to simulate observations of the random phenomenon*: You can use random numbers. Consider the digits 0 to 9. The probability of a successful appeal is 0.20, so we let digits 1 and 2 (20% of the digits) represent success and 3, 4, 5, 6, 7, 8, 9, 0 represent failure. With a sequence of randomly generated digits, you keep track of the cash total until you have $10,000 or are broke. For example, a possible sequence of random digits is

Random digit:	9	8	3	5	1	7	9
Cash flow (thousands):	4	3	2	1	2	1	0

The first random digit is 9, not 1 or 2, so the first company did not agree to back you. You lose the $1000 spent to woo that company, leaving you with $4000 (shown on the next line). Likewise, the second random digit is 8, not 1 or 2, so now you are down to $3000. Only the fifth appeal was successful (a random digit of 1, and your capital increased at that stage from $1000 to $2000). You went broke after the seventh appeal.

■ *Carry out the simulation many times*: Let's carry out this simulation 10 times, with the cash flow in parentheses after each random digit. The first case repeats the results shown above.

Simulation #	Sequence of Outcomes	Result
1	9 (4) 8 (3) 3 (2) 5 (1) 1 (2) 7 (1) 9 (0)	broke
2	8 (4) 6 (3) 3 (2) 3 (1) 7 (0)	broke
3	7 (4) 5 (3) 2 (4) 1 (5) 1 (6) 4 (5) 2 (6)	
	1 (7) 4 (6) 8 (5) 9 (4) 4 (3) 5 (2) 7 (1) 3 (0)	broke
4	8 (4) 6 (3) 9 (2) 6 (1) 4 (0)	broke
5	6 (4) 5 (3) 1 (4) 5 (3) 6 (2) 8 (1) 9 (0)	broke
6	3 (4) 7 (3) 7 (2) 8 (1) 0 (0)	broke
7	3 (4) 1 (5) 1 (6) 7 (5) 8 (4) 0 (3) 2 (4)	
	7 (3) 0 (2) 3 (1) 5 (0)	broke
8	2 (6) 8 (5) 9 (4) 3 (3) 3 (2) 4 (1) 2 (2) 5 (1)	
	6 (0)	broke
9	1 (6) 1 (7) 8 (6) 4 (5) 2 (6) 5 (5) 0 (4) 6 (3)	
	1 (4) 1 (5) 5 (4) 5 (3) 0 (2) 1 (3) 4 (2) 4 (1)	
	1 (2) 1 (3) 1 (4) 5 (3) 9 (2) 2 (3) 2 (4) 3 (3)	
	6 (2) 3 (1) 6 (0)	broke
10	8 (4) 7 (3) 8 (2) 7 (1) 2 (2) 1 (3) 7 (2) 5 (1)	
	1 (2) 5 (1) 6 (0)	broke

■ *Summarize results and state the conclusion*: You went broke in all 10 simulations. With this strategy, the estimated probability of success is $0/10 = 0$. The bolder strategy seems to be the better one.

Insight

You can't estimate a probability well with only 10 simulations. In practice, it is common to carry out at least 1000 simulations, to get a better sense of what happens in the long run. Carry out 10 simulations on your own, and (if possible) combine them with results from your entire class. How would you now estimate this probability?

If many, many simulations were performed, which can be easily done with a computer, you would discover that the probability of success with the more cautious strategy is only 0.001. The bolder strategy is better.

TO PRACTICE THIS CONCEPT, TRY EXERCISE 5.57

Probability Is the Key to Statistical Inference

The concepts of probability hold the key to methods for conducting statistical inference–making conclusions about populations using sample data. To help preview this connection, let's consider an opinion poll. Suppose a poll indicates that 45% of those sampled favor legalized gambling. What's the probability that this sample percentage falls within a certain margin of error, say plus or minus 3%, of the true population percentage? The next chapter will build on the probability foundation of this chapter and enable us to answer such a question. We will study how to evaluate the probabilities of all the possible outcomes in a survey sample or an experiment. This will then be the basis of the statistical inference methods we'll learn about in Chapters 7 and 8.

5.43 **Birthdays of Presidents:** Of the first 43 presidents of
[TRY] the United States (George Washington through George
W. Bush), two had the same birthday (Polk and Harding).
Is this highly coincidental? Answer by finding the proba-
bility of at least one birthday match among 43 people.

5.44 **Matching your birthday:** You consider your birth date
to be special since it falls on January 1. Suppose your
class has 25 students.

a. Is the probability of finding at least one student with
a birthday that matches yours greater, the same, or
less than the probability found in Example 13 of a
match for at least two students? Explain.

b. Find that probability.

5.45 **Lots of pairs:** Show that with 25 students, there are 300
pairs of students who can have the same birthday. So it's
really not so surprising if at least two students have the
same birthday. (*Hint:* You can pair 24 other students with
each student, but how can you make sure you don't count
each pair twice?)

5.46 **Horrible 11 on 9/11:** The digits in 9/11 add up to
11 (9 + 1 + 1), American Airlines flight 11 was the
first to hit the World Trade Towers (which took the
form of the number 11), there were 92 people on
board (9 + 2 = 11), September 11 is the 254th day of
the year (2 + 5 + 4 = 11), and there are 11 letters in
Afghanistan, New York City, the Pentagon, and
George W. Bush (see article by L. Belkin, *New York
Times*, August 11, 2002). How could you explain to
someone who has not studied probability that, because
of the way we look for patterns out of the huge number of things that happen, this is not necessarily an
amazing coincidence?

5.47 **Coincidence in your life:** State an event that has happened to you or to someone you know that seems highly
coincidental (such as seeing a friend while on vacation).
Explain why that event may not be especially surprising,
once you think of all the similar types of events that
could have happened to you or someone that you know,
over the course of several years.

5.48 **Monkeys typing Shakespeare:** Since events of low
probability eventually happen if you observe enough trials, a monkey randomly pecking on a typewriter could
eventually write a Shakespeare play just by chance. Let's
see how hard it would be to type the title of *Macbeth*
properly. Assume 50 keys for letters and numbers and
punctuation. Find the probability that the first seven letters that a monkey types are macbeth. (Even if the monkey can type 60 strokes a minute and never sleeps, if we
consider each sequence of 7 keystrokes as a trial, we
would wait on the average over 100,000 years before seeing this happen!)

5.49 **A *true* coincidence at DisneyWorld:** Wisconsin has 5.4
million residents. On any given day, the probability is

1/5000 that a randomly selected Wisconsin resident decides to visit DisneyWorld in Florida.

a. Find the probability that they all will decide to go tomorrow, in which case DisneyWorld has more than 5.4
million people in line when it opens in the morning.

b. What assumptions did your solution in (a) make? Are
they realistic? Explain.

5.50 **Rosencrantz and Guildenstern:** In the opening scene
[TRY] of Tom Stoppard's play *Rosencrantz and Guildenstern are
Dead*, about two Elizabethan contemporaries of Hamlet,
Guildenstern flips a coin 91 times and gets a head each
time. Suppose the coin was balanced.

a. Specify the sample space for 91 coin flips, such that
each outcome in the sample space is equally likely.
How many outcomes are in the sample space?

b. Show Guildenstern's outcome for this sample space.
Show the outcome in which only the second flip is
a tail.

c. What's the probability of the event of getting a head
91 times in a row?

d. What's the probability of at least one tail, in the
91 flips?

e. State the probability model on which your solutions
in (c) and (d) are based.

5.51 **Mammogram diagnostics:** Breast cancer is the most
[TRY] common form of cancer in women, affecting about 10%
of women at some time in their lives. The chance of
breast cancer increases as a woman ages, and the American Cancer Society recommends an annual mammogram
after age 40 to test for its presence. Of the women who
undergo mammograms at any given time, about 1% are
typically estimated to actually have breast cancer. The
likelihood of a false test result varies according to the
breast density and the radiologist's level of experience.
For use of the mammogram to detect breast cancer, typical values reported are sensitivity = 0.86 and
specificity = 0.88.

a. Construct a tree diagram in which the first set of
branches shows whether a woman has breast cancer
and the second set of branches shows the mammogram result. At the end of the final set of branches,
show that $P(S \text{ and } POS) = 0.01 \times 0.86 = 0.0086$,
and report the other intersection probabilities also.

b. Restricting your attention to the two routes that have
a positive test result, show that $P(POS) = 0.1274$.

c. Of the women who receive a positive mammogram
result, what proportion actually have breast cancer?

d. The tree diagram shown here illustrates how
$P(S|POS)$ can be so small, using a typical group of
100 women who have a mammogram. Explain how
to get the frequencies shown on the branches, and
explain why this suggests that $P(S|POS)$ is only
about 0.08.

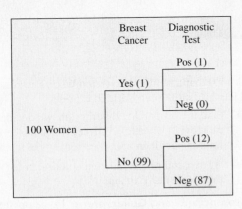

Typical results of mammograms for 100 women

5.52 **More screening for breast cancer:** Refer to the previous exercise. For young women, the prevalence of breast cancer is lower. Suppose the sensitivity is 0.86 and the specificity is 0.88, but the prevalence is only 0.001.

a. Given that a test comes out positive, find the probability that the woman truly has breast cancer.

b. Show how to use a tree diagram with frequencies for a typical sample of 1000 women to explain to someone who has not studied statistics why the probability found in (a) is so low.

c. Of the cases that are positive, explain why the proportion in error is likely to be larger for a young population than for an older population.

5.53 **Convicted by mistake:** In a jury trial, suppose the probability the defendant is convicted, given guilt, is 0.95, and the probability the defendant is acquitted, given innocence, is 0.95. Suppose that 90% of all defendants truly are guilty.

a. Given that the defendant is convicted, find the probability he or she was actually innocent. Draw a tree diagram or construct a contingency table to help you answer.

b. Repeat (a), but under the assumption that 50% of all defendants truly are guilty.

5.54 **Was OJ actually guilty?:** Former pro football star O.J. Simpson was accused of murdering his wife. In the trial, a defense attorney pointed out that although Simpson had been guilty of earlier spousal abuse, annually less than 40 women are murdered per 100,000 incidents of partner abuse. This means that P(murdered by partner | partner abuse) = 40/100,000, but more relevant is P(murdered by partner | partner abuse and women murdered). Every year it is estimated that 5 of every 100,000 women in the United States who are murdered are killed by someone other than their partner (Gigerenzer, 2002, p. 144). Part of a tree diagram is shown starting with 100,000 women.

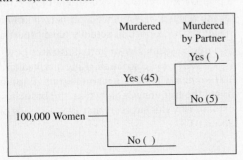

a. Based on the results stated, explain why the numbers 45 and 5 are entered as shown on two of the branches.

b. Fill in the two blanks shown in the tree diagram.

c. Conditional on partner abuse and the woman being murdered (by someone), explain why the probability the woman was murdered by her partner is 40/45. Why is this so dramatically different from P(murdered by partner | partner abuse) = 40/100,000?

5.55 **DNA evidence compelling?:** DNA evidence can be extracted from biological traces such as blood, hair, and saliva. "DNA fingerprinting" is increasingly used in the courtroom as well as in paternity testing. Given that a person is innocent, suppose that the probability of their DNA matching that found at the crime scene is only 0.000001, one in a million. Further, given that a person is guilty, suppose that the probability of their DNA matching that found at the crime scene is 0.99. Jane Doe's DNA matches that found at the crime scene.

a. Find the probability that Jane Doe is actually innocent, if unconditionally her probability of innocence is 0.50. Interpret. Show your solution by introducing notation for events, specifying probabilities that are given, and using a tree diagram to find your answer.

b. Repeat (a) if the unconditional probability of innocence is 0.99. Compare results.

c. Explain why it is very important for a defense lawyer to explain the difference between P(DNA match | person innocent) and P(person innocent | DNA match).

5.56 **Triple blood test:** Example 8 about the Triple Blood Test for Down syndrome found the results shown in the table below.

| Down | Blood Test | | |
	POS	NEG	Total
Yes	48	6	54
No	1307	3921	5228
Total	1355	3927	5282

a. Estimate the probability that Down syndrome occurs (Down = Yes).

b. Find the estimated (i) sensitivity, (ii) specificity.

c. Find the estimated (i) P(Yes | POS), (ii) P(No | NEG).

d. Explain how the probabilities in (b) and (c) give four ways of describing the probability that a diagnostic test makes a correct decision.

5.57 **Simulating donations to local blood bank:** The director of a local blood bank is in need of a donor with type AB blood. The distribution of blood types for Americans is estimated by the American Red Cross to be

Type	A	B	O	AB
Probability	40%	10%	45%	5%

The director decides that if more than 20 donors are required before the first donor with AB blood appears, she will need to issue a special appeal for AB blood.

a. Conduct a simulation 10 times, using the *random numbers* applet on the text CD or a calculator or software, to estimate the probability this will happen. Show all steps of the simulation, including any assumptions that you make. Refer to Example 16 as a model for carrying out this simulation.

b. In practice, you would do at least 1000 simulations to estimate this probability well. You'd then find that the probability of exceeding 20 donors before finding a type AB donor is 0.36. Actually, simulation is not needed. Show how to find this probability using the methods of this chapter.

CHAPTER REVIEW

Answers to the Chapter Figure Questions

Figure 5.1: The cumulative proportion after each roll is evaluated as the frequency of six's rolled through that trial (roll) number divided by the total number of trials (rolls). The cumulative proportions are $1/1 = 1$ for roll 1, $1/2 = 0.50$ for roll 2, $2/3 = 0.67$ for roll 3, and $2/4 = 0.50$ for roll 4.

Figure 5.2: It should be very, very close to 1/2.

Figure 5.3: $2 \times 2 \times 2 \times 2 = 16$ branches.

Figure 5.4: Venn diagrams will vary. One possible Venn diagram is

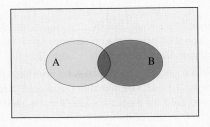

Figure 5.5: The favorable outcomes to the event that the student answers the first question correctly are {CCC, CCI, CIC, CII}. This event is not disjoint from either of the two labeled events in the Venn diagram.

Figure 5.6: P(A or B) = P(A) + P(B) − P(A and B)

Figure 5.7: P(A or B) = P(A) + P(B) when the events A and B are disjoint.

Figure 5.8: We would not expect the trials to be independent if the student is not guessing. If the student is using prior knowledge, the probability of getting a correct answer on a question, such as question 3, may be more likely if the student answers questions 1 and 2 correctly.

Figure 5.9: P(A | B) is not necessarily equal to P(B | A). They have the same numerator, P(A and B), but different denominators, P(B) and P(A) respectively, which may not be equal.

Figure 5.11: As the prevalence rate increases, of the individuals that test positive, a higher percentage of positive test results will occur with individuals that actually use drugs. For example, if 5% of the population uses drugs, of the 115 individuals out of 1000 that would be expected to test positive, 48 would be actual drug users while 67 would be non drug users.

CHAPTER SUMMARY

In this chapter, we've seen how to quantify uncertainty and randomness. With many independent trials of random phenomena, outcomes do show certain regularities. The proportion of times an outcome occurs, in the long run, is its **probability**.

The **sample space** is the set of all possible outcomes of a random phenomenon. An **event** is a subset of the sample space. Two events A and B are **disjoint** if they have no common elements.

We find probabilities using basic rules:

- The **probability** of each individual outcome falls between 0 and 1, and the total of all the individual probabilities equals 1.
- The **probability of an event** is the sum of the probabilities of individual outcomes in that event.

- For an event A and its **complement A^c** (the outcomes not in A),

$$P(A^c) = 1 - P(A).$$

- P(A or B) = P(A) + P(B) − P(A and B). This simplifies to P(A or B) = P(A) + P(B) when the events are disjoint.

- P(A and B) = P(A | B) × P(B), where P(A | B) denotes the **conditional probability** of event A, given that event B occurs. Equivalently, the conditional probability satisfies

$$P(A \mid B) = \frac{P(A \text{ and } B)}{P(B)}.$$

Likewise, P(A and B) = P(B | A)P(A), and P(B | A) = P(A and B)/P(A).

■ When A and B are **independent**, $P(A|B) = P(A)$ and $P(B|A) = P(B)$. Then, also $P(A \text{ and } B) = P(A) \times P(B)$.

A **probability model** states certain assumptions and, based on them, finds the probabilities of events of interest.

Understanding probability helps us to make informed decisions in our lives. In the next three chapters, we'll see how an understanding of probability gives us the basis for making statistical inferences about a population.

Summary of New Notation in Chapter 5

A, B, C   Events
A^c Complement of event A (the outcomes *not* in A)
$P(A)$ Probability of event A

$P(A|B)$ Conditional probability of event A, given event B (| denotes "given")

CHAPTER PROBLEMS

Practicing the Basics

5.58 Barry Bonds home runs: In his prime, the probability that major league baseball player Barry Bonds hit a home run when he came to bat was about 1/10.

 a. Does this mean that if we watched him at bat 10 times, he would almost certainly get exactly one home run? Explain.

 b. Explain what this probability means in terms of watching him at bat over a longer period, say 1000 times over two seasons of his prime.

5.59 Due for a boy?: A couple has five children, all girls. They are expecting a sixth child. The father tells a friend that by the law of large numbers the chance of a boy is now much greater than 1/2. Comment on the father's statement.

5.60 P (life after death): Explain the difference between the relative frequency and subjective definitions of probability. Illustrate by explaining how it is possible to give a value for (a) the probability of life after death, (b) the probability that in the morning you remember at least one dream that you had in the previous night's sleep, based on what you observe every morning for the next year.

5.61 Choices for lunch: For the set lunch at Amelia's Restaurant, customers can select one meat dish, one vegetable, one beverage, and one dessert. The menu offers two meats (beef and chicken), three vegetables (corn, green beans, or potatoes), three beverages (Coke, ice tea, or coffee), and one dessert (Amelia's apple pie).

 a. Use a tree diagram to list the possible meals and to determine how many there are.

 b. In practice, would it be sensible to treat all the outcomes in the sample space as equally likely for the customer selections we'd observe? Why or why not?

5.62 Caught doctoring the books: After the major accounting scandals with Enron, a large energy company, the question may be posed, "Was there any way to examine

Enron's accounting books to determine if they had been 'doctored'?" One way uses *Benford's Law*. This states that in a variety of circumstances, numbers as varied as populations of small towns, figures in a newspaper or magazine, and tax returns and other business records begin with the digit "1" more often than other digits. This law states that the probabilities for the digits 1 through 9 are approximately:

Digit	1	2	3	4	5	6	7	8	9
Probability	0.30	0.18	0.12	0.10	0.08	0.07	0.06	0.05	0.04

 a. If we were to randomly pick one of the digits between 1 and 9 using a random number table or software, what is the probability for each digit?

 b. When people attempt to fake numbers, there's a tendency to use 5 or 6 as the initial digit more often than predicted by Benford's law. What is the probability of 5 or 6 as the first digit by (i) Benford's law, (ii) random selection?

5.63 Life after death: In a General Social Survey, in response to the question "Do you believe in life after death?" 907 answered yes and 220 answered no.

 a. Based on this survey, estimate the probability that a randomly selected adult in the United States believes in life after death.

 b. A married couple is randomly selected. Estimate the probability that both subjects believe in life after death.

 c. What assumption did you make in answer (b)? Explain why that assumption is probably unrealistic, making this estimate unreliable.

5.64 Death penalty jury: In arguing against the death penalty, Amnesty International has pointed out supposed inequities, such as the many times a black person has been given the death penalty by an all-white jury. If jurors are selected randomly from an adult population, find the probability that all 12 jurors are white when the population is (a) 90% white, (b) 50% white.

5.65 Driver's exam: Three fifteen-year-old friends with no particular background in driver's education decide to take the written part of the Georgia Driver's Exam. Each exam was graded as a pass (P) or a failure (F).

    a. How many outcomes are possible for the grades received by the three friends together? Using a tree diagram, list the sample space.

    b. If the outcomes in the sample space in (a) are equally likely, find the probability that all three pass the exam.

    c. In practice, the outcomes in (a) are *not* equally likely. Suppose that statewide 70% of fifteen-year-olds pass the exam. If these three friends were a random sample of their age group, find the probability that all three pass.

    d. In practice, explain why probabilities that apply to a random sample are not likely to be valid for a sample of three friends.

5.66 Visit France and Italy: Students in a geography class are asked whether they've ever visited France, and whether they've ever visited Italy.

    a. For a randomly selected student, would you expect these events to be independent, or dependent? Explain.

    b. How would you explain to someone who has never studied statistics what it means for these events to be independent?

5.67 Insurance coverage: The Gallup Organization conducted a national poll in June 2003. The adults in the poll were asked the type of health insurance coverage they have, given they had used an emergency room at least once in the last 12 months. The table shows results. For the population who had used an emergency room at least once in the past 12 months:

    a. Estimate the probability that a patient has health insurance.

    b. Given that a subject has health insurance, estimate the probability it is private.

Type of Insurance Coverage	Proportion
Private insurance	0.22
Medicare/Medicaid	0.36
No health insurance	0.27
Other	0.15

5.68 Teens and drugs: In 2000 the Center on Addiction and Substance Abuse (CASA) at Columbia University conducted a survey asking 1000 randomly chosen teenagers questions about their parents and the teenagers' views on the use of illegal substances. Twenty-six percent of the sampled teenagers cited drugs as their biggest concern. Of them, 31% said "drugs can ruin your life and cause harm." Which of these two percentages refers to a conditional probability? For it, identify the event conditioned on and the event to which the probability refers.

5.69 Teens and parents: In the CASA teen survey described in the previous exercise, 27% of the teens said they lived with "hands-on" parents—parents who have established a household culture of rules and expectations for their teen's behav-

ior. An excellent relationship with their fathers was reported by 47% of teens living in "hands-on" households but only 13% of teens living in "hands-off" households.

    a. Which of these percentages estimates a conditional probability? For each that does, identify the event conditioned on and the event to which the probability refers.

    b. For the variables "type of parent" and "whether the teen has an excellent relationship with father," construct a contingency table showing counts for the 1000 subjects.

    c. Using the contingency table, given that a teen has an excellent relationship with the father, estimate the probability that the teenager has "hands off" parents.

5.70 Laundry detergent: A manufacturer of laundry detergent has introduced a new product that it claims to be more environmentally sound. An extensive survey gives the percentages shown in the table.

Result of Advertising for New Laundry Product		
	Tried the New Product	
Advertising Status	Yes	No
Seen the ad	10%	25%
Have not seen ad	5%	60%

    a. Estimate the probability that a randomly chosen consumer would have seen advertising for the new product and tried the product.

    b. Given that a randomly chosen consumer has seen the product advertised, estimate the probability that the person has tried the product.

    c. Let A be the event that a consumer has tried the product. Let B be the event that a consumer has seen the product advertised. Express the probabilities found in (a) and (b) in terms of A and B.

    d. Are A and B independent? Justify your answer.

5.71 Board games and dice: A board game requires players to roll a pair of balanced dice for each player's turn. Denote the outcomes by (die 1, die 2), such as (5, 3) for a 5 on die 1 and a 3 on die 2.

    a. List the sample space of the 36 possible outcomes for the two dice.

    b. Let A be the event that you roll doubles (that is, each die has the same outcome). List the outcomes in A, and find its probability.

    c. Let B be the event that the sum on the pair of dice is 7. Find $P(B)$.

    d. Find the probability of (i) A and B, (ii) A or B, (iii) B given A.

    e. Are events A and B independent, or disjoint, or neither? Explain.

5.72 Roll two more dice: Refer to the previous exercise. Define event D as rolling an odd sum with two dice.

    a. B denotes a sum of 7 on the two dice. Find $P(B$ and $D)$. When an event B is contained within an event D, as here, explain why $P(B$ and $D) = P(B)$.

b. Find P(B or D). When an event B is contained within an event D, explain why P(B or D) = P(D).

5.73 Conference dinner: Of the participants at a conference, 50% attended breakfast, 90% attended dinner, and 40% attended both breakfast and dinner. Given that a participant attended breakfast, find the probability that she also attended dinner.

5.74 Waste dump sites: A federal agency is deciding which of two waste dump projects to investigate. A top administrator estimates that the probability of federal law violations is 0.30 at the first project and 0.25 at the second project. Also, he believes the occurrences of violations in these two projects are disjoint.

a. What is the probability of federal law violations in the first project or in the second project?

b. Given that there is not a federal law violation in the first project, find the probability that there is a federal law violation in the second project.

c. In reality, the administrator confused disjoint and independent, and the events are actually independent. Answer (a) and (b) with this correct information.

5.75 A dice game: Consider a game in which you roll two dice, and you win if the total is 7 or 11 and you lose if the total is 2, 3, or 12. You keep rolling until one of these totals occurs. Using conditional probability, find the probability that you win.

5.76 No coincidences: Over time, you have many conversations with a friend about your favorite actress, favorite musician, favorite book, favorite TV show, and so forth for 100 separate topics. On any given topic, there's only a 0.02 probability that you agree. If you did agree on a topic, you would consider it to be coincidental.

a. If whether you agree on any two different topics are independent events, find the probability that you and your friend *never* have a coincidence on these 100 topics.

b. Find the probability that you have a coincidence on at least one topic.

5.77 Amazing roulette run?: A roulette wheel in Monte Carlo has 18 even-numbered slots, 18 odd-numbered slots, a slot numbered zero, and a double zero slot. On August 18, 1913, it came up even 26 times in a row.[7] As more and more evens occurred, the proportion of people betting on an odd outcome increased, as they figured it was "due."

a. Comment on this strategy.

b. Find the probability of 26 evens in a row if each slot is equally likely.

c. Suppose that over the past 100 years there have been 1000 roulette wheels, each being used hundreds of times a day. Is it surprising if sometime in the previous 100 years one of these wheels had 26 evens in a row? Explain.

---

[7]*What Are the Chances?* by B. K. Holland
(Johns Hopkins University Press, 2002, p. 10).

5.78 Death penalty and false positives: For the decision about whether to convict someone charged with murder and give them the death penalty, consider the variables "reality" (defendant innocent, defendant guilty) and "decision" (convict, acquit).

a. Explain what the two types of errors are in this context.

b. Jurors are asked to convict a defendant if they feel the defendant is guilty "beyond a reasonable doubt." Suppose this means that given the defendant is executed, the probability that he or she truly was guilty is 0.99. For the 873 people put to death from the time the death penalty was reinstated in 1977 until September 2003, find the probability that (i) they were all truly guilty, (ii) at least one of them was actually innocent.

c. How do the answers in (b) change if the probability of true guilt is actually 0.95?

5.79 Screening smokers for lung cancer: An article about using a diagnostic test (helical computed tomography) to screen adult smokers for lung cancer warned that a negative test may cause harm by providing smokers with false reassurance, and a false positive test results in an unnecessary operation opening the smoker's chest (Mahadevia et al., *JAMA*, vol. 289, pp. 313–322, 2003). Explain what false negatives and false positives mean in the context of this diagnostic test.

5.80 Screening for heart attacks: Biochemical markers are used by emergency room physicians to aid in diagnosing patients who have suffered acute myocardial infarction (AMI), or what's commonly referred to as a heart attack. One type of biochemical marker used is creatine kinase (CK). Based on a review of published studies on the effectiveness of these markers (by E. M. Balk et al., *Annals of Emergency Medicine*, vol. 37, pp. 478–494, 2001), CK had an estimated sensitivity of 37% and specificity of 87%. Consider a population having a prevalence rate of 25%.

a. Explain in context what is meant by the sensitivity equaling 37%.

b. Explain in context what is meant by the specificity equaling 87%.

c. Construct a tree diagram for this diagnostic test. Label the branches with the appropriate probabilities.

5.81 Screening for colorectal cancer: Gigerenzer (2002, p. 105) reported that on the average, "Thirty out of every 10,000 people have colorectal cancer. Of these 30 people with colorectal cancer, 15 will have a positive hemoccult test. Of the remaining 9,970 people without colorectal cancer, 300 will still have a positive hemoccult test."

a. Sketch a tree diagram or construct a contingency table to display the counts.

b. Of the 315 people mentioned above who have a positive hemoccult test, what proportion actually have colorectal cancer? Interpret.

5.82 Color blindness: For genetic reasons, color blindness is more common in men than women: 5 in 100 men and 25 in 10,000 women suffer from color blindness.

   a. Defining events, identify these proportions as conditional probabilities.

   b. If the population is half male and half female, what proportion of the population is color blind? Use a tree diagram or contingency table with frequencies to portray your solution, showing what you would expect to happen with 20,000 people.

   c. Given that a randomly chosen person is color blind, what's the probability that person is female? Use the tree diagram or table from (b) to find the answer.

5.83 HIV testing: For a combined ELISA–Western blot blood test for HIV positive status, the sensitivity is about 0.999 and the specificity is about 0.9999 (Gigerenzer 2002, pp. 124, 126).

   a. Consider a high-risk group in which 10% are truly HIV positive. Construct a tree diagram to summarize this diagnostic test.

   b. Display the intersection probabilities from (a) in a contingency table.

   c. A person from this high-risk group has a positive test result. Using the tree diagram or the contingency table, find the probability that this person is truly HIV positive.

   d. Explain why a positive test result is more likely to be in error when the prevalence is lower. Use tree diagrams or contingency tables with frequencies for 10,000 people with 10% and 1% prevalence rates to illustrate your arguments.

5.84 Prostate cancer: A study of the PSA blood test for diagnosing prostate cancer in men (by R. M. Hoffman et al., *BMC Family Practice*, vol. 3, p. 19, 2002) used a sample of 2620 men who were 40 years and older. When a positive diagnostic test result was defined as a PSA reading of at least 4, the sensitivity was estimated to be 0.86 but the specificity only 0.33.

   a. Suppose that 10% of those who took the PSA test truly had prostate cancer. Given that the PSA was positive, use a tree diagram and/or contingency table to estimate the probability that the man truly had prostate cancer.

   b. Illustrate your answer in (a) by using a tree diagram or contingency table with frequencies showing what you would expect for a typical sample of 1000 men.

   c. Lowering the PSA boundary to 2 for a positive result changed the sensitivity to 0.95 and the specificity to 0.20. Explain why, intuitively, if the cases increase for which a test is positive, the sensitivity will go up but the specificity will go down.

Concepts and Investigations

5.85 Simulate law of large numbers: Using the *simulating the probability of a head with a fair coin* applet on the text CD or other software, simulate the flipping of a balanced coin.

   a. Report the cumulative proportion of heads after (i) 10 flips, (ii) 100 flips, (iii) 1000 flips, (iv) 10,000 flips. Explain how the results illustrate the law of large numbers and the long-run relative frequency definition of probability.

   b. Using the *simulating the probability of rolling a 3 or 4* applet or other software, simulate the roll of a die with a success defined as rolling a 3 or 4, using (i) 10 rolls, (ii) 100 rolls, (iii) 1000 rolls, (iv) 10,000 rolls. Summarize results.

5.86 Illustrate probability terms with scenarios:

   a. What is a sample space? Give an example of a sample space for a scenario involving (i) a designed experiment, (ii) an observational study.

   b. What are disjoint events? Give an example of two events that are disjoint.

   c. What is a conditional probability? Give an example of two events in your everyday life that you would expect to be (i) independent, (ii) dependent.

5.87 Short term vs. long run: Short-term aberrations do not affect the long run. To illustrate, suppose that you flip a coin 10 times and you get 10 heads. Find the cumulative proportion of heads, including these first ten flips, if (a) in the 100 flips after those 10 you get 50 heads (so there are now 60 heads in 110 flips), (b) in the 1000 flips after those 10 you get 500 heads, (c) in the 10,000 flips after those 10 you get 5000 heads. What number is the cumulative proportion tending toward as *n* increases?

5.88 Risk of space shuttle: After the *Columbia* space shuttle disaster, a former NASA official who faulted the way the agency deals with safety risk warned (in an AP story, March 7, 2003) that NASA workers believed, "If I've flown 20 times, the risk is less than if I've flown just once."

   a. Explain why it would be reasonable for someone to form this belief, from the way we use empirical evidence and margins of error to estimate an unknown probability.

   b. Explain a criticism of this belief, using coin flipping as an analogy.

5.89 Mrs. Test: "Mrs. Test" (see www.mrstest.com) sells diagnostic tests for various conditions. Their Web site gives only imprecise information about the accuracy of the tests. The test for pregnancy is said to be "over 99% accurate." Describe at least four different probabilities to which this could refer.

5.90 Marijuana leads to heroin?: Nearly all heroin addicts have used marijuana sometime in their lives. So, some argue that marijuana should be illegal because marijuana users are likely to become heroin addicts. Use a Venn diagram to illustrate the fallacy of this argument, by sketching sets for M = marijuana use and H = heroin use such that $P(M \mid H)$ is close to 1 but $P(H \mid M)$ is close to 0.

5.91 Stay in school: Suppose 80% of students finish high school. Of them, 50% finish college. Of them, 20% get a masters' degree. Of them, 30% get a Ph.D.

  a. What percentage of students get a Ph.D.?

  b. Explain how your reasoning in (a) used a multiplication rule with conditional probabilities.

  c. Given that a student finishes college, find the probability of getting a Ph.D.

5.92 How good is a probability estimate?: In Example 8 about Down syndrome, we estimated the probability of a positive test result (predicting that Down syndrome is present) to be $P(POS) = 0.257$, based on observing 1355 positive results in 5282 observations. How good is such an estimate? From Section 4.2, $1/\sqrt{n}$ is an approximate margin of error in estimating a proportion with n observations.

  a. Find the approximate margin of error to describe how well this proportion estimates the true probability, $P(POS)$.

  b. The *long run* in the definition of probability refers to letting n get very large. What happens to this margin of error formula as n keeps growing, eventually toward infinity? What's the implication of this?

5.93 Protective bomb: Before the days of high security at airports, there was a legendary person who was afraid of traveling by plane because someone on the plane might have a bomb. He always brought a bomb himself on any plane flight he took, believing that the chance would be astronomically small that two people on the same flight would both have a bomb. Explain the fallacy in his logic, using ideas of independent events and conditional probability.

5.94 Streak shooter: Sportscaster Maria Coselli claims that players on the New York Knicks professional basketball team are streak shooters. To make her case, she looks at the statistics for all the team's players over the past three games and points out that one of them (Joe Smith) made 6 shots in a row at one stage. Coselli argues, "Over a season, Smith makes only 45% of his shots. The probability that he would make 6 in a row if the shots were independent is $(0.45)^6 = 0.008$, less than 1 in a hundred. This would be such an unusual occurrence that we can conclude that Smith is a streak shooter."

  a. Explain the flaws in Coselli's logic.

  b. Use this example to explain how some things that look highly coincidental may not be so when viewed in a wider context.

Select the appropriate answer(s) to multiple-choice Exercises 5.95–5.99. Some exercises may have more than one correct answer.

5.95 MC1: For two events A and B, $P(A) = 0.5$ and $P(B) = 0.2$. Then $P(A \text{ or } B)$ equals

  a. 0.10, if A and B are independent

  b. 0.70, if A and B are independent

  c. 0.60, if A and B are independent

  d. 0.70, if A and B are disjoint

5.96 MC2: Which of the following is always true?

  a. If A and B are independent, then they are also disjoint.

  b. $P(A \mid B) + P(A \mid B^c) = 1$

  c. If $P(A \mid B) = P(B \mid A)$, then A and B are independent.

  d. If A and B are disjoint, then A and B cannot occur at the same time.

5.97 Coin flip: A balanced coin is flipped 100 times. By the law of large numbers:

  a. There will almost certainly be *exactly* 50 heads and 50 tails.

  b. If we got 100 heads in a row, almost certainly the next flip will be a tail.

  c. For the 100 flips, the probability of getting 100 heads equals the probability of getting 50 heads.

  d. It is absolutely impossible to get a head every time.

  e. None of the above.

5.98 Dreams come true: You have a dream in which you see your favorite movie star in person. The very next day, you are visiting Manhattan and you see her walking down Fifth Avenue.

  a. This is such an incredibly unlikely coincidence that you should report it to your local newspaper.

  b. This is somewhat unusual, but given the many dreams you could have in your lifetime about people you know or know of, it is not an incredibly unlikely event.

  c. If you had not had the dream, you would definitely not have seen the film star the next day.

  d. This proves the existence of ESP.

5.99 Comparable risks: Mammography is estimated to save about 1 life in every 1000 women. "Participating in annual mammography screening ... has roughly the same effect on life expectancy as reducing the distance one drives each year by 300 miles" (Gigerenzer 2002, pp. 60, 73). Which of the following do you think has the closest effect on life expectancy as that of smoking throughout your entire life?

  a. Taking a commercial airline flight once a year

  b. Driving about ten times as far every year as the average motorist

  c. Drinking a cup of coffee every day

  d. Eating a fast-food hamburger once a month

  e. Never having a mammogram (if you are a woman)

Select true or false for Exercises 5.100–5.101.

5.100 TF1: When you flip a coin ten times, you are more likely to get the sequence HTHTHTHTHT than the sequence HHHHHTTTTT.

5.101 TF2: When you flip a balanced coin twice, there are three things that can happen: 0 heads, 1 head, or 2 heads.

Since there are three possible outcomes, they each have probability 1/3. (This was claimed by the French mathematician, Jean le Rond d' Alembert, in 1754. To reason this, write down the sample space for the possible sequence of results for the two flips.)

5.102 Driving vs. flying: In the U.S. in 2002, about 43,000 people died in auto crashes and 0 people died in commercial airline accidents. G. Gigerenzer (2002, p. 31) states, "The terrorist attack on September 11, 2001 cost the lives of some 3,000 people. The subsequent decision of millions to drive rather than fly may have cost the lives of many more." Explain the reasoning behind this statement.

5.103 Prosecutor's fallacy: An eyewitness to the crime says that the person who committed it was male, between 15 and 20 years old, Hispanic, and drove a blue Honda. The prosecutor points out that a proportion of only 0.001 people living in that city match all those characteristics, and one of them is the defendant. Thus, the prosecutor argues that the probability that the defendant is not guilty is only 0.001. Explain what is wrong with this logic. (*Hint*: Is $P(\text{match}) = P(\text{not guilty} \mid \text{match})$?)

◆◆ **5.104 Generalizing the addition rule:** For events A, B, and C such that each pair of events is disjoint, use a Venn diagram to explain why

$$P(A \text{ or } B \text{ or } C) = P(A) + P(B) + P(C).$$

◆◆ **5.105 Generalizing the multiplication rule:** For events A, B, and C, explain why $P(A \text{ and } B \text{ and } C) =$

$$P(A) \times P(B \mid A) \times P(C \mid A \text{ and } B).$$

◆◆ **5.106 Bayes's rule:** Suppose we know $P(A)$, $P(B \mid A)$, and $P(B^c \mid A^c)$, but we want to find $P(A \mid B)$.

a. Using the definition of conditional probability for $P(A \mid B)$ and for $P(B \mid A)$, explain why $P(A \mid B) = P(A \text{ and } B)/P(B) = [P(B \mid A)P(A)]/P(B)$.

b. Splitting the event that B occurs into two parts, according to whether or not A occurs, explain why $P(B) = P(B \text{ and } A) + P(B \text{ and } A^c)$.

c. Using (b) and the definition of conditional probability, explain why

$$P(B) = P(B \mid A)P(A) + P(B \mid A^c)P(A^c).$$

d. Combining what you have shown in (a)-(c), reason that

$$P(A \mid B) = \frac{P(B \mid A)P(A)}{P(B \mid A)P(A) + P(B \mid A^c)P(A^c)}.$$

This formula is called **Bayes rule**. It is named after a British clergyman who discovered the formula in 1763.

Class Explorations

5.107 Simulating matching birthdays: Do you find it hard to believe that the probability of at least one birthday match in a class of 25 students is 0.57? Let's simulate the answer. Using the *random numbers* applet on the text CD, each student in the class should simulate 25 numbers between 1 and 365. Observe whether there is at least one match of two numbers out of the 25 simulated numbers.

a. Combine class results. What proportion of the simulations had at least one birthday match?

b. Repeat (a) for a birthday match in a class of 50 students. Again, combine class results. What is the simulated probability of at least one match?

5.108 Simulate table tennis: In table tennis, the first person to get at least 21 points while being ahead of the opponent by at least two points wins the game. In games between you and an opponent, suppose successive points are independent, and suppose the probability of your winning any given point is 0.40.

a. Do you have any reasonable chance to win a game? Showing all steps, simulate two games, using the *random numbers* applet on the text CD or other software. (*Hint:* Let the integers 0–3 represent winning a point and 4–9 represent losing a point.)

b. Combining results for all students in the class, estimate the probability of winning.

5.109 Which tennis strategy is better?: A tennis match can consist of the best of three sets (that is, the winner is the first to win two sets) or the best of five sets (the winner is the first to win three sets). Which would you be better off playing if you are the weaker player and have probability 0.40 of winning any particular set?

a. Simulate 10 matches of each type, using *the random numbers* applet on the text CD or other software. Show the steps of your simulation, and specify any assumptions.

b. Combining results for all students in the class, estimate the probability of the weaker player winning the match under each type of match.

BIBLIOGRAPHY

Against All Odds: Inside Statistics (1989). Produced by Consortium for Mathematics and its Applications, for PBS. See www.pbs.org/als/against_odds

Gigerenzer, G. (2002). *Calculated Risks*. New York: Simon & Schuster.

Wilson, R., and E. A. C. Crouch (2001). *Risk-Benefit Analysis*. Cambridge, MA: Harvard Univ. Press.

6 | Probability Distributions

EXAMPLE 1

How Can We Predict Election Results Using Exit Polls?

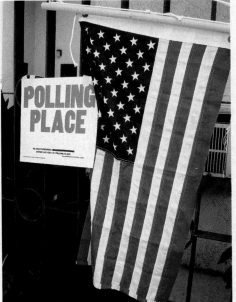

Picture the Scenario

In the United States, Canada, the United Kingdom, Australia, and many other countries, news organizations declare the outcomes of many election contests well before all the votes have been counted. They develop this projection with a technique known as "exit polling," an opinion poll in which voters are randomly sampled after leaving the voting booth. Using an exit poll, a polling organization predicts a winner after learning how a small number of people voted, often only a few thousand out of possibly millions of voters. What amazes many people is that these predictions almost always turn out to be correct. (A famous exception will be discussed at the end of the chapter.)

In California in October 2003, a special election was held to determine whether or not Governor Gray Davis should be recalled from office. In the same election, Californians voted for the candidate who would replace Davis as governor if he were recalled. The four candidates included the Hollywood actor Arnold Schwarzenegger. If a majority voted yes for the recall, the candidate who received the most votes would be the new governor. The exit poll on which TV networks relied for their projections found, after sampling 3160 voters, that 54% of the sample voted to recall Gray Davis and 46% voted not to recall him (www.cnn.com/ELECTION/2003/recall). At the time of the exit poll, the percentage of the entire voting population (nearly 8 million people) that voted in favor of the recall was unknown. In determining whether they could predict a winner, the TV networks had to figure whether the exit poll data gave enough evidence to predict that the population percentage voting in favor of recalling Davis was above 50%.

Questions to Explore

■ How close can we expect a sample percentage to be to the population percentage? For instance, if 54% of 3160 sampled voters supported the recall, how close to 54% is the percentage of the entire population of 8 million voters who voted for the recall of Governor Gray Davis?

■ How does the sample size influence our analyses? For instance, could we sample 100 voters instead of 3160 voters and make an accurate inference about the population percentage favoring recall of Davis, or do we perhaps need tens of thousands in our sample?

Thinking Ahead

In this chapter, we'll apply the probability tools of the previous chapter to analyze how likely it is that sample results will be "close" to population values. For example, in Examples 16 and 17, we'll see why these results for 3160 voters allow us to make a prediction about the outcome for the entire population of nearly 8 million voters. This will be our first example of the use of *inferential statistics*.

Recall

Section 1.2 defined a **parameter** to be a numerical summary of the population, such as a population mean or a population proportion. ■

Having learned the basics of probability in Chapter 5, we'll next study how probability provides the basis for making statistical inferences. As Chapter 1 explained, inferential statistical methods use sample data to make decisions and predictions about a population. The population summaries are called **parameters**. Inferential methods are the main focus of the rest of this book.

Before starting to address inferential statistics later in this chapter, we need to learn a few more fundamentals about probability. We'll learn how to summarize possible outcomes and their probabilities with a **probability distribution**. We'll then study two very commonly used probability distributions—the **normal** and the **binomial**. We'll see in the rest of the book that the normal distribution, which has a bell-shaped graph, plays a key role in statistical inference.

6.1 | How Can We Summarize Possible Outcomes and Their Probabilities?

Recall

The characteristics we measure are called **variables** because their values vary from subject to subject. Section 2.1 introduced **discrete** and **continuous** variables. If the possible outcomes are a set of separate numbers, the variable is discrete (for example, a variable expressed as "the number of ..." with possible values 0, 1, 2, ...). If the variable can take any value in an interval, such as the proportion of a tree leaf that shows a fungus (any real number between 0 and 1), it is continuous. ■

With proper methods of gathering data, the numerical values that a variable assumes are the result of some random phenomenon. For example, the randomness may involve selecting a random sample from a population or performing a randomized experiment. In such cases, to emphasize that the outcome of the variable is *random*, the variable is called a **random variable**.

Random Variable

A **random variable** is a numerical measurement of the outcome of a random phenomenon. Often, the randomness results from the use of random sampling or a randomized experiment to gather the data.

We've used letters near the end of the alphabet, such as x, to symbolize variables. We'll also use letters such as x for the possible value of a random variable. When we want to refer to the random variable itself, rather than a particular value, we'll use a capital letter, such as X. For instance, X = number of heads in three flips of a coin is a random variable, whereas $x = 2$ is one of its possible values.

Because a random variable refers to the outcome of a random phenomenon, each possible outcome has a specific probability of occurring. The **probability distribution** of a random variable specifies its possible values and their probabilities. An advantage of a variable being also a "random variable" is that it's possible to specify such probabilities. Without randomness, it's just not possible to know the probabilities of the possible outcomes.

Probability Distributions for Discrete Random Variables

When a random variable has separate possible values, such as 0, 1, 2, 3 for the number of heads in three flips of a coin, it is called **discrete**. The **probability distribution** of a discrete random variable assigns a probability to each possible value. Each probability falls between 0 and 1, and the sum of the probabilities of all possible values equals 1. We let P(x) denote the probability of a possible value x, such as P(2) for the probability the random variable takes the value 2.

IN WORDS

From Chapter 5, a **probability** is a long-run **proportion**. So, it can take any value in the interval from 0 up to 1. The probabilities for all the possible outcomes add up to 1.0, so that's the sum of the probabilities in a probability distribution.

Probability Distribution of a Discrete Random Variable

A **discrete** random variable X takes a set of separate values (such as 0, 1, 2, ...). Its **probability distribution** assigns a probability P(x) to each possible value x.

■ For each x, the probability P(x) falls between 0 and 1.

■ The sum of the probabilities for all the possible x values equals 1.

For instance, let X be a random digit selected using software for random numbers as part of the process of identifying subjects to include in a random sample. The possible values for X are $x = 0, 1, 2, \ldots, 8, 9$. Each digit is equally likely, so the probability distribution is

$$P(0)=P(1)=P(2)=P(3)=P(4)=P(5)=P(6)=P(7)=P(8)=P(9)=0.10.$$

Each probability falls between 0 and 1, and the probabilities add up to 1.0. If X is the outcome of rolling a balanced die, then the possible values for X are $x = 1, 2, 3, 4, 5, 6$, and $P(1) = P(2) = P(3) = P(4) = P(5) = P(6) = 1/6$.

Random variables can also be **continuous**, having possible values that are an interval rather than a set of separate numbers. We'll learn about continuous random variables and their probability distributions later in this section.

EXAMPLE 2

How Many Home Runs Will the Red Sox Hit in a Game?

Picture the Scenario

One of the authors is a long-suffering fan of the Boston Red Sox baseball team. Finally in 2004, the Red Sox won the World Series (the crowning achievement in major league baseball) for the first time since 1918. This ended many years of frustration, which many attributed to a curse on the team for trading Babe Ruth to the New York Yankees in 1920.

In much of their history, the Red Sox have been a good-hitting team. In a given Boston Red Sox game, what can we expect for X = number of home runs the Red Sox hit? Table 6.1 shows an estimate[1] of the probability distribution that was valid for the team in 2004 for X.

TABLE 6.1: Probability Distribution of Number of Home Runs in a Game for Boston Red Sox

Number of Home Runs	Probability
0	0.23
1	0.38
2	0.22
3	0.13
4	0.03
5	0.01
6 or more	0.00

Question to Explore

a. Show this satisfies the two properties needed for a probability distribution.

b. What is the estimated probability of at least three home runs?

---

[1] Source: www.usatoday.com/sports/baseball/al/redsox/home.htm, based on 2004 regular season. For instance, they hit 0 home runs in 38 of the 162 games, so the probability of 0 is estimated to be 38/162 = 0.23.

Recall

P(3) denotes the probability that the random variable takes the value 3, that is, the probability that the number of home runs in a game equals 3. ■

Think It Through

a. In Table 6.1 the two conditions in the definition of a probability distribution are satisfied:

- For each x, the probability $P(x)$ falls between 0 and 1.
- The sum of the probabilities for all the possible x values equals 1.

b. The estimated probability of at least 3 home runs is

$$P(3) + P(4) + P(5) = 0.13 + 0.03 + 0.01 = 0.17.$$

Insight

The probability distribution of a discrete random variable can also be displayed with a *graph* and sometimes with a *formula*. Figure 6.1 displays a graph for the probability distribution in Table 6.1. It is somewhat skewed to the right.

▲ **FIGURE 6.1: MINITAB Histogram for Probability Distribution of X = Number of Home Runs by the Boston Red Sox in a Game.** The stem over a possible value has height equal to its probability. **Question:** How would you describe the shape of the distribution?

TO PRACTICE THIS CONCEPT, TRY EXERCISE 6.1 AND 6.2

As in describing sample data, a graphical display is useful for revealing the three key components of a distribution: center, spread, and shape. How do we find summary measures of the center and spread of a probability distribution?

The Mean of a Probability Distribution

To describe characteristics of a probability distribution, we can use any of the numerical summaries defined in Chapter 2. These include the mean, median, quartiles, and standard deviation. It is most common to use the *mean* to describe the center and the *standard deviation* to describe the spread.

Recall that numerical summaries of populations are called **parameters**. You can think of a **population distribution** as merely being a type of probability distribution—one that applies for selecting a subject at random from a population. Like numerical summaries of populations, numerical summaries of probability distributions are also referred to as **parameters**. Most parameters are denoted by Greek letters. The mean of a probability distribution is denoted by μ and the standard deviation is denoted by σ.

Suppose we repeatedly observe values of a random variable, such as repeatedly noting the outcome when we roll a die. The mean μ of the probability distribution for that random variable is the value we would get, in the long run, for the average of those values. This "long run" interpretation parallels the interpretation (in Section 5.1) of probability itself as a summary of the long-run behavior of a random phenomenon.

EXAMPLE 3

What's the Expected Number of Home Runs in a Baseball Game?

Picture the Scenario

Let's refer back to the probability distribution for X = number of home runs the Boston Red Sox baseball team hit in a game. The table is shown again in the margin.

Recall

Table 6.1 shows

Number of Home Runs	Probability
0	0.23
1	0.38
2	0.22
3	0.13
4	0.03
5	0.01

Question to Explore

Find the mean of this probability distribution, and interpret it.

Think It Through

Since $P(0) = 0.23$, over the long run you expect $x = 0$ (that is, no home runs) in 23% of the games. Likewise, you expect $x = 1$ home run 38% of the time, and so forth. In 100 games, for example, you expect $x = 0$ about 23 times, $x = 1$ about 38 times, and so forth. Since the mean equals the total of the observations divided by the sample size, for 100 games you expect a mean of about

$$\mu = \frac{\overset{\overset{\text{23 terms}}{\downarrow}}{(0 + 0 + \cdots + 0)} + \overset{\overset{\text{38 terms}}{\downarrow}}{(1 + 1 + \cdots + 1)} + \overset{\overset{\text{22 terms}}{\downarrow}}{(2 + \cdots + 2)} + \overset{\overset{\text{13 terms}}{\downarrow}}{(3 + \cdots + 3)} + (4 + 4 + 4) + 5}{100}$$

$$= \frac{0\overset{\overset{\text{23 times}}{\downarrow}}{(23)} + 1\overset{\overset{\text{38 times}}{\downarrow}}{(38)} + 2\overset{\overset{\text{22 times}}{\downarrow}}{(22)} + 3\overset{\overset{\text{13 times}}{\downarrow}}{(13)} + 4(3) + 5(1)}{100} = \frac{138}{100} = 1.38$$

Note that rather than adding a number 38 times, you can multiply the number by 38, such as 1(38). If this probability distribution applies in a large number of games, you'd expect the mean number of home runs hit by the Red Sox to be 1.38, more than one a game on the average, but not much higher than one. Across all teams in major league baseball, in 2004 the mean number of home runs per game was 1.12.

Insight

Since 23/100 = 0.23, 38/100 = 0.38, and so forth, the above calculation has the form

$$\mu = \frac{0(23) + 1(38) + 2(22) + 3(13) + 4(3) + 5(1)}{100}$$

$$= 0(0.23) + 1(0.38) + 2(0.22) + 3(0.13) + 4(0.03) + 5(0.01)$$

$$= 0 \times P(0) + 1 \times P(1) + 2 \times P(2) + 3 \times P(3) + 4 \times P(4) + 5 \times P(5).$$

Each possible value x is multiplied by its probability. In fact, for any discrete random variable, the mean of its probability distribution results from multiplying each possible value x by its probability $P(x)$, and then adding.

TO PRACTICE THIS CONCEPT, TRY EXERCISE 6.3

IN WORDS

To get the **mean** of a probability distribution, multiply each possible value of the random variable by its probability, and then add all these products.

IN WORDS

A **weighted average** is used when each x value is not equally likely. If a particular x value is more likely to occur, it has a larger influence on the mean, which is the balance point of the distribution.

Recall

Section 2.3 showed how to find the **sample mean** to describe the center of quantitative data. ■

Mean of a Discrete Probability Distribution

The **mean of a probability distribution** for a discrete random variable is

$$\mu = \Sigma x P(x),$$

where the sum is taken over all possible values of x.

For Boston Red Sox home runs in baseball, for example,

$$\mu = \Sigma x P(x) = 0P(0) + 1P(1) + 2P(2) + 3P(3) + 4P(4) + 5P(5)$$
$$= 0(0.23) + 1(0.38) + 2(0.22) + 3(0.13) + 4(0.03) + 5(0.01) = 1.38.$$

The mean $\mu = \Sigma x P(x)$ is called a **weighted average**: Values x that are more likely receive greater weight $P(x)$. It does not make sense to take a simple average of the possible values of x, $(0 + 1 + 2 + 3 + 4 + 5)/6 = 2.5$, because some outcomes are much more likely than others.

Consider the special case in which the outcomes are *equally likely*. Suppose there are n such possible outcomes, each with probability $1/n$, such as $n = 6$ outcomes for rolling a die, each with probability $1/6$. Then

$$\mu = \Sigma x P(x) = \Sigma x(1/n) = (\Sigma x)/n.$$

The formula $(\Sigma x)/n$ is the same as the ordinary one for a sample mean. So the formula $\mu = \Sigma x P(x)$ generalizes the ordinary formula to allow different outcomes not to be equally likely and to apply to probability distributions as well as to sample data.

The mean of the probability distribution of a random variable X is also called the **expected value of X**. The expected value reflects not what we'll observe in a *single* observation, but rather what we expect for the *average* in a long run of observations. In the preceding example, the expected value of the number of Boston Red Sox home runs in a game is $\mu = 1.38$. As with means of sample data, the mean of a probability distribution doesn't have to be one of the possible values for the random variable. We will not see 1.38 home runs in a game, but the long-run average of observing some games with 0 home runs, some with 1, some with 2, and so forth, is 1.38.

EXAMPLE 4

Are You a Risk Taker or Risk Averse?

Picture the Scenario

Are you "risk averse," a person who prefers the sure thing to a risky action that could give you a better or a worse outcome? Or are you "risk taking," willing to gamble in hopes of achieving the better outcome?

Questions to Explore

a. You are given $1000 to invest. You must choose between (i) a sure gain of $500 and (ii) a 0.50

chance of a gain of $1000 and a 0.50 chance to gain nothing. What is the expected gain with each strategy? Which do you prefer?

b. You are given $2000 to invest. You must choose between (i) a sure loss of $500 and (ii) a 0.50 chance of losing $1000 and a 0.50 chance to lose nothing. What is the expected loss with each strategy? Which do you prefer?

Think It Through

a. The expected gain is $500 with the sure strategy (i), since that strategy has gain $500 with probability 1.0. With the risk-taking strategy (ii), the probability distribution is shown in the margin. The expected gain is $\mu = \Sigma x P(x) = \$0(0.50) + \$1000(0.50) = \$500$.

b. The expected loss is $500 with the sure strategy (i). With the risk-taking strategy (ii), the expected loss is $\mu = \Sigma x P(x) = \$0(0.50) + \$1000(0.50) = \$500$.

Insight

In each case, the expected values are the same with each strategy. Yet most people prefer the sure gain strategy (i) in case (a) but the risk-taking strategy (ii) in case (b). They are risk averse in case (a) but risk taking in case (b). This was explored in research by Daniel Kahneman and Amos Tversky, for which Kahneman won the Nobel Prize in 2002.

TO PRACTICE THIS CONCEPT, TRY EXERCISE 6.7

Risk-Taking Probability Distribution	
Gain x	P(x)
0	0.50
1000	0.50

Summarizing the Spread of a Probability Distribution

As with distributions of sample data, it's useful to summarize both the *center* and the *spread* of a probability distribution. For instance, suppose two different investment strategies have the same expected payout. Does one strategy have more variability in its payoffs?

The **standard deviation** of a probability distribution, denoted by σ, measures its spread. Larger values for σ correspond to greater spread. Roughly, σ describes how far the random variable falls, on the average, from the mean of its distribution. We won't deal with the formula for calculating σ until we study particular types of probability distributions in Sections 6.3–6.6 (See also Exercise 6.119.)

EXAMPLE 5

Risk Taking
Entails More Variability

Picture the Scenario

Let's revisit the first scenario in Example 4. You are given $1000 to invest and must choose between (i) a sure gain of $500 and (ii) a 0.50 chance of a gain of $1000 and a 0.50 chance to gain nothing. Table 6.2 shows the probability distribution of the gain X for the two strategies.

TABLE 6.2: Probability Distribution of $X =$ Gain.

The sure strategy has a guaranteed gain of $500. With the risk-taking strategy you have a 0.50 probability of a gain of $1000 and a 0.50 probability of gaining nothing.

Sure Strategy		Risk Taking	
x	P(x)	x	P(x)
500	1.0	0	0.50
		1000	0.50

Question to Explore

Example 4 showed that both strategies have the same mean, namely $500. Which of the two probability distributions would have the larger standard deviation?

Think It Through

There is *no* variability for the sure strategy. The standard deviation of this probability distribution is 0, the smallest possible value. With the risk-taking strategy, no matter what the outcome (either $0 or $1000) the result will be $500 from the mean of $500. In fact, the standard deviation of the probability distribution for the risk-taking strategy is $500. With that strategy, there's much more variability in what can happen.

Insight

In practice, different investment strategies are often compared by their variability. For example, an article in the March 2005 issue of *T. Rowe Price Investor* notes that its Capital Appreciation Fund has been less volatile than most multi-cap value funds. For the five years 2000–2004, the standard deviation of the fund's annual return was 9.6, compared to 16.1 for the Lipper Multi-Cap Value Funds Average. What's the practical implication of this? The T. Rowe Price fund showed less variability in its yield from year to year.

TO PRACTICE THIS CONCEPT, TRY EXERCISE 6.9

Probability Distribution for a Categorical Variable

In examples so far, variables have been quantitative rather than categorical. Partly this is because a random variable is defined to be a *numerical* measurement of the outcome of a random phenomenon. However, we'll see later in this chapter that for categorical variables having only two categories, it's useful to represent the two possible outcomes by the numerical values 0 and 1.

For example, suppose you've conducted a marketing survey of many potential customers to estimate the probability a customer would buy a new product you are developing and plan to sell. Your study estimates that the probability is 0.20. If you denote the possible outcomes (success, failure) for whether a customer buys the product by (1, 0), then the probability distribution of X for the outcome (shown in the margin figure) is

x	P(x)
0	0.80
1	0.20

The mean of this probability distribution is

$$\mu = \Sigma x P(x) = 0(0.80) + 1(0.20) = 0.20.$$

The mean is the probability of success. *For random variables that have possible values 0 and 1, the mean is the probability of the outcome designated by 1.* We'll find this result to be quite useful later in the chapter.

Recall

We saw in the Insight for Example 12 of Section 2.3 that the proportion is a special case of a mean calculated for observations that equal 1 for the outcome of interest and 0 otherwise.■

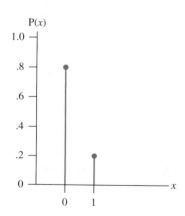

Recall

Section 2.1 defined and gave examples of **continuous** variables. They have an infinite continuum of possible values in an interval. Examples are time, age, and size measures such as height and weight.■

Probability Distributions for Continuous Random Variables

A random variable is called **continuous** when its possible values form an interval. For instance, a recent study by the U.S. Census Bureau analyzed the time that people take to commute to work. Commuting time can be measured with real number values, such as between 0 minutes and 150 minutes.

Many Intervals

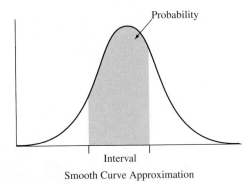

Probability

Interval

Smooth Curve Approximation

Probability distributions of continuous random variables assign probabilities to any interval of the possible values. For instance, a probability distribution for commuting time provides the probability that the travel time is less than 15 minutes or the probability that the travel time is between 30 and 60 minutes. The probability that a random variable falls in any particular interval is between 0 and 1, and the probability of the entire interval containing all the possible values equals 1.

When a random variable is continuous, the intervals of values for the bars of a histogram for probabilities can be chosen as desired. For instance, one possibility for commuting time is {0 to 30, 30 to 60, 60 to 90, 90 to 120, 120 to 150}, quite wide intervals. By contrast, using {0 to 1, 1 to 2, 2 to 3, ..., 149 to 150} gives lots of very narrow intervals. As the number of intervals increases, with their width narrowing, the shape of the histogram gradually approaches a smooth curve. We'll use such curves to portray probability distributions of continuous random variables. See the graphs in the margin.

Probability Distribution of a Continuous Random Variable

A **continuous** random variable has possible values that form an interval. Its **probability distribution** is specified by a curve that determines the probability that the random variable falls in any particular interval of values.

- Each interval has probability between 0 and 1. This is the area under the curve, above that interval.

- The interval containing all possible values has probability equal to 1, so the total area under the curve equals 1.

Figure 6.2 shows a graph for a probability distribution of X = commuting time for workers in the U.S. who commute to work. Historically, in surveys about travel to work, a commute of 45 minutes has been thought of as the maximum time that people would be willing to spend. However, a recent U.S. Census Bureau report[2] based on the 2000 census suggested that of the 97% of U.S. workers who do not work at home, 15% of the commuters have a commuting time above 45 minutes. The shaded area in Figure 6.2 refers to the probability of values higher than 45.0, for those who travel to work. This area equals 15% of the total area under the curve. So the probability is 0.15 of spending more than 45 minutes commuting to work. The report stated that the probability distribution is skewed to the right, with a mean of 25.5 minutes.

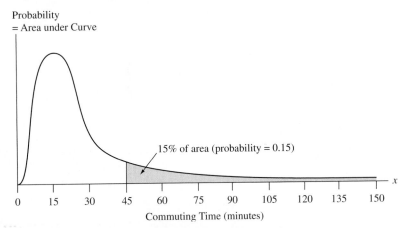

▲ **FIGURE 6.2: Probability Distribution of Commuting Time.** The area under the curve for values higher than 45 is 0.15. **Question:** Identify the area under the curve represented by the probability that commuting time is less than 15 minutes, which equals 0.29.

---

[2]*Journey to Work: 2000*, issued March 2004 by U.S. Census Bureau.

Did You Know?

For continuous random variables, the area above a single value equals 0. For example, the probability that the commuting time equals 23.7693611045 ... minutes is 0, but the probability that commuting time is between 23.5 and 24.5 minutes is positive. ■

For continuous random variables, in practice we need to round off with our measurements. Probabilities are given for *intervals* of values rather than individual values. In measuring commuting time, the U.S. Census Bureau asked, "How many minutes did it usually take to get from home to work last week?", and it gave as possible responses 0, 1, 2, 3, ... Then, for instance, a commuting time of 24 minutes actually means the interval of real numbers that round to the integer value 24. This is the area under the curve for the interval between 23.50 and 24.50. In practice, the probability that the commuting time falls in some given interval, such as above 45 minutes, or between 30 and 60 minutes, is of greater interest than the probability it equals some particular single value.

IN PRACTICE: Continuous variables are measured in a discrete manner

In practice, **continuous** variables are measured in a **discrete** manner because of rounding. With rounding, a continuous random variable can take on a large number of separate values. A probability distribution for a continuous random variable is used to approximate the probability distribution for the possible rounded values.

EXAMPLE 6

Probability Distribution for Height

Picture the Scenario

Figure 6.3 shows a histogram of heights of females at the University of Georgia, with a smooth curve superimposed.

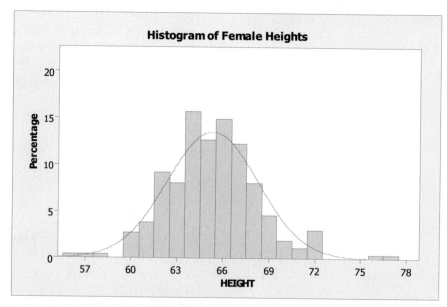

Histogram of Female Heights

▲ **FIGURE 6.3: Histogram of Female Student Heights with Bell-Shaped Curve Superimposed.** Height is continuous but is measured as discrete by rounding to the nearest inch. The smooth curve approximates the probability distribution for height, treating it as a continuous random variable. **Question:** How would you describe the shape, center, and spread of the distribution?

Question to Explore

What does the smooth curve represent?

Think It Through

In theory, height is a continuous random variable. In practice, it is measured here by rounding to the nearest inch. For instance, the bar of the histogram above 64 represents heights between 63.5 and 64.5 inches, which were rounded to 64 inches. The histogram gives the probability distribution for the discrete way height is actually measured. The smooth curve approximates it and approximates the probability distribution that height would have if we could actually measure it precisely as a continuous random variable. The area under this curve between two points approximates the probability that height falls between those points.

Insight

The smooth curve has a bell shape. In the next section we'll study a probability distribution that has this shape and we'll learn how to find probabilities for it.

TO PRACTICE THIS CONCEPT, TRY EXERCISE 6.11

Sometimes we can use a sample space to help us find a probability distribution, as you'll see in Exercises 6.1 and 6.2 and examples later in the chapter. Sometimes the probability distribution is available with a *formula* or a *table* or a *graph*, as we'll see in the next two sections. Sometimes, though, it's necessary to use simulation to approximate the distribution, as shown in Activity 2 at the end of the chapter.

6.1 Practicing the Basics

6.1 **Rolling dice:**

   a. State the probability distribution for the outcome of rolling a balanced die. (This is called the **uniform distribution** on the integers $1, 2, \ldots, 6$.)

   b. Two balanced dice are rolled. Show that the probability distribution for $X =$ total on the two dice is as shown in the figure. (*Hint*: First construct the sample space of the 36 equally likely outcomes you could get. For example, you could denote the six outcomes where you get a 1 on the first die by $(1, 1), (1, 2), (1, 3), (1, 4), (1, 5), (1, 6)$, where the second number in a pair is the number you get on the second die.)

   c. Show that the probabilities in (b) satisfy the two conditions for a probability distribution.

6.2 **Move first in Monopoly:** In *Monopoly*, dice are first
TRY used to determine which player gets to move first. Suppose there are two players in the game. Each player rolls a die and the player with the higher number gets to move first. If the numbers are the same, the players roll again.

    a. Using the sample space {(1, 1), (1, 2), (1, 3), (1, 4), (1, 5), (1, 6), (2, 1)...(6, 5), (6, 6)} of the 36 equally likely outcomes for the two die, show that the probability distribution for the maximum of the two numbers is as shown in the table.

Probability Distribution of X = Maximum Number for Rolls of Two Dice

x	P(x)
1	1/36
2	3/36
3	5/36
4	7/36
5	9/36
6	11/36

    b. Show that the two conditions in the definition of a probability distribution are satisfied.

6.3 **Boston Red Sox hitting:** In 2004, the table shows the
TRY probability distribution of the number of bases for a randomly selected time at bat for a Boston Red Sox player (excluding times when the player got on base because of a "walk" or being "hit by pitch"). In 71.8% of the at-bats the player was out, 17.4% of the time the player got a "single" (one base), 6.5% of the time the player got a "double" (two bases), 0.4% of the time the player got a "triple," and 3.9% of the time the player got a home run.

    a. Verify that the probabilities give a legitimate probability distribution.

    b. Find the mean of this probability distribution. Interpret it.

Boston Red Sox hitting

Number of Bases	Probability
0	0.718
1	0.174
2	0.065
3	0.004
4	0.039

6.4 **Grade distribution:** An instructor always assigns final grades such that 20% are A, 40% are B, 30% are C, and 10% are D. The grade-point scores are 4 for A, 3 for B, 2 for C, and 1 for D.

    a. Specify the probability distribution for the grade point of a randomly selected student of this instructor.

    b. Find the mean of this probability distribution.

6.5 **Bilingual Canadians?:** Let X = number of languages in which a person is fluent. According to Statistics Canada, for residents of Canada this has probability distribution $P(0) = 0.02$, $P(1) = 0.81$, $P(2) = 0.17$, with negligible probability for higher values of x.

    a. Explain why it does not make sense to compute the mean of this probability distribution as $(0 + 1 + 2)/3 = 1.0$.

    b. Find the correct mean.

6.6 **Playing the lottery:** The state of Ohio has several statewide lottery options (www.ohiolottery.com). One of the simpler ones is a "Pick 3" game in which you pick one of the 1000 three-digit numbers between 000 and 999. The lottery selects a three-digit number at random. With a bet of $1, you win $500 if your number is selected and nothing ($0) otherwise. The Florida lottery has the same game (called CASH 3, with the STRAIGHT play option) and the same winnings structure.

    a. With a single $1 bet, what is the probability that you win $500?

    b. Let X denote your winnings for a $1 bet, so $x = \$0$ or $x = \$500$. Construct the probability distribution for X.

    c. Show that the mean of the distribution equals 0.50, corresponding to an expected return of 50 cents for the dollar paid to play.

6.7 **Which wager do you prefer?:** You are given $100 and
TRY told that you must pick one of two wagers, for an outcome based on flipping a fair coin:

    1. You win $200 if it comes up heads and lose $50 if it comes up tails.

    2. You win $350 if it comes up heads and lose your original $100 if it comes up tails.

    a. Without doing any calculation, which wager would you prefer? (There is no "correct" answer. Just specify which you'd choose. Most people pick strategy 1, because of the smaller possible loss.)

    b. Find the expected outcome for each wager. Which wager is better in this sense?

6.8 **Ohio's irresistible lottery:** In Ohio's "Pick 4" lottery, you pick one of the 10,000 four-digit numbers between 0000 and 9999 and (with a $1 bet) win $5000 if you get it correct. In terms of your expected winnings, with which game are you better off—playing Pick 4, or playing Pick 3 in which you win $500 for a correct choice of a three-digit number? Justify your answer.

6.9 **Ideal number of children:** Let X denote the response of
TRY a randomly selected person to the question, "What is the ideal number of children for a family to have?" The probability distribution of X in the United States is approximately[3] as shown in the table, according to the gender of the person asked the question.

Probability Distribution of X = Ideal Number of Children

x	P(x) Females	P(x) Males
0	0.00	0.04
1	0.03	0.03
2	0.63	0.57
3	0.23	0.23
4	0.11	0.13

---

[3] These probability distributions are merely approximations, suggested by results from a recent General Social Survey.

a. The means are similar, 2.42 for females and 2.38 for males. Which probability distribution would you expect to have the larger standard deviation? Why? (*Hint:* Which is more likely to take a relatively small or relatively large value?)

b. Explain why a practical implication of the answer in (a) is that females hold slightly more consistent views than males about the ideal family size.

6.10 **Uniform distribution:** A random number generator is used to generate a real number at random between 0 and 1, equally likely to fall anywhere in this interval of values. (For instance, 0.3794259832... is a possible outcome.)

a. Sketch a curve of the probability distribution of this random variable, which is the continuous version of the **uniform distribution** (see Exercise 6.1).

b. What is the mean of this probability distribution?

c. Find the probability that this random variable falls between 0.25 and 0.75.

6.11 **TV watching:** A social scientist uses the General Social Survey to study how much time per day people spend watching TV. The variable denoted by TVHOURS at the GSS Web site measures this using the values 0, 1, 2, ..., 24.

a. Explain how, in theory, TV watching is a continuous random variable.

b. An article about the study shows two histograms, both skewed to the right, to summarize TV watching for females and males. Since TV watching is in theory continuous, why were histograms used instead of curves?

c. If the article instead showed two curves, explain what they would represent.

6.2 | How Can We Find Probabilities for Bell-Shaped Distributions?

Some probability distributions merit special attention because they are useful for many applications. They have formulas or tables that provide probabilities of the possible outcomes. We next learn about a probability distribution, called the **normal distribution**, that is very commonly used for continuous random variables. It is characterized by a particular[4] symmetric, bell-shaped curve with two parameters—the mean μ and the standard deviation σ.

The Normal Distribution: A Probability Distribution with Bell-Shaped Curves

Figure 6.3 at the end of the previous section showed that heights of female students at the University of Georgia have approximately a bell-shaped distribution. The approximating curve describes a probability distribution that has a mean of 65.0 and a standard deviation of 3.5. In fact, adult female heights in North America have approximately a normal distribution with $\mu = 65.0$ inches and $\sigma = 3.5$ inches. Adult male heights have approximately a normal distribution with $\mu = 70.0$ inches and $\sigma = 4.0$ inches. The adult males tend to be a bit taller (since $70 > 65$) with heights a bit more spread out (since $4.0 > 3.5$). See Figure 6.4.

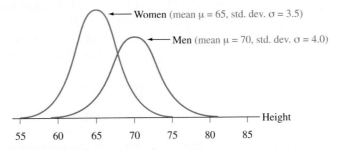

▲ **FIGURE 6.4: Normal Distributions for Women's Height and Men's Height.** For each different combination of μ and σ values, there is a normal distribution with mean μ and standard deviation σ. **Question:** According to $\mu = 70$ and $\sigma = 4$, within what interval do almost all of the men's heights fall?

For any real number for the mean μ and any positive number for the standard deviation σ, there is a normal distribution with that mean and standard deviation.

---

[4]A mathematical formula specifies *which* bell-shaped curve is the normal distribution, but it is complex and we'll not use it in this text. In the next chapter, we'll learn about another bell-shaped distribution, one with "thicker tails" than the normal has.

Normal Distribution

The **normal distribution** is symmetric, bell shaped, and characterized by its mean μ and standard deviation σ. The probability within any particular number of standard deviations of μ is the same for all normal distributions. This probability equals 0.68 within 1 standard deviation, 0.95 within 2 standard deviations, and 0.997 within 3 standard deviations. Figure 6.5 illustrates.

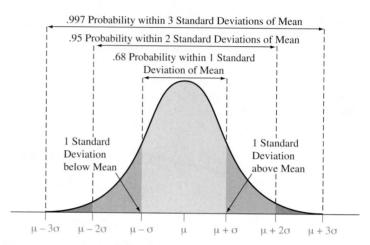

▲ **FIGURE 6.5: The Normal Distribution.** The probability equals 0.68 within one standard deviation of the mean, 0.95 within two standard deviations, and 0.997 within three standard deviations. **Question:** How do these probabilities relate to the Empirical Rule?

To illustrate, for adult female height, $\mu = 65.0$ and $\sigma = 3.5$ inches. Since

$$\mu - 2\sigma = 65.0 - 2(3.5) = 58.0 \text{ and } \mu + 2\sigma = 65.0 + 2(3.5) = 72.0,$$

about 95% of the female heights fall between 58 inches and 72.0 inches (6 feet). For adult male height, $\mu = 70.0$ and $\sigma = 4.0$ inches. About 95% fall between $\mu - 2\sigma = 70.0 - 2(4.0) = 62$ inches and $\mu + 2\sigma = 70.0 + 2(4.0) = 78$ inches (6 1/2 feet).

The property of the normal distribution in the definition above tells us probabilities within 1, 2, and 3 standard deviations of the mean. The multiples 1, 2, and 3 of the number of standard deviations from the mean are denoted by the symbol z in general. For instance, $z = 2$ for two standard deviations. For each fixed number z, the probability within z standard deviations of the mean is the area under the normal curve between $\mu - z\sigma$ and $\mu + z\sigma$. This is shown in Figure 6.6. For every normal distribution, this probability is 0.68 for $z = 1$, so 68% of the area (probability) of a normal distribution falls between $\mu - \sigma$ and $\mu + \sigma$. Similarly, this probability is 0.95 for $z = 2$, and nearly 1.0 for $z = 3$ (that is, between $\mu - 3\sigma$ and $\mu + 3\sigma$). The total probability for any normal distribution equals 1.0.

Recall

From Section 2.5, the **z-score** for an observation is the number of standard deviations that it falls from the mean. ■

▶ **FIGURE 6.6: The Probability between $\mu - z\sigma$ and $\mu + z\sigma$.** This is the area highlighted under the curve. It is the same for every normal distribution and depends only on the value of z. Figure 6.5 showed this for z = 1, 2, and 3, but z can be any other number also.

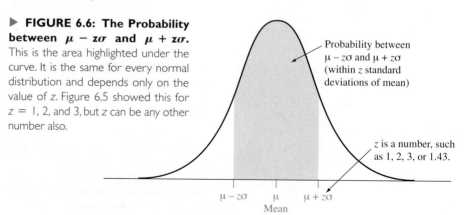

The normal distribution is the most important one in statistics. It's important partly because many variables have approximate normal distributions, as illustrated in Figure 6.3 with female heights. It's important also because it approximates well many discrete distributions when there are a large number of possible outcomes. The main reason for its prominence, however, is that many statistical methods use the normal distribution even when the data are not bell shaped. We'll see why later in this chapter.

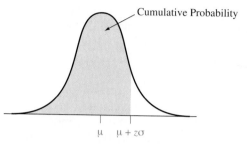

Cumulative Probability

$\mu \quad \mu + z\sigma$

Finding Probabilities for the Normal Distribution

As we'll discuss below, the probabilities 0.68, 0.95, and 0.997 within 1, 2, and 3 standard deviations of the mean are no surprise, because of the Empirical Rule. But what if we wanted to find the probability within, say, 1.43 standard deviations?

Table A at the end of the text enables us to find normal probabilities. It tabulates the normal **cumulative probability** falling *below* the point $\mu + z\sigma$ (see the margin figure). The leftmost column of Table A lists the values for z to one decimal point, with the second decimal place listed above the columns. Table 6.3 shows a small excerpt from Table A. The tabulated probability for $z = 1.43$ falls in the row labeled 1.4 and in the column labeled 0.03. It equals 0.9236. For every normal distribution, the probability that falls below $\mu + 1.43\sigma$ equals 0.9236. Figure 6.7 illustrates.

TABLE 6.3: Part of Table A for Normal Cumulative (Left-Tail) Probabilities.

The top of the table gives the second digit for z. The table entry is the probability falling below $\mu + z\sigma$, for instance, 0.9236 below $\mu + 1.43\sigma$ for z = 1.43.

z					Second Decimal Place of z					
	.00	.01	.02	.03	.04	.05	.06	.07	.08	.09
0.0	.5000	.5040	.5080	.5120	.5160	.5199	.5239	.5279	.5319	.5359
...										
1.3	.9032	.9049	.9066	.9082	.9099	.9115	.9139	.9147	.9162	.9177
1.4	.9192	.9207	.9222	.9236	.9251	.9265	.9278	.9292	.9306	.9319
1.5	.9332	.9345	.9357	.9370	.9382	.9394	.9406	.9418	.9429	.9441

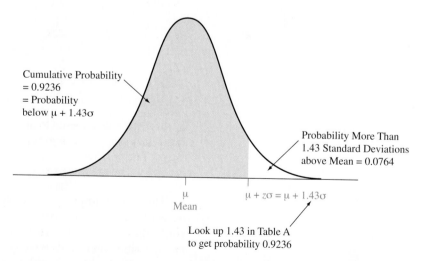

Cumulative Probability
= 0.9236
= Probability
below $\mu + 1.43\sigma$

Probability More Than
1.43 Standard Deviations
above Mean = 0.0764

μ
Mean

$\mu + z\sigma = \mu + 1.43\sigma$

Look up 1.43 in Table A
to get probability 0.9236

▲ **FIGURE 6.7: The Normal Cumulative Probability, Less than z Standard Deviations above the Mean.** Table A lists a cumulative probability of 0.9236 for $z = 1.43$, so 0.9236 is the probability less than 1.43 standard deviations above the mean of any normal distribution (that is, below $\mu + 1.43\sigma$). The complement probability of 0.0764 is the probability *above* $\mu + 1.43\sigma$ in the right tail.

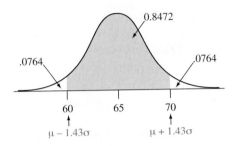

Since an entry in Table A is a probability *below* $\mu + z\sigma$, one minus that probability is the probability *above* $\mu + z\sigma$. For example, the right-tail probability above $\mu + 1.43\sigma$ equals $1 - 0.9236 = 0.0764$. By the symmetry of the normal curve, this probability also refers to the left tail below $\mu - 1.43\sigma$, which you'll find in Table A by looking up $z = -1.43$. The negative z-scores in the table refer to cumulative probabilities for random variable values *below* the mean.

Since the probability is 0.0764 in each tail, the total probability *more than* 1.43 standard deviations from the mean equals $2(0.0764) = 0.1528$. The total probability equals 1, so the probability falling *within* 1.43 standard deviations of the mean equals $1 - 0.1528 = 0.8472$, about 85%. For instance, 85% of women in North America have height between $\mu - 1.43\sigma = 65.0 - 1.43(3.5) = 60$ inches and $\mu + 1.43\sigma = 65 + 1.43(3.5) = 70$ inches (that is, between 5 feet and 5 feet, 10 inches).

Table A lists cumulative probabilities, rather than central probabilities within z standard deviations of the mean. This is for convenience, because cumulative probabilities and their complements in the right tail are more useful in later chapters. Normal cumulative probabilities are also available in statistical software and on some calculators.

Normal Probabilities and the Empirical Rule

The Empirical Rule (Section 2.4) states that for an approximately bell-shaped distribution, about 68% of observations fall within one standard deviation of the mean, 95% within two standard deviations, and all or nearly all within three. In fact, those percentages came from probabilities calculated for the normal distribution.

For instance, a value that is two standard deviations below the mean has $z = -2.00$. The cumulative probability below $\mu - 2\sigma$ listed in Table A opposite $z = -2.00$ is 0.0228. The right-tail probability above $\mu + 2\sigma$ also equals 0.0228, by symmetry. See Figure 6.8. The probability falling more than two standard deviations from the mean in either tail is $2(0.0228) = 0.0456$. Thus, the probability that falls *within* two standard deviations of the mean equals $1 - 0.0456 = 0.9544$. When a variable has a normal distribution, 95.44% of the distribution (95%, rounded) falls within two standard deviations of the mean.

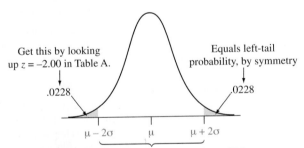

▲ **FIGURE 6.8: Normal Probability within Two Standard Deviations of the Mean.** Probabilities in one tail determine probabilities in the other tail by symmetry. Subtracting the total two-tail probability from 1.0 gives probabilities within a certain distance of the mean. **Question:** Can you do the analogous calculation for *three* standard deviations?

The approximate percentages that the Empirical Rule lists are the actual percentages for the normal distribution, rounded. For instance, you should verify that the probability within one standard deviation of the mean of a normal distribution equals 0.68. (*Hint:* Get the left-tail probability by looking up $z = -1.00$, double it for the two-tail probability, and then subtract from 1.) The probability within three standard deviations of the mean equals 0.997, or 1.00 rounded off. The Empirical Rule stated the probabilities as being *approximate* rather than *exact* because that rule referred to *all approximately* bell-shaped distributions, not just the normal.

How Can We Find the Value of z for a Certain Cumulative Probability?

In practice, we'll sometimes need to find the value of z that corresponds to a certain normal cumulative probability. How can we do this? To illustrate, let's find the value of z for a cumulative probability of 0.025. We look up the cumulative probability of 0.025 in the body of Table A. It corresponds to $z = -1.96$, since it is in the row labeled -1.9 and in the column labeled 0.06. So a probability of 0.025 lies below $\mu - 1.96\sigma$. Likewise, a probability of 0.025 lies above $\mu + 1.96\sigma$. A total probability of 0.050 lies more than 1.96σ from μ. Precisely 95.0% of a normal distribution falls within 1.96 standard deviations of the mean. We've seen in the previous subsection that 95.44% falls within 2.00 standard deviations, and we now see that precisely 95.0% falls within 1.96 standard deviations.

95.0%
(probability
0.95)

0.025   0.025

$\mu - 1.96\sigma$ μ $\mu + 1.96\sigma$

EXAMPLE 7

What IQ Do You Need to Get into MENSA?

Picture the Scenario

Mensa[5] is a society of high-IQ people whose members have a score on an IQ test at the 98th percentile or higher. The Stanford-Binet IQ scores that are used as the basis for admission into Mensa are approximately normally distributed with a mean of 100 and a standard deviation of 16.

Questions to Explore

a. How many standard deviations above the mean is the 98th percentile?

b. What is the IQ score for that percentile?

Think It Through

a. For a value to represent the 98th percentile, its cumulative probability must equal 0.98, by the definition of a percentile. See Figure 6.9.

▶ **FIGURE 6.9: The 98th Percentile for a Normal Distribution.** This is the value such that 98% of the distribution falls below it and 2% falls above. **Question:** Where is the 2nd percentile located?

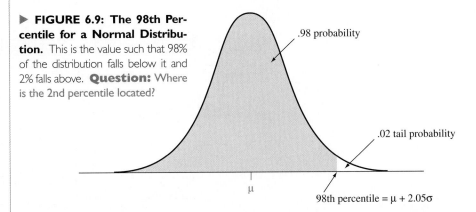

.98 probability

.02 tail probability

μ

98th percentile = $\mu + 2.05\sigma$

The cumulative probability of 0.980 in the body of Table A corresponds to $z = 2.05$. The 98th percentile is 2.05 standard deviations above the mean, at $\mu + 2.05\sigma$.

Recall

The **percentile** was defined in Section 2.5. For the 98th percentile, 98% of the distribution falls below that point. ▪

---

[5]See www.mensa.org.

$$ S = \sqrt{\frac{\Sigma(x - \bar{x})^2}{N - 1}} $$

b. Since $\mu = 100$ and $\sigma = 16$, the 98th percentile of IQ scores equals

$$\mu + 2.05\sigma = 100 + 2.05(16) = 133.$$

In summary, 98% of the IQ scores fall below 133, and an IQ score of at least 133 is required to join Mensa.

Insight

About 2% of IQ scores are higher than 133. By symmetry, about 2% of IQ scores are lower than $\mu - 2.05\sigma = 100 - 2.05(16) = 67$. This is the 2nd percentile. The remaining 96% of the IQ scores fall between 67 and 133, which is the region within 2.05 standard deviations of the mean.

It's also possible to use software to find normal probabilities or z-scores. The margin shows a screen shot using the TI-83+/84 to find the 98th percentile of IQ.

```
invNorm(.98,100,
16)
              132.86
■
```

TI-83+/84 output

TO PRACTICE THIS CONCEPT, TRY EXERCISES 6.16 AND 6.23(B)

Using z = Number of Standard Deviations to Find Probabilities

Recall

Section 2.5 showed that for sample data, the number of standard deviations that a value x falls from the mean of the sample is

$$z = \frac{\text{observed value} - \text{mean}}{\text{standard deviation}}. \ ■$$

We've used the symbol z to represent the *number of standard deviations* a value falls from the mean. If we have a value x of a random variable, how can we figure out the number of standard deviations it falls from the mean μ of its probability distribution? The difference between x and μ equals $x - \mu$. The **z-score** expresses this difference as a number of standard deviations, using $z = (x - \mu)/\sigma$.

> **z-Score for a Value of a Random Variable**
>
> The **z-score** for a value x of a random variable is the number of standard deviations that x falls from the mean μ. It is calculated as
>
> $$z = \frac{x - \mu}{\sigma}.$$

The formula for the z-score is useful when we are given the value of x for some normal random variable and need to find a probability relating to that value. We convert x to a z-score and then use a normal table to find the appropriate probability. The next two examples illustrate.

EXAMPLE 8

Finding Your Relative Standing on the SAT

Picture the Scenario

Scores on the verbal or math portion of the Scholastic Aptitude Test (SAT), a college entrance examination, are approximately normally distributed with mean $\mu = 500$ and standard deviation $\sigma = 100$. The scores range from 200 to 800.

Questions to Explore

a. If one of your SAT scores was $x = 650$, how many standard deviations from the mean was it?

b. What percentage of SAT scores was higher than yours?

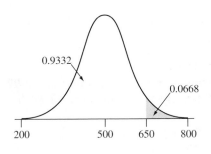

Think It Through

a. The SAT score of 650 has a z-score of $z = 1.50$ because 650 is 1.50 standard deviations above the mean. In other words, $x = 650 = \mu + z\sigma = 500 + z(100)$, where $z = 1.50$. We can find this directly using the formula

$$z = \frac{x - \mu}{\sigma} = \frac{650 - 500}{100} = 1.50.$$

b. The percentage of SAT scores that were higher than 650 is the right-tail probability above 650, for a normal random variable with mean $\mu = 500$ and standard deviation $\sigma = 100$. From Table A, the z-score of 1.50 has cumulative probability 0.9332. That's the probability *below* 650, so the right-tail probability above it is $1 - 0.9332 = 0.0668$. Only about 7% of SAT test scores fall above 650. In summary, a score of 650 was well above average, in the sense that relatively few students scored higher.

Insight

Positive z-scores occur when the value x falls *above* the mean μ. Negative z-scores occur when x falls *below* the mean. For instance, SAT = 350 has a z-score of

$$z = \frac{x - \mu}{\sigma} = \frac{350 - 500}{100} = -1.50.$$

The SAT score of 350 is 1.50 standard deviations *below* the mean. The probability that an SAT score falls below 350 is also 0.0668. Figure 6.10 illustrates.

TO PRACTICE THIS CONCEPT, TRY EXERCISE 6.21

▶ **FIGURE 6.10: Normal Distribution for SAT.** The SAT scores of 650 and 350 have z-scores of 1.50 and −1.50 because they fall 1.50 standard deviations above and below the mean. **Question:** Which SAT scores have $z = 3.0$ and $z = -3.0$?

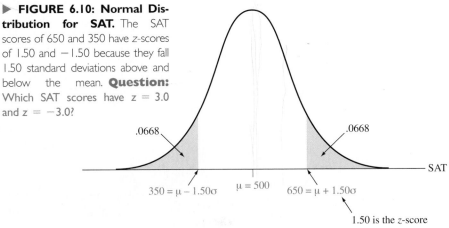

EXAMPLE 9

What Proportion of Students Get a Grade of B?

Picture the Scenario

On the midterm exam in introductory statistics, an instructor always gives a grade of B to students who score between 80 and 90.

Question to Explore

One year, the scores on the exam have approximately a normal distribution with mean 83 and standard deviation 5. About what proportion of students get a B?

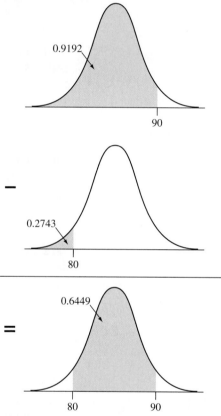

normalcdf(80,90,
83,5)
            .645

TI-83 +/84 output

Think It Through

A midterm exam score of 90 has a z-score of

$$z = \frac{x - \mu}{\sigma} = \frac{90 - 83}{5} = 1.40.$$

Its cumulative probability of 0.9192 (from Table A) means that about 92% of the exam scores were below 90. Similarly, an exam score of 80 has a z-score of

$$z = \frac{x - \mu}{\sigma} = \frac{80 - 83}{5} = -0.60.$$

Its cumulative probability of 0.2743 means that about 27% of the exam scores were below 80. See the normal curves below. It follows that about $0.9192 - 0.2743 = 0.6449$, or about 64%, of the exam scores were in the B range.

Insight

Here, we took the difference between two separate cumulative probabilities to find a probability between two points.

TO PRACTICE THIS CONCEPT, TRY EXERCISE 6.26

Here's a summary of how we've used z-scores so far:

Using z-Scores to Find Normal Probabilities or Random Variable x Values:

■ If we're given a value x and need to find a probability, convert x to a z-score using $z = (x - \mu)/\sigma$, use a table of normal probabilities to get a cumulative probability, and then convert it to the probability of interest.

■ If we're given a probability and need to find the value of x, convert the probability to the related cumulative probability, and find the z-score using a normal table and then evaluate $x = \mu + z\sigma$.

In Example 7, we used the equation $x = \mu + z\sigma$ to find a percentile score (namely, 98th percentile $= \mu + 2.05\sigma = 100 + 2.05(16) = 133$). In Examples 8 and 9, we used the equation $z = (x - \mu)/\sigma$ to determine how many standard deviations certain scores fell from the mean, which enabled us to find probabilities relating to those scores. Of course, in using a normal table, we need to keep in mind that it provides cumulative probabilities and make any needed adjustments, as in these examples.

Another use of z-scores is for comparing observations from different normal distributions in terms of their relative distances from the mean.

EXAMPLE 10

How Can We Compare Test Scores That Use Different Scales?

Picture the Scenario

There are two primary standardized tests used by college admissions, the SAT and the ACT.[6]

Question to Explore

When you applied to college, you scored 650 on an SAT exam, which had mean $\mu = 500$ and standard deviation $\sigma = 100$. Your friend took the comparable ACT in 2001, scoring 30. That year, the ACT had $\mu = 21.0$ and $\sigma = 4.7$. How can we compare these scores to tell who did better?

Think It Through

The test scores of 650 and 30 are not directly comparable because the SAT and ACT have different means and different standard deviations. But we can convert them to z-scores and analyze how many standard deviations each falls from the mean.

With $\mu = 500$ and $\sigma = 100$, we saw in Example 8 that a SAT test score of $x = 650$ converts to a z-score of $z = (x - \mu)/\sigma = (650 - 500)/100 = 1.50$. With $\mu = 21.0$ and $\sigma = 4.7$, an ACT score of 30 converts to a z-score of

$$z = \frac{x - \mu}{\sigma} = \frac{30 - 21}{4.7} = 1.91.$$

The ACT score of 30 is a bit higher than the SAT score of 650, since ACT = 30 falls 1.91 standard deviations above its mean, whereas SAT = 650 falls 1.50 standard deviations above its mean. In this sense, the ACT score is better even though its numerical value is smaller.

Insight

The SAT and ACT tests both have approximately normal distributions. From Table A, $z = 1.91$ (for the ACT score of 30) has a cumulative probability of 0.97. Of all students who took the ACT, only about 3% scored above 30. From Table A, $z = 1.50$ (for the SAT score of 650) has a cumulative probability of 0.93. Of all students who took the SAT, about 7% scored above 650.

TO PRACTICE THIS CONCEPT, TRY EXERCISE 6.27

The Standard Normal Distribution has Mean 0 and Standard Deviation 1

Many statistical methods refer to a particular normal distribution called the **standard normal distribution**.

---

[6]See www.sat.org and www.act.org.

Standard Normal Distribution

The **standard normal distribution** is the normal distribution with mean $\mu = 0$ and standard deviation $\sigma = 1$. It is the distribution of normal z-scores.

For the standard normal distribution, the number falling z standard deviations above the mean is $\mu + z\sigma = 0 + z(1) = z$, simply the z-score itself. For instance, the value of 2.0 is two standard deviations above the mean, and the value of -1.3 is 1.3 standard deviations below the mean. As Figure 6.11 shows, the original values are the same as the z-scores, since

$$z = \frac{x - \mu}{\sigma} = \frac{x - 0}{1} = x.$$

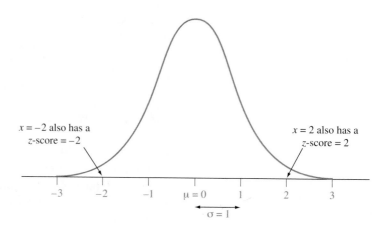

▲ **FIGURE 6.11: The Standard Normal Distribution.** This has mean 0 and standard deviation 1. The random variable value x is the same as its z-score. **Question:** What are the limits within which almost all its values fall?

Examples 8 and 10 dealt with SAT scores, having $\mu = 500$ and $\sigma = 100$. Suppose we convert each SAT score x to a z-score by using $z = (x - \mu)/\sigma = (x - 500)/100$. Then $x = 650$ converts to $z = 1.50$, and $x = 350$ converts to $z = -1.50$. When the values for a normal distribution are converted to z-scores, those z-scores have a mean of 0 and have a standard deviation of 1. That is, the entire set of z-scores has the standard normal distribution.

z-Scores and the Standard Normal Distribution

When a random variable has a normal distribution and its values are converted to z-scores by subtracting the mean and dividing by the standard deviation, the z-scores have the **standard normal** distribution (mean 0, standard deviation 1).

We'll find this result to be quite useful for statistical inference in coming chapters.

6.2 Practicing the Basics

6.12 Probabilities in tails: For a normal distribution, find the probability that an observation is

   a. at least one standard deviation above the mean,

   b. at least one standard deviation below the mean.

   c. In each case, sketch a curve and show the tail probability.

6.13 Tail probability in graph: For the normal distribution shown below, find the probability that an observation falls in the shaded region.

μ $\mu + .67\sigma$

6.14 Empirical Rule: Verify the Empirical Rule by showing that for a normal distribution, the probability (rounded to two decimal places) within

   a. 1 standard deviation of the mean equals 0.68.

   b. 2 standard deviations of the mean equals 0.95.

   c. 3 standard deviations of the mean is very close to 1.00.

6.15 Central probabilities: For a normal distribution, verify that the probability (rounded to two decimal places) within

   a. 1.64 standard deviations of the mean equals 0.90.

   b. 2.58 standard deviations of the mean equals 0.99.

   c. Find the probability that falls within 0.67 standard deviations of the mean.

   d. Sketch these three cases on a single graph.

6.16 z-score for given probability in tails: For a normal
[TRY] distribution,

   a. Find the z-score for which a total probability of 0.02 falls more than z standard deviations (in either direction) from the mean.

   b. For this z, explain why the probability more than z standard deviations above the mean equals 0.01.

   c. Explain why $\mu + 2.33\sigma$ is the 99th percentile.

6.17 Probability in tails for given z-score: For a normal distribution,

   a. Show that a total probability of 0.01 falls more than $z = 2.58$ standard deviations from the mean.

   b. Find the z-score for which the two-tail probability that falls more than z standard deviations from the mean in either direction equals (a) 0.05, (b) 0.10. Sketch the two cases on a single graph.

6.18 z-score for right-tail probability:

   a. For the normal distribution shown below, find the z-score.

   b. Find the value of z (rounding to two decimal places) for right-tail probabilities of (i) 0.05, (ii) 0.005.

μ $\mu + z\sigma$

6.19 z-score and central probability: Find the z-score such that the interval within z standard deviations of the mean for a normal distribution contains

   a. 50% of the probability.

   b. 90% of the probability.

   c. Sketch the two cases on a single graph.

6.20 Female heights: The normal distribution for women's height in North America has $\mu = 65$ inches, $\sigma = 3.5$ inches. Most major airlines have height requirements for flight attendants (www.cabincrewjobs.com). Although exceptions are made, the minimum height requirement is 62 inches. What proportion of adult females in North America are not tall enough to be a flight attendant?

6.21 Blood pressure: A World Health Organization study
[TRY] (the MONICA project) of health in various countries reported that in Canada, systolic blood pressure readings have a mean of 121 and a standard deviation of 16. A reading above 140 is considered to be high blood pressure.

   a. What is the z-score for a blood pressure reading of 140?

   b. If systolic blood pressure in Canada has a normal distribution, what proportion of Canadians suffers from high blood pressure?

   c. What proportion of Canadians has systolic blood pressures in the range from 100 to 140?

6.22 Working hours: According to *Current Population Reports*, self-employed individuals in the United States work an average of 44.6 hours per week, with a standard deviation of 14.5. If this variable is approximately normally distributed, find the proportion of the self-employed who work more than 40 hours per week. Sketch a graph, and mark off on it 44.6, 40, and the region to which the answer refers.

6.23 MDI: The Mental Development Index (MDI) of the Bayley
[TRY] Scales of Infant Development is a standardized measure used in observing infants over time. It is approximately normal with a mean of 100 and a standard deviation of 16.

a. What proportion of children has MDI of (i) at least 120? (ii) at least 80?

TRY b. Find the MDI score that is the 99th percentile.

c. Find the MDI score such that only 1% of the population has MDI below it.

6.24 Quartiles and outliers: Refer to the previous exercise.

a. Find the z-score corresponding to the lower quartile (Q1) of a normal distribution.

b. Find and interpret the lower quartile and upper quartile of the MDI.

c. Find the interquartile range (IQR) of MDI scores.

d. Section 2.5 defined an observation to be a potential outlier if it is more than $1.5 \times$ IQR below Q1 or above Q3. Find the intervals of MDI scores that would be considered potential outliers.

6.25 Murder rates: In 1999, the murder rates (per 100,000 residents) for the 50 states and the District of Columbia (D.C.) had a mean of 5.9 and a standard deviation of 6.3 (*Statistical Abstract of the United States*, 2001).

a. D.C. had a murder rate of 46.4. Find its z-score. If the distribution were roughly normal, would this be unusually high? Explain.

b. Based on the mean and standard deviation, do you think that the distribution of murder rates is approximately normal? Why or why not?

6.26 How many get a C?: Final exam scores have approximately a normal distribution with mean 76 and standard deviation 8. The instructor gives a C to students who score between 70 and 80. About what proportion of students get a C?

6.27 SAT versus ACT: For an SAT distribution ($\mu = 500$, $\sigma = 100$) and an ACT distribution ($\mu = 21, \sigma = 4.7$), which score is relatively higher, an SAT of 600 or an ACT of 25? Explain.

6.28 Relative height: Refer to the normal distributions for women's height ($\mu = 65, \sigma = 3.5$) and men's height ($\mu = 70, \sigma = 4.0$). A man's height of 75 inches and a woman's height of 70 inches are both 5 inches above their means. Which is relatively taller? Explain why.

6.3 | How Can We Find Probabilities When Each Observation Has Two Possible Outcomes?

We next study the most important probability distribution for discrete random variables. Learning about it helps us answer the questions we asked at the beginning of the chapter—for instance, whether results of an exit poll allow us to predict the outcome of an election.

The Binomial Distribution: Probabilities for Counts with Binary Data

In many applications, each observation is **binary**: It has one of two possible outcomes. For instance, a person may

■ accept, or decline, an offer from a bank for a credit card

■ have, or not have, health insurance

■ vote yes or no in a referendum, such as whether to recall a governor from office

With a sample, we summarize such variables by counting the *number* or the *proportion* of cases with an outcome of interest. For instance, with a sample of size $n = 5$, let the random variable X denote the number of people who vote "yes" about some issue in a referendum. The possible values for X are 0, 1, 2, 3, 4, and 5. Under certain conditions, a random variable X that counts the number of observations of a particular type has a probability distribution called the **binomial**.

Consider n cases, called **trials**, in which we observe a binary random variable. This is a *fixed number*, such as $n = 5$ for a sample of 5 voters. The number X of trials in which the outcome of interest occurs can take any one of the integer values $0, 1, 2, \ldots, n$. The binomial distribution gives probabilities for these possible values of X when the following three conditions hold:

Conditions for Binomial Distribution

- Each of n trials has two possible outcomes. The outcome of interest is called a "success" and the other outcome is called a "failure."
- Each trial has the same probability of a success. This is denoted by p.
- The probability of a failure is denoted by $1 - p$.
- The n trials are independent. That is, the result for one trial does not depend on the results of other trials.

The **binomial random variable** X is the number of successes in the n trials.

Flipping a coin is a prototype for when the conditions for the binomial apply:

- Each trial is a flip of the coin. There are two possible outcomes for each flip, head or tail. Let's identify (arbitrarily) head as "success."
- The probability p of a head equals 0.50 for each flip if head and tail are equally likely.
- The flips are independent, since the result for the next flip does not depend on the outcomes of previous flips.

The binomial random variable X counts the number of heads (the outcome of interest) in the n flips. With $n = 3$ coin flips, X = number of heads could equal 0, 1, 2, or 3.

EXAMPLE 11

Finding Binomial Probabilities for an ESP Experiment

Picture the Scenario

John Doe claims to possess extrasensory perception (ESP). An experiment is conducted in which a person in one room picks one of the integers 1, 2, 3, 4, 5 at random and concentrates on it for one minute. In another room, John Doe identifies the number he believes was picked. The experiment is done with three trials. After the third trial, the random numbers are compared with John Doe's predictions. Doe got the correct result twice.

Question to Explore

If John Doe does not actually have ESP and is merely guessing the number, what is the probability that he'd make a correct guess on two of the three trials?

Think It Through

Let X = number of correct guesses in $n = 3$ trials. Then $X = 0, 1, 2,$ or 3. Let p denote the probability of a correct guess for a given trial. If Doe is guessing, $p = 0.2$ for Doe's prediction of one of the five possible integers. Also, $1 - p = 0.8$ is the probability of an incorrect prediction on a given trial. Denote the outcome on a given trial by S or F, representing "success" or "failure" for whether Doe's guess was correct or not. Table 6.4 shows the eight outcomes in the sample space for this experiment. For instance, FSS represents a correct guess on the second and third trials. It also shows their probabilities, by using the multiplication rule for independent events.

Recall

From Section 5.2, for independent events, $P(A \text{ and } B) = P(A)P(B)$.
Thus, $P(FSS) = P(F)P(S)P(S)$
$= 0.8 \times 0.2 \times 0.2.$ ◾

TABLE 6.4: Sample Space and Probabilities for Three Guesses.

The probability of a correct guess is 0.2 on each of the three trials, if John Doe does not have ESP.

Outcome	Probability	Outcome	Probability
SSS	$0.2 \times 0.2 \times 0.2 = (0.2)^3$	SFF	$0.2 \times 0.8 \times 0.8 = (0.2)^1(0.8)^2$
SSF	$0.2 \times 0.2 \times 0.8 = (0.2)^2(0.8)^1$	FSF	$0.8 \times 0.2 \times 0.8 = (0.2)^1(0.8)^2$
SFS	$0.2 \times 0.8 \times 0.2 = (0.2)^2(0.8)^1$	FFS	$0.8 \times 0.8 \times 0.2 = (0.2)^1(0.8)^2$
FSS	$0.8 \times 0.2 \times 0.2 = (0.2)^2(0.8)^1$	FFF	$0.8 \times 0.8 \times 0.8 = (0.8)^3$

The three ways John Doe could make two correct guesses in three trials are SSF, SFS, and FSS. Each of these has probability equal to $(0.2)^2(0.8) = 0.032$. The total probability of two correct guesses is

$$3(0.2)^2(0.8) = 3(0.032) = 0.096.$$

Insight

In terms of the probability $p = 0.2$ of a correct guess on a particular trial, the solution $3(0.2)^2(0.8)$ for $x = 2$ correct in $n = 3$ trials equals $3p^2(1-p)^1 = 3p^x(1-p)^{n-x}$. The multiple of 3 represents the number of ways that 2 successes can occur in 3 trials (SSF or SFS or FSS). You can use similar logic to evaluate the probability that $x = 0$, or 1, or 3. Try $x = 1$, for which you should get $P(1) = 0.384$.

TO PRACTICE THIS CONCEPT, TRY EXERCISE 6.29, PART (A)

The formula for binomial probabilities When the number of trials n is large, it's tedious to write out all the possible outcomes in the sample space. But there's a formula you can use to find binomial probabilities for *any n*.

Probabilities for a Binomial Distribution

Denote the probability of success on a trial by p. For n independent trials, the probability of x successes equals

$$P(x) = \frac{n!}{x!(n-x)!} p^x (1-p)^{n-x}, \quad x = 0, 1, 2, \ldots, n.$$

Did You Know?

The term with factorials at the start of the binomial formula is

$$\binom{n}{x} = \frac{n!}{x!(n-x)!},$$

which is also called the **binomial coefficient.** It is the number of combinations of n things taken x at a time. ◾

The symbol $n!$ is called **n factorial**. It represents $n! = 1 \times 2 \times 3 \times \cdots \times n$, the product of all integers from 1 to n. That is, $1! = 1, 2! = 1 \times 2 = 2$, $3! = 1 \times 2 \times 3 = 6, 4! = 1 \times 2 \times 3 \times 4 = 24$, and so forth. Also, 0! is defined to be 1. For given values for p and n, you can find the probabilities of the possible outcomes by substituting values for x into the binomial formula.

Let's use this formula to find the answer for Example 11 above about ESP:

- The random variable X represents the number of correct guesses (successes) in $n = 3$ trials of the ESP experiment.

- The probability of a correct guess in a particular trial is $p = 0.2$.

- The probability of exactly 2 correct guesses is the binomial probability with $n = 3$ trials, $x = 2$ correct guesses, and $p = 0.2$ probability of a correct guess for a given trial,

$$P(2) = \frac{n!}{x!(n-x)!} p^x(1-p)^{n-x} = \frac{3!}{2!1!}(0.2)^2(0.8)^1 = 3(0.04)(0.8) = 0.096.$$

What's the role of the different terms in this binomial formula?

■ The factorials term tells us the number of possible outcomes that have $x = 2$ successes. Here, $[3!/(2!1!)] = (3 \times 2 \times 1)/(2 \times 1)(1) = 3$ tells us there were 3 possible outcomes with 2 successful guesses, namely SSF, SFS, and FSS.

■ The term $(0.2)^2(0.8)^1$ with the power exponents gives the probability for each such sequence. Here, the probability is $(0.2)^2(0.8) = 0.032$ for each of the three sequences having $x = 2$ successful guesses, for a total probability of $3(0.032) = 0.096$.

Try to calculate P(1) by letting $x = 1$ in the binomial formula with $n = 3$ and $p = 0.2$. You should get 0.384. You'll see that there are again 3 possible sequences, now each with probability $(0.2)^1(0.8)^2 = 0.128$. Table 6.5 summarizes the calculations for all four possible x values. You can also find binomial probabilities using statistical software, such as MINITAB or a calculator with statistical functions.

TABLE 6.5: The Binomial Distribution for $n = 3$, $p = 0.20$.

When $n = 3$, the binomial random variable X can take any integer value between 0 and 3. The total probability equals 1.0.

x	$P(x) = [n!/(x!(n - x)!)]p^x(1 - p)^{n-x}$
0	$0.512 = [3!/(0!3!)](0.2)^0(0.8)^3$
1	$0.384 = [3!/(1!2!)](0.2)^1(0.8)^2$
2	$0.096 = [3!/(2!1!)](0.2)^2(0.8)^1$
3	$0.008 = [3!/(3!0!)](0.2)^3(0.8)^0$

EXAMPLE 12

Are Women Passed Over for Managerial Training?

Picture the Scenario

While the binomial distribution is useful for finding probabilities for such trivial pursuits as flipping a coin or checking ESP claims, it's also an important tool in helping us understand more serious issues. An example is the presence of bias in hiring, promotion, and daily life. Take, for example, the case of a large supermarket chain in Florida. The chain occasionally selects employees to receive management training. A group of women there claimed that female employees were passed over for this training in favor of their male colleagues.[7] The company denied this claim. A similar claim of gender bias was recently made about promotions and pay for the 1.6 million women who work or who have worked for Wal-Mart.[8]

Question to Explore

Suppose the large employee pool (more than a thousand people) that can be tapped for management training is half female and half male. Since this program began, none of the 10 employees chosen for management training were female. How can we investigate statistically the women's assertion of gender bias?

---

[7] *Tampa Tribune*, April 6, 1996.

[8] *New York Times*, June 25, 2004.

Think It Through

Other factors being equal, at each choice the probability of selecting a female equals 0.50 and the probability of selecting a male equals 0.50. If the employees are selected randomly in terms of gender, about half of the employees picked should be females and about half should be male. Due to ordinary sampling variation, however, it need not happen that *exactly* 50% of those selected are female. A simple analogy is in flipping a coin 10 times. We won't necessarily see exactly 5 heads. The number of heads is a random variable with a binomial distribution.

Since none of the 10 employees chosen for management training has been female, we might be inclined to support the women's claim. However, would these results be all that unlikely, if there were no gender bias? Equivalently, if we flip a coin 10 times, would it be very surprising if we got 0 heads?

Let X denote the number of females selected for management training in a random sample of 10 employees. Then, the possible values for X are $0, 1, \ldots, 10$, and X has the binomial distribution with $n = 10$ and $p = 0.50$. For each x between 0 and 10, we can find the probability that x of the 10 people selected are female. We use the binomial formula for that x value with $n = 10$ and $p = 0.50$, namely

$$P(x) = \frac{n!}{x!(n-x)!}p^x(1-p)^{n-x} =$$

$$\frac{10!}{x!(10-x)!}(0.50)^x(0.50)^{10-x}, x = 0, 1, 2, \ldots, 10.$$

The probability that no females are chosen ($x = 0$) equals

$$P(0) = \frac{10!}{0!10!}(0.50)^0(0.50)^{10} = (0.50)^{10} = 0.001.$$

Any number raised to the power of 0 equals 1. Also, $0! = 1$, and the 10! terms in the numerator and denominator cancelled, leaving $P(0) = (0.50)^{10}$. If the employees were chosen randomly, it is very unlikely (one chance in a thousand) that none of the 10 selected for management training would have been female.

Insight

In summary, because this probability is so small, seeing no women chosen would make us highly skeptical that the choices were random with respect to gender.

TO PRACTICE THIS CONCEPT, TRY EXERCISES 6.37

In this example, we found the probability that $x = 0$ female employees would be chosen for management training. To get a more complete understanding of just which outcomes are likely, you can find the probabilities of *all* the possible x values. The table in the margin lists the entire binomial distribution for $n = 10$, $p = 0.50$. In the table, the least likely values for x are 0 and 10. If the employees were randomly selected, it is highly unlikely that 0 females or 10 females would be selected.

If your statistical software provides binomial probabilities, see if you can verify that $P(0) = 0.001$ and $P(1) = 0.010$. The margin also shows a screen shot of the TI-83+ calculator providing the binomial probability of $x = 6$. Figure 6.12 is a graph of the binomial distribution with $n = 10$ and $p = 0.50$. It has a symmetric appearance around $x = 5$. For instance, $x = 10$ has the same probability as $x = 0$. The binomial distribution is perfectly symmetric only when $p = 0.50$.

Binomial Probability Distribution for n = 10 and p = 0.5

x	P(x)	x	P(x)
0	0.001	6	0.205
1	0.010	7	0.117
2	0.044	8	0.044
3	0.117	9	0.010
4	0.205	10	0.001
5	0.246		

```
binompdf(10,.5,6
)
             .205
```

TI-83 +/84 output

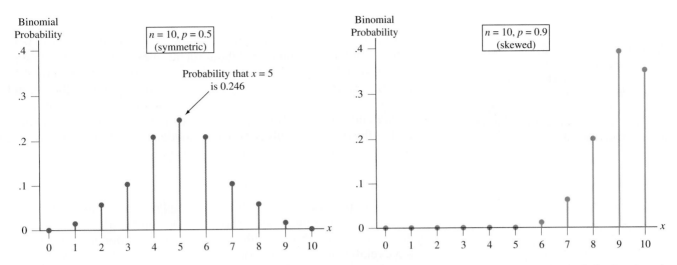

▲ **FIGURE 6.12: Binomial Distributions when $n = 10$ for $p = 0.5$ and for $p = 0.9$.** The binomial probability distribution is symmetric when $p = 0.5$, but it can be quite skewed for p near 0 or near 1. **Question:** How do you think the distribution would look if $p = 0.1$?

When $p \neq 0.50$, the binomial distribution has a skewed appearance. The degree of skew increases as p gets closer to 0 or 1. To illustrate, Figure 6.12 also shows the binomial distribution for $n = 10$ when $p = 0.9$. If 90% of the potential trainees were female, it would not be especially surprising to observe 10, 9, 8, or even 7 females in the sample, but the probabilities drop sharply for smaller x-values.

Do the Binomial Conditions Apply?

Before you use the binomial distribution, check that its three conditions apply. These are (1) binary data (success or failure), (2) the same probability of success for each trial (denoted by p), and (3) independent trials. To judge this, ask yourself whether the observations resemble coin flipping, the simple prototype for the binomial. For instance, for Example 12 on gender bias in selecting employees for management training, this does seem plausible. In this instance,

- The data are binary because (female, male) plays the role of (head, tail).
- If employees are randomly selected, the probability p of selecting a female on any given trial is 0.50.
- With random sampling, the outcome for one trial is separate and independent from the outcome of another trial.

EXAMPLE 13

Binomial Sampling for Managerial Training?

Picture the Scenario

Consider the gender bias investigation in Example 12. Suppose the population of potential trainees contained only four people, two men and two women (instead of more than 1000 people, as in Example 12), and the number chosen was $n = 2$.

Question to Explore

Do the binomial conditions apply for calculating the probability, under random sampling, of selecting 0 women trainees in the two choices for management training?

Recall

As Section 5.3 discussed, if once subjects are selected from a population they are not eligible to be selected again, this is called *sampling without replacement*. Example 10 there showed the effect on sampling from a small population versus a large population. ■

Think It Through

For the first trainee selected, the probability of a woman is $2/4 = 0.50$. The usual sampling is "sampling without replacement," in which the first person selected is no longer in the pool for future selections. So, given that the first person selected was male, the conditional probability that the second person selected is male equals $1/3$, since the pool of potential trainees now has 1 man and 2 women. So, the outcome of the second selection *depends* on that of the first. The trials are *not* independent, which the binomial requires. In summary, the binomial conditions do not apply.

Insight

This example suggests a caution with applying the binomial to a random sample from a population: For trials to be sufficiently "close" to independent with common probability p of success, the population size must be large relative to the sample size.

TO PRACTICE THIS CONCEPT, TRY EXERCISE 6.40

The margin shows a guideline about the relative sizes of the sample and population for which the binomial formula works well:

Suppose your school has 4000 students. Then the binomial formula is adequate as long as the sample size is less than 10% of 4000, which is 400. Why? Because the probability of success for any one observation will be similar regardless of what happens on other observations. Likewise, Example 12 dealt with the selection of 10 employees for management training when the employee pool for such training was more than 1000 people. Again, the sample size was less than 10% of the population size, so using the binomial is valid.

IN PRACTICE: Population and Sample Sizes to Use the Binomial

For sampling n separate subjects from a population (that is, sampling without replacement), the exact probability distribution of the number of successes is too complex to discuss in this text, but the binomial distribution approximates it well when n is less than 10% of the population size. In practice, sample sizes are usually small compared to population sizes, and this guideline is satisfied.

Mean and Standard Deviation of the Binomial Distribution

Example 12 applied the binomial distribution for the number of women selected for management training when $n = 10$ and $p = 0.50$. If p truly equals 0.50, out of 10 selections, what do you expect for the number of women selected for management training?

As with any discrete probability distribution, we can use the formula $\mu = \Sigma x P(x)$ to find the mean. However, finding the mean μ and standard deviation σ is actually simpler for the binomial distribution. There are special formulas based on the number of trials n and the probability p of success on each trial.

Binomial Mean and Standard Deviation

The binomial probability distribution for n trials with probability p of success on each trial has mean μ and standard deviation σ given by

$$\mu = np, \quad \sigma = \sqrt{np(1 - p)}.$$

The formula for the mean makes sense. If the probability of success is p for a given trial, then we expect about a proportion p of the n trials to be successes, or about np total. If we sample $n = 10$ people from a population in which half are female, then we expect that about $np = 10(0.50) = 5$ in the sample will be female.

When the number of trials n is large, it is tedious to calculate binomial probabilities of all the possible outcomes. Often, it's adequate merely to use the mean and standard deviation to describe where most of the probability falls. The binomial distribution has a bell shape when n is large, so we can use the normal distribution to approximate that nearly all the probability falls between $\mu - 3\sigma$ and $\mu + 3\sigma$.

EXAMPLE 14

How Can We Check for Racial Profiling?

Picture the Scenario

In the 1990s, the U.S. Justice Department and other groups studied pos-

sible abuse by Philadelphia police officers in their treatment of minorities. One study, conducted by the American Civil Liberties Union,[9] analyzed whether African-American drivers were more likely than others in the population to be targeted by police for traffic stops. They studied the results of 262 police car stops during one week in 1997. Of those, 207 of the drivers were African American, or 79% of the total.

Question to Explore

At that time, Philadelphia's population was 42.2% African American. Does the number of African Americans stopped suggest possible bias, being higher than we would expect if we take into account ordinary random variation?

Think It Through

We'll treat the 262 police car stops as $n = 262$ trials. Suppose that the percentage of drivers in Philadelphia during that week who were African-American was the same as their representation in the population (42.2%). Then, if there is no bias, the chance for any given police car stop that the driver is African American is $p = 0.422$ (other things being equal, such as the rate of violating traffic laws). Suppose also that successive police car stops are independent. (They would not be, for example, if once a car was stopped, the police followed that car and stopped it repeatedly.) Under these assumptions, for the 262 police car stops, the number of African Americans stopped has a binomial distribution with $n = 262$ and $p = 0.422$.

The binomial distribution with $n = 262$ and $p = 0.422$ has

$$\mu = np = 262(0.422) = 111,$$

$$\sigma = \sqrt{np(1 - p)} = \sqrt{262(0.422)(0.578)} = 8.$$

We'll see below that this binomial distribution is approximated reasonably well by the normal distribution. So, the probability within 3 standard deviations of the mean is close to 1.0. This is the interval between

$$\mu - 3\sigma = 111 - 3(8) = 87 \quad \text{and} \quad \mu + 3\sigma = 111 + 3(8) = 135.$$

If no racial profiling is happening, we would not be surprised if between about 87 and 135 of the 262 people stopped were African Americans. See the smooth curve approximation for the binomial in the margin. However, the actual number stopped (207) is well above these values. This suggests that the number of African Americans stopped may be too high, even taking into account random variation.

Recall

From Section 6.2, when a distribution has a normal distribution, nearly 100% of it falls within 3 standard deviations of the mean. ▪

87   111   135   207

---

[9]www.archive.aclu.org/profiling/report

Insight

By this approximate analysis, we would not expect to see 207 African Americans stopped if there were truly a 0.422 chance of finding an African-American behind the wheel on each stop. If we were to use software to do a more precise analysis by calculating the binomial probabilities of *all* possible values $0, 1, 2, \ldots, 262$ when $n = 262$ and $p = 0.422$, we'd find that the probability of getting 207 or a larger value out in the right tail of the distribution is 0 to the precision of many decimal places (that is, $0.00000000 \ldots$). However, a limitation of both this analysis and the one based on the normal approximation is that people do different amounts of driving, so we don't really know that 42.2% of the *potential* stops were African American. For instance, many suburban residents (who are predominantly white) drive into Philadelphia for work, so that percentage may be less than 42.2%.

TO PRACTICE THIS CONCEPT, TRY EXERCISE 6.36

IN PRACTICE: When Is the Binomial Distribution Approximately Normal?

The binomial distribution can be approximated well by the normal distribution when n is large enough that the expected number of successes, np, and the expected number of failures, $n(1 - p)$, are both at least 15.

The solution in Example 14 treated the binomial distribution as having approximately a normal distribution. This holds when n is sufficiently large. The margin shows a guideline.[10]

In Example 14, of those stopped, the expected number who were African-Americans was $np = 262(0.422) = 111$. The expected number who were *not* African-American was $n(1 - p) = 262(0.578) = 151$. Both exceed 15, so this binomial distribution has approximately a normal distribution.

---

[10]A lower bound of 15 is actually a bit conservative. The binomial is bell shaped even when both np and $n(1 - p)$ are about 10. We use 15 here because it ties in better with a guideline in the next chapter for using the normal distribution for inference about proportions.

6.3 Practicing the Basics

6.29 **ESP:** Jane Doe claims to possess extrasensory perception (ESP). She says she can guess more often than not the outcome of a flip of a balanced coin in another room. In an experiment, a coin is flipped three times. If she does not actually have ESP, find the probability distribution of the number of her correct guesses.

  a. Do this by constructing a sample space, finding the probability for each point, and using them to construct the probability distribution.

  b. Do this using the formula for the binomial distribution.

6.30 **More ESP:** In Example 11 on ESP, John Doe had to predict which of five numbers was chosen in each of three trials. Doe did not actually have ESP. Explain why this experiment satisfies the three conditions for the binomial distribution.

  a. For the analogy with coin flipping, what plays the role of (head, tail)?

  b. Explain why it is sensible to assume the same probability of a correct guess on each trial.

  c. Explain why it is sensible to assume independent trials.

6.31 **Symmetric binomial:** Construct a table with the possible values and their probabilities for a binomial distribution, using the binomial formula or software.

  a. Do this for $n = 3$ and $p = 0.5$. Construct a graph, and note that it is symmetric, which is always true when $p = 0.5$.

  b. Do this for $n = 3$ and $p = 0.2$. Is its graph symmetric?

6.32 **Number of girls in a family:** Each newborn baby has a probability of approximately 0.49 of being female and 0.51 of being male. For a family with four children, let X = number of children who are girls.

  a. Explain why the three conditions are satisfied for X to have the binomial distribution.

  b. Identify n and p for the binomial distribution.

  c. Find the probability that the family has two girls and two boys.

6.33 **NBA shooting:** In the National Basketball Association, the top free-throw shooters usually have probability of about 0.90 of making any given free throw.

  a. During a game, one such player (Dolph Schayes) shot 10 free throws. Let X = number of free throws made. What must you assume in order for X to have a binomial distribution? (Studies have shown that such assumptions are well satisfied for this sport.)

b. Specify the values of n and p for the binomial distribution of X in (a).

c. Find the probability that he made (i) all ten free throws, (ii) 9 free throws.

6.34 **Season performance:** Refer to the previous exercise. Over the course of a season, this player shoots 400 free throws.

a. Find the mean and standard deviation of the probability distribution of the number of free throws he makes.

b. By the normal distribution approximation, within what range would you expect the number made to almost certainly fall? Why?

c. Within what range would you expect the *proportion* made to fall?

6.35 **Is the die balanced?:** A balanced die with 6 sides is rolled 60 times.

a. For the binomial distribution of X = number of 6s, what is n and what is p?

b. Find the mean and the standard deviation of the distribution of X. Interpret.

c. If you observe $x = 0$, would you be skeptical that the die is balanced? Explain why, based on the mean and standard deviation of X.

d. Show that the probability that $x = 0$ is 0.0000177.

6.36 **Exit poll:** Example 1 discussed an exit poll of 3160 voters for the election about the recall of the California governor. Let X denote the number who voted for the recall. Suppose that in the population, exactly 50% voted for it.

a. Explain why this scenario would seem to satisfy the three conditions needed to use the binomial distribution. Identify n and p for the binomial.

b. Find the mean and standard deviation of the probability distribution of X.

c. Using the normal distribution approximation, give an interval in which you would expect X almost certainly to fall, if truly $p = 0.50$. (*Hint*: You can follow the reasoning of Example 14 on racial profiling.)

d. Now, suppose that the exit poll had $x = 1706$. What would this suggest to you about the actual value of p?

6.37 **Jury duty:** The juror pool for the upcoming murder trial of a celebrity actor contains the names of 100,000 individuals in the population who may be called for jury duty. The proportion of the available jurors on the population list who are Hispanic is 0.40. A jury of size 12 is selected at random from the population list of available jurors. Let X = the number of Hispanics selected to be jurors for this jury.

a. Is it reasonable to assume that X has a binomial distribution? If so, identify the values of n and p. If not, explain why not.

b. Find the probability that no Hispanic is selected.

c. If no Hispanic is selected out of a sample of size 12, does this cast doubt on whether the sampling was truly random? Explain.

6.38 **Euro:** *Eurobarometer* has tracked opinions of Europeans about the common currency (the euro) now used in many European countries. When it was introduced in January 2002, 67% of adult residents of the affected countries indicated they were happy that the euro had arrived. Suppose a poll of size 1000 is planned next year to estimate the percentage of people who now approve of the common currency. Suppose the population proportion equals 0.67.

a. With a random sample, explain why it is reasonable to use the binomial for the probability distribution of the number of the 1000 who will indicate approval of the euro. For the binomial, what is n and what is p?

b. Find the mean and standard deviation of the probability distribution in (a).

c. Suppose the poll is taken and that 800 people indicate approval. Does this give strong evidence that the percentage approving of the euro has gone up? Explain.

6.39 **Checking guidelines:** For Example 12 on the gender distribution of management trainees, the population size was more than one thousand, half of whom were female. The sample size was 10.

a. Check that the guideline was satisfied about the relative sizes of the population and the sample, thus allowing you to use the binomial for the probability distribution for the number of females selected.

b. Check that the guideline was satisfied for this binomial distribution to be approximated well by a normal distribution.

6.40 **Class sample:** Four of the 20 students (20%) in a class are fraternity or sorority members. Five students are picked at random. Does X = the number of students in the sample who are fraternity or sorority members have the binomial distribution with $n = 5$ and $p = 0.20$? Explain why or why not.

6.41 **Binomial needs fixed n:** For the binomial distribution, the number of trials n is a fixed number. Let X denote the number of girls in a randomly selected family in Canada that has three children. Let Y denote the number of girls in a randomly selected family in Canada (that is, the number of children could be any number). A binomial distribution approximates well the probability distribution for one of X and Y, but not for the other.

a. Explain why.

b. Identify the case for which the binomial applies, and identify n and p.

6.42 **Binomial assumptions:** For the following random variables, explain why at least one condition needed to use the binomial distribution is unlikely to be satisfied.

a. X = number of people in a family of size 4 who go to church on a given Sunday, when any one of them goes 50% of the time in the long run (binomial, $n = 4$, $p = 0.50$). (*Hint*: Is the independence assumption plausible?)

b. X = number voting for the Democratic candidate out of the 100 votes in the first precinct that reports results, when 60% of the population voted for the Democrat (binomial, $n = 100$, $p = 0.60$). (*Hint*: Is the first precinct a random sample?)

c. X = number of females in a random sample of 4 students from a class of size 20, when half the class is female (binomial, $n = 4$, $p = 0.50$).

Recall

From Sec. 1.2, a **statistic** is a numerical summary of sample data, such as a sample proportion or a sample mean. A **parameter** is a numerical summary of a population, such as a population proportion or a population mean. ■

6.4 | How Likely Are the Possible Values of a Statistic? The Sampling Distribution

In this chapter, we've used probability distributions with known parameter values (such as $p = 0.50$ for the binomial) to find probabilities of possible outcomes of a random variable. In practice, we seldom know the values of parameters. They are estimated using sample data from observational studies or from experiments.

Let's consider the California recall election mentioned in Example 1. Before all the votes were counted, the proportion of the population of voters who voted in favor of recalling Governor Gray Davis was an unknown parameter. An exit poll of 3160 voters reported that the sample proportion who voted in favor of recall was 0.54. This statistic estimates the population proportion. How do we know it's a good estimate? The total number of voters was nearly eight million, and the poll sampled a minuscule portion of them. This section introduces a type of probability distribution that helps us determine how close to the population parameter a statistic is likely to fall.

Representing Sampling Variability by a Sampling Distribution

For an exit poll of voters, a person's vote is a variable because the outcome varies from voter to voter. A statistic such as a sample proportion can also be regarded as a variable, because its value varies from sample to sample. For example, before the exit-poll sample was selected, the value of the sample proportion who voted to recall Governor Davis was unknown. It was a random variable. The proportion of surveyed voters who voted for recall in one poll probably differed somewhat from the proportion for another sample of voters in another poll.

Suppose the *population* proportion is 0.56. Would we see very similar sample proportions in different polls, such as 0.54 in one, 0.57 in another, 0.55 in another? Or would they tend to be quite different, such as 0.40 in one, 0.59 in another, and 0.73 in another? To answer, we need to learn about a probability distribution that provides the possible values for the sample proportion and their probabilities. A probability distribution for a statistic (such as a sample proportion) is called a **sampling distribution**.

IN WORDS

Different polls take different random samples, which have different sample proportion values. The **sampling distribution** specifies the possible sample proportion values and their probabilities.

Sampling Distribution

The **sampling distribution** of a statistic is the probability distribution that specifies probabilities for the possible values the statistic can take.

For sampling 3000 voters in an exit poll, imagine *all* the distinct samples of 3000 voters you could possibly get. Each such sample has a value for the sample proportion voting a certain way (for instance, yes on a referendum). Suppose you could look at each possible sample, find the sample proportion for each one, and then construct the frequency distribution of those sample proportion values. That would be precisely the sampling distribution of the sample proportion.

A sampling distribution is merely a type of probability distribution. Rather than giving probabilities for an observation for an individual subject, it gives probabilities for the value of a statistic for a sample of subjects, such as a sample proportion. The next example illustrates. Example 15 goes through the steps of finding the sampling distribution of a sample proportion.

EXAMPLE 15

Which Brand of Pizza Do You Prefer?

Picture the Scenario

The North End of Boston is a neighborhood that has, for many years, housed a large number of Italian immigrants to the United States. The neighborhood supports an impressive number of Italian restaurants and cafés. Aunt Erma's Restaurant in the North End specializes in pizza baked in a wood-burning oven. The owners plan an advertising campaign with the claim that more people prefer the taste of their pizza (which we'll denote by A) than the current leading fast-food chain selling pizza (which we'll denote by D). To support their claim, they plan to randomly sample n people in Boston. Each person is asked to taste a slice of pizza A and a slice of pizza D. Subjects are blindfolded so they cannot see the pizza when they taste it, and the order of giving them the two slices is randomized. They are then asked which pizza tastes better.

Questions to Explore

a. In the entire Boston population, suppose that exactly half would prefer pizza A and half would prefer pizza D in such a taste experiment. Construct the sampling distribution of the sample proportion who prefer Aunt Erma's pizza, when $n = 3$.

b. In practice, we don't know the population proportion preferring each pizza type. Do the results in (a) suggest that we can gauge Boston's tastes well by sampling only 3 people?

Think It Through

a. Let's use a symbol with three entries to represent the responses for each possible sample of size $n = 3$. For instance, ADD represents a sample in which the first subject sampled preferred pizza A, and the second and third subjects preferred pizza D. The eight possible samples of size 3, and the number and the sample proportion of each sample that preferred pizza A, are

Sample	Number Who Prefer Pizza A	Proportion Who Prefer Pizza A
AAA	3	1
AAD	2	2/3
ADA	2	2/3
DAA	2	2/3
ADD	1	1/3
DAD	1	1/3
DDA	1	1/3
DDD	0	0

The sample proportion that prefers pizza A can be 0, 1/3, 2/3, or 1.0. When the population proportion preferring pizza A is $p = 0.50$, the 8 samples are equally likely, each occurring with probability 1/8. The sample proportion 0 occurs for only 1 of the 8 possible samples, namely DDD, so its probability equals 1/8. The sample proportion 1/3 occurs for 3 samples, ADD, DAD, DDA, so its probability equals 3/8. Similarly, one can find the probability for each possible sample proportion value. Table 6.6 shows the entire sampling distribution.

TABLE 6.6: Sampling Distribution of Sample Proportion.

The random sample size is $n = 3$ from a population with population proportion $p = 0.50$. The eight possible samples provide probabilities for the four possible sample proportion values. **Question:** How can we determine the probabilities when $p \neq 0.50$?

Sample Proportion	Probability
0	1/8
1/3	3/8
2/3	3/8
1	1/8

---

b. Figure 6.13 portrays the sampling distribution of the sample proportion for $n = 3$, when the population proportion is $p = 0.50$. With a very small sample such as $n = 3$, the sample proportion need not fall near the population proportion. Aunt Erma's restaurant would not be able to gauge Boston's tastes well by sampling only three people.

▶ **FIGURE 6.13: Sampling Distribution of Sample Proportion.** The sample size is $n = 3$ from a population with proportion of 0.50. **Question:** How could you find this sampling distribution of the sample proportion with $n = 3$ if 60% (rather than 50%) of the population prefers Aunt Erma's pizza?

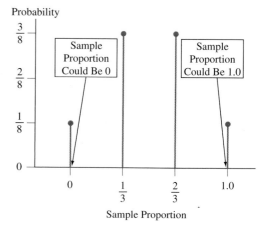

Insight

Why would we care about finding a sampling distribution? Because it shows us whether a sample proportion is likely to be near the population proportion we're really interested in. In practice, of course, samples are almost always much larger than $n = 3$. We used a small n in this example merely so it was easy to find and show all the probabilities in the sampling distribution.

Since the population is quite large compared to the sample size, we can also find the probabilities for each possible sample proportion value using the binomial distribution. The *number* of people preferring pizza A in the sample is a binomial random variable with $n = 3$ and $p = 0.50$. For instance, the probability of a sample proportion of 1/3 equals the probability of 1 person out of 3 preferring pizza A. This equals

$$P(1) = \frac{n!}{x!(n - x)!} p^x (1 - p)^{n-x} = (3!/(1!2!))(0.50)^1(0.50)^2 = 3/8.$$

It's especially helpful to use the binomial formula when p differs from 0.50, since then the eight possible samples listed above would not be equally likely.

TO PRACTICE THIS CONCEPT, TRY EXERCISE 6.43

Every sample statistic has a sampling distribution. Besides the sample proportion, there is a sampling distribution of a sample mean, a sampling distribution of a sample median, a sampling distribution of a sample standard deviation, and so forth.

Sampling distribution of sample proportion is a sampling distribution of a sample mean

In Example 15 above, suppose that for each person we defined a binary random variable, X, to represent their preference, as

$$x = 1, \text{ subject prefers Aunt Erma's pizza A}$$
$$x = 0, \text{ subject prefers pizza D}$$

For example, the sample DAD in which only the second subject sampled preferred pizza A is represented numerically by $(0, 1, 0)$. When we denote the two possible outcomes by 0 and 1, the proportion of times that 1 occurs is the sample mean. For the sample $(0, 1, 0)$, for example, the sample mean equals $(0 + 1 + 0)/3 = 1/3$. This is the sample proportion of the 3 subjects that preferred pizza A.

This example shows that *the sampling distribution of a sample proportion is actually a special case of a sampling distribution of a sample mean.* We'll study the sampling distribution of the sample mean in Section 6.5.

Describing the Sampling Distribution of a Sample Proportion

The binomial probability distribution, introduced in Section 6.3, is an example of a sampling distribution. It is the sampling distribution for the *number* of successes in n independent trials. In practice, it's often preferable to report the sample *proportion* of successes. The proportion is simpler to interpret, because the possible values fall between 0 and 1 regardless of the value of n. As Example 15 on the pizza preference study showed, there is a close connection between the two.

Consider a binomial random variable X for $n = 3$ trials, such as the *number* of people who preferred Aunt Erma's pizza in the pizza taste comparison. The number who preferred it can equal $x = 0, 1, 2, 3$. These correspond to sample *proportions* of 0, 1/3, 2/3, 1. The proportion values are less spread out, by a factor of $1/n = 1/3$, going from 0 to 1 instead of from 0 to 3. This is because the proportion divides the count by 3. The smaller spread for the proportion values is illustrated by the sampling distributions shown in Figure 6.14.

The formulas for the mean and the standard deviation of the sampling distribution of the *proportion* of successes are the formulas for the mean and standard deviation of the *number* of successes divided by n.

Recall

From the Insight for Example 12 of Section 2.3, the **sample proportion** is a special case of a sample mean calculated for observations that equal 1 for the outcome of interest and 0 otherwise. In Section 6.1 we saw that for binary random variables that take the possible values 0 and 1, the mean is the probability of the outcome designated by 1. ▪

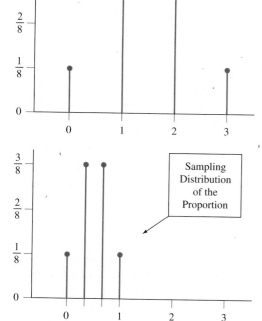

Sampling Distribution of the Number

Sampling Distribution of the Proportion

▲ **FIGURE 6.14: The Sampling Distribution of the Sample *Proportion* Is less Spread Out than the Sampling Distribution of the *Number* of Successes.**

Recall

For a binomial random variable
X = number of successes in n trials,
Section 6.3 gave the formulas mean = np
and standard deviation = $\sqrt{np(1 - p)}$. ▪

Mean and Standard Deviation of Sampling Distribution of a Proportion

For a binomial random variable with n trials and probability p of success for each, the sampling distribution of the *proportion* of successes has

$$\text{Mean} = p \text{ and standard deviation} = \sqrt{\frac{p(1 - p)}{n}}.$$

To obtain these values, take the mean np and standard deviation $\sqrt{np(1 - p)}$ for the binomial distribution of the *number* of successes and divide by n.

Example 15 considered a random sample of $n = 3$ people from a population in which the proportion who prefer pizza A equals $p = 0.50$. The sample proportion preferring pizza A can take values 0, 1/3, 2/3, 1. The sampling distribution, shown in the second part of Figure 6.14, has a mean of $p = 0.50$ and a standard deviation of $\sqrt{p(1 - p)/n} = \sqrt{(0.50)(0.50)/3} = 0.29$.

EXAMPLE 16

Exit Poll of California Voters Revisited

Picture the Scenario

Example 1 discussed an exit poll of 3160 voters for the election to decide

whether then California governor Gray Davis should be recalled from office. If the recall were successful, Davis would be replaced as governor by the candidate who received the most votes in the election. The winning candidate was Arnold Schwarzenegger.

Questions to Explore

Suppose that exactly 50% of the population of all voters voted in favor of the recall. Describe the mean and standard deviation of the sampling distribution of

a. the *number* in the sample who voted in favor of it.

b. the *proportion* in the sample who voted for it.

Think It Through

a. Each voter had two options (voting yes or voting no about the recall), so we have binary data. The *number* of people in the exit-poll sample of 3160 voters who voted for it had possible values $0, 1, 2, 3, \ldots, 3159, 3160$. If the recall actually had 50% support in the population of voters, the outcome is a binomial random variable with $n = 3160$ and $p = 0.50$. That random variable has mean $\mu = np = 3160(0.50) = 1580$. We expect about 1580 to vote for the recall and about 1580 to vote against it. The standard deviation is
$$\sigma = \sqrt{np(1-p)} = \sqrt{3160(0.50)(0.50)} = 28.1.$$ There is variability around the mean, and we would probably not observe *exactly* 1580 voters supporting the recall. For example, we might observe 1550, or perhaps 1610, both values falling about a standard deviation away from the mean.

b. The *proportion* of people in the sample who voted for the recall is the binomial random variable (for the *count*) divided by 3160. It had possible values $0, 1/3160, 2/3160, 3/3160, \ldots, 3159/3160, 1$. If the population proportion $p = 0.50$, then the sampling distribution of the sample proportion has

$$\text{Mean} = p = 0.50$$
$$\text{Standard deviation} = \sqrt{\frac{p(1-p)}{n}} = \sqrt{\frac{(0.50)(0.50)}{3160}} = \sqrt{0.000079} = 0.0089.$$

There is variability around the mean, and we would probably not observe *exactly* a sample proportion of 0.50 supporting the recall. For example, we might observe a sample proportion of 0.49, or 0.51, both values falling about a standard deviation from the mean.

Insight

The standard deviation of the sampling distribution of the sample *proportion* is very small (0.0089). This tells us that with $n = 3160$, the sample proportion will probably fall close to the population proportion of 0.50.

TO PRACTICE THIS CONCEPT, TRY EXERCISE 6.47

IN WORDS

The **standard error** describes how much a statistic varies from sample to sample. For instance, the standard error of a sample proportion describes the spread of the sample proportion values from all the possible samples.

The Standard Error

To distinguish the standard deviation of a *sampling* distribution from the standard deviation of an ordinary probability distribution, in the rest of the text we shall refer to the standard deviation of a sampling distribution as a **standard error**.

Standard Error

The standard deviation of a sampling distribution is called a **standard error**.

For instance, the standard deviation $\sqrt{p(1-p)/n}$ of the sampling distribution of the sample proportion is called the **standard error of the sample proportion**.

EXAMPLE 17

Can You Predict the Outcome of the Special Election?

Picture the Scenario

Let's now conduct an analysis that uses the actual exit poll of 3160 voters for the California election about recalling Governor Gray Davis. In that exit poll, the recall of Davis was supported by 54% of the 3160 voters sampled.

Questions to Explore

a. If the population proportion supporting recall was 0.50, would it have been unlikely to observe the exit-poll sample proportion of 0.54?

b. Based on the answer to (a), would you be willing to predict that Davis would be recalled from office?

Think It Through

a. We use the fact that the binomial distribution for the *number* voting for recall has approximately a normal distribution (see the Recall box). When this happens, so does the sampling distribution of the sample *proportion*, which we get by dividing the number by 3160. Figure 6.15 shows a normal

Recall

By the guideline in Section 6.3, the binomial distribution is approximately normal when the expected numbers of successes and failures, np and $n(1-p)$, are both at least 15. Here, $np = n(1-p) = 3160(0.50) = 1580$ are both much larger than 15. ▪

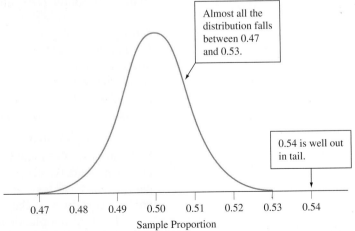

Almost all the distribution falls between 0.47 and 0.53.

0.54 is well out in tail.

Sample Proportion

▲ **FIGURE 6.15: Sampling Distribution of Sample Proportion Voting for Recall in California Special Election.** This shows where we expect sample proportions to fall for random samples of size $n = 3160$, when a population proportion $p = 0.50$ voted for the recall. Nearly all the distribution falls between 0.47 and 0.53. The observed sample proportion of 0.54 is far out in the right tail. This would be very unlikely to happen if $p = 0.50$. Observing a sample proportion of 0.54 makes us doubt that p actually equals 0.50 or any number less than that.

approximation for the sampling distribution of the sample proportion. When we assume $p = 0.50$, this distribution has a mean of $p = 0.50$ and standard error of $\sqrt{p(1 - p)/n} = \sqrt{0.50(0.50)/3160} = 0.0089$. Nearly the entire distribution falls within three standard errors of the mean, or within about $3(0.0089) = 0.03$ of the mean of 0.50.

So, if $p = 0.50$, it is very likely that the sample proportion falls within 0.03 of p, that is, between 0.47 and 0.53. It would be surprising to observe a sample proportion of 0.54. Converting the sample proportion value of 0.54 to a z-score,

$$z = (0.54 - 0.50)/0.0089 = 4.5.$$

This tells us that the sample proportion of 0.54 is more than four standard errors from the expected value of 0.50. In summary, this analysis suggests that if the population support were $p = 0.50$, then the sample proportion of 0.54 voting for recall would be very unlikely.

b. Figure 6.15 suggests that the probability is high that the sample proportion falls within 0.03 of the population proportion. Since the actual sample proportion was 0.54, this suggests that values for p between about 0.51 and 0.57 (that is, within 0.03 of 0.54) are more plausible than 0.50 or numbers below that. Because the sample proportion of 0.54 would be extremely unusual if $p = 0.50$ (or less than 0.50), we have strong evidence that actually p was larger than 0.50. In summary, we would predict that Governor Gray Davis would be recalled from office.

Insight

In this special election, when all 7,974,834 votes were tallied, 55.4% voted for the recall and 44.6% voted against it. The prediction, based on the exit poll, that Governor Davis would be recalled was correct. He was removed from office, and Arnold Schwarzenegger became the next California governor.

If the sample proportion favoring the recall had been closer to 0.50, we would have been unwilling to make a prediction. For instance, a sample proportion of 0.51 for a sample of size 3160 would not have been especially unlikely if truly $p = 0.50$ or is slightly below 0.50.

TO PRACTICE THIS CONCEPT, TRY EXERCISES 6.47 AND 6.48

In the previous example, we evaluated a region over which nearly all of the sampling distribution concentrates under the stated condition and then checked whether the sample proportion falls within that region. We did not need to use the formula for binomial probabilities. Indeed, calculating the probabilities would have been tedious for the 3160 possible values $(0, 1, 2, \ldots, 3160)$ that the number voting for recall could equal. The probability of *any* single value can be extremely small when n is large.

A Sampling Distribution Shows How a Statistic Would Vary with Many Similar Studies

Sampling distributions describe the variability that occurs in collecting data and using statistics to estimate population parameters. If different polling organizations each take a sample and estimate the population proportion who voted a certain way, they will get different estimates, because their samples have different people. Figure 6.15 portrayed the variability for different samples of size $n = 3160$.

Sampling distributions are the foundation of inferential statistics. They help us to predict how close a statistic falls to the parameter it estimates. This accuracy of prediction depends on the sample size. For instance, for a random sample of size 3160, we've seen that the probability is high that a sample proportion falls within 0.03 of the population proportion. The sampling distribution (shown in Figure 6.14) for a sample of size 3 is much more spread out, and such accuracy does not occur.

Here's a summary of what we've learned so far about the sampling distribution of a sample proportion:

Summary of Sampling Distribution of a Sample Proportion

For a random sample of size n from a population with proportion p, the **sampling distribution of the sample proportion** has

$$\text{mean} = p, \quad \text{standard error} = \sqrt{\frac{p(1 - p)}{n}}$$

If n is sufficiently large that the expected numbers of outcomes of the two types, np and $n(1 - p)$, are both at least 15, then this sampling distribution is approximately normal.

To get a feel for the important concept of a sampling distribution, it's helpful to simulate a sampling distribution by repeatedly taking samples of a particular size. That's the purpose of the following activity.

ACTIVITY 1: Simulating a Sampling Distribution

Let's continue to explore how a sample proportion may vary from sample to sample. We'll do this using random numbers and then using the *sampling distributions* applet, under the assumption that the population proportion voting for recall of Governor Gray Davis in the California special election was $p = 0.50$. You can use random numbers and the sampling distribution applet with other values for p as well.

How can we simulate the sampling distribution of sample proportions for random samples of size $n = 1000$? How would you describe the shape, center, and variability?

We can simulate selecting a voter at random from a population by picking a random number. To simulate random sampling from a population in which exactly 50% of all voters voted for each candidate, we identify the 50 two-digit numbers between 00 and 49 as voting "yes" about the recall and the 50 two-digit numbers between 50 and 99 as voting "no." That is,

00 01 02 03 04 ... 49	50 51 52 53 54 ... 99
Vote "yes" on recall	Vote "no" on recall

Then each of "yes" and "no" has a 50% chance of selection on each choice. To simulate random sampling when 45% of the population vote for recall, we would use only the digits 00 to 44 for a yes vote and 45 to 99 for a no vote.

The first two digits of the first column of the random number table in Chapter 4 (Table 4.1, part of the first column of which is shown below) provide the random numbers 10, 22, 24, 42, 37, 77, and so forth. The first number, 10, falls between 00 and 49, so in this simulation the first person sampled voted for the recall. Of the first 6 people selected, 5 voted for the recall (they have numbers between 00 and 49). Selecting 1000 two-

digit random numbers simulates the votes for randomly sampling 1000 residents of the much larger population.

Random Numbers
10480
22368
24130
42167
37570
77921

Rather than laboriously using random numbers, you can use the *sampling distributions* applet on the text CD to generate randomly a binomial random variable for a particular n and p. Select *binary* for the parent population, setting the population proportion as $p = 0.50$. Select for the sample size $n = 1000$. For the third graph, select the option *count* from the menu. For the fourth graph, select the option *proportion*. For one of our simulations, we got $x = 523$ voting "yes" and 477 voting "no" on recall. The sample proportion of yes votes was $523/1000 = 0.523$, near the "true" proportion of 0.50. This particular estimate was good. Were we simply lucky? We repeated the process and used this applet to generate another binomial with $n = 1000$ and $p = 0.50$. This time the sample proportion of yes votes was $491/1000 = 0.491$, different but also quite good.

Now, you use the applet to take a random sample of size 1000. Did you get a sample proportion close to 0.50? Perform this simulation of a random sample of size 1000 ten times, each time observing from the graphs the counts and the corresponding sample proportion of yes votes.

We programmed software to perform this process of picking 1000 people a million separate times, so we could search for a pattern in the results. Figure 6.16 shows a histogram of the million outcomes for the sample proportion. The simulated sample proportions fell in a bell shape around the true population value of $p = 0.50$. Nearly all the sample proportions fell between 0.45 and 0.55, that is, within 0.05 of the true population value of 0.50. Did the ten sample proportion values in your simulation all fall between 0.45 and 0.55?

This result is not a surprise. We know that the standard error of this simulated sampling distribution is

$$\text{Standard error} = \sqrt{p(1 - p)/n} = \sqrt{(0.50)(0.50)/1000}$$
$$= \sqrt{0.00025} = 0.0158.$$

Because the simulated sampling distribution is approximately normal, the probability is very close to 1.0 that the sample proportions would fall within 3 standard errors of 0.50, that is, within $3(0.0158) = 0.047$.

Return to the *sampling distributions* applet and perform 1000 simulations by setting $N = 1000$ with a sample size of $n = 1000$ each, for $p = 0.50$. Compare the sampling distribution in graph 4 of the 1000 sample proportions to Figure 6.16. Are the results similar?

This bar, with height 0.08, means 8% of the one million samples had sample proportion between .47 and .48

◀ **FIGURE 6.16: Results of Simulating the Sample Proportion Voting "Yes" about the Recall of Governor Gray Davis of California.** One million random samples of 1000 subjects each were simulated from a population in which half favor each option. **Question:** For $n = 1000$, explain why it is very likely that the sample proportion falls within 0.05 of the population value of 0.50.

TO PRACTICE THIS ACTIVITY, TRY EXERCISE 6.52

6.4 Practicing the Basics

6.43 **Pizza:** Refer to Example 15, "Which brand of pizza **TRY** do you prefer?" Now use sample size $n = 4$, when the population proportion who preferred Aunt Erma's pizza is 0.50.

   a. Enumerate the 16 possible samples and the possible values for the sample proportion preferring Aunt Erma's pizza.

   b. Using (a), construct the sampling distribution of the sample proportion of people who prefer Aunt Erma's pizza. That is, set up a table (like Table 6.6) showing the sample proportion values and their probabilities.

   c. In practice, you would not know the population proportion value. Based on (b), if you sample only four people, can you be confident that the sample proportion is close to the population proportion? Explain.

6.44 **Pizza using the binomial:** Refer to the previous exercise.

   a. Explain how to use the binomial distribution to find the probability that two of the four people preferred Aunt Erma's pizza.

   b. Suppose now that the population proportion that prefers Aunt Erma's pizza is 0.60. Use the binomial to find the probability that two people prefer Aunt Erma's pizza.

   c. For (b), explain why the sample space approach would no longer have equally likely points.

6.45 **Relative frequency of heads:** Construct the sampling distribution of the sample proportion of heads obtained in the experiment of flipping a balanced coin

   a. Once.

   b. Twice. (*Hint:* The possible samples are {HH, HT, TH, TT}.)

c. Three times. (*Hint:* There are eight possible samples.)

d. Refer to (a)–(c). Describe how the shape of the sampling distribution is changing as the number of flips *n* increases.

6.46 Migraine headaches: For the population of people who suffer occasionally from migraine headaches, suppose *p* = 0.60 is the proportion who get some relief from taking ibuprofen. For a particular subject, let *x* = 1 if they get relief and *x* = 0 if they do not. For a random sample of 24 people who suffer from migraines:

a. State the probability distribution for each observation.

b. Find the mean of the sampling distribution of the sample proportion who get relief.

c. Find the standard error of the sampling distribution of the sample proportion.

d. Explain what the standard error describes.

6.47 Other scenario for exit poll: Refer to Examples 16 and [TRY] 17 about the exit poll, for which the sample size was 3160. Suppose that 55% of *all* voters voted for the recall (which was actually the case).

a. Define a binary random variable taking values 0 and 1 that represents the vote for a particular voter (1 = vote for recall). State its probability distribution.

b. Identify *n* and *p* for the binomial distribution that is the sampling distribution of the *number* in the sample who voted for the recall out of the sample of size 3160.

c. Find the mean and standard deviation of the binomial random variable in (b).

d. Find the mean and standard error of the sampling distribution of the *proportion* of the 3160 people in the sample who voted for the recall.

6.48 Exit poll and n: Refer to the previous exercise.

[TRY] a. In (d), if the mean was 0.55 and the standard error was 0.0089, based on the approximate normality of the sampling distribution, give an interval of values within which the sample proportion will almost certainly fall.

b. Based on the result in (a), if you take an exit poll and observe a sample proportion of 0.60, would you be safe in concluding that the population proportion probably exceeds 0.55? Why?

6.49 Rolling a die: Two balanced dice are rolled.

Probability Distribution of X = Sum of Two Dice			
x	p(x)	x	p(x)
2	1/36	8	5/36
3	2/36	9	4/36
4	3/36	10	3/36
5	4/36	11	2/36
6	5/36	12	1/36
7	6/36		

a. Exercise 6.1 found the probability distribution for the sum of the two dice, shown again above. Using this distribution, construct the sampling distribution for the sample mean $\bar{x}$ of the two numbers rolled.

b. Show that the mean of the probability distribution of each die's outcome is 3.5.

c. By the symmetric form of the sampling distribution in (a), show that its mean is 3.5. (We'll see in the next section that the mean of the sampling distribution of $\bar{x}$ equals the mean of the probability distribution for each observation.)

6.50 Effect of n on sample proportion: The figure illustrates two sampling distributions for sample proportions when the population proportion *p* = 0.50.

a. Find the standard error for the sampling distribution of the sample proportion with (i) *n* = 100, (ii) *n* = 1000.

b. Explain why the sample proportion would be very likely (as the figure suggests) to fall (i) between 0.35 and 0.65 when *n* = 100, and (ii) between 0.45 and 0.55 when *n* = 1000.

c. Explain how the results in (b) indicate that the sample proportion tends to more precisely estimate the population proportion when the sample size is larger.

6.51 Purpose of sampling distribution: You'd like to estimate the proportion of all students in your school who are fluent in more than one language. You poll a random sample of students and get a sample proportion of 0.12. Explain why the standard error of the sampling distribution of the sample proportion gives you useful information for helping you gauge how close this sample proportion is to the unknown population proportion.

6.52 Simulating the exit poll: Simulate an exit poll of [TRY] 100 voters, using the *sampling distributions* applet on the text CD, assuming that the population proportion is 0.50. Refer to Activity 1 for guidance on using the applet.

a. What sample proportion did you get? Why do you not expect to get exactly 0.50?

b. Simulate this exit poll 10 times. Keep the sample size at *n* = 100 and *p* = 0.50. Describe the 10 sample proportion values. If you took a huge number of samples of size 100 each, what would you predict for the value of the standard deviation of the sample proportions?

c. Now change the population proportion to 0.70, keeping the sample size *n* = 100. Simulate the exit poll ten times. How would you say the results differ from those in (b)?

d. Using the appropriate formulas from this section, for part (c) find the mean and the standard error of the sampling distribution of the sample proportion.

6.5 | How Close Are Sample Means to Population Means?

The previous section introduced the *sampling distribution*. This is a probability distribution for the possible values of a statistic. For instance, we learned how much a sample proportion can vary among different random samples. As we've seen, when we denote the two possible outcomes by 0 and 1, the sample proportion is merely a sample mean $\bar{x}$. Because the sample mean is so commonly used, we'll now give its sampling distribution special attention. Using it, we'll see how to predict how close a particular sample mean $\bar{x}$ falls to the population mean μ.

There are two main results about the sampling distribution of the sample mean: One result provides formulas for its mean and standard deviation (the "standard error"). The other indicates that its shape is often approximately a normal distribution, as we observed in the previous section for the sample proportion.

The Mean and Standard Error of the Sampling Distribution of $\bar{x}$

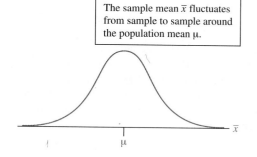

The sample mean $\bar{x}$ fluctuates from sample to sample around the population mean μ.

Before we select a sample, the sample mean $\bar{x}$ is a random variable—we don't know what value it will take. The value varies from sample to sample. By contrast, the *population* mean μ is a single fixed number. At a given time for a particular scenario, there can be many different samples but there's a single population. For random sampling, the sample mean $\bar{x}$ can fall either above or below the population mean μ. In fact, *the mean of the sampling distribution of the sample mean $\bar{x}$ equals the population mean μ.* The margin figure illustrates. It shows the sample mean $\bar{x}$ as a random variable with distribution concentrated around the value of the population mean μ.

How can we describe the spread of the sampling distribution of $\bar{x}$? As we mentioned in the previous section, we use the standard deviation, and to distinguish the standard deviation of a sampling distribution from other standard deviations, we call it a **standard error**.

Standard Error of the Sample Mean $\bar{x}$

The standard deviation of the sampling distribution of the sample mean $\bar{x}$ is called the **standard error of $\bar{x}$.**

Here's a way to think of the standard error: Suppose each of several researchers was studying some topic, independently of each other. Each researcher gathers a random sample of the same size n and finds a sample mean to estimate a population mean μ. If we could collect all their $\bar{x}$-values, those values would fluctuate around that unknown μ. Moreover, the standard deviation of all those $\bar{x}$-values would approximate the standard error. So, the standard error describes the variability in results we can expect from different studies about the same topic.

Do you think that sample mean values tend to vary more, or vary less, than individual observations? For example, suppose each year the job placement center at your school selects a random sample of 100 graduating seniors who have accepted jobs and records the starting salary for each. It then reports the sample mean of the 100 starting salaries. When this is done for several years, would the sample means vary more, or vary less, than the individual salaries?

For random sampling, the standard error of $\bar{x}$ depends on the sample size n and the spread of the probability distribution from which the observations are sampled—the *population distribution*. It can be shown to equal

$$\text{Standard error} = \frac{\text{Population standard deviation}}{\text{Square root of sample size}} = \frac{\sigma}{\sqrt{n}}$$

We'll shortly explain why and how the sample size n affects the standard error. But you can see from the formula that the standard error is less than the population standard deviation, σ: *Means of samples tend to vary less than individual observations.* The graduating senior salaries (in thousands of dollars) might look like

25, 39, 15, 62, 44, 18, 30, 72, 51, 18, ... (standard deviation of about 20, say)

whereas the sample mean salaries of 100 seniors from year to year might look like

35, 38, 35, 40, 36, ... (standard deviation of about 2, say).

The individual salaries vary much more than the sample mean salaries do.

In summary, the following result describes the center and spread of the sampling distribution of the sample mean $\bar{x}$.

Mean and Standard Error of Sampling Distribution of Sample Mean $\bar{x}$

For a random sample of size n from a population having mean μ and standard deviation σ, the **sampling distribution** of the sample mean $\bar{x}$ has

center described by the **mean μ** (the same as the mean of the population)

spread described by the **standard error**, which equals $\sigma/\sqrt{n}$ (the population standard deviation divided by the square root of the sample size).

EXAMPLE 18

How Much Do Mean Sales Vary from Week to Week?

Picture the Scenario

The sales of food and drink in Aunt Erma's pizza restaurant vary from day to day. The daily sales figures fluctuate independently from a probability distribution having a mean of $\mu = \$900$ and a standard deviation of $\sigma = \$300$.

Questions to Explore

The mean sales for the seven days in a week are computed each week. The weekly means are plotted over time to help detect substantial increases or decreases in sales.

a. What would the weekly mean sales figures fluctuate around?

b. How much variability would you expect in the weekly sample mean sales figures? Find the standard error, and interpret.

Think It Through

a. The sample means fluctuate around the mean of the probability distribution for each observation. This is $\mu = \$900$.

b. The sampling distribution of the sample mean for $n = 7$ has mean \$900. Its standard error equals

$$\frac{\sigma}{\sqrt{n}} = \frac{300}{\sqrt{7}} = 113.$$

If we were to observe the sample mean sales for several weeks, the sample means would vary around \$900, with a spread described by the standard error of 113.

Insight

Figure 6.17 portrays a possible population distribution for daily sales. It also shows the sampling distribution of the mean sales $\bar{x}$ for $n = 7$. There is less variability from week to week in the *mean* sales than there is from day to day in the *daily* sales.

The assumptions for the probability model in this example are a bit unrealistic. Some days of the week (such as weekends) may be busier than others, and certain periods of the year may be busier (such as the summer, or New England's leaf-viewing season in the fall). Also, in practice we

Sampling Distribution
(standard error = 113)

Population Distribution
($\sigma = 300$)

0   900

▲ **FIGURE 6.17: A Population Distribution for Daily Sales and the Sampling Distribution of Weekly Mean Sales $\bar{x}$ ($n = 7$).** There is more variability day-to-day in the daily sales than week-to-week in the weekly mean sales.

would not know values of population parameters such as the population mean and standard deviation. However, this example shows that even under some simplifying assumptions, we'd still expect to see quite a bit of variability from week to week in the sample mean sales. We should not be surprised if the mean is $800 this week, $1000 next week, and so forth. Knowing how to find a standard error gives us a mechanism for understanding how much variability to expect in sample statistics.

TO PRACTICE THIS CONCEPT, TRY EXERCISES 6.53 AND 6.54

Effect of *n* on the Standard Error

Let's consider again the formula $\sigma/\sqrt{n}$ for the standard error of the sample mean. Notice that as the sample size n increases, the denominator increases, so the standard error decreases. This has an important practical implication: With larger samples, the sample mean is more likely to fall close to the population mean.

EXAMPLE 19

What's the Long-Run Consequence to Your Finances of Playing a Lottery?

Picture the Scenario

Many U.S. states, Canadian provinces, and countries have lottery games. Most lotteries use "numbers games." They are usually several-digit generalizations of the following game: You bet a dollar on a number between 0 and 9 that you pick at random. If you are correct, you win $5, and otherwise, you win nothing.

X = winnings in one game	
x	**P(x)**
0	0.90
5	0.10

Let X denote your winnings for a single play of this lottery. The possible values for X are $0 and $5. The probability distribution of X is $P(0) = 0.90$ and $P(5) = 0.10$, since there is a 1 in 10 chance that you pick the correct number and win $5. This probability distribution has mean

$$\mu = \Sigma xP(x) = 0(0.90) + 5(0.10) = 0.50,$$

that is, half a dollar. This is the expected value of the outcome X for a single play. The expected return on the dollar bet is only 50 cents. This does not seem attractive, but it is a typical return. (Some are even worse; see the New Jersey lottery in Exercise 6.68.) Naturally, the expected return is less than you pay to play, because the government wants to make money! It can be shown that the standard deviation of this probability distribution is $\sigma = 1.50$.

Questions to Explore

Find the mean and standard error of the sampling distribution of your mean winnings if you play this game (a) once a *week* for the next year (that is, 52 times), (b) once a *day* for the next year (that is, 365 times).

Think It Through

The sampling distribution of the sample mean winnings $\bar{x}$ has the same mean as the probability distribution of X for a single play, namely, $\mu = 0.50$. For playing the lottery n times, its standard error is

$$\sigma/\sqrt{n} = 1.50/\sqrt{n}.$$

a. If you play $n = 52$ times, the standard error equals $1.50/\sqrt{52} = 0.21$.

b. If you're really serious and play 365 times over the next year, the standard error equals $\sigma/\sqrt{n} = 1.50/\sqrt{365} = 0.08$.

Insight

Figure 6.18 portrays the dependence of the sampling distribution on the sample size. As n increases, the standard error gets smaller, so the sampling distribution gets narrower. Thus, the sample mean winnings tend to get closer to the mean of the probability distribution for each game, 0.50, which is the expected value of X. The *law of large numbers* states that as n increases, the sample mean tends to get closer to the expected value of X. As you play the lottery more, the mean of your winnings tends to get closer to 0.50. This is bad news (for you, not the lottery) when you pay $1 each time to play.

▲ **FIGURE 6.18: The Sampling Distribution of $\bar{x}$ when $n = 52$ and when $n = 365$.** The standard error is smaller when n is larger. **Question:** What is the practical implication of the effect of n on the standard error?

TO PRACTICE THIS CONCEPT, TRY EXERCISE 6.55(B) AND CONDUCT THE SIMULATION STUDY IN EXERCISE 6.105 TO OBSERVE HOW SAMPLING DISTRIBUTIONS DEPEND ON n

This example reinforces what we observed in the previous section about the effect of increased sample size on the precision of a sample proportion in estimating a population proportion. The larger the sample size, the smaller is the standard error and, thus, the smaller is the potential error in estimating a population parameter.

Sample Means Have Normal Sampling Distributions for Large *n*

Example 15 constructed the sampling distribution for the sample proportion preferring Aunt Erma's pizza in a taste test for a sample of size $n = 3$. The sampling distribution had somewhat of a mound shape. See the margin. A simulation in Activity 1 for the sample mean with $n = 1000$ showed even more of a bell shape. Is it typical for sampling distributions to have bell shapes?

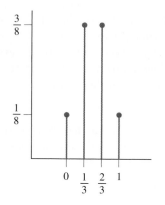

Let's look at another example, this time one in which the probability distribution of X is skewed. Exercise 6.2 derived the probability distribution of X = maximum value on two rolls of a die. The figure below displays the

distribution. It is skewed to the left with a mean of $\mu = 4.5$ and a standard deviation of $\sigma = 1.4$. Suppose we repeatedly sample from this distribution with random sample sizes of $n = 30$. Figure 6.19 displays the histogram of the sample means for a huge number of simulated samples from this skewed distribution. (In theory, the sampling distribution refers to an infinite number of samples of size 30 each; using software, we took 10,000 samples of size 30 each, which gives similar results.)

Even though the probability distribution from which we took the samples (shown above) is highly skewed, this sampling distribution has a bell shape. The histogram in Figure 6.19 has a normal distribution superimposed that has the same mean as the sampling distribution ($\mu = 4.5$) and the same standard deviation as the standard error of the sampling distribution, which is $\sigma/\sqrt{n} = 1.4/\sqrt{30} = 0.26$.

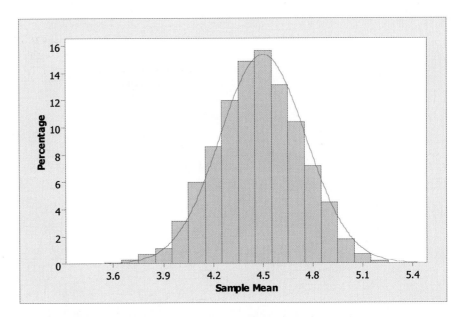

▲ **FIGURE 6.19: Histogram of Simulated Sample Means with *n* = 30.** Each sample was simulated from a discrete probability distribution (shown on previous page) that is skewed to the left with $\mu = 4.5$ and $\sigma = 1.4$. The superimposed normal distribution has mean 4.5 and standard deviation 0.26. **Question:** How does this sampling distribution relate with respect to shape, center, and spread to the probability distribution from which the samples were taken?

We've seen here a quite surprising result: Even when a probability distribution is not at all bell-shaped, the sampling distribution of the sample mean $\bar{x}$ can have a bell shape. This is a consequence of the second main result of this section. It states that the sampling distribution of $\bar{x}$ often has approximately a normal distribution. This result is called the **central limit theorem**. (It's called a "theorem" because it can be proved mathematically.)

Central Limit Theorem: Approximately Normal Sampling Distribution for Sample Mean $\bar{x}$

For random sampling with a large sample size n, the sampling distribution of the sample mean $\bar{x}$ is approximately a normal distribution.

This result applies *no matter what the shape* of the probability distribution from which the samples are taken. This probability distribution is called the *population distribution*, because in practice it describes a population from which we take a sample. *For relatively large sample sizes, the sampling distribution is bell shaped even if the population distribution is highly discrete or highly skewed.* This is remarkable. We observed this in Figure 6.19 with a skewed, highly discrete population, using $n = 30$, which is not all that "large" but is the size of sample we sometimes see in practice.

Figure 6.20 displays sampling distributions of the sample mean $\bar{x}$ for four different shapes for the population distribution from which samples are taken. The population shapes are shown at the top of the figure. Below them are portrayed the sampling distributions of $\bar{x}$ for random sampling of sizes $n = 2, 5,$ and 30. Even if the population distribution itself is uniform (column 1 of the figure) or U-shaped (column 2) or skewed (column 3), the sampling distribution has approximately a bell shape when n is at least about 30, and sometimes for n as small as 5. In addition, the spread of the sampling distribution noticeably decreases as n increases, because the standard error decreases.

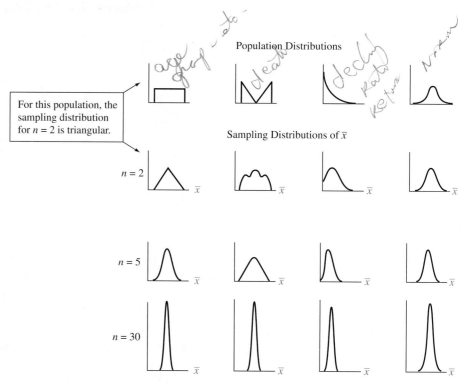

For this population, the sampling distribution for $n = 2$ is triangular.

▲ **FIGURE 6.20: Four Population Distributions and the Corresponding Sampling Distributions of x̄.** Regardless of the shape of the population distribution, the sampling distribution becomes more bell shaped as the random sample size n increases. You can use the *sampling distribution* applet on the text CD to simulate how the first population distribution shown has a sampling distribution for the sample mean that becomes more nearly normal as n increases. See Exercise 6.105 for details.

If the population distribution is approximately normal, then the sampling distribution is approximately normal for *all* sample sizes. The rightmost column of Figure 6.20 shows this case.

> **IN PRACTICE: Sampling distribution of x̄ more bell shaped as *n* increases**
>
> The sampling distribution of x̄ takes more of a bell shape as the random sample size n increases. The more skewed the population distribution, the larger n must be before the shape is close to normal. In practice, the sampling distribution is usually close to normal when the sample size n is at least about 30.

How Does the Central Limit Theorem Help Us Make Inferences?

The central limit theorem and the formula for the standard error of the sample mean x̄ have many implications. These include the following:

Recall

From Section 6.2, about 95% of a normal distribution falls within 2 standard deviations of the mean, and 99.7% within 3 standard deviations. ■

- When the sampling distribution of the sample mean x̄ is approximately normal, x̄ falls within 2 standard errors of the population mean μ with probability close to 0.95, and x̄ almost certainly falls within 3 standard errors of μ. Results of this type are vital to inferential methods that predict how close sample statistics fall to unknown population parameters.

- For *large n*, the sampling distribution is approximately normal even if the population distribution is not. This enables us to make inferences about population means regardless of the shape of the population distribution. This is helpful in practice, since usually we don't know the shape of the population distribution. Often, it is quite skewed.

EXAMPLE 20

What's The Probability You Come Out Ahead in Playing the Lottery?

Picture The Scenario

In Example 19, we considered the following lottery game: You bet a dollar on a number between 0 and 9 that you pick at random. If you are correct, you win $5, and otherwise, you win nothing. Let X denote your winnings. Then, the possible values for X are $0 or $5. From Example 19, the probability distribution is $P(0) = 0.90$ and $P(5) = 0.10$, which has mean $\mu = 0.50$ (50 cents) and standard deviation $\sigma = 1.50$.

Questions to Explore

Let's see how your probability of coming out ahead (winning more than you pay to play) depends on how often you play. What's the probability that you come out ahead if you play this game (a) once, (b) once a week for the next year, (c) once a day for the next year?

Think It Through

You come out ahead if your mean winnings $\bar{x}$ exceed $1, the amount you pay to play each time.

a. In playing $n = 1$ time, you come out ahead only if you win. The probability of this is the probability that you win in playing once, which is 0.10.

b. If you play once a week for the next year, then $n = 52$. We need to find the probability that the sample mean exceeds $1. By the central limit theorem, for large n, the sampling distribution of $\bar{x}$ is approximately normal. This is true even though the population distribution is very nonnormal. In fact, the population distribution here is binary and highly skewed, consisting of a large probability at $x = 0$ and a small probability at $x = 5$. The sampling distribution of $\bar{x}$ is approximately normal with mean $\mu = 0.50$ and standard error $\sigma/\sqrt{n} = 1.50/\sqrt{52} = 0.208$. Figure 6.21 shows this curve.

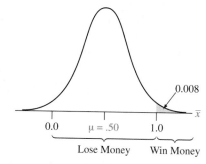

▲ **FIGURE 6.21: Approximate Sampling Distribution of the Sample Mean $\bar{x}$ when $\mu = 0.50$ and Standard Error = 0.208.** We use this to approximate the probability that $\bar{x}$ falls above $1 and you win more money than you pay to play the lottery 52 times. **Question:** Does this have the same shape as the probability distribution for each game? Explain.

The probability that you come out ahead after playing 52 times (that is, the sample mean exceeds $1) is approximately the probability above 1.0 in the right tail of this figure. So we need to find the probability that a normal random variable with mean 0.50 and standard error 0.208 takes a value above 1. The value 1.0 has a z-score of

$$z = (\text{value} - \text{mean})/(\text{standard error}) = (1.0 - 0.50)/0.208 = 2.40.$$

TI-83+/84 output

TI-83+/84 output

From Table A, this has a cumulative probability of 0.992. Thus, the right-tail probability is $1 - 0.992 = 0.008$. There is roughly 1 chance in a hundred that you'll win more than you paid to play, if you play 52 times (Exercise 6.111 gives a more precise answer).

c. What happens if you're really serious and play once a day for the next year, for a total of 365 times? Then, the standard error equals $\sigma/\sqrt{n} = 1.50/\sqrt{365} = 0.0785$. A sample mean winnings of 1 then has a z-score of

$$z = (1.0 - 0.50)/0.0785 = 6.4.$$

You won't find this in Table A, but software tells us that the probability in the right tail is only 0.0000000001, which is 1 in ten billion. This chance of coming out ahead is less than the chance that you would be chosen if one person of everyone in the entire world were picked at random!

Insight
Over the course of a year, it's nearly impossible to win more than you paid to play. With large n, it's hard for gamblers to come out ahead (unless they can cheat).

TO PRACTICE THIS CONCEPT, TRY EXERCISES 6.55(C) AND 6.56

6.5 Practicing the Basics

6.53 **Education of the self-employed:** According to a recent *Current Population Reports*, the population distribution of number of years of education for self-employed individuals in the United States has a mean of 13.6 and a standard deviation of 3.0.

a. Identify the random variable X whose distribution is described here.

b. Find the mean and standard error of the sampling distribution of $\bar{x}$ for a random sample of size 100. Interpret them.

c. Repeat (b) for $n = 400$. Describe the effect of increasing n.

6.54 **Rolling two dice:** Let X denote the outcome of rolling a die. From Exercise 6.49, the sampling distribution for the sample mean outcome from rolling two dice is $P(1) = 1/36, P(1.5) = 2/36, P(2) = 3/36, \ldots,$ $P(3.5) = 6/36, P(4) = 5/36, \ldots, P(6) = 1/36.$

a. Construct a graph of the (i) probability distribution of X, (ii) sampling distribution of the sample mean for $n = 2$. (The first column of Figure 6.20 portrays a case like this, and shows how the sampling distribution of the sample mean becomes more bell-shaped for $n = 5$ rolls and $n = 30$ rolls.)

b. The probability distribution of X has mean 3.50 and standard deviation 1.71. Find the mean and standard error of the sampling distribution of the sample mean for (i) $n = 2$, (ii) $n = 30$. What is the effect of n on the sampling distribution?

6.55 **Playing roulette:** A roulette wheel in Las Vegas has 38 slots. If you bet a dollar on a particular number, you'll win $35 if the ball ends up in that slot and $0 otherwise. Roulette wheels are calibrated so that each outcome is equally likely.

a. Let X denote your winnings when you play once. State the probability distribution of X. It has mean 0.921 and standard deviation 5.603.

b. You decide to play once a minute for 12 hours a day for the next week, a total of 5040 times. Show that the sampling distribution of your sample mean winnings has mean = 0.921 and standard error = 0.079.

c. Refer to (b). Using the central limit theorem, find the probability that with this amount of roulette playing, your mean winnings is at least $1, so that you have not lost money after this week of playing. (*Hint*: Find the probability that a normal random variable with mean 0.921 and standard deviation 0.079 exceeds 1.0.)

6.56 **Canada lottery:** In one lottery option in Canada [TRY] (www.lotterycanada.com), you bet on a six-digit number between 000000 and 999999. For a $1 bet, you win $100,000 if you are correct. The mean and standard deviation of the probability distribution for the lottery winnings are $\mu = 0.100$ (that is, 10 cents) and $\sigma = 31.623$. Joe figures that if he plays enough times every day, eventually he will strike it rich, by the law of large numbers. Over the course of several years, he plays one million times. Let $\bar{x}$ denote his average winnings.

   a. Find the mean and standard error of the sampling distribution of $\bar{x}$.

   b. About how likely is it that Joe's average winnings exceed $1, the amount he paid to play each time? Use the central limit theorem in answering.

6.57 **Income of farm workers:** For the population of farm workers in New Zealand, suppose that weekly income has a distribution that is skewed to the right with a mean of $\mu = \$500$ (N.Z.) and a standard deviation of $\sigma = \$160$. A researcher, unaware of these values, plans to randomly sample 100 farm workers and use the sample mean annual income $\bar{x}$ to estimate μ.

   a. Show that the standard error of $\bar{x}$ equals 16.0.

   b. Explain why it is almost certain that the sample mean will fall within $48 of $500.

   c. The sampling distribution of $\bar{x}$ provides the probability that $\bar{x}$ falls within a certain distance of μ, regardless of the value of μ. Show how to calculate the probability that $\bar{x}$ falls within $20 of μ for all such workers. (*Hint:* Using the standard error, convert the distance 20 to a z-score for the sampling distribution.)

6.58 **Restaurant profit?:** Jan's All You Can Eat Restaurant charges $8.95 per customer to eat at the restaurant. They find that their expense per customer, based on how much the customer eats and the expense of labor, has a distribution that is skewed to the right with a mean of $8.20 and a standard deviation of $3.

   a. If the 100 customers on a particular day have the characteristics of a random sample from their cus-

tomer base, find the mean and standard error of the sampling distribution of their sample mean expense per customer.

   b. Find the probability that the restaurant makes a profit that day, with the sample mean expense being less than $8.95. (*Hint:* Apply the central limit theorem to the sampling distribution in (a).)

6.59 **Survey accuracy:** A study investigating the relationship between age and annual medical expenses randomly samples 100 individuals in Springfield, Illinois. It is hoped that the sample will have a similar mean age as the entire population.

   a. If the standard deviation of the ages of all individuals in Springfield is $\sigma = 15$, find the probability that the mean age of the individuals sampled is within 2 years of the mean age for all individuals in Springfield. (*Hint:* Find the sampling distribution of the mean age and use the central limit theorem. You don't have to know the mean to answer this, but if it makes it easier for you, use a value such as $\mu = 30$.)

   b. Would the probability be larger, or smaller, if $\sigma = 10$? Why?

6.60 **Blood pressure:** Vincenzo De Cerce was diagnosed with high blood pressure. He was able to keep his blood pressure in control for several months by taking blood pressure medicine (amlodipine besylate). De Cerce's blood pressure is monitored by taking three readings a day, in early morning, at midday, and in the evening.

   a. During this period, the probability distribution of his systolic blood pressure reading had a mean of 130 and a standard deviation of 6. If the successive observations behave like a random sample from this distribution, find the mean and standard error of the sampling distribution of the sample mean for the three observations each day.

   b. Suppose that the probability distribution of his blood pressure reading is normal. What is the shape of the sampling distribution? Why?

   c. Refer to (b). Find the probability that the sample mean exceeds 140, which is considered problematically high. (*Hint:* Use the sampling distribution, not the probability distribution for each observation.)

6.6 | How Can We Make Inferences About a Population?

Sampling distributions are fundamental to statistical inference—making decisions and predictions about populations based on sample data. We'll end the chapter with a preview of this connection, which we'll explore in detail in the next two chapters. But first, let's summarize three quite distinct types of distributions we've studied: the **population distribution**, the **data distribution**, and the **sampling distribution of a statistic**. Here is a capsule description:

Population Distribution, Data Distribution, Sampling Distribution

- **Population distribution:** This is the probability distribution from which we take the sample. Values of its *parameters*, such as the population proportion p for a categorical variable and the population mean μ for a quantitative variable, are usually unknown. They're what we'd like to learn about.

- **Data distribution:** This is the distribution of the sample data. It's the distribution we actually see in practice. It's described by sample *statistics*, such as a sample proportion or a sample mean. With random sampling, the larger the sample size n, the more closely it resembles the population distribution. With larger n, the higher the probability that a sample *statistic* (such as a sample proportion or the sample mean) falls close to the population parameter (such as the population proportion p or the population mean μ).

- **Sampling distribution:** This is the probability distribution of a sample statistic, such as a sample proportion or sample mean. With random sampling, it provides probabilities for all the possible values of the statistic. The sampling distribution provides the key for telling us how close a sample statistic falls to the unknown parameter we'd like to make an inference about. For large n, it's approximately a normal distribution, by the central limit theorem. Its standard deviation is called the *standard error*.

For *binary* data, the sampling distribution of the sample proportion has mean equal to the population proportion p, and

$$\text{standard error} = \sqrt{p(1 - p)/n}.$$

For a *quantitative* variable, the sampling distribution of the sample mean $\bar{x}$ has mean equal to the population mean μ, and

$$\text{standard error} = \sigma/\sqrt{n},$$

where σ is the population standard deviation.

EXAMPLE 21

Population, Data, and Sampling Distributions for Family Size

Picture the Scenario

The distribution of family size in a particular tribal society is skewed to the right, with population mean $\mu = 5.2$ and population standard deviation $\sigma = 3.0$. These values are not known to an anthropologist who samples families in this society to estimate mean family size. For a random sample of 36 families, she gets a mean of 4.6 and a standard deviation of 3.2.

Questions to Explore

a. Identify the *population distribution*, and state its mean and standard deviation.

b. Identify the *data distribution*, and state its mean and standard deviation.

c. Identify the *sampling distribution* of $\bar{x}$, and state its mean and standard error.

Think It Through

a. The *population distribution* is the set of family-size values for everyone in the population of this tribal society. It is skewed to the right and described by the population mean of 5.2 and the population standard deviation of 3.0.

b. The *data distribution* is the collection of family-size values for the 36 families in the sample. It is described by the sample mean of 4.6 and the sample standard deviation of 3.2. It would probably look like the population distribution, being skewed to the right.

c. Consider all the possible samples of size 36 that could be gathered from this population. If we could find the sample mean $\bar{x}$ for each, then the frequency distribution of all their values would make up the *sampling distribution* of $\bar{x}$. That is, this is the collection of all the possible sample mean values and their probabilities. Since n is relatively large ($n = 36$), the sampling distribution is approximately a normal distribution, by the central limit theorem. The mean of this distribution is the same as the population mean, $\mu = 5.2$. The standard error is $\sigma/\sqrt{n} = 3.0/\sqrt{36} = 3.0/6 = 0.50$. Unlike the population and data distributions, the sampling distribution is bell shaped, and it is much less spread out. Sample means vary much less than individual observations in the sample or population.

Insight

In practice, the anthropologist does not know the population mean. How can she determine how near her sample mean of 4.6 falls to the population mean? The next chapter shows how she could use the sample data distribution to estimate the standard error. This tells her how spread out the sampling distribution is. As we've seen, knowing the standard error helps us predict how close the sample mean falls to the population mean.

TO PRACTICE THIS CONCEPT, TRY EXERCISES 6.61 AND 6.62

This example showed the distinction among these three distributions for a quantitative variable. Let's now look at an example for a categorical variable.

EXAMPLE 22

Population, Data, and Sampling Distributions for an Exit Poll

Picture the Scenario

The 2000 U.S. senatorial election in New York pitted Hillary Clinton against Rick Lazio. The exit poll on which TV networks relied for their projections found after sampling 2232 voters that 55.7% of the sample (1244 voters) voted for Clinton and 44.3% (988 voters) voted for Lazio. When all 6.2 million votes were tallied, 56.0% voted for Clinton and 44.0% voted for Lazio.

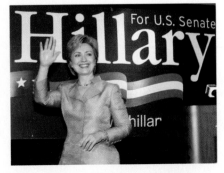

Questions to Explore

Let X = vote outcome, with $x = 1$ for Hillary Clinton and $x = 0$ for Rick Lazio. What are the (a) population distribution, (b) data distribution, and (c) sampling distribution of the sample proportion for this scenario?

Think It Through

a. The *population distribution* is the 6.2 million values of the x vote variable, 44.0% of which are 0 and 56.0% of which are 1. That is, if we look at the votes for all subjects

Voter	Vote	x
1	Clinton	1
2	Lazio	0
3	Lazio	0
. . .		
6,200,000	Clinton	1

56.0% of the time we'd see $x = 1$. The mean of all the 0 and 1 observations would be $p = 0.560$, the population proportion supporting Clinton. This value $p = 0.560$ is the parameter value describing this distribution. The distribution is described by a graph consisting of two stems, with mean 0.560, as the first plot in Figure 6.22 shows. It's not at all bell shaped.

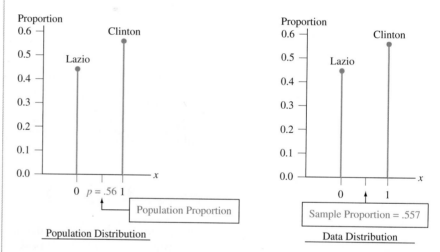

▲ **FIGURE 6.22:** The population (6.2 million voters) and data ($n = 2232$) distributions of candidate preference (0 = Lazio, 1 = Clinton). **Question:** Why do these look so similar?

b. The exit poll sampled $n = 2232$ voters. A graph of the 2232 votes in the sample provides a *data distribution*. We'd have a data file like the one shown above, but for 2232 voters rather than 6.2 million voters. Figure 6.22 also shows this distribution. Like the population distribution, it concentrates at $x = 0$ and $x = 1$. It is described by sample statistics. The sample proportion who voted for Clinton was $1244/2232 = 0.557$. The sample data are summarized by the sample proportion of 0.557. So 55.7% of the data values are 1, and the height of the spike for Clinton in the data distribution figure in Figure 6.22 is 0.557. The larger the sample size, the more the data distribution tends to resemble the population distribution, since the data we observe are a subset of the population values. Here, they are very similar, since n is quite large.

Recall

The number of successes has
mean $= np$
std. dev. $= \sqrt{np(1-p)}$.

The proportion of successes has
mean $= p$
std. error $= \sqrt{p(1-p)/n}$. ▪

c. The *sampling distribution* of the sample proportion, by the central limit theorem, is approximately normal. Its mean is $p = 0.560$ (the population proportion) with

$$\text{Standard error} = \sqrt{p(1-p)/n} = \sqrt{0.560(1-0.560)/2232} = 0.011.$$

With $n = 2232$, the sampling distribution has very little spread. Figure 6.23 portrays this sampling distribution, relative to the population distribution.

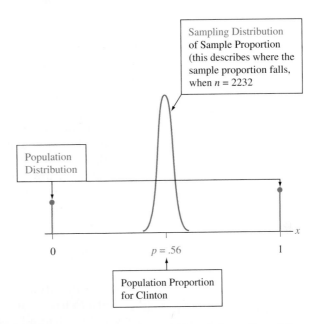

▲ **FIGURE 6.23: The Population Distribution of Votes (1 = Clinton, 0 = Lazio) and the Sampling Distribution of the Sample Proportion of Votes for Clinton.** Both distributions have mean 0.560, namely the population proportion, but the sampling distribution is approximately normal. It looks quite different from the population distribution, and it gets narrower and more bell shaped as *n* increases. **Question:** Which of these two distributions resembles the data distribution?

Insight

The *population distribution* and the *data distribution* are not bell shaped. They are discrete, concentrated at only the two values 0 and 1. They describe the relative number of 0 and 1 values (votes for Lazio and votes for Clinton), in the population and in the sample.

By contrast, the *sampling distribution* describes the scatter of possible sample proportion values voting for Clinton when $n = 2232$. It looks very different from the population and data distributions. It is bell shaped and much less spread out. You can think of the sampling distribution curve as describing what would happen if we could look at *every* possible sample of 2232 people from this population of voters, find the sample proportion voting for Clinton for each such sample, and then plot the distribution of those sample proportions. In practice, of course, we'd never do this, but the curve helps us to visualize how much variation there can be among different exit polls in what they report.

IN PRACTICE: Exit polls use a multistage type of random sampling

Exit polls use a multistage type of random sampling, not simple random sampling. The standard error formulas in this chapter apply for simple random sampling. However, they provide good approximations for the standard errors with many multistage samples and are usually used for such sampling schemes also.

TO PRACTICE THIS CONCEPT, TRY EXERCISE 6.63

How Far Off Will a Sample Prediction Be?

In the Clinton-Lazio exit-poll example, the sampling distribution of the sample proportion for $n = 2232$ is approximately a normal distribution. Its mean is 0.560 (the population proportion), and its standard error is 0.011. Based on the spread of the sampling distribution, what could we have concluded if we did *not* know the population proportion but did have the results of the exit poll and had to decide whether to predict a winner?

The sample proportion of the 2232 voting for Hillary Clinton was 0.557. Because the sampling distribution is approximately normal, we know that the sample proportion was very likely to fall within three standard errors of the population proportion, that is, within $3(0.011) = 0.033$ of p. Thus, using the sample proportion of 0.557, we would predict that the population proportion voting for Clinton was between about $0.557 - 0.033 = 0.524$ and $0.557 + 0.033 = 0.590$. Since all the values in the interval from 0.524 to 0.590 are above 0.50, this suggests that the population proportion was larger than 0.50. Based on results for these 2232 voters, we would predict that Hillary Clinton won the election.

This reasoning is similar to what we'll use in coming chapters to make inferences about population parameters. Fundamental to the logic is the central limit theorem result—that, with large random samples, sample proportions and sample means have approximately normal sampling distributions. *In fact, most statistics used to estimate population parameters have approximately normal sampling distributions, for large random samples. This is the primary reason for the key role of the normal distribution in statistics.*

We'll see in the next chapter that not all bell-shaped curves represent normal distributions. The normal distribution refers to a particular bell-shaped curve—the curve that comes from generating sampling distributions for many statistics with large random samples.

You may have noticed a slight flaw in the logic above about predicting the outcome of the Clinton-Lazio election. In practice, because we don't know the value of the parameter p, we also would not know the standard error, $\sqrt{p(1 - p)/n}$. We'd have to approximate the standard error by substituting the sample proportion, for instance using $\sqrt{0.557(0.443)/2232} = 0.0105$. For most inference methods, we'll use standard error formulas that substitute sample values for unknown parameter values.

IN PRACTICE: Standard errors are estimated

Standard errors have exact values depending on parameter values, such as $\sqrt{p(1 - p)/n}$ for a sample proportion and $\sigma/\sqrt{n}$ for a sample mean. In practice, these are unknown, so inference methods use standard errors that substitute sample values (such as sample proportions and sample standard deviations) in the exact formulas. These *estimated* standard errors are the numbers which we'll use in the next two chapters and that software reports for standard errors.

A final caveat: If the sampling is not random, the results we've used for sampling distributions no longer apply. For instance, in predicting the winner of an election, we should not merely use the first 3000 votes reported to a network, because they would not be a random sample of all voters. (They might all come from one county, for instance.)

Likewise, sometimes nonstatistical factors adversely affect the accuracy of exit-poll predictions. A famous case occurred in the 2000 presidential election. About the time the polls closed in Florida, the major networks predicted that Al Gore would win that state based on exit polling. Later, a more complete count put

George W. Bush ahead by 537 votes, which resulted in Bush gaining enough electoral college votes to become president—even though he received half a million fewer votes than Gore nationwide. But, this was not necessarily the fault of the exit poll. In the 2000 presidential election, about 200,000 marked ballots in Florida were thrown out as invalid either because the tabulating machine could not read them or because the voter marked two names rather than one, often because of a confusing ballot design (see Example 6 in Chapter 3 and Exercise 2.83 in Chapter 2). Of the ballots marked with at least two names, Al Gore was marked on 84,197 and George W. Bush on 37,731. This source of error was so large that it negated the precision of a statistical prediction in a close election.

6.6 Practicing the Basics

6.61 Household size: In 2000, according to the U.S. Census Bureau, the number of people in a household had a mean of 2.6 and a standard deviation of 1.5. This is based on census information for the population. Suppose the Census Bureau instead had estimated this mean using a random sample of 225 homes. Suppose the sample had a sample mean of 2.4 and standard deviation of 1.4.

a. Identify the random variable X. Indicate whether it is quantitative or categorical.

b. Describe the center and spread of the population distribution.

c. Describe the center and spread of the data distribution.

d. Describe the center and spread (standard error) of the sampling distribution of the sample mean for 225 homes.

6.62 Sunshine City: Sunshine City was designed to attract retired people, and its current population of 50,000 residents has a mean age of 60 years and a standard deviation of 16 years. The distribution of ages is skewed to the left. A random sample of 100 residents of Sunshine City has $\bar{x} = 58.3$ and $s = 15.0$.

a. Describe the center and spread of the population distribution.

b. Describe the center and spread of the data distribution. What shape does it probably have?

c. Describe the center and spread of the sampling distribution of the sample mean for $n = 100$. What shape does it have?

d. Explain why it would not be unusual to observe a person of age under 40 in Sunshine City, but it would be highly unusual to observe a *sample mean* under 40 for a random sample size of 100 people.

6.63 Gender distributions: At a university, 60% of the 7400 students are female. The student newspaper reports results of a survey of a random sample of 50 students about various topics involving alcohol abuse, such as participation in binge drinking. They report that their sample contained 26 females.

a. Explain how you can set up a binary random variable X to represent gender.

b. Identify the population distribution of gender at this university.

c. Identify the data distribution of gender for this sample.

d. Identify the sampling distribution of the sample proportion of females in the sample. State its mean and standard error for a random sample of size 50.

6.64 Shapes of distributions:

a. With random sampling, does the shape of the data distribution tend to resemble more closely the sampling distribution or the population distribution? Explain.

b. Explain carefully the difference between a *data distribution* of sample observations and the *sampling distribution* of the sample proportion for a binary variable that can take values of only 0 and 1.

6.65 Predicting the Florida senate race: In the 2000 U.S. senatorial election in Florida, an exit poll indicated that 901 voted for Democrat Bill Nelson and 832 voted for Republican Bill McCollum. Would you have been willing to predict the winner? Use the reasoning shown in Example 22. That is, justify your answer by finding the sample proportion voting for Nelson and showing what you would expect for that proportion if the population proportion voting for him was $p = 0.50$, based on the standard error of the sample proportion for $n = 1733$ when $p = 0.50$.

6.66 Did Schwarzenegger win?: In the 2003 California special election, of the candidates running to replace Gray Davis as governor, an exit poll indicated that 1081 voted for Arnold Schwarzenegger and 817 voted for Cruz Bustamonte. Would you predict a winner? Justify your answer by finding the sample proportion voting for Schwarzenegger and showing what you would expect for it if the population proportion voting for him was 0.50, based on the standard error of the sample proportion for the sample size this exit poll had.

Answers to the Chapter Figure Questions

Figure 6.1: The shape of the distribution is skewed to the right.

Figure 6.2: The area under the curve from 0 to 15 should be shaded.

Figure 6.3: The shape of the distribution is approximately symmetric and bell shaped. The center of the distribution is about 65 inches. The spread as described by the standard deviation is about 3.5 inches.

Figure 6.4: We evaluate $70 \pm 3(4)$, giving as an interval for the men's heights from 58 inches to 82 inches.

Figure 6.5: These probabilities are similar to the percentages stated for the Empirical Rule. The Empirical Rule states that for an approximately bell-shaped distribution, approximately 68% of the observations fall within one standard deviation of the mean, 95% within two standard deviations of the mean, and nearly all within three standard deviations of the mean.

Figure 6.8: For three standard deviations, the probability in one tail is 0.0013. By symmetry, this is also the probability in the other tail. The total tail area is 0.0026, which subtracted from 1 gives an answer of 0.997.

Figure 6.9: The 2nd percentile is the value located on the left side of the curve such that 2% of the distribution falls below it and 98% falls above.

Figure 6.10: The SAT score with $z = -3$ is $500 - 3(100) = 200$. The SAT score with $z = 3$ is $500 + 3(100) = 800$.

Figure 6.11: The values of -3 and 3.

Figure 6.12: The distribution with $p = 0.1$ is skewed to the right.

Figure 6.13: We can find the probabilities for each possible sample proportion value using the binomial distribution. The binomial formula

is useful when p differs from 0.50. Then the eight possible samples we found would not each have probability 1/8.

Figure 6.16: The standard error is
$$\sqrt{p(1 - p)/n} = \sqrt{(0.50)(0.50)/1000} = \sqrt{0.00025} = 0.0158.$$
Since the simulated sampling distribution is bell shaped, the probability is close to 1.0 that the sample proportion falls within three standard errors of 0.50. Three standard errors equals $3(0.0158) = 0.047$, or approximately 0.05.

Figure 6.18: The larger the sample size, the smaller the standard error and thus, the closer the sample mean tends to fall to the population mean.

Figure 6.19: The probability distribution is highly skewed left. However, the sampling distribution of sample means from this probability distribution is bell shaped. The mean of the sampling distribution is similar to the mean of the probability distribution, 4.5. The standard error of the sampling distribution is smaller than the standard deviation of the probability distribution. The standard error equals the standard deviation of the probability distribution divided by the square root of the sample size n, that is, $1.4/\sqrt{30} = 0.26$.

Figure 6.21: The probability distribution and the sampling distribution do not have the same shape. The sampling distribution is bell shaped. The probability distribution is discrete, concentrated at two values, $0 with height of 0.9 and $5 with height of 0.1.

Figure 6.22: The larger the sample size, the more we expect the data distribution to resemble the population distribution.

Figure 6.23: The population distribution resembles the data distribution.

CHAPTER SUMMARY

- A **random variable** is a numerical measurement of the outcome of a random phenomenon. As with ordinary variables, random variables can be **discrete** (taking separate values) or **continuous** (taking an interval of values).

- A **probability distribution** specifies probabilities for the possible values of a random variable. Probability distributions have summary measures of the center and the spread, such as the mean μ and standard deviation σ. The mean (also called **expected value**) for a discrete random variable is

$$\mu = \Sigma x P(x),$$

where P(x) is the probability of the outcome x and the sum is taken over all possible outcomes.

- The **normal distribution** is the probability distribution of a continuous random variable that has a symmetric bell-shaped graph specified by mean (μ) and standard deviation (σ) parameters. For any z, the probability within z standard deviations of μ is the same for every normal distribution.

- The **z-score** for an observation x equals
$$z = (x - \mu)/\sigma.$$

It measures the number of standard deviations that x falls from the mean μ. For a normal distribution, the z-scores have the **standard normal distribution**, which has mean $= 0$ and standard deviation $= 1$.

■ The **binomial distribution** is the probability distribution of the discrete random variable that measures the number of successes X in n independent trials, with probability p of a success on a given trial. It has

$$P(x) = \frac{n!}{x!(n-x)!} p^x (1-p)^{n-x}, \quad x = 0, 1, 2, \ldots, n.$$

The mean is $\mu = np$ and standard deviation is

$$\sigma = \sqrt{np(1-p)}.$$

■ A **sampling distribution** is a probability distribution of a sample statistic, such as the sample proportion or the sample mean. It specifies probabilities for the values of the statistic for all the possible random samples.

■ The **sampling distribution of the sample mean** $\bar{x}$ is centered at the population mean μ. Its spread is described by its standard deviation, called the **standard error**. This is related to the standard deviation of the population by standard error $= \sigma/\sqrt{n}$. For binary data (0 or 1), the standard error simplifies to $\sqrt{p(1-p)/n}$, where p is the population proportion.

■ The **central limit theorem** states that for random samples of sufficiently large size (at least about 30 is usually enough), the sampling distribution of the sample mean is approximately normal. This holds *no matter what the shape of the population distribution*. It applies to sample proportions also, since the sample proportion is a sample mean of 0 and 1 values. In that case, the sampling distribution of the sample proportion is approximately normal whenever n is large enough that both np and $n(1-p)$ are at least 15.

The bell-shaped appearance of the sampling distributions for most statistics is the main reason for the importance of the normal distribution.

Summary of New Notation in Chapter 6

$P(x)$	Probability that a random variable takes value x
μ	Mean of the probability distribution or population distribution

σ	Standard deviation of the probability distribution or population distribution
$p, 1-p$	Probabilities of the two possible outcomes of a binary variable

CHAPTER PROBLEMS

Practicing the Basics

6.67 **Grandparents:** Let X = the number of living grandparents that a randomly selected adult American has. According to recent General Social Surveys, its probability distribution is approximately $P(0) = 0.71$, $P(1) = 0.15, P(2) = 0.09, P(3) = 0.03, P(4) = 0.02$.

   a. Does this refer to a discrete, or a continuous, random variable? Why?

   b. Show that the probabilities satisfy the two conditions for a probability distribution.

   c. Find the mean of this probability distribution.

6.68 **Playing the lottery in NJ:** New Jersey has a "Pick 3" game in which you pick one of the 1000 3-digit numbers between 000 and 999, receiving $275 for a winning $1 bet and nothing otherwise.

   a. Construct the probability distribution of the random variable X = winnings.

   b. Find the mean of the probability distribution.

   c. Based on the mean in (b) and the $1 cost to play the game, on average, how much can you expect to lose each time you play this lottery?

6.69 **More NJ Lottery:** New Jersey has a "six-way combination" lottery game in which you pick three digits (each can be any of 0, 1, 2, ...,9) and you win if the digits the lottery picks are the same as yours in any of the six possible orders. For a dollar bet, a winner receives $45.50. Two possible strategies are (a) to pick three different digits, or (b) to pick the same digit three times. Let X = winnings for a single bet.

   a. Find the probability distribution of X and its mean for strategy (a).

   b. Find the probability distribution of X and its mean for strategy (b).

   c. Which is the better strategy? Why?

6.70 **Are you risk averse?** You need to choose between two alternative programs for dealing with the outbreak of a deadly disease. In program 1, 200 people are saved. In program 2, there is a 2/3 chance that no one is saved and a 1/3 chance that 600 people are saved.

   a. Find the expected number of lives saved with each program.

b. Now you need to choose between program 3, in which 400 people will die, and program 4, in which there is a 1/3 chance that no one will die and a 2/3 chance that 600 people would die. Find the expected number of deaths with each program.

c. Explain why programs 1 and 3 are similar, and 2 and 4 are similar. (If you had to choose, would you be like most people and be risk averse in (a), choosing program 1, and risk taking in (b), choosing program 4?)

6.71 **Flyers' insurance:** An insurance company sells a policy to airline passengers for $1. If a flyer dies on a given flight (from a plane crash), the policy gives $100,000 to the chosen beneficiary. Otherwise, there is no return. Records show that a passenger has about a 1 in a million chance of dying on any given flight. You buy a policy for your next flight.

a. Specify the probability distribution of the amount of money the beneficiary makes from your policy.

b. Find the mean of the probability distribution in (a). Interpret.

c. Explain why the company is very likely to make money in the long run.

6.72 **Normal probabilities:** For a normal distribution, find the probability that an observation is

a. Within 1.96 standard deviations of the mean.

b. More than 2.33 standard deviations from the mean.

6.73 **z-scores:** Find the z-score such that the interval within z standard deviations of the mean contains probability (a) 0.95, (b) 0.99 for a normal distribution. Sketch the two cases on a single graph.

6.74 **z-score and tail probability:**

a. Find the z-score for the number that is less than only 1% of the values of a normal distribution. Sketch a graph to show where this value is.

b. Find the z-scores corresponding to the (a) 90th, (b) 99th percentiles of a normal distribution.

6.75 **Quartiles:** If z is the positive number such that the interval within z standard deviations of the mean contains 50% of a normal distribution, then

a. Explain why this value of z is about 0.67.

b. Explain why the first and third quartiles equal $\mu - 0.67\sigma$ and $\mu + 0.67\sigma$.

c. The interquartile range, IQR, relates to σ by IQR = $2 \times 0.67\sigma$. Explain why.

6.76 **Cholesterol:** The American Heart Association reports that a total cholesterol score of 240 or higher represents high risk of heart disease. A study of postmenopausal women (*Clin. Drug. Invest.*, 2000, vol. 20, pp. 207–214) reported a mean of 220 and standard deviation of 40. If the total cholesterol scores have a normal distribution, what proportion of the women fall in the high-risk category?

6.77 **Female height:** For the normal height distribution of adult females in North America ($\mu = 65.0$ inches, $\sigma = 3.5$ inches), find the proportion who are

a. under five feet (60 inches).

b. over six feet (72 inches).

c. between 60 and 70 inches.

6.78 **Male height:** Repeat the previous exercise using the normal height distribution of adult males in North America ($\mu = 70.0, \sigma = 4.0$).

6.79 **Gestation times:** For 5459 pregnant women using Aarhus University Hospital in Denmark in a two-year period who reported information on length of gestation until birth, the mean was 281.9 days, with standard deviation 11.4 days. A baby is classified as premature if the gestation time is 258 days or less (*British Medical Journal*, July 24, 1993, p. 234). If gestation times are normally distributed, what's the proportion of babies who are born prematurely?

6.80 **Energy use:** An energy study in Gainesville, Florida, found that household use of electricity has a mean of 1200 kWh (kilowatt-hours) and a standard deviation of 500 kWh in a typical summer month. In June 2002, to encourage conservation, Gainesville Regional Utilities announced that the base cost would be $0.05 per kWh up to 1500 kWh of use and then $0.06 above that.

a. If the distribution were approximately normal, about what proportion of users would have their summer bill computed entirely based on the lower rate?

b. Suppose that actually for summer months, $\mu = 1000$ and $\sigma = 1000$. Do you think that this distribution of use is truly normal? Why or why not?

6.81 **Winter energy use:** Refer to the previous exercise. Electricity use is much less in Florida in the winter. To make electricity use values comparable from different seasons, in a given month each home's use is converted to a z-score. For each home with a z-score greater than 1.5 (electricity use more than 1.5 standard deviations above the mean), their bill contains a note suggesting a reduction in electricity use, to conserve resources. If the distribution of electricity use is normal, what proportion of households receives this note?

6.82 **Global warming:** Suppose that weekly use of gasoline for motor vehicle travel by adults in North America has approximately a normal distribution with a mean of 20 gallons and a standard deviation of 6 gallons. Many people who worry about global warming believe that Americans should pay more attention to energy conservation. Assuming that the standard deviation and the normal shape are unchanged, to what level must the mean reduce so that 20 gallons per week is the third quartile rather than the mean?

6.83 **Fast-food profits:** Mac's fast-food restaurant finds that its daily profits have a normal distribution with mean $140 and standard deviation $80.

a. Find the probability that the restaurant loses money on a given day (that is, daily profit less than 0).

b. Find the probability that the restaurant makes money for the next seven days in a row. What assumptions must you make for this calculation to be valid? (*Hint:* Use the binomial distribution.)

6.84 Metric height: A Dutch researcher reads that male height in the Netherlands has a normal distribution with $\mu = 72.0$ inches and $\sigma = 4.0$ inches. She prefers to convert this to the metric scale (1 inch = 2.54 centimeters). The mean and standard deviation then have the same conversion factor.

   a. In centimeters, would you expect the distribution still to be normal? Explain.

   b. Find the mean and standard deviation in centimeters. (*Hint*: What does 72.0 inches equal in centimeters?)

   c. Find the probability that height exceeds 200 centimeters.

6.85 Manufacturing tennis balls: According to the rules of tennis, a tennis ball is supposed to weigh between 56.7 grams (2 ounces) and 58.5 grams (2 1/16 ounces). A machine for manufacturing tennis balls produces balls with a mean of 57.6 grams and a standard deviation of 0.3 grams, when it is operating correctly. Suppose that the distribution of the weights is normal.

   a. If the machine is operating properly, find the probability that a ball manufactured with this machine satisfies the rules.

   b. After the machine has been used for a year, the process still has a mean of 57.6, but because of wear on certain parts the standard deviation increases to 0.6 grams. Find the probability that a manufactured ball satisfies the rules.

6.86 Bride's choice of surname: *Time* magazine reported (June 17, 2002) that 80% of brides take the surname of their new husband. Ann notes that of her 4 best friends who recently married, none kept her own name. If they had been a random sample of brides, how likely would this have been to happen?

6.87 Football wins: A football team has probability 0.50 of winning any particular game. In a season, it plays six conference games.

   a. Using the binomial distribution, identify n and p, and find the probability that the team wins 3 of its conference games.

   b. What assumptions does the calculation in (a) require? Ordinarily, would you expect that a team would have exactly the same probability of winning in each game?

6.88 Weather: A weather forecaster states, "The chance of rain is 50% on Saturday and 50% again on Sunday. So there's a 100% chance of rain sometime over the weekend." If whether or not it rains on Saturday is independent of whether or not it rains on Sunday, find the actual probability of rain *at least once* during the weekend.

   a. Answer using methods from Chapter 5, such as by listing equally likely sample points or using the formula for the probability of a union of two events.

   b. Answer using the binomial distribution.

6.89 Dating success: Based on past experience, Julio believes he has a 60% chance of success when he calls a woman and asks for a date.

   a. State assumptions needed for the binomial distribution to apply to the number of times he is successful on his next five requests.

   b. If he asks the same woman each of the five times, is it sensible to treat these requests as independent trials?

   c. Under the binomial assumptions, state n and p and the mean of the distribution.

6.90 Revisiting Canada Lottery: In one Canadian lottery option, you bet on one of the million six-digit numbers between 000000 and 999999. For a $1 bet, you win $100,000 if you are correct. In playing n times, let X be the number of times you win.

   a. Find the mean of the distribution of X, in terms of n.

   b. How large an n is needed for you to expect to win once (that is, $np = 1$)?

   c. If you play the number of times n that you determined in (b), show that your expected winnings is $100,000 but your expected profit is $-$900,000.

6.91 Female driving deaths: In a given year, the probability that an adult American female dies in a motor vehicle accident equals 0.0001 (*Statistical Abstract of the United States*, 2001).

   a. In a city having 1 million adult American females, state assumptions for a binomial distribution to apply to $X =$ the number of them who die in the next year from motor vehicle accidents. Identify n and p for that distribution.

   b. If the binomial applies, find the mean and standard deviation of X.

   c. Based on the normal distribution, find the interval of possible outcomes within three standard deviations of the mean that is almost certain to occur.

   d. Refer to the assumptions for the analysis in (a). Explain at least one way they may be violated.

6.92 Males drive more poorly?: Refer to the previous exercise. The probability of a motor vehicle death for adult American males is 0.0002. Repeat (b) and (c) for a city having 1 million of them, and compare results to those for females.

6.93 Terrorism: In June 2002, the Gallup organization asked a sample of 1000 Americans, "What do you think is the most important problem facing this country today?" A proportion of 0.46 answered terrorism. The Gallup report indicated that this had a margin of error of plus or minus 0.03 as an estimate of the population proportion. Usually the margin of error is defined as a distance of two standard errors on each side of the sample proportion, using the sample proportion in place of the unknown population proportion in the standard error formula. Show how Gallup got 0.03 as the margin of error (which they expressed as plus or minus 3%).

6.94 Governor exit poll: In a governor's race, the winner receives a population proportion of $p = 0.70$ of the votes. On the day of the election, an exit poll is taken of 1000 randomly selected voters. Consider the sampling distribution of the sample proportion of these 1000 voters who voted for the winner.

   a. Find the mean and standard error of the sampling distribution of this proportion.

   b. What do you expect for the shape of the sampling distribution? Why?

c. If the population proportion is truly $p = 0.70$, would it be surprising if the exit poll had a sample proportion of 0.60? Justify your answer by using the central limit theorem to approximate the probability of a sample proportion of 0.60 or less.

6.95 Exit poll: In the New York exit poll of Example 22, it is possible that all 2232 voters sampled happened to be Hillary Clinton supporters. Investigate how surprising this would be, if actually 56% of the population voted for Clinton, by

a. Finding the probability that all 2232 people voted for Clinton. (*Hint*: Use the binomial.)
b. Finding the number of standard errors that the sample proportion of 1.0 falls from the population proportion of 0.56.

6.96 Physicians' assistants: The 2000 American Academy of Physician Assistants (AAPA) (www.aapa.org) survey of the population of 1434 physicians assistants who were working full-time after graduating in 1999 reported a mean annual income of $58,000 and standard deviation of $11,000.

a. Suppose the AAPA had randomly sampled 100 physicians' assistants instead of collecting data for all 1434 of them. Describe the mean, standard error, and shape of the sampling distribution of the sample mean.
b. Using this sampling distribution, find the z-score for a sample mean of $60,000.
c. Using (b), find the probability that the sample mean would fall within $2000 of the population mean.

6.97 Bank machine withdrawals: An executive in an Australian savings bank decides to estimate the mean amount of money withdrawn in bank machine transactions. From past experience, she believes that $50 (Australian) is a reasonable guess for the standard deviation of the distribution of withdrawals. She would like her sample mean to be within $10 of the population mean. Estimate the probability that this happens if she randomly samples 100 withdrawals. (*Hint*: How many standard errors of the sample mean is $10?)

6.98 PDI: The scores on the Psychomotor Development Index (PDI), a scale of infant development, are approximately normally distributed with mean 100 and standard deviation 15. An infant is selected at random.

a. Find the probability that the infant's PDI score exceeds 103.
b. Find the probability that PDI is between 97 and 103.
c. Find the z-score for a PDI value of 90.

6.99 Mean PDI: Refer to the previous exercise. A study uses a random sample of 225 infants. Using the sampling distribution of the sample mean PDI:

a. Find the probability that the sample mean exceeds 103. (*Hint*: You need to find the standard error of the sampling distribution of the sample mean.)
b. Find the probability that the sample mean falls between 97 and 103.

c. Find the z-score from the sampling distribution corresponding to a sample mean of 90. A PDI value of 90 is not surprising, but would you be surprised to observe a sample mean PDI score of 90? Why?
d. Compare the results of parts (a)–(c) with those in the previous exercise, and interpret the differences.

6.100 Number of sex partners: According to recent General Social Surveys, in the United States, the distribution for adults of X = number of sex partners in the past 12 months has a population mean of about 1.0 and a standard deviation of about 1.0.

a. Does X have a normal distribution? Explain.
b. For a random sample of 100 adults, describe the sampling distribution of $\bar{x}$ and give its mean and standard error. What is the effect of X not having a normal distribution?

◆◆6.101 Using control charts to assess quality: In many industrial production processes, measurements are made periodically on critical characteristics to ensure that the process is operating properly. Observations vary from item to item being produced, perhaps reflecting variability in material used in the process and/or variability in the way a person operates machinery used in the process. There is usually a target mean for the observations, which represents the long-run mean of the observations when the process is operating properly. There is also a target standard deviation for how observations should vary around that mean if the process is operating properly. A **control chart** is a method for plotting data collected over time to monitor whether the process is operating within the limits of expected variation. A control chart that plots *sample means* over time is called an $\bar{x}$**-chart**.

As shown below, the horizontal axis is the time scale and the vertical axis shows possible sample mean values. The horizontal line in the middle of the chart shows the target for the true mean. The upper and lower lines are called the **upper control limit** and **lower control limit**, denoted by **UCL** and **LCL**. These are usually drawn three standard errors above and below the target value. The region between the LCL and UCL contains the values that theory predicts for the sample mean when the process is in control. When a sample mean falls above the UCL or below the LCL, it indicates that something may have gone wrong in the production process.

$\bar{x}$-chart

a. Walter Shewhart invented this method in 1924 at Bell Labs. He suggested using three standard errors in setting the UCL and LCL to achieve a balance between having the chart fail to diagnose a problem and having it indicate a problem when none actually existed. If the process is working properly ("in statistical control") and if n is large enough that $\bar{x}$ has approximately a normal distribution, what is the probability that it indicates a problem when none exists? (That is, what's the probability a sample mean will be at least three standard errors from the target, when that target is the true mean?)

b. What would the probability of falsely indicating a problem be if we used two standard errors instead for the UCL and LCL?

6.102 Too little or too much Coke?: Refer to the previous exercise. When a machine for dispensing a cola drink into bottles is in statistical control, the amount dispensed has a mean of 500 ml (milliliters) and a standard deviation of 4 ml.

a. In constructing a control chart to monitor this process with periodic samples of size 4, how would you select the target line and the upper and lower control limits?

b. If the process actually deteriorates and operates with a mean of 491 ml and a standard deviation of 6 ml, what is the probability that the next value plotted on the control chart indicates a problem with the process, falling more than three standard errors from the target? What do you assume in making this calculation?

6.103 Lack of control: Exercise 6.101 introduced control charts. When about nine sample means in a row fall on the same side of the target for the mean in a control chart, this is an indication of a potential problem, such as a shift up or a shift down in the true mean relative to the target value. If the process is actually in control and has a normal distribution around that mean, what is the probability that the next nine sample means in a row would (a) all fall above the mean, (b) all fall above or all fall below the mean? (*Hint*: Use the binomial distribution, treating the successive observations as independent.)

Concepts and Investigations

6.104 Simulating pizza preference: Use the *sampling distributions* applet on the text CD.

a. For Example 15, in which $n = 3$ people indicate their preferred pizza, simulate the sampling distribution of the *count* preferring pizza A when $p = 0.50$, by taking 1000 samples of size 3. Sketch a plot of the sampling distribution that you generate.

b. Simulate the sampling distribution of sample *proportions* for Example 15 with $n = 3$ subjects, by taking 1000 samples of size 3. Sketch a plot of the sampling distribution that you generate. Compare this generated sampling distribution to the sampling distribution of counts generated in part (a). Comment on the shape, means, and standard deviations. Does your generated sampling distribution for sample proportions look roughly like Figure 6.13 (shown again here)?

c. Repeat parts (a) and (b) using a sample size $n = 3$ and $p = 0.80$. Compare the sampling distribution that you generated to the sampling distributions generated in parts (a) and (b). Comment on similarities and differences.

6.105 CLT through simulation: Use the *sampling distributions* applet on the text CD to see how the first population distribution shown in Figure 6.20 (reproduced here) has a more nearly normal sampling distribution for the mean as n increases and how the spread of that sampling distribution changes as n increases. You will need to select *uniform* for the parent population and select your range of values for this population distribution.

a. Use the applet to create sampling distributions for the sample mean using different sample sizes n. Carry out each simulation with a different sample size taking $N = 1000$ repeated samples. Write a short report in which you describe the investigation you carried out, and the results you obtained.

b. Repeat part (a) using the fourth population distribution shown in Figure 6.20. This is a normal distribution. You will need to select *bell shaped* for the parent population. The population will default to mean = 25 and standard deviation = 8.23. In your report, compare the effect of the sample size n on the sampling distribution shape based on whether the parent population is a normal distribution or a non-normal distribution (such as the uniform distribution).

6.106 Family size in Gaza: The Palestinian Central Bureau of Statistics (www.pcbs.org) asked mothers of age 20–24 about the ideal number of children. For those living on the Gaza Strip, the probability distribution is approximately $P(1) = 0.01$, $P(2) = 0.10$, $P(3) = 0.09$, $P(4) = 0.31$, $P(5) = 0.19$, and $P(6$ or more$) = 0.29$. Because the last category is open-ended, it is not possible to calculate the mean exactly. Explain why you can find the *median* of the distribution, and find it.

6.107 Longest streak made: In basketball, when the probability of making a free throw is 0.50 and successive shots are independent, the probability distribution of the longest streak of shots made has $\mu = 4$ for 25 shots, $\mu = 5$ for 50 shots, $\mu = 6$ for 100 shots, and $\mu = 7$ for 200 shots.

  a. How does the mean change for each doubling of the number of shots taken? Interpret.

  b. What would you expect for the longest number of consecutive shots made in a sequence of (i) 400 shots, (ii) 3200 shots?

  c. For a long sequence of shots, the probability distribution of the longest streak is approximately bell shaped and σ equals approximately 1.9, no matter how long the sequence (Schilling, 1990). Explain why the longest number of consecutive shots made has more than a 95% chance of falling within about 4 of the mean, whether we consider 400 shots, 3200 shots, or a million shots.

6.108 Stock market randomness: Based on the previous exercise and what you have learned in this and the previous chapter (for example, Exercise 5.12), if you are a serious investor, explain why you should not get too excited if sometime in the next year the stock market goes up for seven days in a row.

6.109 Airline overbooking: For the Boston to Chicago route, an airline flies a Boeing 737–800 with 170 seats. Based on past experience, the airline finds that people who purchase a ticket for this flight have 0.80 probability of showing up for the flight. They routinely sell 190 tickets for the flight, claiming it is unlikely that more than 170 people show up to fly.

  a. Provide statistical reasoning they could use for this decision.

  b. Describe a situation in which the assumptions on which their reasoning is based may not be satisfied.

6.110 Babies in China: The sex distribution of new babies is close to 50% each, with the percentage of males usually being just slightly higher. In China in recent years, the percentage of female births seems to have dropped, a combination of policy limiting family size, the desire of most families to have at least one male child, the possi-

bility of determining sex well before birth, and the availability of abortion. Suppose that historically 49% of births in China were female but birth records in a particular town for the past year show 800 females born and 1200 males. Conduct an investigation to determine if the current probability of a female birth in this town is less than 0.49, by using the mean and standard deviation of the probability distribution of what you would observe with 2000 births if it were still 0.49.

6.111 Come out ahead in lottery: Example 20 used the central limit theorem to approximate the probability of coming out ahead if you play a lottery 52 times. Each game cost $1, with winnings equal to $5 with probability 0.1 and $0 with probability 0.9.

  a. Explain why you would have to win at least 11 times to come out ahead after playing 52 times.

  b. Explain why the exact probability in (a) equals the binomial probability of at least 11 successes in 52 trials, with success probability 0.1. (This equals 0.013.)

6.112 True or false: As the sample size increases, the standard error of the sampling distribution of $\bar{x}$ increases. Explain your answer.

Select the correct response in the following multiple-choice questions (6.113–6.116).

6.113 Guess answers: A question has four possible answers, only one of which is correct. You randomly guess the correct response. With 20 such questions, the distribution of the number of incorrect answers

  a. is binomial with $n = 20$ and $p = 0.25$.

  b. is binomial with $n = 20$ and $p = 0.50$.

  c. has mean equal to 10.

  d. has probability $(.75)^{20}$ that all 20 guesses are incorrect.

6.114 Terrorist coincidence?: On 9/11/2002, the first anniversary of the terrorist destruction of the World Trade Center in New York City, the winning three-digit New York State Lottery number came up 9-1-1. The probability of this happening was

  a. 1/1000.

  b. $(1/1000)^2 = 0.000001$.

  c. 1 in a billion.

  d. 3/10.

6.115 Standard error: Which of the following is *not* correct? The standard error of a statistic describes

  a. the standard deviation of the sampling distribution of that statistic.

  b. the standard deviation of the sample data measurements.

  c. how close that statistic falls to the parameter that it estimates.

  d. the variability in the values of the statistic for repeated random samples of size n.

6.116 CLT: The central limit theorem implies

  a. All variables have approximately bell-shaped data distributions if a random sample contains at least about 30 observations.

b. Population distributions are normal whenever the population size is large.

c. For sufficiently large random samples, the sampling distribution of $\bar{x}$ is approximately normal, regardless of the shape of the population distribution.

d. The sampling distribution of the sample mean looks more like the population distribution as the sample size increases.

♦♦6.117 SAT and ethnic groups: Lake Wobegon Junior College admits students only if they score above 800 on the sum of their verbal and math SAT scores. Applicants from ethnic group A have a mean of 1000 and a standard deviation of 200 on this test, and applicants from ethnic group B have a mean of 900 and a standard deviation of 200. Both distributions are approximately normal.

a. Find the proportion not admitted for each ethnic group.

b. Both ethnic groups have the same size. Of the students who are not admitted, what proportion is from group B?

c. A state legislator proposes that the college lower the cutoff point for admission to 600, thinking that of the students who are not admitted, the proportion from ethnic group B would decrease. If this policy is implemented, determine the effect on the answer to (b), and comment.

♦♦6.118 Sample = population: Let X = GPA for students in your school.

a. What would the sampling distribution of the sample mean look like if you sampled *every* student in the school, so the sample size equals the population size? (*Hint:* The sample mean then equals the population mean.)

b. How does the sampling distribution compare to the population distribution if we take a sample of size $n = 1$?

♦♦6.119 Standard deviation of a discrete probability distribution: The **variance** of a probability distribution of a random variable is a weighted average of its squared distances from the mean μ. For discrete random variables, it equals

$$\sigma^2 = \sum (x - \mu)^2 P(x).$$

Multiply each possible squared deviation $(x - \mu)^2$ by its probability $P(x)$, and then add. The **standard deviation** σ is the positive square root of the variance.

a. Suppose $x = 1$ with probability p and $x = 0$ with probability $(1 - p)$, so that $\mu = p$. Since $(x - \mu)^2$ equals $(0 - p)^2 = p^2$ when $x = 0$ and $(1 - p)^2$ when $x = 1$, derive that $\sigma^2 = p(1 - p)$ and $\sigma = \sqrt{p(1 - p)}$, the special case of the binomial σ with $n = 1$.

b. With n trials, using the formula $\sigma/\sqrt{n}$ for a standard error, explain why the standard error of a sample proportion equals $\sqrt{p(1 - p)/n}$.

♦♦6.120 Binomial probabilities: Justify the $p^x(1 - p)^{n-x}$ part of the binomial formula for the probability $P(x)$ of a particu-

lar sequence with x successes, using what you learned in Section 5.2 about probabilities for intersections of independent events.

♦♦6.121 Waiting time for doubles: Most discrete random variables can take on a finite number of values. Let X = the number of rolls of two dice necessary until doubles (the same number on each die) first appears. The possible values for this discrete random variable (called the **geometric**) are $1, 2, 3, 4, 5, 6, 7$, and so on, still separate values (and discrete) but now an infinite number of them.

a. Using intersections of independent events, explain why $P(1) = 1/6$, $P(2) = (5/6)(1/6)$, and $P(3) = (5/6)^2(1/6)$.

b. Find $P(4)$, and explain how to find $P(x)$ for an arbitrary positive integer x.

♦♦6.122 Finite populations: The formula $\sigma/\sqrt{n}$ for the standard error of $\bar{x}$ actually is an approximation that treats the population size as *infinitely* large relative to the sample size n. The exact formula for a *finite* population size N is

$$\text{Standard error} = \sqrt{\frac{N - n}{N - 1}} \frac{\sigma}{\sqrt{n}}$$

The term $\sqrt{(N - n)/(N - 1)}$ is called the **finite population correction**.

a. When $n = 300$ students are selected from a college student body of size $N = 30,000$, show that the exact standard error equals $0.995\sigma/\sqrt{n}$. (When n is small compared to the population size N, the approximate formula works very well.)

b. If $n = N$ (that is, we sample the entire population), show that the exact standard error equals 0. In other words, no sampling error occurs, since $\bar{x} = \mu$ in that case.

Class Explorations

6.123 Best of 7 games: In professional baseball, basketball, and hockey in North America, the final two teams in the playoffs play a "best of seven" series of games. The first team to win four games is the champion. Use simulation with the *random numbers* applet on the text CD to approximate the probability distribution of the number of games needed to declare a champion, when (a) the teams are evenly matched, (b) the better team has probability 0.90 of winning any particular game. In each case, conduct 10 simulations. Then combine results with other students in your class and estimate the mean number of games needed in each case. In which case does the series tend to be shorter?

6.124 Simulate a sampling distribution: The table below provides the ages of all 50 heads of households in a small Nova Scotian fishing village. The distribution of these ages is characterized by $\mu = 47.18$ and $\sigma = 14.74$.

Name	Age	Name	Age	Name	Age	Name	Age
Alexander	50	Griffith	66	McTell	49	Shindell	33
Bell	45	Grosvenor	51	MacLeod	30	Staines	36
Black	23	Ian	57	Mayo	28	Stewart	25
Bok	28	Jansch	40	McNeil	31	Thames	29
Clancy	67	Kagan	36	Mitchell	45	Thomas	57
Cochran	62	Lavin	38	Morrison	43	Todd	39
Fairchild	41	Lunny	81	Muir	43	Travers	50
Finney	68	MacColl	27	Renbourn	54	Trickett	64
Fisher	37	Mallett	37	Rice	62	Tyson	76
Fraser	60	McCusker	56	Rogers, G.	67	Watson	63
Fricker	41	Macdonald	71	Rogers, S.	48	Young	29
Gaughan	70	McDonald	39	Rowan	32		
Graham	47	McTell	46	Rusby	42		

a. Each student or the class should construct a graphical display (stem-and-leaf plot, dot plot, or histogram) of the population distribution of the ages.

b. Each student should select nine random numbers between 01 and 50, with replacement. Using these numbers, each student should sample nine heads of households and find their sample mean age. Collect all sample mean ages. Using technology, construct a graph (using the same graphical display as in part (a)) of the simulated sampling distribution of the $\bar{x}$-values for all the student samples. Compare it to the distribution in (a).

c. Find the mean of the $\bar{x}$-values in (b). How does it compare to the value you would expect in a long run of repeated samples of size 9?

d. Find the standard deviation of the $\bar{x}$-values in (b). How does it compare to the value you would expect in a long run of repeated samples of size 9?

6.125 Coin-tossing distributions: For a single toss of a balanced coin, let $x = 1$ for a head and $x = 0$ for a tail.

a. Construct the probability distribution for x, and calculate its mean.

b. The coin is flipped ten times, yielding six heads and four tails. Construct the data distribution.

c. Each student in the class should flip a coin ten times and find the proportion of heads. Collect the sample proportion of heads from each student. Summarize the simulated sampling distribution by constructing a plot of all the proportions obtained by the students. Describe the shape and spread of the sampling distribution compared to the distributions in (a) and (b).

d. If you performed the experiment in (c) a huge number of times, what would you expect to get for the (i) mean, (ii) standard error of the sample proportions?

6.126 Sample vs. sampling: Each student should bring ten coins to class. For each coin, observe its age, the difference between the current year and the year on the coin.

a. Using all the students' observations, the class should construct a histogram of the sample ages. What is its shape?

b. Now each student should find the mean for that student's ten coins, and the class should plot the means of all the students. What type of distribution is this, and how does it compare to the one in (a)? What concepts does this exercise illustrate?

ACTIVITY 2: What "Hot Streaks" Should We Expect in Basketball?

In basketball games, TV commentators and media reporters often describe a player as being "hot" if he or she makes several shots in a row. Yet statisticians have shown that for players at the professional level, the frequency and length of streaks of good (or poor) shooting are similar to what we'd expect if the success of the shots were random, with the outcome of a particular shot being independent of previous shots.[11]

Shaquille O'Neal is one of the top players in the National Basketball Association, but he is a poor free-throw shooter. He makes about 50% of his free-throw attempts over the course of a season. Let's suppose that he has probability 0.50 of making any particular free throw that he takes in a game. Suppose also whether or not he makes any particular free throw is independent of his previous ones. He takes 20 free throws during the game. Let X denote the longest streak he makes in a row during the game. This is a discrete random variable, taking the possible values $x = 0, 1, 2, 3, \ldots, 20$. Here, $x = 0$ if he makes none of the 20, $x = 1$ if he never makes more than 1 in a row, $x = 2$ if his longest streak is 2 in a row, and so forth.

[11]For instance, see articles by A. Tversky and T. Gilovich, *Chance*, vol. 2 (1989), pp. 16–21 and 31–34.

What is the probability distribution of X? This is difficult to find using probability rules, so let's approximate it by simulating 10 games with $n = 20$ free throws in each, using either coin flipping or the *random numbers* applet on the text CD. Representing each of O'Neal's shots by the flip of a coin, we treat a head as making a shot and a tail as missing it. We simulate O'Neal's 20 free throws by flipping a coin 20 times. We did this and got

TTHHHTHTTTHTHHTHTTTHT

This corresponds to missing shots 1 and 2, then making shots 3, 4, and 5, missing shot 6, and so forth. The simulated value of X = the longest streak of shots made is the longest sequence of heads in a row. In the sequence just shown, the longest streak made is $x = 3$, corresponding to the heads on flips 3, 4, and 5.

You would do the 20 coin flips 10 separate times to simulate what would happen in 10 games with 20 free throws in each. This would be a bit tedious. Instead, we suggest that you use your own judgment to write down quickly 10 sets of 20 H and T symbols (as shown above for one set) on a sheet of paper to reflect the sort of results you would expect for 10 games with 20 free throws in each. After doing this, find the 10 values of X = longest streak of H's, using each set of 20 symbols. Do you think your instructor would be able to look at your 200 H's and T's and figure out that you did not actually flip the coin?

A more valid way to do the simulation uses software, such as the *random numbers* applet on the text CD. For each digit, we could let 0–4 represent making a shot (H) and 5–9 as missing it (T). Use this to simulate 10 games with $n = 20$ free throws in each. When we did this, we got the following results for the first three games:

Coin flips	x
HTHHTHTTTHTHHHHTTHTT	4
HTTTHTHHTHHHTTHTHHTH	3
TTHTHHHTHTTHTTTTHTHH	3

In practice, to get accurate results you have to simulate a *huge* number of games (at least a thousand), for each set of 20 free throws observing the longest streak of successes in a row. Your results for the probability distribution would then approximate[12] those shown in the table.

Probability Distribution of X = *Longest Streak of Successful Free Throws*. The distribution refers to 20 free throws with a 0.50 chance of success for each. All potential x values higher than 9 had a probability of 0.00 to two decimal places and a total probability of only 0.006.

x	$P(x)$	x	$P(x)$
0	0.00	5	0.13
1	0.02	6	0.06
2	0.20	7	0.03
3	0.31	8	0.01
4	0.23	9	0.01

The probability that O'Neal never makes more than 4 free throws in a row equals $P(0) + P(1) + P(2) + P(3) + P(4) = 0.76$. This would usually be the case. The mean of the probability distribution is $\mu = 3.7$.

The longest streak of successful shots tends to be longer, however, with a larger number of total shots. With 200 shots, the distribution of the longest streak has a mean of $\mu = 7$. Although it *would* be a bit unusual for O'Neal to make his *next* 7 free throws in a row, it would not be at all unusual if he makes 7 in a row *sometime* in his next 200 free throws. Making 7 in a row is then not really a hot streak, but merely what we expect by random variation. Sports announcers often get excited by streaks that they think represent a "hot hand" but merely represent random variation.

With 200 shots, the probability is only 0.03 that the longest streak equals 4 or less. Look at the 10 sets of 20 H and T symbols that you wrote on a sheet of paper to reflect the results you expected for 10 games of 20 free throws each. Find the longest streak of H's out of the string of 200 symbols. Was your longest streak 4 or less? If so, your instructor could predict that you faked the results rather than used random numbers, because there's only a 3% chance of this if they were truly generated randomly. Most students underestimate the likelihood of a relatively long streak.

This concept relates to the discussion in Section 5.4 about coincidences. By itself, an event may seem unusual. But when you think of all the possible coincidences and all the possible times they could happen, it is probably not so unusual.

---

[12]Schilling, M. F. (1990). The longest run of heads. *The College Mathematics Journal*, vol. 21, pp. 196–207.

7 | Statistical Inference: Confidence Intervals

EXAMPLE 1

Analyzing Data from the General Social Survey

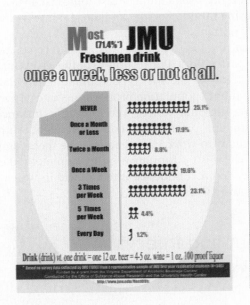

Picture the Scenario

For more than 30 years, the National Opinion Research Center at the University of Chicago (www.norc.uchicago.edu) has conducted each year or every other year an opinion survey called the General Social Survey (GSS). The survey randomly samples about 2000 adult Americans. In a 90-minute in-person interview, the interviewer asks a long list of questions about opinions and behavior for a wide variety of issues. Other nations have similar surveys. For instance, every five years Statistics Canada conducts its own General Social Survey. *Eurobarometer* regularly samples about 1000 people in each country in the European Union.

Analyzing data from such databases helps researchers to learn about how people think and behave at a given time and to track opinions over time. Activity 1 in Chapter 1 showed how to access the GSS data at http://sda.berkeley.edu/GSS. Some examples and exercises in previous chapters have used GSS data.

Questions to Explore

Based on data from a recent GSS, how can you make an inference about

- The proportion of Americans who are willing to pay higher prices in order to protect the environment?

- The proportion of Americans who believe a wife should sacrifice her career for her husband's?

- The mean number of hours that Americans watch TV per day?

Thinking Ahead

We will analyze data from the GSS in examples and exercises throughout this chapter. For instance, in Example 3 we'll see how to estimate the proportion of Americans who are "environmentally friendly" in terms of being willing to pay higher prices in order to protect the environment. We haven't shown any data yet, but do you have a guess for how high this proportion is?

We'll answer the other two questions above in Examples 2 and 6, and in exercises we'll explore opinions on such issues as whether it should or should not be the government's responsibility to reduce income differences between the rich and the poor, whether a preschool child is likely to suffer if his or her mother works, and how conservative or liberal Americans are (on the average) in political ideology.

Recall

A **statistic** describes a **sample**. Examples are the sample mean $\bar{x}$ and standard deviation s.

A **parameter** describes a **population**. Examples are the population mean μ and standard deviation σ.

Statistical inference uses sample statistics to make decisions and predictions about population parameters. ■

Recall

Sections 6.4–6.6 introduced the **sampling distribution**, which specifies the possible values a statistic can take and their probabilities. ■

A sample of about 2000 people (as the GSS takes) is usually relatively small. For instance, in the United States, a survey of this size gathers data for less than 1 out of every 100,000 people. How can we possibly make reliable predictions about the entire population with so few people?

We now have the tools to see how this is done. We're ready to learn about a powerful use of statistics: **statistical inference** about population parameters using sample data. Inference methods help us to predict how close a sample statistic falls to the population parameter. We can make decisions and predictions about populations even if we have data for relatively few subjects from that population. A preview of statistical inference at the end of the previous chapter showed it's often possible to predict the winner of an election in which millions of people voted, knowing only how a couple of thousand people voted.

For statistical inference methods, you may wonder, what's the relevance of the material you've been learning about the role of randomization in gathering data (Chapter 4), concepts of probability (Chapter 5), and the normal distribution and its use as a sampling distribution (Chapter 6)? They're important because of two key aspects of statistical inference:

■ *Statistical inference methods use probability calculations that assume that the data were gathered with a random sample or a randomized experiment.*

■ *The probability calculations refer to a sampling distribution of a statistic, which is often approximately a normal distribution.*

In other words, statistical inference uses sampling distributions of statistics calculated from data gathered using randomization, and those sampling distributions are often approximately normal.

The next two chapters present the two types of statistical inference methods—**estimation** of population parameters and **testing hypotheses** about the parameter values. This chapter discusses the first type, estimating population parameters. We'll learn how to estimate population proportions for categorical variables and population means for quantitative variables. For instance, a study dealing with how college students pay for their education might estimate the proportion of college students who work part time and the mean annual income for those who do work. The most informative estimation method constructs an interval of numbers, called a **confidence interval**, within which the unknown parameter value is believed to fall.

7.1 | What Are Point and Interval Estimates of Population Parameters?

Population parameters have two types of estimates:

Point Estimate and Interval Estimate

A **point estimate** is a *single number* that is our "best guess" for the parameter.

An **interval estimate** is an *interval of numbers* within which the parameter value is believed to fall.

For example, the 1998 General Social Survey asked, "Do you believe in Hell?" From the sample data, the point estimate for the proportion of adult Americans who would respond "yes" equals 0.74—more than 7 out of 10. The adjective "point" in point estimate refers to using a single number or *point* as the parameter estimate.

An interval estimate, found with a method introduced in the next section, predicts that the proportion of *all* adult Americans who believe in Hell falls between 0.71 and 0.77. That is, it predicts that the sample point estimate of 0.74 falls within a *margin of error* of 0.03 of the population proportion. Figure 7.1 illustrates.

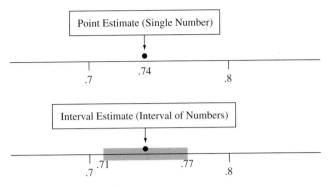

▲ **FIGURE 7.1:** A **point estimate** predicts a parameter by a single number. An **interval estimate** is an interval of numbers that are believable values for the parameter. **Question:** Why is a point estimate alone not sufficiently informative?

A point estimate by itself is not sufficient. It doesn't tell us *how close* the estimate is likely to be to the parameter. An interval estimate is more useful. It incorporates a margin of error, so it helps us to gauge the accuracy of the point estimate.

Point Estimation: How Do We Make a Best Guess for a Population Parameter?

Once we've collected the data, how do we find a point estimate, representing our "best guess" for a parameter value? The answer is straightforward—we can simply use an appropriate sample statistic. For example, for a population mean μ, the sample mean $\bar{x}$ is a point estimate of μ. For the population proportion, the sample proportion is a point estimate.

Point estimates are the most common form of inference reported by the mass media. For example, every month the Gallup organization conducts a survey to estimate the U.S. president's popularity, and the mass media report results. In April 2005, this survey reported that 45% of the American public approved of President Bush's performance in office. By contrast, surveys in Canada and in European countries gave him lower approval ratings, for instance, 32% in Canada, 30% in Great Britain, 19% in France, 19% in Spain, and 17% in Germany.[1] These were *point estimates* rather than parameters, since they used a sample of about 1000 people in each country rather than the entire populations. For simplicity, we'll usually use the term "estimate" in place of "point estimate" when there is no risk of confusing it with an interval estimate.

Properties of point estimators For any particular parameter, there are several possible point estimates. For a normal distribution, for instance, the center is the mean μ and the median, since that distribution is symmetric. So, with sample data from a normal distribution, two possible estimates of that center value are the sample mean and the sample median. What makes a particular estimate better than others? *A good estimator* of a parameter has two desirable properties:

> **Property 1:** A good estimator has a sampling distribution that is centered at the parameter, in the sense that the parameter is the mean of that sampling distribution. An estimator with this property is said to be **unbiased**.

Did You Know?

Mass media, such as newspapers and television networks, often report sample percentages as **point estimates** of population percentages. ■

---

[1] Polling done for Associated Press by Ipsos, an international polling company (reported Dec. 14, 2004).

From Section 6.5, we know that for random sampling the mean of the sampling distribution of the sample mean $\bar{x}$ equals the population mean μ. So, the sample mean $\bar{x}$ is an unbiased estimator of μ. Figure 7.2 recalls this result from Section 6.5.

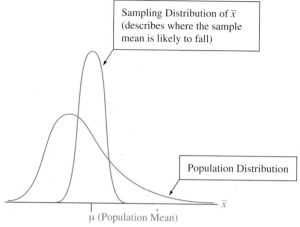

Sampling Distribution of $\bar{x}$ (describes where the sample mean is likely to fall)

Population Distribution

μ (Population Mean)

$\bar{x}$

▲ **FIGURE 7.2: The Sample Mean $\bar{x}$ Is an Unbiased Estimator.** Its sampling distribution is centered at the parameter it estimates—the population mean μ. **Question:** When is the sampling distribution bell shaped, as it is in this figure?

Recall

From Sections 6.4–6.6, a **standard error** of a statistic is the standard deviation of the sampling distribution of the statistic. It describes the variability in the possible values of the statistic, for the given sample size. It also tells us how much the statistic would vary from sample to sample of that size. ■

The sample proportion is an unbiased estimator of a population proportion. In polling about the president's popularity, we don't know whether any particular sample proportion supporting the president falls below or above the actual population proportion. We do know, however, that with random sampling the sample proportions tend to fall *around* the population proportion.

Property 2: A good estimator has a *small standard error* compared to other estimators. This means it tends to fall closer than other estimates to the parameter. For example, for estimating the center of a normal distribution, the sample mean has smaller standard error than the sample median (Exercise 7.117). The sample mean is a better estimator of this parameter. In this text, we'll use estimators that are unbiased (or nearly so, in practical terms) and that have relatively small standard errors.

Interval Estimation: Constructing an Interval that Contains the Parameter (We Hope!)

A recent survey[2] of new college graduates estimated the mean salary for those who had taken a full-time job after graduation to equal $37,000. Does $37,000 seem plausible to you? Too high? Too low? Any one point estimate may or may not be close to the parameter it estimates. For the estimate to be useful, we need to know how close it is likely to fall to the actual parameter value. Is the estimate of $37,000 likely to be within $1000 of the actual population mean? Within $5,000? Within $10,000? Inference about a parameter should provide not only a point estimate but should also indicate its likely precision.

An **interval estimate** indicates precision by giving an interval of numbers around the point estimate. The interval is made up of numbers that are the most believable values for the unknown parameter, based on the data observed. For instance, perhaps a survey of new college graduates predicts that the mean salary of all the graduates working fulltime falls somewhere between $35,000 and $39,000, that is, within a *margin of error* of $2000 of the point estimate of $37,000. An interval estimate is designed to contain the parameter with some chosen probability, such as 0.95. Because interval estimates contain the parameter with a certain degree of confidence, they are referred to as **confidence intervals**.

---

[2] By the National Association of Colleges and Employers, www.naceweb.org

Confidence Interval

A **confidence interval** is an interval containing the most believable values for a parameter. The probability that this method produces an interval that contains the parameter is called the **confidence level**. This is a number chosen to be close to 1, most commonly 0.95.

The interval from $35,000 to $39,000 is an example of a confidence interval. It was constructed using a confidence level of 0.95. This is often expressed as a percentage, and we say that we have "95% confidence" that the interval contains the parameter. It is a **95% confidence interval**.

How can we construct a confidence interval? The key is the *sampling distribution* of the point estimate. This distribution tells us the probability that the point estimate will fall within any certain distance of the parameter.

What's the Logic behind Constructing a Confidence Interval?

To construct a confidence interval, we'll put to work some results about sampling distributions that we learned in the previous chapter. Let's do this for estimating a proportion. We saw that the sampling distribution of a sample proportion:

■ Gives the possible values for the sample proportion and their probabilities

■ Is approximately a normal distribution, for large random samples, because of the central limit theorem

■ Has mean equal to the population proportion

■ Has standard deviation called the **standard error**

Let's use these results to construct a 95% confidence interval for a population proportion. From Chapter 6, approximately 95% of a normal distribution falls within two standard deviations of the mean. More precisely, we saw in Section 6.2 that the mean plus and minus 1.96 standard deviations includes exactly 95% of a normal distribution. Since the sampling distribution of the sample proportion is approximately normal, with probability 0.95 the sample proportion falls within about 1.96 standard errors of the population proportion (see the margin figure). The distance of 1.96 standard errors is the **margin of error**. We've been using this term since Chapter 4. Let's take a closer look at how it's calculated:

Margin of Error

The **margin of error** measures how accurate the point estimate is likely to be in estimating a parameter. It is a multiple of the standard error of the sampling distribution of the estimate, such as 1.96 × (standard error) when the sampling distribution is a normal distribution.

Once the sample is selected, if the sample proportion *does* fall within 1.96 standard errors of the population proportion, then the interval from

[sample proportion − 1.96(standard error)] to

[sample proportion + 1.96(standard error)]

contains the population proportion. In other words, with probability about 0.95, a sample proportion value occurs such that the interval

sample proportion ± 1.96(standard error)

contains the unknown population proportion. This interval of numbers is an approximate **95% confidence interval** for the population proportion.

Sampling Distribution of Sample Proportion

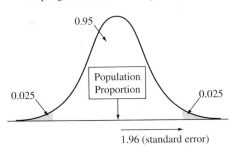

Recall

Using normal cumulative probabilities (Table A or a calculator), $z = 1.96$ has cumulative probability 0.975, right-tail probability 0.025, two-tail probability 0.05, and central probability 0.95. ■

Recall

Example 6 in Chapter 4 showed how the **margin of error** is reported in practice for a sample proportion. We approximated it there by $1/\sqrt{n}$. This is a rough approximation for 1.96 (standard error), as shown following Example 9 in Section 7.4. ■

IN WORDS

"sample proportion ± 1.96 (standard error)" represents taking the sample proportion and adding and subtracting 1.96 standard errors.

EXAMPLE 2

Should a Wife Sacrifice Her Career for Her Husband's?

Picture the Scenario

One question on the General Social Survey asks whether you agree or disagree with the following statement: "It is more important for a wife to help her husband's career than to have one herself." The last time this was asked (in 1998), 19% of 1823 respondents agreed. So the sample proportion agreeing was 0.19. From a formula in the next section, we'll see that this estimate has standard error = 0.01.

Questions to Explore

a. Find and interpret the margin of error for a 95% confidence interval for the population proportion who agreed with the statement about a woman's role.

b. Construct the 95% confidence interval, and interpret it in context.

Think It Through

a. The margin of error for a 95% confidence interval for a population proportion equals $1.96 \times$ (standard error), or $1.96(0.01) = 0.02$. This means that with probability 0.95, the error in using the point estimate to predict the population proportion (in 1998) is no greater than 0.02.

b. The approximate 95% confidence interval is

$$\text{Sample proportion} \pm 1.96(\text{standard error}),$$

which is $0.19 \pm 1.96(0.01)$, or 0.19 ± 0.02.

This gives the interval of proportions from 0.17 to 0.21, denoted by (0.17, 0.21). In summary, using this 95% confidence interval, we predict that the population proportion (in 1998) who believed it is more important for a wife to help her husband's career than to have one herself is somewhere between 0.17 and 0.21.

Insight

In 1977 when this question was first asked on the GSS, the point estimate was 0.57 and the 95% confidence interval was (0.55, 0.59). The proportion of Americans who agree with this statement decreased considerably during the 1980s and 1990s.

TO PRACTICE THIS CONCEPT, TRY EXERCISE 7.7

In summary, *a confidence interval is constructed by adding and subtracting a margin of error to a given point estimate. The margin of error is based on the standard error of the sampling distribution of that point estimate.* When the sampling distribution is approximately normal, a 95% confidence interval has margin of error equal to 1.96 standard errors.

The sampling distribution of most point estimates is approximately normal when the random sample size is relatively large. Thus, this logic of taking the margin of error for a 95% confidence interval to be approximately two standard errors applies with large random samples, such as those found in the General Social Survey. The next two sections show more precise details for estimating proportions and means.

ACTIVITY 1: Download Data from the General Social Survey

We saw in Chapter 1 that it's easy to download data from the GSS. Let's recall how.

■ Go to the Web site sda.berkeley.edu/GSS/. Click on the link there for the GSS. In the *Select an action* menu, select the *Frequencies and crosstabulation* option and click on *Start*.

■ The GSS name for the variable in Example 2 (whether a wife should sacrifice her career for her husband's) is FEHELP. Type FEHELP as the *Row variable* name. Click on *Run the table*. Now you'll see category counts and the percentages for all the years combined in which this question was asked.

■ To download results only for the year 1998, go back to the previous menu and enter YEAR (1998) in the *Selection Filter* space. When you click again on *Run the table*, you'll see that in 1998 2.4% strongly agreed and 16.8% agreed, a total of about 19% (a proportion of 0.19) in the two agree categories.

Now, create other results. At the page at which you *Select an action*, click on *Open extra codebook window* and then *Standard codebook* and you will see indexes for the variables. Look up a subject that interests you and find the GSS code name for a variable. For example, to find the percentage who believe in hell, enter HELL as the code name.

7.1 Practicing the Basics

7.1 Health care: A study dealing with health-care issues plans to take a sample survey of 1500 Americans to estimate the proportion who have health insurance and the mean dollar amount that Americans spent on health care this past year.

   a. Identify two population parameters that this study will estimate.

   b. Identify two statistics that can be used to estimate these parameters.

7.2 Types of estimates: Explain the difference between a point estimate and an interval estimate.

■ **7.3 Believe in hell?:** Using the General Social Survey Web site sda.berkeley.edu/GSS, find the point estimate of the population proportion of Americans who would have answered "yes, definitely" in 1998 when asked whether they believe in hell (variable HELL).

7.4 Help the poor?: One question (coded EQUALIZE) on the 1998 General Social Survey asked, "Do you think that it should be government's responsibility to reduce income differences between the rich and the poor?" Of the possible responses, 548 picked "definitely or probably should be," and 620 picked "probably or definitely should not be."

   a. Find the point estimate of the population proportion who would answer "definitely or probably should be."

   b. The margin of error of this estimate is 0.03. Explain what this represents.

7.5 Watching TV: In response to the GSS question in 2002 about the number of hours daily spent watching TV, the responses by the seven subjects who identified themselves as Buddhists were 2, 2, 1, 3, 2, 3, 2.

   a. Find a point estimate of the population mean for Buddhists.

   b. The margin of error for this point estimate is 0.64. Explain what this represents.

7.6 Starting salaries: The starting salaries of a random sample of three students who graduated from North Carolina State University last year with majors in the mathematical sciences are $45,000, $35,000, and $55,000.

   a. Find a point estimate of the population mean of the starting salaries of all math science graduates at that university last year.

   b. A method that we'll study in Section 7.3 provides a margin of error of $24,800 for a 95% confidence interval for the population mean starting income. Construct that interval.

   c. Use this example to explain why a point estimate alone is usually insufficient for statistical inference.

7.7 Believe in heaven?: When the 1998 GSS asked 1158 subjects, "Do you believe in Heaven?" (coded HEAVEN), the proportion who answered yes was 0.86. From results in the next section, the standard error of this estimate is 0.01.

   a. Find and interpret the margin of error for a 95% confidence interval for the population proportion of Americans who believe in heaven.

   b. Construct the 95% confidence interval. Interpret it in context.

■ **7.8 Feel lonely often?:** The 1996 GSS asked 1450 subjects, "On how many days in the past 7 days have you felt lonely?" At sda.berkeley.edu/GSS, enter LONELY as the variable, check *Statistics* in the menu of *Table Options*, and click on *Run the Table* to see the responses.

a. Report the percentage making each response and the mean and standard deviation of the responses. Interpret.

b. Note that the output also reports a standard error value of 0.06. What does this mean?

7.9 **CI for loneliness:** Refer to the previous exercise. The margin of error for a 95% confidence interval for the population mean is 0.12. Construct that confidence interval, and interpret it.

7.10 **Newspaper article:** Conduct a search to find a newspaper or magazine article that reported a margin of error. If you prefer, use the Internet for the search.

a. Specify the population parameter, the value of the sample statistic, and the size of the margin of error.

b. Explain how to interpret the margin of error.

7.2 | How Can We Construct a Confidence Interval to Estimate a Population Proportion?

Let's now see the details of how to construct a confidence interval for a population proportion. We'll apply the ideas just discussed at the end of Section 7.1. The data are categorical, specifically *binary* (two categories): Each observation either falls or does not fall in the category of interest. We'll use the generic terminology "success" and "failure" for these two possible outcomes (as in the discussion of the binomial distribution in Section 6.3). We summarize the data by the sample proportion of successes and construct a confidence interval for the population proportion.

Finding the 95% Confidence Interval for a Population Proportion

The 2000 General Social Survey asked respondents whether they would be willing to pay much higher prices in order to protect the environment. Of $n = 1154$ respondents, 518 said yes. The sample proportion of yes responses was $518/1154 = 0.45$, less than half. How can we construct a confidence interval for the population proportion that would respond yes?

We symbolize a population proportion by p. The point estimate of the population proportion is the *sample proportion.*

We symbolize the sample proportion by $\hat{p}$, called "p-hat."

In statistics, the circumflex ("hat") symbol over a parameter symbol represents a point estimate of that parameter. Here, the sample proportion $\hat{p}$ is a point estimate, such as $\hat{p} = 0.45$ for the proportion willing to pay much higher prices to protect the environment.

For large random samples, the central limit theorem tells us that the sampling distribution of the sample proportion $\hat{p}$ is approximately normal. As discussed in the previous section, the z-score for a 95% confidence interval with the normal sampling distribution is 1.96. So there is about a 95% chance that $\hat{p}$ falls within 1.96 standard errors of the population proportion p. A 95% confidence interval uses margin of error = 1.96(standard error). The formula

[point estimate ± margin of error] becomes $\hat{p} \pm 1.96$(standard error).

The exact standard error of a sample proportion equals $\sqrt{p(1-p)/n}$. This formula depends on the unknown population proportion, p. In practice, we don't know p, and we need to estimate the standard error.

IN WORDS

It is traditional in statistics to use Greek letters for most parameters. Some books use the Greek letter π ("pi") for a proportion. Since π already has its own meaning in math and statistics (3.14 . . .), we use the symbol p. This is also the parameter for the probability of success in the binomial distribution (Section 6.3). The sample proportion $\hat{p}$ is read as "p-hat."

Recall

The **standard error** of a sample proportion was introduced in Section 6.4 as the standard deviation of the sampling distribution of the sample proportion. ▪

The exact values of standard errors depend on parameter values. In practice, these are unknown, so we find standard errors by substituting estimates of parameters. The term "standard error" is then used as short-hand for what is actually an "estimated standard deviation of a sampling distribution." Beginning here, we'll use **standard error** to refer to this estimated value, since that's what we'll use in practice. When we need to refer to the exact value based on the actual parameter value, we'll call it the "exact standard error."

In summary, in practice *a standard error is an estimated standard deviation of a sampling distribution*. We will use *se* as shorthand for *standard error*. For example, for finding a confidence interval for a population proportion p, the standard error is

$$se = \sqrt{\hat{p}(1 - \hat{p})/n}.$$

IN WORDS

se = standard error

IN WORDS

To find the **95% confidence interval**, you take the sample proportion and add and subtract 1.96 standard errors.

A 95% confidence interval for a population proportion p is

$$\hat{p} \pm 1.96(se), \text{ with } se = \sqrt{\hat{p}(1 - \hat{p})/n},$$

where $\hat{p}$ denotes the sample proportion based on n observations.

Figure 7.3 shows the sampling distribution of $\hat{p}$ and how there's about a 95% chance that $\hat{p}$ falls within 1.96(se) of the population proportion p. This confidence interval is designed for large samples. We'll be more precise about what "large" means after the following example.

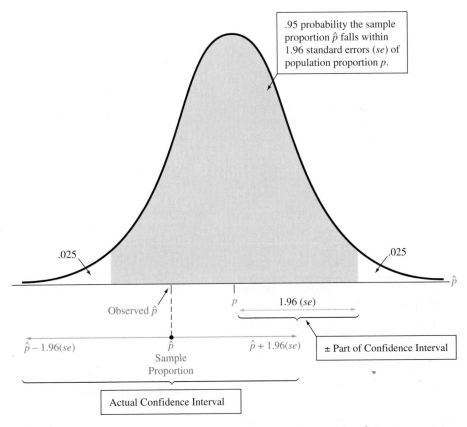

▲ FIGURE 7.3: Sampling Distribution of Sample Proportion $\hat{p}$. For large random samples, the sampling distribution is normal around the population proportion p, so $\hat{p}$ has probability 0.95 of falling within 1.96(se) of p. As a consequence, $\hat{p} \pm 1.96$(se) is a 95% confidence interval for p. **Question:** Why is the confidence interval $\hat{p} \pm 1.96$(se) instead of $p \pm 1.96$(se)?

Would You Pay Higher Prices to Protect the Environment?

Picture the Scenario

Many people consider themselves "green," meaning they are supportive (in theory) when it comes to environmental issues. But how do they act in practice? For instance, Americans' per capita use of energy is roughly double that of Western Europeans. If you live in North America, would you be willing to pay the same price for gas that Europeans do (often about $6 or more per gallon), if the government proposed a significant price hike as an incentive for conservation and for driving more fuel-efficient cars to reduce air pollution and its impact on global warming?

Questions to Explore

In 2000, the GSS asked subjects whether they would be willing to pay much higher prices in order to protect the environment. Of $n = 1154$ respondents, 518 indicated a willingness to do so.

a. Find a 95% confidence interval for the population proportion of adult Americans willing to do so at the time of that survey.

b. Interpret that interval.

Think It Through

a. The sample proportion that estimates the population proportion p is $\hat{p} = 518/1154 = 0.45$. The standard error of the sample proportion $\hat{p}$ equals

$$se = \sqrt{\hat{p}(1 - \hat{p})/n} = \sqrt{(0.45)(0.55)/1154} = \sqrt{0.0002144} = 0.015.$$

Using this se, a 95% confidence interval for the population proportion is

$$\hat{p} \pm 1.96(se), \text{ which is } 0.45 \pm 1.96(0.015)$$
$$= 0.45 \pm 0.03, \text{ or}(0.42, 0.48).$$

Recall

$se = 0.015$ means that if many random samples of 1154 people each were taken to gauge their opinion about this issue, the standard deviation of the sample proportions would be about 0.015. They would vary relatively little. ■

b. At the 95% confidence level, we estimate that the population proportion of adult Americans willing to pay much higher prices to protect the environment was at least 0.42 but no more than 0.48, that is, between 42% and 48%. The point estimate of 0.45 has a margin of error of 0.03. None of the numbers in the confidence interval (0.42, 0.48) fall above 0.50. So we infer that less than half the population was willing to pay much higher prices to protect the environment.

Insight

As usual, results depend on the question's wording. For instance, when asked whether the government should impose strict laws to make industry do less damage to the environment, a 95% confidence interval for the population proportion responding yes is (0.87, 0.90). (See Exercise 7.14.)

TO PRACTICE THIS CONCEPT, TRY EXERCISE 7.13

Recall

From Section 6.3, the **binomial** random variable X counts the number of successes in n observations, and the sample proportion equals x/n, for instance, $518/1154 = 0.45$. ▪

Table 7.1 shows how the MINITAB software reports the data summary and confidence interval. Here, X represents the number that *support* paying much higher prices. The notation reflects that this is the outcome of a random variable, specifically a binomial random variable. The heading "Sample p" stands for the sample proportion $\hat{p}$, and "CI" stands for confidence interval. In reporting results from such output, you should use only the first two or three significant digits. Report the confidence interval as (.42, .48) or (.420, .478) rather than (.420176, .477571). Software's extra precision provides accurate calculations in finding *se* and the confidence interval. However, the extra digits are distracting when reported to others and do not tell them anything extra in a practical sense about the population proportion. *Likewise, if you do a calculation with a hand calculator, don't round off while doing the calculation or your answer may be affected, but do round off when you report the final answer.*

TABLE 7.1: MINITAB Output for 95% Confidence Interval for a Proportion for Example 3

The italicized lines are added annotation to explain what's shown.

X	N	Sample p	95.0% CI
518	1154	0.448873	(.420176, 0.477571)
↑	↑	↑	↑
Category Count	*Sample Size*	*Sample Proportion*	*Endpoints of Confidence Interval*

Sample Size Needed for Validity of Confidence Interval for a Proportion

The confidence interval formula $\hat{p} \pm 1.96(se)$ applies with *"large" random samples*. This is because the sampling distribution of the sample proportion $\hat{p}$ is then approximately normal and the *se* estimate also tends to be good. This is what allows us to use the z-score of 1.96 from the normal distribution.

In practice, "large" means that *you should have at least 15 "successes" and at least 15 "failures"* for the binary outcome.[3] At the end of Section 7.4, we'll see how the large-sample method can fail when this guideline is not satisfied.

Recall

From Section 6.3, the binomial distribution is bell shaped when the expected counts np and $n(1 - p)$ of successes and failures are both at least 15. Here, we don't know p, and we use the guideline with the observed counts of successes and failures. ▪

Sample Size Needed for Large-Sample Confidence Interval for a Proportion

For the 95% confidence interval $\hat{p} \pm 1.96(se)$ for a proportion p to be valid, you should have at least 15 successes and 15 failures. This can also be expressed as

$$n\hat{p} \geq 15 \text{ and } n(1 - \hat{p}) \geq 15.$$

This guideline was easily satisfied in Example 3. The binary outcomes had counts 518 willing to pay much higher prices and 636 ($= 1154 - 518$) unwilling, both much larger than 15.

---

[3]Many statistics texts use 5 or 10 as the minimum instead of 15, but recent research suggests that's too small (e.g., L. Brown et al., *Statistical Science*, vol. 16, pp. 101–133, 2001).

EXAMPLE 4

What Proportion *Won't* Pay Higher Prices to Protect the Environment?

Picture the Scenario

Example 3 found a confidence interval for the population proportion p that will pay higher prices to support the environment. The population proportion that *won't* pay higher prices to support the environment is then $1 - p$.

Question to Explore

How can we construct a 95% confidence interval for the population proportion $1 - p$?

Think It Through

We can use the formula for a 95% confidence interval, because the GSS uses random sampling and because there's at least 15 successes and 15 failures (the counts are 518 and 636, with $n = 1154$). The estimate of $1 - p$ is $636/1154 = 0.55$, which is $1 - \hat{p} = 1 - 0.45$. When $n = 1154$ and $\hat{p} = 0.45$, we saw that the standard error of $\hat{p}$ is $se = \sqrt{\hat{p}(1 - \hat{p})/n} = 0.015$. Similarly, the standard error for $1 - \hat{p} = 0.55$ is

$$\sqrt{\text{proportion}(1 - \text{proportion})/n} = \sqrt{(0.55)(0.45)/1154} = 0.015.$$

This is necessarily the same as the standard error of $\hat{p}$. A 95% confidence interval for the population proportion that won't pay higher prices is

$$\text{sample proportion} \pm 1.96(se), \text{ or } 0.55 \pm 1.96(0.015),$$
$$\text{which is } 0.55 \pm 0.03, \text{ or} (0.52, 0.58).$$

In summary, we can be 95% confident that the population proportion that won't pay higher prices to support the environment falls between 0.52 and 0.58.

Insight

Now $0.52 = 1 - 0.48$ and $0.58 = 1 - 0.42$, where $(0.42, 0.48)$ is the 95% confidence interval for the population proportion who *will* pay higher prices. For a binary variable, inferences about the second category (*won't* pay higher prices) follow directly from those for the first category (*will* pay higher prices) by subtracting each endpoint of the confidence interval from 1.0. We do not need to construct the confidence interval separately for both categories. The confidence interval for one determines the confidence interval for the other.

TO PRACTICE THIS CONCEPT, TRY EXERCISE 7.16

How Can We Use a Confidence Level Other than 95%?

So far we've used a confidence level of 0.95, that is, "95% confidence." This means that there's a 95% chance that a sample proportion value $\hat{p}$ occurs such that the confidence interval $\hat{p} \pm 1.96(se)$ contains the unknown value of the population proportion p. With probability 0.05, however, the method produces a confidence interval that misses p. The population proportion then does *not* fall in the interval, and the inference is incorrect.

In practice, the confidence level 0.95 is the most common choice. But some applications require greater confidence. This is often true in medical research, for example. For estimating the probability p that a new treatment for a deadly

Recall

99% confidence interval: From Section 6.2, for central probability 0.99, you look up the cumulative probability of 0.005 or $1 - 0.005 = 0.995$ in Table A, or use software or a calculator, to find $z = 2.58$. ■

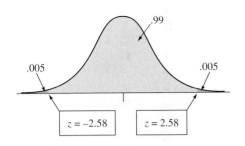

disease works better than the treatment currently used, we would want to be extremely confident about any inference we make. To increase the chance of a correct inference (that is, having the interval contain the parameter value), we use a larger confidence level, such as 0.99.

Now, 99% of the normal sampling distribution for the sample proportion $\hat{p}$ occurs within 2.58 standard errors of the population proportion p. So, with probability 0.99, $\hat{p}$ falls within 2.58(se) of p. A 99% confidence interval for p is $\hat{p} \pm 2.58(se)$.

EXAMPLE 5

Should a Husband Be Forced to Have Children?

Picture the Scenario

A recent GSS asked "If the wife in a family wants children, but the husband decides that he does not want any children, is it all right for the husband to refuse to have children?" Of 598 respondents, 366 said yes and 232 said no.

Questions to Explore

a. Find a 99% confidence interval for the population proportion who would say yes.

b. How does it compare to the 95% confidence interval?

Think It Through

a. The assumptions for the method are satisfied in that the GSS sample was randomly selected, and there were at least 15 successes and 15 failures (366 yes and 232 no). The sample proportion who said yes is $\hat{p} = 366/598 = 0.61$. Its standard error is

$$se = \sqrt{\hat{p}(1 - \hat{p})/n} = \sqrt{(0.61)(0.39)/598} = 0.020.$$

The 99% confidence interval is

$$\hat{p} \pm 2.58(se), \text{ or } 0.61 \pm 2.58(0.020),$$
$$\text{which is } 0.61 \pm 0.05, \text{ or}(0.56, 0.66).$$

In summary, we can be 99% confident that between 56% and 66% of the U.S. adult population agree that it's all right for the husband to refuse to have children. Statistical software or a statistical calculator can calculate confidence intervals. Screen shots from the TI-83+/84 are shown in the margin.

b. The 95% confidence interval is $0.61 \pm 1.96(0.020)$, which is 0.61 ± 0.04, or (0.57, 0.65). This is a bit narrower than the 99% confidence interval of (0.56, 0.66). Figure 7.4 illustrates.

```
1-PropZInt
x:366
n:598
C-Level:.99
Calculate
```

```
1-PropZInt
(.56,.66)
p̂=.61
n=598.00
```

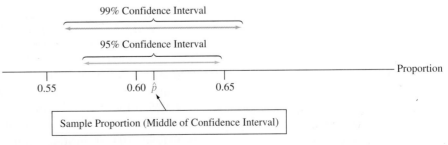

▲ **FIGURE 7.4: A 99% Confidence Interval Is Wider than a 95% Confidence Interval.**
Question: If you want greater confidence, why would you expect a wider interval?

Insight

The inference with a 99% confidence interval is less precise, with margin of error equal to 0.05 instead of 0.04. Having a greater margin of error is the sacrifice for gaining greater assurance (99% rather than 95%) of correctly inferring where p falls. A medical study that uses a higher confidence level will be less likely to make an incorrect inference, but it will not be able to "narrow in" as well on where the true parameter value falls.

TO PRACTICE THIS CONCEPT, TRY EXERCISE 7.20

Why settle for anything less than 100% confidence? To be absolutely 100% certain of a correct inference (that is, of capturing the parameter value inside the confidence interval), the confidence interval must contain *all* possible values for the parameter. For example, a 100% confidence interval for the population proportion believing it is all right for a husband to refuse to have children goes from 0.0 to 1.0. This inference would tell us only that some number between 0.0% and 100.0% of Americans feel this way. This is obviously not helpful. In practice, we settle for a little less than perfect confidence so we can estimate the parameter value more precisely. It is far more informative to have 99% confidence that the population proportion is between 0.56 and 0.66 than to have 100% confidence that it is between 0.0 and 1.0.

In using confidence intervals, *we must compromise between the desired margin of error and the desired confidence of a correct inference.* As one gets better, the other gets worse. This is why you would probably not use a 99.9999% confidence interval. It would usually have too large a margin of error to tell you much about where the parameter falls (its z-score is 4.9). In practice, 95% confidence intervals are the most common.

IN PRACTICE: Margin of error refers to 95% confidence

When the news media report a **margin of error**, it is the margin of error for a 95% confidence interval.

What Is the Error Probability for the Confidence Interval Method?

The general formula for the confidence interval for a population proportion is

$$\text{sample proportion} \pm (z\text{-score from normal table})(\text{standard error}),$$
$$\text{which in symbols is } \hat{p} \pm z(se).$$

The z-score depends on the confidence level. Table 7.2 shows the z-scores for the confidence levels usually used in practice. There is no need to memorize them. You can find them yourself, using Table A or a calculator or software.

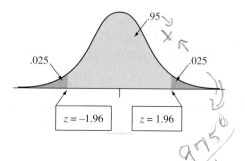

TABLE 7.2: z-Scores for the Most Common Confidence Levels

The large-sample confidence interval for the population proportion is $\hat{p} \pm z(se)$.

Confidence Level	Error Probability	z-Score	Confidence Interval
0.90	0.10	1.645	$\hat{p} \pm 1.645(se)$
0.95	0.05	1.96	$\hat{p} \pm 1.96(se)$
0.99	0.01	2.58	$\hat{p} \pm 2.58(se)$

Let's review how by finding the z-score for a 95% confidence interval. When 0.95 probability falls within z standard errors of the mean, then 0.05 probability falls in the two tails and $0.05/2 = 0.025$ falls in each tail. Looking up 0.025 in the body of Table A, we find $z = -1.96$, or we find $z = 1.96$ if we look up the cumulative probability $1 - 0.025 = 0.975$ corresponding to tail probability 0.025 in the right tail. The 95% confidence interval equals $\hat{p} \pm 1.96(se)$. Try this again for 99% confidence, using either the table or a calculator or software (you should get $z = 2.58$).

Table 7.2 contains a column labeled **error probability**. This is the probability that the method results in an incorrect inference, namely, that the data generates a confidence interval that does *not* contain the population proportion. The error probability equals 1 minus the confidence level. For example, when the confidence level equals 0.95, the error probability equals 0.05. The error probability is the two-tail probability under the normal curve for the given z-score. Half the error probability falls in each tail. For 95% confidence with its error probability of 0.05, the z-score of 1.96 is the one with probability $0.05/2 = 0.025$ in each tail.

Summary: Confidence Interval for a Population Proportion p

A confidence interval for a population proportion p, using the sample proportion $\hat{p}$ and the standard error $se = \sqrt{\hat{p}(1 - \hat{p})/n}$ for sample size n, is

$$\hat{p} \pm z(se), \text{ which is } \hat{p} \pm z\sqrt{\hat{p}(1 - \hat{p})/n}.$$

For 95% and 99% confidence intervals, z equals 1.96 and 2.58. To use this method, you need

- Data obtained by randomization (such as a random sample)
- A large enough sample size n so that the number of successes and the number of failures, that is, $n\hat{p}$ and $n(1 - \hat{p})$, are both at least 15.

What's the Effect of the Sample Size?

We'd expect that estimation should be more precise with larger sample sizes. With more data, we know more about the population. The margin of error is $z(se) = z\sqrt{\hat{p}(1 - \hat{p})/n}$. This margin decreases as the sample size n increases, for a given value of $\hat{p}$. The larger the value of n, the narrower is the interval.

To illustrate, in Example 5, suppose the survey result of $\hat{p} = 0.61$ for the proportion believing a husband can refuse to have children had resulted from a sample of size $n = 150$, only one-fourth the actual sample size of $n = 598$. Then the standard error would be

$$se = \sqrt{\hat{p}(1 - \hat{p})/n} = \sqrt{0.61(0.39)/150} = 0.040,$$

twice as large as the $se = 0.020$ we got for $n = 598$. The 99% confidence interval would be

$$\hat{p} \pm 2.58(se) = 0.61 \pm 2.58(0.040), \text{ which is } 0.61 \pm 0.10, \text{ or } (0.51, 0.71).$$

The margin of error of 0.10 is twice as large as the margin of error of 0.05 in the 99% confidence interval with the sample size $n = 598$ in Example 5. The confidence interval with larger n is narrower.

Because the standard error has the square root of n in the denominator, and because $\sqrt{4n} = 2\sqrt{n}$, *quadrupling* the sample size *halves* the standard error. That is, we must quadruple n, rather than double it, to halve the margin of error.

In summary, we've observed the following properties of a confidence interval:

Summary: Effects of Confidence Level and Sample Size on Margin of Error

The **margin of error** for a confidence interval:

- Increases as the confidence level increases
- Decreases as the sample size increases.

For instance, a 99% confidence interval is wider than a 95% confidence interval, and a confidence interval with 200 observations is narrower than one with 100 observations. These properties apply to *all* confidence intervals, not just the one for the population proportion.

Interpretation of the Confidence Level

In Example 3, the 95% confidence interval for the population proportion p willing to pay higher prices to protect the environment was (0.42, 0.48). The value of p is unknown to us, so we don't know that it actually falls in that interval. It may actually equal 0.41 or 0.49, for example.

So what does it mean to say that we have "95% confidence"? The meaning refers to a *long-run* interpretation—how the method performs when used over and over with many different random samples. If we used the 95% confidence interval method over time to estimate many population proportions, then *in the long run about 95% of those intervals would give correct results, containing the population proportion*. This happens because 95% of the sample proportions would fall within 1.96(se) of the population proportion. A graphical example is the $\hat{p}$ in line 1 of Figure 7.5.

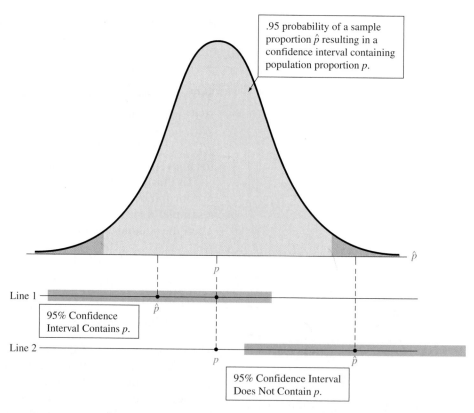

▲ **FIGURE 7.5: The Sampling Distribution of the Sample Proportion $\hat{p}$.** The lines below it show two possible $\hat{p}$ values and the corresponding 95% confidence intervals for the population proportion p. The interval on line 1 provides a correct inference, but the one on line 2 provides an incorrect inference. **Question:** Can you identify on the figure all the $\hat{p}$ values for which the 95% confidence interval would not contain p?

Saying that a particular interval contains p with "95% confidence" signifies that in the long run, 95% of such intervals would provide a correct inference (containing the actual parameter value). On the other hand, in the long run, 5% of the time the sample proportion $\hat{p}$ *does not* fall within 1.96(se) of p. If that happens, then the confidence interval *does not* contain p, as is seen for $\hat{p}$ in line 2 of Figure 7.5.

By our choice of the confidence level, we can control the chance that we make a correct inference. If an error probability of 0.05 makes us too nervous, we can instead form a 99% confidence interval. This is in error only 1% of the time in the long run. But then we must settle for a wider confidence interval and less precision.

Long Run versus Subjective Probability Interpretation of Confidence

You might be tempted to interpret a statement such as "We can be 95% confident that the population proportion p falls between 0.42 and 0.48" as meaning that the *probability* is 0.95 that p falls between 0.42 and 0.48. However, probabilities apply to statistics (such as in sampling distributions of the sample proportion), not to parameters. The estimate $\hat{p}$, not the parameter p, is the random variable having a sampling distribution and probabilities. The 95% confidence refers not to a probability for the population proportion p but rather to a probability that applies to the confidence interval *method* in its relative frequency sense: If we use it over and over for various samples, in the long run we make correct inferences 95% of the time.

Section 5.1 mentioned a *subjective* definition of probability that's an alternative to the relative frequency definition. This approach treats the *parameter* as a random variable. Statistical inferences based on the subjective definition of probability *do* make probability statements about parameters. For instance, with it you *can* say that the probability is 0.95 that p falls between 0.42 and 0.48. Statistical inference based on the subjective definition of probability is called **Bayesian statistics**.[4] It has gained in popularity in recent years, but it is beyond the scope of this text.

---

[4]The name refers to **Bayes theorem**, which can generate probabilities of parameter values, given the data, from probabilities of the data, given the parameter values.

ACTIVITY 2: **Let's Simulate the Performance of Confidence Intervals**

Let's get a feel for how confidence intervals sometimes provide an incorrect inference. To do this, for a given population proportion value we will simulate taking many samples and forming a confidence interval for each sample. We can then check how often the intervals provide an incorrect inference. We can conduct the simulation using statistical software (such as MINITAB) or using an applet in which we can control the parameter value, the sample size, and the confidence level.

Try this by going to the *confidence intervals for a proportion* applet on the text CD. We'll set the population $p = 0.50$, and see what happens when we take samples of size 50 and form 95% and 99% confidence intervals. At the menu, set the proportion value to $p = 0.5$ and set the sample size to $n = 50$. Click *Simulate* and 100 samples will be generated. The applet calculates the resulting 95%

and 99% confidence intervals using the sample proportion from each sample. The applet summaries the number of intervals that captured the true population proportion p. How many confidence intervals out of 100 intervals captured $p = 0.5$ at the 95% confidence level? At the 99% confidence level? How many would you expect at each confidence level to capture p, giving us a correct inference? How many intervals would you expect at each confidence level to provide an incorrect inference?

To get a feel for what happens "in the long run," do this simulation 10,000 times by clicking the *Simulate* button multiple times until you have a cumulative total of 10,000 confidence intervals. How many confidence intervals out of 10,000 intervals captured $p = 0.5$ at the 95% confidence level? At the 99% confidence level?

TO PRACTICE THIS ACTIVITY, TRY EXERCISES 7.23 AND 7.24

7.11 **Drug abuse:** In 2004 the National Center on Addiction and Substance Abuse at Columbia University (www.casacolumbia.org) took a random telephone survey of 1000 students age 12–17. Among other things, students were asked, "What is the most important problem facing people your age?" with options including drugs, crime, peer pressure, sexuality, and the environment. The most common response was drugs, cited by 29%. A story about the survey stated, "It has a margin of error of plus or minus 3 percentage points." Explain how they got this.

7.12 **Crime victims:** A recent General Social Survey asked, "During the last year, did anyone take something from you by using force—such as a stickup, mugging, or threat?" Of 987 subjects, 17 answered yes and 970 answered no.

   a. Find the point estimate of the proportion of the population who were victims.

   b. Find the standard error of this estimate.

   c. Find the margin of error for a 95% confidence interval.

   d. Construct the 95% confidence interval for the population proportion. Can you conclude that fewer than 10% of all adults in the United State were victims?

7.13 **How green are you?:** When the 2000 GSS asked subjects (variable GRNSOL) whether they would be willing to accept cuts in their standard of living to protect the environment, 344 of 1170 subjects said yes.

   a. Estimate the population proportion who would answer yes.

   b. Find the margin of error for this estimate.

   c. Find a 95% confidence interval for that proportion. What do the numbers in this interval represent?

   d. State and check the assumptions needed for the interval in (c) to be valid.

7.14 **Make industry help environment?:** When the GSS recently asked subjects whether it should or should not be the government's responsibility to impose strict laws to make industry do less damage to the environment (variable GRNLAWS), 1093 of 1232 subjects said yes.

   a. What assumptions are made to construct a 95% confidence interval for the population proportion who would say yes? Do they seem satisfied here?

   b. Construct the 95% confidence interval. Interpret in context. Can you conclude whether a majority or minority of the population would answer yes?

7.15 **Favor death penalty:** In the 2000 General Social Survey, respondents were asked whether they favored or opposed the death penalty for people convicted of murder. Software shows results

X	N	Sample p	95.0% CI
1764	2565	0.6877	(0.670, 0.706)

Here, X refers to the number of the respondents who were in favor.

   a. Show how to obtain the value reported under "Sample p."

   b. Interpret the confidence interval reported, in context.

   c. Explain what the "95% confidence" refers to, by describing the "long-run" interpretation.

   d. Can you conclude that more than half of all American adults were in favor? Why?

7.16 **Oppose death penalty:** Refer to the previous exercise. Show how you can get a 95% confidence interval for the proportion of American adults who were *opposed* to the death penalty from the confidence interval stated in the previous exercise for the proportion in favor. (*Hint*: The proportion opposed is 1 minus the proportion in favor.)

7.17 **Reincarnation:** A Harris poll (www.harrisinteractive.com) of a random sample of 2201 adults in the United States in 2003 reported that 90% believe in God, 82% believe in heaven, 69% believe in hell (although only 2% of those who did so expected to go there), and 51% believe in ghosts. The results on belief in reincarnation, using MINITAB software, are

X	N	Sample p	95% CI
594	2201	0.269877	(0.251333, 0.288422)

Here, X denotes the number who believed in reincarnation. Explain how to interpret "Sample p" and "95% CI" on this printout.

7.18 **z-score and confidence level:** Which z-score is used in a (a) 90%, (b) 98%, (c) 99.9% confidence interval for a population proportion?

7.19 **Fear of breast cancer:** A recent survey of 1000 American women between the ages of 45 and 64 asked them what medical condition they most feared. Of those sampled, 61% said breast cancer, 8% said heart disease, and the rest picked other conditions. By contrast, currently about 3% of female deaths are due to breast cancer, whereas 32% are due to heart disease.[5]

   a. Construct a 90% confidence interval for the population proportion of women who most feared breast cancer. Interpret.

   b. Indicate the assumptions you must make for the inference in (a) to be valid.

7.20 **Wife doesn't want kids:** The 1996 GSS asked, "If the husband in a family wants children, but the wife decides that she does not want any children, is it all right for the wife to refuse to have children?" Of 708 respondents, 576 said yes.

   a. Find a 99% confidence interval for the population proportion who would say yes. Can you conclude that the population proportion exceeds 75%? Why?

   b. Without doing any calculation, explain whether the interval in (a) would be wider or narrower than a 95% confidence interval for the population proportion who would say yes.

[5]See B. Lomborg, *The Skeptical Environmentalist,* Cambridge University Press, 2001, p. 222.

7.21 Exit-poll predictions: A national television network takes an exit poll of 1400 voters after each has cast a vote in a state gubernatorial election. Of them, 660 say they voted for the Democratic candidate and 740 say they voted for the Republican candidate.

  a. Treating the sample as a random sample from the population of all voters, would you predict the winner? Base your decision on a 95% confidence interval.

  b. Base your decision on a 99% confidence interval. Explain why you need stronger evidence to make a prediction when you want greater confidence.

7.22 Exit poll with smaller sample: In the previous exercise, suppose the same proportions resulted from $n = 140$ (instead of 1400), with counts 66 and 74.

  a. Now does a 95% confidence interval allow you to predict the winner? Explain.

  b. Explain why the same proportions but with smaller samples provide less information. (*Hint*: What effect does n have on the standard error?)

 7.23 Simulating confidence intervals: Repeat the simulation **TRY** activity at the end of the section, but this time forming 1000 confidence intervals at the 95% and 99% confidence levels. What percentage of the 1000 confidence intervals contained $p = 0.50$? What percentage did you expect?

 7.24 Simulating poor confidence intervals: Using the **TRY** *confidence interval for a proportion* applet on the text CD, let's check that the large-sample confidence interval for a proportion may work poorly with small samples. Set $n = 10$ and $p = 0.10$. Generate 100 random samples, each of size 10, and for each one, form a 95% confidence interval for p.

  a. How many of the intervals fail to contain the true value, $p = 0.10$?

  b. How many would you expect not to contain the true value? What does this suggest?

  c. To see that this is not a fluke, now take 1000 samples and see what percentage of 95% confidence intervals contain 0.10. (*Note*: For every interval formed, the number of successes is smaller than 15, so the large-sample formula is not adequate.)

  d. Using the *sampling distribution* applet, generate 1000 random samples of size 10 when $p = 0.10$. Plot the empirical sampling distribution of the sample proportion values. Is it bell shaped and symmetric? Use this to help explain why the large-sample confidence interval performs poorly in this case.

7.3 | How Can We Construct a Confidence Interval to Estimate a Population Mean?

We've learned how to construct a confidence interval for a population proportion—a summary parameter for a categorical variable. Next we'll learn how to construct a confidence interval for a population mean—a summary parameter for a quantitative variable. We'll analyze GSS data to estimate the mean number of hours per day that Americans watch television. The method resembles that for a proportion. The confidence interval again has the form

$$\text{point estimate} \pm \text{margin of error.}$$

The margin of error again equals a multiple of a standard error. To apply this formula, what do you think plays the role of the point estimate and the role of the standard error (se)?

How to Construct a Confidence Interval for a Population Mean

Recall

Section 6.5 introduced the standard error of the sampling distribution of $\bar{x}$, which describes how much the sample mean varies from sample to sample for a given size n. ▪

The sample mean $\bar{x}$ is the point estimate of the population mean μ. In Section 6.5, we learned that the exact standard error of the sample mean equals $\sigma/\sqrt{n}$, where σ is the population standard deviation. Like the exact standard error of the sample proportion, the exact standard error of the sample mean depends on a parameter whose value is unknown, in this case σ. In practice, we estimate σ by the sample standard deviation s. So, the standard error used in confidence intervals is

$$se = s/\sqrt{n}.$$

We get a confidence interval by taking the sample mean and adding and subtracting the margin of error. We'll see the details following the next example.

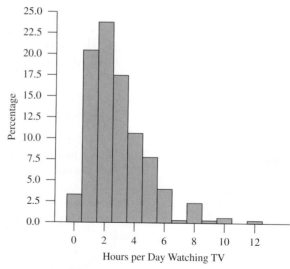

EXAMPLE 6

On the Average, How Much Television Do You Watch?

Picture the Scenario

Are you a couch potato? How much of the typical person's day is spent in front of the TV set? In recent years excessive TV watching has been claimed to be one factor, other than diet, for the increasing proportion of Americans who suffer from obesity. A recent General Social Survey asked respondents, "On the average day, about how many hours do you personally watch television?" A computer printout (from MINITAB) summarizes the results for the GSS variable, TV:

Variable	N	Mean	StDev	SE Mean	95.0% CI
TV	905	2.983	2.361	.0785	(2.83, 3.14)

We see that the sample size was 905, the sample mean was 2.98, the sample standard deviation was 2.36, the standard error of the sample mean was 0.0785, and the 95% confidence interval for the population mean time μ spent watching TV goes from 2.83 to 3.14 hours per day.

Questions to Explore

a. What do the sample mean and standard deviation suggest about the likely shape of the population distribution?

b. How did the software get the standard error? What does it mean?

c. Interpret the 95% confidence interval reported by software.

Think It Through

a. For the sample mean of $\bar{x} = 2.98$ and standard deviation of $s = 2.36$, the lowest possible value of 0 falls only a bit more than one standard deviation below the mean. This information suggests that the population distribution of TV watching may be skewed to the right. Figure 7.6

Percentage (y-axis): 0.0, 2.5, 5.0, 7.5, 10.0, 12.5, 15.0, 17.5, 20.0, 22.5, 25.0

Hours per Day Watching TV (x-axis): 0, 2, 4, 6, 8, 10, 12

▲ **FIGURE 7.6: Histogram of Number of Hours a Day Watching Television.**
Question: Does the skew affect the validity of a confidence interval for the population mean?

shows a histogram of the data, which suggests the same thing. The median was 2, the lower and upper quartiles were 1 and 4, the 95th percentile was 7, yet some subjects reported much higher values.

b. With sample standard deviation $s = 2.36$ and sample size $n = 905$, the standard error of the sample mean is

$$se = s/\sqrt{n} = 2.36/\sqrt{905} = 0.0785 \text{ hours.}$$

If many studies were conducted about TV watching, with $n = 905$ for each, the sample mean would not vary much among those studies.

c. A 95% confidence interval for the population mean μ of TV watching in the United States is $(2.83, 3.14)$ hours. We can be 95% confident that Americans watch between 2.83 and 3.14 hours of TV a day, on the average.

Insight

Because the sample size was relatively large, the estimation is precise and the confidence interval is quite narrow. The larger the sample size, the smaller the standard error and the subsequent margin of error.

TO PRACTICE THIS CONCEPT, TRY EXERCISE 7.25

We have not yet seen how software found the margin of error for the confidence interval in Example 6. As with the proportion, the margin of error for a 95% confidence interval is roughly two standard errors. However, we need to introduce a new distribution similar to the normal distribution to give us a more precise margin of error. We'll find the margin of error by multiplying *se* by a score that is a bit larger than the z-score when n is small but very close to it when n is large.

The t Distribution and Its Properties

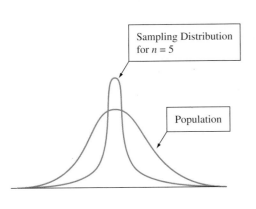

Sampling Distribution for $n = 5$

Population

We'll now learn about a confidence interval that applies even for small sample sizes. A disadvantage is that it makes the assumption that the *population distribution of the variable is normal*. In that case, the sampling distribution of $\bar{x}$ is normal even for small sample sizes. (The right panel of Figure 6.20, which showed sampling distributions for various shapes of population distributions, illustrated this, as does the figure here in the margin.) When the population distribution is normal, the sampling distribution of $\bar{x}$ is normal for all n, not just large n.

Suppose we knew the exact standard error, $\sigma/\sqrt{n}$, of the sample mean. Then, with the additional assumption that the population is normal, with small n we could use the formula $\bar{x} \pm z(se)$ with that exact standard error, for instance with $z = 1.96$ for 95% confidence. In practice, we don't know the population standard deviation σ. Substituting the sample standard deviation s for σ to get $se = s/\sqrt{n}$ then introduces extra error. This error can be sizeable when n is small. To account for this increased error, we must replace the z-score by a slightly larger score, called a *t*-score. The confidence interval is then a bit wider. *The t-score is like a z-score but it comes from a bell-shaped distribution that has slightly thicker tails than a normal distribution.* This distribution is called the *t* **distribution**.

IN PRACTICE: The t distribution adjusts for estimating σ

In practice, we estimate the standard error of the sample mean by $se = s/\sqrt{n}$. Then we multiply *se* by a *t*-score from the **t distribution** to get the margin of error for a confidence interval for the population mean.

Recall

The t distribution resembles the *standard normal* distribution, being bell shaped around a mean of 0. Its standard deviation is a bit larger than 1, the precise value depending on what is called the **degrees of freedom**, denoted by **df**. For inference about a population mean, the degrees of freedom equal $df = n - 1$, one less than the sample size. Before presenting this confidence interval for a mean, we list the major properties of the t distribution.

Properties of the t distribution

■ The t distribution is bell shaped and symmetric about 0.

■ The probabilities depend on the degrees of freedom, df. The t distribution has a slightly different shape for each distinct value of df, and different t-scores apply for each df value.

■ The t distribution has thicker tails and is more spread out than the standard normal distribution. The larger the df value, however, the closer it gets to the standard normal. Figure 7.7 illustrates this point. When df is about 30 or more, the two distributions are nearly identical.

■ A t-score multiplied by the standard error gives the margin of error for a confidence interval for the mean.

Table B at the end of the text lists t-scores from the t distribution for the right-tail probabilities of 0.100, 0.050, 0.025, 0.010, 0.005, and 0.001. The table labels these by $t_{.100}$, $t_{.050}$, $t_{.025}$, $t_{.010}$, $t_{.005}$, and $t_{.001}$. For instance, $t_{.025}$ has probability 0.025 in the right tail, a two-tail probability of 0.05, and is used in 95% confidence intervals. Statistical software reports t-scores for any tail probability.

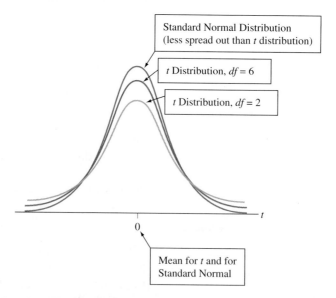

Standard Normal Distribution
(less spread out than t distribution)

t Distribution, $df = 6$

t Distribution, $df = 2$

Mean for t and for
Standard Normal

▲ **FIGURE 7.7: The t Distribution Relative to the Standard Normal Distribution. The t distribution gets closer to the standard normal as the degrees of freedom (df) increase. The two are practically identical when df ≥ 30. Question:** Can you find z-scores (such as 1.96) for a normal distribution on the t table (Table B)?

Table 7.3 is an excerpt from the t table (Table B). To illustrate its use, suppose the sample size is 7. Then the degrees of freedom $df = n - 1 = 6$. Row 6 of the t table shows the t-scores for $df = 6$. The column labeled $t_{.025}$ contains t-scores with right-tail probability equal to 0.025. With $df = 6$, this t-score is $t_{.025} = 2.447$. This means that 2.5% of the t distribution falls in the right tail above 2.447. By symmetry, 2.5% also falls in the left tail below $-t_{.025} = -2.447$. Figure 7.8 illustrates. When $df = 6$, the probability equals 0.95 between -2.447 and 2.447. This is the t-score for a 95% confidence interval when $n = 7$. The confidence interval is $\bar{x} \pm 2.447(se)$.

TABLE 7.3: Part of Table B Displaying t-Scores

The scores have right-tail probabilities of 0.100, 0.050, 0.025, 0.010, 0.005, and 0.001. When $n = 7$, $df = 6$, and $t_{.025} = 2.447$ is the t-score with right-tail probability = 0.025 and two-tail probability = 0.05. It is used in a 95% confidence interval, $\bar{x} \pm 2.447(se)$.

	Confidence Level					
	80%	**90%**	**95%**	**98%**	**99%**	**99.8%**
df	$t_{.100}$	$t_{.050}$	$t_{.025}$	$t_{.010}$	$t_{.005}$	$t_{.001}$
1	3.078	6.314	12.706	31.821	63.657	318.3
...						
6	1.440	1.943	2.447	3.143	3.707	5.208
7	1.415	1.895	2.365	2.998	3.499	4.785

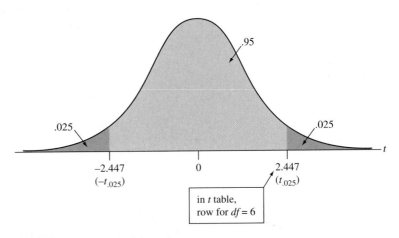

▲ **FIGURE 7.8: The t Distribution with df = 6.** 95% of the distribution falls between -2.447 and 2.447. These t-scores are used with a 95% confidence interval when n = 7. **Question:** Which t-scores contain the middle 99% of a t distribution (for a 99% confidence interval)?

Using the t Distribution to Construct a Confidence Interval for a Mean

The confidence interval for a mean has margin of error that equals a t-score times the standard error.

Summary: 95% Confidence Interval for a Population Mean

A 95% confidence interval for the population mean μ is

$$\bar{x} \pm t_{.025}(se), \text{ where } se = s/\sqrt{n}.$$

Here, $df = n - 1$ for the t-score $t_{.025}$ that has right-tail probability 0.025 (total probability 0.05 in the two tails and 0.95 between $-t_{.025}$ and $t_{.025}$). To use this method, you need

- Data obtained by randomization (such as a random sample)
- An approximately normal population distribution.

EXAMPLE 7

eBay Auctions of Palm Handheld Computers

Picture the Scenario

eBay is a popular Internet company for personal auctioning of just about anything. When you list an item to sell on eBay, there is an online auction format in which the product sells for the highest price bid over a set period of time (1, 3, 5, 7, or 10 days). In addition, you can offer potential buyers a "buy-it-now" option, whereby they can buy the product immediately at a fixed price that you set.

Do you tend to get a higher, or a lower, price if you give bidders the "buy-it-now" option? Let's consider some data from sales of the Palm M515 PDA (personal digital assistant), a popular handheld computer, during the first week of May 2003. During that week, 25 of these handheld computers were auctioned off, 7 of which had the "buy-it-now" option. Here are the final prices (in dollars) at which the item sold:

Buy-it-now option: 235, 225, 225, 240, 250, 250, 210

Bidding only: 250, 249, 255, 200, 199, 240, 228, 255, 232, 246, 210, 178, 246, 240, 245, 225, 246, 225

Questions to Explore

a. Use numerical and graphical descriptive statistics to summarize selling prices for the two types of auctions.

b. Consider the probability distribution of selling prices with the buy-it-now option. State and check the assumptions for using these data to find a 95% confidence interval for the mean of that distribution.

c. Find the 95% confidence interval for the buy-it-now option, and interpret it. How does it compare to the 95% confidence interval for the mean sales price for the bidding only option, which in Exercise 7.29 you'll see is (220.70, 242.52)?

Think It Through

a. When we use MINITAB and request descriptive statistics, we get the mean and standard deviation and the five-number summary using quartiles:

buy_now	N	Mean	StDev	Minimum	Q1	Median	Q3	Maximum
no	18	231.61	21.94	178.0	221.25	240.0	246.75	255.0
yes	7	233.57	14.64	210.0	225.00	235.0	250.00	250.0

The sample mean selling price is a bit higher for items with the buy-it-now option, $233.57 compared to $231.61, but the medians have the reverse ordering. There's a bit less variability in the selling prices with the buy-it-now option.

Figure 7.9 shows a MINITAB dot plot for each set of selling prices. The plot shows the greater variability for the 18 observations not having the buy-it-now option as well as some evidence of skew to the left.

b. The confidence interval using the t distribution assumes a random sample from an approximately normal population distribution of selling prices. Unlike a survey such as the GSS, with the eBay selling prices, the sample and population distinction is not so clear. We'll regard the observed selling prices as independent observations from a probability distribution for selling price at that point in time (May 2003). That probability distribution reflects the variability in selling price from auction to auction. Inference treats that probability distribution like a population distribution and treats the observed selling prices as a random sample from it.

We can get some information about the shape of the probability distribution of selling price by looking at Figure 7.9. With only 7 observations for the buy-it-now option, it's hard to tell much. A bell-shaped, discrete population distribution could generate sample data such as shown in the second line of Figure 7.9. Later in this section, we'll discuss this assumption further.

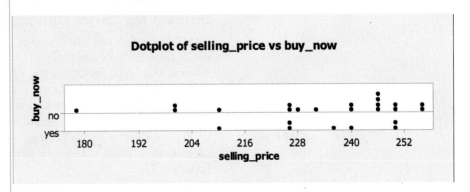

▲ **FIGURE 7.9: MINITAB Dot Plot for eBay Selling Prices for Palm Handheld Computers.** Data are shown for 7 observations in which buyers had the buy-it-now option and 18 observations in which they did not have this option. **Question:** What does this plot tell you about how the selling prices compare for the two cases?

c. Let μ denote the population mean for the buy-it-now option. The estimate of μ is the sample mean of 235, 225, 225, 240, 250, 250, 210, which is $\bar{x}$ = \$233.57. This and the sample standard deviation of s = \$14.64 are reported in the above table. The standard error of the sample mean is

$$se = s/\sqrt{n} = 14.64/\sqrt{7} = 5.53.$$

Since n = 7, the degrees of freedom are $df = n - 1 = 6$. For a 95% confidence interval, from Table 7.3 we use $t_{.025}$ = 2.447. The 95% confidence interval is

$$\bar{x} \pm t_{.025}(se) = 233.57 \pm 2.447(5.53),$$
which is 233.57 ± 13.54, or (220.03, 247.11).

With 95% confidence, the range of believable values for the mean selling price for this probability distribution is \$220.03 to \$247.11. In the margin, a screen shot shows how the TI-83+/84 reports this confidence interval. Table 7.4 shows the way MINITAB reports it, with "SE Mean" being the standard error of the sample mean.

TABLE 7.4: MINITAB Output for 95% Confidence Interval for Mean

Variable	N	Mean	StDev	SE Mean	95.0% CI
selling_price	7	233.57	14.64	5.53	(220.03, 247.11)

```
TInterval
 Inpt:Data Stats
 x:233.57
 Sx:14.64
 n:7
 C-Level:.95
 Calculate
```

```
Interval
(220.03,247.11)
x=233.57
Sx=14.64
n=7.00
```

TI-83+/84 output

This confidence interval is not much different from the confidence interval (220.70, 242.52) for the mean without the buy-it-now option. The intervals overlap a lot. There's not enough information for us to conclude that one probability distribution clearly has a higher mean than the other.

Insight

With small samples, we usually must sacrifice precision. In Exercise 7.33 we'll make a similar comparison using a larger data set, and we will be able to infer that one option is better. In Chapter 9 we'll learn about inferential methods directed specifically toward comparing population means or proportions.

TO PRACTICE THIS CONCEPT, TRY EXERCISE 7.31

How Do We Find a t Confidence Interval for Other Confidence Levels?

The 95% confidence interval uses $t_{.025}$, the t-score for a right-tail probability of 0.025, since 95% of the probability falls between $-t_{.025}$ and $t_{.025}$. For 99% confidence, the error probability is 0.01, the probability is $0.01/2 = 0.005$ in each tail, and the appropriate t-score is $t_{.005}$. The top margin of Table B and Table 7.3 show both the t subscript notation and the confidence level.

For instance, the buy-it-now sample in the above example has $n = 7$, so $df = 6$ and from Table 7.3, $t_{.005} = 3.707$. A 99% confidence interval for the mean selling price is (again using $\bar{x} = 233.57$ and $se = 5.53$, as we found in Example 7)

$$\bar{x} \pm t_{.005}(se) = 233.57 \pm 3.707(5.53),$$
$$\text{which is } 233.57 \pm 20.50, \text{ or } (213.07, 254.07).$$

With a larger confidence level, the confidence interval is wider than the 95% confidence interval of (220.03, 247.11).

If the Population Is Not Normal, Is the Method "Robust"?

A basic assumption of the confidence interval using the t distribution is that the population distribution is normal. This is worrisome, because many variables have distributions that are far from a bell shape. How problematic is it if we use the t confidence interval even if the population distribution is not normal? For large random samples, it's not problematic, because of the central limit theorem. The sampling distribution is bell shaped even when the population distribution is not. But what about for small n?

For the confidence interval in Example 7 with $n = 7$ to be valid, we must assume that the probability distribution of selling price is normal. Does this assumption seem plausible? A dot plot, histogram, or stem-and-leaf plot gives us some information about the population distribution, but it is not precise when n is small and it tells us little when $n = 7$. Fortunately, the confidence interval using the t distribution is a **robust** method in terms of the normality assumption.

Robust Statistical Method

A statistical method is said to be **robust** with respect to a particular assumption if it performs adequately even when that assumption is violated.

Even if the population distribution is not normal, confidence intervals using t-scores usually work quite well. The actual probability that the 95% confidence interval method provides a correct inference is close to 0.95 and gets closer as n increases.

Recall

Section 2.5 identified an observation as a potential **outlier** if it falls more than $1.5 \times IQR$ below the first quartile or above the third quartile, or if it falls more than 3 standard deviations from the mean. ■

The most important case when the t confidence interval method does *not* work well is *when the data contain extreme outliers*. Partly this is because of the effect on the method, but also because the mean itself may not then be a representative summary of the center. In Example 7 with the 7 observations in the "buy-it-now" sample, you can check that no potential outliers are identified. Another case that calls for caution is with binary data, in which case the mean is a proportion. Section 7.2 presented a separate method for binary data.

The t confidence interval method is not robust to violations of the random sampling assumption. The t method, like all inferential statistical methods, has questionable validity if the method for producing the data did not use randomization.

IN PRACTICE: Assumptions are rarely perfectly satisfied

Knowing that a statistical method is **robust** (that is, it still performs adequately) even when a particular assumption is violated is important, because in practice assumptions are rarely perfectly satisfied. Confidence intervals for a mean using the **t distribution** are robust against most violations of the normal population assumption. However, you should check the data graphically to identify **outliers** that could affect the validity of the mean or its confidence interval. Also, unless the data production used **randomization**, statistical inference may be inappropriate.

The Standard Normal Distribution Is the t Distribution with $df = \infty$

Look at the table of t-scores (Table B in the Appendix), part of which is shown in Table 7.5 here. As df increases, you move down the table. The t-score decreases toward the z-score for a standard normal distribution. For instance, when df increases from 1 to 100 in Table B, the t-score $t_{.025}$ that has right-tail probability equal to 0.025 decreases from 12.706 to 1.984. This reflects the t distribution becoming less spread out and more similar in appearance to the standard normal distribution as df increases. The z-score with right-tail probability of 0.025 for the standard normal distribution is $z = 1.96$. When df is above about 30, the t-score is similar to this z-score. For instance, they both round to 2.0. The t-score gets closer and closer to the z-score as df keeps increasing. *You can think of the standard normal distribution as a t distribution with $df = infinity$.*

TABLE 7.5: Part of Table B Displaying t-Scores for Large df Values

The z-score of 1.96 is the t-score $t_{.025}$ with right-tail probability of 0.025 and $df = \infty$.

	Confidence Level					
	80%	90%	95%	98%	99%	99.8%
	Right-Tail Probability					
df	$t_{.100}$	$t_{.050}$	$t_{.025}$	$t_{.010}$	$t_{.005}$	$t_{.001}$
1	3.078	6.314	12.706	31.821	63.657	318.3
30	1.310	1.697	2.042	2.457	2.750	3.385
40	1.303	1.684	2.021	2.423	2.704	3.307
50	1.299	1.676	2.009	2.403	2.678	3.261
60	1.296	1.671	2.000	2.390	2.660	3.232
80	1.292	1.664	1.990	2.374	2.639	3.195
100	1.290	1.660	1.984	2.364	2.626	3.174
∞	1.282	1.645	1.960	2.326	2.576	3.090

Did You Know?

z-score = t-score with $df = \infty$ (infinity) ■

Recall

The reason we use a t-score instead of a z-score in the confidence interval for a mean is that it accounts for the extra error due to estimating σ by s.■

Table 7.5 shows the first row and the last several rows of Table B. The last row lists the z-scores for various confidence levels, opposite $df = \infty$ (infinity). The t-scores are not printed for $df > 100$, but they are close to the z-scores. For instance, to get the confidence interval about TV watching in Example 6, for which the GSS sample had $n = 905$, software uses the $t_{.025}$ score for $df = 905 - 1 = 904$, which is 1.963. This is nearly identical to the z-score of 1.960 from the standard normal distribution.

You can get t-scores for *any df* value using software and many calculators, so you are not restricted to Table B. (For instance, MINITAB provides percentile scores for various distributions under the *CALC* menu.) If you don't have access to software, you won't be far off if you use a z-score instead of a t-score for df values larger than shown in Table B (above 100). For a 95% confidence interval you will then use

$$\bar{x} \pm 1.96(se) \text{ instead of } \bar{x} \pm t_{.025}(se).$$

You will not get *exactly* the same result that software would give, but it will be very, very close.

IN PRACTICE: Use t for inference about μ whenever you estimate σ

Statistical software and calculators use the t distribution for *all* cases when the sample standard deviation s is used to estimate the population standard deviation σ. The normal population assumption is mainly relevant for small n, but even then the t confidence interval is a robust method, working well unless there are extreme outliers or the data are binary.

On the Shoulders of . . . William S. Gosset

How do you find the best way to brew beer if you have only small samples?

The statistician and chemist William S. Gosset was employed as a Brewer in charge of the experimental unit of Guinness Breweries in Dublin, Ireland. The search for a better stout in 1908 led him to the discovery of the t distribution. Only small samples were available from several of his experiments pertaining to the selection, cultivation, and treatment of barley and hops for the brewing process. The established statistical methods at that time relied on large samples and the normal distribution. Because of company policy forbidding the publication of company work in one's own name, Gosset used the pseudonym "Student" in articles he wrote about his discoveries. The t distribution became known as *Student's t* distribution, a name sometimes still used today.

7.25 Females' ideal number of children: The 2002 General Social Survey asked, "What do you think is the ideal number of children for a family to have?" The 497 females who responded had a median of 2, mean of 3.02, and standard deviation of 1.81.

a. What is the point estimate of the population mean?

b. Find the standard error of the sample mean.

c. The 95% confidence interval is (2.89, 3.21). Interpret.

d. Is it plausible that the population mean $\mu = 2$? Explain.

7.26 Males' ideal number of children: Refer to the previous exercise. For the 397 males in the sample, the mean was 2.89 and the standard deviation was 1.77.

a. Find the point estimate of the population mean, and show that its standard error is 0.089.

b. The 95% confidence interval is (2.72, 3.06). Explain what "95% confidence" means for this interval.

7.27 Using t-table: Using Table B or software or a calculator, report the t-score which you multiply by the standard error to form the margin of error for a

a. 95% confidence interval for a mean with 5 observations.

b. 95% confidence interval for a mean with 15 observations.

c. 99% confidence interval for a mean with 15 observations.

7.28 Anorexia in teenage girls: A study[6] compared various therapies for teenage girls suffering from anorexia, an eating disorder. For each girl, weight was measured before and after a fixed period of treatment. The variable measured was the change in weight, X = weight at the end of the study minus weight at the beginning of the study. The therapies were designed to aid weight gain, corresponding to positive values of X. For the sample of 17 girls receiving the family therapy, the changes in weight during the study were

11, 11, 6, 9, 14, −3, 0, 7, 22, −5, −4, 13, 13, 9, 4, 6, 11.

a. Plot these with a dot plot or box plot, and summarize.

b. Verify that the weight changes have $\bar{x} = 7.29$ and $s = 7.18$ pounds.

c. Verify that the standard error of the sample mean was $se = 1.74$.

d. To use the t distribution, explain why $df = 16$ and a 95% confidence interval uses the t-score equal to 2.120.

e. Let μ denote the population mean change in weight for this therapy. Verify that the 95% confidence interval for μ is (3.6, 11.0). Explain why this suggests that

the true mean change in weight is positive, but possibly quite small.

7.29 eBay without buy-it-now option: From Example 7, selling prices of the Palm M515 PDA for a one-week selling period in May 2003 when the buy-it-now option was not used were

250, 249, 255, 200, 199, 240, 228, 255, 232, 246, 210, 178, 246, 240, 245, 225, 246, 225.

These data have $\bar{x} = 231.61$, $s = 21.94$, Q1 = 221.25, Median = 240.0, Q3 = 246.75.

a. What assumptions are needed to construct a 95% confidence interval for μ? Point out any that seem questionable, based on a histogram or the dot plot in Figure 7.9.

b. Show that the 95% confidence interval is (220.70, 242.52). Interpret it in context.

c. Check whether this data set has any potential outliers according to the criterion of (i) 1.5 × IQR below Q1 or above Q3, (ii) 3 standard deviations from the mean.

d. Figure 7.9 shows that the $178 observation is quite a bit lower than the others. Deleting this observation, find the mean and standard deviation, and construct the 95% confidence interval for μ. How does it compare to the 95% confidence interval (220.70, 242.52) using all the data?

7.30 New graduates' income: Exercise 7.6 reported annual incomes of $45,000, $35,000, and $55,000 for three new college graduates with math majors.

a. Using software or a calculator, verify that the 95% confidence interval for the population mean goes from $20,159 to $69,841.

b. Name two things you could do to get a narrower interval than the one in (a).

c. Construct a 99% confidence interval. Why is it wider than the 95% interval?

d. On what assumptions is the interval in (a) based? Explain how important each assumption is.

7.31 TV watching for Muslims: Having estimated the mean amount of time spent watching TV in Example 6, we might want to estimate the mean for various groups, such as different religious groups. Let's consider Muslims. A recent GSS had responses on TV watching from seven subjects who identified their religion as Muslim.

Their responses on the number of hours of TV watching were 0, 0, 2, 2, 2, 4, 6, shown also in the accompanying MINITAB dot plot.

---

[6]Data courtesy of Prof. Brian Everitt, Institute of Psychiatry, London.

TV watching for Muslims in GSS sample

a. What must we assume to use these data to find a 95% confidence interval for the mean amount of TV watched by the population of American Muslims?

b. The table shows the way MINITAB reports results. Explain how to interpret the 95% confidence interval in context.

Variable	N	Mean	StDev	SE Mean	95.0% CI
television	7	2.286	2.138	0.808	(0.308, 4.264)

c. What is the main factor causing the confidence interval to be so wide?

7.32 Wage discrimination?: According to a union agreement, the mean income for all senior-level assembly-line workers in a large company equals $500 per week. A representative of a women's group decides to analyze whether the mean income for female employees matches this norm. For a random sample of nine female employees, using software she obtains a 95% confidence interval of (371, 509). Explain what is wrong with each of the following interpretations of this interval.

a. We infer that 95% of the women in the population have income between $371 and $509 per week.

b. If random samples of 9 women were repeatedly selected, then 95% of the time the sample mean income would be between $371 and $509.

c. We can be 95% confident that $\bar{x}$ is between $371 and $509.

d. If we repeatedly sampled the entire population, then 95% of the time the population mean would be between $371 and $509.

7.33 More eBay selling prices: Example 7 analyzed selling prices of the Palm M515 PDA for a one-week selling period in May 2003. When prices are analyzed for a three-month period in 2003, MINITAB shows the results in the table.

Variable	Buy now	N	Mean	SE Mean	StDev	Median
Price	No	136	223.36	2.04	23.74	225.01
	Yes	132	235.45	1.79	20.57	239.99

The 95% confidence intervals for the mean are (231.91, 238.99) with the buy-it-now option and (219.32, 227.40) without it. Interpret each of these intervals, and explain what you learn by comparing them.

7.34 How often read a newspaper?: For the "FL student survey" data file on the text CD, software reports the results for responses on the number of times a week the subject reads a newspaper:

Variable	N	Mean	Std Dev	SE Mean	95.0% CI
News	60	4.1	3.0	0.387	(3.325, 4.875)

a. Is it plausible that $\mu = 7$, where μ is the population mean for all Florida students? Explain.

b. Suppose that the sample size had been 240, with $\bar{x} = 4.1$ and $s = 3.0$. Find a 95% confidence interval, and compare it to the one reported. Describe the effect of sample size on the margin of error.

c. Does it seem plausible that the population distribution of this variable is normal? Why?

d. Explain the implications of the term "robust" regarding the normality assumption made to conduct this analysis.

7.35 Political views: The General Social Survey asks respondents to rate their political views on a seven-point scale, where 1 = extremely liberal, 4 = moderate, and 7 = extremely conservative. A researcher analyzing data from the 2002 GSS gets SPSS output:

Variable	N	Mean	StDev	SE Mean	95% CI
Polviews	1331	4.12	1.39	0.0381	(4.05, 4.19)

a. Show how to construct the confidence interval from the other information provided.

b. Can you conclude that the population mean is higher than the moderate score of 4.0? Explain.

c. Would the confidence interval be wider, or narrower, (i) if you constructed a 99% confidence interval, (ii) if $n = 500$ instead of 1331?

7.36 Length of hospital stays: A hospital administrator wants to estimate the mean length of stay for all inpatients using that hospital. Using a random sample of 100 records of patients for the previous year, she reports that "The sample mean was 5.3. In repeated random samples of this size, the sample mean could be expected to fall within 1.0 of the true mean about 95% of the time." Explain the meaning of this sentence from the report, showing what it suggests about the 95% confidence interval.

7.37 Effect of n: Find the margin of error for a 95% confidence interval when the sample standard deviation equals 100, with a sample size of (a) 400, (b) 1600. What is the effect of the sample size?

7.38 Effect of confidence level: Find the margin of error when the sample standard deviation equals 100 for a sample size of 400, using confidence level (a) 95%, (b) 99%. What is the effect of the choice of confidence level?

7.39 Catalog mail-order sales: A company that sells its products through mail-order catalogs wants information about the success of its most recent catalog. The company decides to estimate the mean dollar amount of items ordered from those who received the catalog. For a random sample of 100 customers from their files, only 5 made an order, so 95 of the response values were $0. The overall mean of all 100 orders was $10, with a standard deviation of $10.

a. Is it plausible that the population distribution is normal? Explain, and discuss how much this affects the validity of a confidence interval for the mean.

b. Find a 95% confidence interval for the mean dollar order for the population of all customers who received this catalog. Normally, the mean of their sales per catalog is about $15. Can we conclude that it declined with this catalog? Explain.

■ **7.40   Number of children:** For the question, "How many children have you ever had?" use the GSS Web site sda.berkeley.edu/GSS with the variable CHILDS to find the sample mean and standard deviation for the 2000 survey.

a. Show how to get the standard error of 0.03 that the GSS reports for a random sample of 2801 adults.

b. Construct a 95% confidence interval for the population mean. Can you conclude that the population mean is less than 2.0? Explain.

c. Discuss the assumptions for the analysis in (b) and whether that inference seems to be justified.

■ **7.41   Simulating the confidence interval:** Go to the *confidence intervals for a mean* applet on the text CD. Choose the sample size of 50 and the skewed population distribution with $\mu = 100$. Generate 100 random samples, each of size 50, and for each one form a 95% confidence interval for the mean.

a. How many of the intervals fail to contain the true value?

b. How many would you expect not to contain the true value?

c. Now repeat the simulation using 1000 random samples of size 50. Why do close to 95% of the intervals contain μ, even though the population distribution is quite skewed?

7.4 | How Do We Choose the Sample Size for a Study?

Have you ever wondered how the sample sizes are determined for polls? How does a polling organization know whether it needs 10,000 people, or only 100 people, or some odd number such as 745? The simple answer is that this depends on how much precision is needed, as measured by the margin of error. The smaller the margin of error, the larger the sample size must be. We'll next learn how to determine which sample size has the desired margin of error. For instance, we'll find out how large an exit poll must be so that a 95% confidence interval for the population proportion voting for a candidate has a certain margin of error, such as 0.04.

The key results for finding the sample size for a random sample are as follows:

■ The *margin of error* depends on the *standard error* of the sampling distribution of the point estimate.

■ The *standard error* itself depends on the *sample size*.

So one of the main components of the margin of error is the sample size n. Once we specify a margin of error with a particular confidence level, we can determine the value of n that has a standard error giving that margin of error.

Choosing the Sample Size for Estimating a Population Proportion

How large should n be to estimate a population proportion? First we must decide on the desired *margin of error*—how close the sample proportion should be to the population proportion. Second, we must choose the *confidence level* for achieving that margin of error. In practice, 95% confidence intervals are most common. If we specify a margin of error of 0.04, this means that a 95% confidence interval should equal the sample proportion plus and minus 0.04.

EXAMPLE 8

What Sample Size Do You Need for an Exit Poll?

Picture the Scenario

A television network plans to predict the outcome of an election between Jacalyn Levin and Roberto Sanchez. They will do this with an exit poll that randomly samples voters on election day. They want a reasonably accurate estimate of the population proportion that voted for Levin. The final poll a week before election day estimated her to be well ahead, 58% to 42%, so they do not expect the outcome to be close. Since their finances for this project are limited, they don't want to collect a large sample if they don't need it. They decide to use a sample size for which the margin of error is 0.04, rather than their usual margin of error of 0.03.

Question to Explore

What is the sample size for which a 95% confidence interval for the population proportion has margin of error equal to 0.04?

Recall

From Section 7.2, a 95% confidence interval for a population proportion p is

$$\hat{p} \pm 1.96(se),$$

where $\hat{p}$ denotes the sample proportion and the standard error (se) is

$$se = \sqrt{\hat{p}(1 - \hat{p})/n}. \blacksquare$$

Think It Through

The 95% confidence interval for a population proportion p is $\hat{p} \pm 1.96(se)$. So, if the sample size is such that $1.96(se) = 0.04$, then the margin of error will be 0.04. See Figure 7.10.

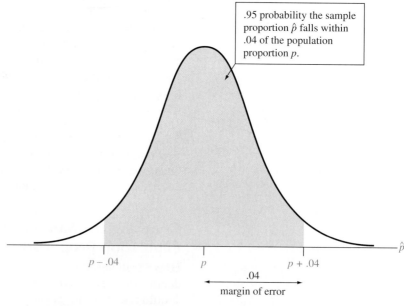

.95 probability the sample proportion $\hat{p}$ falls within .04 of the population proportion p.

$p - .04$ p $p + .04$

.04

margin of error

▲ **FIGURE 7.10: Sampling Distribution of Sample Proportion $\hat{p}$ such that a 95% Confidence Interval Has Margin of Error 0.04.** We need to find the value of n that has this margin of error. **Question:** What must we assume for this distribution to be approximately normal?

Let's find the value of the sample size n for which $0.04 = 1.96(se)$. For a confidence interval for a proportion, the standard error is $\sqrt{\hat{p}(1 - \hat{p})/n}$. So the equation $0.04 = 1.96(se)$ becomes

$$0.04 = 1.96\sqrt{\hat{p}(1 - \hat{p})/n}.$$

To find the answer, we solve algebraically for *n*. Using a bit of algebra, you can check that

$$n = (1.96)^2 \hat{p}(1 - \hat{p})/(0.04)^2.$$

(If your algebra is rusty, don't worry—we'll show a general formula below.)

Now, we face a problem. We're doing this calculation *before* gathering the data, so we don't yet have a sample proportion $\hat{p}$. The formula for *n* depends on $\hat{p}$ because the standard error depends on it. Since $\hat{p}$ is unknown, we must substitute an educated guess for what we'll get once we gather the sample and analyze the data.

Since the latest poll *before* election day predicted that 58% of the voters preferred Levin, it is sensible to substitute 0.58 for $\hat{p}$ in this equation. Then we find

$$n = (1.96)^2 \hat{p}(1 - \hat{p})/(0.04)^2 = (1.96)^2 (0.58)(0.42)/(0.04)^2 = 584.9.$$

In summary, a random sample of size about *n* = 585 should give a margin of error of about 0.04 for a 95% confidence interval for the population proportion.

Insight

Sometimes, we may have no idea what to expect for $\hat{p}$. We may prefer not to guess the value it will take, as we did in this example. We'll next learn what we can do in those situations.

TO PRACTICE THIS CONCEPT, TRY EXERCISE 7.43

How Can We Select a Sample Size without Guessing a Value for $\hat{p}$?

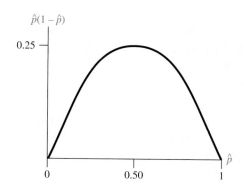

In the previous example for determining the sample size *n* for estimating a proportion, the solution for *n* was proportional to $\hat{p}(1 - \hat{p})$. The figure in the margin shows how that product depends on the value of $\hat{p}$. The largest possible value for $\hat{p}(1 - \hat{p})$ is 0.25, which occurs when $\hat{p} = 0.50$. You can check by plugging values of $\hat{p}$ into $\hat{p}(1 - \hat{p})$ that this product is near 0.25 unless $\hat{p}$ is quite far from 0.50. For example, $\hat{p}(1 - \hat{p}) = 0.24$ when $\hat{p} = 0.40$ or $\hat{p} = 0.60$.

In the formula for determining *n*, setting $\hat{p} = 0.50$ gives the largest value for *n* out of all the possible values to substitute for $\hat{p}$. So doing this is the "safe" approach that guarantees we'll have enough data. In the election exit-poll example above, for a margin of error of 0.04 we then get

$$n = (1.96)^2 \hat{p}(1 - \hat{p})/(0.04)^2$$
$$= (1.96)^2 (0.50)(0.50)/(0.04)^2 = 600.$$

This compares to *n* = 585 from guessing that $\hat{p} = 0.58$. Using the slightly larger value of *n* = 600 ensures that the margin of error for a 95% confidence interval will not exceed 0.04, no matter what value $\hat{p}$ takes once we collect the data.

This safe approach is not always sensible, however. Substituting $\hat{p} = 0.50$ gives us an overly large solution for *n* if $\hat{p}$ actually falls far from 0.50. Suppose that based on other studies we expect $\hat{p}$ to be about 0.10. Then an adequate sample size to achieve a margin of error of 0.04 is

$$n = (1.96)^2 \hat{p}(1 - \hat{p})/(0.04)^2$$
$$= (1.96)^2 (0.10)(0.90)/(0.04)^2 = 216.$$

A sample size of 600 would be much larger and more costly than needed.

General sample size formula for estimating a population proportion

A general formula exists for determining the sample size, based on solving algebraically for n by setting the margin of error formula equal to the desired value. Let m denote the desired margin of error. This is $m = 0.04$ in the above example. The general formula also uses the z-score for the confidence level.

Summary: Sample Size for Estimating a Population Proportion

The random sample size n for which a confidence interval for a population proportion p has margin of error m (such as $m = 0.04$) is

$$n = \frac{\hat{p}(1 - \hat{p})z^2}{m^2}.$$

The z-score is based on the confidence level, such as $z = 1.96$ for 95% confidence. You either guess the value you'd get for the sample proportion $\hat{p}$ based on other information or take the safe approach of setting $\hat{p} = 0.5$.

EXAMPLE 9

How Large Should an Exit Poll Be If a Race Is Close?

Picture the Scenario

An election is expected to be close. The pollsters who are planning an exit poll decide that a margin of error of 0.04 is too large.

Question to Explore

How large should the sample size be for the margin of error of a 95% confidence interval to equal 0.02?

Think It Through

Since the election is expected to be close, we expect $\hat{p}$ to be near 0.50. In the formula, we set $\hat{p} = 0.50$ to be safe. We also set the margin of error $m = 0.02$ and use $z = 1.96$ for a 95% confidence interval. The required sample size is

$$n = \frac{\hat{p}(1 - \hat{p})z^2}{m^2} = \frac{(0.50)(0.50)(1.96)^2}{(0.02)^2} = 2401.$$

The sample size of about 2400 is four times the sample size of 600 necessary to guarantee a margin of error of $m = 0.04$. *Reducing the margin of error by a factor of one-half requires quadrupling n.*

Insight

Example 22 in Chapter 6 described an exit poll for the U.S. senatorial race in 2002 between Hillary Clinton and Rick Lazio. That exit poll used a similar sample size ($n = 2232$) to the value of 2401 just determined. Did they achieve a margin of error of about 0.02? From that example, the sample proportion who voted for Clinton was 0.557. A 95% confidence interval for the population proportion p who voted for Clinton is $\hat{p} \pm 1.96(se)$, with $se = \sqrt{\hat{p}(1 - \hat{p})/n} = \sqrt{(0.557)(0.443)/2232} = 0.0105$. This gives the interval

$$0.557 \pm 1.96(0.0105), \text{ which is } 0.557 \pm 0.021, \text{ or } (0.536, 0.578).$$

In fact, the margin of error *was* 0.02 (rounded off). This was small enough to predict (correctly) that Clinton would win, since the confidence interval values (0.536, 0.578) all fell above 0.50.

TO PRACTICE THIS CONCEPT, TRY EXERCISE 7.42

Samples taken by polling organizations typically contain 1000–2000 subjects. This is large enough to estimate a population proportion with a margin of error of about 0.02 or 0.03. At first glance, it seems astonishing that a sample of this size from a population of perhaps many millions is adequate for predicting outcomes of elections, summarizing opinions on controversial issues, showing relative sizes of television audiences, and so forth. The basis for this inferential power lies in the standard error formulas for the point estimates, with random sampling. Good estimates result no matter how large the population size.[7]

Revisiting the approximation $1/\sqrt{n}$ for the margin of error Chapter 4 introduced a margin of error approximation of $1/\sqrt{n}$ for estimating a population proportion. What's the connection between this approximation and the more exact margin of error we've used in this chapter? Let's take the margin of error $1.96\sqrt{\hat{p}(1-\hat{p})/n}$ for a 95% confidence interval, round the z-score to 2 and replace the sample proportion by the value of 0.50 that gives the maximum possible standard error. Then we get margin of error

$$2\sqrt{0.50(0.50)/n} = 2(0.50)\sqrt{1/n} = 1/\sqrt{n}.$$

So for a 95% confidence interval, the margin of error is approximately $1/\sqrt{n}$ when $\hat{p}$ is near 0.50.

Choosing the Sample Size for Estimating a Population Mean

To derive the sample size for estimating a population mean, you set the margin of error equal to its desired value and solve for n. Recall that a 95% confidence interval for the population mean is

$$\bar{x} \pm t_{.025}(se), \text{ where } se = s/\sqrt{n}$$

and s is the sample standard deviation. If you don't know n, you also don't know the degrees of freedom and the t-score. However, we saw in Table B that when $df > 30$, the t-score for 95% confidence intervals is about 2.0, eventually growing closer to the z-score of 1.96 as it keeps growing. Setting $2.0(s/\sqrt{n})$ equal to a desired margin of error m and solving for n yields the following result:

Sample Size for Estimating a Population Mean

The random sample size n for which a 95% confidence interval for a population mean has margin of error approximately equal to m is

$$n = \frac{4s^2}{m^2}.$$

To use this formula, you guess the value you'll get for the sample standard deviation s (perhaps based on other studies).

In practice, since you don't yet have the data, you don't know the value of the sample standard deviation s. You must substitute an educated guess for s. Sometimes you can use the sample standard deviation from a similar study already conducted. The next example shows another sort of reasoning to form an educated guess.

---

[7]In fact, the mathematical derivations of these methods treat the population size as infinite. See Exercise 6.122.

EXAMPLE 10

Finding *n* to Estimate Mean Education in South Africa

Picture the Scenario

A social scientist plans a study of adult South Africans living in townships on the outskirts of Cape Town, to investigate educational attainment in the black community. Educational attainment is the number of years of education completed. Many of the study's potential subjects were forced to leave Cape Town in 1966 when the government passed a law forbidding blacks to live in the inner cities. Under the apartheid system, black South African children were not required to attend school, so some residents had very little education.

Question to Explore

How large a sample size is needed so that a 95% confidence interval for the mean number of years of attained education has margin of error equal to 1 year?

Think It Through

No prior information is stated about the standard deviation of educational attainment for the township residents. As a crude approximation, we might guess that the sample education values will fall within a range of about 18 years, such as between 0 and 18 years. If the data distribution is bell-shaped, the range from $\bar{x} - 3s$ to $\bar{x} + 3s$ will contain all or nearly all the distribution. Since the distance from $\bar{x} - 3s$ to $\bar{x} + 3s$ equals $6s$, the range of 18 years would equal about $6s$. Then, solving $18 = 6s$ for s, $18/6 = 3$ is a crude guess for s. So we'd expect a sample standard deviation value of about $s = 3$.

The desired margin of error is $m = 1$ year. The required sample size is

$$n = \frac{4s^2}{m^2} = \frac{4(3^2)}{1^2} = 36.$$

We need to randomly sample about 36 subjects for a 95% confidence interval for mean educational attainment to have a margin of error of 1 year.

Insight

A more cautious approach would select the *largest* value for the standard deviation that is plausible. This will give the largest sensible guess for how large *n* needs to be. For example, we could reasonably predict that *s* will be no greater than 4, since a range of six standard deviations then extends over 24 years. Then we get $n = 4(4^2)/(1^2) = 64$. If we collect the data and the sample standard deviation is actually less than 4, we will have more data than we need. The margin of error will be even less than 1.0.

Recall

Section 2.4 noted that for an approximately symmetric, bell-shaped distribution, we can approximate the standard deviation by roughly a sixth of the range. ■

TO PRACTICE THIS CONCEPT, TRY EXERCISE 7.48

If we plan to use a confidence level other than 95%, we'd replace 4 in the formula for n by the square of the approximate t-score for that confidence. For large n, recall that the t-score is approximately equal to the z-score in the last row of Table B.

What Other Factors Affect the Choice of the Sample Size?

We've looked at two factors that play a role in determining a study's sample size.

- The first is the desired *precision*, as measured by the *margin of error m*.
- The second is the *confidence level*, which determines the z-score or t-score in the sample size formulas.

  Other factors also play a role.

- A third factor is the *variability* in the data.

  Let's look at the formula $n = 4s^2/m^2$ for the sample size for estimating a mean. The greater the value expected for the standard deviation s, the larger the sample size needed. If subjects have little variation (that is, s is small), we need fewer data than if they have substantial variation. Suppose a study plans to estimate the mean level of education in several countries. Western European countries have relatively little variation, as students are required to attend school until the middle teen years. To estimate the mean to within a margin of error of $m = 1$, we need fewer observations than in South Africa.

- A fourth factor is *financial*.

  Larger samples are more time consuming to collect. They may be more expensive than a study can afford. Cost is often a major constraint. You may need to ask, "Should we go ahead with the smaller sample that we can afford, even though the margin of error will be greater than we would like?"

What If You Have to Use a Small n?

Sometimes, because of financial or ethical reasons, it's just not possible to take as large of a sample as we'd like. For example, each observation may result from an expensive experimental procedure. A consumer group that estimates the mean repair cost after a new-model automobile crashes into a concrete wall at 30 miles per hour would probably not want to crash a large sample of cars so they can get a narrow confidence interval!

If n must be small, how does that affect the validity of the confidence interval methods? The t methods for a mean are valid *for any n*. When n is small, though, you need to be extra cautious to look for extreme outliers or great departures from the normal population assumption (such as implied by highly skewed data). These can affect the results and the validity of using the mean as a summary of center.

For the confidence interval formula for a proportion, we've seen that we need at least 15 successes and at least 15 failures. Why? Otherwise, the central limit theorem no longer applies. The numbers of successes and failures must both be at least 15 for the normal distribution to approximate well the binomial distribution or the sampling distribution of a sample proportion. An even more serious difficulty is that the exact standard error depends on the parameter we're trying to estimate. If the estimate $\hat{p}$ is far from p, as often happens for small n, then so may be the estimate $se = \sqrt{\hat{p}(1 - \hat{p})/n}$ of the exact standard error. As a result, the confidence interval formula works poorly, as we'll see in the next example.

EXAMPLE 11

What Proportion of Students Are Vegetarians?

Picture the Scenario

In recent years, vegetarianism has gone from being viewed as a slightly eccentric fringe practice to receiving mainstream attention. Seemingly more and more people cite health, animal rights, and religious reasons for becoming vegetarians. Yet a cover story of the subject in *Time* magazine (July 15, 2002) quoted a poll of 10,000 Americans in which only 4% said they were vegetarians.

For a recent class project, a student randomly sampled 20 fellow students at the University of Florida (UF) to estimate the proportion at that university who were vegetarians. Of the 20 students she sampled, none were vegetarians.

Question to Explore

What does a 95% confidence interval tell her about the proportion of vegetarians among all 49,000 students at UF?

Think It Through

Let p denote the population proportion of vegetarians at the university. The sample proportion was $\hat{p} = 0/20 = 0.0$. When $\hat{p} = 0.0$, then $se = \sqrt{\hat{p}(1 - \hat{p})/n} = \sqrt{0.0(1.0)/20} = 0.0$. The 95% confidence interval for the population proportion of vegetarians is

$$\hat{p} \pm z(se) = 0.0 \pm 1.96(0.0), \text{ which is } 0.0 \pm 0.0, \text{ or } (0.0, 0.0).$$

This investigation told the student that she can be 95% confident that p falls between 0 and 0, that is, that $p = 0$. She was surprised by this result. It seemed unrealistic to conclude that *no* student at the university is a vegetarian.

Insight

Do you trust this inference? Just because no one in a small sample is a vegetarian, would you conclude that no one in the much larger *population* is a vegetarian? We doubt it. Perhaps the population proportion p is close to 0, but it is almost surely not exactly equal to 0.

The confidence interval formula $\hat{p} \pm z\sqrt{\hat{p}(1 - \hat{p})/n}$ is valid only if the sample has at least 15 vegetarians and at least 15 nonvegetarians. The sample did not have at least 15 vegetarians, so we have to use a different method.

TO PRACTICE THIS CONCEPT, TRY EXERCISE 7.51(A)-(C)

Confidence Interval for a Proportion with Small Samples

Constructing a Small-Sample Confidence Interval for a Proportion p

Suppose a random sample does *not* have at least 15 successes and 15 failures. The confidence interval formula $\hat{p} \pm z\sqrt{\hat{p}(1 - \hat{p})/n}$ still is valid if we use it after adding 2 to the original number of successes and 2 to the original number of failures. This results in adding 4 to the sample size n.

The sample of size $n = 20$ in Example 11 had 0 vegetarians and 20 non-vegetarians. We can apply the confidence interval formula with $0 + 2 = 2$ vegetarians and $20 + 2 = 22$ nonvegetarians. The value of the sample size for the formula is then $n = 24$. *Now we can use the formula, even though we don't have at least 15 vegetarians.* We get

$$\hat{p} = 2/24 = 0.083, \, se = \sqrt{\hat{p}(1 - \hat{p})/n} = \sqrt{(0.083)(0.917)/24} = 0.056.$$

The resulting 95% confidence interval is

$$\hat{p} \pm 1.96(se), \text{ which is } 0.083 \pm 1.96(0.056), \text{ or } (-0.03, 0.19).$$

A proportion cannot be negative, so we report the interval as $(0.0, 0.19)$. We can be 95% confident that the proportion of vegetarians at the University of Florida is no greater than 0.19.

This approach enables us to use a large-sample method even when the sample size is small. With it, the point estimate moves the sample proportion a bit toward $1/2$ (e.g., from 0.0 to 0.083). This is particularly helpful when the ordinary sample proportion is 0 or 1, which we would not usually expect to be a believable estimate of a population proportion. Why do we add 2 to the counts of the two types? The reason is that the confidence interval then approximates one based on a more complex method (described in Exercise 7.115) that does not require estimating the exact standard error.[8]

TO PRACTICE THIS CONCEPT, TRY EXERCISE 7.51(D)

[8]See article by A. Agresti and B. Coull (who proposed this small-sample confidence interval), *The American Statistician*, vol. 52, pp. 119–126, 1998.

7.4 Practicing the Basics

7.42 South Africa study: The researcher planning the study in South Africa also will estimate the population proportion having at least a high school education. No information is available about its value. How large a sample size is needed to estimate it to within 0.07 with 95% confidence?

7.43 Binge drinkers: A study in 2001 at the Harvard School of Public Health found that 44% of 10,000 sampled college students were binge drinkers. A student at the University of Minnesota plans to estimate the proportion of college students at that school who are binge drinkers. How large a random sample would she need to estimate it to within 0.05 with 95% confidence, if before conducting the study she uses the Harvard study results as a guideline?

7.44 Abstainers: The Harvard study mentioned in the previous exercise estimated that 19% of college students abstain from drinking alcohol. To estimate this proportion in your school, how large a random sample would you need to estimate it to within 0.05 with probability 0.95, if before conducting the study

  a. You are unwilling to predict the proportion value at your school.
  b. You use the Harvard study as a guideline.

  c. Use the results from (a) and (b) to explain why strategy (a) is inappropriate if you are quite sure you'll get a sample proportion that is far from 0.50.

7.45 How many businesses fail?: A study is planned to estimate the proportion of businesses started in the year 2000 that had failed within five years of their startup. How large a sample size is needed to guarantee estimating this proportion correct to within

  a. 0.10 with probability 0.95?
  b. 0.05 with probability 0.95?
  c. 0.05 with probability 0.99?
  d. Compare sample sizes for parts (a) and (b), and (b) and (c), and summarize the effects of decreasing the margin of error and increasing the confidence level.

7.46 Canada and the death penalty: A poll in Canada in 1998 indicated that 48% of Canadians favor imposing the death penalty (Canada does not have it). A report by Amnesty International on this and related polls (www.amnesty.ca) did not report the sample size but stated "Polls of this size are considered to be accurate within 2.5 percentage points 95% of the time." About how large was the sample size?

7.47 **Farm size:** An estimate is needed of the mean acreage of farms in Ontario, Canada. A 95% confidence interval should have a margin of error of 25 acres. A study ten years ago in this province had a sample standard deviation of 200 acres for farm size.

a. About how large a sample of farms is needed?

b. A sample is selected of the size found in (a). However, the sample has a standard deviation of 300 acres, rather than 200. What is the margin of error for a 95% confidence interval for the mean acreage of farms?

7.48 **Income of Native Americans:** How large a sample size do we need to estimate the mean annual income of Native Americans in Onondaga County, New York, correct to within $1000 with probability 0.99? No information is available to us about the standard deviation of their annual income. We guess that nearly all of the incomes fall between $0 and $120,000 and that this distribution is approximately bell shaped.

7.49 **Population variability:** Explain the reasoning behind the following statement: In studies about a very diverse population, large samples are often necessary, whereas for more homogeneous populations smaller samples are often adequate. Illustrate for the problem of estimating mean income for all medical doctors in the United States compared to estimating mean income for all entry-level employees at McDonald's restaurants in the United States.

7.50 **Web survey to get large *n*:** A newspaper wants to gauge public opinion about legalization of marijuana. The sample size formula indicates that they need a random sample of 875 people to get the desired margin of error. But surveys cost money, and they can only afford to randomly sample 100 people. Here's a tempting alternative: If they place a question about that issue on their Web site, they will get more than 1000 responses within a day at little cost. Are they better off with the random sample of 100 responses or the Website volunteer sample of more than 1000 responses? (*Hint*: Think about the issues discussed in Section 4.2 about proper sampling of populations.)

7.51 **Do you like tofu?** You randomly sample five students at your school to estimate the proportion of students who like tofu. All five students say they like it.

a. Find the sample proportion who like it.

b. Find the standard error. Does its usual interpretation make sense?

c. Find a 95% confidence interval, using the large-sample formula. Is it sensible to conclude that *all* students at your school like tofu?

d. Why is it not appropriate to use the ordinary large-sample confidence interval in (c)? Use a more appropriate approach, and interpret the result.

7.52 **Alleviate PMS?:** A pharmaceutical company proposes a new drug treatment for alleviating symptoms of PMS (premenstrual syndrome). In the first stages of a clinical trial, it was successful for 7 out of 10 women.

a. Construct a 95% confidence interval for the population proportion.

b. Is it plausible that it's successful for only half the population? Explain.

7.53 **Accept a credit card?:** A bank wants to estimate the proportion of people who would agree to take a credit card they offer if they send a particular mailing advertising it. For a trial mailing to a random sample of 100 potential customers, 0 people accept the offer. Can they conclude that fewer than 10% of their population of potential customers would take the credit card? Answer by finding a 95% confidence interval.

7.5 | How Do Computers Make New Estimation Methods Possible?

We've seen how to construct point and interval estimates of a population proportion and a population mean. Confidence intervals are relatively simple to construct for these parameters. For some parameters, it's not so easy because it's difficult to derive the sampling distribution or the standard error of a point estimate. We'll now introduce a relatively new simulation method for constructing a confidence interval that statisticians often use for such cases.

The Bootstrap: Using Simulation to Construct a Confidence Interval

When it is difficult to derive a standard error or a confidence interval formula that works well, you can "pull yourself up by your bootstraps" to attack the problem without using mathematical formulas. A recent computational invention, actually called the **bootstrap**, does just that.

The bootstrap is a simulation method that resamples from the observed data. It treats the data distribution as if it were the population distribution. You resample, *with replacement, n* observations from the data distribution. Each of the original n data points has probability $1/n$ of selection for each "new" observation. For this new sample of size n, you construct the point estimate of the parameter. You then resample another set of n observations from the original data distribution and construct another value of the point estimate. In the same way, you repeat this resampling process (using a computer) from the original data distribution a very large number of times, for instance, selecting 10,000 separate samples of size n and calculating 10,000 corresponding values of the point estimate.

This spread of resampled point estimates provides information about the accuracy of the original point estimate. For instance, a 95% confidence interval for the parameter is the 95% central set of the resampled point estimate values. These are the ones that fall between the 2.5th percentile and 97.5th percentile of those values. This takes a lot of computation, but it is simple with modern computing power.

EXAMPLE 12

How Variable Are Your Weight Readings on a Scale?

Picture the Scenario

Instruments used to measure physical characteristics such as weight and blood pressure do not give the same value every time they're used in a given situation. The measurements vary. One of the authors recently bought a scale (called "Thinner") that is supposed to give precise weight readings. To investigate how much the weight readings tend to vary, he weighed himself ten times, taking a 30-second break after each trial to allow the scale to reset. He got the values (in pounds):

160.2, 160.8, 161.4, 162.0, 160.8, 162.0, 162.0, 161.8, 161.6, 161.8.

These ten trials have a mean of 161.44 and standard deviation of 0.63.

Since weight varies from trial to trial, it has a probability distribution. You can regard this distribution as describing "long-run population" values that you would get if you could conduct a huge number of weight trials. The sample mean and standard deviation estimate the center and spread of the distribution. Ideally, you would like the scale to be precise and give the same value every time. Then the population standard deviation would be 0.0, but it's not that precise in practice.

How could we get a confidence interval for the population standard deviation? The sample standard deviation s has an approximate normal distribution for very large n. However, its standard error is highly sensitive to any assumption we make about the shape of the population distribution. It is safer to use the bootstrap method to construct a confidence interval.

Question to Explore

Using the bootstrap method, find a 95% confidence interval for the population standard deviation.

Think It Through

We sample from a distribution that has probability 1/10 at each of the values in the sample. For each new observation this corresponds to selecting a random digit and making the observation 160.2 if we get 0, 160.8 if we get 1, ..., 161.8 if we get 9. This can be done with software. The bootstrap with 100,000 re-samples of the data uses the following steps:

■ Randomly sample 10 observations from this sample data distribution. We did this and got the 10 new observations

160.8, 161.4, 160.2, 161.8, 162.0, 161.8, 161.6, 161.8, 161.6, 160.2.

For the 10 new observations, the sample standard deviation is $s = 0.67$.

■ Proceed as in the preceding step, taking 100,000 separate re-samples of size 10. This gives us 100,000 values of the sample standard deviation. Figure 7.11 shows a histogram of their values.

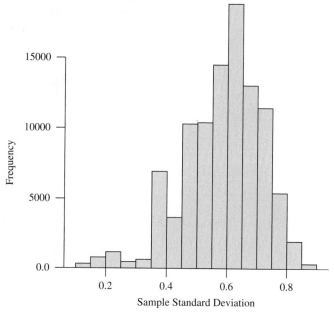

▲ **FIGURE 7.11: A Bootstrap Frequency Distribution of Standard Deviation Values.** These were obtained by taking 100,000 samples of size 10 each from the sample data distribution. **Question:** What is the practical reason for using the bootstrap method?

■ Now identify the middle 95% of these 100,000 sample standard deviation values. For the 100,000 samples we took, the 2.5th percentile was 0.26 and the 97.5th percentile was 0.80. In other words, 95% of the re-samples had sample standard deviation values between 0.26 and 0.80. This is our 95% bootstrap confidence interval for the population standard deviation.

In summary, the 95% confidence interval for σ is (0.26, 0.80). A typical deviation of a weight reading from the mean might be rather large, nearly a pound.

Insight

Figure 7.11 is skewed and is quite irregular. This appearance is because of the small sample size ($n = 10$). Such simulated distributions take a more regular shape when n is large, usually becoming symmetric and bell shaped when n is sufficiently large.

TO PRACTICE THIS CONCEPT, TRY EXERCISE 7.55

The bootstrap method was invented by Brad Efron, a statistician at Stanford University, in 1979. It is now widely used, taking advantage of modern computing power. New statistical methods continue to be developed today. (The small-sample confidence interval for a proportion at the end of Section 7.4 was developed in 1998 by one of the authors with then Ph.D. student Brent Coull.) The ever-increasing power of computers is making new methods feasible that were inconceivable when confidence interval methods were first developed in the 1920s and 1930s.

Finally, a word of caution: In the estimation of parameters, "margin of error" refers to the size of error resulting from having data from a random sample rather than the population—what's called *sampling error*. This is the error that the sampling distribution describes in showing how close the estimate is likely to fall to the parameter. But that's not the only source of potential error. Data may be missing for a lot of the target sample, some observations may be recorded incorrectly by the data collector, and some subjects may not tell the truth. When errors like these occur, the actual confidence level may be much lower than advertised. Be skeptical about a claimed margin of error and confidence level unless you know that the study was well conducted and these other sources of error are negligible.

On the Shoulders of . . . Ronald A. Fisher

How do you conduct scientific inquiry, whether it be developing methods of experimental design, finding the best way to estimate a parameter, or answering specific questions such as which fertilizer works best?

R. A. Fisher. Fisher was the statistician most responsible for the statistical methods used to analyze data today.

Compared with other mathematical sciences, statistical science is a mere youth. The most important contributions were made between 1920 and 1940 by the British statistician Ronald A. Fisher (1890–1962). While working at an agricultural research station north of London, Fisher was the first statistician to show convincingly the importance of randomization in designing experiments. He also developed the theory behind point estimation and proposed several new statistical methods.

Fisher was involved in a wide variety of scientific questions, ranging from finding the best ways to plant crops to controversies (in his time) about whether smoking is harmful. He also did fundamental work in genetics and is regarded as a giant in that field as well. Fisher showed the importance of natural selection, above and beyond its manifestation in evolution. For example, he showed that correlations between relatives could be used to make inference about genetic properties such as dominance.

Fisher had strong disagreements with others about the way statistical inference should be conducted. One of his main adversaries was Karl Pearson. Fisher corrected a major error Pearson made in proposing methods for contingency tables, and he also criticized Pearson's son's work on developing the theory of confidence intervals in the 1930s. Although Fisher often disparaged the ideas of other statisticians, he reacted strongly if anyone criticized him in return. Writing about Pearson, Fisher once said, "If peevish intolerance of free opinion in others is a sign of senility, it is one which he had developed at an early age."

7.5 Practicing the Basics

7.54 Why bootstrap?: Explain the purpose of using the bootstrap method.

7.55 Estimating variability: Refer to Example 12 about
[TRY] weight readings of a scale. For 10 successive trials on the next day, the weight values were

$$159.8, 159.8, 159.6, 159.0, 158.4,$$
$$159.2, 158.8, 158.4, 158.8, 159.0.$$

Explain the steps of how you could use the bootstrap method to get a 95% confidence interval for a "long-run" standard deviation of such values.

7.56 Bootstrap the proportion: We want a 95% confidence interval for the population proportion of students in a high school in Dallas, Texas who can correctly find Iraq on an unlabelled globe. For a random sample of size 50, 10 get the correct answer.

a. Using software or the *sampling distributions* applet on the text CD, set the population menu to "binary" and treat the sample proportion as the population proportion by setting the proportion parameter to $0.20 = 10/50$. Take a random sample of size 50, and find the sample proportion of correct answers.

b. Take 100 resamples like the one in (a), each time calculating the sample proportion. Take one sample at a time, recording each sample proportion. Now, construct a 90% confidence interval by identifying the 5th and 95th percentiles of the sample proportions. This is the 90% bootstrap confidence interval.

c. Explain why the sample proportion does not fall exactly in the middle of the bootstrap confidence interval. (*Hint:* Is the sampling distribution symmetric or skewed?)

CHAPTER REVIEW

Answers to the Chapter Figure Questions

Figure 7.1: A point estimate alone will not tell us how close the estimate is likely to be to the parameter.

Figure 7.2: When the random sample size is relatively large, by the central limit theorem.

Figure 7.3: We don't know the value of p, the population proportion, to form the interval $p \pm 1.96(se)$. The population proportion p is what we're trying to estimate.

Figure 7.4: Having greater confidence means that we want to have greater assurance of a correct inference. Thus, it is natural that we would expect a wider interval of believable values for the population parameter.

Figure 7.5: The $\hat{p}$ values falling in the darker shaded left and right tails of the bell-shaped curve.

Figure 7.6: This skew should not affect the validity of the confidence interval for the mean because of the large sample size ($n = 905$).

Figure 7.7: Using Table B or Table 7.5, $t = 1.96$ when $df = \infty$ with right-tail probability $= 0.025$.

Figure 7.8: $t = -3.707$ and $t = 3.707$.

Figure 7.9: The do not buy-it-now option has greater variability and more skew than the buy-it-now option. This is largely influenced by the value $178. The centers of the two distributions seem to be similar.

Figure 7.10: The sample size n is sufficiently large such that $np \geq 15$ and $n(1 - p) \geq 15$.

Figure 7.11: The bootstrap method is used when it is difficult to derive a standard error or confidence interval formula by using mathematical techniques.

CHAPTER SUMMARY

We've now learned how to **estimate** the population proportion p for categorical variables and the population mean μ for quantitative variables.

- A **point estimate** (or "**estimate**," for short) is our best guess for the unknown parameter value. An estimate of the population mean μ is the sample mean $\bar{x}$. An estimate of the population proportion p is the sample proportion $\hat{p}$.

- A **confidence interval** contains the most plausible values for a parameter. Confidence intervals for most parameters have the form

  Estimate ± margin of error,

  which is estimate ± (z- or t- score) × (se),

where se is the standard error of the estimate. For the proportion, the score is a z-score from the normal distribution. For the mean, the score is a *t*-score from the **t distribution** with degrees of freedom **df = n − 1.** The *t*-score is similar to a *z*-score when $df \geq 30$. Table 7.6 summarizes the point and interval estimation methods.

■ The *z*- or *t*-score depends on the **confidence level,** the probability that the method produces a confidence interval that contains the population parameter value. For a proportion, for instance, since a probability of 0.95 falls within 1.96 standard errors of the center of the normal sampling distribution, we use $z = 1.96$ for 95% confidence. *To achieve greater confidence, we make the sacrifice of a larger margin of error and wider confidence interval.*

■ For estimating a **mean,** the **t distribution** accounts for the extra variability due to using the sample standard deviation *s* to estimate the population standard deviation in finding a standard error. The *t* method assumes that the population distribution is normal. This ensures that the sampling distribution of $\bar{x}$ is bell shaped. This assumption is mainly important for small *n*, because

when *n* is large the central limit theorem guarantees that the sampling distribution is bell shaped.

■ For estimating a **proportion,** the formulas rely on the central limit theorem. For large random samples, this guarantees that the sample proportion has a normal sampling distribution. For estimating a proportion with small samples (fewer than 15 successes or fewer than 15 failures), the confidence interval formula $\hat{p} \pm z(se)$ still applies if we use it after adding 2 successes and 2 failures (and add 4 to *n*).

■ Before conducting a study, we can **determine the sample size n** having a certain margin of error. Table 7.6 shows the sample size formulas. To use them, we must (1) select the margin of error *m*, (2) select the confidence level, which determines the *z*-score or *t*-score, and (3) guess the value the data will have for the sample standard deviation *s* (to estimate a population mean) or the sample proportion $\hat{p}$ (to estimate a population proportion). In the latter case, substituting $\hat{p} = 0.50$ guarantees that the sample size is large enough regardless of the value the sample has for $\hat{p}$.

Table 7.6: Estimation Methods for Means and Proportions

Parameter	Point Estimate	Standard Error	Confidence Interval	Sample Size for Margin of Error *m*
Proportion *p*	$\hat{p}$	$se = \sqrt{\hat{p}(1-\hat{p})/n}$	$\hat{p} \pm z(se)$	$n = [\hat{p}(1-\hat{p})z^2]/m^2$
Mean μ	$\bar{x}$	$se = s/\sqrt{n}$	$\bar{x} \pm t(se)$	$n = (4s^2)/m^2$

Note: The *z*- or *t*-score depends on the confidence level. The *t*-score has $df = n − 1$. The sample size formula for a mean applies for a 95% confidence interval.

SUMMARY OF NOTATION

se = standard error
$\hat{p}$ = sample proportion
m = margin of error

$t_{.025}$ = *t*-score with right-tail probability 0.025
df = degrees of freedom
$\quad$ ($= n − 1$ for inference about a mean)

CHAPTER PROBLEMS

Practicing the Basics

7.57 **Divorce and age of marriage:** A July 2002 report published by the Centers for Disease Control estimated that for non-Hispanic white women, the probability that their first marriage breaks up within 10 years is 0.50 for those whose marriage occurred before their 18th birthday, but only 0.22 for those whose marriage occurred at age 25 and over.

    a. Are these point estimates or interval estimates?

    b. Is the information given here sufficient to allow you to construct confidence intervals? Why or why not?

7.58 **Snoring and insomnia:** In 2002, the National Sleep Foundation (www.sleepfoundation.org) conducted 1010 phone interviews using a random sampling of phone numbers. Of the respondents, 37% reported problems with snoring and 58% reported problems with insomnia. One way to report the latter result says, "It is estimated that 58% of people have problems with insomnia. This estimate has a margin of error of plus or minus 3%." How could you explain what this means to someone who has not taken a statistics course?

7.59 **British monarchy:** In February 2002, the Associated Press quoted a survey of 3000 British residents conducted by YouGov.com. It stated, "Only 21 percent wanted to see the monarchy abolished, but 53% felt it should become more democratic and approachable. No margin of error was given." If the sample was random, find the 95% margin of error for each of these estimated proportions.

7.60 **Born again:** A poll of a random sample of $n = 2000$ Americans by the Pew Research Center (www.people-press.org) in March 2002 indicated that 36% considered themselves "born-again" or evangelical Christians. How would you explain to someone who has not studied statistics:

    a. What it means to call this a *point estimate.*

    b. Why this does not mean that *exactly* 36% of *all* Americans are born-again or evangelical Christians.

7.61 **Life after death:** The variable POSTLIFE in the 2002 General Social Survey asked "Do you believe in life after death?" Of 1211 respondents, 975 answered yes. A report based on these data stated that "81% of Americans believe in life after death. The margin of error for this result is plus or minus 2%." Explain how you could form a 95% confidence interval using this information, and interpret that confidence interval in context.

7.62 Female belief in life after death: Refer to the previous exercise. The following printout shows results for the females in the sample, where X = the number answering yes. Explain how to interpret each item, in context.

```
Sample     X      N    Sample p      95.0% CI
Females   550    648    0.84844    (0.821, 0.876)
```

7.63 Vegetarianism: *Time* magazine (July 15, 2002) quoted a poll of 10,000 Americans in which only 4% said they were vegetarians.

  a. What has to be assumed about this sample in order to construct a confidence interval for the population proportion of vegetarians?

  b. Construct a 99% confidence interval for the population proportion. Explain why the interval is so narrow, even though the confidence level is high.

  c. In interpreting this confidence interval, can you conclude that fewer than 10% of Americans are vegetarians? Explain your reasoning.

7.64 Alternative therapies: The Department of Public Health at the University of Western Australia conducted a survey in which they randomly sampled general practitioners in Australia.[9] One question asked whether the GP had ever studied alternative therapy, such as acupuncture, hypnosis, homeopathy, and yoga. Of 282 respondents, 132 said yes. Is the interpretation "We are 95% confident that the percentage of all GPs in Australia who have ever studied alternative therapy equals 46.8%" correct or incorrect? Explain.

7.65 Population data: You would like to find the proportion of bills passed by Congress that were vetoed by the president in the last congressional session. After checking congressional records, you see that for the population of all 40 bills passed, 15 were vetoed. Does it make sense to construct a confidence interval using these data? Explain. (*Hint*: Identify the sample and population.)

■ 7.66 Wife supporting husband: Consider the statement that it is more important for a wife to help her husband's career than to have one herself, from the GSS (variable denoted FEHELP).

  a. Go to the Web site, sda.berkeley.edu/GSS. In 1998, find the number who agreed with that statement and the sample size.

  b. Show that the sample proportion was 0.192, with standard error 0.0092.

  c. Show that a 99% confidence interval for the population proportion who would agree is 0.192 ± 0.024, or (0.168, 0.216), and interpret it.

■ 7.67 Legalize marijuana? The General Social Survey has asked respondents, "Do you think the use of marijuana should be made legal or not?" Go to the GSS Web site, sda.berkeley.edu/GSS. For the 2002 survey with variable GRASS:

  a. Of the $n = 851$ respondents, how many said "legal" and how many said "not legal"? Report the sample proportions.

  b. Is there enough evidence to conclude whether a majority or a minority of the population support legalization? Explain your reasoning.

  c. Now look at the data on this variable for all years by entering YEAR as the column variable. Describe any trend you see over time in the proportion favoring legalization.

7.68 Smoking: A report in 2001 by the U.S. National Center for Health Statistics provided an estimate of 23% for the percentage of Americans over the age of 18 who were currently smokers. The sample size was 38,633. Assuming that this sample has the characteristics of a random sample, a 99.999999% confidence interval for the proportion of the population who were smokers is (0.23, 0.24). When the sample size is extremely large, explain why even confidence intervals with large confidence levels are narrow.

7.69 Nondrinkers: Refer to the previous exercise. The same study provided the following results for estimating the proportion of adult Americans who have had fewer than 12 drinks in their lifetime.

```
Sample      X        N      Sample p     95.0% CI
Nodrink   8654    38633    0.2240    (0.220, 0.228)
```

Explain how to interpret all results on this printout, in context.

7.70 Travel to work: As part of the 2000 census, the Census Bureau surveyed 700,000 households to study transportation to work. They reported that 76.3% drove alone to work, 11.2% carpooled, 5.1% took mass transit, 3.2% worked at home, 0.4% bicycled, and 3.8% took other means.

  a. With such a large survey, explain why the margin of error for any of these values is extremely small.

  b. The survey also reported that the mean travel time to work was 24.3 minutes, compared to 22.4 minutes in 1990. Explain why this is not sufficient information to construct a confidence interval for the population mean. What else would you need?

7.71 t-scores:

  a. Show how the *t*-score for a 95% confidence interval changes as the sample size increases from 10 to 20 to 30 to infinity.

  b. What does the answer in (a) suggest about how the *t* distribution compares to the standard normal distribution?

7.72 Buddhists watching TV: In a recent GSS, the responses about the number of hours daily spent watching TV for the five subjects who identified themselves as Buddhists were

0, 5, 0, 1, 2.

---

⁹This was reported at www.internethealthlibrary.com/Surveys/

a. Find the mean, standard deviation, and standard error.

b. Construct a 95% confidence interval for the population mean.

c. Specify the assumptions for the method. What can you say about their validity for these data?

7.73 Psychologists' income: In 1999, the American Psychological Association conducted a survey of a random sample of psychologists to estimate mean incomes for psychologists of various types (research.apa.org). Of the 10 clinical psychologists with 5–9 years experience who were in a medical psychological group practice, the mean income was $78,500 with a standard deviation of $33,287.

a. Construct a 95% confidence interval for the population mean. Interpret.

b. What assumption about the population distribution of psychologists' incomes does the confidence interval method make?

c. If the assumption about the shape of the population distribution is not valid, does this invalidate the results? Explain.

7.74 More psychologists: Refer to the previous exercise. Interpret each item on the following printout that software reports for the clinical psychologists with 5–9 years experience who worked in a rehabilitation facility:

```
Variable N  Mean  StDev  SE Mean    95.0% CI
income   8 57625 12614  4459.7  (47079.4, 68170.6)
```

7.75 How long lived in town?: The General Social Survey has asked subjects, "How long have you lived in the city, town or community where you live now?" The responses of 1415 subjects in one survey had a mode of less than one year, a median of 16 years, a mean of 20.3 and a standard deviation of 18.2.

a. Do you think that the population distribution is normal? Why or why not?

b. Based on your answer in (a), can you construct a 95% confidence interval for the population mean? If not, explain why not. If so, do so and interpret.

7.76 How often do women feel sad?: A recent GSS asked, "How many days in the past 7 days have you felt sad?" The 816 women who responded had a median of 1, mean of 1.81, and standard deviation of 1.98. The 633 men who responded had a median of 1, mean of 1.42, and standard deviation of 1.83.

a. Find a 95% confidence interval for the population mean for women. Interpret.

b. Do you think that this variable has a normal distribution? Does this cause a problem with the confidence interval method in (a)? Explain.

7.77 How often feel sad?: Refer the previous exercise. This question was asked of 10 students in a class at the University of Wisconsin recently. The responses were

$$0, 0, 1, 0, 7, 2, 1, 0, 0, 3.$$

Find and interpret a 90% confidence interval for the population mean, and indicate what you would have to assume for this inference to apply to the population of all University of Wisconsin students.

■ 7.78 Happy often?: The 1996 GSS asked, "How many days in the past 7 days have you felt happy?"

a. Using the GSS variable HAPFEEL, verify that the sample had a mean of 5.27 and a standard deviation of 2.05. What was the sample size?

b. Find the standard error for the sample mean.

c. Stating assumptions, construct and interpret a 95% confidence interval for the population mean. Can you conclude that the population mean is at least 5.0?

■ 7.79 Revisiting mountain bikes: Use the "Mountain Bike" data file from the text CD, shown also below.

a. Form a 95% confidence interval for the population mean price of all mountain bikes. Interpret.

b. What assumptions are made in forming the interval in part (a)? State at least one important assumption that does not seem to be satisfied, and indicate its impact on this inference.

Mountain Bikes	
Brand and Model	**Price($)**
Trek VRX 200	1000
Cannondale SuperV400	1100
GT XCR-4000	940
Specialized FSR	1100
Trek 6500	700
Specialized Rockhop	600
Haro Escape A7.1	440
Giant Yukon SE	450
Mongoose SX 6.5	550
Diamondback Sorrento	340
Motiv Rockridge	180
Huffy Anorak 36789	140

■ 7.80 eBay selling prices: For eBay auctions of the Palm M515 PDA on the final day for which data were available for auctions of this product, the selling prices were

$$\$212, \$249, \$250, \$240, \$210, \$230, \$195, \$193.$$

a. Explain what a parameter μ might represent that you could estimate with these data.

b. Find the point estimate of μ.

c. Find the standard deviation of the data and the standard error of the sample mean. Interpret.

d. Find the 95% confidence interval for μ. Interpret the interval in context.

■ 7.81 Income for families in public housing: A survey is taken to estimate the mean annual family income for families living in public housing in Chicago. For a random sample of 29 families, the annual incomes (in hundreds of dollars) are as follows:

90 77 100 83 64 78 92 73 122 96 60 85 86 108
70 139 56 94 84 111 93 120 70 92 100 124 59 112 79

a. Construct a box plot of the incomes. What do you predict about the shape of the population distribution? Does this affect the possible inferences?

b. Using software, find point estimates of the mean and standard deviation of the family incomes of all families living in public housing in Chicago.

c. Obtain and interpret a 95% confidence interval for the population mean.

7.82 Females watching TV: The GSS asked in 2002 "On the average day about how many hours do you personally watch television?" Software reports the results for females,

```
Variable  N   Mean  StDev  SE Mean   95.0% CI
TV       506  3.061 2.122  0.0944  (2.88, 3.25)
```

a. Would you expect that TV watching has a normal distribution? Why or why not?

b. On what assumptions is the confidence interval shown based? Are any of them violated here? If so, is the reported confidence interval invalid? Explain.

c. What's wrong with the interpretation, "In the long run, 95% of the time females watched between 2.88 and 3.25 hours of TV a day."

7.83 Males watching TV: Refer to the previous exercise. The 399 males had a mean of 2.885 and a standard deviation of 2.633. The 95% confidence interval for the population mean is (2.63, 3.14). Interpret in context.

7.84 Working mother: In response to the statement on a recent General Social Survey, "A preschool child is likely to suffer if his or her mother works," the response categories (Strongly agree, Agree, Disagree, Strongly disagree) had counts (91, 385, 421, 99). Scores (2, 1, −1, −2) were assigned to the four categories, to treat the variable as quantitative. Software reported

```
Variable   N    Mean   StDev  SE Mean     95.0% CI
Response  996  −0.052  1.253  0.0397  (−0.130, 0.026)
```

a. Explain what this choice of scoring assumes about relative distances between categories of the scale.

b. Based on this scoring, how would you interpret the sample mean of −0.052?

c. Explain how you could also make an inference about proportions for these data.

7.85 Highest grade completed: The 2002 GSS asked, "What is the highest grade that you finished and got credit for?" (variable EDUC). Of 2753 respondents, the mean was 13.4, the standard deviation was 3.0, and the proportion who gave responses below 12 (i.e., less than a high school education) was 0.16. Explain how you could analyze these data by making inferences about a population mean or about a population proportion, or both. Show how to implement one of these types of inference with these data.

7.86 Interpreting an interval for μ: Refer to the previous exercise. For the 429 African Americans in the 2000 GSS, a 99% confidence interval for the mean of EDUC is (12.1, 12.6). Explain why the following interpretation is incorrect: 99% of all African Americans have completed grades between 12.1 and 12.6.

7.87 Sex partners in previous year: The 2002 General Social Surveys asked respondents how many sex partners they had in the previous 12 months (variable PARTNERS). Software summarizes the results of the responses by

```
Variable    N    Mean  StDev  SE Mean    95.0% CI
partners  2242  1.08   1.08   0.023   (1.03, 1.13)
```

a. Based on the reported sample size and standard deviation, verify the reported value for the standard error.

b. Based on these results, explain why the distribution was probably skewed to the right.

c. Explain why the skew need not cause a problem with constructing a confidence interval for the population mean, unless there are extreme outliers such as a reported value of 1000.

7.88 Men don't go to the doctor: A survey of 1084 men age 18 and older in 1998 for the Commonwealth Fund (www.cmwf.org) indicated that more than half did not have a physical exam or a blood cholesterol test in the past year. A medical researcher plans to sample men in her community randomly to see if similar results occur. How large a random sample would she need to estimate this proportion to within 0.05 with probability 0.95?

7.89 Driving after drinking: In December 2004, a report based on the National Survey on Drug Use and Health estimated that 20% of all Americans of ages 16 to 20 drove under the influence of drugs or alcohol in the previous year (AP, Dec. 30, 2004). A public health unit in Wellington, New Zealand, plans a similar survey for young people of that age in New Zealand. They want a 95% confidence interval to have a margin of error of 0.04.

a. Find the necessary sample size if they expect results similar to those in the United States.

b. Suppose that in determining the sample size, they use the safe approach that sets $\hat{p} = 0.50$ in the formula for n. Then, how many records need to be sampled? Compare this to the answer in (a). Explain why it is better to make an educated guess about what to expect for $\hat{p}$, when possible.

7.90 Changing views of United States: The June 2003 report on *Views of a Changing World*, conducted by the Pew Global Attitudes Project (www.people-press.org), discussed changes in views of the United States by other countries. In the largest Muslim nation, Indonesia, a poll conducted in May 2003 after the Iraq war began reported that 83% had an unfavorable view of America, compared to 36% a year earlier. The 2003 result was claimed to have a margin of error of 3 percentage points. How can you approximate the sample size the study was based on?

7.91 Mean property tax: A tax assessor wants to estimate the mean property tax bill for all homeowners in Madison, Wisconsin. A survey ten years ago got a sample mean and standard deviation of $1400 and $1000.

a. How many tax records should the tax assessor randomly sample in order for a 95% confidence interval for the mean to have a margin of error equal to $100? What assumption does your solution make?

b. In reality, suppose that they'd now get a standard deviation equal to $1500. Using the sample size you derived in (a), without doing any calculation, explain whether the margin of error for a 95% confidence interval would be less than $100, equal to $100, or more than $100.

c. Refer to (b). Would the probability that the sample mean falls within $100 of the population mean be less than 0.95, equal to 0.95, or greater than 0.95? Explain.

7.92 Kicking accuracy: A football coach decides to estimate the kicking accuracy of a player who wants to join the team. Of 10 extra point attempts, the player makes all ten.

a. Find a 95% confidence interval for the probability that the player makes any given extra point attempt.

b. What's the lowest value that you think is plausible for that probability?

c. How would you interpret the random sample assumption in this context? Describe a scenario such that it would not be sensible to treat these 10 kicks as a random sample.

Concepts and Investigations

7.93 Religious beliefs: A column by *New York Times* columnist Nicholas Kristof (August 15, 2003) discussed results of polls indicating that religious beliefs in the United States tend to be quite different from those in other Western nations. He quoted recent Gallup and Harris polls of random samples of about 1000 Americans estimating that 83% believe in the Virgin Birth of Jesus but only 28% believe in evolution. A friend of yours is skeptical, claiming that it's impossible to predict beliefs of over 200 million adult Americans with interviewing only 1000 of them. Write a one-page report in this context to show how you could explain about random sampling, the margin of error, and how a margin of error depends on the sample size.

7.94 TV watching and race: For the number of hours of TV watching, the 2000 GSS reported a mean of 2.75 for the 1435 white subjects, with a standard deviation of 2.25. The mean was 4.14 for the 305 black subjects, with a standard deviation of 3.56. Analyze these data, preparing a short report in which you mention the methods used and the assumptions on which they are based, and summarize and interpret your findings.

7.95 Housework and gender: Using data from the National Survey of Families and Households, a study (by S. South and G. Spitze, *American Sociological Review*, Vol. 59, 1994, pp. 327–347) reported the descriptive statistics in the following table for the hours spent in housework. Analyze these data. Summarize results in a short report, including assumptions you made to perform the inferential analyses.

Gender	Sample Size	Mean	Std Dev
Men	4252	18.1	12.9
Women	6764	32.6	18.2

7.96 Women's role opinions: When subjects in a recent GSS were asked whether they agreed with the following statements, the (yes, no) counts under various conditions were as follows:

- Women should take care of running their homes and leave running the country up to men: (275, 1556).

- It is better for everyone involved if the man is the achiever outside the home and the woman takes care of the home and the family: (627, 1208).

- A preschool child is likely to suffer if his or her mother works: (776, 1054).

Analyze these data. Prepare a one-page report stating assumptions, showing results of description and inference, and summarizing conclusions.

7.97 Types of estimates: An interval estimate for a mean is more informative than a point estimate, because with an interval estimate you can figure out the point estimate, but with the point estimate alone you have no idea how wide the interval estimate is. Explain why this statement is correct, illustrating using the reported 95% confidence interval of (4.0, 5.6) for the mean number of dates in the previous month based on a sample of women at a particular college.

7.98 Width of a confidence interval: Why are confidence intervals wider when we use larger confidence levels but narrower when we use larger sample sizes, other things being equal?

7.99 99.9999% confidence: Explain why confidence levels are usually large, such as 0.95 or 0.99, but not extremely large, such as 0.999999. (*Hint:* What impact does the extremely high confidence level have on the margin of error?)

7.100 Need 15 successes and 15 failures: To use the large-sample confidence interval for p, you need at least 15 successes and 15 failures. Show that the smallest value of n for which the method can be used is (a) 30 when $\hat{p} = 0.50$, (b) 50 when $\hat{p} = 0.30$, (c) 150 when $\hat{p} = 0.10$. That is, the overall n must increase as $\hat{p}$ moves toward 0 or 1. (When the true proportion is near 0 or 1, the sampling distribution can be highly skewed unless n is quite large.)

7.101 Outliers and CI: For the observations 0,0,2,2,2,4,6 on TV watching for the seven Muslims in the sample in 1998, a 95% confidence interval for the population mean for that group is (0.3, 4.3). Suppose the observation of 6 for the seventh subject was incorrectly recorded as 60. What would have been obtained for the 95% confidence interval? Compare to the interval (0.3, 4.3). How does this warn you about potential effects of outliers when you construct a confidence interval for a mean?

7.102 What affects n?: Using the sample size formula $n = [\hat{p}(1 - \hat{p})z^2]/m^2$ for a proportion, explain the effect on n of (a) increasing the confidence level, (b) decreasing the margin of error.

Select the best response in Exercises 7.103–7.106.

7.103 CI property: Increasing the confidence level causes the margin of error of a confidence interval to (a) increase, (b) decrease, (c) stay the same.

7.104 CI property 2: Other things being equal, increasing n causes the margin of error of a confidence interval to (a) increase, (b) decrease, (c) stay the same.

7.105 Number of close friends: Based on responses of 1467 subjects in a General Social Survey, a 95% confidence interval for the mean number of close friends equals (6.8, 8.0). Which two of the following interpretations are correct?

   a. We can be 95% confident that $\bar{x}$ is between 6.8 and 8.0.

   b. We can be 95% confident that μ is between 6.8 and 8.0.

   c. Ninety-five percent of the values of X = number of close friends (for this sample) are between 6.8 and 8.0.

   d. If random samples of size 1467 were repeatedly selected, then 95% of the time $\bar{x}$ would be between 6.8 and 8.0.

   e. If random samples of size 1467 were repeatedly selected, then in the long run 95% of the confidence intervals formed would contain the true value of μ.

7.106 Why z?: The reason we use a z-score from a normal distribution in constructing a large-sample confidence interval for a proportion is that

   a. For large random samples the sampling distribution of the sample proportion is approximately normal.

   b. The population distribution is normal.

   c. For large random samples the data distribution is approximately normal.

   d. For any n we use the t distribution to get a confidence interval, and for large n the t distribution looks like the standard normal distribution.

7.107 Mean age at marriage: A random sample of 50 records yields a 95% confidence interval of 21.5 to 23.0 years for the mean age at first marriage of women in a certain county. Explain what is wrong with each of the following interpretations of this interval.

   a. If random samples of 50 records were repeatedly selected, then 95% of the time the sample mean age at first marriage for women would be between 21.5 and 23.0 years.

   b. Ninety-five percent of the ages at first marriage for women in the county are between 21.5 and 23.0 years.

   c. We can be 95% confident that $\bar{x}$ is between 21.5 and 23.0 years.

   d. If we repeatedly sampled the entire population, then 95% of the time the population mean would be between 21.5 and 23.5 years.

7.108 Interpret CI: For the previous exercise, provide the proper interpretation.

7.109 True or false 1: Suppose a 95% confidence interval for the population proportion of students at your school who regularly drink alcohol is (0.61, 0.67). The inference is that you can be 95% confident that the sample proportion falls between 0.61 and 0.67.

7.110 True or false 2: The confidence interval for a mean with a random sample of size $n = 2000$ is invalid if the population distribution is bimodal.

7.111 True or false 3: If you have a volunteer sample instead of a random sample, then a confidence interval for a parameter is still completely reliable as long as the sample size is larger than about 30.

7.112 Women's satisfaction with appearance: A special issue of *Newsweek* in March 1999 on women and their health reported results of a poll of 757 American women aged 18 or older. When asked, "How satisfied are you with your overall physical appearance?" 30% said very satisfied, 54% said somewhat satisfied, 13% said not too satisfied, and 3% said not at all satisfied. True or false: Since all these percentages are based on the same sample size, they all have the same margin of error.

◆◆7.113 Opinions over time about the death penalty: For many years, the General Social Survey has asked respondents whether they favor the death penalty for persons convicted of murder. Support has been quite high in the United States, one of few Western nations that currently has the death penalty. The following figure uses the 20 General Social Surveys taken between 1974 and 2000 and plots the 95% confidence intervals for the population proportion in the United States who supported the death penalty in each of the 20 years of these surveys.

Twenty 95% confidence intervals for the population proportions supporting the death penalty.

a. When we say we have "95% confidence" in the interval for a particular year, what does this mean?

b. For 95% confidence intervals constructed using data for 20 years, let X = the number of the intervals that contain the true parameter values. Find the probability that $x = 20$, that is, all 20 inferences are correct. (*Hint:* You can use the binomial distribution to answer this.)

c. Refer to (b). Find the mean of the probability distribution of X.

d. What could you do differently so it is more likely that all 20 inferences are correct?

◆◆ **7.114 Why called "degrees of freedom"?:** You know the sample mean $\bar{x}$ of n observations. Once you know $(n - 1)$ of the observations, show that you can find the remaining one. In other words, for a given value of $\bar{x}$, the values of $(n - 1)$ observations determine the remaining one. In summarizing scores on a quantitative variable, having $(n - 1)$ *degrees of freedom* means that only that many observations are independent. (If you have trouble with this, try to show it for $n = 2$, for instance showing that if you know that $\bar{x} = 80$ and you know that one observation is 90, then you can figure out the other observation. The *df* value also refers to the divisor in $s^2 = \Sigma(x - \bar{x})^2/(n - 1)$.)

◆◆ **7.115 Estimating p without estimating se:** The large-sample confidence interval for a proportion substitutes $\hat{p}$ for the unknown value of p in the exact standard error of $\hat{p}$. A less approximate 95% confidence interval has endpoints determined by the p values that are 1.96 standard errors from the sample proportion, without estimating the standard error. To do this, you solve for p in the equation

$$|\hat{p} - p| = 1.96\sqrt{p(1 - p)/n}.$$

a. For Example 11 with no vegetarians in a sample of size 20, substitute $\hat{p}$ and n in this equation and show that the equation is satisfied at $p = 0$ and at $p = 0.161$. So the confidence interval is $(0, 0.161)$, compared to $(0, 0)$ with $\hat{p} \pm 1.96(se)$.

b. Which confidence interval seems more believable? Why?

◆◆ **7.116 *m* and *n*:** Consider the sample size formula $n = [\hat{p}(1 - \hat{p})z^2]/m^2$ for estimating a proportion. When $\hat{p}$ is close to 0.50, for 95% confidence explain why this formula gives roughly $n = 1/m^2$.

◆◆ **7.117 Median as point estimate:** When the population distribution is normal, the population mean equals the population median. How good is the sample median as a point estimate of this common value? For a random sample, the estimated standard error of the sample median equals $1.25(s/\sqrt{n})$. If the population is normal, explain why the sample mean tends to be a better estimate than the sample median.

Class Explorations

◆◆ **7.118 Randomized response:** To encourage subjects to make honest responses on sensitive questions, the method of *randomized response* is often used. Let's use your class to estimate the proportion who have had alcohol at a party. Before carrying out this method, the class should discuss what they would guess for the value of the proportion of students in the class who have had alcohol at a party. Come to a class consensus. Now each student should flip a coin, in secret. If it is a head, toss the coin once more and report the outcome, head or tails. If the first flip is a tail, report instead the response to whether you have had alcohol, reporting the response *head* if the true response is *yes* and reporting the response *tail* if the true response is *no*. Let p denote the true probability of the *yes* response on the sensitive question.

a. Explain why the numbers in the following table are the probabilities of the four possible outcomes.

b. Let $\hat{q}$ denote the sample proportion of subjects who report *head* for the second response. Explain why we can set $\hat{q} = 0.25 + p/2$ and hence use $\hat{p} = 2\hat{q} - 0.5$ to estimate p.

c. Using this approach with your class, estimate the probability of having had alcohol at a party. Is it close to the class guess?

Table for Randomized Response

FIRST COIN	SECOND RESPONSE	
	Head	Tail
Head	0.25	0.25
Tail	$p/2$	$(1 - p)/2$

7.119 GSS project: The instructor will assign the class a theme to study. Download recent results for variables relating to that theme from sda.berkeley.edu/GSS. Find and interpret confidence intervals for relevant parameters. Prepare a two-page report summarizing results.

EXAMPLE 1

Are Astrology Predictions Better than Guessing?

Picture the Scenario

Astrologers believe that the positions of the planets and the moon at the moment of your birth determine your personality traits. But have you ever seen any scientific evidence that astrology works? One scientific "test of astrology" used the following experiment[1]: Each of 116 volunteers were asked to give their dates and times of birth. From this information, an astrologer prepared each subject's horoscope based on the positions of the planets and the moon at the moment of birth. Each volunteer also filled out a California Personality Index survey. Then the birth data and horoscope for one subject, together with the results of the personality survey for that individual and for two other participants randomly selected from the experimental group, were given to an astrologer. The astrologer was asked to predict which of the three personality charts matched the birth data and horoscope for the subject.

Let p denote the probability of a correct prediction by an astrologer. Suppose an astrologer actually has no special predictive powers, as would be expected by those who view astrology as "quack science." The predictions then merely correspond to random guessing, that is, picking one of the three personality charts at random, so $p = 1/3$. However, the participating astrologers claimed that $p > 1/3$. They felt they could predict better than with random guessing.

Questions to Explore

- How can we use data from such an experiment to summarize the evidence about the claim by the astrologers?
- How can we decide, based on the data, whether or not the claims are believable?

Thinking Ahead

In this chapter we'll learn how to use inferential statistics to answer these questions. An inferential method called a **significance test** analyzes the evidence that the data provide in favor of the astrologers' claim. We'll analyze data from the astrology experiment in Examples 3, 5, and 13.

Therapeutic touch is another practice that many believe is "quack science." Its practitioners claim to be able to heal many medical conditions by using their hands to manipulate a "human energy field" they perceive above the patient's skin. We'll analyze data from an experiment to investigate this claim in Examples 2, 6, and 14.

[1] S. Carlson, *Nature*, vol. 318, pp. 419–425, 1985.

The significance test is the second major method for making statistical inference about a population. Like a confidence interval for estimating a parameter (the first major method), this method uses probability to determine the likely parameter values while controlling the chance of an incorrect inference. With this method, we'll be able to use data to answer questions about astrology, therapeutic touch, and many other topics, such as the following:

■ Does a proposed diet truly result in weight loss, on the average?

■ Is there evidence of discrimination against women in promotion decisions?

■ Does one advertising method result in better sales, on the average, than another advertising method?

8.1 | What Are the Steps for Performing a Significance Test?

The main goal in many research studies is to check whether the data support certain statements or predictions. These statements are **hypotheses** about a population. They are usually expressed in terms of population parameters for variables measured in the study.

Hypothesis

In statistics, a **hypothesis** is a statement about a population, usually of the form that a parameter takes a particular numerical value or falls in a certain range of values.

For instance, the parameter might be a population proportion or a probability. Here's an example of a hypothesis, for the astrology experiment in Example 1:

Hypothesis: Using a person's horoscope, the probability p that an astrologer can correctly predict which of three personality charts applies to that person equals 1/3. That is, astrologers' predictions correspond to random guessing.

A **significance test** (or "test" for short) is a method of using data to summarize the evidence about a hypothesis. For instance, if a high proportion of the astrologers' predictions are correct, the data might provide strong evidence against the hypothesis that $p = 1/3$ in favor of an alternative hypothesis representing the astrologers' claim that $p > 1/3$.

Before conducting a significance test, we identify the variable measured and the population parameter of interest. For a categorical variable the parameter is a proportion and for a quantitative variable the parameter is a mean. Section 8.2 shows the details for tests about proportions. Section 8.3 presents tests about means.

A significance test has five steps. We'll now introduce these steps.

Step 1: Assumptions
Each significance test makes certain assumptions or has certain conditions under which it applies. Foremost, a test assumes that the data production used randomization. Other assumptions may include ones about the sample size or about the shape of the population distribution.

Step 2: Hypotheses
Each significance test has two hypotheses about a population parameter.

IN WORDS

H_0: null hypothesis (read as "H zero" or "H naught")
H_a: alternative hypothesis (read as "H a")
In everyday English, null is an adjective meaning "of no consequence or effect, amounting to nothing."

Null Hypothesis, Alternative Hypothesis

The **null hypothesis** is a statement that the parameter takes a particular value.

The **alternative hypothesis** states that the parameter falls in some alternative range of values.

The value in the null hypothesis usually represents *no effect*. The value in the alternative hypothesis then represents an effect of some type.

The symbol H_0 denotes **null hypothesis** and the symbol H_a denotes **alternative hypothesis**.

For the astrology experiment in Example 1, consider the hypothesis, "Based on any person's horoscope, the probability p that an astrologer can correctly predict which of three personality charts applies to that person equals 1/3." This hypothesis states that there is *no effect* in the sense that an astrologer's predictive power is no better than random guessing. This is a *null hypothesis*. It is symbolized by H_0: $p = 1/3$. If it is true, any difference that we observe between the sample proportion of correct guesses and 1/3 is due merely to ordinary sampling variability. The *alternative hypothesis* states that there *is* an effect—an astrologer's predictions are *better* than random guessing. This is H_a: $p > 1/3$.

A null hypothesis has a *single* parameter value, such as H_0: $p = 1/3$. An alternative hypothesis has a *range* of values that are alternatives to the one in H_0, such as H_a: $p > 1/3$ or H_a: $p \neq 1/3$. You formulate the hypotheses for a significance test *before* viewing or analyzing the data.

EXAMPLE 2

Hypotheses for a Therapeutic Touch Study

Picture the Scenario

Therapeutic touch (TT) practitioners claim to improve or heal many medical conditions by using their hands to manipulate a "human energy field" they perceive above the patient's skin. (The patient does not have to be touched.) A test investigating this claim used the following experiment:[2] A TT practitioner was blind-folded. In each trial, the researcher placed her hand over either the right or left hand of the TT practitioner, the choice being determined by flipping a coin. The TT practitioner was asked to identify whether his or her right or left hands was closer to the hand of the researcher. Let p denote the probability of a correct prediction by a TT practitioner. With random guessing, $p = 1/2$. However, the TT practitioners claimed that they could do better than random guessing. They claimed that $p > 1/2$.

---

[2]Rosa, L. et al., *JAMA*, vol. 279, pp. 1005–1010, 1998.

Questions to Explore

Consider the hypothesis, "In any given trial, the probability p of guessing the correct hand is larger than 1/2." This is the TT practitioners' claim.

a. Is this a null or an alternative hypothesis?

b. How can we express the hypothesis that being a TT practitioner has "no effect" on the probability of a correct guess?

Think It Through

a. The hypothesis states that $p > 1/2$. It has a range of parameter values, so it is an alternative hypothesis, symbolized by H_a: $p > 1/2$.

b. The quoted hypothesis is an alternative to the null hypothesis that a TT practitioner's predictions are equivalent to random guessing. This "no effect" hypothesis is H_0: $p = 1/2$.

TO PRACTICE THIS CONCEPT, TRY EXERCISES 8.1 AND 8.2

In a significance test, the null hypothesis is presumed to be true unless the data give strong evidence against it. The "burden of proof" falls on the alternative hypothesis. In the TT study, we assume that $p = 0.50$ unless the data provide strong evidence against it and in favor of the TT practitioners' claim that $p > 0.50$. An analogy may be found in a legal trial in a courtroom, in which a jury must decide the guilt or innocence of a defendant. The null hypothesis, corresponding to "no effect," is that the defendant is innocent. The alternative hypothesis is that the defendant is guilty. The jury presumes the defendant is innocent unless the prosecutor can provide strong evidence that the defendant is guilty "beyond a reasonable doubt." The "burden of proof" is on the prosecutor to convince the jury that the defendant is guilty.

Step 3: Test Statistic

The parameter to which the hypotheses refer has a point estimate. A **test statistic** describes how far that estimate falls from the parameter value given in the null hypothesis. Usually this is measured by the number of standard errors between the two.

For instance, consider the null hypothesis H_0: $p = 1/3$ that, based on a person's horoscope, the probability p that an astrologer can correctly predict which of three personality charts applies to that person equals 1/3. For the experiment described in Example 1, 40 of 116 predictions were correct. The estimate of the probability p is the sample proportion, $\hat{p} = 40/116 = 0.345$. The test statistic compares this to the value in the null hypothesis ($p = 1/3$), using a z-score that measures the number of standard errors that the estimate falls from 1/3. The next section shows the details.

Step 4: P-value

To interpret a test statistic value, we use a probability summary of the evidence against the null hypothesis, H_0. Here's how we get it: We presume that H_0 is true, since the "burden of proof" is on the alternative, H_a. Then we consider the sorts of values we'd expect to get for the test statistic, according to its sampling distribution. If the sample test statistic falls well out in a tail of the sampling distribution, it is far from what H_0 predicts. If H_0 were true, such a value would be unusual. When there are a large number of possible outcomes, any single one may be unlikely, so we summarize how far out in the tail the test statistic falls by the tail probability of that value and values even more extreme. See Figure 8.1. This probability is called a **P-value**. The smaller the P-value, the stronger the evidence is against H_0.

Recall

The **standard error** of an estimate describes the spread of the sampling distribution of that estimate (Sections 6.4 and 7.2). Denoted by se, it is used both in confidence intervals and in significance tests. ■

IN WORDS

The **P-value** is a tail probability beyond the observed test statistic value. Smaller P-values provide stronger evidence against the null hypothesis.

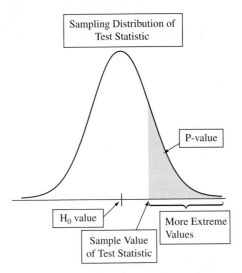

▲ **FIGURE 8.1: Suppose H_0 Were True. The P-value Is the Probability of a Test Statistic Value Like the Observed One or Even More Extreme.** This is the shaded area in the tail of the sampling distribution. **Question:** Which gives stronger evidence against the null hypothesis, a P-value of 0.20 or of 0.01? Why?

In the astrology study, suppose a P-value is small, such as 0.01. This means that if H_0 were true (that an astrologer's predictions correspond to random guessing), it would be unusual to get sample data such as we observed. Such a P-value provides strong evidence against the null hypothesis of random guessing and in support of the astrologers' claim. On the other hand, if the P-value is not near 0, the data are consistent with H_0. For instance, a P-value such as 0.26 or 0.63 indicates that if the astrologer were actually randomly guessing, the observed data would not be unusual.

P-value

The **P-value** is the probability that the test statistic equals the observed value or a value even more extreme. It is calculated by presuming that the null hypothesis H_0 is true.

Step 5: Conclusion

The conclusion of a significance test reports the P-value and *interprets* what it says about the question that motivated the test. Sometimes this includes a decision about the validity of the null hypothesis H_0. For instance, based on the P-value, can we reject H_0 and conclude that astrologers' predictions are better than random guessing? As we'll discuss in the next section, we can reject H_0 in favor of H_a when the P-value is very small, such as 0.05 or less.

SUMMARY: The Five Steps of a Significance Test

1. Assumptions

Specify the variable and parameter. Assumptions commonly pertain to method of data production (randomization), sample size, shape of population distribution.

2. Hypotheses

Null hypothesis, H_0 (a single parameter value, usually "no effect")
Alternative hypothesis, H_a (a set of alternative parameter values)

3. Test statistic

Measures distance between point estimate of parameter and its null hypothesis value, usually by the number of standard errors between them.

4. P-value

Presume H_0 to be true. The P-value is the probability the test statistic takes the observed value or a value more extreme. Smaller P-values represent stronger evidence against H_0.

5. Conclusion

Report and interpret the P-value in the context of the study. Based on the P-value, make a decision about H_0 if one is needed.

8.1 **H₀ or Hₐ?:** For (a) and (b), is the statement a null hypothesis, or an alternative hypothesis?

TRY

   a. In Canada, the proportion of adults who favor legalized gambling equals 0.50.

   b. The proportion of all Canadian college students who are regular smokers is less than 0.24, the value it was ten years ago.

   c. Introducing notation for a parameter, state the hypotheses in (a) and (b) in terms of the parameter values.

8.2 **H₀ or Hₐ?:** For each of the following, is the statement a null hypothesis, or an alternative hypothesis? Why?

TRY

   a. The mean IQ of all students at Lake Wobegon High School is larger than 100.

   b. The probability of rolling a 6 with a particular die equals 1/6.

   c. The proportion of all new business enterprises that remain in business for at least five years is less than 0.50.

8.3 **Burden of proof:** For a new pesticide, should the Environmental Protection Agency (EPA) have the burden of proof to show that it is harmful to the environment, or should the producer of the pesticide have the burden of proof to show that it is not harmful to the environment? Give the analog of the null hypothesis and the alternative hypothesis if the burden of proof is on the EPA to show the new pesticide is harmful.

8.4 **Iowa GPA:** Suppose the mean GPA of all students graduating from the University of Iowa in 1975 was 2.80. The registrar plans to look at records of students graduating in 2005 to see if mean GPA has changed. Define notation and state the null and alternative hypotheses for this investigation.

8.5 **Low-carbohydrate diet:** A study plans to have a sample of obese adults follow a proposed low-carbohydrate diet for three months. The diet imposes limited eating of starches (such as bread and pasta) and sweets, but otherwise no limit on calorie intake. Consider the hypothesis,

TRY

The population mean of the values of weight change = weight at start of study − weight at end of study is a positive number.

   a. Is this a null or an alternative hypothesis? Explain your reasoning.

   b. Define a relevant parameter, and express the hypothesis that the diet has "no effect" in terms of that parameter. Is it a null or alternative hypothesis?

8.6 **Examples of hypotheses:** Give an example of a null hypothesis and an alternative hypothesis about a (a) population proportion, (b) population mean.

8.7 **z-test statistic:** To test H₀: $p = 0.50$ that a population proportion equals 0.50, the test statistic is a z-score that measures the number of standard errors between the sample proportion and the H₀ value of 0.50. If $z = 3.6$, do the data support the null hypothesis, or do they give strong evidence against it? Explain.

8.8 **P-value:** Indicate whether each of the following P-values gives strong evidence or not especially strong evidence against the null hypothesis.

   a. 0.38

   b. 0.001

8.2 | Significance Tests About Proportions

For categorical variables, the parameters of interest are the population proportions in the categories. We'll use the astrology study to illustrate a significance test for population proportions.

EXAMPLE 3

Are Astrologers' Predictions Better than Guessing?

Picture the Scenario

Many people take astrological predictions seriously, but there has never been any scientific evidence that astrology works. One scientific "test of astrology" used the experiment mentioned in Example 1: For each of 116 adult volunteers, an astrologer prepared a horoscope based on the positions of the planets and the moon at the moment of the person's birth. Each adult subject also filled out a California Personality Index (CPI) survey. For a given adult, his or her birth data and horoscope were shown to an astrologer together with the results of the personality survey for that adult and for two other adults randomly selected from the experimental group. The astrologer was asked which personality chart of the three subjects was the correct one for that adult, based on their horoscope.

The 28 participating astrologers were randomly chosen from a list prepared by the National Council for Geocosmic Research (NCGR), an organization dealing with astrology and respected by astrologers worldwide. The NCGR sampling frame consisted of astrologers with some background in psychology who were familiar with the CPI and who were held in high esteem by their astrologer peers. The experiment was double-blind: Each subject was identified by a random number, and neither the astrologers nor the experimenter knew which number corresponded to which subject. The chapter of the NCGR that recommended the participating astrologers claimed that the probability of a correct guess on any given trial in the experiment was larger than 1/3, the value for random guessing. (In fact, they felt that it would exceed 1/2.)

Question to Explore

Put this investigation in the context of a significance test by stating null and alternative hypotheses.

Think It Through

The variable specifying the outcome of any given trial in the experiment is categorical. The categories are "correct prediction" and "incorrect prediction." For each person, let p denote the probability of a correct prediction by the astrologer. We can regard this as the population proportion of correct guesses for the population of people and population of astrologers from which the study participants were sampled. Hypotheses refer to the probability p of a correct prediction. With random guessing, $p = 1/3$. To test the hypothesis of random guessing against the astrologers' claim that $p > 1/3$, we would test H_0: $p = 1/3$ against H_a: $p > 1/3$.

Insight

In the experiment, the astrologers were correct with 40 of their 116 predictions. We'll see how to use these data to test these hypotheses as we work through the five steps of a significance test for a proportion in the next subsection.

TO PRACTICE THIS CONCEPT, TRY EXERCISES 8.9 AND 8.10

What Are the Steps of a Significance Test about a Population Proportion?

This section presents a significance test about a population proportion that applies with relatively large samples. Here are the five steps of the test:

Step 1: Assumptions

■ The variable is categorical.

■ The data are obtained using randomization (such as a random sample).

■ The sample size is sufficiently large that the sampling distribution of the sample proportion $\hat{p}$ is approximately normal. The approximate normality happens when the *expected numbers of successes and failures are both at least 15, at the null hypothesis value for p.*

The sample size guideline is the one we used in Chapter 6 for judging when the normal distribution approximates the binomial distribution well. For the astrology experiment with $n = 116$ trials, when H_0 is true that $p = 1/3$, we expect $116(1/3) = 38.7 \approx 39$ correct guesses and $116(2/3) = 77.3 \approx 77$ incorrect guesses. Both of these are above 15, so the sample size guideline is satisfied.

As with other statistical inference methods, without randomization the validity of the results is questionable. A survey should use random sampling. An experiment should use principles of randomization and blinding with the

Recall

Section 6.4 introduced the sampling distribution of a sample proportion. It has exact standard error

$$\sqrt{p(1 - p)/n}$$

and is well approximated by the normal distribution when

$$np \geq 15 \text{ and } n(1 - p) \geq 15. ■$$

study subjects, as was done in the astrology study. In that study, the astrologers were randomly selected, but the subjects evaluated were people (mainly students) who volunteered for the study. Because of this, any inference applies to the population of astrologers but only to the particular subjects in the study. If the study could have randomly chosen the subjects as well, then the inference would extend more broadly to *all* people.

Step 2: Hypotheses

The null hypothesis of a test about a proportion has the form

$$H_0: p = p_0,$$

where p_0 represents a particular proportion value between 0 and 1. In Example 3, the null hypothesis of no effect states that the astrologers' predictions correspond to random guessing. This is $H_0: p = 1/3$. The null hypothesis value p_0 is 1/3.

The alternative hypothesis refers to alternative parameter values from the number in the null hypothesis. One possible alternative hypothesis has the form

$$H_a: p > p_0.$$

This is used when a test is designed to detect whether p is *larger* than the number in the null hypothesis. In the astrology experiment, the astrologers claimed they could predict *better* than by random guessing. Their claim corresponds to $H_a: p > 1/3$. This is called a **one-sided** alternative hypothesis, because it has values falling only on one side of the null hypothesis value. We'll use this H_a below. The other possible one-sided alternative hypothesis is $H_a: p < p_0$, such as $H_a: p < 1/3$.

A **two-sided** alternative hypothesis has the form

One-Sided $H_a: p > 1/3$

$$H_a: p \neq p_0.$$

It includes *all* the other possible values, both below and above the value p_0 in H_0. It states that the population proportion *differs* from the number in the null hypothesis. An example is $H_a: p \neq 1/3$. This states that the population proportion equals some number other than 1/3.

In summary, for Example 3 we'll use $H_0: p = 1/3$ and the one-sided $H_a: p > 1/3$.

Two-Sided $H_a: p \neq 1/3$

Step 3: Test Statistic

The test statistic measures how far the sample proportion $\hat{p}$ falls from the null hypothesis value p_0, relative to what we'd expect if H_0 were true. The sampling distribution of the sample proportion has mean equal to the population proportion p and exact standard error equal to $\sqrt{p(1 - p)/n}$. When H_0 is true, $p = p_0$, so the sampling distribution has mean p_0 and exact standard error $se_0 = \sqrt{p_0(1 - p_0)/n}$. (We use the zero subscript here on se to reflect using the value the exact se takes when H_0 is true. We substitute p_0 rather than $\hat{p}$ for p in the standard error because we're presuming H_0 to be true in conducting the test.) The test statistic is

$$z = \frac{\hat{p} - p_0}{se_0} = \frac{\hat{p} - p_0}{\sqrt{\dfrac{p_0(1 - p_0)}{n}}} =$$

$$\frac{\text{sample proportion} - \text{null hypothesis proportion}}{\text{standard error when null hypothesis is true}}.$$

This z-score measures the number of standard errors between the sample proportion $\hat{p}$ and the null hypothesis value p_0.

In testing $H_0: p = 1/3$ for the astrology experiment with $n = 116$ trials, the standard error is $\sqrt{p_0(1 - p_0)/n} = \sqrt{[(1/3)(2/3)]/116} = 0.0438$. The astrologers were correct with 40 of their 116 predictions, a sample proportion of

$\hat{p} = 0.345$. The test statistic is

$$z = \frac{\hat{p} - p_0}{se} = \frac{\hat{p} - p_0}{\sqrt{\dfrac{p_0(1 - p_0)}{n}}} = \frac{0.345 - 1/3}{0.0438} = 0.26.$$

The sample proportion of 0.345 is only 0.26 standard errors above the null hypothesis value of 1/3.

Step 4: P-value

Does $z = 0.26$ give much evidence against H_0: $p = 1/3$ and in support of H_a: $p > 1/3$? The P-value summarizes the evidence. It describes how unusual the data would be if H_0 were true, that is, if the probability of a correct prediction were 1/3. The P-value is the probability that the test statistic takes a value like the observed test statistic or even more extreme, if actually $p = 1/3$.

Figure 8.2 shows the approximate sampling distribution of the z test statistic when H_0 is true. This is the **standard normal distribution**. For the astrology study, $z = 0.26$. Values even farther out in the right tail, above 0.26, are even more extreme.

When a random variable has a large number of possible values, the probability of any single value is usually very small. (For the normal distribution, the probability of a single value, such as $z = 0.26$, is *zero*, because the area under

▲ **FIGURE 8.2: Calculation of P-value, when z = 0.26, for Testing H₀: p = 1/3 against Hₐ: p > 1/3.** Presuming that H_0 is true, the P-value is the right-tail probability of a test statistic value even more extreme than observed. **Question:** Logically, why are the *right-tail* z-scores considered to be the *more extreme* values for testing H_0 against H_a: $p > 1/3$?

the curve above a single point is 0.) So the P-value is taken to be the probability of a *region* of values, specifically the more extreme values, $z > 0.26$. From Table A or using software, the right-tail probability above 0.26 is 0.40. This P-value of 0.40 tells us that if H_0 were true, the probability would be 0.40 that the test statistic would be more extreme than the observed value.

Step 5: Conclusion

We summarize the test by reporting and interpreting the P-value. The P-value of 0.40 is not especially small. It does *not* provide strong evidence against H_0: $p = 1/3$ and in favor of the astrologers' claim that $p > 1/3$. The sample data are the sort we'd expect to see if $p = 1/3$ (that is, if astrologers were randomly guessing the personality type). We would not conclude that astrologers have special predictive powers.

TO PRACTICE THIS CONCEPT, TRY EXERCISE 8.14

How Do We Interpret the P-value?

A significance test analyzes the strength of the evidence against the null hypothesis, H_0. We start by presuming that H_0 is true, putting the *burden of proof* on H_a. The approach taken is the indirect one of *proof by contradiction*. To convince ourselves that H_a is true, we must show the data contradict H_0, by showing they'd be unusual if H_0 were true. We analyze whether the data would be unusual if H_0 were true by finding the P-value. If the P-value is small, the data contradict H_0 and support H_a.

In the astrology study, the P-value for H_a: $p > 1/3$ is the right-tail probability of 0.40 from the sampling distribution of the z statistic. This P-value also approximates the probability that the sample proportion $\hat{p}$ takes a value that is as least as far above the null hypothesis value of 1/3 as the observed value of $\hat{p} = 0.345$ (See the margin figure). Since the P-value is not small, if truly $p = 1/3$, it would not be unusual to observe $\hat{p} = 0.345$. Based on the data, it is believable that the astrologers' predictions merely correspond to random guessing.

Why do we find the total probability in the *right tail*? Because the alternative hypothesis H_a: $p > 1/3$ has values *above* (that is, to the right of) the null hypothesis value of 1/3. It's the relatively *large* values of $\hat{p}$ that support this alternative hypothesis.

Why do smaller P-values indicate stronger evidence against H_0? Because the data would then be more unusual if H_0 were true. For instance, if we got a P-value of 0.01, then we might be more impressed by the astrologers' claims. When H_0 is true, the P-value is roughly equally likely to fall anywhere between 0 and 1. By contrast, when H_0 is false, the P-value is more likely to be near 0 than near 1.

Sampling Distribution of $\hat{p}$

P-value 0.40

1/3

$\hat{p} = 0.345$

H_0 Value

Two-Sided Significance Tests

Sometimes we're interested in investigating whether a proportion falls above or below some point. For instance, can we conclude whether the population proportion who voted for a particular candidate is above 1/2, or below 1/2? We then use a *two-sided* alternative hypothesis. This has the form H_a: $p \neq p_0$, such as H_a: $p \neq 1/2$.

For two-sided tests, the values that are more extreme than the observed test statistic value are ones that fall farther in the tail in *either* direction. The P-value is the *two-tail* probability under the standard normal curve, because these are the test statistic values that provide even stronger evidence in favor of H_a: $p \neq p_0$ than the observed value. We calculate this by finding the tail probability in a single tail and then doubling it, since the distribution is symmetric. See Figure 8.3.

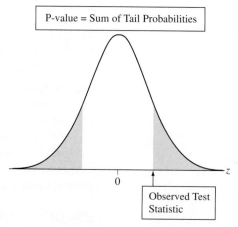

P-value = Sum of Tail Probabilities

0

Observed Test Statistic

▲ **FIGURE 8.3: For the Two-Sided Alternative H_a: $p \neq p_0$, the P-value Is a Two-Tail Probability. Question:** Logically, why are both the left-tail values and the right-tail values the more extreme values for H_a: $p \neq p_0$?

EXAMPLE 4

Dr. Dog: Can Dogs Detect Cancer by Smell?

Picture the Scenario

A recent study investigated whether dogs can be trained to distinguish a patient with bladder cancer by smelling certain compounds released in the patient's urine.[3] Six dogs of varying breeds were trained to discriminate between urine from patients with bladder cancer and urine from control patients without it. The dogs were taught to indicate which among several specimens was from the bladder cancer patient by lying beside it.

An experiment was conducted to analyze how the dogs' ability to detect the correct urine specimen compared to what would be expected with random guessing. Each of the six dogs was tested with nine trials. In each trial, one urine sample from a bladder cancer patient was randomly placed among six control urine samples. In the total of 54 trials with the six dogs, the dogs made the correct selection 22 times.

Let p denote the probability that a dog makes the correct selection on a given trial. Since the urine from the bladder cancer patient was one of seven specimens, with random guessing $p = 1/7$.

Question to Explore

Did this study provide strong evidence that the dogs' predictions were better or worse than with random guessing? Specifically, is there strong evidence that $p > 1/7$, with dogs able to select better than with random guessing, or that $p < 1/7$, with dogs' selections being poorer than random guessing?

Think It Through

The outcome of each trial is binary. The categories are correct selection and incorrect selection. Since we want to test whether the probability of a correct selection differs from random guessing, the hypotheses are

$$H_0: p = 1/7 \text{ and } H_a: p \neq 1/7.$$

The null hypothesis represents *no effect*, the selections being like random guessing. The alternative hypothesis says there is an effect, the selections differing from random guessing. You might instead use $H_a: p > 1/7$, if you expect the dogs' predictions to be better than random guessing. Most medical studies, however, use two-sided alternative hypotheses. This represents an open-minded research approach that recognizes that if an effect exists, it could be negative rather than positive.

The sample proportion of correct selections by the dogs was $\hat{p} = 22/54 = 0.407$, for the sample size $n = 54$. The null hypothesis value is $p_0 = 1/7$. When $H_0: p = 1/7$ is true, the expected counts are $np_0 = 54(1/7) = 7.7$ correct selections and $n(1 - p_0) = 54(6/7) = 46.3$ incorrect selections. The first of these is not larger than 15, so according to the sample size guideline for step 1 of the test, n is not large enough to use the large-sample test. Later in this section, we'll see that the *two-sided* test is robust when this assumption is not satisfied. So we'll use the large-sample test, under the realization that the P-value is a good approximation for the P-value of a small-sample test mentioned later.

Recall

From Section 7.3, a method is **robust** with respect to a particular assumption if it works well even when that assumption is violated. ■

[3]Article by C. M. Willis et al., *British Medical Journal*, vol. 329, September 25, 2004.

The standard error is $se_0 = \sqrt{p_0(1 - p_0)/n} = \sqrt{(1/7)(6/7)/54} = 0.0476$. The test statistic for H_0: $p = 1/7$ ($= 0.143$) equals

$$z = \frac{\hat{p} - p_0}{se_0} = \frac{0.407 - (1/7)}{0.0476} = 5.6.$$

The sample proportion, 0.407, falls more than five standard errors above the null hypothesis value of 1/7.

Figure 8.4 shows the approximate sampling distribution of the z test statistic when H_0 is true. The test statistic value of 5.6 is well out in the right tail. Values farther out in the tail, above 5.6, are even more extreme. The P-value is the total two-tail probability of the more extreme outcomes, above 5.6 or below -5.6. From software, the cumulative probability in the left tail below $z = -5.6$ is 0.000 to three decimal places (in fact, it is 0.00000001), and the probability in the two tails equals $2(0.000) = 0.000$. This P-value provides extremely strong evidence against H_0: $p = 1/7$.

▲ **FIGURE 8.4: Calculation of P-value, when z = 5.6, for Testing H_0: p = 1/7 against H_a: p ≠ 1/7.** Presuming H_0 is true, the P-value is the two-tail probability of a test statistic value even more extreme than observed. **Question:** Is the P-value of 0.000 strong evidence supporting H_0, or strong evidence against H_0?

When the P-value in a two-sided test is small, the point estimate tells us the direction in which the parameter appears to differ from the null hypothesis value. In summary, since the P-value is very small and $\hat{p} > 1/7$, the evidence strongly suggests that the dogs' selections are *better* than random guessing.

Insight

Recall that one assumption for this significance test is randomization for obtaining the data. This study is like most medical studies in that its subjects were a *convenience sample* rather than a random sample from some population. It is not practical that a study can identify the population of all people who have bladder cancer and then randomly sample them for an experiment. Likewise, the dogs were not randomly sampled. Because of this, any inferential predictions are tentative. They are valid only to the extent that the patients and the dogs in the experiment are representative of their populations. The predictions become more conclusive if similar results occur in other studies with other samples. In medical studies, even though the sample is not random, it is important to employ randomization in any experimentation, for instance in the placement of the bladder cancer patient's urine specimen among the six control urine specimens.

Recall

See Section 4.2 for potential difficulties with convenience samples, which use subjects who are conveniently available. ■

TO PRACTICE THIS CONCEPT, TRY EXERCISE 8.16

Summary of How the Alternative Hypothesis Determines the P-value

The P-value is the probability of the values that are more extreme than the observed test statistic value. What is "more extreme" depends on the alternative hypothesis. For the two-sided alternative hypothesis, more extreme values fall in each direction, so the P-value is a two-tail probability. A one-sided alternative states that the parameter falls in a particular direction relative to the null hypothesis value. The P-value then uses only the tail in that direction. For instance, in the astrology study, the alternative $H_a: p > 1/3$ used only the *right* tail. For the alternative $H_a: p < 1/3$ in the other direction, we would have used the probability to the *left* of the observed test statistic value. See Figure 8.5. The P-values for the two one-sided alternative hypotheses add up to 1.0.

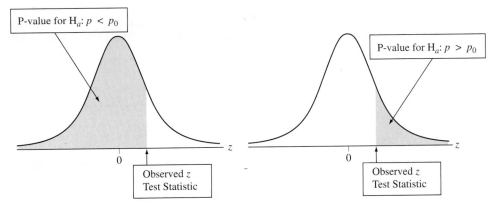

▲ **FIGURE 8.5: Calculation of P-value for One-Sided Alternative Hypotheses.** **Question:** For a given one-sided H_a, how do we know which tail to use for finding the tail probability?

Here's a summary of how we find the P-value:

Summary of P-values for Different Alternative Hypotheses

Alternative Hypothesis	P-value
$H_a: p > p_0$	*Right-tail probability*
$H_a: p < p_0$	*Left-tail probability*
$H_a: p \neq p_0$	*Two-tail probability*

The Significance Level Tells Us How Strong the Evidence Must Be

Sometimes we need to make a decision about whether the data provide sufficient evidence to reject H_0. Before seeing the data, we decide how small the P-value would need to be to reject H_0. For example, we might decide that that we will reject H_0 if the P-value ≤ 0.05. The cutoff point of 0.05 is called the **significance level**. It is shown in Figure 8.6.

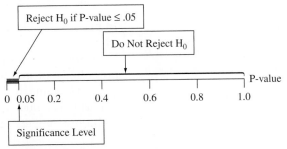

▲ **FIGURE 8.6: The Decision in a Significance Test.** Reject H_0 if the P-value is less than or equal to a chosen **significance level,** usually 0.05.

Significance Level

The **significance level** is a number such that we reject H_0 if the P-value is less than or equal to that number. In practice, the most common significance level is 0.05.

Table 8.1 summarizes the two possible outcomes for a test decision when the significance level is 0.05. We either "reject H_0" or "do not reject H_0." If the P-value if larger than 0.05, the data do not contradict H_0 sufficiently for us to reject it. Then, H_0 is still believable to us, and we "do not reject H_0." If the P-value is ≤ 0.05, the data provide enough evidence to "reject H_0." Recall that H_a had the "burden of proof," and in this case we feel that the proof is sufficient. When we reject H_0, we say the results are **statistically significant**.

TABLE 8.1: Possible Decisions in a Test with Significance Level = 0.05

P-value	Decision About H_0
≤ 0.05	Reject H_0
> 0.05	Do not reject H_0

EXAMPLE 5

What's the Decision for the Astrology Study?

Picture the Scenario

Let's continue our analysis of the astrology study from Example 3. The parameter p is the probability that an astrologer picks the correct one of three personality charts, based on the horoscope and birth data provided. We tested H_0: $p = 1/3$ against H_a: $p > 1/3$ and got a P-value of 0.40.

Questions to Explore

What decision would we make for a significance level of (a) 0.05? (b) 0.50?

Think It Through

a. For a significance level of 0.05, the P-value of 0.40 is *not* less than 0.05. So we do not reject H_0. The evidence is not strong enough to conclude that the astrologers' predictions are better than random guessing.

b. For a significance level of 0.50, the P-value of 0.40 *is* less than 0.50. So we reject H_0 in favor of H_a. We conclude that this result provides sufficient evidence to support $p > 1/3$. The results are statistically significant at the 0.50 significance level.

Insight

In practice, significance levels are normally close to 0, such as 0.05 or 0.01. The reason is that the significance level is also a type of error probability. We'll see in Section 8.4 that when H_0 is true, the significance level is the probability of making an error by rejecting H_0. We used the significance level of 0.50 in (b) above for illustrative purposes, but the value of 0.05 used in part (a) is much more typical. The astrologers' predictions are consistent with random guessing, and we would not reject H_0 for these data.

TO PRACTICE THIS CONCEPT, TRY EXERCISE 8.17

IN PRACTICE: Report the P-value

Report the P-value, rather than merely indicating whether the results are statistically significant. Learning the actual P-value is more informative than learning only whether the test is "statistically significant at the 0.05 level." The P-values of 0.01 and 0.049 are both statistically significant in this sense, but the first P-value provides much stronger evidence than the second.

Now that we've studied all five steps of a significance test about a proportion, let's use them in a new example, much like those you'll see in the exercises. In this example and in the exercises, it may help you to refer to the following summary box:

SUMMARY: Steps of a Significance Test for a Population Proportion p

1. Assumptions

- Categorical variable, with population proportion p defined in context
- Randomization, such as a simple random sample, for gathering data
- n large enough to expect at least 15 successes and 15 failures under H_0 (that is $np_0 \geq 15$ and $n(1 - p_0) \geq 15$). This is mainly important for one-sided tests.

2. Hypotheses

Null: $H_0: p = p_0$, where p_0 is the hypothesized value, such as
$\qquad H_0: p = 0.50$

Alternative: $H_a: p \neq p_0$, such as $H_a: p \neq 0.50$ (two-sided) or
$\qquad H_a: p < p_0$ (one-sided) or $H_a: p > p_0$ (one-sided)

3. Test statistic

$$z = \frac{\hat{p} - p_0}{se_0} \text{ with } se_0 = \sqrt{p_0(1 - p_0)/n}$$

4. P-value

Alternative hypothesis	P-value	
$H_a: p > p_0$	Right-tail probability	
$H_a: p < p_0$	Left-tail probability	
$H_a: p \neq p_0$	Two-tail probability	

5. Conclusion

Smaller P-values give stronger evidence against H_0. If a decision is needed, reject H_0 if the P-value is less than or equal to the preselected significance level (such as 0.05). Relate the conclusion to the context of the study.

EXAMPLE 6

Can TT Practitioners Detect a Human Energy Field?

Picture the Scenario

Let's revisit the therapeutic touch (TT) experiment from Example 2. Each trial investigated whether a TT practitioner could correctly identify (while blind-folded) which of their hands was closer to the hand of a researcher. The researcher determined hand placement by flipping a coin. In a set of 150 trials with 15 TT practitioners (10 trials each), the TT practitioners were correct with 70 of their 150 predictions.

Did You Know?

The TT experiment was a science fair project of a nine-year old girl. Her project was published in a major medical journal, *Journal of the American Medical Association (JAMA)*. What inspiration for future science fair researchers! ■

Recall

H_0 contains a single number, whereas H_a has a range of values. ■

Sampling Distribution of z Test Statistic (Standard Normal)

Area 0.21

P-value = 0.79

−0.82   0

```
1-PropZTest
 Prop>.50
 z=-.82
 P=.79
 p̂=.47
 n=150.00
```

TI-83 +/84 *output*

Questions to Explore

How strong is the evidence to support the TT practitioners' claim that they can predict the correct hand better than with random guessing? What decision would be made for a 0.05 significance level?

Think It Through

Let's follow the five steps of a significance test to organize our response to these questions:

1. Assumptions:

■ The response is categorical, with outcomes "correct" and "incorrect" for a TT practitioner's prediction. Let p denote the probability of a correct prediction.

■ We'll treat the 150 trials as a random sample of the possible trials that could occur in such an experiment with these TT practitioners. The inference will apply to the 15 particular TT practitioners used in this experiment. If the study randomly selected them from the population of TT practitioners, then the inference would apply to the entire population of TT practitioners.

■ We'll discuss the sample size requirement in the next step.

2. Hypotheses:

If a TT practitioner merely guesses, then $p = 0.50$. However, the TT practitioners claimed that $p > 0.50$. To check their claim, we'll test H_0: $p = 0.50$ against H_a: $p > 0.50$. Supposing H_0 is true, of 150 trials we expect $150(0.50) = 75$ correct predictions and $150(0.50) = 75$ incorrect ones. These both exceed 15. The sample size is large enough to use the large-sample significance test.

3. Test statistic:

The sample estimate of the probability p of a correct decision is $\hat{p} = 70/150 = 0.467$. The standard error of $\hat{p}$ when H_0: $p = 0.50$ is true is

$$se_0 = \sqrt{p_0(1 - p_0)/n} = \sqrt{0.50(0.50)/150} = 0.0408.$$

The value of the test statistic is

$$z = \frac{\hat{p} - p_0}{se_0} = \frac{0.467 - 0.50}{0.0408} = -0.82.$$

The sample proportion is 0.82 standard errors below the null hypothesis value.

4. P-value:

For testing H_0: $p = 0.50$ against H_a: $p > 0.50$, the P-value is the *right-tail* probability above $z = -0.82$ in the standard normal distribution (see the figure in the margin). From software or a normal table, this is $1 - 0.21 = 0.79$. TI-83+/84 output is also provided in the margin.

5. Conclusion:

The P-value of 0.79 is not small. If the null hypothesis H_0: $p = 0.50$ were true, the data we observed would not be unusual. With a 0.05 significance level, the evidence is not strong enough to reject H_0. It seems plausible that $p = 0.50$. In summary, this experiment does not suggest that the TT practitioners can predict better than random guessing.

Insight

The P-value is larger than 1/2 because the data do not provide evidence against H_0: $p = 0.50$ in favor of H_a: $p > 0.50$. The sample proportion of 0.467 actually falls in the direction of the *other* tail, $p < 0.50$. In this experiment, the TT practitioners predicted *worse* than what we'd expect for random guessing.

TO PRACTICE THIS CONCEPT, TRY EXERCISE 8.15

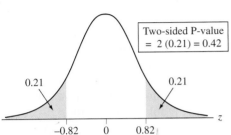

The one-sided alternative H_a: $p < 0.50$ states that the probability of a correct prediction is *worse* than with random guessing. Then the P-value is a *left-tail* probability, everything to the left of $z = -0.82$ under the standard normal curve. The P-value is 0.21. See the margin figure. The sum of the P-values for the one-sided alternatives always equals 1.0. The P-value = 0.79 for H_a: $p > 0.50$, the P-value = 0.21 for H_a: $p < 0.50$, and $0.79 + 0.21 = 1.0$.

In practice, many researchers would take an objective view and allow for predictions to possibly be better or possibly be worse than random guessing, by using the two-sided H_a: $p \neq 0.50$. The P-value is then the two-tail probability, $2(0.21) = 0.42$. See the second figure in the margin.

IN PRACTICE: Using software for significance tests

In practice, you should pick H_a before seeing the data, and you can use software to do the test. For the one-sided test of Example 6 with H_a: $p > 0.50$, MINITAB reports the results shown in Table 8.2. Unless requested otherwise, by default software reports the two-sided P-value.

Some software reports the P-value to several decimals, such as 0.792892. We recommend rounding it to two or three decimal places, for instance, to 0.79, before reporting it. Reporting a P-value as 0.792892 suggests greater accuracy than actually exists, since the normal sampling distribution is only *approximate*.

TABLE 8.2: MINITAB Output for One-Sided Test in Example 6

```
Test of p = 0.5 vs p > 0.5
     X           N        Sample p      Z-Value       P-Value
    70         150        0.466667       -0.82          0.793
```

Category Count	Sample Size	Sample Proportion	Test Statistic	

"Do Not Reject H_0" Does Not Mean "Accept H_0"

A small P-value means that the sample data would be unusual if H_0 were true. If the P-value is not small, such as 0.79 for the one-sided test in Example 6 on TT practitioners, the null hypothesis is plausible. In this case, the conclusion is reported as "Do not reject H_0," since the data do not contradict H_0.

"*Do not reject H_0*" *is not the same as saying "accept H_0.*" The population proportion has many plausible values besides the number in the null hypothesis. For instance, consider Example 6, with p the probability of a correct prediction by a TT practitioner. We did not reject H_0: $p = 0.50$. Thus, p may equal 0.50, but other values are also believable. A 95% confidence interval for p is

$$\hat{p} \pm 1.96\sqrt{\hat{p}(1-\hat{p})/n}, \text{ or } 0.467 \pm 1.96\sqrt{(0.467)(0.533)/150},$$

which equals $(0.39, 0.55)$. Even though insufficient evidence exists to reject H_0, it is improper to accept it and conclude that $p = 0.50$. It is plausible that p is as low as 0.39 or as high as 0.55.

An analogy here is that of a legal trial in a courtroom. The null hypothesis is that the defendant is innocent. The alternative hypothesis is that the defendant is guilty. If the jury acquits the defendant, this does not mean that it *accepts* the defendant's claim of innocence. It merely means that innocence is plausible because guilt has not been established *beyond a reasonable doubt*.

We can "Accept H_a" when it contains all the plausible values The null hypothesis contains a single possible value for the parameter. Saying "Do not reject H_0" instead of "Accept H_0" emphasizes that that value is merely one of many plausible ones. Because of sampling error, there is a range of believable values besides the H_0 value. "Accept H_a" *is* permissible for the alternative hypothesis. When the P-value is sufficiently small, the entire range of believable values falls within the range of numbers contained in H_a.

Use p_0 instead of $\hat{p}$ in the standard error for significance tests In calculating the standard error in the test statistic, we substituted the null hypothesis value p_0 for the population proportion p in the formula $\sqrt{p(1-p)/n}$ for the exact standard error. *The parameter values for sampling distributions in significance tests are those from H_0, since the P-value is calculated presuming that H_0 is true.* So we use p_0 in standard errors for tests. This practice differs from confidence intervals, in which the sample proportion $\hat{p}$ substitutes for p (which is unknown) in the exact standard error. The confidence interval method[4] does not have a hypothesized value for p, so that method substitutes the point estimate $\hat{p}$ for p. To differentiate between the two cases, we've denoted $se_0 = \sqrt{p_0(1-p_0)/n}$ and $se = \sqrt{\hat{p}(1-\hat{p})/n}$.

How Do We Decide between a One-Sided and a Two-Sided Test?

In practice, two-sided tests are more common than one-sided tests. Even if we think we know the direction of an effect, two-sided tests can also detect an effect that falls in the opposite direction. For example, in a medical study, even if we think a drug will perform better than a placebo, using a two-sided alternative allows us to detect if the drug is actually worse, perhaps because of bad side effects. However, as we saw with the astrology and TT examples, in some scenarios a one-sided test is natural.

Here are some things to consider in deciding on the alternative hypothesis:

■ In deciding between one-sided and two-sided alternative hypotheses in a particular exercise or in practice, *consider the context of the real problem.*

For instance, in the astrology experiment, to test whether someone can guess *better* than with random guessing, we used the values $p > 1/3$ in the alternative hypothesis corresponding to that possibility. An exercise that says "Test whether the population proportion *differs* from 0.50" suggests a two-sided alternative, H_a: $p \neq 0.50$, to allow for p to be larger or smaller than 0.50. "Test whether the population proportion is *larger* than 0.50" suggests the one-sided alternative, H_a: $p > 0.50$.

■ In most research articles, significance tests use two-sided P-values.

Partly this reflects an objective approach to research that recognizes that an effect could go in either direction. Using an alternative hypothesis in which an effect can go in either direction is regarded as the most even-handed way to perform the test. In using two-sided P-values, researchers avoid the suspicion that they chose H_a when they saw the direction in which the data occurred. That is not ethical and would be cheating. (The P-value would then need to be based on a different sampling distribution that is conditioned on that extra information.)

■ Confidence intervals are two-sided.

The practice of using a two-sided test coincides with the ordinary approach for confidence intervals, which are two-sided, obtained by adding and subtracting some quantity from the point estimate. There is a way to construct one-sided confidence intervals, for instance, concluding that a population proportion is *at least* equal to 0.70. In practice, though, two-sided confidence intervals are much more common.

---

[4]If we instead conduct the test by substituting the sample proportion in the standard error, the standard normal approximation for the sampling distribution of z is much poorer.

As you specify the hypotheses, also keep in mind three basic facts:

- The null hypothesis has an equal sign (such as $H_0: p = 0.5$), but the alternative hypothesis does not.
- You shouldn't pick H_a based on looking at the data.
- The hypotheses always refer to population parameters, not sample statistics.

So *never* express a hypothesis using sample statistic notation, such as $H_0: \hat{p} = 0.5$. There is no need to conduct inference about statistics such as the sample proportion $\hat{p}$, because you can find their values exactly from the data.

The Binomial Test for Small Samples

The test about a proportion applies normal sampling distributions for the sample proportion $\hat{p}$ and the z test statistic. Therefore, it is a *large-sample* test, because the central limit theorem implies approximate normality of the sampling distribution for large random samples. The guideline is that the expected numbers of successes and failures should be at least 15, when H_0 is true; that is, $np_0 \geq 15$ and $n(1 - p_0) \geq 15$.

In practice, the large-sample z test performs well for *two-sided* alternatives even for small samples. When p_0 is below 0.3 or above 0.7 and n is small, the sampling distribution is quite skewed. However, a tail probability that is smaller than the normal probability in one tail is compensated by a tail probability that is larger than the normal probability in the other tail. Because of this, the P-value from the two-sided test using the normal table approximates well a P-value from a small-sample test.[5]

For one-sided tests, when p_0 differs from 0.5, the large-sample test does not work well when the sample size guideline ($np_0 \geq 15$ and $n(1 - p_0) \geq 15$) is violated. In that case, you should use a small-sample test. This test uses the binomial distribution with parameter value p_0 to find the exact probability of the observed value and all the more extreme values, according to the direction in H_a. Since one-sided tests with small n are not common in practice, we will not study the binomial test here. Exercises 8.23 and 8.24 show how to do it.

---

[5]With small samples, this test works much better than the confidence interval of Section 7.2. Having an exact, rather than estimated, standard error, makes a great difference.

8.2 Practicing the Basics

8.9 **Psychic:** A person who claims to be psychic says that the probability p that he can correctly predict the outcome of the roll of a die in another room is greater than 1/6, the value that applies with random guessing. If we want to test this claim, we could use the data from an experiment in which he predicts the outcomes for n rolls of the die. State hypotheses for a significance test, letting the alternative hypothesis reflect the psychic's claim.

8.10 **Believe in astrology?:** You plan to apply significance testing to your own experiment for testing astrology, in which astrologers have to guess which of four personality profiles is the correct one for someone who has a particular horoscope. Define notation and state hypotheses, letting one hypothesis reflect the possibility that the astrologers' predictions could be better than random guessing.

8.11 **Get P-value from z:** For a test of $H_0: p = 0.50$, the z test statistic equals 1.04.

a. Find the P-value for $H_a: p > 0.50$.

b. Find the P-value for $H_a: p \neq 0.50$.

c. Find the P-value for $H_a: p < 0.50$.
   (*Hint:* The P-values for the two possible one-sided tests must sum to 1.)

d. Do any of the P-values in (a), (b), or (c) give strong evidence against H_0? Explain.

8.12 **Get more P-values from z:** Refer to the previous exercise. Suppose $z = 2.50$ instead of 1.04.

a. Find the P-value for (i) $H_a: p > 0.50$, (ii) $H_a: p \neq 0.50$, (iii) $H_a: p < 0.50$.

b. Do any of the P-values in (a) provide strong evidence against H_0? Explain.

8.13 **Find test statistic and P-value:** For a test of $H_0: p = 0.50$, the sample proportion is 0.35 based on a sample size of 100.

a. Show that the test statistic is $z = -3.0$.

b. Find the P-value for H_a: $p < 0.50$.

c. Does the P-value in (b) give much evidence against H_0? Explain.

8.14 **Another test of astrology:** Examples 1, 3, and 5 referred to a study about astrology. Another part of the study used the following experiment: Professional astrologers prepared horoscopes for 83 adults. Each adult was shown three horoscopes, one of which was the one an astrologer prepared for them and the other two were randomly chosen from ones prepared for other subjects in the study. Each adult had to guess which of the three was theirs. Of the 83 subjects, 28 guessed correctly.

a. Defining notation, set up hypotheses to test that the probability of a correct prediction is 1/3 against the astrologers' claim that it exceeds 1/3.

b. Show that the sample proportion = 0.337, the standard error of the sample proportion for the test is 0.052, and the test statistic is $z = 0.08$.

c. Find the P-value. Would you conclude that people are more likely to select their "correct" horoscope than if they were randomly guessing, or are results consistent with random guessing?

8.15 **Another test of therapeutic touch:** Examples 2 and 6 described a study about therapeutic touch (TT). A second run of the same experiment in the study used 13 TT practitioners who had to predict the correct hand in each of 10 trials.

a. Defining notation, set up hypotheses to test that the probability of a correct guess is 0.50 against the TT practitioners' claim that it exceeds 0.50.

b. The 130 trials had 53 correct guesses. Find and interpret the test statistic.

c. Report the P-value. (*Hint*: Since the sample proportion is in the opposite direction from the practitioners' claim, you should get a P-value > 0.50.) Indicate your decision, in the context of this experiment, using a 0.05 significance level.

d. Check whether the sample size was large enough to make the inference in (c). Indicate what assumptions you would need to make for your inferences to apply to all TT practitioners and to all subjects.

8.16 **Testing a headache remedy:** Studies that compare treatments for chronic medical conditions such as headaches can use the same subjects for each treatment. With a crossover design, each person crosses over from using one treatment to another during the study. One such study considered a drug (a pill called Sumatriptan) for treating migraine headaches in children.[6] The study observed each of 30 children at two times when he or she had a migraine headache. The child received the drug at one time and a placebo at the other time. The order of treatment was randomized and the study was double-blind. For each child, the response was whether the drug or the placebo provided better pain relief. Let p denote the proportion of children having better pain relief with the drug, in the population of children who suffer periodically from migraine headaches. Can you conclude that $p > 0.50$, with more than half of the population getting better pain relief with the drug, or that $p < 0.50$, with less than half getting better pain relief with the drug (i.e., the placebo being better)? Of the 30 children, 22 had more pain relief with the drug and 8 had more pain relief with the placebo.

a. For testing H_0: $p = 0.50$ against H_a: $p \neq 0.50$, show that the test statistic $z = 2.56$.

b. Show that the P-value is 0.0104. Interpret.

c. Check the assumptions needed for this test, and discuss the limitations due to using a convenience sample.

8.17 **Gender bias in selecting managers:** For a large supermarket chain in Florida, a women's group claimed that female employees were passed over for management training in favor of their male colleagues. The company denied this claim, saying they picked the employees from the eligible pool at random to receive this training. Statewide, the large pool of more than 1000 eligible employees who can be tapped for management training is 40% female and 60% male. Since this program began, 28 of the 40 employees chosen for management training were male and 12 were female.

a. In a significance test, the random sampling assumption is the claim of the company. In defining hypotheses, explain why this company's claim of a lack of gender bias is a "no effect" hypothesis. State the null and alternative hypotheses for a test to investigate the strength of evidence to support the women's claim.

b. The table shows results of using MINITAB to do a large-sample analysis. Explain why the large-sample analysis is justified, and show how software obtained the test statistic value.

```
Test of p = 0.60 vs. not = 0.60
   X    N   Sample p     95.0% CI     Z-Value  P-Value
  28   40   0.70000   (0.558,0.842)    1.29    0.1967
```

c. To what alternative hypothesis does the P-value in the table refer? Use it to find the P-value for the alternative hypothesis you specified in (a), and interpret it.

d. What decision would you make for a 0.05 significance level? Interpret.

---

[6]Data based on those in a study by M. L. Hamalainen et al., reported in *Neurology*, vol. 48, pp. 1100–1103, 1997.

8.18 **Gender discrimination:** Refer to the previous exercise.

    a. Explain why the alternative hypothesis of bias against males is H_a: $p < 0.60$.

    b. Show that the P-value for testing H_0: $p = 0.60$ against H_a: $p < 0.60$ equals 0.90. Interpret. Why is it large?

8.19 **Garlic to repel ticks:** A recent study (*J. Amer. Med. Assoc.*, vol. 284, p. 831, 2000) considered whether daily consumption of 1200 mg of garlic could reduce tick bites. The study used a crossover design with a sample of Swedish military conscripts, half of whom used placebo first and garlic second and half the reverse. The authors did not present the data, but the effect they described is consistent with garlic being more effective with 37 subjects and placebo being more effective with 29 subjects. Does this suggest a real difference between garlic and placebo, or are the results consistent with random variation? Answer by

    a. Identifying the relevant variable and parameter.

    b. Stating hypotheses for a large-sample two-sided test and checking that sample size guidelines are satisfied for that test.

    c. Finding the test statistic value.

    d. Finding and interpreting a P-value and stating the conclusion in context.

8.20 **Exit-poll predictions:** According to an exit poll in the 2000 New York senatorial election, 55.7% of the sample of size 2232 reported voting for Hillary Clinton. Is this enough evidence to predict who won? Test that the population proportion who voted for Clinton was 0.50 against the alternative that it differed from 0.50. Answer by

    a. Identifying the variable and parameter, and defining notation.

    b. Stating hypotheses and checking assumptions for a large-sample test.

    c. Finding the test statistic for the test.

    d. Reporting the P-value and interpreting it.

    e. Explaining how to make a decision for the significance level of 0.05.

8.21 **Which cola?:** The 49 students in a class at the University of Florida made blinded evaluations of pairs of cola drinks. For the 49 comparisons of Coke and Pepsi, Coke was preferred 29 times. In the population that this sample represents, is this strong evidence that a majority prefers one of the drinks? Refer to the following MINITAB printout.

```
Test of p = 0.50 vs. not = 0.50
  X    N   Sample p    95.0% CI     Z-Value  P-Value
  29   49   0.5918   (0.454, 0.729)  1.286    0.1985
```

    a. Explain how to get the test statistic value that MINITAB reports.

    b. Explain how to get the "P-value." Interpret it.

    c. Based on the result in (b), does it make sense to "accept H_0"? Explain.

    d. What does the 95% confidence interval tell you that the test does not?

8.22 **How to sell a burger:** A fast-food chain wants to compare two ways of promoting a new burger (a turkey burger). One way uses a coupon available in the store. The other way uses a poster display outside the store. Before the promotion, their marketing research group matches 50 pairs of stores. Each pair has two stores with similar sales volume and customer demographics. The store in a pair that uses coupons is randomly chosen, and after a month-long promotion, the increases in sales of the turkey burger are compared for the two stores. The increase was higher for 28 stores using coupons and higher for 22 stores using the poster. Is this strong evidence to support the coupon approach, or could this outcome be explained by chance? Answer by performing all five steps of a two-sided significance test.

♦♦8.23 **A binomial headache:** A null hypothesis states that the population proportion p of headache sufferers who have more pain relief with aspirin than with another pain reliever equals 0.50. For a crossover study with 10 subjects, all 10 have more relief with aspirin. If the null hypothesis were true, by the binomial distribution the probability of this sample result equals $(0.50)^{10} = 0.001$. In fact, this is the small-sample P-value for testing H_0: $p = 0.50$ against H_a: $p > 0.50$. Does this P-value give (a) strong evidence in favor of H_0, or (b) strong evidence against H_0? Explain why.

♦♦8.24 **P-value for small samples:** Example 4, on whether dogs can detect bladder cancer by selecting the correct urine specimen (out of 7), used the normal sampling distribution to find the P-value. The normal distribution P-value approximates a P-value using the binomial distribution. That binomial P-value is more appropriate when either expected count is less than 15. In Example 4, n was 54, and 22 of the 54 selections were correct.

    a. Supposing H_0: $p = 1/7$ is true, X = number of correct selections has the binomial distribution with $n = 54$ and $p = 1/7$. Why?

    b. For H_a: $p > 1/7$, with $x = 22$, the P-value using the binomial is $P(22) + P(23) + \cdots + P(54)$, where $P(x)$ denotes the binomial probability of outcome x with $p = 1/7$. (This equals 0.0000019.) Why would the P-value be this sum rather than just $P(22)$?

8.3 | Significance Tests About Means

For quantitative variables, significance tests usually refer to the population mean μ. We illustrate the significance test about means with the following example.

EXAMPLE 7

Mean Weight Change in Anorexic Girls

Picture the Scenario

A recent study compared different psychological therapies for teenage girls suffering from anorexia, an eating disorder that can cause them to become dangerously underweight.[7] Each girl's weight was measured before and after a period of therapy. The variable of interest was the weight change, defined as weight at the end of the study minus weight at the beginning of the study. The weight change was positive if the girl gained weight and negative if she lost weight. The therapies were designed to aid weight gain.

In this study, 29 girls received the cognitive behavioral therapy. This form of psychotherapy stresses identifying the thinking that causes the undesirable behavior and replacing it with thoughts designed to help to improve this behavior. Table 8.3 shows the data. The weight changes for the 29 girls had a sample mean of $\bar{x} = 3.00$ pounds and standard deviation of $s = 7.32$ pounds.

TABLE 8.3: Weights of Anorexic Girls (in Pounds) before and after Treatment

Example 7 uses the weight change as the variable of interest.

Girl	Weight Before	After	Change	Girl	Weight Before	After	Change	Girl	Weight Before	After	Change
1	80.5	82.2	1.7	11	85.0	96.7	11.7	21	83.0	81.6	−1.4
2	84.9	85.6	0.7	12	89.2	95.3	6.1	22	76.5	75.7	−0.8
3	81.5	81.4	−0.1	13	81.3	82.4	1.1	23	80.2	82.6	2.4
4	82.6	81.9	−0.7	14	76.5	72.5	−4.0	24	87.8	100.4	12.6
5	79.9	76.4	−3.5	15	70.0	90.9	20.9	25	83.3	85.2	1.9
6	88.7	103.6	14.9	16	80.6	71.3	−9.3	26	79.7	83.6	3.9
7	94.9	98.4	3.5	17	83.3	85.4	2.1	27	84.5	84.6	0.1
8	76.3	93.4	17.1	18	87.7	89.1	1.4	28	80.8	96.2	15.4
9	81.0	73.4	−7.6	19	84.2	83.9	−0.3	29	87.4	86.7	−0.7
10	80.5	82.1	1.6	20	86.4	82.7	−3.7				

Question to Explore

How could we frame this investigation in the context of a significance test that can detect a positive or negative effect of the therapy? State the null and alternative hypotheses for that test.

---

[7]The data are courtesy of Prof. Brian Everitt, Institute of Psychiatry, London.

Think It Through

The response variable, weight change, is quantitative. Hypotheses refer to the population mean weight change μ, namely whether it is 0 (the "no effect" value) or differs from 0. To use a significance test to analyze the strength of evidence about the therapy's effect, we'll test H_0: $\mu = 0$ against H_a: $\mu \neq 0$.

Insight

We'll see how to test these hypotheses as we work through the five steps of a test about means in the next subsection. We'll analyze whether the difference between the sample mean weight change of 3.0 and the no-effect value of 0 can be explained by random variability.

TO PRACTICE THIS CONCEPT TRY EXERCISE 8.5 (B)

What Are the Steps of a Significance Test about a Population Mean?

A significance test about a mean has the same five steps as a test about a proportion: Assumptions, hypotheses, test statistic, P-value, and conclusion. We will mention modifications as they relate to the test for a mean. However, the reasoning is the same as that for the significance test for a proportion.

Step 1: Assumptions
The three basic assumptions of a test about a mean are as follows:

■ The variable is quantitative.

■ The data production employed randomization.

■ The *population distribution* is approximately normal. This is most crucial when n is small and H_a is one-sided, as discussed later in the section.

The anorexia study is like the 'dogs detecting cancer' study in Example 4 in that its subjects were a convenience sample. Inferences are tentative. They are more convincing if researchers can argue that the girls in the sample are representative of the population of girls who suffer from anorexia. The study did employ randomization in assigning girls to one of three therapies, only one of which (cognitive behavioral) is considered in this example.

Step 2: Hypotheses
The null hypothesis in a test about a population mean has form

$$H_0: \mu = \mu_0,$$

where μ_0 denotes a particular value for the population mean. The two-sided alternative hypothesis

$$H_a: \mu \neq \mu_0$$

includes values both below and above the number μ_0 listed in H_0. Also possible are the one-sided alternative hypotheses,

$$H_a: \mu > \mu_0 \text{ or } H_a: \mu < \mu_0.$$

For instance, let μ denote the mean weight change for the population represented by the sample of anorexic girls. If the therapy has no effect, then $\mu = 0$. If the therapy has a beneficial effect on weight, as the study expected, then $\mu > 0$. To test that the therapy has no effect against the alternative that it has a beneficial effect, we test H_0: $\mu = 0$ against H_a: $\mu > 0$. In practice, the two-sided alternative H_a: $\mu \neq 0$ is more common, to take an objective approach that can detect either a positive or a negative effect of the therapy.

Recall

From Section 6.5, the exact standard error of the sample mean is $\sigma/\sqrt{n}$, where σ = population standard deviation. In practice, σ is unknown, so we estimate the standard error by

$$se = s/\sqrt{n}. ■$$

Step 3: Test Statistic

The test statistic is the distance between the sample mean $\bar{x}$ and the null hypothesis value μ_0, as measured by the number of standard errors between them. This is measured by

$$\frac{(\bar{x} - \mu_0)}{se} = \frac{\text{sample mean} - \text{null hypothesis mean}}{\text{standard error of sample mean}}.$$

In practice, as in forming a confidence interval for a mean (Section 7.3), the standard error is $se = s/\sqrt{n}$. The test statistic is

$$t = \frac{(\bar{x} - \mu_0)}{se} = \frac{(\bar{x} - \mu_0)}{s/\sqrt{n}}.$$

In the anorexia study, the sample mean $\bar{x} = 3.00$ and the sample standard deviation $s = 7.32$. The standard error $se = s/\sqrt{n} = 7.32/\sqrt{29} = 1.36$. The test statistic equals

$$t = (\bar{x} - \mu_0)/se = (3.00 - 0)/1.36 = 2.21.$$

Recall

You can review the *t distribution* in Section 7.3. We used it there to form a confidence interval for a population mean μ. ■

We use the symbol t rather than z for the test statistic because, as in forming a confidence interval, using s to estimate σ introduces additional error: The sampling distribution is more spread out than the standard normal. When H_0 is true, the t test statistic has approximately the *t distribution*. The t distribution is specified by its degrees of freedom, which equal $n - 1$ for inference about a mean. This test statistic is called a *t statistic*.

Figure 8.7 shows the t sampling distribution. The farther $\bar{x}$ falls from the null hypothesis mean μ_0, the farther out in a tail the t test statistic falls, and the stronger the evidence is against H.

▲ **FIGURE 8.7: The t Distribution of the t Test Statistic.** There is stronger evidence against H_0 when the t test statistic falls farther out in a tail. The P-value for a two-sided H_a is a two-tail probability (shaded in figure). **Question:** Why is it that t-scores farther out in the tails provide stronger evidence against H_0?

Step 4: P-Value

The P-value is a single tail or a two-tail probability depending on whether the alternative hypothesis is one-sided or two-sided.

Alternative Hypothesis	P-value
$H_a: \mu \neq \mu_0$	*Two-tail probability from t distribution*
$H_a: \mu > \mu_0$	*Right-tail probability from t distribution*
$H_a: \mu < \mu_0$	*Left-tail probability from t distribution*

For the anorexia study with $H_a: \mu \neq 0$, the P-value is the two-tail probability of a test statistic value farther out in either tail than the observed value of 2.21. See Figure 8.7. This probability is double the single-tail probability. Table 8.4 shows the way MINITAB software reports results. Since $n = 29$, $df = n - 1 = 28$. The P-value is 0.036, or 0.04 rounded to two decimal places. This is the two-tail probability of the t test statistic values below -2.21 and above $+2.21$ when $df = 28$.

TABLE 8.4: MINITAB Output for Analyzing Data from Anorexia Study

```
One-Sample T: wt_change

Test of mu = 0 vs not = 0

Variable    N   Mean   StDev  SE Mean      95% CI         T     P
wt_change  29  3.000  7.3204  1.3594  (0.2155, 5.7845)  2.21  0.036
```

Step 5: Conclusion

The conclusion of a significance test reports the P-value and *interprets* what it says about the question that motivated the test. Sometimes this includes a decision about the validity of H_0. We reject the null hypothesis when the P-value is less than or equal to the pre-selected significance level. In the anorexia study, the small P-value of 0.04 provides considerable evidence against the null hypothesis that the therapy has no effect. If we had pre-selected a significance level of 0.05, this would be enough evidence to reject H_0: $\mu = 0$ in favor of H_a: $\mu \neq 0$.

The alternative hypothesis states that the population mean weight change μ is not equal to zero. The positive value for the sample mean ($\bar{x} = 3.0$) suggests that $\mu > 0$. The cognitive behavioral therapy seems to be beneficial. The effect may be small in practical terms, however. The 95% confidence interval shown in Table 8.4 predicts that μ falls between 0.2 and 5.8 pounds.

TO PRACTICE THIS CONCEPT, TRY EXERCISE 8.30

When research scientists conduct a significance test, their report would not show all results of a printout but would instead present a simple summary such as, "The diet had a statistically significant positive effect on weight (mean change $= 3$ pounds, $n = 29$, $t = 2.21$, P-value $= 0.04$)." Often results are summarized even further, such as "The diet had a significant positive effect on weight (P-value < 0.05)." This brief a summary is undesirable, because it does not show the estimated size of the effect (3 pounds) or the actual P-value. Don't condense your conclusions this much.

The t Statistic and z Statistic Have the Same Form

As you read the examples in this section, notice the parallel between each step of the test for a mean and the test for a proportion. For instance, the t test statistic for a mean has the same form as the z test statistic for a proportion, namely,

Form of Test Statistic

$$\frac{\text{Estimate of parameter} - H_0 \text{ value of parameter}}{\text{standard error of estimate}}$$

For the test about a mean, the estimate $\bar{x}$ of the population mean μ replaces the estimate $\hat{p}$ of the population proportion p, the H_0 mean μ_0 replaces the H_0 proportion p_0, and the standard error of the sample mean replaces the standard error of the sample proportion. The box summarizes tests for population means.

SUMMARY: Steps of a Significance Test for a Population Mean μ

1. Assumptions

- Quantitative variable, with population mean μ defined in context
- Data are obtained using randomization, such as a simple random sample
- Population distribution is approximately normal

2. Hypotheses

Null: H_0: $\mu = \mu_0$, where μ_0 is the hypothesized value (such as H_0: $\mu = 0$)

Alternative: H_a: $\mu \neq \mu_0$ (two-sided) or

$\qquad\qquad$ H_a: $\mu < \mu_0$ (one-sided) or H_a: $\mu > \mu_0$ (one-sided)

3. Test statistic

$$t = \frac{(\bar{x} - \mu_0)}{se}, \text{ where } se = s/\sqrt{n}$$

4. P-value   Use t distribution (Table B) with $df = n - 1$

Alternative Hypothesis	P-value
H_a: $\mu \neq \mu_0$	*Two-tail probability*
H_a: $\mu > \mu_0$	*Right-tail probability*
H_a: $\mu < \mu_0$	*Left-tail probability*

5. Conclusion

Smaller P-values give stronger evidence against H_0 and supporting H_a. If using a significance level to make a decision, reject H_0 if P-value is less than or equal to the significance level (such as 0.05). Relate the conclusion to the context of the study.

How Can We Perform a One-Sided Test about a Population Mean?

One-sided alternative hypotheses apply for a prediction that μ differs from the null hypothesis value in a certain direction. For example, H_a: $\mu > 0$ predicts that the true mean is *larger* than the null hypothesis value. Its P-value is the probability of a t value *larger* than the observed value, that is, in the *right* tail. Likewise, for H_a: $\mu < 0$ the P-value is the *left*-tail probability. In each case, again $df = n - 1$.

EXAMPLE 8

Does a Low-Carbohydrate Diet Work?

Picture the Scenario

In a recent study,[8] 41 overweight subjects were placed on a low carbohydrate diet but given no limit on caloric intake. The prediction was that subjects on such a diet would lose weight, on the average. After 16 weeks, their weight change averaged -9.7 kg (that is, a *decrease* in weight, on the average), with a standard deviation of 3.4 kg.

---

[8]*Am. J. Med.*, vol. 113, pp. 30–36, 2002

Question to Explore

Find and interpret the P-value for testing $H_0: \mu = 0$ against $H_a: \mu < 0$.

Think It Through

The standard error $se = s/\sqrt{n} = 3.4/\sqrt{41} = 0.53$. The test statistic equals

$$t = (\bar{x} - \mu_0)/se = (-9.7 - 0)/0.53 = -18.3,$$

with $df = n - 1 = 40$. Since H_a predicts that the mean is *below* 0, the P-value is the left-tail probability *below* the test statistic value of -18.3. This is zero to many decimal places. See Figure 8.8. The P-value of 0.000 (rounded to three decimal places) means that if $H_0: \mu = 0$ were true, it would be practically impossible to observe a t statistic of -18.3 or even larger in the negative direction. A P-value of 0.000 is very strong evidence against $H_0: \mu = 0$ and in favor of $H_a: \mu < 0$.

▲ **FIGURE 8.8: P-value for Testing $H_0: \mu = 0$ against $H_a: \mu < 0$. Question:** Why are large negative, rather than large positive, t test statistic values the ones that support $H_a: \mu < 0$?

In summary, on the average, people on this diet lose weight. This conclusion is tentative, since this study used a convenience sample rather than a random sample. No information was given about the shape of the distribution of weight changes, but with n this large ($n = 41$) the normal population assumption is not usually crucial because the sampling distribution will be roughly bell shaped.

Insight

The P-value is the probability of t values below the observed t test statistic value of -18.3. Equivalently, it is the probability of a sample mean weight change $\bar{x} < -9.7$ (the observed value), if H_0 were true that $\mu = 0$. See the first figure in the margin.

TO PRACTICE THIS CONCEPT, TRY EXERCISE 8.28

We've seen that if $\mu = 0$, it would be extremely unusual to get a sample mean of -9.7. If in fact μ were a number greater than 0, it would be even more unusual. For example, a sample value of $\bar{x} = -9.7$ is even more unusual when $\mu = 10$ than when $\mu = 0$, since -9.7 is farther out in the tail of the sampling distribution of $\bar{x}$ when $\mu = 10$ than when $\mu = 0$. See the second figure in the margin.

Thus, when we reject $H_0: \mu = 0$ in favor of $H_0: \mu < 0$, we can also reject the broader null hypothesis of $H_0: \mu \geq 0$,

In other words, we conclude that $\mu = 0$ is false and that $\mu > 0$ is false. However, statements of null hypotheses use a *single* number in the null hypothesis, because a single number is entered in the test statistic to compare to the sample mean.

How Can We Use the t Table to Approximate a P-Value?

Statistical software and many calculators can find the P-value for you, as illustrated in Table 8.4 for the anorexia study of Example 7. If you have the test statistic value but are not using software, you can use the *t* table (Table B in the Appendix). Table B is not detailed enough to provide an *exact* tail probability, but it provides enough information to determine whether a one-tail probability is greater than or less than 0.100, 0.050, 0.025, 0.010, 0.005, or 0.001.

Let's see how to do this for the *t* test statistic from the anorexia study (Example 7), $t = 2.21$ based on $df = 28$. Look at the row of Table B for $df = 28$. You will see

df	$t_{.100}$	$t_{.050}$	$t_{.025}$	$t_{.010}$	$t_{.005}$	$t_{.001}$
28	1.313	1.701	2.048	2.467	2.763	3.408

Note that $t = 2.21$ falls between 2.048 and 2.467. Now, the value $2.048 = t_{.025}$ has a right-tail probability of 0.025 and the value $2.467 = t_{.010}$ has a right-tail probability of 0.010. So the right-tail probability for $t = 2.21$ falls between 0.010 and 0.025. Figure 8.9 illustrates.

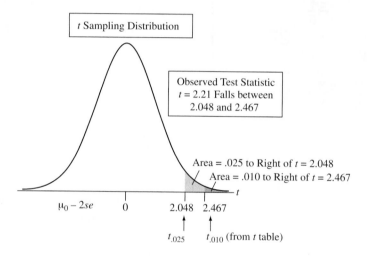

▲ **FIGURE 8.9: For df = 28, t = 2.21 has a Right-Tail Probability between 0.010 and 0.025.** The two-sided P-value falls between 2(0.010) = 0.02 and 2(0.025) = 0.05. **Question:** Using software or a calculator, can you show that the actual right-tail probability equals 0.018 and that the two-sided P-value equals 0.036?

To get the P-value for the two-sided H_a, double the bounds, because you want a two-tail probability. For these data, we double 0.01 and 0.025 to report $0.02 <$ P-value < 0.05. So, we can reject H_0 at the 0.05 significance level. In fact, software told us (Table 8.4) that the actual P-value $= 0.036$.

Results of Two-Sided Tests and Results of Confidence Intervals Agree

For the anorexia study, we got a P-value of 0.036 for testing $H_0: \mu = 0$ against $H_a: \mu \neq 0$ for the mean weight change with the cognitive behavioral therapy. With the 0.05 significance level, we would reject H_0. Table 8.4 showed that a 95% confidence interval for the population mean weight change μ is (0.2, 5.8) pounds. The confidence interval shows just how different from 0 the population mean weight change is likely to be. It is estimated to fall between 0.2 and 5.8 pounds. We infer that the population mean weight change μ is positive, because all the numbers in this interval are greater than 0, but the effect of the therapy may be very small.

Both the significance test and the confidence interval suggested that μ differs from 0. In fact, conclusions about means using two-sided significance tests are consistent with conclusions using confidence intervals. If a two-sided test says you can reject the hypothesis that $\mu = 0$, then 0 is not in the corresponding confidence interval.

If P-value ≤ 0.05 in a two-sided test, a 95% confidence interval does not contain the H$_0$ value   In the anorexia study of Example 7, the two-sided test of H$_0$: $\mu = 0$ has P-value = $0.04 \leq 0.05$. This small P-value says we can reject H$_0$ at the 0.05 significance level. The 95% confidence interval for μ is $(0.2, 5.8)$. The interval does not contain 0.0, so this method also suggests that μ is not exactly equal to 0.0.

By contrast, suppose that the P-value > 0.05 in a two-sided test of H$_0$: $\mu = 0$, so we cannot reject H$_0$ at the 0.05 significance level. Then, a 95% confidence interval for μ *does* contain 0. According to both methods, the value of 0 is a plausible one for μ.

Why are confidence intervals and two-sided tests about means consistent?   Figure 8.10 illustrates why decisions from two-sided tests about means are consistent with confidence intervals.

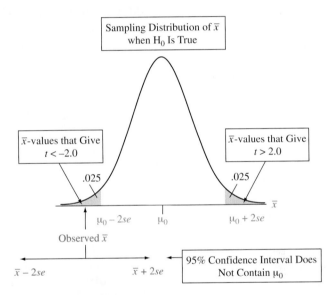

▲ **FIGURE 8.10: Relation between Confidence Interval and Significance Test.** With large samples, if the sample mean falls more than about two standard errors from μ_0, then μ_0 does not fall in the 95% confidence interval and also μ_0 is rejected in a test at the 0.05 significance level. **Question:** Inference about proportions does not have an *exact* equivalence between confidence intervals and tests. Why? (*Hint:* Are the exact same standard error values used in the two methods?)

With large samples, the t-score for a 95% confidence interval is approximately 2, so the confidence interval is roughly $\bar{x} \pm 2(se)$. If this interval does not contain a particular value μ_0, this means that the sample mean $\bar{x}$ falls more than 2 standard errors from μ_0. But this means that the test statistic $t = (\bar{x} - \mu_0)/se$ is larger than 2 in absolute value. But when this happens, the two-tail P-value is less than 0.05, so we would reject the hypothesis that $\mu = \mu_0$.

What If the Population Does Not Satisfy the Normality Assumption?

For the test about a mean, the third assumption states that the population distribution should be approximately normal. This ensures that the sampling distribution of the sample mean $\bar{x}$ is normal and, after using s to estimate σ, the

Recall

From Section 6.5, with random sampling the **sampling distribution of the sample mean is approximately normal** for large n, by the central limit theorem. ■

test statistic has the *t* distribution. For large samples (roughly about 30 or higher), this assumption is usually not important. Then, an approximate normal sampling distribution occurs for $\bar{x}$ regardless of the population distribution. (Remember the central limit theorem?) Chapter 14 presents a type of statistical method, called *nonparametric*, that does not require the normal assumption. However, the normal assumption made with small samples for the *t* test is not crucial when we use a two-sided H_a.

Recall

Section 7.3 defined a statistical method to be **robust** with respect to a particular assumption if it performs adequately even when that assumption is violated. ■

IN PRACTICE: Two-sided inferences are robust

Two-sided inferences using the *t* distribution are *robust* against violations of the normal population assumption. They still usually work well if the actual population distribution is not normal. The test does not work well for a one-sided test with small *n* when the population distribution is highly skewed.

Checking for normality in the anorexia study   Figure 8.11 shows a histogram and a box plot of the data from the anorexia study. With small *n*, such plots are very rough estimates of the population distribution. It can be difficult to determine whether the population distribution is approximately normal. However, Figure 8.11 does suggest skew to the right, with a small proportion of girls having considerable weight gains.

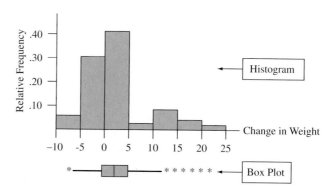

▲ **FIGURE 8.11: Histogram and Box Plot of Weight Change for Anorexia Sufferers. Question:** What do these plots suggest about the shape of the population distribution of weight change?

As just mentioned, a two-sided *t* test still works quite well even if the population distribution is skewed. So, we feel comfortable with the two-sided test in Example 7. However, this plot makes us a bit wary about using a one-sided test for these data. The sample size is not large ($n = 29$), and the histogram in Figure 8.11 shows substantial skew, with the box plot highlighting six quite large weight change values.

Regardless of robustness, look at the data   Whether *n* is small or large, you should look at the data to check for severe skew or for severe outliers that occur primarily in one direction. They could cause the sample mean to be a misleading measure. For the anorexia data, the median weight change is only 1.4 pounds, somewhat less than the mean of 3.0 because of the skew to the right. Even though the significance test indicated that the population mean weight change was positive, the sample median is another indication that the size of the effect could be small. You also need to be cautious about any conclusion if it changes after removing an extreme outlier from the data set (see Exercise 8.36).

What Effect Does the Sample Size Have on P-values?

The result of a significance test can depend strongly on how much data you have. For a given sample effect, the test statistic tends to be larger as the sample size increases, as the next example shows.

EXAMPLE 9

Testing Whether Software Operates Properly

Picture the Scenario

Random numbers are used for conducting simulations and for identifying subjects to be chosen for a random sample. A difficult programming challenge is writing a computer program that can generate random numbers properly. Each digit must be equally likely to be 0, 1, 2, ..., 9, and the digits in the sequence must be independent: The probability of each digit is 0.10, regardless of which digits were generated in the past. One test of software for generating random numbers checks whether the mean of the generating process differs from the mean of 4.50 that holds if each digit 0, 1, ..., 9 truly has probability 0.10 of occurring each time.

Questions to Explore

Consider the test of $H_0: \mu = 4.50$ against $H_a: \mu \neq 4.50$ when a sequence of random digits has sample mean 4.40 and standard deviation 2.90, if (i) $n = 100$, (ii) $n = 10,000$. Table 8.5 shows software output for the two cases.

a. Show how the *se* values and test statistic values were obtained in the two cases.

b. Explain the practical implications about the effect of *n* on results of a test.

TABLE 8.5: Effect on a Significance Test of Increasing the Sample Size *n*

For a given size of effect (such as $\bar{x} - \mu_0 = -0.10$), as *n* increases, the test statistic increases in absolute value and the P-value decreases.

Test of mu = 4.50 vs. not = 4.50

(i) $n = 100 \rightarrow$

Variable	N	Mean	StDev	SE Mean	T	P
RanDigit	100	4.40	2.90	0.290	−0.345	0.731

(ii) $n = 10,000 \rightarrow$

Variable	N	Mean	StDev	SE Mean	T	P
RanDigit	10,000	4.40	2.90	0.029	−3.45	0.0006

Think It Through

a. In both cases, $\bar{x} = 4.40$, $s = 2.90$, and the sample effect as measured by $\bar{x} - \mu_0 = 4.40 - 4.50 = -0.10$ is the same. However, the standard error is smaller with a larger sample size. The *se* values are

$$\text{(i) } se = s/\sqrt{n} = 2.90/\sqrt{100} = 0.29,$$

$$\text{(ii) } se = s/\sqrt{n} = 2.90/\sqrt{10,000} = 0.029.$$

The test statistics then are also quite different:

$$\text{(i) } t = \frac{(\bar{x} - \mu_0)}{se} = \frac{4.40 - 4.50}{0.29} = -0.345,$$

$$\text{(ii) } t = \frac{(\bar{x} - \mu_0)}{se} = \frac{4.40 - 4.50}{0.029} = -3.45.$$

From Table 8.5, the P-values are dramatically different, 0.73 compared to 0.0006. The same effect, $\bar{x} - \mu_0 = -0.10$, but based on a larger sample size, results in a much smaller P-value.

b. For a given sample effect, larger sample sizes produce larger test statistics (in absolute value). Why does this happen? As n increases, the standard error in the denominator of the t statistic decreases. So the t statistic itself increases. The two-sided P-value is then smaller. This makes sense: We can be more certain that a given sample effect reflects a true population effect if the sample size is large than if it is small.

Insight

An implication of this result is that, for large n, statistical significance may not imply an important result in practical terms. For instance, you can get a small P-value even if the sample mean falls quite near the null hypothesis value. We'll discuss this further in Section 8.5.

TO PRACTICE THIS CONCEPT, TRY EXERCISE 8.27

8.3 Practicing the Basics

8.25 Which t has P-value = 0.05?: A t test for a mean uses a sample of 15 observations. Find the t test statistic value that has a P-value of 0.05 when the alternative hypothesis is (a) $H_a: \mu \neq 0$, (b) $H_a: \mu > 0$, (c) $H_a: \mu < 0$?

8.26 Practice mechanics of a t test: A study has a random sample of 20 subjects. The test statistic for testing $H_0: \mu = 100$ is $t = 2.40$. Find the approximate P-value for the alternative, (a) $H_a: \mu \neq 100$, (b) $H_a: \mu > 100$, (c) $H_a: \mu < 100$.

8.27 Effect of n: Refer to the previous exercise. If the same
TRY sample mean and standard deviation had been based on $n = 5$ instead of $n = 20$, the test statistic would have been $t = 1.20$.

a. Would the P-value for $H_a: \mu \neq 0$ be larger, or smaller, than when $t = 2.40$? Why?

b. Other things being equal, explain why larger sample sizes result in smaller P-values.

8.28 More on the anorexia study: Suppose the study about
TRY the anorexia therapy (Example 7) had predicted that the therapy induces a *positive* mean weight change. We could then test $H_0: \mu = 0$ against $H_a: \mu > 0$. In Example 7, we saw that $\bar{x} = 3.0$ for $n = 29$ subjects, and $t = 2.21$, which has P-value $= 0.036$ for $H_a: \mu \neq 0$.

a. What is the P-value for testing $H_0: \mu = 0$ against $H_a: \mu > 0$?

b. How is the P-value interpreted?

c. Explain why rejecting $H_0: \mu = 0$ in favor of $H_a: \mu > 0$ also inherently rejects the broader null hypothesis of $H_0: \mu \leq 0$. (*Hint:* If $\bar{x} = 3.0$ and

$t = 2.21$ would be unlikely if $\mu = 0$, would they be even more unlikely if $\mu < 0$?)

8.29 Lake pollution: An industrial plant claims to discharge no more than 1000 gallons of wastewater per hour, on the average, into a neighboring lake. An environmental action group decides to monitor the plant, in case this limit is being exceeded. Doing so is expensive, and only a small sample is possible. A random sample of four hours is selected over a period of a week. The observations are

2000, 1000, 3000, 2000.

a. Show that $\bar{x} = 2000$, $s = 816.5$, and standard error $= 408.25$.

b. To test $H_0: \mu = 1000$ vs. $H_a: \mu > 1000$, show that the test statistic equals 2.45.

c. Using Table B or software, show that the P-value is less than 0.05, so there is enough evidence to reject the null hypothesis at the 0.05 significance level.

d. Explain how your one-sided analysis in (b) implicitly tests the broader null hypothesis that $\mu \leq 1000$.

8.30 Weight change for controls: A disadvantage of the exper-
TRY imental design in Example 7 on weight change in anorexic girls is that girls could change weight merely from participating in a study. In fact, girls were randomly assigned to receive a therapy or to serve in a control group, so it was possible to compare weight change for the therapy group to the control group. For the 26 girls in the control group, the weight change had $\bar{x} = -0.5$ and $s = 8.0$. Repeat all five steps of the test of $H_0: \mu = 0$ against $H_a: \mu \neq 0$ for this group, and interpret the P-value.

8.31 Crossover study: A crossover study of 13 children suffering from asthma (*Clinical and Experimental Allergy*, vol. 20, pp. 429–432, 1990) compared single inhaled doses of formoterol (F) and salbutamol (S). The outcome measured was the child's peak expiratory flow (PEF) eight hours following treatment. The data on PEF follow:

Child	F	S	Child	F	S	Child	F	S	Child	F	S
1	310	270	5	410	380	9	330	365	13	220	90
2	385	370	6	370	300	10	250	210			
3	400	310	7	410	390	11	380	350			
4	310	260	8	320	290	12	340	260			

Let μ denote the population mean of the difference between the PEF values for the F and S treatments. Use a calculator or software for the following analyses:

a. Form the 13 difference scores, for instance $310 - 270 = 40$ for child 1 and $330 - 365 = -35$ for child 9, always taking $F - S$. Construct a dot plot or a box plot. Describe the sample data distribution.

b. Carry out the five steps of the significance test for a mean of the difference scores, using $H_0: \mu = 0$ and $H_a: \mu \neq 0$.

c. Discuss whether the assumptions seem valid for this example. What is the impact of using a convenience sample?

8.32 Too little or too much wine?: Hosmer Winery on Cayuga Lake, New York has a machine for dispensing their wine into bottles, which are advertised to contain 750 ml. When the machine is in statistical control, the amount dispensed has a mean of 752 ml. Four observations are taken each day, to plot a daily mean over time on a control chart to check for irregularities. The most recent day's observations were

$$745, 740, 750, 745.$$

Could the difference between the sample mean and the target value be due to random variation, or can you conclude that the true mean is now different from 752? Answer by showing the five steps of a significance test, making a decision using a 0.05 significance level.

8.33 Selling a burger: In Exercise 8.22, a fast-food chain compared two ways of promoting a turkey burger. In a separate experiment with 10 pairs of stores, the difference in the month's increased sales between the store that used coupons and the store with the outside poster had a mean of $3000. Does this indicate a true difference between mean sales for the two advertising approaches? Answer by using the output shown to test that the population mean difference is 0, carrying out the five steps of a significance test. Make a decision using a 0.05 significance level.

```
Test of mu = 0 vs. mu not = 0

Variable    N      Mean     StDev    SE Mean
Opinion    10      3000     4000     1264.91

Variable       95.0% CI        T        P
Opinion    (138.8, 5861.2)   2.372    0.04177
```

8.34 Assumptions important?: Refer to the previous exercise.

a. Explain how the result of the 95% confidence interval shown in the table agrees with the test decision using the 0.05 significance level.

b. Suppose you instead wanted to perform a one-sided test, because the study predicted that the increase in sales would be higher with coupons. Explain why the normal population assumption may possibly be problematic.

8.35 Anorexia in teenage girls: Example 7 described a study about various therapies for teenage girls suffering from anorexia. For each of 17 girls who received the family therapy, the changes in weight were

$$11, 11, 6, 9, 14, -3, 0, 7, 22, -5, -4, 13, 13, 9, 4, 6, 11.$$

a. Plot these with a dot plot or box plot, and summarize.

b. Verify that the weight changes have $\bar{x} = 7.29$, $s = 7.18$, and $se = 1.74$ pounds.

c. Give all steps of a significance test about whether the population mean was 0, against an alternative designed to see if there is any effect.

8.36 Sensitivity study: Ideally, results of a statistical analysis should not depend greatly on a single observation. To check this, it's a good idea to conduct a **sensitivity study**. This entails redoing the analysis after deleting an outlier from the data set or changing its value to a more typical value and checking whether results change much. If results change little, this gives us more faith in the conclusions that the statistical analysis reports. For the weight changes in Table 8.3 from the anorexia study (shown again below and also in the "anorexia" data file on the text CD), the greatest reported value of 20.9 pounds was a severe outlier. Suppose this observation was actually 2.9 pounds but was incorrectly recorded. Redo the two-sided test of that example, and summarize how the results differ. Does the ultimate conclusion depend on that single observation?

Weight Changes in Anorexic Girls		
1.7	11.7	−1.4
0.7	6.1	−0.8
−0.1	1.1	2.4
−0.7	−4.0	12.6
−3.5	20.9	1.9
14.9	−9.3	3.9
3.5	2.1	0.1
17.1	1.4	15.4
−7.6	−0.3	−0.7
1.6	−3.7	

8.37 Test and CI: Results of 99% confidence intervals are consistent with results of two-sided tests with which significance level? Explain the connection.

8.4 | Decisions and Types of Errors in Significance Tests

In significance tests, the P-value summarizes the evidence about H_0. A P-value such as 0.001 casts strong doubt on H_0 being true, because if it were true the observed data would be very unusual.

When we need to decide whether the evidence is strong enough to reject H_0, we've seen that the key is whether the P-value falls below a pre-specified **significance level**. The significance level is usually denoted by the Greek letter α (alpha). In practice, $\alpha = 0.05$ is most common: We reject H_0 if the P-value ≤ 0.05. We do not reject H_0 if the P-value > 0.05. The smaller α is, the stronger the evidence must be to reject H_0. To avoid bias, we select α *before* looking at the data.

Two Potential Types of Errors in Test Decisions

Because of sampling variability, decisions in significance tests always have some uncertainty. A decision can be in error. For instance, in the anorexia study of Example 7, we got a P-value of 0.04 for testing $H_0: \mu = 0$ of no weight change, on the average. With significance level $\alpha = 0.05$, we rejected H_0 and concluded that anorexic girls have a positive mean weight change when undergoing the therapy. If the therapy truly has a positive effect, this is a correct decision. In reality, though, perhaps the therapy has no effect, and the population mean weight change (unknown to us) is actually 0.

Tests have two types of potential errors, called **Type I** and **Type II errors**.

Type I and Type II Errors

When H_0 is true, a **Type I error** occurs when H_0 is rejected.
When H_0 is false, a **Type II error** occurs when H_0 is not rejected.

If actually the anorexia therapy has no effect (that is, if $\mu = 0$), we've made a Type I error. H_0 was actually true, but we rejected it.

Consider the experiment about astrology in Example 5. In that study $H_0: p = 1/3$ corresponded to random guessing by the astrologers. We got a P-value of 0.40. With significance level = 0.05, we do not reject H_0. If truly $p = 1/3$, this is a correct decision. However, if astrologers actually can predict better than random guessing (so that $p > 1/3$), we've made a Type II error, failing to reject H_0 when it is false.

When we make a decision, there are four possible results. These refer to the two possible decisions combined with the two possibilities for whether H_0 is true. Table 8.6 summarizes these four possibilities. In practice, when we plan to make a decision in a significance test, it's important to know the probability of an incorrect decision.

TABLE 8.6: The Four Possible Results of a Decision in a Significance Test

Type I and Type II errors are the two possible incorrect decisions. We make a correct decision if we do not reject H_0 when it is true or if we reject it when it is false.

Reality About H_0	Decision Do not reject H_0	Reject H_0	
H_0 true	Correct decision	Type I error ←	Type I error occurs if reject H_0 when it is actually true.
H_0 false	Type II error	Correct decision	

An analogy: Decision errors in a legal trial   These two types of errors can occur with any decision having two options, one of which is incorrect. For instance, consider a decision in a legal trial. The null hypothesis tested is the defendant's claim of innocence. The alternative hypothesis is that the defendant is guilty. The jury rejects H_0 if it decides that the evidence is sufficient to convict. The defendant is then judged guilty. A Type I error, rejecting a true null hypothesis, occurs in convicting a defendant who is actually innocent. Not rejecting H_0 means the defendant is acquitted (judged not guilty). A Type II error, not rejecting H_0 even though it is false, occurs in acquitting a defendant who is actually guilty. See Table 8.7.

TABLE 8.7: Possible Results of a Legal Trial

Defendant	Legal Decision	
	Acquit	Convict
Innocent (H_0)	Correct decision	Type I error
Guilty (H_a)	Type II error	Correct decision

The Significance Level Is the Probability of a Type I Error

When H_0 is actually true, let's see how to find the probability of a Type I error. This is the probability of rejecting H_0, even though it is actually true. We'll find this for the two-sided test about a proportion.

With the $\alpha = 0.05$ significance level, we reject H_0 if the P-value ≤ 0.05. For two-sided tests about a proportion, the two-tail probability that forms the P-value is ≤ 0.05 whenever the test statistic z satisfied $|z| \geq 1.96$. The collection of test statistic values for which a test rejects H_0 is called the **rejection region**. These are the z test statistic values that occur when the sample proportion falls at least 1.96 standard errors from the null hypothesis value. They are the values we'd least expect to observe if H_0 were true. Figure 8.12 illustrates.

▶ **FIGURE 8.12: The Rejection Region is the Set of Test Statistic Values that Reject H_0.** For a two-sided test about a proportion with significance level 0.05, these are the z test statistic values that exceed 1.96 in absolute value. **Question:** If instead we use a one-sided H_a: $p > p_0$, what is the rejection region?

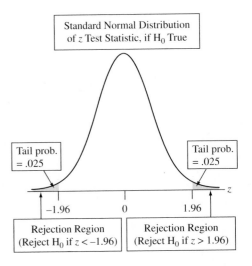

Now, if H_0 is actually true, the sampling distribution of the z test statistic is the standard normal. Therefore, the probability of rejecting H_0, which is the probability that $|z| \geq 1.96$, is exactly 0.05. But this is precisely the significance level.

P(Type I error) = Significance level α

Suppose H_0 is true. The probability of rejecting H_0, thereby making a Type I error, equals the significance level for the test.

IN PRACTICE: We don't know if a decision is correct

In practice, we don't know whether a decision in a significance test is correct, just as we don't know whether a particular confidence interval truly contains an unknown parameter value. However, we can control the *probability* of an incorrect decision for either type of inference.

We can control the probability of a Type I error by our choice of the significance level. The more serious the consequences of a Type I error, the smaller α should be. In practice, $\alpha = 0.05$ is most common, just as a probability of error of 0.05 in interval estimation is most common (that is, 95% confidence intervals). However, this may be too high when a decision has serious implications. For example, suppose a convicted defendant gets the death penalty. Then, if a defendant is actually innocent, we would hope that the probability of conviction is smaller than 0.05.

Although we don't know whether the decision in a particular test is correct, we justify the method in terms of the long-run proportions of Type I and Type II errors. We'll learn how to calculate P(Type II error) later in the chapter.

EXAMPLE 10

Can We Eliminate Type I Errors from Legal Verdicts?

Picture the Scenario

In an ideal world, Type I or Type II errors would not occur. In practice, however, whether in significance tests or in applications such as courtroom trials or medical diagnoses, errors do happen. It can be surprising and disappointing how often they do, as we saw in the diagnostic testing examples of Chapter 5. Likewise, we've all read about defendants who were given the death penalty but were later determined to be innocent, but we don't have reliable information about how often this occurs.

Question to Explore

When we make a decision, why don't we use an extremely small probability of Type I error, such as $\alpha = 0.000001$? For instance, why don't we make it almost impossible to convict someone who is really innocent?

Think It Through

When we make α smaller in a significance test, we need a smaller P-value to reject H_0. It then becomes harder to reject H_0. But this means that it will also be harder even if H_0 is false. The stronger the evidence that is required to convict someone, the more likely it becomes that we will fail to convict defendants who are actually guilty. In other words, the smaller we make the probability of Type I error, the *larger* the probability of Type II error becomes (that is, failing to reject H_0 even though it is false).

If we tolerate only an extremely small chance of a Type I error (such as $\alpha = 0.000001$), then the test may be unlikely to reject the null hypothesis even if it is false—for instance, unlikely to convict someone even if they are guilty. Some of our laws are set up to make Type I errors very unlikely, and as a consequence some truly guilty individuals are not punished for their crimes.

Insight

This reasoning reflects a fundamental relation between the probabilities of the two types of errors, for a given sample size n:

As P(Type I Error) Goes Down, P(Type II Error) Goes Up

The two probabilities are inversely related.

TO PRACTICE THIS CONCEPT, TRY EXERCISE 8.43 PART (C)

Except in the final section of this chapter, we will not calculate the probability of a Type II error. This calculation can be complex. In practice, to make a decision in a test, we only need to set the probability of Type I error, which is the significance level α.

These days, most research articles merely report the P-value rather than a decision about whether to reject H_0. From the P-value, readers can see the strength of evidence against H_0 and make their own decisions, if they want to.

ACTIVITY 1:   Why is 0.05 Commonly Used as a Significance Level?

A crossover study compares a drug with a placebo for children who suffer from migraine headaches. The study observed each child at two times when he or she had a migraine headache. The child received the drug at one time and a placebo at the other time. Let p denote the probability that the pain relief is better with the drug. You are going to decide whether you can reject H_0: $p = 0.50$ in favor of H_a: $p \neq 0.50$. Ahead of time, you have no idea whether the drug will be better, or worse, than the placebo.

■ The first child does better with the placebo. Would you reject H_0?

■ The second child also does better with the placebo. Would you now reject H_0?

■ The third child also does better with the placebo. Would you now reject H_0?

■ The fourth child also does better with the placebo. Would you now reject H_0?

■ The fifth child also does better with the placebo. Are you ready yet to reject H_0?

If you are like many people, by the time you see the fifth straight success for the placebo over the drug, you are willing to predict that the placebo is better. If the null hypothesis that $p = 0.50$ is actually true, then by the binomial distribution the probability this happens is $(0.50)^5 = 1/32 = 0.03$. For a two-sided test, this result gives a P-value $= 2(0.03) = 0.06$, close to 0.05. So, for many people, it takes a P-value near 0.05 before they feel there is enough evidence to reject a null hypothesis. This may be one reason the significance level of 0.05 has become common over the years in a wide variety of disciplines that use significance tests.

TO PRACTICE THIS CONCEPT, TRY EXERCISE 8.119

8.4   Practicing the Basics

8.38   **Dr. Dog:** In the experiment in Example 4, we got a P-value $= 0.000$ for testing H_0: $p = 1/7$ about dogs' ability to diagnose urine from bladder cancer patients.

a. For the significance level 0.05, what decision would you make?

b. If you made an error in (a), what type of error was it? Explain what the error means in context of the Dr. Dog experiment.

8.39   **Error probability:** A significance test about a proportion is conducted using a significance level of 0.05. The test statistic equals 2.58. The P-value is 0.01.

a. If H_0 were true, for what probability of a Type I error was the test designed?

b. If this test resulted in a decision error, what type of error was it?

8.40   **Astrology errors:** Example 3, in testing H_0: $p = 1/3$ against H_a: $p > 1/3$, analyzed whether astrologers could predict the correct personality chart (out of three possible ones) for a given horoscope better than by random

guessing. In the words of that example, what would be a (a) Type I error, (b) Type II error?

8.41   **Anorexia errors:** Example 7 tested a therapy for anorexia, using hypotheses H_0: $\mu = 0$ and H_a: $\mu \neq 0$ about the population mean weight change μ. In the words of that example, what would be a (a) Type I error, (b) Type II error?

8.42   **Anorexia decision:** Refer to the previous exercise. When we test H_0: $\mu = 0$ against H_a: $\mu > 0$, we get a P-value of 0.02.

a. What would the decision be for a significance level of 0.05? Interpret in context.

b. If the decision in (a) is in error, what type of error is it?

c. Suppose the significance level were instead 0.01. What decision would you make, and if it is in error, what type of error is it?

8.43   **Decision errors in medical diagnostic testing:**
TRY   Consider medical diagnostic testing, such as using a mammogram to detect whether a woman may have breast

cancer. Identify the null hypothesis of no effect with no disease. Identify rejecting H_0 with a positive diagnostic test, which means that the test predicts that the person has the disease. See the table for a summary of the possible outcomes:

Medical Diagnostic Testing

Disease	Medical Diagnosis	
	Negative	Positive
No (H_0)	Correct	Type I error
Yes (H_a)	Type II error	Correct

a. When a radiologist interprets a mammogram, explain why a Type I error is a "false positive," predicting that a woman has breast cancer when actually she does not.

b. A Type II error is a "false negative." What does this mean, and what is the consequence of such an error to the woman?

c. [TRY] A radiologist wants to decrease the chance of telling a woman that she may have breast cancer when actually she does not. To do this, a positive test result will be reported only when there is *extremely* strong evidence that breast cancer is present. What is the disadvantage of this approach?

8.44 **Detecting prostate cancer:** Refer to the previous exercise about medical diagnoses. A *New York Times* article (February 17, 1999) about the PSA blood test for detecting prostate cancer stated: "The test fails to detect prostate cancer in 1 in 4 men who have the disease."

a. For the PSA test, explain what a Type I error is, and explain the consequence to a man of this type of error.

b. For the PSA test, what is a Type II error? What is the consequence to a man of this type of error?

c. To which type of error does the probability of 1 in 4 refer?

d. The article also stated that "as many as two-thirds of the men tested receive false-positive results." That is, given that you receive a positive result, the probability that you do not actually have prostate cancer is 2/3. Explain the difference between this and the conditional probability of a Type I error, given that you do not actually have prostate cancer.

8.45 **Which error is worse?:** Which error, Type I or Type II, would usually be considered more serious for decisions in the following tests? Explain why.

a. A trial to test a murder defendant's claimed innocence, when conviction results in the death penalty.

b. A medical diagnostic procedure, such as a mammogram.

8.5 | Limitations of Significance Tests

Chapters 7 and 8 have presented the two primary methods of statistical inference—confidence intervals and significance testing. We will use both of these methods throughout the rest of the book. Of the two methods, confidence intervals can be more useful, for reasons we'll discuss in this section. Significance tests have more potential for misuse. We'll now summarize their major limitations.

Statistical Significance Does Not Mean Practical Significance

When we conduct a significance test, its main relevance is studying whether the true parameter value is

■ above, or below, the value in H_0, and

■ sufficiently different from the value in H_0 to be of practical importance.

A significance test gives us information about whether the parameter differs from the H_0 value and its direction from that value, but we'll see now that it does not tell us about practical importance.

There is an important distinction between *statistical significance* and *practical significance*. A small P-value, such as 0.001, is highly statistically significant, giving strong evidence against H_0. It does not, however, imply an *important* finding in any practical sense. The small P-value means that if H_0 were true, the observed data would be unusual. It does not mean that the true value of the parameter is far from the null hypothesis value in practical terms. In particular, whenever the sample size is large, small P-values can occur when the point estimate is near the parameter value in H_0, as the following example shows.

EXAMPLE 11

Political Conservatism and Liberalism in America

Picture the Scenario

Where do Americans say they fall on the conservative–liberal political

spectrum? The General Social Survey asks, "I'm going to show you a seven-point scale on which the political views that people might hold are arranged from extremely liberal,

point 1, to extremely conservative, point 7. Where would you place yourself on this scale?" Table 8.8 shows the scale and the distribution of 2644 responses for the GSS in 2000.

TABLE 8.8: Responses of 2644 Subjects on a Seven-Point Scale of Political Views

Category	Count
1. Extremely liberal	107
2. Liberal	308
3. Slightly liberal	285
4. Moderate, middle of road	1054
5. Slightly conservative	390
6. Conservative	411
7. Extremely conservative	89

IN WORDS

An **ordinal variable** is a categorical variable for which the categories are ordered from low to high in some sense. Table 8.8 lists categories of political views from low to high in terms of conservatism.

This categorical variable has seven categories that are ordered in terms of degree of liberalism or conservatism. Categorical variables that have *ordered* categories are called **ordinal variables**. Sometimes we treat an ordinal variable in a quantitative manner by assigning scores to the categories and summarizing the data by the mean. This summarizes whether observations gravitate toward the conservative or the liberal end of the scale. For the category scores of 1 to 7, as in Table 8.8, a mean of 4.0 corresponds to the moderate outcome. A mean below 4 shows a propensity toward liberalism, and a mean above 4 shows a propensity toward conservatism. The 2644 observations in Table 8.8 have a mean of 4.10 and a standard deviation of 1.41.

Questions to Explore

a. Do these data indicate that the population has a propensity toward liberalism or toward conservatism? Answer by conducting a significance test that compares the population mean to the moderate value of 4.0, by testing $H_0: \mu = 4.0$ against $H_a: \mu \neq 4.0$.

b. Does this test show (i) statistical significance? (ii) practical significance?

Think It Through

a. For the sample data, the standard error $se = s/\sqrt{n} = 1.41/\sqrt{2644} = 0.0274$. The test statistic for $H_0: \mu = 4.0$ equals

$$t = (\bar{x} - \mu_0)/se = (4.10 - 4.00)/0.0274 = 3.65.$$

Its two-sided P-value is 0.0003. There is *extremely* strong evidence that the true mean exceeds 4.0, with the true mean falling on the conservative side of moderate.

b. Although the P-value is very small, on a scale of 1 to 7, the sample mean of 4.10 is very close to the value of 4.0 in H_0. It is only one tenth of the distance from the moderate score of 4.0 to the slightly conservative score of 5.0. Although the difference of 0.10 between the sample mean of 4.10 and the null hypothesis mean of 4.0 is highly significant statistically, this difference is small in practical terms. In practice, we'd regard a mean of 4.10 as "moderate" on this 1 to 7 scale. In summary, there's statistical significance but not practical significance.

Insight

As we also saw in Example 9, with large samples P-values can be small even when the sample estimate falls near the parameter value in H_0. The P-value measures the extent of evidence about H_0, not how far the true parameter value happens to be from H_0. Always inspect the difference between the sample estimate and the null hypothesis value to gauge the practical implications of a test result.

TO PRACTICE THIS CONCEPT, TRY EXERCISE 8.47

Significance Tests Are Less Useful than Confidence Intervals

Although significance tests are useful, most statisticians believe that this method has been overemphasized in research.

■ A significance test merely indicates whether the particular parameter value in H_0 (such as $\mu = 0$) is plausible.

When a P-value is small, the significance test indicates that the hypothesized value is not plausible, but it tells us little about which potential parameter values *are* plausible.

■ A confidence interval is more informative, because it displays the entire set of believable values.

A confidence interval shows how badly H_0 may be false by showing whether the values in the interval are far from the H_0 value. It helps us to determine whether the difference between the true value and the H_0 value has practical importance.

Let's illustrate with Example 11 and the 1 to 7 scale for political beliefs. A 95% confidence interval for μ is $\bar{x} \pm 1.96(se) = 4.10 \pm 1.96(0.0274)$, or (4.05, 4.15). We can conclude that the difference between the population mean and the moderate score of 4.0 is small. Figure 8.13 illustrates. Although the P-value of 0.0003 provided extremely strong evidence against $H_0: \mu = 4.0$, in practical terms the confidence interval shows that H_0 is not wrong by much.

By contrast, if $\bar{x}$ had been 6.10 (instead of 4.10), the 95% confidence interval would equal (6.05, 6.15). This indicates a substantial difference from 4.0, the mean response being near the conservative score rather than near the moderate score.

▶ **FIGURE 8.13: Statistical Significance but not Practical Significance.** In testing H_0: $\mu = 4.0$, the P-value = 0.0003, but the confidence interval of (4.05, 4.15) shows that μ is very close to the H_0 value of 4.0. **Question:** For H_0: $\mu = 4.0$, does a sample mean of 6.10 and confidence interval of (6.05, 6.15) indicate (a) statistical significance? (b) practical significance?

Recall

$P(A|B)$ denotes the conditional probability of event A, given event B. To find a P-value, we condition on H_0 being true (that is, we presume it is true), rather than find the probability it is true. ▪

Misinterpretations of Results of Significance Tests

Unfortunately, results of significance tests are often misinterpreted. Here are common misinterpretations, some of which we've already discussed:

▪ **"Do not reject H_0" does not mean "Accept H_0."** If you get a P-value above 0.05 when the significance level is 0.05, you cannot conclude that H_0 is correct. We can never accept a single value, which H_0 contains, such as $p = 0.50$ or $\mu = 0$. A test merely indicates whether a particular parameter value is plausible. A confidence interval shows that there is a *range* of plausible values, not just a single one.

▪ **Statistical significance does not mean practical significance.** A small P-value does not tell us whether the parameter value differs by much in practical terms from the value in H_0.

▪ **The P-value cannot be interpreted as the probability that H_0 is true.** The P-value is

**P(test statistic takes observed value or beyond in tails | H_0 true),
not P(H_0 true | observed test statistic value).**

We've been calculating probabilities about test statistic values, not about parameters. It makes sense to find the probability that a test statistic takes a value in the tail, but the probability that a population mean = 0 does not make sense, because probabilities do not apply to parameters. The null hypothesis H_0 is either true or not true, and we simply do not know which.[9]

▪ **It is misleading to report results only if they are "statistically significant."** Some research journals have the policy of publishing results of a study only if the P-value ≤ 0.05. Here's a danger of this policy: If there truly is no effect, but 20 researchers independently conduct studies about it, we would expect about 20(0.05) = 1 of them to obtain significance at the 0.05 level merely by chance. (When H_0 is true, about 5% of the time we get a P-value below 0.05 anyway.) If that researcher then submits results to a journal but the other 19 researchers do not, the article published will be a Type I error—reporting an effect when there really is not one. The popular media may then also report on this study, causing the general public to hear about an effect that does not actually exist.

▪ **Some tests may be statistically significant just by chance.** Related to the last comment, you should never scan pages and pages of computer output for results that are statistically significant and report only those. If you run 100 tests, even if all the null hypotheses are correct, you would expect to get P-values of 0.05 or less about 100(0.05) = 5 times. Keep this in mind and be skeptical of reports of significance that might merely reflect ordinary random variability. For instance, suppose an article reports an unusually high rate of a rare type of cancer in your town. It could be due to some cause such as air or water pollution. However, if researchers found this by looking at data for *all* towns and cities nationwide, it could also be due to random variability. Determining which is true may not be easy.

---

[9]It is possible to find probabilities about parameter values using an approach called **Bayesian statistics**, but this requires extra assumptions and is beyond the scope of this text.

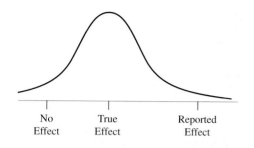

No Effect   True Effect   Reported Effect

■ **True effects may not be as large as initial estimates reported by the media.** Even if a statistically significant result is a true effect, the true effect may be smaller than suggested in the first article about it. For instance, often several researchers perform similar studies, but the results that get attention are the most extreme ones. This sensationalism may come about because the researcher who is the first to publicize the result is the one who got the most impressive sample result, perhaps way out in the tail of the sampling distribution of all the possible results. Then the study's estimate of the effect may be greater than later research shows it to be. See the margin figure and the next example.

EXAMPLE 12

Are Many Medical "Discoveries" Really Type I Errors?

Picture the Scenario

What can be done with heart attack victims to increase their chance of survival? In 1992, trials of a clot-busting drug called anistreplase suggested that it doubled the chance of survival. Likewise, a 1993 study estimated that injections of magnesium could double the chance of survival. However, a much larger study in 1995 of heart attack survival rates among 58,000 patients indicated that this optimism was premature. The actual effectiveness of anistreplase seemed to be barely half that estimated by the original trial, and magnesium injections seemed to have no effect at all.[10] The anistreplase finding is apparently an example of a true effect not being as large as the initial estimate, and the report from the original magnesium study may well have been a Type I error.

Question to Explore

In medical studies, suppose that a true effect exists only 10% of the time. Suppose also that when an effect truly exists, there's a 50% chance of making a Type II error and failing to detect it. These were the hypothetical percentages used in a recent article in a medical journal.[11] The authors noted that many medical studies have a high Type II error rate because they are not able to use a large sample size. Assuming these rates, could a substantial percentage of medical "discoveries" actually be Type I errors? Given that H_0 is rejected, approximate the P(Type I error) by considering a tree diagram of what you would expect to happen with 1000 medical studies that test various hypotheses.

Think It Through

Figure 8.14 is a tree diagram of what's expected. A true effect exists 10% of the time, or in 100 of the 1000 studies. We do not get a small enough P-value to detect this true effect 50% of the time, that is, in 50 of these 100 studies. An effect will be reported for the other 50 of the 100 that do truly have an effect. Now, for the 900 cases in which there truly is no effect, with the usual significance level of 0.05 we expect 5% of the 900 studies (that is, 45 studies) to incorrectly reject the null hypothesis and predict that there is an effect. So, of the 1000 studies, we expect 50 to report an effect that is truly there, but we also expect 45 to report an effect that does not really exist. If the assumptions are reasonable, then a pro-

---

[10]"The great health hoax," by R. Matthews, in *The Sunday Telegraph*, September 13, 1998.

[11]By J. Sterne, G. Smith, and D. R. Cox, *BMJ*, vol. 322, pp. 226–231 (2001).

portion of $45/(45 + 50) = 0.47$ of medical studies that report effects (that is, reject H_0) are actually reporting Type I errors. In summary, nearly half the time when an effect is reported, there actually is no effect in the population!

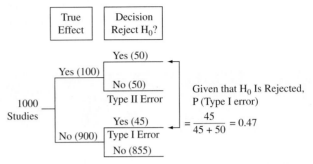

▲ FIGURE 8.14: Tree Diagram of Results of 1000 Hypothetical Medical Studies. This assumes that a true effect exists 10% of the time and that there's a 50% chance of a Type II error when an effect truly does exist.

Insight

Be skeptical when you hear reports of new medical advances. The true effect may be weaker than reported, or there may actually be no effect at all.

TO PRACTICE THIS CONCEPT, TRY EXERCISE 8.51

On the Shoulders of . . . Jerzy Neyman and the Pearsons

How can you build a framework for making decisions about everything from whether a roulette wheel is balanced to whether cloud-seeding results in rain?

The methods of confidence intervals and hypothesis testing were introduced in a series of articles beginning in 1928 by Jerzy Neyman (1894–1981) and Egon Pearson (1895–1980). Neyman emigrated from Poland to England and then to the United States, where he established a top-notch Statistics Department at the University of California at Berkeley. He helped develop the theory of statistical inference, and he applied the theory to scientific questions in a variety of areas, such as agriculture, astronomy, biology, medicine, and weather modification. For instance, late in his career, Neyman's analysis of data from several randomized experiments showed that cloud-seeding can have a considerable effect on rainfall.

Much of Neyman's theoretical research was done with Egon Pearson, a professor at University College, London. Pearson's father, Karl Pearson, had developed one of the first statistical tests in 1900 to study various hypotheses, including whether the outcomes on a roulette wheel were equally likely. (We'll study his test, the **chi-squared test**, in Chapter 10.) Neyman and the younger Pearson developed the decision-making framework that introduced the two types of errors and the most powerful significance tests for various hypotheses.

◀ JERZY NEYMAN AND EGON PEARSON.
Neyman and Pearson developed statistical theory for making decisions about hypotheses.

Finally, remember that tests and confidence intervals use *sample* data to make inferences about *populations*. If you have data for an entire population, statistical inference is not relevant. For instance, if your college reports that the class of entering freshmen had a mean high school GPA of 3.60, there is no need to perform a test or construct a confidence interval. You already know the mean for that population.

8.5   Practicing the Basics

8.46 Misleading summaries?: Two researchers conduct separate studies to test H_0: $p = 0.50$ against H_a: $p \neq 0.50$, each with $n = 400$. Researcher A gets 220 observations in the category of interest, and $\hat{p} = 220/400 = 0.550$. Researcher B gets 219 in the category of interest, and $\hat{p} = 219/400 = 0.5475$.

  a. Show that $z = 2.00$ and P-value $= 0.046$ for Researcher A. Show that $z = 1.90$ and P-value $= 0.057$ for Researcher B.

  b. Using $\alpha = 0.05$, indicate in each case whether the result is "statistically significant." Interpret.

  c. From (a) and (b), explain why important information is lost by reporting the result of a test as "P-value ≤ 0.05" versus "P-value > 0.05," or as "reject H_0" versus "Do not reject H_0," instead of reporting the actual P-value.

  d. Show that the 95% confidence interval for p is (0.501, 0.599) for Researcher A and (0.499, 0.596) for Researcher B. Explain how this method shows that, in practical terms, the two studies had very similar results.

8.47 Practical significance: A study considers whether the mean score on a college entrance exam for students in 2005 is any different from the mean score of 500 for students who took the same exam in 1975. Let μ represent the mean score for all students who took the exam in 2005.

  a. Find the P-value for testing H_0: $\mu = 500$ against H_a: $\mu \neq 500$, based on a random sample of 40,000 students who took the exam in 2005, getting $\bar{x} = 498$ and $s = 100$.

  b. Explain why the test result is statistically significant but not practically significant.

8.48 Effect of n: Example 11 analyzed political conservatism and liberalism in America. Suppose that the sample mean of 4.10 and sample standard deviation of 1.41 were from a sample size of only 25, rather than 2644.

  a. Find the test statistic.

  b. Find the P-value for testing H_0: $\mu = 4.0$ against H_a: $\mu \neq 4.0$. Interpret.

  c. Show that a 95% confidence interval for μ is (3.5, 4.7).

  d. Together with the results of Example 11, explain what this illustrates about the effect of sample size on (i) the size of the P-value (for a given mean and standard deviation), (ii) the width of the confidence interval.

8.49 Fishing for significance: A marketing study conducts 60 significance tests about means and proportions for several groups. Of them, 3 tests are statistically significant at the 0.05 level. The study's final report stresses only the tests with significant results, not mentioning the other 57 tests. What is misleading about this?

8.50 Selective reporting: In 2004, New York Attorney General Eliot Spitzer filed a lawsuit against GlaxoSmithKline pharmaceutical company, claiming that the company failed to publish results of one of their studies that showed that an antidepressant drug (Paxil) may make adolescents more likely to commit suicide. Partly as a consequence, editors of 11 medical journals agreed to a new policy to make researchers and companies register all clinical trials when they begin, so that negative results cannot later be covered up. The *International Journal of Medical Journal Editors* wrote, "Unfortunately, selective reporting of trials does occur, and it distorts the body of evidence available for clinical decision-making." Explain why this controversy relates to the argument that it is misleading to report results only if they are "statistically significant." (*Hint:* See the subsection of this chapter on misinterpretations of significance tests.)

8.51 How many medical discoveries are Type I errors?: Refer to Example 12. Using a tree diagram, given that H_0 is rejected, approximate P(Type I error) under the assumption that a true effect exists 20% of the time and that there's a 30% chance of a Type II error.

8.52 Interpret medical research studies:

  a. An advertisement by Schering Corp. in 1999 for the allergy drug Claritin mentioned that in a clinical trial, the proportion who showed symptoms of nervousness was not significantly greater for patients taking Claritin than for patients taking placebo. Does this mean that the population proportion having nervous symptoms is exactly the same using Claritin and using placebo? How would you explain this to someone who has not studied statistics?

  b. An article in the medical journal *BMJ* (by M. Petticrew et al., published 11/02) found no evidence to back the commonly held belief that a positive attitude can lengthen the lives of cancer patients. The authors noted that the studies that had indicated a benefit from some coping strategies tended to be smaller studies with weaker designs. Using this example and the text discussion, explain why you need to have some skepticism when you hear that new research suggests that some therapy or drug has an impact in treating a disease.

8.6 | How Likely is a Type II Error (Not Rejecting H_0, Even though It's False)?

The probability of a Type I error is the significance level α of the test. Given that H_0 is true, when $\alpha = 0.05$, the probability of rejecting it equals 0.05.

Given that H_0 is false, a Type II error results from *not* rejecting H_0. This probability has more than one value, because H_a contains a range of possible values for the parameter. Each value in H_a has its own probability of a Type II error. Let's see how to find the probability of a Type II error at a particular value.

EXAMPLE 13

Finding P(Type II Error) as Part of a Study Design

Picture the Scenario

Examples 1, 3, and 5 discussed an experiment to test astrologers' predictions. For each person's horoscope, an astrologer must predict which of three personality charts is the actual one. Let p denote the probability of a correct prediction. Consider the test of $H_0: p = 1/3$ (astrologers' predictions are like random guessing) against $H_a: p > 1/3$ (better than random guessing), using the 0.05 significance level. Suppose an experiment plans to use $n = 116$ people, as this experiment did.

Questions to Explore

a. For what values of the sample proportion can we reject H_0?

b. The National Council for Geocosmic Research claimed that p would be 0.50 or higher. If truly $p = 0.50$, for what values of the sample proportion would we make a Type II error, failing to reject H_0 even though it's false?

c. If truly $p = 0.50$, what is the probability that a significance test based on this experiment will make a Type II error?

Think It Through

a. For testing $H_0: p = 1/3$ with $n = 116$, the standard error for the test statistic is

$$se_0 = \sqrt{p_0(1 - p_0)/n} = \sqrt{[(1/3)(2/3)]/116} = 0.0438.$$

For $H_a: p > 1/3$, a test statistic of $z = 1.645$ has a P-value (right-tail probability) of 0.05. If $z \geq 1.645$, the P-value is ≤ 0.05 and we can reject H_0. That is, we reject H_0 when $\hat{p}$ falls at least 1.645 standard errors above $p_0 = 1/3$,

$$\hat{p} \geq 1/3 + 1.645(se) = 1/3 + 1.645(0.0438) = 0.405.$$

Figure 8.15 shows the sampling distribution of $\hat{p}$ and this rejection region. The figure is centered at 1/3, because the test statistic is calculated and the rejection region is formed presuming that H_0 is correct.

b. When H_0 is false, a Type II error occurs when we fail to reject H_0. From part (a) and Figure 8.15, we do not reject H_0 if $\hat{p} < 0.405$.

0.05

0   1.645   z

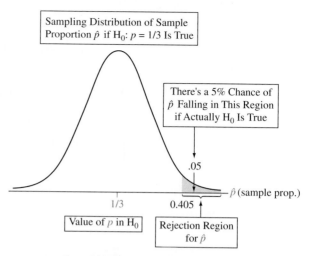

▲ **FIGURE 8.15: For Sample Proportion $\hat{p}$ above 0.405, Reject H_0: $p = 1/3$ against H_a: $p > 1/3$ at the 0.05 Significance Level.** When the true $p > 1/3$, a Type II error occurs if $\hat{p} < .405$, since then the P-value > 0.05 and we do not reject H_0. **Question:** Why does each possible value of p from H_a have a separate probability of a Type II error?

c. If the true value of p is 0.50, then the true sampling distribution of $\hat{p}$ is centered at 0.50, as Figure 8.16 shows. The probability of a Type II error is the probability that $\hat{p} < 0.405$ when $p = 0.50$. When $p = 0.50$, the standard error of $\hat{p}$ for a sample size of 116 is $\sqrt{0.50(0.50)/116} = 0.0464$. (Note that this differs a bit from the standard error for the test statistic, which uses 1/3 instead of 0.50 for p.)

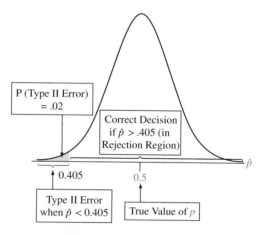

▲ **FIGURE 8.16: Calculation of P(Type II Error) when True Parameter $p = 0.50$.** A Type II error occurs if the sample proportion $\hat{p} < 0.405$. **Question:** If the value of p decreases to 0.40, will the probability of Type II error decrease, or increase?

For a normal sampling distribution with mean 0.50 and standard error 0.0464, the $\hat{p}$ value of 0.405 has a z-score of

$$z = (0.405 - 0.50)/0.0464 = -2.04.$$

Using Table A or software, the left-tail probability below -2.04 for the standard normal distribution equals 0.02. In summary, when $p = 0.50$ the probability of making a Type II error and failing to reject H_0: $p = 1/3$ is only 0.02.

Figure 8.17 shows the two figures together that we used in our reasoning. The normal distribution with mean 1/3 was used to find the rejection region, based on what we expect for $\hat{p}$ when H_0: $p = 1/3$ is true. The normal distribution with mean 0.50 was used to find the probability that $\hat{p}$ fails to fall in the rejection region even though $p = 0.50$ (that is, a Type II error occurs).

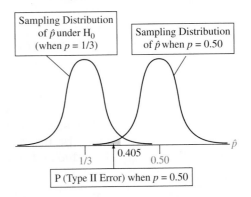

▲ **FIGURE 8.17: Sampling Distributions of $\hat{p}$ when H_0: $p = 1/3$ is true and when $p = 0.50$. Question:** Why does the shaded area under the left tail of the curve for the case $p = 0.50$ represent P(Type II error) for that value of p?

Insight

If astrologers truly had the predictive power claimed by the National Council for Geocosmic Research, the experiment would have been unlikely to fail to detect it.

TO PRACTICE THIS CONCEPT, TRY EXERCISE 8.54

The probability of a Type II error increases when the true parameter value moves closer to H_0. To verify this, try to find the probability of a Type II error when $p = 0.40$ instead of 0.50. You should get P(Type II error) = 0.54. (Remember to recompute the exact standard error, now using 0.40 for p.) If an astrologer can predict better than random guessing but not *much* better, we may not detect it with this experiment. Figure 8.18 plots P(Type II error) for various values of p above 1/3. The farther the parameter value falls from the number in H_0, the less likely a Type II error.

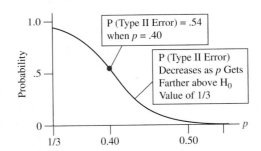

▲ **FIGURE 8.18: Probability of a Type II Error for Testing H_0: $p = 1/3$ against H_a: $p > 1/3$.** This is plotted for the values of p in H_a, when the significance level = 0.05. **Question:** If the sample size n increases, how do you think this curve would change?

For a fixed significance level α, P(Type II error) decreases

■ as the parameter value moves farther into the H_a values and away from the H_0 value.

■ as the sample size increases.

Also, recall that P(Type II error) increases as α decreases. One reason that extremely small values, such as $\alpha = 0.001$, are not common is that P(Type II error) is then too high. We may be unlikely to reject H_0, even though it is false.

Before conducting a study, researchers should find P(Type II error) for the size of effect they want to be able to detect. If P(Type II error) is high, it may not be worth conducting the study unless they can use a larger sample size and lower the probability. They won't know the value of the parameter, so they won't know the actual P(Type II error). It may be large if n is small and if the true parameter value is not far from the value in H_0. This may be the reason a particular significance test does not obtain a small P-value and reject H_0.

The Power of a Test

When H_0 is false, you want the probability of rejecting it to be high. The probability of rejecting H_0 is called the **power** of the test. For a particular value of the parameter from the range of alternative hypothesis values,

$$\boxed{\textbf{Power} = 1 - \textbf{P(Type II error)}}$$

In Example 13, for instance, P(Type II error) = 0.02 when $p = 0.50$. Therefore, the power of the test at $p = 0.50$ is $1 - 0.02 = 0.98$. The higher the power, the better, so a test power of 0.98 is quite good.

In practice, it is ideal for studies to have high power while using a relatively small significance level, such as 0.05 or 0.01. Before granting research support, many agencies (such as the National Institutes of Health) expect research scientists to show that for the planned study reasonable power (usually, at least 0.80) exists at values of the parameter that are considered practically significant.

EXAMPLE 14

What Was the Power of the Test in the TT Experiment?

Picture the Scenario

In Example 6 on the therapeutic touch (TT) experiment, the data did not support the TT practitioners' claim to be able to detect a human energy field. The P-value was not small for testing H_0: $p = 0.50$ against H_a: $p > 0.50$, where p is the probability of a correct prediction about which hand was near the researcher's hand. The medical journal article about the study stated, "The statistical power of this experiment was sufficient to conclude that if TT practitioners could reliably detect a human energy field, the study would have demonstrated this." For the test of H_0: $p = 0.50$ with one of the sets of trials, the power was reported as 0.96 if actually $p = 2/3$.

Questions to Explore

a. How should the probability of 0.96 be interpreted?

b. In context, what is a Type II error for this experiment?

c. If $p = 2/3$, what is the probability of committing a Type II error?

Think It Through

a. The power of 0.96 is the probability of correctly rejecting H_0 when it is false. If the actual probability of correct predictions by TT practitioners was 2/3, there was a 96% chance of data such that the significance test performed would reject H_0.

b. A Type II error occurs if we do not reject H_0: $p = 0.50$, when actually TT practitioners *can* predict correctly more than half the time. The consequence would be to question the truthfulness of what they have been practicing for over 30 years, when they actually do have some ability.

c. If $p = 2/3$, the value of P(Type II error) is $1 - $ (Power at $p = 2/3$). This is $1 - 0.96 = 0.04$. A Type II error was unlikely if truly $p = 2/3$.

Insight

Based on the data, the researchers remained skeptical that TT practitioners could detect a human energy field. Because of the strong power at a value of p that TT practitioners claim is plausible, the researchers felt justified in concluding that "TT claims are groundless and further use of TT by health professionals is unjustified."

TO PRACTICE THIS CONCEPT, TRY EXERCISE 8.58

ACTIVITY 2: Let's Simulate the Performance of Significance Tests

Let's get a feel for the two possible errors in significance tests. To do this, for a given population proportion value you can simulate many samples and perform a significance test for each. You can then check how often the tests make an incorrect inference. You can conduct the simulation using statistical software (such as MINITAB) or using a statistical applet that allows you to control the null hypothesis value, the true parameter value, the sample size, and the significance level.

Try this by going to the *hypothesis tests for a proportion* applet on the text CD. Set the null hypothesis as $p = 1/3 = 0.33$, for a one-sided test ($p > 0.33$) with sample size $n = 116$, a case Example 3 on the astrology experiment considered. At the menu, set the true proportion value to $p = 0.33$. Click *Simulate* and 100 samples will be generated. The applet calculates the resulting z-statistic and P-value for the sample proportion from each sample. The applet summaries the number of significance tests that reject the null hypothesis at the 0.05 and 0.01 significance levels. How many

significance tests out of 100 tests rejected the null at the 0.05 significance level? At the 0.01 significance level? How many would you expect at each significance level to reject the null, giving us an incorrect inference, a Type I error?

To get a feel for what happens "in the long run," do this simulation 10,000 times by clicking the *Simulate* button multiple times until you have a cumulative total of 10,000 significance tests. How many significance tests out of 10,000 tests incorrectly rejected the null at the 0.05 significance level? At the 0.01 significance level?

Next, change the value of p to 0.50, but keep the null hypothesis as $p = 0.33$, so H_0 is actually false. Perform a single significance test. Now if you get a P-value above 0.05 or 0.01 and fail to reject H_0, you made a Type II error, since H_0 is actually false. Perform the significance test 10,000 times. What percentage of times did you make a Type II error using a significance level of 0.05? By Example 13, this should happen only about 2% of the time.

TO PRACTICE THIS CONCEPT, TRY EXERCISE 8.59

8.53 **Two sampling distributions:** A study is designed to test H_0: $p = 0.50$ against H_a: $p > 0.50$, taking a random sample of size $n = 100$, using significance level 0.05.

  a. Show that the rejection region consists of values of $\hat{p} > 0.582$.

  b. Sketch a single picture that shows (i) the sampling distribution of $\hat{p}$ when H_0 is true and (ii) the sampling distribution of $\hat{p}$ when $p = 0.60$. Label each sampling distribution with its mean and standard error, and highlight the rejection region.

  c. Find P(Type II error) when $p = 0.60$.

8.54 **Gender bias in selecting managers:** Exercise 8.17 **TRY** tested the claim that female employees were passed over for management training in favor of their male colleagues. Statewide, the large pool of more than 1000 eligible employees who can be tapped for management training is 40% female and 60% male. Let p be the probability of selecting a female for any given selection. For testing H_0: $p = 0.40$ against H_a: $p < 0.40$ based on a random sample of 50 selections, using the 0.05 significance level, verify that

  a. A Type II error occurs if the sample proportion falls less than 1.645 standard errors below the null hypothesis value, which means that $\hat{p} > 0.286$.

  b. When $p = 0.20$, a Type II error has probability 0.064.

8.55 **Balancing Type I and Type II errors:** Recall that the smaller the probability of a Type I error, α, the larger the P(Type II error). Let's check this for Example 13. There we found P(Type II error) for testing H_0: $p = 1/3$ (astrologers randomly guessing) against H_a: $p > 1/3$ when actually $p = 0.50$, with $n = 116$. If we use $\alpha = 0.01$, verify that

  a. A Type II error occurs if the sample proportion falls less than 2.326 standard errors above the null hypothesis value, which means $\hat{p} < 0.435$.

  b. When $p = 0.50$, a Type II error has probability 0.08. (By comparison, Example 13 found P(Type II error) = 0.02 when $\alpha = 0.05$, so we see that P(Type II error) increased when P(Type I error) decreased.)

8.56 **P(Type II error) large when p close to H_0:** For testing H_0: $p = 1/3$ (astrologers randomly guessing) against H_a: $p > 1/3$ with $n = 116$, Example 13 showed that P(Type II error) = 0.02 when $p = 0.50$. Now suppose that $p = 0.35$.

  a. Show that P(Type II error) = 0.89.

  b. Explain intuitively why P(Type II error) is large when the parameter value is close to the value in H_0 and decreases as it moves farther from that value.

8.57 **Type II error with two-sided H_a:** In Example 13 for testing H_0: $p = 1/3$ (astrologers randomly guessing) with $n = 116$ when actually $p = 0.50$, suppose we used H_a: $p \neq 1/3$. Then show that:

  a. A Type II error occurs if $0.248 < \hat{p} < 0.419$.

  b. The probability is 0.00 that $\hat{p} < 0.248$ and 0.96 that $\hat{p} > 0.419$.

  c. P(Type II error) = 0.04.

8.58 **Power of TT:** Consider Example 14 about the power **TRY** of the test used in the TT experiment for testing H_0: $p = 0.50$ against H_a: $p > 0.50$, where p is the probability of a correct prediction about which hand was nearer the researcher's hand. In a significance test planned for a second set of trials, the power was reported as 0.98 at $p = 2/3$.

  a. How should the probability of 0.98 be interpreted?

  b. In context, what is a Type II error for this experiment?

  c. What is the probability of committing a Type II error if $p = 2/3$?

8.59 **Simulating Type II errors:** Refer to the simulation in **TRY** Activity 2 at the end of the section.

  a. Repeat the simulation, now assuming that actually $p = 0.45$. In a large number of simulations, what proportion of the time did you make a Type II error at the significance level of 0.05? What does theory predict for this proportion?

  b. Do you think that the proportion of Type II errors will increase, or decrease, if actually $p = 0.50$? Check your intuition by conducting another simulation.

  c. Refer to (a). Do you think that the proportion of Type II errors will increase, or decrease, if the sample size in each simulation is 50 instead of 116? Check your intuition by conducting another simulation.

CHAPTER REVIEW

CHAPTER SUMMARY

A significance test helps us to judge whether a particular value for a parameter is believable. Each significance test has five steps:

1. **Assumptions:** The most important assumption for any significance test is that the data result from a *random sample* or *randomized experiment*. Specific significance tests make other assumptions, such as summarized in the box below for significance tests about means and proportions.

2. **Null and alternative hypotheses about the parameter:** Null hypotheses have the form H_0: $p = p_0$ for a proportion and H_0: $\mu = \mu_0$ for a mean, where p_0 and μ_0 denote particular values, such as $p_0 = 0.5$ and $\mu_0 = 0$. The most common alternative hypothesis is **two-sided**, such as H_a: $p \neq 0.5$. **One-sided** hypotheses such as H_a: $p > 0.5$ and H_a: $p < 0.5$ are also possible.

3. **Test statistic:** This measures how far the sample estimate of the parameter falls from the null hypothesis value. The z statistic for proportions and the t statistic for means have the form

   $$\text{Test statistic} = \frac{\text{parameter estimate} - \text{null hypothesis value}}{\text{standard error}}.$$

   This measures the number of standard errors that the parameter estimate ($\hat{p}$ or $\overline{x}$) falls from the null hypothesis value (p_0 or μ_0).

4. **P-value:** This is a probability summary of the evidence that the data provide about the null hypothesis. It equals the probability that the test statistic takes a value like the observed one or even more extreme.

   ■ It is calculated by presuming that H_0 is true.

■ The test statistic values that are "more extreme" depend on the alternative hypothesis. When H_a is two-sided, the P-value is a two-tail probability. When H_a is one-sided, the P-value is a one-tail probability.

■ When the P-value is small, the observed data would be unusual if H_0 were true. The smaller the P-value, the stronger the evidence against H_0.

5. **Conclusion:** A test concludes by interpreting the P-value in the context of the study. Sometimes a decision is needed,

using a fixed **significance level** α, usually $\alpha = 0.05$. Then we reject H_0 if the P-value $\leq \alpha$. Two types of error can occur:

■ A **Type I error** results from rejecting H_0 when it is true. When H_0 is true, the significance level = P(Type I error).

■ When H_0 is false, a **Type II error** results from failing to reject H_0.

SUMMARY: Significance Tests for Population Proportions and Means

	Parameter	
	Proportion	**Mean**
1. Assumptions	Categorical variable	Quantitative variable
	Randomization	Randomization
	Expected numbers of successes and failures ≥ 15	Approximately normal population
2. Hypotheses	$H_0: p = p_0$ $H_a: p \neq p_0$ (two-sided) $H_a: p > p_0$ (one-sided) $H_a: p < p_0$ (one-sided)	$H_0: \mu = \mu_0$ $H_a: \mu \neq \mu_0$ (two-sided) $H_a: \mu > \mu_0$ (one-sided) $H_a: \mu < \mu_0$ (one-sided)
3. Test statistic	$z = \dfrac{\hat{p} - p_0}{se}$ $(se = \sqrt{p_0(1 - p_0)/n})$	$t = \dfrac{\bar{x} - \mu_0}{se}$ $(se = s/\sqrt{n})$
4. P-value	Two-tail ($H_a: p \neq p_0$) or right tail ($H_a: p > p_0$) or left tail ($H_a: p < p_0$) probability from standard normal distribution	Two-tail ($H_a: \mu \neq \mu_0$) or right tail ($H_a: \mu > \mu_0$) or left tail ($H_a: \mu < \mu_0$) probability from t distribution $(df = n - 1)$
5. Conclusion	Interpret P-value in context Reject H_0 if P-value $\leq \alpha$	Interpret P-value in context Reject H_0 if P-value $\leq \alpha$

Where We're Going

To introduce statistical inference, Chapters 7 and 8 presented methods for a single proportion or mean. In practice, inference is used more commonly to compare parameters for different groups (for example, females and males). The next chapter shows how to compare proportions and how to compare means for two groups.

SUMMARY OF NOTATION

H_0 = null hypothesis, H_a = alternative hypothesis
p_0 = null hypothesis value of proportion,
μ_0 = null hypothesis value of mean

α = significance level
= probability of Type I error
(usually 0.05; the P-value must be $\leq \alpha$ to reject H_0)

CHAPTER PROBLEMS

Practicing the Basics

8.60 **H$_0$ or H$_a$?:** For each of the following hypotheses, explain whether it is a null hypothesis or an alternative hypothesis:

a. For females, the population mean on the political ideology scale is equal to 4.0.

b. For males, the population proportion who support the death penalty is larger than 0.50.

c. The diet has an effect, the population mean change in weight being less than 0.

d. For all Subway (submarine sandwich) stores worldwide, the difference between sales this month and in the corresponding month a year ago has a mean of 0.

8.61 **ESP:** A person who claims to possess extrasensory perception (ESP) says she can guess more often than not the outcome of a flip of a balanced coin. Out of 20 flips, she guesses correctly 12 times. Would you conclude that she truly has ESP? Answer by reporting all five steps of a significance test of the hypothesis that each of her guesses has probability 0.50 of being correct against the alternative that corresponds to her having ESP.

8.62 **Free-throw accuracy:** Consider all cases in which a pro basketball player shoots two free throws and makes one and misses one. Which do you think is more common: making the first and missing the second, or missing the first and making the second? One of the best shooters was Larry Bird of the Boston Celtics. During 1980–1982 he made only the first free throw 34 times and made only the second 48 times (A. Tversky and T. Gilovich, *Chance*, vol. 2, pp. 16–21, 1989). Does this suggest that one sequence was truly more likely than the other sequence for Larry Bird? Answer by conducting a significance test, (a) defining notation and specifying assumptions and hypotheses, (b) finding the test statistic, and (c) finding the P-value and interpreting in this context.

8.63 **Box or Draper?:** A mayoral election in Madison, Wisconsin, has two candidates, Box and Draper.

a. For a random sample of 400 voters in an exit poll, 230 voted for Box and 170 for Draper. Conduct all five steps of a test of H$_0$: $p = 0.50$ against H$_a$: $p \neq 0.50$, where p denotes the probability that a randomly selected voter prefers Box. Are you willing to predict the outcome of the election? Explain how to make a decision, using a significance level of 0.05.

b. Suppose the sample size had been 40 voters, of whom 23 voted for Box. Show that the sample proportion is the same as in (a), but the test statistic and P-value are very different. Are you now willing to predict the outcome of the election?

c. Using (a) and (b), explain how results of a significance test can depend on the sample size.

8.64 **Protecting the environment?:** When the 2000 General Social Survey asked, "Would you be willing to pay much higher taxes in order to protect the environment?" (variable GRNTAXES), 369 people answered yes and 483 answered no. (We exclude those who made other responses.) Let p denote the population proportion who would answer yes. MINITAB shows the following results to analyze whether a majority or minority of Americans would answer yes:

```
Test of p = 0.5 vs p not = 0.5

X    N    Sample p      95% CI        Z-Value   P-Value
369  852  0.433099  (0.3998, 0.4663)   -3.91     0.000
```

a. What are the assumptions for the significance test? Do they seem to be satisfied for this application? Explain.

b. For this printout, specify the null and alternative hypotheses that are tested, and report the point estimate of p and the value of the test statistic.

c. Report and interpret the P-value in context.

d. According to the P-value, is it plausible that $p = 0.50$? Explain.

e. Explain an advantage of the confidence interval shown over the significance test.

8.65 **Affirmative action:** A Pew Research Center poll (May 14, 2003) of 1201 adults asked the question, "All in all, do you think affirmative action programs designed to increase the number of black and minority students on college campuses are a good thing or a bad thing?" 60% said good, 30% said bad, and 10% said don't know. Let p denote the population proportion who said it is good. Conduct all five steps of a test of H$_0$: $p = 0.50$ against H$_a$: $p \neq 0.50$. Can you reject H$_0$ using a significance level of 0.05?

8.66 **Plant inheritance:** In an experiment on chlorophyll inheritance in maize (corn), of the 1103 seedlings of self-fertilized green plants, 854 seedlings were green and 249 were yellow. Theory predicts the ratio of green to yellow is 3 to 1. Show all five steps of a test of the hypothesis that 3 to 1 is the true ratio. Interpret the P-value in context.

8.67 **Ellsberg paradox:** You are told that a ball will be randomly drawn from one of two boxes (A and B), both of which contain black balls and red balls, and if a red ball is chosen, you will win $100. You are also told that Box A contains half black balls and half red balls, but you are not told the proportions in Box B. Which box would you pick?

a. Set up notation and specify hypotheses to test whether the population proportion who would pick Box A is 0.50.

b. For a random sample of 40 people, 36 pick Box A. Can you make a conclusion about whether the proportion for one option is higher in the population? Explain all steps of your reasoning. (Logically those who picked Box A would seem to think that Box B has greater chance of a black ball. However, a paradox first discussed by Daniel Ellsberg predicts that if they were now told that they would instead receive $100 if a black ball is chosen, they would overwhelmingly pick Box A again, because they prefer definite information over ambiguity.)

8.68 **Start a hockey team:** A fraternity at a university lobbies the administration to start a hockey team. To bolster its case, it reports that of a simple random sample of 100 students, 83% support starting the team. Upon further investigation, their sample has 80 males and 20 females. Should you be skeptical of whether the sample was random, if you know that 55% of the student body population was male? Answer this by performing all steps of a two-sided significance test.

8.69 Interest charges on credit card: A bank wants to evaluate which credit card would be more attractive to its customers: One with a high interest rate for unpaid balances but no annual cost, or one with a low interest rate for unpaid balances but an annual cost of $40. For a random sample of 100 of its 52,000 customers, 40 say they prefer the one that has an annual cost. Software reports the following:

```
Test of p = 0.50 vs p not = 0.50

X   N  Sample p   95.0% CI    Z-Value  P-Value

40 100 0.40000 (0.304,0.496)  -2.00  0.04550
```

Explain how to interpret all results on the printout. What would you tell the company about what the majority of its customers prefer?

8.70 Jurors and gender: A jury list contains the names of all individuals who may be called for jury duty. The proportion of the available jurors on the list who are women is 0.53. If 40 people are selected to serve as candidates for being picked on the jury, show all steps of a significance test of the hypothesis that the selections are random with respect to gender.

a. Set up notation and hypotheses, and specify assumptions.

b. 5 of the 40 selected were women. Find the test statistic.

c. Report the P-value, and interpret.

d. Explain how to make a decision using a significance level of 0.01.

8.71 Type I and Type II errors: Refer to the previous exercise.

a. Explain what Type I and Type II errors mean in the context of that exercise.

b. If you made an error with the decision in (d), is it a Type I or a Type II error?

8.72 Levine = author?: The authorship of an old document is in doubt. A historian hypothesizes that the author was a journalist named Jacalyn Levine. Upon a thorough investigation of Levine's known works, it is observed that one unusual feature of her writing was that she consistently began 10% of her sentences with the word "whereas." To test the historian's hypothesis, it is decided to count the number of sentences in the disputed document that begin with the word *whereas*. Out of the 300 sentences in the document, none begin with that word. Let p denote the probability that any one sentence written by the unknown author of the document begins with the word "whereas."

a. Conduct a test of the hypothesis $H_0: p = 0.10$ against $H_a: p \neq 0.10$. What conclusion can you make, using a significance level of 0.05?

b. What assumptions are needed for that conclusion to be valid? (F. Mosteller and D. Wallace conducted an investigation similar to this to determine whether Alexander Hamilton or James Madison was the author of 12 of the *Federalist Papers*. See *Inference and Disputed Authorship: The Federalist*, Addison-Wesley, 1964.)

8.73 Practice steps of test for mean: For a quantitative variable, you want to test $H_0: \mu = 0$ against $H_a: \mu \neq 0$. The 10 observations are

$$3, 7, 3, 3, 0, 8, 1, 12, 5, 8.$$

a. Show that (i) $\bar{x} = 5.0$, (ii) $s = 3.71$, (iii) standard error = 1.17, (iv) test statistic = 4.26, (v) $df = 9$.

b. The P-value is 0.002. Interpret, and make a decision using a significance level of 0.05. Interpret.

c. If you had instead used $H_a: \mu > 0$, what would the P-value be? Interpret it.

d. If you had instead used $H_a: \mu < 0$, what would the P-value be? Interpret it. (*Hint*: Recall that the two one-sided P-values should sum to 1.)

8.74 Two ideal children?: Is the ideal number of children equal to 2, or higher or lower than that? For testing that the mean response from a recent GSS equals 2.0 for the question, "What do you think is the ideal number of children to have?", software shows results:

```
Test of mu = 2.0 vs not = 2.0
Variable  N   Mean  StDev  SE Mean   T      P
Children 1302 2.490 0.850  0.0236  20.80  0.0000
```

a. Report the test statistic value, and show how it was obtained from other values reported in the table.

b. Explain what the P-value represents, and interpret its value.

8.75 Hours at work: When the subjects in the 2002 GSS who were working full or part-time were asked how many hours they worked in the previous week at all jobs (variable HRS1), software produced the following analyses:

```
Test of mu = 40 vs not = 40
 N   Mean   StDev  SE Mean      95% CI        T     P
1729 41.777 14.623 0.3517 (41.087,42.467)  5.05 0.000
```

For this printout,

a. State the hypotheses.

b. Explain how to interpret the values of (i) SE Mean, (ii) T, (iii) P.

c. Show the correspondence between a decision in the test using a significance level of 0.05 and whether 40 falls in the confidence interval reported.

8.76 Females liberal or conservative?: Example 11 compared mean political beliefs (on a 1 to 7 point scale) to the moderate value of 4.0, using GSS data. Test whether the population mean equals 4.00 for females, for whom the sample mean was 4.02 and standard deviation was 1.40 for a sample of size 1504. Carry out the five steps of a significance test, reporting and interpreting the P-value in context.

8.77 Blood pressure: When Vincenzo Baranello's blood pressure is in control, his systolic blood pressure reading has a mean of 130. For the last six times he has monitored his blood pressure, he has obtained the values

$$140, 150, 155, 155, 160, 140.$$

a. Does this provide strong evidence that his true mean has changed? Carry out the five steps of the significance test, interpreting the P-value.

b. Review the assumptions that this method makes. For each assumption, discuss it in context.

8.78 Increasing blood pressure: In the previous exercise, suppose you had predicted that if the mean changed, it would have increased above the control value. State the alternative hypothesis for this prediction, and report and interpret the P-value.

■ **8.79 Tennis balls in control?:** When it is operating correctly, a machine for manufacturing tennis balls produces balls with a mean weight of 57.6 grams. The last eight balls manufactured had weights

$$57.3, 57.4, 57.2, 57.5, 57.4, 57.1, 57.3, 57.0$$

a. Using a calculator or software, find the test statistic and P-value for a test of whether the process is in control against the alternative that the true mean of the process now differs from 57.6.

b. For a significance level of 0.05, explain what you would conclude. Express your conclusion so it would be understood by someone who never studied statistics.

8.80 Catalog sales: A company that sells products through mail-order catalogs wants to evaluate whether the mean sales for their most recent catalog were higher or lower than the mean of $15 from past catalogs. For a random sample of 100 customers, the mean sales were $10, with a standard deviation of $10. Find a P-value to provide the extent of evidence that the mean differed with this catalog. Interpret.

8.81 Wage claim false?: Management claims that the mean income for all senior-level assembly-line workers in a large company equals $500 per week. An employee decides to test this claim, believing that it is actually less than $500. For a random sample of nine employees, the incomes are

$$430, 450, 450, 440, 460, 420, 430, 450, 440.$$

Conduct a significance test of whether the population mean income equals $500 per week against the alternative that it is less. Include all assumptions, the hypotheses, test statistic, and P-value, and interpret the result in context.

8.82 CI and test: Refer to the previous exercise.

a. For which significance levels can you reject H_0? (i) 0.10, (ii) 0.05, (iii) 0.01.

b. Based on the answers in (a), for which confidence coefficients would the confidence interval contain 500? (i) 0.90, (ii) 0.95, (iii) 0.99.

c. Use (a) and (b) to illustrate the correspondence between results of significance tests and results of confidence intervals.

8.83 CI and test connection: The P-value for testing $H_0: \mu = 100$ against $H_a: \mu \neq 100$ is 0.043.

a. What decision is made using a 0.05 significance level?

b. If the decision in (a) is in error, what type of error is it?

c. Does a 95% confidence interval for μ contain 100? Explain.

8.84 Religious beliefs: A journal article that deals with changes in religious beliefs over time states, "For these subjects, the difference in their responses on the scale of religiosity between age 16 and the current survey was statistically significant (P-value < 0.05)."

a. How would you explain to someone who has never taken a statistics course what it means for the result to be "statistically significant" with a P-value < 0.05?

b. Explain why it would have been more informative if the authors provided the actual P-value rather than merely indicating that it is below 0.05.

c. Can you conclude that a practically *important* change in religiosity has occurred between age 16 and the time of the current survey? Why or why not?

8.85 How to reduce chance of error?: In making a decision in a significance test, a researcher worries about rejecting H_0 when it may actually be true.

a. Explain how the researcher can control the probability of this type of error.

b. Why should the researcher probably not set this probability equal to 0.00001?

8.86 Legal trial errors: Consider the analogy (discussed in Section 8.4) between making a decision about a null hypothesis in a significance test and making a decision about the innocence or guilt of a defendant in a criminal trial.

a. Explain the difference between Type I and Type II errors in the trial setting.

b. In this context, explain intuitively why decreasing the chance of Type I error increases the chance of Type II error.

8.87 P(Type II error) with smaller *n*: Consider Example 13 about testing $H_0: p = 1/3$ against $H_a: p > 1/3$ for the astrology study, with $n = 116$. Find P(Type II error) for testing $H_0: p = 1/3$ against $H_a: p > 1/3$ when actually $p = 0.50$, if the sample size is 60 instead of 116. Do this by showing that

a. The standard error is 0.061 when H_0 is true.

b. The rejection region consists of $\hat{p}$ values above 0.433.

c. When $p = 0.50$, the probability that $\hat{p}$ falls below 0.433 is the left-tail probability below -1.03 under a standard normal curve. What is the answer? Why would you expect P(Type II error) to be larger when *n* is smaller?

Concepts and Investigations

■ **8.88 Student data:** Refer to the "FL student survey" data file on the text CD. Test whether the (a) population mean political ideology (on a scale of 1 to 7, where 4 = moderate) equals or differs from 4.0, (b) population proportion favoring affirmative action equals or differs from 0.50. For each part, write a one-page report showing all five steps of the test, including what you must assume for each inference to be valid.

■ **8.89 Class data:** Refer to the data file your class created in Activity 3 at the end of Chapter 1. For a variable chosen by your instructor, conduct inferential statistical analyses. Prepare a report, summarizing and interpreting your findings. In this report, also use graphical and numerical methods presented earlier in this text to describe the data.

8.90 Gender of best friend: A GSS question asked the gender of your best friend. Of 1381 people interviewed, 147 said their best friend had the opposite gender, and 1234 said their best friend had the same gender. Prepare a short report in which you analyze these data using a confidence interval and a significance test. Which do you think is more informative? Why?

8.91 Baseball home team advantage: In major league baseball's 2004 season, the home team won 1299 games and the away team won 1129 (www.mlb.com).

a. Although these games are not a random sample, explain how you could think of these data as giving you information about some long-run phenomenon.

b. Analyze these data using a (i) significance test and (ii) confidence interval. Which method is more informative? Why?

8.92 Statistics and scientific objectivity: The President of the American Statistical Association recently stated, "Statistics has become the modern-day enforcer of scientific objectivity. Terms like randomization, blinding, and 0.05 significance wield a no-doubt effective objectivity nightstick." He also discussed how learning what effects *aren't* in the data is as important as learning what effects *are* significant. In this vein, explain how statistics provides an objective framework for testing the claims of what many believe to be "quack science," such as astrology and therapeutic touch.

(Source: Bradley Efron, Amstat News, July 2004, p. 3.)

8.93 Two-sided or one-sided?: A medical researcher gets a P-value of 0.056 for testing $H_0: \mu = 0$ against $H_a: \mu \neq 0$. Since he believes that the true mean is positive and is worried that his favorite journal will not publish the results because they are not "significant at the 0.05 level," he instead reports in his article the P-value of 0.028 for $H_a: \mu > 0$. Explain what is wrong with

a. Reporting the one-sided P-value after seeing the data.

b. The journal's policy to publish results only if they are statistically significant.

8.94 Interpret P-value: An article in a marketing journal states that "no statistically significant difference was found between men and women in the proportion who said they would consider buying a hybrid car (P-value = 0.63)." In practical terms, how would you explain to someone who has not studied statistics what this means?

8.95 Subgroup lack of significance: A crossover study on comparing a magnetic device to placebo for reducing pain in 54 people suffering from low back or knee pain (www.holcombhealthcare.com/reports/pub-bio.html) reported a significant result overall, the magnetic device being preferred to placebo. However, when the analysis was done separately for the 27 people with shorter illness duration and the 27 people with longer illness duration, results were not significant. Explain how it might be possible that an analysis could give a P-value below 0.05 using an entire sample but not with subgroups, even if the subgroups have the same effects. (*Hint*: What is the impact of the smaller sample size for the subgroups?)

8.96 Birth control pills and breast cancer: A medical study in 1996 concluded that women who take birth control pills have a higher chance of getting breast cancer. A later study by scientists at the NIH and CDC of 9200 women reported no effect (*New England Journal of Medicine*, 6/27/2002). Discuss the factors that can cause different medical studies to come to different conclusions.

8.97 Overestimated effect: When medical stories in the mass media report dangers of certain agents (e.g., coffee drinking), later research often suggests that the effects may not exist or are weaker than first believed. Explain how this could happen if some journals tend to publish only statistically significant results.

8.98 Choosing α: An alternative hypothesis states that a newly developed drug is better than the one currently used to treat a serious illness. If we reject H_0, the new drug will be prescribed instead of the current one.

a. Why might we prefer to use a smaller significance level than 0.05, such as 0.01?

b. What is a disadvantage of using $\alpha = 0.01$ instead of 0.05?

8.99 Why not accept H_0?: Explain why the terminology "do not reject H_0" is preferable to "accept H_0."

8.100 Report P-value: It is more informative and potentially less misleading if you conclude a test by reporting and interpreting the P-value rather than by merely indicating whether or not you reject H_0 at the 0.05 significance level. One reason is that a reader can then tell whether the result is significant *at any significance level*. Give another reason.

8.101 Significance: Explain the difference between *statistical significance* and *practical significance*. Make up an example to illustrate your reasoning.

8.102 More doctors recommend: An advertisement by Company A says that three of every four doctors recommend pain reliever A than all other brands combined.

a. If the company based this claim on interviewing a random sample of doctors, explain how they could use a significance test to back up the claim.

b. Explain why this claim would be more impressive if it is based on a (i) random sample of 40 doctors than if it is based on a random sample of 4 doctors, (ii) random sample of 40 doctors nationwide than the sample of all 40 doctors who work in a particular hospital.

8.103 Medical diagnosis error: Consider the medical diagnosis of breast cancer with mammograms. An AP story (September 19, 2002) said that a woman has about a 50% chance of having a false positive diagnosis over the course of 10 annual mammography tests. Relate this result to the chance of eventually making a Type I error if you do lots of significance tests.

8.104 Bad P-value interpretations: A random sample of size 1000 has $\bar{x} = 104$. The P-value for testing $H_0: \mu = 100$ against $H_a: \mu \neq 100$ is 0.057. Explain what is incorrect about each of the following interpretations of this P-value, and provide a proper interpretation.

a. The probability that the null hypothesis is correct equals 0.057.

b. The probability that $\bar{x} = 104$ if H_0 is true equals 0.057.

c. If in fact $\mu \neq 100$ so H_0 is false, the probability equals 0.057 that the data would show at least as much evidence against H_0 as the observed data.

d. The probability of a Type I error equals 0.057.

e. We can accept H_0 at the $\alpha = 0.05$ level.

f. We can reject H_0 at the $\alpha = 0.05$ level.

8.105 Interpret P-value: One interpretation for the P-value is that it is the smallest value for the significance level α for which we can reject H_0. Illustrate using the P-value of 0.057 from the previous exercise.

8.106 Incorrectly posed hypotheses: What is wrong with expressing hypotheses about proportions and means in a form such as $H_0: \hat{p} = 0.50$ and $H_0: \bar{x} = 0$?

Select the correct response in multiple-choice Exercises 8.107–8.110.

8.107 Small P-value: The P-value for testing $H_0: \mu = 100$ against $H_a: \mu \neq 100$ is 0.001. This indicates that

a. There is strong evidence that $\mu = 100$.

b. There is strong evidence that $\mu \neq 100$, since if μ were equal to 100, it would be unusual to obtain data such as those observed.

c. The probability that $\mu = 100$ is 0.001.

d. The probability that $\mu = 100$ is the significance level, usually taken to be 0.05.

8.108 Probability of P-value: When H_0 is true in a t test with significance level 0.05, the probability that the P-value falls ≤ 0.05

a. equals 0.05.

b. equals 0.95.

c. equals 0.05 for a one-sided test and 0.10 for a two-sided test.

d. can't be specified, because it depends also on P(Type II error).

8.109 Pollution: Exercise 8.29 concerned an industrial plant that may be exceeding pollution limits. An environmental action group took four readings to analyze whether the true mean discharge of wastewater per hour exceeded the company claim of 1000 gallons. When we make a decision in the one-sided test using $\alpha = 0.05$:

a. If the plant is not exceeding the limit, but actually $\mu = 1000$, there is only a 5% chance that we will conclude that they are exceeding the limit.

b. If the plant is exceeding the limit, there is only a 5% chance that we will conclude that they are not exceeding the limit.

c. The probability that the sample mean equals exactly the observed value would equal 0.05 if H_0 were true.

d. If we reject H_0, the probability that it is actually true is 0.05.

e. All of the above.

8.110 Interpret P(Type II error): For a test of $H_0: \mu = 0$ against $H_a: \mu > 0$ based on $n = 30$ observations and using $\alpha = 0.05$ significance level, P(Type II error) $= 0.36$ at $\mu = 4$. Identify the response that is *incorrect*.

a. At $\mu = 5$, P(Type II error) < 0.36.

b. If $\alpha = 0.01$, then at $\mu = 4$, P(Type II error) > 0.36.

c. If $n = 50$, then at $\mu = 4$, P(Type II error) > 0.36.

d. The power of the test is 0.64 at $\mu = 4$.

8.111 True or false 1: It is always the case that P(Type II error) $= 1 - $ P(Type I error).

8.112 True or false 2: If we reject $H_0: \mu = 0$ in a study about change in weight on a new diet using $\alpha = 0.01$, then we also reject it using $\alpha = 0.05$.

8.113 True or false 3: A study about the change in weight on a new diet reports P-value $= 0.043$ for testing $H_0: \mu = 0$ against $H_a: \mu \neq 0$. If the authors had instead reported a 95% confidence interval for μ, then the interval would have contained 0.

8.114 True or false 4: A 95% confidence interval for $\mu = $ population mean IQ is (96, 110). So, in the test of $H_0: \mu = 100$ against $H_a: \mu \neq 100$, the P-value > 0.05.

8.115 True or false 5: For a fixed significance level α, the probability of a Type II error increases when the sample size increases.

8.116 True or false 6: The P-value is the probability that H_0 is true.

♦♦8.117 Standard error formulas: Suppose you wanted to test $H_0: p = 0.50$, but you had 0 successes in n trials. If you had found the test statistic using the $se = \sqrt{\hat{p}(1 - \hat{p})/n}$ designed for confidence intervals, show what happens to the test statistic. Explain why $se = \sqrt{p_0(1 - p_0)/n}$ is a more appropriate *se* for tests.

♦♦8.118 Rejecting true H_0?: A medical researcher conducts a significance test whenever she analyzes a new data set. Over time, she conducts 100 independent tests.

a. Suppose the null hypothesis is true in every case. What is the distribution of the number of times she rejects the null hypothesis at the 0.05 level?

b. Suppose she rejects the null hypothesis in five of the tests. Is it plausible that the null hypothesis is correct in every case? Explain.

Class Explorations

8.119 Refer to the activity at the end of Section 8.4 (pg. 403). Each **TRY** student should indicate how many successive "wins" by the (a) placebo over the drug would be necessary before they would feel comfortable rejecting $H_0: p = 0.5$ in favor of $H_a: p \neq 0.5$ and concluding that $p < 0.5$, (b) drug over the placebo would be necessary before they would feel comfortable rejecting $H_0: p = 0.5$ in favor of $H_a: p \neq 0.5$ and concluding that $p > 0.5$. The instructor will compile a "distribution of significance levels" for the two cases. Are they the same? In principle, should they be?

8.120 Refer to Exercise 7.118, "Randomized response," in Chapter 7 (pg. 365). Before carrying out the method described there, the class was asked to hypothesize or predict what they believe is the value for the population proportion of students that have had alcohol at a party. Use the class estimate for $\hat{p}$ to carry out the significance test for testing this hypothesized value. Discuss whether to use a one-sided or two-sided alternative hypothesis. Describe how the confidence interval formed in Exercise 7.118 relates to the significance test results.

9 | Comparing Two Groups

EXAMPLE 1

Making Sense of Newspaper Articles

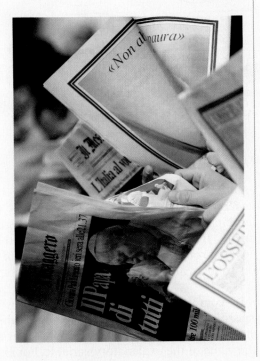

Picture the Scenario

Here are some headlines of recent news stories:

"Males more likely than females to be binge drinkers"

"A 20-year-old obese person can lose 20 years of life"

"Religion more important among Americans than Europeans"

"Drinking tea may help heart attack victims"

"TV violence linked to aggressive behavior"

Pick up a major newspaper any day and you're likely to see a story with a headline like these. Such stories usually refer to research studies that used statistical methods to make inferences. Most commonly the studies involve a comparison of two groups or two treatments (such as males vs. females, obese vs. nonobese, drinking tea vs. not drinking it).

Questions to Explore

Many medical studies in recent years have explored the potential benefits of regular intake of a simple drug—aspirin. For instance, one study in which subjects were randomly allocated to take either aspirin or placebo regularly showed a lower heart attack rate for the subjects who were taking aspirin. The sample proportions over a five-year period were 0.009 for aspirin and 0.017 for placebo, based on sample sizes of about 11,000 each.

- How can you construct a confidence interval to compare the population proportions of heart attack for those taking aspirin to those taking placebo?

- How can you tell whether the sample difference in heart attack rates is statistically significant? Does it truly help if you take aspirin regularly?

Thinking Ahead

This chapter shows how to **compare two groups** on a categorical or quantitative outcome. To compare groups, we'll use the inferential statistical methods that the previous two chapters introduced—confidence intervals and significance tests.

For *categorical* variables, the inferences compare *proportions*. In Examples 2–4 we'll compare aspirin and placebo treatments on the proportions having a heart attack for the study mentioned above.

For *quantitative* variables, the inferences compare *means*. In Examples 6–8 we'll compare the extent of nicotine addiction for female teenage smokers and male teenage smokers and we'll compare nicotine addiction for teenagers who smoke with teenagers who have smoked in the past but no longer do so.

In reading this book, you are becoming an "educated consumer" of information based on statistics. By now, you know to be skeptical of studies that do not or could not employ randomization in the sampling procedure or the experimental design. By the end of this chapter, you'll know about other potential pitfalls. You'll be better able to judge how much credence you can give the claims made in newspaper stories. Such stories nearly always report only "statistically significant" results. Occasionally such a report may be a Type I error, claiming an effect that actually does not exist in the population. Some may predict that effects are larger than they truly are in that population. Does drinking tea really help heart attack victims as much as reported? Or, as Figure 9.1 suggests, is this merely today's random medical news?

Reprinted with special permission of King Features Syndicate.

▲ **FIGURE 9.1: Today's Random Medical News**

Bivariate Analyses: A Response Variable and a Binary Explanatory Variable

Consider a study that compares female and male college students on the proportion who say they have participated in binge drinking. The two groups being compared, females and males, are the categories of a *binary* variable—gender. Methods for comparing two groups are special cases of **bivariate** statistical methods: There are two variables, an outcome variable and the binary variable that specifies the groups.

The outcome variable on which comparisons are made is the **response variable**. The binary variable that specifies the groups is the **explanatory variable**. Statistical methods analyze how the outcome on the response variable *depends on* or is *explained by* the value of the explanatory variable. For example, participation in binge drinking (yes or no) is a response variable and gender is an explanatory variable. Our interest is in studying how binge drinking depends on gender, not how gender depends on binge drinking.

Dependent and Independent Samples

Most comparisons of groups use **independent samples** from the groups: The observations in one sample are *independent* of those in the other sample. For instance, randomized experiments that randomly allocate subjects to two treatments have independent samples. An example is the study mentioned in Example 1 comparing heart attack rates for aspirin and placebo groups: The observation for any subject taking aspirin is independent of the observation for any subject taking placebo. Another way independent samples occur is

Recall

A **binary** variable has two possible outcomes, such as gender—either female or male. See the beginning of Chapter 3 to review the distinction between **response** and **explanatory** variables. ■

when an observational study separates subjects into groups according to their value for an explanatory variable, such as smoking status. If the overall sample was randomly selected, then the groups (for instance, smokers and nonsmokers) are independent random samples.

Dependent samples result when the data are **matched pairs**—each subject in one sample is matched with a subject in the other sample. An example is a set of married couples, the men being in one sample and the women in the other. Also, dependent samples occur when each subject is observed at two times, so the two samples have the same people. An example is a diet study in which subjects' weights are measured before and after participating in the diet. A particular subject's weight before and after the diet, such as (144 lb., 127 lb.), form a matched pair. Data from dependent samples need different statistical methods than data from independent samples. We'll study them in Section 9.4. Sections 9.1–9.3 show how to analyze independent samples, first for a categorical response variable and then for a quantitative response variable.

IN WORDS

When the two samples have the same subjects, they are **dependent**. They're also dependent if each subject in one sample is naturally paired with a subject in the other (such as a husband in one sample with his wife in the other sample).

9.1 | Categorical Response: How Can We Compare Two Proportions?

For a *categorical* response variable, inferences compare groups in terms of their population proportions in a particular category. Let p_1 represent the population proportion for the first group and p_2 the population proportion for the second group. We can compare the groups by their difference, $(p_1 - p_2)$. This is estimated by the difference of the sample proportions, $(\hat{p}_1 - \hat{p}_2)$. Let n_1 and n_2 denote the sample sizes for the two groups.

EXAMPLE 2

Aspirin, the Wonder Drug

Picture the Scenario

Here are some recent titles of newspaper articles about beneficial effects of aspirin:

"Aspirin cuts deaths after heart attack"

"Aspirin could lower risk of ovarian cancer"

"New study finds a daily aspirin lowers the risk of colon cancer"

"Aspirin may lower the risk of Hodgkin's"

Most of us think of aspirin as a simple pill that helps relieve pain. In recent years, though, researchers have been on the lookout for new ways that aspirin may be helpful. Studies have shown that taking aspirin regularly may possibly forestall Alzheimer's disease and may increase the chance of survival for a person who has suffered a heart attack. Other studies have suggested that aspirin may protect against cancers of the pancreas, colon, ovaries, and prostate. As of 2003, the National Cancer Institute had 15 studies on the use of aspirin-like drugs to prevent cancer.[1]

Increasing attention has focused on aspirin since a landmark study about whether regular aspirin intake reduces deaths from heart disease. The Physicians Health Study Research Group at Harvard Medical School conducted a five-year randomized study about this. The study subjects were 22,071 male physicians. Every other day, study participants took either an

[1] *New York Times* story by G. Kolata, March 11, 2003.

Recall

Section 4.3 discussed the importance of **randomization** in experimental design and introduced double blinding. Exercises 4.63–4.65 in that chapter show other data from this study. ■

aspirin tablet or a placebo tablet. The physicians were randomly assigned to the aspirin or to the placebo group. The study was double-blind: The physicians did not know which pill they were taking, nor did those who evaluated their responses.

Table 9.1 shows the study results. This is a **contingency table**, a data summary for categorical variables introduced in Section 3.1. Of the 11,034 physicians who took the placebo, 189 suffered heart attacks during the study. Of the 11,037 physicians who took aspirin, 104 had heart attacks.

TABLE 9.1: Whether Subject Had a Heart Attack, for Placebo and Aspirin Treatment Groups

| Group | Heart Attack | | Total |
	Yes	No	
Placebo	189	10,845	11,034
Aspirin	104	10,933	11,037

Questions to Explore

a. What is the response variable, and what are the groups to compare?

b. What are the two population parameters to compare? Estimate the difference between them using the data in Table 9.1.

Think it Through

a. In Table 9.1, the response variable is whether the subject had a heart attack, with categories yes and no. Group 1 is the physicians who took placebo and Group 2 is the physicians who took aspirin. They are the categories of the explanatory variable.

b. For the population from which this sample was taken, the proportion who would have a heart attack if they participated in this experiment is represented by p_1 for taking placebo and p_2 for taking aspirin. The sample proportions of heart attack were

$$\hat{p}_1 = 189/11034 = 0.017$$

for the $n_1 = 11,034$ subjects in the placebo group and

$$\hat{p}_2 = 104/11037 = 0.009$$

for the $n_2 = 11,037$ subjects in the aspirin group. Since $(\hat{p}_1 - \hat{p}_2) = 0.017 - 0.009 = 0.008$, the proportion who had a heart attack was 0.008 higher for those who took placebo. In percentage terms, the difference was $1.7\% - 0.9\% = 0.8\%$, less than 1 percent.

Insight

The sample proportion of subjects who had a heart attack was smaller for the aspirin treatment group. But we really want to know if this is true also for the population. To make an inference about the difference of population proportions, $(p_1 - p_2)$, we need to learn how much the difference $(\hat{p}_1 - \hat{p}_2)$ between the sample proportions would tend to vary from study to study. This is described by the standard error of the sampling distribution for the difference between the sample proportions.

TO PRACTICE THIS CONCEPT, TRY EXERCISES 9.2 AND 9.3(A)

What Is the Standard Error for Comparing Two Proportions?

Just as a sample proportion has a standard error that describes how well it estimates a population proportion, so does the difference $(\hat{p}_1 - \hat{p}_2)$ between

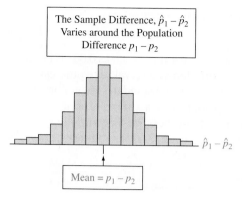

The Sample Difference, $\hat{p}_1 - \hat{p}_2$ Varies around the Population Difference $p_1 - p_2$

$\hat{p}_1 - \hat{p}_2$

Mean $= p_1 - p_2$

Recall

From Section 7.2, for using a confidence interval to estimate a proportion p, the **standard error** of a sample **proportion** $\hat{p}$ is

$$\sqrt{\frac{\hat{p}(1 - \hat{p})}{n}}.\,\blacksquare$$

two sample proportions have a standard error. For instance, if other studies like the Physicians Health Study had been conducted at the same time, $(\hat{p}_1 - \hat{p}_2)$ would not have equaled 0.008 for each of them. This estimate would vary from study to study. The standard error describes the spread of the sampling distribution of the estimate, and thus how much it would vary. It is interpreted as the standard deviation of the estimates $(\hat{p}_1 - \hat{p}_2)$ from different randomized experiments of this size.

The formula for the standard error of $(\hat{p}_1 - \hat{p}_2)$ is

$$se = \sqrt{\frac{\hat{p}_1(1 - \hat{p}_1)}{n_1} + \frac{\hat{p}_2(1 - \hat{p}_2)}{n_2}}.$$

We'll see where this formula comes from at the end of the section. For now, notice that if you ignore one of the two samples (and half of this formula), you get the usual standard error for a proportion shown in the Recall box.

EXAMPLE 3

How Well Can We Compare Heart Attack Rates for Aspirin and Placebo?

Picture the Scenario

In Example 2, the sample proportions suffering a heart attack were $\hat{p}_1 = 189/11{,}034 = 0.017$ for placebo (group 1) and $\hat{p}_2 = 104/11{,}037 = 0.009$ for aspirin (group 2). The estimated difference was $\hat{p}_1 - \hat{p}_2 = 0.017 - 0.009 = 0.008$.

Questions to Explore

a. What is the standard error of this estimate?

b. What is the interpretation of this standard error?

Think It Through

a. Using the standard error formula given above with the values just recalled,

$$se = \sqrt{\frac{\hat{p}_1(1 - \hat{p}_1)}{n_1} + \frac{\hat{p}_2(1 - \hat{p}_2)}{n_2}} =$$

$$\sqrt{\frac{(0.017)(0.983)}{11034} + \frac{(0.009)(0.991)}{11037}} = 0.0015.$$

b. Consider all the possible random samples of size about 11,000 for each group that could participate in this five-year study. The difference $(\hat{p}_1 - \hat{p}_2)$ between the sample proportions of heart attack would not always equal 0.008 but would vary from sample to sample. The standard deviation of the $(\hat{p}_1 - \hat{p}_2)$ values for the different possible samples would equal about 0.0015.

Insight

From the *se* formula, we see that *se* decreases as n_1 and n_2 increase. The standard error is very small for these data because the sample sizes were so large. This means that the $(\hat{p}_1 - \hat{p}_2)$ values would be very similar from study to study. It also implies that $(\hat{p}_1 - \hat{p}_2) = 0.008$ is quite precise as an estimate of the difference of population proportions.

TO PRACTICE THIS CONCEPT, TRY EXERCISE 9.3(B)

Recall

You can review Section 7.2 for a discussion of the **confidence interval** for a single **proportion**. This has form

$$\hat{p} \pm z(se),$$

with $z = 1.96$ for 95% confidence. ■

Confidence Interval for the Difference between Two Population Proportions

The standard error helps us predict how close an estimate such as 0.008 is likely to be to the population value $(p_1 - p_2)$. A confidence interval takes the estimate and adds and subtracts a margin of error based on the standard error. For a large random sample, recall that the sampling distribution of a sample proportion is approximately normal. This is due to the central limit theorem. The sampling distribution of $(\hat{p}_1 - \hat{p}_2)$ is also approximately normal when each sample size is large, as specified in the summary box below. See Figure 9.2.

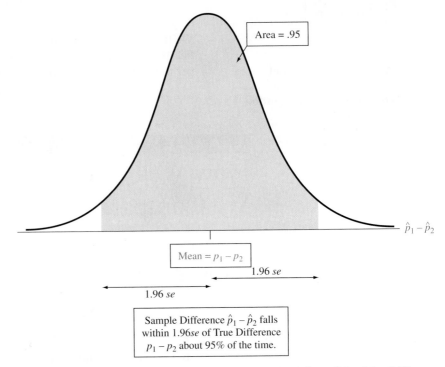

Area = .95

Mean = $p_1 - p_2$

1.96 *se*

1.96 *se*

Sample Difference $\hat{p}_1 - \hat{p}_2$ falls within 1.96*se* of True Difference $p_1 - p_2$ about 95% of the time.

$\hat{p}_1 - \hat{p}_2$

▲ **FIGURE 9.2: Sampling Distribution of the Estimate ($\hat{p}_1 - \hat{p}_2$) of the Difference between Two Population Proportions.** With large random samples, this is approximately normal about the difference in the populations, $(p_1 - p_2)$. **Questions:** For the aspirin and heart attack study (Examples 2 and 3), to what do $(\hat{p}_1 - \hat{p}_2)$ and $(p_1 - p_2)$ refer? What is the mean of this sampling distribution if heart attack rates are identical for aspirin and placebo?

As in the single proportion case, to get the margin of error, we multiply the standard error by a z-score from the normal distribution table. The confidence interval has the form

$$(\hat{p}_1 - \hat{p}_2) \pm z(se).$$

Confidence Interval for the Difference between Two Population Proportions

A confidence interval for the difference $(p_1 - p_2)$ between two population proportions is

$$(\hat{p}_1 - \hat{p}_2) \pm z(se), \text{ where } se = \sqrt{\frac{\hat{p}_1(1 - \hat{p}_1)}{n_1} + \frac{\hat{p}_2(1 - \hat{p}_2)}{n_2}}.$$

The z-score depends on the confidence level, such as $z = 1.96$ for 95% confidence. To use this method, you need

■ Independent random samples for the two groups.
■ Large enough sample sizes n_1 and n_2 so that, in each sample, there are at least 10 "successes" and at least 10 "failures."

Recall

The confidence interval for a single proportion required at least 15 successes and 15 failures. Here, the method works well with slightly smaller samples, at least 10 of each type in each group. ■

Confidence Interval Comparing Heart Attack Rates for Aspirin and Placebo

Picture the Scenario

In the Physicians Health Study, the estimated difference between placebo and aspirin in the proportions suffering heart attacks was $\hat{p}_1 - \hat{p}_2 = 0.017 - 0.009 = 0.008$. In Example 3 we found that this estimate has a standard error of 0.0015.

Question to Explore

What can we say about the difference of proportions for the population of male physicians? Construct a 95% confidence interval for $(p_1 - p_2)$, and interpret.

Think It Through

From Table 9.1, shown again in the margin, the two outcome counts were at least 10 for each group, so the large-samples confidence interval method is valid. A 95% confidence interval for $(p_1 - p_2)$ is

$$(\hat{p}_1 - \hat{p}_2) \pm 1.96(se), \text{ or } 0.008 \pm 1.96(0.0015),$$
$$\text{which is } 0.008 \pm 0.003, \text{ or } (0.005, 0.011).$$

Suppose this experiment could be conducted with the entire population. The inference that $(p_1 - p_2)$ is between 0.005 and 0.011 means that the population proportion p_1 of heart attacks for those taking placebo would be between 0.005 higher and 0.011 higher than the population proportion p_2 of heart attacks for those taking aspirin. Since both endpoints of the confidence interval (0.005, 0.011) for $(p_1 - p_2)$ are positive, we infer that $(p_1 - p_2)$ is positive. That is, it seems that p_1 is larger than p_2: The population proportion of heart attacks is larger when subjects take the placebo than when they take aspirin. Table 9.2 shows the way MINITAB reports this result and the margin shows screen shots from the TI-83+/84.

	Heart Attack	
Group	**Yes**	**No**
Placebo	189	10,845
Aspirin	104	10,933

IN WORDS

The interval (0.005, 0.011) for $(p_1 - p_2)$ predicts that the population proportion for the first group is between 0.005 and 0.011 larger than the population proportion for the second group.

```
2-PropZInt
 x1:189
 n1:11034
 x2:104
 n2:11037
 C-Level:.95
 Calculate
```

```
2-PropZInt
 (.005,.011)
 p1=.017
 p2=.009
 n1=11034.000
 n2=11037.000
 ■
```

TABLE 9.2: MINITAB Output for Confidence Interval Comparing Proportions

	Number of heart attacks		$\hat{p}$ values
Sample	X	N	Sample p
1	189	11034	0.017129
2	104	11037	0.009423

Difference = p(1) − p(2)

Estimate for difference: 0.00771   ← This is $(\hat{p}_1 - \hat{p}_2)$

95% CI for difference: (0.00469, 0.01072)

Insight

All the numbers in the confidence interval fall near 0. This suggests that the population difference $(p_1 - p_2)$ is small. However, this difference, small as it is, may be important in public health terms. For instance, projected over a population of 200 million adults (as in the United States or in Western Europe), a decrease over a five-year period of 0.01 in the proportion of people suffering heart attacks would mean two million fewer people having heart attacks.

The Physicians Health Study used male doctors in the United States. So, strictly speaking, we can apply this inference only to the U.S. population of male doctors. Before concluding that aspirin benefits a larger population, we'd want to see results of studies with more diverse groups. In fact, that has since happened. A similar benefit was observed in a large-scale study of female nurses and in studies of populations in other countries.

This example shows how the use of statistics can result in conclusions that help to benefit public health. A recent article[2] about proper and improper scientific methodology stated. "The most important discovery of modern medicine is not vaccines or antibiotics, it is the randomized double-blind study, by means of which we know what works and what doesn't."

TO PRACTICE THIS CONCEPT, TRY EXERCISE 9.6

Recall

For **99% confidence** we use $z = 2.58$, because 99% of the standard normal distribution falls between -2.58 and 2.58. ■

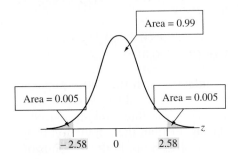

As in the one-sample case, we can use a higher confidence level, but the interval is then wider. For instance, a 99% confidence interval uses $z = 2.58$ and equals

$$(\hat{p}_1 - \hat{p}_2) \pm 2.58(se), \text{ or } 0.008 \pm 2.58(0.0015),$$
$$\text{which is } 0.008 \pm 0.004, \text{ or}(0.004, 0.012).$$

We can be 99% confident that the population proportion of heart attacks is between 0.004 higher and 0.012 higher for the placebo treatment than for the aspirin treatment. This is slightly wider than the 95% confidence interval of $(0.005, 0.011)$. With very large sample sizes, it is often worth the slight widening, to obtain greater confidence.

Key Points in Interpreting a Confidence Interval Comparing Proportions

Example 4 illustrated how to interpret a confidence interval for $(p_1 - p_2)$ in context. Whether a particular group is called group 1 or group 2 is completely arbitrary. If we reverse the labels, each endpoint of the confidence interval reverses sign. For instance, if we instead call aspirin group 1 and placebo group 2, then $(\hat{p}_1 - \hat{p}_2) = 0.009 - 0.017 = -0.008$. The 95% confidence interval is then $(-0.011, -0.005)$ instead of $(0.005, 0.011)$. Since this confidence interval contains entirely negative numbers, we infer that $(p_1 - p_2)$ is negative, that is, p_1 is smaller than p_2. Since group 1 is now aspirin, the interval $(-0.011, -0.005)$ predicts that the population proportion of heart attacks is between 0.011 less and 0.005 less for aspirin than for placebo. (The negative signs translate to the first proportion being *less* than the second.)

Another key point to keep in mind is that if the confidence interval contains 0, then it is plausible that $(p_1 - p_2) = 0$, that is, $p_1 = p_2$. The population proportions might be equal. In such a case insufficient evidence exists to infer which of p_1 or p_2 is larger. See Figure 9.3.

▲ **FIGURE 9.3: Three Confidence Intervals for Difference between Proportions,** $p_1 - p_2$. When the confidence interval for $p_1 - p_2$ contains 0 (the middle interval above), the population proportions may be equal. When it does not contain 0, we can infer which population proportion is larger. **Question:** For the two confidence intervals that do not contain 0, how can we tell which population proportion is predicted to be larger?

[2]By R. L. Park, *The Chronicle of Higher Education*, 1/31/2003.

For instance, Exercise 9.6 summarizes a Swedish study comparing population proportions of heart attack for a placebo group and an aspirin group. The study had a much smaller sample size than the Physicians Health Study, so its standard error was larger. Its 95% confidence interval for $(p_1 - p_2)$ is $(-0.005, 0.034)$. Since the interval contains 0, the population proportions might well be equal. This interval infers that the population proportion p_1 of heart attacks for the placebo group is as much as 0.005 lower or as much as 0.034 higher than the population proportion p_2 for the aspirin group. A *negative* value for $(p_1 - p_2)$, such as -0.005, means that p_1 could be *below* p_2, whereas a *positive* value for $(p_1 - p_2)$, such as $+.034$, means that p_1 could be *above* p_2.

SUMMARY: Interpreting a Confidence Interval for a Difference of Proportions

- Check whether 0 falls in the confidence interval. If so, it is plausible (but not necessary) that the population proportions are equal.

- If all values in the confidence interval for $(p_1 - p_2)$ are positive, you can infer that $(p_1 - p_2) > 0$, that is, $p_1 > p_2$. The interval shows just how much larger p_1 might be. If all values in the confidence interval are negative, you can infer that $(p_1 - p_2) < 0$, that is, that $p_1 < p_2$.

- The magnitude of values in the confidence interval tells you how *large* any true difference is. If all values in the confidence interval are near 0, the true difference may be relatively small in practical terms.

Significance Tests Comparing Population Proportions

Recall

Section 8.2 presented the **significance test** for a single **proportion**. The z test statistic measures the number of standard errors that the sample proportion falls from the value in the null hypothesis. We call H$_0$ the "null" hypothesis because it usually represents no effect. ▪

Another way to compare two population proportions p_1 and p_2 is with a significance test. The null hypothesis is H$_0$: $p_1 = p_2$, the population proportion taking the same value for each group. In terms of the difference of proportions, this is H$_0$: $(p_1 - p_2) = 0$, *no difference*, or *no effect*.

Under the presumption for H$_0$ that $p_1 = p_2$, we estimate the common value of p_1 and p_2 by the proportion of the *total* sample in the category of interest. We denote this by $\hat{p}$. For example, if group 1 had 7 successes in $n_1 = 20$ observations and group 2 had 5 successes in $n_2 = 10$ observations, then $\hat{p}_1 = 7/20 = 0.35$, $\hat{p}_2 = 5/10 = 0.50$, and $\hat{p} = (7 + 5)/(20 + 10) = 12/30 = 0.40$. The proportion $\hat{p}$ is called a **pooled estimate**, because it pools together the total number of successes and total number of observations from the two samples. Whenever the sample sizes n_1 and n_2 are roughly equal, it falls about half way between $\hat{p}_1$ and $\hat{p}_2$. Otherwise, it falls closer to the sample proportion that has the larger sample size.

The test statistic measures the number of standard errors that the sample estimate $(\hat{p}_1 - \hat{p}_2)$ of $(p_1 - p_2)$ falls from its null hypothesis value of 0. It is

$$z = \frac{\text{Estimate} - \text{Null hypothesis value}}{\text{Standard error}} = \frac{(\hat{p}_1 - \hat{p}_2) - 0}{se_0}.$$

The standard error for the test, denoted by se_0, is based on the presumption stated in H$_0$ that $p_1 = p_2$. It uses the pooled estimate $\hat{p}$ to estimate each population proportion. This standard error is

Recall

A **P-value** is the probability that the test statistic equals the observed value or a value even more extreme (in one tail for a one-sided H$_a$ or both tails for a two-sided H$_a$). It is calculated presuming that H$_0$ is true. Smaller P-values provide stronger evidence against H$_0$. ▪

$$se_0 = \sqrt{\frac{\hat{p}(1 - \hat{p})}{n_1} + \frac{\hat{p}(1 - \hat{p})}{n_2}} = \sqrt{\hat{p}(1 - \hat{p})\left(\frac{1}{n_1} + \frac{1}{n_2}\right)}.$$

In practice, when the sample proportions are close, se_0 is very close to se used in a confidence interval, which does not presume equal proportions.

As usual, the P-value for H$_0$: $p_1 = p_2$ depends on whether the alternative hypothesis is two-sided, H$_a$: $p_1 \neq p_2$, or one-sided, H$_a$: $p_1 > p_2$ or H$_a$: $p_1 < p_2$. When it is two-sided, the P-value is the two-tail probability beyond the observed *z* test statistic value from the standard normal distribution. This is the probability, presuming H_0 to be true, of obtaining results more extreme than observed in either direction.

The box outlines the steps of two-sided tests comparing proportions.

Two-Sided Significance Test for Comparing Two Population Proportions

1. Assumptions

- Categorical response variable for two groups
- Independent random samples
- n_1 and n_2 are large enough that there are at least five successes and five failures in each group

2. Hypotheses

Null H_0: $p_1 = p_2$ (that is, $p_1 - p_2 = 0$)

Alternative H_a: $p_1 \neq p_2$ (one-sided H_a also possible; see after Example 5)

3. Test Statistic

$$z = \frac{(\hat{p}_1 - \hat{p}_2) - 0}{se_0} \text{ with } se_0 = \sqrt{\hat{p}(1 - \hat{p})\left(\frac{1}{n_1} + \frac{1}{n_2}\right)}$$

4. P-value

P-value = Two tail probability from standard normal distribution (Table A) of values even more extreme than observed *z* test statistic

5. Conclusion

Smaller P-values give stronger evidence against H_0 and supporting H_a. Interpret the P-value in context. If a decision is needed, reject H_0 if P-value ≤ significance level (such as 0.05).

EXAMPLE 5

Is TV Watching Associated with Aggressive Behavior?

Picture the Scenario

Is there a link between television violence and aggressive behavior by those who watch a lot of TV? Various studies have investigated this. A recent study[3] considered whether greater levels of TV watching by teenagers were associated with a greater likelihood of committing aggressive acts over the years. The researchers randomly sampled 707 families in two counties in northern New York State and made follow-up observations over 17 years. Table 9.3 shows some results about whether a sampled teenager later conducted any aggressive act against another person, according to a self report by that person or by their mother.

---

[3]by J. G. Johnson et al., *Science*, vol. 295, March 29, 2002, pp. 2468–2471.

TABLE 9.3: TV Watching by Teenagers and Later Aggressive Acts

TV Watching	Aggressive Act		
	Yes	No	Total
Less than 1 hour per day	5	83	88
At least 1 hour per day	154	465	619

We'll identify Group 1 as those who watched less than 1 hour of TV per day, on the average, as teenagers. Group 2 is those who averaged at least 1 hour of TV per day, as teenagers. Denote the population proportion committing aggressive acts by p_1 for the lower level of TV watching and by p_2 for the higher level of TV watching.

Questions to Explore

a. Find and interpret the P-value for testing $H_0: p_1 = p_2$ against $H_a: p_1 \neq p_2$.

b. Make a decision about H_0 using the significance level of 0.05.

Think It Through

a. The study used random sampling and then classified the teenagers by the level of TV watching, so the samples were independent random samples. Each count in Table 9.3 is at least 5, so we can use a large-sample test. The sample proportions of aggressive acts were $\hat{p}_1 = 5/88 = 0.057$ for the lower level of TV watching and $\hat{p}_2 = 154/619 = 0.249$ for the higher level. Under the null hypothesis presumption that $p_1 = p_2$, the pooled estimate of the common value p is $\hat{p} = (5 + 154)/(88 + 619) = 159/707 = 0.225$.

The standard error for the test is

$$se_0 = \sqrt{\hat{p}(1 - \hat{p})\left(\frac{1}{n_1} + \frac{1}{n_2}\right)} = \sqrt{(0.225)(0.775)\left(\frac{1}{88} + \frac{1}{619}\right)} = 0.0476.$$

$.174375(.01236$
$.00162$

The test statistic for $H_0: p_1 = p_2$ is

$$z = \frac{(\hat{p}_1 - \hat{p}_2) - 0}{se_0} = \frac{(0.057 - 0.249) - 0}{0.0476} = \frac{-0.192}{0.0476} = -4.04.$$

For the two-sided alternative hypothesis, the P-value is the two-tail probability from the standard normal distribution. A z-score of -4.04 is far out in the left tail. From tables (such as Table A) or software it has a P-value $= 2(0.00003) = 0.00006$, or 0.0001 rounded to four decimal places. Extremely strong evidence exists against the null hypothesis that the population proportions committing aggressive acts are the same for the two levels of TV watching. The study provides strong evidence in support of H_a.

b. Since the P-value is less than 0.05, we can reject H_0. We accept $H_a: p_1 \neq p_2$ and conclude that the population proportions of aggressive acts differ for the two groups. The sample values suggest that the population proportion is higher for the higher level of TV watching. The final row of Table 9.4 shows the way MINITAB reports this result and the margin of the next page shows screen shots from the TI-83+/84.

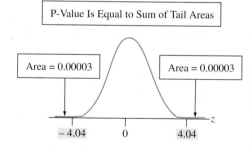

P-Value Is Equal to Sum of Tail Areas

Area = 0.00003   Area = 0.00003

−4.04   0   4.04 z

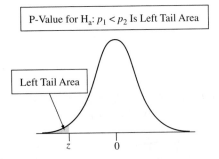

Recall

An **observational** study merely observes the explanatory variable, such as TV watching. An **experimental** study randomly allocates subjects to its levels. Sections 4.1 and 4.4 discussed limitations of observational studies, such as effects that lurking variables may have on the association. ■

P-Value for H_a: $p_1 < p_2$ Is Left Tail Area

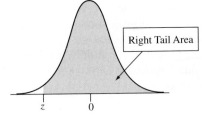

P-Value for H_a: $p_1 > p_2$ Is Right Tail Area

TABLE 9.4: MINITAB Output for Example 5 on TV Watching and Aggression			
Sample	X	N	Sample p
1	5	88	0.056818
2	154	619	0.248788

Difference = p(1) − p(2)

Estimate for difference: −0.191970

95% CI for difference: (−0.251124, −0.132816)

Test for difference = 0 (vs not = 0): z = −4.04 P-Value = 0.000

Insight

This was an observational study. In practice, it would be impossible to conduct an experimental study by randomly assigning teenagers to watch little TV or watch much TV over several years. Also, just because a person watches more TV does not mean they watch more violence. We need to be cautious of effects of lurking variables when we make conclusions. It is not proper to conclude that greater levels of TV watching *cause* later aggressive behavior. For instance, perhaps those who watched more TV tended to have lower education levels, and perhaps lower education tends to be associated with a greater likelihood of aggressive acts.

TO PRACTICE THIS CONCEPT, TRY EXERCISE 9.8

What if the alternative hypothesis is one-sided?   Example 5 requested a test for the two-sided alternative, H_a: $p_1 \neq p_2$. Two-sided P-values are normally reported in journals. However, suppose the researchers specifically predicted that greater levels of TV watching by teenagers were associated with greater likelihood of committing aggressive acts years later. Then they may have preferred to use the one-sided alternative, H_a: $p_1 < p_2$. This states that the population proportion of aggressive acts is higher at the higher level of TV watching. Equivalently, it is H_a: $(p_1 - p_2) < 0$.

This one-sided H_a predicts that $(\hat{p}_1 - \hat{p}_2) < 0$ and hence that the test statistic $z < 0$. The P-value is then a left-tail probability below the observed test statistic, $z = -4.04$. From Table A or software, this is tiny, about 0.00003. The conclusion is again that there was significantly greater probability of aggression for those who had watched more TV.

The one-sided H_a: $p_1 > p_2$ predicts that $(\hat{p}_1 - \hat{p}_2) > 0$ and hence that the test statistic $z > 0$. The P-value is then a right-tail probability. This alternative was not considered relevant for this study.

What Is the Standard Error for Comparing Two Statistics?

Now that we've seen how to compare proportions inferentially, let's learn where the *se* formulas given in this section come from. Whenever we estimate a difference between two population parameters, a general rule specifies the standard error.

Standard Error of the Difference between Two Estimates

For two estimates from independent samples, the standard error is

$$se(\text{estimate 1} - \text{estimate 2}) = \sqrt{[se(\text{estimate 1})]^2 + [se(\text{estimate 2})]^2}$$

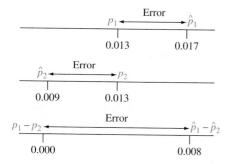

Each estimate has sampling error, and the variabilities add together to determine the standard error of the difference of the estimates.

Notice that we *add* squared standard errors under the square root sign, rather than subtract them. For example, if we're comparing two proportions, this means that the standard error of the difference is *larger* than the standard error for either sample proportion alone. Why is this? In practical terms, $(\hat{p}_1 - \hat{p}_2)$ is often farther from $(p_1 - p_2)$ than $\hat{p}_1$ is from p_1 or $\hat{p}_2$ is from p_2. For instance, in the aspirin and heart attacks study, suppose that

$$p_1 = p_2 = 0.013 \text{ (unknown to us)},$$

but the sample proportions were

$$\hat{p}_1 = 0.017 \text{ and } \hat{p}_2 = 0.009,$$

as was actually observed for the sample data. Then the errors of estimation were

$$\hat{p}_1 - p_1 = 0.017 - 0.013 = 0.004 \text{ and } \hat{p}_2 - p_2 = 0.009 - 0.013 = -0.004,$$

each estimate being off by a distance of 0.004. But then $(\hat{p}_1 - \hat{p}_2) = 0.017 - 0.0009 = 0.008$. That is, the estimate is 0.008 from $(p_1 - p_2) = 0$, larger than the error for either proportion individually. See the margin figure.

Let's apply the formula in the previous formula box to the comparison of two proportions. The standard error of a single sample proportion $\hat{p}$ is $se = \sqrt{\hat{p}(1 - \hat{p})/n}$. Then, for two samples, the boxed formula tells us that the *se* of $(\hat{p}_1 - \hat{p}_2)$ is

$$se = \sqrt{[se(\text{estimate 1})]^2 + [se(\text{estimate 2})]^2} = \sqrt{\frac{\hat{p}_1(1 - \hat{p}_1)}{n_1} + \frac{\hat{p}_2(1 - \hat{p}_2)}{n_2}}.$$

We used this *se* formula to construct a confidence interval for the difference between two population proportions.

Small-Sample Inference for Comparing Proportions

The confidence interval for a difference of proportions specifies that each sample should have at least 10 outcomes of each type. For smaller sample sizes the method may not work well: The sampling distribution of $(\hat{p}_1 - \hat{p}_2)$ may not be close to normal, and the estimate of the standard error may be poor. We won't cover this case here, but it is briefly discussed in Exercise 9.103. There are also small-sample significance tests for comparing proportions. We will discuss one, called *Fisher's exact test*, in the next chapter in a more general context. Such tests have their own disadvantages, however, and for two-sided alternatives the large-sample test usually performs quite well for small samples also.

9.1 Practicing the Basics

9.1 **Wealth gap:** A family's net worth accounts for the values of items such as a home and car, checking and savings accounts and stocks, minus debts such as mortgage, car loans and credit card bills. An Associated Press (AP) story (October 18, 2004) about a Census Bureau survey reported that in recent years the net worth of white households in the United States has increased while that of black households has decreased. In 2002 the median net worth was estimated to be about $89,000 for white households and $6000 for black households.

a. Identify the response variable and the explanatory variable.

b. Identify the two groups that are the categories of the explanatory variable.

c. The estimated medians were based on a sample of households. Were the samples of white households and black households independent samples, or dependent samples? Explain.

9.2 **Sampling sleep:** The 2002 "Sleep in America" poll of 1010 adults reported that respondents slept an average of 6.9 hours on weekdays and 7.5 hours on weekends, and that 30% of respondents got 8 or more hours of sleep on weekdays whereas 52% got 8 or more hours of sleep on weekends (www.sleepfoundation.org).

a. To compare the means or the percentages using inferential methods, should you treat the samples on weekdays and weekends as independent samples, or as dependent samples? Explain.

b. To compare these results to polls of other people taken in previous years, should you treat the samples in the two years as independent samples, or as dependent samples? Explain.

9.3 **Binge drinking:** The College Alcohol Study at the Harvard School of Public Health has interviewed random samples of students at 119 4-year colleges several times since 1993 (see *Journal of American College Health*, vol. 50, pp. 203–217, 2002). Of the students who reported drinking alcohol, the percentage who reported that drinking "to get drunk" is an important reason for drinking was 39.9% of 12,708 students in 1993 and 48.2% of 8783 students in 2001.

a. Estimate the difference between the proportions in 2001 and 1993, and interpret.

b. Find the standard error for this difference. Interpret it.

c. Construct and interpret a 95% confidence interval to estimate the true change, explaining how your interpretation reflects whether the interval contains 0. What is the main factor that causes the confidence interval to be so narrow?

d. State and check the assumptions for the confidence interval in (c) to be valid.

9.4 **Less smoking now?** The National Health Interview Survey conducted of 33,326 adults by the U.S. National Center for Health Statistics in 2003 indicated that 21.4% of adults were current smokers. A similar study conducted in 1991 of 42,000 adults indicated that 25.6% were current smokers.

a. Find and interpret a point estimate of the difference between the proportion of current smokers in 1991 and the proportion of current smokers in 2003.

b. A 99% confidence interval for the true difference is (0.034, 0.050). Interpret.

c. What assumptions must you make for the interval in (b) to be valid?

9.5 **Do you believe in miracles?** Let p_1 and p_2 denote the population proportions of males and females in the United States who answer, "yes, definitely" when asked whether they believe in miracles. Estimate these by going to the General Social Survey web site (sda.berkeley.edu/GSS), selecting the frequencies and crosstabulation option, and then enter the variable MIRACLES for the rows, SEX for the columns, and YEAR (1998) as the selection filter (1998 was the most recent year this was asked).

a. Report point estimates of p_1 and p_2 (you should get $\hat{p}_2 = 0.563$).

b. Construct a 95% confidence interval for $(p_1 - p_2)$, specifying the assumptions you make to use this method. Interpret.

c. Based on the interval in (b), explain why the proportion believing in miracles may have been quite a bit larger for females, or it might have been only slightly larger.

9.6 **Aspirin and heart attacks in Sweden:** A Swedish study used 1360 patients who had suffered a stroke. The study randomly assigned each subject to an aspirin treatment or a placebo treatment.[4] The table shows MINITAB output, where X is the number of deaths due to heart attack during a follow-up period of about 3 years. Sample 1 received the placebo and sample 2 received aspirin.

a. Explain how to obtain the values labeled "Sample p".

b. Explain how to interpret the value given for "estimate for difference".

c. Explain how to interpret the confidence interval, indicating the relevance of 0 falling in the interval.

d. If we instead let sample 1 refer to the aspirin treatment and sample 2 the placebo treatment, explain how the estimate of the difference and the 95% confidence interval would change. Explain how then to interpret the confidence interval.

Deaths due to heart attacks in Swedish study:

Sample	X	N	Sample p
1	28	684	0.040936
2	18	676	0.026627

Difference = p(1) − p(2)
Estimate for difference: 0.0143085
95% CI for difference: (−0.00486898, 0.0334859)
Test for difference = 0 (vs not = 0):
Z = 1.46 P-Value = 0.144

9.7 **Swedish study test:** Refer to the previous exercise.

a. State the hypotheses that were tested.

b. Explain how to interpret the P-value for the test.

c. Report the P-value for the one-sided alternative hypothesis that the chance of death due to heart attack is lower for the aspirin group.

d. Even though the difference between the sample proportions was larger than in the Physicians Health Study (Examples 2–4), the results are less statistically significant. Explain how this could be. (*Hint*: How do the sample sizes compare for the two studies? How does this affect the standard error and thus the test statistic and P-value?)

9.8 **Significance test for aspirin and heart attacks study:** For the Physicians Health Study, consider the null hypothesis that the population proportion of heart attacks p_1 for placebo is the same as the population proportion p_2 for aspirin. The sample proportions were $\hat{p}_1 = 189/11{,}034 = 0.017$ and $\hat{p}_2 = 104/11{,}037 = 0.0094$.

a. For testing H_0: $p_1 = p_2$ against H_a: $p_1 \neq p_2$, show that the pooled estimate of the common value p under H_0 is $\hat{p} = 0.013$ and the standard error is 0.0015.

[4]Based on results described in *Lancet*, vol. 338, pp. 1345–1349 (1991).

b. Show that the test statistic is $z = 5.0$.

c. Find and interpret the P-value in context.

9.9 **Drinking and unplanned sex:** In the study of binge drinking mentioned in Exercise 9.3, the percent who said they had engaged in unplanned sexual activities because of drinking alcohol was 19.2% in 1993 and 21.3% in 2001. Is this change statistically significant at the 0.05 significance level?

a. Specify assumptions, notation, and hypotheses for a two-sided test.

b. Show how to find the pooled estimate of the proportion to use in a test. (*Hint*: You need the count in each year rather than the proportion. In 1993, note this is $(0.192)12,708 = 2440$.) Interpret this estimate.

c. Find the standard error for the difference used in the test. Interpret it.

d. Find the test statistic and P-value. Make a decision using a significance level of 0.05.

e. Some might argue that the result in (d) reflects statistical significance but not practical significance. Explain the basis of this argument.

9.10 **Comparing marketing commercials:** Two TV commercials are developed for marketing a new product. A volunteer test sample of 200 people is randomly split into two groups of 100 each. In a controlled setting, Group A watches commercial A and Group B watches commercial B. In Group A, 25 say they would buy the product. In group B, 20 say they would buy the product. The marketing manager who devised this experiment concludes that commercial A is better. Is this conclusion justified? Analyze the data, if you prefer using software (such as MINITAB) for which you can enter summary counts.

a. Show all steps of your analysis, including assumptions.

b. Comment on the manager's conclusion, and indicate limitations of the experiment.

9.11 **Hormone therapy for menopause:** The Women's Health Initiative conducted a randomized experiment to see if hormone therapy was helpful for post-menopausal women. The women were randomly assigned to receive the estrogen plus progestin hormone therapy or a placebo. After 5 years, 107 of the 8506 on the hormone therapy developed cancer and 88 of the 8102 in the placebo group developed cancer. Is this a significant difference?

a. Set up notation, and state assumptions and hypotheses.

b. Find the test statistic and P-value, and interpret. (If you prefer, use software, such as MINITAB, for which you can conduct the analysis by entering summary counts.)

c. What is your conclusion, for a significance level of 0.05? (The study was planned to be 8 years long but was stopped after 5 years because of increased heart and cancer problems for the therapy group. This shows a benefit of doing two-sided tests, as results sometimes have the opposite direction from the expected one.)

9.2 | Quantitative Response: How Can We Compare Two Means?

We can compare two groups on a *quantitative* response variable by comparing their means. What does the difference between the sample means tell us about the difference between the population means? We'll find out in this section.

EXAMPLE 6

Teenagers Hooked on Nicotine

Picture the Scenario

A 30-month study evaluated the degree of addiction that teenagers form to nicotine once they begin experimenting with smoking.[5] Random numbers were used to sample 679 seventh-grade students in two Massachusetts cities. Of them, the 332 students who had ever used tobacco by the start of the study were the subjects evaluated. The response variable was constructed using a questionnaire developed for the study, called the Hooked on Nicotine Checklist (HONC). This is a list of ten questions such as "Have you ever tried to quit but couldn't?", "Do

[5]J. DiFranza et al., *Archives of Pediatric and Adolescent Medicine*, vol. 156, pp. 397–403, 2002.

you ever have strong cravings to smoke?", and "Is it hard to keep from smoking in places where you are not supposed to, like school?" The HONC score is the total number of questions to which a student answered yes during the study. Each student's HONC score falls between 0 and 10. The higher the score, the more hooked on nicotine a student is judged to be.

The study considered explanatory variables, such as gender, that might be associated with the HONC score. Figure 9.4, taken from the journal article about this study, shows the sample data distributions of the HONC scores for females and males who had ever smoked. Table 9.5 reports some descriptive statistics.

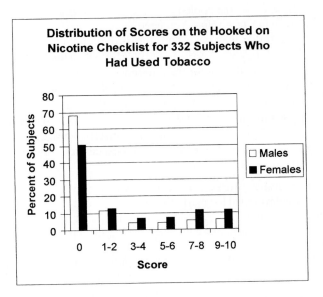

▲ **FIGURE 9.4: Sample Data Distribution of Hooked on Nicotine Checklist (HONC) Scores for Teenagers Who Have Smoked. Question:** Which group seems to have greater nicotine dependence, females or males? Explain why.

TABLE 9.5: Summary of Hooked on Nicotine Checklist (HONC) Scores, by Gender

| Group | Sample Size | HONC Score | |
		Mean	Standard Deviation
Females	150	2.8	3.6
Males	182	1.6	2.9

Question to Explore

How can we compare the sample HONC scores for females and males?

Think It Through

From Figure 9.4, a substantial proportion of teenagers who had ever smoked show no nicotine dependence. For both females and males the sample data distribution is highly skewed to the right. Because of the skew, we could use the median to summarize the sample HONC scores, but that has limited usefulness for such highly discrete data: The median is 0 both for boys and girls. We will estimate the population means and the difference between them. For Table 9.5, let's identify females as group 1 and males as group 2. Then $\bar{x}_1 = 2.8$ and $\bar{x}_2 = 1.6$. We estimate $(\mu_1 - \mu_2)$ by $(\bar{x}_1 - \bar{x}_2) = 2.8 - 1.6 = 1.2$. On the average, females answered yes to about one more question on the HONC scale than males did.

Insight

This analysis uses descriptive statistics only. We'll next see how to make inferences about the difference between population means. You can conduct an inferential comparison of HONC scores by gender in Exercise 9.21. How do you think the nonnormality of HONC distributions affects inference about the population means?

TO PRACTICE THIS CONCEPT, TRY EXERCISE 9.19(A) AND (B)

Standard Error for Comparing Two Means

Recall

From the formula box at the end of Section 9.1, for two estimates from independent samples,

$$se(\text{estimate 1} - \text{estimate 2}) = \sqrt{[se(\text{est. 1})]^2 + [se(\text{est. 2})]^2},$$

the square root of the sum of squared standard errors of the two estimates. ■

How well does the difference $(\bar{x}_1 - \bar{x}_2)$ between two sample means estimate the difference between two population means? This is described by the standard error of the sampling distribution of $(\bar{x}_1 - \bar{x}_2)$.

We can find the standard error of $(\bar{x}_1 - \bar{x}_2)$ using the general rule in the formula box at the end of Section 9.1. The standard error of a single sample mean is $se = s/\sqrt{n}$, where s is the sample standard deviation. Let s_1 and s_2 denote the sample standard deviations for the first and second samples. Then, for independent samples, the standard error of $(\bar{x}_1 - \bar{x}_2)$ is

$$se = \sqrt{[se(\bar{x}_1)]^2 + [se(\bar{x}_2)]^2} = \sqrt{\frac{s_1^2}{n_1} + \frac{s_2^2}{n_2}}.$$

EXAMPLE 7

How Can We Compare Nicotine Dependence for Smokers and Ex-Smokers?

Picture the Scenario

Another explanatory variable in the teenage smoking study was whether a subject was still a smoker when the study ended. The study had 75 smokers and 257 ex-smokers at the end of the study. The HONC means describing nicotine addiction were 5.9 ($s = 3.3$) for the smokers and 1.0 ($s = 2.3$) for the ex-smokers.

Questions to Explore

a. Compare smokers and ex-smokers on their mean HONC scores.

b. What is the standard error of the difference in sample mean HONC scores? How do you interpret that se?

Think It Through

a. Let's regard the smokers as group 1 and ex-smokers as group 2. Then, $\bar{x}_1 = 5.9$ and $\bar{x}_2 = 1.0$. Since $(\bar{x}_1 - \bar{x}_2) = 5.9 - 1.0 = 4.9$, on the average smokers answered yes to nearly five more questions than ex-smokers did on the ten-question HONC scale. That's a large sample difference.

b. Applying the formula for the se of $(\bar{x}_1 - \bar{x}_2)$ given above to these data,

$$se = \sqrt{\frac{s_1^2}{n_1} + \frac{s_2^2}{n_2}} = \sqrt{\frac{(3.3)^2}{75} + \frac{(2.3)^2}{257}} = 0.41.$$

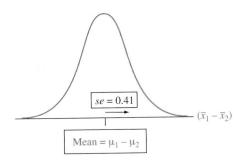

This describes the spread of the sampling distribution of $(\bar{x}_1 - \bar{x}_2)$, as shown in the margin figure. If other random samples of this size were taken from the study population, the difference between the sample means would vary from study to study. The standard deviation of the values for $(\bar{x}_1 - \bar{x}_2)$ would equal about 0.41.

Insight

This standard error will be used in confidence intervals and in significance tests for comparing means.

TO PRACTICE THIS CONCEPT, TRY EXERCISE 9.12(B)

Confidence Interval for the Difference between Two Population Means

As usual, a confidence interval takes the estimate and adds and subtracts a margin of error. For large random samples

- The sampling distribution of a sample mean is approximately normal, by the central limit theorem.
- Likewise, $(\bar{x}_1 - \bar{x}_2)$ has a sampling distribution that is approximately normal.
- For 95% confidence, the margin of error is about two standard errors, so the confidence interval for the difference $(\mu_1 - \mu_2)$ between the population means is approximately

$$(\bar{x}_1 - \bar{x}_2) \pm 2(se).$$

More precisely, the multiple of se is a t-score from a table of t distribution values, because of the extra variability that results from needing to estimate parameters in finding the se.

The degrees of freedom for the t-score multiple of se depends on the sample standard deviations and the sample sizes. The formula is messy and does not give insight into the method, so we relegate it to a footnote.[6] If $s_1 = s_2$ and $n_1 = n_2$, it simplifies to $df = (n_1 + n_2 - 2)$. This is the sum of the df values for single-sample inference about each group, that is, $df = (n_1 - 1) + (n_2 - 1) = n_1 + n_2 - 2$. *Generally, df falls somewhere between $n_1 + n_2 - 2$ and the minimum of $(n_1 - 1)$ and $(n_2 - 1)$.*

> **IN PRACTICE: Software finds *df* for comparing means**
>
> Software calculates the *df* value for comparing means for you. When s_1 and s_2 are similar and n_1 and n_2 are close, *df* is close to $n_1 + n_2 - 2$. Otherwise, if you're not using software, you can take *df* to be the smaller of $(n_1 - 1)$ and $(n_2 - 1)$ and this will be "safe," as the *t*-score will be larger than you actually need.

When either sample size is small (roughly $n_1 < 30$ or $n_2 < 30$), we cannot rely on the central limit theorem. In that instance, the method makes the assumption that the population distributions are normal, so that $(\bar{x}_1 - \bar{x}_2)$ has a bell-shaped sampling distribution. In practice, the method is *robust*, and it works quite well even if the distributions are not normal. This is subject to the usual caveat: We need to be on the lookout for outliers that might affect the means or their usefulness as a summary measure.

Recall

See Section 7.3 to review the **confidence interval** for a single **mean**, the ***t* distribution** and *t*-scores from it. The 95% confidence interval is

$$\bar{x} \pm t_{.025}(se)$$

where $se = s/\sqrt{n}$. Recall that $df = n - 1$ for inference about a single mean.■

Recall

From Section 7.3, a method is **robust** with respect to a particular assumption if it works well even when that assumption is violated.■

[6]It is $df = \dfrac{\left(\dfrac{s_1^2}{n_1} + \dfrac{s_2^2}{n_2} \right)^2}{\dfrac{1}{n_1 - 1}\left(\dfrac{s_1^2}{n_1} \right)^2 + \dfrac{1}{n_2 - 1}\left(\dfrac{s_2^2}{n_2} \right)^2}$, called the Welch-Satterthwaite formula.

Confidence Interval for Difference between Population Means

For two samples with sizes n_1 and n_2 and standard deviations s_1 and s_2, a 95% confidence interval for the difference $(\mu_1 - \mu_2)$ between the population means is

$$(\bar{x}_1 - \bar{x}_2) \pm t_{.025}(se), \text{ with } se = \sqrt{\frac{s_1^2}{n_1} + \frac{s_2^2}{n_2}}.$$

Software provides $t_{.025}$, the t-score with right-tail probability 0.025 (total probability $= 0.95$ between $-t_{.025}$ and $t_{.025}$). This method assumes

- Independent random samples from the two groups
- An approximately normal population distribution for each group (This is mainly important for small sample sizes, and even then the method is robust to violations of this assumption.)

EXAMPLE 8

Nicotine—How Much More Addicted Are Smokers than Ex-Smokers?

Picture the Scenario

For the teenage smoking study, let's make an inference comparing the population mean nicotine addiction (as summarized by the HONC score) for smokers and ex-smokers. From Example 7, $\bar{x}_1 = 5.9$ with $s_1 = 3.3$ for $n_1 = 75$ smokers, and $\bar{x}_2 = 1.0$ with $s_2 = 2.3$ for $n_2 = 257$ ex-smokers.

Questions to Explore

a. Were the HONC sample data distributions for smokers and ex-smokers approximately normal? How does this affect inference?

b. Software reports a 95% confidence interval of (4.1, 5.7) for the difference between the population mean HONC score for smokers and ex-smokers. Show how it obtained this confidence interval, and interpret it.

Think It Through

a. For the ex-smokers (group 2), the values $\bar{x}_2 = 1.0$ and $s_2 = 2.3$ suggest that the HONC distribution is far from bell-shaped, because the lowest possible HONC score of 0 is less than one standard deviation below the mean. This is not problematic. With large samples, the confidence interval method does not require normal population distributions, because the central limit theorem implies that the sampling distribution is approximately normal. (Recall that even for small samples the method is robust when the population distributions are not normal.)

b. For these sample sizes and s_1 and s_2 values, software reports that $df = 95$. The t-score for a 95% confidence interval is then $t_{.025} = 1.985$. (As an approximation, Table B reports the value $t_{.025} = 1.984$ for $df = 100$. Also, the smaller of $(n_1 - 1)$ and $(n_2 - 1)$ is 74, so if you did not have software to find df for you, to be safe you could use the t-score with $df = 74$.)

Denote the population mean HONC score by μ_1 for smokers and μ_2 for ex-smokers. From Example 7, $(\bar{x}_1 - \bar{x}_2) = 4.9$ has a standard error of $se = 0.41$. The 95% confidence interval for $(\mu_1 - \mu_2)$ is

$$(\bar{x}_1 - \bar{x}_2) \pm 1.985(se), \text{ or } 4.9 \pm 1.985(0.41),$$
$$\text{which equals } 4.9 \pm 0.8, \text{ or } (4.1, 5.7).$$

We can be 95% confident that $(\mu_1 - \mu_2)$, the difference between the population mean HONC scores for smokers and for ex-smokers, falls between 4.1 and 5.7. We can infer that the population mean for the smokers is between 4.1 higher and 5.7 higher than for the ex-smokers. The margin shows screen shots from the TI-83+/84.

Insight

We are 95% confident that the smokers answer "yes" to about 4 to 6 more of the 10 questions, on the average, than the ex-smokers. For the 10-point HONC scale, this is quite a substantial difference.

> **TO PRACTICE THIS CONCEPT, TRY EXERCISE 9.19**

How Can We Interpret a Confidence Interval for a Difference of Means?

Besides interpreting the confidence interval comparing two means in context, you can judge the implications using the same criteria as in comparing two proportions:

- *Check whether 0 falls in the interval.* When it does, 0 is a plausible value for $(\mu_1 - \mu_2)$, meaning that possibly $\mu_1 = \mu_2$. Figure 9.5 illustrates. For example, suppose the confidence interval for $\mu_1 - \mu_2$ equals $(-4.1, 5.7)$. Then, the population mean for smokers may be as much as 4.1 lower or as much as 5.7 higher than the population mean for ex-smokers.

▲ **FIGURE 9.5: Three Confidence Intervals for the Difference between Two Means.** When the interval contains 0, it is plausible that the population means may be equal. Otherwise, we can predict which is larger. **Question:** Why do positive numbers in the confidence interval for $(\mu_1 - \mu_2)$ suggest that $\mu_1 > \mu_2$?

- *A confidence interval for $(\mu_1 - \mu_2)$ that contains only positive numbers suggests that $(\mu_1 - \mu_2)$ is positive. We then infer that μ_1 is larger than μ_2.* The 95% confidence interval in Example 8 is $(4.1, 5.7)$. Since $(\mu_1 - \mu_2)$ is the difference between the mean for smokers and the mean for ex-smokers, we infer that the population mean is higher for smokers.

- *A confidence interval for $(\mu_1 - \mu_2)$ that contains only negative numbers suggests that $(\mu_1 - \mu_2)$ is negative. We then infer that μ_1 is smaller than μ_2.*

- *Which group is labeled 1 and which is labeled 2 is arbitrary.* If you change this, the confidence interval has the same endpoints but with different sign. For instance, $(4.1, 5.7)$ becomes $(-5.7, -4.1)$.

Recall

Section 8.3 presented the **significance test** for a single **mean**. The t test statistic measures the number of standard errors that the sample mean falls from the value in the null hypothesis. ■

Significance Tests Comparing Population Means

Another way to compare two population means is with a significance test of the null hypothesis $H_0: \mu_1 = \mu_2$ of equal means. The assumptions of the test are the same as for a confidence interval—independent random samples and approximately normal population distributions for each group. When n_1 and n_2 are at least about 30 each, the normality assumption is not important, because of the central limit theorem. For two-sided alternatives, the test is robust against violations of the normal assumption even when the sample sizes are small. However, one-sided t tests are not trustworthy if a sample size is below 30 and the population distribution is highly skewed.

The test uses the usual form for a test statistic,

$$\frac{\text{Estimate of parameter} - \text{null hypothesis value of parameter}}{\text{Standard error of estimate}}.$$

Treating the difference $(\mu_1 - \mu_2)$ as the parameter, we test that $(\mu_1 - \mu_2) = 0$, that is, the null hypothesis value of the parameter $(\mu_1 - \mu_2)$ is 0. The estimate of the parameter $(\mu_1 - \mu_2)$ is $(\bar{x}_1 - \bar{x}_2)$. The standard error of $(\bar{x}_1 - \bar{x}_2)$ is the same for a significance test as for a confidence interval. The test statistic is

$$t = \frac{(\bar{x}_1 - \bar{x}_2) - 0}{se}, \text{ where } se = \sqrt{\frac{s_1^2}{n_1} + \frac{s_2^2}{n_2}}.$$

When H_0 is true, this statistic has approximately a t sampling distribution. Software can determine the df value.

The P-value for the test depends on whether the alternative hypothesis is two-sided, $H_a: \mu_1 \neq \mu_2$, or one-sided, $H_a: \mu_1 > \mu_2$ or $H_a: \mu_1 < \mu_2$. For the two-sided alternative, the P-value is the two-tail probability beyond the observed t value. That is, we find the probability of results more extreme in either direction, under the presumption that H_0 is true. Software reports this for us. The box summarizes the steps of the two-sided test.

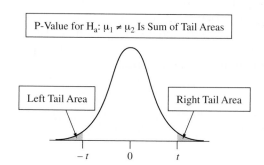

P-Value for $H_a: \mu_1 \neq \mu_2$ Is Sum of Tail Areas

Left Tail Area   Right Tail Area

$-t \quad 0 \quad t$

Two-Sided Significance Test for Comparing Two Population Means

1. Assumptions

- Quantitative response variable for two groups
- Independent random samples
- Approximately normal population distribution for each group. (This is mainly important for small sample sizes, and even then the two-sided test is robust to violations of this assumption.)

2. Hypotheses

$H_0: \mu_1 = \mu_2$
$H_a: \mu_1 \neq \mu_2$ (one-sided $H_a: \mu_1 > \mu_2$ or $H_a: \mu_1 < \mu_2$ also possible)

3. Test Statistic

$$t = \frac{(\bar{x}_1 - \bar{x}_2) - 0}{se} \text{ where } se = \sqrt{\frac{s_1^2}{n_1} + \frac{s_2^2}{n_2}}$$

4. P-value

P-value = Two-tail probability from t distribution of values even more extreme than observed t test statistic, with df given by software

5. Conclusion

Smaller P-values give stronger evidence against H_0 and supporting H_a. Interpret the P-value in context, and if a decision is needed, reject H_0 if P-value $\leq$ significance level (such as 0.05).

EXAMPLE 9

Does Cell Phone Use while Driving Impair Reaction Times?

Picture the Scenario

Chapter 4 discussed three research studies about whether cell phone use may cause cancer. Most scientists think that greater potential harm results from the use of cell phones while driving. A recent experiment[7] investigated whether cell phone use impairs drivers' reaction times, using a sample of 64 students from the University of Utah. Students were randomly assigned to a cell phone group or to a control group, 32 to each. On a machine that simulated driving situations, at irregular periods a target flashed red or green. Participants were instructed to press a "brake button" as soon as possible when they detected a red light. The control group listened to a radio broadcast or to books-on-tape while they performed the simulated driving. The cell phone group carried out a conversation about a political issue on the cell phone with someone in a separate room.

For each subject, the experiment measured their mean response time over all the trials. The purpose was to analyze whether the population mean response time would differ for the two groups. Averaged over all trials and subjects, the mean response time was 585.2 milliseconds (a bit over half a second) for the cell-phone group and 533.7 milliseconds for the control group. Figure 9.6 shows box plots of the responses for the two groups.

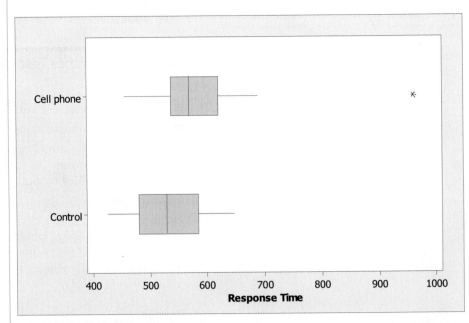

▲ **FIGURE 9.6: MINITAB Box Plots of Response Times for Cell Phone Study.**
Question: Does either box plot show any irregularities that could affect the analysis?

---

[7]Data courtesy of David Strayer, Univ. of Utah. Example based on Experiment 1a in article by D. Strayer and W. Johnston, *Psych. Science*, vol. 21, pp. 462–466, 2001.

```
2-SampTTest
 Inpt:Data Stats
 x̄1:585.2
 Sx1:89.6
 n1:32
 x̄2:533.7
 Sx2:65.3
↓n2:32■
```

```
2-SampTTest
↑n1:32
 x̄2:533.7
 Sx2:65.3
 n2:32
 µ1:≠µ2 <µ2 >µ2
 Pooled:No Yes
 Calculate Draw
```

```
2-SampTTest
 µ1≠µ2
 t=2.628
 p=.011
 df=56.685
 x̄1=585.200
↓x̄2=533.700
```

Denote the population mean response time by μ_1 for the cell phone group and by μ_2 for the control group. Table 9.6 shows how MINITAB reports inferential comparisons of those two means. The margin contains screen shots from the TI-83+/84.

TABLE 9.6: MINITAB Output Comparing Mean Response Times for Cell Phone and Control Groups

	N	Mean	StDev
Cell phone	32	585.2	89.6
Control	32	533.7	65.3

Difference = mu(Cell phone) − mu(Control)

Estimate for difference: 51.5172

95% CI for difference: (12.2393, 90.7951)

T-Test of difference = 0 (vs not =):

T-Value = 2.63 P-Value = 0.011 DF = 56

Questions to Explore

a. Show how MINITAB got the test statistic for testing $H_0: \mu_1 = \mu_2$ against $H_a: \mu_1 \neq \mu_2$.

b. Report and interpret the P-value, and state the decision you would make for a 0.05 significance level.

c. What do the box plots tell us about the suitability of these analyses? What effect does the outlier for the cell phone group have on the analysis?

Think It Through

a. We estimate the difference $\mu_1 - \mu_2$ by $(\bar{x}_1 - \bar{x}_2) = 585.2 - 533.7 = 51.5$, as reported in Table 9.6. The standard error of this estimate is

$$se = \sqrt{\frac{s_1^2}{n_1} + \frac{s_2^2}{n_2}} = \sqrt{\frac{(89.6)^2}{32} + \frac{(65.3)^2}{32}} = 19.6.$$

The test statistic for $H_0: \mu_1 = \mu_2$ equals

$$t = \frac{(\bar{x}_1 - \bar{x}_2)}{se} = \frac{51.5}{19.6} = 2.63.$$

b. The P-value is the two-tail probability from a t distribution. Table 9.6 reports $df = 56$ for that distribution and a P-value = 0.01. If H_0 were true, the probability would be 0.01 of getting a t test statistic this large or even larger in either tail. The P-value is less than 0.05, so we can reject H_0. There is enough evidence to conclude that the population mean response times differ between the cell phone and control groups. The sample means suggest that the population mean is higher for the cell phone group.

c. The t inferences assume normal population distributions. The box plots do not show any substantial skew, but there is an extreme outlier for the cell phone group. One subject in that group had a very slow mean reaction time. Because this observation is so far from the others in that group, it's a good idea to make sure the results of the analysis aren't affected too strongly by that single observation (for instance, in case it was recorded improperly).

If we delete the extreme outlier for the cell phone group, software reports

```
                N      Mean      StDev
Cell phone  31     573.1     58.9
Control      32     533.7     65.3
Estimate for difference: 39.4265
95% CI for difference: (8.1007, 70.7522)
T-Test of difference = 0 (vs not =):
T-Value = 2.52   P-Value = 0.015   DF = 60
```

The mean and standard deviation for the cell phone group now decrease substantially. However, the t test statistic is not much different, and the P-value is still small, 0.015, leading to the same conclusion.

Insight

Even though the difference between the sample means decreased from 51.5 to 39.4 when we deleted the outlier, the standard error also got smaller (you can check that it equals 15.7) because of the smaller standard deviation for the cell phone group after removing the outlier. That's why the t test statistic did not change much. In practice, you should not delete outliers from a data set without sufficient cause (for example, if it seems the observation was incorrectly recorded). However, it's a good idea to check for sensitivity of an analysis to an outlier, as we did here, by repeating the analysis without it. If the results change much, it means that the inference including the outlier is on shaky ground.

TO PRACTICE THIS CONCEPT, TRY EXERCISE 9.21

Example 9 used a two-sided alternative, which is the way that research results are usually reported in journal articles (as this study was). But the researchers thought that the mean response time would be *greater* for the cell phone group than for the control. So, for their own purposes, they could find the P-value for the one-sided $H_a: \mu_1 > \mu_2$ (that is, $\mu_1 - \mu_2 > 0$), which predicts a higher mean response time for the cell phone group. For the analysis without the outlier, the P-value is the probability to the *right* of $t = 2.52$. This is half the two-sided P-value, namely $0.015/2 = 0.008$. There is very strong evidence in favor of this alternative.

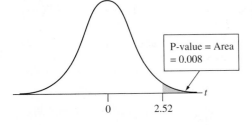

P-value = Area = 0.008

Connection between Confidence Intervals and Tests

We learn even more by constructing a confidence interval for $(\mu_1 - \mu_2)$. Let's do this for the data set without the outlier, for which $df = 60$ and the difference of sample means of 39.4 had a standard error of 15.7. The t-score with $df = 60$ for a 95% confidence interval is $t_{.025} = 2.000$. The confidence interval is

$$(\bar{x}_1 - \bar{x}_2) \pm t_{.025}(se), \text{ which is } 39.4 \pm 2.000(15.7), \text{ or } (8.1, 70.8).$$

This is also shown on the printout above. It is more informative than the test, because it shows the range of realistic values for the difference between the population means. The interval (8.1, 70.8) for $\mu_1 - \mu_2$ is quite wide. It tells us that the population means could be similar, or the mean response may be as much as about 71 milliseconds higher for the cell phone group. This is nearly a tenth of a second, which could be crucial in a practical driving situation.

The confidence interval (8.1, 70.8) does not contain 0. This inference agrees with the significance test that the population mean response times differ. Recall that Section 8.3 showed that the result of a two-sided significance test about a mean is consistent with a confidence interval for that mean. The same is true in comparing two means. If a test rejects H_0: $\mu_1 = \mu_2$, then the confidence interval for $(\mu_1 - \mu_2)$ having the same error probability does not contain 0.

In Example 9 about cell phone use without the outlier, the P-value $= 0.015$, so we rejected H_0: $\mu_1 = \mu_2$ at the 0.05 significance level. Likewise, the 95% confidence interval for $(\mu_1 - \mu_2)$ of (8.1, 70.8) does not contain 0, the null hypothesis value.

By contrast, when a 95% confidence interval for $(\mu_1 - \mu_2)$ *contains* 0, then 0 is a plausible value for $(\mu_1 - \mu_2)$. In a test of H_0: $\mu_1 = \mu_2$ against H_a: $\mu_1 \neq \mu_2$, the P-value would be >0.05. The test would not reject H_0 at the 0.05 significance level and would conclude that the population means may be equal.

9.2 Practicing the Basics

9.12 Housework for women and men: Do women tend to spend more time on housework than men? If so, how much more? Based on data from the National Survey of Families and Households, one study (by S. South and G. Spitze, *American Sociological Review*, vol. 59, 1994, pp. 327–347) reported the results in the table for the number of hours spent in housework per week.

		HOUSEWORK HOURS	
Gender	Sample Size	Mean	Standard Deviation
Women	6764	32.6	18.2
Men	4252	18.1	12.9

a. Show that this study estimated that, on the average, women spend 14.5 more hours a week on housework than men.

b. Find the standard error for comparing the means. What factor causes the standard error to be small compared to the sample standard deviations for the two groups?

c. A 95% confidence interval comparing the population means for women and men is (13.9, 15.1). Interpret, indicating the relevance of 0 not falling in the interval.

d. State the assumptions upon which the interval in (c) is based.

9.13 More confident about housework: Refer to part (c) in the previous exercise.

a. Show that a 99% confidence interval is 14.5 ± 0.8, or (13.7, 15.3). (*Hint*: For such large sample sizes, the t-score is practically identical to a z-score.)

b. Explain why this interval is wider than the 95% confidence interval.

9.14 Employment by gender: The study described in the previous exercise also evaluated the weekly time spent in employment. Software shows the results.

```
Two-sample T for Employ
Gender    N     Mean   StDev   SE Mean
Men      4252   31.8   22.6    0.3466
Women    6764   18.4   20.0    0.2432
Difference = mu(Men) - mu(Women)
95% CI for difference: (12.57, 14.23)
T-Test of difference = 0 (vs not =):
T-Value = 31.65  P-Value = 0.000
```

a. Does it seem plausible that employment has a normal distribution for each gender? Explain.

b. What effect does the answer to (a) have on inference comparing population means? What assumptions are made for the inferences in this table?

c. Explain how to interpret the confidence interval.

d. Refer to (c). Do you think that the population means are equal? Explain.

9.15 Ideal number of children: The web site sda.berkeley.edu/GSS for the General Social Survey enables you to compare means for groups, by clicking on "Comparison of means" at the homepage. Compare female and male responses on the question, "What is the ideal number of children for a family to have?" in 1994 (the last year this was asked). Identify NUMKIDS as the "dependent variable," SEX as the "row variable," use YEAR(1994) as the "selection variable," and request standard deviations and standard errors.

a. The 772 females who responded had a mean of 2.49 and standard deviation of 0.81. What were the results for males?

b. Report the 95% confidence interval for the difference between the population means for females and males. Interpret.

9.16 eBay auctions: Example 7 in Chapter 7 showed final prices at which the Palm M515 PDA (personal digital assistant), a popular handheld computer, sold. Some auctions used a "buy-it-now" option, whereby a bidder can buy the product immediately at a fixed price set by the seller. Here are the data from sales during the first week of May 2003.

1. Buy-it-now option: 235, 225, 225, 240, 250, 250, 210
2. Bidding only: 250, 249, 255, 200, 199, 240, 228, 255, 232, 246, 210, 178, 246, 240, 245, 225, 246, 225

Construct a 95% confidence interval for comparing the population means. Interpret, and explain what (if any) effect the type of sale method has on the mean sales price.

9.17 Test comparing auction prices: Refer to the previous exercise.

   a. Plot the data to compare prices for type of sale. What do you learn from the plots?

   b. Find the test statistic and P-value for $H_0: \mu_1 = \mu_2$ and $H_a: \mu_1 \neq \mu_2$. Interpret.

   c. State the assumptions for the method in (b). Is there any evidence any of them are violated for these data? If so, how, if at all, does this affect the validity of the analysis in (b)?

9.18 Chelation useless?: Chelation is an alternative therapy for heart disease that uses repeated intravenous administration of a human-made amino acid in combination with oral vitamins and minerals. Practitioners believe it removes calcium deposits from buildup in arteries. However, the evidence for a positive effect is anecdotal or comes from nonrandomized, uncontrolled studies. A double-blind randomized clinical trial comparing chelation to placebo used a treadmill test in which the response was the length of time until a subject experienced ischemia (lack of blood flow and oxygen to the heart muscle).

   a. After 27 weeks of treatment, the sample mean time for the chelation group was 652 seconds. A 95% confidence interval for the population mean for chelation minus the population mean for placebo was −53 to 36 seconds. Explain how to interpret the confidence interval.

   b. A significance test comparing the means had P-value = 0.69. Specify the hypotheses for this test, which was two-sided.

   c. The authors concluded from the test, "There is no evidence to support a beneficial effect of chelation therapy" (M. Knudtson et al., *JAMA*, vol. 287, p. 481, 2002). Explain how this conclusion agrees with inference based on the values in the confidence interval.

9.19 Some smoked but didn't inhale: Refer to Examples 6–8 on nicotine dependence for teenage smokers. Another explanatory variable was whether a subject reported inhaling when smoking. The table reports descriptive statistics.

| Group | HONC SCORE | | |
	Sample Size	Mean	Standard Deviation
Inhalers	237	2.9	3.6
Noninhalers	95	0.1	0.5

   a. Explain why (i) the overwhelming majority of noninhalers must have had HONC scores of 0, (ii) on the average, those who reported inhaling answered yes

to nearly three more questions than those who denied inhaling.

   b. Might the HONC scores have been approximately normal for each group? Why or why not?

   c. Find the standard error for the estimate $(\bar{x}_1 - \bar{x}_2) = 2.8$. Interpret.

   d. The 95% confidence interval for $(\mu_1 - \mu_2)$ is (2.3, 3.3). What can you conclude about the population means for inhalers and noninhalers?

9.20 Inhaling affect HONC?: Refer to the previous exercise.

   a. Show that the test statistic for $H_0: \mu_1 = \mu_2$ equals $t = 11.7$. If the population means were equal, explain why it would be nearly impossible by random variation to observe this large a test statistic.

   b. What decision would you make about H_0, at common significance levels? Can you conclude which group had higher mean nicotine dependence? How?

   c. State the assumptions for the inference in this exercise.

9.21 Females or males more nicotine dependent?: Refer to Example 6, "Teenagers Hooked on Nicotine." Of those who had tried tobacco, the mean HONC score was 2.8 ($s = 3.6$) for the 150 females and 1.6 ($s = 2.9$) for the 182 males.

   a. Find a standard error for comparing the sample means. Interpret.

   b. Find the test statistic and P-value for $H_0: \mu_1 = \mu_2$ and $H_a: \mu_1 \neq \mu_2$. Interpret, and explain what (if any) effect gender has on the mean HONC score.

   c. Do you think that the HONC scores were approximately normal for each gender? Why or why not? How does this affect the validity of the analysis in (b)?

9.22 Female and male monthly smokers: Refer to the previous exercise. A subject was called a "monthly smoker" if he or she had smoked cigarettes over an extended period of time. The 74 female monthly smokers had a mean HONC score of 5.4 ($s = 3.5$), and the 71 male monthly smokers had a mean HONC score of 3.9 ($s = 3.6$). Using software (such as MINITAB) that can conduct analyses using summary statistics, repeat steps (b) and (c) of the previous exercise.

9.23 TV watching and gender: For the 2002 General Social Survey, a comparison of females and males on the number of hours a day that the subject watched TV gave

Group	n	Mean	StDev	SE Mean	95.0% CI
Females	506	3.06	2.12	0.0942	(2.87, 3.25)
Males	399	2.88	2.63	0.1317	(2.62, 3.14)

   a. Set up the hypotheses of a significance test to analyze whether the population means differ for females and males.

   b. Using software (such as MINITAB) in which you can use summarized data, conduct all parts of the significance test. Interpret the P-value, and report the conclusion for a significance level of 0.05.

   c. If you were to construct a 95% confidence interval comparing the means, would it contain 0? Answer based on the result of (b), without actually finding the interval.

d. Do you think that the distribution of TV watching is approximately normal? Why or why not? Does this affect the validity of your inferences? What assumptions do the methods make?

9.24 **Student survey:** Refer to the "FL student survey" data file on the text CD. Identify number of times reading a newspaper as the response variable and gender as the explanatory variable. The observations are as follows:

Females: 5, 3, 6, 3, 7, 1, 1, 3, 0, 4, 7, 2, 2, 7, 3, 0, 5, 0, 4, 4, 5, 14, 3, 1, 2, 1, 7, 2, 5, 3, 7

Males: 0, 3, 7, 4, 3, 2, 1, 12, 1, 6, 2, 2, 7, 7, 5, 3, 14, 3, 7, 6, 5, 5, 2, 3, 5, 5, 2, 3, 3

Using software,

a. Construct and interpret a plot comparing responses by females and males.

b. Construct and interpret a 95% confidence interval comparing population means for females and males.

c. Show all five steps of a significance test comparing the population means.

d. State and check the assumptions for (b) and (c).

9.25 **Normal assumption:** The methods of this section make the assumption of a normal population distribution. Why do you think this is more relevant for small samples than for large samples?

9.3 Other Ways of Comparing Means and Comparing Proportions

We've now learned about the primary methods for comparing two proportions or two means. We'll next study an alternative method for comparing means that is useful when it is reasonable to assume that the two groups have similar variability on the response variable. This method, like those of Sections 9.1 and 9.2, focuses on inference about the *difference* between the parameters. Often it's useful to estimate a *ratio* of parameters instead. We'll also illustrate this, as an alternative way of comparing two proportions or two means.

Comparing Means, Assuming Equal Population Standard Deviations

An alternative *t* method to the one described in Section 9.2 is sometimes used when, under the null hypothesis, it is reasonable to expect the *variability* as well as the mean to be the same. For example, consider a study comparing a drug to a placebo treatment. If the drug has no effect, then we expect the entire distributions of the response variable to be identical for the two groups, not just the mean. This method requires an extra assumption in addition to the usual ones of independent random samples and approximately normal population distributions:

The population standard deviations are equal, that is, $\sigma_1 = \sigma_2$ (see Figure 9.7).

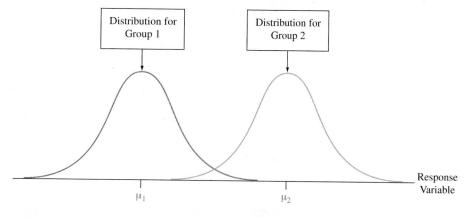

▲ **FIGURE 9.7: Two Groups with Equal Population Standard Deviations. Question:** What would a graph of the sample data look like to make you doubt this assumption?

This alternative method estimates the common value σ of σ_1 and σ_2 by

$$s = \sqrt{\frac{(n_1 - 1)s_1^2 + (n_2 - 1)s_2^2}{n_1 + n_2 - 2}}.$$

The estimate s, called the **pooled standard deviation**, combines information from the two samples to provide a single estimate of spread. The term inside the square root is a weighted average of the two sample squared standard deviations. When $n_1 = n_2$, it's the ordinary average, the sum of the squared standard deviations divided by 2. The estimate s falls between s_1 and s_2. The degrees of freedom for this method are $df = n_1 + n_2 - 2$, which appears in the denominator of the formula for s.

Comparing Population Means, Assuming Equal Population Standard Deviations

Using the pooled standard deviation estimate s of $\sigma = \sigma_1 = \sigma_2$, the standard error of $(\bar{x}_1 - \bar{x}_2)$ simplifies to

$$se = \sqrt{\frac{s^2}{n_1} + \frac{s^2}{n_2}} = s\sqrt{\frac{1}{n_1} + \frac{1}{n_2}}.$$

Otherwise, inference formulas look the same as those that do not assume $\sigma_1 = \sigma_2$:

- A 95% confidence interval for $(\mu_1 - \mu_2)$ is $(\bar{x}_1 - \bar{x}_2) \pm t_{.025}(se)$.
- The test statistic for H_0: $\mu_1 = \mu_2$ is $t = (\bar{x}_1 - \bar{x}_2)/se$.

These methods have $df = n_1 + n_2 - 2$. To use them, you assume

- Independent random samples from the two groups
- An approximately normal population distribution for each group (This is mainly important for small sample sizes, and even then the confidence interval and two-sided test are usually robust to violations of this assumption.)
- $\sigma_1 = \sigma_2$ (In practice, they are not usually relied on if one *sample* standard deviation is more than double the other one.)

IN PRACTICE: Robustness of two-sided inferences

Confidence intervals and two-sided tests using this alternative method are robust. They work well even when the population distributions are not normal and when the population standard deviations are not exactly equal. This is particularly true when the sample sizes are similar and not extremely small. In practice, however, these alternative methods are not usually used if one *sample* standard deviation is more than double the other one.

EXAMPLE 10

Is Arthroscopic Surgery Better than Placebo?

Picture the Scenario

Arthroscopic surgery is often used when medical therapy fails to relieve pain from osteoarthritis of the knee. In the United States, more than 650,000 arthroscopic surgeries are performed yearly, at a cost of about $5000 each. It is quite a common procedure with athletes, especially basketball players. About half of the time patients report relief from pain, but there is no evidence that arthroscopy provides a cure. A recent study[8] conducted a randomized trial to assess the usefulness of arthroscopic surgery.

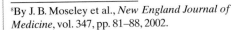

[8]By J. B. Moseley et al., *New England Journal of Medicine*, vol. 347, pp. 81–88, 2002.

Over a three-year period, patients suffering from osteoarthritis who had at least moderate knee pain were recruited from a medical center in Houston. Patients were randomly assigned to one of three groups, to receive one of two types of arthroscopic surgeries or a surgery that was actually a placebo procedure. In the arthroscopic surgeries, the "lavage group" had the joint flushed with fluid but instruments were not used to remove tissue, whereas the "debridement group" had tissue removal as well. In the placebo procedure, the same incisions were made in the knee as with surgery and the surgeon manipulated the knee as if surgery was being performed, but none was actually done. The study was double-blind: Both the patients and the assessors were blinded to the treatment assignments.

A knee-specific pain scale was created for the study. Administered two years after the surgery, it ranged from 0 to 100, with higher scores indicating more severe pain. Table 9.7 shows summary statistics. The pain scale was the response variable, and the treatment group was the explanatory variable.

TABLE 9.7: Summary of Knee Pain Scores

The descriptive statistics compare lavage and debridement arthroscopic surgery to a placebo (fake surgery) treatment.

| | | KNEE PAIN SCORE | |
Group	Sample Size	Mean	Standard Deviation
1. Placebo	60	51.6	23.7
2. Arthroscopic—lavage	61	53.7	23.7
3. Arthroscopic—debridement	59	51.4	23.2

Denote the population mean of the pain scores by μ_1 for the placebo group and by μ_2 for the lavage arthroscopic group. Most software gives you the option of assuming equal population standard deviations. Table 9.8 is MINITAB output for the two-sample t inferences that do this. (We'll consider the debridement arthroscopic group in an exercise.)

TABLE 9.8: MINITAB Output for Comparing the Mean Knee Pain for Placebo and Arthroscopic Surgery Groups

This analysis assumes equal population standard deviations.

```
  Sample    N     Mean    StDev   SE Mean
    1       60    51.6    23.7     3.1
    2       61    53.7    23.7     3.0
Difference = mu(1) - mu(2)
Estimate for difference: -2.10000
95% CI for difference: (-10.63272, 6.43272)
T-Test of difference = 0 (vs not =):
T-Value = -0.49  P-Value = 0.627  DF = 119
Both use Pooled StDev = 23.7000
```

Questions to Explore

a. Does the t test inference in Table 9.8 seem appropriate for these data?

b. Show how to find the pooled standard deviation estimate of σ, the standard error, the test statistic, and its df.

c. Identify on the printout the P-value for testing $H_0: \mu_1 = \mu_2$ against $H_a: \mu_1 \neq \mu_2$. With the 0.05 significance level, can you reject H_0? Interpret.

Think It Through

a. If the arthroscopic surgery has no real effect, the population distribution of pain score should be the same for it as for the placebo. Not only will the population means be equal, but so will the population standard deviations. So, in testing whether this surgery has "no effect," it is sensible to use a method that assumes equal population standard deviations. Table 9.8 shows that the sample standard deviations are identical. This is mere coincidence, and we would not normally expect this even if the population standard deviations were identical. But this suggests it should be fine to use the method assuming equal population standard deviations.

b. The pooled standard deviation estimate of the common value σ of $\sigma_1 = \sigma_2$ is

$$s = \sqrt{\frac{(n_1 - 1)s_1^2 + (n_2 - 1)s_2^2}{n_1 + n_2 - 2}} =$$

$$\sqrt{\frac{(60 - 1)23.7^2 + (61 - 1)23.7^2}{60 + 61 - 2}} = 23.7,$$

shown at the bottom of Table 9.8. We estimate the difference $\mu_1 - \mu_2$ by $(\bar{x}_1 - \bar{x}_2) = 51.6 - 53.7 = -2.1$, as reported in Table 9.8. The standard error of $(\bar{x}_1 - \bar{x}_2)$ equals

$$se = s\sqrt{\frac{1}{n_1} + \frac{1}{n_2}} = 23.7\sqrt{\frac{1}{60} + \frac{1}{61}} = 4.31.$$

The test statistic for H_0: $\mu_1 = \mu_2$ equals

$$t = \frac{(\bar{x}_1 - \bar{x}_2)}{se} = \frac{51.6 - 53.7}{4.31} = -0.49.$$

Its $df = (n_1 + n_2 - 2) = (60 + 61 - 2) = 119$. When H_0 is true, the test statistic has the t distribution with $df = 119$.

c. From Table 9.8, the two-sided P-value equals 0.63. So, if the true population means were equal, by random variation it would not be surprising to observe a test statistic of this size. The P-value is larger than 0.05, so we cannot reject H_0. In summary, we don't have enough evidence to conclude that the mean pain level differs for the placebo treatment and the arthroscopic surgery treatment.

Insight

Table 9.8 reports a 95% confidence interval for $(\mu_1 - \mu_2)$ of $(-10.6, 6.4)$. Since the interval contains 0, it is plausible that there is no difference between the population means. This is the same conclusion as made with the significance test. We infer that the population mean for the knee pain score could be as much as 10.6 lower or as much as 6.4 higher for the placebo treatment than for the arthroscopic surgery. On the pain scale with range 100, this is a relatively small difference in practical terms.

TO PRACTICE THIS CONCEPT, TRY EXERCISE 9.28

Suppose that the researchers specifically predicted that the lavage surgery would give *better* pain relief than the placebo. Then they would find the P-value for the one-sided H_a: $\mu_1 > \mu_2$ (that is, $\mu_1 - \mu_2 > 0$), which predicts higher mean pain with the placebo. The P-value is the probability to the *right*

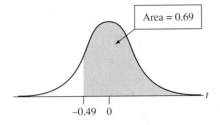

Area = 0.69

−0.49 0

of $t = -0.49$. Software reports this as 0.69. There is no evidence in favor of this alternative. The P-value exceeds 0.50 because the sample mean pain was actually *lower* for the placebo, not higher.

Comparing Population Standard Deviations

Many texts and most software present a statistic denoted by F for testing the hypothesis that the population standard deviations are equal. This test assumes that the population distributions are normal. Unfortunately, this F-test is *not* a robust method.

IN PRACTICE: *F* test for comparing standard deviations is not robust

The **F test for comparing standard deviations** of two populations performs poorly if the populations are not close to normal. Statisticians do not recommend it for general use because of this. If the data show evidence of a potentially large difference in standard deviations, with one of the sample standard deviations being at least double the other, it is better to use the two-sample t-inferences (of Section 9.2) that do not have this extra assumption.

The Ratio of Proportions: The Relative Risk

	Heart Attack		Prop.
Group	Yes	No	**Yes**
Placebo	189	10,845	.0171
Aspirin	104	10,933	.0094

Examples 2–4 discussed data from the Physicians Health Study comparing the proportions of heart attacks between a group who took placebo for five years and a group that took aspirin. See the margin table. The difference $\hat{p}_1 - \hat{p}_2 = 0.0171 - 0.0094 = 0.008$ was small. However, this inequality seems more substantial when viewed by their *ratio*, which is a statistic commonly reported in medical journals.

Ratio of Proportions (Relative Risk)

The **ratio of proportions** for two groups is $\hat{p}_1/\hat{p}_2$. In medical applications for which the proportion refers to a category that is an undesirable outcome, such as death or having a heart attack, this ratio is called the **relative risk**.

The ratio describes the sizes of the proportions *relative* to each other. For the Physicians Health Study, $\hat{p}_1 = 0.0171$ for placebo, $\hat{p}_2 = 0.0094$ for aspirin, and

$$\text{sample relative risk} = \hat{p}_1/\hat{p}_2 = 0.0171/0.0094 = 1.82.$$

This means that the proportion of the placebo group who had a heart attack was 1.82 times the proportion of the aspirin group who had a heart attack.

The population relative risk is p_1/p_2. When $p_1 = p_2$, the relative risk equals 1. The sample relative risk of 1.82 for the Physicians Health Study is quite far from 1, indicating a practically significant effect.

Recall that the labeling of the groups is arbitrary. If we had labeled group 1 as those taking aspirin, then the relative risk is $0.0094/0.0171 = 0.55$. The proportion of the aspirin group who had a heart attack was 0.55 times the proportion of the placebo group who had a heart attack. But $0.55 = 1/1.82$. *Whether the relative risk equals a particular number or its reciprocal merely depends on the group labeling.*

Journal articles about medical studies often report the relative risk, especially when both proportions are close to 0. Software can form a confidence interval[9] for a population relative risk. It is common for the articles to report this also.

---

[9]The formula for the confidence interval is complex, and we do not cover it in this text.

EXAMPLE 11

Does Alcohol Consumption Help Reduce the Risk of Stroke?

Picture the Scenario

Since results of the Physicians Health Study were first reported, that study has continued to follow the sampled doctors. One analysis considered whether subjects who regularly consumed some alcohol might have a reduced risk of stroke. An article in a medical journal[10] stated, "Compared with participants who had less than one drink per week, those who drank more had a reduced overall risk of stroke (relative risk, 0.79; 95% confidence interval, 0.66 to 0.94)."

Question to Explore

How do you interpret the relative risk value and the reported confidence interval?

Think It Through

Let $\hat{p}_1$ = sample proportion of the regular alcohol drinkers who had strokes and $\hat{p}_2$ = sample proportion of the light or nondrinkers who had strokes. Then the sample relative risk of 0.79 means that $\hat{p}_1/\hat{p}_2 = 0.79$. So, $\hat{p}_1 = 0.79\hat{p}_2$. The proportion of the regular alcohol drinkers who had strokes was 0.79 times the proportion of the light or nondrinkers who had strokes. Since the relative risk was less than 1.0, the proportion of strokes was smaller for the regular alcohol drinkers.

The 95% confidence interval for the population relative risk of stroke was (0.66, 0.94). Since all numbers in the interval are less than 1.0, we can infer that $p_1/p_2 < 1.0$ (see the margin figure). That is, it seems that $p_1 < p_2$. We can infer that the population proportion of strokes is smaller for those who drink alcohol regularly.

Insight

Regular alcohol drinking seems to correspond to a reduction in the incidence of stroke. Since the upper endpoint of the confidence interval is 0.94, however, the population relative risk could be close to 1.0. So, the effect of drinking might be weak. Another reason to be cautious is that this was an observational study, rather than a randomized clinical trial such as was used to compare aspirin and placebo. (The study did not randomly pick half the subjects to be the ones who drank alcohol regularly!) In any case, the conclusion does not imply that it's good to drink a lot of alcohol every day. High alcohol consumption can lead to its own medical problems, such as alcoholism and liver damage.

TO PRACTICE THIS CONCEPT, TRY EXERCISE 9.33

[10]*New England Journal of Medicine*, vol. 341, pp. 1557–1564 (1999).

The Ratio of Means

You can also use the ratio to compare means. For example, the National Center for Health Statistics recently reported that the mean weight for adult American women was 140 pounds in 1962 and 164 pounds in 2002. Since 164/140 = 1.17, the mean in 2002 was 1.17 times the mean in 1962. This can also be expressed by saying that the mean increased by 17% over this 40-year period.

9.3 Practicing the Basics

9.26 Bulimia CI: A study of bulimia among college women (J. Kern and T. Hastings, *Journal of Clinical Psychology*, vol. 51, 1995, p. 499) studied the connection between childhood sexual abuse and a measure of family cohesion (the higher the score, the greater the cohesion). The sample mean on the family cohesion scale was 2.0 for 13 sexually abused students ($s = 2.1$) and 4.8 for 17 non-abused students ($s = 3.2$). Assuming equal population standard deviations:

a. Find the standard error for comparing the means.

b. Construct a 95% confidence interval for the difference between the mean family cohesion for sexually abused students and non-abused students. Interpret.

9.27 Bulimia test: Refer to the previous exercise.

a. Find the P-value for testing whether the population means are equal. Use a two-sided alternative.

b. Summarize assumptions for the analysis in (a). Do you think the normality assumption is justified? If not, what is the consequence of violating it?

9.28 Surgery vs. placebo for knee pain: Refer to Example 10, "Is Arthroscopic Surgery Better than Placebo?" Here we show MINITAB output comparing mean knee pain scores for the placebo (group 1) to debridement arthroscopic surgery (group 2).

a. State and interpret the result of the confidence interval.

b. State all steps and interpret the result of the significance test.

c. Based on the confidence interval and test, would you conclude that the arthroscopic surgery works better than placebo? Explain.

```
Sample    N    Mean    StDev   SE Mean
  1       60   51.6    23.7    3.1
  2       59   51.4    23.2    3.0
Difference = mu(1) − mu(2)
Estimate for difference: 0.200000
95% CI for difference: (−8.316130, 8.716130)
T-Test of difference = 0 (vs not =):
T-Value = 0.05  P-Value = 0.963  DF = 117
Both use Pooled StDev = 23.4535
```

▉ 9.29 Comparing clinical therapies: A clinical psychologist wants to choose between two therapies for treating severe cases of mental depression. She selects six patients who are similar in their depressive symptoms and in their overall quality of health. She randomly selects three of the patients to receive therapy 1, and the other three receive therapy 2. She selects small samples for ethical reasons—if her experiment indicates that one therapy is superior, she will use that therapy on all her other depression patients. After one month of treatment, the improvement in each patient is measured by the change in a score for measuring severity of mental depression. The higher the score, the better. The improvement scores are

Therapy 1: 30, 45, 45

Therapy 2: 10, 20, 30

Analyze these data (you can use software, if you wish), assuming equal population standard deviations.

a. Show that $\bar{x}_1 = 40$, $\bar{x}_2 = 20$, $s = 9.35$, $se = 7.64$, $df = 4$, and a 95% confidence interval comparing the means is $(−1.2, 41.2)$.

b. Explain how to interpret what the confidence interval tells you about the therapies. Why do you think that it is so wide?

c. When the sample sizes are very small, it may be worth sacrificing some confidence to achieve more precision. Show that a 90% confidence interval is $(3.7, 36.3)$. At this confidence level, can you conclude that therapy 1 is better?

9.30 Clinical therapies 2: Refer to the previous exercise.

a. For the null hypothesis, $H_0: \mu_1 = \mu_2$, show that $t = 2.62$ and the two-sided P-value $= 0.059$. Interpret.

b. What decision would you make in the test, using a (i) 0.05, (ii) 0.10 significance level? Explain what this means in the context of the study.

c. Suppose the researcher had predicted ahead of time that therapy 1 would be better. To which H_a does this correspond? Report the P-value for it, and make a decision with significance level 0.05.

9.31 Vegetarians more liberal?: When a sample of social science graduate students at the University of Florida gave their responses on political ideology (ranging from 1 = very liberal to 7 = very conservative), the mean was 3.18 ($s = 1.72$) for the 51 nonvegetarian students and 2.22 ($s = 0.67$) for the 9 vegetarian students. Software for comparing the means provides the printout, which shows results first for inferences that assume equal population standard deviations and then for inferences that allow them to be unequal.

```
Sample    N     Mean    StDev   SE Mean
  1      51     3.18    1.72     0.24
  2       9     2.22    0.67     0.22
Difference = mu(1) - mu(2)
95% CI for difference: (-0.209716, 2.129716)
T-Test of difference = 0 (vs not =):
T-Value = 1.64  P-Value = 0.106  DF = 58
Both use Pooled StDev = 1.6162

95% CI for difference: (0.289196, 1.630804)
T-Test of difference = 0 (vs not =):
T-Value = 2.92  P-Value = 0.007  DF = 30
```

a. Explain why the results of the two approaches differ so much. Which do you think is more reliable?

b. State your conclusion about whether the true means are plausibly equal.

9.32 Claritin and nervousness: An advertisement by Schering Corp. for the allergy drug Claritin mentioned that in a pediatric randomized clinical trial, symptoms of nervousness were shown by 4 of 188 patients on Claritin and 2 of 262 patients taking placebo. Find and interpret the sample relative risk.

9.33 Fish and heart disease: The Nurses' Health Study has followed about 85,000 female nurses over time. Compared with women who ate fish less than once a month, a medical article (F. Hu et al., *JAMA*, vol. 287, pp. 1815–1821, 2002) reported "the relative risk of coronary heart disease was 0.66 (95% CI: 0.50 to 0.89) for fish consumption 5 or more times per week." Explain how to interpret the (a) reported relative risk, (b) confidence interval.

9.34 Aspirin and heart attacks: For the Swedish study described in Exercise 9.6, during the follow-up period 28 of 684 taking placebo died from a heart attack and 18 of 676 taking aspirin died from a heart attack.

a. Show that the relative risk is 1.54. Interpret.

b. Software reports a 95% confidence interval of (0.86, 2.75) for the population relative risk. Interpret the lower endpoint of 0.86.

c. Explain the relevance of the confidence interval containing 1.0.

9.35 Obesity and cancer: A recent study (by E. Calle et al., *New England Journal of Medicine*, vol. 348, pp. 1625–1638, April 24, 2003) suggested that obesity may be linked to 1 in 6 cancer deaths. The article reported that, compared to women of normal weight, the relative percent increase in death rate from cancer for women with BMI [body mass

index = BMI = 703 (weight in pounds)/(height in inches)^2] of 40 or higher was 62%. This means that the relative risk, defined as the probability of death for those with high BMI divided by the probability of death for those with low BMI, equals 1.62. The articles' abstract reported a 95% confidence interval of the relative risk for women of (1.40, 1.87). Fill in the blanks in the interpretation, "In the population of women, we are 95% confident that the increase in the death rate for the highly obese group was between _____% and _____%.

9.36 Reducing risk summaries and interpretations:

a. An AP story (October 24, 2002) about new research reported in the *New England Journal of Medicine* stated, "Those getting aspirin within 48 hours of bypass surgery were only one-third as likely to die in the hospital as the others" (not receiving aspirin). Which descriptive measure does "one-third" refer to? Interpret.

b. An AP story (September 10, 2003) stated, "Brisk walking for just an hour or two weekly can help older women reduce their risk of breast cancer by 18 percent," according to an analysis of 74,171 women in the Women's Health Initiative study. Explain how to form a sample relative risk measure that equals 0.82.

c. A Knight Ridder story (October 17, 2003) stated, "Taking an aspirin a day can reduce the risk of a first heart attack by 32 percent." State this result by defining and interpreting a relative risk.

d. An AP story (August 1, 2002) reported "A study of 5,881 men and women published in today's *New England Journal of Medicine* showed that the risk of heart failure is double in obese people and 34% higher in those overweight, compared with those of normal weight." Report the relative risk values in this statement, and interpret.

9.37 We're getting heavier: According to the National Center for Health Statistics, men in the United States weighed an average of 166 pounds in 1962 and 191 pounds in 2002. Compare these using the ratio of means, and interpret.

9.38 Comparing net worths: An AP story (see also Exercise 9.1) reported that in the United States in 2002 the median net worth was about $89,000 for white households and $8000 for Hispanic households. The story reported that the median net worth of white households was about 11 times that of Hispanic households.

a. Show how the researchers got the value 11, and explain what summary measure is estimated by this value.

b. Why do you think the Census Bureau used the median instead of the mean for this comparison?

9.4 | How Can We Analyze Dependent Samples?

With **dependent samples**, each observation in one sample has a matched observation in the other sample. The observations are called **matched pairs**. We've already seen examples of matched-pairs data. In Example 7 of Chapter 8, for instance, each of a sample of anorexic girls had her weight measured before and after participating in a treatment for anorexia. The weights before the treatment form one group of observations. The weights after the treatment form the other group. The same girls were in each sample, so the samples were dependent.

Matched Pairs Design for Cell Phones and Driving Study

Picture the Scenario

Example 9 analyzed whether the use of cell phones impairs reaction times in a driving skills test. The analysis used independent samples—one group used cell phones and a separate control group did not use them. An alternative design uses the same subjects for both groups. Reaction times are measured when subjects performed the driving task without using cell phones and then again while the same subjects used cell phones. In fact, the study quoted in Example 9 also did this. Before observations were taken for these two conditions, subjects had a warm-up time interval to become familiar with the task, so reaction times would not tend to change from one condition to the other merely because of increased familiarity.

Table 9.9 shows the mean of the reaction times (in milliseconds) for each subject under each condition. Figure 9.8 shows box plots of the data for the two conditions.

TABLE 9.9: Reaction Times on Driving Skills Task before and while Using Cell Phone

The difference score is the reaction time using the cell phone minus the reaction time not using it, such as $636 - 604 = 32$ milliseconds.

	Using cell phone?				Using cell phone?		
Student	**No**	**Yes**	**Difference**	**Student**	**No**	**Yes**	**Difference**
1	604	636	32	17	525	626	101
2	556	623	67	18	508	501	−7
3	540	615	75	19	529	574	45
4	522	672	150	20	470	468	−2
5	459	601	142	21	512	578	66
6	544	600	56	22	487	560	73
7	513	542	29	23	515	525	10
8	470	554	84	24	499	647	148
9	556	543	−13	25	448	456	8
10	531	520	−11	26	558	688	130
11	599	609	10	27	589	679	90
12	537	559	22	28	814	960	146
13	619	595	−24	29	519	558	39
14	536	565	29	30	462	482	20
15	554	573	19	31	521	527	6
16	467	554	87	32	543	536	−7

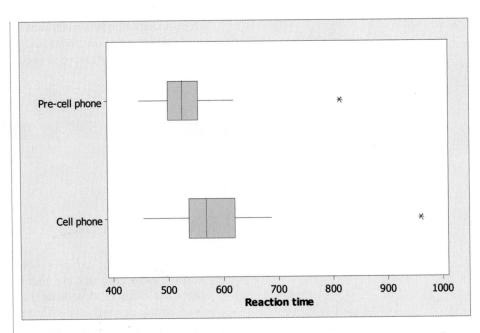

▲ **FIGURE 9.8: Box Plots of Observations on Reaction Times. Question:** From the data file, what do the outliers for the two distributions have in common?

Question to Explore

a. Summarize the sample data distributions for the two conditions.

b. To compare the mean response times using statistical inference, should we treat the samples as independent, or dependent?

Think It Through

a. The box plots show that the reaction times tend to be larger for the cell phone condition. But each sample data distribution has one extremely large outlier.

b. The pre–cell phone responses in Table 9.9 were made by the same subjects as the responses using cell phones. These are matched pairs data, because each control observation (sample 1) pairs with a cell phone observation (sample 2). Because this part of the study used the same subjects for each sample, the samples are dependent.

Insight

The data file shows that subject number 28 had a large reaction time in each case. This one subject had a very slow reaction time, regardless of the condition.

TO PRACTICE THIS CONCEPT, TRY EXERCISE 9.39(A)

Why would we use dependent samples instead of independent samples? A major benefit is that many sources of potential bias are controlled so we can make a more accurate comparison. Using matched pairs keeps many other factors fixed that could affect the analysis. Often this results in the benefit of smaller standard errors.

From Figure 9.8, for instance, the sample mean was higher when subjects used cell phones. This did not happen because the subjects using cell phones tended to be older than the subjects not using them. Age was not a lurking variable, because each sample had the same subjects. When we used independent samples (different subjects) for the two conditions in Example 9, the two samples could differ somewhat on characteristics that might affect the results, such as physical fitness or gender or age. With independent samples, studies ideally use randomization to assign subjects to the two groups to try to minimize the extent to which this happens. This was done in Example 9. With observational studies, however, this is not an option.

To Compare Means with Matched Pairs, Use Paired Differences

For each matched pair in Table 9.9, we construct a new variable consisting of a difference score (d for difference),

d = reaction time using cell phone − reaction time without cell phone.

For subject 1, for example, $d = 636 - 604 = 32$. Table 9.9 also shows these 32 difference scores. The sample mean of these difference scores, which we denote by $\bar{x}_d$, is

$$\bar{x}_d = (32 + 67 + 75 + \cdots - 7)/32 = 50.6.$$

The sample mean of the difference scores necessarily equals the difference between the means for the two samples. In Table 9.9, the mean reaction time without the cell phone is $(604 + 556 + 540 + \cdots + 543)/32 = 534.6$, the mean reaction time using the cell phone is $(636 + 623 + 615 + \cdots + 536)/32 = 585.2$, and the difference between these means equals 50.6. This is also the mean of the differences, $\bar{x}_d$.

For Dependent Samples, Mean of Differences = Difference of Means

For dependent samples, the difference $(\bar{x}_1 - \bar{x}_2)$ between the means of the two samples equals the mean $\bar{x}_d$ of the difference scores for the matched pairs.

IN WORDS

For **dependent** samples, we calculate the **difference** scores and then use the one-sample methods of Chapters 7 and 8 with them. We don't need separate two-sample methods for dependent samples.

Likewise, the difference $(\mu_1 - \mu_2)$ between the population means is identical to the parameter μ_d that is the population mean of the difference scores. So the sample mean $\bar{x}_d$ of the differences not only estimates μ_d, the population mean difference, but also the difference $(\mu_1 - \mu_2)$. We can base inference about $(\mu_1 - \mu_2)$ on inference about the population mean of the difference scores. *This simplifies the analysis, since it reduces a two-sample problem to a one-sample analysis using the difference scores.*

Let n denote the number of observations in each sample. This equals the number of difference scores. The 95% confidence interval for the population mean difference has the form introduced in Section 7.3. It is

$$\bar{x}_d \pm t_{.025}(se) \text{ with } se = s_d/\sqrt{n},$$

where $\bar{x}_d$ is the sample mean of the difference scores and s_d is their standard deviation. The t-score comes from the t table with $df = n - 1$.

Likewise, to test the hypothesis $H_0: \mu_1 = \mu_2$ of equal means, we can conduct the single-sample test of $H_0: \mu_d = 0$ with the difference scores. The test statistic is

$$t = \frac{\bar{x}_d - 0}{se} \text{ with } se = s_d/\sqrt{n}.$$

Recall

You can review **confidence intervals for a mean** in Section 7.3 and **significance tests for a mean** in Section 8.3. ▪

This compares the sample mean of the differences to the null hypothesis value of 0. The standard error is the same as for the confidence interval. The $df = n - 1$. Since this test uses the difference scores for the pairs of observations, it is called a **paired-difference t test**. Software can do the computations for us.

Because these paired-difference inferences are special cases of single-sample inferences about a population mean, they make the same assumptions:

■ The sample of difference scores is a random sample from a population of such difference scores.

■ The difference scores have a population distribution that is approximately normal. This is mainly important for small samples (less than about 30) and for one-sided inferences.

Confidence intervals and two-sided tests are **robust**: They work quite well even if the normality assumption is violated. One-sided tests do not work well when the sample size is small and the distribution of differences is highly skewed.

EXAMPLE 13

Matched Pairs Analysis of Cell Phone Impact on Driver Reaction Time

Picture the Scenario

The matched pairs data in Table 9.9 showed the reaction times for the sampled subjects using and not using cell phones. Figure 9.9 is a box plot of the $n = 32$ difference scores. Table 9.10 shows MINITAB output for these data. The first part shows the mean, standard deviation, and standard error for each sample and for the differences. The second part shows inference about the mean of the differences.

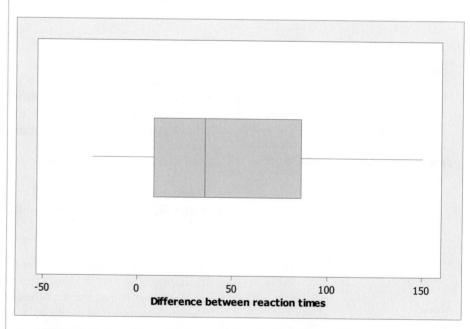

▲ **FIGURE 9.9: Box Plot of Difference Scores from Table 9.9. Question:** How is it that some of the scores plotted here are negative?

```
T-Test
 Inpt:Data Stats
 μ0:0
 x̄:50.625
 Sx:52.4858
 n:32
 μ:≠μ0 <μ0 >μ0
Calculate Draw
```

```
T-Test
 μ≠0.000
 t=5.456
 p=5.803E-6
 x̄=50.625
 Sx=52.486
 n=32.000
```

```
TInterval
 Inpt:Data Stats
 x̄:50.625
 Sx:52.4858
 n:32
 C-Level:.95
Calculate
```

```
TInterval
 (31.702,69.548)
 x̄=50.625
 Sx=52.486
 n=32.000
```

TABLE 9.10: Software Output for Matched-Pairs Analyses with Table 9.9

The margin shows screen shots from the TI-83+/84.

Paired T for Cell phone - Pre-cell phone

	N	Mean	StDev	SE Mean
Cell phone	32	585.188	89.646	15.847
Pre-cell phone	32	534.563	66.448	11.746
Difference	32	50.6250	52.4858	9.2783

95% CI for mean difference: (31.7019, 69.5481)

T-Test of mean difference = 0 (vs not = 0):

T-Value = 5.46   P-Value = 0.000

Questions to Explore

a. How can you conduct and interpret the significance test reported in this table?

b. How can you construct and interpret the confidence interval reported in this table?

Think It Through

a. The box plot shows skew to the right for the difference scores. Two-sided inference is robust to violations of the assumption of a normal population distribution. The box plot does not show any severe outliers that would raise questions about the validity of using the mean to summarize the difference scores.

The printout reports the sample mean difference, $\bar{x}_d = 50.6$, and the standard deviation of the difference scores, $s_d = 52.5$. The standard error is $se = s_d/\sqrt{n} = 52.5/\sqrt{32} = 9.28$, also reported on the printout. The t test statistic for the significance test of $H_0: \mu_d = 0$ (and hence equal population means for the two conditions) against $H_a: \mu_d \neq 0$ is

$$t = \bar{x}_d/se = 50.6/9.28 = 5.46.$$

With 32 difference scores, $df = n - 1 = 31$. Table 9.10 reports the two-sided P-value of 0.000. There is extremely strong evidence that the population mean reaction times are different.

b. For a 95% confidence interval for $\mu_d = \mu_1 - \mu_2$, with $df = 31$, $t_{.025} = 2.040$. We can use $se = 9.28$ from part (a). The confidence interval equals

$$\bar{x}_d \pm t_{.025}(se), \text{ or } 50.6 \pm 2.040(9.28),$$
$$\text{which equals } 50.6 \pm 18.9, \text{ or } (31.7, 69.5),$$

as reported on the printout. We infer that the population mean when using cell phones is between about 32 and 70 milliseconds higher than when not using cell phones. The confidence interval does not contain 0, so we can infer that the population mean reaction time is greater when using a cell phone. The confidence interval *is* more informative than the significance test, since it predicts just how large the difference might be.

The sample was taken from the population of students at the University of Utah. The article about the study did not indicate whether the subjects were randomly selected, so inferential conclusions are tentative.

Insight

The study also showed that reaction times tended to be similar with hands-free as with hand-held cell phones. It also showed that the probability of missing a simulated traffic signal doubled when students used cell phones. A later related study[11] showed that the mean reaction time of students using cell phones was similar to that of elderly drivers not using cell phones. The study concluded, "The net effect of having younger drivers converse on a cell phone was to make their average reactions equivalent to those of older drivers who were not using a cell phone."

TO PRACTICE THIS CONCEPT, TRY EXERCISES 9.39 AND 9.40

In summary, comparing means with matched-pairs data is easy. We merely use methods we've already learned about in Chapters 7 and 8 for single samples.

SUMMARY: Comparing Means of Dependent Samples

To compare means with dependent samples, construct confidence intervals and significance tests using the single sample of difference scores,

$$d = \text{observation in sample 1} - \text{observation in sample 2}.$$

The 95% confidence interval $\bar{x}_d \pm t_{.025}(se)$ and the test statistic $t = (\bar{x}_d - 0)/se$ are the same as for a single sample. The assumptions are also the same: A random sample and a normal population distribution of difference scores.

Comparing Proportions with Dependent Samples

We next present methods for comparing *proportions* with dependent samples. We'll again be able to do this by conducting single-sample inference with difference scores.

EXAMPLE 14

Comparing Beliefs in Heaven and in Hell

Picture the Scenario

A recent General Social Survey asked subjects whether they believed in heaven and whether they believed in hell. For the 1120 subjects who responded, Table 9.11 shows the data in contingency table form. The rows of Table 9.11 are the response categories for belief in heaven. The columns are the same categories for belief in hell. In the U.S. adult population, let p_1 denote the proportion who believe in heaven and let p_2 denote the proportion who believe in hell.

---

[11]D. L. Strayer and F. A. Drews, *Human Factors*, vol. 46, pp. 640–649, 2004.

TABLE 9.11: Opinions about Existence of Heaven and Hell

Belief in Heaven	Belief in Hell Yes	No	Total
Yes	833	125	**958**
No	2	160	**162**
Total	**835**	**285**	**1120**

Questions to Explore

a. How can we estimate p_1, p_2, and their difference?

b. Are the samples used to estimate p_1 and p_2 independent, or dependent, samples?

c. How can we use a sample mean of difference scores to estimate $(p_1 - p_2)$?

Think It Through

a. The counts in the two margins of the table summarize the responses. Of the 1120 subjects, 958 said they believed in heaven, so $\hat{p}_1 = 958/1120 = 0.86$. Of the same 1120 subjects, 835 said they believed in hell, so $\hat{p}_2 = 835/1120 = 0.75$. Since $(\hat{p}_1 - \hat{p}_2) = 0.86 - 0.75 = 0.11$, we estimate that 11% more people believe in heaven than in hell.

b. Each sample of 1120 responses refers to the same 1120 subjects. Any given subject's response on belief in heaven can be matched with that subject's response on belief in hell. So the samples for these two proportions are dependent.

c. Recall that a proportion is a mean when we code the responses by 1 and 0. For belief in heaven or belief in hell, let 1 = yes and 0 = no. Then, for belief in heaven, 958 responses were 1 and 162 responses were 0. The mean of the 1120 observations was $[958(1) + 162(0)]/1120 = 958/1120 = 0.86$, which is $\hat{p}_1$. For belief in hell, 835 responses were 1 and 285 responses were 0, and the mean was $\hat{p}_2 = 835/1120 = 0.75$.

Table 9.12 shows the possible paired binary responses, for the 0 and 1 coding. The difference scores for the heaven and hell responses are $d = 1, 0$, or -1, as shown. The sample mean of the 1120 difference scores equals

$$[0(833) + 1(125) - 1(2) + 0(160)]/1120 = 123/1120 = 0.11.$$

This equals the difference of proportions, $\hat{p}_1 - \hat{p}_2 = 0.86 - 0.75 = 0.11$. In summary, the mean of the difference scores equals the difference between the sample proportions.

TABLE 9.12: Using 0 and 1 Responses for Binary Matched Pairs Data

Heaven	Hell	Interpretation	Difference d	Frequency
1	1	believe in heaven and in hell	$1 - 1 = 0$	833
1	0	believe in heaven but not in hell	$1 - 0 = 1$	125
0	1	believe in hell but not in heaven	$0 - 1 = -1$	2
0	0	do not believe in heaven or hell	$0 - 0 = 0$	160

Insight

We've converted two samples with 1120 binary observations each to a single sample of 1120 difference scores. We can use single-sample methods with the differences, as we just did for matched pairs analysis of means.

TO PRACTICE THIS CONCEPT, TRY EXERCISE 9.46

Recall

Section 2.3 (Example 12) and Section 6.4 (following Example 15) showed that a **proportion** is a **mean** when we code the two possible outcomes by 1 and 0. ■

Confidence Interval Comparing Proportions with Matched Pairs Data

To conduct inference about the difference $(p_1 - p_2)$ between the population proportions, we can use the fact (shown above) that the sample difference $(\hat{p}_1 - \hat{p}_2)$ is the mean of difference scores when we code the responses by 1 and 0. We can find a confidence interval for $(p_1 - p_2)$ by finding a confidence interval for the population mean of difference scores.

EXAMPLE 15

Inference Comparing Beliefs in Heaven and in Hell

Picture the Scenario

We continue our analysis of the data on belief in heaven and/or hell. We use the data in the binary form summarized in Table 9.12. When software analyzes the 1120 paired difference scores, we get the results in Table 9.13:

TABLE 9.13: Software Output for Analyzing Difference Scores from Table 9.12 to Compare Beliefs in Heaven and Hell

	N	Mean	StDev	SE Mean
Difference	1120	0.1098	0.3185	0.00952

95% CI for mean difference: (0.091, 0.128)

Questions to Explore

a. Explain how software got the confidence interval from the other results shown.

b. Interpret the reported 95% confidence interval.

Think It Through

a. We've already seen that the sample mean of the 1120 difference scores is $(\hat{p}_1 - \hat{p}_2) = 0.11$. Table 9.13 reports that the standard error of the sample mean difference is $se = 0.0095$. For $n = 1120$, $df = 1119$, and $t_{.025} = 1.962$. The population mean difference is $(p_1 - p_2)$, the difference between the population proportion p_1 who believe in heaven and the population proportion p_2 who believe in hell. A 95% confidence interval for $(p_1 - p_2)$ equals

Recall

With such a large *df*, you could use the *z*-score of 1.96 if your software or calculator cannot find the *t*-score for you. For such a very large *n*, the nonnormality of the data is not problematic. ■

$$\text{Sample mean difference} \pm t_{.025}(se), \text{ which is}$$
$$0.11 \pm 1.962(0.0095), \text{ or } (0.09, 0.13).$$

b. We can be 95% confident that the population proportion p_1 believing in heaven is between 0.09 higher and 0.13 higher than the population proportion p_2 believing in hell. Since the interval contains only positive values, we infer that $p_1 > p_2$.

Insight

As in the matched pairs inferences comparing two means, we conducted the inference comparing two proportions by using inference for a single parameter—namely, the population mean of the differences.

TO PRACTICE THIS CONCEPT, TRY EXERCISE 9.47

$(\hat{p}_1 - \hat{p}_2) = 0$			
Heaven	Hell	d	Freq.
1	1	0	833
1	0	1	63
0	1	−1	63
0	0	0	160

Belief in Heaven	Belief in Hell	
	Yes	No
Yes	833	125
No	2	160

$(p_1 - p_2) = 0$ if cells with counts 125 and 2 have equal population counts.

McNemar Test Comparing Proportions with Matched Pairs Data

For the difference scores on the binary response, $(\hat{p}_1 - \hat{p}_2) = 0$ if the number of subjects having $d = 1$ equals the number of subjects having $d = -1$. Then, the sample mean of the differences equals 0. See the margin table for an example. Equivalently, the number believing in heaven but *not* in hell equals the number believing in hell but *not* in heaven. Similarly, in the population, if $H_0: p_1 = p_2$ is true, then the population mean of the difference scores is 0. So, we can test $H_0: (p_1 - p_2) = 0$ by testing whether the population mean of difference scores is 0.

For testing $H_0: p_1 = p_2$ with two dependent samples, there's actually a simple way of calculating a test statistic. It uses the two counts in the contingency table that have the differences of 1 and −1, that is, the frequencies of "yes" on one response and "no" on the other. This test statistic equals their difference divided by the square root of their sum. For Table 9.11 (shown again in the margin), for instance,

$$z = \frac{125 - 2}{\sqrt{125 + 2}} = 10.9.$$

We use the symbol z, because it has an approximate standard normal distribution when $H_0: p_1 = p_2$ is true.

For $z = 10.9$, the two-sided P-value equals 0.000000.... The sample has *extremely* strong evidence that the proportion p_1 believing in heaven is higher than the proportion p_2 believing in hell. The inference that $p_1 > p_2$ agrees with the conclusion in Example 15 based on the confidence interval for $(p_1 - p_2)$. The advantage of the confidence interval is that it indicates *how different* the proportions are likely to be. In percentage terms, we infer that the belief in heaven is about 9% to 13% higher.

The z test comparing proportions with dependent samples is often called **McNemar's test**, in honor of a psychologist (Quinn McNemar) who proposed it in 1947. It is a large-sample test. It applies when the sum of the two counts used in the test is at least 30 (the sum is $125 + 2 = 127$ above). For two-sided tests, however, the test is robust and works well even for small samples.

SUMMARY: McNemar Test Comparing Proportions from Dependent Samples

Hypotheses: $H_0: p_1 = p_2$, H_a can be two-sided or one-sided.

Test Statistic: For the two counts for the frequency of "yes" on one response and "no" on the other, the z test statistic equals their difference divided by the square root of their sum. The sum of the counts should be at least 30, but in practice the two-sided test works well even if this is not true.

P-value: For $H_a: p_1 \neq p_2$, two-tail probability of z test statistic values more extreme than observed z, using standard normal distribution.

9.4 Practicing the Basics

9.39 **Does exercise help blood pressure?** Several recent studies have suggested that people who suffer from abnormally high blood pressure can benefit from regular exercise. A medical researcher decides to test her belief that walking briskly for at least half an hour a day has the effect of lowering blood pressure. She conducts a small pilot study. If results from it are supportive, she will apply for funding for a larger study. She randomly samples three of her patients who have high blood pressure. She measures their systolic blood pressure initially and then again a month later after they participate in her exercise program. The table shows the results.

Subject	Before	After
1	150	130
2	165	140
3	135	120

a. Explain why the three "before" observations and the three "after" observations are dependent samples.

b. Find the sample mean of the before scores, the sample mean of the after scores, and the sample mean of d = before − after. How are they related?

c. Find a 95% confidence interval for the difference between the population means of subjects before and after going through such a study. Interpret.

9.40 Test for blood pressure: Refer to the previous exercise. The output shows some results of using software to analyze the data with a significance test.

```
Paired T for Before—After
              N     Mean    StDev    SE Mean
Before        3     150.0   15.0     8.660
After         3     130.0   10.0     5.774
Difference    3     20.0     5.0     2.887
T-Test of mean difference = 0 (vs not= 0):
T-Value = 6.93  P-Value = 0.020
```

a. State the hypotheses to which the reported P-value refers.

b. Explain how to interpret the P-value. Does the exercise program seem beneficial to lowering blood pressure?

c. What are the assumptions on which this analysis is based?

9.41 Social activities for students: As part of her class project, a student at the University of Florida randomly sampled 10 fellow students to investigate their most common social activities. As part of the study, she asked the students to state how many times they had done each of the following activities during the previous year: going to a movie, going to a sporting event, or going to a party. The table shows the data.

Frequency of attending movies, sports events, and parties

	ACTIVITY		
Student	Movies	Sports	Parties
1	10	5	25
2	4	0	10
3	12	20	6
4	2	6	52
5	12	2	12
6	7	8	30
7	45	12	52
8	1	25	2
9	25	0	25
10	12	12	4

a. To compare the mean movie attendance and mean sports attendance using statistical inference, should we treat the samples as independent or dependent? Why?

b. The figure is a dot plot of the $n = 10$ difference scores for movies and sports. Does this show any irregularities that would make statistical inference unreliable?

Dot plot of difference scores for Exercise 9.41

c. Using the MINITAB output shown for these data, show how the 95% confidence interval was obtained from the other information given in the printout. Interpret the interval.

d. Show how the test statistic shown on the printout was obtained from the other information given. Report the P-value, and interpret in context.

MINITAB Output for Inferential Analyses:
Paired T for movies - sports

```
              N     Mean      StDev      SE Mean
movies        10    13.0000   13.1740    4.1660
sports        10     9.0000    8.3799    2.6499
Difference    10     4.00000  16.16581   5.11208
95% CI for mean difference: (−7.56432, 15.56432)
T-Test of mean difference = 0 (vs not = 0):
T-Value = 0.78  P-Value = 0.454
```

9.42 More social activities: Refer to the previous exercise. The output shows the result of comparing the mean responses on parties and sports.

```
Paired T for parties - sports
              N     Mean      StDev      SE Mean
parties       10    21.8000   18.5760    5.8742
sports        10     9.0000    8.3799    2.6499
Difference    10    12.8000   22.5477    7.1302
95% CI for mean difference: (−3.3297, 28.9297)
T-Test of mean difference = 0 (vs not = 0):
T-Value = 1.80  P-Value = 0.106
```

a. Explain how to interpret the reported 95% confidence interval.

b. State the hypotheses to which the P-value refers, and interpret its value.

c. Explain the connection between the results of the test and the confidence interval.

d. What assumptions are necessary for these inferences to be appropriate?

9.43 Movies vs. parties: Refer to the previous two exercises. Using software, compare the responses on movies and parties using (a) all steps of a significance test, (b) a 95% confidence interval. Interpret results in context.

9.44 "Freshman 15" a myth?: The "freshman 15" is the name of a common belief that college students, particularly women, gain an average of 15 pounds during their first year of college. In one study of 110 female first-year college students (*Psychology of Women Quarterly*, vol. 17, pp. 119–126, 1993), the mean weight was 130.6 pounds at the beginning of the study and 131.4 pounds six months later. The standard deviation of weights was about 20 pounds at each time.

a. Estimate the change in the mean weight over this six-month period.

b. Is this sufficient information to find a confidence interval or conduct a test about the change in the mean over the six months? If not, what else do you need to know?

9.45 **Checking for "freshman 15":** Refer to the previous exercise. Suppose that the change in weight scores for the 110 freshmen had a standard deviation of 2.0 pounds.

a. Explain how this standard deviation could be so much less than the standard deviation of 20 for the weight scores at each time.

b. Is 15 pounds a plausible mean weight change in the population of freshman women? Answer by constructing a 95% confidence interval for the population mean change in weight or conducting a significance test of the hypothesis that the mean weight change in the population equals 15. Interpret.

c. What assumptions are necessary for the inference in (b)?

9.46 **Obesity now and in 20 years:** Many medical studies
TRY have used a large sample of subjects from Framingham, Massachusetts who have been followed since 1948. A recent study (*Annals of Internal Medicine*, vol. 138, pp. 24–32, 2003) gave the contingency table shown for weight at a baseline time and then 20 years later.

| Baseline | 20 Years After Baseline | |
	Normal	Overweight
Normal	695	368
Overweight	87	827

a. Find the sample proportion with normal weight at (i) baseline, (ii) 20 years later.

b. Identify the two samples and whether they are independent or dependent. Explain.

9.47 **Change coffee brand?:** A study was conducted to see if
TRY an advertisement campaign would increase market share for Sanka instant decaffeinated coffee (R. Grover and V. Srinivasan, *J. Marketing Research*, vol. 24, pp. 139–153, 1987). Subjects sampled who use instant decaffeinated coffee were asked which brand they bought last. They were asked this before the campaign and after it. The results are shown in the table.

a. Estimate the population proportion choosing Sanka for the (i) first purchase, (ii) second purchase.

b. Explain how each proportion in (a) can be found as a sample mean and how the estimated difference of population proportions is a difference of sample means.

c. The table also shows results of a confidence interval for the difference between the population proportions. Explain how to interpret it.

Two Purchases of Coffee

| First Purchase | Second Purchase | |
	Sanka	Other brand
Sanka	155	49
Other brand	76	261

	N	Mean	StDev	SE Mean
Difference	541	0.050	0.478	0.021

95% CI for mean difference: (0.010, 0.090)

9.48 **President's popularity:** Last month a random sample of 1000 subjects was interviewed and asked whether they think the president is doing a good job. This month the same subjects were asked this again. The results are: 450 said "yes" each time, 450 said "no" each time, 60 said "yes" on the first survey and "no" on the second survey, and 40 said "no" on the first survey and "yes" on the second survey.

a. Form a contingency table showing these results.

b. Estimate the proportion giving a favorable rating (i) last month, (ii) this month.

c. Show how each of the proportions in (b) can be obtained as a sample mean.

d. Find the test statistic and P-value for applying McNemar's test that the population proportion was the same each month. Interpret.

9.49 **Heaven and hell:** Results of polls about belief in heaven and hell depend strongly on question wording. For instance, a survey conducted by the International Social Science Program at NORC asked Americans whether they "definitely believe in heaven" or "definitely believe in hell" (see religioustolerance.com). The percentage who answered yes was 63.1% for heaven and 49.6% for hell.

a. Estimate the difference between the population proportions believing in heaven and believing in hell.

b. In this survey, suppose that 65 subjects believed in heaven but not in hell, whereas 35 believed in hell but not in heaven. Report the (i) assumptions, (ii) hypotheses, (iii) test statistic, (iv) P-value, (v) conclusion for testing that the probability of belief is the same for heaven and for hell.

9.50 **Heaven and hell around the world:** Refer to the previous exercise. Results in this poll also depended strongly on the country in which the question was asked. For instance, the percentages believing in (heaven, hell) were (52%, 26%) in Ireland, (32%, 19%) in New Zealand, (28%, 22%) in Italy, (25%, 13%) in Great Britain, (21%, 11%) in the Netherlands, and (18%, 9%) in Germany. If for a given country you know the sample size was 1000, does this give you enough information to compare the two proportions inferentially for that country? Explain.

9.5 | How Can We Adjust for Effects of Other Variables?

When we find a practically significant difference between two groups, can we identify a reason for the difference? As Chapters 3 and 4 explained, an association may be due to a *lurking variable* not measured in the study. Example 5 in Section 9.1 indicated that teenagers who watch more TV have a tendency later

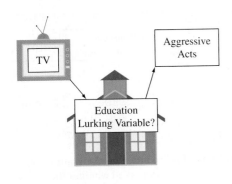

in life to commit more aggressive acts. Why is this? Might education be a lurking variable? Perhaps teenagers who watch more TV tend to attain lower educational levels, and perhaps lower education tends to be associated with higher levels of aggression.

To investigate why differences occur between groups, we must measure potential lurking variables and use them in the statistical analysis. For instance, suppose we planned to study TV watching and aggression, and we thought that educational level was a potential lurking variable. Then, in the study we would want to measure educational level. Suppose we did so using categories (did not complete high school, completed high school but did not attend college, attended college). We could then evaluate TV watching and aggressive behavior separately for subjects in these three educational levels. For instance, we might obtain the results in Table 9.14. At each educational level, we still find higher aggression for those who watched more TV.

TABLE 9.14: Percentage Conducting Aggressive Act, according to Level of Teenage TV Watching, Controlling for Educational Level

For example, for those who attended college who watched TV less than 1 hour per day, 2% committed an aggressive act.

	Educational Level		
TV Watching	Less than High school	High school	College
Less than 1 hour per day	8%	4%	2%
At least 1 hour per day	30%	20%	10%

This analysis uses three variables—level of TV watching, whether commit aggressive acts, and educational level. This takes us from a **bivariate** analysis (two variables) to a **multivariate** analysis (more than two variables): Whether the subject has committed aggressive acts is the response variable, the level of TV watching is the explanatory variable, and the educational level is called a **control variable**.

Control variable

A **control variable** is a variable that is held constant in a multivariate analysis.

Table 9.14 treated educational level as a control variable. Within each of the three education groups, educational level is approximately constant.

Statistical Control

To analyze whether an association can be explained by a third variable (such as educational level), we treat that third variable as a control variable. We conduct the ordinary bivariate analysis, such as comparing two proportions, while holding that control variable constant at fixed values. Then, whatever association occurs cannot be due to effects of the control variable, because in each part of the analysis it is not allowed to vary.

In Table 9.14, at each educational level, the percentage committing an aggressive act is higher for those who watched more TV. If these were the actual results, then the association observed in Example 5 between TV watching and aggressive acts was not because of education. In each part of Table 9.14, educational level is approximately constant, so changes in aggression as TV watching changes are not due to changes in educational level. To conduct inference, we could use the methods of this chapter to compare the population proportions of aggression at each education level.

Conducting a statistical analysis while holding a control variable constant has an analogy in scientific laboratory experiments. Such experiments try to hold certain variables constant so they do not affect results. For instance, a chemistry experiment might control temperature by holding it constant throughout the experiment. Otherwise, a sudden increase or decrease in the temperature might cause a reaction that would otherwise not happen. Unlike experiments, observational studies cannot assign subjects to particular values of variables that we want to control, such as educational level. But we can approximate an experimental control by grouping together observations with similar values on the control variable, as just described. This is a **statistical control**, rather than an experimental control.

The next example shows that after we control for a variable, the results can look quite different from those that fail to take that control variable into account.

EXAMPLE 16

Is the Death Penalty Associated with Race?

Picture the Scenario

The United States is one of only a few Western nations that impose the

death penalty. Surveys suggest that it has strong support, but there has been increasing attention on potential inequities. Are those convicted of murder more likely to get the death penalty if they are black than if they are white, or if they are poor than if they are wealthy?

Table 9.15 comes from one of the first studies into potential racial inequities of the death penalty.[12] The 326 subjects were defendants convicted of homicide in murder trials in the state of Florida. The variables are the defendant's race and whether the defendant received the death penalty. The contingency table shows that about 12% of white defendants and about 10% of black defendants received the death penalty.

In this study, the difference between the percentages of white defendants and black defendants who received the death penalty is small (1.7%), but the percentage was lower for black defendants. This may surprise you, since most media stories about racial inequities have reported that black defendants seem to be *more* likely to receive the death penalty.

TABLE 9.15: Defendant's Race and Death Penalty Verdict for Homicide Cases

Defendant's Race	Death Penalty		Total	Percentage Yes
	Yes	No		
White	19	141	160	11.9
Black	17	149	166	10.2

Is there some explanation for these results? Does a control variable lurk that explains why relatively fewer black defendants got the death penalty? Researchers who study the death penalty stress that the *victim's* race is often an important factor. So, let's control for victim's race. That is, we'll construct a table like Table 9.15 *separately* for cases in which the victim was white and for cases in which the victim was black. Table 9.16 shows this three-variable table.

---

[12]M. Radelet, *American Sociological Review*, vol. 46, pp. 918–927 (1981).

TABLE 9.16: Defendant's Race and Death Penalty Verdict, Controlling for Victim's Race

Victim's Race	Defendant's Race	Death Penalty Yes	No	Total	Percentage Yes
White	White	19	132	151	12.6
	Black	11	52	63	17.5
Black	White	0	9	9	0.0
	Black	6	97	103	5.8

Table 9.16 shows the effect of defendant's race on the death penalty verdict while controlling for victim's race, that is, when victim's race is kept constant at white or kept constant at black. This table decomposes Table 9.15 into two separate contingency tables, one for each victim's race. You can recover Table 9.15 by summing corresponding entries, $19 + 0 = 19$, $132 + 9 = 141$, $11 + 6 = 17$, and $52 + 97 = 149$.

Questions to Explore

a. Summarize the association between defendant's race and the death penalty, after controlling for victim's race.

b. Describe the difference in the results between ignoring and controlling victim's race.

c. Why does the control for victim's race make such a difference?

Think It Through

a. The top part of Table 9.16 lists cases in which the victim's race was white. Relatively more black defendants than white defendants got the death penalty. The difference between the percentages is $17.5 - 12.6 = 4.9$. The bottom part of the table lists cases in which the victim's race was black. Again, relatively more black defendants got the death penalty. The difference between the percentages is $5.8 - 0.0 = 5.8$. In summary, controlling for victim's race, more black defendants than white defendants received the death penalty.

b. Table 9.15 showed that a *larger* percentage of white defendants than black defendants got the death penalty. By contrast, controlling for victim's race, Table 9.16 showed that a *smaller* percentage of white defendants than black defendants got the death penalty. This was true for each victim's race category.

c. In Table 9.16, look at the percentages who got the death penalty. When the victim was white they are quite a bit larger (12.6 and 17.5) than when the victim was black (0.0 and 5.8). That is, defendants who killed a white person were more likely to get the death penalty. Now look at the totals for the four combinations of victim's race and defendant's race. The most common cases are white defendants having white victims (151 times) and black defendants having black victims (103 times). In summary, white defendants usually had white victims and black defendants usually had black victims, and killing a white person was more likely to result in the death penalty than killing a black person. These two factors operating together produce an overall association that shows (in Table 9.15) that a higher percentage of white defendants than black defendants got the death penalty.

Insight

Overall, relatively more white defendants got the death penalty. Controlling for victim's race, however, relatively more black defendants got the

Recall

Section 3.4 first introduced **Simpson's paradox**. It is named after a British statistician who wrote an article in 1951 about mathematical conditions under which the association can change direction when you control for a variable.■

Recall

From the end of Section 3.4, we say that defendant's race and victim's race are *confounded* in their effects on the death penalty verdict.■

death penalty. The effect changes direction after we control for victim's race. This shows that the association at each level of a control variable can have a different direction than overall when that third variable is ignored instead of controlled. This phenomenon is called **Simpson's paradox**.

In Table 9.16, the case having the highest percentage of times in which the death penalty was imposed was when a black defendant had a white victim. By contrast, it was never imposed when a white defendant had a black victim. Similar results have occurred in other studies of the death penalty. See, for instance, Exercises 9.52 and 3.50.

> **TO PRACTICE THIS CONCEPT, TRY EXERCISE 9.52**

With statistical control imposed for a variable, the results tend to change considerably when that control variable has a strong association both with the response variable and the explanatory variable. In Example 16, for instance, victim's race was the control variable. Victim's race had a noticeable association both with the death penalty response (the death penalty was given more often when the victim was white) and with defendant's race (defendants and victims usually had the same race).

> **IN PRACTICE: Control for variables associated both with the response and explanatory variables**
>
> In determining which variables to control in observational studies, *researchers choose variables that they expect to have a practically significant association with both the response variable and the explanatory variable.* A statistical analysis that controls such variables can have quite different results than when those variables are ignored.

Statistical control is also relevant when the response variable is quantitative. The difference between means for two groups can change substantially after controlling for a third variable. See, for example, Exercise 9.54.

ACTIVITY 1: Making Sense of Newspaper Articles?

This chapter began by stating that many newspaper articles describe research studies that used the statistical methods of this chapter. At this stage of your Statistics course, you should be able to read and understand those articles much better than you could before you knew the basis of such terms as "margin of error" and "statistical significance" and "randomized clinical trial".

For the next three days, read a daily newspaper such as *The New York Times*. Make a copy of any article you see that refers to a study that apparently used statistics to help make it conclusions. For one article, prepare a one-page report that you can read in your class that answers the following questions:

■ What was the purpose of the study?

■ Was the study an experiment or an observational? Can you provide any details about the study design and sampling method?

■ Identify explanatory and response variables.

■ Can you tell whether the statistical analysis used (1) independent samples or dependent samples, (2) a comparison of proportions or a comparison of means?

■ Can you think of any lurking variables that could have affected the results?

■ Are there limitations you can see that would make you skeptical to put much faith in the conclusions of this study?

You may be able to find additional information to answer these questions by browsing the web or going to the research journal that published the results.

9.51   Benefits of drinking: A *USA Today* story (September 27, 2004) about the medical benefits of moderate drinking of alcohol stated, "While wine may protect a 65-year-old man (from heart disease), it may offer no help to a 22-year-old. The young have a low risk of heart disease."

  a. Explain how this refers to an analysis of three variables. Identify those variables.

  b. Suppose age is treated as a control variable when we compare moderate drinkers to abstainers in their heart attack rates. Explain how this analysis shows that an effect of an explanatory variable on a response variable can change at different values of a control variable.

9.52   Death penalty in Kentucky: A study of the death penalty in Kentucky reported the results shown in the table. (T. Keil and G. Vito, *Amer. J. Criminal Justice*, vol. 20, pp. 17–36, 1995).

TRY

  a. Find and compare the percentage of white defendants with the percentage of black defendants who received the death penalty, when the victim was
  (i) white, (ii) black.

  b. In the analysis in (a), identify the response variable, explanatory variable, and control variable.

  c. Construct the summary 2 × 2 table that ignores, rather than controls, victim's race. Compare the overall percentages of white defendants and black defendants who got the death penalty (ignoring, rather than controlling, victim's race). Compare to (a).

  d. Do these data satisfy Simpson's paradox? If not, explain why not. If so, explain what is responsible for Simpson's paradox occurring.

  e. Explain, without doing the calculations, how you could inferentially compare the proportions of white and black defendants who get the death penalty (i) ignoring victim's race, (ii) controlling for victim's race.

Victim's Race	Defendant's Race	Death Penalty Yes	No	TOTAL
White	White	31	360	391
	Black	7	50	57
Black	White	0	18	18
	Black	2	106	108

9.53   Basketball paradox: The table summarizes shooting percentage in the 2001–2002 season in the National Basketball Association by Brent Barry and Shaquille O'Neal.

Type of Shot	Brent Barry	Shaquille O'Neal
2-point shots	58.8% (237/403)	58.0% (712/1228)
3-point shots	42.4% (164/387)	0% (0/1)
Overall	50.8% (401/790)	57.9% (712/1229)

Source: www.nba.com and article by T. P. Ryan, *Chance*, vol. 16, p. 16, 2003.

  a. Treating the type of shot as a control variable, whether a shot is made as the response variable, and the player as the explanatory variable, explain how this table illustrates Simpson's paradox.

  b. Explain how O'Neal could have the higher overall percentage, even though he made a lower percentage of each type of shot.

9.54   Teacher salary, gender, and academic level: The table lists the mean salary, in thousands of dollars, of full-time instructional faculty on nine-month contracts in U.S. institutions of higher education in 2000, by gender and academic level.[13] Regard salary as the response variable, gender as the explanatory variable, and academic level as the control variable.

Mean Salary (Thousands of Dollars) for Men and Women Faculty Members

GENDER	ACADEMIC LEVEL Professor	Associate	Assistant	Instructor	Overall
Men	76.5	55.9	46.4	35.9	60.1
Women	67.1	52.1	43.4	34.2	49.0

  a. Find the difference between men and women faculty members on their mean salary (i) overall, (ii) after controlling for academic rank.

  b. The overall difference between the mean salary of men and women faculty members was larger than the difference at each academic level. What could be a reason for this? (Simpson's paradox does not hold here, but the gender effect does weaken when we control for academic level.)

9.55   Family size in Canada: The table shows the mean number of children in Canadian families, classified by whether the family was English speaking or French speaking and by whether the family lived in Quebec or in another province.

Mean Number of Children in Canada

PROVINCE	English Speaking	French Speaking
Quebec	1.64	1.80
Other	1.97	2.14
Overall	1.95	1.85

  a. Overall, compare the mean number of children for English speaking and French speaking families.

  b. Compare the means, controlling for province (Quebec, Others).

  c. How is it possible that for each level of province the mean is higher for French speaking families, yet overall the mean is higher for English speaking families? Which paradox does this illustrate?

9.56   Heart disease and age: In the United States, the median age of residents is lowest in Utah. At each age level, the death rate from heart disease is higher in Utah than in Colorado. Overall, the death rate from heart disease is lower in Utah than Colorado. Are there any contradictions here, or is this possible? Explain.

9.57   Breast cancer over time: The percentage of women who get breast cancer sometime during their lifetime is higher now than in 1900. Suppose that breast cancer incidence tends to increase with age, and suppose that women tend to live longer lives now than in 1900. Explain why a comparison of breast cancer rates now with the rate in 1900 could show different results if we control for the age of the woman.

---

[13]*Source:* U.S. Department of Education.

Answers to the Chapter Figure Questions

Figure 9.2: $\hat{p}_1 - \hat{p}_2$ represents the difference between the sample proportions suffering heart attack with placebo and with aspirin. $p_1 - p_2$ represents the difference between the population proportions. The mean of this sampling distribution is zero if heart attack rates are identical for aspirin and placebo.

Figure 9.3: If the interval contains only negative values, $p_1 - p_2$ is predicted to be negative, so p_1 is predicted to be smaller than p_2. If the interval contains only positive values, $p_1 - p_2$ is predicted to be larger than p_2.

Figure 9.4: Females seem to have the greater nicotine dependence. The female percentages are higher for scores 1 through 10 (a higher score indicates higher nicotine dependence) and the male percentage is higher for score 0.

Figure 9.5: Positive numbers in the confidence interval for $(\mu_1 - \mu_2)$ suggest that $\mu_1 - \mu_2 > 0$, and thus that $\mu_1 > \mu_2$.

Figure 9.6: The box plot for the cell phone group indicates an extreme outlier. It's important to check that the results of the analysis aren't affected too strongly by that single observation.

Figure 9.7: One sample distribution will be much more spread out than the other.

Figure 9.8: The outliers for the two distributions are both from subject number 28.

Figure 9.9: This box plot represents a student's difference score in reaction time using the cell phone minus the reaction time not using it. Some students had a slower reaction time when not using the cell phone, resulting in a negative difference.

CHAPTER SUMMARY

This chapter introduced inferential methods for comparing two groups.

- For categorical data, we compare the *proportions* of individuals in a particular category. Confidence intervals and significance tests apply to the difference between the population proportions, $(p_1 - p_2)$, for the two groups. The test of $H_0: p_1 = p_2$ analyzes whether the population proportions are equal. If the test has a small P-value, or if the confidence interval for $(p_1 - p_2)$ does not contain 0, we conclude that they differ.

- An alternative way of comparing proportions uses the **relative risk**, which is the *ratio* of proportions. The population relative risk p_1/p_2 is 1.0 when the population proportions are equal.

- For quantitative data, we compare the *means* for the two groups. Confidence intervals and significance tests apply to $(\mu_1 - \mu_2)$.

Table 9.17 summarizes two-sided estimation and testing methods for large, **independent random samples**. This is the most common case in practice.

TABLE 9.17: Comparing Two Groups, for Large, Independent Random Samples

TYPE OF RESPONSE VARIABLE

	Categorical	Quantitative
Estimation		
1. Parameter	$p_1 - p_2$	$\mu_1 - \mu_2$
2. Point estimate	$(\hat{p}_1 - \hat{p}_2)$	$\bar{x}_1 - \bar{x}_2$
3. Standard error (se)	$\sqrt{\dfrac{\hat{p}_1(1-\hat{p}_1)}{n_1} + \dfrac{\hat{p}_2(1-\hat{p}_2)}{n_2}}$	$\sqrt{\dfrac{s_1^2}{n_1} + \dfrac{s_2^2}{n_2}}$
4. 95% confidence int.	$(\hat{p}_1 - \hat{p}_2) \pm 1.96(se)$	$(\bar{x}_1 - \bar{x}_2) \pm t_{.025}(se)$
Significance Test		
1. Assumptions	Random sampling, at least 5 of each type for each group	Random sampling Normal population (test robust)
2. Hypotheses	$H_0: p_1 = p_2 \ (p_1 - p_2 = 0)$ $H_a: p_1 \neq p_2$	$H_0: \mu_1 = \mu_2 \ (\mu_1 - \mu_2 = 0)$ $H_a: \mu_1 \neq \mu_2$
3. Test statistic	$z = (\hat{p}_1 - \hat{p}_2)/se_0$ $\left[se_0 = \sqrt{\hat{p}(1-\hat{p})\left(\dfrac{1}{n_1} + \dfrac{1}{n_2}\right)}\right.$ with $\hat{p}$ = pooled proportion]	$t = (\bar{x}_1 - \bar{x}_2)/se$ [can also use $se = s\sqrt{\dfrac{1}{n_1} + \dfrac{1}{n_2}}$ with s = pooled standard deviation]
4. P-value	Two-tail probability from standard normal distribution	Two-tail probability from t distribution
5. Conclusion	Interpret P-value in context Reject H_0 if P-value $\leq \alpha$	Interpret P-value in context Reject H_0 if P-value $\leq \alpha$

Both for differences of proportions and differences of means, confidence intervals have the form

$$\text{Estimate} \pm (z\text{- or } t\text{-score})(\text{standard error}),$$

such as for 95% confidence intervals,

$$(\hat{p}_1 - \hat{p}_2) \pm 1.96(se), \text{ and } (\overline{x}_1 - \overline{x}_2) \pm t_{.025}(se).$$

For significance tests, the test statistic equals the estimated difference divided by the standard error, $z = (\hat{p}_1 - \hat{p}_2)/se_0$ and $t = (\overline{x}_1 - \overline{x}_2)/se$.

With **dependent samples**, each observation in one sample is matched with an observation in the other sample. We compare means or proportions by analyzing the mean of the differences between the paired observations. The confidence interval and test procedures can apply one-sample methods to the difference scores.

At this stage, you may feel confused about which method to use for any given exercise. It may help if you use the following checklist. Ask yourself, do you have

■ Means or proportions (quantitative or categorical variable)?
■ Independent samples or dependent samples?
■ Confidence interval or significance test?
■ Large n_1 and n_2 or not?

In practice, most applications have large, independent samples, and confidence intervals are more useful than tests. So the following cases are especially important: confidence intervals comparing proportions with large independent samples, and confidence intervals comparing means with large independent samples.

SUMMARY OF NOTATION

A subscript index identifies the group number: For instance,

n_1 = sample size for group 1, n_2 = sample size for group 2

Likewise for population means μ_1, μ_2, sample means $\overline{x}_1$, $\overline{x}_2$, sample standard deviations s_1, s_2, population proportions p_1, p_2, sample proportions $\hat{p}_1$, $\hat{p}_2$. For dependent samples, $\overline{x}_d$ and μ_d are sample mean and population mean of difference scores, and s_d is standard deviation of difference scores.

CHAPTER PROBLEMS

Practicing the Basics

9.58 Pick the method: Steve Solomon, the owner of Leonardo's Italian restaurant, wonders whether a redesigned menu will increase, on the average, the amount that customers spend in the restaurant. For the following scenarios, pick a statistical method from this chapter that would be appropriate for analyzing the data, indicating whether the samples are independent or dependent, which parameter is relevant, and what inference method you would use:

a. Solomon records the mean sales the week before the change and the week after the change and then wonders whether the difference is "statistically significant."

b. Solomon randomly samples 100 people and shows them each both menus, asking them to give a rating between 0 and 10 for each menu.

c. Solomon randomly samples 100 people and shows them each both menus, asking them to give an overall rating of "positive" or "negative" to each menu.

d. Solomon randomly samples 100 people and randomly separates them into two groups of 50 each. He asks those in group 1 to give a rating to the old menu and those in group 2 to give a rating to the new menu, using a 0 to 10 rating scale.

9.59 Opinion about America: A report (December 4, 2002) by the Pew Research Center on *What the World Thinks in 2002* reported that "The American public is strikingly at odds with publics around the world in its views about the U.S. role in the world and the global impact of American actions." Conclusions were based on polls in several countries in 1999/2000 and in 2002. One of the largest changes reported was in Pakistan, where the percentage of interviewed subjects who had a favorable view of the United States was 23% in 1999/2000 and 10% in 2002. The sample sizes were apparently about 1000 for the first survey and 2032 for the second.

a. Identify the response variable and the explanatory variable.

b. If the two surveys used separate samples of subjects, should we treat these samples as dependent or as independent in order to conduct inference? Explain.

c. The same report indicated that in the 1999/2000 survey the percentage in Pakistan who thought the spread of American ideas and customs was good was 2% ($n = 2032$). To compare this to the percentage who had a favorable view of the U.S. in that survey, should you treat the samples as dependent, or as independent? Why?

9.60 More obese over time: The Centers for Disease Control (www.cdc.gov) periodically takes large randomized surveys to track health of Americans. In a survey of 11,207 adults in 1976–1980, 47% were overweight (body mass index BMI ≥ 25). In a survey of 3,601 adults in 1999–2000, 64% were overweight.

a. Estimate the change in the population proportion who are obese, and interpret.

b. The standard error for estimating this difference equals 0.009. What is the main factor that causes *se* to be so small?

c. The 95% confidence interval comparing the population proportions is (0.15, 0.19). Interpret, taking into account whether 0 is in this interval.

9.61 Marijuana and gender: In a survey conducted by Wright State University, senior high school students were asked if they had ever used marijuana. The table shows results of one analysis, where X is the count who said yes. Assuming these observations can be treated as a random sample from a population of interest,

a. Interpret the reported estimate and the reported confidence interval. Explain how to interpret the fact that 0 is not in the confidence interval.

b. Explain how the confidence interval would change if males were group 1 and females were group 2. Express an interpretation for the interval in that case.

```
Sample       X        N      Sample p
1. Female    445     1120     0.3973
2. Male      515     1156     0.4455
estimate for p(1) - p(2): -0.0482
95% CI for p(1) - p(2): (-0.0887, -0.0077)
```

9.62 Gender and belief in afterlife: The table shows results from the 1998 General Social Survey on gender and whether one believes in an afterlife.

| | BELIEF IN AFTERLIFE | | |
GENDER	Yes	No	TOTAL
Female	509	116	625
Male	398	104	502

a. Denote the population proportion who believe in an afterlife by p_1 for females and by p_2 for males. Estimate p_1, p_2, and $(p_1 - p_2)$.

b. Find the standard error for the estimate of $(p_1 - p_2)$. Interpret.

c. Construct a 95% confidence interval for $(p_1 - p_2)$. Can you conclude which of p_1 and p_2 is larger? Explain.

d. Suppose that, unknown to us, $p_1 = 0.81$ and $p_2 = 0.74$. Does the confidence interval in (d) contain the parameter it is designed to estimate? Explain.

9.63 Belief depend on gender?: Refer to the previous exercise.

a. Find the standard error of $(\hat{p}_1 - \hat{p}_2)$ for a test of $H_0: p_1 = p_2$.

b. For a two-sided test, find the test statistic and P-value, and make a decision using significance level 0.05. Interpret.

c. Suppose that actually $p_1 = 0.81$ and $p_2 = 0.74$. Was the decision in (c) in error?

d. State the assumptions on which the methods in this exercise are based.

9.64 Females or males have more close friends?: A recent GSS reported that the 486 females had a mean of 8.3 close friends ($s = 15.6$) and the 354 males had a mean of 8.9 close friends ($s = 15.5$).

a. Estimate the difference between the population means for males and females.

b. The 95% confidence interval for the difference between the population means is 0.6 ± 2.1. Interpret.

c. For each gender, does it seem like the distribution of number of close friends is normal? Why? How does this affect the validity of the confidence interval in (b)?

9.65 Heavier horseshoe crabs more likely to mate?: A study of a sample of horseshoe crabs on a Florida island (J. Brockmann, *Ethology*, vol. 102, pp. 1–21, 1996) investigated the factors that were associated with whether female crabs had a male crab mate. Basic statistics, including the five-number summary on weight (kg) for the 111 female crabs who had a male crab nearby and for the 62 female crabs who did not have a male crab nearby, are given in the table. Assume that these horseshoe crabs have the properties of a random sample of all such crabs.

Summary Statistics for Weights of Horseshoe Crabs

	n	Mean	Std. Dev.	Min	Q1	Med	Q3	Max
Mate	111	2.6	0.6	1.5	2.2	2.6	3.0	5.2
No Mate	62	2.1	0.4	1.2	1.8	2.1	2.4	3.2

a. Sketch box plots for the weight distributions of the two groups. Interpret by comparing the groups with respect to shape, center, and spread.

b. Estimate the difference between the mean weights of female crabs who have mates and female crabs who do not have mates.

c. Find the standard error for the estimate in (a).

d. Construct a 90% confidence interval for the difference between the population mean weights, and interpret.

9.66 TV watching and race: The 2002 GSS asked about the number of hours you watch TV per day. An analysis that evaluates this by race shows the results:

```
Race    N      Mean   StDev  SE Mean
Black   305    4.14   3.56   0.2038
White   1435   2.75   2.25   0.0594
Difference = mu (Black) - mu (White)
Estimate for difference:1.39
95% CI for difference:(0.97, 1.81)
T-Test of difference = 0 (vs not =):
T-Value = 6.55 P-Value = 0.000
```

a. Do you believe that TV watching has a normal distribution for each race? Why or why not? What effect does this have on inference comparing population means?

b. Explain how to interpret the reported confidence interval. Can you conclude that one population mean is higher? If so, which one? Explain.

c. On what assumptions is this inference based?

9.67 Test TV watching by race: Refer to the previous exercise.

a. Specify the hypotheses that are tested in the output shown.

b. Report the value of the test statistic and the P-value. Interpret.

c. Make a decision, using the 0.05 significance level.

d. Explain the connection between the result of this significance test in (c) and the result of the confidence interval in the previous exercise.

9.68 How often do you feel sad?: A recent General Social Survey asked, "How many days in the past 7 days have you felt sad?" Software comparing results for men and women who responded showed the following results.

```
Gender    N    Mean  StDev  SE Mean
Female   816   1.81  1.98   .06931
Male     633   1.42  1.83   .07274
Difference = mu (Female) — mu (Male)
Estimate for difference:0.39
95% CI for difference:(0.232, 0.548)
T-Test of difference = 0 (vs not =):
T-Value = 4.84   P-Value = 0.000
```

a. Explain how to interpret the P-value. Do you think that the population means may be equal? Why?

b. Explain how to interpret the confidence interval. What do you learn from the confidence interval that you cannot learn from the test?

c. What assumptions are made for these inferences? Can you tell from the summary statistics on the printout whether any of them is badly violated? What's the effect?

9.69 Parental support and household type: A recent study interviewed youths with a battery of questions that provides a summary measure of perceived parental support. This measure had sample means of 46 ($s = 9$) for the single-mother households and 42 ($s = 10$) for the households with both biological parents. One conclusion in the study stated, "The mean parental support was 4 units higher for the single-mother households. If the true means were equal, a difference of at least this size could be expected only 2% of the time. For samples of this size, 95% of the time one would expect this difference to be within 3.4 of the true value." Explain how this conclusion refers to the results of (a) a confidence interval, (b) a significance test.

9.70 Car bumper damage: An automobile company compares two different types of front bumpers for their new model by driving sample cars into a concrete wall at 20 miles per hour. The response is the amount of damage to the car, as measured by the repair costs, in hundreds of dollars. Due to the costs, the study uses only six cars, obtaining results for three bumpers of each type. The results are in the table. Conduct statistical inference (95% confidence interval or significance test with significance level 0.05) about the difference between the population means, using software if you wish. Can you conclude that the mean is higher for one bumper type?

Bumper A	Bumper B
11	1
15	3
13	4

9.71 Teenage anorexia: Example 7 in Section 8.3 described a study that used a cognitive behavioral therapy to treat a sample of teenage girls who suffered from anorexia. The study observed the mean weight change after a period of treatment. Studies of that type also usually have a control group that receives no treatment or a standard treatment. Then researchers can analyze how the change in weight compares for the treatment group to the control group. In fact, the anorexia study had a control group that received a standard treatment. Teenage girls in the study were randomly assigned to the cognitive behavioral treatment (Group 1) or to the control group (Group 2). The figure shows box plots of the weight changes for the two groups (displayed vertically). The output shows how MINITAB reports inferential comparisons of those two means.

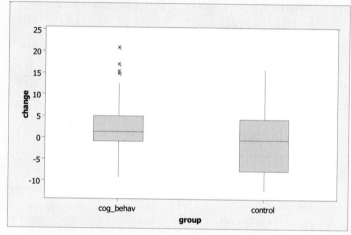

Box plots of weight change for anorexia study.

MINITAB output comparing mean weight changes for Exercise 9.71

```
group       N    Mean   StDev  SE Mean
cog_behav   29   3.01   7.31   1.4
control     26  -0.45   7.99   1.6
Difference = mu (cog_behav) — mu (control)
Estimate for difference: 3.45690
95% CI for difference:(-0.70643, 7.62022)
T-Test of difference = 0 (vs not =):
T-Value = 1.67   P-Value = 0.102   DF = 50
```

a. Report and interpret the P-value for testing $H_0: \mu_1 = \mu_2$ against $H_a: \mu_1 \neq \mu_2$.

b. Summarize the assumptions needed for the analysis in (a). Based on the box plots, would you be nervous if you had to perform a one-sided test instead? Why?

c. The reported 95% confidence interval tells us that if the population mean weight change is less for the cognitive behavioral group than for the control group, it is just barely less (less than 1 pound), but if the population mean change is greater, it could be nearly 8 pounds greater. Explain how to get this interpretation from the interval reported.

d. Explain the correspondence between the confidence interval and the decision in the significance test for a 0.05 significance level.

9.72 Anorexia paradox?: Refer to the previous exercise. For the control group, the 26 changes in weights have a mean of −0.45. From Example 7 in Chapter 8, for the therapy group the sample mean change in weight of 3.0 pounds was significantly different from 0. However, the previous exercise showed it is not significantly different from the mean change for the control group, even though that group had a *negative* sample mean change. How do you explain this paradox?(*Hint*: From Section 9.2, how does the *se* value for estimating a *difference* between two means compare to the *se* value for estimating a single mean?)

9.73 Surgery vs. placebo for knee pain: Refer to Example 10 on whether arthroscopic surgery is better than placebo. The table below shows the pain scores one year after surgery. Using software (such as MINITAB) that can conduct analyses using summary statistics, compare the placebo to the debridement group using a 95% confidence interval. Use the method that assumes equal population standard deviations. Explain how to interpret the interval.

Knee Pain Score

GROUP	SAMPLE SIZE	MEAN	STANDARD DEVIATION
Placebo	60	48.9	21.9
Arthroscopic—lavage	61	54.8	19.8
Arthroscopic—debridement	59	51.7	22.4

9.74 More knee pain: Refer to the previous exercise. Compare the placebo to the debridement group using a significance test. State the assumptions, and explain how to interpret the P-value.

9.75 Anorexia again: Refer to Exercise 9.71, comparing mean weight changes in anorexic girls for cognitive behavioral therapy and a control group. The MINITAB output shows results of doing analyses while assuming equal population standard deviations.

MINITAB output for comparing mean weight changes

```
group       N     Mean   StDev   SE Mean
cog_behav  29     3.01    7.31     1.4
control    26    -0.45    7.99     1.6
Difference = mu (cog_behav) - mu (control)
Estimate for difference: 3.45690
95% CI for difference:(-0.68014, 7.59393)
T-Test of difference = 0(vs not =):
T-Value = 1.68  P-Value = 0.100  DF = 53
Both use Pooled StDev = 7.6369
```

a. Interpret the reported confidence interval.

b. Interpret the reported P-value.

c. What would be the P-value for $H_a: \mu_1 > \mu_2$? Interpret it.

d. What assumptions do these inferences make?

9.76 Breast-feeding helps IQ?: A Danish study of individuals born at a Copenhagen hospital between 1959 and 1961 reported higher mean IQ scores for adults who were breast-fed for longer lengths of time as babies (E. Mortensen et al., *JAMA*, vol. 287, pp. 2365–2371, 2002). The mean IQ score was 98.1 ($s = 15.9$) for the 272 sub-

jects who had been breast-fed for no longer than a month and 108.2 ($s = 13.1$) for the 104 subjects who had been breast-fed for 7–9 months.

a. With software that can analyze summarized data, use an inferential method to analyze whether the corresponding population means differ, assuming the population standard deviations are equal. Interpret.

b. Was this an experimental or an observational study? Can you think of a potential lurking variable?

9.77 Eat tomatoes: The Health Professional Follow-Up Study followed the health history and dietary habits of 47,000 men from 1986 to 1998, over which time 2481 of the men developed prostate cancer. An article about the study (E. Giovannucci et al., *Journal of the National Cancer Institute*, vol. 94, pp. 391–398, 2002) reported that "intake of tomato sauce was associated with a reduction in prostate cancer risk (relative risk for at least 2 servings per week versus less than 1 serving per month = 0.77, 95% CI = 0.66 to 0.90, P-value < .001)." Explain how to interpret the (a) relative risk value, (b) confidence interval.

9.78 Drink tea: A recent study followed patients for four years after they had a heart attack (K. Mukamal et al., *Circulation*, vol. 105, pp. 2476–2481, 2002). Patients were interviewed about their weekly caffeinated tea consumption during the year before the heart attack. The relative risk of another heart attack during the study period for those who reported drinking at least 14 cups a week, compared to those who reported not drinking tea, was 0.56, with a 95% confidence interval of 0.37 to 0.84.

a. Interpret the confidence interval for the population relative risk.

b. According to these results, can we conclude that in a corresponding population, tea drinking reduces the chance of another heart attack? Explain.

9.79 Improving employee evaluations: Each of a random sample of ten customer service representatives from a large department store chain answers a questionnaire about how they respond to various customer complaints. Based on the responses, a summary score measures how positively the employees react to complaints. This is measured both before and after employees undergo an intensive training course designed to improve such scores. A report about the study states, "The mean was significantly higher after taking the training course [$t = 3.40$ ($df = 9$), P-value < 0.05]." Explain how to interpret this to someone who has never studied statistics. What else should have been reported, to make this more informative?

9.80 Which tire is better?: A tire manufacturer believes that a new tire it is introducing (Brand A) will have longer wear than the comparable tire (Brand B) sold by their main competitor. To get evidence to back up their claim in planned advertising, they conduct a study. On each of four cars they use a tire of Brand A on the left front and a tire of Brand B on the right front. The response is the number of thousands of miles until a tire wears out, according to a tread marking on the tire. The sample mean response is 50 for Brand A and 40 for Brand B.

a. Show a pattern of four pairs of observations with these means for which you think Brand A would be judged better according to statistical inference. You do not need to actually conduct the inference. (*Hint*: What affects the value of the test statistic, other than the sample means and *n*?)

b. How could the design of this study be improved?

9.81 TV or rock music a worse influence?: In a recent General Social Survey, subjects were asked to respond to the following: "Children are exposed to many influences in their daily lives. What kind of influence does each of the following have on children? 1. Programs on network television, 2. rock music." The possible responses were (very negative, negative, neutral, positive, very positive). The responses for 12 of the sampled subjects, using scores $(-2, -1, 0, 1, 2)$ for the possible responses are given in the table:

Subject	TV	Rock	Subject	TV	Rock	Subject	TV	Rock
1	0	-1	5	-1	-1	9	-1	-1
2	0	0	6	-2	-2	10	0	1
3	1	-2	7	-1	0	11	1	-1
4	0	1	8	1	-1	12	-1	-2

a. When you compare responses for TV and Rock, are the samples independent or dependent? Why?

b. Use software to construct a 95% confidence interval for the population difference in means. Interpret in the context of the variables studied.

c. Use software to conduct a significance test of equality of the population means for TV and Rock. Interpret the P-value.

9.82 Influence of TV and movies: Refer to the previous exercise. The GSS also asked about the influence of Movies. The responses for these 12 subjects were $-1, 1, 0, 2, 0, -2, -1, 0, -1, 1, 1, -1$. The results of using MINITAB to compare the influence of Movies and TV are shown below. Explain how to interpret (a) the confidence interval, (b) the significance test results.

```
Paired T for movies - tv
              N     Mean      StDev    SE Mean
movies       12  -0.083333  1.164500  0.336162
tv           12  -0.250000  0.965307  0.278660
Difference   12   0.166667  0.937437  0.270615

95% CI for mean difference:
(-0.428952, 0.762286)

T-Test of mean difference = 0 (vs not = 0):
T-Value = 0.62   P-Value = 0.551
```

9.83 Crossover study: The table summarizes results of a crossover study to compare results of low-dose and high-dose analgesics for relief of menstrual bleeding (B. Jones and M. Kenward, *Statistics in Medicine*, vol. 6, pp. 555–564, 1987).

a. Find the sample proportion of successes for each dose of the analgesic.

b. Find the P-value for testing that the probability of success is the same with each dose, for a two-sided alternative. Interpret. What assumptions does this inference make?

	High Dose	
Low Dose	Success	Failure
Success	53	8
Failure	16	9

9.84 Belief in ghosts and in astrology: A poll by Louis Harris and Associates of 1249 Americans indicated that 36% believe in ghosts and 37% believe in astrology.

a. Is it valid to compare the proportions using inferential methods for independent samples? Explain.

b. Do you have enough information to compare the proportions using methods for dependent samples? If yes, do so. If not, explain what else you would need to know.

9.85 Fast food, TV, and obesity: A March, 2003 article in the *Gainesville Sun* about a study at Boston's Children's Hospital had the headline, "New study: Lots of fast food and TV triple risk of obesity." The response variable was binary, namely whether a child was obese. Define, in terms of this study, the sample statistic to which the "triple risk" refers.

9.86 Seat belts help?: For automobile accidents in a recent year in Maine, injuries occurred to 3865 of 30,902 subjects not wearing seat belts, and to 2409 of 37,792 subjects wearing seat belts. Find and interpret the relative risk for these data.

9.87 Death penalty paradox: Exercise 3.50 showed results of another study about the death penalty and race. The data are shown below.

a. Treating victim's race as the control variable, show that Simpson's paradox occurs.

b. Explain what causes the paradox to happen.

		Death Penalty	
Vic Race	**Def Race**	Yes	No
W	W	53	414
	B	11	37
B	W	0	16
	B	4	139

9.88 Death rate paradoxes: The crude death rate is the number of deaths in a year, per size of the population, multiplied by 1000.

a. According to the U.S. Bureau of the Census, in 1995 Mexico had a crude death rate of 4.6 (i.e., 4.6 deaths per 1000 population) while the United States had a crude death rate of 8.4. Explain how this overall death rate could be higher in the United States even if the United States had a lower death rate than Mexico for people of each specific age.

b. For each age level, the death rate is higher in South Carolina than in Maine. Overall, the death rate is higher in Maine (H. Wainer, *Chance*, vol. 12, p. 44, 1999). Explain how this could be possible.

9.89 Income and gender: For a particular Big Ten university, the mean income for male faculty is $8000 higher than the mean income for female faculty. Explain how this difference could disappear:

a. Controlling for number of years since received highest degree. (*Hint*: What if relatively few female professors were hired until recent years?)

b. Controlling for college of employment.

Concepts and Investigations

9.90 Student survey: Refer to the "FL student survey" data file on the text CD. Using software, prepare a short report summarizing the use of confidence intervals and significance tests (including checking assumptions) to compare males and females in terms of opinions about whether there is life after death.

9.91 More eBay selling prices: Exercise 9.16 showed selling prices of the Palm M515 PDA for a one-week selling period in May 2003. When prices are analyzed for a three-month period in 2003, we get the results shown in the table.

Variable	Buy now	n	Mean	StDev
Price	No	136	223.36	23.74
	Yes	132	235.45	20.57

Analyze these data. Prepare a one-page summary of your analyses, highlighting your conclusion about whether it is good to use the buy it now option.

9.92 Attractiveness and getting dates: The results in the table are from a study of physical attractiveness and subjective well-being (E. Diener et al., *Journal of Personality and Social Psychology*, vol. 69, 1995, pp. 120–129). As part of the study, college students in a sample were rated by a panel on their physical attractiveness. The table presents the number of dates in the past three months for students rated in the top or bottom quartile of attractiveness. Analyze these data. Write a short report that summarizes the analyses and makes interpretations.

	NO. DATES, MEN			NO. DATES, WOMEN		
ATTRACTIVENESS	Mean	StdDev	n	Mean	StdDev	n
More	9.7	10.0	35	17.8	14.2	33
Less	9.9	12.6	36	10.4	16.6	27

9.93 Pay discrimination against women?: A *Time Magazine* article titled "Wal-Mart's Gender Gap" (July 5, 2004) stated that in 2001 women managers at Wal-Mart earned $14,500 less than their male counterparts.

a. If these data are based on a random sample of managers at Wal-Mart, what more would you need to know about the sample to determine whether this is a "statistically significant" difference?

b. If these data referred to *all* the managers at Wal-Mart and if you can get the information specified in (a), is it relevant to conduct a significance test? Explain.

In multiple-choice exercises 9.94–9.96, select the correct response.

9.94 Alcoholism and gender: Suppose that a 99% confidence interval for the difference $p_1 - p_2$ between the proportions of men and women in California who are alcoholics equals $(0.02, 0.09)$.

a. We are 99% confident that the proportion of alcoholics is between 0.02 and 0.09.

b. We are 99% confident that the proportion of men in California who are alcoholics is between 0.02 and 0.09 larger than the proportion of women in California who are.

c. We can conclude that the population proportions may be equal.

d. We are 99% confident that a minority of California residents are alcoholics.

e. Since the confidence interval does not contain 0, it is impossible that $p_1 = p_2$.

9.95 Comparing mean incomes: A study compares the population mean annual incomes for Hispanics (μ_1) and for whites (μ_2) having jobs in construction, using a 95% confidence interval for $\mu_1 - \mu_2$.

a. If the confidence interval is $(-3000, -6000)$, then at this confidence level we conclude that the mean income for whites is less than for Hispanics.

b. If the confidence interval is $(-3000, 1000)$, then the test of $H_0: \mu_1 = \mu_2$ against $H_a: \mu_1 \neq \mu_2$ with significance level 0.05 rejects H_0.

c. If the confidence interval is $(-3000, 1000)$, then we can conclude that $\mu_1 = \mu_2$.

d. If the confidence interval is $(-3000, 1000)$ then we are 95% confident that the population mean annual income for Hispanics is between 3000 less and 1000 more than the population mean annual income for whites.

9.96 Sample size and significance: If the sample proportions in Example 4 comparing heart attack rates for aspirin and placebo had sample sizes of only 1000 each, rather than about 11,000 each, then the 95% confidence interval for $(p_1 - p_2)$ would be $(-0.002, 0.018)$ rather than $(0.005, 0.011)$. This reflects that

a. When an effect is small, it may take very large samples to have much power for establishing statistical significance.

b. Smaller sample sizes are preferable, because there is more of a chance of capturing 0 in a confidence interval.

c. Confidence intervals get wider when sample sizes get larger.

d. The confidence interval based on small sample sizes must be in error, because it is impossible for the parameter to take a negative value.

9.97 Positive values in CI: True or false? If a 95% confidence interval for $(\mu_1 - \mu_2)$ contains only positive numbers, then we can conclude that both μ_1 and μ_2 are positive.

9.98 Afford food?: True or false? A 2003 survey by the Pew Research Center asked whether there have been times in the past year the respondent has been unable to afford food. Of advanced economies, the country with the highest response was the United States, 15%. Worldwide, the highest response was in Angola, 86%. Because the same question was asked in both countries, the samples are dependent.

9.99 Control for clinic: Suppose there are more successes with treatment A than with treatment B at a clinic in Rochester, and there are more successes with treatment A than with treatment B at a clinic in Syracuse. True or false?: For the overall sample (combining results for the two cities), there must be more successes with treatment A than with treatment B.

◆◆9.100 Guessing on test: A test consists of 100 true-false questions. Joe did not study, and on each question he randomly guesses the correct response. Jane studied a little and has a 0.60 chance of a correct response for each question.

    a. Approximate the probability that Jane's score is nonetheless lower that Joe's.(*Hint*: Use the sampling distribution of the difference of sample proportions.)

    b. Intuitively, do you think that the probability answer to (a) would decrease or would increase if the test had only 50 questions? Explain.

◆◆9.101 Standard error of difference: From the box formula for the standard error at the end of Section 9.1,

$$se(\text{estimate 1} - \text{estimate 2}) =$$
$$\sqrt{[se(\text{estimate 1})]^2 + [se(\text{estimate 2})]^2},$$

if you know the *se* for each of two independent estimates, you can find the *se* of their difference. This is useful, because often articles report a *se* for each sample mean or proportion, but not the *se* or a confidence interval for their difference. Exercise 9.46 referred to a large sample of subjects from Framingham, Massachusetts, who have been followed in a study since 1948. That study recently estimated the number of years of life lost by being obese and a smoker. For females of age 40, adjusting for other factors, the number of years of life left were estimated to have a mean of 46.3 ($se = 0.6$) for nonsmokers of normal weight and a mean of 33.0 ($se = 1.8$) for smokers who were obese. Construct a 95% confidence interval for the population mean number of years lost. Interpret.

◆◆9.102 Lots of variation in the reported importance of religion:

    a. For comparisons of groups in which $n_1 = n_2$, with common value denoted by n, use the fact that the largest possible value of $\hat{p}(1 - \hat{p})$ occurs at $\hat{p} = 0.5$ to show that the margin of error for a large-sample 95% confidence interval for $(p_1 - p_2)$ can be no greater than $2\sqrt{0.5/n} = \sqrt{2/n}$.

    b. A 2003 report by the Pew Research Center reported that the percent of people who find religion very

important in their life is 59% in the United States, 30% in Canada, 33% in Great Britain, 27% in Italy, 21% in Germany, and 11% in France. Assuming that each country had a random sample of 1000 people, use the $\sqrt{2/n}$ bound to identify any pairs of countries for which the true percentages might not be different.

◆◆9.103 Small-sample CI: The small-sample confidence interval for comparing two proportions is a simple adjustment of the large-sample one. Recall that for a small-sample confidence interval for a single proportion, we used the ordinary formula after adding 4 observations, 2 of each type (see the end of Section 7.4). In the two-sample case we also add 4 observations, 2 of each type, and then use the ordinary formula, by adding 1 observation of each type to each sample.[14] When the proportions are near 0 or near 1, results are more sensible than with the ordinary formula. Suppose there are no successes in either group, with $n_1 = n_2 = 10$.

    a. With the ordinary formula, show that (i) $\hat{p}_1 = \hat{p}_2 = 0$, (ii) $se = 0$, and (iii) the 95% confidence interval for $p_1 - p_2$ is 0 ± 0, or $(0, 0)$. Obviously, it is too optimistic to predict that the true difference is exactly equal to 0.

    b. Find the 95% confidence interval using the small-sample method described here. Are the results more plausible?

◆◆9.104 Graphing Simpson's paradox: The figure illustrates Simpson's paradox for Example 16 on the death penalty. For each defendant's race, the figure plots the percentage receiving the death penalty. Each percentage is labeled by a letter symbol giving the category of the victim's race. Surrounding each observation is a circle of an area representing the number of observations on which that percentage is based. For instance, the W in the largest circle

Percentage Receiving Death Penalty by Defendant's Race, Controlling Victim's Race (labels in center of circles) and Ignoring Victim's Race (overall values marked by × for white defendants and black defendants).

---

[14]This is a new method. It was first proposed by A. Agresti and B. Caffo in the journal, *The American Statistician*, vol. 54, pp. 280–288 (2000).

represents a percentage of 12.6 receiving the death penalty for cases with white defendants and white victims. That circle is largest because the number of cases at that combination (151) is largest. When we add results across victim's race to get a summary result ignoring that variable, the largest circles, having the greater number of cases, have greater influence. Thus, the summary percentages for each defendant's race, marked on the figure by an ×, fall closer to the center of the larger circles than to the center of the smaller circles.

a. Explain why a comparison of a pair of circles having the same letter in the middle indicates that the death penalty was more likely for black defendants, when we control for victim's race.

b. Explain why a comparison of the × marks shows that overall the death penalty was more likely for white defendants than black defendants.

c. For white defendants, why is the overall percentage who got the death penalty so close to the percentage for the case with white victims?

Class Explorations

9.105 **Reading the medical literature:** Refer to Activity 2 below of reading an article in a medical journal. Your instructor will pick a recent article at the Web site for the *British Medical Journal.* Prepare a short report in which you summarize the main conclusions of the article and explain how those conclusions depended on statistical analyses. After you have done this, your entire class will discuss this article and the statistical analyses.

9.106 **Review the medical literature:** Your instructor will pick a medical topic of interest to the class. Find a recent article of a medical journal that reports results of a research study on that topic. Describe the statistical analyses that were used in that article. Did the article use (a) descriptive statistics, (b) confidence intervals, (c) significance tests? If so, explain how these methods were used. Prepare a one-page summary of your findings that you will present to your class.

ACTIVITY 2: Can You Understand the Statistics in a Medical Journal?

Medical research makes frequent use of the statistical methods of this chapter. If you pick up nearly any medical journal, you will see results of statistical analyses in its articles. For instance, you can read the *British Medical Journal* (*BMJ*) at bmj.com. On the Internet, pull up the April 19, 2003 issue of the journal. Browse through the article on "Behavioural counseling to increase consumption of fruit and vegetables in low income adults: randomized trial" by A. Steptoe et al. You will see a variety of statistical analyses used. See if you can answer these questions:

■ What was the objective of the study?

■ Was the study an experiment or an observational study? Summarize the design and the subjects used in the study.

■ Identify response and explanatory variables.

■ According to Table 2 in the article, the 135 subjects in the nutrition counseling group had a baseline mean of 3.67 and standard deviation of 2.00 for the number of portions per day they ate of fruits and vegetables. Over the 12 months of the study, the mean increased by 0.99. Report the 95% confidence interval for the change in the mean, and interpret. Report the corresponding results for the behavioral counseling group. Were the groups in this analysis independent samples or dependent samples?

■ Adjusted for confounding variables, the difference between the change in the mean for the behavioral counseling group and the nutrition counseling group was 0.62. Report and interpret the 95% confidence interval for the true difference, and report the P-value for testing that the difference was 0. Were the groups in this analysis independent samples, or dependent samples?

■ According to Table 2, at the baseline for the nutrition counseling group, the proportion of subjects who ate at least five portions a day of fruits and vegetables was 0.267, and this increased by 0.287 over the 12 months. Report and interpret the 95% confidence interval for the change in the true proportion. Was this analysis for independent samples, or for dependent samples? Was this change in the proportion significantly larger, or smaller, than the change for the behavioral counseling group?

■ What were the primary conclusions of the study?

■ Describe two limitations of the study, as explained in the sections on "Representativeness of the sample" and "Limitations of the study."

Your instructor may have the class discuss some of these issues.

EXAMPLE I

What's Associated with Happiness?

Picture the Scenario

On what does your overall happiness depend? Whether you have a partner in your life? How many friends you have? Whether you like your job? How much money you make? How often you attend religious services? How good your health is? How attractive you are? The country you live in?

To investigate which variables are associated with happiness, we can use data from the General Social Survey (GSS). In each survey, the GSS asks, "Taken all together, would you say that you are very happy, pretty happy, or not too happy?" Table 10.1 uses the 2002 survey to cross-tabulate happiness with family income, here measured as the response to the question, "Compared with American families in general, would you say that your family income is below average, average, or above average?"

TABLE 10.1: Happiness and Family Income, from 2002 General Social Survey			
	HAPPINESS		
INCOME	Not too Happy	Pretty Happy	Very Happy	**Total**
Above average	21	159	110	290
Average	53	372	221	646
Below average	94	249	83	426

Questions to Explore

- How can you determine if there is an association between happiness and family income in the population of all adult Americans?

- If there is an association, what is its nature? For example, do people with above average family income tend to be happier than people with below average family income?

- Can you think of a variable that might have a stronger association with happiness than family income does?

Thinking It Through

The variables in Table 10.1 are both categorical. The focus of this chapter is learning how to detect and how to describe associations between categorical variables. To do this, we'll continue to apply the basic tools of statistical inference. These inferential methods help us answer questions such as "Do people with higher family incomes tend to be happier?" and "Are married people happier than unmarried people?" We'll analyze Table 10.1 in Examples 3 and 4 and other data on happiness in Example 8 and in many of the exercises.

Recall

In the introduction to Chapter 3, we stated that two variables have an **association** if a particular value for one variable is more likely to occur with certain values of the other variable—for example, if being very happy is more likely to happen if a person has an above average income. Section 3.1 introduced the analysis of association for categorical variables. ■

Association between Variables

Let's recap where we are at this stage of our study. Chapter 6 introduced the fundamental concept of a *sampling distribution*. Chapter 7 showed how the sampling distribution is the basis of estimation, using *confidence intervals*, and Chapter 8 showed how it is the basis of *significance testing*. Those chapters focused on inference for a single proportion or a single mean.

Chapter 9 introduced methods for two variables—a response variable and a binary explanatory variable for which the two categories define two groups to compare (such as females and males). For a binary response variable, we compared two proportions, and for a quantitative response variable, we compared two means. Other methods apply when (a) both variables are quantitative or (b) both variables are categorical with possibly several categories. We'll study these cases in this chapter and in Chapter 11.

This chapter presents methods for investigations in which both variables are categorical. Table 10.1 shows an example of such data. We'll begin by being more precise about what we mean by an association or lack of association, in the context of contingency tables such as this one.

10.1 | What Is Independence and What Is Association?

As a first step, we identify the response variable and the explanatory variable. In Table 10.1 it's more natural to study how happiness depends on income than how income depends on happiness. We'll treat happiness as the response variable and income as the explanatory variable.

Comparing Percentages

Table 10.1 is easier to digest if we convert the frequencies to percentages for the response categories. Within each category of income, Table 10.2 shows the percentages for the three categories of happiness. For example, of the 290 subjects who reported their family income as above average, 110 identified themselves as very happy. This is a proportion of 110/290 = 0.38, or 38%. By contrast, 83 out of the 426 below average income subjects said they were very happy, a percentage of 19%. The row totals in Table 10.1 are the basis of these percentage calculations. Within each row the percentages sum to 100%.

TABLE 10.2: Conditional Percentages for Happiness Categories, Given Each Category of Family Income

The percentages are computed by dividing the counts in Table 10.1 by their row totals, and then multiplying by 100.

	HAPPINESS			
INCOME	Not too Happy	Pretty Happy	Very Happy	**TOTAL**
Above average	7%	55%	38%	290 (100%)
Average	8%	58%	34%	646 (100%)
Below average	22%	59%	19%	426 (100%)

Recall

Section 3.1 introduced **conditional proportions**, which are proportions for the categories of a categorical response variable that are calculated **conditional** upon (that is, given) the value of another variable. ■

The three percentages in a particular row are called **conditional percentages**. They refer to a sample data distribution of happiness, *conditional* on the category for family income. They form the **conditional distribution** for happiness, given a particular income level. For those who reported above-average family income, the conditional distribution of happiness is the set of percentages (7%, 55%, 38%) for the responses (not too happy, pretty happy, very happy). The proportions (0.07, 0.55, 0.38) are the estimated **conditional probabilities** of happiness, given above-average family income. For instance, given that a subject reported above-average family income, the estimated probability of being very happy is 0.38.

Let's compare the happiness percentages for the subjects who reported above average income to those who reported below average income. From Table 10.2, the higher-income subjects were more likely to be very happy (38% vs. 19%) and less likely to be not too happy (7% vs. 22%). Figure 10.1 uses a bar graph to portray the three conditional distributions.

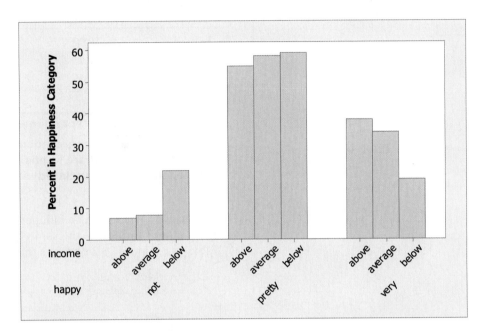

▲ **FIGURE 10.1: MINITAB Bar Graph of Conditional Distributions of Happiness, Given Income, from Table 10.2.** For each happiness category, you can compare the three categories of income on the percent in that happiness category. **Question:** Based on this graph, how would you say that happiness depends on income?

We'll use these guidelines when constructing tables with conditional distributions:

■ Make the response variable the column variable. The response categories are shown across the top of the table.

■ Compute conditional proportions for the response variable within each row by dividing each cell frequency by the row total. Those proportions multiplied by 100 are the conditional percentages.

■ Include the total sample sizes on which the percentages are based. That way, readers can reconstruct the cell counts, if they would like to.

Independence versus Dependence (Association)

Table 10.2 shows conditional distributions for a sample. We'll use sample conditional distributions to make inferences about the corresponding population conditional distributions.

For instance, to investigate how happiness compares for females and males, we'd make inferences about the population conditional distributions of happiness for females and for males. Suppose that those population distributions are as shown in Table 10.3. (In fact, the percentages in Table 10.3 are consistent with data from the GSS. See Exercise 10.4.) We see that the percentage classified as very happy is the same for females and males, 32%. Similarly, the percentage classified as pretty happy (58%) and the percentage classified as not too happy

IN WORDS

Happiness and gender are **independent** if (as in Table 10.3) the probability of a particular response on happiness is the same for females and males.

(10%) is the same for females and for males. The probability that a person is classified in a particular happiness category is the same for females and males. The categorical variables happiness and gender are then said to be **independent**.

TABLE 10.3: Population Conditional Distributions Showing Independence

The conditional distribution of happiness is the same for each gender, namely (10%, 58%, 32%).

| GENDER | HAPPINESS | | | TOTAL |
	Not too Happy	Pretty Happy	Very Happy	
Females	10%	58%	32%	100%
Males	10%	58%	32%	100%
Overall	10%	58%	32%	100%

Independence and Dependence (Association)

Two categorical variables are **independent** if the population conditional distributions for one of them are identical at each category of the other. The variables are **dependent** (or **associated**) if the conditional distributions are not identical.

For two variables to be independent, the population percentage in any category of one variable is the same for all categories of the other variable, as in Table 10.3. Figure 10.2 portrays this for Table 10.3.

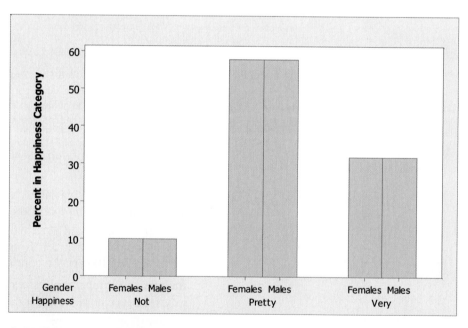

▲ **FIGURE 10.2: MINITAB** Output of Population Conditional Distributions Showing Independence. **Question:** What feature of the bar graph reflects the independence?

Recall

This is an additional use of the term **independent**.

Section 5.3 in Chapter 5 defined A and B to be **independent events** if

$$P(A \mid B) = P(A).$$

Chapter 9 used **independent samples** to mean that observations in one sample are independent of those in the other sample. ■

This definition extends independence of **events** to independence of **variables**. Recall that two *events* A and B are independent if P(A) is the same as the conditional probability of A, given B, denoted by $P(A \mid B)$. When two *variables* are independent, any event about one variable is independent of any event about the other variable. In Table 10.3, $P(A \mid B) = P(A)$ if A is an event about happiness and B is an event about gender. For example if A = very happy and B = female, then $P(A \mid B) = 0.32$. From the margin of the table for the "overall" distribution, this also equals P(A).

EXAMPLE 2

Is Belief in Life after Death Associated with Race?

Picture the Scenario

Table 10.4 cross-tabulates belief in life after death with race, using data

from the 2000 GSS. The table also shows sample conditional distributions for life after death, given race, and the overall percentages for the life after death categories.

TABLE 10.4: Sample Conditional Distributions for Belief in Life After Death, Given Race

| RACE | BELIEF IN LIFE AFTER DEATH | | TOTAL |
	Yes	No	
White	1339 (81.7%)	300 (18.3%)	1639 (100%)
Black	260 (82.5%)	55 (17.5%)	315 (100%)
Other	88 (80.0%)	22 (20.0%)	110 (100%)
Overall	1687 (81.7%)	377 (18.3%)	2064 (100%)

Question to Explore

Are race and belief in life after death independent or dependent?

Think It Through

The conditional distributions in Table 10.4 are similar but not exactly identical. So it is tempting to conclude that the variables are dependent. For example, if A = "yes" for belief in life after death, we estimate $P(A | \text{white}) = 1339/1639 = 0.817$, $P(A | \text{black}) = 0.825$, $P(A | \text{other}) = 0.800$, whereas we estimate that $P(A) = 1687/2064 = 0.817$. However, the definition of independence between variables refers to the *population*. Since Table 10.4 refers to a *sample* rather than a population, it provides evidence but it does not definitively tell us whether these variables are independent or dependent.

Even if the variables *were* independent, we would not expect the *sample* conditional distributions to be identical. Because of sampling variability, each sample percentage typically differs somewhat from the true population percentage. We would expect to observe *some* differences between sample conditional distributions such as we see in Table 10.4 even if *no* differences exist in the population.

Insight

If the observations in Table 10.4 were the entire population, the variables would be dependent. But the association would not be practically important, because the percentages are so similar from one category to another of race. The next section presents a significance test of the hypothesis that the variables are independent.

TO PRACTICE THIS CONCEPT, TRY EXERCISE 10.2

10.1 Gender gap in politics?: In the United States, is there a "gender gap" in political beliefs? That is, do women and men tend to differ in their political thinking and voting behavior? The table taken from the 2002 GSS relates gender and political party identification. Subjects indicated whether they identified more strongly with the Democratic or Republican party or as Independents.

POLITICAL PARTY IDENTIFICATION

GENDER	Democrat	Independent	Republican	Total
Females	567	534	395	1496
Males	356	460	369	1185

a. Identify the response variable and the explanatory variable.

b. Construct a table that shows the conditional distributions of the response variable. Interpret.

c. Give a hypothetical example of population conditional distributions for which these variables would be independent.

d. Sketch bar graphs to portray the distributions in (b) and in (c).

10.2 Beliefs of college freshmen: Every year, a large-scale poll of college freshmen conducted by the Higher Education Research Institute at UCLA asks their opinions about a variety of issues. In 2002, 46% of men and 35% of women in the survey of 283,000 college freshmen indicated support for legalization of marijuana.

a. If results for the population of college freshmen were similar to these, would gender and opinion about legalizing marijuana be independent or dependent?

b. Give an example of hypothetical population percentages for which these variables would be independent.

10.3 FBI statistics: Data posted at the FBI Web site indicated that of all blacks slain in 2002, 92% were slain by blacks, and of all whites slain in 2002, 85% were slain by whites. Let y denote race of victim and x denote race of murderer.

a. Which conditional distributions do these statistics refer to, those of y at given categories of x, or those of x at given categories of y? Set up a table with race of murderer as rows and race of victim as columns, showing the two conditional distributions.

b. Are x and y independent or dependent? Explain. (*Hint*: These results refer to the population.)

10.4 Happiness and gender: The contingency table shown relates happiness and gender for the 2000 GSS.

HAPPINESS

GENDER	Not too Happy	Pretty Happy	Very Happy	Total
Females	163	898	502	1563
Males	130	705	379	1214

a. Identify the response variable and the explanatory variable.

b. Construct a table or graph showing the conditional distributions. Interpret.

c. Give an example of population conditional distributions that would seem to be consistent with this sample and for which happiness and gender would be independent.

10.5 Marital happiness and income: In the GSS, subjects who were married were asked the happiness of their marriage, the variable coded as HAPMAR.

a. Go to the GSS Web site sda.berkeley.edu/GSS/ and construct a contingency table for 2002 relating HAPMAR to family income as measured in this section, by entering FINRELA (r: 1-2; 3; 4-5) as the row variable and YEAR(2002) in the selection filter.

b. Construct a table or graph that shows the conditional distributions of marital happiness, given family income. How would you describe the association?

c. Compare the conditional distributions to those in Table 10.2. For a given family income, what tends to be higher, general happiness or marital happiness for those who are married? Explain.

10.6 What is independent of happiness?: Which one of the following variables would you think most likely to be independent of happiness: belief in an afterlife, family income, quality of health, region of the country in which you live, satisfaction with job? Explain the basis of your reasoning.

10.7 Sample evidence about independence: Refer to the previous exercise. Go to the GSS Web site and construct a table relating happiness (HAPPY) to the variable you chose (AFTERLIF, FINRELA, HEALTH, REGION, or JOBSAT). Inspect the conditional distributions and indicate whether independence seems plausible, with the sample conditional distributions all being quite similar.

10.2 | How Can We Test Whether Categorical Variables Are Independent?

How can we judge whether two categorical variables are independent or dependent? The definition of independence refers to the *population*. Could the observed *sample* association be due to sampling variation? Or would the observed results be unusual if the variables were truly independent?

A significance test can answer these questions. The hypotheses for the test are

H_0: The two variables are independent.

H_a: The two variables are dependent (associated).

The test assumes random sampling and a large sample size. Specific guidelines are discussed later in the section. The test compares the cell counts in the contingency table with counts we would expect to see if the null hypothesis of independence were true.

What Do We Expect for Cell Counts If the Variables Are Independent?

The count in any particular cell is a random variable: Different samples have different values for the count. The mean of its distribution is called an **expected cell count**. This is found under the presumption that H_0 is true. That is, the expected cell counts are values that satisfy the null hypothesis of independence.

Table 10.5 shows the observed cell counts from Table 10.1, with the expected cell counts below them. How do we find the expected cell counts? Recall that when two variables are independent, any event about one variable is independent of any event about the other variable. Consider the events A = "not too happy" and B = "above average income." From the margins of Table 10.5 we estimate P(A) by 168/1362 and P(B) by 290/1362. The event "A and B" (not too happy *and* above average income) corresponds to the first cell in the table, with cell count of 21. Under the presumption of independence,

$$P(A \text{ and } B) = P(A) \times P(B),$$

which we estimate by (168/1362)(290/1362). For the $n = 1362$ subjects, the *number* of them we expect in this cell is 1362 times the cell's probability. So we estimate the **expected cell count** (presuming H_0 to be true) by

$$1362\left(\frac{168}{1362}\right)\left(\frac{290}{1362}\right) = \frac{168 \times 290}{1362} = 35.8.$$

If happiness were independent of income, we would expect about 36 subjects in that first cell. Since the observed cell count is 21, fewer people were of above average income and not too happy than we would expect if the variables were independent.

Recall

From Section 5.2, two events A and B can also be defined as **independent** if

$$P(A \text{ and } B) = P(A)P(B). \blacksquare$$

TABLE 10.5: Happiness by Family Income, Showing Observed and Expected Cell Counts

We use the highlighted totals to get the expected count of 35.8 = (290 × 168)/1362 in the first cell.

INCOME	HAPPINESS			Total
	Not too happy	Pretty happy	Very happy	
Above average	21	159	110	290
	35.8	166.1	88.1	
Average	53	372	221	646
	79.7	370.0	196.4	
Below average	94	249	83	426
	52.5	244.0	129.5	
Total	168 (12.3%)	780 (57.3%)	414 (30.4%)	1362 (100%)

This discussion illustrates the general rule for calculating expected cell counts.

Expected Cell Count

For a particular cell, the **expected cell count** equals

$$\text{Expected cell count} = \frac{(\text{Row total}) \times (\text{Column total})}{\text{Total sample size}}.$$

For instance, the first cell in Table 10.5 has expected cell count = $(290 \times 168)/1362 = 35.8$, the product of the row total (290) for that cell by the column total (168) for that cell, divided by the overall sample size (1362).

Let's see more about why this rule makes sense. For A = "not too happy" we estimated $P(A)$ by $168/1362 = 0.123$. If the variables were independent, we would expect this probability to be the same if we condition on any event about income. For instance, for B = "above average income," we expect $P(A|B) = P(A)$. Thus, 12.3% of the 290 subjects who report above average income should be classified in the not too happy category. The expected cell count is then $0.123(290) = (168/1362)290 = 35.8$. But this is the column total of 168 times the row total of 290 divided by the overall sample size of 1362.

Table 10.5 shows that for the overall sample, the percentages in the three happiness categories are (12.3%, 57.3%, 30.4%). You can check that the *expected cell counts* have these percentages in each of the three rows. For instance, for the first cell in row 1, $35.8/290 = 0.123$. In fact, *the expected frequencies are values that have the same row and column totals as the observed counts, but for which the conditional distributions are identical.*

The Chi-Squared Test Statistic

The test statistic for the test of independence summarizes how close the observed cell counts fall to the expected cell counts. Symbolized by X^2, it is called the **chi-squared statistic**, taking the name of its (approximate) sampling distribution. It is the oldest test statistic in use today. Introduced by the British statistician Karl Pearson in 1900, it is sometimes also called "Pearson's chi-squared statistic."

IN WORDS

The Greek letter chi is written as χ and pronounced "ki" (k + eye). Later in this section, we'll see that chi-squared (χ^2) is the name of the approximate sampling distribution of the statistic denoted here by X^2. The Roman letter denotes the statistic and the Greek letter denotes the sampling distribution.

Chi-Squared Statistic

The **chi-squared statistic** summarizes how far the observed cell counts in a contingency table fall from the expected cell counts for a null hypothesis. Its formula is

$$X^2 = \sum \frac{(\text{observed count} - \text{expected count})^2}{\text{expected count}}.$$

For each cell, square the difference between the observed count and expected count, and then divide that square by the expected count. After calculating this term for every cell, sum the terms to find X^2.

When H_0: independence is true, the observed and expected cell counts tend to be close for each cell. Then, X^2 has a relatively small value. If H_0 is false, at least some observed counts and expected counts tend to be far apart. For a cell where this happens, its value of $[(\text{observed count} - \text{expected count})^2/\text{expected count}]$ tends to be large. Then, X^2 has a relatively large value.

EXAMPLE 3

Chi-Squared for Happiness and Family Income

Picture the Scenario

Table 10.5 showed the observed and expected cell counts for the data on family income and happiness. Table 10.6 shows MINITAB output for the chi-squared test. The margin on the next page shows screen shots from the TI-83+/84. The matrix [B] referred to there (screen shot not shown) contains the expected cell counts.

```
MATRIX[A] 3 x3
[ 21    159    110  ]
[ 53    372    221  ]
[ 94    249    83   ]
```

```
X²-Test
  Observed:[A]
  Expected:[B]
  Calculate Draw
```

```
X²-Test
  X²=73.352
  P=4.444E⁻15
  df=4.000
```

ACTIVITY I

For the software you use, try to re-create a table like Table 10.6. First create a data file with nine rows, such as shown below, and then choose an option to create a contingency table and find appropriate statistics

income	happy	count
above	not	21
above	pretty	159
above	very	110
average	not	53
average	pretty	372
average	very	221
below	not	94
below	pretty	249
below	very	83

TABLE 10.6: Chi-Squared Test of Independence of Happiness and Family Income

This table shows how MINITAB reports, for each cell, the observed and expected cell counts and the contribution [(observed − expected)^2/expected] to X^2, as well as the overall X^2 value and its P-value for testing H_0: independence.

```
Rows: income              C₁      C₂      Columns: C₃ happy
                         not    pretty        very
above  R₁                 21     159          110   ← observed cell count
                        35.77   166.08       88.15  ← expected cell count
                        6.099    0.302        5.416 ← contribution to X²

average  R₂               53     372          221
                        79.68   369.96       196.36
                        8.935    0.011        3.092

below  R₃                 94     249          83
                        52.55   243.96       129.49
                        32.703   0.104        16.690

Cell contents:       Count
                     Expected count
                     Contribution to Chi-square
Pearson Chi-Square = 73.4, DF = 4, P-Value = 0.000
```

Questions to Explore

a. State the null and alternative hypotheses for this test.

b. Report the X^2 statistic and explain how it was calculated.

Think It Through

a. The hypotheses for the chi-squared test are

H_0: Happiness and family income are independent.

H_a: Happiness and family income are dependent (associated).

The alternative hypothesis is that there's an association between happiness and family income.

b. Finding X^2 involves a fair amount of computation, and it's best to let software do the work for us. Beneath the observed and expected cell counts in a cell, MINITAB reports the contribution of that cell to the X^2 statistic. For the first cell (above average income and not too happy), for instance,

$$(\text{observed count} - \text{expected count})^2/\text{expected count} =$$
$$(21 - 35.8)^2/35.8 = 6.1.$$

For all nine cells,

$$X^2 = \frac{(21 - 35.8)^2}{35.8} + \frac{(159 - 166.1)^2}{166.1} +$$

$$\frac{(110 - 88.1)^2}{88.1} + \frac{(53 - 79.7)^2}{79.7} + \frac{(372 - 370.0)^2}{370.0}$$

$$+ \frac{(221 - 196.4)^2}{196.4} + \frac{(94 - 52.5)^2}{52.5} + \frac{(249 - 244.0)^2}{244.0} + \frac{(83 - 129.5)^2}{129.5}$$

$$= 6.1 + 0.3 + 5.4 + 8.9 + 0.1 + 3.1 + 32.7 + 0.1 + 16.7 = 73.4.$$

Insight

Table 10.6 has some large differences between observed and expected cell counts, so X^2 is large. The larger the X^2 value, the greater the evidence against H_0: independence and in support of the alternative hypothesis that happiness and income are associated. We next learn how to interpret the magnitude of the X^2 test statistic, so we know what is "small" and what is "large" and how to find the P-value.

TO PRACTICE THIS CONCEPT, TRY EXERCISE 10.11

The Chi-Squared Distribution

To convert the X^2 test statistic to a P-value, we use the sampling distribution of the X^2 statistic. For large sample sizes, this sampling distribution is well approximated by the **chi-squared probability distribution**. Figure 10.3 shows several chi-squared distributions.

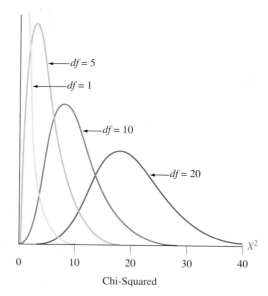

▲ **FIGURE 10.3: The Chi-Squared Distribution.** The curve has larger mean and standard deviation as the degrees of freedom increase. **Question:** Why can't the chi-squared statistic be negative?

The main properties of the chi-squared distribution are as follows:

■ It falls on the positive part of the real number line. The X^2 test statistic cannot be negative, since it sums squared differences divided by positive expected frequencies. The minimum possible value, $X^2 = 0$, would occur if observed count = expected count in each cell.

■ The precise shape of the distribution depends on the **degrees of freedom** (*df*). For testing independence in a table with r rows and c columns (called an $r \times c$ table), the formula for the degrees of freedom is

$$df = (r - 1)(c - 1).$$

For example, Table 10.5 has $r = 3$ and $c = 3$, so $df = (3 - 1)(3 - 1) = 2 \times 2 = 4$. We'll see the reason behind this formula later in the section.

■ The mean of the distribution equals the *df* value. Since $df = (r - 1)(c - 1)$, larger numbers of rows and columns produce larger *df* values. Since larger tables have more terms in the summation for the X^2 statistic, the X^2 values also tend to be larger.

- It is skewed to the right. As df increases, the skew lessens and the chi-squared curve becomes more bell-shaped and spread out.

- The larger the X^2 value, the greater the evidence against H_0: independence. Thus, the P-value equals the right-tail probability. It measures the probability that the X^2 statistic for a random sample would be larger than the observed value, if the variables are truly independent. Figure 10.4 depicts the P-value.

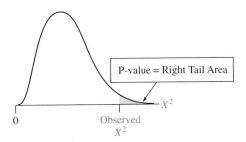

▲ **FIGURE 10.4: The P-value for the Chi-Squared Test of Independence.** This is the right-tail probability, above the observed value of the X^2 test statistic. **Question:** Why do we not also use the left tail in finding the P-value?

Table C at the back of the text lists values from the chi-squared distribution for various right-tail probabilities. These are X^2 test statistic values that have P-values equal to those probabilities. Table 10.7 shows an excerpt from Table C for small df values. For example, a 3×3 table has $df = 4$, for which Table 10.7 reports that an X^2 value of 9.49 has P-value = 0.05. In practice, software provides the P-value.

TABLE 10.7: Some Rows of Table C Displaying Chi-Squared Values

The values have right-tail probabilities between 0.250 and 0.001. For a table with $r = 3$ rows and $c = 3$ columns, $df = (r - 1)(c - 1) = 4$, and 9.49 is the chi-squared value with a right-tail probability of 0.05.

df	\multicolumn{7}{c}{RIGHT-TAIL PROBABILITY}
	.250	.100	.050	.025	.010	.005	.001
1	1.32	2.71	3.84	5.02	6.63	7.88	10.83
2	2.77	4.61	5.99	7.38	9.21	10.60	13.82
3	4.11	6.25	7.81	9.35	11.34	12.84	16.27
4	5.39	7.78	9.49	11.14	13.28	14.86	18.47

EXAMPLE 4

Are Happiness and Income Independent?

Picture the Scenario

For testing the null hypothesis of independence of happiness and family income, Table 10.6 in Example 3 reported a test statistic value of $X^2 = 73.4$.

Questions to Explore

a. What is the P-value for the chi-squared test of independence for these data?

b. State your decision for a significance level of 0.05 and interpret in context.

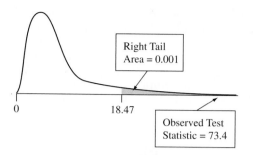

Think It Through

a. Since the table has $r = 3$ rows and $c = 3$ columns, $df = (r - 1)(c - 1) = 4$. In Table 10.7, for $df = 4$ the largest chi-squared value shown is 18.47. It has tail probability $= 0.001$ (see the margin figure). Since $X^2 = 73.4$ falls well above this, it has a smaller right-tail probability. Thus, the P-value is <0.001. The actual P-value would be 0 to many decimal places. MINITAB reports P-value $= 0.000$, as Table 10.6 shows.

b. Since the P-value is below 0.05, we can reject H_0. In the population, we conclude that an association exists between happiness and family income.

Insight

The extremely small P-value provides very strong evidence against H_0: independence. If the variables were independent, it would be highly unusual for a random sample to have this large a chi-squared statistic.

> **TO PRACTICE THIS CONCEPT, TRY EXERCISE 10.10**

How Large a Sample Size Is Needed for the Chi-Squared Test?

Presuming H_0 is true, the sampling distribution of the X^2 test statistic gets closer to the chi-squared distribution as the sample size increases. The approximation is good when each expected cell count exceeds about 5. Section 10.5 discusses the sample size issue further and presents a small-sample test.

The box summarizes the steps of the chi-squared test of independence.

The Five Steps of the Chi-Squared Test of Independence

1. Assumptions: Two categorical variables

Randomization, such as random sampling

Expected count ≥ 5 in all cells (otherwise, use small-sample test in Section 10.5)

2. Hypotheses:

H_0: The two variables are independent.

H_a: The two variables are dependent (associated).

3. Test statistic:

$$X^2 = \sum \frac{(\text{observed count} - \text{expected count})^2}{\text{expected count}},$$

where expected count $=$ (row total $\times$ column total)/total sample size

4. P-value: Right-tail probability above observed X^2 value, for the chi-squared distribution with $df = (r - 1)(c - 1)$

5. Conclusion: Report P-value and interpret in context. If a decision is needed, reject H_0 when P-value $\leq$ significance level (such as 0.05).

As with other inferential methods, the chi-squared test assumes randomization, such as simple random sampling which many surveys use. As explained in Section 4.4, the General Social Survey uses a rather complex multistage random sampling design. Its characteristics are similar, however, to those of a simple random sample. Inferences that assume simple random sampling, such as the chi-squared test, are routinely used with GSS data. The chi-squared test is also valid with randomized experiments. Then subjects are randomly assigned to different treatments, and the contingency table summarizes the subjects' responses on a categorical response variable. If the table compares groups, the random samples from those groups must be independent random samples. The test works best with large sample sizes.

IN PRACTICE: Chi-squared is also used as a "test of homogeneity"

The **chi-squared test** does not depend on which is the response variable and which is the explanatory variable (if either). The steps of the test and the results are identical either way. When a response variable *is* identified and the population conditional distributions are identical, they are said to be **homogeneous**. The chi-squared test of independence is then often referred to as a **test of homogeneity**.

Chi-Squared and the Test Comparing Proportions in 2 × 2 Tables

In practice, contingency tables of size 2 × 2 are very common. These occur in summarizing responses of two groups on a binary response variable, such as belief in life after death (yes, no). For convenience, here, as in the discussion of the binomial distribution in Section 6.2, we label the two possible outcomes by the generic labels *success* and *failure*.

Denote the population proportion of success by p_1 in group 1 and p_2 in group 2. Then $(1 - p_1)$ and $(1 - p_2)$ are the population proportions of failures. Table 10.8 displays the notation. The rows are the groups to be compared and the columns are the response categories.

TABLE 10.8: A 2 × 2 Table Compares Two Groups on a Binary Response Variable

The proportions in a row form a population conditional distribution.

GROUP	POPULATION PROPORTION		
	Success	Failure	Total
1	p_1	$1 - p_1$	1.0
2	p_2	$1 - p_2$	1.0

If the response variable is independent of the group, then $p_1 = p_2$, so the conditional distributions are identical. H_0: independence is, equivalently, the *homogeneity* hypothesis H_0: $p_1 = p_2$ of equality of population proportions.

Section 9.1 presented a z test of H_0: $p_1 = p_2$. The test statistic is the difference between the sample proportions divided by its standard error, that is,

$$z = (\hat{p}_1 - \hat{p}_2)/se_0,$$

where se_0 is the standard error estimated under the presumption that H_0 is true. In fact, using algebra, it can be shown that the chi-squared statistic for 2 × 2 tables is related to this z statistic by

$$X^2 = z^2.$$

Square the z statistic, and you get the X^2 statistic. For H_a: $p_1 \neq p_2$, the z test has exactly the same P-value as the chi-squared test.

EXAMPLE 5

Aspirin and Heart Attacks Revisited

Picture the Scenario

Examples 2–4 in Chapter 9 discussed the Physicians Health Study, in which doctors were randomly assigned to take aspirin or placebo regularly. That example analyzed a 2×2 contingency table that compared the proportions of heart attacks during a five-year period for those who took placebo and for those who took aspirin. Table 10.9 shows the data again with the expected cell counts, as part of MINITAB output for a chi-squared test of independence.

TABLE 10.9: Annotated MINITAB Output for Chi-Squared Test of Independence of Group (Placebo, Aspirin) and Whether Subject Had a Heart Attack

The same P-value results as with a two-sided z test comparing the two population proportions.

```
Rows: group     Columns: heart

                yes       no

placebo         189       10845  ← cell counts

                146       10888  ← expected cell counts

aspirin         104       10933

                147       10890

Cell Contents:            Count

                          Expected count

Pearson Chi-Square = 25.01, DF = 1, P-Value = 0.000
```

Denote the population proportion of heart attacks by p_1 for the placebo treatment and by p_2 for the aspirin treatment. For H_0: $p_1 = p_2$, the z test of Section 9.1 has a test statistic value of $z = 5.0$ (Exercise 9.8). Its P-value is 0.000 for H_a: $p_1 \neq p_2$.

Questions to Explore

a. What are the hypotheses for the chi-squared test for these data?

b. Report the test statistic and P-value for the chi-squared test. How do these relate to results from the z test comparing the proportions?

Think It Through

a. The null hypothesis is that whether a doctor has a heart attack is independent of whether he takes placebo or aspirin. This is equivalent to H_0: $p_1 = p_2$, the population proportion of heart attacks being the same for each group. The alternative hypothesis is that there's an association. This is equivalent to H_a: $p_1 \neq p_2$.

b. From Table 10.9, more doctors taking placebo had heart attacks than we would expect if the variables were independent (189 observed vs. 146 expected). Fewer doctors taking aspirin had heart attacks than we would expect if the variables were independent (104 observed vs. 147 expected). Table 10.9 reports $X^2 = 25.0$. This is the square of the z test statistic, $X^2 = z^2 = (5.0)^2 = 25.0$. Table 10.9 reports a chi-squared P-value of 0.000. This is very strong evidence that the population proportions of heart attacks differed for those taking aspirin and for those taking placebo. The sample proportions indicate that the aspirin group had a lower rate of heart attacks than the placebo group.

Insight

For 2×2 tables, $df = (r - 1)(c - 1) = (2 - 1)(2 - 1) = 1$. *Whenever a statistic has a standard normal distribution, the square of that statistic has a chi-squared distribution with $df = 1$.* The chi-squared test and the two sided z test necessarily have the same P-value and the same conclusion. Here, both the z and X^2 statistics show extremely strong evidence against the null hypothesis of equal population proportions. An advantage of the z test over X^2 is that it also can be used with one-sided alternative hypotheses. The direction of the effect is lost in squaring z and using X^2.

> **TO PRACTICE THIS CONCEPT, TRY EXERCISE 10.18**

For 2×2 tables, we don't really need the chi-squared test, because we can use the z test. Why then can't we use a z statistic to test independence for larger tables, for which $df > 1$? The reason is that a z statistic can be used only to compare a *single* estimate to a *single* null hypothesis value. Examples are a z statistic for comparing a sample proportion to a null hypothesis proportion, or a difference of sample proportions to a null hypothesis value of 0 for $p_1 - p_2$.

When a table is larger than 2×2 and thus $df > 1$, we need more than one difference parameter to describe the association. For instance, suppose Table 10.9 had three groups: placebo, regular-dose aspirin, and low-dose (children's) aspirin. Then the null hypothesis of independence corresponds to $p_1 = p_2 = p_3$, where p_3 is the population proportion of heart attacks for the low-dose aspirin group. The comparison parameters are $(p_1 - p_2)$, $(p_1 - p_3)$, and $(p_2 - p_3)$. We could use a z statistic for each comparison, but not a single z statistic for the overall test of independence.

	Heart Attack	
Group	Yes	No
Placebo	p_1	$1 - p_1$
Regular dose	p_2	$1 - p_2$
Low dose	p_3	$1 - p_3$

Interpretation of Degrees of Freedom

The *df* value in a chi-squared test indicates how many parameters are needed to determine all the comparisons for describing the contingency table. For instance, we've just seen that a 3×2 table for comparing three treatments (placebo, regular aspirin, low-dose aspirin) on whether a subject had a heart attack would have three parameters for making comparisons: $(p_1 - p_2)$, $(p_1 - p_3)$, and $(p_2 - p_3)$. We need to know only two of these to figure out the third. For instance, if we know $(p_1 - p_2)$ and $(p_2 - p_3)$, then

$$(p_1 - p_3) = (p_1 - p_2) + (p_2 - p_3).$$

A 3×2 table can use two parameters to determine all the comparisons, so its $df = 2$.

The *df* term also has the following interpretation: Given the row and column marginal totals in an $r \times c$ contingency table, the cell counts in a rectangular block of size $(r - 1) \times (c - 1)$ determine all the other cell counts. We illustrate using Table 10.10 from the GSS, which cross-tabulates political views by whether the subject would ever vote for a female for president. The table shows results for those subjects who called themselves extremely liberal, moderate, or extremely conservative.[1] For this 3×2 table, suppose we know the two counts in the upper left-hand $(3 - 1) \times (2 - 1) = 2 \times 1$ part of the table. Table 10.10 shows the

---
[1]Data from 1998 GSS, the most recent one to ask the question about a female president.

cell counts in this block. Given these and the row and column totals, we can determine all the other cell counts.

TABLE 10.10: Illustration of Degrees of Freedom

A block of $(r - 1)(c - 1)$ cell counts determine the others, given the marginal totals. Here, a $(3 - 1)(2 - 1) = 2 \times 1$ block of cell counts in the first two rows and first column determines the others.

VOTE FOR FEMALE PRESIDENT

POLITICAL VIEWS	Yes	No	TOTAL
Extremely liberal	42	2	44
Moderate	594	35	629
Extremely conservative	35	10	45
Total	671	47	718

	Vote female?		
Political	Yes	No	**Total**
Liberal	42	2	44
Moderate	594	35	629
Conserv.	35	10	45
Total	671	47	718

For instance, the marginal total in row 1 is 44 and the first cell count is 42, so the second cell has $44 - 42 = 2$ subjects. Likewise, there are

$629 - 594 = 35$ subjects in the second cell in the second row, and

$671 - (42 + 594) = 35$ subjects in the third cell in the first column.

From these and the fact that the third row has 45 observations, there must be $45 - 35 = 10$ in the final cell in the third row and second column. To check that you understand this, try reconstructing the table if you know only the cell counts in the two cells that are in the second and third rows and second column.

Once the marginal totals are fixed in a contingency table, a block of only $(r - 1) \times (c - 1)$ cell counts is free to vary, since these cell counts determine the remaining ones. The degrees of freedom value equals the number of cells in this block, or $df = (r - 1)(c - 1)$.

Limitations of the Chi-Squared Test

The chi-squared test of independence, like other significance tests, provides limited information. If the P-value is very small, strong evidence exists against the null hypothesis of independence. We can infer that the variables are associated. The chi-squared statistic and that P-value tell us nothing, however, about the nature or the strength of the association. We know there is *statistical* significance, but the test alone does not indicate whether there is *practical* significance as well. A large X^2 means *strong evidence* of association, but not necessarily a strong association. We'll see how to investigate the strength of association in the next section.

The chi-squared test is often misused. Some common misuses are applying it

- When some of the expected frequencies are too small.
- When separate rows or columns are *dependent* samples,[2] such as when each row of the table has the same subjects.
- To data that do not result from a random sample or randomized experiment.
- By classifying quantitative variables into categories. This results in a loss of information. It is usually more appropriate to analyze the data with methods for quantitative variables, such as the next chapter presents.

Recall

From Section 8.5, a small P-value does not imply an important result in practical terms. Statistical significance is not the same as *practical* significance. ■

---

[2]With dependent samples, McNemar's test (Section 9.4) is the appropriate test with binary variables, and extensions of it handle categorical variables with more than two categories.

ACTIVITY 2: Analyzing Contingency Tables of GSS Data

It's easy to analyze associations between categorical variables measured by the General Social Survey. Let's see how.

- Go to the GSS Web site, sda.berkeley.edu/GSS/. After selecting the most recent GSS data set, in the "select an action" menu choose the "frequencies and crosstabulation" option and click on START.

- The GSS names for Table 10.1 are HAPPY (for general happiness) and FINRELA (for family income in relative terms). Enter HAPPY in the column space and FINRELA in the row space to construct a table with these as the column and row variables. Entering YEAR (2002) in the "selection filter" restricts the search to GSS data from the survey in 2002.

- Put a check in the ROW box for the Percentaging option, and in the STATISTICS box for "table options," and click on RUN THE TABLE.

You'll now get a contingency table relating happiness to family income for the 2002 survey. The table shows the cell counts and sample conditional distributions. It treats as the response variable the one you check for the Percentaging option. The output also shows some statistics (such as Chisq(P) for the Pearson chi-squared statistic) that we're using in this chapter to analyze contingency tables.

The family income variable has five categories (far below average, below average, average, above average, far above average). Table 10.1 combined categories 1 and 2 into a single category (for "below average") and categories 4 and 5 into a single category (for "above average"). To create this collapsed table, type the family income variable as FINRELA(r:1-2;3; 4-5) instead of FINRELA.

Now, create a table relating another variable to happiness. At the page at which you "select an action," click on "open extra codebook window" and then "standard codebook" and you will see indexes for the variables. Look up a subject that interests you and find the GSS code name for a variable. For the table you create

- Find the chi-squared statistic and *df* value.
- Report the P-value and interpret results in context.

10.2 Practicing the Basics

10.8 What gives P-value = 0.05?: How large a X^2 test statistic value provides a P-value of 0.05 for testing independence for the following table dimensions?

- **a.** 2×2
- **b.** 2×3
- **c.** 2×5
- **d.** 5×5
- **e.** 3×9

10.9 Happiness and gender: For a 2×3 table on gender and happiness in Exercise 10.4 (shown again below), software tells us that $X^2 = 0.27$ and the P-value = 0.87. Show the five steps of the chi-squared test for these data.

	Happiness		
Gender	**Not**	**Pretty**	**Very**
Females	163	898	502
Males	130	705	379

10.10 Marital happiness and income: In Exercise 10.5 when you used the GSS to download a 3×3 table for marital happiness and family income in 2002, you should have obtained the table shown below.

	Marital happiness		
Income	**Not**	**Pretty**	**Very**
Above	6	43	75
Average	6	113	178
Below	6	57	117

- **a.** State the null and alternative hypotheses for the test.
- **b.** How large a X^2 value would give a P-value of exactly 0.05?
- **c.** The chi-squared statistic for the table equals 4.2. Find and interpret the P-value.
- **d.** State your decision for a significance level of 0.05, and interpret in context.

10.11 Life after death and gender: In the 2002 GSS, 425 of 563 males and 550 of 648 females indicated a belief in life after death.

- **a.** Construct a 2×2 contingency table relating gender (categories male and female) as the rows to belief about life after death (categories yes and no) as the columns. Check to make sure that the row totals are 563 and 648.
- **b.** Find the four expected cell counts for the chi-squared test. Compare them to the observed cell counts, identifying cells having more observations than expected.
- **c.** The data have $X^2 = 16.9$. Set up its calculation by showing how to substitute the observed and expected cell counts you found into its formula.

10.12 Basketball shots independent?: In pro basketball games during 1980–1982, when Larry Bird of the Boston Celtics missed his first free throw, 48 out of 53 times he made the second one, and when he made his first free throw, 251 out of 285 times he made the second one.

- **a.** Form a 2×2 contingency table that cross-tabulates the outcome of the first free throw (with categories made and missed) and the outcome of the second free throw (made and missed).

b. When we use MINITAB to analyze the contingency table, we get the result

```
Pearson Chi-Square = 0.273, DF = 1,
        P-Value = 0.602
```

Does it seem as if his success on the second shot depends on whether he made the first? Explain how to interpret the result of the chi-squared test.

10.13 Cigarettes and marijuana: The table below refers to a survey[3] in which senior high school students in Dayton, Ohio were randomly sampled. It cross-tabulates whether a student had ever smoked cigarettes and whether a student had ever used marijuana. Analyze these data by (a) finding and interpreting conditional distributions with marijuana use as the response variable, (b) reporting all five steps of the chi-squared test of independence.

	Marijuana	
Cigarettes	Yes	No
Yes	914	581
No	46	735

10.14 Smoking and alcohol: Refer to the previous exercise. A similar table relates cigarette use to alcohol use. Below is MINITAB output for the chi-squared test.

a. True or false? The chi-squared statistic and P-value shown refer to the column variable (alcohol use) being the response variable, and if we instead use cigarette use as the response variable we'll get different values.

b. Explain what value you would get for the z statistic and P-value if you conducted a significance test of H_0: $p_1 = p_2$ against H_a: $p_1 \neq p_2$, where p_1 is the population proportion of non-cigarette-users who have drunk alcohol and p_2 is the population proportion of cigarette users who have drunk alcohol.

Dayton student survey for Exercise 10.14

```
Row: cigarette Columns: alcohol
          no     yes
no       281     500
yes       46    1449
Pearson Chi-Square = 451.404,
DF = 1, P-Value = 0.000
```

10.15 Help the environment: In 2000 the GSS asked whether a subject is willing to accept cuts in the standard of living to help the environment (GRNSOL), with categories (vw = very willing, fw = fairly willing, nwu = neither willing nor unwilling, nvw = not very willing, nw = not at all willing). When this was cross-tabulated with gender (SEX), as shown below, $X^2 = 8.0$.

a. What are the hypotheses for the test to which X^2 refers?

b. Report r and c and the df value on which X^2 is based.

c. Is the P-value less than (i) 0.05? (ii) 0.10? Explain.

d. What conclusion would you make, using a significance level of (i) 0.05, (ii) 0.10? State your conclusion in the context of this study.

	Help the Environment				
Sex	vw	fw	nwu	nvw	nw
F	34	149	160	142	168
M	30	131	152	98	106

10.16 What predicts being green?: Refer to the previous exercise. What is associated with environment opinion? Go to the GSS Web site. sda.berkeley.edu/GSS/

a. Find a variable that is significantly associated with GRNSOL, using the 0.05 significance level. (*Hint:* You could first try some demographic variables, such as gender (variable SEX) or race (RACE).)

b. For the variable you found, in which cells are there considerably (i) more, (ii) fewer, subjects than you would expect if the variables were independent?

10.17 Aspirin and heart attacks: Exercise 9.6 in Chapter 9 reported a Swedish study in which heart attacks were suffered by 28 of the 684 subjects taking placebo and 18 of the 676 subjects taking aspirin.

a. Report the data in the form of a 2 × 2 contingency table.

b. Show how to carry out all five steps of the null hypothesis that having a heart attack is independent of whether one takes placebo or aspirin. (You should get a chi-squared statistic equal to 2.1.) Interpret.

10.18 z-test for heart attack study: Refer to the previous exercise. The printout for Exercise 9.6 in Chapter 9 reported

```
Test for difference = 0 (vs not = 0):
        Z = 1.46 P-Value = 0.144
```

a. Define population proportions p_1 and p_2 and state the hypotheses for that test.

b. Explain how the result of the chi-squared test in (b) in the previous exercise corresponds to this z test result.

10.19 Happiness and life after death: Are people happier who believe in life after death?

a. Go to the GSS Web site sda.berkeley.edu/GSS and download the contingency table for the 2002 survey relating happiness and whether you believe in life after death (variables denoted HAPPY and POSTLIFE). Construct the conditional distributions, using happiness as the response variable, and interpret.

b. Check the statistics facility in GSS to show that $X^2 = 0.97$. What does the GSS report for the P-value? Interpret.

c. Based on the test in (b), can you conclude that happiness and opinion about life after death are independent? That is, can you accept the null hypothesis? Explain.

10.20 What is independent of happiness?: Refer to Exercises 10.6 and 10.7. For the variables that you thought might be independent,

a. At the GSS Web site, conduct all five steps of the chi-squared test.

b. Based on (a), which inference is most appropriate?: (i) We accept the hypothesis that the variables are independent; (ii) The variables may be independent; (iii) The variables are associated.

---

[3]*Source*: Personal communication from Harry Khamis, Wright State University.

10.3 | How Strong Is the Association?

Three questions are normally addressed in analyzing contingency tables:

■ **Is there an association?** The chi-squared test of independence addresses this. When the P-value is small, we infer that the variables are associated.

■ **How do the cell counts differ from what independence predicts?** To answer this question, we compare each observed cell count to the corresponding expected cell count. We'll say more about this in Section 10.4.

■ **How strong is the association?** Analyzing the *strength* of the association reveals whether the association is an important one, or if it is statistically significant but weak and unimportant in practical terms. We now focus on this third question.

Measures of Association

For recent GSS data, Table 10.11 shows associations between opinion about the death penalty and two explanatory variables—gender and race. Both tables have large chi-squared statistics and small P-values for the test of independence. We can conclude that both gender and race are associated with the death penalty opinion.

TABLE 10.11: GSS Data (from 2000) Showing Race and Gender as Explanatory Variables for Opinion about the Death Penalty

	OPINION				OPINION		
RACE	Favor	Oppose	**N**	**GENDER**	Favor	Oppose	**N**
White	74%	26%	2050	Male	75%	25%	1149
Black	43%	57%	383	Female	63%	37%	1416
Chi-squared = 144.6				Chi-squared = 42.2			
$df = 1$, P-value = 0.000				$df = 1$, P-value = 0.000			

The X^2 test statistic is larger for the first table. Does this mean that opinion is more strongly associated with race than with gender? How can we quantify the strength of the association in a contingency table?

Measure of Association

A **measure of association** is a statistic or a parameter that summarizes the strength of the dependence between two variables.

A measure of association takes a range of values from one extreme to another as data range from the weakest to the strongest association. It is useful for *comparing* associations, to determine which is stronger. Later in this section we'll see that although the chi-squared test tells us how strong the *evidence* is that an association truly exists in the population, it does not describe the strength of that association.

Difference of Proportions

An easily interpretable measure of association is the *difference between the proportions* making a particular response. The population difference of proportions is 0 whenever the conditional distributions are identical, that is, when the variables are independent. The difference of proportions falls between -1 and $+1$, where 1 and -1 represent the strongest possible association. Whether you get a negative or a positive value merely reflects how the rows are ordered. Section 9.1 showed how to use a sample difference of proportions value to conduct inference (such as a confidence interval) about a population value.

Table 10.12 shows two hypothetical contingency tables relating a subject's income and whether the subject responds to a promotional mailing from a bank offering a special credit card. Case (a) exhibits the weakest possible association—*no* association. Both the high-income and the low-income subjects have 60% rejecting the credit card. The difference of proportions is 0. By contrast, case (b) exhibits the strongest possible association. All those with high income accept the card, whereas all those with low income decline it. The difference of proportions is −1.

TABLE 10.12: Decision about Accepting a Credit Card Cross-Tabulated by Income

In the first table, the conditional distributions are the same in each row, so the difference of proportions is 0. In the second table the acceptance decision is completely dependent on income level, and the difference of proportions is −1.

	(A) ACCEPT CREDIT CARD			(B) ACCEPT CREDIT CARD		
INCOME	No	Yes	TOTAL	No	Yes	TOTAL
High	240 (60%)	160 (40%)	400 (100%)	0 (0%)	400 (100%)	400 (100%)
Low	360 (60%)	240 (40%)	600 (100%)	600 (100%)	0 (0%)	600 (100%)

	Opinion	
Race	Favor	Oppose
White	74%	26%
Black	43%	57%
Gender		
Male	75%	25%
Female	63%	37%

In practice, we don't expect data to follow either of these extremes, but *the stronger the association, the larger the absolute value of the difference of proportions.* For Table 10.11, shown again in the margin, the difference of proportions is $0.74 - 0.43 = 0.31$ for whites and blacks, and the difference of proportions is $0.75 - 0.63 = 0.12$ for males and females. The difference of 0.31 for whites and blacks is much larger than the difference of 0.12 for males and females. If similar results hold in the entire population, then opinion about the death penalty is more strongly associated with race than with gender.

EXAMPLE 6

Do Student Stress and Depression Depend on Gender?

Picture the Scenario

Every year, the Higher Education Research Institute at UCLA conducts a large-scale survey of college freshmen on a variety of issues. From the survey of 283,000 freshmen in 2002, Table 10.13 compares females and males on the percent who reported feeling frequently stressed (overwhelmed by all they have to do) and the percent who reported feeling frequently depressed during the past year. P-values for chi-squared tests of independence were 0.000 (rounded) for each data set.

TABLE 10.13: Conditional Distributions of Stress and Depression, by Gender

	STRESS			DEPRESSION			
GENDER	Yes	No	TOTAL	GENDER	Yes	No	TOTAL
Female	35%	65%	100%	Female	8%	92%	100%
Male	16%	84%	100%	Male	6%	94%	100%

Question to Explore

Which response variable, stress or depression, has the stronger sample association with gender?

Think It Through

The difference of proportions between females and males was $0.35 - 0.16 = 0.19$ for feeling stressed. It was $0.08 - 0.06 = 0.02$ for feeling depressed. Since 0.19 is much larger than 0.02, there is a greater difference between reports of females and males on their feeling stress than on depression. In the sample, stress has the stronger association with gender.

Insight

Although the difference of proportions can be as large as 1 or −1, in practice in comparisons of groups, it is rare to find differences near these limits.

TO PRACTICE THIS CONCEPT, TRY EXERCISE 10.22

Recall

Section 9.3 introduced the **relative risk**. It showed that software can report a confidence interval for a population relative risk. ■

The Ratio of Proportions: Relative Risk

Another measure of association, introduced in Section 9.3, is the *ratio* of two proportions, p_1/p_2. In medical applications in which the proportion refers to an adverse outcome, it is called the **relative risk**.

EXAMPLE 7

Relative Risk for Seat Belt Use and Outcome of Auto Accidents

Picture the Scenario

Based on records of automobile accidents in a recent year, the Department

of Highway Safety and Motor Vehicles in Florida reported Table 10.14 on the counts of non-fatalities and fatalities. The data have $X^2 = 2338$, with P-value = 0.00000....

TABLE 10.14: Outcome of Auto Accident by Whether or Not Subject Wore Seat Belt

	OUTCOME		
Wore Seat Belt	Survived	Died	Total
Yes	412,368	510	412,878
No	162,527	1601	164,128

Question to Explore

Treating the auto accident outcome as the response variable, find and interpret the relative risk.

Think It Through

In Table 10.14, the adverse outcome is death, rather than survival. The relative risk is formed for that outcome. For those who wore a seat belt, the proportion who died equaled $510/412,878 = 0.00124$. For those who did not wear a seat belt, the proportion who died was $1601/164,128 = 0.00975$. The relative risk is the ratio, 0.00124/0.00975 = 0.127. The proportion of subjects wearing a seat belt who died was 0.127 times the proportion of subjects not wearing a seat belt who died.

Equivalently, since 0.00975/0.00124 = 1/0.127 = 7.9, the proportion of subjects *not wearing* a seat belt who died was nearly 8 times the proportion of subjects *wearing* seat belts who died. This reciprocal value is the relative risk when the rows are interchanged, with "no" for seat-belt use in row 1.

Insight

The ordering of rows is arbitrary. Many find it easier to interpret the relative risk for the ordering for which its value is *above* 1.0. A relative risk of 7.9 represents a strong association. This is far from the value of 1.0 that would occur if the proportion of deaths were the same for each group. Wearing a seat belt has a practically significant effect in enhancing the chance of surviving an auto accident.

TO PRACTICE THIS CONCEPT, TRY EXERCISES 10.25, PART (C), AND 10.28

In Example 7 the proportion of interest was close to 0 for each group, but the relative risk was far from 1. When this happens, an association may be practically important even though the difference of proportions is near 0 (it was 0.0085 in Example 7). If the chance of death is only a bit higher when one does not wear a seat belt, but that chance is nearly 8 times the chance of death when wearing a seat belt, the association is practically significant. The relative risk is often more informative than the difference of proportions for comparing proportions that are both close to 0.

Properties of the Relative Risk

- The relative risk can equal any nonnegative number.
- When $p_1 = p_2$, the variables are independent and relative risk = 1.0.
- Values farther from 1.0 (in either direction) represent stronger associations. Two values for the relative risk represent the same strength of association, but in opposite directions, when one value is the reciprocal of the other. See Figure 10.5.

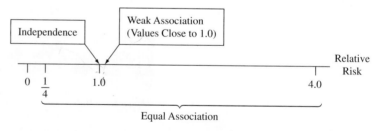

▲ **FIGURE 10.5: Values of the Relative Risk Farther from 1.0 Represent Stronger Association. Question:** Why do relative risks that take reciprocal values, such as 4 and $\frac{1}{4}$, represent the same strength of association?

A relative risk of 8 is farther from independence and represents a stronger association than a relative risk of 4, and a relative risk of 1/8 is farther from independence than a relative risk of $\frac{1}{4}$. But relative risk = 4.0 and relative risk = 1/4.0 = 0.25 represent the same strength of association. When relative risk $p_1/p_2 = 0.25$, then p_1 is 0.25 times p_2. Equivalently, p_2 is 1/0.25 = 4.0 times p_1. Whether we get 4.0 or 0.25 for the relative risk depends merely on which group we call group 1 and which we call group 2.

Association in $r \times c$ Tables

The difference of proportions and the relative risk are defined for 2×2 tables. For $r \times c$ tables, they can compare proportions in a particular response category for various pairs of rows.

EXAMPLE 8

General Happiness and Marital Happiness

Picture the Scenario

Earlier in the chapter we studied the association between happiness and family income. For those subjects who are married, a possible predictor of happiness is their reported happiness with their marriage. Table 10.15 cross-tabulates these variables, using data from the 2002 GSS. It also reports conditional distributions with happiness as the response variable. The data have chi-squared statistic $X^2 = 164.2$, with $df = 4$ and a P-value of 0.0000....

TABLE 10.15: Happiness and Marital Happiness

The percentages are the conditional distributions of happiness in each row. When percentages are compared in rows 1 and 3, the highlighted cells reveal a strong association.

MARITAL HAPPINESS	HAPPINESS			
	Not too Happy	Pretty Happy	Very Happy	TOTAL
Not too Happy	7 (38.9%)	9 (50.0%)	2 (11.1%)	18 (100%)
Pretty Happy	17 (7.9%)	173 (80.5%)	25 (11.6%)	215 (100%)
Very Happy	14 (3.8%)	136 (36.8%)	220 (59.5%)	370 (100%)

Question to Explore

Describe the association between marital happiness and general happiness, using the "not too happy" and "very happy" categories.

Think It Through

Let's consider the "Very happy" category of happiness. The difference between the proportions in this category for the "Not too happy" and "Very happy" categories of marital happiness (that is, rows 1 and 3) is $0.111 - 0.595 = -0.48$. Those who are not too happy with their marriage are much less likely to be very happy overall. Consider next the "Not too happy" category of happiness. The difference between the proportions in this category for the "Not too happy" and "Very happy" categories of marital happiness is $0.389 - 0.038 = 0.35$. This is also a substantial difference.

Likewise, we could use the *ratio* of proportions. For instance, consider the "Not too happy" category of happiness as an undesirable outcome. The relative risk of this outcome, comparing the "Not too happy" and "Very happy" categories of marital happiness, is $0.389/0.038 = 10.2$. The estimated proportion who are not too happy is about 10 times larger for those who have "not too happy" marriages than for those who have "very happy" marriages. This ratio of 10.2 is very far from 1.0, indicating that this part of the table reveals an extremely strong association.

Recall

From Table 10.1,

Income	Not too happy	n
Below average	94	426
Above average	21	290 ■

Insight

It is unusual to find such strong associations. For instance, family income also has a substantial association with happiness, but it is not as strong. Using Table 10.1 (partly shown in the margin), you can check that the relative risk of being "Not too happy" equals 3.0 when you compare the below average and above average levels of income. This is much less than the relative risk of 10.2 and is a weaker association.

> **TO PRACTICE THIS CONCEPT, TRY EXERCISES 10.24 AND 10.25**

Chi-Squared and Association

You may be tempted to regard a large value of the X^2 statistic in the test of independence as indicating a strong association. This is a misuse of chi-squared.

> **IN PRACTICE: Large X^2 does not mean there's a strong association**
>
> A large value for X^2 provides strong evidence that the variables are associated. It does not imply, however, that the variables have a strong association. This statistic merely indicates (through its P-value) *how certain* we can be that the variables are associated, not how *strong* that association is. For a given strength of association, larger X^2 values occur for larger sample sizes. Large X^2 values can occur with weak associations, if the sample size is sufficiently large.

For example, cases A, B, and C in Table 10.16 are hypothetical tables relating gender and whether or not one attends religious services weekly. The association in each case is weak—the conditional distribution for females (51% yes, 49% no) is nearly identical to the conditional distribution for males (49% yes, 51% no). All three cases show the same weak strength of association: The difference between the proportions of females and males who say "yes" is $0.51 - 0.49 = 0.02$ in each.

TABLE 10.16: Response on Attending Religious Services Weekly by Gender, Showing Weak but Identical Associations

In each case the difference between proportions is small, only 0.02.

A.	Yes	No	n	B.	Yes	No	n	C.	Yes	No	n
Female	51%	49%	100		51%	49%	200		51%	49%	10,000
Male	49%	51%	100		49%	51%	200		49%	51%	10,000
Chi-squared = 0.08				Chi-squared = 0.16				Chi-squared = 8.0			
P-value = 0.78				P-value = 0.69				P-value = 0.005			

For the sample of size 200 in case A, $X^2 = 0.08$, which has a P-value of 0.78. Case B has twice as large a sample size, with the cell counts doubling. For its sample of size 400, $X^2 = 0.16$, for which the P-value $= 0.69$. So, when the cell counts double, X^2 doubles. Similarly, for the sample size of 20,000 (100 times as large as $n = 200$) in case C, $X^2 = 8.0$ (100 times as large as $X^2 = 0.08$). Then the P-value $= 0.005$.

In summary, for fixed conditional distributions, the value of X^2 is directly proportional to the sample size—larger values occur with larger sample sizes. Like other test statistics, the larger the X^2 statistic, the smaller the P-value and the stronger the evidence against the null hypothesis. However, a small P-value results from even a weak association when the sample size is large, as case C shows.

10.21 Party ID and race: The table shows 2000 GSS data on race and political party identification.

PARTY IDENTIFICATION

Race	Democrat	Independent	Republican	*n*
Black	251	151	23	425
White	615	931	642	2188

a. The chi-squared test of independence has $X^2 = 187.2$. Carry out the five steps of the test, and interpret in the context of these variables.

b. Estimate the difference between black subjects and white subjects in the proportion who identify themselves as Democrats. Interpret. Does this seem to be a weak (practically insignificant), or a strong (practically significant), association? Explain.

10.22 Party ID, race and gender: Refer to the previous exercise and to the data in the table below from Exercise 10.1. Which has a stronger association with whether one identifies as a Democrat—race or gender? Justify your answer by comparing the difference between blacks and whites in the proportion identifying as Democrat to the difference between females and males in the proportion identifying as Democrat.

Data from Exercise 10.1

Gender	Democrat	*n*
Female	567	1496
Male	356	1185

10.23 Happiness and highest degree: The table shows 2002 GSS data on happiness and the highest degree attained.

HIGHEST DEGREE	HAPPINESS Not too Happy	Pretty Happy	Very Happy
< High school	36	113	54
HS, junior college	105	482	237
College, grad school	29	188	124

a. The chi-squared test of independence has $X^2 = 15.2$. What conclusion would you make using a significance level of 0.05? Interpret.

b. Does this large chi-squared value mean there is a strong association between happiness and highest degree? Explain.

c. Estimate the difference between the lowest and highest education groups in the proportion who report being not too happy. Interpret.

d. Find and interpret the relative risk of being not too happy, comparing the lowest and highest education groups. Interpret.

10.24 Income and highest degree: The table shows 2002 GSS data on family income and the subject's highest degree attained, for which $X^2 = 158.0$ (P-value = 0.000).

GSS data for Exercise 10.24

HIGHEST DEGREE	FAMILY INCOME Below average	Average	Above average
< High school	101	85	14
HS, junior college	274	414	133
College, grad school	50	147	144

a. Estimate the difference between the lowest and highest education groups in the proportion who report (i) below average income, (ii) above average income. Interpret. Does this seem to be a practically significant, or practically insignificant, association?

b. Find and interpret the relative risk of a below average income, comparing the lowest and highest education groups. Interpret.

10.25 Smoking and alcohol: The table refers to a survey[4] of senior high school students in Dayton, Ohio. It cross-tabulates whether a student had ever smoked cigarettes and whether a student had ever drunk alcohol. The table also shows MINITAB output of conditional distributions.

a. Describe the strength of association using the difference between users and nonusers of cigarettes in the proportions who have used alcohol. Interpret.

b. Describe the strength of association using the difference between users and nonusers of alcohol in the proportions who have used cigarettes. Interpret.

c. Describe the strength of association using the relative risk of using cigarettes, comparing those who have used or not used alcohol. Interpret.

Data from Dayton student survey for Exercise 10.25

```
Rows:  cigarette      Columns:  alcohol

             no        yes

no          281        500

          35.98      64.02        ◄── For row conditional distribution

          85.93      25.65        ◄── For column conditional distribution

yes          46       1449

           3.08      96.92

          14.07      74.35

Cell Contents:           Count
                         % of Row
                         % of Column
```

10.26 Sex of victim and offender: For murders in the United States in 2002, the table cross-tabulates the sex of the victim by the sex of the offender, using data from

---

[4]*Source*: Harry Khamis, Wright State University.

www.fbi.gov. Find and interpret a measure of association, treating sex of victim as the response variable.

Data for Exercise 10.26

SEX OF OFFENDER	SEX OF VICTIM	
	Female	Male
Female	183	528
Male	1778	4328

10.27 Prison and gender: According to the U.S. Department of Justice, in 2002 the incarceration rate in the nation's prisons was 832 per 100,000 male residents, and 58 per 100,000 female residents.

a. Find the relative risk of being incarcerated, comparing males to females. Interpret.

b. Find the difference of proportions of being incarcerated. Interpret.

c. Which measure do you think is more appropriate for these data? Why?

10.28 Homicide victims by race: According to recent data

TRY from the U.S. National Center for Health Statistics, the annual probability that a male between the ages of 20 and 24 is a homicide victim is about 0.00164 for blacks and 0.00015 for whites.

a. Compare these rates using the difference of proportions, and interpret.

b. Compare these rates using the relative risk, and interpret.

c. Which of the two measures seems more useful when both proportions are very close to 0? Explain.

10.29 Death penalty associations: Table 10.11, summarized again here, showed the associations between death penalty opinion and gender and race.

Race	Opinion	
	Favor	Oppose
White	74%	26%
Black	43%	57%
Gender		
Male	75%	25%
Female	63%	37%

a. True or false? The table with the larger X^2 statistic necessarily has the stronger association. Explain.

b. To make an inference about the strength of association in the population, you can construct confidence intervals around the sample differences of proportions. The 95% confidence intervals are (0.26, 0.37) comparing whites and blacks and (0.08, 0.15) comparing males and females in the proportions favoring the death penalty. In the *population*, can you make a conclusion about which variable is more strongly associated with the death penalty opinion? Explain.

10.30 Chi-squared vs. measuring association: For Table 10.15 on general happiness and marital happiness, the chi-squared statistic equals 164.2 ($df = 4$, P-value = 0.000). Explain the difference between the purpose of the chi-squared test and the descriptive analysis in Example 8 comparing conditional distributions using measures of association. (*Hint*: Is chi-squared a descriptive, or inferential, analysis?)

10.4 | How Can Residuals Reveal the Pattern of Association?

The chi-squared test and measures of association such as $p_1 - p_2$ and p_1/p_2 are fundamental methods for analyzing contingency tables. The P-value for X^2 summarizes the strength of evidence against H_0: independence. If the P-value is small, then we conclude that *somewhere* in the contingency table the population cell proportions differ from independence. The chi-squared test does not indicate, however, whether *all* cells deviate greatly from independence or perhaps only some of them do so.

Recall

Section 3.3 introduced a **residual** for quantitative variables as the difference between an observed response and a predicted response for a regression equation. For categorical variables, a residual compares an observed count to an expected count that represents what H_0 predicts. ■

Residual Analysis

A cell-by-cell comparison of the observed counts with the counts that are expected when H_0 is true reveals the nature of the evidence. The difference between an observed and expected count in a particular cell is called a **residual**.

For example, Table 10.5 reported observed and expected cell counts for happiness and family income. The first column of this table is summarized in the margin, with the expected cell counts in parentheses. The first cell reports 21 subjects with above average income who are not too happy. The expected

Income	Not too happy
Above average	21 (35.8)
Average	53 (79.7)
Below average	94 (52.5)

count is 35.8. The residual is $21 - 35.8 = -14.8$. The residual is *negative* when, as in this cell, *fewer* subjects are in the cell than expected under H_0. The residual is *positive* when *more* subjects are in the cell than expected, as in the cell having observed count = 94 (below average income subjects who are not too happy) and residual $94 - 52.5 = 41.5$.

How do we know whether a residual is large enough to indicate strong evidence of a deviation from independence in that cell? To answer this, we use an *adjusted* form of the residual that behaves like a z-score. It is called the **standardized residual**.

Standardized Residual

The **standardized residual** for a cell equals

$$(\text{observed count} - \text{expected count})/se.$$

Here, se denotes a standard error for the sampling distribution of (observed count − expected count), when the variables are independent.

---

observed count − expected

oc − exo

A standardized residual reports the number of standard errors that an observed count falls from its expected count. The se describes how much the (observed − expected) difference would tend to vary in repeated random sampling if the variables were independent. Its formula is complex, and we'll not give it here. Software can easily find standardized residuals for us.

When H_0: independence is true, the standardized residuals have approximately a standard normal distribution: They fluctuate around a mean of 0, with a standard deviation of 1. There is about a 5% chance that any particular standardized residual exceeds 2 in absolute value, and less than a 1% chance that it exceeds 3 in absolute value. See the margin figure. A large standardized residual provides evidence against independence in that cell. A value exceeding about 3 in absolute value provides strong evidence. A value above $+3$ suggests that the population proportion for that cell is *higher* than what independence predicts. A value below -3 indicates that the population proportion for that cell is *lower* than what independence predicts.

When we inspect many cells in a table, some standardized residuals could be large just by random variation. Values below -3 or above $+3$, however, are quite convincing evidence of a true effect in that cell.

Area = about 95%

Standardized Residuals

EXAMPLE 9

Standardized Residuals for Religiosity and Gender

Picture the Scenario

Table 10.16 near the end of the previous section showed some artificial tables comparing females and males on their religious attendance. Let's look at some actual data on religiosity. Table 10.17 displays observed and expected cell counts and the standardized residuals for the association between gender and response to the question, "To what extent do you consider yourself a religious person?" The possible responses were (very religious, moderately religious, slightly religious, not religious at all). The data are from the 1998 GSS, the most recent time the religiosity question was asked. The table is in the form provided by MINITAB (which actually calls the standardized residuals "adjusted residuals").

TABLE 10.17: Religiosity by Gender, with Expected Counts and Standardized Residuals

Large *positive* standardized residuals (in green) indicate strong evidence of a *higher* population cell proportion than expected under independence. Large *negative* standardized residuals (in red) indicate strong evidence of a *lower* population proportion than expected.

Rows: gender   Columns: religiosity

	Very	Moderately	Slightly	Not	All
Female	170	340	174	95	779 ← Counts
	146.3	330.8	182.9	119.0	← Expected counts
	3.226	0.988	−1.114	−3.548	← Standardized residuals
Male	98	266	161	123	648
	121.7	275.2	152.1	99.0	
	−3.226	−0.988	1.114	3.548	
All	268	606	335	218	1427

Cell Contents: Count
                  Expected count
                  Adjusted residual ← This is what **Minitab** calls standardized residuals.

Questions to Explore

a. How would you interpret the standardized residual of 3.226 in the first cell?

b. Interpret the standardized residuals in the entire table.

Think It Through

a. For the cell for females who are very religious, observed count = 170, expected count = 146.3, and the standardized residual = 3.226. The difference between the observed and expected counts is more than three standard errors. Since the standardized residual exceeds 3.0 in absolute value, this cell shows a much greater discrepancy than we'd expect if the variables were truly independent. There's strong evidence that the population proportion for that cell (female and very religious) is *higher* than independence predicts.

b. Table 10.17 exhibits large positive residuals above 3.0 in two cells: females who are very religious and males who are not at all religious. These are the two cells in which the observed count is much *larger* than the expected count. Those cells had strong evidence of more frequent occurrence than if religiosity and gender were independent. The table exhibits large negative residuals for the cells for females who are not at all religious and males who are very religious. In these cells, the observed count is much *smaller* than the expected count. There's strong evidence that the population has fewer subjects in these cells than if the variables were independent.

Insight

The standardized residuals help to describe the pattern of this association: Compared to what we'd expect if religiosity and gender were independent, there are more females who are very religious and more males who are not at all religious.

TO PRACTICE THIS CONCEPT, TRY EXERCISE 10.31

10.4 Practicing the Basics

10.31 Standardized residuals for happiness and income:
TRY The table displays the observed and expected cell counts and the standardized residuals for testing independence of happiness and family income, for GSS data.

Happiness by family income, with expected counts and standardized residuals

```
Rows: income   Columns: happiness
          not    pretty    very    All
above      21       159     110    290
         35.8     166.1    88.1
        -2.973   -0.947   3.144
average    53       372     221    646
         79.7     370.0   196.4
        -4.403    0.224   2.907
below      94       249      83    426
         52.5     244.0   129.5
         7.368    0.595  -5.907
All       168       780     414   1362
Cell Contents:  Count
                Expected count
                Adjusted residual
```

a. How would you interpret the standardized residual of 3.144?

b. Interpret the standardized residuals highlighted in green.

c. Interpret the standardized residuals highlighted in red.

10.32 Happiness and religious attendance: The table shows MINITAB output for data from the 2000 GSS on happiness and frequency of attending religious services ($1 = $ at most several times a year, $2 = $ once a month to nearly every week, $3 = $ every week to several times a week).

```
Rows: religion   Columns: happiness
   Not too happy  Pretty happy  Very happy
1      189            908          382
      3.899          4.393       -7.252
2       53            311          180
     -0.788         -0.227        0.763
3       46            335          294
     -3.750         -4.838        7.625
Cell Contents: Count
               Adjusted residual
Pearson Chi-Square = 73.460, DF = 4,
P-Value = 0.000
```

a. Based on the chi-squared statistic and P-value, give a conclusion about the association between the variables.

b. The numbers below the counts in the table are standardized residuals. Which cells have strong evidence that in the population there are more subjects than if the variables were independent?

c. Which cells have strong evidence that in the population there are fewer subjects than if the variables were independent?

10.33 Marital happiness and general happiness: Table 10.15 showed the association between general happiness and marital happiness. The table shown here gives the standardized residuals for those data, in parentheses.

a. Explain what a small standardized residual such as -0.2 in the second cell represents.

b. Identify the cells in which you'd infer that the population has more cases than would occur if happiness and marital happiness were independent.

MARITAL		HAPPINESS	
HAPPINESS	Not too Happy	Pretty Happy	Very Happy
Not too Happy	7 (5.8)	9 (−0.2)	2 (−2.6)
Pretty Happy	17 (1.2)	173 (10.1)	25 (−10.9)
Very Happy	14 (−3.2)	136 (−9.9)	220 (11.6)

10.34 Happiness and marital status: For the 2000 GSS, the table shows cell counts and standardized residuals (in parentheses) for happiness and marital status. Summarize the association by indicating which marital statuses have strong evidence of (i) more, (ii) fewer people in the population in the "very happy" category than if the variables were independent.

Marital		Happiness	
Status	Very Happy	Pretty Happy	Not too Happy
Married	555	650	67
	(12.4)	(−6.5)	(−8.3)
Widowed	66	161	42
	(−2.7)	(0.7)	(2.8)
Divorced	93	276	65
	(−5.0)	(2.7)	(3.3)
Separated	17	65	26
	(−3.6)	(0.5)	(4.7)
Never Married	150	450	93
	(−6.6)	(4.4)	(2.8)

10.35 Gender gap?: The table in Exercise 10.1 on gender and party identification is shown again below. The standardized residuals that exceeded 2.0 in absolute value were 4.25 for females who identified as Democrats, −4.25 for males who identified as Democrats, −2.70 for females who identified as Republicans, and 2.70 for males who identified as Republicans. Interpret.

Sex	Political Party		
	Dem	Indep	Repub
F	567	534	395
M	356	460	369

■ 10.36 Ideology and political party: Go to the GSS web site sda.berkeley.edu/GSS/. For the 2002 survey, construct the 7 × 8 contingency table relating political ideology (POLVIEWS) and party identification (PARTYID).

a. Summarize results of carrying out the chi-squared test.

b. Find the standardized residuals (by checking "Show *z* statistic"). What do you learn from them that you do not learn from the chi-squared test?

10.5 | What if the Sample Size Is Small? Fisher's Exact Test

The chi-squared test of independence, like one- and two-sample z tests for proportions, is a large-sample test. When the expected frequencies are small, any of them being less than about 5, small-sample tests are more appropriate than the chi-squared test. For 2 × 2 contingency tables, **Fisher's exact test** is a small-sample test of independence.

Fisher's exact test of the null hypothesis that two variables are independent was proposed by the eminent British statistician and geneticist, Sir Ronald Fisher. (If you've surmised by now that the British have had a major impact on how we perform modern statistics, you are correct.) The test is called *exact* because it uses a sampling distribution that gives exact probability calculations rather than the approximate ones that use the chi-squared distribution.

The calculations for Fisher's exact test are complex and beyond the scope of this text. The principle behind the test is straightforward (see Exercise 10.73), however, and statistical software provides its P-value. The smaller the P-value, the stronger is the evidence that the variables are associated.

EXAMPLE 10

Tea Tastes Better with Milk Poured First?

Picture the Scenario

In introducing his small-sample test of independence in 1935, Fisher described the following experiment: Fisher was fond of the British tradition of taking a break in the afternoon for tea. One day, his colleague Dr. Muriel Bristol claimed that when drinking tea she could tell whether the milk or the tea had been added to the cup first (she preferred milk first). To test her claim, Fisher asked her to taste eight cups of tea, four of which had the milk added first and four of which had the tea added first. She was told there were four cups of each type and was asked to indicate which four had the milk added first. The order of presenting the cups to her was randomized.

Table 10.18 shows the result of the experiment. Dr. Bristol predicted three of the four cups correctly that had milk poured first.

TABLE 10.18: Result of Tea-Tasting Experiment

The table cross-tabulates what was actually poured first (milk or tea) by what Dr. Bristol predicted was poured first. She had to indicate which four of the eight cups had the milk poured first.

ACTUAL	PREDICTION		
	Milk	Tea	Total
Milk	3	1	4
Tea	1	3	4
Total	4	4	8

For those cases where milk was poured first (row 1), denote the probability of guessing that milk was added first by p_1. For those cases where tea was poured first, denote the probability of guessing that milk was added first by p_2. If she really could predict when milk was poured first better than with random guessing, then $p_1 > p_2$. So we shall test H_0: $p_1 = p_2$ (prediction independent of actual pouring order) against H_a: $p_1 > p_2$.

Questions to Explore

a. Show the sample space of the five possible contingency table outcomes that could have occurred for this experiment.

b. Table 10.19 shows the result of using SPSS software to conduct the chi-squared test of the null hypothesis that her predictions were independent of the actual order of pouring. Is this test and its P-value appropriate for these data? Why or why not?

TABLE 10.19: Result of Fisher's Exact Test for Tea-Tasting Experiment

The chi-squared P-value is listed under "Asymp. Sig." and the Fisher's exact test P-values are listed under "Exact Sig." "Sig" is short for *significance* and "asymp." is short for *asymptotic*.

	Value	df	Asymp. Sig. (2-sided)	Exact Sig. (2-sided)	Exact Sig. (1-sided)
Pearson Chi-Square	2.000	1	.157		
Fisher's Exact Test				.486	.243

4 cells (100.0%) have expected count less than 5.

c. Table 10.19 also shows the result for Fisher's exact test of the same null hypothesis. The one-sided version of the test pertains to the alternative that her predictions are *better* than random guessing. How does the one-sided P-value relate to the probabilities for the five tables in (a)? Does this P-value suggest that she had the ability to predict better than random guessing?

IN WORDS

In statistics, **asymptotic** means "large sample." Mathematically, the theory behind large-sample statistical methods applies as the sample size grows toward infinity "asymptotically."

Think It Through

a. The possible sample tables have four observations in each row, because there were four cups with milk poured first and four cups with tea poured first. The tables also have four observations in each column, because Dr. Bristol was told there were four cups of each type and was asked to predict which four had the milk added first. Of the four cups where milk was poured first (row 1 of the table), the number she could predict correctly is 4, 3, 2, 1, or 0. These outcomes correspond to the sample tables:

	Guess Poured First									
Poured First	Milk	Tea	Milk	Tea	Milk	Tea	Milk	Tea	Milk	Tea
Milk	4	0	3	1	2	2	1	3	0	4
Tea	0	4	1	3	2	2	3	1	4	0

These are the five possible tables with totals of 4 in each row and 4 in each column.

b. The cell counts in Table 10.18 are small. Each cell has expected count $(4 \times 4)/8 = 2$. Since they are less than 5, the chi-squared test is not appropriate. In fact, Table 10.19 warns us that all four cells have expected counts less than 5. The table reports $X^2 = 2.0$ with P-value $= 0.157$, but software provides results whether or not they are valid, and it's up to us to know when we can use a method.

c. The one-sided P-value reported by SPSS is for the alternative $H_a: p_1 > p_2$. It equals 0.243. This is the probability, presuming H_0 to be true, of the observed table (Table 10.18) and the more extreme table giving even more evidence in favor of $p_1 > p_2$. That more extreme table is the one with counts $(4, 0)$ in the first row and $(0, 4)$ in the second row, corresponding to guessing all four cups correctly. The P-value of 0.243 does not give much evidence against the null hypothesis.

Insight

This experiment did not establish an association between the actual order of pouring and Dr. Bristol's predictions. If she had predicted all four cups correctly, the one-sided P-value would have been 0.014. We might then believe her claim. With a larger sample, we would not need such extreme results to get a small P-value. According to a biography of Fisher by one of his daughters, in reality Dr. Bristol did eventually convince Fisher of her tea-tasting predictive ability.

TO PRACTICE THIS CONCEPT, TRY EXERCISE 10.39

Recall

Expected cell count = (row total × column total) divided by total sample size. ■

	Guess		
Actual	Milk	Tea	**Total**
Milk	3	—	4
Tea	—	—	4
Total	4	4	8

In part (a) above, notice that for the given row and column marginal totals of four each, the count in the first cell determines the other three cell counts. For example, for the actual data (Table 10.18), knowing she guessed three of the four cups correctly that actually had milk poured first determines the other three cells (see the margin). Because the count in the first cell determines the others, that cell count is regarded as the test statistic for Fisher's exact test. The P-value finds the probability of that count and of the other possible counts for that cell that provide even more evidence in support of H_a.

Summary of Fisher's Exact Test of Independence for 2 × 2 Tables

1. Assumptions:

Two binary categorical variables

Randomization, such as random sampling

2. Hypotheses:

H_0: The two variables are independent (H_0: $p_1 = p_2$)

H_a: The two variables are associated

$$(H_a: p_1 \neq p_2 \text{ or } H_a: p_1 > p_2 \text{ or } H_a: p_1 < p_2).$$

3. Test statistic: First cell count (this determines the others, given the margins).

4. P-value: Probability that the first cell count equals the observed value or a value even more extreme than observed in the direction predicted by H_a.

5. Conclusion: Report P-value and interpret in context. If a decision is needed, reject H_0 when P-value ≤ significance level (such as 0.05).

The main use of Fisher's exact test is for small sample sizes, because the chi-squared test is not then valid. However, Fisher's exact test can be used with *any* sample sizes. It also extends to tables of arbitrary size $r \times c$. The computations are complex, and unfortunately most software currently can perform it only for 2 × 2 tables.

Using the Ordering of Categories

Some categorical variables have a natural ordering of the categories. There is a "low" end and a "high" end of the categorical scale. Examples are education attained (less than high school, high school, some college, college degree, advanced degree), appraisal of a company's inventory level (too low, about right, too high), and the happiness and income variables used in this chapter. Such variables are called **ordinal** variables. The chi-squared test of independence treats the categories as unordered. It takes the same value no matter in what order the categories are listed in the table. This is because with any reordering of the categories, the set of row and column totals does not change, and so the expected frequency does not change that goes with a particular observed cell count.

When there is only a weak association, tests that use the ordering information are often more powerful than chi-squared, giving smaller P-values and stronger evidence against H_0. One way to do this treats the two categorical variables as *quantitative*: Assign scores to the rows and to the columns, and then use the *correlation* to describe the strength of the association. A significance test can be based on the size of the correlation. This is beyond our scope here, but the next chapter presents inference methods for two quantitative variables.

10.37 Keeping old dogs mentally sharp: In an experiment with 7–11 year-old beagles, the dogs had to try to learn how to find a treat under a certain black-colored block and then relearn that task with a white-colored block. The control group of dogs received standard care and diet. The diet and exercise group were given dog food fortified with vegetables and citrus pulp and vitamin E and C supplements plus extra exercise and social play. All 12 dogs in the diet and exercise group were able to solve the entire task, but only 2 of the 8 dogs in the control group could do so (*Source:* N. W. Milgram et al., *Neurobiology of Aging*, vol. 26, pp. 77–90, 2005).

    a. Show how to summarize the results in a contingency table.

    b. Conduct all steps of Fisher's exact test of the hypothesis that whether a dog can solve the task is independent of the treatment group. Use the two-sided alternative hypothesis. Interpret.

    c. Why is it improper to conduct the chi-squared test for these data?

10.38 Tea-tasting results: Consider the tea-tasting experiment of Example 10 and Table 10.18. Consider the possible sample table in which all four of her predictions about the cups that had milk poured first are correct. Using software, find the P-value for the one-sided alternative. Interpret the P-value.

10.39 Claritin and nervousness: An advertisement by Schering Corporation in 1999 for the allergy drug Claritin mentioned that in a pediatric randomized clinical trial, symptoms of nervousness were shown by 4 of 188 patients on Claritin and 2 of 262 patients taking placebo. Denote the population proportion who would show such symptoms by p_1 for Claritin and by p_2 for placebo. The computer printout shows results of significance tests for H_0: $p_1 = p_2$.

    a. Report the P-value for the small-sample test, with H_a: $p_1 \neq p_2$. Interpret in the context of this study.

    b. Is it appropriate to conduct the chi-squared test for these data? Why or why not?

Analyses of Claritin data for Exercise 10.39

```
Rows: treatment Columns: nervousness
               yes      no
Claritin        4       184
Placebo         2       260


Statistic                        P-Value
Fisher's Exact Test (2-Tail)  0.24
Chi-squared = 1.55               0.21
```

10.40 AIDS and condom use: Chatterjee et al. (1995, p. 132) described a study about the effect of condoms in reducing the spread of AIDS. This two-year Italian study followed heterosexual couples where one partner was infected with the HIV virus. Of 171 couples who always used condoms, 3 partners became infected with HIV, while of 55 couples who did not always use condoms, 8 partners became infected. Test whether the rates are significantly different,

    a. Define p_1 and p_2 in this context, and specify the null and two-sided alternative hypotheses.

    b. For these data, software reports

```
Pearson Chi-Square = 14.704, DF = 1,
P-Value = 0.0001

* Note * 1 cell with expected count less than 5

Fisher's exact test: P-Value = 0.0007
```

Report the result of the test that you feel is most appropriate for these data. For the test you chose, report the P-value and interpret in the context of this study.

CHAPTER REVIEW

Answers to the Chapter Figure Questions

Figure 10.1: The higher-income subjects were more likely to be very happy and less likely to be not too happy.

Figure 10.2: The height of the bars is the same for males and females at each level of happiness.

Figure 10.3: The chi-squared statistic can't be negative because it sums squared differences divided by positive expected frequencies.

Figure 10.4: We use only the right tail because larger chi-square values provide greater evidence against H_0. Chi-squared values in the left tail represent small differences between observed and expected frequencies and do not provide evidence against H_0.

Figure 10.5: Whether we get $\frac{1}{4}$ or 4 for the relative risk depends merely on which group we call group 1 and which we call group 2.

CHAPTER SUMMARY

This chapter showed how to analyze the association between two categorical variables:

- By *describing the counts* in contingency tables using the **conditional distributions** across the categories of the response variable. If the population conditional distributions are identical, the two variables are **independent**.

- By using the **chi-squared** statistic to **test the null hypothesis of independence**.

- By *describing the strength of association with a* **measure of association** such as the **difference of proportions** and the **ratio of proportions** (the **relative risk**). When there is independence, a population difference of proportions equals 0 and a population relative risk equals 1. The stronger the association, the farther the measures fall from these baseline values.

- By *describing the pattern of association* by comparing observed and expected cell counts using **standardized residuals**. A standardized residual reports the number of standard errors that an observed count falls from an expected count. A value larger than about 3 in absolute value indicates that the cell provides strong evidence of association.

The **expected cell counts** are values with the same margins as the observed cell counts but that satisfy the null hypothesis of independence. The **chi-squared test statistic** compares the observed cell counts to the expected cell counts, using

$$X^2 = \sum \frac{(\text{observed count} - \text{expected count})^2}{\text{expected count}}.$$

Under the null hypothesis, the X^2 test statistic has a large-sample **chi-squared distribution**. The **degrees of freedom** depend on the number of rows r and the number of columns c through $df = (r - 1)(c - 1)$. The P-value is the right-tail probability above the observed value of X^2.

The expected cell counts for the chi-squared test should be at least 5. **Fisher's exact test** does not have a sample size restriction. It is used to test independence with samples that are too small for the chi-squared test.

The next chapter introduces methods for describing and making inferences about the association between two *quantitative* variables. We'll learn more about the correlation and regression methods that Chapter 3 introduced.

SUMMARY OF NOTATION

X^2 = chi-squared statistic for testing independence of two categorical variables

CHAPTER PROBLEMS

Practicing the Basics

10.41 Female for President?: When recent General Social Surveys have asked, "If your party nominated a woman for President, would you vote for her if she were qualified for the job?", about 94% of females and 94% of males answered yes, the rest answering no.

   a. For males and for females, report the conditional distributions on this response variable in a 2 × 2 table, using outcome categories (yes, no).

   b. If results for the entire population are similar to these, does it seem possible that gender and opinion about having a woman President are independent? Explain.

10.42 Down syndrome diagnostic test: The table shown, from Example 8 in Chapter 5, cross-tabulates whether a fetus has Down syndrome by whether or not the triple blood diagnostic test for Down syndrome is positive (that is, indicates that the fetus has Down syndrome).

   a. Tabulate the conditional distributions for the blood test result, given the true Down syndrome status.

   b. For the Down cases, what percentage was diagnosed as positive by the diagnostic test? For the unaffected

cases, what percentage got a negative test result? Does the diagnostic test appear to be a good one?

   c. Construct the conditional distribution on Down syndrome status, for those who have a positive test result. (*Hint*: You condition on the first column total and find proportions in that column.) Of those cases, what percentage truly have Down syndrome? Is the result surprising? Explain why this probability is small.

Down Syndrome Status	Blood Test Result Positive	Negative	Total
D (Down)	48	6	54
D^c (unaffected)	1307	3921	5228
Total	1355	3927	5282

10.43 Down and chi-squared: For the data in the previous example, $X^2 = 114.4$. Show all steps of the chi-squared test of independence.

10.44 What is *df*?: The contingency table shown below has $df = 4$. Show this, by filling in the missing cell counts.

Table for Exercise 10.44

	A	B	C	D	E	Total
	24	21	12	10	—	100
	—	—	—	—	—	100
Total	40	40	40	40	40	200

10.45 Echinacea no better than placebo?: Echinacea is a widely used herbal remedy for treatment of upper respiratory tract infections (URIs). However, in the primary analyses in a randomized, double-blind study of children, it did not do better than placebo in relieving cold symptoms (article by J. Taylor et al., *JAMA*, vol. 290, pp. 2824–2830, 2003). For one of the secondary analyses, the article stated that "Of the children who had at least 1 URI, 64.4% of children in the placebo group had more than 1 URI compared with 52.3% of children receiving Echinacea (P-value = .015 by chi-squared test)."

a. Identify the response variable and the explanatory variable and their categories for the 2 × 2 contingency table that provided this particular analysis.

b. How would you explain to someone who has never studied statistics how to interpret the parenthetical part of the quoted sentence?

10.46 Happiness and number of friends: The table shows GSS data on happiness and the number of close friends, with expected cell counts given underneath the observed counts.

Number of Close Friends	Not too happy	Pretty happy	Very happy	Total
0–1	34	92	43	169
	19.2	95.1	54.7	
2–5	83	418	200	701
	79.5	394.6	226.9	
6 or more	47	304	225	576
	65.3	324.3	186.4	
Total	164	814	468	1446

(Happiness is the column header spanning Not too happy, Pretty happy, Very happy.)

a. Suppose the variables were independent. Explain what this means in this context.

b. Explain what is meant by an expected cell count. Show how to get the expected cell count for the first cell, for which the observed count is 34.

c. Compare the expected cell frequencies to the observed counts. Based on this, what is a profile of subjects who tend to be (i) more happy than independence predicts, (ii) less happy than independence predicts.

10.47 Gender gap?: Exercise 10.1 showed a 2 × 3 table relating gender and political party identification, shown again below. The chi-squared statistic for these data equals 18.8. Conduct all five steps of the chi-squared test.

Sex	Dem	Indep	Repub
F	567	534	395
M	356	460	369

(Political Party is the column header spanning Dem, Indep, Repub.)

10.48 Job satisfaction and income: A recent GSS was used to cross-tabulate income (<15 thousand, 15–25 thousand, 25–40 thousand, >40 thousand) in dollars with job satisfaction (very dissatisfied, little dissatisfied, moderately satisfied, very satisfied) for 96 subjects.

a. For these data, $X^2 = 6.0$. What is its df value, and what is its approximate sampling distribution, if H_0 is true?

b. For this test, the P-value is 0.74. Interpret in the context of these variables.

c. What decision would you make with a 0.05 significance level? Can you accept H_0 and conclude that job satisfaction is independent of income?

10.49 Aspirin and heart attacks: For the Physicians' Health Study described in Example 5, the table shows a finer breakdown in terms of whether a subject suffered a fatal heart attack, nonfatal heart attack, or no heart attack (from *N. Engl. J. Med.*, vol. 318, pp. 262–264, 1988).

Physicians health study data for Exercise 10.49

	MYOCARDIAL INFARCTION		
Group	Fatal Attack	Nonfatal Attack	No Attack
Placebo	18	171	10,845
Aspirin	5	99	10,933

a. Use software to test independence. Show (i) assumptions, (ii) hypotheses, (iii) test statistic, (iv) P-value, (v) conclusion in the context of this study.

b. Describe the association by finding and interpreting the relative risk for the fatal attack category.

10.50 Women's role: For a recent GSS and the statement, "Women should take care of running their homes and leave running the country up to men," 14.8% of the male respondents agreed. Of the female respondents, 15.9% agreed. Of respondents having less than a high school education, 39.0% agreed. Of respondents having at least a high school education, 11.7% agreed.

a. Report the difference between the proportion of males and the proportion of females who agree.

b. Report the difference between the proportion at the low education level and the proportion at the high education level who agree.

c. Which variable, gender or educational level, seems to have the stronger association with opinion? Explain your reasoning.

10.51 Seat belt helps?: The table refers to passengers in autos and light trucks involved in accidents in the state of Maine in a recent year.

a. Use the difference of proportions to describe the strength of association. Interpret.

b. Use the relative risk to describe the strength of association. Interpret.

Maine accident data for Exercise 10.51

	Injury	
Seat Belt	Yes	No
No	3865	27,037
Yes	2409	35,383

Source: Dr. Cristanna Cook, Medical Care Development, Augusta, Maine.

10.52 Race and party ID: The table shows an SPSS printout for some analyses of 2000 GSS data on race and party ID.

a. Interpret the expected count for the first cell.

b. Interpret the standardized residual of 12.4 for the first cell (SPSS calls it an adjusted residual).

c. How would you summarize to someone who has never studied statistics what you learn from the standardized residuals given in the four corner cells on this printout?

race * party_ID Crosstabulation

			party ID			
			democrat	independent	republican	Total
race	black	Count	251	151	23	425
		Expected Count	140.9	176.0	108.2	425.0
		Adjusted Residual	12.4	-2.7	-10.4	
	white	Count	615	931	642	2188
		Expected Count	725.1	906.0	556.8	2188.0
		Adjusted Residual	-12.4	2.7	10.4	
Total		Count	866	1082	665	2613
		Expected Count	866.0	1082.0	665.0	2613.0

10.53 Happiness and sex: A contingency table from the 2002 GSS relating happiness to number of sex partners in the previous year (0, 1, at least 2) had standardized residuals as shown in the table. Interpret the highlighted standardized residuals.

Results on happiness and sex for Exercise 10.53

```
Rows: partners Columns: happy

        not       pretty       very

0       40         152          61

        1.938      0.812       -2.267

1       76         371         240

       -1.600     -3.332        4.744

2       20         117          29

       -0.106      3.571       -3.778
```

10.54 Education and religious beliefs: When data from a recent GSS were used to form a 3 × 3 table that cross-tabulates highest degree (1 = less than high school, 2 = high school or junior college, 3 = bachelor or graduate) with religious beliefs (F = fundamentalist, M = moderate, L = liberal), the three largest standardized residuals were −6.8 for the cell (3, F), 6.3 for the cell (3, L), and 4.5 for the cell (1, F). Summarize how to interpret these three standardized residuals.

10.55 TV and aggression: From a study described in Example 5 in the previous chapter, the proportion of males committing aggressive acts was 4 out of 45 for those who watched less than 1 hour of TV a day, and 117 out of 315 for those who watched at least 1 hour per day. MINITAB reports the printout shown for conducting Fisher's exact test (two-sided) for these data.

a. State hypotheses for the test, defining the notation.

b. Report the P-value that software reports, and interpret in the context of this study.

Data on TV watching and aggression for Exercise 10.55

```
Rows: TV Columns: aggression
            no          yes
high       198          117
low         41            4
Fisher's exact test:
P-Value = 0.0000751
```

10.56 Botox side effects: An advertisement for Botox Cosmetic by Allergan, Inc. for treating wrinkles appeared in several magazines. The back page of the ad showed the results of a randomized clinical trial to compare 405 people receiving Botox injections to 130 people receiving placebo in terms of the frequency of various side effects. The side effect of pain in the face was reported by 9 people receiving Botox and 0 people receiving placebo. Use software to conduct a two-sided test of the hypothesis that the probability of this side effect is the same for each group. Report all steps of the test, and interpret results.

Concepts and Applications

10.57 Student data: Refer to the "FL student survey" data file on the text CD. Using software, create and analyze descriptively and inferentially the contingency table relating religiosity and belief in life after death. Summarize your analyses in a short report.

10.58 Marital happiness decreasing?: At sda.berkeley.edu/GSS, cross-tabulate HAPMAR with YEAR, so you can see how conditional distributions on marital happiness in the GSS have changed since 1973. Using conditional distributions and standardized residuals, explain how these results show a very slight trend over time for fewer people to report being very happy. (You can get the standardized residuals by checking "show z statistic" at the web site.)

10.59 Another predictor of happiness?: Go to sda.berkeley.edu/GSS and find a variable that is associated with happiness, other than variables used in this chapter. Use methods of this chapter to describe and make inferences about the association, in a one-page report.

10.60 Pregnancy associated with contraceptive use?: Whether a young married woman becomes pregnant in the next year is a categorical variable with categories (yes, no), and whether she and her partner use contraceptives is another categorical variable with categories (yes, no). Would you expect these variables to be independent, or associated? Explain.

10.61 Babies and gray hair: A young child wonders what causes women to have babies. For each woman who lives on her block, she observes whether her hair is gray and whether she has young children, with the results shown in the table below.

a. Construct the 2 × 2 contingency table that cross-tabulates "gray hair" (yes, no) with "has young children" (yes, no) for these nine women.

b. Treating "has young children" as the response variable, obtain the conditional distributions for those women who have gray hair and for those who do not. Does there seem to be an association?

c. Noticing this association, the child concludes that not having gray hair is what causes women to have children. Use this example to explain why association does not necessarily imply causation.

Woman	Gray Hair	Young Children
Andrea	No	Yes
Mary	Yes	No
Linda	No	Yes
Jane	No	Yes
Maureen	Yes	No
Judy	Yes	No
Margo	No	Yes
Carol	Yes	No
Donna	No	Yes

10.62 When is chi-squared not valid?: Give an example of a contingency table for which the chi-squared test of independence should not be used.

10.63 Gun homicide in United States and Britain:
According to recent United Nations figures, the annual gun homicide rate is 62.4 per one million residents in the United States and 1.3 per one million residents in Britain.

a. Show how to compare the proportion of residents of the two countries killed annually by guns using the difference of proportions. Show how the results differ according to whether the United States or Britain is identified as Group 1.

b. Show how to compare the proportion of residents of the two countries using the relative risk. Show how the results differ according to whether the United States or Britain is identified as Group 1.

c. When both proportions are very close to 0, as in this example, which measure do you think is more useful for describing the strength of association? Why?

10.64 Aspirin and colon cancer: An Associated Press story (April 8, 2002) summarized a Dartmouth Medical School study on colon cancer. The research was done on 1121 people who had polyps removed during routine screening. They were randomly assigned to take baby aspirin or placebo daily. Three years later, 38% of those taking baby aspirin had new polyps, compared with 47% of those taking placebo. The description of the results stated that taking aspirin reduces the risk of new polyps by 19%. Explain how they got this estimate. (*Hint*: It refers to a difference *and* a ratio.)

10.65 $X^2 = 0$: True or false? The null hypothesis for the test of independence between two categorical variables is H_0: $X^2 = 0$, for the sample chi-squared statistic X^2. (*Hint*: Do hypotheses refer to a sample or the population?)

10.66 Group 1 becomes group 2: True or false? Interchanging two rows or interchanging two columns in a contingency table has no effect on the value of the X^2 statistic.

10.67 Statistical but not practical significance: True or false? Even when the sample conditional distributions in a contingency table are only slightly different, when the sample size is very large it is possible to have a large X^2 statistic and a very small P-value for testing H_0: independence.

10.68 Statistical vs. practical significance: In any significance test, when the sample size is very large, we have not necessarily established an important result when we obtain statistical significance. Explain what this means in the context of analyzing contingency tables with a chi-squared test.

10.69 Normal and chi-squared with $df = 1$: When $df = 1$, the P-value from the chi-squared test of independence is the same as the P-value for the two-sided test comparing two proportions with the z test statistic. This is because of a direct connection between the standard normal distribution and the chi-squared distribution with $df = 1$: Squaring a z-score yields a chi-squared value with $df = 1$ having chi-squared right-tail probability equal to the two-tail normal probability for the z-score.

a. Illustrate this with $z = 1.96$, the z-score with a two-tail probability of 0.05. Using the chi-squared table or software, show that the square of 1.96 is the chi-squared score for $df = 1$ with a P-value of 0.05.

b. Show the connection between the normal and chi-squared values with P-value = 0.01.

◆◆ **10.70 Multiple response variables:** Each subject in a sample of 100 men and 100 women is asked to indicate which of the following factors (one or more) are responsible for increases in crime committed by teenagers: A—the increasing gap in income between the rich and poor, B—the increase in the percentage of single-parent families, C—insufficient time that parents spend with their children. To analyze whether responses differ by gender of respondent, we cross-classify the responses by gender, as the table shows.

a. Is it valid to apply the chi-squared test of independence to these data? Explain.

b. Explain how this table actually provides information needed to cross-classify gender with each of three variables. Construct the contingency table relating gender to opinion about whether factor A is responsible for increases in teenage crime.

Three factors for explaining teenage crime

Gender	A	B	C
Men	60	81	75
Women	75	87	86

◆◆ **10.71 Standardized residuals for 2 × 2 tables:** The table below shows the standardized residuals in parentheses for GSS data about the statement, "Women should take care of running their homes and leave running the country up to men." The absolute value of the standardized residual is 13.2 in every cell. For chi-squared tests with 2 × 2 tables, since $df = 1$, only one nonredundant piece of information exists about whether an association exists. If observed count > expected count in one cell, observed count < expected count in the other cell in that row or column. Explain why, using the fact that observed and expected counts have the same row and column totals. (In fact, in 2 × 2 tables, all four standardized residuals have absolute value equal to the square root of the X^2 test statistic.)

Year	Agree	Disagree
1974	509 (13.2)	924 (−13.2)
1998	280 (−13.2)	1534 (13.2)

◆◆ 10.72 Spread of chi-squared: For the chi-squared distribution, the <u>mean</u> equals df and the standard deviation equals $\sqrt{2(df)}$.

a. Explain why, as a rough approximation, for a large df value, 95% of the chi-squared distribution falls with-in <u>$df \pm 2\sqrt{2(df)}$</u>.

b. With $df = 8$, show that $df \pm 2\sqrt{2(df)}$ gives the interval $(0, 16)$ for *approximately* containing 95% of the distribution. Using the chi-squared table, show that *exactly* 95% of the distribution actually falls between 0 and 15.5.

◆◆ 10.73 Explaining Fisher's exact test: A pool of six candidates for three managerial positions includes three females and three males. Denote the three females by F1, F2, F3 and the three males by M1, M2, M3. The result of choosing three individuals for the managerial positions is (F2, M1, M3).

a. Identify the 20 possible samples that could have been selected. Explain why the contingency table relating gender to whether chosen for the observed sample is

	CHOSEN FOR POSITION	
GENDER	Yes	No
Male	2	1
Female	1	2

b. Let $\hat{p}_1$ denote the sample proportion of males selected and $\hat{p}_2$ the sample proportion of females. For the observed table, $\hat{p}_1 - \hat{p}_2 = (2/3) - (1/3) = 1/3$. Of the 20 possible samples, show that 10 have $\hat{p}_1 - \hat{p}_2 \geq 1/3$.

Thus, if the three managers were randomly selected, the probability would equal $10/20 = 0.50$ of obtaining $\hat{p}_1 - \hat{p}_2 \geq 1/3$. This is the reasoning that provides the one-sided P-value for Fisher's exact test with H_a: $p_1 > p_2$.

◆◆ 10.74 Likelihood-ratio chi-squared: For testing independence, most software also reports another chi-squared statistic, called **likelihood-ratio chi-squared**. It equals

$$G^2 = 2\sum \text{observed count} \times \log\left(\frac{\text{observed count}}{\text{expected count}}\right).$$

It has similar properties as the X^2 statistic, such as $df = (r - 1)(c - 1)$.

a. Show that $G^2 = X^2 = 0$ when each observed count = expected count.

b. Explain why in practice you would not expect to get *exactly* $G^2 = X^2 = 0$, even if the variables are truly independent.

Class Explorations

10.75 Conduct a research study using the GSS: Go to the GSS codebook at sda.berkeley.edu/GSS. Your instructor will assign a categorical response variable. Conduct a research study in which you find at least two other categorical variables that have both a statistically significant and a practically significant association with that response variable, using data for the most recent year in which the response variable was surveyed. Analyze each association using the methods of this chapter. Write a two-page report, describing your analyses and summarizing the results of your research study. Be prepared to discuss your results in a class discussion, because the class will discuss which variables seemed to be especially relevant.

BIBLIOGRAPHY

Agresti, A. (1996). *An Introduction to Categorical Data Analysis*. New York: Wiley.

Chatterjee et al. (1995) A Case Book for *A First Course in Statistics and Data Analysis*. New York. Wiley.

11 Analyzing the Association Between Quantitative Variables: Regression Analysis

EXAMPLE 1

How Can You Estimate a Person's Strength?

Picture the Scenario

How can you measure a person's strength? One way is to find the *maximum* number of pounds that the individual can bench press. However, this technique can be risky for people who are unfamiliar with proper lifting techniques or who are inexperienced in using a bench press. Is there a variable that is easier to measure yet is a good predictor of the maximum bench press?

There have been many studies about strength using males but relatively few using females. One exception was a recent study of 57 female athletes in a high school in Augusta, Georgia. Among the variables measured were ones that are easier and safer to assess than maximum bench press but that are thought to correlate highly with it. One such variable is the number of times that a girl can lift a bench press set at only 60 pounds (a relatively low weight) before she becomes too fatigued to lift it again. The data are in the "high school female athletes" data file on the text CD. Let x = number of 60-pound bench presses performed (before fatigue) and let y = maximum bench press.

Questions to Explore

- How well can we predict an athlete's maximum bench press from knowing the number of 60-pound bench presses that she can perform?

- What can we say about the association between these variables in the population?

Thinking Ahead

The bench press variables are quantitative. This chapter presents methods for analyzing the association between two quantitative variables. Like the methods presented in the previous chapter for categorical variables, these methods enable us to answer questions such as

- Could the variables x and y realistically be independent (in the population), or can we conclude that there is an association between them?

- If the variables are associated, how strong is the association?

- How does the outcome for the response variable depend on the value of the explanatory variable, and which observations are unusual?

The analyses that address these questions are collectively called a **regression analysis**. We'll conduct regression analyses of the data from the female athlete strength study in several examples in this chapter (Examples 2, 3, 6, 9, 11, 12, 15, and 16).

Recall

From Section 3.3, a **regression line** is a straight line that predicts the value of a response variable y from the value of an explanatory variable x. From Section 3.2, the **correlation** is a summary measure of association that falls between -1 and $+1$. ▪

Chapter 3 introduced regression analysis. We'll next review and learn more about using a **regression line** to predict the response variable y and the **correlation** to describe the strength of association. We'll then learn how to make inferences about the regression line for a population. We'll also learn how the variability of the data points around the regression line helps us to detect unusual observations and to predict how far from the line a value of y is likely to fall. Finally, we'll study an alternative regression analysis that is useful when two quantitative variables have a curved rather than a straight-line relationship.

11.1 | How Can We "Model" How Two Variables Are Related?

IN WORDS

y = response variable
Graph on vertical (y) axis

x = explanatory variable
Graph on horizontal (x) axis

Regression uses x to predict y

As in most statistical analyses, the first step of a regression analysis is to identify the response and explanatory variables. In Example 1 we want to know how well we can predict an athlete's maximum bench press, so that's the response variable. The predictor, the number of 60-pound bench presses that an athlete can manage, is the explanatory variable. As in Chapter 3, we use y to denote the response variable and x to denote the explanatory variable.

The Scatterplot: Is There a Straight-Line Trend?

Is there an association between the number of 60-pound bench presses and the maximum bench press? The first step in answering this question is to look at the data. Recall that a **scatterplot** shows, for each observation, a point representing its value on the two variables. The points are shown relative to the x (horizontal) and y (vertical) axes. The scatterplot shows whether there is roughly a straight-line trend, in which case the relationship between x and y is said to be approximately **linear**.

EXAMPLE 2

What Do We Learn from a Scatterplot in the Strength Study?

Picture the Scenario

The strength study discussed in Example 1 measured the maximum bench press by giving a girl a 60-pound weight to lift and then increasing the weight in 5-pound increments until the girl could no longer lift it. The response outcome is the maximum weight that the girl bench pressed. Let's look at the "high school female athletes" data file on the text CD. The first four entries of the 57 lines for x = the number of 60-pound bench presses and y = maximum bench press show the values

x	y
10	80
12	85
20	85
5	65

For the 57 girls in this study, these variables are summarized by

x: mean = 11.0, standard deviation = 7.1
y: mean = 79.9 lbs., standard deviation = 13.3 lbs.

Question to Explore

Using the entire data file for the 57 athletes, construct a scatterplot. Interpret it.

Think It Through

With any statistical software, you can construct a scatterplot after identifying in the data file the variables that play the role of x and y. Figure 11.1 shows the scatterplot that MINITAB produces. It shows that female athletes with higher numbers of 60-pound bench presses also tended to have higher values for the maximum bench press. The data points follow roughly an increasing linear trend. The margin shows a screen shot of the scatterplot from the TI-/83+84.

TI-83+/84 *Output*

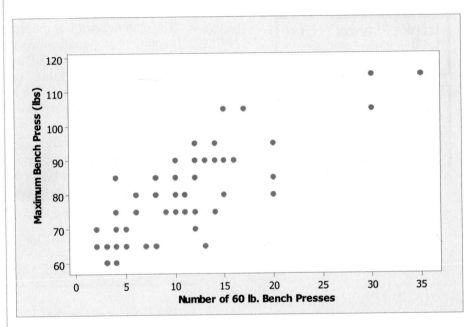

▲ **FIGURE 11.1: Scatterplot for y = Maximum Bench Press and x = Number of 60-Pound Bench Presses. Question:** How can you tell from this plot that (a) two athletes had y = maximum bench press of only 60 pounds, the weight used for finding x, (b) five-pound increments were used in determining maximum bench press—60, 65, 70, and so on?

Insight

The three data points at the upper right represent subjects who could do a large number of 60-pound bench presses. Their maximum bench presses were also high.

TO PRACTICE THIS CONCEPT, TRY EXERCISE 11.7A

The Regression Line Equation Uses x to Predict y

When the scatterplot shows a linear trend, a straight line fitted through the data points describes that trend. As in Chapter 3, we use the notation

$$\hat{y} = a + bx$$

for this line, called the **regression line**. The symbol $\hat{y}$ represents the *predicted value* of the response variable y. In the formula, a is the **y-intercept** and b is the **slope**.

EXAMPLE 3

Which Regression Line Predicts Maximum Bench Press?

Picture the Scenario

Let's continue our analysis of the "high school female athletes" data file.

Questions to Explore

a. Using software, find the regression line for y = maximum bench press and x = number of 60-pound bench presses.

b. Interpret the slope by comparing the predicted maximum bench press for subjects at the highest and lowest levels of x in the sample (35 and 2).

Think It Through

a. The data file denotes the maximum bench press variable by BP and the number of 60-pound bench presses by BP_60. Using software, we pick the regression option with BP as the response variable and BP_60 as the explanatory variable. Table 11.1 shows some output using MINITAB. We'll interpret some of this, such as standard errors and t statistics, later in the chapter. The margin shows TI-83+/84 output.

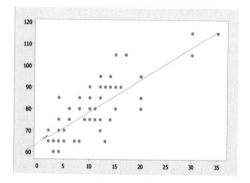

```
LinReg(a+bx) L₁,
L₂■
```

```
LinReg
 y=a+bx
 a=63.537
 b=1.491
```

TABLE 11.1: MINITAB Printout for Regression Analysis of y = Maximum Bench Press (BP) and x = Number of 60-Pound Bench Presses (BP_60)

The regression equation is BP = 63.5 + 1.49 BP_60

Predictor	Coef	SE Coef	T	P
Constant	63.537	1.956	32.48	0.000
BP_60	1.4911	0.1497	9.96	0.000

The output tells us that $\hat{y}$ = predicted maximum bench press (BP) relates to x = number of 60-pound bench presses (BP_60) by

BP = 63.5 + 1.49 BP_60, that is, $\hat{y}$ = 63.5 + 1.49x.

The y-intercept is 63.5 and the slope is 1.49. These are also shown in the column labeled "Coef," an abbreviation for *coefficient*. The slope appears opposite the variable name for which it is the coefficient, "BP_60." The predictor "Constant" refers to the y-intercept. The margin figure plots the regression line through the scatterplot.

b. The slope of 1.49 tells us that predicted maximum bench press $\hat{y}$ increases by about $1\frac{1}{2}$ pounds for every additional 60-pound bench press an athlete can do. The impact on $\hat{y}$ of a 33-unit change in x, from the sample minimum of x = 2 to the maximum of x = 35, is 33(1.49) = 49.2 pounds. An athlete who can do 35 60-pound bench presses has a predicted maximum bench press nearly 50 pounds higher than an athlete who can do only 2 60-pound bench presses. Those predicted values are $\hat{y}$ = 63.5 + 1.49(2) = 66.5 pounds at x = 2 and $\hat{y}$ = 63.5 + 1.49(35) = 115.7 pounds at x = 35.

Insight

The slope of 1.49 is positive: As x increases, the predicted value $\hat{y}$ increases. The association is *positive*. When the association is *negative*, the predicted value $\hat{y}$ *decreases* as x increases. When the slope = 0, the regression line is horizontal.

Recall

The direction of the association (positive or negative) refers to the sign of the slope and whether the line slopes upward or downward. ■

TO PRACTICE THIS CONCEPT, TRY EXERCISE 11.1

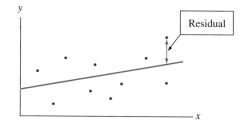

One reason for plotting the data before you do a regression analysis is to check for outliers. The regression line can be pulled toward an outlier and away from the general trend of points. We observed in Section 3.4 that an observation can be **influential** in affecting the regression line when two things happen:

■ Its x value is low or high compared to the rest of the data

■ It does not fall in the straight-line pattern that the rest of the data have.

In Figure 11.1, three subjects had quite large values of x. However, their values of y fit in with the trend exhibited by the other data, so those points are not influential.

Residuals Are Prediction Errors for the Least Squares Line

The regression equation is often called a **prediction equation**, because substituting a particular value of x into the equation provides a prediction for y at that value of x. For instance, the third athlete in the data file could do $x = 20$ sixty-pound bench presses. Her predicted maximum bench press is

$$\hat{y} = 63.5 + 1.49x = 63.5 + 1.49(20) = 93.3 \text{ pounds.}$$

The difference $y - \hat{y}$ between an observed outcome y and its predicted value $\hat{y}$ is the *prediction error*, called a **residual**. For instance, the third athlete in the data file had a maximum bench press of $y = 85$ pounds. Since her $\hat{y} = 93.3$ pounds, the prediction error is $y - \hat{y} = 85 - 93.3 = -8.3$.

Here's a summary of some things we learned about residuals in Section 3.3:

■ *Each observation has a residual.* Some are positive, some are negative, and their average equals 0.

■ In the scatterplot, *a residual is the vertical distance between the data point and the regression line.* The smaller the distance, the better the prediction.

■ We can summarize how near the regression line the data points fall by

$$\text{sum of squared residuals} = \sum (\text{residual})^2 = \sum (y - \hat{y})^2.$$

■ The regression line has a smaller sum of squared residuals than any other line. It is called the **least squares** line, because of this property.

Regression Model: A Line Describes How the *Mean* of y Depends on x

At a given value of x, the equation $\hat{y} = a + bx$ predicts a single value $\hat{y}$ of the response variable. However, we should not expect all subjects at that value of x to have the same value of y. *Variability* occurs in their y values.

For example, let $x =$ number of years of education and $y =$ annual income (in dollars), for the adult residents in the work force of your hometown. For a random sample, suppose you find $\hat{y} = -20,000 + 4000x$. Those workers with $x = 12$ years of education have predicted annual income

$$\hat{y} = -20,000 + 4000(12) = 28,000.$$

It's not the case that every worker with 12 years of education would have annual income $28,000. Income is not completely dependent upon education. One person may have income $35,000, another $15,000, and so forth. Instead, you can think of $\hat{y} = 28,000$ as estimating the *mean* annual income for all workers with $x = 12$.

Likewise, there's a mean of the annual income values at *each* separate education value. For those residents with 13 years of education, the estimated mean annual income is $-20,000 + 4000(13) = 32,000$. The slope is 4000, so the estimated mean goes up by $4000 (from $28,000 to $32,000) for this 1-year increase in education (from 12 to 13).

The regression line connects the estimated means of y at the various x values

In summary, $\hat{y} = a + bx$ describes the relationship between x and the *estimated means* of y at the various values of x.

Recall

Parameters describe the population. They are often denoted by Greek symbols. Here, α is *alpha,* β is *beta,* and μ_y is *mu-sub-y.*■

A similar equation describes the relationship *in the population* between x and the means of y. This **population regression equation** is denoted by

$$\mu_y = \alpha + \beta x.$$

Here, α is a population y-intercept and β is a population slope. These are parameters, so in practice their values are unknown. The parameter μ_y denotes the population mean of y for all the subjects at a particular value of x. It takes a different value at each separate x value. In practice, we estimate the population regression equation using the prediction equation for the sample data. We'll see how to conduct inference about the unknown parameters in Sections 11.3 and 11.4.

A straight line is the simplest way to describe the relationship between two quantitative variables. In practice, any given relationship is not *exactly* linear. The equation $\mu_y = \alpha + \beta x$ merely *approximates* the actual relationship between x and the population means of y. It is a **model**.

Model

A **model** is a simple approximation for how variables relate in a population.

Figure 11.2 shows how a regression model can approximate the true relationship between x and the mean of y. The true relationship is unlikely to be *exactly* a straight line. That's not a problem, as long as a straight line provides a reasonably good approximation, as we expect of a *model*.

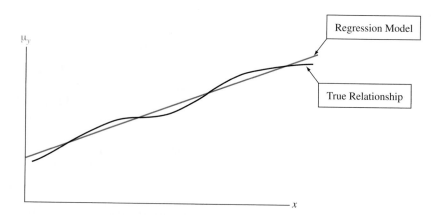

▲ **FIGURE 11.2: The Regression Model** $\mu_y = \alpha + \beta x$ **for the Means of y Is a Simple Approximation for the True Relationship. Question:** Can you sketch a true relationship for which this model is a very *poor* approximation?

If the true relationship is far from a straight line, this regression model may be a poor one. Figure 11.3 illustrates. This figure shows a U-shaped relationship. It requires a parabolic curve rather than a straight line to describe it well. To check the straight-line assumption, you should always construct a scatterplot.

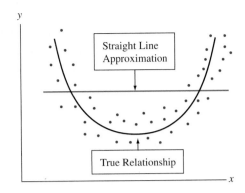

▲ **FIGURE 11.3: The Straight-Line Regression Model Provides a Poor Approximation when the Actual Relationship Is Highly Nonlinear. Question:** What type of mathematical function might you consider using for a regression model in this case?

How could you get into trouble by using a straight-line regression model even when the true relationship is U-shaped? First, predictions about y would be poor. Second, inference about the association could be misleading. The variables are associated, since the mean of y changes as x does. The slope (and correlation) might not detect this, however, since the prediction line would be close to horizontal even though a substantial association exists.

The Regression Model Also Allows Variability about the Line

At each fixed value of x, variability occurs in the y values around their mean μ_y. For instance, at education = 12 years, annual income varies around the mean annual income for all workers with 12 years of education. The probability distribution of y values at a fixed value of x is a **conditional distribution**. At each value of x, there is a conditional distribution of y values. A regression model also describes these distributions. An additional parameter σ describes the standard deviation of each conditional distribution.

Recall

Chapter 10 used a **conditional distribution** to describe percentages in categories of a categorical response variable, at each value of an explanatory variable. Each conditional distribution refers to possible values of the response variable at a fixed value of the explanatory variable. ▪

IN WORDS

A **regression model** uses a formula (usually a straight line) to approximate how the expected value for y changes at different values of x. It also describes the spread of observations on y around the line.

SUMMARY: Regression Model

A **regression model** describes how the population mean μ_y of each conditional distribution for the response variable depends on the value x of the explanatory variable. A **straight-line regression model** uses the line $\mu_y = \alpha + \beta x$ to connect the means. The model also has a parameter σ that describes variability of observations around the mean of y at each x value.

EXAMPLE 4

Regression Model for Income and Education

Picture the Scenario

As described above, suppose the regression line $\mu_y = -20,000 + 4000x$ models the relationship for the population of working adults in your hometown between x = number of years of education and the mean of y = annual income. This tells us that income goes up as education does, on the average, but how much variability is there? Suppose also that the conditional distribution of annual income at each value of x is modeled by a normal distribution, with standard deviation $\sigma = 13,000$.

Question to Explore

Use this regression model to describe the mean and the variability around the mean for the conditional distributions of income at education values 12 and 16 years.

Think It Through

This regression model states that for workers with x years of education, their annual incomes have a normal distribution with a mean of $\mu_y = -20,000 + 4000x$ and a standard deviation of $\sigma = 13,000$. For those having a high school education ($x = 12$), the mean annual income is $\mu_y = -20,000 + 4000(12) = 28,000$ and the standard deviation is 13,000. Those with a college education ($x = 16$) have a mean annual income of $\mu_y = -20,000 + 4000(16) = 44,000$ and a standard deviation of 13,000.

Since the conditional distributions are modeled as normal, the model predicts that nearly all of the y values fall within three standard deviations of the mean. For instance, for those with 16 years of education, $\mu_y = 44,000$ and $\sigma = 13,000$. Since

$$44,000 - 3(13,000) = 5000 \text{ and } 44,000 + 3(13,000) = 83,000,$$

the model predicts that nearly all of their annual incomes fall between $5000 and $83,000.

Insight

Figure 11.4 portrays this regression model. It plots the regression equation and the conditional distributions of y = income at x = 12 years and at x = 16 years of education. We'll see how to use data to estimate the variability σ later in the chapter.

TO PRACTICE THIS CONCEPT, TRY EXERCISE 11.4

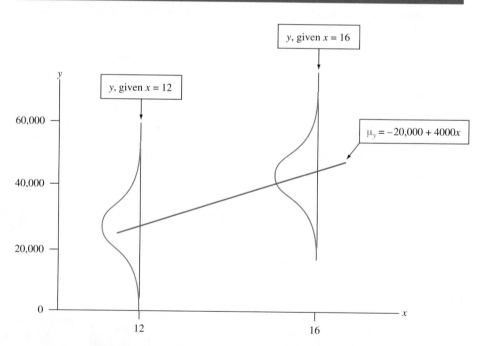

▲ **FIGURE 11.4: The Regression Model** $\mu_y = -20,000 + 4000x$, **with** $\sigma = 13,000$, **Relating the Means of y = Annual Income to x = Years of Education. Question:** What do the bell-shaped curves around the line at $x = 12$ and at $x = 16$ represent?

In Figure 11.4, each conditional distribution is normal, and each has the same standard deviation, $\sigma = 13,000$. But a *model* merely *approximates* reality. In practice, the means of the conditional distributions would not *perfectly* follow a straight line. Those conditional distributions would not be *exactly* normal. The standard deviation would not be *exactly* the same for each conditional distribution. For instance, the conditional distributions might be somewhat skewed to the right, and incomes might vary more for college graduates than for high school graduates.

IN PRACTICE: A model is a simple *approximation* for reality

A statistical **model** never holds *exactly* in practice. It is merely a simple approximation for reality. But even though it does not describe reality exactly, a model is useful if the true relationship is close to what the model predicts.

11.1 Practicing the Basics

11.1 **Car mileage and weight:** The "Car weight and mileage" data file on the text CD shows the weight (in pounds) and mileage (miles per gallon) of 25 different model autos.
TRY
   a. Identify the natural response variable and explanatory variable.

   b. The regression of mileage on weight has MINITAB regression output

```
Predictor   Coef     SE Coef     T       P
Constant   45.645     2.603    17.54  0.000
weight  -0.0052220 0.0006271  -8.33  0.000
```

   State the prediction equation and report the y-intercept and slope.

   c. Interpret the slope in terms of a 1000-pound increase in the vehicle weight.

   d. Does the y-intercept have any contextual meaning for these data? (*Hint:* The weight values range between 2460 and 6400 pounds.)

11.2 **Predicting car mileage:** Refer to the previous exercise.

   a. Find the predicted mileage for the Toyota Corolla, which weighs 2590 pounds.

   b. Find the residual for the Toyota Corolla, for which observed mileage equals 38.

   c. Sketch a graphical representation of the residual in (b).

11.3 **Children of working females:** For recent data on several nations, the prediction equation relating y = fertility rate (mean number of children per adult woman) to x = female economic activity (the female labor force as a percentage of the male labor force) is $\hat{y} = 5.2 - 0.044x$.

   a. Find the predicted fertility for Saudi Arabia, which had the lowest value of $x = 28$.

   b. Find the predicted fertility for Vietnam, which had the highest value of $x = 91$.

   c. Find the residual for Vietnam, which had $y = 2.3$.

   d. Suppose this equation applies to all x between 0 and 100. Interpret the y-intercept. Use the slope to describe how predicted fertility changes as x increases from 0 to 100.

11.4 **Income higher with age:** Suppose the regression line
TRY
   $\mu_y = -10,000 + 1000x$ models the relationship for the

population of working adults in Canada between x = age and the mean of y = annual income (in Canadian dollars). The conditional distribution of y at each value of x is modeled as normal, with $\sigma = 5000$. Use this regression model to describe the mean and the variability around the mean for the conditional distribution at age (a) 20 years, (b) 50 years.

11.5 **Mu, not y:** For a population regression equation, why is it more sensible to write $\mu_y = \alpha + \beta x$ instead of $y = \alpha + \beta x$? Explain with reference to the variables x = height and y = weight for the population of girls in elementary schools in your hometown.

11.6 **Parties and dating:** Let y = number of parties attended in the past month and x = number of dates in the past month, measured for all single students at your school. Explain the mean and variability aspects of the regression model $\mu_y = \alpha + \beta x$, in the context of these variables. In your answer, explain why (a) it is more sensible to use a straight line to model the *means* of the conditional distributions rather than individual observations, (b) the model needs to allow variation around the mean.

11.7 **Predicting college GPA:** For the "Georgia student survey" data file on the text CD, look at college GPA and
TRY
   high school GPA.

   a. Identify the response and explanatory variables and construct a scatterplot. What is the effect on this plot of several students having high school GPAs of exactly 4.0?

   b. Find the sample prediction equation. Predict the college GPA for students having high school GPA of (i) 3.0, (ii) 4.0. Explain how the slope relates to the answers.

11.8 **Exercise and watching TV:** For the "Georgia student survey" file on the text CD, let y = exercise and x = watch TV (minutes per day).

   a. Construct a scatterplot. Identify an outlier that could have an impact on the fit of the regression model. What would you expect its effect to be on the slope?

   b. Fit the model with and without that observation. Summarize its impact.

11.2 | How Can We Describe Strength of Association?

In the straight-line regression model, the slope indicates whether the association is positive or negative and describes the trend. The slope does not, however, describe the *strength* of the association. Chapter 3 showed that the **correlation** does this.

Recall

You can review the **correlation** r and its properties in Section 3.2. Figure 3.7 displayed scatterplots having a variety of values for r. ■

The Correlation Measures the Strength of Linear Association

The correlation, denoted by r, describes *linear association* between two variables. Here's a summary of its basic properties:

■ The correlation r has the same sign as the slope b. Thus, $r > 0$ when the points in the scatterplot have an upward trend and $r < 0$ when the points have a downward trend.

■ The correlation r always falls between -1 and $+1$, that is, $-1 \leq r \leq 1$.

■ The larger the absolute value of r, the stronger the linear association, with $r = \pm 1$ when the data points all fall exactly on the regression line.

When you have several quantitative variables, software can provide a **correlation matrix**. This is a square table, listing the variables in the rows and again in the columns and showing the correlation between each pair. This table does not differentiate between response and explanatory variables, because *the correlation when we use y to predict x is the same as the correlation when we use x to predict y.*

EXAMPLE 5

Correlations with Worldwide Internet Use

Picture the Scenario

Table 3.4 in Chapter 3 showed recent data for 39 nations on several variables, from the "Human development" data file on the text CD. One variable in the file is *Internet use*, the percentage of adult residents who use the Internet. Some of its values are shown in the table in the margin on the next page. Which variables are strongly associated with Internet use? Let's consider four possibilities.

GDP: Gross domestic product, per capita, in thousands of U.S. dollars (a measure of a nation's economic development)

Cellular: Percentage of adults who are cellular-phone subscribers

Fertility: Mean number of children per adult woman

CO2: Carbon-dioxide emissions, per capita (a measure of air pollution)

Table 11.2 displays a correlation matrix for these five variables.

Nation	Internet
Australia	37.1
Brazil	4.7
China	2.6
Egypt	0.9
France	26.4
Germany	37.4
India	0.7
Ireland	23.3
Israel	27.7
Japan	38.4
Mexico	3.6
New Zealand	46.1
Russia	2.9
Sweden	51.6
United Kingdom	33.0
United States	50.2

TABLE 11.2: Correlation Matrix for Internet Use, Gross Domestic Product (GDP), Carbon Dioxide Emissions (CO2), Cellular-Phone Use, and Fertility

These are computed for the data for 39 nations in Table 3.4. Intersect the row for one variable with the column for a second variable to find their correlation.

	INTERNET	GDP	CELLULAR	FER
GDP	0.888			
CELLULAR	0.818	0.874		
FER	−0.551	−0.615	−0.607	
CO2	0.680	0.786	0.563	−0.464

Questions to Explore

a. Which variable has the strongest linear association with Internet use?

b. Which variable is negatively associated with the others?

Think It Through

a. The correlations with Internet use appear in the first column. The variable most strongly linearly associated with Internet use is the one having the largest correlation, in absolute value. This is GDP. Its correlation with Internet use is $r = 0.888$, a very strong positive association.

b. From the row and column for fertility (the mean number of children per adult woman), we see it is negatively correlated with each of the other four variables. For instance, $r = -0.551$ between fertility and Internet use. There is a moderate tendency for Internet use to decrease as fertility increases.

Insight

Why are Internet use and fertility negatively correlated? Is there a causal connection, whereby spending more time having children reduces the chance to use the Internet? Or, might a confounding variable explain the association? The association might be explained by the correlations these variables have with GDP. Nations that are more economically advanced tend to have lower fertility ($r = -0.615$) and also tend to have higher Internet use ($r = 0.888$). As GDP goes up, fertility goes down and Internet use goes up. This may induce a negative correlation between fertility and Internet use.

TO PRACTICE THIS CONCEPT, TRY EXERCISE 11.10

Finding the Correlation and the Prediction Equation

Software uses the sample data to find the prediction equation $\hat{y} = a + bx$ and the correlation r. Sections 3.2 and 3.3 in Chapter 3 gave the formulas. We saw there that $r > 0$ and $b > 0$ when most observations are in the quadrants where x and y are both above their means or both below their means. See Figure 11.5. By contrast, $r < 0$ and $b < 0$ when subjects who are above the mean on one variable tend to be below the mean on the other variable.

The formulas for the slope b and the y-intercept a (shown again in the margin) use the correlation r and the sample means $\bar{x}$ and $\bar{y}$ of the x values and the y values and the sample standard deviations s_x and s_y. The slope b is proportional to the correlation r, and we'll discuss their connection below.

Recall

From Section 3.3,

$$b = r\left(\frac{s_y}{s_x}\right),$$

$$a = \bar{y} - b\bar{x}. ■$$

▲ **FIGURE 11.5: The Correlation *r* and the Slope *b* are Positive When Most Data Points Are above the Mean on Both Variables or Below the Mean on Both.** These are the areas shaded in this figure. **Question:** Where do most data points usually fall when the slope and correlation are negative?

IN PRACTICE: Check the model graphically; use software for computations

Check whether it is sensible to do a regression analysis by looking at a scatterplot. If you see an approximately linear relationship, you can let technology do the computational work of finding the correlation, slope, and y-intercept.

The formula $a = \bar{y} - b\bar{x}$ for the y-intercept can be rewritten as $\bar{y} = a + b\bar{x}$. This shows that if we substitute $x = \bar{x}$ into the regression equation $\hat{y} = a + bx$, then the predicted outcome is $\hat{y} = a + b\bar{x} = \bar{y}$. In words, *a subject who is average on x is predicted to be average on y.*

These formulas help to show what's going on when you do a regression analysis. Software can do the computational work for you.

The Correlation Is a Standardized Slope

Why do we need the correlation? Why can't we use the *slope* to describe the strength of the association? The reason is that the slope's numerical value depends on the units of measurement. Example 3 found a slope of 1.49 pounds in predicting y = maximum bench press using x = number of 60-pound bench presses. Suppose we instead measure y in kilograms (kg). Since 1 kg = 2.2 pounds, a slope of 1.49 pounds is equivalent to a slope of 1.49/2.2 = 0.68 kilograms. In *grams*, the slope would be 680. Whether a slope is a small number or a large number merely depends on the units of measurement.

The correlation is a *standardized* version of the slope. Unlike the slope b, the correlation does not depend on units of measurement. It takes the same value regardless of whether maximum bench press is measured in pounds, kilograms, or grams. The standardization adjusts the slope b for the way it depends on the standard deviations of x and y. Since the correlation r and slope b are related by $b = r(s_y/s_x)$, equivalently,

$$r = b\left(\frac{s_x}{s_y}\right).$$

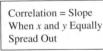

Suppose the data are equally spread out for each variable, with $s_x = s_y$. Then, $r = b$: The correlation and the slope are the same. So, in summary,

■ The correlation r does not depend on the units of measurement.

■ The correlation represents the value that the slope equals if the two variables are equally spread out.

For instance, suppose the standard deviation equals 10 both for x = midterm exam score and also for y = final exam score of your statistics course. Then, if $\hat{y} = 24.0 + 0.70x$ (that is, slope = 0.70), the correlation is also 0.70 between the two exam scores. In practice, the standard deviations are not usually identical. However, this tells us that the correlation represents what we would get for the slope of the regression line if the two variables were equally spread out.

EXAMPLE 6

What's the Correlation for Predicting Strength?

Picture the Scenario
For the female athlete strength study (Examples 1–3), x = number of 60-pound bench presses and y = maximum bench press had

> x: mean = 11.0, standard deviation = 7.1
>
> y: mean = 79.9, standard deviation = 13.3 (both in pounds)
>
> regression equation: $\hat{y} = 63.5 + 1.49x$.

Questions to Explore
a. Find the correlation r between these two variables.
b. Show that r does not change value if you measure y in kilograms.

Think It Through
a. The slope of the regression equation is $b = 1.49$. Since $s_x = 7.1$ and $s_y = 13.3$,

$$r = b\left(\frac{s_x}{s_y}\right) = 1.49\left(\frac{7.1}{13.3}\right) = 0.80.$$

The variables have a strong, positive association.

b. If y = had been measured in kilograms, the y values would have been divided by 2.2, since one kg = 2.2 pounds. For instance, subject 1 had y = 80 pounds, which is 80/2.2 = 36.4 kg. Likewise, the standard deviation s_y of 13.3 in pounds would have been divided by 2.2 to get 13.3/2.2 = 6.05 in kg. The slope of 1.49 would have been divided by 2.2, giving 0.68, since 1.49 pounds = 0.68 kg. Then

$$r = b(s_x/s_y) = 0.68(7.1/6.05) = 0.80.$$

The correlation is the same (0.80) if we measure y in pounds or in kilograms.

Insight
Now if we change units from kilograms to *grams*, s_y changes from 6.05 to 6050, b changes from 0.68 to 680, but again $r = 0.80$ because r *does not depend on the units*.

TO PRACTICE THIS CONCEPT, TRY EXERCISE 11.13

Regression toward the Mean

An important property of the correlation is that at any particular x value, the predicted value of y is relatively closer to its mean than x is to its mean.

> **If an x value is a certain number of standard deviations from its mean, then the predicted y is r times that many standard deviations from its mean.**

Let's illustrate for $r = 0.50$ and a subject who is one standard deviation above the mean of x, that is with x-value, $x = \bar{x} + s_x$. The predicted value of y is 0.50 standard deviations above the mean, $\bar{y}$. That is, at $x = \bar{x} + s_x$, the predicted outcome is $\hat{y} = \bar{y} + 0.50s_y$. The predicted y is relatively closer to its mean. See Figure 11.6, which also recalls that for subjects at the mean $\bar{x}$ of x, their predicted value of y is $\bar{y}$.

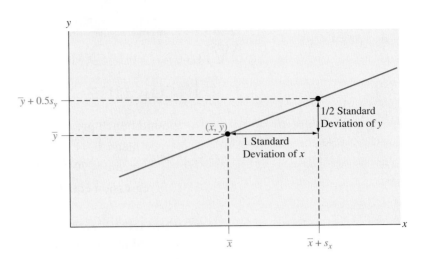

▲ **FIGURE 11.6: The Predicted Value of y Is Closer to Its Mean $\bar{y}$ than x Is to Its Mean $\bar{x}$ (in Number of Standard Deviations).** When x is one standard deviation above $\bar{x}$, the predicted y value is r standard deviations above $\bar{y}$ (shown in the figure for r = 1/2). **Question:** When x is two standard deviations above $\bar{x}$, how many standard deviations does the predicted y value fall above $\bar{y}$ (if r = $\frac{1}{2}$)?

EXAMPLE 7

Very Tall Parents Tend to Have Tall, but Less Tall, Children

Picture the Scenario

Francis Galton was a British scientist who discovered the basic ideas of regression and correlation in the 1880s. His discovery occurred after he observed that very tall parents tended to have tall children, but on the average not quite so tall. For instance, for all fathers with height 7 feet, perhaps their sons average 6 feet 5 inches when fully grown—taller than average, but not so extremely tall. Likewise, for fathers with height 5 feet, perhaps their sons average 5 feet 5 inches—shorter than average, but not so extremely short.

In his research, Galton accounted for gender height differences by multiplying each female height by 1.08, so heights of mothers and daughters had about the same mean as heights of fathers and sons. Then, for each son or daughter, he summarized their father's and mother's heights by parents' height = (father's height + mother's height)/2, the mean of their heights.

Question to Explore

Galton found a correlation of 0.5 between x = parents' height and y = child's height. How does his observation about very tall or very short parents tending to have children who are not so very tall or so very short relate to the property about the correlation that a predicted value of y is relatively closer to its mean than x is to its mean?

Think It Through

From the property of r with r = 0.5, when x = parents' height is a certain number of standard deviations from its mean, then y = child's predicted height is *half* as many standard deviations from its mean. For example, if the parents' height is 2 standard deviations above the mean, the child is predicted to be 1 standard deviation above the mean (half as far, in relative terms, when r = 0.5). At $x = \bar{x} + 3s_x$, we predict $\hat{y} = \bar{y} + 1.5s_y$. In each case, on the average a child's height is above the mean, but only half as far above the mean. Likewise, at $x = \bar{x} - 3s_x$, we predict $\hat{y} = \bar{y} - 1.5s_y$.

Insight

The correlation r is no greater than 1, in absolute value. So, a y value is predicted to be fewer standard deviations from its mean than x is from its mean.

TO PRACTICE THIS CONCEPT, TRY EXERCISE 11.22

IN WORDS

In English, "regression" means a going back, or returning. Here, the predicted value of y is going back toward the mean.

In summary, the predicted y is relatively closer to its mean than x is to its mean. Because of this, there is said to be **regression toward the mean**. This is the origin of the name that Francis Galton chose for regression analysis.

For all cases in which parents' height is *extremely* tall, say 3 standard deviations above the mean (essentially at the upper limit of observed heights), it's no surprise that their children would tend to be shorter. For all cases in which parents' height is extremely short, say 3 standard deviations below the mean, it's no surprise that their children would tend to be taller. In both these cases, we'd expect regression toward the mean. What's interesting and perhaps surprising is that regression of y toward the mean happens not only at the very extreme values of x. Exercise 11.98 shows why regression toward the mean happens.

EXAMPLE 8

The Placebo Effect, or Regression toward the Mean?

Picture the Scenario

A clinical trial admits subjects suffering from high blood cholesterol counts (over 225 mg/dl). The subjects are randomly assigned to take either a placebo or a drug that is being tested for reducing cholesterol levels. After the three-month study, the mean cholesterol level for subjects taking the drug drops from 270 to 230. However, the researchers are surprised to see that the mean cholesterol level for the placebo group also drops, from 270 to 250.

Question to Explore

Explain how this placebo effect could merely reflect regression toward the mean.

Think It Through

For a group of people, a person's cholesterol reading at one time would likely be positively correlated with their reading three months later. So, for all people who are not taking the drug, a subject with relatively high cholesterol at one time would also tend to have relatively high cholesterol three months later. By regression toward the mean, however, subjects who are relatively high at one time will, on the average, be lower at a later time. So, if a study gives placebo to people with relatively high cholesterol (that is, in the right-hand tail of the blood cholesterol distribution), on the average we expect their values three months later to be lower.

Insight

Regression toward the mean is pervasive. In sports, excellent performance tends to be followed by good, but less outstanding, performance. A football team that wins all its games in the first half of its schedule will probably not win all its games in the second half. A baseball player who hits 0.400 in the first month will probably not be hitting that high at the end of the season.

By contrast, the good news about regression toward the mean is that very poor performance tends to be followed by improved performance. If you got the worst score in your statistics class on the first exam, you probably did not do so poorly on the second exam (but you were probably still below the mean).

> **TO PRACTICE THIS CONCEPT, TRY EXERCISE 11.19**

The Squared Correlation (*r*-squared) Describes Predictive Power

If you know how many times a person can bench press 60 pounds, can you predict well their maximum bench press? Another way to describe the strength of association refers to how close predictions for y tend to be to observed y values. The variables are strongly associated if you can predict y much better by substituting x values into the prediction equation $\hat{y} = a + bx$ than by merely using the sample mean $\bar{y}$ and ignoring x.

For a given person, the prediction error is the difference between the observed and predicted values of y.

- The error using the regression line to make a prediction is $y - \hat{y}$.
- The error using $\bar{y}$ to make a prediction is $y - \bar{y}$.

For each potential predictor ($\hat{y}$ and $\bar{y}$), some errors are positive, some errors are negative, and the sum of the errors for the sample equals 0. You can summarize the sizes of the errors by the sum of their squared values,

$$\text{Error summary} = \sum(\text{observed } y \text{ value} - \text{predicted } y \text{ value})^2.$$

When we predict y using $\bar{y}$ (that is, ignoring x), the error summary equals

$$\sum(y - \bar{y})^2.$$

Recall

$\sum(y - \bar{y})^2$ is the numerator of the variance of the y values. $\sum(y - \hat{y})^2$ is what's minimized in finding the least squares estimates for the regression equation. ■

This is called the **total sum of squares**. When we predict y using x with the regression equation, the error summary equals

$$\sum(y - \hat{y})^2.$$

This is called the **residual sum of squares**, since it sums the squared residuals.

When a strong linear association exists, the regression equation predictions ($\hat{y}$) tend to be much better than $\bar{y}$. Then, $\sum(y - \hat{y})^2$ is much less than $\sum(y - \bar{y})^2$. The difference between the two error summaries depends on the units of measure: It's different, for instance, with weight measured in kilograms than in pounds. We can eliminate this dependence on units by converting the difference to a proportion, by dividing by $\sum(y - \bar{y})^2$. This gives the summary measure of association,

$$r^2 = \frac{\sum(y - \bar{y})^2 - \sum(y - \hat{y})^2}{\sum(y - \bar{y})^2}.$$

This measure falls on the same scale as a proportion, 0 to 1.

The measure r^2 is interpreted as the **proportional reduction in error**. If $r^2 = 0.40$, for instance, the error using $\hat{y}$ to predict y is 40% smaller than the error using $\bar{y}$ to predict y. We use the notation r^2 for this measure because, in fact, it can be shown that this measure equals the square of the correlation r.

IN PRACTICE: Get r^2 by squaring the correlation

If you know the correlation r, it is simple to calculate r^2, by squaring the correlation. The formula shown above for r^2 is useful for interpretation of r^2, but it's not needed for calculation.

EXAMPLE 9

What Does r^2 Tell Us in the Strength Study?

Picture the Scenario

For the female athlete strength study, Example 6 showed that x = number of 60-pound bench presses and y = maximum bench press had a correlation of 0.80.

Question to Explore

Find and interpret r^2.

Think It Through

Since the correlation $r = 0.80$, $r^2 = (0.80)^2 = 0.64$. For predicting maximum bench press, the regression equation has 64% less error than $\bar{y}$ has. "Error" here refers to the summary given by the sum of squared prediction errors.

Insight

Since $r^2 = 0.64$ is quite far from 0, we can predict y much better using the regression equation than using $\bar{y}$. In this sense, the association is quite strong.

TO PRACTICE THIS CONCEPT, TRY EXERCISE 11.14(C)

It's also possible to calculate r^2 directly from the definition. Software for regression routinely provides tables of sums of squares. Table 11.3 is an example. The heading SS stands for "sum of squares." From Table 11.3, the residual sum of squared errors is $\sum (y - \hat{y})^2 = 3522.8$, and the total sum of squares is $\sum (y - \bar{y})^2 = 9874.6$. Thus,

$$r^2 = \frac{\sum (y - \bar{y})^2 - \sum (y - \hat{y})^2}{\sum (y - \bar{y})^2} = \frac{9874.6 - 3522.8}{9874.6} = 0.643.$$

TABLE 11.3: Annotated MINITAB Table Showing Sums of Squares and r^2 for Strength Study

Under the heading for Source, "Total" refers to $\sum (y - \bar{y})^2$ and "Residual Error" refers to $\sum (y - \hat{y})^2$. The "Regression" sum of squares equals their difference, which is the numerator of r^2.

Source	SS	
		◄— SS stands for "Sum of squares"
Regression	6354.8	◄— This is 9874.6 – 3522.8, the numerator of r^2
Residual Error	3522.8	◄— $\Sigma(y - \hat{y})^2$
Total	9874.6	◄— $\Sigma(y - \bar{y})^2$
R-Sq = 64.3%		◄— This is r^2, in percentage terms

Normally, it is unnecessary to perform this computation, since most software reports r or r^2 or both.

A table that reports the sums of squares used in regression analysis is called an **analysis of variance table**, or **ANOVA table** for short. We'll discuss the ANOVA table further in Section 11.4.

Properties of r-squared

The r^2 measure, like the correlation r, measures the strength of *linear* association. We emphasize *linear* because r^2 compares predictive power of the *straight-line* regression equation to $\bar{y}$. The main properties of r^2 are as follows:

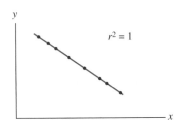

- Since $-1 \le r \le 1$, r^2 falls between 0 and 1.

- $r^2 = 1$ when $\sum(y - \hat{y})^2 = 0$, which happens only when all the data points fall exactly on the regression line. There is then no prediction error using x to predict y (that is, $y = \hat{y}$ for each observation). This corresponds to $r = \pm 1$.

- $r^2 = 0$ when $\sum(y - \hat{y})^2 = \sum(y - \bar{y})^2$. This happens when the slope $b = 0$, in which case each $\hat{y} = \bar{y}$. The regression line and $\bar{y}$ then give the same predictions.

- The closer r^2 is to 1, the stronger the linear association: The more effective the regression equation $\hat{y} = a + bx$ then is compared to $\bar{y}$ in predicting y.

Because the calculation of prediction error in r^2 refers to *squared* distances, it relates to comparing the overall *variance* of the y-values around $\bar{y}$ to the variance around the regression line. For example, $r^2 = 0.64$ means that the estimated variance around the regression line is 64% less than the overall variance of the y values. This is often phrased in the style, "64% of the variability in y is explained by x." This interpretation has the weakness, however, that variability is summarized by the *variance*. Some statisticians find r^2 to be less useful than r, because (being based on sums of squares) it uses the square of the original scale of measurement. It's easier to interpret the original scale than a squared scale. This is a strong advantage of the standard deviation over the variance.

SUMMARY: Correlation r and Its Square r^2

Both the correlation r and its square r^2 describe the strength of association. They have different interpretations. The correlation falls between -1 and $+1$. It represents the *slope* of the regression line when x and y have equal standard deviations. It governs the extent of "regression toward the mean." The r^2 measure falls between 0 and 1. It summarizes the reduction in sum of squared errors in *predicting* y using the regression line instead of using the mean of y.

What Factors Affect the Correlation?

Section 3.4 showed that certain "regression outliers" can unduly influence the slope and the correlation. A single observation can have a large influence if its x-value is unusually large or unusually small and if it falls quite far from the trend of the rest of the data.

Besides being influenced by outliers, the size of the correlation (and r^2) depends strongly on two other factors.

- First, if the subjects are *grouped* for the observations, such as when the data refer to county summaries instead of individual people, the correlation tends to increase in magnitude.

Suppose you want to find the correlation between number of years of education completed and annual income. You could measure these variables for a sample of individuals. Or you could use summary education and income measures such as means or medians for counties (or states or provinces). The scatterplot in Figure 11.7 shows that at the individual level, the correlation could be much weaker. Lots of variability in income exists for individuals, but not much variability in income (mean or median) exists for counties. The summary values for counties fall closer to a straight line.

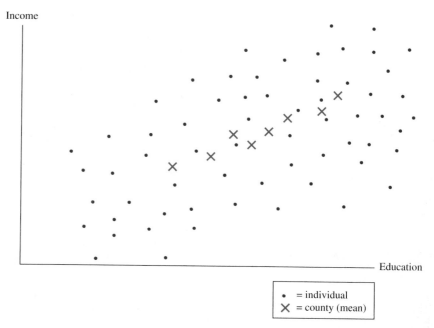

▲ **FIGURE 11.7: The Effect of Grouping of Subjects on the Correlation.** There is a stronger linear trend for the countywide data than for the data on individuals. **Question:** Why would you expect much more variability for income of individuals than for mean income of counties?

Because correlations can change dramatically when data are grouped, it is misleading to take results for groupings such as counties and extend them to individuals. If income and education are strongly correlated when we measure the summary means for each county, this does not imply they will be strongly correlated for measurements on individuals. Making predictions about individuals based on the summary results for groups is known as the **ecological fallacy** and should be avoided.

▪ Second, the size of the correlation depends on the range of x values sampled: The correlation tends to be smaller when we sample only a restricted range of x values than when we use the entire range.

For the regression of y = Internet use on x = GDP for 39 nations in the "Human development" data file, the correlation is 0.89. When we use only the nations that are above the median on GDP, however, the correlation is 0.44. When we use only the countries below the median on GDP, the correlation is 0.48. We get a misleading indication of the correlation between Internet use and GDP if we use a sample having only a restricted range of the x values.

By contrast, the *prediction equation* remains valid when we observe a restricted range of x values. The slope is usually not much affected by the range of x values, and we simply limit our predictions to that range. The correlation makes most sense, however, when x takes its full range of values. Otherwise, it can be misleading.

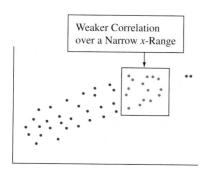

Weaker Correlation over a Narrow *x*-Range

EXAMPLE 10

How Well Does the SAT Predict College GPA?

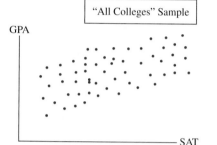

Picture the Scenario

Consider the correlation between score on the SAT college entrance exam and later performance in college as measured by GPA at the time of graduation.

Question to Explore

For which group would these variables have a stronger correlation: All students who graduate from Harvard University this year, or all students who graduate from college somewhere in the United States this year?

Think It Through

The correlation depends in magnitude on the variability in SAT scores. For Harvard students, the SAT scores will concentrate narrowly at the upper end of the scale. So, the correlation would probably be weak. By contrast, for all students who finish college, SAT scores would range from very low to very high. We would likely see a stronger correlation for them.

Insight

What reason explains this property of the correlation? Recall the formula, $r = b(s_x/s_y)$. When we use a much wider range of x values, s_x increases a lot. So, if the slope does not change much and if the variability of the y values is not much larger with the expanded sample, r will tend to increase because it is proportional in value to s_x.

TO PRACTICE THIS CONCEPT, TRY EXERCISE 11.25

On the Shoulders of: Francis Galton (1857–1936)

Heights and Peas – What do they have in common?

Francis Galton, like his cousin Charles Darwin, was interested in genetics. In his work on inheritance, he collected data on parents' and children's heights. He found that, *on the average*, the heights of children were not as extreme as those of the parents and tended toward the mean height. There was *regression toward the mean*. This refers to this *average*, not to every observation. It does not imply that over the course of many generations, heights will *all* get less and less spread out.

Galton also conducted experiments examining the weight of sweet peas. The mean weight of offspring peas was closer than the weight of the parent group to the mean for all peas. Again, Galton observed regression toward the mean.

Galton discovered a way to summarize linear associations numerically. He named this numerical summary the "coefficient of correlation." In 1896 the British statistician Karl Pearson (who proposed the chi-squared test for contingency tables in 1900) derived the current method for estimating the correlation using sample data.

11.9 Dollars and thousands of dollars: If a slope is 1.50 when x = income in thousands of dollars, then what is the slope when x = income in dollars? (*Hint:* A one-dollar change has only 1/1000 of the impact of a one-thousand-dollar change?)

11.10 Comparing slopes and correlations: Suppose we want
TRY to describe whether Internet use is more strongly associated with GDP or with fertility.

   a. Can we compare the slopes when GDP and fertility each predict Internet use in separate regression equations? Why or why not?

   b. According to the correlation matrix in Table 11.2, part of which is shown below, which is more strongly associated with Internet use? Why?

	INTERNET
GDP	0.888
CELLULAR	0.818
FER	−0.551
CO2	0.680

11.11 When can you compare slopes?: Although the slope does not measure association, it *is* useful for comparing effects for two variables that have the *same* units. For the "Human development" data file of 39 nations on the text CD, let x = GDP (thousands of dollars per capita). For predicting y = percentage using cell phones, the prediction equation is $\hat{y} = -0.13 + 2.62x$. For predicting y = percentage using Internet, the prediction equation is $\hat{y} = -3.61 + 1.55x$.

   a. Explain how to interpret the two slopes.

   b. For these nations, explain why a 1-unit increase in GDP has a slightly greater impact on the percentage using cell phones than on the percentage using the Internet.

11.12 Sketch scatterplot: Sketch a scatterplot, identifying quadrants relative to the sample means as in Figure 11.5, for which (a) the slope and correlation would be negative, (b) the slope and correlation would be approximately zero.

11.13 Sit-ups and the 40-yard dash: Is there a relationship
TRY between x = how many sit-ups you can do and y = how fast you can run 40 yards (in seconds)? The MINITAB output of a regression analysis for the female athlete strength study is shown here.

Predictor	Coef	SE Coef	T	P
Constant	6.7065	0.1779	37.70	0.000
SIT-UPS	−0.024346	0.006349	−3.83	0.000
R-Sq = 21.1%				

   a. Find the predicted time in the 40-yard dash for a subject who can do (i) 10 sit-ups, (ii) 40 sit-ups. (The minimum and maximum in the study were 10 and 39.) Relate the difference in predicted times to the slope.

   b. The number of sit-ups had mean 27.175 and standard deviation 6.887. The time in the 40-yard dash had mean 6.045 and standard deviation 0.365. Show how the correlation relates to the slope, and find its value.

11.14 Weight, height, and fat: For the female athlete strength
TRY study, the output shows a correlation matrix for height, weight, body mass index (BMI), and percentage of body fat (BF%).

	HT (IN)	WT (lbs)	BF%
WT (lbs)	0.553		
BF%	0.216	0.871	
BMI	0.153	0.898	0.945

   a. Which pair of variables has the (i) strongest association, (ii) weakest association.

   b. Interpret the sign and the strength of the correlation between height and weight.

   c. Find and interpret r^2 for height and weight.

   d. If height and weight were measured instead with metric units, would any results differ in (a)–(c)? Explain.

11.15 Student ideology: For the "FL student survey" data file on the text CD, the sample correlation between y = political ideology (scored 1 to 7, with 7 the most conservative) and x = number of times a week reading a newspaper is $r = -.066$.

   a. Interpret the sign of the correlation.

   b. Find and interpret r^2. Is the sample association strong, or weak? Explain.

11.16 Verbal and math SAT: All students who attend Lake Wobegon College must take the math and verbal SAT exams. Both exams have a mean of 500 and a standard deviation of 100. The regression equation relating y = math SAT score and x = verbal SAT score is $\hat{y} = 250 + 0.5x$.

   a. Find the predicted math SAT score for a student who has the mean verbal SAT score of 500. (Note: At the x value equal to $\bar{x}$, the predicted value of y equals $\bar{y}$.)

   b. Show how to find the correlation. Interpret its value as a standardized slope. (*Hint:* Both standard deviations are equal.)

   c. Find r^2 and interpret its value.

11.17 SAT regression toward mean: Refer to the previous exercise.

   a. Predict the math SAT score for a student who has a verbal SAT = 800.

   b. The correlation is 0.5. Interpret the prediction in (a) in terms of regression toward the mean.

11.18 GPAs and TV watching: For the "Georgia student survey" data file on the text CD, the table shows the correlation matrix for college GPA, high school GPA, and daily time spent watching TV.

	HSGPA	CGPA
CGPA	0.505	
WatchTV	−0.333	−0.353

   a. Interpret r and r^2 between time spent watching TV and college GPA.

   b. One student is two standard deviations above the mean on high school GPA. How many standard deviations would you expect that student to be above the mean on college GPA? Use your answer to explain "regression toward the mean."

11.19 Placebo helps cholesterol?: A clinical trial admits subjects suffering from high cholesterol, who are then randomly assigned to take a drug or a placebo for a twelve-week study. For the population, without taking any drug, the correlation between the cholesterol readings at times twelve weeks apart is 0.70. The mean cholesterol reading at any given time is 200, with the same standard deviation at each time. Consider all the subjects with a cholesterol level of 300 at the start of the study, who take placebo.

a. What would you predict for their mean cholesterol level at the end of the study?

b. Does this suggest that placebo is effective for treating high cholesterol? Explain.

11.20 Does tutoring help?: For a class of 100 students, the teacher takes the 10 students who performed poorest on the midterm exam and enrolls them in a special tutoring program. Both the midterm and final have a class mean of 70 with standard deviation 10, and the correlation is 0.50 between the two exam scores. The mean for the specially tutored students increases from 50 on the midterm to 60 on the final. Can we conclude that the tutoring program was successful? Explain, identifying the response and explanatory variables and the role of regression toward the mean.

11.21 What's wrong with your stock fund?: Last year you looked at all the financial firms that had stock growth funds. You picked the growth fund that had the best performance last year (ranking at the 99th percentile on performance) and invested all your money in it this year. This year, with their new investments, they ranked only at the 65th percentile on performance. Your friend suggests that their stock picker became complacent or was burned out. Can you give another explanation?

11.22 Golf regression: In the first round of a golf tournament, five players tied for the lowest round, at 65. The mean score of all players was 75. If the mean score of all players is also 75 in the second round, what does regression toward the mean suggest about how well we can expect the five leaders to do, on the average, in the second round? (*Hint*: Suppose the standard deviation is also the same in each round.)

11.23 Car weight and mileage: The "Car weight and mileage" data file on the text CD shows the weight and the mileage per gallon of gas of 25 cars of various models. The regression of mileage on weight has $r^2 = 0.75$. Explain how to interpret this in terms of how well you can predict a car's mileage if you know its weight.

11.24 Food and drink sales: The owner of Bertha's Restaurant is interested in whether an association exists between the amount spent on food and the amount spent on drinks for the restaurant's customers. She decides to measure each variable for every customer in the next month. Each day she also summarizes the mean amount spent on food and the mean amount spent on drinks. Which correlation between amounts spent on food and drink do you think would be higher, the one computed for the 2500 customers in the next month, or the one computed using the means for the 30 days of the month? Why? Sketch a sample scatterplot showing what you expect for each case as part of your answer.

11.25 Yale and UConn: For which student body do you think the correlation between high-school GPA and college GPA would be higher: Yale University or the University of Connecticut? Explain why.

11.26 Violent crime and single-parent families: Use software to analyze the "U.S. statewide crime" data file on the text CD on y = violent crime rate and x = percentage of single parent families.

a. Construct a scatterplot. What does it show?

b. One point is quite far removed from the others, having a much higher value on both variables than the rest of the sample, but it fits in well with the linear trend exhibited by the rest of the points. Show that the correlation changes from 0.77 to 0.59 when you delete this observation. Why does it drop so dramatically?

11.27 Correlations for the strong and for the weak: Refer to the "High school female athlete" data file on the text CD.

a. Find the correlation between x = number of 60-pound bench presses before fatigue and y = maximum bench press. Interpret.

b. Find the correlation using only the x values (i) below the median of 10, (ii) above the median of 10. Compare to the correlation in (a). Why are they so different?

11.3 | How Can We Make Inferences About the Association?

Section 11.1 showed how a regression line *models a relationship* between two quantitative variables, when the means of y at various values of x follow approximately a straight line. Section 11.2 showed how the correlation r and its square, r^2, describe the *strength of the association*.

The sample regression equation, r, and r^2 are *descriptive* parts of a regression analysis. The *inferential* parts of regression use the tools of confidence intervals and significance tests. They provide inference about the regression equation and the correlation and r-squared in the *population* of interest. For instance, in studying ways to measure strength, we're interested not just in the 57 female athletes in the Georgia study described in Example 1 but in a population of female athletes.

Assumptions for Regression Analysis

Some assumptions are needed to use regression in a descriptive manner. Extra assumptions are needed to make inferences. Let's start with the descriptive part.

Basic assumption for using regression line for description:

The population means of y at different values of x have a straight-line relationship with x, that is, $\mu_y = \alpha + \beta x$.

This assumption states that a *straight-line* regression model is valid. You can check it by constructing a scatterplot. Statistical inferences in regression analysis make two additional assumptions:

Extra assumptions for using regression to make statistical inference:

- The data were gathered using randomization, such as random sampling or a randomized experiment.
- The population values of y at each value of x follow a normal distribution, with the same standard deviation at each x value.

The randomness assumption applies for any statistical inference. For inferences to be valid, the sample must be representative of the population.

As in the case of inference about a mean, the assumption about normality is what leads to t test statistics having t distributions. This assumption is never *exactly* satisfied in practice. However, the closer reality resembles this ideal, the more appropriate are the confidence interval and test procedures for the regression model. This assumption is less important than the other two, especially when the sample size is large. In that case, an extended central limit theorem implies that sample slopes have bell-shaped sampling distributions, no matter what shape the population distribution of y has at each value of x.

Testing Independence between Quantitative Variables

Under the assumptions for inference, suppose that the slope β of the regression line $\mu_y = \alpha + \beta x$, equals 0, as Figure 11.8 shows. Then, the mean of y is identical at each x value. In fact, the two quantitative variables are *statistically independent*: The outcome for y does not depend on the value of x. It does not help us to know the value of x if we want to predict the value of y.

Recall

By definition, **models** merely approximate the true relationship. A relationship will not be exactly linear, with exactly normal distributions for y at each x and with exactly the same standard deviation of y values at each x value. A model is useful as long as the approximation is reasonably good. ■

Recall

Section 11.1 showed that a regression model allows **variability** around a line $\mu_y = \alpha + \beta x$, that describes how the mean of y changes as x changes. ■

▶ **FIGURE 11.8: Quantitative Variables x and y Are Statistically Independent when the True Slope $\beta = 0$.** Each normal curve shown here represents the variability in y values at a particular value of x. When $\beta = 0$, the normal distribution of y is the same at each value of x. **Question:** How can you express the null hypothesis of independence between x and y in terms of a parameter from the regression model?

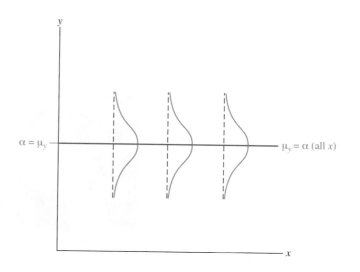

The null hypothesis that x and y are statistically independent is $H_0: \beta = 0$. Its significance test has the same purpose as the chi-squared test of independence for categorical variables (Chapter 10). It investigates how much evidence there is that the variables truly are associated. The smaller the P-value, the greater is the evidence.

Usually, the alternative hypothesis for the test of independence is two-sided, $H_a: \beta \neq 0$. Also possible is a one-sided alternative, $H_a: \beta > 0$ or $H_a: \beta < 0$, when you predict the direction of the association. The test statistic in each case equals

$$t = \frac{b - 0}{se},$$

where b is the sample slope and se denotes its standard error. Software supplies the slope b, the se, the t test statistic, and its P-value. The formula for se is rather complex, and we won't need it here (it's in Exercise 11.100.) The form of the test statistic is the usual one for a t or z test. We take the estimate (b) of the parameter (β), subtract the null hypothesis value ($\beta = 0$), and divide by the standard error. The se describes the spread of the sampling distribution that measures how that estimate varies from sample to sample of size n (see the margin figure).

The symbol t is used for this test statistic because, under the assumptions, it has the t sampling distribution. The P-value for $H_a: \beta \neq 0$ is the two-tail probability from the t distribution (Table B). The degrees of freedom are $df = n - 2$. Software provides the P-value for any value of df.

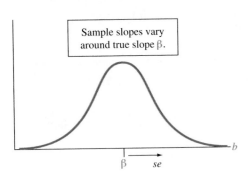

Sample slopes vary around true slope β.

Recall

Sections 7.3 and 8.3 used $df = n - 1$ for inference using the t distribution, so why is $df = n - 2$ here? We'll see that in regression, $df = n -$ number of parameters in equation for mean, μ_y.

The equation $\mu_y = \alpha + \beta x$ has two parameters (α and β) so $df = n - 2$. The inferences in Chapters 7 and 8 had one parameter (μ), so $df = n - 1$. More complex models in Chapter 12 with several explanatory variables need the more general formula. ■

Steps of Two-Sided Significance Test about a Population Slope β

1. **Assumptions:** (1) Population satisfies regression line $\mu_y = \alpha + \beta x$, (2) data gathered using randomization, (3) population y values at each x value have normal distribution, with same standard deviation at each x value.

2. **Hypotheses:** $H_0: \beta = 0$, $H_a: \beta \neq 0$

3. **Test statistic:** $t = (b - 0)/se$, where software supplies sample slope b and its se.

4. **P-value:** Two-tail probability of t test statistic value more extreme than observed, using t distribution with $df = n - 2$.

5. **Conclusions:** Interpret P-value in context. If decision needed, reject H_0 if P-value $\leq$ significance level (such as 0.05).

EXAMPLE 11

Is Strength Associated with 60-Pound Bench Presses?

Picture the Scenario

One purpose of the strength study introduced in Example 1 was to analyze whether $x =$ number of times an athlete can lift a 60-pound bench press helps us predict $y =$ maximum number of pounds the athlete can bench press. Table 11.4 shows the regression analysis for the 57 female athletes, with x denoted by BP_60 and y denoted by BP. The margin of the next page shows screen shots from the TI-83+/84.

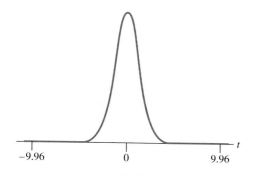

TI-83/84+ *Output*

TABLE 11.4: MINITAB Printout for Regression Analysis of y = Maximum Bench Press (BP) and x = Number of 60-Pound Bench Presses (BP_60)

The regression equation is BP = 63.5 + 1.49 BP_60

Predictor	Coef	SE Coef	T	P
Constant	63.537	1.956	32.48	0.000
BP_60	1.4911	0.150	9.96	0.000

R-Sq = 64.3%

Questions to Explore

a. Conduct a two-sided significance test of the null hypothesis of independence.

b. Report the P-value for the alternative hypothesis of a positive association, which is a sensible one for these variables. Interpret the results in context.

Think It Through

a. From Table 11.4, the prediction equation is $\hat{y} = 63.5 + 1.49x$. Here are the steps of a significance test of the null hypothesis of independence.

1. Assumptions: The scatterplot in Figure 11.1, shown again in the margin, revealed a linear trend, with scatter of points having similar spread at different x values. The straight-line regression model $\mu_y = \alpha + \beta x$ seems appropriate. The 57 female athletes were *all* the female athletes at a particular high school. This was a convenience sample rather than a random sample of the population of female high school athletes. Although the goal was to make inferences about that population, inferences are tentative because the sample was not random.

2. Hypotheses: The null hypothesis that the variables are independent is $H_0: \beta = 0$. The two-sided alternative hypothesis of dependence is $H_a: \beta \neq 0$.

3. Test statistic: For the sample slope $b = 1.49$, Table 11.4 reports standard error, $se = 0.150$. This is listed under "SE Coef," in the row of the table for the predictor (BP_60). For testing $H_0: \beta = 0$, the t test statistic is

$$t = (b - 0)/se = 1.49/0.150 = 9.96.$$

This is extremely large. The sample slope falls nearly 10 standard errors above the null hypothesis value. The t test statistic appears in Table 11.4 under the column labeled "T." The sample size equals $n = 57$, so $df = n - 2 = 55$.

4. P-value: The P-value, listed in Table 11.4 under the heading "P," is 0.000 rounded to three decimal places. Software reports the P-value for the two-sided alternative, $H_a: \beta \neq 0$. It is the two-tailed probability of "more extreme values" above 9.96 and below −9.96. See the margin figure.

5. Conclusion: If H_0 were true that the population slope β equals 0, it would be extremely unusual to get a sample slope as far from 0 as $b = 1.49$. The P-value gives very strong evidence against H_0. We conclude that an association exists between the number of 60-pound bench presses and maximum bench press.

b. The one-sided alternative $H_a: \beta > 0$ predicts a positive association. For it, the P-value is halved, because it is then the right-tail probability of $t > 9.96$. This also equals 0.000, to three decimal places. On the average, we infer that maximum bench press increases as the number of 60-pound bench presses increases.

Insight

In practice, studies must often rely on convenience samples. Results may be biased if the study subjects differ in an important way from those in the population of interest. Here, inference is reliable only to the extent that the sample is representative of the population of female high school athletes. This is a common problem with studies of this type, in which it would be difficult to arrange for a random sample of all subjects but a sample is conveniently available locally. We can place more faith in the inference if similar results occur in other studies.

The printout in Table 11.4 also contains a standard error and t statistic for testing that the population y-intercept α equals 0. This information is usually not of interest. Rarely is there any reason to test the hypothesis that a y-intercept equals 0.

TO PRACTICE THIS CONCEPT, TRY EXERCISE 11.29(A)

You probably did not expect maximum bench press and number of 60-pound bench presses to be independent. The test confirms that they are positively associated. As usual, a confidence interval is more informative. It will help us learn about the actual *size* of the slope.

A Confidence Interval Tells Us How Closely We Can Estimate the Slope

Recall

For **95% confidence**, the error probability is 0.05, and the t-score with half this probability in the right tail is denoted $t_{.025}$ in the t table. ■

A small P-value in the significance test of $H_0: \beta = 0$ suggests that the population regression line has a nonzero slope. To learn just how far the slope β falls from 0, we construct a confidence interval. A 95% confidence interval for β has the formula

$$b \pm t_{.025}(se).$$

The t-score is found by software or in a t table (such as Table B) with $df = n - 2$. This inference makes the same assumptions as the significance test.

EXAMPLE 12

Estimating the Slope for Predicting Maximum Bench Press

Picture the Scenario

For the female athlete strength study, the sample regression equation is $\hat{y} = 63.5 + 1.49x$. From Table 11.4, the sample slope $b = 1.49$ has standard error $se = 0.150$.

Questions to Explore

a. Construct a 95% confidence interval for the population slope β.

b. What are the plausible values for the increase in maximum bench press, on the average, for each additional 60-pound bench press that a female athlete can do?

Think It Through

a. For a 95% confidence interval, the $t_{.025}$ value for $df = n - 2 = 55$ is $t_{.025} = 2.00$. (If your software does not supply t-scores, you can find or approximate $t_{.025}$ from Table B.) The confidence interval is

$$b \pm t_{.025}(se) = 1.49 \pm 2.00(0.150),$$
$$\text{which is } 1.49 \pm 0.30 \text{ or} (1.2, 1.8).$$

b. We can be 95% confident that the population slope β falls between 1.2 and 1.8. On average, the maximum bench press increases by between 1.2 and 1.8 pounds for each additional 60-pound bench press that an athlete can do.

Insight

Confidence intervals and two-sided significance tests about slopes are consistent: When a two-sided test has P-value below 0.05, casting doubt on $\beta = 0$, the 95% confidence interval for β does not contain 0.

TO PRACTICE THIS CONCEPT, TRY EXERCISE 11.29(B)

A confidence interval for β may not have a useful interpretation if a one-unit increase in x is relatively small (or large) in practical terms. We can then instead estimate the effect for an increase in x that is a more relevant portion of the actual range of x values. For instance, an increase of 10 units in x has change 10β in the mean of y. To find the confidence interval for 10β, multiply the endpoints of the confidence interval for β by 10.

In the strength study, x = number of 60-pound bench presses varied between 2 and 35, with a standard deviation of 7.1. A change of 1 is very small. Let's estimate the effect of a 10-unit increase in x. Since the 95% confidence interval for β is (1.2, 1.8), the 95% confidence interval for 10β has endpoints $10(1.2) = 12$ and $10(1.8) = 18$. On the average, we infer that the maximum bench press increases by at least 12 pounds and at most 18 pounds, for an increase of 10 in the number of 60-pound bench presses. For instance, the interval from 12 to 18 is also a 95% confidence interval for the difference between the mean maximum bench press for athletes who can do $x = 25$ bench presses and athletes who can do $x = 15$ (since $25 - 15 = 10$).

Inferences Also Apply to the Correlation

Since $r = b(s_x/s_y)$, the sample correlation $r = 0$ whenever the sample slope $b = 0$. Likewise, the population correlation equals 0 precisely when the population slope $\beta = 0$. We don't need to test whether the population correlation equals 0, because its t statistic is identical to the one for testing H_0: $\beta = 0$ using the sample slope b.

Software that reports the correlation matrix gives the option of getting the two-sided P-value for testing the significance of the correlation. Table 11.5 illustrates the way MINITAB reports the correlation matrix for the variables in the strength study (x = BP_60 and y = BP), including also two other variables that measure the maximum leg press (LP) and the number of 200-pound leg presses that the athlete can do (LP_200). For instance, the correlation between maximum bench press (BP) and the number of 60-pound bench presses (BP_60) equals 0.802. Its P-value for testing that the population correlation equals zero against the two-sided alternative is 0.000. This must be the same as the P-value (in Table 11.4) for the test for the slope. Table 11.5 lists the P-value underneath the correlation value. In fact, the P-value is 0.000 for the correlation for each of the six pairs of variables in this table.

TABLE 11.5: MINITAB Printout of Correlation Matrix for Strength Study

The value below a correlation is the P-value for testing that the population correlation equals 0.

	BP_60	BP	LP_200	
BP	0.802			
	0.000			
LP_200	0.611	0.578		
	0.000	0.000		
LP	0.669	0.501	0.793	← correlation
	0.000	0.000	0.000	← P-value

11.3 Practicing the Basics

11.28 t-score?: A regression analysis is conducted with 25 observations.

  a. What is the *df* value for inference about the slope β?

  b. Which two *t* test statistic values would give a P-value of 0.05 for testing H_0: $\beta = 0$ against H_a: $\beta \neq 0$?

  c. Which *t*-score would you multiply the standard error by in order to find the margin of error for a 95% confidence interval for β?

11.29 Predicting house prices: For the "house selling prices" data file on the text CD, MINITAB results of a regression analysis are shown for 100 homes relating y = selling price (in dollars) to x = the size of the house (in square feet).

  a. Show all steps of a two-sided significance test of independence. Could the sample association between these two variables be explained by random variation?

  b. Show that a 95% confidence interval for the population slope is (64, 90).

  c. A builder had claimed that selling price increases $100, on the average, for every extra square foot. Based on (b), what would you conclude about this claim?

House selling prices and size of home

Predictor	Coef	SE Coef	T	P
Constant	9161	10760	0.85	0.397
size	77.008	6.626	11.62	0.000

11.30 House prices in bad part of town: Refer to the previous exercise. Of the 100 homes, 25 were in a part of town considered less desirable. For a regression analysis using y = selling price and x = size of house for these 25 homes,

  a. You plan to test H_0: $\beta = 0$ against H_a: $\beta > 0$. Explain what H_0 means, and explain why a data analyst might choose a one-sided H_a for this test.

  b. For this one-sided alternative hypothesis, how large would the *t* test statistic need to be in order to get a P-value equal to (i) 0.05, (ii) 0.01?

11.31 Strength as leg press: The female athletes strength study also considered prediction of y = maximum leg press (LP) using x = number of 200-pound leg presses (LP_200). MINITAB results of a regression analysis are shown.

```
The regression equation is LP = 234 + 5.27 LP_200
Predictor    Coef      SE Coef    T       P
Constant     233.89    13.06      17.90   0.000
LP_200       5.2710    0.5469     9.64    0.000
R-Sq = 62.8%
```

  a. Show all steps of a two-sided significance test of the hypothesis of independence.

  b. Find a 95% confidence interval for the true slope. What do you learn from this that you cannot learn from the significance test in (a)?

11.32 More boys are bad?: A study of 375 women who lived in pre-industrial Finland (by S. Helle et al., *Science*, vol. 296, p. 1085, 2002) using Finnish church records from 1640 to 1870 found that there was roughly a linear relationship between y = lifelength (in years) and x = number of sons the woman had, with a slope estimate of $-0.65(se = 0.29)$.

  a. Interpret the sign of the slope. Is the effect of having more boys good, or bad?

  b. Show all steps of the test of the hypothesis that lifelength is independent of number of sons, for the two-sided alternative hypothesis. Interpret the P-value.

  c. Construct a 95% confidence interval for the true slope. Interpret. Is it plausible that the effect is relatively weak, with true slope near 0?

11.33 More girls are good?: Repeat the previous exercise using x = number of daughters the woman had, for which the slope estimate was 0.44 ($se = 0.29$).

11.34 CI and two-sided tests correspond: Refer to the previous two exercises. Using significance level 0.05, what decision would you make? Explain how that decision is in agreement with whether 0 falls in the confidence interval. Do this for the data for (a) girls, (b) boys.

11.35 Advertising and sales: Each month, the owner of Café Gardens restaurant records y = monthly total sales receipts and x = amount spent that month on advertising, both in thousands of dollars. For the first four months of operation, the observations are as shown in the table. The correlation equals 0.857.

Advertising	Sales
0	4
1	7
2	8
5	9

   a. Find the mean and standard deviation for each variable.

   b. Using the formulas for the slope and the y-intercept or software, find the regression line.

   c. The se of the slope estimate is 0.364. Test the null hypothesis that these variables are independent, using a significance level of 0.05.

■ **11.36 Student GPAs:** Refer to the "Georgia student survey" data file on the text CD. Treat college GPA as the response variable and high school GPA as the explanatory variable, and suppose these students are a random sample of all University of Georgia students.

   a. Can you conclude that these variables are associated in that population? Show all steps of the relevant significance test with significance level 0.05, and interpret.

   b. Find a 95% confidence interval for the population slope. Interpret the endpoints, and explain the correspondence with the result of the significance test in (a).

11.4 | What Do We Learn from How the Data Vary Around the Regression Line?

We've used regression to describe and to make inferences about the relationship between two quantitative variables. Now, we'll see what we can learn from the variability of the data around the regression line. We'll see that a type of residual helps us detect unusual observations. We'll also see that the sizes of the residuals affect how well we can predict y or the mean of y at any given value of x.

Recall

A **residual** is a prediction error—the difference $y - \hat{y}$ between an observed outcome y and its predicted value $\hat{y}$. For contingency tables, Section 10.4 constructed a **standardized residual** by dividing the difference between a cell count and its expected value by a standard error. In regression, an observation y takes the place of the cell count and the predicted value $\hat{y}$ takes the place of the expected cell count. ■

Standardized Residuals

The magnitude of the residuals depends on the units of measurement for y. If we measure the bench press in kilograms instead of pounds, we'll get different residuals. A *standardized* version of the residual does not depend on the units. It equals the residual divided by a standard error that describes the sampling variability of the residuals. This ratio is called a **standardized residual**,

$$\text{Standardized residual} = \frac{y - \hat{y}}{se(y - \hat{y})}.$$

The se formula is complex, but we can rely on software to find this.

   A standardized residual behaves like a z-score. It indicates how many standard errors a residual falls from 0. If the relationship is truly linear and if the standardized residuals have approximately a bell-shaped distribution, absolute values larger than about 3 should be quite rare. Often, observations with standardized residuals larger than 3 in absolute value represent outliers—observations that are far from what the model predicts.

EXAMPLE 13

Detecting an Underachieving College Student

Picture the Scenario

Two of the variables in the "Georgia student survey" data file on the text CD are college GPA and high school GPA (variables CGPA and HSGPA). These were measured for a sample of 59 students at the University of Georgia. Identifying y = CGPA and x = HSGPA, we find $\hat{y} = 1.19 + 0.64x$.

Question to Explore

MINITAB highlights observations that have standardized residuals with absolute value larger than 2 in a table of "unusual observations." Table 11.6 shows this table for these data. Interpret the results for observation 59.

TABLE 11.6: Observations with Large Standardized Residuals in Student GPA Regression Analysis, as Reported by MINITAB

Obs	HSGPA	CGPA	Fit	Residual	St Resid	← standardized residuals
14	3.30	2.60	3.29	−0.69	−2.26R	
28	3.80	2.98	3.61	−0.63	−2.01R	
59	3.60	2.50	3.48	−0.98	−3.14R	

R denotes an observation with a large standardized residual.

Think It Through

Observation 59 is a student who had high school GPA $x = 3.60$, college GPA $y = 2.50$, predicted college GPA $\hat{y} = 3.48$ (the "fit"), and residual $= y - \hat{y} = -0.98$. The reported standardized residual of -3.14 indicates that the residual is 3.14 standard errors below 0. This student's actual college GPA is quite far below what the regression line predicts.

Insight

Based on his or her high school GPA and predicted college GPA, this student with an actual college GPA of 2.50 seems to be an underachiever in college.

TO PRACTICE THIS CONCEPT, TRY EXERCISE 11.37

When a standardized residual is larger than about 3 in absolute value, check out whether the observation is unusual in some way. Does it fall well away from the linear trend that the other points follow? Does it have too much influence on the results? As we've observed, a severe regression outlier can substantially affect the fit, especially when the value of the explanatory variable is relatively high or low and when the overall sample size is not very large.

Keep in mind that *some* large standardized residuals may occur just because of ordinary random variability. Table 11.6 reports three observations with standardized residuals having absolute value above 2 (labeled by R in the table). Even if the model is perfect we'd expect about 5% of the standardized residuals to have absolute value above 2 just by chance. So it's not at all surprising to find 3 such values out of 59 observations. This does not suggest that the model fits poorly.

Checking the Response Distribution with a Histogram of Residuals

To detect unusual observations, it's helpful to construct a histogram of the residuals. This also helps us to check the inference assumption that the conditional distribution of y is normal. If this assumption is true, the residuals should have approximately a bell-shaped histogram. To check this, software can construct a histogram of the residuals or the standardized residuals.

Recall

About 5% of a normal distribution is more than 2 standard deviations from the mean. ■

EXAMPLE 14

What Do Residuals for Predicting College GPA Tell Us?

Picture the Scenario

For the regression model of Example 13 predicting

college GPA from high school GPA, Figure 11.9 is a MINITAB histogram of the standardized residuals. The margin shows a MINITAB boxplot of the standardized residuals.

Question to Explore

What does Figure 11.9 tell you?

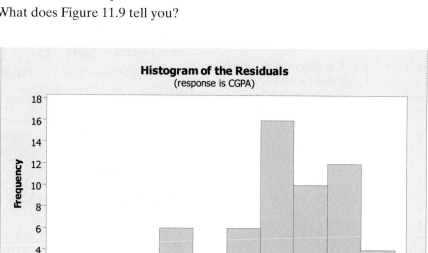

▲ **FIGURE 11.9: Histogram of Standardized Residuals for Regression Model Predicting College GPA.** **Question:** How many observations do the three left-most bars represent?

Think It Through

The standardized residuals show some skew to the left. It may be that for a fixed value of high school GPA, college GPA tends to be skewed to the left, with some students doing much more poorly than the regression model would predict. The large negative standardized residual of -3.14 in Example 13, summarized by the left-most bar in Figure 11.9, may merely reflect skew in the distribution of college GPA. However, each of the three bars farthest to the left represents only a single observation, so this conclusion is tentative and requires a larger sample to check more thoroughly.

Insight

The sample size was 59. When n is not especially large, a graph like Figure 11.9 is an imprecise estimate of a corresponding graph for the population. Although this graph shows some evidence of skew, much of this evidence is based on only 3 observations. This is not strong evidence that the conditional distribution is highly nonnormal. In viewing such graphs, we need to be careful not to let a few observations influence our judgment too much. We're mainly looking for dramatic departures from the assumptions.

TO PRACTICE THIS CONCEPT, TRY EXERCISE 11.39

If the distribution of the residuals is not normal, two-sided inferences about the slope parameter still work quite well. The *t* inferences are *robust*. The normality assumption is not as important as the assumption that the regression equation approximates well the true relationship between the predictor and the mean of *y*. If the model assumes a linear effect but the effect is actually U-shaped, for instance, descriptive and inferential statistics will be seriously faulty.

We'll study residuals in more detail in the next chapter. We'll learn there how to plot them to check other model assumptions.

Recall

From Section 11.1, σ **describes the spread** of *y* values at each fixed *x* value, that is, the spread of a **conditional distribution** of *y* values. ▪

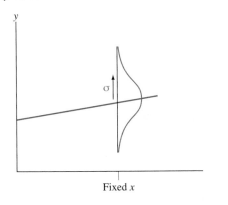

The Residual Standard Deviation and Mean Square Error (MSE)

Recall that the sample prediction equation $\hat{y} = a + bx$ estimates a population regression equation, $\mu_y = \alpha + \beta x$. For statistical inference, the regression model also assumes that the conditional distribution of *y* at a fixed value of *x* is normal, with the same standard deviation at each *x*. This standard deviation, denoted by σ, refers to the variability of *y* values for all subjects with the same value of *x*. This is a parameter that also can be estimated from the data.

The estimate of σ uses $\sum (y - \hat{y})^2$, the *residual sum of squares*, which summarizes sample variability about the regression line. The estimate, called the **residual standard deviation**, is

$$ s = \sqrt{\frac{\sum (y - \hat{y})^2}{n - 2}}. $$

It describes the typical size of the residuals. The $n - 2$ term in the denominator is the *df* value of the *t* distribution used for inference about β in the previous section.

EXAMPLE 15

How Variable Are the Athletes' Strengths?

Picture the Scenario

Let's return to the analysis of *y* = maximum bench press and *x* = number of 60-pound bench presses, for 57 female high school athletes. The prediction equation is $\hat{y} = 63.5 + 1.49x$. In Table 11.8 below, we'll see that the residual sum of squares equals 3522.8.

Questions to Explore

a. Find the residual standard deviation of *y*.

b. Interpret the value you obtain in context at the sample mean value for *x* of 11.

Think It Through

a. Since $n = 57$, $df = n - 2 = 55$. The residual standard deviation of *y* values is

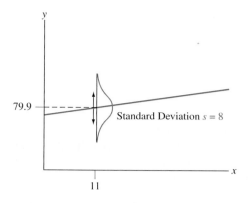

79.9 ----

Standard Deviation $s = 8$

11

Recall

$$s_y = \sqrt{\frac{\sum(y - \bar{y})^2}{n - 1}}$$ ▪

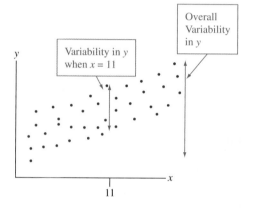

Overall Variability in y

Variability in y when $x = 11$

11

$$s = \sqrt{\frac{\sum(y - \hat{y})^2}{n - 2}} = \sqrt{\frac{3522.8}{55}} = \sqrt{64.1} = 8.0.$$

b. At any fixed value x of number of 60-pound bench presses, the model estimates that the maximum bench press values vary around a mean of $\hat{y} = 63.5 + 1.49x$ with a standard deviation of 8.0. See the figure in the margin.

At $\bar{x} = 11$, the predicted maximum bench press is $\hat{y} = 63.5 + 1.49(11) = 79.9$. For female high school athletes who can do 11 60-pound bench presses, we estimate that the maximum bench press values have a mean of about 80 pounds and a standard deviation of 8.0 pounds.

Insight

Why does the residual standard deviation ($s = 8.0$) differ from the standard deviation ($s_y = 13.3$) of the 57 y values in the sample? The reason is that s_y refers to the variability of *all* the y values around their mean, not just those at a fixed x value. That is, $s_y = 13.3$ describes variability about the overall mean of $\bar{y} = 80$ for *all* 57 high school female athletes, whereas $s = 8.0$ describes variability at a fixed x value such as $x = 11$. See the margin figure.

When the correlation is strong, at a fixed value of x we see less variability than the overall sample has. For instance, at the fixed value $x = 11$, we describe the variability in maximum bench press values by $s = 8.0$, whereas overall we describe variability in maximum bench press values by $s_y = 13.3$.

TO PRACTICE THIS CONCEPT, TRY EXERCISE 11.43(A)

How Can Intervals Predict y Values and Their Mean at a Given Value of x?

Consider all female high school athletes who can manage $x = 11$ 60-pound bench presses before fatigue. Based on the results for the sample, which do you think you can estimate more precisely: $y =$ maximum bench press for a randomly chosen *one* of them, or $\mu_y =$ the *mean* of the y values for *all* of them? We'll see that the estimate is the same in each case, but the margins of error are quite different.

For the straight-line regression model, we estimate μ_y, the population mean of y at a given value of x, by $\hat{y} = a + bx$. How good is this estimate? We can use its *se* to construct a 95% **confidence interval for μ_y.** This interval is

$$\hat{y} \pm t_{.025} \, (se \text{ of } \hat{y}).$$

Again, the *t*-score has $df = n - 2$. The standard error formula is complex (we'll show a simple approximation below), and in practice we rely on software.

The estimate $\hat{y} = a + bx$ for the *mean* of y at a fixed value of x is also a prediction for the outcome on y for a *particular* subject at that value. With most regression software you can form an interval within which an outcome y itself is likely to fall. The interval is called a **prediction interval for y.**

What's the difference between the prediction interval for y and the confidence interval for μ_y? The *prediction interval for y* is an inference about where *individual observations* fall, whereas the *confidence interval for μ_y* is an inference about where a *population mean* falls. Use a prediction interval for y if you want to predict where a single observation on y will fall. Use a confidence interval for μ_y if you want to estimate the mean of y for everyone having a particular x value. These inferences make the same assumptions as the regression inferences of the previous section.

EXAMPLE 16

Predicting Maximum Bench Press and Estimating Its Mean

Picture the Scenario

We've seen that the equation $\hat{y} = 63.5 + 1.49x$ predicts y = maximum bench press using x = number of 60-pound bench presses. For $x = 11$, Table 11.7 shows how MINITAB reports a confidence interval (CI) for the population mean of y and a prediction interval (PI) for a single y value. The predicted value itself is reported under the heading "Fit."

TABLE 11.7: MINITAB Output for Confidence Interval (CI) and Prediction Interval (PI) on Maximum Bench Press for Athletes Who Do 11 60-Pound Bench Presses before Fatigue

```
Predicted Values for New Observations
New Obs     Fit   SE Fit      95% CI            95% PI
    1     79.94   1.06   (77.81, 82.06)   (63.76, 96.12)
Values of Predictors for New Observations
New Obs     BP_60
    1       11.0
```

Questions to Explore

a. Using $\hat{y}$ and its se reported in Table 11.7, find and interpret a 95% *confidence interval* for the population mean of the maximum bench press values for all female high school athletes who can do $x = 11$ sixty-pound bench presses.

b. Report and interpret a 95% *prediction interval* for a single new observation on maximum bench press, for a randomly chosen female high school athlete with $x = 11$.

Think It Through

a. At $x = 11$, the predicted maximum bench press is $\hat{y} = 63.5 + 1.49(11) = 79.94$ pounds, the "fit." Table 11.7 reports a standard error for this estimate, $se = 1.06$, under the heading "SE Fit". With $n = 57$ we've seen that $df = n - 2 = 55$ and the t-score is $t_{.025} = 2.00$. So, the 95% *confidence interval for the population mean* of maximum bench press values at $x = 11$ is

$$\hat{y} \pm t_{.025}(se \text{ of } \hat{y}),$$

which is $79.94 \pm 2.00(1.06)$, or 79.94 ± 2.12, that is $(77.8, 82.1)$.

This is labeled as "95% CI" on the MINITAB printout. For *all* female high school athletes who can do 11 sixty-pound bench presses, we are 95% confident that the *mean* of their maximum bench press values falls between about 78 and 82 pounds.

b. MINITAB reports the 95% *prediction interval* (63.8, 96.1) under the heading "95% PI." This predicts where maximum bench press y will fall for a randomly chosen female high school athlete having $x = 11$. Equivalently, this refers to where 95% of the corresponding population values fall. For *all* female high school athletes who can do 11 sixty-pound bench presses, we predict that 95% of them have maximum bench press between about 64 and 96 pounds. Look at Figure 11.1, reproduced in the margin. Locate $x = 11$. Of all possible data points at that x value, we predict that 95% of them would fall between about 64 and 96.

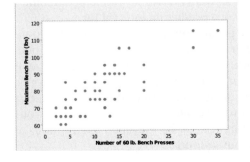

Insight

The 95% *prediction interval* (63.8, 96.1) predicts the maximum bench press *y* for a randomly chosen female high school athlete having *x* = 11. The 95% *confidence interval* (77.8, 82.1) estimates the *mean* of such *y* values for *all* female high school athletes having *x* = 11. The prediction interval for a *single* observation *y* is much wider than the confidence interval for the *mean* of *y*. In other words, you can estimate a population mean more precisely than you can predict a single observation.

TO PRACTICE THIS CONCEPT, TRY EXERCISE 11.41

The margins of error for these intervals use the residual standard deviation *s*. For an approximately normal conditional distribution for *y*, about 95% of the observations fall within about two standard deviations of the true mean μ_y at a particular value of *x*. For large *n*, $\hat{y}$ is close to μ_y, especially near the mean of the *x* values. Also, for large *n* the residual standard deviation *s* estimates the standard deviation σ well. So we could predict that about 95% of the *y* values would fall within $\hat{y} \pm 2s$. In fact, this is roughly what software does to form a prediction interval, when *n* is large and *x* is at or near the mean. The margin of error is approximately a *t*-score times *s* for predicting an individual observation and a *t*-score times $s/\sqrt{n}$ for estimating a mean.

Prediction Interval for *y* and Confidence Interval for μ_y at fixed value of *x*:

For large samples with an *x* value equal to or close to the mean of *x*,

■ The 95% **prediction interval** for *y* is approximately $\hat{y} \pm 2s$.

■ The 95% **confidence interval** for μ_y is approximately

$$\hat{y} \pm 2(s/\sqrt{n}),$$

where *s* is the residual standard deviation. Software uses *exact* formulas. We show these *approximate* formulas here merely to give a sense of what these intervals do. *In practice, use software.*

For instance, for female student athletes, about 95% of the observations fall within about 2*s* = 2(8.0) = 16 of the true mean at a particular value of *x*. At *x* = 11, which is the mean of *x* for the sample, $\hat{y}$ = 79.94 ≈ 80. So we predict that about 95% of the maximum bench press values would fall within $\hat{y} \pm 2s$, which is 80 ± 16, or between 64 and 96. This is the 95% PI result in Table 11.7. This formula is only an approximate one (because of using $\hat{y}$ as an approximation for μ_y and 2 as an approximation for the *t*-score), and you should rely on software for more precise results.

Use these confidence intervals and prediction intervals with caution. For them to be valid, the true relationship must be close to linear, with about the same variability of *y* values at each fixed *x* value. You can get a rough visual check from the scatterplot. The variability of points around the regression line should be similar for the various possible *x* values. If it isn't, don't use these confidence and prediction intervals.

Because the model assumptions never hold exactly in practice, these inferences are sometimes inaccurate. For instance, suppose the variability in *y* values is small at small values of *x* and large at large values of *x*. The prediction interval will have similar widths in each case. See Figure 11.10, which shows curves for the lower and upper endpoints of prediction intervals at all the possible *x* values. At a fixed *x* value, there should be a 95% chance a data point falls between the curves. In case (a), prediction intervals are justified. In case (b), they are not. At a large value of x in case (b), a 95% prediction interval for y actually has a much smaller than 95% chance (perhaps even less than 50%) of containing a value of y.

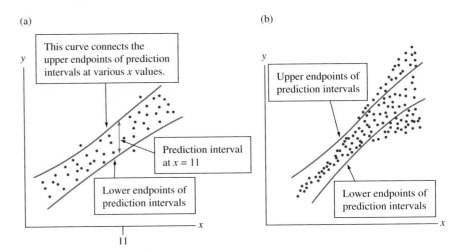

▲ **FIGURE 11.10: When a Regression Model Assumption is Badly Violated, Prediction Intervals May Perform Poorly.** The curves plot how the endpoints of a 95% prediction interval change as x changes. For case (a), inferences are valid. For case (b), a 95% prediction interval has actual probability of containing y above 0.95 for small x and below 0.95 for large x. **Question:** How can you tell from the figure that case (b) will lead to inaccurate inferences?

The Analysis of Variance (ANOVA) Table Summarizes Variability

Software reports the sums of squares (SS) used in estimating standard deviations and variances (their squares) in an **analysis of variance (ANOVA) table**. Table 11.8 shows the ANOVA table that MINITAB software reports for the strength study.

TABLE 11.8: Analysis of Variance (ANOVA) Table for the Strength Study

Analysis of Variance

Source	DF	SS	MS	F	P
Regression	1	6351.8	6351.8	99.17	0.000
Residual Error	55	3522.8	64.1		
Total	56	9874.6			

In the ANOVA table, the *total sum of squares* is $\sum(y - \bar{y})^2$. It summarizes the total variability of the y values. The ANOVA table breaks the total SS into two parts. One part is the *residual sum of squares*, $\sum(y - \hat{y})^2$. It represents the error in using the regression line to predict y. The other part is called the *regression sum of squares*. Its formula is

regression SS $= \sum(\hat{y} - \bar{y})^2$, and it equals $\sum(y - \bar{y})^2 - \sum(y - \hat{y})^2$.

It summarizes how much less error there is in predicting y using the regression line compared to using $\bar{y}$. Recall that this difference is the numerator of r^2. The sum

Regression SS + Residual SS = Total SS.

In Table 11.8, regression SS + residual SS $= 6351.8 + 3522.8 = 9874.6$, the total SS. Each sum of squares in the ANOVA table has an associated degrees

of freedom value. For instance, *df* for the residual sum of squares equals
$n - 2 = 57 - 2 = 55$. The ratio of a sum of squares to its *df* value is called a
mean square. It is listed in the ANOVA table under the heading MS.

In Table 11.8 the ratio of $\sum(y - \hat{y})^2 = 3522.8$ to its $df = 55$ gives 64.1.
This ratio is the **mean square error**, often abbreviated by **MS error** or by **MSE**.
It is roughly a mean of the squared errors in using the regression line to predict *y*. Equaling $s^2 = \sum(y - \hat{y})^2/(n - 2)$, it estimates the *variance* of the
conditional distributions. In practice, it's easier to interpret its square root, *s*,
the residual standard deviation of *y*.

The ANOVA *F* Statistic Also Tests for an Association

The ANOVA table contains values labeled *F* and P. The *F* value is the ratio of
the mean squares,

$$F = \frac{\text{Mean square for regression}}{\text{Mean square error}}.$$

This is an alternative test statistic for testing $H_0: \beta = 0$ against $H_a: \beta \neq 0$. In
fact, it can be shown that *F is the square of the t statistic* for this null hypothesis.
Using the sampling distribution of *F* values, which we'll discuss in the next
chapter, we get exactly the same P-value as with the *t* test for the two-sided alternative hypothesis.

In Table 11.8, the *F* statistic equals the ratio of MS values, $6351.8/64.1 =$
99.17. Table 11.4 listing the parameter estimates and their standard errors
showed the *t* test:

Predictor	Coef	SE Coef	T	P
BP_60	1.4911	0.150	9.96	0.000

The *F* statistic of 99.17 here is the square of the *t* statistic value $t = 9.96$. The P-
value, labeled as P on Table 11.8, is 0.000 to three decimal places, necessarily
the same as the P-value for the two-sided *t* test.

Why do we need the *F* statistic if it does the same thing as a *t* statistic? In
the next chapter, which generalizes regression to handle multiple predictors,
we'll see that the main use of an *F* statistic is for testing effects of *several* predictors at once. We can use the *t* distribution to test the effect of only a single
predictor in a given test.

11.4 Practicing the Basics

11.37 Poor predicted strengths: The MINITAB output shows
[TRY] the large standardized residuals for the female athlete
strength study.

Large standardized residuals for strength study:

Obs	BP_60	BP	Fit	Residual	St Resid
10	15.0	105.00	85.90	19.10	2.41R
25	17.0	105.00	88.88	16.12	2.04R
37	13.0	65.00	82.92	−17.92	−2.26R

a. Explain how to interpret all the entries in the row of
the output for athlete 10.

b. Out of 57 observations, is it surprising that 3 observations would have standardized residuals with absolute
value above 2.0? Explain.

11.38 Loves TV and exercise: For the "Georgia student survey" file on the text CD, let *y* = exercise and

x = watch TV. One student reported watching TV an average of 180 minutes a day and exercising 60 minutes a
day. This person's residual was 48.8 and standardized
residual was 6.41.

a. Interpret the residual, and use it to find the predicted
value of exercise.

b. Interpret the standardized residual.

11.39 Bench press residuals: The figure is a histogram of the
[TRY] standardized residuals for the regression of maximum
bench press on number of 60-pound bench presses, for
the high school female athletes.

a. Which distribution does this figure provide information about?

b. What would you conclude based on this figure?

Histogram of the Residuals
response is BP

11.40 Predicting house prices: The "house selling prices" data file on the text CD has several predictors of house selling prices. The table here shows the ANOVA table for a regression analysis of y = selling price (thousands of dollars) and x = size of house (thousands of square feet). The prediction equation is $\hat{y} = 9.2 + 77.0x$.

ANOVA table for selling price and size of house:

Source	DF	SS	MS	F	P
Regression	1	182220	182220	135.07	0.000
Residual					
Error	98	132213	1349		
Total	99	314433			

a. What was the sample size? (*Hint:* You can figure it out from the residual *df*.)

b. The sample mean house size was 1.53 thousand square feet. What was the sample mean selling price? (*Hint:* What does $\hat{y}$ equal when $x = \bar{x}$?)

c. Estimate the standard deviation of the selling prices for homes that have $x = 1.53$. Interpret.

d. Report an approximate prediction interval within which you would expect about 95% of the selling prices to fall, for homes of size $x = 1.53$.

11.41 Predicting clothes purchases: For a random sample TRY of children from a school district in South Carolina, a regression analysis is conducted of y = amount spent on clothes in the past year (dollars) and x = year in school. MINITAB reports the tabulated results for observations at $x = 12$.

```
Predicted Values for New Observations
New Obs   Fit   SE Fit  95% CI      95% PI
      1   448.0  10.6   (427, 469)  (101, 795)
```

a. Interpret the value listed under "Fit."

b. Interpret the interval listed under "95% CI."

c. Interpret the interval listed under "95% PI."

11.42 CI vs. PI: Using the context of the previous exercise, explain the difference between the purpose of a 95% prediction interval (PI) for an observation and a 95% confidence interval (CI) for the mean of y at a given value of x. Why would you expect the PI to be wider than the CI?

11.43 ANOVA table for leg press: Exercise 11.31 referred to an TRY analysis of leg strength for 57 female athletes, with y = max-imum leg press and x = number of 200-pound leg presses until fatigue, for which $\hat{y} = 233.89 + 5.27x$. The table shows ANOVA results from SPSS for the regression analysis.

ANOVA[b]

Model		Sum of Squares	df	Mean Square	F	Sig.
1	Regression	121082.4	1	121082.400	92.875	.000 [a]
	Residual	71704.442	55	1303.717		
	Total	192786.8	56			

a. Show that the residual standard deviation is 36.1. Interpret it.

b. For this sample, $\bar{x} = 22.2$. For female athletes with $x = 22$, what would you estimate the variability to be of their maximum leg press values? If the y values are approximately normal, find an interval within which about 95% of them would fall.

11.44 Predicting leg press: Refer to the previous exercise. MINITAB reports the tabulated results for observations at $x = 25$.

```
Predicted Values for New Observations
New
Obs  Fit    SE Fit  95% CI            95% PI
 1  365.66  5.02   (355.61,375.72)   (292.61,438.72)
```

a. Show how MINITAB got the "Fit" of 365.66.

b. Using the predicted value and *se* value, explain how MINITAB got the interval listed under "95% CI." Interpret this interval.

c. Interpret the interval listed under "95% PI."

11.45 Variability and F: Refer to the previous two exercises.

a. In the ANOVA table, show how the Total SS breaks into two parts, and explain what each part represents.

b. From the ANOVA table, explain why the overall sample standard deviation of y values is $s_y = \sqrt{192787/56} = 58.7$. Explain the difference between the interpretation of this standard deviation and the residual s of 36.1.

c. Exercise 11.31 reported a t statistic of 9.64 for testing independence of these variables. Report the F test statistic from the table here, and explain how it relates to that t statistic.

11.46 Assumption violated: For prediction intervals, an important inference assumption is a constant residual standard deviation of y values at different x values. In practice, the residual standard deviation often tends to be larger when μ_y is larger.

a. Sketch a hypothetical scatterplot for which this happens, using observations for the previous year on x = family income and y = amount donated to charity.

b. Explain why a 95% prediction interval would not work well at very small or at very large x values.

11.47 Understanding an ANOVA table: For a random sample of U.S. counties, the ANOVA table shown refers to hypothetical data on x = percentage of the population aged over 50 and y = per capita expenditure (dollars) on education.

a. Fill in the blanks in the table.

b. For what hypotheses can the F test statistic be used?

Source	DF	SS	MS	F
Model	1	200000	___	___
Error	31	700000	___	
Total	32	900000		

■ **11.48 Predicting GPA:** Refer to the "Georgia student survey" data file on the text CD. Regress y = college GPA on x = high school GPA.

a. Stating the necessary assumptions, find a 95% confidence interval for the mean college GPA for all University of Georgia students who have high school GPA = 3.6.

b. Find a 95% prediction interval for college GPA of a randomly chosen student having high school GPA = 3.6. Interpret.

c. Explain the difference between the purposes of the intervals in (a) and (b).

■ **11.49 GPA ANOVA:** Report the ANOVA table for the previous exercise.

a. Show how the Total SS breaks into two parts, and explain what each part represents.

b. Find the estimated residual standard deviation of y. Interpret it.

c. Find the sample standard deviation s_y of y values. Explain the difference between the interpretation of this standard deviation and the residual standard deviation in (b).

11.5 | Exponential Regression: A Model for Nonlinearity

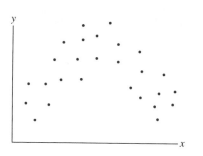

The straight line is by far the most common equation for a regression model. Sometimes, however, a scatterplot indicates substantial curvature in a relationship. Then equations that provide curvature are more appropriate.

Occasionally a scatterplot has a parabolic appearance: As x increases, y tends to go up and then it goes back down (as shown in the margin figure), or the reverse. Then the regression model can use an equation giving the shape of a parabola. More often, y tends to continually increase or to continually decrease, but the trend shows curvature. Let's see a mechanism for how this can happen.

EXAMPLE 17

Exponential Growth in Population Size

Picture the Scenario

The population size of the United States has been growing rapidly in recent years, much of it due to immigration. According to the 2000 census, the population size was about 280 million on January 1, 2000.

Questions to Explore

a. Suppose that the rate of growth after 2000 is 2% a year. That is, the population is 2% larger at the end of each year than it was at the beginning of the year. Find the population size after (i) 1 year, (ii) 2 years, (iii) 10 years.

b. Give a formula for the population size in terms of x = number of years since 2000.

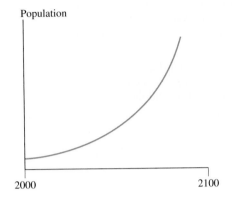

Population

2000   2100

Think It Through

a. With a 2% growth rate, the population size one year after 2000 is $280 \times 1.02 = 285.6$ million. That is, increasing the population size by 2% corresponds to multiplying it by 1.02. The population size after 2 years is 2% higher than this, or

$$285.6 \times 1.02 = (280 \times 1.02) \times 1.02 =$$
$$280 \times (1.02)^2 = 291.3 \text{ million.}$$

After 10 years, the population size is $280 \times 1.02 \times 1.02 \times 1.02 \times \cdots \times 1.02 = 280 \times (1.02)^{10} = 341.3$ million.

b. Can you see the pattern? For each additional year, we mulitply by another factor of 1.02. After x years, the population size is 280×1.02^x million.

Insight

The population size formula, 280×1.02^x, is called **exponential growth**. Plotted, the response goes up faster than a straight line. See the margin figure. The amount of change in y per unit change in x increases as x increases. After 100 years (that is, in the year 2100), taking $x = 100$, population size $= 280 \times (1.02)^{100} = 2{,}028.5$ million, more than 2 billion people!

TO PRACTICE THIS CONCEPT, TRY EXERCISE 11.50

A regression function that uses x in the *exponent* is called **exponential regression**.

Exponential Regression Model

An **exponential regression model** has the formula

$$\mu_y = \alpha\beta^x$$

for the mean μ_y of y at a given value of x, where α and β are parameters.

The formula 280×1.02^x for U.S. population size in Example 17 has the form $\alpha\beta^x$, with $\alpha = 280$, $\beta = 1.02$, and $x =$ number of years since 2000. Exponential regression can model quantities that tend to increase by increasingly large amounts over time.

In the exponential regression equation, the explanatory variable x appears as the exponent of a parameter. Unlike with straight-line regression, the mean μ_y and the effect parameter β can take only positive values. As x increases, the mean μ_y continually increases when $\beta > 1$. It continually decreases when $0 < \beta < 1$. Figure 11.11 shows the shape for the two cases. We provide interpretations for the model parameters later in this section.

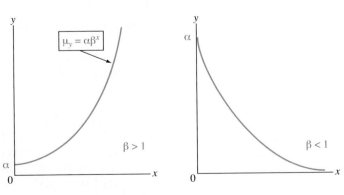

▲ **FIGURE 11.11: The Exponential Regression Curve for $\mu_y = \alpha\beta^x$. Question:** Why does μ_y decrease if $\beta = 0.5$, even though $\beta > 0$?

For exponential regression, the *logarithm* of the mean is a linear function of x. When the exponential regression model holds, a plot of the log of the y values versus x should show an approximate straight-line relation with x. Don't worry if you have forgotten your high school math about logarithms. You will not need to use logarithms to understand how to fit the model. Software can do it. It's more important to know when it is appropriate and to be able to interpret the model fit.

EXAMPLE 18

Explosion in Number of People Using the Internet

Picture the Scenario

Table 11.9 shows the number of people worldwide using the Internet (in millions) starting in 1995. Figure 11.12 plots these values. They increase over time, and the amount of increase from one year to the next seems to itself increase over time.

	No. Years Since 1995	Number People	Log No. People	Predicted Number
YEAR	x	y	$log(y)$	$\hat{y}$
1995	0	16	1.20	20.4
1996	1	36	1.56	36.1
1997	2	76	1.88	63.9
1998	3	147	2.17	113.2
1999	4	195	2.29	200.4
2000	5	369	2.57	354.9
2001	6	513	2.71	628.4

TABLE 11.9: Number of People Worldwide Using Internet (in Millions)

Source: "The Roads and Crossroads of Internet History," by Gregory Gromov; see www.netvalley.com/intvalstat.html

▲ **FIGURE 11.12: Plot of Number of People Using Internet since 1995.**

Table 11.9 also shows the logarithm of the Internet-user counts, using base-10 logs. (The model makes sense with logs to any base.) Figure 11.13 plots these log values over time. They appear to grow approximately linearly. In fact, the correlation between the log of the population size and the year number is 0.99, a very strong linear association. This suggests that growth in Internet users over this time period was approximately exponential.

▲ **FIGURE 11.13: Plot of Log of Number of Internet Users since 1995.** When the log of the response has a straight-line relationship with the explanatory variable, the exponential regression model is appropriate.

Question to Explore

Let x denote the number of years since 1995. That is, 1995 is $x = 0$, 1996 is $x = 1$, and so forth up to 2001 which is $x = 6$. The exponential regression model fitted to y = number of Internet users and x gives

$$\hat{y} = 20.38 \times 1.7708^x.$$

What does this equation predict for the number of Internet users in 1995 and in 2000?

Think It Through

In 1995, $x = 0$, so the predicted number of Internet users is

$$\hat{y} = 20.38 \times 1.7708^0 = 20.38 \text{ million}.$$

For the year 2000, $x = 5$, and the predicted number is $\hat{y} = 20.38 \times 1.7708^5 = 354.9$ million. Table 11.9 shows the predicted value for each year between 1995 and 2001.

Insight

Comparing the observed values in Table 11.9 to the predicted values for this model, we see that the model fits the data pretty well. However, there is some indication that the actual growth was slowing a bit in 2001, since the observation (513) that year was quite a bit less than the predicted value (628.4).

TO PRACTICE THIS CONCEPT, TRY EXERCISE 11.52

Interpreting Exponential Regression Models

From what we've just seen in the previous example, here's how to interpret parameters in the exponential regression model, $\mu_y = \alpha\beta^x$: The parameter α represents the mean of y when $x = 0$, since $\beta^0 = 1$. The parameter β represents the *multiplicative* effect on the mean of y for a one-unit increase in x. The mean of y at $x = 1$ equals the mean of y at $x = 0$ *multiplied* by β. For instance, for the equation $\hat{y} = 20.38 \times 1.77^x$ with $\beta = 1.77$, the predicted number of Internet users is 20.38 million in 1995 (for which $x = 0$), and in 1996 (for which $x = 1$) it equals 20.38 million *times* 1.77.

By contrast, the parameter β in the straight-line model $\mu_y = \alpha + \beta x$ represents an *additive* effect on the mean of y for a one-unit increase in x. The mean of y at $x = 1$ equals the mean of y at $x = 0$ *plus* β. The straight-line model fitted to the Internet users data had fit $\hat{y} = -50.7 + 81.3x$. This model predicts that the number of Internet users increases by 81.3 million people every year. The model is not appropriate for these data. For example, it gives an absurd prediction at 1995: Plugging in $x = 0$, it predicts $\hat{y} = -50.7 + 81.3(0) = -50.7$ million Internet users.

For the straight-line model, μ_y changes by the same *quantity* for each one-unit increase in x, whereas for the exponential model, μ_y changes by the same *percentage* for each 1-unit increase. For the exponential regression model with Table 11.9, the predicted number of Internet users multiplies by 1.77 each year. This equation corresponds to a predicted 77% growth per year. With this exponential regression equation, over time the actual quantity of growth tends to get larger every year, as Figure 11.12 showed.

11.5 Practicing the Basics

11.50 Savings grow exponentially: You invest $100 in a savings account with interest compounded annually at 10%.

a. How much money does the account have after 1 year?

b. How much money does the account have after 5 years?

c. How much money does the account have after x years?

d. How many years does it take until your savings more than double in size?

11.51 Growth by year vs. decade: You want your savings to double in a decade.

a. Explain why 7.2% interest a year would do this. (*Hint:* What does $(1.072)^{10}$ equal?)

b. You might think that 10% interest a year would give 100% interest (that is, double your savings) over a decade. Explain why interest of 10% a year would actually cause your savings to multiply by 2.59 over a decade.

11.52 U.S. population growth: The table shows the U.S. population size (in millions) at 10-year intervals beginning in 1890. Let x denote the number of decades since 1890. That is, 1890 is $x = 0$, 1900 is $x = 1$, and so forth. The exponential regression model fitted to $y = $ population size and x gives $\hat{y} = 68.33 \times 1.1418^x$.

U.S. population sizes (in millions) from 1890 to 2000

Year	Population Size	Year	Population Size	Year	Population Size
1890	63.0	1930	122.8	1970	203.3
1900	76.0	1940	131.7	1980	226.5
1910	92.0	1950	151.3	1990	248.7
1920	105.7	1960	179.3	2000	280.0

Source: U.S. Census Bureau

a. Show that the predicted population sizes are 68.3 million in 1890 and 293.8 million in 2000.

b. Explain how to interpret the value 1.1418 in the prediction equation.

c. The correlation equals 0.996 between the log of the population size and the year number. What does this suggest about whether the exponential regression model is appropriate for these data?

11.53 Future shock: Refer to the previous exercise, for which predicted population growth was 14.18% per decade. Suppose the growth rate is now 15% per decade. Explain why the population size will (a) double after five decades, (b) quadruple after 100 years (10 decades), (c) be 16 times its original size after 200 years. (The exponentially increasing function has the property that its doubling time is constant. This is a fast increase, even though the annual rate of growth seems small.)

11.54 Age and death rate: Let x denote a person's age and let y be the death rate, measured as the number of deaths per thousand individuals of a fixed age within a period of a year. For men in the United States, these variables follow approximately the equation $\hat{y} = 0.32(1.078)^x$.

  a. Interpret 0.32 and 1.078 in this equation.

  b. Find the predicted death rate when age is (i) 20, (ii) 50, (iii) 80.

  c. In every how many years does the death rate double? (*Hint:* What is x such that $(1.078)^x = 2$?)

■ **11.55 Leaf litter decay:** Ecologists believe that organic material decays over time according to an *exponential decay* model. This is the case $0 < \beta < 1$ in the exponential regression model, for which μ_y decreases over time. The rate of decay is determined by a number of factors, including composition of material, temperature, and humidity. In an experiment carried out by researchers at the University of Georgia Ecology Institute, leaf litter was allowed to sit for a 20-week period in a bag in a moderately forested area. Initially, the total weight of the organic mass in the bag was 75.0 kg. Each week, the remaining amount (y) was measured. The table shows the weight y by x = number of weeks of time that have passed.

x	y	x	y	x	y	x	y	x	y	x	y
0	75.0	1	60.9	2	51.8	3	45.2	4	34.7	5	34.6
6	26.2	7	20.4	8	14.0	9	12.3	11	8.2	15	3.1
20	1.4										

  a. Construct a scatterplot. Why is a straight-line model inappropriate?

  b. Show that the ordinary regression model gives the fit, $\hat{y} = 54.98 - 3.59x$. Find the predicted weight after $x = 20$ weeks. Does this prediction make sense? Explain.

  c. Plot the log of y against x. Does a straight-line model now seem appropriate?

  d. The exponential regression model has prediction equation $\hat{y} = 80.6(0.813)^x$. Find the predicted weight (i) initially, (ii) after 20 weeks.

  e. Interpret the coefficient 0.813 in the prediction equation.

11.56 More leaf litter: Refer to the previous exercise.

  a. The correlation equals -0.890 between x and y, and -0.997 between x and $\log(y)$. What does this tell you about which model is more appropriate?

  b. The half-life is the time for the weight remaining to be one-half of the original weight. Use the equation $\hat{y} = 80.6(0.813)^x$ to predict the half-life of this organic material. (*Hint:* By trial and error, find the value of x for which $(0.813)^x$ is about $1/2$.)

CHAPTER REVIEW

Answers to the Chapter Figure Questions

Figure 11.1: (a) There are two dots at the value of 60 pounds on the vertical axis (representing maximum bench press) (b) All dots in the scatterplot are in "rows" corresponding to the values 60, 65, 70, etc., on the vertical axis.

Figure 11.2: Answers will vary. One possible sketch is Figure 11.3.

Figure 11.3: A mathematical function that has a parabolic shape.

Figure 11.4: The bell-shaped curves represent the conditional distributions of income at $x = 12$ years and $x = 16$ years. Each conditional distribution is assumed to be normal and to have the same standard deviation, $\sigma = \$13,000$.

Figure 11.5: This happens when most observations are in the quadrants where data points are above the mean on one variable and below the mean on the other variable, the upper-left and lower-right quadrants.

Figure 11.6: One standard deviation.

Figure 11.7: The spread of the sampling distribution of a sample mean is much less than the spread of the distribution of individual observations.

Figure 11.8: H_0: $\beta = 0$.

Figure 11.9: Three observations, one for each bar.

Figure 11.10: The curves that represent the lower and upper endpoints of the prediction interval are too wide for the x values on the lower end (the 95% prediction interval for y has a greater than 95% chance of containing a value of y) and too narrow for the x values on the upper end (a 95% prediction interval for y has a much smaller than 95% chance of containing a value of y).

Figure 11.11: When x increases by 1, the mean μ_y is multiplied by β. When $0 < \beta < 1$, multiplying by β causes the mean to decrease.

CHAPTER SUMMARY

In Chapters 9–11 we've learned how to detect and describe *association between two variables*. Chapter 9 showed how to compare means or proportions for two groups. Chapter 10 dealt with *association between two categorical variables*. This chapter showed how to analyze *association between two quantitative variables*.

A regression analysis investigates the relationship between a quantitative explanatory variable x and a quantitative response variable y.

■ At each value of x, there is a conditional distribution of y values that summarizes how y varies at that value. We denote the population mean of the conditional distribution of y by μ_y. The **regression model** $\mu_y = \alpha + \beta x$ with y-intercept α and slope β uses a straight line to approximate the relationship between x and the mean μ_y of the conditional distribution of y at the different possible values of x. The *sample* **prediction equation** $\hat{y} = a + bx$ predicts y and estimates the mean of y at the fixed value of x.

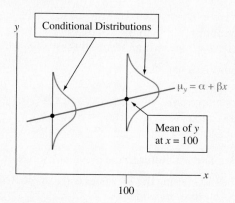

■ The **correlation** r describes the strength of linear association. It has the same sign as the slope b but falls between -1 and $+1$. The weaker the correlation, the greater the *regression toward the mean*, with the y values tending to fall closer to their mean (in terms of the number of standard deviations) than the x values fall to their mean. The squared correlation, r^2, describes the proportional reduction in the sum of squared errors using the prediction equation $\hat{y} = a + bx$ to predict y compared to using the sample mean $\bar{y}$ to predict y.

■ **Inference:** A significance test of H_0: $\boldsymbol{\beta = 0}$ for the population slope β tests **statistical independence** of x and y. It has test statistic
$$t = (b - 0)/se,$$
for the sample slope b and its standard error. Interval estimation is also useful for

■ A confidence interval for β

■ A confidence interval for μ_y, the mean of y at a given value of x.

■ A **prediction interval** for a value of y at a given value of x

These inferences all use the t distribution with $df = n - 2$. Their basic assumptions are

1. The population mean of y has a straight-line relationship with x.
2. The data were gathered using randomization.
3. The distribution of y at each value of x is normal, with the same standard deviation at each x value.

■ An **ANOVA table** displays sums of squares, their df values, and an **F statistic** (which equals the square of the t statistic) that also tests H_0: $\beta = 0$. In this table, the **residual sum of squares** $\sum(y - \hat{y})^2$ takes each **residual** (prediction error) $y - \hat{y}$ and then squares and adds them. The residual SS divided by its df value of $n - 2$ is the **mean square error**. Its square root is the **residual standard deviation** estimate s of the spread σ of the conditional distribution of y at each fixed x. A residual divided by its se is a **standardized residual**, which measures the number of standard errors that a residual falls from 0. It helps us identify unusual observations.

■ The **exponential regression** model $\mu_y = \alpha\beta^x$ has an increasing or a decreasing curved shape. For it, a 1-unit increase in x has a *multiplicative* effect of β on the mean, rather than an *additive* effect as when $\mu_y = \alpha + \beta x$.

SUMMARY OF NOTATION

μ_y = population mean of conditional distribution of y values at fixed value of x

$\mu_y = \alpha + \beta x$ = population straight-line regression equation

$\hat{y} = a + bx$ = sample prediction equation that estimates the population regression equation

r^2 = proportional reduction in prediction error = square of correlation r

$\sum(y - \hat{y})^2$ = residual sum of squares, which summarizes how well $\hat{y}$ predicts y. This is the numerator of the mean squared error, and its square root is the numerator of the residual standard deviation s, and it is used in finding r^2.

$\sum(y - \bar{y})^2$ = total sum of squares, which summarizes how well $\bar{y}$ predicts y. This is used in finding r^2.

$\mu_y = \alpha\beta^x$ = exponential regression model: mean $= \alpha$ when $x = 0$ and the mean multiplies by β for each 1-unit increase in x.

CHAPTER PROBLEMS

Practicing the Basics

11.57 Parties and sports: Let y = number of parties attended in the past month and x = number of sports events watched in the past month, measured for all students at your school. Explain the mean and variability about the mean aspects of the regression model $\mu_y = \alpha + \beta x$, in the context of these variables. In your answer, explain why (a) it is more sensible to let $\alpha + \beta x$ represent the *means* of the conditional distributions rather than individual observations, (b) the model allows variation around the mean with its σ parameter.

11.58 Verbal-math correlation: A report summarizing scores for students on a verbal aptitude test x and a mathematics aptitude test y states that $\bar{x} = 480, \bar{y} = 500, s_x = 80$, $s_y = 120$, and $r = 0.60$.

a. Find the slope of the regression line, based on its connection with the correlation.

b. Find the y-intercept of the regression line (*Hint:* Use its formula in Section 11.2), and state the prediction equation.

c. Find the prediction equation for predicting verbal test result using math test result.

11.59 Short people: Do very short parents tend to have children who are even shorter, or short but not as short as they are? Explain, identifying the response and explanatory variables and the role of regression toward the mean.

11.60 Income and education in Florida: The "Florida crime" data file on the text CD contains data for all counties in Florida on y = median annual income (thousand of dollars) for residents of the county and x = percent of residents with at least a high school education. The table shows some summary statistics and results of a regression analysis.

Income and education for Florida counties

Variable	Mean	Std Dev	Predictor	Parameter Estimate
INCOME	24.51	4.69	INTERCEPT	−4.60
EDUC	69.49	8.86	EDUC	0.42

a. Find the correlation. Interpret using (i) the sign, (ii) the magnitude.

b. Find the predicted median income for a county that is one standard deviation above the mean on x, and use it to explain the concept of regression toward the mean.

c. Suppose you instead sampled individuals (rather than counties) in Florida and observed y = annual income and x = whether have at least a high school education (1 = yes, 0 = no). Do you think their correlation would be larger, or smaller, than the correlation in (a)? Why?

11.61 Fertility and contraception: In the *Human Development Report*, one variable measured was x = percentage of adults who use contraceptive meth-

ods. Part of a MINITAB printout is shown for a regression analysis using y = fertility (mean number of children per adult woman), for 22 nations listed in that report. For those nations x had a mean of 60.0 and standard deviation of 20.6.

a. Report the prediction equation and find the predicted fertility when (i) $x = 0$, (ii) $x = 100$. Show how the difference between these can be obtained using the slope.

b. For Belgium, $x = 51.0$ and $y = 1.3$. Find the predicted fertility and the residual. (This is the observation with the largest residual.)

c. The standardized residual for Belgium was −2.97. Interpret it.

Fertility and contraception around the world

Predictor	Coef	SE Coef	T	P
Constant	6.6633	0.4771	13.97	0.000
CONTRA	−0.064843	0.007539	−8.60	0.000

Analysis of Variance

Source	DF	SS
Regression	1	37.505
Residual Error	20	10.138
Total	21	47.644

11.62 SS, r, and r^2: Refer to the previous exercise.

a. Find r^2 using the SS values listed in the ANOVA table. Interpret its value.

b. Using r^2 and the sign of the slope, find the correlation. Interpret its value.

11.63 Bedroom residuals: For the "house selling prices" data set on the text CD, when we regress y = selling price (in dollars) on x = number of bedrooms, we get the results shown in the printout.

a. One home with 3 bedrooms sold for $338,000. Find the residual, and interpret.

b. The home in (a) had a standardized residual of 4.02. Interpret.

House selling prices and number of bedrooms

Variable	Mean	StDev	Minimum	Q1	Median	Q3	Maximum
price	126698	56357	21000	86625	123750	155625	338000
Bedrooms	2.990	0.659	1.000	3.000	3.000	3.000	5.000

Predictor	Coef	SE Coef	T	P
Constant	33778	24637	1.37	0.173
Bedrooms	31077	8049	3.86	0.000

S = 52771.5   R-Sq = 13.2%

11.64 Bedrooms affect price?: Refer to the previous exercise.

a. Explain what the regression parameter β means in this context.

b. Construct and interpret a 95% confidence interval for β.

c. Use the result of (b) to form a 95% confidence interval for the difference in the mean selling prices for homes with $x = 4$ bedrooms and with $x = 2$ bedrooms. (*Hint*: How does this difference in means relate to the slope?)

11.65 Types of spread: Refer to the previous two exercises.

a. Explain the difference between the residual standard deviation of 52,771.5 and the standard deviation of 56,357 reported for the selling prices.

b. Since they're not much different, explain why this means that number of bedrooms is not strongly associated with selling price. Support this by reporting and interpreting the value of r^2.

11.66 Predicting body fat: Part of a MINITAB printout is shown for a regression of $y = $ percent of body fat (BF%) on $x = $ body mass index (BMI) for the 57 high school female athletes. For these data, BF% has mean $= 25.1$ and standard deviation $= 5.0$, and BMI has mean $= 22.9$ and standard deviation $= 3.5$.

Predictor	Coef	SE Coef	T	P
Constant	−5.556	1.453	−3.82	0.000
BMI	1.33812	0.06264	21.36	0.000

S= 1.6404   R-Sq = 89.2%

Analysis of Variance

Source	DF	SS	MS	F	P
Regression	1	1228.0	1228.0	456.32	0.000
Residual Error	55	148.0	2.7		
Total	56	1376.0			

a. Report and interpret r^2. Is the association strong, or weak?

b. Interpret the P-value for the test of independence between BF% and BMI.

c. The 95% confidence interval for the true slope is (1.21, 1.46). Explain how this is in agreement with the result of the t test in (b).

d. Find and interpret the residual standard deviation.

11.67 Body weight and lean body mass: The "college athletes" data set on the text CD comes from a study of University of Georgia Division 1 female athletes from nine sports. SPSS reports the results in the table of regressing TBW = total body weight (in pounds) on LBM = lean body mass, which equals TBW $\times$ (1 − proportion of body fat):

a. To what does the reported T value of 17.91 refer?

b. Explain how to interpret its associated P-value.

Coefficients [a]

Model		Unstandardized Coefficients		t	Sig.
		B	Std. Error		
1	(Constant)	.415	7.452	.056	.956
	LBM	1.225	.068	17.913	.000

[a.] Dependent Variable: TBW

■ **11.68 Exercise and college GPA:** For the "Georgia student survey" file on the text CD, let $y = $ exercise and $x = $ college GPA.

a. Construct a scatterplot. Identify an outlier that could influence the regression line. What would you expect its effect to be on the slope and the correlation?

b. Fit the model. Find the standardized residual for that observation. Interpret.

c. Fit the model without the outlying observation. Summarize its impact.

11.69 Bench press predicting leg press: For the study of high school female athletes, when we use
$x = $ maximum bench press (BP) to predict
$y = $ maximum leg press (LP), we get the results shown below. The sample mean of BP was 80.

Predictor	Coef	SE Coef	T	P
Constant	173.98	41.75	4.17	0.000
BP	2.2158	0.5155	4.30	0.000

Predicted Values for New Observations

New Obs	Fit	SE Fit	95% CI	95% PI
1	351.25	6.78	(337.65,364.84)	(247.70,454.80)

Values of Predictors for New Observations

New Obs	BP
1	80.0

a. Interpret the confidence interval listed under "95% CI."

b. Interpret the interval listed under "95% PI." What's the difference between the purpose of the 95% PI and the 95% CI?

11.70 Leg press ANOVA: The analysis in the previous exercise has the ANOVA table shown.

Analysis of Variance

Source	DF	SS	MS	F	P
Regression	1	48483	48483	18.48	0.000
Residual Error	55	144304	2624		
Total	56	192787			

a. For those female athletes who had maximum bench press equal to the mean of 80 pounds, what is the estimated standard deviation of their maximum leg press values?

b. Assuming that maximum leg press has a normal distribution, show how to find an approximate 95% prediction interval for the y-values at $x = 80$.

11.71 Savings grow: You invest $1000 in an account having interest such that your principal doubles every 10 years.

a. How much money would you have after 50 years?

b. If you were still alive in 100 years, show that you'd be a millionaire.

c. Give the equation that relates $y = $ principal to $x = $ number of decades for which your money has been invested.

11.72 Florida population: The population size of Florida (in thousands) since 1830 has followed approximately the exponential regression $\hat{y} = 46(1.036)^x$. Here, $x = $ year − 1830 (so, $x = 0$ for 1830 and $x = 170$ for the year 2000).

a. What has been the approximate rate of growth per year?

b. Find the predicted population size in (i) 1830, (ii) 2000.

c. Find the predicted population size in 2100. Do you think that this same formula will continue to hold between 2000 and 2100? Why?

11.73 World population growth: The table shows the world population size (in billions) since 1900.

World population sizes (in billions)

Year	Population	Year	Population
1900	1.65	1965	3.33
1910	1.75	1970	3.69
1920	1.86	1975	4.07
1930	2.07	1980	4.43
1940	2.30	1985	4.83
1950	2.52	1990	5.26
1955	2.76	1995	5.67
1960	3.02	2000	6.07

Source: United Nations

a. Let x denote the number of years since 1900. The exponential regression model fitted to y = population size and x gives $\hat{y} = 1.4193 \times 1.014^x$. Show that the predicted population sizes are 1.42 billion in 1900 and 5.70 billion in 2000.

b. Explain why the fit of the model corresponds to a rate of growth rate of 1.4% per year.

c. For this model fit, explain why the predicted population size (i) doubles after 50 years, (ii) quadruples after 100 years.

d. The correlation equals 0.948 between the population size and the year number and it equals 0.985 between the log of the population size and the year number. What does this suggest about whether the straight-line regression model or the exponential regression model is more appropriate for these data?

11.74 Match the scatterplot: Match each of the following scatterplots to the description of its regression and correlation. The plots are the same except for a single point. Justify your answer for each scatterplot. (*Hint*: Think about the possible effect of an outlier in the x-direction and an outlier in the y-direction relative to the regression line for the rest of the data.)

a. $r = -0.46$ $\hat{y} = 142 - 15.6x$
b. $r = -0.86$ $\hat{y} = 182 - 25.8x$
c. $r = -0.74$ $\hat{y} = 165 - 21.8x$

Scatterplot 1

Scatterplot 2

Scatterplot 3

Concepts and Investigations

11.75 Softball data: The "softball" data file on the text CD contains the records of a University of Georgia coed intramural softball team for 277 games over a 20-year period. (The players changed, but the team continued.) The variables include, for each game, the team's number of runs scored (RUNS), number of hits (HIT), number of errors (ERR), and the difference (DIFF) between the number of runs scored by the team and by the other team. Let DIFF be the response variable. Note that DIFF > 0 means the team won and DIFF < 0 means the team lost.

a. Construct a modified boxplot of DIFF. What do the three outlying observations represent?

b. Find the prediction equation relating DIFF to RUNS. Show that the team is predicted to win when RUNS = 8 or more.

c. Construct the correlation matrix for RUNS, HIT, ERR, and DIFF, and interpret.

d. Conduct statistical inference about the slope of the relationship.

11.76 Runs and hits: Refer to the previous exercise. Conduct a regression analysis of y = RUNS and x = HIT. Does a straight-line regression model seem appropriate? Prepare a report

a. Using graphical ways of portraying the individual variables and their relationship.

b. Interpreting descriptive statistics for the individual variables and their relationship.

c. Viewing standardized residuals to find games with unusual results.

d. Conducting statistical inference about the slope of the relationship.

11.77 GPA and TV watching: Using software with the "FL student survey" data file on the text CD, conduct regression analyses relating y = high school GPA and x = hours of TV watching. Prepare a two-page report, showing descriptive and inferential methods for analyzing the relationship.

11.78 Female athletes' speed: For the "high school female athletes" data set on the text CD, conduct a regression analysis using the time for the 40-yard dash as the response variable and weight as the explanatory variable. Prepare a two-page report, indicating why you conducted each analysis and interpreting the results.

11.79 Sports and regression: One of your relatives is a big sports fan but has never taken a statistics course. Explain how you could describe the concept of "regression toward the mean" in terms of a sports application, without using technical jargon.

11.80 Regression toward the mean paradox: Does regression toward the mean imply that, over many generations, there are fewer and fewer very short people and very tall people? Explain your reasoning. (*Hint:* What happens if you look backward in time in doing the regressions?)

11.81 Height and weight: Suppose the correlation between height and weight is 0.50 for a sample of males in elementary school, and also 0.50 for a sample of males in middle school. If we combine the samples, explain why the correlation will probably be larger than 0.50.

11.82 Income and education: Explain why the correlation between x = number of years of education and y = annual income is likely to be smaller if we use a random sample of adults who have a Ph.D. than if we use a random sample of all adults.

11.83 Dollars and pounds: Annual income, in dollars, was the response variable in a regression analysis. For a British version of a written report about the analysis, all responses were converted to British pounds sterling (1 pound equaled $2.00, when this was done).

a. How, if at all, does the slope of the prediction equation change?

b. How, if at all, does the correlation change?

c. How, if at all, does the t statistic change for testing the effect of a predictor?

11.84 All models are wrong: The statistician George Box, who had an illustrious academic career at the University of Wisconsin, is often quoted as saying, "All models are wrong, but some models are useful." Why do you think that, in practice,

a. All models are wrong?

b. Some models are *not* useful?

11.85 *df* for *t* tests in regression: In regression modeling, for t tests about regression parameters, $df = n$ − number of parameters in equation for the mean.

a. Explain why $df = n - 2$ for the model $\mu_y = \alpha + \beta x$.

b. Chapter 7 discussed how to estimate a single mean μ. Treating this as the parameter in a simpler regression model, $\mu_y = \mu$, with a single parameter, explain why $df = n - 1$ for inference about a single mean.

11.86 Assumptions: What assumptions are needed to use the regression equation $\mu_y = \alpha + \beta x$ (a) to *describe* the relationship between two variables and (b) to make *inferences* about the relationship. In case (b), which assumption is least critical?

11.87 Assumptions fail?: Refer to the previous exercise. In view of these assumptions, indicate why such a model would or would not be good in the following situations:

a. x = year (from 1900 to 2005), y = percentage unemployed workers in the United States. (*Hint:* Does y continually tend to increase or decrease?)

b. x = age of subject, y = subject's annual medical expenses. (*Hint:* Suppose expenses tend to be relatively high for the newborn and for the elderly.)

c. x = per capita income, y = life expectancy, for nations. (*Hint:* The increasing trend eventually levels off.)

11.88 Lots of standard deviations: Explain carefully the interpretations of the standard deviations (a) s_y, (b) s_x, (c) residual standard deviation s, (d) se of slope estimate b.

11.89 Growth in Florida: A newspaper article quoted a planner in a Florida city as saying, "This city has been growing at the rate of 4.2% per year. That's not slow growth by any means. It corresponds to 42% growth per decade."

a. Explain what is incorrect about this statement.

b. If, in fact, the current population size of the city is 100,000 and in each of the next ten years the city increases in size by 4.2% relative to the previous year, then what is the population size after a decade? What percentage growth occurs for the decade?

11.90 Population growth: Exercise 11.52 about U.S. population growth showed a predicted growth rate of 14.18% per decade.

a. Show that this is equivalent to a 1.33% predicted growth *per year*.

b. Explain why the predicted U.S. population size (in millions) x years after 1890 is $68.3(1.0133)^x$.

Select the best response in multiple-choice exercises 11.91–11.95.

11.91 Interpret *r*: One can interpret $r = 0.30$ or the corresponding $r^2 = 0.09$ as follows:

a. A 30% reduction in error occurs in using x to predict y.

b. A 9% reduction in error occurs in using x to predict y compared to using $\bar{y}$ to predict y.

c. 9% of the time $\hat{y} = y$.

d. y changes 0.3 unit for every one-unit increase in x.

e. x changes 0.3 standard deviations when y changes one standard deviation.

11.92 Correlation invalid: The correlation is appropriate for describing association between two quantitative variables

a. When different people measure the variables using different units (e.g., kilograms vs. pounds).

b. When the relationship is highly nonlinear.

c. When the slope of the regression equation is 0 using nearly all the data, but a couple of outliers are extremely high on y at the high end of the x-scale.

d. When the sample has a much narrower range of x values than does the population.

e. When the response variable and explanatory variable are both categorical.

11.93 Slope and correlation: The slope of the least squares regression equation and the correlation are similar in the sense that

a. They both must fall between -1 and $+1$.

b. They both describe the strength of association.

c. Their squares both have proportional reduction in error interpretations.

d. They have the same t statistic value for testing H_0: Independence.

e. They both are unaffected by severe outliers.

11.94 Income and height: University of Rochester economist Steven Landsburg surveyed economic studies in England and the United States that showed a positive correlation between height and income. The article stated that in the United States each one-inch increase in height was worth about $1500 extra earnings a year, on the average (*Toronto Globe and Mail*, 4/1/2002). The regression equation that links y = annual earnings (in thousands of dollars) to x = height (in inches)

a. has y-intercept = $1500.

b. has slope 1.5.

c. has slope 1/1500.

d. has slope 1500.

e. has correlation 0.150.

11.95 Regress x on y: The regression of y on x has a prediction equation of $\hat{y} = -2.0 + 5.0x$ and a correlation of 0.3. Then, the regression of x on y

a. also has a correlation of 0.3.

b. could have a negative slope.

c. has $r^2 = \sqrt{0.3}$.

d. equals $1/(-2.0 + 5.0y)$.

11.96 True-false questions: The variables y = annual income (thousands of dollars), x_1 = number of years of education, and x_2 = number of years experience in job are measured for all the employees having city-funded jobs, in Knoxville, Tennessee. Suppose that the following regression equations and correlations apply:

i. $\hat{y} = 10 + 1.0x_1, \quad r = 0.30.$

ii. $\hat{y} = 14 + 0.4x_2, \quad r = 0.60.$

The correlation is -0.40 between x_1 and x_2. Which of the following statements are true, and which are false?

a. The strongest sample association is between y and x_2.

b. A standard deviation increase in education corresponds to a predicted increase of 0.3 standard deviations in income.

c. There is a 30% reduction in error in using education, instead of $\bar{y}$, to predict y.

d. When x_1 is the predictor of y, the sum of squared residuals is larger than when x_2 is the predictor of y.

e. If $s = 8$ for the model using x_1 to predict y, then it is not unusual to observe an income of $100,000 for an employee who has 10 years of education.

◆◆ 11.97 Golf club velocity and distance: A study about the effect of the swing on putting in golf (by C. M. Craig et al., *Nature*, vol. 405, pp. 295–296, 2000) showed a very strong linear relationship between y = putting distance and the square of x = club's impact velocity (r^2 is in the range 0.985 to 0.999).

a. For the model $\mu_y = \alpha + \beta x^2$, explain why it is sensible to set $\alpha = 0$.

b. If the appropriate model is $\mu_y = \beta x^2$, explain why if you double the velocity of the swing, you can expect to quadruple the putting distance.

◆◆ 11.98 Why is there regression toward the mean?: Refer to the relationship $r = (s_x/s_y)b$ between the slope and correlation, which is equivalently $s_x b = r s_y$.

a. Explain why an increase in x of s_x units relates to a change in the predicted value of y of $s_x b$ units. (For instance, if $s_x = 10$, it corresponds to a change in $\hat{y}$ of $10b$.)

b. Based on (a), explain why an increase of one standard deviation in x corresponds to a change of only r standard deviations in the predicted y-variable.

◆◆ 11.99 r² and variances: Suppose $r^2 = 0.30$. Since $\sum(y - \bar{y})^2$ is used in estimating the overall variability of the y values and $\sum(y - \hat{y})^2$ is used in estimating the residual variability at any fixed value of x, explain why approximately the estimated variance of the conditional distribution of y for a given x is 30% smaller than the estimated variance of the marginal distribution of y.

◆◆ 11.100 Standard error of slope: The formula for the standard error of the sample slope b is $se = s/\sum(x - \bar{x})^2$, where s is the residual standard deviation of y.

a. Show that the smaller the value of s, the more precisely b estimates β.

b. Explain why a small s occurs when the data points show little variability about the prediction equation.

c. Explain why the standard error of b decreases as the sample size increases and when the x values are more highly spread out.

◆◆ 11.101 Regression with an error term[1]: An alternative to the regression formula $\mu_y = \alpha + \beta x$ expresses each y value, rather than the mean of the y values, in terms of x. This approach models an observation on y as

$$y = \text{mean} + \text{error} = \alpha + \beta x + \varepsilon$$

where the mean $\mu_y = \alpha + \beta x$ and the error $= \varepsilon$.

[1]This formula is less useful than the one for the mean because it does not apply to regression models when y is not assumed to be normal, such as logistic regression for binary data.

The **error term** denoted by ε (the Greek letter epsilon) represents the deviation of the observation from the mean, that is, $\varepsilon = \text{error} = y - \text{mean}$.

a. If an observation has $\varepsilon > 0$, explain why the observation falls above the mean.

b. What does ε equal when the observation falls exactly at the mean? The ε term represents the error that results from using the mean value $(\alpha + \beta x)$ of y at a certain value of x for the prediction of the individual observation on y.

c. For the sample data and their prediction equation $\hat{y} = a + bx$, explain why an analogous equation to the population equation $y = \alpha + \beta x + \varepsilon$ is $y = a + bx + e$, where e is the residual, $e = y - \hat{y}$. (The residual e estimates ε. We can interpret e as a sample residual and ε as a **population residual**.)

d. Explain why it does not make sense to use the simpler model, $y = \alpha + \beta x$, which does not have the error term.

◆◆ **11.102** **Rule of 72:** You invest $1000 at 6% compound interest a year. How long does it take until your investment is worth $2000?

a. Based on what you know about exponential regression, explain why the answer is the value of x for which $1000(1.06)^x = 2000$.

b. Using the property of logarithms that $\log(a^x) = x \log(a)$, show that the answer x satisfies $x[\log(1.06)] = \log(2)$, or $x = \log(2)/\log(1.06) = 12$.

c. The "rule of 72" says that if you divide 72 by the interest rate, you will find approximately how long it takes your money to double. According to this rule, about how long does it take your money to double at an interest rate of (i) 1%, (ii) 18%?

Class Explorations

■ **11.103** **Analyze your data:** Refer to the data file you created in Activity 3 at the end of Chapter 1. For variables chosen by your instructor, conduct a regression and correlation analysis. Report both descriptive and inferential statistical analyses, interpreting and summarizing your findings, and prepare to discuss results in class.

ACTIVITY 1: SAT French Exam and Years of Study

In 2003, high school students taking the SAT French II language exam after 2–4 years of study scored, on average, 35 points higher for each additional year of study (*Source*: http://www.collegeboard.com/). The scores of students with 2 years of study averaged 505. For each number of years of study, the standard deviation was approximately 100. Let $y =$ French II exam score and $x =$ number of years of study beyond 2 years (so $x = 0, 1,$ or 2 for 2, 3, and 4 years of study).

1. Explain why the regression equation $\mu_y = \alpha + \beta x$ is $\mu_y = 505 + 35x$.

2. Suppose y has a normal conditional distribution. Using software, randomly select a student who has 2 years of French study, a student who has 3 years, and a student who has 4 years, by simulating from the appropriate normal distributions. Record each exam score.

3. Do step 2 at least 100 times. Store your results in a data file. You now have 100 values of y at each value of x.

4. Construct parallel dot plots and/or box plots for the sample conditional distributions. Compare the mean of the simulated distribution at each x to the actual mean, based on $\mu_y = 505 + 35x$. Is the variability similar for the three sample conditional distributions?

5. Fit a regression line, $\hat{y} = a + bx$, to the simulated values. How does this line compare to the actual regression line, $\mu_y = 505 + 35x$?

The simulated results in (4) and (5) are not *exactly* equal to those for the model because they use a sample, whereas $\mu_y = 505 + 35x$ refers to a population.

EXAMPLE 1

How Can You Predict the Selling Price of a House?

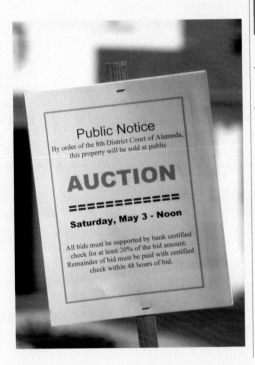

Picture the Scenario

You are saving to buy a house or condo and you wonder how much it will cost. Depending on what you want and its location, the price will vary dramatically. The "house selling prices" data file on the text CD has observations on 100 home sales in Gainesville, Florida, in November, 2003. Table 12.1 shows data for two homes.

TABLE 12.1: Selling Prices and Related Factors for a Sample of Home Sales

Home	Selling Price	House Size	Number of Bedrooms	Number of Bathrooms	Lot Size	Real Estate Tax	NW
1	145,000	1240	3	2	18,000	1360	yes
2	69,000	1120	3	2	17,000	830	no

Variables listed are the selling price (in dollars), house size (in square feet), number of bedrooms, number of bathrooms, the lot size (in square feet), the annual real estate tax (in dollars), and whether the house is in the northwest (NW) quadrant of town, which is considered the most desirable location. Table 12.2 reports the mean and standard deviation of these variables for all 100 home sales.

TABLE 12.2: Descriptive Statistics for Sales of 100 Homes

	Selling price	House size	Bedrooms	Bathrooms	Lot size	Tax
Mean	126,698	1526	3.0	1.9	19417	1668
Standard deviation	56,357	557	0.66	0.51	8524	1009

Questions to Explore

In your community, if you know values of such variables,

- How can you predict a home's selling price?
- How can you describe the association between selling price and the other variables?
- How can you make inferences about the population based on the sample data?

Thinking Ahead

You can find a regression equation to predict selling price by treating one of the other variables as the explanatory variable. Since there are *several* explanatory variables, perhaps you can make better predictions by using *all* of them at once. That's the idea behind **multiple regression**. It uses *more than one* explanatory variable to predict a response variable. We'll learn how to use multiple regression in this chapter, and we'll apply it to a variety of analyses with these data in Examples 2, 3, 8, 9, and 10.

Besides helping you to better predict a response variable, multiple regression can help you analyze the association between two variables while controlling (keeping fixed) values of other variables. Such adjustment is important, because an effect of an explanatory variable can change considerably after you account for potential lurking variables. A multiple regression analysis provides information not available from simple regression analyses involving only two variables at a time.

Section 12.1 presents the basics of multiple regression. Section 12.2 extends the concepts of correlation and r^2 for describing strength of association to multiple predictors (explanatory variables). Section 12.3 presents inference methods, and Section 12.4 shows how to check the model. Section 12.5 extends the model further to incorporate categorical explanatory variables. The final section presents a regression model for a categorical response variable.

Recall

This chapter builds strongly on the content in Chapter 11 on **regression** and **correlation** methods. You may want to review Sections 11.1 and 11.2 as you read Sections 12.1 and 12.2. ■

12.1 | How Can We Use Several Variables to Predict a Response?

Chapter 11 modeled a quantitative explanatory variable x and the mean μ_y of a quantitative response variable y by the straight-line equation $\mu_y = \alpha + \beta x$. We refer to this model with a *single* predictor as **bivariate** (two-variable) regression, since it contains only two variables (x and y).

Now, suppose there are two predictors, denoted by x_1 and x_2. The bivariate regression equation generalizes to the **multiple regression** equation,

$$\mu_y = \alpha + \beta_1 x_1 + \beta_2 x_2.$$

In this equation, α, β_1, and β_2 are parameters. When we substitute values for x_1 and x_2, the equation specifies the population mean of y for all subjects with those values of x_1 and x_2. When there are additional predictors, each has a βx term.

Multiple Regression Model

The **multiple regression model** relates the mean μ_y of a quantitative response variable y to a set of explanatory variables $x_1, x_2, \ldots$. For three explanatory variables, for example, the multiple regression equation is $\mu_y = \alpha + \beta_1 x_1 + \beta_2 x_2 + \beta_3 x_3$, and the sample prediction equation is $\hat{y} = a + b_1 x_1 + b_2 x_2 + b_3 x_3$.

With sample data, software estimates the multiple regression equation. It uses the method of least squares to find the "best" prediction equation (the one with the smallest possible sum of squared residuals).

EXAMPLE 2

Predicting Selling Price Using House and Lot Sizes

Picture the Scenario

For the house selling price data described in Example 1, MINITAB reports the results in Table 12.3 for a multiple regression analysis with selling price as the response variable and with house size and lot size as explanatory variables.

TABLE 12.3: Regression of Selling Price on House Size and Lot Size

The regression equation is price = −10536 + 53.8 house_size + 2.84 lot_size

Predictor	Coef	SE Coef	T	P
Constant	−10536	9436	−1.12	0.267
House_size	53.779	6.529	8.24	0.000
Lot_size	2.8404	0.427	6.66	0.000

Questions to Explore

a. State the prediction equation.

b. The first home listed in Table 12.1 has house size = 1240 square feet, lot size = 18,000 square feet, and selling price $145,000. Find its predicted selling price and the residual (prediction error). Interpret the residual.

Think It Through

a. The response variable is y = selling price. Let x_1 = house size and x_2 = lot size. From Table 12.3, the prediction equation is $\hat{y} = -10{,}536 + 53.8x_1 + 2.84x_2$.

b. For $x_1 = 1240$ and $x_2 = 18{,}000$, the predicted selling price is

$$\hat{y} = -10{,}536 + 53.8(1240) + 2.84(18{,}000) = 107{,}276, \text{ that is, } \$107{,}276.$$

The residual is the prediction error, $y - \hat{y} = 145{,}000 - 107{,}276 = 37{,}724$ This result tells us that the actual selling price was $37,724 higher than predicted.

Insight

The coefficients of house size and lot size are positive. As these variables increase, the predicted selling price increases, as we would expect.

TO PRACTICE THIS CONCEPT, TRY EXERCISE 12.1

IN PRACTICE: The number of explanatory variables you can use depends on how much data you have

In practice, you should not use many explanatory variables in a multiple regression model unless you have lots of data. A rough guideline is that *the sample size n should be at least 10 times the number of explanatory variables*. For example, to use two explanatory variables, you should have at least $n = 20$.

Plotting Relationships

Always look at the data before doing a multiple regression analysis. Most software has the option of constructing scatterplots on a single graph for each pair of variables. This type of plot is called a **scatterplot matrix**.

Figure 12.1 shows a MINITAB scatterplot matrix for selling price, house size, and lot size. It shows each pair of variables twice. For a given pair, in one plot a variable is on the y-axis and in another it is on the x-axis. For instance, selling price of house is on the y-axis for the plots in the first row, whereas it is on the x-axis for the plots in the first column. Since selling price is the response variable for this example, the plots in the first row (where selling price is on the y-axis) are the ones of primary interest. These graphs show strong positive linear relationships between selling price and both house size and lot size. Because a scatterplot matrix shows each pair of variables twice, you only need to look at the plots in the upper-right triangle.

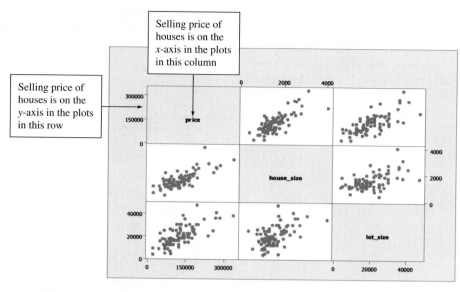

▲ **FIGURE 12.1: Scatterplot Matrix for Selling Price, House Size, and Lot Size.** The middle plot in the top row has house size on the x-axis and selling price on the y-axis. The first plot in the second row reverses this, with selling price on the x-axis and house size on the y-axis. **Questions:** Why are the plots of main interest the ones in the first row?

Each scatterplot portrays only *two* variables. It's a two-dimensional picture. A multiple regression equation, which has *several* variables, is more difficult to portray graphically.

Interpretation of Multiple Regression Coefficients

The simplest way to interpret a multiple regression equation looks at it in *two dimensions* as a function of a *single* explanatory variable. We can look at it this way by fixing values for the other explanatory variable(s). For instance, let's fix x_1 = house size at the value 2000 square feet. Then the prediction equation simplifies to one with x_2 = lot size alone as the predictor,

$$\hat{y} = -10{,}536 + 53.8(2000) + 2.84x_2 = 97{,}022 + 2.84x_2.$$

For 2000 square-foot houses, the predicted selling price relates to lot size by $\hat{y} = 97{,}022 + 2.84x_2$. Since the slope coefficient of x_2 is 2.84, the predicted selling price increases by \$2.84 for every square foot increase in lot size. One square foot is a relatively tiny amount. For a 1000 square-foot increase, the predicted selling price increases by $1000(2.84) = \$2840$.

Likewise, we could fix the lot size, and then describe how the predicted selling price depends on the house size. Let's consider houses with lot size $x_2 = 30{,}000$ square feet. The prediction equation becomes

$$\hat{y} = -10{,}536 + 53.8x_1 + 2.84(30{,}000) = 74{,}676 + 53.8x_1.$$

For houses with this lot size, the predicted selling price increases by \$53.80 for every additional square foot in house size.

In summary, an increase of a square foot in *house* size has a larger impact on the selling price (\$53.80) than an increase of a square foot in *lot* size (\$2.84). We can compare slopes for these explanatory variables because their units of measurement are the same (square feet). Slopes can't be compared when the units differ.

Summarizing the Effect While Controlling for a Variable

The multiple regression model states that *each* explanatory variable has a straightline relationship with the mean of *y*, given fixed values of the other explanatory variables. Specifically, *the model assumes that the slope for a particu-*

IN WORDS

For a rectangular house lot with a width of 100 feet, adding 10 feet of depth to the lot increases the lot size by $100 \times 10 = 1000$ square feet, a more meaningful increase than 1 square foot.

Recall

Section 9.5 showed that in multivariate analyses, we can **control** a variable statistically by keeping its value constant while we study the association between other variables. ■

lar explanatory variable is identical for all fixed values of the other explanatory variables. For instance, the coefficient of x_1 in the prediction equation $\hat{y} = -10,536 + 53.8x_1 + 2.84x_2$ is 53.8 regardless of whether we plug in $x_2 = 10,000$ or $x_2 = 30,000$ or $x_2 = 50,000$ for the lot size. When you fix x_2 at these three levels, you can check that

x_2	$\hat{y} = -10,536 + 53.8x_1 + 2.84x_2$
10,000	$\hat{y} = 17,868 + 53.8x_1$
30,000	$\hat{y} = 74,676 + 53.8x_1$
50,000	$\hat{y} = 131,484 + 53.8x_1$

The slope effect of house size is 53.8 for each equation. Setting x_2 at a variety of values yields a collection of parallel lines, each having slope 53.8. See Figure 12.2.

▲ **FIGURE 12.2: The Relationship between $\hat{y}$ and x_1 for the Multiple Regression Equation $\hat{y} = -10,536 + 53.8x_1 + 2.84x_2$.** This shows how the equation simplifies when lot size $x_2 = 10000$, or $x_2 = 30000$, or $x_2 = 50000$. **Question:** The lines move upward (to higher $\hat{y}$ values) as x_2 increases. How would you interpret this fact?

When we fix the value of x_2 we are holding it constant: We are **controlling** for x_2. That's the basis of the major difference between the interpretation of slopes in multiple regression and in bivariate regression:

■ In multiple regression, a slope describes the effect of an explanatory variable while *controlling* effects of the other explanatory variables in the model.

■ Bivariate regression has only a single explanatory variable. So a slope in bivariate regression describes the effect of that variable while *ignoring* all other possible explanatory variables.

For example, the bivariate regression between $y =$ selling price and $x_1 =$ house size is $\hat{y} = 9161 + 77.0x_1$. In this equation, $x_2 =$ lot size and other possible predictors are *ignored*, not *controlled*. The equation describes the relationship for *all* the house sales in the data set. The equation $\hat{y} = 74,676 + 53.8x_1$ we obtained above by substituting $x_2 = 30,000$ into the multiple regression equation applies only for houses that have that fixed lot size. Lot size is controlled. This is why the slopes are different, 77.0 for bivariate regression and 53.8 for multiple regression.

One of the main uses of multiple regression is to identify potential lurking variables and control for them by including them as explanatory variables in the model. Doing so can have a major impact on a variable's effect. When we control a variable, we keep that variable from influencing the associations among the other variables in the study. As we've seen before, the direction of the effect can even change after we control for a variable. Exercise 12.5 illustrates this for multiple regression modeling.

Multiple regression is a powerful tool. However, cautions we made for bivariate regression apply to multiple regression also. For example, it is very risky to use the prediction equation to extrapolate for predictor values far from the values that you observe.

12.1 **Predicting weight:** For a study of University of Georgia female athletes, the prediction equation relating y = total body weight (in pounds) to x_1 = height (in inches) and x_2 = percent body fat is $\hat{y} = -121 + 3.50x_1 + 1.35x_2$.

a. Find the predicted total body weight for a female athlete at the mean values of 66 and 18 for x_1 and x_2.

b. An athlete with $x_1 = 66$ and $x_2 = 18$ has actual weight $y = 115$ pounds. Find the residual, and interpret it.

12.2 **Does study help GPA?:** For the "Georgia student survey" file on the text CD, the prediction equation relating y = college GPA to x_1 = high school GPA and x_2 = study time (hours per day), is $\hat{y} = 1.13 + 0.643x_1 + 0.0078x_2$.

a. Find the predicted college GPA of a student who has a high school GPA of 3.5 and who studies 3 hours a day.

b. For students with fixed study time, what is the change in predicted college GPA when high school GPA increases from 3.0 to 4.0?

12.3 **Predicting college GPA:** For all students at Walden University, the prediction equation for y = college GPA (range 0–4.0) and x_1 = high school GPA (range 0–4.0) and x_2 = college board score (range 200–800) is $\hat{y} = 0.20 + 0.50x_1 + 0.002x_2$.

a. Find the predicted college GPA for students having (i) high school GPA = 4.0 and college board score = 800, (ii) $x_1 = 2.0$ and $x_2 = 200$.

b. For those students with $x_2 = 500$, show that $\hat{y} = 1.20 + 0.50x_1$.

c. For those students with $x_2 = 600$, show that $\hat{y} = 1.40 + 0.50x_1$. Thus, compared to (b), the slope for x_1 is still 0.50, and increasing x_2 by 100 (from 500 to 600) shifts the intercept upward by $100 \times$ (slope for x_2) $= 100(0.002) = 0.20$ units.

12.4 **Interpreting slopes on GPA:** Refer to the previous exercise.

a. Explain why setting x_2 at a variety of values yields a collection of parallel lines relating $\hat{y}$ to x_1. What is the value of the slope for those parallel lines?

b. Since the slope 0.50 for x_1 is larger than the slope 0.002 for x_2, does this imply that x_1 has a larger effect than x_2 on y in this sample? Explain.

12.5 **Does more education cause more crime?:** The "Florida crime" data file on the text CD has data for the 67 counties in Florida on

y = crime rate: Annual number of crimes in county per 1000 population

x_1 = education: Percentage of adults in county with at least a high school education

x_2 = urbanization: Percentage in county living in an urban environment.

The figure shows a scatterplot matrix. The correlations are 0.47 between crime rate and education, 0.68 between crime rate and urbanization, and 0.79 between education and urbanization. MINITAB multiple regression results are also displayed.

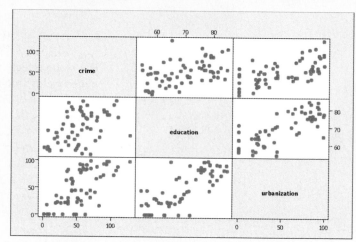

Scatterplot Matrix for Crime Rate, Education, and Urbanization

Multiple regression for y = crime rate, x_1 = education, and x_2 = urbanization.

Predictor	Coef	SE Coef	T	P
Constant	59.12	28.37	2.08	0.041
education	-0.5834	0.4725	-1.23	0.221
urbanization	0.6825	0.1232	5.54	0.000

a. Find the predicted crime rate for a county that has 0% in an urban environment and (i) 70% high school graduation rate, (ii) 80% high school graduation rate.

b. Use results from (a) to explain how education affects the crime rate, controlling for urbanization, interpreting the slope coefficient −0.58 of education.

c. Using the prediction equation, show that the equation relating crime rate and education when urbanization is fixed at (i) 0, (ii) 50, (iii) 100, is as follows:

x_2	$\hat{y} = 59.12 - 0.5834x_1 + 0.6825x_2$
0	$\hat{y} = 59.1 - 0.58x_1$
50	$\hat{y} = 93.2 - 0.58x_1$
100	$\hat{y} = 127.4 - 0.58x_1$

Sketch a plot with these lines and use it to interpret the effect of education on crime rate, controlling for urbanization.

d. The scatterplot matrix shows that education has a *positive* association with crime rate, but the multiple regression equation shows that the association is *negative* when we keep x_2 = urbanization fixed. The reversal in the association is an example of **Simpson's paradox** (See Example 16 in Sec. 3.4 and Example 16 in Sec. 9.5). Consider the hypothetical figure below. Sketch lines that represent (i) the bivariate relationship, ignoring the information on urbanization, (ii) the relationship for counties having urbanization = 50. Use this figure and the correlations provided to explain how Simpson's paradox can happen.

Crime Rate

Hypothetical scatter diagram for crime rate and education, labeling by urbanization.

12.6 Crime rate and income: Refer to the previous exercise. MINITAB reports the results below for the multiple regression of y = crime rate on x_1 = median income (in thousands of dollars) and x_2 = urbanization.

Results of regression analysis

```
Predictor      Coef   SE Coef      T       P
Constant      39.97    16.35     2.44   0.017
income       -0.7906   0.8049   -0.98   0.330
urbanization  0.6418   0.1110    5.78   0.000
```

Correlations: crime, income, urbanization

```
                crime   urbanization
urbanization    0.677
income          0.434      0.731
```

a. Report the prediction equations relating crime rate to income at urbanization levels of (i) 0, (ii) 100. Interpret.

b. For the bivariate model relating y = crime rate to x = income, MINITAB reports

```
crime = -11.6 + 2.61 income
```

Interpret the effect of income, according to the sign of its slope. How does this effect differ from the effect of income in the multiple regression equation?

c. The correlation matrix for these three variables is shown in the table. Use these correlations to explain why the income effect seems so different in the models in (a) and (b).

d. Do these variables satisfy Simpson's paradox? Explain.

12.7 Putting or driving more important in golf?: In the 1997 Masters golf tournament, a multiple regression of total score y of a player in the tournament (for the 72 holes) with explanatory variables DD = player's average driving distance (yards), FWAYS = number of fairways hit in regulation, GIR = number of greens hit in regulation, and PUTTS = total number of putts, gave the re-

sults shown below.[1] Tiger Woods won the tournament with $y = 270$.

Multiple regression for y = total score in Masters golf tournament

```
Predictor   Coef    SE Coef     t        P
Constant  261.78    16.342    16.02   0.0001
DD        -0.0482    0.0306   -1.58   0.123
FWAYS      0.0029    0.1521    0.02   0.985
GIR       -1.307     0.1122  -11.65   0.0001
PUTTS      0.852     0.0960    8.87   0.0001
```

a. Find the predicted total score for a golfer who had DD = 280 yards, FWAYS = 40, GIR = 40, and PUTTS = 110.

b. Explain how to interpret the coefficient for PUTTS.

c. According to the prediction equation, which helps the predicted score decrease the most, for given values of the other variables? (i) increasing DD by 10 yards, (ii) increasing FWAYS by 10, (iii) increasing GIR by 10, (iv) decreasing PUTTS by 10. Justify your answer.

12.8 Comparable lot size and house size effects: In Example 2, the prediction equation between y = selling price and x_1 = house size and x_2 = lot size was $\hat{y} = -10{,}536 + 53.8x_1 + 2.84x_2$.

a. For fixed lot size, how much is the house selling price predicted to increase for each square foot increase in house size? Why?

b. For fixed house size, how much would lot size need to increase to have the same impact?

12.9 Controlling can have no effect: Suppose that the correlation between x_1 and x_2 equals 0. Then, for multiple regression with those predictors, it can be shown that the slope for x_1 is the same as in bivariate regression when x_1 is the only predictor. Explain why you would expect this to be true. (*Hint:* If you don't control x_2, would you expect it to have an impact on how x_1 affects y, if x_1 and x_2 have correlation of 0?)

12.10 House selling prices: Using software with the "house selling prices" data file on the text CD, analyze y = selling price, x_1 = real estate tax and x_2 = lot size.

a. Construct box plots for each variable and a scatterplot matrix or scatter diagrams between y and each of x_1 and x_2. Interpret.

b. Find the multiple regression prediction equation.

c. Interpret the effect of x_2 in the equation in (b).

---

[1]*Source:* J. Earley, Proceedings of SAS Users Group International Conference, 1998.

12.2 Extending the Correlation and *R*-Squared for Multiple Regression

The correlation r and its square, r^2, describe strength of association in a straight-line regression analysis. These measures can also describe association between y and a set of explanatory variables that predict y in a multiple regression model.

Multiple Correlation

To summarize how well a multiple regression model predicts y, we analyze how well the observed y values correlate with the predicted $\hat{y}$ values. As a set, the explanatory variables are strongly associated with y if the correlation between the y and $\hat{y}$ values is strong. Treating the $\hat{y}$ values as a variable gives us a way of summarizing several explanatory variables by *one* variable, for which we can use its ordinary correlation with the y values. The correlation between the observed y values and the predicted $\hat{y}$ values from the multiple regression model is called the **multiple correlation**.

IN WORDS

From Sections 3.2 and 11.2, the **correlation** describes the association between the response variable y and a single explanatory variable x. It is denoted by r (lowercase). The **multiple correlation** describes the association between y and a set of explanatory variables in a multiple regression model. It is denoted by R (uppercase).

> **Multiple Correlation, R**
>
> For a multiple regression model, the **multiple correlation** is the correlation between the observed y values and the predicted $\hat{y}$ values. It is denoted by R.

For each subject, the regression equation provides a predicted value $\hat{y}$. So, each subject has a y value and a $\hat{y}$ value. For the two houses listed in Table 12.1, Table 12.4 shows the actual selling price y and the predicted selling price $\hat{y}$ from the equation $\hat{y} = -10{,}536 + 53.8x_1 + 2.84x_2$ with house size and lot size as predictors. The correlation computed between all 100 pairs of y and $\hat{y}$ values is the multiple correlation, R. Software tells us that this equals 0.84.

Recall

For the first home sale listed in Table 12.1, $y = 145{,}000$, $x_1 = 1240$ and $x_2 = 18{,}000$. ■

TABLE 12.4: Selling Prices and their Predicted Values

These values refer to the two home sales listed in Table 12.1. The predictors are $x_1 =$ house size and $x_2 =$ lot size.

Home	Selling Price	Predicted Selling Price
1	145,000	$\hat{y} = -10{,}536 + 53.8(1240) + 2.84(18{,}000) = 107{,}276$
2	69,000	$\hat{y} = -10{,}536 + 53.8(1120) + 2.84(17{,}000) = 97{,}983$

The larger the multiple correlation, the better are the predictions of y by the set of explanatory variables. For the housing data, $R = 0.84$ indicates a strong association.

The predicted values $\hat{y}$ cannot correlate negatively with y. Otherwise, the predictions would be worse than merely using $\bar{y}$ to predict y. Therefore, R *falls between 0 and 1*. In this way, the multiple correlation R differs from the bivariate correlation r between y and a single variable x, which falls between -1 and $+1$.

R-squared

Recall

Section 11.2 introduced r^2 (**r-squared**) as measuring the proportional reduction in prediction error from using $\hat{y}$ to predict y compared to using $\bar{y}$ to predict y. It falls between 0 and 1, with larger values representing stronger association. ■

For predicting y, the square of R describes the relative improvement from using the prediction equation instead of using the sample mean $\bar{y}$. The error in using the prediction equation to predict y is summarized by the *residual sum of squares*, $\Sigma(y - \hat{y})^2$. The error in using $\bar{y}$ to predict y is summarized by the *total sum of squares*, $\Sigma(y - \bar{y})^2$. The *proportional reduction in error* is

$$R^2 = \frac{\Sigma(y - \bar{y})^2 - \Sigma(y - \hat{y})^2}{\Sigma(y - \bar{y})^2}.$$

The better the predictions are using the regression equation, the larger R^2 is. The bivariate measure r^2 is the special case of R^2 applied to regression with one explanatory variable. As the notation suggests, for multiple regression R^2 is the square of the multiple correlation. Regression software reports R^2, and you can take the positive square root of it to get the multiple correlation.

EXAMPLE 3

How Well Can We Predict House Selling Prices?

Picture the Scenario

For the 100 observations on $y =$ selling price, $x_1 =$ house size, and $x_2 =$ lot size, Table 12.5 shows the ANOVA (analysis of variance) table that MINITAB reports for the multiple regression model.

TABLE 12.5: ANOVA Table and R-Squared for Predicting House Selling Price Using House Size and Lot Size		
R-Sq = 71.1%		
Analysis of Variance		
Source	DF	SS
Regression	2	223676
Residual Error	97	90756
Total	99	314433

Questions to Explore

a. Show how to use the sums of squares in the ANOVA table to find R^2 for this multiple regression model. Interpret.

b. Find and interpret the multiple correlation.

Think It Through

a. From the sum of squares (SS) column, the total sum of squares is $\Sigma(y - \bar{y})^2 = 314{,}433$. The residual sum of squares from using the multiple regression equation to predict y is $\Sigma(y - \hat{y})^2 = 90{,}756$. The value of R^2 is

$$R^2 = \frac{\Sigma(y - \bar{y})^2 - \Sigma(y - \hat{y})^2}{\Sigma(y - \bar{y})^2} = \frac{314{,}433 - 90{,}756}{314{,}433} = 0.711.$$

Using house size and lot size together to predict selling price reduces the prediction error by 71%, relative to using $\bar{y}$ alone to predict selling price. The R^2 statistic appears (in percentage form) in Table 12.5 under the heading "R-sq."

b. The multiple correlation between selling price and the two explanatory variables is $R = \sqrt{R^2} = \sqrt{0.711} = 0.84$. This equals the correlation for the 100 homes between the observed selling prices and the predicted selling prices from multiple regression. There's a strong association between the observed and the predicted selling prices. In summary, house size and lot size very much help us to predict selling prices.

Insight

The bivariate r^2 values for predicting selling price are 0.58 with house size as the predictor and 0.51 with lot size as the predictor. The multiple regression model has $R^2 = 0.71$, so it provides better predictions than either bivariate model.

TO PRACTICE THIS CONCEPT, TRY EXERCISE 12.11 ■

Properties of R^2

The example showed that R^2 for the multiple regression model was larger than r^2 for a bivariate model using only one of the explanatory variables. In fact, *a key property of R^2 is that it cannot decrease when predictors are added to a model.*

The difference $\Sigma(y - \bar{y})^2 - \Sigma(y - \hat{y})^2$ that forms the numerator of R^2 also appears in the ANOVA table. It is called the *regression sum of squares*. A simpler formula for it is

$$\text{Regression SS} = \Sigma(\hat{y} - \bar{y})^2.$$

If each $\hat{y} = \bar{y}$, then the regression equation predicts no better than $\bar{y}$. Then regression SS = 0, and $R^2 = 0$.

The numerical values of the sums of squares reported in Table 12.5 result when y is measured in *thousands* of dollars (for instance, for the first house y is 145.0 rather than 145,000, and $\hat{y}$ is 107.3 rather than 107,300). We truncated the numbers for convenience, because the sums of squares would otherwise be enormous with each one having six more zeros on the end! It makes no difference to the result. Another property of R and R^2 is that, like the correlation, *their values don't depend on the units of measurement.*

In summary, the properties of R^2 are similar to those of r^2 for bivariate models. Here's a list of the main properties:

Properties of R^2

- R^2 falls between 0 and 1. The larger the value, the better the explanatory variables collectively predict y.
- $R^2 = 1$ only when all residuals are 0, that is, when all regression predictions are perfect (each $y = \hat{y}$), so residual SS $= \Sigma(y - \hat{y})^2 = 0$.
- $R^2 = 0$ when each $\hat{y} = \bar{y}$. In that case, the estimated slopes all equal 0, and the correlation between y and each explanatory variable equals 0.
- R^2 gets larger, or at worst stays the same, whenever an explanatory variable is added to the multiple regression model.
- The value of R^2 does not depend on the units of measurement.

Table 12.6 shows how R^2 increases as we add explanatory variables to a multiple regression model to predict $y =$ house selling price. The single predictor in the data set that is most strongly associated with y is the house's real estate tax assessment ($r^2 = 0.679$). When we add house size as a second predictor, R^2 goes up from 0.679 to 0.730. As other predictors are added, R^2 continues to go up, but not by much. Predictive power is not much worse with only real estate tax as a predictor than with all six predictors in the regression model.

TABLE 12.6: R^2 Values for Multiple Regression Models for $y =$ House Selling Price

NW is a geographic variable defined in Section 12.5.

Explanatory Variables in Model	R^2
Tax	0.679
Tax, House size	0.730
Tax, House size, Lot size	0.757
Tax, House size, Lot size, NW	0.769
Tax, House size, Lot size, NW, No. baths	0.773
Tax, House size, Lot size, NW, No. baths, No. bedrooms	0.775

Although R^2 goes up by only small amounts after real estate tax is in the model, this does not mean that the other predictors are only weakly correlated with selling price. For instance, the correlation between lot size and selling price is 0.71. Because the predictors are themselves highly correlated, once one or two of them are in the model, the remaining ones don't help much in adding to the predictive power. For instance, lot size is highly positively correlated with real estate tax and with size of house. So, once real estate tax and size of house are included as predictors in the model, there's not much benefit to including lot size as an additional predictor.

IN PRACTICE: R^2 often does not increase much after a few predictors are in the model

When there are many explanatory variables but the correlations among them are strong, once you have included a few of them in the model, R^2 usually doesn't increase much more when you add additional ones. This does not mean that the additional variables are uncorrelated with the response variable, but merely that they don't add much new power for predicting y, given the values of the predictors already in the model.

As in the bivariate case, a disadvantage of R^2 (compared to the multiple correlation R) is that its units are the *square* of the units of measurement. The R^2 of 0.71 in Example 3 implies that the estimated *variance* of y at fixed values of the predictors is 71% less than the overall variance of y.

12.2 Practicing the Basics

12.11 **Predicting golf scores:** In the 1997 Masters golf tournament, a multiple regression of total score y of a player in the tournament on DD = player's average driving distance (yards), FWAYS = number of fairways hit in regulation, GIR = number of greens hit in regulation, and PUTTS = total number of putts gave the ANOVA table shown[2].

ANOVA table for y = total score in Masters golf tournament

Source	DF	SS	MS	F	P
Regression	4	1599.00	399.75	57.1	0.0001
Residual Error	41	287.00	7.00		
Total	45	1886.00			

S = 2.65   R-Sq = 84.8%

a. Show how R^2 was calculated from the SS values, and report its value.

b. Interpret the R^2 value. Does the multiple regression equation help us predict the total golf score much better than we could without knowing that equation?

c. Find the multiple correlation. Interpret.

12.12 **Predicting weight:** Let's use multiple regression to predict total body weight (in pounds) using data from a study of University of Georgia female athletes.

Possible predictors are HGT = height (in inches), %BF = percent body fat, and age. The display shows correlations among these explanatory variables.

	TBW	HGT	%BF
HGT	0.745		
%BF	0.390	0.096	
AGE	−0.187	−0.120	0.024

a. Which explanatory variable gives by itself the best predictions of weight? Explain.

b. With height as the sole predictor, $\hat{y} = -106 + 3.65$ (HGT) and $r^2 = 0.55$. If you add %BF as a predictor, you know that R^2 will be at least 0.55. Explain why.

c. When you add % body fat to the model, $\hat{y} = -121 + 3.50(\text{HGT}) + 1.35(\%\text{BF})$ and $R^2 = 0.66$. When you add age to the model, $\hat{y} = -97.7 + 3.43(\text{HGT}) + 1.36(\%\text{BF}) - 0.960(\text{AGE})$ and $R^2 = 0.67$. Once you know height and % body fat, does age seem to help you in predicting weight? Explain, based on comparing the R^2 values.

12.13 **When does controlling have little effect?:** Refer to the previous exercise. Height has a similar estimated slope for each of the three models. Why do you think that controlling for % body fat and then age does not change the effect of height much? (*Hint:* How strongly is height correlated with the other two variables?)

---

[2]*Source:* J. Earley, Proceedings of SAS Users Group International Conference, 1998.

12.14 Internet use: For countries listed in the "Human development" data file on the text CD, y = Internet use (percent) is predicted by x_1 = per capita GDP (gross domestic product, in thousands of dollars) with $r^2 = 0.79$. Adding x_2 = percent using cell phones to the model yields the results in the display below.

a. Report R^2, and show how it is determined by SS values in the ANOVA table.

b. Interpret its value as a proportional reduction in prediction error.

Regression of Internet use on GDP and cell-phone use

```
Predictor   Coef    SE Coef    T      P
Constant   -3.601   2.506    -1.44  0.159
GDP         1.2799  0.2703    4.74  0.000
CELLULAR    0.1021  0.0900    1.13  0.264
S = 8.56746      R-Sq = 79.6%
```

Analysis of Variance

```
Source          DF      SS
Regression       2    10316.8
Residual Error  36     2642.5
Total           38    12959.3
```

12.15 More Internet use: In the previous exercise, $r^2 = 0.79$ when x_1 is the predictor and $R^2 = 0.796$ when both x_1 and x_2 are predictors. Why do you think that the predictions of y don't improve much when x_2 is added to the model, even though x_2 is itself highly associated with y (with $r^2 = 0.67$)? (*Hint:* Would you expect x_1 and x_2 to be highly correlated? If so, what's the effect?)

12.16 Softball data: For the "softball" dataset on the text CD, for each game the variables are a team's number of runs scored (RUNS), number of hits (HIT), number of errors (ERR), and the difference (DIFF) between the number of runs scored by that team and by the other team, which is the response variable. MINITAB reports

```
Difference = - 4.03 + 0.0260 Hits + 1.04
Run - 1.22 Errors
```

a. If you know the team's number of runs and number of errors in a game, explain why it does not help much to know how many hits the team has.

b. Explain why the result in (a) is also suggested by knowing that $R^2 = 0.7594$ for this model, whereas $R^2 = 0.7593$ when only runs and errors are the explanatory variables in the model.

12.17 Slopes, correlations, and units: In Example 2 on y = house selling price, x_1 = house size, and x_2 = lot size, $\hat{y} = -10{,}536 + 53.8x_1 + 2.84x_2$, and $R = 0.84$.

a. Interpret the value of the multiple correlation.

b. Suppose house selling prices are changed from dollars to *thousands* of dollars. Explain why each y value in Table 12.1 is divided by 1000, and the prediction equation changes to $\hat{y} = -10.536 + 0.0538x_1 + 0.00284x_2$.

c. In (b), does the multiple correlation change to 0.00084? Justify your answer.

12.18 Predicting college GPA: Using software with the "Georgia student survey" data file from the text CD, find and interpret the multiple correlation and R^2 for the relationship between y = college GPA, x_1 = high school GPA, and x_2 = study time.

12.3 | How Can We Use Multiple Regression to Make Inferences?

We've seen that multiple regression uses more than one explanatory variable to predict a response variable y. With it, we can study an explanatory variable's effect on y while controlling other variables that could affect the results. Now, let's turn our attention to using multiple regression to make inferences about the population.

Inferences require the same assumptions as in bivariate regression:

■ The regression equation truly holds for the population means.

■ The data were gathered using randomization.

■ The response variable y has a normal distribution at each combination of values of the explanatory variables, with the same standard deviation.

The first assumption implies that there is a straight-line relationship between the mean of y and each explanatory variable, with the same slope at each value of the other predictors. We'll see how to check this important assumption in Section 12.4.

Recall

Section 11.3 showed that the *t* **test** statistic for $H_0: \beta = 0$ in the bivariate model is

$$t = \frac{(\text{sample slope} - 0)}{\text{std. error of sample slope}}$$

$$= (b - 0)/se,$$

with $df = n - 2$. ▪

Inferences about Individual Regression Parameters

For the bivariate model, $\mu_y = \alpha + \beta x$, there's a *t* test for the null hypothesis $H_0: \beta = 0$ that *x* and *y* are statistically independent. Likewise, a *t* test applies to any slope parameter in multiple regression. Let's consider a particular parameter, say β_1.

If $\beta_1 = 0$, the mean of *y* is identical for all values of x_1, at fixed values for the other explanatory variables. So, $H_0: \beta_1 = 0$ states that *y and x_1 are statistically independent, controlling for the other variables.* This means that once the other explanatory variables are in the model, it doesn't help to have x_1 in the model. The alternative hypothesis usually is two sided, $H_a: \beta_1 \neq 0$, but one-sided alternative hypotheses are also possible.

The test statistic for $H_0: \beta_1 = 0$ is

$$t = (b_1 - 0)/se,$$

where *se* is the standard error of the slope estimate b_1 of β_1. Software provides the *se* value, the *t* test statistic, and the P-value. If H_0 is true and the inference assumptions hold, the *t* test statistic has the *t* distribution. The degrees of freedom are

$$df = \text{sample size } n - \text{number of parameters in the regression equation.}$$

The bivariate model $\mu_y = \alpha + \beta x$ has two parameters (α and β), so $df = n - 2$, as Section 11.3 used. The model with two predictors, $\mu_y = \alpha + \beta_1 x_1 + \beta_2 x_2$, has three parameters, so $df = n - 3$. Likewise, with three predictors, $df = n - 4$, and so on.

Significance Test about a Multiple Regression Parameter (such as β_1)

1. Assumptions:

- Each explanatory variable has a straight-line relation with μ_y, with the same slope for all combinations of values of other predictors in model

- Data gathered with randomization (such as random sample)

- Normal distribution for *y* with same standard deviation at each combination of values of other predictors in model

2. Hypotheses:

$H_0: \beta_1 = 0$

$H_a: \beta_1 \neq 0$

When H_0 is true, *y* is independent of x_1, controlling for the other predictors.

3. Test statistic: $t = (b_1 - 0)/se$. Software supplies the slope estimate b_1 and its *se*.

4. P-value: Two-tail probability from *t* distribution of values larger than observed *t* test statistic (in absolute value). The *t* distribution has

$$df = n - \text{number of parameters in regression equation}$$

(such as $df = n - 3$ when $\mu_y = \alpha + \beta_1 x_1 + \beta_2 x_2$, which has 3 parameters).

5. Conclusion: Interpret P-value; compare to significance level if decision needed.

EXAMPLE 4

What Helps Predict a Female Athlete's Weight?

Picture the Scenario

The "college athletes" data set on the text CD comes from a study of 64 University of Georgia female athletes who participated in Division I sports. The study measured several physical characteristics, including total body weight in pounds (TBW), height in inches (HGT), the percent of body fat (%BF), and age. Table 12.7 shows summary statistics for these variables.

TABLE 12.7: Summary Statistics for Study of Female College Athletes

Variables are TBW = total body weight, HGT = height, %BF = percent body fat, and AGE.

Variable	Mean	StDev	Minimum	Q1	Median	Q3	Maximum
TBW	133.0	17.2	96.0	119.2	131.5	143.8	185.0
HGT	65.5	3.5	56.0	63.0	65.0	68.2	75.0
%BF	18.4	4.1	11.2	15.2	18.5	21.5	27.6
AGE	20.0	1.98	17.0	18.0	20.0	22.0	23.0

Table 12.8 shows results of fitting a multiple regression model for predicting weight using the other variables. The predictive power is good, with $R^2 = 0.669$.

TABLE 12.8: Multiple Regression Analysis for Predicting Weight

Predictors are HGT = height, %BF = percent body fat, and age of subject.

Predictor	Coef	SE Coef	T	P
Constant	−97.69	28.79	−3.39	0.001
HGT	3.4285	0.3679	9.32	0.000
%BF	1.3643	0.3126	4.36	0.000
AGE	−0.9601	0.6483	−1.48	0.144

R-Sq = 66.9%

Question to Explore

a. Interpret the effect of age on weight in the multiple regression equation.

b. In the population, does age help you to predict weight, if you already know height and percent body fat? Show all steps of a significance test, and interpret.

Think It Through

a. Let $\hat{y}$ = predicted weight, x_1 = height, x_2 = %body fat, and x_3 = age. Then

$$\hat{y} = -97.7 + 3.43x_1 + 1.36x_2 - 0.96x_3.$$

The slope coefficient of age is −0.96. The sample effect of age on weight is negative, which may seem surprising, but in practical terms it is small: For athletes having fixed values of x_1 and x_2, the predicted weight decreases by only 0.96 pounds for a 1-year increase in age, and the ages vary only from 17 to 23.

b. If $\beta_3 = 0$, then x_3 = age has *no* effect on weight in the population, controlling for height and body fat. The hypothesis that age does *not* help us better predict weight, if we already know height and body fat, is $H_0: \beta_3 = 0$. Here are the steps:

1. **Assumptions:** The 64 female athletes were a convenience sample, not a random sample. Although the goal was to make inferences about all female college athletes, inferences are tentative. We'll discuss the other assumptions and learn how to check them in Section 12.4.

2. **Hypotheses:** The null hypothesis is H_0: $\beta_3 = 0$. Since there's no prior prediction about whether the effect of age is positive or negative (for fixed values of x_1 and x_2), we use the two-sided H_a: $\beta_3 \neq 0$.

3. **Test statistic:** Table 12.8 reports a slope estimate of -0.960 for age and a standard error of $se = 0.648$. It also reports the t test statistic of

$$t = (b_3 - 0)/se = -0.960/0.648 = -1.48.$$

Since the sample size equals $n = 64$ and the regression equation has 4 parameters, the degrees of freedom are $df = n - 4 = 60$.

4. **P-value:** Table 12.8 reports P-value $= 0.14$. This is the two-tailed probability of a t statistic below -1.48 or above 1.48, if H_0 were true.

5. **Conclusion:** The P-value of 0.14 does not give much evidence against the null hypothesis that $\beta_3 = 0$. At common significance levels, such as 0.05, we cannot reject H_0. Age does not significantly predict weight if we already know height and percentage of body fat.

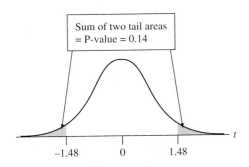

Sum of two tail areas = P-value = 0.14

Insight

By contrast, Table 12.8 shows that $t = 9.3$ for testing the effect of height (H_0: $\beta_1 = 0$) and $t = 4.4$ for testing the effect of %BF (H_0: $\beta_2 = 0$). Both P-values are 0.000. It *does* help us to have each of these variables in the model, given the other two.

TO PRACTICE THIS CONCEPT, TRY EXERCISE 12.19

As usual, a test merely tells us whether the null hypothesis is plausible. In Example 4 we saw that β_3 may equal 0, but what are its other plausible values? Might it actually be far from 0? A confidence interval answers this question.

Confidence Interval for a Multiple Regression β Parameter

A 95% confidence interval for a β slope parameter in multiple regression equals

$$\text{Estimated slope} \pm t_{.025}(se).$$

The t-score has $df = n -$ number of parameters in regression equation, as in the t test. The assumptions are also the same as for the t test.

EXAMPLE 5

What's Plausible for the Effect of Age on Weight?

Picture the Scenario

For the "college athletes" data set of Example 4, consider the multiple regression analysis of y = weight and predictors x_1 = height, x_2 = % body fat, and x_3 = age.

Question to Explore

Find and interpret a 95% confidence interval for β_3, the effect of age while controlling for height and percent of body fat.

Recall

From Table 12.8,

Predictor	Coef	SE
Constant	−97.7	28.8
HGT	3.43	0.368
%BF	1.36	0.313
AGE	−0.96	0.648 ■

Think It Through

From the previous example, $df = 60$. For $df = 60$, $t_{.025} = 2.00$. From Table 12.8 (shown partly in the margin), the estimate of β_3 is -0.96, with $se = 0.648$. The 95% confidence interval equals

$$b_3 \pm t_{.025}(se) = -0.96 \pm 2.00(0.648),$$
$$\text{or } -0.96 \pm 1.30, \text{ roughly } (-2.3, 0.3).$$

At fixed values of x_1 and x_2, we infer that the population mean of weight changes very little (and may not change at all) for a 1-year increase in age.

Insight

The confidence interval contains 0. Age may have *no* effect on weight, once we control for height and % body fat. This is in agreement with not rejecting $H_0: \beta_3 = 0$ in favor of $H_a: \beta_3 \neq 0$ at the $\alpha = 0.05$ level in the significance test.

TO PRACTICE THIS CONCEPT, TRY EXERCISE 12.20

Estimating Variability around the Regression Equation

As in bivariate regression, a standard deviation parameter σ describes variability of the observations around the regression equation. Its sample estimate is

$$s = \sqrt{\frac{\text{Residual SS}}{df}} = \sqrt{\frac{\sum (y - \hat{y})^2}{n - (\text{number of parameters in regression equation})}}.$$

This **residual standard deviation** describes the typical size of the residuals. Its degrees of freedom equal the *df* used in the *t* test and the confidence interval about a slope parameter. Software reports *s* and its square, the **mean square error**.

EXAMPLE 6

Estimating Variability of Female Athletes' Weight

Picture the Scenario

For the "college athletes" data set (Examples 4 and 5), Table 12.9 shows the ANOVA table for the regression of weight on height, % body fat, and age.

TABLE 12.9: ANOVA Table for Multiple Regression Analysis of Athlete Weights

Analysis of Variance

Source	DF	SS	MS	F	P
Regression	3	12407.9	4136.0	40.48	0.000
Residual Error	60	6131.0	102.2		
Total	63	18539.0			

S = 10.1086   R-Sq = 66.9%

Question to Explore

For female athletes at particular values of height, percent of body fat, and age, estimate the standard deviation of their weights.

Recall

Recall that the prediction equation was $\hat{y} = -97.7 + 3.43x_1 + 1.36x_2 - 0.96x_3$, so there were 4 parameters in the regression model.▪

Recall

For bivariate regression, Example 16 in Section 11.4 discussed confidence intervals for the mean and prediction intervals for y at a particular setting of x. These intervals can be formed with multiple regression as well.▪

Think It Through

The SS column tells us that the residual sum of squares is 6131.0. There were $n = 64$ observations and 4 parameters in the regression model, so the DF column reports $df = n - 4 = 60$ opposite the residual SS. The mean square error is

$$s^2 = (\text{residual SS})/df = 6131.0/60 = 102.2.$$

It appears in the mean square (MS) column, in the row labeled "Residual Error." The residual standard deviation is $s = \sqrt{102.2} = 10.1$, identified as S. For athletes with certain fixed values of height, % body fat, and age, the weights vary with a standard deviation of about 10 pounds.

Insight

If the conditional distributions of weight are approximately bell-shaped, about 95% of the weight values fall within about $2s = 20$ pounds of the true regression equation. More precisely, software can report *prediction intervals* within which a response outcome has a certain chance of falling. For instance, at $x_1 = 66$, $x_2 = 18$, and $x_3 = 20$, which are close to the mean predictor values, software reports $\hat{y} = 133.9$ and a 95% prediction interval of 133.9 ± 20.4.

TO PRACTICE THIS CONCEPT, TRY EXERCISE 12.21

In the sample, weight has standard deviation = 17 pounds, describing variability around mean weight = 133 pounds. How is it that the *residual* standard deviation could be only 10 pounds? The residual standard deviation is smaller because it refers to variability at *fixed* values of the predictors. Weight varies less at given values of the predictors than it does overall. If we could predict weight perfectly knowing height, % body fat, and age, the residual standard deviation would be 0.

The ANOVA table also reports another mean square, called a *mean square for regression*, or *regression mean square* for short. We'll next see how to use the mean squares to conduct a significance test about all the slope parameters together.

Do the Explanatory Variables Collectively Have an Effect?

Do the explanatory variables collectively have a statistically significant effect on the response variable y? With 3 predictors in a model, we can check this by testing

$$H_0: \beta_1 = \beta_2 = \beta_3 = 0.$$

This hypothesis states that the mean of y does not depend on *any* of the predictors in the model. That is, *y is statistically independent of all the explanatory variables.* The alternative hypothesis is

$$H_a: \text{At least one } \beta \text{ parameter is not equal to } 0.$$

This states that *at least one* explanatory variable is associated with y.

The null hypothesis that all the slope parameters equal 0 is equivalent to the hypothesis that the population values of the multiple correlation and R^2 equal 0. The equivalence occurs because the population values of R and R^2 equal 0 only in those situations in which all the β parameters equal 0.

The test statistic for H_0 is denoted by F. It equals the ratio of the mean squares from the ANOVA table,

$$F = \frac{\text{Mean square for regression}}{\text{Mean square error}}.$$

We won't need the formulas for the mean squares here, but the value of F is proportional to $R^2/(1 - R^2)$. As R^2 increases, the F test statistic increases.

When H_0 is true, the expected value of the F test statistic is approximately 1. When H_0 is false, F tends to be larger than 1. The larger the F test statistic, the stronger the evidence against H_0. The P-value is the right-tail probability from the sampling distribution of the F test statistic.

From the ANOVA table in Table 12.9 (shown again in the margin) for the regression model predicting weight, the mean square for regression = 4136.0 and the mean square error = 102.2. Then the test statistic

$$F = 4136.0/102.2 = 40.5.$$

This F value is extremely large, and the ANOVA table reports P-value = 0.000.

The *F* Distribution and Its Properties

The sampling distribution of the F test statistic is called the **F distribution**. The symbol for the F statistic and its distribution honors the eminent British statistician R. A. Fisher, who discovered the F distribution in 1922. Like the chisquared distribution, the F distribution can assume only nonnegative values and it is skewed to the right. See Figure 12.3. The mean is approximately 1.

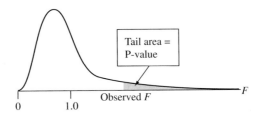

▲ **FIGURE 12.3: The *F* Distribution and the P-Value for *F* tests. Question:** Why does the P-value use the right tail but not the left tail?

The precise shape of the F distribution is determined by two degrees-of-freedom terms, denoted by df_1 and df_2. These are the df values in the ANOVA table for the two mean squares whose ratio equals F. The first one is

$$df_1 = \text{number of explanatory variables in the model.}$$

The second,

$$df_2 = n - \text{number of parameters in regression equation,}$$

is the df for the t tests about the individual regression parameters. Sometimes df_1 is called the **numerator *df***, because it is listed in the ANOVA table next to the mean square for regression, which goes in the numerator of the F test statistic. Likewise, df_2 is called the **denominator *df***, because it is listed in the ANOVA table next to the mean square error, which goes in the denominator of the F test statistic.

Table D at the end of the text lists the F values that have P-value = 0.05, for various df_1 and df_2 values. Table 12.10 shows a small excerpt. For instance, when $df_1 = 3$ and $df_2 = 40$, $F = 2.84$ has P-value = 0.05. When $F > 2.84$, the test statistic is farther out in the tail and the P-value < 0.05. Regression software reports the actual P-value.

ANOVA table from Table 12.9

Source	DF	SS	MS	F	P
Reg.	3	12407.9	4136.0	40.5	.000
Error	60	6131.0	102.2		
Total	63	18539.0			

TABLE 12.10: An Excerpt of Table D Displaying *F* Values

These are the values that have right-tail probability equal to 0.05.

df_2	1	2	3	4	5
30	4.17	3.32	2.92	2.69	2.69
40	4.10	3.23	2.84	2.61	2.45
60	4.00	3.15	2.76	2.52	2.37

Summary of F Test That All the Multiple Regression Beta Parameters = 0

1. **Assumptions:** Multiple regression equation holds, data gathered randomly, normal distribution for y with same standard deviation at each combination of predictors.

2. **Hypotheses:**

   H_0: $\beta_1 = \beta_2 = \ldots = 0$ (all the beta parameters in the model = 0)

   H_a: At least one β parameter differs from 0.

3. **Test statistic:** $F = $ (mean square for regression)/(mean square error)

4. **P-value:** Right-tail probability above observed F test statistic value from F distribution with

   $df_1 = $ number of explanatory variables,

   $df_2 = n - $ (number of parameters in regression equation).

5. **Conclusion:** The smaller the P-value, the stronger the evidence that at least one explanatory variable has an effect on y. If a decision is needed, reject H_0 if P-value≤significance level, such as 0.05. Interpret in context.

EXAMPLE 7

The F Test for Predictors of Athletes' Weight

Picture the Scenario

For the 64 female college athletes, the ANOVA table for the multiple regression predicting $y = $ weight using $x_1 = $ height, $x_2 = \%$ body fat, and $x_3 = $ age shows

Source	DF	SS	MS	F	P
Regression	3	12407.9	4136.0	40.48	0.000
Residual Error	60	6131.0	102.2		

Questions to Explore

a. State and interpret the null hypothesis tested in this table.

b. From the F table, which F value would have a P-value of 0.05 for these data?

c. Report the observed test statistic and P-value. Interpret the P-value, and make a decision for a 0.05 significance level.

Think It Through

a. Since there are 3 explanatory variables, the null hypothesis is H_0: $\beta_1 = \beta_2 = \beta_3 = 0$. It states that weight is independent of height, % body fat, and age.

b. In the DF column, the ANOVA table shows $df_1 = 3$ and $df_2 = 60$. The F table or software indicates that the F value with right-tail probability of 0.05 is 2.76. (See also Table 12.10.)

c. From the ANOVA table, the observed F test statistic value is 40.5. Since this is well above 2.76, the P-value is less than 0.05. The ANOVA table reports P-value = 0.000. If H_0 were true, it would be extremely unusual to get such a large F test statistic. We can reject H_0 at the 0.05 significance level. In summary, we conclude that at least one predictor has an effect on weight.

Insight

The F test tells us that *at least one* explanatory variable has an effect. From the individual t tests in Table 12.8, we've seen that x_1 and x_2 have significant effects, but x_3 does not have a significant effect (controlling for x_1 and x_2).

TO PRACTICE THIS CONCEPT, TRY EXERCISE 12.25

Model Building

Unless the sample size is small and the correlation is weak between y and each explanatory variable, the F test usually has a small P-value. If the explanatory variables are chosen sensibly, at least one should have *some* predictive power. When the P-value is small, we can conclude merely that *at least* one explanatory variable affects y. The more narrowly focused t inferences about individual slopes judge *which* effects are nonzero and estimate their sizes.

IN PRACTICE: The overall F test is done before the individual t inferences

The F test is normally done *first* before looking at the individual t inferences. The F test result tells us whether there is sufficient evidence to make it worthwhile to consider the individual effects. When there are many explanatory variables, doing the F test first provides protection from doing lots of t tests and having one of them be significant merely by random variation when, in fact, there truly are no effects in the population.

After finding the highly significant F test result of Example 7, we would study the individual effects, as we did in Examples 4 and 5. When we look at an individual t inference, suppose we find that the plausible values for a parameter are all relatively near 0. This was the case for the effect of age on weight (controlling for the other variables) as shown by the confidence interval in Example 5. Then, to simplify the model, you can remove that predictor, refitting the model with the other predictors. When we do this for y = weight, x_1 = height, and x_2 = % body fat, we get

$$\hat{y} = -121.0 + 3.50x_1 + 1.35x_2.$$

For the simpler model, $R^2 = 0.66$. This is nearly as large as the value of $R^2 = 0.67$ with age also in the model. The predictions of weight are essentially as good without age in the model.

When you have several potential explanatory variables for a multiple regression model, how do you decide which ones to include? As just mentioned, the lack of statistical and practical significance is one criterion for deleting a term from the model. Many other possible criteria can also be considered. Most regression software has automatic "stepwise" procedures that successively add or delete predictor variables from models according to criteria such as statistical significance. Such methods must be used with great caution, however. There is no guarantee that the final model chosen is sensible. Model selection requires quite a bit of statistical sophistication, and for applications with many potential explanatory variables we recommend that you seek guidance from a statistician.

Recall

The multiple regression equation using all three predictors was

$$\hat{y} = -97.7 + 3.43x_1 + 1.36x_2 - 0.96x_3.$$

The effects of height and % body fat are similar in the equation without x_3 = age. ■

12.3 Practicing the Basics

12.19 Predicting GPA: For the 59 observations in the "Georgia student survey" data file on the text CD, the result of regressing college GPA on high school GPA and study time is shown below.

College GPA, high school GPA, and study time

The regression equation is

CGPA = 1.13 + 0.643 HSGPA + 0.0078 Studytime

Predictor	Coef	SE Coef	T	P
Constant	1.1262	0.5690	1.98	0.053
HSGPA	0.6434	0.1458	4.41	0.000
Studytime	0.00776	0.01614	0.48	0.633

S = 0.3188 R-Sq = 25.8%

a. Explain in nontechnical terms what it means if the population slope coefficient for high school GPA equals 0.

b. Show all steps for testing the hypothesis that this slope equals 0.

12.20 Study time help GPA?: Refer to the previous exercise.

a. Report and interpret the P-value for testing the hypothesis that the population slope coefficient for study time equals 0.

b. Find a 95% confidence interval for the true slope for study time. Explain how the result is in accord with the result of the test in (a).

c. Does the result in (a) imply that in the corresponding population, study time has no association with college GPA? Explain. (*Hint:* What is the impact of also having HSGPA in the model?)

12.21 Variability in college GPA: Refer to the previous two exercises.

a. Report the residual standard deviation. What does this describe?

b. Interpret the residual standard deviation by predicting where approximately 95% of the Georgia college GPAs fall when high school GPA = 3.80 and study time = 5.0 hours per day, which are the means.

12.22 Does leg press help predict body strength?: Chapter 11 analyzed strength data for 57 female high school athletes. Upper body strength was summarized by the maximum number of pounds the athlete could bench press (denoted BP). This was predicted well by the number of times she can do a 60-pound bench press (denoted BP_60). Can we predict BP even better if we also know how many times an athlete can perform a 200-pound leg press? The table shows results after adding this second predictor (denoted LP_200) to the model.

Multiple regression analysis of strength data

Predictor	Coef	SE Coef	T	P
Constant	60.596	2.872	21.10	0.000
BP_60	1.3318	0.1876	7.10	0.000
LP_200	0.2110	0.1519	1.39	0.171

Analysis of Variance

Source	DF	SS	MS	F	P
Regression	2	6473.3	3236.6	51.39	0.000
Res. Error	54	3401.3	63.0		
Total	56	9874.6			

S = 7.93640 R-Sq = 65.6%

a. Does LP_200 have a significant effect on BP, if BP_60 is also in the model? Show all steps of a significance test to answer this.

b. Show that the 95% confidence interval for the slope for LP_200 equals 0.21 ± 0.30, roughly (−0.1, 0.5). Based on this interval, does LP_200 seem to have a strong impact, or a weak impact, on predicting BP, if BP_60 is also in the model?

c. Given that LP_200 is in the model, provide evidence from a significance test that shows why it *does* help to add BP_60 to the model.

12.23 Leg press uncorrelated with strength?: The P-value of 0.17 in (a) of the previous exercise suggests that LP_200 plausibly had no effect on BP, once BP_60 is in the model. Yet when LP_200 is the sole predictor of BP, the correlation is 0.58 and the significance test for its effect has a P-value of 0.000, suggesting very strong evidence of an effect. Explain why this is not a contradiction.

12.24 Interpret strength variability: Refer to the previous two exercises. The sample standard deviation of BP was 13.3. The residual standard deviation of BP when BP_60 and LP_200 are predictors in a multiple regression model is 7.9.

a. Explain the difference between the interpretations of these two standard deviations.

b. If the conditional distributions of BP are approximately bell shaped, explain why most maximum bench press values fall within about 16 pounds of the regression equation, when the predictors BP_60 and LP_200 are near their mean values.

c. At $x_1 = 11$ and $x_2 = 22$, which are close to the mean values, software reports $\hat{y} = 80$ and a 95% prediction interval of 80 ± 16, or (64, 96). Is this interval an inference about where the population BP values fall or where the population *mean* of the BP values fall (for subjects having $x_1 = 11$ and $x_2 = 22$)? Explain.

d. Refer to (c). Would it be unusual for a female athlete with these predictor values to be able to bench press more than 100 pounds? Why?

597

12.25 Any predictive power?: Refer to the previous three
TRY exercises.

a. State and interpret the null hypothesis tested with the
F statistic in the above ANOVA table.

b. From the F table, which F statistic value would have a
P-value of 0.05 for these data?

c. Report the observed F test statistic and its P-value.
Interpret the P-value, and make a decision for a 0.05
significance level. Explain in nontechnical terms what
the result of the test means.

12.26 Predicting pizza revenue: Aunt Erma's Pizza restaurant keeps monthly records of total revenue, amount
spent on TV advertising, and amount spent on newspaper
advertising.

a. Specify notation and formulate a multiple regression equation for predicting the monthly revenue.
Explain how to interpret the parameters in the
equation.

b. State the null hypothesis that you would test if you
want to analyze whether TV advertising is helpful,
for a given amount of newspaper advertising.

c. State the null hypothesis that you would test if
you want to analyze whether *at least one* of the
sources of advertising has some effect on monthly
revenue.

12.27 Regression for mental health: A study in Alachua
County, Florida, investigated an index of mental health
impairment, which had $\bar{y} = 27.3$ and $s = 5.5$. Two explanatory variables were x_1 = life events score
(mean = 44.4, $s = 22.6$) and x_2 = SES (socioeconomic
status, mean = 56.6, $s = 25.3$). Life events is a composite measure of the number and severity of major life
events, such as death in the family, that the subject
experienced within the past three years. SES is a
composite index based on occupation, income, and
education. The table shows data for 6 subjects. The
complete data[3] for a random sample of 40 adults in
the county is in the "mental health" data file on
the text CD. Some regression results are also shown.

a. Find the 95% confidence interval for β_1.

b. Explain why the interval in (a) means that an
increase of 100 units in life events corresponds
to anywhere from a 4 to a 17 unit increase in
mean mental impairment, controlling for SES.
(This lack of precision reflects the small
sample size.)

**y = mental impairment, x_1 = life events index, and
x_2 = socioeconomic status**

y	x_1	x_2	y	x_1	x_2	y	x_1	x_2
17	46	84	26	50	40	30	44	53
19	39	97	26	48	52	31	35	38

```
Predictor Coef     SE Coef      T       P
Constant 28.230    2.174      12.98   0.000
life      0.10326 0.03250      3.18   0.003
ses      -0.09748 0.02908     -3.35   0.002
S = 4.55644 R-Sq = 33.9%

Analysis of Variance
Source        DF    SS      MS     F      P
Regression     2  394.24 197.12  9.49  0.000
Res. Error    37  768.16  20.76
Total         39 1162.40
```

12.28 Mental health again: Refer to the previous exercise.

a. Report the test statistic and P-value for testing
$H_0: \beta_1 = \beta_2 = 0$.

b. State the alternative hypothesis that is supported by
the result in (a).

c. Does the result in (a) imply that necessarily *both* life
events and SES are needed in the model? Explain.

12.29 More predictors for house price: The SPSS table
shows results for predicting y = house selling price using
x_1 = size of home, x_2 = lot size, and x_3 = real estate tax.

**Regression of selling price on house size, lot size, and
real estate tax**

Coefficients[a]

Model		Unstandardized Coefficients		t	Sig.
		B	Std. Error		
1	(Constant)	6305.319	9567.273	.659	.511
	size	34.512	7.543	4.575	.000
	lot	1.594	.491	3.246	.002
	Taxes	22.035	5.195	4.242	.000

a. Dependent Variable: price

a. State the null hypothesis for an F test, in the context
of these variables.

b. The F statistic equals 99.65, with P-value = 0.000.
Interpret.

c. Explain in nontechnical terms what you learn from
the results of the t tests reported in the table for the
three explanatory variables.

12.30 House prices: Use software to do further analyses
with the multiple regression model of y = selling price
of home, x_1 = size of home, and x_2 = lot size, considered in Section 12.1. The data file "house selling price"
is on the text CD.

a. Report the F statistic and state the hypotheses to
which it refers. Report its P-value, and interpret. Why
is it not surprising to get a small P-value for this test?

b. Report and interpret the t statistic and P-value for
testing $H_0: \beta_2 = 0$ against $H_a: \beta_2 > 0$.

c. Construct a 95% confidence interval for β_2, and
interpret. This inference is more informative than the
test in (b). Explain why.

[3]Based on a larger survey reported in a Ph.D. thesis by C. E. Holzer III at University of Florida.

12.4 | Checking a Regression Model Using Residual Plots

In capsule form, recall that the three assumptions for inference with a multiple regression model are that (1) the regression equation approximates well the true relationship between the predictors and the mean of y (that is, straight-line relationships between the mean of y and each explanatory variable, at fixed values of the other explanatory variables), (2) the data were gathered randomly, and (3) y has a normal distribution with the same standard deviation at each combination of predictors. Now, let's see how to check the assumptions about the regression equation and the distribution of y.

For bivariate regression, the scatterplot provides a simple visual check of whether the straight-line model is appropriate. For multiple regression, a plot of all the variables at once would require many dimensions. So instead, we study how the predicted values $\hat{y}$ compare to the observed values y. This is done using the residuals, $y - \hat{y}$.

Checking Shape and Detecting Unusual Observations

Consider first the assumption that the conditional distribution of y is normal, at any fixed values of the explanatory variables. If this is true, the residuals should have approximately a bell-shaped histogram. As Section 11.4 noted, software can plot a histogram of the residuals or of the *standardized residuals*, which are the residuals divided by their standard errors. Nearly all the standardized residuals should fall between about −3 and +3. A standardized residual below −3 or above +3 indicates a potential regression outlier—an observation that is far from what the model predicts.

When some observations have large standardized residuals, you should think about whether there is an explanation. Often this merely reflects skew in the conditional distribution of y, with a long tail in one direction. Other times the observations differ from the others on a variable that was not included in the model. Once that variable is added, those observations cease to be so unusual. For instance, suppose the data file from the weight study of Examples 4–7 also contained observations for a few males. Then we might have observed a few y values considerably above the others and with very large positive standardized residuals. These residuals would probably diminish considerably once the model also included gender as a predictor.

EXAMPLE 8

Residuals for House Selling Price

Picture the Scenario

For the house selling price data (Examples 1–3), Figure 12.4 is a MINITAB histogram of the standardized residuals for the multiple regression model predicting selling price by the house size and the lot size.

Question to Explore

What does this figure tell you? Interpret it.

Think It Through

The residuals are roughly bell shaped about 0. They fall between about −3 and +3. The shape suggests that the conditional distribution of the response variable may have a bit of skew, but no severe nonnormality is indicated.

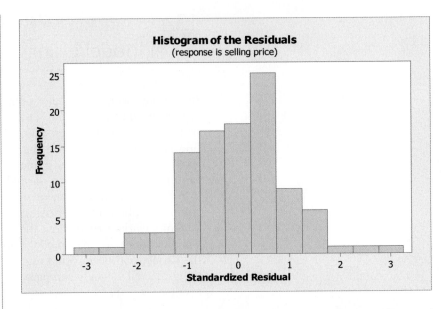

▲ **FIGURE 12.4: Histogram of Standardized Residuals for Multiple Regression Model Predicting Selling Price. Question:** Give an example of a shape for this histogram that would indicate that a few observations are highly unusual.

Insight

When *n* is small, don't make the mistake of "reading too much" into such plots. We're mainly looking for dramatic departures from the assumptions and highly unusual observations that "stand apart" from the others.

TO PRACTICE THIS CONCEPT, TRY EXERCISE 12.31

Two-sided inferences about slope parameters are *robust*. The normality assumption is not as important as the assumption that the regression equation approximates well the true relationship between the predictors and the mean of *y*. We consider that assumption next.

Plotting Residuals against Each Explanatory Variable

Plots of the residuals against each explanatory variable help us check for potential problems with the regression model. We discuss this in terms of x_1, but you should view a plot for each predictor. Ideally, the residuals should fluctuate randomly about 0, as in Figure 12.5a. There should be no obvious change in trend or change in variation as the values of x_1 increase.

By contrast, a scattering of the residuals as in Figure 12.5c suggests that *y* is actually *nonlinearly* related to x_1. It suggests that *y* tends to be *below* $\hat{y}$ for very small and very large x_1 values (giving negative residuals) and *above* $\hat{y}$ for medium-sized x_1 values (giving positive residuals).

▲ **FIGURE 12.5: Possible Patterns for Residuals, Plotted Against an Explanatory Variable. Question:** Why does the pattern in (c) suggest that the effect of x_1 is not linear?

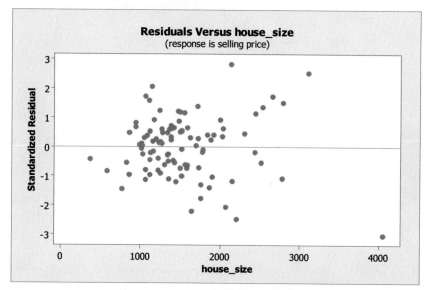

A common occurrence is that residuals have increased variation as x_1 increases, as seen in Figure 12.5b. A pattern like this indicates that the residual standard deviation of y is not constant—the data are more spread out at larger values of x_1 (see the margin figure). Two-sided inferences about slope parameters still perform pretty well. However, ordinary prediction intervals are invalid. The width of such intervals is similar at relatively high and relatively low x_1 values, but in reality, observations on y are much more spread out at high x_1 values than at low x_1 values.

IN PRACTICE: Use caution in interpreting residual patterns

Residual patterns are often not as neat as the ones in Figure 12.5. Be careful not to let a single outlier or ordinary sampling variability overly influence your reading of a pattern from the plot.

EXAMPLE 9

Another Residual Plot for House Selling Prices

Picture the Scenario

For the house selling price data, Figure 12.6 is a residual plot for the multiple regression model relating selling price to house size and to lot size. It plots the standardized residuals against house size.

▲ **FIGURE 12.6: Standardized Residuals of Selling Price Plotted against House Size, for Model with House Size and Lot Size as Predictors. Questions:** How does this plot suggest that selling price has more variability at higher house size values, for given lot sizes? You don't see lot size on the plot, so how do its values affect the analysis?

Question to Explore

Does this plot suggest any irregularities with the model?

Think It Through

As house size increases, the variability of the standardized residuals seems to increase. This suggests more variability in selling prices when house size is larger, for a given lot size. We must be cautious, though, because the single point with a large negative residual for the largest house and the two points with large positive residuals may catch our eyes more than the others. Generally, it's not a good idea to let just a few points overly influence your judgment about the shape of a residual pattern. However, there is definite evidence that the variability increases as house size increases. A larger data set would provide more evidence about this.

Insight

Nonconstant variability does not invalidate the use of multiple regression. It would, however, make us cautious about using prediction intervals. We would expect predictions about selling price to tend to have smaller prediction errors when house size is small than when house size is large.

TO PRACTICE THIS CONCEPT, TRY EXERCISE 12.32

12.4 Practicing the Basics

12.31 Body weight residuals: Examples 4–7 used multiple regression to predict total body weight of college athletes. The figure shows the standardized residuals for another multiple regression model for predicting weight.

   a. About which distribution do these give you information—the overall distribution of weight, or the conditional distribution of weight at fixed values of the predictors?

   b. What does the histogram suggest about the likely shape of this distribution? Why?

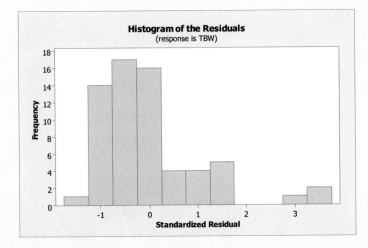

12.32 Strength residuals: In Chapter 11, we analyzed strength data for a sample of female high school athletes. The figure below is a residual plot for the multiple regression model relating the maximum number of pounds the athlete could bench press (BP) to the number of 60-pound bench presses (BP_60) and the number of 200-pound leg presses (LP_200). It plots the standardized residuals against the values of LP_200.

   a. You don't see BP_60 on the plot, so how do its values affect the analysis?

   b. Explain how the plot might suggest less variability at the lower values of LP_200.

   c. Suppose you remove the three points with standardized residuals around −2. Then is the evidence about variability in (b) so clear? What does this suggest about cautions in looking at residual plots?

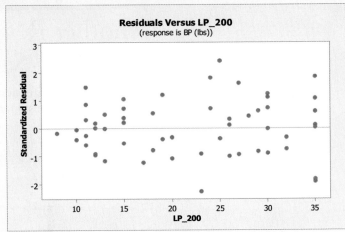

12.33 More residuals for strength: Refer to the previous exercise. The figure below is a residual plot for the model relating maximum bench press to LP_200 and BP_60. It plots the standardized residuals against the values of BP_60. Does this plot suggest any irregularities with the model? Explain.

12.34 Nonlinear effects of age: Suppose you fit a straight-line regression model to y = amount of time sleeping per day and x = age of subject. Values of y in the sample tend to be quite large for young children and for elderly people, and they tend to be lower for other people. Sketch what you would expect to observe for (a) the scatterplot of x and y, (b) a plot of the residuals against the values of age.

12.35 Driving accidents: Suppose you fit a straight-line regression model to x = age of subjects and y = driving accident rate. Sketch what you would expect to observe for (a) the scatterplot of x and y, (b) a plot of the residuals against the values of age.

12.36 Why inspect residuals?: When we use multiple regression, what's the purpose of doing a residual analysis? Why can't we just construct a single plot of the data for all the variables at once in order to tell whether the model is reasonable?

12.37 College athletes: The "college athletes" data set on the text CD comes from a study of University of Georgia female athletes. The response variable BP = maximum bench press has explanatory variables LBM = lean body mass (which is weight times 1 minus the proportion of body fat) and REP_BP = number of repetitions before fatigue with a 70-pound bench press. Let's look at all the steps of a regression analysis for these data.

a. The first figure shows a scatterplot matrix. Which two plots in the figure describe the associations with BP as a response variable? Describe those associations.

b. Results of a multiple regression analysis are shown below. Write down the prediction equation, and interpret the coefficient of REP_BP.

c. Report R^2, and interpret its value in the context of these variables.

d. Based on the value of R^2, report and interpret the multiple correlation.

e. Interpret results of the F test that BP is independent of these two predictors. Show how to obtain the F statistic from the mean squares in the ANOVA table.

f. Once REP_BP is in the model, does it help to have LBM as a second predictor? Answer by showing all steps of a significance test for a regression parameter.

g. Examine the histogram shown of the residuals for the multiple regression model. What does this describe, and what does it suggest?

h. Examine the plot shown of the residuals plotted against values of REP_BP. What does this describe, and what does it suggest?

i. From the plot in (h), can you identify a subject whose BP value was considerably lower than expected based on the predictor values? Identify by indicating the approximate values of REP_BP and the standardized residual for that subject.

Regression of maximum bench press on LBM and REP_BP

```
Predictor     Coef     SE Coef    T       P
Constant      55.012   7.740      7.11    0.000
LBM           0.16676  0.07522    2.22    0.030
REP_BP        1.6575   0.1090     15.21   0.000

S = 7.11957  R-Sq = 83.2%

Analysis of Variance

Source       DF    SS        MS       F        P
Regression   2     15283.0   7641.5   150.75   0.000
Res.         61    3092.0    50.7
Total        63    18375.0
```

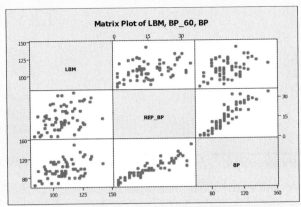

Scatterplot matrix for Exercise 12.37.

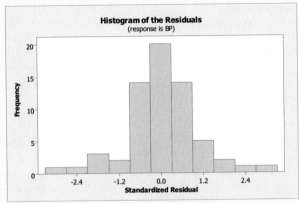

Residual plot for Exercise 12.37, part (g).

Residual plot for Exercise 12.37, part (h).

12.38 House prices: Use software with the "house selling price" data file on the text CD to do residual analyses with the multiple regression model for y = house selling price, x_1 = real estate tax, and x_2 = lot size.

a. Find a histogram of the standardized residuals. What assumption does this check? What do you conclude from the plot?

b. Plot the standardized residuals against the lot size. What does this check? What do you conclude from the plot?

12.39 Selling prices level off: In the previous exercise, suppose house selling price tends to increase with a straight-line trend for small to medium size lots, but then levels off as lot size gets large, for a fixed value of real estate tax. Sketch the pattern you'd expect to get if you plotted the residuals against lot size.

12.5 | How Can Regression Include Categorical Predictors?

So far, we've studied regression models for quantitative variables. Next, we'll learn how to include a *categorical explanatory* variable. The final section shows how to perform regression for a *categorical response* variable.

Indicator Variables

Regression models specify categories of a categorical explanatory variable using artificial variables, called **indicator variables**. The indicator variable for a particular category is binary. It equals 1 if the observation falls into that category and it equals 0 otherwise.

In the house selling prices data set, the city region in which a house is located is a categorical variable. It was measured with categories (NW, Other), the northwest being considered the most desirable area. The SW, SE, and NE regions were combined to form the "other" category. The indicator variable x for region is

$$x = 1 \text{ if house in NW}$$
$$x = 0 \text{ otherwise.}$$

This indicator variable indicates whether or not a home is in the NW. Let's see how an indicator variable works in a regression equation. To start, suppose region is the only predictor of selling price. The regression model is then $\mu_y = \alpha + \beta x$, with x as just defined. Substituting the possible values 1 and 0 for x,

$$\mu_y = \alpha + \beta(1) = \alpha + \beta, \text{ if house is in NW (so } x = 1)$$
$$\mu_y = \alpha + \beta(0) = \alpha, \qquad \text{if house is not in NW (so } x = 0).$$

The difference between the mean selling price in the NW and in other regions is

$$(\mu_y \text{ for NW}) - (\mu_y \text{ for other regions}) = (\alpha + \beta) - \alpha = \beta.$$

The coefficient β of the indicator variable x is the difference between the mean selling prices for homes in the NW and for homes not in the NW.

EXAMPLE 10

Including Region in Regression for House Selling Price

Picture the Scenario

Let's now fit a regression model for y = selling price of home using x_1 = house size and x_2 = region. Table 12.11 shows MINITAB output.

TABLE 12.11: Regression Analysis of y = Selling Price Using x_1 = House Size and x_2 = Indicator Variable for Region (Whether House Is in NW)

```
The regression equation is
price = − 15258 + 78.0 House_size + 30569 Region
```

Predictor	Coef	SE Coef	T	P
Constant	−15258	11908	−1.28	0.203
House_size	77.985	6.209	12.56	0.000
Region	30569	7949	3.85	0.000

S = 34390.2   R-Sq = 63.5%

Questions to Explore

a. Find and plot the lines showing how predicted selling price varies as a function of house size, for homes in the NW and for homes not in the NW.

b. Interpret the coefficient of the indicator variable for region.

Think It Through

a. Table 12.11 reports the prediction equation,

$$\hat{y} = -15{,}258 + 78.0x_1 + 30{,}569x_2.$$

For homes not in the NW, $x_2 = 0$. The prediction equation for $y = $ selling price using $x_1 = $ house size then simplifies to

$$\hat{y} = -15{,}258 + 78.0x_1 + 30{,}569(0) = -15{,}258 + 78.0x_1.$$

For homes in the NW, $x_2 = 1$. The prediction equation then simplifies to

$$\hat{y} = -15{,}258 + 78.0x_1 + 30{,}569(1) = 15{,}311 + 78.0x_1.$$

Both lines have the same slope, 78. For homes in the NW and for homes not in the NW, the predicted selling price increases by \$78 for each square-foot increase in house size. Figure 12.7 plots the two prediction equations. The quantitative explanatory variable, house size, is on the x-axis. The figure portrays a separate line for each category of region (NW, other).

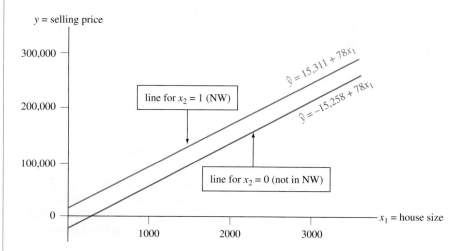

▲ **FIGURE 12.7: Plot of Equation Relating $\hat{y} = $ Predicted Selling Price to $x_1 = $ House Size, According to $x_2 = $ Region (1 = NW, 0 = other). Question:** Why are the lines parallel?

b. At a fixed value of $x_1 = $ house size, the difference between the predicted selling prices for homes in the NW and homes not in the NW is

$$(15{,}311 + 78.0x_1) - (-15{,}258 + 78.0x_1) = 30{,}569.$$

This is precisely the coefficient of the indicator variable, x_2. For any fixed value of house size, we predict that the selling price is \$30,569 higher for homes in the NW.

Insight

Since the two lines have the same slope, they are parallel. The line for homes in the NW is above the other line because its y-intercept is larger. This means that for any fixed value of house size, the predicted selling price is higher for homes in the NW. The P-value of 0.000 for the test for the coefficient of the indicator variable suggests that this difference is statistically significant.

TO PRACTICE THIS CONCEPT, TRY EXERCISE 12.40

A categorical variable having more than two categories uses an additional indicator variable for each category. For instance, if we want to use all four categories (NW, SW, NE, SE) for region in the model, we need three indicator variables:

$$x_1 = 1 \text{ for houses in NW, and } x_1 = 0 \text{ otherwise,}$$
$$x_2 = 1 \text{ for houses in SW, and } x_2 = 0 \text{ otherwise,}$$
$$x_3 = 1 \text{ for houses in NE, and } x_3 = 0 \text{ otherwise.}$$

If $x_1 = x_2 = x_3 = 0$, the house is in the SE. We don't need a separate indicator variable for SE, as it would be redundant. We can tell whether a house is in the SE merely from seeing the values of x_1, x_2, and x_3. Generally, *a categorical explanatory variable in a regression model uses one fewer indicator variable than the number of categories.* For instance, with the two categories (NW, other), we needed only a single indicator variable.

Why can't we specify the four categories merely by setting up a variable x that equals 1 for homes in the NW, 2 for the SW, 3 for the NE, and 4 for the SE? Because this would treat region as *quantitative* rather than *categorical*. It would treat region as if different categories corresponded to different *amounts* of the variable. But the variable measures *which* region, not *how much* region. Treating it as quantitative is inappropriate.

Is There Interaction?

In Example 10, the regression equation simplified to two straight lines:

$$\hat{y} = 15{,}311 + 78.0x_1 \text{ for homes in the NW,}$$
$$\hat{y} = -15{,}258 + 78.0x_1 \text{ for homes in other regions.}$$

Both equations have the same slope. The model forces the effect of x_1 = house size on selling price to be the same for both regions.

Likewise, in a multiple regression model, *the slope of the relationship between the population mean of y and each explanatory variable is identical for all values of the other explanatory variables.* Such models are sometimes too simple. The effect of an explanatory variable may *change* considerably as the value of another explanatory variable in the model changes. The multiple regression model we've studied assumes this does not happen. When it does happen, there is **interaction**.

Interaction

For two explanatory variables, **interaction** exists between them in their effects on the response variable when the slope of the relationship between μ_y and one of them changes as the value of the other changes.

Suppose the *actual* population relationship between x_1 = house size and the mean selling price is

$$\mu_y = 15{,}000 + 100x_1 \text{ for homes in the NW,}$$

and

$$\mu_y = 12{,}000 + 25x_1 \text{ for homes in other regions.}$$

The slope for the effect of x_1 differs (100 vs. 25) for the two regions. There is then interaction between house size and region in their effects on selling price. See Figure 12.8.

How can you allow for interaction when you do a regression analysis? With two explanatory variables, one quantitative and one categorical (such as in Example 10), you can fit a separate line between the two quantitative variables for each category of the categorical variable.

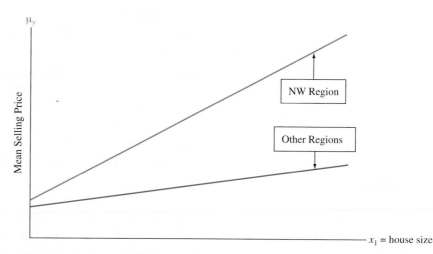

▲ **FIGURE 12.8: An Example of Interaction.** There's a larger slope between selling price and house size for homes in the NW than in other regions.

EXAMPLE 11

Comparing Winning High Jumps for Men and Women

Recall

These data were first analyzed in Example 11 in Chapter 3.■

Picture the Scenario:

Men have competed in the high jump in the Olympics since 1896 and women since 1928. Figure 12.9 shows how the winning high jump in the Olympics has changed over time for men and women. The "high jump" data file on the text CD contains the winning heights for each year. A multiple regression analysis of y = winning height (in meters) as a function of x_1 = number of years since 1928 (when women first participated in the high jump) and x_2 = gender (1 = male, 0 = female) gives $\hat{y} = 1.6 + 0.0069x_1 + 0.34x_2$.

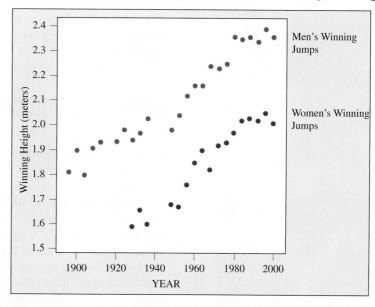

▲ **FIGURE 12.9: Scatterplot for the Winning High Jumps (in Meters) in the Olympic Games.**

Questions to Explore

a. Interpret the coefficient of year and the coefficient of gender in the equation.

b. To allow interaction, we can fit equations separately to the data for males and the data for females. We then get $\hat{y} = 1.98 + 0.0057x_1$ for males and $\hat{y} = 1.58 + 0.0070x_1$ for females. Describe the interaction by comparing slopes.

c. Describe the interaction allowed in (b) by comparing predicted winning high jumps for males and females in 1928 and in 2000.

Think It Through

a. For the prediction equation $\hat{y} = 1.6 + 0.0069x_1 + 0.34x_2$, the coefficient of year is 0.0069. For each gender, the predicted winning high jump increases by 0.0069 meters per year. This seems small, but over a hundred years it projects to an increase of $100(0.0069) = 0.69$ meters, about 27 inches. The model does not allow interaction, as it assumes that the slope of 0.0069 is the same for each gender. The coefficient of gender is 0.34. In a given year, the predicted winning high jump for men is 0.34 meters higher than for women. Because this model does not allow interaction, the predicted difference between men and women is the same each year.

b. The slope of 0.0070 for females is higher than the slope of 0.0057 for males. So the predicted winning high jump increases a bit more for females than for males over this time period.

c. In 1928, $x_1 = 0$, and the predicted winning high jump was $\hat{y} = 1.98 + 0.0057(0) = 1.98$ for males and $\hat{y} = 1.58 + 0.0070(0) = 1.58$ for females, a difference of 0.40 meters. In 2000, $x_1 = 2000 - 1928 = 72$, and the predicted winning high jump was $1.98 + 0.0057(72) = 2.39$ for males and $1.58 + 0.0070(72) = 2.08$ for females, a difference of 0.31 meters. The predicted difference between the winning high jumps of males and females decreased a bit between 1928 and 2000.

Insight

When we allow interaction, the estimated slope is a bit higher for females than for males. This is what caused the difference in predicted winning high jumps to be less in 2000 than in 1928. However, the slopes were not dramatically different, as Figure 12.9 shows that the points go up at similar rates for the two genders. The sample degree of interaction was not strong.

TO PRACTICE THIS CONCEPT, TRY EXERCISES 12.46 AND 12.47

How do we know whether the interaction shown by the sample is sufficiently large to indicate that there is interaction in the population? There is a significance test for checking this, but it is beyond the scope of this book. In practice, it's usually adequate to investigate interaction informally using graphics. For example, suppose there are two predictors, one quantitative and one categorical. Then you can plot y against the quantitative predictor, identifying the data points by the category of the categorical variable, as we did in Example 11. Do the points seem to go up or go down at quite different rates, taking into account sampling error? If so, it's safer to fit a separate regression line for each category, which then allows different slopes.

12.40 Winning high jump: Refer to Example 11 on winning Olympic high jumps. The prediction equation relating y = winning height (in meters) as a function of x_1 = number of years since 1928 and x_2 = gender (1 = male, 0 = female) is $\hat{y} = 1.587 + 0.00686x_1 + 0.341x_2$.

a. Using this equation, find the prediction equations relating winning height to year, separately for males and for females.

b. Find the predicted winning height in 2008 for (i) females, (ii) males, and show how the difference between them relates to a parameter estimate for the model.

12.41 eBay selling prices: Example 7 in Chapter 7 showed selling prices on eBay of the Palm M515 PDA, a popular handheld computer. When y = selling price is regressed on x_1 = number of bids placed and x_2 = whether the seller offered buyers the "buy-it-now" option (1 = yes, 0 = no), $\hat{y} = 238 - 0.723x_1 - 2.77x_2$.

a. Interpret the estimated effect of the number of bids placed.

b. Interpret the estimated effect of using the "buy-it-now" option.

12.42 Predict using lot size and NW: For the "house selling prices" data set, when we regress y = selling price on x_1 = lot size and x_2 = region (1 = NW, 0 = other), we get the results shown.

Regression of selling price on lot size and region (1 = NW, 0 = other)

Predictor	Coef	SE Coef	T	P
Constant	32139	11102	2.89	0.005
lot_size	4.6534	0.4821	9.65	0.000
region	5604	9443	0.59	0.554

S = 39803.6   R-Sq = 51.1%

a. Report the regression equation. Find and interpret the separate lines relating predicted selling price to lot size for homes in the NW and for homes not in the NW.

b. Sketch how selling price varies as a function of lot size, for homes in the NW and for homes not in the NW.

c. Estimate the difference between the mean selling price in the NW and in other regions, controlling for lot size.

12.43 Quality and productivity: The table shows data from 27 automotive plants on y = number of assembly defects per 100 cars and x = time (in hours) to assemble each vehicle. The data are in the "quality and productivity" file on the text CD.

Number of defects in assembling 100 cars and time to assemble each vehicle

Plant	Defects	Time	Plant	Defects	Time	Plant	Defects	Time
1	39	27	10	89	17	19	69	54
2	38	23	11	48	20	20	79	18
3	42	15	12	38	26	21	29	31
4	50	17	13	68	21	22	84	28
5	55	12	14	67	26	23	87	44
6	56	16	15	69	30	24	98	23
7	57	18	16	69	32	25	100	25
8	56	26	17	70	31	26	140	21
9	61	20	18	68	37	27	170	28

Source: S. Chatterjee, M. Handcock, and J. Simonoff, *A Casebook for a First Course in Statistics and Data Analysis* (Wiley, 1995); based on graph in *The Machine that Changed the World,* by J. Womack, D. Jones, and D. Roos (Macmillan, 1990).

a. The prediction equation is $\hat{y} = 61.3 + 0.35x$. Find the predicted number of defects for a car having assembly time (i) 12 hours (the minimum), (ii) 54 hours (the maximum).

b. The first 11 plants were Japanese facilities and the rest were not. Let x_1 = time to assemble vehicle and x_2 = whether facility is Japanese (1 = yes, 0 = no). The fit of the multiple regression model is $\hat{y} = 105.0 - 0.78x_1 - 36.2x_2$. Interpret the coefficients that estimate the effect of x_1 and the effect of x_2.

c. Explain why (a) and (b) indicate that Simpson's paradox has occurred.

d. Explain *how* Simpson's paradox occurred. To do this, construct a scatterplot between y and x_1 in which points are identified by whether the facility is Japanese. Note that the Japanese facilities tended to have low values for both x_1 and x_2.

12.44 Predicting hamburger sales: A chain restaurant that specializes in selling hamburgers wants to analyze how y = sales for a customer (the total amount spent by a customer on food and drinks, in dollars) depends on the location of the restaurant, which is classified as inner city, suburbia, or at an interstate exit.

a. Construct indicator variables x_1 for inner city and x_2 for suburbia so you can include location in a regression equation for predicting the sales.

b. For (a), suppose $\hat{y} = 5.8 - 0.7x_1 + 1.2x_2$. Find the difference between the estimated mean sales in suburbia and at interstate exits.

12.45 Houses, tax, and NW: Use the "house selling prices" data file at the text CD to regress selling price on real estate tax and whether the home is in the NW region (1 = NW, 0 = other).

a. Report the prediction equation. Find and interpret the equations predicting selling price using real estate tax, for homes in the NW and for homes not in the NW.

b. How do you interpret the coefficient of the indicator variable for NW?

12.46 Tax and region interact?: Refer to the previous exercise.

[TRY] **a.** Explain what the "no interaction" assumption means for this model.

b. Sketch a hypothetical scatter diagram, showing points identified by region, suggesting that there is actually a substantial degree of interaction.

12.47 Equal high jump for men and women: Refer to Example 11 and Exercise 12.40, with $\hat{y}$ = predicted winning high jump and x_1 = number of years since 1928. When equations are fitted *separately* for males and for females, we get $\hat{y} = 1.98 + 0.00570x_1$ for males and $\hat{y} = 1.58 + 0.00698x_1$ for females.

a. In allowing the lines to have different slopes, the overall model allows for _____ between gender and year in their effects on the winning high jump. (Fill in the correct word.)

b. Show that both equations yield the same predicted winning high jump when $x_1 = 312$ (that is, 312 years after 1928, or in 2240).

c. Is it sensible to use this model to predict that in the year 2240 men and women will have about the same winning high jump? Why or why not?

12.48 Comparing sales: You own a gift shop that has a campus location and a shopping mall location. You want to compare the regressions of y = daily total sales on x = number of people who enter the shop, for total sales listed by day at the campus location and at the mall location. Explain how you can do this using regression modeling

a. With a single model, having an indicator variable for location, that assumes the slopes are the same for each location.

b. With separate models for each location, permitting the slopes to be different.

12.6 | How Can We Model a Categorical Response?

The regression models studied so far are designed for a quantitative response variable y. When y is categorical, a different regression model applies, called **logistic regression**. For instance, logistic regression can model

■ A voter's choice in an election (Democrat or Republican), with explanatory variables annual income, political ideology, religious affiliation, and race.

■ Whether a credit card holder pays their bill on time (yes or no), with explanatory variables family income and the number of months in the past year that the customer paid the bill on time.

Logistic regression is used in many business applications, such as for modeling whether a customer buys a particular product, or for modeling whether a credit-card holder pays a bill on time. It's also used a lot in medical applications, such as for modeling whether a patient is cured or not. We'll study logistic regression in this section for the special case of **binary** y.

The Logistic Regression Model

Recall

Section 2.3 (Example 12) and Section 6.4 (following Example 15) showed that **a proportion for a category is a mean calculated for a binary variable** that has score 1 for that category and score 0 otherwise. ■

Denote the possible outcomes for y by 0 and 1. We'll use the generic terms *failure* and *success* for these outcomes, as with trials for the binomial distribution in Section 6.3. The population mean of the 0 and 1 scores equals the population *proportion* of 1 outcomes (successes) for the response variable. That is, $\mu_y = p$, where p denotes the population proportion of successes. This proportion also represents the *probability* that a randomly selected subject has a success outcome. The model describes how p depends on the values of the explanatory variables.

For a single explanatory variable, x, the straight-line regression model is

$$p = \alpha + \beta x.$$

As Figure 12.10 shows, this model implies that p falls below 0 or above 1 for sufficiently small or large x values. However, a proportion must fall between 0 and 1. Although the straight-line model may be valid over a restricted range of x values, it is usually inadequate when there are multiple explanatory variables.

Figure 12.10 also shows a more realistic model. It has a curved, S-shape instead of a straight-line trend.

▲ **FIGURE 12.10: Two Possible Regressions for a Probability _p_ of a Binary Response Variable.** A straight line is usually less appropriate than an S-shaped curve. **Question:** Why is the straight-line regression model for a binary response variable often poor?

IN WORDS

In math, the symbol _e_ represents a number that, like π (pi), occurs often. It equals 2.7183.... In Section 11.5, we saw that a number raised to a power (such as _x_) represents an **exponential** function.

Logistic regression is used when the response variable has only two possible outcomes (it's **binary.**)

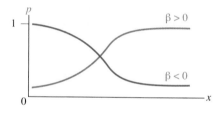

Logistic Regression

A regression equation for an S-shaped curve for the probability of success _p_ is

$$p = \frac{e^{\alpha + \beta x}}{1 + e^{\alpha + \beta x}}.$$

This equation for _p_ is called the **logistic regression** equation.

Here, _e_ raised to a power represents the _exponential function_ evaluated at that number. Most calculators have an e^x key that provides values of _e_ raised to a power. The model has two parameters, α and β. Since the numerator of the formula for _p_ is smaller than the denominator, the model forces _p_ to fall between 0 and 1. With a regression model having this S-shape, the probability _p_ falls between 0 and 1 for all possible _x_ values.

Like the slope of a straight line, the parameter β in this model refers to whether the mean of _y_ increases or decreases as _x_ increases. When $\beta > 0$, the probability _p_ increases as _x_ increases. When $\beta < 0$, the probability _p_ decreases as _x_ increases. See the margin figures. If $\beta = 0$, _p_ does not change as _x_ changes, so the curve flattens to a horizontal straight line. The steepness of the curve increases as the absolute value of β increases. However, unlike in the straight-line model, β is _not_ a slope—the change in the mean per 1-unit change in _x_. For this S-shaped curve, the rate at which the curve climbs or descends _changes_ according to the value of _x_.

Software can estimate the parameters α and β in the logistic regression model. Calculators or software can find estimated probabilities $\hat{p}$ based on the model fit.

EXAMPLE 12

Annual Income and Having a Travel Credit Card

Picture the Scenario

An Italian study with 100 randomly selected Italian adults considered factors that are associated with whether a person possesses at least one travel credit card, such as American Express or Diners Club. Table 12.12 shows results for the first 15 people on this response variable and on the person's annual income, in thousands of euros.[4] (As of 2005, 1 Euro equals about $1.30 U.S.) The complete data set is in the "Credit card and income" data file on the text CD. Let _x_ = annual income and let _y_ = whether the person possesses a travel credit card (1 = yes, 0 = no).

---

[4]The data were originally recorded in Italian lira but have been changed to euros and adjusted for inflation.

TABLE 12.12: Annual Income (in Thousands of Euros) and Whether Possess a Travel Credit Card

The response y equals 1 if a person has a travel credit card and equals 0 otherwise. The complete data set is on the text CD.

Income	y	Income	y	Income	y
12	0	14	1	15	0
13	0	14	0	15	1
14	1	14	0	15	0
14	0	14	0	15	0
14	0	14	0	15	0

Source: R. Piccarreta, Bocconi University, Milan (personal communication)

Questions to Explore

Table 12.13 shows part of the printout that software provides for conducting a logistic regression analysis.

TABLE 12.13: Results of Logistic Regression for Italian Credit Card Data

Predictor	Coef	SE Coef	Z	P
Constant	−3.5180	0.71034	−4.95	0.000
income	0.1054	0.02616	4.03	0.000

a. State the prediction equation for the probability of owning a travel credit card, and explain whether annual income has a positive or a negative effect.

b. Find the estimated probability of possessing a travel credit card at the lowest and highest annual income levels in the sample, which were $x = 12$ and $x = 65$.

Think It Through

a. Substituting the α and β estimates from Table 12.13 into the logistic regression model formula, $p = e^{\alpha + \beta x}/(1 + e^{\alpha + \beta x})$, we get the equation for the estimated probability $\hat{p}$ of possessing a travel credit card,

$$\hat{p} = \frac{e^{-3.52 + 0.105x}}{1 + e^{-3.52 + 0.105x}}.$$

Since the estimate 0.105 of β (the coefficient of x) is positive, this sample suggests that annual income has a positive effect: The estimated probability of possessing a travel credit card is higher at higher levels of annual income.

b. For subjects with income $x = 12$ thousand euros, the estimated probability of possessing a travel credit card equals

$$\hat{p} = \frac{e^{-3.52 + 0.105(12)}}{1 + e^{-3.52 + 0.105(12)}} = \frac{e^{-2.26}}{1 + e^{-2.26}} = \frac{0.104}{1.104} = 0.09.$$

For $x = 65$, the highest income level in this sample, you can check that the estimated probability equals 0.97.

Insight

There is a strong effect. The estimated probability of having a travel credit card changes from 0.09 to 0.97 (nearly 1.0) as annual income changes over its range.

Using software, we could also fit the straight-line regression model. Its prediction equation is $\hat{p} = -0.159 + 0.0188x$. However, its $\hat{p}$ predictions are quite different at the low end and at the high end of the annual income scale. At $x = 65$, for instance, it provides the prediction $\hat{p} = 1.06$. This is a poor prediction, because we know that a proportion must fall between 0 and 1.

TO PRACTICE THIS CONCEPT, TRY EXERCISE 12.50

Ways to Interpret the Logistic Regression Model

The sign of β in the logistic regression model tells us whether the probability p increases or decreases as x increases. How else can we interpret the model parameters and their estimates? Here, we'll describe three ways.

First, to describe the effect of x, you can compare estimates of p at two different values of x. One possibility is to compare $\hat{p}$ found at the minimum and maximum values of x, as we did in Example 12. There we saw that the estimated probability changed from 0.09 to 0.97, a considerable change. An alternative is to use instead values of x to evaluate this probability that are not as affected by outliers, such as the first and third quartiles.

Second, it's often informative to know the value of x at which $p = 0.50$, that is, that value of x for which each outcome is equally likely. This depends on the logistic regression parameters α and β. It can be shown to equal $x = -\alpha/\beta$ (Exercise 12.94).

Finally, the simplest way to use the logistic regression parameter β to interpret the *steepness* of the curve uses a straight-line approximation. Because the logistic regression formula is a curve rather than a straight line, β itself is no longer the ordinary slope. At the x value where $p = 0.50$, it can be shown that the line drawn tangent to the logistic regression curve has slope $\beta/4$. See Figure 12.11. The value $\beta/4$ represents the approximate change in the probability p for a one-unit increase in x, when x is close to the value at which $p = 0.50$. Tangent lines at other points have weaker slopes (Exercise 12.93), close to 0 when p is near 0 or 1.

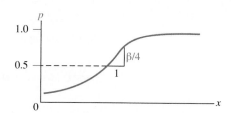

▲ **FIGURE 12.11: When the Probability $p = 0.50$, for a 1-unit Change in x, p Changes by about $\beta/4$.** Here, β is the coefficient of x from the logistic regression model.

EXAMPLE 13

Interpreting Effect of Income on Credit Card Use

Picture the Scenario

For the Italian travel credit card study, Example 12 found that the equation

$$\hat{p} = \frac{e^{-3.52+0.105x}}{1 + e^{-3.52+0.105x}}$$

estimates the probability p of having such a credit card as a function of annual income x.

Questions to Explore

Interpret this equation by (a) finding the income value at which the estimated probability of having a credit card equals 0.50, (b) finding the approximate rate of change in the probability at that income value.

Think It Through

a. The estimate of α is -3.52 and the estimate of β is 0.105. Substituting the estimates into the expression $-\alpha/\beta$ for the value of x at which $p = 0.50$, we get $x = (3.52)/(0.105) = 33.5$. The estimated probability of possessing a travel credit card equals 0.50 when annual income equals 33.5 thousand euros.

b. A line drawn tangent to the logistic regression curve at the point where $p = 0.50$ has slope equal to $\beta/4$. The estimate of this slope is $0.105/4 = 0.026$. For each increase of one thousand euros in annual income near the income value of 33.5 thousand euros, the estimated probability of possessing a travel credit card increases by approximately 0.026.

Insight

Figure 12.12 shows the estimated logistic regression curve, highlighting what we've learned in Examples 12 and 13: The estimated probability increases from 0.09 to 0.97 between the minimum and maximum income values, and it equals 0.50 at an annual income of 33.5 thousand euros.

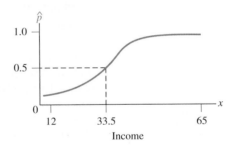

▲ **FIGURE 12.12: Logistic Regression Curve Relating Estimated Probability $\hat{p}$ of Having a Travel Credit Card to Annual Income x.**

TO PRACTICE THIS CONCEPT, TRY EXERCISE 12.54

Inference for Logistic Regression

Software also reports a z test statistic for the hypothesis $H_0: \beta = 0$. When $\beta = 0$, the probability of possessing a travel credit card is the same at all income levels. The test statistic equals the ratio of the estimate b of β divided by its standard error. From Table 12.13 (shown again in the margin), the z test statistic equals

$$z = (b - 0)/se = (0.105 - 0)/0.0262 = 4.0.$$

This has a P-value of 0.000 for $H_a: \beta \neq 0$. Since the sample slope is positive, for the population of adult Italians, there is strong evidence of a positive association between annual income and having a travel credit card.

The result of this test is no surprise. We would expect people with higher annual incomes to be more likely to have travel credit cards. Some software can construct a confidence interval for the probability p at various x-levels. A 95% confidence interval for the probability of possessing a credit card equals $(0.04, 0.19)$ at the lowest sample income level of 12 thousand euros, and $(0.78, 0.996)$ at the highest sample income level of 65 thousand euros.

Recall

Table 12.13 was

Predictor	Coef	SE	Z	P
Constant	−3.518	.710	−4.95	.000
income	0.105	.0262	4.03	.000

Sections 7.2 and 8.2 showed that **inference about proportions uses z test statistics** rather than t test statistics. ■

Multiple Logistic Regression

Just as ordinary regression extends to handle several explanatory variables, so does logistic regression. Also, logistic regression can include categorical explanatory variables using indicator variables.

EXAMPLE 14

Estimating Proportion of Students Who've Used Marijuana

Picture the Scenario

Table 12.14 is a three-variable contingency table from a survey conducted by Wright State University. The survey asked senior high-school students in a nonurban area near Dayton, Ohio, whether they had ever used each of alcohol, cigarettes, or marijuana. We'll treat marijuana use as the response variable and cigarette use and alcohol use as explanatory variables.

TABLE 12.14: Alcohol, Cigarette, and Marijuana Use for High School Seniors

Alcohol Use	Cigarette Use	Marijuana Use Yes	No
Yes	Yes	911	538
	No	44	456
No	Yes	3	43
	No	2	279

Source: Professor Harry Khamis, Wright State University (personal communication).

Questions to Explore

Let y indicate marijuana use, coded $1 =$ yes, $0 =$ no. Let x_1 be an indicator variable for alcohol use ($1 =$ yes, $0 =$ no), and let x_2 be an indicator variable for cigarette use ($1 =$ yes, $0 =$ no). Table 12.15 shows MINITAB output for a logistic regression model.

TABLE 12.15: MINITAB Output for Estimating the Probability of Marijuana Use Based on Alcohol Use and Cigarette Use

Predictor	Coef	SE Coef	Z	P
Constant	−5.30904	0.475190	−11.17	0.000
alcohol	2.98601	0.464671	6.43	0.000
cigarettes	2.84789	0.163839	17.38	0.000

a. Report the prediction equation, and interpret.
b. Find the estimated probability $\hat{p}$ of having used marijuana (i) for students who have not used alcohol or cigarettes, (ii) for students who have used both alcohol and cigarettes.

Think It Through

a. From Table 12.15, the logistic regression prediction equation is

$$\hat{p} = \frac{e^{-5.31+2.99x_1+2.85x_2}}{1 + e^{-5.31+2.99x_1+2.85x_2}}.$$

The coefficient of alcohol use (x_1) is positive (2.99). The indicator variable x_1 equals 1 for those who've used alcohol. Thus, the alcohol users have a higher estimated probability of using marijuana, controlling for whether they used cigarettes. Likewise, the coefficient of cigarette use (x_2) is positive (2.85), so the cigarette users have a higher estimated probability of using marijuana, controlling for whether they used alcohol. Table 12.15 tells us that for each predictor, the test statistic is very large. In other words, the estimated effect is a large number of standard errors from 0. The P-values are both 0.000, so there is strong evidence that the corresponding population effects are positive also.

b. For those who have not used alcohol or cigarettes, $x_1 = x_2 = 0$. For them, the estimated probability of marijuana use is

$$\hat{p} = \frac{e^{-5.31+2.99(0)+2.85(0)}}{1 + e^{-5.31+2.99(0)+2.85(0)}} = \frac{e^{-5.31}}{1 + e^{-5.31}} = \frac{0.0049}{1.0049} = 0.005.$$

For those who have used alcohol and cigarettes, $x_1 = x_2 = 1$. For them, the estimated probability of marijuana use is

$$\hat{p} = \frac{e^{-5.31+2.99(1)+2.85(1)}}{1 + e^{-5.31+2.99(1)+2.85(1)}} = 0.628.$$

In summary, the probability that students have tried marijuana seems to depend greatly on whether they've used alcohol and cigarettes.

Insight

Likewise, you can find the estimated probability of using marijuana for those who have used alcohol but not cigarettes (let $x_1 = 1$ and $x_2 = 0$) and for those who have not used alcohol but have used cigarettes. Table 12.16 summarizes results. We see that marijuana use is unlikely unless a student has used both alcohol and cigarettes.

TABLE 12.16: Estimated Probability of Marijuana Use, by Alcohol Use and Cigarette Use, based on Logistic Regression Model

The sample proportions of marijuana use are shown in parentheses for the four cases.

Alcohol Use	Cigarette Use Yes	No
Yes	0.628 (0.629)	0.089 (0.088)
No	0.079 (0.065)	0.005 (0.007)

TO PRACTICE THIS CONCEPT, TRY EXERCISES 12.56 AND 12.57

Recall

Table 12.14 was

Alcohol	Cigarette	Marijuana Yes	No
Yes	Yes	911	538
	No	44	456
No	Yes	3	43
	No	2	279 ■

Checking the Logistic Regression Model

How can you check whether a logistic regression model fits the data well? When the explanatory variables are categorical, you can find the sample proportions for the outcome of interest. Table 12.16 also shows these for the marijuana use example. For instance, for those who had used alcohol and cigarettes, from Table 12.14 (shown again in the margin) the sample proportion who had used marijuana is 911/(911 + 538) = 0.629. Table 12.16 shows that the sample proportions are close to the estimated proportions generated by the model. The model seems to fit well. When at least one explanatory variable is quantitative, model checking is more complex and is beyond the scope of this text.

Table 12.16 may suggest the question, why bother to fit the model? Why not merely inspect a table of sample proportions? One important reason is that the model enables us to easily conduct inferences about the effects of explanatory variables, while controlling for other variables. For instance, if we want to test the effect of alcohol use on marijuana use, controlling for cigarette use, we can test $H_0: \beta_1 = 0$ in the model with alcohol use and cigarette use as predictors.

12.6 Practicing the Basics

12.49 Income and credit cards: Example 12 used logistic regression to estimate the probability of having a travel credit card when x = annual income (in thousands of euros). At the mean income of 25 thousand euros, show that the estimated probability of having a travel credit card equals 0.29.

12.50 Baseball pitching: In the first nine decades of the twentieth century in baseball's National League, George Will reported (*Newsweek*, April 10, 1989) that the proportion of times the starting pitcher pitched a complete game were 0.727 (1900–1909), 0.634, 0.500, 0.443, 0.416, 0.328, 0.272, 0.225, 0.133 (1980–1989). Let $x = 1, 2 \ldots, 9$ denote time for the nine decades. A logistic regression model fitted to describe the time trend in these data gives prediction equation

$$\hat{p} = \frac{e^{1.148-0.315x}}{1 + e^{1.148-0.315x}}.$$

a. Use this to predict the proportion of complete games in the decade $x = 12$, which is 2010–2019.

b. Straight-line regression gives $\hat{p} = 0.76 - 0.07x$. Find its prediction for $x = 12$. Is it realistic? Indicate a limitation of that model in describing proportions.

12.51 Horseshoe crabs: A study of horseshoe crabs by the zoologist Dr. Jane Brockmann at the University of Florida used logistic regression to predict the probability that a female crab had a male partner nesting nearby. One explanatory variable was x = weight of the female crab (in kilograms). The results were

Predictor	Coef
Constant	−3.695
Weight	1.815

The quartiles for weight were Q1 = 2.00, Q2 = 2.35, and Q3 = 2.85.

a. Find the estimated probability of a male partner at Q1 and at Q3.

b. Interpret the effect of weight by estimating how much the probability increases over the middle half of the sampled weights, between Q1 and Q3.

12.52 More crabs: Refer to the previous exercise. For what weight values do you estimate that a female crab has probability (a) 0.50, (b) greater than 0.50, (c) less than 0.50, of having a male partner nesting nearby?

12.53 Voting and income: A logistic regression model describes how the probability of voting for the Republican candidate in a presidential election depends on x, the voter's total family income (in thousands of dollars) in the previous year. The prediction equation for a particular sample is

$$\hat{p} = \frac{e^{-1.00+0.02x}}{1 + e^{-1.00+0.02x}}.$$

Find the estimated probability of voting for the Republican candidate when (a) income = 10 thousand dollars, (b) income = 100 thousand dollars. Describe how the probability seems to depend on income.

12.54 Equally popular candidates: Refer to the previous exercise.

a. At which income level is the estimated probability of voting for the Republican candidate equal to 0.50?

b. Over what region of income values is the estimated probability of voting for the Republican candidate (a) greater than 0.50, (b) less than 0.50?

c. At the income level for which $\hat{p} = 0.50$, give a linear approximation for the change in the probability for each thousand-dollar increase in income.

12.55 Many predictors of voting: Refer to the previous two exercises. When the explanatory variables are x_1 = family income, x_2 = number of years of education, and x_3 = gender (1 = male, 0 = female), suppose a logistic regression reports

Predictor	Coef	SE Coef
Constant	−2.40	0.12
income	0.02	0.01
education	0.08	0.05
gender	0.20	0.06

For this sample, x_1 ranges from 6 to 157 with a standard deviation of 25, and x_2 ranges from 7 to 20 with a standard deviation of 3.

a. Interpret the effects using the sign of the coefficient for each predictor.

b. Illustrate the gender effect by finding and comparing the estimated probability of voting Republican for (a) a man with 16 years of education and income 40 thousand dollars, (b) a woman with 16 years of education and income 40 thousand dollars.

12.56 Athletes' graduation: The U.S. National Collegiate Athletic Association conducted a study of student athlete graduation rates. The table shows the data.

Athlete graduation

Group	Sample Size	Graduates
White females	796	498
White males	1625	878
Black females	143	54
Black males	660	197

Source: J. J. McArdle and F. Hamagami, *J. Amer. Statist. Assoc.,* vol. 89 (1994), pp. 1107–1123.

  a. Identify the response variable.
  b. Express the data in the form of a three-variable contingency table that cross-classifies whether graduated $(1 = \text{yes}, 0 = \text{no})$, race, and gender.
  c. When we use indicator variables for race $(1 = \text{white}, 0 = \text{black})$ and for gender $(1 = \text{female}, 0 = \text{male})$, the coefficients of those predictors in the logistic regression model are 1.015 for race and 0.352 for gender. Based on these estimates, which race and gender combination has the highest estimated probability of graduation? Why?

12.57 Death penalty and race: The three-dimensional contingency table shown is from a study of the effects of racial characteristics on whether or not individuals convicted of homicide receive the death penalty. The subjects classified were defendants in indictments involving cases with multiple murders in Florida between 1976 and 1987.

Death penalty verdict by defendant's race and victims' race

Defendant's Race	Victims' Race	Death Penalty Yes	Death Penalty No	Percent Yes
White	White	53	414	11.3
	Black	0	16	0.0
Black	White	11	37	22.9
	Black	4	139	2.8

Source: M. L. Radelet and G. L. Pierce, *Florida Law Rev.,* vol. 43, 1991, pp. 1–34.

  a. Based on the percentages shown, controlling for victims' race, for which defendant's race was the death penalty more likely?
  b. Let $y = $ death penalty verdict $(1 = \text{yes}, 0 = \text{no})$, let d be an indicator variable for defendant's race $(1 = \text{white}, 0 = \text{black})$, and let v be an indicator variable for victims' race $(1 = \text{white}, 0 = \text{black})$. The logistic regression prediction equation is

$$\hat{p} = \frac{e^{-3.596 - 0.868d + 2.404v}}{1 + e^{-3.596 - 0.868d + 2.404v}}.$$

According to this equation, for which of the four groups is the death penalty most likely? Explain your answer.

12.58 Death penalty probabilities: Refer to the previous exercise.

  a. Based on the prediction equation, when the defendant is black and the victims were white, show that the estimated death penalty probability is 0.233.
  b. The model-estimated probabilities are 0.011 when the defendant is white and victims were black, 0.113 when the defendant and the victims were white, and 0.027 when the defendant and the victims were black. Construct a table cross-classifying defendant's race by victims' race, and show the estimated probability of the death penalty for each cell. Use this to interpret the effect of defendant's race.
  c. Collapse the contingency table over victims' race, and show that (ignoring victims' race) white defendants were more likely than black defendants to get the death penalty. Comparing this with what happens when you control for victims' race, explain how Simpson's paradox occurs.

CHAPTER REVIEW

Answers to the Chapter Figure Questions

Figure 12.1: Selling price is the response variable, and the graphs that show it as the response variable are in the first row.

Figure 12.2: The coefficient (2.84) of $x_2 = $ lot size is positive, so increasing it has the effect of increasing the predicted selling price (for fixed house size).

Figure 12.3: We use only the right tail because larger F values provide greater evidence against H_0.

Figure 12.4: A histogram that shows a few observations that have extremely large or extremely small standardized residuals, well removed from the others.

Figure 12.5: The pattern suggests that y tends to be below $\hat{y}$ for very small and large x_1 values and above $\hat{y}$ for medium-sized x_1 values.

The effect of x_1 appears to be better modeled by a mathematical function that has a parabolic shape.

Figure 12.6: As house size increases, the variability of the standardized residuals appears to increase, suggesting more variability in selling prices when house size is larger, for a given lot size. The lot size values affect the analysis because they have an effect on the predicted values for the model, on which the residuals are based.

Figure 12.7: Each line has the same slope.

Figure 12.10: A straight line implies that p falls below 0 or above 1 for sufficiently small or large x values. This is a problem, because a proportion must fall between 0 and 1.

CHAPTER SUMMARY

This chapter generalized regression to include more than one explanatory variable in the model. The **multiple regression model** relates the mean μ_y of a response variable y to several explanatory variables, for instance,

$$\mu_y = \alpha + \beta_1 x_1 + \beta_2 x_2$$

for two predictors. Advantages over bivariate regression include better predictions of y and being able to study the effect of an explanatory variable on y while *controlling* (keeping fixed) values of other explanatory variables in the model.

- The **regression parameters** (the betas) are slopes that represent effects of the explanatory variables while controlling for the other variables in the model. For instance, β_1 represents the change in the mean of y for a one-unit increase in x_1, at fixed values of the other explanatory variables.

- The **multiple correlation** R and its square (R^2) describe the predictability of the response variable y by the set of explanatory variables. The multiple correlation R equals the correlation between the observed and predicted y values. Its square, R^2, is the proportional reduction in error from predicting y using the prediction equation instead of using $\overline{y}$ (and ignoring x). Both R and R^2 fall between 0 and 1, with larger values representing stronger association.

- An **F statistic** tests $H_0: \beta_1 = \beta_2 = \cdots = 0$, which states that y is independent of all the explanatory variables in the model. The F test statistic equals a ratio of mean squares. A small P-value suggests that at least one explanatory variable affects the response.

- Individual **t tests** and **confidence intervals** for each β parameter analyze separate population effects of each explanatory variable, controlling for the other variables in the model.

- Categorical explanatory variables can be included in a regression model using **indicator variables**. With two categories, the indicator variable equals 1 when the observation is in the first category and 0 when it is in the second.

- For binary response variables, the **logistic regression model** describes how the probability of a particular category depends on the values of explanatory variables. An S-shaped curve describes how the probability changes as the value of a quantitative explanatory variable increases.

Table 12.17 summarizes the basic properties and inference methods for multiple regression and those that Chapter 11 introduced for bivariate regression, with quantitative variables.

TABLE 12.17: Summary of Bivariate and Multiple Regression

	Bivariate Regression	Multiple Regression	
Model formula	$\mu_y = \alpha + \beta x$	$\mu_y = \alpha + \beta_1 x_1 + \beta_2 x_2 + \cdots$	
		Simultaneous Effect	Separate Effects
Properties of measure	$\beta = $ slope		$\beta_1, \beta_2, \ldots$ are slopes,
	$r = $ sample correlation	$R = $ multiple correlation	for each predictor at fixed
	$-1 \leq r \leq 1$	$0 \leq R \leq 1$	values of others
	r^2 is proportional reduction in error (PRE)	R^2 is PRE	
Hypotheses of No effect	$H_0: \beta = 0$	$H_0: \beta_1 = \beta_2 = \ldots = 0$	$H_0: \beta_1 = 0,$ $H_0: \beta_2 = 0, \ldots$
Test statistic	$t = (b - 0)/se$ $df = n - 2$	$F = $ (MS for regression)/MSE $df_1 = $ no. predictors $df_2 = n - $ no. regression para.	$t = (b_i - 0)/se, \ldots$ $df = n - $ no. regression para.

SUMMARY OF NOTATION

$x_1, x_2, \ldots = $ explanatory variables in multiple regression model

$R = $ multiple correlation $= $ correlation between observed y and predicted y values

$R^2 = $ proportional reduction in prediction error, for multiple regression

$F = $ test statistic for testing that all β parameters $= 0$ in regression model

$e^{\alpha + \beta x} = $ exponential function used in numerator and denominator of logistic regression equation, which models the probability p of a binary outcome by

$$p = e^{\alpha + \beta x}/(1 + e^{\alpha + \beta x}).$$

CHAPTER PROBLEMS

Practicing the Basics

12.59 House prices: This chapter has considered many aspects of regression analysis. Let's consider several of them at once by using software with the "house selling price" data file to conduct a multiple regression analysis of $y = $ selling price of home, $x_1 = $ size of home, $x_2 = $ number of bedrooms, $x_3 = $ number of bathrooms.

    a. Construct a scatterplot matrix. Identify the plots that pertain to selling price as a response variable. Interpret, and explain how the highly discrete nature of x_2 and x_3 affects the plots.

    b. Fit the model. Write down the prediction equation, and interpret the coefficient of size of home by its effect when x_2 and x_3 are fixed.

    c. Show how to calculate R^2 from SS values in the ANOVA table. Interpret its value in the context of these variables.

    d. Find and interpret the multiple correlation.

    e. Show all steps of the F test that selling price is independent of these predictors. Explain how to obtain the F statistic from the mean squares in the ANOVA table.

f. Report the t statistic for testing H_0: $\beta_2 = 0$. Report its P-value for H_a: $\beta_2 > 0$, and interpret. Why do you think this effect is not significant? Does this mean that the number of bedrooms is not associated with selling price?

g. Construct and examine the histogram of the residuals for the multiple regression model. What does this describe, and what does it suggest?

h. Construct and examine the plot shown of the residuals plotted against size of home. What does this describe, and what does it suggest?

12.60 Predicting body strength: In Chapter 11, we analyzed strength data for a sample of female high school athletes. When we predict the maximum number of pounds the athlete can bench press using the number of times she can do a 60-pound bench press (BP_60), we get $r^2 = 0.643$. When we add the number of times an athlete can perform a 200-pound leg press (LP_200) to the model, we get $\hat{y} = 60.6 + 1.33(BP\_60) + 0.21(LP\_200)$ and $R^2 = 0.656$.

a. Find the predicted value and residual for an athlete who has BP = 85, BP_60 = 10, and LP_200 = 20.

b. Find the prediction equation for athletes who have LP_200 = 20, and explain how to interpret the slope for BP_60.

c. Note that $R^2 = 0.656$ for the multiple regression model is not much larger than $r^2 = 0.643$ for the bivariate model. What does this suggest?

12.61 Softball data: Refer to the "softball" dataset on the text CD. Regress the difference (DIFF) between the number of runs scored by that team and by the other team on the number of hits (HIT) and the number of errors (ERR).

a. Report the prediction equation, and interpret the slopes.

b. From part (a), approximately how many hits does the team need so that the predicted value of DIFF is positive (corresponding to a predicted win), if they can play error-free ball (ERR = 0)?

12.62 Violent crime: A MINITAB printout is provided from fitting the multiple regression model to "U.S. statewide crime" data on the text CD for the 50 states (excluding Washington, D.C.), on y = violent crime rate, x_1 = poverty rate, and x_2 = percent living in urban areas.

a. Predict the violent crime rate for Massachusetts, which has violent crime rate = 476, poverty rate = 10.2%, and urbanization = 92.1%. Find the residual, and interpret.

b. Interpret the effects of the predictors by showing the prediction equation relating y and x_1 for states with (i) $x_2 = 0$, (ii) $x_2 = 100$. Interpret.

Regression of violent crime rate on poverty rate and urbanization

```
Predictor      Coef     SE Coef     T        P
Constant      -270.7    121.1     -2.23    0.030
poverty        28.334     7.249    3.91    0.000
urbanization    5.416     1.035    5.23    0.000
```

12.63 Effect of poverty on crime: Refer to the previous exercise. Now we add x_3 = percentage of single-parent families to the model. The SPSS table shows results. Without x_3 in the model, poverty has slope 28.33, and when x_3 is added,

poverty has slope 14.95. Explain the differences in the interpretations of these two slopes.

Violent crime predicted by poverty rate, urbanization, and single-parent family

Coefficients[a]

Model		Unstandardized Coefficients		t	Sig.
		B	Std. Error		
1	(Constant)	-631.700	149.604	-4.222	.000
	poverty	14.953	7.540	1.983	.053
	urbanization	4.406	.973	4.528	.000
	singleparent	25.362	7.220	3.513	.001

[a.] Dependent Variable: violentcrimerate

12.64 Correlates of fertility: For data on 39 nations in the "Human development" data file, let y = fertility (FER, the mean number of children per adult woman), x_1 = female economic activity (FEC, the female labor force as a percentage of the male labor force), and x_2 = gross domestic product (GDP). The correlation matrix for these variables is

```
           FER       FEC
FEC      -0.592
GDP      -0.615     0.502
```

a. For the multiple regression model, the multiple correlation $R = 0.70$. Interpret.

b. From looking at the correlation matrix, how is it that you know that R cannot be smaller than 0.615?

c. Report and interpret the value of R^2.

12.65 Predicting pollution: Using the "human development" data file, regressing $CO2$ use (a measure of air pollution) on gross domestic product (GDP) gives a correlation of 0.786. When we add life expectancy as a second explanatory variable, the multiple correlation is $R = 0.787$.

a. Explain how to interpret the multiple correlation.

b. For predicting $CO2$, did it help much to add life expectancy to the model? Explain.

c. Refer to (b). Does this mean that life expectancy is very weakly correlated with $CO2$? Explain.

d. Explain why a large positive standardized residual means that a country has a relatively large $CO2$ value, for its GDP and life expectancy. The two biggest "energy hogs"—the countries with the largest positive standardized residuals—are Australia (2.84) and the United States (2.28). Interpret the standardized residual for Australia.

12.66 Modeling fertility: For the "Human development" data file for several nations, a MINITAB printout is shown below from fitting a multiple regression model to y = fertility (FER), x_1 = literacy (LIT), and x_2 = female economic activity (FEC). Report the value of each of the following:

a. r between y and x_1

b. R^2

c. Total sum of squares

d. Residual sum of squares

e. Standard deviation of y

f. Residual standard deviation of y

g. Test statistic value for H_0: $\beta_1 = 0$

h. P-value for H_0: $\beta_1 = \beta_2 = 0$.

Analysis of fertility, literacy, and female economic activity

```
Variable Mean StDev Minimum Q1    Median Q3     Maximum
FER      2.38 1.24  1.10   1.60  2.00   2.70   7.00
LIT     86.33 20.78 26.90 78.70 96.90 100.00 100.00
FEC     63.85 16.58 28.00 50.00 65.00  80.00  91.00
```

Correlation matrix

```
        FEC    FER
FER    -0.592
LIT     0.638 -0.836
```

```
Predictor   Coef       SE Coef    T       P
Constant    6.8492     0.5129     13.35   0.000
LIT        -0.046313   0.007053   -6.57   0.000
FEC        -0.007384   0.008839   -0.84   0.409
S = 0.695606 R-Sq = 70.4%
```

Analysis of Variance

```
Source          DF     SS       MS      F       P
Regression       2    41.484   20.742  42.87   0.000
Residual Error  36    17.419    0.484
Total           38    58.904
```

12.67 Significant fertility prediction?: Refer to the previous exercise.

a. Show how to construct the F statistic for testing $H_0: \beta_1 = \beta_2 = 0$ from the reported mean squares, report its P-value, and interpret.

b. If these are the only nations of interest to us for this study, rather than a random sample of such nations, is this significance test relevant? Explain.

12.68 eBay selling prices and power sellers: Example 7 in Chapter 7 showed selling prices on eBay of the Palm M515 PDA in May 2003. When y = selling price is regressed on x_1 = number of bids placed and x_2 = whether the seller was classified as a "power seller" (1 = yes, 0 = no), $\hat{y} = 235 - 0.611x_1 + 2.20x_2$.

a. Find the equation relating predicted selling price to number of bids for (i) nonpower sellers, (ii) power sellers. Sketch the two equations on a single plot.

b. Estimate the mean selling prices for the nonpower sellers and for the power sellers, when number of bids equals 10. How does the difference between them relate to the coefficient of x_2 in the multiple regression equation?

12.69 Mother working and child's education: A study of mother's occupational status and children's schooling used logistic regression to estimate the probability that the child obtains a high school degree, using explanatory variables father's education, mother's education, father's occupational level, mother's occupational level, whether mother employed (1 = yes, 0 = no), and number of siblings. Why do you think that logistic regression was used instead of ordinary regression?

12.70 Horseshoe crabs and width: A study of horseshoe crabs found a logistic regression equation for predicting the probability that a female crab had a male partner nesting nearby using x = width of the carapace shell of the female crab (in centimeters). The results were

```
Predictor    Coef
Constant    -12.351
Width         0.497
```

a. For width, Q1 = 24.9 and Q3 = 27.7. Find the estimated probability of a male partner at Q1 and at Q3. Interpret the effect of width by estimating the increase in the probability over the middle half of the sampled widths.

b. At which carapace shell width level is the estimated probability of a male partner (i) equal to 0.50? (ii) greater than 0.50? (iii) less than 0.50?

12.71 AIDS and AZT: In a study (reported in *New York Times,* February 15, 1991) on the effects of AZT in slowing the development of AIDS symptoms, 338 veterans whose immune systems were beginning to falter after infection with the AIDS virus were randomly assigned either to receive AZT immediately or to wait until their T cells showed severe immune weakness. The study classified the veterans' race, whether they received AZT immediately, and whether they developed AIDS symptoms during the three-year study. Let x_1 denote whether used AZT (1 = yes, 0 = no) and let x_2 denote race (1 = white, 0 = black). A logistic regression analysis for the probability of developing AIDS symptoms gave the prediction equation

$$\hat{p} = \frac{e^{-1.074 - 0.720x_1 + 0.056x_2}}{1 + e^{-1.074 - 0.720x_1 + 0.056x_2}}.$$

a. Interpret the sign of the effect of AZT.

b. Show how to interpret the AZT effect for a particular race by comparing the estimated probability of AIDS symptoms for black veterans with and without immediate AZT use.

c. The *se* value was 0.279 for the AZT use effect. Does AZT use have a significant effect, at the 0.05 significance level? Show all steps of a test to justify your answer.

12.72 Factors affecting first home purchase: The table summarizes results of a logistic regression model for predictions about first home purchase by young married households. The response variable is whether the subject owns a home (1 = yes, 0 = no). The explanatory variables are husband's income, wife's income (each in ten-thousands of dollars), the number of years the respondent has been married, the number of children aged 0–17 in the household, and an indicator variable that equals 1 if the subject's parents owned a home in the last year the subject lived in the parental home.

a. Explain why, other things being fixed, the probability of home ownership increases with husband's earnings, wife's earnings, the number of children, and parents' home ownership.

b. From the table, explain why the number of years married seems to show little evidence of an effect, given the other variables in the model.

Results of logistic regression for probability of home ownership		
Variable	**Estimate**	**Std. Error**
Husband earnings	0.569	0.088
Wife earnings	0.306	0.140
No. years married	-0.039	0.042
No. children	0.220	0.101
Parents' home ownership	0.387	0.176

Source: J. Henretta, *Social Forces,* Vol. 66, 1987, pp. 520–536.

Concepts and Investigations

12.73 Student data: Refer to the "FL student" data file on the text CD. Using software, conduct a regression analysis using y = college GPA and predictors high school GPA and sports (number of weekly hours of physical exercise). Prepare a report, summarizing your graphical analyses, bivariate models and interpretations, multiple regression models and interpretations, inferences, checks of effects of outliers, and overall summary of the relationships.

12.74 Why regression?: In 100–200 words, explain to someone who has never studied statistics what multiple regression does and how it can be useful.

12.75 Modeling salaries: The table shows results of fitting a regression model to data on salaries (in dollars) of about 35,000 college professors. Four predictors are categorical (binary), with indicator variable defined in parentheses. Prepare a one-page report, interpreting the effects of the predictors.

Modeling professor salaries

Variable	Estimate	Std. Error
Intercept	10,033.2	
Years of Experience	162.7	2.0
Terminal degree (Ph.D. = 1)	264.0	40.1
Degree quality	616.8	34.5
Research productivity	616.5	7.7
Marital status (married = 1)	850.3	51.2
Race (nonwhite = 1)	−310.3	107.0
Gender (female = 1)	−1863.7	62.1

Source: N. Langton and J. Pfeffer, *Amer. Sociolog. Rev.*, Vol. 59 (1994), p. 236.

Choose the correct response for multiple-choice exercises 12.76–12.78.

12.76 Interpret parameter: If $\hat{y} = 2 + 3x_1 + 5x_2 - 8x_3$, then controlling for x_2 and x_3, the change in the estimated mean of y when x_1 is increased from 10 to 20

  a. equals 30.

  b. equals 0.3.

  c. Cannot be given—depends on specific values of x_2 and x_3.

  d. Must be the same as when we ignore x_2 and x_3.

12.77 Interpret indicator: In the model $\mu_y = \alpha + \beta_1 x_1 + \beta_2 x_2$, suppose that x_2 is an indicator variable for gender, equaling 1 for females and 0 for males.

  a. We set $x_2 = 0$ if we want a predicted mean without knowing gender.

  b. The slope effect of x_1 is β_1 for males and β_2 for females.

  c. β_2 is the difference between the population mean of y for females and for males.

  d. β_2 is the difference between the population mean of y for females and males, for all those subjects having x_1 fixed, such as $x_1 = 10$.

12.78 Regression effects: Multiple regression is used to model y = annual income using x_1 = number of years of education and x_2 = number of years employed in current job.

  a. It is possible that the coefficient of x_2 is positive in a bivariate regression but negative in multiple regression.

  b. It is possible that the correlation between y and x_1 is 0.30 and the multiple correlation between y and x_1 and x_2 is 0.26.

  c. If the F statistic for H_0: $\beta_1 = \beta_2 = 0$ has a P-value = 0.001, then we can conclude that *both* predictors have an effect on annual income.

  d. If $\beta_2 = 0$, then annual income is independent of x_2 in bivariate regression.

12.79 True or false about R and R^2: For each of the following statements, indicate whether it is true or false. If false, explain why it is false.

  a. The multiple correlation equals the ordinary correlation computed between the values of the response variable and the values $\hat{y}$ predicted by the regression model.

  b. The multiple correlation is like the ordinary correlation in that it falls between −1 and 1.

  c. R^2 describes how well you can predict y using x_1 when you control for the other variables in the multiple regression model.

  d. It's impossible for R^2 to go down when you add explanatory variables to a regression model.

12.80 True or false in regression: For each of the following statements, indicate whether it is true or false. If false, explain why it is false. In regression analysis

  a. The estimated coefficient of x_1 can be positive in the bivariate model but negative in a multiple regression model.

  b. When a model is refitted after y = income is changed from dollars to euros, R^2, the correlation between y and x_1, the F statistics and t statistics will not change.

  c. If $r^2 = 0.6$ between y and x_1 and if $r^2 = 0.6$ between y and x_2, then for the multiple regression model with both predictors $R^2 = 1.2$.

  d. The multiple correlation between y and $\hat{y}$ can equal −0.40.

12.81 True or false about slopes: For data on y = college GPA, x_1 = high school GPA, and x_2 = average of mathematics and verbal entrance exam score, we get $\hat{y} = 2.70 + 0.45x_1$ for bivariate regression and $\hat{y} = 0.3 + 0.40x_1 + 0.003x_2$ for multiple regression. For each of the following statements, indicate whether it is true or false. Give a reason for your answer.

  a. The correlation between y and x_1 is positive.

  b. A one-unit increase in x_1 corresponds to a change of 0.45 in the predicted value of y, controlling for x_2.

  c. Controlling for x_1, a 100-unit increase in x_2 corresponds to a predicted increase of 0.30 in college GPA.

12.82 Scores for religion: You want to include religious affiliation as a predictor in a regression model, using the categories Protestant, Catholic, Jewish, other. You set up a variable x_1 that equals 1 for Protestants, 2 for Catholics, 3 for Jewish, and 4 for other, using the model $\mu_y = \alpha + \beta x_1$. Explain why this is inappropriate.

12.83 Lurking variable: Give an example of three variables for which you expect $\beta \neq 0$ in the model $\mu_y = \alpha + \beta x_1$ but $\beta_1 = 0$ in the model $\mu_y = \alpha + \beta_1 x_1 + \beta_2 x_2$. (*Hint:* The bivariate effect of x_1 could be completely due to a lurking variable, x_2.)

12.84 Properties of R^2: Using its definition in terms of SS values, explain why $R^2 = 1$ only when all the residuals are 0, and $R^2 = 0$ when each $\hat{y} = \bar{y}$. Explain what this means in practical terms.

12.85 Why an F test?: When a model has a very large number of predictors, even when none of them truly have an effect in the population, one or two may look significant in t tests merely by random variation. Explain why performing the F test first can safeguard against getting such false information from t tests.

12.86 Multicollinearity: For the "high school female athletes" data file, regress the maximum bench press on weight and % body fat.
   a. Show that the F test is statistically significant at the 0.05 significance level.
   b. Show that the P-values are both larger than 0.35 for testing the individual effects with t tests. (It seems like a contradiction when the F test tells us that at least one predictor has an effect but the t tests indicate that neither predictor has a significant effect. This can happen when the predictor variables are highly correlated, so a predictor has little impact when the other predictors are in the model. Such a condition is referred to as **multicollinearity**. In this example, the correlation is 0.871 between weight and % body fat.)

12.87 Logistic vs. linear: For binary response variables, one reason that logistic regression is usually preferred over straight-line regression is that a fixed change in x often has a smaller impact on a probability p when p is near 0 or near 1 than when p is near the middle of its range. Let y refer to the decision to rent or to buy a home, with $p =$ the probability of buying, and let $x =$ weekly family income. In which case do you think an increase of $100 in x has greater effect: when $x = 50,000$ (for which p is near 1), when $x = 0$ (for which p is near 0), or when $x = 500$? Explain how your answer relates to the choice of a linear versus logistic regression model.

12.88 Adjusted R^2: When we use R^2 for a random sample to estimate a population R^2, it's a bit biased. It tends to be a bit too large, especially when n is small. Some software also reports

$$\text{Adjusted } R^2 = R^2 - \{p/[n - (p + 1)]\}(1 - R^2),$$

where $p =$ number of predictor variables in the model. This is slightly smaller than R^2 and is less biased. Suppose $R^2 = 0.500$ for a model with $p = 2$ predictors. Calculate adjusted R^2 for the following sample sizes: 10, 100, 1000. Show that the difference between adjusted R^2 and R^2 diminishes as n increases.

12.89 R can't go down: The least squares prediction equation provides predicted values $\hat{y}$ with the strongest possible correlation with y, out of all possible prediction equations of that form. Based on this property, explain why the multiple correlation R cannot decrease when you add a variable to a multiple regression model. (*Hint*: The prediction equation for the simpler model is a special case of a prediction equation for the full model that has coefficient 0 for the added variable.)

12.90 Indicator for comparing two groups: Chapter 9 presented methods for comparing means for two groups. Explain how it's possible to perform a significance test of equality of two population means as a special case of a regression analysis. (*Hint*: The regression model then has a single explanatory variable—an indicator variable for the two groups being compared. What does $\mu_1 = \mu_2$ correspond to in terms of a value of a parameter in this model?)

12.91 Simpson's paradox: Let $y =$ death rate and $x =$ average age of residents, measured for each county in Louisiana and in Florida. Draw a hypothetical scatterplot, identifying points for each state, such that the mean death rate is higher in Florida than in Louisiana when x is ignored, but lower when it is controlled. (*Hint*: When you fit a line for each state, the line should be higher for Louisiana, but the y-values for Florida should have an overall higher mean.)

12.92 Parabolic regression: A regression formula that gives a parabolic shape instead of a straight line for the relationship between two variables is

$$\mu_y = \alpha + \beta_1 x + \beta_2 x^2.$$

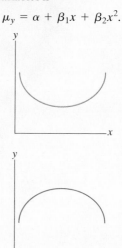

   a. Explain why this is a multiple regression model, with x playing the role of x_1 and x^2 (the square of x) playing the role of x_2.
   b. For x between 0 and 5, sketch the prediction equation (i) $\hat{y} = 10 + 2x + 0.5x^2$, (ii) $\hat{y} = 10 + 2x - 0.5x^2$. This shows how the parabola is bowl-shaped or mound-shaped, depending on whether the coefficient of x^2 is positive or negative.

12.93 Logistic slope: At the x value where the probability of success is some value p, the line drawn tangent to the logistic regression curve has slope $\beta p(1 - p)$.
   a. Explain why the slope is $\beta/4$ when $p = 0.5$.
   b. Show that the slope is weaker at other p values by evaluating this at $p = 0.1, 0.3, 0.7$, and 0.9. What does the slope approach as p gets closer and closer to 0 or 1? Sketch a curve to illustrate.

12.94 When is $p = 0.50$?: When $\alpha + \beta x = 0$, so that $x = -\alpha/\beta$, show that the logistic regression equation $p = e^{\alpha + \beta x}/(1 + e^{\alpha + \beta x})$ gives $p = 0.50$.

Class Explorations

12.95 Class data: Refer to the data file your class created in Activity 3 in Chapter 1. For variables chosen by your instructor, fit a multiple regression model and conduct descriptive and inferential statistical analyses. Interpret and summarize your findings, and prepare to discuss these in class.

13

Comparing Groups: Analysis of Variance Methods

EXAMPLE 1

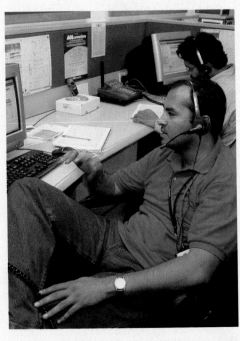

Investigating Customer Satisfaction

Picture the Scenario

In recent years, many companies have increased the attention paid to measuring and analyzing customer satisfaction. Here are examples of two recent studies of customer satisfaction:

- A company that makes personal computers has provided a toll-free telephone number for owners of their PCs to call and seek technical support. For years the company had two service centers for these calls: in San Jose, California, and in Toronto, Canada. Recently the company "outsourced" many of the customer service jobs to a new center in Bangalore, India, because employee salaries are much lower there. The company wanted to compare customer satisfaction at the three centers.

- An airline has a toll-free telephone number that potential customers can call to make flight reservations. Usually the call volume is heavy and callers are placed on hold until an agent is free to answer. Researchers working for the airline recently conducted a randomized experiment to analyze whether callers would tend to remain on hold longer if they heard (a) an advertisement about the airline and its current promotions, (b) recorded Muzak ("elevator music"), (c) recorded, classical music. Currently, messages are five minutes long and then repeated, and the researchers also wanted to find out if it would make a difference if messages instead were 10 minutes long before repeating.

Questions to Explore

In the second study, the company's executive officer (CEO) had some familiarity with statistical methods, based on a course he took in college. He asked the researchers:

- In this experiment, are the sample mean times that callers stayed on hold before hanging up significantly different for the three recorded message types?

- What conclusions can you make if you take into account both the type of recorded message and whether it was repeated every five minutes or every ten minutes?

Thinking Ahead

Chapter 9 showed how to compare two means. In practice, there may be *several* means to compare, such as in the second example. This chapter shows how to use statistical inference to compare several means. We'll see how to determine whether a set of sample means is significantly different and how to estimate the differences among corresponding population means. To illustrate, we'll analyze data from the second study in Examples 2–4 and 7.

The methods introduced in this chapter apply when a quantitative response variable has a categorical explanatory variable. The categories of the explanatory variable identify the groups to be compared in terms of their means on the response variable. For example, the first study above compared mean customer satisfaction for three groups—the customers who call the service centers at the three locations. The response variable is customer satisfaction (on a scale of 0 to 10) and the explanatory variable is the service center location.

The inferential method for comparing means of several groups is called **analysis of variance**, denoted **ANOVA**. Section 13.1 shows that the name "analysis of variance" refers to the significance test's focus on two types of variability in the data. Section 13.2 shows how to construct confidence intervals comparing the means. It also shows that ANOVA methods are special cases of a multiple regression analysis.

Categorical explanatory variables in multiple regression and in ANOVA are often referred to as **factors**. ANOVA with a single factor, such as service center location, is called **one-way ANOVA**. Section 13.3 introduces ANOVA for two factors, called **two-way ANOVA**. For instance, for the second study above two-way ANOVA analyzes how the mean telephone holding time varies across categories of recorded message type and categories defined by the length of time before repeating the message.

13.1 | How Can We Compare Several Means?: One-Way ANOVA

The **analysis of variance** method compares means of *several* groups. Let g denote the number of groups. Each group has a corresponding population of subjects. The means of the response variable for the g populations are denoted by $\mu_1, \mu_2, \ldots, \mu_g$.

Hypotheses and Assumptions for the ANOVA Test Comparing Means

The analysis of variance is a significance test of the null hypothesis of equal population means,

$$H_0: \mu_1 = \mu_2 = \cdots = \mu_g.$$

An example is $H_0: \mu_1 = \mu_2 = \mu_3$ for testing population mean satisfaction at $g = 3$ service center locations. The alternative hypothesis is

$$H_a: \text{at least two of the population means are unequal.}$$

If H_0 is false, perhaps all the population means differ, but perhaps merely one mean differs from the others. The test analyzes whether the differences observed among the *sample* means could have reasonably occurred by chance, if the null hypothesis of equal *population* means were true.

The assumptions for the ANOVA test comparing population means are as follows:

- The population distributions of the response variable for the g groups are normal, with the same standard deviation for each group.

- Randomization: In a survey sample, independent random samples are selected from the g populations. In an experiment, subjects are randomly assigned separately to the g groups.

We'll discuss these assumptions in some detail later in the section. Under the first assumption, when the population means are equal, the population distribution of the response variable is the same for each group. The population distribution does not depend on the group to which a subject belongs. *The ANOVA test is a test of independence between the quantitative response variable and the group factor.*

How Long Will You Tolerate Being Put on Hold?

Picture the Scenario

Let's refer back to the second scenario in Example 1. An airline has a toll-free telephone number for reservations. Often the call volume is heavy, and callers are placed on hold until a reservation agent is free to answer. The airline hopes a caller remains on hold until the call is answered, so as not to lose a potential customer.

The airline recently conducted a randomized experiment to analyze whether callers would remain on hold longer, on the average, if they heard (a) an advertisement about the airline and its current promotions, (b) Muzak, or (c) classical music (Vivaldi's *Four Seasons*). The company randomly selected one out of every 1000 calls in a particular week. For each call, they randomly selected one of the three recorded messages to play and then measured the number of minutes that the caller remained on hold before hanging up (these calls were purposely not answered). The total sample size was 15. The company kept the study small, hoping it could make conclusions without alienating too many potential customers! Table 13.1 shows the data. It also shows the mean and standard deviation for each recorded message type.

TABLE 13.1: Telephone Holding Times by Type of Recorded Message

Each observation is the number of minutes a caller remained on hold before hanging up, rounded to the nearest minute.

Recorded Message	Holding Time Observations	Sample Size	Mean	Standard Deviation
Advertisement	5, 1, 11, 2, 8	5	5.4	4.2
Muzak	0, 1, 4, 6, 3	5	2.8	2.4
Classical	13, 9, 8, 15, 7	5	10.4	3.4

Questions to Explore

a. What are the hypotheses for the ANOVA test?

b. Figure 13.1 displays the sample means. Since these means are quite different, is there sufficient evidence to conclude that the population means differ?

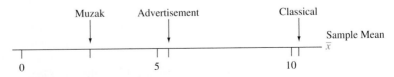

▲ **FIGURE 13.1: Sample Means of Telephone Holding Times for Callers Who Hear One of Three Recorded Messages. Question:** Since the sample means are quite different, can we conclude that the population means differ?

Think It Through

a. Denote the holding time means for the population that these three random samples represent by μ_1 for the advertisement, μ_2 for Muzak, and μ_3 for classical music. ANOVA tests whether these are equal. The null hypothesis is $H_0: \mu_1 = \mu_2 = \mu_3$. The alternative hypothesis is that at least two of the population means are different.

b. The sample means are quite different. But even if the population means are equal, we expect the sample means to differ, because of sampling variability. So this alone is not sufficient evidence to enable us to reject H_0.

Insight

The strength of evidence against H_0 will also depend on the sample sizes and the variability of the data. We'll next study how to test these hypotheses.

TO PRACTICE THIS CONCEPT, TRY EXERCISE 13.1 PARTS (A) AND (B)

Variability between Groups and within Groups Is the Key to Significance

The ANOVA method is used to compare population means. So, why is it called analysis of *variance*? The reason is that the test statistic uses evidence about two types of variability. Rather than presenting a formula now for this test statistic, which is rather complex, we'll discuss the reasoning behind it, which is quite simple.

Table 13.1 listed data for three groups. Figure 13.2a shows the same data with dot plots. Suppose the sample data were different, as shown in Figure 13.2b. The data in 13.2b have the same means as the data in Table 13.1 but have smaller standard deviations. Which case do you think gives stronger evidence against $H_0: \mu_1 = \mu_2 = \mu_3$?

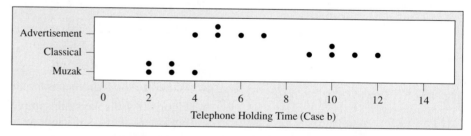

▲ **FIGURE 13.2: Data from Table 13.1 (Case a) and Hypothetical Data (Case b) that Have the Same Means but Less Variability within Groups. Question:** Do the data in case b give stronger, or weaker, evidence against $H_0: \mu_1 = \mu_2 = \mu_3$ than the data in case a? Why?

What's the difference between the data in these two cases? The variability *between* pairs of sample means is the same in each case, because the sample means are the same. However, the variability *within* each sample is much smaller in Figure 13.2b than in Figure 13.2a. The sample standard deviation is about 1.0 for each sample in Figure 13.2b whereas it is between 2.4 and 4.2 for the samples in Figure 13.2a. We'll see that the evidence against $H_0: \mu_1 = \mu_2 = \mu_3$ is stronger when the variability *within* each sample is smaller. Case b has less variability within each sample than case a, and it gives stronger evidence against H_0. The evidence against H_0 is also stronger when the variability *between* sample means increases (that is, when the sample means are farther apart) and as the sample sizes increase.

ANOVA F Test Statistic

The **analysis of variance (ANOVA) F test statistic** summarizes

$$F = \frac{\text{Between-groups variability}}{\text{Within-groups variability}}.$$

The larger the variability *between* groups relative to the variability *within* groups, the larger the F test statistic tends to be. For instance, $F = 6.4$ for the data in Figure 13.2, case a, whereas $F = 67.8$ for the data in Figure 13.2, case b. The between-groups variability is the same in each figure, but Figure 13.2b has much less within-groups variability and thus a larger test statistic value. Later in the section we'll see the two types of variability described in the F test statistic are measured by estimates of *variances* (squared standard deviations).

The Test Statistic for Comparing Means Has the F Distribution

When H_0 is true, the F test statistic has the F sampling distribution. The formula for the F test statistic is such that when H_0 is true, the F distribution has a mean of approximately 1. When H_0 is false, the F test statistic tends to be larger than 1, more so as the sample sizes increase. The larger the F test statistic value, the stronger the evidence against H_0.

Recall that we used the F distribution in the F test that the slope parameters of a multiple regression model are all zero (Section 12.3). As in that test, the P-value here is the probability (presuming that H_0 is true) that the F test statistic is larger than the observed F value. That is, it is the right-hand tail probability, as shown in the margin figure, representing results even more extreme than observed. The larger the F test statistic, the smaller the P-value.

In Section 12.3 we learned that the F distribution has two *df* values. For ANOVA with g groups and total sample size for all groups combined of $N = n_1 + n_2 + \cdots + n_g$,

$$df_1 = g - 1 \text{ and } df_2 = N - g.$$

Table D in the appendix reports F values having P-value of 0.05, for various df_1 and df_2 values. For any given F value and df values, software provides the P-value. The next box summarizes the steps of an ANOVA F test.

Steps of ANOVA F Test for Comparing Population Means of Several Groups

1. **Assumptions:** Independent random samples, normal population distributions with equal standard deviations

2. **Hypotheses:** H_0: $\mu_1 = \mu_2 = \cdots = \mu_g$ (Equal population means for g groups),
   H_a: at least two of the population means are unequal.

3. **Test statistic:** $F = \dfrac{\text{Between-groups variability}}{\text{Within-groups variability}}.$

   F sampling distribution has $df_1 = g - 1$, $df_2 = N - g$ = total sample size − no. groups

4. **P-value:** Right-tail probability above observed F value

5. **Conclusion:** Interpret in context. If decision needed, reject if P-value $\leq$ significance level (such as 0.05).

EXAMPLE 3

ANOVA for Customers' Telephone Holding Times

Picture the Scenario

Examples 1 and 2 discussed a study, using 15 subjects, of the length of time that callers to an airline's toll-free telephone number remain on hold before hanging up. The study compared three recorded messages: An advertisement about the airline, Muzak, and classical music. Let μ_1, μ_2, and μ_3 denote the population mean telephone holding times for the three messages.

Questions to Explore

a. For testing $H_0: \mu_1 = \mu_2 = \mu_3$ based on this experiment, what value of the F test statistic would have a P-value of 0.05?

b. For the data in Table 13.1, we'll see that software reports $F = 6.4$. Based on the answer to (a), will the P-value be larger, or smaller, than 0.05?

c. Can you reject H_0, using a significance level of 0.05? What can you conclude from this?

Think It Through

a. With $g = 3$ groups and a total sample size of $N = 15$ (5 in each group), the test statistic has

$$df_1 = g - 1 = 2 \text{ and } df_2 = N - g = 15 - 3 = 12.$$

From Table D (see the excerpt in the margin), with these df values an F test statistic value of 3.88 has a P-value of 0.05.

b. Since the F test statistic of 6.4 is farther out in the tail than 3.88, the right-tail probability above 6.4 is less than 0.05. So, the P-value is less than 0.05.

c. Since P-value < 0.05, there is sufficient evidence to reject $H_0: \mu_1 = \mu_2 = \mu_3$. We conclude that a difference exists among the three types of messages in the population mean amount of time that customers are willing to remain on hold.

Insight

We'll see that software reports a P-value $= 0.013$. This is quite strong evidence against H_0. If H_0 were true, there'd be only about a 1% chance of getting an F test statistic value larger than the observed F value of 6.4.

TO PRACTICE THIS CONCEPT, TRY EXERCISE 13.2

F values with P-value = 0.05			
		df_1	
df_2	1	2	3
6	5.99	5.14	4.76
12	4.75	3.88	3.49
18	4.41	3.55	3.16
24	4.26	3.40	3.01
30	4.17	3.32	2.92
120	3.92	3.07	2.68

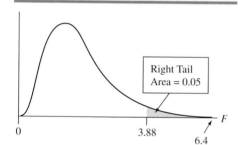

The Variance Estimates and the ANOVA Table

Now let's take a closer look at the F test statistic. Denote the group sample means by $\bar{y}_1, \bar{y}_2, \ldots, \bar{y}_g$. (We use y rather than x because the quantitative variable is the response variable when used in a corresponding regression analysis.) We'll see that the F test statistic depends on these sample means and also on the sample standard deviations, $s_1, s_2, \ldots, s_g$, for the g groups and the sample sizes.

Let σ denote the standard deviation for each of the g population distributions. One assumption for the ANOVA F test is that each population has the same standard deviation. The F test statistic for $H_0: \mu_1 = \mu_2 = \cdots = \mu_g$ is the ratio of two estimates of σ^2, the population *variance* for each group. Since we won't usually do computations by hand, we'll show formulas merely to give a

Recall

From Section 2.4, a **variance** is a squared standard deviation. ■

better sense of what the F test statistic represents. The formulas are simplest when the group sample sizes are equal (as in Example 3), the case we'll show.

The estimate of σ^2 in the *denominator* of the F test statistic uses the variability *within* each group. The sample standard deviations $s_1, s_2, \ldots, s_g$ summarize the variation of the observations within the groups around their means.

■ With equal sample sizes, the **within-groups estimate** of the variance σ^2 is the mean of the g sample variances for the g groups,

$$\text{Within-groups variance estimate } s^2 = \frac{s_1^2 + s_2^2 + \cdots + s_g^2}{g}.$$

Did You Know?

Without n in the formula, the between-groups estimate is the sample variance of the g sample means. That sample variance of $\{\bar{y}_1, \bar{y}_2, \ldots, \bar{y}_g\}$ estimates the variance of the sampling distribution of each sample mean, which is σ^2/n. Multiplying by n then gives an estimate of σ^2 itself. See Exercise 13.68. ■

When the sample sizes are not equal, the within-groups estimate is a *weighted average* of the sample variances, with greater weight given to samples with larger sample sizes. In either case, this estimate is unbiased: Its sampling distribution has σ^2 as its mean, regardless of whether H_0 is true.

The estimate of σ^2 in the *numerator* of the F test statistic uses the variability *between* each sample mean and the overall sample mean $\bar{y}$ for all the data.

■ With equal sample sizes, n in each group, the **between-groups estimate** of the variance σ^2 is

$$\text{Between-groups variance estimate} = \frac{n[(\bar{y}_1 - \bar{y})^2 + (\bar{y}_2 - \bar{y})^2 + \cdots + (\bar{y}_g - \bar{y})^2]}{g - 1}.$$

If H_0 is true, this estimate is also unbiased. We expect this estimate to take a similar value as the within-groups estimate, apart from sampling error. If H_0 is false, however, the population means differ and the sample means tend to differ more greatly. Then, the between-groups estimate tends to overestimate σ^2.

The F test statistic is the ratio of these two estimates of the population variance,

$$F = \frac{\text{Between-groups estimate of } \sigma^2}{\text{Within-groups estimate of } \sigma^2}.$$

Recall

In Sections 11.4 and 12.3, we saw that a **mean square** (MS) is a ratio of a sum of squares (SS) to its *df* value. ■

Computer software displays the two estimates in an ANOVA table similar to tables displayed in regression. Table 13.2 shows the basic format, illustrating for the test in Example 3

■ The MS column contains the two estimates, which are called **mean squares**.

■ The ratio of the two mean squares is the F test statistic, $F = 74.6/11.6 = 6.4$.

This F test statistic has P-value = 0.013, also shown in Table 13.2.

TABLE 13.2: ANOVA Table for F Test Using Data from Table 13.1

Source	DF	SS	MS	F	P
Group	2	149.2	74.6	6.43	0.013
Error	12	139.2	11.6		
Total	14	288.4			

Between groups → Group

Within groups → Error

F test statistic = ratio of the MS values

The mean square in the Group row of the ANOVA table is based on variability *between* the groups. It is the *between-groups* estimate of the population variance σ^2. This is 74.6 in Table 13.2, listed under MS. The mean square error is the *within-groups* estimate s^2 of σ^2. This is 11.6 in Table 13.2. The "Error" label for this MS refers to the fact that it summarizes the error from not being able to predict subjects' responses exactly if we know only the group to which they belong.

Each mean square equals a sum of squares (in the SS column) divided by a degrees of freedom value (in the DF column). The *df* values for Group and Error in Table 13.2 are the *df* values for the *F* distribution, $df_1 = g - 1 = 3 - 1 = 2$ and $df_2 = N - g = 15 - 3 = 12$.

The sum of the between-groups sum of squares and the within-groups (Error) sum of squares is the **total sum of squares**. This is the sum of squares of the combined sample of *N* observations around the overall sample mean. The analysis of variance partitions the total sum of squares into two independent parts, the between-groups SS and the within-groups SS. It can be shown that the total sum of squares equals

$$\text{Total SS} = \sum (y - \bar{y})^2 = \text{between-groups SS} + \text{within-groups SS}.$$

In Table 13.2, for example,

$$\text{Total SS} = 288.4 = 149.2 + 139.2.$$

The total SS divided by $N - 1$ (the *df* for the total SS) is the sample variance when the data for the *g* groups are combined and treated as a single sample. The margin shows the TI-83+/84 output for this one-way ANOVA.

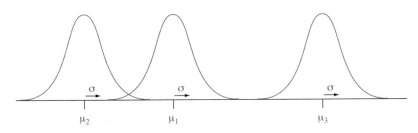

Assumptions and the Effects of Violating Them

The assumptions for the ANOVA *F* test comparing population means are (1) that the population distributions of the response variable for the *g* groups are normal, (2) those distributions have the same standard deviation σ, and (3) the data resulted from randomization. Figure 13.3 portrays the population distribution assumptions.

▲ **FIGURE 13.3: The Assumptions about the Population Distributions: Normal with Equal Standard Deviations. Question:** In practice, what types of evidence would suggest that these assumptions are badly violated?

The assumptions that the population distributions are normal with identical standard deviations seem stringent. They are never satisfied exactly in practice. Moderate violations of the normality assumption are not serious, however. The *F* sampling distribution still provides a reasonably good approximation to the actual sampling distribution of the *F* test statistic. This becomes even more the case as the sample sizes increase, because the sampling distribution then has weaker dependence on the form of the population distributions.

Moderate violations of the equal population standard deviation assumption are also not serious. When the sample sizes are identical for the groups, the *F* test still works well even with severe violations of this assumption. When the sample sizes are not equal, *the F test works quite well as long as the largest group standard deviation is no more than about twice the smallest group standard deviation.*

You can construct box plots or dot plots for the sample data distributions to check for extreme violations of normality. Misleading results may occur with the F test if the distributions are highly skewed and the sample size N is small, or if there are relatively large differences among the standard deviations (the largest sample standard deviation being more than double the smallest one) and the sample sizes are unequal. When the distributions are very highly skewed, the mean may not even be an especially relevant summary measure.[1]

IN PRACTICE Robustness of ANOVA *F* test

Since the ANOVA F test is robust to moderate breakdowns in the population normality and equal standard deviation assumptions, in practice it is used unless (i) graphical methods show extreme skew for the response variable or (ii) the largest group standard deviation is more than about double the smallest group standard deviation and the sample sizes are unequal.

EXAMPLE 4

Checking Assumptions for Telephone Holding Time Study

Picture the Scenario

Let's check the assumptions for the F test on telephone holding times (Example 3).

Question to Explore

Is it appropriate to apply ANOVA to the data in Table 13.1 to compare mean telephone holding times for three message types?

Think It Through

Subjects were selected randomly for the experiment and assigned randomly to the three recorded messages. From Table 13.1 (summarized in the margin), the largest sample standard deviation of 4.2 is less than twice the smallest standard deviation of 2.4 (In any case, the sample sizes are equal, so this assumption is not crucial). The sample sizes in Table 13.1 are small, so it is difficult to make judgments about shapes of population distributions. However, the dot plots in Figure 13.2 (case a) did not show evidence of severe nonnormality. Thus, ANOVA is suitable for these data.

Insight

As in other statistical inferences, the method used to gather the data is the most crucial factor. Inferences have greater validity when the data came from an experimental study that randomly assigned subjects to groups or from an observational study that used random sampling.

TO PRACTICE THIS CONCEPT, TRY EXERCISE 13.11, PART (B)

Recall

Table 13.1 showed the following:

Recorded Message	Sample Size	Mean	Std. Dev.
Advert.	5	5.4	4.2
Muzak	5	2.8	2.4
Classical	5	10.4	3.4

■

Recall

See the beginning of Chapter 9 to review the distinction between **independent samples** and **dependent samples**. ■

The ANOVA methods in this chapter are designed for *independent* samples. Recall that for independent samples, the subjects in one sample are distinct from those in other samples. Separate methods, beyond the scope of this text, handle *dependent* samples. An example is when the same group of subjects is observed at different times, such as in an experiment before and after exposure to some treatment.

---

[1]Chapter 14 discusses methods that apply when these assumptions are badly violated.

Recall

To compare two means while assuming equal population standard deviations, Section 9.3 showed the test statistic is

$$t = \frac{\bar{y}_1 - \bar{y}_2}{s\sqrt{\dfrac{1}{n_1} + \dfrac{1}{n_2}}},$$

where the standard deviation s pools information from within both samples. It has $df = n_1 + n_2 - 2 = N - g$ for $N = n_1 + n_2$ and $g = 2$. ■

Why Not Use Several t Tests to Compare the Means?

With two groups, Section 9.3 showed how a t test can compare the means under the assumption of equal population standard deviations. That test also uses between-groups and within-groups variation. The t test statistic has the difference *between* the two group means in the numerator and a denominator based on pooling variability *within* the two groups. See the formula in the margin recall box. In fact, if we apply the ANOVA F test to data from $g = 2$ groups, it can be shown that the F test statistic equals the square of this t test statistic. The P-value for the F test is exactly the same as the two-sided P-value for the t test. We can use either test to conduct the analysis.

When there are several groups, instead of using the F test why not use a t test to compare each pair of means? One reason for doing the F test is that using a *single* test rather than multiple tests enables us to control the probability of a type I error. With a significance level of 0.05 in the F test, for instance, the probability of incorrectly rejecting a true H_0 is fixed at 0.05. When we do a separate t test for each pair of means, by contrast, a type I error probability applies for *each* comparison. We are not then controlling the *overall* type I error rate for all the comparisons.

But the F test has its own disadvantages. With a small P-value, we can conclude that the population means are not identical. However, the result of the F test does not tell us *which* groups are different or *how* different they are. In the example above, we have not concluded whether one recorded message works significantly better than the other two at keeping potential customers on the phone. We can address these issues using confidence intervals, as the next section shows.

13.1 Practicing the Basics

13.1 **Hotel satisfaction:** The CEO of a company that owns
TRY five resort hotels wants to evaluate and compare satisfaction with the five hotels. The company's research department randomly sampled 125 people who had stayed at any of the hotels during the past month and asked them to rate their expectations of the hotel before their stay and to rate the quality of the actual stay at the hotel. Both observations used a rating scale of 0–10, with $0 =$ very poor and $10 =$ excellent. The researchers compared the hotels on the gap between prior expectation and actual quality, using the difference score, $y =$ performance gap = (prior expectation score – actual quality score).

  a. Identify the response variable, the factor, and the categories that form the groups.

  b. State the null and alternative hypotheses for conducting an ANOVA.

  c. Explain why the df values for this ANOVA are $df_1 = 4$ and $df_2 = 120$.

  d. How large an F test statistic is needed to get a P-value = 0.05 in this ANOVA?

13.2 **Satisfaction with banking:** A bank conducts a survey in
TRY which it randomly samples 400 of its customers. The survey asks the customers which way they use the bank the most: (1) interacting with a teller at the bank, (2) using

ATMs, or (3) using the bank's Internet banking service. It also asks their level of satisfaction with the service they most often use (on a scale of 0 to 10 with $0 =$ very poor and $10 =$ excellent). Does mean satisfaction differ according to how they most use the bank?

  a. Identifying notation, state the null and alternative hypotheses for conducting an ANOVA with data from the survey.

  b. Report the df values for this ANOVA. About what range of F test statistic values give a P-value below 0.05?

  c. For the data, $F = 0.46$ and the P-value equals 0.63. What can you conclude?

  d. What were the assumptions on which the ANOVA was based? Which assumption is the most important?

13.3 **What's the best way to learn French?:** The table below shows scores on the first quiz (maximum score 10 points) for eighth-grade students in an introductory-level French course. The instructor grouped the students in the course as follows:

  Group 1: Never studied foreign language before but have good English skills

  Group 2: Never studied foreign language before and have poor English skills

  Group 3: Studied at least one other foreign language

French scores on the quiz

	Group 1	Group 2	Group 3
	4	1	9
	6	5	10
	8		5
Mean	6.0	3.0	8.0
Std. Dev.	2.000	2.828	2.646

Source	DF	SS	MS	F	P
Group	2	30.00	15.00	2.50	0.177
Error	5	30.00	6.00		
Total	7	60.00			

a. Defining notation and using results obtained with software, also shown in the table, report the five steps of the ANOVA test.

b. The sample means are quite different, but the P-value is not small. Name one important reason for this. (*Hint*: For given sample means, how do the results of the test depend on the sample sizes?)

c. Was this an experimental study, or an observational study? Explain how a lurking variable could be responsible for Group 3 having a larger mean than the others. (Thus, even if the P-value were small, it is inappropriate to assume that having studied at least one foreign language causes one to perform better on this quiz.)

13.4 What affects the F value?: Refer to the previous exercise.

a. Suppose that the first observation in the second group was actually 9, not 1. Then the standard deviations are the same as reported in the table, but the sample means are 6, 7, and 8 rather than 6, 3, and 8. Do you think the F test statistic would be larger, the same, or smaller? Explain your reasoning, without doing any calculations.

b. Suppose you had the same means as shown in the table but the sample standard deviations were 1.0, 1.8, and 1.6, instead of 2.0, 2.8, and 2.6. Do you think the F test statistic would be larger, the same, or smaller? Explain your reasoning.

c. Suppose you had the same means and standard deviations as shown in the table but the sample sizes were 30, 20, and 30, instead of 3, 2, and 3. Do you think the F test statistic would be larger, the same, or smaller? Explain your reasoning.

d. In (a), (b), and (c), would the P-value be larger, the same, or smaller? Why?

13.5 Outsourcing: Example 1 at the beginning of this chapter mentioned a study to compare customer satisfaction at service centers in San Jose, California; Toronto, Canada; and Bangalore, India. Each center randomly sampled 100 people who called during a two-week period. Callers rated their satisfaction on a scale of 0 to 10, with higher scores representing greater satisfaction. The sample means were 7.6 for San Jose, 7.8 for Toronto, and 7.1 for Bangalore. The table shows the results of conducting an ANOVA.

a. Define notation and specify the null hypothesis tested in this table.

b. Explain how to obtain the F test statistic value reported in the table from the MS values shown, and report the values of df_1 and df_2 for the F distribution.

c. Interpret the P-value reported for this test. What conclusion would you make using a 0.05 significance level?

Customer satisfaction with outsourcing

Source	DF	SS	MS	F	P
Group	2	26.00	13.00	27.6	0.000
Error	297	140.00	0.47		
Total	299	60.00			

13.6 Good friends and astrological sign: The 1998 General Social Survey asked subjects how many good friends they have. Is this associated with the respondent's astrological sign (the 12 symbols of the Zodiac)? The ANOVA table for the GSS data reports $F = 0.61$ based on $df_1 = 11$, $df_2 = 813$.

a. Introduce notation, and specify the null hypothesis and alternative hypothesis for the ANOVA.

b. Based on what you know about the F distribution, would you guess that the test statistic value of 0.61 provides strong evidence against the null hypothesis? Explain.

c. Software reports a P-value of 0.82. Explain how to interpret this.

13.7 How many kids to have?: A recent General Social Survey asked subjects "What is the ideal number of kids for a family?" Do responses tend to depend on the subjects' religious affiliation? Results of an ANOVA are shown in the printout, for religion categories (Protestant, Catholic, Jewish, Other, or none).

a. Define notation and specify the null hypothesis tested in this printout.

b. Summarize the assumptions made to conduct this test.

c. Report the F test statistic value and the P-value for this test. Interpret the P-value.

d. Based on (c), can you conclude that *every* pair of religious affiliations has different population means for ideal family size? Explain.

Ideal number of kids in a family by religious affiliation

Source	DF	SS	MS	F	P
Religion	3	11.72	3.91	5.48	0.001
Error	1295	922.82	0.71		
Total	1298	934.54			

13.8 Smoking and personality: A study about smoking and personality (by A. Terracciano and P. Costa, *Addiction*, vol. 99, pp. 472–481, 2004) used a sample of 1638 adults in the Baltimore Longitudinal Study on Aging. The subjects formed three groups according to smoking status (never, former, current). Each subject completed a personality questionnaire that provided scores on various personality scales designed to have overall means of about 50 and standard deviations of about 10. The table shows some results for three traits, giving the means with standard deviations in parentheses.

	Never smokers (n = 828)	Former smokers (n = 694)	Current smokers (n = 116)	F
Neuroticism	46.7 (9.6)	48.5 (9.2)	51.9 (9.9)	17.77
Extraversion	50.4 (10.3)	50.2 (10.0)	50.9 (9.4)	0.24
Conscientiousness	51.8 (10.1)	48.9 (9.7)	45.6 (10.3)	29.42

a. For the F test for the extraversion scale, using the 0.05 significance level, what conclusion would you make?

b. Refer to (a). Does this mean that the population means are necessarily equal?

■ **13.9 Florida student data:** Refer to the "FL student survey" data file on the text CD. For the response variable, use the number of weekly hours engaged in sports and other physical exercise. For the explanatory variable, use gender.

a. Using software, for each gender construct box plots and find the mean and standard deviation of the response variable.

b. Conduct an ANOVA. Report the hypotheses, F test statistic value, P-value, and interpret results.

c. To compare the means, suppose you instead used the two-sided t test from Section 9.3, which assumes equal population standard deviations. How would the t test statistic and P-value for that test relate to the F test statistic and P-value for ANOVA?

■ **13.10 Software and French ANOVA:** Refer to Exercise 13.3 above. Using software,

a. Create the data file and find the sample means and standard deviations.

b. Find and report the ANOVA table. Interpret the P-value.

c. Change an observation in Group 2 so that the P-value will be smaller. Specify the value you changed, and report the resulting F test statistic and the P-value. Explain why the value you changed would have this effect.

■ **13.11 Comparing therapies for anorexia:** The "anorexia"
TRY data file on the text CD shows weight change for 72 anorexic teenage girls who were randomly assigned to one of three psychological treatments. Use software to analyze these data. (The change scores are given in the file for the control and cognitive therapy groups. You can create the change scores for the family therapy group.)

a. Construct box plots for the three groups. Use these and sample summary means and standard deviations to describe the three samples.

b. For the one-way ANOVA comparing the three mean weight changes, report the test statistic and P-value. Explain how to interpret.

c. State and check the assumptions for the test in (b).

13.2 | How Should We Follow Up an ANOVA F Test?

When an analysis of variance F test has a small P-value, the test does not specify *which* means are different or *how* different they are. In practice, we can estimate differences between population means with confidence intervals.

Confidence Intervals Comparing Pairs of Means

Recall

s^2 is an unbiased estimator of σ^2, regardless of whether H_0 is true in the ANOVA F test. ■

Let s denote the residual standard deviation. This is the square root of the within-groups variance estimate s^2 of σ^2 that is the denominator of the F test statistic in ANOVA. It is also the square root of the mean square error that software reports (the MS in the row of the ANOVA table for Error).

Confidence interval comparing means

For two groups i and j, with sample means $\bar{y}_i$ and $\bar{y}_j$ having sample sizes n_i and n_j, the 95% confidence interval for $\mu_i - \mu_j$ is

$$(\bar{y}_i - \bar{y}_j) \pm t_{.025}s\sqrt{\frac{1}{n_i} + \frac{1}{n_j}}.$$

The t-score from the t table has $df = N - g =$ total sample size – no. groups.

The df value of $N - g$ for the t-score is also df_2 for the F test. This is the df for the MS error. For $g = 2$ groups, $N = n_1 + n_2$ and $df = N - g = (n_1 + n_2 - 2)$. This confidence interval is then identical to the one introduced in Section 9.3 for $(\mu_1 - \mu_2)$ based on a pooled standard deviation. In the context of follow-up of the ANOVA F test by forming this confidence interval to compare a pair of means, some software (such as MINITAB) refers to this as the **Fisher method**.

When the confidence interval does not contain 0, we can infer that the population means are different. The interval shows just how different they may be. Software can also report confidence intervals for the individual means.

EXAMPLE 5

Number of Good Friends and Happiness

Picture the Scenario

Chapter 10 investigated the association between happiness and several categorical variables, using data from the General Social Survey. The respondents indicated whether they were very happy, pretty happy, or not too happy. Is happiness associated with having lots of friends? A recent GSS asked, "About how many good friends do you have?" Here, we could treat either happiness (variable HAPPY in GSS) or number of good friends (NUMFREND in GSS) as the response variable. If we choose number of good friends, then we are in the ANOVA setting, having a quantitative response variable and a categorical explanatory variable (happiness).

For each happiness category, Table 13.3 shows the sample mean and standard deviation for the number of good friends. It also shows the ANOVA table for the *F* test comparing the population means. The small P-value of 0.03 suggests that at least two of the three population means are different.

TABLE 13.3: Summary of ANOVA for Comparing Mean Number of Good Friends for Three Happiness Categories

The analysis is based on GSS data.

	Very happy	Pretty happy	Not too happy
Mean	10.4	7.4	8.3
Standard deviation	17.8	13.6	15.6
Sample size	276	468	87

Source	DF	SS	MS	F	P
Group	2	1626.8	813.4	3.47	0.032
Error	828	193900.9	234.2		
Total	830	195527.7			

Question to Explore

Use 95% confidence intervals to compare the population mean number of good friends for the three pairs of happiness categories—very happy with pretty happy, very happy with not too happy, and pretty happy with not too happy.

Think It Through

From Table 13.3, the MS error is 234.2. The residual standard deviation is $s = \sqrt{234.2} = 15.3$, with $df = 828$ (listed in the row for the MS error). For a 95% confidence interval with $df = 828$, software reports $t_{.025} = 1.963$. For comparing the very happy and pretty happy categories, the confidence interval for $\mu_1 - \mu_2$ is

$$(\bar{y}_1 - \bar{y}_2) \pm t_{.025}\, s\sqrt{\frac{1}{n_1} + \frac{1}{n_2}} = (10.4 - 7.4) \pm 1.963(15.3)\sqrt{\frac{1}{276} + \frac{1}{468}}$$

which is 3.0 ± 2.3, or $(0.7, 5.3)$.

We infer that the population mean number of good friends is between about 1 and 5 higher for those who are very happy than for those who are pretty happy. Since the confidence interval contains only positive numbers, this suggests that $\mu_1 - \mu_2 > 0$; that is, μ_1 exceeds μ_2. On the average, people who are very happy have more good friends than people who are pretty happy.

For the other comparisons, you can find

Very happy, not too happy: 95% CI for $\mu_1 - \mu_3$ is
$(10.4 - 8.3) \pm 3.7$, or $(-1.6, 5.8)$.

Pretty happy, not too happy: 95% CI for $\mu_2 - \mu_3$ is
$(7.4 - 8.3) \pm 3.5$, or $(-4.4, 2.6)$.

These two confidence intervals contain 0. So, there's not enough evidence to conclude that a difference exists between μ_1 and μ_3 or between μ_2 and μ_3.

Insight

The confidence intervals are quite wide, even though the sample sizes are fairly large. This is because the sample standard deviations (and hence s) are large. Table 13.3 reports that the sample standard deviations are larger than the sample means, suggesting that the three distributions are skewed to the right. The margin figure shows a box plot of the overall sample data distribution of number of good friends, except for many large outliers. It would also be sensible to compare the median responses, but these are not available at the GSS website. Do nonnormal population distributions invalidate this inferential analysis? We'll discuss this next.

1 3 4   8   16
Number of Good Friends

TO PRACTICE THIS CONCEPT, TRY EXERCISE 13.12

Recall

From Section 9.2, a 95% confidence interval for $\mu_2 - \mu_1$ using separate standard deviations s_1 and s_2 is

$$(\bar{y}_1 - \bar{y}_2) \pm t_{.025}\sqrt{\frac{s_1^2}{n_1} + \frac{s_2^2}{n_2}}.$$

Software supplies the df value and the confidence interval. ■

What Are the Effects of Violating Assumptions?

The t confidence intervals have the same assumptions as the ANOVA F test: (1) normal population distributions with (2) identical standard deviations and (3) data that resulted from randomization. These inferences also are not highly dependent on the normality assumption, especially when the sample sizes are large, such as in Example 5 above. When the standard deviations are quite different, with the ratio of the largest to smallest exceeding about 2, it is preferable to use the confidence interval formula from Section 9.2 (see the margin Recall box) that uses *separate* standard deviations for the groups rather than a single pooled value s. That approach does not assume equal standard deviations.

In Example 5, the sample standard deviations are not very different (ranging from 13.6 to 17.8), and the GSS is a multistage random sample with properties similar to a simple random sample. Thus, assumptions (2) and (3) are reasonably well satisfied. The sample sizes are fairly large, so assumption (1) of normality is not crucial. It's justifiable to use ANOVA and follow-up confidence intervals with these data.

Controlling Overall Confidence with Many Confidence Intervals

With g groups, there are $g(g - 1)/2$ pairs of groups to compare. With $g = 3$, for instance, there are $g(g - 1)/2 = 3(2)/2 = 3$ comparisons: Group 1 with group 2, group 1 with group 3, and group 2 with group 3.

The confidence interval method just discussed is mainly used when *g* is small or when only a few comparisons are of main interest. When there are many groups, the number of comparisons can be large. For example, when $g = 10$, there are $g(g - 1)/2 = 45$ pairs of means to compare. If we plan to construct 95% confidence intervals for these comparisons, an error probability of 0.05 applies to *each* comparison. On the average, $45(0.05) = 2.25$ of the confidence intervals would *not* contain the true differences of means.

For 95% confidence intervals, the confidence level of 0.95 is the probability that *any particular* confidence interval that we plan to construct will contain the parameter. The probability that *all* the confidence intervals will contain the parameters is considerably smaller than the confidence level for any particular interval. How can we construct the intervals so that the 95% confidence extends to the *entire set* of intervals rather than to *each single* interval? Methods that control the probability that *all* confidence intervals will contain the true differences in means are called **multiple comparison methods**. For these methods, *all* intervals are designed to contain the true parameters *simultaneously* with an overall fixed probability.

Multiple comparisons for comparing all pairs of means

Multiple comparison methods compare pairs of means with a confidence level that applies simultaneously to the entire set of comparisons rather than to each separate comparison.

The simplest multiple comparison method uses the confidence interval formula from the beginning of this section. However, for each interval it uses a *t*-score for a more stringent confidence level. This ensures that the overall confidence level is acceptably high. The desired overall error probability is split into equal parts for each comparison. Suppose we want a confidence level of 0.95 that *all* confidence intervals will be simultaneously correct. If we plan five confidence intervals comparing means, then the method uses error probability $0.05/5 = 0.01$ for each one; that is, a 99% confidence level for each separate interval. This approach ensures that the overall confidence level is *at least* 0.95. (It is actually slightly larger.) Called the **Bonferroni** method, it is based on a special case of a probability theorem shown by an Italian probabilist, Carlo Bonferroni, in 1936 (Exercise 13.69).

We shall instead use the **Tukey method**. It is designed to give overall confidence level *very close* to the desired value (such as 0.95), and it has the advantage that its confidence intervals are slightly narrower than the Bonferroni intervals. The Tukey method is more complex, using a sampling distribution pertaining to the difference between the largest and smallest of the *g* sample means. We do not present its formula, but it is easy to obtain with software.

Recall

John Tukey was responsible for many statistical innovations, including box plots and other methods of **exploratory data analysis**. ■

EXAMPLE 6

Multiple Comparison Intervals for Number of Good Friends

Picture the Scenario

Example 5 compared the population mean numbers of good friends, for three levels of reported happiness. There, we constructed a *separate* 95% confidence interval for the difference between each pair of means. Table 13.4 displays these three intervals. It also displays the confidence intervals that software reports for the Tukey multiple comparison method.

TABLE 13.4: Multiple Comparisons of Mean Number of Good Friends for Three Happiness Categories			

An asterisk[*] indicates a significant difference, with the confidence interval not containing 0.

Groups	Difference of means	Separate 95% CIs	Tukey 95% Multiple Comparison CIs
(Very happy, Pretty happy)	$\mu_1 - \mu_2$	$(0.7, 5.3)^*$	$(0.3, 5.7)^*$
(Very happy, Not too happy)	$\mu_1 - \mu_3$	$(-1.6, 5.8)$	$(-2.6, 6.5)$
(Pretty happy, Not too happy)	$\mu_2 - \mu_3$	$(-4.4, 2.6)$	$(-5.1, 3.3)$

Question to Explore

a. Explain how the Tukey multiple comparison confidence intervals differ from the separate confidence intervals in Table 13.4.

b. Summarize results shown for the Tukey multiple comparisons.

Think It Through

a. The Tukey confidence intervals hold with *overall* confidence level about 95%. This confidence applies to the entire set of three intervals. The Tukey confidence intervals are wider than the separate 95% confidence intervals because the multiple comparison approach uses a higher confidence level for each separate interval to ensure achieving the overall confidence level of 95% for the entire set of intervals.

b. The Tukey confidence interval for $\mu_1 - \mu_2$ contains only positive values, so we infer that $\mu_1 > \mu_2$. The mean number of good friends is higher, although perhaps barely so, for those who are very happy than for those who are pretty happy. The other two Tukey intervals contain 0, so we cannot infer that those pairs of means differ.

Insight

Figure 13.4 summarizes the three Tukey comparisons from Table 13.4. The intervals have different lengths because the group sample sizes are different.

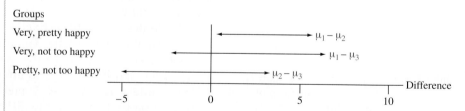

▲ **FIGURE 13.4: Summary of Tukey Comparisons of Pairs of Means.**

TO PRACTICE THIS CONCEPT, TRY EXERCISE 13.15

ANOVA and Regression

ANOVA can be presented as a special case of multiple regression. The factor enters the regression model using *indicator variables*. Each indicator variable takes only two values, 0 and 1, and indicates whether an observation falls in a particular group.

With three groups, we need two indicator variables to indicate the group membership. The first indicator variable is

$$x_1 = 1 \text{ for observations from the first group}$$
$$= 0 \text{ otherwise.}$$

Recall

You can review **indicator variables** in Section 12.5. We used them there to include a categorical explanatory variable in a regression model. ■

The second indicator variable is

$$x_2 = 1 \text{ for observations from the second group}$$
$$= 0 \text{ otherwise.}$$

The indicator variables identify the group to which an observation belongs as follows:

$$\text{Group 1: } x_1 = 1 \text{ and } x_2 = 0$$
$$\text{Group 2: } x_1 = 0 \text{ and } x_2 = 1$$
$$\text{Group 3: } x_1 = 0 \text{ and } x_2 = 0.$$

We don't need a separate indicator variable for the third group. We know an observation is in that group if $x_1 = 0$ and $x_2 = 0$.

With these indicator variables, the multiple regression equation for the mean of y is

$$\mu_y = \alpha + \beta_1 x_1 + \beta_2 x_2.$$

For observations from the third group, $x_1 = x_2 = 0$, and the equation reduces to

$$\mu_y = \alpha + \beta_1(0) + \beta_2(0) = \alpha.$$

So the parameter α represents the population mean of the response variable y for the last group. For observations from the first group, $x_1 = 1$ and $x_2 = 0$, so

$$\mu_y = \alpha + \beta_1(1) + \beta_2(0) = \alpha + \beta_1$$

equals the population mean μ_1 for that group. Similarly, $\alpha + \beta_2$ equals the population mean μ_2 for the second group (let $x_1 = 0$ and $x_2 = 1$).

Since $\alpha + \beta_1 = \mu_1$ and $\alpha = \mu_3$, the difference between the means

$$\mu_1 - \mu_3 = (\alpha + \beta_1) - \alpha = \beta_1.$$

That is, the coefficient β_1 of the first indicator variable represents the difference between the first mean and the last mean. Likewise, $\beta_2 = \mu_2 - \mu_3$. The beta coefficients of the indicator variables represent differences between the mean of each group and the mean of the last group. Table 13.5 summarizes the parameters of the regression model and their correspondence with the three population means.

TABLE 13.5: Interpretation of Coefficients of Indicator Variables in Regression Model $\mu_y = \alpha + \beta_1 x_1 + \beta_2 x_2$

The indicator variables represent a categorical predictor with three categories specifying three groups.

Group	Indicator x_1	x_2	Mean of y	Interpretation of β
1	1	0	$\mu_1 = \alpha + \beta_1$	$\beta_1 = \mu_1 - \mu_3$
2	0	1	$\mu_2 = \alpha + \beta_2$	$\beta_2 = \mu_2 - \mu_3$
3	0	0	$\mu_3 = \alpha$	

How Can We Use Regression for the ANOVA Comparison of Means?

For three groups, the null hypothesis for the ANOVA *F* test is $H_0: \mu_1 = \mu_2 = \mu_3$. If H_0 is true, then $\mu_1 - \mu_3 = 0$ and $\mu_2 - \mu_3 = 0$. In the multiple regression model

$$\mu_y = \alpha + \beta_1 x_1 + \beta_2 x_2$$

Recall

Section 12.3 introduced the *F* test that all the beta parameters in a multiple regression model equal 0. ■

with indicator variables, recall that $\mu_1 - \mu_3 = \beta_1$ and $\mu_2 - \mu_3 = \beta_2$. Therefore, the ANOVA null hypothesis $H_0: \mu_1 = \mu_2 = \mu_3$ is equivalent to $H_0: \beta_1 = \beta_2 = 0$ in the regression model. If the beta parameters in the regression model all equal 0, then the mean of the response variable equals α for each group. We can perform the ANOVA test comparing means using the F test of $H_0: \beta_1 = \beta_2 = 0$ for this regression model.

EXAMPLE 7

Regression Analysis of Telephone Holding Times

Picture the Scenario

Let's return to the data we analyzed in Examples 1–4 on telephone holding times for callers to an airline for which the recorded message is an advertisement, Muzak, or classical music.

Questions to Explore

a. Set up indicator variables to use regression to model the mean holding times with the type of recorded message as explanatory variable.

b. Table 13.6 shows a portion of a MINITAB printout for fitting this model. Use it to find the estimated mean holding time for the advertisement recorded message.

c. Use Table 13.6 to conduct the ANOVA F test (which Example 3 had shown).

TABLE 13.6: Printout for Regression Model $\mu_y = \alpha + \beta_1 x_1 + \beta_2 x_2$ for Telephone Holding Times and Type of Recorded Message

The indicator variables are x_1 for the advertisement and x_2 for Muzak.

Predictor	Coef	SE Coef	T	P
Constant	10.400	1.523	6.83	0.000
x1	−5.000	2.154	−2.32	0.039
x2	−7.600	2.154	−3.53	0.004

Analysis of Variance

Source	DF	SS	MS	F	P
Regression	2	149.20	74.60	6.43	0.013
Residual Error	12	139.20	11.60		
Total	14	288.40			

Think It Through

a. The factor (type of recorded message) has three categories—advertisement, Muzak, and classical music. We set up indicator variables x_1 and x_2 with

$$x_1 = 1 \text{ for the advertisement (and 0 otherwise),}$$
$$x_2 = 1 \text{ for Muzak (and 0 otherwise),}$$
$$\text{so } x_1 = x_2 = 0 \text{ for classical music.}$$

The regression model for the mean of $y =$ telephone holding time is then

$$\mu_y = \alpha + \beta_1 x_1 + \beta_2 x_2.$$

b. From Table 13.6, the prediction equation is

$$\hat{y} = 10.4 - 5.0x_1 - 7.6x_2.$$

For the advertisement, $x_1 = 1$ and $x_2 = 0$, so the estimated mean is $\hat{y} = 10.4 - 5.0(1) - 7.6(0) = 5.4$. This is the sample mean for the five subjects in that group.

c. From Table 13.6, the *F* test statistic for testing

$$H_0: \beta_1 = \beta_2 = 0$$

is $F = 6.43$, with $df_1 = 2$ and $df_2 = 12$. This null hypothesis is equivalent to

$$H_0: \mu_1 = \mu_2 = \mu_3.$$

Table 13.6 reports a P-value of 0.013. The regression approach provides the same *F* test statistic and P-value as the ANOVA did in Table 13.2 in Example 3.

Insight

Testing that the beta coefficients equal zero is a way of testing that the population means are equal. Likewise, so do confidence intervals for those coefficients give us confidence intervals for differences between means. For instance, since $\beta_1 = \mu_1 - \mu_3$, a confidence interval for β_1 is also a confidence interval comparing μ_1 and μ_3. From Table 13.6, the estimate -5.0 of β_1 has $se = 2.154$. Since $df = 12$ for that interval, $t_{.025} = 2.179$, and a 95% confidence interval for $\mu_1 - \mu_3$ is

$$-5.0 \pm 2.179(2.154), \text{ or } -5.0 \pm 4.7, \text{ which is } (-9.7, -0.3).$$

This agrees with the 95% confidence interval you would obtain using the difference between the sample means and its standard error.

TO PRACTICE THIS CONCEPT, TRY EXERCISE 13.17

Table 13.2: ANOVA Table

Source	DF	SS	MS	F	P
Group	2	149.2	74.6	6.4	0.013
Error	12	139.2	11.6		
Total	14	288.4			

Table 13.6: ANOVA Table for Regression Model

Source	DF	SS	MS	F	P
Regression	2	149.2	74.6	6.4	.013
Resid.	12	139.2	11.6		
Total	14	288.4			

Notice the similarity between the ANOVA table for comparing means (Table 13.2) and the ANOVA table for regression (Table 13.6), both shown again in the margin. The "between-groups sum of squares" for ordinary ANOVA is the "regression sum of squares" for the regression model. This is the variability explained by the indicator variables for the groups. The "error sum of squares" for ordinary ANOVA is the residual error sum of squares for the regression model. This represents the variability within the groups. This sum of squares divided by its degrees of freedom is the mean square error (MS in the error row), which is also the within-groups estimate = 11.6 of the variance of the observations within the groups. The regression mean square is the between-groups estimate = 74.6. The ratio of the regression mean square to the mean square error is the *F* test statistic ($F = 6.4$).

So far, this chapter has shown how to compare groups for a single factor. This is **one-way ANOVA**. Sometimes the groups to compare are the cells of a cross-classification of two or more factors. For example, the four groups (employed men, employed women, unemployed men, unemployed women) result from cross classifying employment status and gender. The next section presents **two-way ANOVA**, the procedure for comparing the mean of a quantitative response variable across categories of each of two factors.

13.12 House prices and region: For the "house selling prices" data file on the text CD, the output shows the result of conducting an ANOVA comparing mean house selling prices (in $1000) by the region (NW, NE, SW, SE) of the city (Gainesville, Florida) in which it is located. It also shows a summary table of means and standard deviations of the selling prices, by region. (There is no standard deviation reported for the SE region, because you cannot find a standard deviation when there is only one observation.) Consider the NW and the NE, where 92 of the 100 observations fell.

a. Using information given in the tables, show how to construct a 95% confidence interval comparing the corresponding population means.

b. Interpret the confidence interval.

Region	N	Mean	StDev
NE	17	103.21	55.62
NW	75	133.32	54.93
SE	1	22.50	*
SW	7	127.70	54.95

Source	DF	SS	MS	F	P
Region	3	23534	7845	2.59	0.057
Error	96	290899	3030		
Total	99	314433			

13.13 Religiosity and number of good friends: In 1998 the GSS asked, "How often do you attend religious services?" The display shows results of ANOVA for comparing three levels of this variable on the mean number of good friends. The number of times attending religious services was at least two to three times a month for the "high" group, several times a year to about once a month for the "medium" group, and at most once a year for the "low" group. Construct a 95% confidence interval to compare the population mean number of good friends for the high and low religiosity groups. Interpret the interval.

Summary of ANOVA for mean number of good friends and religiosity

	High	Medium	Low
Mean	12.1	6.4	6.2
Sample size	337	159	330

Source	DF	SS	MS	F	P
Group	2	6748.2	3374.1	14.2	0.000
Error	823	196136.9	238.3		
Total	825	202885.1			

13.14 Comparing telephone holding times: Examples 2 and 3 analyzed whether telephone callers to an airline would stay on hold different lengths of time, on the average, if they heard (a) an advertisement about the airline, (b) Muzak, (c) classical music. The sample means were 5.4, 2.8, and 10.4, with $n_1 = n_2 = n_3 = 5$. The ANOVA test had $F = 74.6/11.6 = 6.4$ and a P-value of 0.013.

a. Show that a 95% confidence interval comparing the population mean times that callers are willing to remain on hold for classical music and Muzak is (2.9, 12.3).

b. The margin of error was 4.7 for this comparison. Without doing a calculation, explain why the margin of error is 4.7 for comparing *each* pair of means.

c. The 95% confidence intervals are (0.3, 9.7) for $\mu_3 - \mu_1$ and (−2.1, 7.3) for $\mu_1 - \mu_2$. Interpret the three confidence intervals, and summarize what the airline company learned from this study.

d. The confidence intervals are wide. In the design of this experiment, what could you change to estimate the differences in means more precisely?

13.15 Tukey holding time comparisons: Refer to the previous exercise. We could instead use the Tukey method to construct multiple comparison confidence intervals. The Tukey confidence intervals having *overall* confidence level 95% have margins of error of 5.7, compared to 4.7 for the separate 95% confidence intervals in the previous exercise.

a. According to this method, which groups are significantly different?

b. Why are the margins of error larger than with the separate 95% intervals?

13.16 REM sleep: A psychologist compares the mean amount of time of rapid-eye movement (REM) sleep for subjects under three conditions. She randomly assigns 12 subjects to the three groups, 4 per group. The sample means for the three groups were 18, 15, and 12. The table shows the ANOVA table from SPSS.

REM sleep

Source	DF	SS	MS	F	P
Group	2	72.00	36.00	0.79	0.481
Error	9	408.00	45.33		
Total	11	480.00			

a. Report and interpret the P-value for the ANOVA *F* test.

b. For the Tukey 95% multiple comparison confidence intervals comparing each pair of means, the margin of error for each interval is 13.3. True or false: Since all the confidence intervals contain 0, it is plausible that all three population means equal 0.

c. Would the margin of error for each separate 95% confidence interval be less than 13.3, equal to 13.3, or larger than 13.3? Explain why.

13.17 REM regression: Refer to the previous exercise.

a. Set up indicator variables for a regression model so that an *F* test for the regression parameters is equivalent to the ANOVA test comparing the three means.

b. Express the null hypothesis both in terms of population means and in terms of regression parameters for the model in (a).

c. The sample means were 18, 15, and 12. Explain why the prediction equation is $\hat{y} = 12 + 6x_1 + 3x_2$. Interpret the three parameter estimates in this equation.

13.18 Outsourcing satisfaction: Exercise 13.5 showed an ANOVA for comparing mean customer satisfaction scores for three service centers. The sample means on a scale of 0 to 10 were 7.60 in San Jose, 7.80 in Toronto, and 7.10 in Bangalore. Each sample size = 100, MS error = 0.47, and the F test statistic = 27.6 has P-value < 0.001.

a. Explain why the margin of error for separate 95% confidence intervals is the same for comparing the population means for each pair of cities. Show that this margin of error is 0.19.

b. Find the 95% confidence interval for the difference in population means for each pair of service centers. Interpret.

c. The margin of error for Tukey 95% multiple comparison confidence intervals for comparing the service centers is 0.23. Construct the intervals. Interpret.

d. Why are the confidence intervals different in (b) and in (c)? What is an advantage of using the Tukey intervals?

13.19 Regression for outsourcing: Refer to the previous exercise.

a. Set up indicator variables to represent the three service centers.

b. The prediction equation is $\hat{y} = 7.1 + 0.5x_1 + 0.7x_2$. Show how the terms in this equation relate to the sample means of 7.6 for San Jose, 7.8 for Toronto, and 7.1 for Bangalore.

13.20 Advertising effect on sales: Each of 100 restaurants in a fast-food chain is randomly assigned one of four media for an advertising campaign: A = radio, B = TV, C = newspaper, D = mailing. For each restaurant, the observation is the change in sales, defined as the difference between the sales for the month during which the advertising campaign took place and the sales in the same month a year ago (in thousands of dollars).

a. By creating indicator variables, write a regression equation for the analysis to compare mean change in sales for the four media.

b. Explain how you could use the regression model to test the null hypothesis of equal population mean change in sales for the four media.

c. The prediction equation is $\hat{y} = 35 + 5x_1 - 10x_2 + 2x_3$. Estimate the difference in mean change in sales for media (i) A and D, (ii) A and B.

13.21 French ANOVA: Refer to Exercise 13.3, with data shown again below. Using software,

a. Compare the three pairs of means with separate 95% confidence intervals. Interpret.

b. Compare the three pairs of means with Tukey 95% multiple comparison confidence intervals. Interpret, and explain why the intervals are different than in (a).

Group 1	Group 2	Group 3
4	1	9
6	5	10
8		5

13.3 | What if There Are Two Factors?: Two-Way ANOVA

Recall

A **factor** is a categorical explanatory variable. **One-way ANOVA** has a single factor. ▪

One-way ANOVA is a *bivariate* (two-variable) method. It analyzes the relationship between the mean of a quantitative response variable and the groups that are categories of a factor. ANOVA extends to handle two or more factors. With multiple factors, the analysis is *multivariate*. We'll illustrate for the case of two factors. This extension is **two-way ANOVA**. It enables us to study the effect of one factor at a fixed level of a second factor.

The great British statistician R. A. Fisher (see *On the Shoulders Of R.A. Fisher* at the end of Section 7.5) developed ANOVA methods in the 1920s. Agricultural experiments were the source of many of the early ANOVA applications. For instance, ANOVA has often been used to compare the mean yield of a crop for different fertilizers.

EXAMPLE 8

How Does Corn Yield Depend on Amounts of Fertilizer and Manure?

Did You Know?

A **metric ton** is 1000 kilograms, which is about 2200 pounds. A **hectare** is 10,000 square meters (e.g., 100 meters by 100 meters), which is about 2.5 acres. ■

Picture the Scenario

This example presents a typical ANOVA application, based on a recent study at Iowa State University.[2] A large field was portioned into 20 equal-size plots. Each plot was planted with the same amount of corn seed, using a fixed spacing pattern between the seeds. The goal was to study how the yield of corn later harvested from the plots (in metric tons) depended on the levels of use of nitrogen-based fertilizer and manure. Each factor was measured in a binary manner. The fertilizer level was low = 45 kg per hectare or high = 135 kg per hectare. The manure level was low = 84 kg per hectare or high = 168 kg per hectare.

Questions to Explore

a. What are four treatments you can compare with this experiment?

b. What comparisons are relevant when you control for (keep fixed) manure level?

Think It Through

a. Four treatments result from cross-classifying the two binary factors: fertilizer level and manure level. Table 13.7 shows the four treatments, defined for the $2 \times 2 = 4$ combinations of categories of the two factors (fertilizer level and manure level).

TABLE 13.7: Four Groups for Comparing Mean Corn Yield

These result from the two-way cross classification of fertilizer level with manure level.

Manure	Fertilizer	
	Low	High
Low		
High		

b. We can compare the mean corn yield for the two levels of fertilizer, controlling for manure level (that is, at a fixed level of manure use). For fields in which manure level was *low*, we can compare the mean yields for the two levels of fertilizer use. These refer to the first row of Table 13.7. Likewise, for fields in which manure level was *high*, we can compare the mean yields for the two levels of fertilizer use. These refer to the second row of Table 13.7.

Insight

Among the questions we'll learn how to answer in this section are: Does the mean corn yield depend significantly on the fertilizer level? Does it depend on the manure level? Does the effect of fertilizer depend on the manure level?

TO PRACTICE THIS CONCEPT, TRY EXERCISE 13.22

---

[2]Thanks to Dan Nettleton, Iowa State University, for data on which this example is based.

Inference about Effects in Two-Way ANOVA

In two-way ANOVA, a null hypothesis states that the population means are the same in each category of one factor, at each fixed level of the other factor. For example, we could test

H_0: Mean corn yield is equal for plots at the low and high levels of fertilizer, for each fixed level of manure.

Table 13.8a displays a set of population means satisfying this null hypothesis of "no effect of fertilizer level."

TABLE 13.8: Population Mean Corn Yield Satisfying Null Hypotheses: (a) No Effect of Fertilizer Level, (b) No Effect of Manure Level

		Fertilizer				Fertilizer	
	Manure	Low	High		Manure	Low	High
(a)	Low	10	10	(b)	Low	10	20
	High	20	20		High	10	20

We could also test

H_0: Mean corn yield is equal for plots at the low and high levels of manure, for each fixed level of fertilizer.

Table 13.8b displays a set of population means satisfying this null hypothesis of "no effect of manure level." The effects of individual factors tested with these two null hypotheses are called **main effects**. We'll discuss a third null hypothesis later in the section.

As in one-way ANOVA, the F tests of hypotheses in two-way ANOVA assume that

■ The population distribution for each group is normal.

■ The population standard deviations are identical.

■ The data result from a random sample or randomized experiment.

Here, each group refers to a cell in the two-way classification of the two factors. ANOVA procedures still usually work quite well if the population distributions are not normal with identical standard deviations. As in other ANOVA inferences, the quality of the sample is the most important assumption.

The test statistics have complex formulas. We'll rely on software. As in one-way ANOVA, the test for a factor uses two estimates of the variance for each group. These estimates appear in the mean square (MS) column of the ANOVA table.

F test statistics in two-way ANOVA

For testing the main effect for a factor, the test statistic is the ratio of mean squares,

$$F = \frac{\text{MS for the factor}}{\text{MS error}}.$$

The MS for the factor is a variance estimate based on between-groups variation for that factor. The MS error is a within-groups variance estimate that is always unbiased.

When the null hypothesis of equal population means for the factor is true, the F test statistic values tend to fluctuate around 1. When it is false, they tend to be larger. The P-value is the right-tail probability above the observed F value. That is, it is the probability (presuming H_0 is true) of even more extreme results than we observed in the sample.

EXAMPLE 9

Testing the Main Effects for Corn Yield

Picture the Scenario

Let's analyze the relationship between corn yield and the two factors. Table 13.9 shows the data and the sample mean and standard deviation for each group.

TABLE 13.9: Corn Yield by Fertilizer Level and Manure Level

Fertilizer Level	Manure Level	1	2	3	4	5	Sample Size	Mean	Std. Dev.
High	High	13.7	15.8	13.9	16.6	15.5	5	15.1	1.3
High	Low	16.4	12.5	14.1	14.4	12.2	5	13.9	1.7
Low	High	15.0	15.1	12.0	15.7	12.2	5	14.0	1.8
Low	Low	12.4	10.6	13.7	8.7	10.9	5	11.3	1.9

Questions to Explore

a. Summarize the factor effects as shown by the sample means.

b. Table 13.10 is an ANOVA table for two-way ANOVA. Specify the two hypotheses tested, give the test statistics and P-values, and interpret.

TABLE 13.10: Two-Way ANOVA for Corn Yield Data in Table 13.9

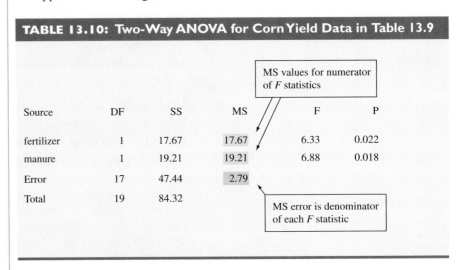

Source	DF	SS	MS	F	P
fertilizer	1	17.67	17.67	6.33	0.022
manure	1	19.21	19.21	6.88	0.018
Error	17	47.44	2.79		
Total	19	84.32			

MS values for numerator of F statistics

MS error is denominator of each F statistic

Think It Through

a. Table 13.9 (with means summarized in the margin) shows that for each manure level, the sample mean yield is higher for the plots using more fertilizer. For each fertilizer level, the sample mean yield is higher for the plots using more manure.

b. First, consider the hypothesis

H_0: Mean corn yield is equal for plots at the low and high levels of fertilizer, for each fixed level of manure

For the fertilizer main effect, Table 13.10 reports that the between-groups estimate of the variance is 17.67. This is the mean square (MS) for fertilizer in the table. The within-group estimate is the MS error, or 2.79. The F test statistic is the ratio,

$$F = 17.67 / 2.79 = 6.33.$$

Recall:

Means from Table 13.9

	Fertilizer	
Manure	Low	High
Low	11.3	13.9
High	14.0	15.1

■

From Table 13.10, the df values are 1 and 17 for the two estimates. From the F distribution with $df_1 = 1$ and $df_2 = 17$, the P-value is 0.022, also reported in Table 13.10. If the population means were equal at the two levels of fertilizer, the probability of an F test statistic value larger than 6.33 would be only 0.022. There is strong evidence that the mean corn yield depends on fertilizer level.

Next, consider the hypothesis

H_0: Mean corn yield is equal for plots at the low and high levels of manure, for each fixed level of fertilizer.

For the manure main effect, the F test statistic is $F = 19.21 / 2.79 = 6.88$. From Table 13.10, $df_1 = 1$ and $df_2 = 17$, and the P-value is 0.018. There is strong evidence that the mean corn yield also depends on the manure level.

Insight

As with any significance test, the information gain is limited. We do not learn *how large* the fertilizer and manure effects are on the corn yield. We can use confidence intervals to investigate the sizes of the main effects. We'll now learn how to do this by using regression modeling with indicator variables.

TO PRACTICE THIS CONCEPT, TRY EXERCISE 13.27

Regression Model with Indicator Variables for Two-Way ANOVA

Let f denote an indicator variable for fertilizer level and let m denote an indicator variable for manure level. Specifically,

$$f = 1 \text{ for plots with high fertilizer level}$$
$$= 0 \text{ for plots with low fertilizer level}$$
$$m = 1 \text{ for plots with high manure level}$$
$$= 0 \text{ for plots with low manure level}$$

The multiple regression model for the mean corn yield with these two indicator variables is

$$\mu_y = \alpha + \beta_1 f + \beta_2 m.$$

To find the population means for the four groups, we substitute the possible combinations of values for the indicator variables. For example, for plots that have high fertilizer level ($f = 1$) and low manure level ($m = 0$), the mean corn yield is

$$\mu_y = \alpha + \beta_1(1) + \beta_2(0) = \alpha + \beta_1.$$

Table 13.11 shows the four means. The difference between the means at the high and low levels of fertilizer equals β_1 for each manure level. That is, the coefficient β_1 of the indicator variable f for fertilizer level equals the difference between the means at its high and low levels, controlling for manure level. The null hypothesis of no fertilizer effect is H_0: $\beta_1 = 0$. Likewise, β_2 is the difference between the means at the high and low levels of manure, for each fertilizer level.

TABLE 13.11: Population Mean Corn Yield for Fertilizer and Manure Levels

Fertilizer	Manure	Indicator variables f	m	Mean of y
High	High	1	1	$\alpha + \beta_1 + \beta_2$
High	Low	1	0	$\alpha + \beta_1$
Low	High	0	1	$\alpha + \beta_2$
Low	Low	0	0	α

We do not need to use regression modeling to conduct the ANOVA F tests. They're easy to do using software. But the modeling approach helps us to focus on estimating the means and the differences among them. We can compare means using an ordinary confidence interval for the regression parameter that equals the difference between those means. The 95% confidence interval has the usual form of

$$\text{parameter estimate} \pm t_{0.025}(se).$$

The df for the t-score is the df value for the MS error.

EXAMPLE 10

Regression Modeling to Estimate and Compare Mean Corn Yields

Picture the Scenario

Table 13.12 shows the result of fitting the regression model for predicting corn yield with indicator variables for fertilizer level and manure level.

TABLE 13.12: Estimates of Regression Parameters for Two-Way ANOVA of the Mean Corn Yield by Fertilizer Level and Manure Level (1 = high, 0 = low)

Predictor	Coef	SE Coef	T	P
Constant	11.6500	0.6470	18.01	0.000
fertilizer	1.8800	0.7471	2.52	0.022
manure	1.9600	0.7471	2.62	0.018

Questions to Explore

a. Use the prediction equation to estimate the mean corn yield for each group.

b. Use a parameter estimate from the prediction equation to compare mean corn yields for the high and low levels of fertilizer, at each manure level.

c. Find a 95% confidence interval comparing the mean corn yield at the high and low levels of fertilizer, controlling for manure level. Interpret it.

Think It Through

a. From Table 13.12, the prediction equation is (rounding to one decimal place)

$$\hat{y} = 11.6 + 1.9f + 2.0m.$$

The y-intercept equals 11.6. This is the estimated mean yield (in metric tons per hectare) when both indicator variables equal 0, that is, with fertilizer and manure at the low levels. The estimated means for the other cases result from substituting values for the indicator variables. For instance, at fertilizer level = high and manure level = low, $f = 1$ and $m = 0$, so the estimated mean yield is $\hat{y} = 11.6 + 1.9(1) + 2.0(0) = 13.5$. Doing this for all four groups, we get

Manure	Fertilizer	
	Low	High
Low	11.6	$11.6 + 1.9 = 13.5$
High	$11.6 + 2.0 = 13.6$	$11.6 + 1.9 + 2.0 = 15.5$

b. The coefficient of the fertilizer indicator variable f is 1.9. This is the estimated difference in mean corn yield between the high and low levels of fertilizer, for each level of manure (for instance, $13.5 - 11.6 = 1.9$ when manure level = low).

c. The estimate of the fertilizer effect is 1.9. Its standard error, reported in Table 13.12, is 0.747. From Table 13.10, the df for the MS error is 17. From Table B, $t_{.025} = 2.11$ when $df = 17$. The 95% confidence interval is

$$1.9 \pm 2.11(0.747), \text{ which is } (0.3, 3.5).$$

At each manure level, we estimate that the mean corn yield is between 0.3 and 3.5 metric tons per hectare higher at the high fertilizer level than at the low fertilizer level. The confidence interval contains only positive values, reflecting the conclusion that the mean yield is significantly higher at the higher level of fertilizer. This agrees with the P-value falling below 0.05 in the test for the fertilizer effect.

Insight

The *estimated* means are not the same as the *sample* means in Table 13.9. (Both sets are shown again in the margin.) The model *smooths* the sample means so that the difference between the estimated means for two categories of a factor is *exactly* the same at each category of the other factor. For example, from the estimated group means found above, the fertilizer effect of $1.9 = 13.5 - 11.6 = 15.5 - 13.6$.

Estimated means

Manure	Fertilizer	
	Low	High
Low	11.6	13.5
High	13.6	15.5

Table 13.9: Sample Means

Manure	Fertilizer	
	Low	High
Low	11.3	13.9
High	14.0	15.1

TO PRACTICE THIS CONCEPT, TRY EXERCISE 13.28

The regression model assumes that the difference between means for the two categories for one factor is the same in each category of the other factor. The next section shows how to check this assumption. When it is reasonable, we can use a single comparison, rather than a separate one at each category of the other variable. In Example 10, we estimated the difference in mean corn yield between the pair of fertilizer levels, a single comparison (with estimate 1.9) holding at each level of manure.

Recall

Section 12.5 introduced the concept of **interaction** between two explanatory variables in their effects on a response variable. In a regression context, **no interaction** implied parallel lines (common slopes). ■

Exploring Interaction between Factors in Two-Way ANOVA

Investigating whether **interaction** occurs is important whenever we analyze multivariate relationships. *No interaction* between two factors means that the effect of either factor on the response variable is the same at each category of the other factor. The above regression model and the ANOVA tests of main effects assume there is no interaction between the factors. What does this mean in this context?

From Example 10, the estimated mean corn yields for the regression model having a indicator variable for fertilizer level and an indicator variable for manure level are

	Fertilizer	
Manure	Low	High
Low	11.6	13.5
High	13.6	15.5

What pattern do these show? Let's plot the means for the two fertilizer levels, within each level of manure. Figure 13.5 shows a plot in which the *y*-axis gives estimated mean corn yields, and points are shown for the four fertilizer-manure combinations. The horizontal axis is not a numerical scale but merely lists the two fertilizer levels. The drawn lines connect the means for the two fertilizer levels, for a given manure level. The absence of interaction is indicated by the *parallel lines*.

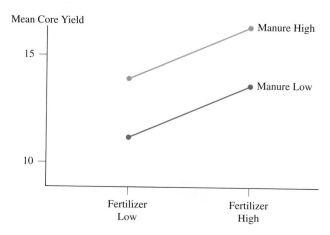

▲ **FIGURE 13.5: Mean Corn Yield, by Fertilizer and Manure Levels, Showing No Interaction.** The parallel lines reflect an absence of interaction. This implies that the difference in estimated means between the two fertilizer levels is the same for each manure level. **Question:** Is it also true that the difference in estimated means between the two manure levels is the same for each fertilizer level?

The parallelism of lines occurs because the difference in the estimated mean corn yield between the high and low levels of fertilizer is the same for each manure level. The difference equals 1.9. Also, the difference between the high and low levels of manure in the estimated mean corn yield is 2.0 for each fertilizer level.

By contrast, Table 13.13 and Figure 13.6 show a set of means for which there is interaction. The difference between the high and low levels of fertilizer in the mean corn yield is $14 - 10 = 4$ for low manure and $12 - 16 = -4$ for high manure. Here, the difference in means depends on the manure level: According to these means, it's better to use a high level of fertilizer when the manure level is low but it's better to use a low level of fertilizer when the manure level is high. Similarly, the manure effect differs at the two fertilizer levels; for the low level, it is $16 - 10 = 6$ whereas at the high level, it is $12 - 14 = -2$. The lines in Figure 13.6 are not parallel.

TABLE 13.13: Means that Show Interaction between the Factors in their Effects on the Response

The effect of fertilizer differs according to whether manure level is low or high. See Figure 13.6.

| | **Fertilizer** | |
Manure	Low	High
Low	10	14
High	16	12

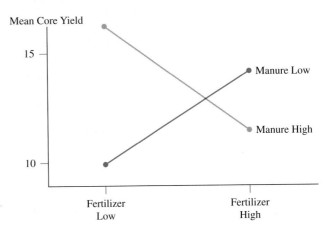

▲ **FIGURE 13.6: Mean Corn Yield, by Fertilizer and Manure Levels, Displaying Interaction. Question:** What aspect of the plot reflects the interaction?

In Table 13.13, suppose the numbers of observations are the same for each group. Then the overall mean corn yield, ignoring manure level, is 13 for each fertilizer level (the average of 10 and 16, and the average of 14 and 12). The overall difference in means between the two fertilizer levels then equals 0. A one-way analysis of mean corn yield by fertilizer level would conclude that fertilizer level has no effect. However, a two-way analysis that allows for interaction would detect that fertilizer has an effect, but that effect differs according to the manure level.

Testing for Interaction

In conducting a two-way ANOVA, *before testing the main effects, it is customary to test a third null hypothesis stating that there is no interaction* between the factors in their effects on the response. The test statistic providing the sample evidence of interaction is

$$F = \frac{\text{MS for interaction}}{\text{MS error}}.$$

When H_0 is false, the F statistic tends to be large. So, as usual, the P-value is the right-tail probability.

EXAMPLE 11

Testing for Interaction with Corn Yield Data

Picture the Scenario

Let's return to our analysis of the corn yield data, summarized in the margin table on the next page. In Example 10 we analyzed these data assuming no interaction. Let's see if that analysis is valid. Table 13.14 is an ANOVA table for a model that allows interaction in assessing the effects of fertilizer level and manure level on the mean corn yield.

Corn Yields			
Fertilizer Level	Manure Level	Mean	Std. Dev.
High	High	15.1	1.3
High	Low	13.9	1.7
Low	High	14.0	1.8
Low	Low	11.3	1.9

TABLE 13.14: Two-Way ANOVA of Mean Corn Yield by Fertilizer Level and Manure Level, Allowing Interaction

Source	DF	SS	MS	F	P
fertilizer	1	17.67	17.67	6.37	0.023
manure	1	19.21	19.21	6.92	0.018
Interaction	1	3.04	3.04	1.10	0.311
Error	16	144.40	2.78		
Total	19	84.32			

F statistic for test of no interaction

Question to Explore

Give the result of the test of H_0: no interaction, and interpret.

Think It Through

The test statistic for H_0: no interaction is

$$F = (\text{MS for interaction})/(\text{MS error}) = 3.04/2.78 = 1.10.$$

Based on the F distribution with $df_1 = 1$ and $df_2 = 16$ for these two mean squares, the ANOVA table reports P-value = 0.31. This is not much evidence of interaction. We would not reject H_0 at the usual significance levels, such as 0.05.

Insight

Since there is not much evidence of interaction, we are justified in conducting the simpler two-way ANOVA about main effects. The tests presented previously in Table 13.12 for effects of fertilizer and manure on mean corn yield are valid.

TO PRACTICE THIS CONCEPT, TRY EXERCISE 13.29

It is not meaningful to test the main effects hypotheses when there is interaction. A small P-value in the test of H_0: no interaction suggests that each factor has an effect, but the size of effect for one factor varies according to the category of the other factor. Then, you should investigate the nature of the interaction by plotting the sample cell means, using a plot like Figure 13.6. You should compare categories of one factor separately at different levels of the other factor.

IN PRACTICE: Check interaction before main effects

In practice, in two-way ANOVA you should *first test the hypothesis of no interaction.* If the evidence of interaction is not strong (that is, if the P-value is not small), then test the main effects hypotheses and/or construct confidence intervals for those effects. But if important evidence of interaction exists, plot and compare the cell means for a factor separately at each category of the other factor.

Recall

From the box at the beginning of Section 13.2, the 95% confidence interval comparing two means is

$$(\bar{y}_i - \bar{y}_j) \pm t_{.025}se, \text{ where } se = s\sqrt{\frac{1}{n_i} + \frac{1}{n_j}},$$

s is the square root of the MS error, and *df* for the *t* distribution is *df* for MS error. ■

For comparing means in two cells using a confidence interval, use the formula from the box at the beginning of Section 13.2, shown again in the margin. Substitute the cell sample sizes for n_i and n_j and use the MS error for the two-way ANOVA that allows interaction.

EXAMPLE 12

Comparing Mean Political Ideology by Gender and Race

Picture the Scenario

In most years, the General Social Survey asks subjects to report their political ideology, measured with seven categories in which 1 = extremely liberal, 4 = moderate, 7 = extremely conservative. Table 13.15 shows results from the 2000 General Social Survey on mean political ideology classified by gender and by race.

TABLE 13.15: Mean Political Ideology by Gender and by Race

Gender	Race	
	Black	White
Female	4.06 ($n = 256$)	4.04 ($n = 1144$)
Male	3.74 ($n = 139$)	4.25 ($n = 973$)

For the test of H_0: no interaction, software reports an F test statistic of 21.7 with $df_1 = 1$ and $df_2 = 2508$, for a P-value of 0.001. So, in comparing females and males on their mean political ideology, we should do it separately by race. The MS error for the model allowing interaction equals 1.996, so the residual standard deviation is $s = \sqrt{1.996} = 1.41$.

Questions to Explore

a. Interpret the significant interaction by comparing sample mean political ideology of females and males for each race descriptively and using 95% confidence intervals.

b. Interpret the confidence intervals derived in (a).

Think It Through

a. The sample means show that for black subjects, females are more conservative (have the higher mean). By contrast, for white subjects, males are more conservative.

For a confidence interval comparing mean political ideology for females and males who are black, the standard error (using the sample sizes reported in the table) is

$$se = s\sqrt{\frac{1}{n \text{ for black females}} + \frac{1}{n \text{ for black males}}} =$$

$$1.41\sqrt{\frac{1}{256} + \frac{1}{139}} = 0.149.$$

The 95% confidence interval is

$$(4.06 - 3.74) \pm 1.96(0.149), \text{ which is } 0.32 \pm 0.29, \text{ or } (0.03, 0.61).$$

Likewise, you can find that the 95% confidence interval comparing mean political ideology for females and males who are white is -0.21 ± 0.12, or $(-0.33, -0.09)$.

b. Since all values in the confidence interval for black subjects are positive, we can infer that black females are *more* conservative than black males in the population. For white subjects, however, all values in the interval are negative. We infer that white females are *less* conservative than white males in the population.

> **Insight**
>
> Both confidence intervals have endpoints that are close to 0. So, the true gender effect in the population could be small for each race.

TO PRACTICE THIS CONCEPT, TRY EXERCISE 13.33

Why Not Instead Perform Two Separate One-Way ANOVAs?

When you have two factors, rather than performing a two-way ANOVA, why not instead perform two separate one-way ANOVAs? For instance, in Example 12 you could compare the mean political ideology for females and males using a one-way ANOVA, ignoring the information about race. Likewise, you could perform a separate one-way ANOVA to compare the means for blacks and whites, ignoring the information about gender.

The main reason is that you learn more from two-way ANOVA. The two-way ANOVA indicates whether there is interaction. When there is, as in Example 12, it is more informative to compare levels of one factor *separately* at each level of the other factor. This enables us to investigate how the effect depends on that other factor. For instance, a one-way ANOVA of mean political ideology by gender might show *no* gender effect, whereas the two-way ANOVA has shown that there *is* a gender effect but it varies by race.

Similarly, in experimental studies, rather than carrying out one experiment to investigate the effect of one factor and a separate experiment to investigate the effect of a second factor, it is better to carry out a *single* experiment and study both factors at the same time. Besides the advantage of being able to check for interaction, this is more cost effective. If we have funds to experiment with 100 subjects, we can use a sample size of 100 for studying each factor with a two-way ANOVA, rather than use 50 subjects in one experiment about one factor and 50 subjects in a separate experiment about the other factor.

Yet another benefit of a two-way ANOVA is that the residual variability, which affects the MS error and the denominators of the F test statistics, tends to decrease. When we use two factors to predict a response variable, we usually tend to get better predictions (that is, less residual variability) than when we use one factor. With less residual (within-groups) variability, we get larger test statistics, and hence greater power for rejecting false null hypotheses.

Factorial ANOVA

The methods of two-way ANOVA extend to the analysis of several factors. A multifactor ANOVA with observations from all combinations of the factors is called **factorial ANOVA**. For example, with three factors, **three-way ANOVA** considers main effects for those factors as well as possible interactions.

IN PRACTICE: Use regression with categorical and quantitative predictors

With several explanatory variables, usually some are categorical and some are quantitative. Then, it is sensible to build a multiple regression model containing both types of predictor. That's what we did in Example 10 in the previous chapter when we modeled house selling prices in terms of house size and the region of town in which the house was located.

13.3 Practicing the Basics

13.22 Reducing cholesterol: An experiment randomly assigns 100 subjects suffering from high cholesterol to one of four groups: low-dose Lipitor, high-dose Lipitor, low-dose Pravachol, high-dose Pravachol. After three months of treatment, the change in cholesterol level is measured.

TRY

a. Identify the response variable and the two factors.

b. What are four treatments that can be compared with this experiment?

c. What comparisons are relevant when we control for dose level?

13.23 Drug main effects: For the previous exercise, show a hypothetical set of population means for the four groups that would have

a. A dose effect but no drug effect.

b. A drug effect but no dose effect.

c. A drug effect and a dose effect.

d. No drug effect and no dose effect.

13.24 Reasons for divorce: The 26 students in a statistics class at the University of Florida were surveyed about their attitudes toward divorce. Each received a response score according to how many from a list of ten possible reasons were regarded as legitimate for a woman to seek a divorce. The higher the score, the more willing the subject is to accept divorce as an option in a variety of circumstances. The students were also asked whether they were "fundamentalist" or "nonfundamentalist" in their religious beliefs and whether their church attendance frequency was "frequent" (more than once a month) or "infrequent." The table displays the data and the results of a two-way ANOVA.

a. State the null hypothesis to which the F test statistic in the Religion row refers.

b. Show how to use mean squares to construct the F test statistic for the religion main effect, report its P-value, and interpret.

		RELIGION	
		Fundamentalist	Non-fundamentalist
CHURCH	Frequent	0, 3, 4, 0, 3, 2, 0, 1, 1, 3	2, 5, 1, 2, 3, 3
ATTENDANCE	Infrequent	4, 3, 4	6, 8, 6, 4, 6, 3, 7, 4

Source	DF	SS	MS	F	P
Religion	1	11.07	11.07	5.20	0.032
Church_attend	1	36.84	36.84	17.32	0.000
Error	23	48.93	2.13		
Total	25	117.12			

13.25 House prices, region, and bedrooms: For the "house selling prices" data file on the text CD, the output shows the result of conducting a two-way ANOVA of house selling prices by the region (NW, NE, SW, SE) of the city (Gainesville, Florida) and the number of bedrooms in the house (1, 2, 3, 4, or 5).

a. For testing the main effect of region, report the F test statistic value, and show how it was formed from other values reported in the ANOVA table.

b. Report the P-value for the main effect test for region, and interpret.

Source	DF	SS	MS	F	P
Region	3	27384	9128	3.59	0.017
Bedrooms	4	57024	14256	5.61	0.000
Error	92	233875	2542		
Total	99	314433			

13.26 Corn and manure: In Example 10, the coefficient of the manure-level indicator variable m is 1.96.

a. Explain why this coefficient is the estimated difference in mean corn yield between the high and low levels of manure, for each level of fertilizer.

b. Explain why the 95% confidence interval for the difference in mean corn yield between the high and low levels of manure is $1.96 \pm 2.11(0.747)$.

13.27 Hang up if message repeated?: Example 2 described an experiment in which telephone callers to an airline were put on hold with an advertisement, Muzak, or classical music in the background. Each caller who was chosen was also randomly assigned to a category of a second factor: Whether the message played was five minutes long or ten minutes long. (In each case, it was repeated at the end.) The table shows the data classified by both factors and the results of a two-way ANOVA.

TRY

Telephone holding times by type of recorded message and repeat time

	Repeat Time	
Message	Ten minutes	Five minutes
Advertisement	8, 11, 2	5, 1
Muzak	1, 4, 3	0, 6
Classical	13, 8, 15	7, 9

Source	DF	SS	MS	F	P
Message	2	149.20	74.60	7.09	0.011
Repeat	1	23.51	23.51	2.24	0.163
Error	11	115.69	10.52		
Total	14	288.40			

a. State the null hypothesis to which the F test statistic in the Message row refers.

b. Show how to use mean squares to construct the F test statistic for the message main effect, report its P-value, and interpret.

c. On what assumptions is this analysis based?

13.28 Regression for telephone holding times: Refer to the previous exercise. Let $x_1 = 1$ for the advertisement and 0 otherwise, $x_2 = 1$ for Muzak and 0 otherwise, and $x_1 = x_2 = 0$ for classical music. Likewise, let $x_3 = 1$ for repeating in 10-minute cycles and $x_3 = 0$ for repeating in 5-minute cycles. The display shows results of a regression of the telephone holding times on these indicator variables.

TRY

Regression for telephone holding times

```
Predictor  Coef    SE Coef   T      P
Constant   8.867   1.776     4.99   0.000
x1         -5.000  2.051    -2.44   0.033
x2         -7.600  2.051    -3.71   0.003
x3          2.556  1.709     1.50   0.163
```

a. State the prediction equation. Interpret the parameter estimates.

b. Find the estimated means for the six groups in the two-way cross-classification of message type and repeat time.

c. Find the estimated difference between the mean holding times for 10-minute repeats and 5-minute repeats, for a fixed message type. How can you get this estimate from a coefficient of an indicator variable in the prediction equation?

13.29 Interaction between message and repeat time?:

TRY Refer to the previous two exercises. When we allow interaction, two of the rows in the ANOVA table are

```
Source        DF  SS      MS     F     P
Group*Repeat  2   15.02   7.51   0.67  0.535
Error         9   100.67  11.19
```
where Group*Repeat denotes the interaction effect.

a. Show the hypotheses, test statistic, and P-value for the test of H_0: no interaction.

b. What is the implication of the result of this test? For instance, the analyses in previous exercises assumed a lack of interaction. Was it valid to do so?

13.30 Income by gender and job type: In 2000, the population mean hourly wage for males was $22 for white-collar jobs, $11 for service jobs, and $14 for blue-collar jobs. For females the means were $15 for white-collar jobs, $8 for service jobs, and $10 for blue-collar jobs.[3]

a. Identify the response variable and the two factors.

b. Show these means in a two-way classification of the two factors.

c. Compare the differences between males and females for (i) white-collar jobs (ii) blue-collar jobs. Explain why there is interaction, and describe it.

d. Show a set of six population mean incomes that would satisfy H_0: no interaction.

13.31 Ideology by gender and race: Refer to Example 12, the sample means from which are shown again below.

Mean political ideology

	Race	
Gender	Black	White
Female	4.06	4.04
Male	3.74	4.25

a. Explain how to obtain the following interpretation for the interaction from the sample means: "For females there is no race effect on ideology. For males, whites are more conservative by about half an ideology category, on the average."

b. Suppose that instead of the two-way ANOVA, you performed a one-way ANOVA with gender as the predictor and a separate one-way ANOVA with race as the predictor. Suppose the ANOVA for gender does not show a significant effect. Explain how this could happen, even though the two-way ANOVA showed a gender effect for each race. (Hint: Will the overall sample means for females and males be more similar than they are for each race?)

c. Refer to (b). Summarize what you would learn about the gender effect from a two-way ANOVA that you would fail to learn from a one-way ANOVA.

13.32 Attractiveness and getting dates: The results in the table are from a study of physical attractiveness and subjective well-being (E. Diener et al., *Journal of Personality and Social Psychology*, Vol. 69, 1995, pp. 120–129). A panel rated a sample of college students on their physical attractiveness. The table presents the number of dates in the past three months for students rated in the top or bottom quartile of attractiveness.

a. Identify the response variable and the factors.

b. Do these data appear to show interaction? Explain.

c. Based on the results in the table, specify one of the ANOVA assumptions that these data violate. Is this the most important assumption?

Dates and attractiveness

ATTRACTIVENESS	NO. DATES, MEN			NO. DATES, WOMEN		
	Mean	Std. Dev	n	Mean	Std. Dev	n
More	9.7	10.0	35	17.8	14.2	33
Less	9.9	12.6	36	10.4	16.6	27

■ 13.33 Diet and weight gain: A randomized experiment[4]

TRY measured weight gain (in grams) of male rats under six diets varying by source of protein (beef, cereal, pork) and level of protein (high, low). Ten rats were assigned to each diet. The data are shown in the table below and are also available in the "Protein and weight gain" data file on the book's CD.

a. Conduct a two-way ANOVA that assumes a lack of interaction. Report the F test statistic and the P-value for testing the effect of the protein level. Interpret.

b. Now conduct a two-way ANOVA that also considers potential interaction. Report the hypotheses, test statistic and P-value for a test of no interaction. What do you conclude at the 0.05 significance level? Explain.

c. Refer to (b). Allowing interaction, construct a 95% confidence interval to compare the mean weight gain for the two protein levels, for the beef source of protein.

---

[3] *Source: The State of Working America 2000–2001*, Economic Policy Institute.

[4] From p. 347 of *Statistical Methods*, 6th edition, by G. Snedecor and W. Cochran (Iowa State Univ. Press, 1967).

Weight gain by source of protein and by level of protein

	High Protein	Low Protein
Beef	73, 102, 118, 104, 81,	90, 76, 90, 64, 86,
	107, 100, 87, 117, 111	51, 72, 90, 95, 78
Cereal	98, 74, 56, 111, 95,	107, 95, 97, 80, 98,
	88, 82, 77, 86, 92	74, 74, 67, 89, 58
Pork	94, 79, 96, 98, 102,	49, 82, 73, 86, 81,
	102, 108, 91, 120, 105	97, 106, 70, 61, 82

■ **13.34 Regression of weight gain on diet:** Refer to the previous exercise.

a. Set up indicator variables for protein source and for protein level, and specify a regression model with effects both of protein level and protein source on weight gain.

b. Fit the model in (a), and explain how to interpret the parameter estimate for the protein level indicator variable.

c. Show how you could test a hypothesis about beta parameters in the model in (a) to analyze the effect of protein source on weight gain.

d. Using the fit of the model, find the estimated mean for each of the six diets. Explain what it means when we say that these estimated means do not allow for interaction between protein level and source in their effects on weight loss.

CHAPTER REVIEW

Answers to the Chapter Figure Questions

Figure 13.1: No, even if the population means are equal, we expect sample means to differ, because of sampling variability. Differences among sample means is not sufficient evidence to conclude that the population means differ.

Figure 13.2: The data in case *b* give stronger evidence against H_0 because case *b* has less variability within each sample than case *a*.

Figure 13.3: Evidence suggesting the normal with equal standard deviation assumptions are violated would be (1) graphical methods showing extreme skew for the response variable or (2) the largest group standard deviation is more than about double the smallest group standard deviation when the sample sizes are unequal.

Figure 13.5: Yes. The parallelism of lines implies that no interaction holds no matter which factor we choose for making comparisons of means.

Figure 13.6: The lines are not parallel.

CHAPTER SUMMARY

Analysis of variance (ANOVA) methods compare several groups according to their means on a quantitative response variable. The groups are categories of categorical explanatory variables. A categorical explanatory variable is also called a **factor**.

■ The **one-way ANOVA** *F* test compares means for a single factor. The groups are specified by categories of a single categorical explanatory variable.

■ **Two-way ANOVA** methods compare means across categories of each of two factors, at fixed levels of the other factor. When there is **interaction** between the factors in their effects, differences between response means for categories of one factor change according to the category of the other factor.

■ **Multiple comparison methods** such as the **Tukey method** compare means for each pair of groups while controlling the overall confidence level.

■ Analysis of variance methods can be conducted by using multiple regression models. The regression model uses **indicator variables** as explanatory variables to represent the factors. Each indicator variable equals 1 for observations from a particular category and 0 otherwise.

■ The ANOVA methods assume randomization and that the population distribution for each group is normal, with the same standard deviation for each group. In practice, *the randomness assumption is the most important*, and ANOVA methods are *robust* to moderate violations of the other assumptions.

CHAPTER PROBLEMS

Practicing the Basics

13.35 Good friends and marital status: Is the number of good friends associated with marital status? For GSS data with marital status measured with the categories (married, widowed, divorced, separated, never married), an ANOVA table reports $F = 0.80$ based on $df_1 = 4, df_2 = 612$.

a. Introduce notation, and specify the null hypothesis and the alternative hypothesis for the ANOVA F test.

b. Based on what you know about the F distribution, would you guess that the test statistic value of 0.80 provides strong evidence against the null hypothesis? Explain.

c. Software reports a P-value of 0.53. Explain how to interpret it.

13.36 Political ideology and party affiliation: From Example 12, the GSS measures political ideology with scores such as 1 = extremely liberal, 4 = moderate, 7 = extremely conservative. In the 2002 GSS, when responses were classified by political party affiliation, results were obtained as shown in the table.

Political ideology by party

Group (Party)	n	Political Ideology Mean	Std. Dev.
Democrat	596	3.65	1.28
Independent	238	3.92	1.33
Republican	465	4.83	1.23

a. Identify the response variable and the factor.

b. Provide notation, and state the hypotheses for applying ANOVA to these data.

c. The ANOVA F statistic equals 113.9, and its P-value is 0.000. Explain how to interpret the result of the test.

13.37 MS error for ideology: The ANOVA table for the analysis in the previous exercise reported a mean square error of 1.62.

a. If we assume that each party has the same variability in political ideology scores, what is the estimated standard deviation for each group?

b. From the table in the previous exercise, explain why the assumption in (a) seems reasonable.

c. Show that the 95% confidence interval for $\mu_3 - \mu_1$ is (1.03, 1.33). Interpret.

13.38 Going to bars and having friends: Do people who go to bars and pubs a lot tend to have more friends? In 1998 the GSS asked, "How often do you go to a bar or tavern?" The table shows results of ANOVA for comparing the mean number of good friends at three levels of this variable. The "very often" group reported going to a bar at least several times a week. The "occasional" group reported going occasionally, but not as often as several times a week.

Summary of ANOVA for mean number of good friends and going to bars

	Very often	Occasional	Never
Mean	12.1	6.4	6.2
Standard dev.	21.3	10.2	14.0
Sample size	41	166	215

Source	DF	SS	MS	F	P
Group	2	1116.8	558.4	3.03	0.049
Error	419	77171.8	184.2		
Total	421	78288.5			

a. State the (i) hypotheses, (ii) test statistic value, (iii) P-value for the significance test displayed in this table. Interpret the P-value.

b. Based on the assumptions you need to make to use the method in (a), is there any aspect of the data summarized here that suggests that the ANOVA test and follow-up confidence intervals may not be appropriate? Explain.

13.39 TV watching: The 2000 General Social Survey asked 1829 subjects how many hours per day they watched TV, on the average. Are there differences in population means according to the race of the subject (white, black, other)? The sample means were 2.75 for whites ($n = 1435$), 4.14 for blacks ($n = 305$), and 2.61 for other ($n = 89$). In a one-way ANOVA, the between-groups estimate of the variance is 249.93 and the within-groups estimate is 6.41.

a. Conduct the ANOVA test and make a decision using 0.05 significance level.

b. The 95% confidence interval comparing the population means is (1.1, 1.7) for blacks and whites, $(-0.4, 0.7)$ for whites and the other category, and (0.9, 2.1) for blacks and the other category. Based on the three confidence intervals, indicate which pairs of means are "significantly different." Interpret.

c. Based on the information given, show how to construct the confidence interval that compares the population mean TV watching for blacks and whites.

d. Refer to (c). Would the corresponding interval formed with the Tukey method be wider, or narrower? Explain why.

13.40 Comparing auto bumpers: A consumer organization compares the sturdiness of three types of front bumpers. In the study, a particular brand of car is driven into a concrete wall at 15 miles per hour. The response is the amount of damage, as measured by the repair costs, in hundreds of dollars. Due to the potentially large costs, the study conducts only two tests with each bumper type, using six cars. The table shows the data and some ANOVA results. Show the (a) assumptions, (b) hypotheses, (c) test statistic and df values, (d) P-value, and (e) interpretation for testing the hypothesis that the true mean repair costs are the same for the three bumper types.

Bumper A		Bumper B		Bumper C
1		2		11
3		4		15

Source	DF	SS	MS	F	P
Bumper	2	148.00	74.00	18.50	0.021
Error	3	12.00	4.00		
Total	5	160.00			

13.41 Compare bumpers: Refer to the previous exercise.

a. Find the margin of error for constructing a 95% confidence interval for the difference between any pair of the true means. Interpret by showing which pairs of bumpers (if any) are significantly different in their true mean repair costs.

b. For Tukey 95% multiple comparison confidence intervals, the margin of error is 8.4. Explain the difference between confidence intervals formed with this method and separate confidence intervals formed with the method in (a).

c. Set up indicator variables for a multiple regression model including bumper type.

d. The prediction equation for (c) is $\hat{y} = 13 - 11x_1 - 10x_2$. Explain how to interpret the three parameter estimates in this model, and show how these estimates relate to the sample means for the three bumpers.

13.42 Segregation by region: Studies of the degree of residential racial segregation often use the *segregation index*. This is the percentage of nonwhites who would have to change the block on which they live in order to produce a fully nonsegregated city—one in which the percentage of nonwhites living in each block is the same for all blocks in the city. This index can assume values ranging from 0 to 100, with higher values indicating greater segregation. (The national average in 2000 for all metropolitan areas considered was 65.5.) The table shows the index for a sample of cities in 2000, classified by region.

Segregation index

Northeast		North Central		South		West	
Buffalo	77	Minneapolis	58	New Orleans	68	San Francisco	60
Newark	80	Detroit	85	Tampa	63	Dallas	59
Philadelphia	72	Chicago	80	Miami	69	Los Angeles	66
Pittsburgh	67	Milwaukee	82	Atlanta	65	Houston	66

Source: Racial and Ethnic Residential Segregation in the United States, 1980–2000, U.S. Census Bureau Series CENSR-3, 2002.

a. Report the mean and standard deviation for each of the four regions.

b. Define notation, and state the hypotheses for one-way ANOVA.

c. Report the F test statistic and its P-value. What do you conclude about the mean segregation indices for the four regions?

d. Suppose we took these data from the Census Bureau report by choosing only the cities in which we know people. Is the ANOVA valid? Explain.

13.43 Compare segregation means: Refer to the previous exercise.

a. Using software, find the margin of error that pertains to each comparison using the Tukey method for 95% multiple comparison confidence intervals.

b. Using (a), determine which pairs of means, if any, are significantly different.

13.44 Georgia political ideology: The "Georgia student survey" file on the text CD asked students their political party affiliation (1 = Democrat, 2 = Republican, 3 = Independent and their political ideology (on a scale from 1 = very liberal to 7 = very conservative). The table shows results of an ANOVA, with political ideology as the response variable.

Georgia political ideology and party affiliation

Affiliation	N	Political ideology	
		Mean	StDev
Democrat	8	2.6250	1.0607
Republican	36	5.5000	1.0000
Independent	15	3.4667	0.9155

Source	DF	SS	MS	F	P
PoliticalAff	2	79.629	39.814	40.83	0.000
Error	56	54.608	0.975		
Total	58	134.237			

a. Does the ANOVA assumption of equal population standard deviations seem plausible, or is it so badly violated that ANOVA is inappropriate?

b. Are the population distributions normal? Why or why not? Which is more important, the normality assumption or the assumption that the groups are random samples from the population of interest?

c. The next table shows 95% confidence intervals comparing pairs of means. Interpret the confidence interval comparing Republicans and Democrats.

Affiliations	95% CI		
	Lower	Center	Upper
Republican—Democrat	2.10	2.87	3.65
Independent—Democrat	−0.02	0.84	1.71
Republican—Independent	1.43	2.03	2.64

d. Explain how you would summarize the results of the ANOVA F test and the confidence intervals to someone who has not studied statistics.

13.45 Comparing therapies for anorexia: The "anorexia" data file on the text CD shows weight change for 72 anorexic teenage girls who were randomly assigned to one of three psychological treatments.

a. Show how to construct 95% confidence intervals to investigate how the population means differ. Interpret them.

b. Report the Tukey 95% multiple comparison confidence intervals. Interpret, and explain why they are wider than the confidence intervals in (a).

13.46 Lot size varies by region?: A geographer compares residential lot sizes in four quadrants of a city. To do this, she randomly samples 300 records from a city file on home residences and records the lot sizes (in thousands of square feet) by quadrant. The ANOVA table, shown in the table below, refers to a comparison of mean lot sizes for the northeast (NE), northwest (NW), southwest (SW), and southeast (SE) quadrants of the city. Fill in all the blanks in the table.

Lot sizes by quadrant of city

Source	DF	SS	MS	F	P-value
Quadrant	___	___	___	___	0.000
Error	296	1480	___		
Total	___	4180			

13.47 House location: Refer to the "house selling price" data file on the text CD.

a. Set up an indicator variable for whether a house is in the NW region. Using software, put this as the sole predictor of house selling price in a regression model. Report the prediction equation, and interpret the intercept and slope estimates.

b. For the model fitted in (a), conduct the t test for the effect of the indicator variable in the regression analysis (that is, test H_0: $\beta = 0$). Interpret.

c. Use software to conduct the F test for the analysis of variance comparing the mean selling prices of homes in the NW and in other regions. Interpret.

d. Explain the connection between the value of t in (b) and the value of F in (c).

13.48 Ideal number of kids by gender and race: The GSS asks, "What is the ideal number of kids for a family?" When we use a recent GSS to evaluate how the mean response depends on gender and race, we get the results shown in the ANOVA table.

a. Identify the response variable and the factors.

b. Explain what it would mean if there were no interaction between gender and race in their effects. Show a hypothetical set of population means that would show a strong race effect and a weak gender effect and no interaction.

c. Using the table, specify the null and alternative hypotheses, test statistic, and P-value for the test of no interaction. Interpret the result.

ANOVA of ideal number of kids by gender and race

Source	DF	SS	MS	F	P
Gender	1	0.25	0.25	0.36	0.546
Race	1	16.98	16.98	24.36	0.000
Interaction	1	0.95	0.95	1.36	0.244
Error	1245	867.72	0.70		
Total	1248	886.12			

13.49 Regress kids on gender and race: Refer to the previous exercise. Let $f = 1$ for females and 0 for males, and let $r = 1$ for blacks and 0 for whites. The regression model for predicting y = ideal number of kids is $\hat{y} = 2.42 + 0.04f + 0.37r$.

a. Interpret the coefficient of f. What is the practical implication of this estimate being so close to 0?

b. Find the estimated mean for each of the four combinations of gender and race.

c. Summarize what you learn about the effects based on the analyses in this and the previous exercise.

13.50 Florida student data: For the "FL student survey" data file on the text CD, we use as the response variable sports (the number of weekly hours engaged in sports and other physical exercise). For the explanatory variables, we use gender and whether the student is a vegetarian. The output shows results of a two-way ANOVA.

a. State the hypotheses for the vegetarian main effect.

b. Show how the F test statistic for (a) was obtained from mean squares reported in the ANOVA table.

c. Report and interpret the P-value of the test for the vegetarian main effect.

Source	DF	SS	MS	F	P
vegetarian	1	7.89	7.89	0.56	0.457
gender	1	64.35	64.35	4.59	0.037
Error	57	799.80	14.03		
Total	59	878.98			

13.51 Regress TV watching on gender and religion: When we use the 2002 GSS and regress y = number of hours per day watching TV on g = gender (1 = male, 0 = female) and religious affiliation ($r1 = 1$ for Protestant, $r2 = 1$ for Catholic, $r3 = 1$ for Jewish, $r1 = r2 = r3 = 0$ for none or other), we get the prediction equation $\hat{y} = 2.4 + 0.2g + 0.5r1 + 0.8r2 - 0.1r3$.

a. Interpret the gender effect.

b. Interpret the coefficient of $r1$.

c. State a corresponding model for the population, and indicate which parameters would need to equal zero for the response variable to be independent of religious affiliation, for each gender.

13.52 Income, gender, and race: According to the U.S. Census Bureau, in 2000, the population median income was estimated to be $29,661 for white females, $25,736 for black females, $40,350 for white males, $30,886 for black males (*Source:* www.census.gov/hhes/www/income00.html).

a. Identify the response variable and the two factors.

b. Show these medians in a two-way classification of the factors.

c. Compare the differences between males and females for (i) white subjects (ii) black subjects. If these are close estimates of the population medians, explain why there is interaction in terms of the median summary of center. Describe its nature.

13.53 Birth weight, age of mother, and smoking: A study about maternal age, smoking, and birth weight of children (by S. Bonnellie, *Pediatric and Perinatal Epidemiology*, vol. 15, pp. 19–26, 2001) indicated that mean birth weight was significantly lower for babies born to mothers who smoked during pregnancy. It also showed that the difference between the means for non-smokers and smokers increased with the age of the mother. Explain how this suggests interaction between smoking status and age of mother in their effects on birth weight of the child.

13.54 TV watching by gender and race: When we use the 2000 GSS to evaluate how the mean number of hours a day watching TV depends on gender and race, we get the results shown in the ANOVA table below.

a. Identify the response variable and the factors.

b. From the table, specify the test statistic and P-value for the test of no interaction. Interpret the result.

c. Is there a significant (i) gender effect, (ii) race effect? Explain.

d. The sample means were 2.71 for white females, 2.79 for white males, 4.13 for black females, and 4.16 for black males. Explain how these results are compatible with the results of the tests discussed in (c).

ANOVA of TV watching by gender and race

Source	DF	SS	MS	F	P
Gender	1	2.22	2.22	0.35	0.555
Race	1	489.65	489.65	76.62	0.000
Interaction	1	0.12	0.12	0.02	0.892
Error	1736	11094.04	6.39		
Total	1739	11583.81			

13.55 Salary and gender: A regression analysis of college faculty salaries (M. Bellas, *Amer. Sociolog. Rev.*, Vol. 59, 1994, p. 807) included an indicator variable for gender (male = 1, female = 0) and an indicator variable for race (nonwhite = 1, white = 0). For annual income measured in thousands of dollars, the estimated coefficients were 0.8 for gender and 0.6 for race.

a. Interpret the coefficient for gender.

b. At particular settings of the other predictors, the estimated mean salary for white females was 30.2 thousand dollars. Using the estimated coefficients, find the estimated means for (i) white males, (ii) black females.

13.56 Political ideology by religion and gender: The table shown summarizes responses on political ideology (where 1 = extremely liberal and 7 = extremely conservative) in the 2000 General Social Survey by religion and gender. The P-value is 0.03 for testing the null hypothesis of no interaction. Explain what this means in the context of this example, and indicate one place in the table that may be responsible for the small P-value. (*Hint:* Is the difference between males and females in sample mean ideology about the same for each religion?)

Political ideology by religion and gender

Religion		Mean	Std Dev.		Mean	Std. Dev.
Protestant	Female	4.23	1.34	Male	4.44	1.41
Catholic	Female	4.10	1.33	Male	4.22	1.42
Jewish	Female	2.59	1.22	Male	3.82	1.68
None	Female	3.34	1.34	Male	3.58	1.42

Concepts and Applications

13.57 Number of friends and degree: Using the GSS Web site sda.berkeley.edu/GSS, analyze whether the number of good friends (the variable NUMFREND) depends on the subject's highest degree (the variable DEGREE). To do so, at this web address, pick "comparison of means" and then enter the variable names and check the "ANOVA stats" box. Prepare a short report summarizing your analysis and its interpretation.

13.58 Sketch within- and between-groups variability: Sketch a dot plot of data for 10 observations in each of three groups such that

a. You believe the P-value would be very small for a one-way ANOVA. (You do not need to do the ANOVA; merely show points for which you think this would happen.)

b. The P-value in one-way ANOVA would not be especially small.

13.59 A = B and B = C, but A ≠ C?: In multiple comparisons following a one-way ANOVA with equal sample sizes, the margin of error with a 95% confidence interval for comparing each pair of means equals 10. Give three sample means illustrating that it is possible that group A is not significantly different from group B and group B is not significantly different from group C, yet group A is significantly different from group C.

13.60 Multiple comparison confidence: For four groups, explain carefully the difference between a confidence level of 0.95 for a single comparison of two means and a confidence level of 0.95 for a multiple comparison of all six pairs of means.

13.61 Another Simpson paradox: The 25 women faculty members in the humanities division of a college have a mean salary of 56,000, whereas the five women faculty in the science division have a mean salary of 70,000. The 20 men in the humanities division have a mean salary of 55,000, and the 30 men in the science division have a mean salary of 69,000.

a. Construct a table of sample mean incomes for the 2 × 2 table of gender and division of college. Use weighted averages to find the overall means for men and women.

b. Discuss how the results of a one-way comparison of mean incomes by gender would differ from the results of a two-way comparison of mean incomes by gender, controlling for division of college. (*Note:* This reversal of which gender has the higher mean salary, according to whether one controls division of college, illustrates *Simpson's paradox*.)

Multiple choice: Select the correct response in Exercises 13.62–13.65.

13.62 ANOVA/regression similarities: Analysis of variance and multiple regression have many similarities. Which of the following is *not* true?

a. The response variable is quantitative for each.

b. They both have F tests for testing that the response variable is statistically independent of the explanatory variable(s).

c. For inferential purposes, they both assume that the response variable y is normally distributed with the same standard deviation at all combinations of levels of the explanatory variable(s).

d. They are both appropriate when the main focus is on describing straight-line effects of quantitative explanatory variables.

13.63 ANOVA variability: One-way ANOVA provides relatively more evidence that $H_0: \mu_1 = \cdots = \mu_g$ is false:

a. The smaller the between-groups variation and the larger the within-groups variation.

b. The smaller the between-groups variation and the smaller the within-groups variation.

c. The larger the between-groups variation and the smaller the within-groups variation.

d. The larger the between-groups variation and the larger the within-groups variation.

13.64 Multiple comparisons: For four means, it is planned to construct Tukey 95% multiple comparison confidence intervals for the differences between the six pairs.

a. For each confidence interval, there is a 0.95 chance that the interval will contain the true difference.

b. The probability that all six confidence intervals will contain the true differences is 0.70.

c. The probability that all six confidence intervals will contain the true differences is about 0.95.

d. The probability that all six confidence intervals will contain the true differences is $(0.95)^6$.

13.65 Interaction: There is interaction in a two-way ANOVA model when

a. The two factors are associated.

b. Both factors have significant effects in the model without interaction terms.

c. The difference in true means between two categories of one factor varies among the categories of the other factor.

d. The mean square for interaction is about the same size as the mean square error.

13.66 Interaction true/false: For subjects aged under 50, there is little difference in mean annual medical expenses for smokers and non-smokers. For subjects aged over 50, there is a large difference. True or false: There is interaction between smoking status and age in their effects on annual medical expenses.

◆◆ 13.67 What causes large or small F?: An experiment used four groups of five individuals each. The overall sample mean was 60.

a. What did the sample means look like if the one-way ANOVA for comparing the means had test statistic $F = 0$? (Hint: What would have to happen in order for the between-groups variability to be 0?)

b. What did the data look like in each group if $F = $ infinity? (*Hint*: What would have to happen in order for the within-groups variability to be 0?)

◆◆ 13.68 Between-subjects estimate: This exercise motivates the formula for the between-subjects estimate of the variance in one-way ANOVA. Suppose each population mean equals μ (that is, H_0 is true) and each sample size equals n. Then the sampling distribution of each $\bar{y}_i$ has mean μ and variance σ^2/n, and the sample mean of the $\bar{y}_i$ values is the overall sample mean, $\bar{y}$.

a. Treating the g sample means as g observations having sample mean $\bar{y}$, explain why $\Sigma(\bar{y}_i - \bar{y})^2/(g - 1)$ estimates the variance σ^2/n of the distribution of the $\{\bar{y}_i\}$ values.

b. Using (a), explain why the between-groups estimate $\Sigma n(\bar{y}_i - \bar{y})^2/(g - 1)$ estimates σ^2. (For the unequal sample size case, the formula replaces n by n_i.)

◆◆ 13.69 Bonferroni multiple comparisons: The Bonferroni theorem states that the probability that *at least* one of a set of events occurs can be no greater than the sum of the separate probabilities of the events. For instance, if the probability of an error for each of five separate confidence intervals equals 0.01, then the probability that *at least* one confidence interval will be in error is no greater than $(0.01 + 0.01 + 0.01 + 0.01 + 0.01) = 0.05$.

a. Following Example 10, construct a confidence interval for each factor and guarantee that they both hold with overall confidence level at least 95%. [*Hint*: Each interval should use $t_{.0125} = 2.46$, and you should get $(0.06, 3.74)$ for fertilizer use and $(0.16, 3.84)$ for manure use.]

b. Exercise 13.8 referred to a study comparing three groups (smoking status never, former, or current) on various personality scales. The study measured 35 personality scales and reported an F test comparing the three smoking groups for each scale. The researchers mentioned doing a Bonferroni correction for the 35 F tests. If the nominal *overall* probability of Type I error was 0.05 for the 35 tests, how small did the P-value have to be for a particular test to be significant? (*Hint*: What should the Type I error probability be for each of 35 tests in order for the overall Type I error probability to be no more than 0.05?)

◆◆ 13.70 Independent confidence intervals: You plan to construct a 95% confidence interval in five different situations with independent data sets.

a. Assuming that the results of the confidence intervals are statistically independent, find the probability that *all* five confidence intervals will contain the parameters they are designed to estimate. (*Hint*: Use the binomial distribution.)

b. Which confidence level should you use for each confidence interval so that the probability that all five intervals contain the parameters equals *exactly* 0.95?

◆◆ 13.71 **Regression or ANOVA?:** You want to analyze $y =$ house selling price and $x =$ number of bathrooms (1, 2, or 3) by testing whether x and y are independent.

a. You could conduct a test of independence using (i) the ANOVA F test for a multiple regression model with two indicator variables, or (ii) a regression t test for the coefficient of the number of bathrooms when it is treated as a quantitative predictor in a straight-line regression model. Explain the difference between these two ways of treating the number of bathrooms in the analysis.

b. What do you think are the advantages and disadvantages of the straight-line regression approach to conducting the test?

c. Give an example of three population means for which the straight-line regression model would be less appropriate than the model with indicator variables.

◆◆ 13.72 **Three factors:** An experiment analyzed how the mean corn yield varied according to three factors: nitrogen-based fertilizer, phosphate-based fertilizer, and potash (potassium chloride) based fertilizer, each applied at low and at high levels.

a. How many groups result from the different combinations of the three factors?

b. Defining indicator variables, state the regression model that corresponds to an ANOVA assuming a lack of interaction.

c. Give possible estimates of the model parameters in (b) for which the estimated corn yield would be highest at the high level of nitrogen fertilizer and phosphate fertilizer and the low level of potash.

Class Investigations

13.73 **Student survey data:** Refer to the student survey data file that your class created with Activity 3 in Chapter 1. For variables chosen by your instructor, use ANOVA methods and related inferential statistical analyses. Interpret and summarize your findings in a short report, and prepare to discuss your findings in class.

EXAMPLE 1

How to Get a Better Tan

Picture the Scenario

During an introductory statistics class, all students were asked to design an experiment about a subject of interest and then to conduct the experiment and analyze the data. One student, Allison, wanted to compare ways of getting a tan without exposure to the sun, because sun exposure has risks such as skin cancer. She decided to investigate which of two treatments would give a better tan—an "instant bronze sunless tanner and bronzer" lotion, applied twice over a two-day period, or a tanning studio at which you lie on a Plexiglas surface and expose your body to longwave ultraviolet (UV) lights from above and below. We'll refer to these treatments as "tanning lotion" and "tanning studio."

The tanning lotion is much less expensive, but Allison predicted that the tanning studio would give a better tan. To investigate this hypothesis she recruited five untanned female friends to participate in an experiment. A friend in the class used a random number generator to pick three of the girls to use the tanning lotion. The other two girls used the tanning studio. After three days, Allison evaluated the tans produced. She was blinded to the treatment allocation, not knowing which girls used which tanning method. Allison ranked the girls in terms of the quality of their tans. The ranks went from 1 to 5, with 1 = best, 5 = worst.

Questions to Explore

- Once Allison ranked the five tanned girls, how could she summarize the evidence in favor of one treatment over the other?

- How can Allison find a P-value to determine if one treatment is significantly better than the other?

Thinking Ahead

You learned in Sections 9.2 and 9.3 how to compare means for two treatments using t tests. The tests assume a *normal* distribution for a quantitative response variable. The t tests are *robust*, usually working quite well even when population distributions are *not* normal. An exception is when the distribution is quite skewed, the sample sizes are small, and the alternative hypothesis is one-sided.

To use the t test, suppose Allison created a quantitative variable by assigning a score between 0 and 10 for each girl to describe the quality of tan. With such small sample sizes (only 2 and 3), she would not be able to assess whether quality of tan is approximately normal. Moreover, her prediction that the studio gives a better tan than the lotion was one-sided. In any case, Allison found it easier to rank the girls than to create a quantitative variable. For these reasons, then, it's not appropriate for her to use a t test to compare the tanning methods.

We'll now learn about an alternative way to compare treatments without having to assume a normal distribution for the response variable. **Nonparametric statistical methods** provide statistical inference without this assumption. They use solely the *ranking* of the subjects on the response variable. In Example 2 we'll use the best known nonparametric method, the **Wilcoxon test**, to analyze the ranking of the girls' tans from Allison's experiment.

Nonparametric methods are especially useful in two cases:

■ When the data are ranks for the subjects (as in Example 1) rather than quantitative measurements.

■ When it's inappropriate to assume normality, and the ordinary statistical method is not robust when the normal assumption is violated. We might prefer not to assume normality because we think the distribution will be highly skewed. Or, perhaps we have no idea about the distribution shape and the sample size is too small to give us much information about it.

Statisticians developed the primary nonparametric statistical methods starting in the late 1940s, long after most other methods described in this text. Since then, many nonparametric methods have been devised to handle a wide variety of scenarios. Entire books are devoted to their use. This final chapter is designed to show you the idea behind nonparametric methods. In Section 14.1 we'll learn about the most popular nonparametric test, the **Wilcoxon test** for comparing two groups. Section 14.2 briefly describes a few other popular nonparametric methods. Details about the nonparametric approach may be found in the books cited in the chapter bibliography.

> **Nonparametric statistical methods**
>
> **Nonparametric statistical methods** are inferential methods that do not assume a particular form of distribution, such as the normal distribution, for the population distribution.

14.1 | How Can We Compare Two Groups by Ranking?

In Example 1, we learned about a tanning experiment. Of five girls, three were randomly assigned to use the tanning lotion and two to use a tanning studio. The girls' tans were then ranked from 1 to 5, with 1 representing the best tan. Let's consider all the possible outcomes for this experiment. Each possible outcome divides the ranks of 1, 2, 3, 4, 5 into two groups—three ranks for the tanning lotion group and two ranks for the tanning studio group. Table 14.1 shows the possible rankings. For instance, in the first case, the three girls using the tanning lotion got the three best ranks, (1, 2, 3).

TABLE 14.1: Possible Rankings of Five Girls on Tanning Quality

Each case shows the three ranks for the girls using the tanning lotion and the two ranks for the girls using the tanning studio. It also shows the sample mean ranks and their difference.

Treatment			Ranks		
Lotion	(1,2,3)	(1,2,4)	(1,2,5)	(1,3,4)	(1,3,5)
Studio	(4,5)	(3,5)	(3,4)	(2,5)	(2,4)
Lotion mean rank	2	2.3	2.7	2.7	3
Studio mean rank	4.5	4	3.5	3.5	3
Difference of mean ranks	−2.5	−1.7	−0.8	−0.8	0
Lotion	(2,3,4)	(1,4,5)	(2,3,5)	(2,4,5)	(3,4,5)
Studio	(1,5)	(2,3)	(1,4)	(1,3)	(1,2)
Lotion mean rank	3	3.3	3.3	3.7	4
Studio mean rank	3	2.5	2.5	2	1.5
Difference of mean ranks	0	0.8	0.8	1.7	2.5

We summarize the outcome of the experiment by finding the mean rank for each group and taking their difference. Table 14.1 shows the mean ranks and their differences. For example, in the first case the mean is $(1 + 2 + 3)/3 = 2$ for the tanning lotion and $(4 + 5)/2 = 4.5$ for the tanning studio. The difference is $2 - 4.5 = -2.5$.

Allison predicted that the tanning studio would tend to give better tans. When this happens in the sample, the ranks for the tanning studio are smaller than those for the tanning lotion. Then the mean rank is larger for the tanning lotion, and the difference between the mean ranks is positive, as in the last case in the table.

Comparing Mean Ranks: The Wilcoxon Test

For this experiment, the samples were independent random samples—the responses for the girls using the tanning lotion were independent of the responses for the girls using the tanning studio. Suppose that the two treatments have identical effects, in the sense that a girl's tan would be the same regardless of which treatment she uses. Then, each of the ten possible outcomes shown in Table 14.1 is equally likely. Each has probability 1/10. Using the ten possible outcomes, we can construct a sampling distribution for the difference between the mean ranks. Table 14.2 shows this sampling distribution. For instance, the difference between the sample mean ranks is 0 for two of the ten samples in Table 14.1, so the probability of this outcome is 2/10.

TABLE 14.2: Sampling Distribution of Difference between Sample Mean Ranks

These probabilities apply when the treatments have identical effects. For example, only one of the ten possible samples in Table 14.1 has difference between the mean ranks equal to –2.5, so this value has probability 1/10.

Difference between mean ranks	Probability
–2.5	1/10
–1.7	1/10
–0.8	2/10
0	2/10
0.8	2/10
1.7	1/10
2.5	1/10

Figure 14.1 displays this sampling distribution. It is symmetric around 0. This is the expected value for the difference between the sample mean ranks if the two treatments truly have identical effects.

▲ **FIGURE 14.1: Sampling Distribution of Difference between Sample Mean Ranks.**
This sampling distribution, which is symmetric around 0, applies when the treatments have identical effects. It is used for the significance test of the null hypothesis that the treatments are identical in their tanning quality.

Allison hypothesized that the tanning studio would give a better tan than the tanning lotion. She wanted to test the null hypothesis

H_0: The treatments are identical in tanning quality.

against

H_a: Better tanning quality results with the tanning studio.

This alternative hypothesis is one-sided. Another way to state H_a is, "The expected value of the sample mean rank is smaller for the tanning studio than for the tanning lotion." On the average, we expect the ranks to be smaller (better) for the tanning studio. Thus, we expect the difference between the sample mean rank for the tanning lotion and the sample mean rank for the tanning studio to be positive.

To find the P-value, we presume that H_0 is true. Then, all samples in Table 14.1 are equally likely, and the sampling distribution is the one shown in Table 14.2. The P-value is the probability of a difference between the sample mean rankings like the observed difference or even more extreme, in terms of giving even more evidence in favor of H_a. The test comparing two groups based on the sampling distribution of the difference between the sample mean ranks is called the **Wilcoxon test**. It is named after the statistician, Frank Wilcoxon, who devised it in 1945.

SUMMARY: Wilcoxon Nonparametric Test for Comparing Two Groups

1. **Assumptions:** Independent random samples from two groups.

2. **Hypotheses:**

   H_0: Identical population distributions for the two groups (This implies equal expected values for the sample mean ranks)

   H_a: Different expected values for the sample mean ranks (two-sided), or

   H_a: Higher expected value for the sample mean rank for a specified group (one-sided)

3. **Test statistic:** Difference between sample mean ranks for the two groups (Equivalently, can use sum of ranks for one sample, as discussed after Example 2).

4. **P-value:** One-tail or two-tail probability, depending on H_a, that the difference between the sample mean ranks is as extreme or more extreme than observed.

5. **Conclusion:** Report the P-value and interpret in context. If a decision is needed, reject H_0 if the P-value $\leq$ significance level, such as 0.05.

EXAMPLE 2

Does the UV Tanning Studio Give a Better Tan than the Tanning Lotion?

Picture the Scenario

Example 1 described an experiment through which a student (Allison) hoped to determine whether a tanning lotion or a UV tanning studio produced a better tan. Table 14.1 showed the possible rankings for five girls. Table 14.2 showed the sampling distribution of the difference between the sample mean ranks, presuming the null hypothesis is true that the tanning treatments have identical effects. For Allison's actual experiment, the ranks were (2, 4, 5) for the girls using the tanning lotion and (1, 3) for the girls using the tanning studio.

Questions to Explore

a. Find and interpret the P-value for comparing the treatments, using the one-sided alternative hypothesis that the tanning studio gives a better tan than the tanning lotion. That is, H_a states that the expected mean rank is larger for the tanning lotion than for the tanning studio, so the difference between the two is positive.

b. What's the smallest possible P-value you could get for this experiment?

Think It Through

a. For the observed sample, the mean ranks are $(2 + 4 + 5)/3 = 3.7$ for the tanning lotion and $(1 + 3)/2 = 2$ for the tanning studio. The test statistic is the difference between the sample mean ranks, $3.7 - 2 = 1.7$. The right tail of the sampling distribution in Figure 14.1 has the large positive differences, for which the ranks tended to be better (lower) with the tanning studio. For the one-sided H_a, the P-value is the probability,

$$\text{P-value} = P(\text{difference between sample mean ranks at least as large as observed}).$$

That is, under the presumption that H_0 is true,

$$\text{P-value} = P(\text{difference between sample mean ranks} \geq 1.7).$$

From Table 14.2 (shown again in the margin) or Figure 14.1 (reproduced below), the probability of a sample mean difference of 1.7 or even larger is $1/10 + 1/10 = 2/10 = 0.20$. This is the P-value. It is not very close to 0. Although there is some evidence that the tanning studio gives a better tan (it *did* have a lower sample mean rank), the evidence is not strong. If the treatments had identical effects, the probability would be 0.20 of getting a sample like we observed or even more extreme.

Recall

Sampling distribution of the difference between mean ranks

Difference	Probability
−2.5	1/10
−1.7	1/10
−0.8	2/10
0	2/10
0.8	2/10
1.7	1/10
2.5	1/10

▪

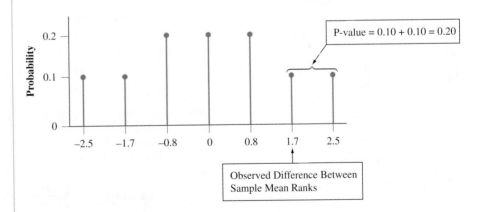

b. In this experiment, suppose the tanning studio gave the two best tans. The ranks would then be $(1, 2)$ for the tanning studio and $(3, 4, 5)$ for the tanning lotion. The difference of sample means then equals $4 - 1.5 = 2.5$. It is the most extreme possible sample, and (from Table 14.2 or Figure 14.1) its tail probability is 0.10. This is the smallest possible one-sided P-value.

Insight

With sample sizes of only 2 for one treatment and 3 for the other treatment, it's not possible to get a very small P-value. If Allison wanted to make a decision using a 0.05 significance level, she would never be able to get strong enough evidence to reject the null hypothesis. To get informative results, she'd need to conduct an experiment with larger sample sizes.

Even if the UV tanning studio were judged better than the tanning lotion, you might think twice about getting a tan this way. See www.ftc.gov/bcp/online /pubs/health/indootan.htm for a discussion of possible dangers from exposure to ultraviolet rays, including premature aging and possible links to malignant melanoma and immune system damage.

TO PRACTICE THIS CONCEPT, TRY EXERCISES 14.1 AND 14.2

Suppose Allison instead used a two-sided H_a, namely that there is a treatment difference without specifying which is better. Then the P-value is a two-tail probability. In Example 2, we would get P-value = 2(0.20) = 0.40.

The Wilcoxon rank sum The Wilcoxon test can, equivalently, use as the test statistic the sum (or mean) of the ranks in just one of the samples. For example, Table 14.3 shows the possible sum of ranks for the tanning lotion. They have the same probabilities as the differences between the sample mean ranks, because the ranks in one sample determine the ranks in the other sample (since the ranks must be the integers 1 through 5) and the mean ranks. For example, the ranks were (2, 4, 5) for the girls using the tanning lotion, for which the sum of ranks is 11. This implies that the ranks were (1, 3) for the girls using the tanning studio, and it implies that the difference between the mean ranks was $(2 + 4 + 5)/3 - (1 + 3)/2 = 3.7 - 2 = 1.7$. The right-tail probability of the observed rank sum of 11 or a more extreme value is again 1/10 + 1/10 = 0.20. Some software reports the sum of ranks as the *Wilcoxon rank sum statistic*, sometimes denoted by W.

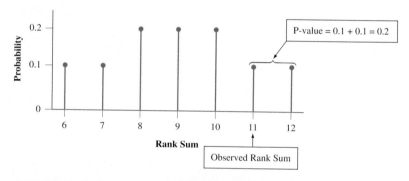

TABLE 14.3: Sampling Distribution of Sum of Ranks

The observed tanning lotion ranks of (2,4,5) have a rank sum of 11. These ranks imply that the tanning studio ranks were (1,3) and that the difference between the sample mean ranks was 1.7.

Tanning lotion ranks	Sum of tanning lotion ranks	Probability
(1, 2, 3)	6	1/10
(1, 2, 4)	7	1/10
(1, 2, 5),(1, 3, 4)	8	2/10
(2, 3, 4),(1, 3, 5)	9	2/10
(2, 3, 5),(1, 4, 5)	10	2/10
(2, 4, 5)	11	1/10
(3, 4, 5)	12	1/10

Large-Sample P-Values Use a Normal Sampling Distribution

With sample sizes of 3 and 2, it was simple to enumerate all the possible outcomes and construct a sampling distribution for the sum of ranks for a sample or for the difference between the mean ranks. With the sample sizes usually used in practice, it is tedious to do this. It's best to use software to get the results. Table 14.4 shows the way SPSS reports results for the Wilcoxon test conducted in Example 2. It shows the P-value of 0.40 for the two-sided alternative hypothesis.

TABLE 14.4: SPSS Output for Wilcoxon Test with Example 2

The "Wilcoxon W" is the sum of ranks (1,3) for the tanning studio. The P-value for the two-sided test is listed opposite "Exact Sig" (where "Sig." is short for significance.)

GROUP	N	Mean Rank	Sum of ranks
lotion	3	3.67	11.00
studio	2	2.00	4.00
Total	5		

Test Statistics

Wilcoxon W	4.000
Z	−1.155
Asymp. Sig. (2-tailed)	.248
Exact Sig. [2*(1-tailed Sig.)]	.400

Two-sided P-value (0.40)

In Example 2 we found the P-value with an exact probability calculation using the actual sampling distribution. With large samples, an alternative approach finds a z test statistic and a P-value based on a normal distribution approximation to the actual sampling distribution. (Some software, such as MINITAB, *only* provides results for the large-sample z test analysis.) For example, in Table 14.4, SPSS reports $z = −1.155$ and "Asymp. Sig. (2-tailed)" = 0.248. These are the results of the large-sample approximate analysis. This analysis is not appropriate for Example 2, because the sample sizes are only 2 and 3 for the two groups.

The large-sample test has a z test statistic because the difference between the sample mean ranks (or the sum of ranks for one sample) has an approximate normal sampling distribution. The z test statistic has the form

$$z = (\text{difference between sample mean ranks})/se.$$

The standard error formula is complex but easily calculated by software. The P-value is then the right-tail, left-tail, or two-tail probability, depending on H_a. *Using the normal distribution for the large-sample test does not mean we are assuming that the response variable has a normal distribution. We are merely using the fact that the sampling distribution is approximately normal.*

Ties Often Occur when We Rank Observations

Often in ranking the observations, some pairs of subjects are tied. They are judged to perform equally well. We then *average the ranks* in assigning them to those subjects. For example, suppose a girl using the tanning studio got the best tan (rank 1), two girls using the tanning lotion got the two worst tans (ranks 4 and 5), but the other girl using the tanning studio and the other girl using the tanning lotion were judged to have equally good tans. Then, those two girls share ranks 2 and 3, and each gets the average rank of 2.5. The ranks are then

Tanning studio: 1, 2.5

Tanning lotion: 2.5, 4, 5

IN WORDS

"Asymp." is short for **asymptotic**, which in statistics means "large-sample" or "as the sample size grows" (mathematically, as it grows toward infinity).

IN PRACTICE: Software and implementing the Wilcoxon test

Software can find tied ranks and then calculate the P-value for the Wilcoxon test. Most software reports only the large-sample approximate P-value when either sample size is larger than some value, typically 20. For example, SPSS reports the small-sample exact P-value only when each sample size is 20 or less.

Using the Wilcoxon Test with a Quantitative Response

In Examples 1 and 2, the response was a rank. When the response variable is quantitative, the Wilcoxon test is applied by converting the observations to ranks. For the combined sample, the observations are ordered from smallest to largest. The smallest observation gets rank 1, the second smallest gets rank 2, and so forth. The test compares the mean ranks for the two samples. Software can implement the test.

EXAMPLE 3

Do Drivers Using Cell Phones Have Slower Reaction Times?

Picture the Scenario

Example 9 in Chapter 9 discussed a recent experiment that investigated whether cell phone use impairs drivers' reaction times. A sample of 64 college students was randomly assigned to a cell phone group or a control group, 32 to each. On a machine that simulated driving situations, participants were instructed to press a "brake button" as soon as possible when they detected a red light. The control group listened to a radio broadcast or to books-on-tape while they performed the simulated driving. The cell phone group carried out a conversation on a cell phone with someone in a separate room.

A subject's reaction time observation is defined to be his or her response time to the red lights (in milliseconds), averaged over all the trials. Figure 14.2 shows box plots of the data for the two groups. Here's some of the data (from the "cell phone" file on the text CD), showing the four smallest observations and the four largest observations for each treatment.

Cell phone:	456	468	482	501	. . . .	672	679	688	960
Control:	426	436	444	449	. . . .	626	626	642	648

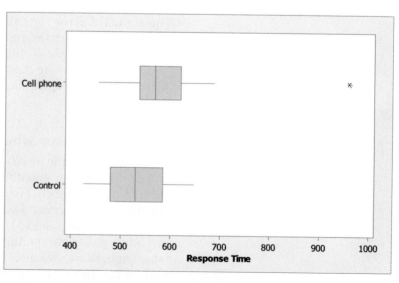

▲ **FIGURE 14.2: Box Plots of Response Times for Cell Phone Study. Question:** Does either box plot show any irregularities that suggest it's safer to use a nonparametric test than a two-sample *t* test?

The t inferences for comparing the treatment means assume normal population distributions. The box plots do not show any substantial skew, but there is an extreme outlier for the cell phone group. One subject in that group had a very slow mean reaction time, 960 milliseconds.

Questions to Explore

a. Explain how to find the ranks for the Wilcoxon test by showing which of the 64 observations get ranks 1, 2, 63, and 64.

b. Table 14.5 shows the SPSS output for conducting the Wilcoxon test. Report and interpret the mean ranks.

TABLE 14.5: SPSS Output for Wilcoxon Test with Data from Cell Phone Study

Ranks

	group	N	Mean Rank	Sum of Ranks
TIME	Control	32	27.03	865.00
	Cell phone	32	37.97	1215.00
	Total	64		

Test Statistics

	TIME
Wilcoxon W	865.000
Z	−2.350
Asymp. Sig. (2-tailed)	.019

c. Report the test statistic and the P-value for the two-sided Wilcoxon test. Interpret.

Think It Through

a. Let's look at the smallest and largest observations for each group that were shown above:

Cell Phone:	456	468	482	501	672	679	688	960
Control:	426	436	444	449	626	626	642	648

We give rank 1 to the *smallest* reaction time, so the value 426 gets rank 1. The second smallest observation is 436, which gets rank 2. The largest of the 64 reaction times, which was 960, gets rank 64. The next largest observation, 688, gets rank 63.

b. Table 14.5 reports mean ranks of 27.03 for the control group and 37.97 for the cell phone group. The smaller mean for the control group suggests that that group tends to have smaller ranks, and thus faster reaction times.

c. The z test statistic takes the difference between the sample mean ranks and divides it by a standard error. Table 14.5 reports $z = -2.35$. The P-value of 0.019, reported as "Asymp. Sig. (2-tailed)," is the two-tail probability for the two-sided H_a. It shows strong evidence against the null hypothesis that the distribution of reaction time is identical for the two treatments. Specifically, the sample mean ranks suggest that reaction times tend to be slower for those using cell phones.

Insight

The observation of 960 would get rank 64 if it were *any* number larger than 688 (the second largest value). So, *the Wilcoxon test is not affected by an outlier*. No matter how far the largest observation falls from the next largest, it still gets the same rank. Likewise, no matter how far the smallest observation is below the next smallest, it still gets the rank of 1.

TO PRACTICE THIS CONCEPT, TRY EXERCISE 14.4

Nonparametric Estimation Comparing Two Groups

Throughout this text, we've seen the importance of *estimating* parameters. We learn more from a confidence interval for a parameter than from a significance test about the value of that parameter. Nonparametric estimation methods, like nonparametric tests, do not require the assumption of normal population distributions.

When the response is quantitative, we can compare a measure of center for the two groups. Chapter 9 did this for the mean. Nonparametric methods are often used when the response distribution may be highly skewed. Then, it can be more informative to summarize each group by the *median*. We then estimate the difference between the population medians for the two groups.

Most software for the Wilcoxon test reports point and interval estimates comparing medians. (Some software, such as MINITAB, refers to the equivalent Mann-Whitney test instead. See Exer. 14.35.) Although this does not require a normal population assumption, it *does* require an extra assumption, namely that *the population distributions for the two groups are symmetric and have the same shape*. Unfortunately, this is a serious restriction, because the need for such inference is mainly when the population distributions are highly skewed and the median is a more representative summary than the mean.

Under the extra assumption, here's how software estimates the difference between the population medians. For each pair of subjects, one from each group, it takes the difference between the response from the first group and the response from the second group. The point estimate is the median of all these differences. Software also reports a confidence interval for the difference between the population medians. The mechanics of this are beyond our scope here.

EXAMPLE 4

Estimating the Difference between Median Reaction Times

Picture the Scenario

Example 3 used the Wilcoxon test to compare reaction time distributions in a simulated driving experiment with occasional red lights, for subjects using cell phones and for a control group. The MINITAB output in Table 14.6 shows results of comparing the distributions using medians. (It uses the Greek letter name *eta*, which is η, to denote the median.)

> ### TABLE 14.6: MINITAB Output for Comparing Medians for Cell Phone Group and Control Group
>
> ```
>                 N     Median
> Cell phone     32     569.00
> Control        32     530.00
>
> Point estimate for ETA1-ETA2 is 44.50 ◄──  ┌─ ETA is MINITAB notation for the median ─┐
> 95.1 Percent CI for ETA1-ETA2 is (8.99, 79.01)
> Test of ETA1 = ETA2 vs. ETA1 not = ETA2 is significant at 0.0184 ◄── ┤ P-value
> ```

Questions to Explore

a. Report the sample medians and the point estimate of the difference between the population medians.

b. Report and interpret the 95% confidence interval for the difference between the population medians.

Think It Through

a. The median reaction times were 569 milliseconds for the cell phone group and 530 milliseconds for the control group. For the cell phone group, for example, half of the reaction times were smaller than 569 milliseconds and half were larger than 569. Table 14.6 reports a point estimate of the difference between the population medians for the two groups of 44.5 milliseconds. (*Note:* This is not the same as the difference between the two sample medians, which is an alternative estimate.)

b. Table 14.6 reports that the 95% confidence interval for the difference between the population medians is (9, 79). We infer that the population median reaction time for the cell phone group is between 9 milliseconds and 79 milliseconds larger than for the control group. This agrees with the conclusion of the Wilcoxon test that the reaction time is significantly larger for the cell phone group (P-value = 0.02).

Insight

Example 9 in Chapter 9 estimated the difference between the population *means* to be 51.5, with a 95% confidence interval of (12, 91). However, those results were influenced by the outlier of 960 for the cell phone group. When estimation focuses on medians rather than means, outliers do not influence the analysis. The lack of an influence of outliers is an advantage of the analysis reported here for the medians.

> ### TO PRACTICE THIS CONCEPT, TRY EXERCISE 14.5

Recall

From the discussion before the example, the **point estimate** of the difference between the population medians equals the median of the differences between responses from the two groups. ∎

The Proportion of Better Responses for a Group

For comparing two groups, here's another way to summarize the results: Look at all the pairs of observations, such that one observation is from one group and the other observation is from the other group. Then find the proportion of the pairs for which the observation from the first group was better.

We illustrate for the tanning experiment described in Examples 1 and 2. The tanning lotion got ranks (2, 4, 5) and the tanning studio got ranks (1, 3). Figure 14.3 depicts all the pairs of girls, such that one girl used the tanning lotion and one girl used the tanning studio. The tanning lotion gave a better tan when we pair the girls who got ranks 2 and 3. That is, the girl using the tanning lotion got the better rank, namely, rank 2. However, for all other pairs the girl who used the tanning studio got the better tan. This is the case when we pair the girls who got ranks 1 and 2, ranks 1 and 4, ranks 1 and 5, ranks 3 and 4, and ranks 3 and 5. With 3 girls using the tanning lotion and 2 girls using the tanning studio, there are $3 \times 2 = 6$ pairs to consider. The tanning studio gave the better tan in 5 of the 6 pairs, a sample proportion of 5/6.

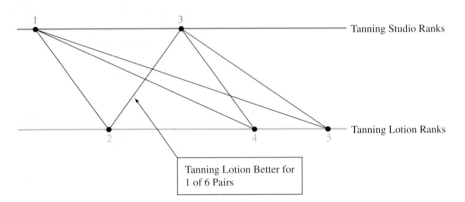

▲ **FIGURE 14.3: Results of Pairing Girls from Two Treatment Groups.** The tanning lotion got ranks (2, 4, 5) and the tanning studio (with the UV tanning bed) got ranks (1, 3). The tanning studio gave a better tan for 5 of the 6 pairs.

This sample proportion is a useful descriptive summary of the effect: "No effect" corresponds to a proportion of 1/2, and greater effects correspond to proportions farther from 1/2. As the proportion gets closer to 0 or to 1, there's a greater difference between the two groups, in the sense that ranks tend to be higher for one group than for the other.

Consider the driving with cell phone experiment analyzed in Examples 3 and 4. For 67% of the pairs the reaction time was greater for the subject using a cell phone, and for 33% of the pairs the reaction time was greater for the subject in the control group. We estimate that the probability is 0.67 of a greater reaction time for the cell phone user.

14.1 Practicing the Basics

14.1 **Tanning experiment:** Suppose the tanning experiment
`TRY` described in Examples 1 and 2 used only 4 girls, 2 for each treatment.

   a. Show the six possible ways the four ranks could be allocated, two to each treatment, with no ties.

   b. For each possible sample, find the mean rank for each treatment and the difference between the mean ranks.

   c. Presuming H_0 is true of identical treatment effects, construct the sampling distribution of the difference between the sample mean ranks for the two treatments.

14.2 **Test for tanning experiment:** Refer to the previous ex-
`TRY` ercise. For the actual experiment, suppose the girls using the tanning studio got ranks 1 and 2 and the girls using the tanning lotion got ranks 3 and 4.

   a. Find and interpret the P-value for the alternative hypothesis that the tanning studio tends to give better tans than the tanning lotion.

   b. Find and interpret the P-value for the alternative hypothesis that the treatments have different effects.

   c. Explain why it is a waste of time to conduct this experiment if you plan to use a 0.05 significance level to make a decision.

14.3 **Comparing clinical therapies:** A clinical psychologist wants to choose between two therapies for treating severe mental depression. She selects six patients who are similar in their depressive symptoms and overall quality of health. She randomly selects three patients to receive therapy 1. The other three receive therapy 2. After one month of treatment, the improvement in each patient is measured by the change in a score for measuring severity of mental depression—the higher the change score, the better. The improvement scores are

Therapy 1: 25, 40, 45

Therapy 2: 10, 20, 30

a. For the possible samples that could have occurred (with no ties), show the possible ways the six ranks could have been allocated to the two treatments.

b. For each possible allocation of ranks, find the mean rank for each treatment and the difference between the mean ranks.

c. Consider the null hypothesis of identical response distributions for the two treatments. Presuming H_0 is true, construct the sampling distribution of the difference between the sample mean ranks for the two treatments.

d. For the actual data shown above, find and interpret the P-value for the alternative hypothesis that the two treatments have different effects.

14.4 Anticipation of hypnosis: The nonparametric statistics textbook by Lehmann (1975) discussed a study on the effect of hypnosis. A group of 16 subjects were randomly split into a treatment group of size 8 to be hypnotized and a control group of size 8. A measure of respiratory ventilation was taken on each subject at the beginning of the experiment, to investigate whether subjects in the treatment group tended to ventilate more in anticipation of the hypnosis. The data were

Controls: 3.99 4.19 4.21 4.54 4.64 4.69 4.84 5.48
Treated: 4.36 4.67 4.78 5.08 5.16 5.20 5.52 5.74

a. State the hypotheses for conducting a one-sided Wilcoxon test.

b. Find the ranks for the two groups and their mean ranks.

c. Software reports a small-sample one-sided P-value of 0.025. Interpret.

14.5 Estimating hypnosis effect: Refer to the previous exercise. For these data, MINITAB reports

```
Point estimate for ETA1-ETA2 is -0.5150
95.9 Percent CI for ETA1-ETA2 is (-1.0100,
0.0196)
W = 49.0
```

Explain how to interpret the reported (a) point estimate, (b) confidence interval.

14.6 eBay auctions: Example 7 in Chapter 7 showed prices at which the Palm M515 PDA, a popular handheld computer, sold on eBay auctions. Some auctions used a "buy-it-now" option, whereby a bidder can buy the product immediately at a fixed price set by the seller. Here are the data from sales during the first week of May 2003.

1. Buy-it-now option: 235, 225, 225, 240, 250, 250, 210
2. Bidding only: 250, 249, 255, 200, 199, 240, 228, 255, 232, 246, 210, 178, 246, 240, 245, 225, 246, 225

Using software,

a. Plot the data. Summarize what the plot shows.

b. State the hypotheses and give the P-value for the Wilcoxon test for comparing the two groups with a two-sided alternative hypothesis.

c. Construct a 95% confidence interval for comparing the population medians. Interpret, and explain what (if any) effect the type of sale method has on the median sales price.

d. State the assumptions for the methods in (b) and (c).

14.7 Teenage anorexia: Example 7 in Section 8.3 and Exercise 9.71 in Chapter 9 described a study that used therapy to treat teenage girls who suffered from anorexia. The girls were randomly assigned to the cognitive behavioral treatment (Group 1) or to the control group (Group 2). The study observed the weight change after a period of time. The output shows results of a nonparametric comparison.

a. Interpret the reported point estimate of the difference between the population medians for the weight changes for the two groups.

b. Interpret the reported confidence interval, and summarize the assumptions on which it is based.

c. Report a P-value for testing the null hypothesis of identical population distributions of weight change. Specify the alternative hypothesis, and interpret the P-value.

MINITAB output comparing weight changes for Exercise 14.7

```
                   N Median
Cognitive_change  29   1.400
Control_change    26  -0.350

Point estimate for ETA1-ETA2 is 3.05
95.0 Percent CI for ETA1-ETA2 is
(-0.60,8.10)
W = 907.0
Test of ETA1 = ETA2 vs ETA1 not = ETA2
is significant at 0.1111
```

14.2 | Nonparametric Methods for Several Groups and for Matched Pairs

This final section describes a few other popular nonparametric methods. We'll first learn about a method that extends the Wilcoxon test to comparisons of mean ranks for *several* groups. We then study two methods for comparing *dependent* samples in the form of *matched pairs* data.

Comparing Mean Ranks of Several Groups: The Kruskal-Wallis Test

Chapter 13 showed that comparisons of means for two groups extend to comparisons for many groups, using the **analysis of variance (ANOVA) F test**. Likewise, the Wilcoxon test for comparing mean ranks of two groups extends to a comparison of mean ranks for several groups. This rank test is called the **Kruskal-Wallis test**, named after the statisticians who proposed it in 1952.

As in the Wilcoxon test, the Kruskal-Wallis test assumes that the samples are independent random samples, and the null hypothesis states that the groups have identical population distributions for the response variable. The test determines the ranks for the entire sample and then finds the sample mean rank for each group. The test statistic is based on the between-groups variability in the sample mean ranks (see the margin comment). Software easily calculates it for us.

The sampling distribution of the test statistic indicates whether the variability among the sample mean ranks is large compared to what's expected under the null hypothesis that the groups have identical population distributions. With g groups, the test statistic has an approximate chi-squared distribution with $g - 1$ degrees of freedom. (Recall that for the chi-squared distribution, the df is the mean.) The approximation improves as the sample sizes increase. The larger the differences among the sample mean ranks, the larger the test statistic and the stronger the evidence against H_0. The P-value is the right-tail probability above the observed test statistic value.

When would we use this test? The ANOVA F test assumes normal population distributions. The Kruskal-Wallis test does not have this assumption. It's a "safer" method to use with small samples when not much information is available about the shape of the distributions. It's also useful when the data are merely ranks and we don't have a quantitative measurement of the response variable. Here's a basic summary of the test:

KRUSKAL-WALLIS STATISTIC

Denote the sample mean rank by $\bar{R}_i$ for group i and by $\bar{R}$ for the combined sample of g groups. The **Kruskal-Wallis** test statistic is

$$\left(\frac{12}{n(n+1)}\right)\Sigma n_i(\bar{R}_i - \bar{R})^2.$$

The constant $12/n(n+1)$ is there so that the test statistic values have approximately a chi-squared sampling distribution.

SUMMARY: Kruskal-Wallis Nonparametric Test for Comparing Several Groups

1. **Assumptions:** Independent random samples from several (g) groups
2. **Hypotheses:**
   H_0: Identical population distributions for the g groups
   H_a: Population distributions not all identical
3. **Test statistic:** Uses between-groups variability of sample mean ranks (see formula in margin box on this page.)
4. **P-value:** Right-tail probability above observed test statistic value from chi-squared distribution with $df = g - 1$.
5. **Conclusion:** Report the P-value and interpret in context.

EXAMPLE 5

Does Heavy Dating Affect College GPA?

Picture the Scenario

In the same statistics class in which Allison chose as a project the experiment comparing tanning methods (Examples 1 and 2), another student in the class (Tim) decided to study whether dating was associated with college GPA. He wondered whether students who date a lot tend to have poorer GPAs. He asked the 17 students in the class to anonymously fill out a short questionnaire in which they were asked to give their college GPA (0 to 4 scale) and to indicate whether during their college careers they had dated regularly, occasionally, or rarely.

Figure 14.4 shows the dot plots that Tim constructed of the GPA data for the three dating groups. Since the dot plots showed evidence of severe skew to the left and since the sample size was small in each group, he felt safer analyzing the data with the Kruskal-Wallis test than with the ordinary ANOVA *F* test.

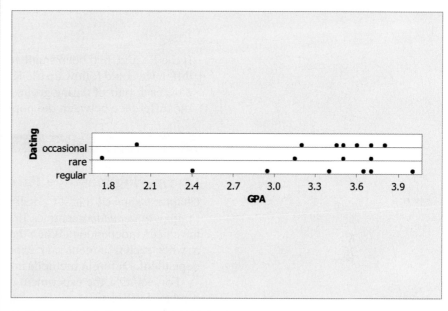

▲ **FIGURE 14.4: Dot Plots of GPA by Dating Group. Question:** Why might we be nervous about using the ordinary ANOVA *F* test to compare mean GPA for the three dating groups?

Table 14.7 shows the data, with the college GPA values ordered from smallest to largest for each dating group. The table in the margin shows the combined sample of 17 observations from the three groups and their ranks. Table 14.7 also shows these ranks as well as the mean rank for each group.

Finding ranks for data in Table 14.7	
GPA	**Rank**
1.75	1
2.00	2
2.40	3
2.95	4
3.15	5
3.20	6
3.40	7
3.44	8
3.50	9.5 ← average of 9 and 10
3.50	9.5
3.60	11
3.67	12
3.68	13
3.70	14
3.71	15
3.80	16
4.00	17

TABLE 14.7: College GPA by Dating Group

Dating Group	GPA Observations	Ranks	Mean rank
Rare	1.75, 3.15, 3.50, 3.68	1, 5, 9.5, 13,	7.1
Occasional	2.00, 3.20, 3.44, 3.50, 3.60, 3.71, 3.80	2, 6, 8, 9.5, 11, 15, 16	9.6
Regular	2.40, 2.95, 3.40, 3.67, 3.70, 4.00	3, 4, 7, 12, 14, 17	9.5

Question to Explore

Table 14.8 shows MINITAB output for the Kruskal-Wallis test. MINITAB denotes the chi-squared test statistic by H. Interpret these results.

TABLE 14.8: Results of Kruskal-Wallis Test for Data in Table 14.7

Kruskal-Wallis Test: GPA versus Dating

```
Dating        N     Median   AVE. Rank
rare          4     3.325      7.1
occasional    7     3.500      9.6
regular       6     3.535      9.5
H = 0.72  DF = 2         P = 0.696 (adjusted for ties)
```

Think It Through

If H_0: identical population distributions for the three groups were true, the Kruskal-Wallis test statistic would have an approximate chi-squared distribution with $df = 2$. Table 14.8 reports that the test statistic is H = 0.72. The P-value is the right-tail probability above 0.72. Table 14.8 reports this as 0.696, about 0.7. It is plausible that GPA is independent of dating group. Table 14.8 shows that the sample median GPAs are not very different, and since the sample sizes are small, these sample medians do not give much evidence against H_0.

Insight

If the P-value had been small, to find out which pairs of groups significantly differ, we could follow up the Kruskal-Wallis test by a Wilcoxon test to compare each pair of dating groups. Or, we could find a confidence interval for the difference between the population medians for each pair.

TO PRACTICE THIS CONCEPT, TRY EXERCISE 14.8

Comparing Matched Pairs: The Sign Test

Chapter 9 showed that it's possible to compare groups using either *independent* samples or *dependent* samples. In the examples so far in this chapter, the samples have been independent. When the subjects in the two samples are matched, such as when each treatment in an experiment uses the same subjects, the samples are dependent. Different methods must then be used.

For example, the experiment described in Examples 1 and 2 to compare tans produced with a tanning lotion and with a tanning studio could have instead used a crossover design: The subjects get a tan using one treatment, and after it wears off they get a tan using the other treatment. The order of using the two treatments is random. For each subject, we observe which treatment gives the better tan. That is, we make comparisons by pairing the two observations for the same subject.

For such a matched pairs experiment, let p denote the population proportion of cases for which a particular treatment does better than the other treatment. Under the null hypothesis of identical treatment effects, $p = 0.50$. That is, each treatment should have the better response outcome about half the time. (We ignore those cases in which each treatment gives the *same* response.) Let n denote the sample number of pairs of observations for which the two responses differ. For large n, we can use the z test statistic to compare the sample proportion $\hat{p}$ to the null hypothesis value of 0.50. (See the margin recall box.) The P-value is based on the approximate standard normal sampling distribution.

A test that compares matched pairs in this way is called a **sign test**. The name refers to how the method evaluates for each matched pair whether the difference between the first and second response is *positive* or *negative*.

Recall

From Section 8.2, to test H_0: $p = 0.50$ with sample proportion $\hat{p}$ when $n \geq 30$, the test statistic is

$$z = (\hat{p} - 0.50)/se,$$

with $se = \sqrt{(0.50)(0.50)/n}$. ■

SUMMARY: Sign Test for Matched Pairs

1. **Assumptions:** Random sample of matched pairs for which we can evaluate which observation in a pair has the better response.

2. **Hypotheses:** H_0: Population proportion $p = 0.50$ who make better response for a particular group
   H_a: $p \neq 0.50$ (two-sided) or H_a: $p > 0.50$ or H_a: $p < 0.50$ (one-sided).

3. **Test statistic:** $z = (\hat{p} - 0.50)/se$, as shown in margin recall box above.

4. **P-value:** For large samples ($n \geq 30$), use tail probabilities from standard normal. For smaller n, use binomial distribution (discussed in Example 7 on page 684).

5. **Conclusion:** Report the P-value and interpret in context.

EXAMPLE 6

Spend More Time Browsing the Internet or Watching TV?

Picture the Scenario

Which do most students spend more time doing—browsing the Internet or watching TV? Let's consider the students surveyed at the University of Georgia whose responses are in the "Georgia student survey" data file. The results for the first three students in the data file (in minutes per day) were

Student	Internet	TV
1	60	120
2	20	120
3	60	90

All three spent more time watching TV. For the entire sample, 35 students spent more time watching TV and 19 students spent more time browsing the Internet. (The analysis ignores the 5 students who reported the same time for each.)

Question to Explore

Let p denote the population proportion who spent more time watching TV. Find the test statistic and P-value for the sign test of H_0: $p = 0.50$ against H_a: $p \neq 0.50$. Interpret.

Think It Through

Here, $n = 35 + 19 = 54$. The sample proportion who spent more time watching TV was $35/54 = 0.648$. For testing that $p = 0.50$, from the previous margin recall box the se of the sample proportion is

$$se = \sqrt{(0.50)(0.50)/n} = \sqrt{(0.50)(0.50)/54} = 0.068.$$

The test statistic is

$$z = (\hat{p} - 0.50)/se = (0.648 - 0.50)/0.068 = 2.18.$$

From the normal distribution table (Table A or software), the two-sided P-value is 0.03. This provides considerable evidence that most students spend more time watching TV than browsing the Internet. The conclusion must be tempered by the fact that the data resulted from a convenience sample (students in a class for a statistics course) rather than a random sample of all college students.

Insight

The sign test uses merely the information about *which* response is higher, not the quantitative information about *how much* higher. This is a disadvantage compared to the corresponding parametric test, the matched pairs t test of Section 9.4, which analyzes the mean of the differences between the two responses. The sign test is most appropriate when we can order the responses but do not have quantitative information, such as in the next example.

TO PRACTICE THIS CONCEPT, TRY EXERCISE 14.10

For small n, we can conduct the sign test by using the *binomial* distribution. The next example illustrates this case as well as a situation in which we can order responses for each pair but do not have quantitative information about how different the responses are.

EXAMPLE 7

A Crossover Experiment Comparing Tanning Methods

Picture the Scenario

When Allison told another girl in the class (Megan) about her planned experiment to compare tanning methods, Megan decided to do a separate tanning experiment. She used a crossover design for a different sample of five girls. The tanning studio gave a better tan than the tanning lotion for four of the five girls.

Question to Explore

Find and interpret the P-value for testing that the population proportion p of girls for whom the tanning studio gives a better tan than the tanning lotion equals 0.50. Use the alternative hypothesis that this population proportion is larger than 0.50, because Megan predicted that the tanning studio would give better tans.

Recall

Section 6.3 presented the **binomial** distribution. With probability p of success on a trial, out of n independent trials the probability of x successes is

$$P(x) = \frac{n!}{x!(n-x)!} p^x (1-p)^{n-x}.$$ ■

Think It Through

The null hypothesis is H_0: $p = 0.50$. For H_a: $p > 0.50$, the P-value is the probability of the observed sample outcome or an even larger one. The sample size ($n = 5$) was small, so we use the binomial distribution rather than its normal approximation to find the P-value.

If $p = 0.50$, from the margin box the binomial probability that $x = 4$ of the $n = 5$ girls would get better tans with the tanning studio is

$$P(4) = \frac{5!}{4!(5-4)!}(0.50)^4(0.50)^1 = 0.156.$$

The more extreme result that all five girls would get better tans with the tanning studio has probability $P(5) = (0.50)^5 = 0.031$. The P-value is the right-tail probability of the observed result and the more extreme one, that is, $0.156 + 0.031 = 0.19$. See the margin figure. In summary, the evidence is not strong that more girls get a better tan from the tanning studio than the tanning lotion.

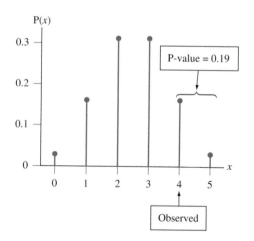

Insight

Megan would instead use the two-sided alternative, H_a: $p \neq 0.50$, if she did not make a prior prediction about which tanning method would be better. The P-value would then be $2(0.19) = 0.38$. With only $n = 5$ observations, the smallest possible two-sided P-value would be $2(0.031) = 0.06$, which occurs when $x = 0$ or when $x = 5$.

TO PRACTICE THIS CONCEPT, TRY EXERCISE 14.12

Ranking Matched Pairs: The Wilcoxon Signed-Ranks Test

With matched pairs data, for each pair the sign test merely observes which treatment does better, but not *how much* better. The **Wilcoxon signed-ranks test** is a nonparametric test designed for cases in which the comparisons of the paired observations can themselves be ranked. For each matched pair of responses, it measures the difference between the responses. It tests the hypothesis,

H_0: population median of difference scores is 0.

The test uses the quantitative information provided by the n difference scores by ranking their magnitudes, in absolute value. (Like the sign test, it ignores observation pairs for which the difference equals 0.) The test statistic is the sum of the ranks for the differences that are positive. Let's illustrate with an example.

EXAMPLE 8

Is a "No Soft Drink Diet" Effective?

Picture the Scenario

A diet for which a person is not allowed to drink soft drinks and other drinks with high sugar content is claimed to promote weight loss. Here are the results after six months, with weights in pounds, for three people who volunteered to try the diet. (The actual sample size was much larger, but it is easier to explain the idea behind the Wilcoxon signed-ranks test when the sample size is very small.)

Subject	Weight before diet	Weight after diet	Weight loss (difference)
1	172	160	12
2	210	195	15
3	168	170	-2

A negative weight loss represents a weight *gain*. Let's test the null hypothesis that the diet has no effect, in the sense that the population median weight loss is 0, against the alternative hypothesis that the population median is positive.

Questions to Explore

a. For ranks applied to the absolute values of the differences, find the rank sum for the differences that were positive.

b. Consider each of the possible ways that positive and negative signs could be assigned to these three differences. For each case, find the rank sum for the positive differences. Create the sampling distribution of this rank sum that applies if the diet truly has no effect.

c. Find the P-value for the Wilcoxon signed-ranks test, using the sampling distribution of the rank sum created in (b).

Think It Through

a. For the data shown above, the differences have absolute values and ranks:

Subject	Weight before	Weight after	Difference	Absolute value	Rank
1	172	160	12	12	2
2	210	195	15	15	3
3	168	170	-2	2	1

The rank sum for the differences that are positive (namely, 12 and 15) is $2 + 3 = 5$.

b. For the absolute difference values of 12, 15, and 2, Table 14.9 shows all the possible ways the differences could have been positive or negative. For each sample, this table also shows the sum of ranks for the positive differences. The observed data are sample 2, which had a rank sum of 5.

TABLE 14.9: Possible Samples with Absolute Difference Values of 12, 15, and 2

Subject	Sample								Rank of absolute value
	1	2	3	4	5	6	7	8	
1	12	12	–12	12	–12	12	–12	–12	2
2	15	15	15	–15	15	–15	–15	–15	3
3	2	–2	2	2	–2	–2	2	–2	1
Sum of ranks for positive differences									
	6	5	4	3	3	2	1	0	

Sampling distribution of rank sum for the positive differences

Rank sum	Probability
0	1/8
1	1/8
2	1/8
3	2/8
4	1/8
5	1/8
6	1/8

If the diet has no effect, then the eight possible samples in Table 14.9 are equally likely. The margin summarizes the sampling distribution of the rank sum for the positive differences, presuming no effect of the diet. For example, the rank sum was 5 for only 1 of the 8 samples, so its probability is 1/8.

c. The larger the sum of ranks for the positive differences, the greater the evidence that the diet has a positive effect. So, the P-value is the probability that this sum of ranks is at least as large as observed. Since 2 of the 8 possible samples had a rank sum for the positive differences of at least 5 (the observed value), the P-value is 2/8 = 0.25.

Insight

Suppose we instead used the sign test. We then observe that 2 of the 3 differences are positive. For the alternative hypothesis that the diet has a positive effect, the P-value is the probability that *at least* 2 of the 3 differences are positive, when the chance is 0.50 that any particular difference is positive. Using the binomial distribution, you can find that the P-value is 0.50 (Exercise 14.12).

The sign test ignores the fact that the two positive differences are larger than the negative difference. The Wilcoxon signed-ranks test uses this information. By taking this extra information into account, its P-value of 0.25 is smaller than the P-value of 0.50 from the sign test. However, the P-value is still not small. With only 3 observations, the one-sided P-value can be no smaller than 1/8, which is the P-value for the largest possible value (which is 6) for the rank sum of positive differences.

TO PRACTICE THIS CONCEPT, TRY EXERCISE 14.14

Although the Wilcoxon signed-ranks test has the advantage compared to the sign test that it can take into account the *sizes* of the differences and not merely their *sign*, it also has a disadvantage. For the possible samples (such as the 8 samples shown in Example 8 above) to be equally likely, it must make an additional assumption: The population distribution of the difference scores must be *symmetric*.

Symmetric Distribution

Difference

SUMMARY: Wilcoxon Signed-Ranks Test for Matched Pairs

1. **Assumptions:** Random sample of matched pairs for which the differences of observations have a symmetric population distribution and can be ranked.

2. **Hypotheses:**

   H_0: Population median of difference scores is 0.

   H_a: Population median of difference scores is not 0 (one-sided also possible).

3. **Test statistic:** Rank the absolute values of the difference scores for the matched pairs and then find the sum of ranks of the differences that were positive.

4. **P-value:** Software can find a P-value based on all the possible samples with the given absolute differences. (For large samples, it uses an approximate normal sampling distribution, as discussed below.)

5. **Conclusion:** Report the P-value and interpret in context.

The symmetry assumption for the Wilcoxon signed-ranks test is a bit weaker than the assumption of a normal population distribution that the matched pairs t test of Section 9.4 makes. However, with small samples, there's not much evidence to check this symmetry assumption. Also, recall that the t test is *robust* for violations of the assumption of normality, especially for two-sided inference. The extra assumption of symmetry for the Wilcoxon signed-ranks test weakens the advantage of using a nonparametric method. Because of this, many statisticians prefer to use the matched pairs t test for such data.

Like the Wilcoxon test for comparing mean ranks for two independent samples, the Wilcoxon signed-ranks test can allow for ties and it has a large-sample z test statistic that has an approximate standard normal sampling distribution. For instance, Example 6 above used the sign test to compare times spent on Internet browsing and watching TV for 54 students. If we create difference scores

$$\text{Difference} = \text{TV watching time} - \text{Internet browsing time}$$

for the students and analyze them with the Wilcoxon signed-ranks test, MINITAB reports the results shown in Table 14.10.

TABLE 14.10: Results of Wilcoxon Signed-Ranks Test for Time Differences between TV Watching and Internet Browsing (from Example 6)

Wilcoxon Signed Ranks Test: difference

Test of median = 0.000000 versus median not = 0.000000

	N	N for Test	Wilcoxon Statistic	P	Estimated Median
difference	59	54	1012.5	0.020	22.50

The Wilcoxon test statistic is the sum of ranks for the positive differences. The two-sided P-value of 0.020 provides strong evidence that the population median of the differences is not 0. There were 59 subjects, but only 54 were used in the test, because 5 difference scores were equal to 0. MINITAB also estimates that the population median difference between the time watching TV and the time browsing the Internet was 22.5 minutes. Separately, it provides a 95% confidence interval for the population median of the differences of 2.5 to 40.0 minutes.

Advantages and Limitations of Nonparametric Statistics

Nonparametric methods make weaker assumptions than the "parametric" methods we've studied in the rest of the text. For comparing two groups, for instance, it's not necessary to assume that the response distribution is normal. This is especially appealing for small samples, because then the normality assumption is more important than it is for larger samples, for which the Central Limit Theorem applies.

Statisticians have shown that nonparametric tests are often very nearly as good as parametric tests even in the exact case for which the parametric tests are designed. To illustrate, suppose two population distributions are normal in shape and have the same standard deviation but different means. Then, the Wilcoxon test is very nearly as powerful in detecting this difference as the t test, even though it uses only the ranks of the observations.

Nonetheless, nonparametric methods have some disadvantages. For example, confidence interval methods have not been as thoroughly developed as significance tests. Also, methods have not yet been well developed for multivariate procedures, such as multiple regression. Finally, recall that in many cases (especially two-sided tests and confidence intervals) most parametric methods are *robust*, working well even if assumptions are somewhat violated.

14.2 Practicing the Basics

14.8 **How long do you tolerate being put on hold?:**
TRY Examples 1–4 and 7 in Chapter 13 referred to the following randomized experiment: An airline analyzed whether telephone callers to their reservations office would remain on hold longer, on the average, if they heard (a) an advertisement about the airline, (b) Muzak, (c) classical music. For 15 callers randomly assigned to these three conditions, the table shows the data. It also shows the ranks for the 15 observations, as well as the mean rank for each group and some results from using MINITAB to conduct the Kruskal-Wallis test.

a. State the null and alternative hypotheses for the Kruskal-Wallis test.

b. Identify the value of the test statistic for the Kruskal-Wallis test, and state its approximate sampling distribution, presuming H_0 is true.

c. Report and interpret the P-value shown for the Kruskal-Wallis test.

d. To find out which pairs of groups significantly differ, how could you follow up the Kruskal-Wallis test?

Telephone holding times by type of recorded message.

Recorded Message	Holding Time Observations	Ranks	Mean rank
Muzak	0, 1, 3, 4, 6	1, 2.5, 5, 6, 8	4.5
Advertisement	1, 2, 5, 8, 11	2.5, 4, 7, 10.5, 13	7.4
Classical	7, 8, 9, 13, 15	9, 10.5, 12, 14, 15	12.1

Kruskal-Wallis Test: Holding Time versus Group

Group	N	Median	Ave Rank
Muzak	5	3.000	4.5
advert	5	5.000	7.4
classical	5	9.000	12.1

H = 7.38 DF = 2 P = 0.025 (adjusted for ties)

14.9 **What's the best way to learn French?:** Exercise 13.3 gave the data in the table for scores on the first quiz for ninth-grade students in an introductory-level French course. The instructor grouped the students in the course as follows:

Group 1: Never studied foreign language before, but have good English skills

Group 2: Never studied foreign language before; have poor English skills

Group 3: Studied at least one other foreign language

The table also shows results of using MINITAB to perform the Kruskal-Wallis test.

a. Find the rank associated with each observation, and show how to find the mean rank for group 1.

b. Report and interpret the P-value for the test.

Scores on the quiz

GROUP 1	GROUP 2	GROUP 3
4	1	9
6	5	10
8		5

```
Kruskal-Wallis Test on response
group   N    Median   Ave Rank
  1     3    6.000      4.3
  2     2    3.000      2.3
  3     3    9.000      6.2
H = 3.13 DF = 2 P = 0.209 (adjusted for ties)
```

14.10 **Sports vs. TV:** Which do students spend more time
TRY doing—playing sports or watching TV? For the students surveyed at the University of Florida whose responses are in the "FL student survey" data file, 24 students spent more time playing sports and 30 stu-

dents spent more time watching TV. Let p denote the corresponding population proportion who spend more time watching TV.

a. Find the test statistic for the sign test of H_0: $p = 0.50$ against H_a: $p \neq 0.50$.

b. Refer to (a). Find and interpret the P-value.

14.11 Cell phones and reaction times: Example 12 in Chapter 9 compared reaction times in a simulated driving test for the same students when they were using a cell phone and when they were not. The table shows data for the first four students. For all 32 students, 26 had faster reaction times when not using the cell phone and 6 had faster reaction times when using it.

a. Are the observations for the two treatments independent samples, or dependent samples? Explain.

b. Let p denote the population proportion who would have a faster reaction time when not using a cell phone. Estimate p based on this experiment.

c. Using all 32 observations, find the test statistic and the P-value for the sign test of H_0: $p = 1/2$ against H_a: $p > 1/2$ Interpret.

d. What is the "parametric" method for comparing the scores? What is an advantage of it over the sign test (*Hint*: Does the sign test use the magnitude of the difference between the two scores, or just its direction?)

Reaction times in cell phone study

Student	Using cell phone? No	Yes	Difference
1	604	636	32
2	556	623	67
3	540	615	75
4	522	672	150

14.12 Sign test for weight loss: Consider Example 8, for
TRY which the changes in weight during three months for three people on a diet were 12, 15, and −2 pounds. Show how to use the sign test to test that the probability that the weight difference is positive equals 0.50 against the alternative hypothesis that it is greater than 0.50.

14.13 Does exercise help blood pressure?: Exercise 9.39 in Chapter 9 discussed a pilot study of people who suffer from abnormally high blood pressure. A medical researcher decides to test her belief that walking briskly for at least half an hour a day has the effect of lowering blood pressure. She randomly samples three of her patients who have high blood pressure. She measures their systolic blood pressure initially and then again a month later after they participate in her exercise program. The table shows the results. Show

how to analyze the data with the sign test. State the hypotheses, find the P-value, and interpret.

SUBJECT	Before	After
1	150	130
2	165	140
3	135	120

14.14 More on blood pressure: Refer to the previous exer-
TRY cise. The analysis there did not take into account the *size* of the change in blood pressure. Show how to do this with the Wilcoxon signed-ranks test.

a. State the hypotheses for that test, for the relevant one-sided alternative hypothesis.

b. Construct the sampling distribution for the rank sum of the positive differences when you consider the possible samples that have absolute differences in blood pressure of 20, 25, and 15.

c. Using the sampling distribution from (b), find and interpret the P-value. (When every difference is positive, or when every difference is negative, this test and the sign test give the same P-value, for a given alternative hypothesis.)

14.15 More on cell phones: Refer to Exercise 14.11 above. That analysis did not take into account the magnitudes of the differences in reaction times. Show how to do this with the Wilcoxon signed-ranks test, illustrating by using only the four observations shown in the table there.

a. State the hypotheses for the relevant one-sided test.

b. Create the sampling distribution of the sum of ranks for the positive differences.

c. Find the P-value, and interpret.

14.16 Use all data on cell phones: Refer to the previous exercise. When we use the data for all 32 subjects, MINITAB reports results for the Wilcoxon signed-ranks test shown below.

Wilcoxon signed-ranks test results for Exercise 14.16

```
Test of median = 0.000000 versus median
not = 0.000000
              N for  Wilcoxon        Estimated
        N    Test   Statistic  P    Median
diff   32    32      490.0    0.000  47.25
```

a. State the null and alternative hypotheses for this test.

b. Explain how MINITAB found the value reported for the Wilcoxon test statistic.

c. Report and interpret the P-value.

d. Report and interpret the estimated median.

CHAPTER REVIEW

Answers to the Chapter Figure Questions

Figure 14.2: Yes, there is an extreme outlier for the cell phone group. This may have a large effect on the mean for that group but will not affect the nonparametric test.

Figure 14.4: The dot plots show evidence of severe skew to the left. Because the sample sizes are very small, severe violation of the normality assumption could invalidate the ANOVA F test.

CHAPTER SUMMARY

Nonparametric statistical methods provide statistical inference without an assumption of normality about the probability distribution of the response variable.

■ The **Wilcoxon test** is a nonparametric test for the null hypothesis of identical population distributions for two groups. It assumes independent random samples and uses the ranks for the combined sample of observations. The P-value is a one- or two-tail probability for the sampling distribution of the difference between the sample mean ranks.

■ Nonparametric methods can also estimate differences between groups, for example to construct a confidence interval for the difference between two population **medians**.

■ Other popular nonparametric methods include

   ■ The **Kruskal-Wallis test** for comparing mean ranks of *several* groups with independent random samples.

   ■ The **sign test** for comparing two groups with *matched pairs* data (dependent samples rather than independent samples) in terms of the proportion of times the response is better for a particular group.

   ■ The **Wilcoxon signed-ranks test** for comparing two groups with *matched pairs data* when the differences can be ranked.

CHAPTER PROBLEMS

Practicing the Basics

14.17 Car bumper damage: An automobile company compares two types of front bumper for their new model by driving sample cars into a concrete wall at 15 miles per hour. The response is the amount of damage to the car, as measured by the repair costs, in dollars. Due to the costs, the study uses only six cars, obtaining results for three bumpers of each type. The results are in the table.

   a. Find the ranks and the mean rank for each bumper type.

   b. Show that there are 20 possible allocations of ranks to the two bumper types.

   c. Explain why the observed ranks for the two groups are one of the two most extreme ways the two groups can differ, for the 20 possible allocations of the ranks.

   d. Explain why the P-value for the two-sided test equals 0.10.

Bumper A	Bumper B
1100	100
1500	300
1300	400

14.18 Comparing more bumpers: Refer to the previous exercise.

   a. Would the results of your analysis change if for the second bumper of type A the repair cost was $9000 instead of $1500? What does this illustrate about the analysis?

   b. Suppose the company actually wanted to compare *three* bumper types. Which significance test could they use to do this if they did not want to assume that the repair costs have a normal distribution for each bumper type?

14.19 Telephone holding times: In Exercise 14.8, the telephone holding times for Muzak and classical music were

   Muzak   0, 1, 4, 6, 3

   Classical   13, 9, 8, 15, 7

   a. For comparing these two groups with the Wilcoxon test, report the ranks and the mean rank for each group.

   b. Two groups of size 5 each have 252 possible allocations of rankings. For a two-sided test of H_0: identical distributions with these data, explain why the P-value is 2/252 = 0.008. Interpret the P-value.

14.20 Treating alcoholics: The nonparametric statistics textbook by Hollander and Wolfe (1999) discussed a study on a social skills training program for alcoholics. A sample of male alcoholics was randomly split into two groups. The control group received traditional treatment. The treatment group received the traditional treatment plus a class in social skills training. Every two weeks for a year after the treatment, subjects indicated the quantity of alcohol they consumed during that period. The summary response was the total alcohol intake for the year (in centiliters).

   a. Suppose the researchers planned to conduct a one-sided test and believed that the response variable could be highly skewed to the right. Why might they prefer to use the Wilcoxon test rather than a two-sample *t* test to compare the groups?

   b. The data were

   Controls: 1042 1617 1180 973 1552 1251
            1151 1511 728 1079 951 1319

   Treated: 874 389 612 798 1152 893 541
           741 1064 862 213

   Show how to find the ranks and the mean ranks for the two groups.

c. MINITAB reports the results shown in the output below. Report and interpret the P-value.

d. Report and interpret the confidence interval shown in the output.

MINITAB output for Exercise 14.20

```
              N      Median
control      12      1165.5
treated      11       798.0
Point estimate for ETA1-ETA2 is 435.5
95.5 Percent CI for ETA1-ETA2 is (186.0,
713.0)
Test of ETA1 = ETA2 vs ETA1 > ETA2 is
significant at 0.0009
```

14.21 Comparing tans: Examples 1 and 2 compared two methods of getting a tan. Suppose Allison conducted an expanded experiment in which nine girls were randomly assigned to one of two brands of tanning lotion or to the tanning studio, 3 girls to each treatment. The nine girls were ranked on their quality of tan.

a. Which nonparametric test could be used to compare the three treatments?

b. Give an example of ranks for the three treatments that would have the largest possible test statistic value and the smallest possible P-value for this experiment. (*Hint*: What allocation of ranks would have the greatest between-groups variation in the mean ranks?)

14.22 Comparing therapies for anorexia: The "anorexia" data file on the text CD shows weight change for 72 anorexic teenage girls who were randomly assigned to one of three psychological treatments. Using software, analyze these data with a nonparametric ANOVA to compare the three weight change distributions.

a. State the hypotheses.

b. Report the test statistic and its sampling distribution.

c. Report the P-value and explain how to interpret it.

14.23 Internet vs. cell phones: For the countries in the "Human development" data file on the text CD, in 4 countries a higher percentage of people used the Internet than used cell phones, while in 35 countries a higher percentage of people used cell phones than the Internet.

a. Show how you could use a nonparametric test to compare Internet use and cell phone use in the population of all countries. State the (i) hypotheses, (ii) test statistic value, and (iii) find and interpret the P-value.

b. Is the analysis in (a) relevant if the 39 countries in the data file are all the countries of interest to you rather than a random sample of countries? Explain.

14.24 Browsing the Internet: Refer to the "Georgia student survey" data file on the text CD. Use a method from this chapter to test whether the amount of time spent browsing the Internet is independent of one's political affiliation. State the (a) hypotheses, (b) test statistic, (c) P-value, and interpret the result.

14.25 GPAs: The "Georgia student survey" data file has data on college GPA and high school GPA for 59 University of Georgia students.

a. If you wanted to use a nonparametric test to check your friend's prediction that high school GPAs tend to be higher than college GPAs, which would you use?

b. What would be a reason for using the nonparametric method to do this?

c. Use the test in (a) to do this analysis, and interpret the result.

14.26 Sign test about the diet: In Exercise 14.12 on the proposed "no soft drink diet," you used the sign test to evaluate the weight differences of 12, 15, and -2. Suppose the weight differences were 120, 15, and -2 instead of 12, 15, and -2.

a. State the hypotheses and find the P-value for the one-sided sign test for evaluating the effect of the diet. Interpret.

b. Compare results to those in Exercise 14.12, and indicate what this tells you about the effect that outliers have on this nonparametric statistical method.

14.27 Wilcoxon signed-ranks test about the diet: Example 8 on the proposed "no soft drink diet" used the Wilcoxon signed-ranks test to evaluate the weight differences of 12, 15, and -2. Suppose the weight differences were 120, 15, and -2 instead of 12, 15, and -2.

a. State the hypotheses and find the P-value for the one-sided Wilcoxon signed-ranks test for evaluating the effect of the diet. Interpret.

b. Compare results to those in Example 8, and indicate what this tells you about the effect that outliers have on this nonparametric statistical method.

Concepts and Investigations

14.28 Student survey: For the "FL student survey" data file on the text CD, we identify the number of times reading a newspaper as the response variable and gender as the explanatory variable. The observations are as follows:

Females: 5, 3, 6, 3, 7, 1, 1, 3, 0, 4, 7, 2, 2, 7, 3, 0, 5, 0, 4, 4, 5, 14, 3, 1, 2, 1, 7, 2, 5, 3, 7

Males: 0, 3, 7, 4, 3, 2, 1, 12, 1, 6, 2, 2, 7, 7, 5, 3, 14, 3, 7, 6, 5, 5, 2, 3, 5, 5, 2, 3, 3

Using software, analyze these data using methods of this chapter. Write a one-page report summarizing your analyses and conclusions.

14.29 Why nonparametrics?: Present a situation for which it's preferable to use a nonparametric method instead of a parametric method, and explain why.

14.30 Why matched pairs?: Refer to Example 7. Describe the advantages of an experiment using a crossover design instead of independent samples to compare the tanning methods.

14.31 Complete the analogy: The *t* test for comparing two means is to the one-way ANOVA *F* test as the Wilcoxon test is to the _____ test.

14.32 Complete the analogy 2: The *t* test for comparing two means is to the Wilcoxon test (for independent samples) as the matched pairs *t* test is to the _____ (for dependent samples in matched pairs).

14.33 True or false?: For a one-sided significance test comparing two means with small samples from highly skewed population distributions, it's safer to use a *t* test than a Wilcoxon test. This is because the Wilcoxon test assumes normal population distributions and is not robust if that assumption is violated.

14.34 Multiple choice: Nonparametric statistical methods are used

  a. Whenever the response variable is known to have a normal distribution.

  b. Whenever the assumptions for a "parametric" method are not *perfectly* satisfied.

  c. When the data are ranks for the subjects rather than quantitative measurements or when it's inappropriate to assume normality and the ordinary statistical method is not robust when the normal assumption is violated.

  d. Whenever we want to compare two methods for getting a good tan.

◆◆ 14.35 Mann-Whitney statistic: For the tanning experiment, Table 14.2 showed the sampling distribution of the difference between the sample mean ranks. Suppose you instead use as test statistic the sample proportion of pairs of girls for which the tanning studio gave a better tan than the tanning lotion. This is the basis of the **Mann-Whitney** test, devised by two statisticians about the same time as Wilcoxon devised his test.

  a. The table shows the proportion of pairs for which the tanning studio gave a better tan, for the possible sample results. Explain how to find these proportions.

  b. Using this table, construct the sampling distribution that applies under the null hypothesis for this proportion.

  c. For the one-sided alternative hypothesis of a better tan with the tanning studio, find the P-value for the observed sample proportion of 5/6. (*Note*: P-values based on this sample proportion are identical to P-values for the Wilcoxon test based on the mean ranks.)

Sample Proportion of Pairs of Girls, One from Each Treatment, for which Tanning Studio Gave Better Tan than Tanning Lotion.

Treatments	Ranks				
Tanning lotion ranks	(1, 2, 3)	(1, 2, 4)	(1, 2, 5)	(1, 3, 4)	(1, 3, 5)
Tanning studio ranks	(4, 5)	(3, 5)	(3, 4)	(2, 5)	(2, 4)
Proportion	0/6	1/6	2/6	2/6	3/6
Tanning lotion ranks	(2, 3, 4)	(1, 4, 5)	(2, 3, 5)	(2, 4, 5)	(3, 4, 5)
Tanning studio ranks	(1, 5)	(2, 3)	(1, 4)	(1, 3)	(1, 2)
Proportion	3/6	4/6	4/6	5/6	6/6

◆◆ 14.36 Rank-based correlation: For data on two quantitative variables, *x* and *y*, an alternative correlation uses the *rankings* of the data. Let *n* denote the number of observations on the two variables. You rank the values of the *x*-variable from 1 to *n* according to their magnitudes, and you separately rank the values of the *y*-variable from 1 to *n*. The correlation computed between the two sets of ranks is called the **Spearman rank correlation**. Like the ordinary correlation, it falls between −1 and +1, with values farther from 0 representing stronger association.

  a. The ordinary correlation can be strongly affected by a regression outlier. Is this true also for the Spearman rank correlation? Why or why not?

  b. If you want to test the null hypothesis of no association, what value for the Spearman rank correlation would go in the null hypothesis?

◆◆ 14.37 Nonparametric regression: Nonparametric methods have also been devised for regression. Here's a simple way to estimate the slope: For each pair of subjects, the slope of the line connecting their two points is the difference between their *y* values divided by the difference between their *x* values. (See the figure below.) With *n* subjects, we can find this slope for each pair of points. (There are $n(n - 1)/2$ pairs of points.) A nonparametric estimate of the slope is the median of all these slopes for the various pairs of points. The ordinary slope (least squares, minimizing the sum of squared residuals) can be strongly affected by a regression outlier. Is this true also for the nonparametric estimate of the slope? Why or why not?

REFERENCES

Hollander, M., and Wolfe, D. (1999). *Nonparametric Statistical Methods*, 2nd edition. Wiley.

Lehmann, E. L. (1975). *Nonparametrics: Statistical Methods Based on Ranks*. Holden-Day.

Appendix A

2-SCORES

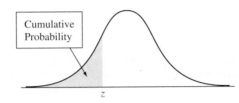

Cumulative Probability

Cumulative probability for z is the area under the standard normal curve to the left of z

z	.00
−5.0	.000000287
−4.5	.00000340
−4.0	.0000317
−3.5	.000233

TABLE A Standard Normal Cumulative Probabilities

z	.00	.01	.02	.03	.04	.05	.06	.07	.08	.09
−3.4	.0003	.0003	.0003	.0003	.0003	.0003	.0003	.0003	.0003	.0002
−3.3	.0005	.0005	.0005	.0004	.0004	.0004	.0004	.0004	.0004	.0003
−3.2	.0007	.0007	.0006	.0006	.0006	.0006	.0006	.0005	.0005	.0005
−3.1	.0010	.0009	.0009	.0009	.0008	.0008	.0008	.0008	.0007	.0007
−3.0	.0013	.0013	.0013	.0012	.0012	.0011	.0011	.0011	.0010	.0010
−2.9	.0019	.0018	.0018	.0017	.0016	.0016	.0015	.0015	.0014	.0014
−2.8	.0026	.0025	.0024	.0023	.0023	.0022	.0021	.0021	.0020	.0019
−2.7	.0035	.0034	.0033	.0032	.0031	.0030	.0029	.0028	.0027	.0026
−2.6	.0047	.0045	.0044	.0043	.0041	.0040	.0039	.0038	.0037	.0036
−2.5	.0062	.0060	.0059	.0057	.0055	.0054	.0052	.0051	.0049	.0048
−2.4	.0082	.0080	.0078	.0075	.0073	.0071	.0069	.0068	.0066	.0064
−2.3	.0107	.0104	.0102	.0099	.0096	.0094	.0091	.0089	.0087	.0084
−2.2	.0139	.0136	.0132	.0129	.0125	.0122	.0119	.0116	.0113	.0110
−2.1	.0179	.0174	.0170	.0166	.0162	.0158	.0154	.0150	.0146	.0143
−2.0	.0228	.0222	.0217	.0212	.0207	.0202	.0197	.0192	.0188	.0183
−1.9	.0287	.0281	.0274	.0268	.0262	.0256	.0250	.0244	.0239	.0233
−1.8	.0359	.0351	.0344	.0336	.0329	.0322	.0314	.0307	.0301	.0294
−1.7	.0446	.0436	.0427	.0418	.0409	.0401	.0392	.0384	.0375	.0367
−1.6	.0548	.0537	.0526	.0516	.0505	.0495	.0485	.0475	.0465	.0455
−1.5	.0668	.0655	.0643	.0630	.0618	.0606	.0594	.0582	.0571	.0559
−1.4	.0808	.0793	.0778	.0764	.0749	.0735	.0721	.0708	.0694	.0681
−1.3	.0968	.0951	.0934	.0918	.0901	.0885	.0869	.0853	.0838	.0823
−1.2	.1151	.1131	.1112	.1093	.1075	.1056	.1038	.1020	.1003	.0985
−1.1	.1357	.1335	.1314	.1292	.1271	.1251	.1230	.1210	.1190	.1170
−1.0	.1587	.1562	.1539	.1515	.1492	.1469	.1446	.1423	.1401	.1379
−0.9	.1841	.1814	.1788	.1762	.1736	.1711	.1685	.1660	.1635	.1611
−0.8	.2119	.2090	.2061	.2033	.2005	.1977	.1949	.1922	.1894	.1867
−0.7	.2420	.2389	.2358	.2327	.2296	.2266	.2236	.2206	.2177	.2148
−0.6	.2743	.2709	.2676	.2643	.2611	.2578	.2546	.2514	.2483	.2451
−0.5	.3085	.3050	.3015	.2981	.2946	.2912	.2877	.2843	.2810	.2776
−0.4	.3446	.3409	.3372	.3336	.3300	.3264	.3228	.3192	.3156	.3121
−0.3	.3821	.3783	.3745	.3707	.3669	.3632	.3594	.3557	.3520	.3483
−0.2	.4207	.4168	.4129	.4090	.4052	.4013	.3974	.3936	.3897	.3859
−0.1	.4602	.4562	.4522	.4483	.4443	.4404	.4364	.4325	.4286	.4247
−0.0	.5000	.4960	.4920	.4880	.4840	.4801	.4761	.4721	.4681	.4641

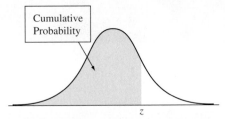

Cumulative Probability

Cumulative probability for z is the area under the standard normal curve to the left of z

TABLE A Standard Normal Cumulative Probabilities (continued)

z	.00	.01	.02	.03	.04	.05	.06	.07	.08	.09
0.0	.5000	.5040	.5080	.5120	.5160	.5199	.5239	.5279	.5319	.5359
0.1	.5398	.5438	.5478	.5517	.5557	.5596	.5636	.5675	.5714	.5753
0.2	.5793	.5832	.5871	.5910	.5948	.5987	.6026	.6064	.6103	.6141
0.3	.6179	.6217	.6255	.6293	.6331	.6368	.6406	.6443	.6480	.6517
0.4	.6554	.6591	.6628	.6664	.6700	.6736	.6772	.6808	.6844	.6879
0.5	.6915	.6950	.6985	.7019	.7054	.7088	.7123	.7157	.7190	.7224
0.6	.7257	.7291	.7324	.7357	.7389	.7422	.7454	.7486	.7517	.7549
0.7	.7580	.7611	.7642	.7673	.7704	.7734	.7764	.7794	.7823	.7852
0.8	.7881	.7910	.7939	.7967	.7995	.8023	.8051	.8078	.8106	.8133
0.9	.8159	.8186	.8212	.8238	.8264	.8289	.8315	.8340	.8365	.8389
1.0	.8413	.8438	.8461	.8485	.8508	.8531	.8554	.8577	.8599	.8621
1.1	.8643	.8665	.8686	.8708	.8729	.8749	.8770	.8790	.8810	.8830
1.2	.8849	.8869	.8888	.8907	.8925	.8944	.8962	.8980	.8997	.9015
1.3	.9032	.9049	.9066	.9082	.9099	.9115	.9131	.9147	.9162	.9177
1.4	.9192	.9207	.9222	.9236	.9251	.9265	.9279	.9292	.9306	.9319
1.5	.9332	.9345	.9357	.9370	.9382	.9394	.9406	.9418	.9429	.9441
1.6	.9452	.9463	.9474	.9484	.9495	.9505	.9515	.9525	.9535	.9545
1.7	.9554	.9564	.9573	.9582	.9591	.9599	.9608	.9616	.9625	.9633
1.8	.9641	.9649	.9656	.9664	.9671	.9678	.9686	.9693	.9699	.9706
1.9	.9713	.9719	.9726	.9732	.9738	.9744	.9750	.9756	.9761	.9767
2.0	.9772	.9778	.9783	.9788	.9793	.9798	.9803	.9808	.9812	.9817
2.1	.9821	.9826	.9830	.9834	.9838	.9842	.9846	.9850	.9854	.9857
2.2	.9861	.9864	.9868	.9871	.9875	.9878	.9881	.9884	.9887	.9890
2.3	.9893	.9896	.9898	.9901	.9904	.9906	.9909	.9911	.9913	.9916
2.4	.9918	.9920	.9922	.9925	.9927	.9929	.9931	.9932	.9934	.9936
2.5	.9938	.9940	.9941	.9943	.9945	.9946	.9948	.9949	.9951	.9952
2.6	.9953	.9955	.9956	.9957	.9959	.9960	.9961	.9962	.9963	.9964
2.7	.9965	.9966	.9967	.9968	.9969	.9970	.9971	.9972	.9973	.9974
2.8	.9974	.9975	.9976	.9977	.9977	.9978	.9979	.9979	.9980	.9981
2.9	.9981	.9982	.9982	.9983	.9984	.9984	.9985	.9985	.9986	.9986
3.0	.9987	.9987	.9987	.9988	.9988	.9989	.9989	.9989	.9990	.9990
3.1	.9990	.9991	.9991	.9991	.9992	.9992	.9992	.9992	.9993	.9993
3.2	.9993	.9993	.9994	.9994	.9994	.9994	.9994	.9995	.9995	.9995
3.3	.9995	.9995	.9995	.9996	.9996	.9996	.9996	.9996	.9996	.9997
3.4	.9997	.9997	.9997	.9997	.9997	.9997	.9997	.9997	.9997	.9998

z	.00
3.5	.999767
4.0	.9999683
4.5	.9999966
5.0	.999999713

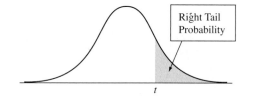

Right Tail Probability

t

TABLE B t Distribution Critical Values

| | Confidence Level | | | | | |
	80%	90%	95%	98%	99%	99.8%
	Right-Tail Probability					
df	$t_{.100}$	$t_{.050}$	$t_{.025}$	$t_{.010}$	$t_{.005}$	$t_{.001}$
1	3.078	6.314	12.706	31.821	63.656	318.289
2	1.886	2.920	4.303	6.965	9.925	22.328
3	1.638	2.353	3.182	4.541	5.841	10.214
4	1.533	2.132	2.776	3.747	4.604	7.173
5	1.476	2.015	2.571	3.365	4.032	5.894
6	1.440	1.943	2.447	3.143	3.707	5.208
7	1.415	1.895	2.365	2.998	3.499	4.785
8	1.397	1.860	2.306	2.896	3.355	4.501
9	1.383	1.833	2.262	2.821	3.250	4.297
10	1.372	1.812	2.228	2.764	3.169	4.144
11	1.363	1.796	2.201	2.718	3.106	4.025
12	1.356	1.782	2.179	2.681	3.055	3.930
13	1.350	1.771	2.160	2.650	3.012	3.852
14	1.345	1.761	2.145	2.624	2.977	3.787
15	1.341	1.753	2.131	2.602	2.947	3.733
16	1.337	1.746	2.120	2.583	2.921	3.686
17	1.333	1.740	2.110	2.567	2.898	3.646
18	1.330	1.734	2.101	2.552	2.878	3.611
19	1.328	1.729	2.093	2.539	2.861	3.579
20	1.325	1.725	2.086	2.528	2.845	3.552
21	1.323	1.721	2.080	2.518	2.831	3.527
22	1.321	1.717	2.074	2.508	2.819	3.505
23	1.319	1.714	2.069	2.500	2.807	3.485
24	1.318	1.711	2.064	2.492	2.797	3.467
25	1.316	1.708	2.060	2.485	2.787	3.450
26	1.315	1.706	2.056	2.479	2.779	3.435
27	1.314	1.703	2.052	2.473	2.771	3.421
28	1.313	1.701	2.048	2.467	2.763	3.408
29	1.311	1.699	2.045	2.462	2.756	3.396
30	1.310	1.697	2.042	2.457	2.750	3.385
40	1.303	1.684	2.021	2.423	2.704	3.307
50	1.299	1.676	2.009	2.403	2.678	3.261
60	1.296	1.671	2.000	2.390	2.660	3.232
80	1.292	1.664	1.990	2.374	2.639	3.195
100	1.290	1.660	1.984	2.364	2.626	3.174
∞	1.282	1.645	1.960	2.326	2.576	3.091

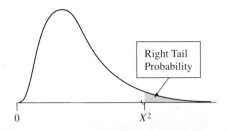

TABLE C Chi-Squared Distribution for Values of Various Right Tail Probabilities

df	Right-Tail Probability						
	0.250	0.100	0.050	0.025	0.010	0.005	0.001
1	1.32	2.71	3.84	5.02	6.63	7.88	10.83
2	2.77	4.61	5.99	7.38	9.21	10.60	13.82
3	4.11	6.25	7.81	9.35	11.34	12.84	16.27
4	5.39	7.78	9.49	11.14	13.28	14.86	18.47
5	6.63	9.24	11.07	12.83	15.09	16.75	20.52
6	7.84	10.64	12.59	14.45	16.81	18.55	22.46
7	9.04	12.02	14.07	16.01	18.48	20.28	24.32
8	10.22	13.36	15.51	17.53	20.09	21.96	26.12
9	11.39	14.68	16.92	19.02	21.67	23.59	27.88
10	12.55	15.99	18.31	20.48	23.21	25.19	29.59
11	13.70	17.28	19.68	21.92	24.72	26.76	31.26
12	14.85	18.55	21.03	23.34	26.22	28.30	32.91
13	15.98	19.81	22.36	24.74	27.69	29.82	34.53
14	17.12	21.06	23.68	26.12	29.14	31.32	36.12
15	18.25	22.31	25.00	27.49	30.58	32.80	37.70
16	19.37	23.54	26.30	28.85	32.00	34.27	39.25
17	20.49	24.77	27.59	30.19	33.41	35.72	40.79
18	21.60	25.99	28.87	31.53	34.81	37.16	42.31
19	22.72	27.20	30.14	32.85	36.19	38.58	43.82
20	23.83	28.41	31.41	34.17	37.57	40.00	45.32
25	29.34	34.38	37.65	40.65	44.31	46.93	52.62
30	34.80	40.26	43.77	46.98	50.89	53.67	59.70
40	45.62	51.80	55.76	59.34	63.69	66.77	73.40
50	56.33	63.17	67.50	71.42	76.15	79.49	86.66
60	66.98	74.40	79.08	83.30	88.38	91.95	99.61
70	77.58	85.53	90.53	95.02	100.4	104.2	112.3
80	88.13	96.58	101.8	106.6	112.3	116.3	124.8
90	98.65	107.6	113.1	118.1	124.1	128.3	137.2
100	109.1	118.5	124.3	129.6	135.8	140.2	149.5

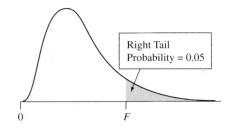

Right Tail Probability = 0.05

TABLE D *F* Distribution for Values of Right-Tail Probability = 0.05

df_2	df_1									
	1	2	3	4	5	6	8	12	24	∞
1	161.45	199.50	215.71	224.58	230.16	233.99	238.88	243.91	249.05	254.31
2	18.51	19.00	19.16	19.25	19.30	19.33	19.37	19.41	19.45	19.50
3	10.13	9.55	9.28	9.12	9.01	8.94	8.85	8.74	8.64	8.53
4	7.71	6.94	6.59	6.39	6.26	6.16	6.04	5.91	5.77	5.63
5	6.61	5.79	5.41	5.19	5.05	4.95	4.82	4.68	4.53	4.37
6	5.99	5.14	4.76	4.53	4.39	4.28	4.15	4.00	3.84	3.67
7	5.59	4.74	4.35	4.12	3.97	3.87	3.73	3.57	3.41	3.23
8	5.32	4.46	4.07	3.84	3.69	3.58	3.44	3.28	3.12	2.93
9	5.12	4.26	3.86	3.63	3.48	3.37	3.23	3.07	2.90	2.71
10	4.96	4.10	3.71	3.48	3.33	3.22	3.07	2.91	2.74	2.54
11	4.84	3.98	3.59	3.36	3.20	3.09	2.95	2.79	2.61	2.40
12	4.75	3.89	3.49	3.26	3.11	3.00	2.85	2.69	2.51	2.30
13	4.67	3.81	3.41	3.18	3.03	2.92	2.77	2.60	2.42	2.21
14	4.60	3.74	3.34	3.11	2.96	2.85	2.70	2.53	2.35	2.13
15	4.54	3.68	3.29	3.06	2.90	2.79	2.64	2.48	2.29	2.07
16	4.49	3.63	3.24	3.01	2.85	2.74	2.59	2.42	2.24	2.01
17	4.45	3.59	3.20	2.96	2.81	2.70	2.55	2.38	2.19	1.96
18	4.41	3.55	3.16	2.93	2.77	2.66	2.51	2.34	2.15	1.92
19	4.38	3.52	3.13	2.90	2.74	2.63	2.48	2.31	2.11	1.88
20	4.35	3.49	3.10	2.87	2.71	2.60	2.45	2.28	2.08	1.84
21	4.32	3.47	3.07	2.84	2.68	2.57	2.42	2.25	2.05	1.81
22	4.30	3.44	3.05	2.82	2.66	2.55	2.40	2.23	2.03	1.78
23	4.28	3.42	3.03	2.80	2.64	2.53	2.37	2.20	2.01	1.76
24	4.26	3.40	3.01	2.78	2.62	2.51	2.36	2.18	1.98	1.73
25	4.24	3.39	2.99	2.76	2.60	2.49	2.34	2.16	1.96	1.71
26	4.23	3.37	2.98	2.74	2.59	2.47	2.32	2.15	1.95	1.69
27	4.21	3.35	2.96	2.73	2.57	2.46	2.31	2.13	1.93	1.67
28	4.20	3.34	2.95	2.71	2.56	2.45	2.29	2.12	1.91	1.65
29	4.18	3.33	2.93	2.70	2.55	2.43	2.28	2.10	1.90	1.64
30	4.17	3.32	2.92	2.69	2.53	2.42	2.27	2.09	1.89	1.62
40	4.08	3.23	2.84	2.61	2.45	2.34	2.18	2.00	1.79	1.51
60	4.00	3.15	2.76	2.53	2.37	2.25	2.10	1.92	1.70	1.39
120	3.92	3.07	2.68	2.45	2.29	2.18	2.02	1.83	1.61	1.25
∞	3.84	3.00	2.60	2.37	2.21	2.10	1.94	1.75	1.52	1.00

TABLE E Table of Random Numbers

Line/Col.	(1)	(2)	(3)	(4)	(5)	(6)	(7)	(8)
1	10480	15011	01536	02011	81647	91646	69179	14194
2	22368	46573	25595	85393	30995	89198	27982	53402
3	24130	48360	22527	97265	76393	64809	15179	24830
4	42167	93093	06243	61680	07856	16376	39440	53537
5	37570	39975	81837	16656	06121	91782	60468	81305
6	77921	06907	11008	42751	27756	53498	18602	70659
7	99562	72905	56420	69994	98872	31016	71194	18738
8	96301	91977	05463	07972	18876	20922	94595	56869
9	89579	14342	63661	10281	17453	18103	57740	84378
10	85475	36857	53342	53988	53060	59533	38867	62300
11	28918	69578	88231	33276	70997	79936	56865	05859
12	63553	40961	48235	03427	49626	69445	18663	72695
13	09429	93969	52636	92737	88974	33488	36320	17617
14	10365	61129	87529	85689	48237	52267	67689	93394
15	07119	97336	71048	08178	77233	13916	47564	81056
16	51085	12765	51821	51259	77452	16308	60756	92144
17	02368	21382	52404	60268	89368	19885	55322	44819
18	01011	54092	33362	94904	31273	04146	18594	29852
19	52162	53916	46369	58586	23216	14513	83149	98736
20	07056	97628	33787	09998	42698	06691	76988	13602

Short Answers to Selected Odd Numbered Exercises

Chapter 1

1.1 (a) The sample, the randomization, and the plan to obtain percentages of each group that have heart attacks **(b)** The actual percentages of the people in the sample who have heart attacks **(c)** The use of statistical methods to predict **1.3** 64.8%; 20.9%; 8.6%; 5.8% **1.5** Responses will differ for each student **1.7 (a)** the new Honda Accord car **(b)** the new Honda Accords that are chosen for the study; **(c)** all new Honda Accords **1.9 (a)** Descriptive statistics; they summarize data from a population. **(b)** Parameters, since refer to population **1.11 (a)** Very short bars toward the top indicate very few old men and women in 1750. **(b)** Bars are much longer for both men and women in 1999 than in 1750. **(c)** The bars for women in their 70's and 80's in 1999 are longer than those for men. **(d)** The bars of people who were born right after World War II are the longest. **1.13 (a)** Yes; **(b)** No. Two random samples may differ by chance. **1.15** The results for this exercise will vary. **1.17** Every person in the population does not have the same chance to be in the sample. **1.19** Answer is data file printed from software. **1.21** With the applet, most samples fall within 14 points of the true population percentage—from about 56 to 84. **1.23 (a)** All American adults **(b)** The sample data are summarized by a proportion, 0.638 **(c)** the population proportion who had at least one such experience **1.25 (a)** sample data; **(b)** percentages reported; **(c)** reporting margin of error for sample percentage **1.27 (a)** Population: all customers; sample: 500 customers **(b)** Descriptive: average sales per person in sample; Inferential: estimate of average sales per person in population. **1.31** Will be different for each student. **1.33** You will get a sample proportion as large as 0.638 when the true proportion is 0.20, only *very* rarely.

Chapter 2

2.1 (a) Categorical: observations belong to one of a set of categories. Quantitative: observations are numerical. **(b)** Examples will differ for each student. **2.3 (a)** quantitative; **(b)** categorical; **(c)** categorical; **(d)** quantitative **2.5 (a)** Discrete: possible variables are separate values, such as 0,1,2,... Continuous: possible variables form an interval. **(b)** Examples will differ for each student. **2.7 (a)** continuous; **(b)** discrete; **(c)** discrete; **(d)** continuous **2.9 (b)** bar chart; **(c)** We get a better sense of the data when we can see the sizes of the various categories. **2.11 (a)** categories of a variable **(b)** frequency and percentage of weather stations in a given region **(c)** bar graph; we can more easily compare the heights of bars. **2.13 (i.)** Graph shows proportions alphabetically; **(ii.)** Pareto chart. The Pareto chart allows us to easily see what outcomes occurred most frequently. **2.15 (b)** Prices fall between $178 and $255, with more PDAs selling at the upper end of this range. **2.17 (a)** 1000 milligrams **(b)** 3000 milligrams **2.19 (a)** 1 to 3, 3 to 5, 5 to 7, 7 to 9, 9 to 11, 11 to 13, 13 to 15, and 15 to 17. **(b)** Bimodal; child cereals, on average, have much more sugar than adult cereals have. **(c)** The dot and stem-and-leaf plots allow us to see all the individual

data points. **(d)** The relative differences among bars would remain the same. **2.21 (a)** Skewed to the right because of some very expensive homes **(b)** Skewed to the right because of the few faculty who overdraw frequently **(c)** Symmetric because most would be in the middle, with some higher and some lower **(d)** Symmetric because most would fall in the middle, with some higher and some lower **2.23 (a)** Discrete; the value for each person would be a whole number. **(b)** (i) 0 ; (ii) 9 ; (iii) 2; (iv) 3 **(c)** unimodal and somewhat skewed to the right. **2.25 (a)** somewhat skewed to the left **(b)** A time plot connects the data points over time in order to show trends. **(c)** A histogram shows us the number of observations at each level more easily then does the time plot. We also can see the shape of the distribution of temperatures from the histogram but not from the time plot. **2.27** Overall, the time plot does seem to show a decrease in temperature over time. **2.29 (a)** mean: 537.14; median: 316 **(b)** Some of these countries have much larger populations than others. **2.31** Both distributions are skewed to the right; when this happens, means are higher than medians. **2.33** The outlier of 0 causes the mean to be lower, but does not affect the median. **2.35 (a)** Mean: 2; median: 0; and mode: 0 **(b)** Mean: 10; median: 0 outlier affects mean but not median **2.37 (a)** The most common score (mode), and the middle score (median), will be zero. The few very expensive hospital stays will inflate the mean. **(b)** number of times arrested in previous year. **2.39** Skewed to the right because the mean is higher than the median. **2.41 (a)** 0.16; **(b)** 0; **(c)** median: 0; mean: 1.95; **(d)** The median is the same for both because the median ignores much of the data. **2.43 (a)** Range = 6; there are six days separating the fewest and most sick days taken. **(b)** s=2.37; the typical distance of an observation from the mean is 2.37. **(c)** Range = 60; standard deviation = 21.06. Both increase when an outlier is added. **2.45** 12 is most realistic. −10 is negative; 0 indicates *no* variability; 3 is too small and 63 is too large for a typical deviation. **2.47 (a)** (i) the right; (ii) middle **(b)** the left because it is bell shaped. **2.49 (a)** 99–167. **(b)** 184. Yes. It is very high; all or nearly all observations will fall within three standard deviations from the mean. **2.51 (a)** One: −0.21 and 0.59; two: −0.61 and 0.99; three: −1.01 and 1.39. There are more observations within one standard deviation of the mean and fewer within two standard deviations than would be predicted by the Empirical Rule. **(b)** This is not a bell-shaped distribution but is highly skewed to the right. **2.53 (a)** Skewed to the right. The standard deviation is larger than the mean and the mean is higher than the median. **(b)** No. The data are not bell shaped. **2.55 (b)** Variable = unemployment, standard deviation = 3.695. This is a typical deviation from the mean unemployment for EU nations. **2.57 (a)** 31; **(b)** 25.5; **(c)** 36; **(d)** 25% of countries have residents who average fewer than 25.5 vacation days, half have residents who average fewer than 31, and 75% of countries have residents who average fewer than 36. **2.59 (a)** One fourth had maximum bench press less than 70 pounds, and one fourth had maximum bench press greater than 90 pounds. **(b)** Symmetric. The mean and median are the about the same, and the first and third quartiles are equidistant from the median. **2.61 (a)** The range is more affected by an outlier, and the standard deviation uses all the data. **(b)** IQR is not be affected by an outlier. **(c)** The standard deviation takes into account the values of *all* observations. **2.63 (a)** Range = 6; there are six days separating the fewest and most sick days taken. **(b)** The range of the middle 50% of scores is 2. **(c)** a. Range: 60; b. IQR: 2. The IQR is least affected by the outlier because it doesn't take the magnitudes of the two extreme scores into account. **2.65 (a)** Skewed to the left because the maximum is closer to the median than is the minimum. **(b)** Numbers are approximate: Min: 50; Q1: 53; Median: 54.2, Q3: 55; Max: 57.3 Skewed to the left because the median is closer to the maximum and Q3 than it is to the minimum and Q1. **2.67 (a)** Minimum: 0; Q1: 0; Median: 0; Q3: 4; Maximum: 10 (part of answer is box plot) **(b)** The middle scores of the whole set of data and of the lower half of the data set are both zero. **(c)** The minimum score also is zero. **2.69 (a)** Outer edges of box: 4.6 and 9.2; ends of whiskers: 3.9, 11.2 **(b)** No; it is only 1.78 standard deviations above the mean. **(c)** 7.1 **2.71 (a)** −3.1 **(b)** The height is below the mean. **(c)** Yes; it is more than three standard deviations from the mean. **2.73 (a)** The distribution depicted in the boxplot (part of answer is box plot) is skewed to the right. One reason that we can tell this is that there are several outliers (indicated by asterisks). **(b)** There are several potential outliers; the 1.5 (IQR) criterion indicates that the data should fall between about −7.5 and 20.5. **2.75 (a)** The percentages do not add up to 100, and Tesco slice seems too large for 27.2%. Also, contiguous colors are similar. **(b)** Bar graph. We would merely have to identify the highest bar. **2.77 (a)** Because the vertical axis does not start at 0, it appears that six times as many people are in the "no, not" column than in the "yes, safer" column. **(b)** With a pie chart, the area of each slice represents the entire percentage who fall in that category. Therefore, the relative sizes of the slices will always represent the relative percentages in each catagory. **2.79 (a)** quantitative; **(b)** quantitative; **(c)** categorical; **(d)** categorical **2.81 (a)** categorical; add percentages: 14.0, 9.6, 36.4, 6.2, 25.5, 8.3, and 100 for total **(b)** These categories are not ordered, and so we would use the mode: Central America. **(c)** The place of birth categories would be organized from highest to lowest percentages. It's much easier to make comparisons and to identify the most common outcomes.

2.83 (a) optical scanning with a two-column ballot **(b)** one column with either optical scanning or votomatic **(c)** bar for each combination (e.g., optical, one column; optical, two columns) **2.85 (c)** 3.5 **(d)** slightly skewed to the right **2.87 (b)** allows one to see the individual scores **(d)** This is a bimodal distribution. **2.89 (a)** skewed to the right, since most values at 0 but some large values **(b)** skewed to the left, since most values at 1 hour or slightly less but some could be quite a bit less **(c)** skewed to the right, since some extremely large values **(d)** skewed to the left, since most values are high but some very young people die **2.91** Larger; right skew would pull the mean higher. **2.93** $1.2 million was the mean because it would be inflated by outliers and severe right skew. **2.95 (a)** 12 is best; −10 is negative (not possible); 1 indicates almost no variability; 60 is almost as large as the whole range. **(b)** −20; standard deviations must be nonnegative.

2.97 (a) Skewed to the right; maximum is much farther from the mean than the minimum is, and the lowest possible value of 0 is only 780/560 = 1.54 standard deviations below the mean. **(b)** Skewed to the right; the standard deviation is almost as large as the mean and the smallest possible value is 0. **2.99 (a)** Nearly all scores would fall between 62.2 and 79.6. **(b)** The center for women is about five inches less than the center for men. The spread, however, is very similar. **(c)** 3.07 **2.101 (a)** Numbers are approximate: Min: 1; Q1: 3; median: 10; Q3: 12; Max: 15 **(b)** Skewed to the left **(c)** −1.58. This sugar value falls 1.58 standard deviations below the mean. **2.103 (a)** Symmetric, although perhaps slightly skewed to the left. The mean and median are almost the same. There is very little spread. (histogram is part of the answer) **(b)** The box plot does not indicate any outliers. (box plot is part of the answer) **(c)** Minimum: 50.0; Q1: 53.1; median: 54.2; Q3: 55.1; Maximum: 57.2 The 5-number summary suggests a roughly symmetric distribution. **(d)** Mean: 54.1; standard deviation: 1.4 **2.105 (a)** Skewed to the right; the distance from the median and from Q3 to the maximum is longer than the distance from the median and from Q1 to the minimum. **(b)** 4 most plausible; −16 is negative; 0 would indicate no variability; 15 is almost as large as the whole range, and 25 is even larger. **2.107 (a)** Range: 14.1; IQR: 6.5 **(b)** There are scores beyond the limit indicated by the criterion of 1.5 (IQR) from Q1 or Q3. **2.109 (a)** The z-score of 1.19 indicates that a person with a blood pressure of 140 falls 1.19 standard deviations above the mean. **(b)** 89 to 153 **2.111 (a)** Neither the minimum nor the maximum score has a z-score more extreme than 3 or −3. **(b)** The maximum is much farther from the mean and median than is the minimum. **(c)** This has a z-score of only −0.62, and so is not unusual. **2.113** The responses will be different for each student depending on the methods used. **2.115 (a)** The midpoint for the adult cereals is 5500 milligrams, with a shape skewed to the right. The midpoint for the child cereals is between 11,000 and 12,000 milligrams with a shape skewed to the left. **(b)** The distributions are fairly similar. (back-to-back stem-and-leaf plot is part of the answer) **2.117 (a)**

$$\bar{x} = \frac{\sum x}{n} = \frac{47(1) + 143(2) + 159(3) + 522(4) + 209(5) + 210(6) + 41(7)}{1,331} =$$

$\frac{5,490}{1,331} = 4.12$ **(b)** 4 **(c)** The median, 4, is the middle score. It is the 666th score. **2.119** (a) **2.121** (a) **2.123** Dividing the frequency for a given category by the total number of subjects would give us the proportion. **2.125** $\bar{x} \pm 3s$ would encompass all or nearly all scores. A range of $3s + 3s = 6s$ contains all or nearly all scores, so approximates the range. **2.127 (a)** With greater spread, numbers tend to be further from the mean. Thus, the absolute values of their deviations from the mean would be larger. When we take the average of all these values, the overall *MAD* is larger than with distributions with less spread. **(b)** Because we square the deviations, a large deviation has greater effect.

Chapter 3

3.1 (a) explanatory: high school GPA; response: college GPA **(b)** explanatory: mother's religion; response: number of children **(c)** explanatory: marital status; response: happiness **3.3 (a)** The response variable is happiness and the explanatory variable is income.

(b) INCOME

	Not Too Happy	Pretty Happy	Very Happy	Total	n
Above average	.11	.57	.32	**1.00**	158
Average	.10	.58	.32	**1.00**	453
Below average	.13	.58	.29	**1.00**	241

There does not seem to be much if any association between these variables. **(c)** .31 **3.5 (a)** response: binge drinking; explanatory: gender **(b)** (i) 1908; (ii) 2854 **(c)** No; these are not proportions of male and female students; these are counts and there are far more females than males in this study.

(d) Gender

	Binge Drinker	Non-Binge Drinker	Total	n
Male	0.49	0.51	**1.00**	3925
Female	0.41	0.59	**1.00**	6979

(e) It appears that men are more likely than are women to be binge drinkers. **3.7 (a)** response: party identification; explanatory: gender **(b)** (i) 0.14; (ii) 0.15 **(c)** (i) 0.44; (ii) 0.28 **(d)** Conditional proportions suggest that more women than men are Democratic and that more men than women are Independent or Republican. **3.9 (a)** positive; as cars age, they tend to have covered more miles. **(b)** negative; as cars age, they tend to be worth less. **(c)** positive; older cars tend to have needed more repairs. **(d)** negative; heavier cars tend to travel fewer miles on a gallon of gas. **3.11 (a)** (box plot is part of answer); skewed to the right **(b)** U.S. (34.3, 19.7) **(c)** strong (the closer a correlation is to 1.0 in absolute value, the stronger it is) **3.13 (a)** GDP **(b)** fertility **(c)** negative correlation, so countries with higher fertility tend to have lower internet use. **3.15 (a)** answer is scatterplot **(b)** 4, 13. **(c)** The value of 13 would have to be changed to 10. **3.17** answer is scatterplot **3.19 (a)** (i) weight; (ii) price. **(b)** Bikes with weights in the middle tend to cost the most. (scatterplot is part of answer) **(c)** negative and fairly small; weight does not appear to affect price strongly in a linear manner. **3.21 (a)** Both box plots indicate that the data are skewed to the right with few counties in the high ranges of vote counts. (two box plots are part of the answer) **(b)** The point close to 3500 on the variable "Buchanan" is an outlier; we were unable to make this comparison from the box plots because there were two separate depictions, one for each candidate. (scatterplot is part of the answer) **(c)** around 1000 **(d)** The point close to 3500 on the variable "Buchanan" is still an outlier. (two box plots and a scatterplot are part of the answer) **3.23 (a)** (i) 6.47; (ii) 5.75 Connect the points ($x = 10, y = 6.47$) and ($x = 40, y = 5.75$). **(b)** The y-intercept indicates that when a person cannot do any sit-ups, she/he would be predicted to run the 40-yard dash in 6.71 seconds. The slope indicates that every increase of one sit-up leads to a decrease in predicted running time of 0.024 seconds. **(c)** The slope indicates a negative correlation. (The slope and correlation have the same sign.) **3.25 (a)** Slope and correlation always have the same sign. **(b)** 46.61%. **(c)** −23.3; it is 23.3% lower than what would be predicted by this equation. **3.27 (a)** (i) 1.5; (ii) 13.1 **(b)** predicted value: 13.1; residual: 6.6; the United States is doing worse than would be predicted in terms of carbon dioxide emissions. **(c)** predicted value: 11.0; residual: −5.3. Switzerland is doing much better than would be predicted with respect to carbon dioxide emissions.

3.29 (a) The software calculates the line for which the sum of squares of the residuals is a minimum. **(b)** The predicted value is 193.9. The residual is −193.9. **(c)** The residual for this point is large relative to the other points on the scatterplot. **(d)** There is an apparent outlier represented by the leftmost bar. **3.31 (a)** answer is a scatterplot **(b)** 1.0; $\hat{y} = 4 + 2x$. **(c)** advertising: 1, 1; sales: 6, 2 **(d)** $\hat{y} = 4 + 2x$; when there is no advertising, it is predicted that sales will be $4,000. For each increase of $1000 in advertising, predicted sales increase by $2000. **3.33 (a)** $\hat{y} = 24 + 0.7x$; predicted final exam score = 80 **3.35. (a)** (answer is a scatterplot) **(b)** From MINITAB: price = 1896 − 40.5 weight. Predicted price decreases by $40.5 for each pound increase in weight. No, because there were no weights around 0. **(c)** $681 **3.37 (a)** From MINITAB: Men_Meters = − 9.01 + 0.00570 Year_Men; for each increase of one year, the predicted amount of meters increased by 0.00570. **(b)** From MINITAB: Women_Meters = − 11.9 + 0.00698 Year_Women; for each increase of one year, the predicted amount of meters increased by 0.00698. **(c)** Men: 2.41; women: 2.09 **3.39 (a)** The observation in the lower left of the scatterplot; it is a regression outlier. **(b)** The prediction from the regression line; it uses the available data to predict. **(c)** No; this data point is very far beyond the data used to determine this line. **3.41 (a)** $\hat{y} = −3.1 + 0.33(15) = 1.85$; $\hat{y} = −3.1 + 0.33(40) = 10.1$ **(b)** $\hat{y} = 8.0 − 0.14(15) = 5.9$; $\hat{y} = 8.0 − 0.14(40) = 2.4$ **(c)** D.C. is a regression outlier because it is well removed from the trend of the rest of the data. **(d)** Because D.C. is so high on both variables, it pulls the line upwards on the right and suggests a positive correlation, when the rest of the data (without D.C.) are negatively correlated. The relationship is best summarized after removing D.C. **3.43 (a)** Yes on all three counts. **(b)** (1) the x value (the United States in this case) is relatively low or high compared to the rest of the data; (2) the observation (the United States) is a regression outlier, falling quite far from the trend that the rest of the data follow. TV watching in the United States is very high despite the very low birth rate. **(c)** The association is (i) very weak without the United States because the six countries, although they vary in birth rates, all have very few televisions, and is (ii) very strong with the United States because the United States is so much higher in number of televisions and so much lower on birth rate that it makes the two variables seem related. **(d)** Because that point has a large effect on pulling the line downward. **3.45 (a)** The point at $x = 0$ has an x value far from all the others and falls far from the trend of the rest of the data. (scatterplot is part of answer) **(b) All data points**: SUGAR(g) = 13.6 − 0.0290 SODIUM(mg); −0.453. **All except the regression outlier**: SUGAR(g) = 18.1 − 0.0503 SODIUM(mg); −0.623 This point lowers the intercept and makes the slope less steep. **3.47 (a)** Not likely. **(b)** Values would be higher on *both* variables as age increases. **(c)** There is an overall positive correlation between height (in inches) and vocabulary (assessed on a scale of 1–10) if we ignore grade; if we look within each grade, we see roughly a horizontal trend and no particular association. (scatterplot part of answer) **3.49 (a)** several possible responses to this exercise (e.g., students who watch TV more may be more likely to see the ads and less likely to spend time with their peers who smoke pot) **(b)** Pot smoking (or the lack thereof) might be caused by many variables (e.g., antidrug ads, television watching, regular school attendance, parental influence, neighborhood type). **3.51 (a)** response: eighth-grade math scores; explanatory: state **(b)** The third variable is race. Nebraska has the overall higher mean because there is a higher percentage of whites and a lower percentage of blacks in Nebraska than in New Jersey, and overall, whites tended to have higher math scores than blacks. **3.53 (a)** response: assessed value; explanatory: square feet **(b)** response: political party; explanatory: gender **(c)** response: income; explanatory: education **(d)** response: pounds lost; explanatory: type of diet

3.55 (a)

Gender	Yes, Definitely	Yes, Probably	No, Probably Not	No, Definitely Not	Total
Male	0.56	0.24	0.11	0.10	**502**
Female	0.61	0.21	0.09	0.09	**625**

(b) Not much, but women are slightly more likely than are men to respond "yes, definitely." **3.57 (a)** income; quantitative **(b)** degree; categorical **(c)** A bar graph could have a separate bar for each degree type. **3.59 (a)** response: gender; explanatory: type of occupation

(b) Executive, administrative, and managerial

Year	Female	Male	Total
1972	0.197	0.803	**1.00**
2002	0.459	0.541	**1.00**

(c) yes; women made up a larger proportion of the executive work force in 2002 than in 1972. **(d)** year and type of occupation. **3.61 (a)** slope: 0.56; As urban nature goes from 0 to 100, predicted crime rate increases by 56. **(b)** Relatively strong, positive relationship **(c)** $b = r\left(\dfrac{s_y}{s_x}\right)$; $0.56 = 0.67(28.3/34.0)$

3.63 (a) response: salary; explanatory: height. **(b)** slope: 789 **(c)** An increase of seven inches (from 5 foot 5 to 6 feet) is worth a predicted $789 per inch, or $5523. **3.65** y-intercept: 0; slope: 1; this means that your predicted college GPA equals your high school GPA. **3.67 (a)** A 1000 increase in x would mean a predicted change in y of -5.2.(poorer mileage). **(b)** 12.32; 4.68. The Hummer gets 4.68 more miles/gallon than one would predict. **3.69 (a)** $4200 **(b)** $r = 0.79$; (i) positive; (ii) strong **3.71 (a)** When economic activity is 0%, predicted birth rate is much higher (36.3) than when economic activity is 100% (6.3). **(b)** women's economic activity; the correlation between birth rate and women's economic activity is bigger in magnitude than the correlation between birth rate and GNP. **3.73 (a)** $-16,000$; **(b)** 3200; **(c)** same (0.50); correlation does not depend on the units used **3.75 (a)** D.C. is the outlier to the far, upper right. This would have an effect on the regression analysis because it is a regression outlier. **(b)** When D.C. is included, the y-intercept decreases and the slope increases. **3.77 (a)** The point with a y-value of around 1500 (scatterplot is part of answer) **(b)** violent crime rate $= 2545 - 24.6$ high school; for each increase of 1% of people with a high school education, the predicted violent crime rate decreases by 24.6. **(c)** violent crime rate $= 2268 - 21.6$ high school; for each increase of 1% of people with a high school education, predicted violent crime rate decreases by 21.6, similar to the slope in (b). **3.79 (a)** negative relationship (scatterplot is part of answer) **(b)** -0.451; negative association. **(c)** Health Ins $= 49.2 - 0.423$

HS Grad Rate; for each increase of one in high school graduation rate, the predicted percentage of individuals without health insurance goes down by 0.423; negative relationship. **3.81 (a)** Men tend to be taller and make more money than women. **(b)** When measured, a lurking variable becomes a confounding variable. **3.83 (a)** As people age, they might both sleep more *and* be more likely to die. **(b)** Subject's age might actually cause people to sleep more and be more likely to die. **3.85** The responses will be different for each student depending on the methods used. **3.87** Temp $= 119 - 0.0294$ Year; The regression line indicates a very slight decrease over time, the opposite of what is indicated by the Central Park data. **3.89** San Francisco could be higher than other cities on lots of variables, but that does not mean those variables cause AIDS, as association does not imply correlation. An alternative explanation is that San Francisco has a relatively high gay population, and AIDS is more common among gays. **3.91 (a)** It's dangerous to extrapolate far beyond one's data. **(b)** Correlation is not causation, for example, there could be a third variable (e.g., income). **(c)** This point would be a regression outlier, and could greatly affect the regression equation. **3.99 (a)** If we convert income from British pounds to dollars, we multiply each score by two. Thus, each y value doubles, the mean of y is now doubled, and the distance of each score from the mean doubles. If this is all so, then the variability (the standard deviation) has now doubled. **(b)** The correlation would not change in value, however, because the correlation is based on standardized versions of the measures—z-scores, rather than raw scores. **3.101 (a)** Algebraically, we can manipulate the formula $a = \bar{y} - b\bar{x}$ to become $\bar{y} = a + b\bar{x}$. We do this simply by adding $b\bar{x}$ to both sides of the equation. Now $\bar{y}$ is alone on one side of the equation because we have both added and subtracted $b\bar{x}$, canceling each other out. And the other side of the equation is $a + b\bar{x}$. The latter formula is very similar to the regression equation, except the generic predicted y, $\hat{y}$, is replaced by the mean of y, $\bar{y}$. Similarly, the generic x is replaced by $\bar{x}$. Thus, a score on any x that is at the mean will predict the mean for y. **(b)** Here are the algebraic steps to go from one formula to the other. *Step 1*: Because we know that $a = \bar{y} - b\bar{x}$, we can replace the a in the regression equation with this formula. $\hat{y} = \bar{y} - b\bar{x} + bx$. *Step 2*: We can now subtract $\bar{y}$ from both sides. It cancels itself out on the right, and is now subtracted from the left. $\hat{y} - \bar{y} = -b\bar{x} + bx$ or $\hat{y} - \bar{y} = bx - b\bar{x}$ (if we switch the two parts of the right-hand side of the equation). *Step 3*: Finally, we can take the b on the right and put it outside of a parenthesis to denote that it is to be multiplied by both variables within the parentheses. $\hat{y} - \bar{y} = b(x - \bar{x})$ This formula tells us that if we figure out how far from the mean our x is, we can multiple that deviation by the slope to figure out how far from the mean the predicted y is.

Chapter 4

4.1 (a) response: whether a subject had brain cancer; explanatory: cell phone use **(b)** observational study because the experimenters did not assign subjects to treatments **4.3** observational study because subjects are not assigned to treatments **4.5** For example, it is possible that schools that institute drug testing are those in higher crime areas than are those that did not choose to use drug testing. **4.7 (a)** response: whether woman's infant had a birth defect; explanatory: whether woman was a hairdresser **(b)** observational study because women were not randomly assigned to be hairdressers or not **(c)** No; it's possible that there's a lurking variable, such as socioeconomic status (e.g., people who are less wealthy or educated tending to have poorer health care). **4.9 (a)** Observational study; **(b)** Observational study; **(c)** Experiment **4.11** The seat belt incident might be the exception, rather than what is typical. Death rates are in fact higher for those who do not wear seat belts. **4.13** Depends on how the first census is defined. By one definition, the first census in Canada was conducted in 1665 (this was a census of "New France"): 3,215. The most recent Canadian census in 2001: 30,007,094. **4.19** Pick five-digit random numbers, ignoring 00000, numbers above 50,000, and duplicates. Keep the first ten numbers that are in range, and find the names associated with those ten numbers. **4.21 (a)** 10 percentage points; **(b)** 5 percentage points; **(c)** 2.5 percentage points. As n increases, the sample becomes a more accurate reflection of the population, and the margin of error decreases. **4.23 (a)** It provides negative information within the question (e.g., "symbol of past slavery"). **(b)** It has neither explicitly negative nor explicitly positive information. **4.25 (a)** all executive recruiters **(b)** intended: 1500; actual: 97; percentage of nonresponse: 93.5% **(c)** Possible sources: nonresponse (those who responded were not representative) and executive recruiters might not be the best source of industries that are hiring. **4.27 (a)** Hong Kong residents **(b)** Subjects were not assigned to have a particular perception. **(c)** This study used a volunteer sample; it may not be representative of the population. **4.29 (a)** Not all parts of the Los Angeles population have representation (because they may not have read the paper a long time). **(b)** They do not use random sampling; it uses the 1000 people who have subscribed the longest. **(c)** Those who take the time to respond might not be representative. There was a high percentage of people who did not respond. **(d)** The question frames the proposal in a positive way, perhaps skewing people's responses. **4.31 (a)** experiment; subjects are randomly assigned to treatments **(b)** among the possible answers: unethical, impossible to ensure that subjects do as assigned and smoke or not smoke according to the assignment, too long to wait for answer **4.33 (a)** experimental units: n rats; explanatory: type of radiation; response: whether tumors develop; treatments: analog cell phone frequency, digital cell phone frequency, and no radiation **(b)** The Australian study just considered brain cancer and the mice were not representative of the overall population of mice (they were specially susceptible to cancer). **4.35** Yes; so that they don't intentionally or unintentionally respond how they think they should **4.37 (a)** Recruit volunteers (the experimental units) with a history of high blood pressure; randomly assign to one of two treatments: the new drug or the current drug; explanatory: treatment type; response: blood pressure after experimental period. **(b)** To make the study double-blind, the two drugs would have to look identical so that neither the subjects nor the experimenters know what drug the subjects are taking. **4.39 (a)** The response: blood pressure, heart rate, and reported stress levels; explanatory: whether caffeine taken, with two treatments: caffeine pill or placebo pill; experimental units: 47 regular coffee drinkers **(b)** crossover

design because all subjects participate in both treatments on different days **4.41 (a)** No; the researchers are not randomly assigning subjects to living situation. **(b)** Yes; those living with smokers and those not living with smokers. **(c)** Yes; randomization of units to treatments occurs within blocks. **4.43 (a)** Number the students from 01 to 05. Pick two-digit numbers randomly. Select the first female student to have her number picked (01 to 03) and the first male student to have his number picked (04 or 05). **(b)** Because every sample of size two does not have an equal chance of being selected (e.g., samples with two males have probability 0 of being selected). **4.45 (a)** Because of the randomness, any number of the accounts sampled could be small. **(b)** Label accounts are 01 to 60. Select five accounts from 01 to 30 and five from 31 to 60. **4.47** It is possible that people with lung cancer had a different diet than did those without. These people might have eaten out at restaurants quite a bit, consuming more fat, and smoking socially. It could have been the fat and not the smoking that caused lung cancer. **4.49** In an observational study, we observe people in the groups they already are in. In an experiment, we actually assign people to the groups of interest. The major weakness of an observational study is that we can't control (such as by balancing through randomization) other possible factors that might influence the outcome variable. **4.51 (a)** explanatory: marriage status; response: health **(b)** People choose their own marital statuses. **(c)** No; one would not be able to assign people randomly to marital status. **4.53** the study; the story of one person is anecdotal and not as strong evidence as a carefully conducted study with a much larger sample size. **4.55 (a)** Restricting the range would not allow us to determine an association between these variables. **(b)** There are several possible responses. For example, we could select a simple random sample, or we could stratify the sample to get equal numbers of students from each year. **4.57 (a)** Pick three random numbers between 0001 nd 3500. **(b)** No; every sample is not equally likely; any possible sample with more than 40 males or fewer that 40 males has probability 0 of being chosen **(c)** stratified random sample; it offers the advantage of having the same numbers of men and women in the study, which would be unlikely if the population had a small proportion of one of these, and this is useful for making comparisons. **4.59 (a)** American adults; sampling bias (Americans who have no telephone or are rarely home would not be interviewed.) **(b)** The margin of error was likely calculated by dividing one by the square root of the sample size of 1000, and multiplying by 100 to get the percentage points. The range of believable values goes from 3 percentage points below to 3 above the reported percentage. **4.61 (a)** explanatory: history of playing video games; response: visual skills **(b)** observational study; because the men were not randomly assigned to treatment **(c)** There are several possible answers e.g., excellent reaction times might make it easier, and therefore more fun, to play video games, and might also lead to better performance on tasks measuring visual skills. **4.63 (a)** response: whether have a heart attack; explanatory: treatment group (aspirin or placebo) **(b)** because physicians were randomly assigned to treatment—either aspirin or placebo **(c)** Because the experiment is randomized, we can assume that the groups are fairly balanced with respect to exercise. **4.65 (a)** Among those in the aspirin group, 0.493 never smoked, 0.397 smoked in the past, and 0.110 are current smokers. Among those in the placebo group, 0.498 never smoked, 0.391 smoked in the past, and 0.111 currently smoke. These proportions are very similar. **(b)** Yes; the treatment groups are balanced in terms of smoking status. **(c)** It does seem that the randomization process did a good job in achieving balanced treatment groups in terms of smoking status. **4.67 (a)** Nicotine patch only: 6.4 percentage points; Zyban only: 6.4 percentage points; Nicotine patch with Zyban: 6.4 percentage points; Placebo only: 7.9 percentage points **(b)** Yes; the margin of error for Zyban only indicates that the low end of believable values is 23.9%, whereas the margin of error for Placebo only indicates that the high end of believable values is 23.5%. **(c)** No; there is substantial overlap between the ranges indicated by the margins of error. **(d)** The results of the study suggest that two of the treatments, Zyban only and Nicotine patch with Zyban, led to higher abstinence percentages than did either of the other two treatments, Nicotine only or Placebo only. **4.69 (a)** response: gas mileage; explanatory: brand of gas; treatments: Brands A and B **(b)** 10 cars would be randomly assigned to Brand A and 10 to Brand B. **(c)** Each car would be a block. It would first use gas from one brand, and then from the other. **(d)** It would reduce the effects of possible lurking variables because the two groups would be identical. **4.71 (a)** clusters; **(b)** because each possible sample does not have an equal chance. For instance, there is no chance of a sample in which just one person from a particular nursing home is sampled. **4.73** This is cluster random sampling. The colleges are the clusters. **4.75 (a)** Retrospective; people are identified based on current disease status and are asked questions about their pasts. Cases: those with brain cancer; controls: those without brain cancer. **(b)** The researchers could compare the cell phone use of the two groups to see if the people with brain cancer had used cell phones more than had the people who did not have brain cancer. **(c)** There could be a lurking variable. People with and without brain cancer might be different, on average, on a variable other than cancer status, such as age. **4.79** No; due to the volunteer nature of the sample. **4.83 (a)** The sample includes only one cluster. **(b)** Sampling just Fridays is an example of sampling bias (values might be higher than usual because days sampled are on the weekend). **4.85** This is not a random sampling method. People who approach the street corner are interviewed as they arrive (and as they agree to the interview!) and may not be representative of their group). **4.87 (a)** Such a study would measure how well a treatment works if patients believe in it, rather than how much a treatment works independent of subjects' beliefs about its efficacy. **(b)** Patients might be reluctant to be randomly assigned to one of the treatments because they might perceive it as inferior to another treatment. **(c)** He or she might feel that *all* patients should get the new treatment. **4.89** Since the disease is rare, this proposed study might not sample anyone that gets it. With the case-control study, one would find a certain number of people who already have the rare disease (the cases), and compare the proportion of these people who had received the vaccine to the proportion in a group of controls who did not have the disease. **4.99 (a)** Although every subject is equally likely to be chosen, every possible sample of 100 is not equally likely. **(b)** The company would determine the first item

using a randomly selected two-digit number between 01 and 50; the item that coincided with that number would be checked, and then every 50th item thereafter also would be checked. **4.101** Consider the hint. The mean population family size is 6. If you choose families, the possible values are 2 and 10, each equally likely, and the sampling is not biased. If you choose individuals, the value 2 has probability 2/12 and the value 10 has probability 10/12. You are likely to overestimate the mean size.

Chapter 5

5.1 The long-run relative frequency definition of probability refers to probability of a particular outcome as the proportion of times that the outcome would occur in a long run of observations. **5.3** No. In the short run, the proportion of a given outcome can fluctuate a lot. **5.5** False. For the typical pro, successive shots are independent, and what happens in shots already taken does not affect much if at all the chance of making future shots. **5.7** False. With a biased sampling design, having a large sample does not remove problems from the sample not being selected to represent the entire population. **5.9** We would be relying on our own judgment rather than objective information such as data, and so would be relying on the subjective definition of probability. **5.11 (b)** about 50 **(c)** The results will depend on the answer to part (a). **(d)** 42% of the answers were "true." We would expect 50%. They are not necessarily identical, because observed percentages of a given outcome can fluctuate in the short run. **(e)** There are some groups of answers that appear nonrandom. For example, there are strings of five "trues" and eight "falses," but this can happen by random variation. **5.13 (a)** Answer is tree diagram with 16 branches at end; **(b)** 64 **5.15 (a)** Answer is tree diagram with 16 possible outcomes. **(b)** The probability of each possible individual outcome is 1/16 =0. 0625. **(c)** 0.3125 **5.17** False. One could have two outcomes that do not have equal chances of occurring. **5.19 (a)** FFFF, FFFM, FFMF, FFMM, FMFF, FMFM, FMMF, FMMM, MMMM, MMMF, MMFM, MMFF, MFMM, MFMF, MFFM, and MFFF. **(b)** 1/16 = 0.0625 **(c)** We have to assume that it is equally likely to have a boy or a girl, and that the gender of each child is independent of the gender of the others. **5.21** 20 heads has probability $(1/2)$ to the 20th power, which is $1/1,048,576 = 0.000001$; Risk of a one in a million death: $1/1,000,000 = 0.000001$ **5.23 (a)** YS; YD; NS; and ND; **(b)** (i) 0.004; (ii)0.284; **(c)** 0.003 **(d)** The answer would have been $P(N \text{ and } D) = P(N) \times P(D) = (0.004)(0.28) = .001$. This indicates that death might depend on seat belt use. **5.25 (a)** There are eight possible outcomes. Answer is tree diagram. **(b)** 0.488 **(c)** They assumed independence, but that might not be so (e.g., three customers are friends, or three members of the same family). **5.27 (a)** $P(B|S)$; **(b)** $P(B^C|S^C)$; **(c)** $P(S^C|B^C)$; **(d)** $P(S|B)$ **5.29 (a)** 0.879; **(b)** 0.139 **5.31 (a)** .004; **(b)** (i) 0.001; (ii) 0.010 **(c)** Neither $P(D|\text{wore seat belt})$ nor $P(D|\text{didn't wear seat belts})$ equals $P(D)$; specifically, 0.001 and 0.010 are different from 0.004. Thus, the events are not independent. **5.33 (a)** $P(D|NEG) = 6/3927 = 0.0015$; **(b)** No; $P(NEG|D) = 6/54 = 0.111$ **5.35 (a)** 0.61; **(b)**(i) 0.68; (ii) 0.54; **(c)** No **5.37(a)** 0.30; **(b)** (i) 0.40; (ii) 0.40 **(c)** No; the probability that he will make the second shot depends on whether he made the first. **5.39 (a)** A: identification of moral values; B: voting for Bush; **(b)** 0.16;

(c) 0.314 **5.41 (a)** P(C|A) = P(C and A)/P(A) = (0.25)/ (0.5) = 0.5 **(b)** These are not independent events because P(C|A), 0.5, is not equal to P(C), 0.25 **5.43** No; 0.92 **5.45** Each student can be matched with 24 other students, for a total of 25(24) pairs. But this considers each pair twice (e.g., student 1 with student 2, and student 2 with student 1), so the answer is 25(24)/2=300. **5.49 (a)** The probability that the first will go multiplied by the probability that the second will go and so on for all 5.4 million, that is (1/5000) taken to the 5.4 million power, which is zero to a huge number of decimal places. **(b)** This solution assumes that each person decides independently of all others. This is not realistic because families and friends often make vacation plans together. **5.51 (a)** The first answer is a tree diagram. **(b)** P(POS) = P(S and POS) + P(S^c and POS) = 0.0086 + 0.1188 = 0.1274 **(c)** .068 **(d)** They are calculated by multiplying the proportion for each branch by the total number. The proportion of positive tests with breast cancer is 1/(1 + 12) = 0.08. **5.53 (a)** P(Innocent|Convicted) = P(Innocent and Convicted)/P(Convicted)=(0.005)/0.86)=0.006 **(b)** P(Innocent|Convicted)= P(Innocent and Convicted)/ P(Convicted) = (0.025)/0.50) = 0.05 **5.55 (a)** Tree diagram is part of answer. P(Innocent|Match) = P(Innocent and Match)/ P(Match) = 0.0000005/0.4950005 = 0.000001 **(b)** Tree diagram is part of answer. P(Innocent|Match) = P(Innocent and Match)/P(Match) = 0.00000099/0.00990099 = 0.00001; When the probability of being innocent is higher, there's a bigger probability of being innocent given a match. **(c)** P(Innocent| Match) can be much different from P(Match|Innocent). **5.57 (b)** (0.95) × (0.95) × (0.95)... × (0.95)—a grand total of twenty 0.95's (or 0.95^{20}) = 0.36. **5.59** The gender of a child is independent of the previous children's genders; the chance is still $^1/_2$. **5.61 (a)** There are 18 possible meals [(2 × 3 × 3 × 1) = 18]. (tree diagram is part of answer) **(b)** No; some menu options would be more popular than are others. **5.63 (a)** 0.805; **(b)** 0.648 **(c)** That the responses of the two subjects are independent; married couples share many beliefs. **5.65 (a)** 8; tree diagram is part of answer; **(b)** 0.125; **(c)** 0.343 **(d)** Three friends are likely to be similar on many characteristics that might affect performance. **5.67 (a)** 0.73; **(b)** 0.30 **5.69 (a)** 47%: conditioned on teen living with "hands-on" parents. 13%: teen living with "hands-off" parents. Both probabilities refer to an excellent relationship with fathers. **(b)** Hands-on: 127 for yes, and 143 for no. Hands off: 95 for yes, and 635 for no. **(c)** a) 0.428. **5.71 (a)** (1, 1); (1, 2); (1, 3); (1, 4); (1, 5); (1, 6); (2, 1); (2, 2); (2, 3); (2, 4); (2, 5); (2, 6); (3, 1); (3, 2); (3, 3); (3, 4); (3, 5); (3, 6); (4, 1); (4, 2); (4, 3); (4, 4); (4, 5); (4, 6); (5, 1); (5, 2); (5, 3); (5, 4); (5, 5); (5, 6); (6, 1); (6, 2); (6, 3); (6, 4); (6, 5); (6, 6) **(b)** (1, 1); (2, 2); (3, 3); (4, 4); (5, 5); and (6, 6); 0.167; **(c)** (1, 6); (2, 5); (3, 4); (4, 3); (5, 2); and (6, 1); 0.167. **(d)** (i) 0; (ii) 0.333; (iii) 0; **(e)** A and B are disjoint. You cannot roll doubles that add up to seven. **5.73** 0.80 **5.75** 0.67 **5.77 (a)** This strategy is a poor one; the chance of an even slot is the same on each spin of the wheel. **(b)** (18/37) to the 26 power, which is essentially 0 **(c)** No; events that seem highly coincidental are often not so unusual when viewed in the context of *all* the possible random occurrences at all times. **5.79** False negatives: test indicates that an adult smoker does not have lung cancer when he or she does have lung cancer. False positives: test indicates the presence of

lung cancer when there is none. **5.81 (b)** 0.048; because so few people have this cancer, most of the positive tests will be false. **5.83 (c)** 0.999 **(d)** A positive result is more likely to be in error when the prevalence is lower as relatively more of the positive results are for people who do not have the condition. Tree diagrams or contingency tables are part of the answer. **5.85 (a)** The cumulative proportion of heads approaches 0.50 with larger numbers of flips, illustrating the law of large numbers and the long-run relative frequency definition of probability. **(b)** The outcome will be similar to that in part (a), with the cumulative proportion of heads approaching one third with larger numbers of flips. **5.87 (a)** 0.545; **(b)** 0.505; **(c)** 0.500; as *n* increases, the cumulative proportion tends toward 0.50. **5.89** (1) specificity: if pregnant, 99% chance of a positive test. (2) sensitivity: if not pregnant, 99% chance of a negative test. (3) 99% chance of pregnancy given a positive test; (4) 99% chance of non-pregnancy given a negative test. **5.91 (a)** 0.024 **(b)** Multiply (1) probability of getting a high school degree by (2) probability of getting a college degree once you had a high school degree by (3) probability of getting a masters' once you had the earlier degrees by (4) probability of getting a Ph.D. once you had the earlier degrees. **(c)** 0.06 **5.93** The event of a person bringing a bomb is independent of the event of any other person bringing a bomb. **5.95** (c) and (d) **5.97** (e) **5.99** (b) **5.101** False **5.103** Being not guilty is a seperate event from the event of matching all the characteristics listed. It might be easiest to illustrate with a contingency table. Suppose there are 100,000 people in the population. We are given that the probability of a match is 0.001. So out of 100,000 people, 100 people would match all the characteristics. Now the question becomes, of those 100 people, what is the probability that a person would not be guilty of the crime? we don't know that probability. However, let's SUPPOSE in the population of 100,000, 5% of the people could be guilty of such a crime, 95% are not guilty. Here's a contingency table that illustrates this concept.

	Guilty	Not Guilty	**Total**
Match		???	**100**
No Match			**99900**
Total	**5000**	**95000**	**100000**

Thus, the P(not guilty|match) = ???/100.
5.105 When two events are not independent, P(A and B) = P(A) × P(B|A); if we think about (A and B), as one event, we can see that P[C and (A and B)] = P(A and B) × P(C|A and B).If we replace P(A and B) with its equivalent, P(A) × P(B|A), we see that P(A and B and C) = P(A) × P(B|A) × P(C|A and B).

Chapter 6

6.1 (a) uniform distribution; P(1) = P(2) = P(3) = P(4) = P(5) = P(6) = 1/6 **(b)** There will be a probability for each *x* from 2 through 12; for example, there are three rolls that add up to four (1,3; 2,2; 3,1); thus, the probability of four is 3/36 = 0.083. **(c)** $0 \le P(x) \le 1$; and $\sum P(x) = 1$. **6.3 (a)** $0 \le P(x) \le 1$; and $\sum P(x) = 1$; **(b)** 0.472 **6.5 (a)** Their corresponding probabilities are not the same; **(b)** 1.15 **6.7 (b)** μ = 0.5(200) + 0.5(−50) = 75; μ = 0.5(350) + 0.5(−100) = 125; wager two is better. **6.9 (a)** Males; the smallest and highest values (0, 4) have higher probabilities for males than females. **(b)** The responses for females tend to be closer to the mean than the responses for males are. **6.11 (a)** Someone could watch exactly one hour of TV or 1.8643 hours of TV. **(b)** Each person reported TV watching by rounding to the nearest whole number. **(c)** These curves would approximate the two histograms to represent the approximate distribution if we could measure TV watching in a continuous manner. Then, the area above an interval represents the proportion of people whose TV watching fell in that interval **6.13** 0.251 **6.15 (a)** 0.9495-0.0505 = 0.899, which rounds to 0.90 **(b)** 0.9951-0.0049 = 0.9902, which rounds to 0.99 **(c)** 0.7486-0.2514 = 0.4972, which rounds to 0.50 **6.17 (a)** Divide by two for amount in each tail, 0.005. Subtract from 1.0 for cumulative probability, 0.995. Look up this probability on Table A to find the z-score of 2.58. **(b)** (a) 1.96; (b) 1.64; graph is part of answer **6.19 (a)** 0.674; **(b)** 1.64 **6.21 (a)** 1.19; **(b)** 0.1170; **(c)** 0.7879 **6.23 (a)** (i) 0.1056; (ii) 0.8944; **(b)** 137.28; **(c)** 62.72 **6.25 (a)** 6.43; yes; it is far more than 3 standard deviations above the mean. **(b)** No; the standard deviation is higher than the mean, so the lowest possible value of 0 is less than one standard deviation below the mean. **6.27** 600; it is further from the mean in terms of the number of standard deviations. **6.29 (a)** Sample space: (SSS, SSF, SFS, SFF, FSS, FSF, FFS, FFF); probability of each is 0.125. The probabilities for $x = 0, 1, 2, 3$ are $P(0) = 0.125, P(1) = 0.375, P(2) = 0.375, P(3) = 0.125$. **(b)** same probabilities as in (a); e.g.,

$$P(0) = \frac{3!}{0!(3-0)!} 0.5^0 (1 - 0.5)^{3-0} = 0.125$$

6.31 (a) $P(0) = 0.125, P(1) = 0.375, P(2) = 0.375, P(3) = 0.125$; answer includes symmetric graph **(b)** $P(0) = 0.512, P(1) = 0.384, P(2) = 0.096, P(3) = 0.008$; answer includes graph that is not symmetric **6.33 (a)** Same probability of success for each free throw, and results of different free throws are independent. Each trial is binary (make or do not make free throw). **(b)** $n = 10$; $p = 0.90$ **(c)** (i) 0.349; (ii) 0.387 **6.35 (a)** $n = 60$; $p = 1/6 = 0.167$ **(b)** $\mu = 10$; $\sigma = 2.89$ **(c)** Yes; 0 is well over three standard deviations from the mean.

(d) $P(0) = \dfrac{60!}{0!(60-0)!} 0.1667^0 \ (1 - 0.1667)^{60-0} =$

$= (1)(1)(0.0000177) = 0.0000177$

6.37 (a) Yes; binary data (Hispanic or not), same probability of success for each trial (i.e., 0.40), and independent trials (whom you pick for the first juror is not likely to affect whom you pick for the others, and $n < 10\%$ of the population size). $n = 12$; $p = 0.40$. **(b)** 0.0022 **(c)** Yes; there is only a 0.22% chance that this would occur if the selection were done randomly. **6.39 (a)** Yes; the sample is less than 10% of the size of the population size. **(b)** No; neither (np) nor $[n(1 - p)]$ is at least 15. Both are 5. **6.41 (a)** The formula relies on a given number of trials, n. **(b)** The binomial applies for X. We only have n for X, not Y. $n = 3$, $p = 0.50$. **6.43 (a)** First six: AAAA, 1; AAAD, 3/4; AADA, 3/4; ADAA, 3/4; DAAA, 3/4; AADD, 1/2 **(b)** proportions and probabilities for table: 0, 1/16; 1/4, 4/16; 1/2, 6/16; 3/4, 4/16; 1, 1/16 **(c)** No; there is probability 6/16 that the sample proportion equals the population proportion of 1/2 but otherwise the sample proportion is far from 1/2. **6.45** Sample proportions and their probabilities are: **(a)** 0, 0; 1,

1, **(b)** 0, 0.25; 1/2, 0.50; 1, 0.25, **(c)** 0, 0.125; 1/3, 0.375; 2/3, 0.375; 1, 0.125. **(d)** The shape begins to take a bell shape. **6.47 (a)** $P(1) = 0.55$ and $P(0) = 0.45$. **(b)** $n = 3160$; $p = 0.55$ **(c)** mean = 1738; $\sigma = 27.97$ **(d)** mean = 0.55; standard error = 0.0089 **6.49 (a)** $P(1) = 1/36, P(1.5) = 2/36, P(2) = 3/36, P(2.5) = 4/36, P(3) = 5/36, P(3.5) = 6/36, P(4) = 5/36, P(4.5) = 4/36, P(5) = 3/36, P(5.5) = 2/36, P(6) = 1/36$. **(b)** Each outcome has a 1/6 chance of occurring. Thus, the mean is $(1 + 2 + 3 + 4 + 5 + 6)/6 = 3.5$. **(c)** For the sampling distribution listed as the solution in part (a), its symmetric form around the value 3.5 suggests that the mean is 3.5. **6.51** The standard error gives a typical value for how far the sample proportion falls from the population proportion. The smaller it is, the better the sample proportion is likely to approximate the population proportion, whose value is typically unknown. **6.53 (a)** education; **(b)** mean = 13.6; standard error = 0.30; **(c)** mean = 13.6; standard error = 0.15. As n increases, mean of the sampling distribution stays the same, and standard error gets smaller. **6.55 (a)** $P(0) = 0.9736$; $P(35) = 0.0263$ **(b)** The mean would be the same as the population mean of 0.921. The standard error would be $\sigma/\sqrt{n} = 5.603/\sqrt{5040} = 0.079$. **(c)** 0.16 **6.57 (a)** Standard error $= \sigma/\sqrt{n} = 160/\sqrt{100} = 16$ **(b)** Three standard errors, or $(3)(16)$, is 48. **(c)** 0.8944 falls below the z-score of 1.25, and 0.1056 falls below -1.25. Thus, $0.8944-0.1056 = 0.7888$ falls between the mean and the z-scores below *and* above it. **6.59 (a)** 0.816; **(b)** smaller because the standard error would be smaller **6.61 (a)** $X =$ number of people in a household, quantitative; **(b)** 2.6, 1.5; **(c)** 2.4, 1.4; **(d)** 2.6, 0.1 **6.63 (a)** by assigning a number, such as 1, to women, and another number, such as 0, to men **(b)** Using 1 for woman, $P(1) = 0.60, P(0) = 0.40$. **(c)** Sample proportions of 0.52 for $x = 1$ (women) and 0.48 for $x = 0$ (men). **(d)** Approximately a normal distribution with mean = 0.60; standard error = 0.069 **6.65** For $n = 1733$, the standard error = 0.012, and we would expect a sample proportion between about 0.464 and 0.536 (that is, 3 standard errors). The sample Nelson proportion of 0.52 is within this range, so we would not have been willing to predict the winner. **6.67 (a)** discrete; you can only have whole numbers of grandparents **(b)** They each fall between 0 and 1, and the sum of the probabilities of all possible values is 1. **(c)** 0.50 **6.69 (a)** Probability distribution: $P(0) = 0.994, P(45.50) = 0.006$ **(b)** Probability distribution: $P(0) = 0.999, P(45.50) = 0.001$ **(c)** The first; because you have a higher chance of winning **6.71 (a)** Probability distribution: $P(0) = 0.999999, P(100,000) = 0.000001$ **(b)** 0.10 **(c)** because the flyer's return on each \$1 spent averages to only \$0.10 **6.73 (a)** 1.96; **(b)** 2.58; answer includes graph **6.75 (a)** 25% between the mean and the positive z-score. Added to 50% below mean, 75% of the normal distribution is below the positive z-score, corresponding to a z-score of 0.67. **(b)** First quartile: 25th percentile; third quartile: 75th percentile. The 75th percentile has a z-score of 0.67. The z-score at the 25th percentile is $z = -0.67$. **(c)** The interquartile range is $Q3-Q1 = (\mu + 0.67\sigma) - (\mu - 0.67\sigma) = 2(0.67)\sigma$. **6.77 (a)** 0.0764; **(b)** 0.0228; **(c)** 0.8472 **6.79** 0.0179 **6.81** 0.0668 **6.83 (a)** 0.04; **(b)** 0.75; each day's take must be independent of every other day's take **6.85 (a)** 0.9974; **(b)** 0.8664 **6.87 (a)** n is 6; p is 0.50; 0.3125 **(b)** The assumptions in (a) are that results of different games are independent and the probability of a win is the same in each game; no. **6.89 (a)** binary data (e.g., yes or no as only options); same probability of success for each call; independent trials (calls) **(b)** no **(c)** $n = 5$; $p = 0.60$; $\mu = 5(0.60) = 3$ **6.91 (a)** binary data (death or no death); same probability of success for each person ($p = 0.0001$); independent outcomes for each woman. $n = 1{,}000{,}000$; $p = 0.0001$ **(b)** mean = 100; standard deviation = 10.0 **(c)** 70.0 to 130.0 **(d)** All women would not have same probability (e.g., some drive much more), and outcomes may not be independent (e.g., one woman may usually ride in the same car as another). **6.93** Standard error = $\sqrt{p(1 - p)/n} = \sqrt{0.46(1 - 0.46)/1000} = 0.016$. If we double this and round it off, we get the margin of error of 0.03. **6.95 (a)** 0.56 to the 2232 power, which is essentially 0; **(b)** 41.9 **6.97** 0.95 **6.99 (a)** 0.0013 **(b)** 0.9974 **(c)** -10; yes; the mean of a sample is unlikely to be 10 standard errors below the mean of its sampling distribution. **(d)** The distribution of sample means (Exer. 6.99) has far less spread than the distribution of scores (Exer. 6.98). **6.101 (a)** 0.003; **(b)** 0.05 **6.103 (a)** 0.00195; **(b)** 0.0039 **6.105** Sampling distribution should become more bell-shaped and the spread smaller as n increases. **6.107 (a)** The mean increases by one for each doubling. **(b)** (i) 8; (ii) 11 **(c)** 4 is a bit more than two standard deviations, and a bell-shaped curve has probability about 0.95 within two standard deviations. **6.109 (a)** The sampling distribution of the sample proportion who show up has mean 0.80, standard error: 0.029. The proportion values within 3 standard errors of the mean are 0.713 to 0.887. The proportion that the plane can handle is 170/190 = 0.895, just above this range. **(b)** Large groups might buy tickets and travel together, violating the assumption of independent trials. **6.111 (a)** You'd spend \$52 to play 52 times. You'd have to win at least 11 times (thus, winning at least \$55) to win more than the \$52 you pay to play. **(b)** If we were to calculate the probability for 11 out of 52 trials, and then 12 out of 52, etc. up to 52 out of 52, we'd get the probability that we'd win more than the \$52 spent. This is 0.013. **6.113** (a) **6.115** (b) **6.117 (a)** 0.1587; 0.3085; **(b)** 0.6603; **(c)** Now, 0.7455 are from group B. This problem is worse! **6.119 (a)** $\sigma^2 = p^2(1 - p) + (1 - p)^2 p = p(1 - p)$. So $\sigma = \sqrt{p(1 - p)}$ **(b)** Since standard deviation of probability distribution equals $\sigma = \sqrt{p(1 - p)}$ and since we must divide by $\sqrt{n}$ to get standard error, we have $\sigma/\sqrt{n} = \sqrt{p(1 - p)/n}$. **6.121 (a)** The probability of rolling doubles is $\frac{1}{6}$. The outcome on the second die has a $\frac{1}{6}$ chance of matching the first. To have doubles occur first on the second roll, you'd have to have no match on the first, a $\frac{5}{6}$ chance. You'd then have to have a match on the second roll, a $\frac{1}{6}$ chance. The probability of both occurring is $\left(\frac{5}{6}\right)\left(\frac{1}{6}\right)$. For no doubles until the third roll, the first and the second roll would not match, a $\frac{5}{6}$ chance for each, followed by doubles on the third, a $\frac{1}{6}$ chance. The probability of all of three events is $\left(\frac{5}{6}\right)^2\left(\frac{1}{6}\right)$. **(b)** By the logic in part (a), $P(4) = \left(\frac{5}{6}\right)\left(\frac{5}{6}\right)\left(\frac{5}{6}\right)\left(\frac{1}{6}\right)$ or $\left(\frac{5}{6}\right)^3\left(\frac{1}{6}\right)$. By extension, we could calculate $P(x)$ for any x by $\left(\frac{5}{6}\right)^{(x-1)}\left(\frac{1}{6}\right)$. **6.125 (a)** Probability distribution $P(0) = 0.5$, $P(1) = 0.5$; mean = 0.5 **(b)** Sample proportion for 0 is 0.4 and for 1 is 0.6 **(d)** (i) mean that's very close to the population proportion; (ii) standard error to approach the standard

error you'd calculate using this formula: $\sqrt{p(1 - p)/n}$. The population proportion is 0.5 and the standard error is 0.158.

Chapter 7

7.1 (a) proportion with health insurance in population; mean amount spent on insurance by population **(b)** sample proportion and sample mean **7.3** 0.554. **7.5 (a)** 2.14; **(b)** the sample point estimate likely falls within 0.64 of the population mean. **7.7 (a)** 0.02. **(b)** 0.84 to 0.88; the interval containing the most believable values for the parameter. **7.9** 1.38 to 1.62 is the range that includes the most believable values for the population mean. **7.11** Margins of error in the media typically are based on a 95% confidence interval. They would have multiplied 1.96, the z-score associated with a 95% confidence interval, by the standard error = $\sqrt{0.29(1 - 0.29)/1000}$, giving a margin of error of 0.03 on the proportion scale, or 3%. **7.13 (a)** 0.294. **(b)** 0.025. **(c)** (0.27, 0.32); the most believable values for the population proportion **(d)** Observations are obtained randomly; number of successes and the number of failures both are greater than 15. Both seem to hold true in this case. **7.15 (a)** Divide the number of those in favor by the total number of respondents, 1764/2565. **(b)** We can be 95% confident that the proportion of the population who are in favor of the death penalty is between 0.670 and 0.706. **(c)** If we use this method over and over for numerous samples, in the long run we make correct inferences (that is, the confidence interval contains the parameter) 95% of the time. **(d)** Yes, all the values in the confidence interval are above 0.50. **7.17** The "sample p" is the proportion of all respondents who believe in reincarnation, 594/2201 = 0.27. We can be 95% confident that the population proportion falls between 0.25 and 0.29. **7.19 (a)** We can be 95% confident that the population proportion falls between 0.585 and 0.635. **(b)** The data must be obtained randomly, and the number of successes and the number of failures both must be greater than 15. **7.21 (a)** Yes; because 0.50 falls outside of the confidence interval of 0.446 to 0.496. **(b)** No; because 0.50 falls in the confidence interval of 0.437 to 0.505. The more confident we are, the wider the confidence interval. **7.23** The percentage we'd expect would be 90%, but the actual value may differ a bit because of sampling variability. **7.25 (a)** 3.02; **(b)** 0.081; **(c)** We're 95% confident the population mean falls between 2.89 and 3.21; **(d)** no; because 2 falls outside the confidence interval. **7.27 (a)** 2.776; **(b)** 2.145; **(c)** 2.977 **7.29 (a)** Data production must have used randomization. The population distribution is assumed to be approximately normal, although this method is robust except in case of extreme outliers or binary data. Here the method is questionable because of an outlier at 178. **(b)** $231.61 - (2.110)(5.1713) = 220.70$; $231.61 + (2.110)(5.1713) = 242.52$. We can be 95% confident that the mean population price is between \$220.70 and \$242.52. **(c)** (i) Scores below 183.00 and above 280.77, such as 178, are potential outliers. **(d)** Mean: 234.76; standard deviation: 17.92; confidence interval: (225.55, 243.97); this confidence interval is narrower than the one using all the data. **7.31 (a)** The data must be produced randomly, and the population distribution should be approximately normal. **(b)** The population mean hours of TV watching is likely to be between 0.3 and 4.3. **(c)** The confidence interval is wide mainly because of the very small sample size. **7.33** The most plausible values for the population mean with the buy-it-now option: $231.91 through $238.99; without the buy-it-now option: $219.32 through $227.40. There is no overlap, so we can infer that the buy-it-now option leads to higher prices. **7.35 (a)** The margin of error equals the standard error, 0.0381, multiplied by the t-score for 1330 degrees of freedom and a 95% confidence interval (1.96). Take the sample mean of 4.12 plus and minus the margin of error. **(b)** Yes; because 4.0 falls below the lowest believable value of the confidence interval. **(c)** (i) wider; (ii) wider **7.37 (a)** 9.8; **(b)** 4.9; as sample size increases, the margin of error is smaller. **7.39 (a)** No, there is a large proportion at the single value of 0. Very little because with large random samples the sampling distribution of the sample mean is approximately normal. **(b)** (8.0, 12.0); yes, because $15 is not in the confidence interval. **7.41 (b)** We would expect 5% of the intervals not to contain the true value. **(c)** With a large sample size, the sampling distribution is approximately normal even when the population distribution is not. The assumption of a normal population distribution becomes less important as n gets larger. **7.43** 379 **7.45 (a)** 96; **(b)** 384; **(c)** 666; **(d)** in both cases we need a larger sample size **7.47 (a)** 256; **(b)** 37.5 **7.49** Diverse populations (medical doctors making a wide range of incomes) have a wider range of observed values, and thus larger standard deviations, than homogeneous populations (entry level McDonald's workers all making close to minimum wage). Larger standard deviations result in larger standard errors and wider confidence intervals. **7.51 (a)** 1.0. **(b)** 0; no; the sampling distribution is likely to have *some* variability, because the true probability (which determines the exact standard error) is positive. **(c)** (1.0, 1.0); no, because this method works poorly in this case. **(d)** We don't have at least 15 outcomes of each type. Using the small-sample method based on adding 2 outcomes of each type, we can be 95% confident that the proportion is between 0.51 and 1.05 (i.e., 1.0, since p cannot exceed 1). **7.53** Yes; 0.10 is not in the confidence interval of 0.00 to 0.04. **7.55** Sample with replacement from the 10 values, taking 10 observations and find the standard deviation. Do this many, many times. The 95% confidence interval would be the values in the middle 95% of the standard deviation values. **7.57 (a)** point estimates; **(b)** no; also need to know sample sizes. **7.59** 0.0145; 0.0178 **7.61** Subtract and add the 2% margin of error from 81%. We are 95% confident that the population percentage who believe in life after death falls between 79 and 81%. **7.63 (a)** Data were obtained randomly. **(b)** (0.035, 0.045); mainly because of the very large sample size **(c)** Yes; 10% falls above the highest believable value in the confidence interval. **7.65** No; because you have data for the entire population; you don't have to estimate it. **7.67 (a)** 306 said "legal." 545 said "not legal." 0.36 and 0.64 **(b)** Minority; 0.50 is above the highest believable value in the 95% confidence interval of 0.33 to 0.39. **(c)** It appears that the proportion favoring legalization is increasing over time. **7.69** Sample: specific sample of non drinkers. X: number who said that they have had fewer than 12 drinks in their lifetime. N: sample size of 38,633. Sample p: proportion of total sample, 0.2240, who said that they had had fewer than 12 drinks in their lifetime. 95.0% CI: 95% confidence interval; we can be 95% confident that the population proportion

falls between 0.220 and 0.228. **7.71 (a)** 2.262, 2.093, 2.045, 1.96 (the z-score associated with a 95% confidence interval). **(b)** The t distribution approaches the standard normal distribution as the sample size gets larger. **7.73 (a)** (54,690, 102,310) **(b)** an approximately normal population distribution **(c)** Not necessarily; the method is robust in terms of the normal distribution assumption. **7.75 (a)** No; the minimum possible value of 0 is barely more than one standard deviation below the mean, and the mean is quite a bit larger than the median. **(b)** Yes, because the normal population assumption is much less important with such a large sample size. (19.4, 21.2) **7.77** We can be 95% confident that the mean for the population of students is between 0.1 and 2.7. Assumptions: data randomly produced; approximately normal population distribution. **7.79 (a)** We can be 95% confident that the population mean price is between $411 and $845. **(b)** Assumptions: data produced randomly; population distributed normally. The population does not seem to be distributed normally (cluster on the low and high ends, with fewer in the middle). The t distribution is fairly robust with respect to the normal distribution assumption. **7.81 (a)** Box plot suggests that the population distribution is skewed. There does not seem to be an extreme outlier, so this should not affect the population inferences. **(b)** 90.2; 21.0 **(c)** We can be 95% confident that population mean income is between $8,256 and $9,823. **7.83** We can be 95% confident that the mean number of hours for the population of males falls between 2.63 and 3.14. **7.85** You could make inferences using the mean of 13.4 (standard deviation of 3.0) or the proportion of 0.16. For the proportion, $se = \sqrt{\hat{p}(1 - \hat{p})/n} = \sqrt{0.16(1 - 0.16)/2753} = 0.007$; $\hat{p} \pm z(se)$ has bounds $0.16 - (1.96)(0.007) = 0.146$; $0.16 + (1.96)(0.007) = 0.174$, giving (0.146, 0.174). **7.87 (a)** $se = s/\sqrt{n} = 1.08/\sqrt{2242} = 0.023$. **(b)** The lowest possible value of 0 is only one standard deviation below the mean. **(c)** Because the method is robust with respect to the normal distribution assumption. **7.89 (a)** 384 **(b)** 600; if we can make an educated guess, we can use a smaller sample size. **7.91 (a)** 400; the solution makes the assumption that the standard deviation will be similar now. **(b)** more than because the standard error will be larger than predicted. **(c)** With a larger margin of error, the 95% confidence interval is wider; thus, the probability that the sample mean is within $100 of the population mean is less than 0.95. **7.95** Report should include confidence intervals (17.7, 18.5); (32.2, 33.0). Because there's no overlap, we conclude that women do more housework than men. **7.97** The mean falls in the middle of the confidence interval of (4.0, 5.6). It equals 4.8. **7.99** An extremely large confidence level makes the confidence interval so wide as to have little use. **7.101** (−10.4, 30.4); An outlier can dramatically affect the standard deviation and standard error. **7.103** (a) **7.105** (b) and (e) **7.107 (a)** The confidence interval refers to the population, not the sample, mean. **(b)** The confidence interval is an interval containing possible means, not possible scores. **(c)** $\bar{x}$ is the sample mean; we know exactly what it is. **(d)** If we sampled the entire population even once, we would know the population mean exactly. **7.109** False **7.111** False **7.113 (a)** that in the long run (if we took many random samples of this size from the population), intervals based on the samples would capture the population proportion 95% of the time.

(b) 0.358 **(c)** 19 **(d)** You could increase the confidence level—to 0.99, for example. **7.115 (a)** $|\hat{p} - p| = 1.96\sqrt{p(1-p)/n}$; $|0 - p| = 1.96\sqrt{p(1-p)/20}$; If we substitute 0, we get 0 on both sides of the equation. If we substitute .161, we get .161 on the left, and on the right. **(b)** This confidence interval seems more believable because it forms an actual interval. **7.117** If the population is normal, the standard error of the median is 1.25 times the standard error of the mean. This means a larger margin of error, and therefore, a wider confidence interval.

Chapter 8

8.1 (a) null; **(b)** alternative; **(c)** H_0: $p = 0.50$; H_a: $p \neq 0.50$; H_0: $p = 0.24$; H_a: $p < 0.24$ **8.3** H_0: The pesticide is not harmful; H_a: The pesticide is harmful. **8.5 (a)** alternative; it has a range of parameter values. **(b)** The relevant parameter is the mean weight change, μ. H_0: $\mu = 0$; this is a null hypothesis. **8.7** strong evidence against; this sample proportion falls over three standard errors from the null hypothesis value. **8.9** Null: psychic will predict the outcome 1/6 of the time; alternative: psychic will predict the outcome more than 1/6 of the time. H_0: $p = 1/6$ and H_a: $p > 1/6$ **8.11 (a)** 0.149; **(b)** 0.298; **(c)** 0.851; **(d)** No; all of them indicate that this null hypothesis is plausible. **8.13 (a)** standard error $= \sqrt{p(1-p)/n} = \sqrt{0.5(1 - 0.5)/100} = 0.05$; $z = (0.35 - 0.50)/0.05 = -3.0$ **(b)** 0.0013 **(c)** Yes; this is a very small P-value; it does not seem plausible that $p = 0.50$. **8.15 (a)** $p =$ population proportion of trials guessed correctly; H_0: $p = 0.50$ and H_a: $p > 0.50$ **(b)** $z = -2.10$; sample proportion is a bit more than 2 standard errors less than expected if the null hypothesis were true. **(c)** .98. Since the P-value is so much greater then $\alpha = 0.05$, there is insufficient evidence that they predict better than by random guessing. **(d)** $np = (130)(0.5) = 65 = n(1 - p)$; the sample size was large enough. We also would need to assume randomly selected practitioners and subjects for this to apply to all practitioners and subjects. **8.17 (a)** H_0: $p = 0.60$ and H_a: $p > 0.60$, where p is probability of selecting a male; the company predicts that the proportion of males chosen for management training is the same as in the eligible pool. **(b)** $np = (40)(0.6) = 24$, and $n(1 - p) = (40)(0.4) = 16$. Software obtained the test statistic by subtracting p predicted by the null from the sample p, then dividing by standard error. **(c)** The P-value in the table refers to the alternative hypothesis that $p \neq 0.60$. For the one-sided alternative, it is half this, or about 0.10. **(d)** For a 0.05 significance level, we would not reject the null hypothesis since the P-value is > 0.05. There is insufficient evidence that there is a greater proportion of male trainees than would be expected. **8.19 (a)** whether garlic or placebo is more effective; $p =$ population proportion of those for whom garlic is more effective **(b)** H_0: $p = 0.50$ and H_a: $p \neq 0.50$; there are at least 15 successes (garlic) and failures (placebo). **(c)** $z = 0.98$. **(d)** 0.33. Since the probability is 0.33 that we would observe our test statistic or one more extreme due to random variation, there is insufficient evidence to support that the population proportion is different from 0.50. **8.21 (a)** Take the difference between sample and null proportions, and divide by standard error. **(b)** The two-tail probability from the standard normal distribution below -1.286 and above 1.286. The P-value of 0.20 tells us that if the null hypothesis were true, a proportion of

0.20 of samples would fall at least this far from the null hypothesis proportion of 0.50. **(c)** No; it is possible that there is a real difference in the population that we did not detect, and we can never accept a null hypothesis. **(d)** The 95% confidence interval tells us the range of plausible values, whereas the test merely tells us that 0.50 is plausible. **8.23** This P-value gives strong evidence against the null hypothesis. It would be very unlikely to have a sample proportion of 1.00 if the actual population proportion were 0.50. **8.25 (a)** 2.145; **(b)** 1.762; **(c)** -1.762 **8.27 (a)** larger; the t value of 1.20 is less extreme. **(b)** because they decrease the standard error (by making its denominator larger) **8.29 (d)** If it would be unusual to get a sample mean of 2000 if the population mean were 1000, it would be even more unusual to get this sample mean if the population mean were less than 1000. **8.31 (a)** 40, 15, 90, 50, 30, 70, 20, 30, -35, 40, 30, 80, 130; positively skewed; box plot part of answer. **(b) 1.** data are quantitative (PEF change scores); data not likely produced using randomization but a convenience sample; test is two-sided, so not likely affected by skewed population distribution. **2.** H_0: $\mu = 0$; H_a: $\mu \neq 0$ **3.** $t = 4.03$ **4.** P-value $= 0.0017$ **5.** strong evidence that PEF levels were lower with salbutamol **(c)** assumptions of random production of data and normal population distribution do not seem valid; convenience sample limits reliability in applying inference to population at large. **8.33 1.** data are quantitative (dollar differences); might have been produced randomly. **2.** H_0: $\mu = 0$; H_a: $\mu \neq 0$ **3.** $t = 2.372$ **4.** P-value $= 0.04$ **5.** Since P-value is $0.04 < 0.05$, sufficient evidence to conclude that the coupons led to higher sales than outside posters **8.35 (a)** The data appear somewhat skewed to left; dot plot part of answer. **(c) 1.** data are quantitative, produced randomly; population distribution approximately normal. **2.** H_0: $\mu = 0$; H_a: $\mu \neq 0$ **3.** $t = 4.19$ **4.** P-value $= 0.001$ **5.** Strong evidence against the null hypothesis that family therapy has no effect. **8.37** 0.01. If the test rejects the null hypothesis with significance level 0.01, then the 99% confidence interval does not contain the value in the null hypothesis. **8.39 (a)** 0.05; **(b)** Type I **8.41 (a)** We rejected the null hypothesis, but therapy actually did not work. **(b)** We failed to reject the null hypothesis, but therapy actually had an effect. **8.43 (a)** We rejected the null hypothesis that there is no disease, but the woman does not have cancer. **(b)** We failed to reject the null hypothesis that there is no disease, but the woman has cancer. The consequence is failing to detect cancer and treat the cancer when it actually exists. **(c)** More women who <u>do</u> have cancer will have false negative tests and not receive treatment. **8.45 (a)** Type I; an innocent man or woman is put to death. **(b)** Type II; someone might not receive life-saving treatment when they need it. **8.47 (a)** 0.00006; **(b)** It is statistically significant because the P-value is very small but it is not practically significant because the sample mean of 498 is very close to the null hypothesis mean of 500. **8.49** We would expect 5% of tests to be significant just by chance if the null hypothesis is true, and for 60 tests this is $0.05(60) = 3$ tests. **8.51** tree diagram is part of answer; there is a 5% chance of a Type I error **8.53 (a)** cutoff z-score of 1.645; standard error $= \sqrt{0.5(1 - 0.5)/100} = 0.050$. The value 1.645 standard errors above 0.50 is $0.50 + 1.645 (0.050) = 0.582$. **(b)** 0.36 **8.55 (a)** cutoff z-score: 2.33; standard error $= 0.0438$. The value

2.33 standard errors above 0.333 is 0.333 + 2.33(0.0438) = 0.435. **(b)** If $p = 0.50$, the z-score for 0.435 in reference to 0.50 is

$$z = \frac{(0.435 - 0.5)}{\sqrt{0.5(1 - 0.5)/116}} = -1.40;$$ proportion of this curve

that is not in the rejection area is 0.08.

8.57 (a) standard error $= \sqrt{\dfrac{p_0(1 - p_0)}{n}} = 0.0438$, where

$p_0 = 1/3$; for H_a, $z = 1.96$ has a P-value of 0.05. We reject H_0 when $|\hat{p} - 1/3| \geq 1.96(se) = 0.086$, hence we need $\hat{p} \geq 0.086 + 1/3 = 0.419$ or $\hat{p} \leq 1/3 - 0.086 = 0.248$; When H_0 is false, a Type II error occurs if $0.248 < \hat{p} < 0.419$. **(b)** $z = (0.248 - 0.50)/0.0464 = -5.43$; the probability that $\hat{p}$ is less than this z-score is 0. $z = (0.419 - 0.50)/0.0464 = -1.75$; the probability that $\hat{p}$ is greater than this z-score is 0.96. **(c)** Probability of Type II error is the portion of the curve (when $p = 0.50$) not over the rejection area. This is 0.04. **8.59 (a)** Theory predicts that P(Type II error) will be larger with a p that is closer to that predicted by the null hypothesis. **(b)** Decrease **(c)** Increase **8.61 1.** data are categorical (correct vs. incorrect guesses); we'll assume the data are obtained randomly. $np = (20)(0.5) < 15$, and $n(1 - p) = (20)(0.5) < 15$ so this test is approximate. **2.** H_0: $p = 0.50$; H_a: $p > 0.50$ **3.** $z = 0.89$ **4.** P-value = 0.19. **5.** There is not strong evidence with a P-value of 0.19 that the probability of a correct guess is higher than 0.50 **8.63 (a) 1.** data are categorical (Box and Draper); we'll assume the data are obtained randomly; $np = (0.5)(400) \geq 15$, and $n(1 - p) = (0.5)(400) \geq 15$ **2.** $p =$ proportion of voters who prefer Box; H_0: $p = 0.50$; H_a: $p \neq 0.50$ **3.** $z = 3.00$ **4.** P-value = 0.003 **5.** Strong evidence to reject null and conclude that proportion of voters who chose Box is different from 0.50 **(b)** $z = \dfrac{0.575 - 0.50}{\sqrt{0.5(1 - 0.5)/40}} = 0.95$; P-value: 0.34. We could not have rejected the null hypothesis. **(c)** As sample size increases, standard error decreases. A smaller standard error leads to a larger z-score and a smaller P-value. **8.65 1.** data are categorical (said good vs. didn't say good); we'll assume the data are obtained randomly; $np = (0.5)(1201) \geq 15$, and $n(1 - p) = (0.5)(1201) \geq 15$ **2.** H_0: $p = 0.50$; H_a: $p \neq 0.50$ **3.** $z = 7.07$ **4.** P-value = 0.000 **5.** Very strong evidence to conclude that more than half of the population think having affirmative action is good. **8.67 (a)** $p =$ proportion of people who would pick Box A; H_0: $p = 0.50$; H_a: $p \neq 0.50$ **(b)** $z = 5.060$; P-value = 0.000. This is very strong evidence to conclude that the population proportion that chooses Box A is not 0.50. Box A seems to be preferred. **8.69** We are testing "p = 0.50 vs not = 0.50." X is 40, the number of people in the sample who preferred the card with the annual cost. 100 is the "N," the size of the whole sample. "Sample p" of 0.40 is the proportion of the sample that preferred the card with the annual cost. "95.0% CI" is the 95% confidence interval. "Z-Value" is the test statistic. "P-Value" tells us that if the null hypothesis were true, the proportion 0.0455 of samples would fall at least this far from the null hypothesis proportion of 0.50. The majority of the customers seem to prefer the card without the annual cost. **8.71 (a)** Type I error would have occurred if we had rejected the null hypothesis, when really women were not being passed over.

A Type II error would occur if we had failed to reject the null, but women really were being picked disproportionate to their representation in the jury pool. **(b)** Type I error **8.73 (b)** Very strong evidence to reject null and conclude that the population mean is not 0 **(c)** 0.001; we have very strong evidence to conclude that the population mean is positive. **(d)** 0.999; insufficient evidence to conclude that the population mean is negative. **8.75 (a)** H_0: $\mu = 40$; H_a: $\mu \neq 40$ **(b)** (i) SE Mean, 0.3517, is standard error; (ii) T, 5.05, is the test statistic; (iii) P-value, 0.000, is probability that we'd get a sample mean at least this far from the value in H_0 if the null hypothesis were true. **(c)** Confidence interval supports the conclusion; 40 falls below the range of plausible values. **8.77 (a) 1.** data are quantitative and have been produced randomly; and have an approximate normal population distribution. **2.** H_0: $\mu = 130$; H_a: $\mu \neq 130$ **3.** The sample mean is 150.0 and standard deviation is 8.37; $t = 5.85$ **4.** P-value = 0.002 **5.** Very strong evidence to conclude that the true mean is different from 130 **8.79 (a)** Software indicates a test statistic of -5.51 and a P-value of 0.001. **(b)** For a significance level of 0.05, we would conclude that the process is not in control. **8.81 1.** data are quantitative and seem to have been produced randomly; also assume approximately normal population distribution. **2.** H_0: $\mu = 500$; H_a: $\mu < 500$ **3.** $t = -13.9$ **4.** P-value = 0.000 **5.** extremely strong evidence that the population mean is less than 500 **8.83 (a)** We can reject the null hypothesis. **(b)** It would be a Type I error. **(c)** No; because the P value is less than 0.05; when a value is rejected by a test at the 0.05 significance level, it does not fall in the 95% confidence interval. **8.85 (a)** by choosing a smaller significance level **(b)** It will be too difficult to reject the null hypothesis, even if the null hypothesis is not true. **8.87 (a)** $\sqrt{0.333(1 - 0.333)/60} = 0.061$ **(b)** When P(Type I error) = significance level = 0.05, $z = 1.645$, the value 1.645 standard errors above 0.333 is 0.433. **(c)** 0.15; Type II error is larger when n is smaller, because a smaller n results in a larger standard error and makes it more difficult to have a sample proportion fall in the rejection region. **8.91 (a)** When we look at many games over time, we see the long run probability of the home team winning. **(b)** $z = 3.45$ for testing that $p = 0.50$ about the probability the home team wins, giving P-value = 0.0006. The confidence interval is (0.515, 0.555). The test merely indicates whether $p = 0.50$ is plausible whereas the confidence interval displays the range of plausible values. **8.93 (a)** After seeing the data, we know the direction of results; it is "cheating" to do a one-tailed test now. **(b)** If there really is no effect, but many studies are conducted, eventually someone will achieve significance, and then the journal will publish a Type I error. **8.95** The subgroups have smaller sample size, so for a particular size of effect will have a smaller test statistic and a larger P-value. **8.97** The studies with the most extreme results will give the smallest P-values and be most likely to be statistically significant. If we could look at how results from all studies vary around a true effect, the most extreme results would be out in a tail, suggesting an effect much larger than it actually is. **8.99** Just because the sample statistic was not extreme enough to conclude that the value at H_0 is unlikely doesn't mean that the value in H_0 is the actual value. A confidence interval would show that there is a whole range of plausible values, not just the null value. **8.101** Statistical significance: strong evidence that the true parameter value is not the value in H_0; practical significance: true parameter is sufficiently different from value in H_0 to be important in practical terms. **8.103** With the probability of a false positive diagnosis being about 50% over the course of 10 mammograms, it would not be unusual for a woman to receive a false positive over the course of having had many mammograms. Likewise, if you conduct 10 significance tests at the 0.05 significance level in 10 cases in which the null hypothesis is actually true, it would not be surprising if you sometime reject the null hypothesis just by chance. **8.105** Suppose P-value = 0.057. If the significance level were any number above this, we would have enough evidence to reject the null; if it were any number below this, there would not be enough evidence. **8.107** (b) **8.109** (a) **8.111** False **8.113** False **8.115** False **8.117** Sample proportion = 0, standard error = 0, test statistic = infinity, which does not make sense. A significance test is conducted by supposing the null is true, so in finding the test statistic we should substitute the null hypothesis value, giving a more appropriate standard error.

Chapter 9

9.1 (a) net worth; race **(b)** white and black households **(c)** independent; no household could be in both samples **9.3 (a)** 0.083; the percentage appeared to have increased between 1993 and 2001. **(b)** 0.0069; standard error is standard deviation of sampling distribution of difference between the sample proportions. **(c)** (0.069, 0.097); it does not contain zero, evidence of an increase in reports of "getting drunk" as an important reason. The interval is narrow because the standard error is small, which is because the sample sizes are so large. **(d)** data are categorical (stated "getting drunk" vs. did not), samples are independent and obtained randomly, sufficiently large sample sizes. **9.5 (a)** 0.454; 0.563 **(b)** We're 95% confident that the population proportion for females falls between 0.050 higher and 0.168 higher than the population proportion for males; 0 is not in this interval, so we conclude that females are more likely to say that they believe in miracles. Assumptions: categorical data, independent and randomly obtained samples, sufficient sample size. **(c)** The confidence interval has a wide range of plausible values, including some (such as 0.05) that indicate a small difference and some (such as 0.16) that indicate a relatively large one. **9.7 (a)** H_0: $p_1 = p_2$; H_a: $p_1 \neq p_2$ **(b)** If the null were true, we would obtain a difference at least this extreme a proportion 0.144 of the time. **(c)** 0.072 **(d)** This study has smaller samples than the Physicians Health Study did. Therefore, its standard error was larger and its test statistic was smaller. A smaller test statistic has a larger P-value. **9.9 (a)** Each sample must have at least five (for a two-sided test) outcomes of each type; data must be categorical; samples must be independent random samples; p is probability that someone says he/she had engaged in unplanned sexual activities because of drinking alcohol; H_0: $p_1 = p_2$; H_a: $p_1 \neq p_2$ **(b)** $\hat{p} = (1871 + 2440)/(8783 + 12708) = 4311/21,491 = 0.201$; proportion of the *total* sample who reported that they had engaged in such activities **(c)** 0.006; interpreted as the standard deviation of the estimates $(\hat{p}_1 - \hat{p}_2)$ from different randomized studies using these sample sizes **(d)** $z = -3.5$; P-value = 0.0005; We have very strong

evidence to conclude that there is a difference in reports between 1993 and 2001. It appears to have increased. **(e)** It is a relatively small increase; a small change can be statistically significant when the sample size is large. **9.11 (a)** Each sample has at least 5 (for two-sided test) outcomes of each type; data are categorical; independent random samples; p: probability someone developed cancer; H_0: $p_1 = p_2$; H_a: $p_1 \neq p_2$ **(b)** The test statistic is 1.03, and the P-value is 0.30. We do not have strong evidence that there are different results. **(c)** We cannot reject the null hypothesis. **9.13 (a)** margin of error: t value of 2.576 multiplied by the standard error of 0.297 gives 0.8. The bounds are 14.5 − 0.8 = 13.7 and 14.5 + 0.8 = 15.3. **(b)** because we have chosen a larger confidence level **9.15 (a)** 2.48; 0.90 **(b)** (−0.09, 0.11) Because 0 is in this interval, there may be no difference in reports of ideal number of children between males and females. **9.17 (a)** Dot plot indicates bidding only method leads to more variable prices. **(b)** 0.26; 0.80; we cannot conclude that there is a difference in mean prices. **(c)** data are quantitative, samples are independent and random, approximately a normal population distribution for each group. The distribution for the bidding only prices does not appear to be normal; the two-sided test, however, is robust with respect to this assumption. **9.19 (a)** (i) because the mean is very close to 0 (ii) Those who reported inhaling had a mean score that was 2.9 − 0.1 = 2.8 higher. **(b)** Not for the noninhalers. The lowest possible value of 0, which was very common, was only a fraction of a standard deviation below the mean. **(c)** 0.239; it's the standard deviation of the difference between sample means from different studies using these sample sizes. **(d)** Because 0 is not in this interval, we can conclude that there is a difference in population mean HONC scores. **9.21 (a)** 0.364 is the standard deviation of the difference between sample means from different studies using these sample sizes. **(b)** $t = 3.30$; P-value = 0.001; very strong evidence that females have higher population mean HONC score. **(c)** No, because in each case the lowest possible value of 0 is less than one standard deviation below the mean. This does not affect the validity of our inference greatly because of the robustness of the two-sided test for the assumption of a normal population distribution for each group. **9.23 (a)** H_0: $\mu_1 = \mu_2$; H_a: $\mu_1 \neq \mu_2$ **(b)** P-value = 0.27; Since the probability is 0.27 of observing our test statistic or one more extreme due to random variation, insufficient evidence to conclude that there is a gender difference in TV watching **(c)** Yes, because according to the test 0 is a plausible value for the difference between the population means. **(d)** No, because in each case the lowest possible value of 0 is only about 1 or 1 1/2 standard deviations below the mean. This does not affect the validity of our inference greatly because of the robustness of the two-sided test for the assumption of a normal population distribution for each group. Inferences assume randomized study, normal population distributions. **9.25** With large samples, the sampling distribution is normal regardless of the shape of the distribution of the population. **9.27 (a)** P-value = 0.011 **(b)** data are quantitative; random samples from two groups; populations have approximately normal distributions; population standard deviations are equal. Normality assumption is likely violated, but we're using a two-sided test, so inferences are robust for that assumption. **9.29 (a)** We can be 95% confident that the true population mean difference between change scores is between −1.2 and 41.2. The therapies may not have different means, but if they do the population mean could be much higher for therapy 1. The confidence interval is wide because the two sample sizes are very small. **(b)** Yes; 0 is no longer in the range of plausible values. **9.31 (a)** The first set of inferences assume equal population standard deviations, but the sample standard deviations suggest this is not plausible. It is more reliable to conduct the second set of inferences, which do not make this assumption. **(b)** It appears that the vegetarian students are more liberal than are the non-vegetarian students. **9.33 (a)** The probability of heart disease for women who ate fish 5 or more times per week was estimated to be 0.66 times the probability of heart disease for women who ate fish less than once a month. **(b)** The confidence interval gives us the range of plausible values for the reported relative risk. Since the interval does not contain 1, in the population we can infer that the probability of heart disease was smaller for women who ate fish 5 or more times per week. **9.35** 40% and 87% **9.37** 191/166 = 1.15. The mean weight of men in 2002 was 1.15 times the mean weight in 1962. The mean increased by 15%. **9.39 (a)** The same patients are in both samples. **(b)** 150; 130; 20; "before" mean minus "after" mean equals mean of the difference scores **(c)** We can be 95% confident that the difference between the population means is between 7.6 and 32.4. **9.41 (a)** dependent; the same students are in both samples. **(b)** No, there is quite a bit of spread, but outlying values appear in both directions. **(c)** The 95% confidence interval was obtained by adding and subtracting the margin of error (the *t*-score of 2.262 for df = 9 times the standard error of 5.11, which is 11.56) from the mean difference score of 4.0. **(d)** The test statistic was obtained as

$$t = \frac{\bar{x}_d - 0}{se} = \frac{4.0 - 0}{5.11} = 0.78;$$ The P-value is 0.45. Since the

P-value is large, we cannot conclude that there is a difference in attendance at movies vs. sports events. **9.43 (a) 1.** Assumptions: the difference scores are a random sample from a population distribution that is approximately normal **2.** $H_0: \mu_1 = \mu_2$ (or population mean of difference scores is 0); $H_a: \mu_1 \neq \mu_2$ **3.** $t = -1.62$ **4.** P-value = 0.14 **5.** Since the probability is 0.14 of observing our test statistic or one more extreme by random variation, insufficient evidence to conclude that the population means are different for movies and parties. **(b)** (−21.1, 3.5); We are 95% confident that the population mean number of times spent attending movies is between 21.1 less and 3.5 higher than the population mean number of times spent attending parties. **9.45 (a)** because even though there is a lot of variability among the initial and final weights of the women, most women do not see a large change in weight over the course of the study, so the weight changes would not vary much. **(b)** no; 95% confidence interval is (0.4, 1.2), so the plausible mean weight change falls in this range. **(c)** data must be quantitative; sample of difference scores must be random; difference scores must have an approximately normal population distribution (particularly with samples of size < 30) **9.47 (a)** 0.377; 0.427. **(b)** Sample mean: for each subject, their coffee score is 1 if they choose Sanka and 0 if not. The mean of the scores is the proportion using Sanka. The estimated difference of population proportions is difference

between the sample means at the two times. **(c)** We can be 95% confident that the population proportion choosing Sanka for the second purchase was between 0.01 and 0.09 higher than the population proportion choosing it for the first purchase. **9.49 (a)** 0.135 **(b)** (ii) $H_0: p_1 = p_2$; $H_a: p_1 \neq p_2$ (iii) z = 3.00 (iv) P-value: 0.0027 (v) strong evidence that there is a difference between the population proportions. **9.51 (a)** response variable: whether have heart disease; explanatory variable: whether drink alcohol moderately; control variable: age **(b)** There is a stronger association between drinking alcohol and whether have heart disease for subjects who are older (in fact, perhaps no association for young subjects). **9.53 (a)** The proportion of shots made is higher for Barry both for 2-point shots and for 3-point shots, but the proportion of shots made is higher for O'Neal overall. **(b)** O'Neal took almost exclusively 2-point shots, where the chance of success is higher. **9.55 (a)** Higher mean for English-speaking families (1.95) than French-speaking families (1.85). **(b)** In each case, higher mean for French-speaking families: Quebec: French (1.80); English (1.64); other provinces: French (2.14); English (1.97). **(c)** There are relatively more English-speaking families in the "other" provinces, and the "other" provinces have a higher mean regardless of language. This illustrates Simpson's paradox. **9.57** There could be no difference in the prevalence of breast cancer now and in 1900 for women of a given age. Overall, the breast cancer rate would be higher now, because more women live to an old age now, and older people are more likely to have breast cancer. **9.59 (a)** response: view of the United States; explanatory: year. **(b)** independent; the two samples are not linked (e.g., not the same people in both groups) **(c)** The two samples include the same subjects so should be treated as dependent. **9.61 (a)** We are 95% confident that the population proportion of females who have used marijuana is at least 0.0077 lower and at most 0.0887 lower than the population proportion of males who have used marijuana. Because 0 is not in the confidence interval, we can conclude that females and males differ with respect to marijuana use. **(b)** The confidence interval would change only in sign. It would now be (0.0077, 0.0887). We are 95% confident that the population proportion of males who have used marijuana is at least 0.0077 higher and at most 0.0887 higher than the population proportion of females who have used marijuana. **9.63 (a)** $se_0 = 0.0238$ **(b)** z = 0.91; P-value: 0.36; we cannot reject the null hypothesis. It is plausible that the population proportion believing in the afterlife is the same for females and males. **(c)** Yes, because there actually was a difference between females and males. **(d)** independent random samples; at least 5 successes and 5 failures in each sample (since two-sided test). **9.65 (a)** Box plots are part of answer. Female crabs have a higher median and a bigger spread if they had a mate (right-skewed distribution) than if they did not have a mate (symmetrical). **(b)** 0.5 **(c)** se = 0.076 **(d)** We can be 90% confident that the difference between the population mean weights of female crabs with and without a mate is between 0.375 and 0.625 kg. **9.67 (a)** $H_0: \mu_1 = \mu_2$; $H_a: \mu_1 \neq \mu_2$ **(b)** 6.55; 0.000. It would be extremely unusual to obtain a test statistic this large if the null hypothesis were true. **(c)** We can reject the null hypothesis. The population mean is higher for blacks. **(d)** When we reject the null hypothesis with significance level 0.05, the 95% confidence interval does not include 0. **9.69 (a)** The 95% confidence interval is 4 plus and minus 3.4, that is (0.6, 7.4). **(b)** It tells us the P-value = 0.02; we would expect a difference of sample means at least this size only 2% of the time if there were truly no difference between the population means. **9.71 (a)** P-value = 0.102, so there is a probability of 0.102 that we would get a test statistic at least this large if the null hypothesis is true that there is no difference between population mean change scores. **(b)** data are quantitative; samples are independent and random; population distributions for each group are approximately normal. Based on the box plots (which show outliers and a skew to the right for the cognitive behavioral group), it would not be a good idea to conduct a one-sided test. It is not as robust as the two-sided test to violations of the normal assumption. **(c)** The lowest plausible difference between means is −0.7, a difference of less than 1 pound. The highest plausible difference between means is 7.6. **(d)** We do not reject the null hypothesis that the difference between the population means is 0, and 0 falls in the 95% confidence interval for the difference between the population means. **9.73** We can be 95% confident that the true population mean difference is between −10.8 and 5.2. **9.75 (a)** We can be 95% confident that the true population mean difference is between −0.7 and 7.6. **(b)** If the null hypothesis were true, there would be probability 0.100 of obtaining a test statistic at least this large (in absolute value). **(c)** P-value = 0.100/2 = 0.050. If the null hypothesis were true, the probability = 0.05 of getting a t statistic of 1.68 or larger. **(d)** quantitative response variable, independent random samples, and approximately normal population distributions for each group **9.77 (a)** For those who ate tomato sauce, the sample proportion of prostate cancer cases was 0.77 times the sample proportion of prostate cancer cases for those who did not. **(b)** Because 1.00 does not fall in the confidence interval, all of the plausible values indicate a decreased risk among those who eat tomato sauce. **9.79** We would explain that there's less than a 5% chance that we'd get a mean at least this much higher after the training course if there were, in fact, no difference. It would have been helpful to have the means, the mean difference, and its standard error (or a confidence interval). **9.81 (a)** dependent; the same people are answering both questions. **(b)** We can be 95% confident that the true population mean difference between ratings of the influence of TV and rock music is between −0.3 and 1.3. **(c)** The P-value indicates that there is a probability of 0.21 that we would obtain a test statistic at least this large if the null hypothesis were true. **9.83 (a)** high dose: 69/86 = 0.802; low dose: 61/86 = 0.709 **(b)** There is a probability of 0.10 that the test statistic would be at least this extreme if the null hypothesis were true. **9.85** Relative risk = 3. The proportion of obesity for those with lots of fast food and TV is three times the proportion of obesity for those who do not. **9.87 (a)** Ignore victim's race: proportion 0.11 of white defendants receive the death penalty, proportion 0.08 of black defendants receive the death penalty. When we take victim's race into account, the direction of the association changes, with black defendant more likely to get the death penalty. Specifically, when victim was white: White defendants: 0.11; black defendants: 0.23; when victim was black: White defendants: 0.00; black defendants: 0.03. **(b)** Death penalty imposed

more when victim was white, and white victims were more common when defendant was white. **9.89 (a)** If most of the female faculty had been hired recently, they would be fewer years from their degree and would have lower incomes. So, overall, mean could be lower for females. **(b)** If more women seek positions in low salary colleges and more men in high salary colleges, there could be no difference in means between males and females in each college yet overall a higher mean for men. **9.91** Summary should include the findings such as: 95% CI for difference: (−17.43, −6.75) T-Test of difference = 0 (vs not =): T-Value = −4.46 P-Value = 0.000. **9.93 (a)** sample standard deviations and sample sizes for the two groups **(b)** No; if we have the information on the entire population, there's no need to make inferences. **9.95** (d) **9.97** False **9.99** False **9.101** We can be 95% confident that the difference between the population means is between 9.6 and 17.0. **9.103 (a)** (i) $\hat{p}_1 = \hat{p}_2 = 0$ because there are no successes in either group (i.e., 0/10 = 0) (ii) $se = 0$ because there is no variability in either group if all responses are the same (and can also see this from formula for se). (iii) The 95% confidence interval would be (0,0) because we'd be adding 0 to 0 (se multiplied by z would always be 0 with se of 0, regardless of the confidence level). **(b)** The new confidence interval, (−0.22, 0.22), is far more plausible than (0, 0).

Chapter 10

10.1 (a) response: political party identification; explanatory: gender **(b)** Women are more likely than are men to be Democrats, whereas men are more likely than are women to be Independents or Republicans. **(c)** Distributions should show percentages in the party categories that are the same for men and women, such as (35%, 31%, 34%). **10.3 (a)** those of x at given categories of y. **(b)** dependent; because the probability of a murderer being of a certain race changes greatly according to the race of the person murdered. **10.5 (a)** Low income people tend to have slightly unhappier marriages than those in the other two brackets; high income people tend to have slightly happier marriages. **(b)** Regardless of income brackets, marital happiness tends to be higher than general happiness.
10.9 1. assumptions: two categorical variables (gender and happiness), randomization, expected count at least five in all cells. 2. H_0: happiness independent of gender; H_a: dependence 3. $X^2 = 0.27$ 4. P-value = 0.87 5. It is plausible that gender and happiness are independent. **10.11 (a)** 453.28; 109.72; 521.72; 126.28; There are more women who believe in the afterlife and men who do not than is expected.
(b) $X^2 = (425 − 453.28)^2/453.28 + (138 − 109.72)^2/109.72 + (550 − 521.72)^2/521.72 + (98 − 126.28)^2/126.28 = 1.76 + 7.29 + 1.53 + 6.33 = 16.9$ **10.13 (a)** (61.1%, 38.9%) and (5.9%, 94.1%); suggests marijuana use much more common for those who have smoked cigarettes than for those who have not. **(b)** 1. assumptions: two categorical variables (cigarette use and marijuana use), randomization, expected count at least five in all cells 2. H_0: marijuana use independent of cigarette use; H_a: dependence 3. $X^2 = 642.0$ 4. P-value: 0.000 5. Extremely strong evidence that marijuana use and cigarette use are associated **10.15 (a)** H_0: opinion about the environment independent of sex; H_a: dependence **(b)** $r = 2$; $c = 5$; df = 4

(c) The P-value is 0.09; (i) no; (ii) yes **(d)** (i) fail to reject the null hypothesis, and conclude that variables may be independent. (ii) reject the null hypothesis, and conclude that opinion depends on gender. **10.17 (b) 1.** Two categorical variables, randomization, expected count at least 5 in all cells **2.** H_0: whether have heart attack independent of treatment; H_a: dependence **3.** $X^2 = 2.1$ **4.** P-value: 0.14 **5.** It is plausible that treatment and heart attack incidence are independent **10.19 (a)** Data do not show a strong association between happiness and belief in life after death. **(b)** 0.62; plausible that happiness and belief in life after death are independent. **(c)** No; although independence is plausible, it is not necessary. We can never accept a null hypothesis, because a confidence interval would show that 0 is not the only plausible value for the difference between two conditional proportions. **10.21 (a) 1.** Two categorical variables, randomization, expected count at least 5 in all cells **2.** H_0: party identification independent of race; H_a: dependence **3.** $X^2 = 187.2$ **4.** P-value: 0.000 **5.** very strong evidence that party identification and race are associated **(b)** 0.31; strong; it is quite far from 0.00, the difference corresponding to no association **10.23 (a)** Reject the null hypothesis, because P-value is less than 0.05. Happiness depends on highest degree. **(b)** No; with a large sample size, even a weak association could be statistically significant. **(c)** 0.092; in the sample, happiness is higher among those who completed college than among those who completed only high school. **(d)** 2.1; college graduates are 2.1 times as likely to be happy as high school graduates **10.25 (a)** The proportion who had used alcohol is 0.33 higher for cigarette users than for non-cigarette users. **(b)** The proportion who had used cigarettes is 0.60 higher for alcohol users than for those who had not used alcohol. **(c)** Alcohol users are 5.3 times as likely to have smoked as are non-alcohol users. **10.27 (a)** 14.3. Men were 14.3 times as likely as women to be incarcerated. **(b)** The proportion of men who are incarcerated is 0.008 higher than the proportion of women who are incarcerated. **(c)** (a); it shows there is a substantial gender effect, which the difference does not show when both proportions are close to 0. **10.29 (a)** False; larger chi-squared might be due to a larger sample size rather than a stronger association. **(b)** Yes; race is more strongly associated. **10.31 (a)** The observed count falls 3.144 standard errors above the expected count. **(b)** They designate conditions in which the observed counts are much higher than the expected counts, relative to what we'd expect due to sampling variability. **(c)** They designate conditions in which the observed counts are much lower than the expected counts, relative to what we'd expect due to sampling variability. **10.33 (a)** observed count for this cell is only 0.2 standard errors from the expected count **(b)** People who are not happy in their marriage are more likely to be not happy in general than would be expected if the variables were independent. People who are pretty happy in their marriage are more likely to be pretty happy overall than would be expected if the variables were independent. Lastly, people who are very happy in their marriage are more likely to be very happy overall than would be expected if the variables were independent. **10.35** Females tend to be Democrat and males tend to be Republican, in greater numbers than one would expect if party ID were independent of gender. **10.37 (b) 1.** There are

two binary categorical variables, and randomization was used. **2.** H_0: $p_1 = p_2$; H_a: $p_1 \neq p_2$ **3.** Test statistic: 2 **4.** P-value: 0.001 **5.** Strong evidence that care and diet are associated with ability to solve a task **(c)** because the expected cell counts are less than 5 for at least some cells **10.39 (a)** 0.24; it is plausible that nervousness and treatment are independent. **(b)** No, because two cells have an expected count of less than five **10.41 (b)** yes; percentages of men and women who would vote for a qualified woman may be the same. **10.43 1.** 2 categorical variables, randomization, expected count at least 5 in all cells **2.** H_0: blood test result independent of Down syndrome status; H_a: dependence **3.** $X^2 = 114.4$, df = 1. **4.** P-value = 0.000 **5.** Very strong evidence of an association between test result and actual status **10.45 (a)** response: URI (more than 1 vs. just 1); explanatory: treatment (placebo vs. Echinacea). **(b)** It seems that number of URI's is associated with treatment. If a child had at least one URI, they were more likely to have another if they only took placebo than if they took Echinacea. **10.47 1.** Two categorical variables, randomization, expected count at least 5 in all cells **2.** H_0: political party independent of sex; H_a: dependence **3.** $X^2 = 18.8$; df = 2 **4.** P-value = 0.000 **5.** very strong evidence that party identification and gender are associated **10.49 (a)** (i) 2 categorical variables, randomization, expected count at least 5 in all cells (ii) H_0: myocardial infarction independent of treatment group; H_a: dependence (iii) $X^2 = 26.9$; df = 2 (iv) P-value = 0.000 (v) very strong evidence of association between myocardial infarction and group **(b)** 3.4. Those on placebo are 3.4 times as likely as those on aspirin to have a fatal heart attack. **10.51 (a)** The proportion who were injured is 0.061 higher for those who did not wear a seat belt than for those who wore a seat belt. **(b)** People were 1.95 times as likely to be injured if they were not wearing a seat belt than if they were wearing a seat belt. **10.53** Those with no partners or two or more partners are less likely to be very happy, and those with one partner are more likely to be very happy, than would be expected if these variables were independent. **10.55 (a)** H_0: $p_1 = p_2$; H_a: $p_1 \neq p_2$; p_1: population proportion of those who are aggressive in the group that watches less than one hour of TV per day; p_2: population proportion of those who are aggressive in the group that watches more than one hour of TV per day **(b)** 0.000075; very strong evidence that TV watching and aggression are associated. **10.57** Report could include: Pearson Chi-Square = 21.4, DF = 6, P-Value = 0.002 **10.61 (b)** yes **(c)** There often are third factors, such as age in this case, that influence an association. **10.63 (a)** Difference of proportions with U.S. as group 1: 0.000061; Britain as group 1: −0.000061 **(b)** United States as group 1: 48; Britain as group 1: 0.021 **(c)** relative risk; the difference between proportions might be very small, even when one is many times larger than the other. **10.65** False **10.67** True **10.69 (a)** The chi squared value for a right-tail probability of 0.05 and df = 1 is 3.84, which is the z value for a two-tail probability of 0.05 squared: (1.96)(1.96) = 3.84. **(b)** The chi-squared value for P-value of 0.01 and df = 1 is 6.64. This is the square of the z value for a two-tailed P-value of 0.01, which is 2.58. **10.71** The two observed values in a given row (or column) must add up to the same total as the two expected values in that same row (or column).

10.73 (a) Here are six of the twenty distinct possible samples that could have been selected: F1, F2, F3; F1, F2, M1; F1, F2, M2; F1, F2, M3; F1, F3, M2. This contingency table shows that one female was chosen (F2) and two were not (F1 and F3). **(b)** Ten samples have a difference greater than or equal to 1/3. A difference greater than or equal to 1/3 indicates more men than women in the sample.

Chapter 11

11.1 (a) Response: mileage; explanatory: weight **(b)** $\hat{y} = 45.6 - 0.0052x$; y-intercept is 45.6; slope is -0.0052. **(c)** For each 1000 pound increase, the predicted mileage will decrease by 5.2 miles per gallon. **(d)** No; the y-intercept is the predicted miles per gallon for a car that weighs 0 pounds. **11.3 (a)** 4.0; **(b)** 1.2; **(c)** 1.1 **(d)** For nations with no female economic activity, the predicted fertility rate is 5.2. As x increases from 0 to 100, the fertility rate decreases from 5.2 to 0.8. **11.5** For a given x-value, there will not be merely one y value — not every elementary schoolgirl in your town who is a given height will weigh the same. **11.7 (a)** Response: college GPA; explanatory: high school GPA; answer includes scatterplot. There is a ceiling effect that limits our ability to predict college GPA. **(b)** From software: CGPA = 1.19 + 0.637 HSGPA; (i) CGPA = 1.19 + 0.637(3.0) = 3.10; (ii) CGPA = 1.19 + 0.637(4.0) = 3.74; predicted college GPA will increase as high school GPA does. **11.9** 0.0015 **11.11 (a)** For a \$1000 increase in GDP, the predicted percentage using cell phones increases by 2.62, and the predicted percentage using the Internet increases by 1.55. **(b)** Because the slope of GDP to cell phone use is larger than that of GDP to Internet use. **11.13 (a)** (i) 6.46; (ii) 5.73; the difference, 0.73, equals the slope multiplied by the difference in numbers of sit-ups: (0.024346)(30) = 0.73.

(b) $r = b\left(\dfrac{s_x}{s_y}\right) = 0.024346\left(\dfrac{6.887}{0.365}\right) = 0.46$ **11.15 (a)** Negative;

reading the newspaper more tends to be associated with being less conservative. **(b)** $r^2 = 0.004$; very weak association **11.17 (a)** 650 **(b)** x = 800 is 3 standard deviations above the mean, so the predicted y value is 0.5(3) = 1.5 standard deviations above the mean. **11.19 (a)** 270; **(b)** No; this decrease could occur merely because of regression to the mean. **11.21** Regression to the mean **11.23** There is a 75% reduction in error in predicting a car's mileage based on knowing the weight, compared to predicting by the mean mileage. This relatively large value means that we can predict a car's mileage quite well if we know its weight. **11.25** University of Connecticut; Yale would have a restricted range of high school GPA values, with nearly all its students clustered very close to the top. **11.27 (a)** 0.80, strong positive association **(b)** (i) 0.67; (ii) 0.475; the correlation usually is smaller in absolute value when the range of predictor values is restricted. **11.29 (a)** (i) Assume randomization, linear trend with normal conditional distribution for y and the same standard deviation at different values of x. (ii) $H_0: \beta = 0$; $H_a: \beta \neq 0$ (iii) $t = 11.6$ (iv) P-value = 0.000 (v) Very strong evidence that an association exists between the size and price of houses; this is extremely unlikely to be due to random variation. **(b)** $b \pm t_{.025}(se) = 77.008 \pm 1.985(6.626)$, which is (64, 90)

(c) An increase of \$100 is outside the confidence interval and so is an implausible value. **11.31 (a)** (i) Assume randomization, linear trend with normal conditional distribution for y and the same standard deviation at different values of x. These data were not gathered using randomization, and so inferences are tentative. (ii) $H_0: \beta = 0$; $H_a: \beta \neq 0$ (iii) $t = 9.6$ (iv) P-value = 0.000 (v) Very strong evidence that an association exists between these variables. **(b)** (4.2, 6.4); it gives us a range of plausible values for the slope of the population. **11.33 (a)** Having more daughters is good. **(b)** (i) Assume that there was roughly a linear relationship between variables, and that the data were gathered using randomization and that the population y values at each x value follow a normal distribution, with roughly the same standard deviation at each x value. (ii) $H_0: \beta = 0$; $H_a: \beta \neq 0$ (iii) $t = 1.52$. (iv) P-value = 0.13. (v) It is plausible that there is no association. **(c)** $(-0.13, 1.01)$. Zero is a plausible value for this slope. **11.35 (a)** Means: advertising is 2, sales is 7. Standard deviations: advertising is 2.16, sales is 2.16. **(b)** $\hat{y} = 5.286 + 0.857x$ **(c)** For $H_0: \beta = 0$; $H_a: \beta \neq 0$, $t = 2.35$, P-value = 0.14, it is plausible that there is no association between advertising and sales. **11.37 (a)** 15.0 = number of 60-pound bench presses and 105.00 = maximum bench press for athlete 10, 85.90 = predicted maximum bench press, 19.10 = difference between the actual and predicted maximums, 2.41 = residual divided by the standard error that describes the sampling variability of the residuals. **(b)** No; we would expect about 5% of standardized residuals to have an absolute value above 2.0. **11.39 (a)** distribution of standardized residuals and hence conditional distribution of maximum bench press. **(b)** The distributions mentioned in (a) seem to be approximately normal. **11.41 (a)** 448 = predicted amount spent on clothes in past year for those in 12th grade of school **(b)** (427, 469): range of plausible values for the population mean of dollars spent on clothes for 12th grade students in the school **(c)** (101, 795): range of plausible values for the individual observations (dollars spent on clothes) for all the 12th grade students at the school **11.43 (a)** The square root of 1303.72 is 36.1. This is the estimated standard deviation of maximum leg presses for female athletes who can do a fixed number of 200-pound leg presses. **(b)** Standard deviation = 36.1; (277.6, 422.0) **11.45 (a)** Total SS = residual SS + regression SS, where residual SS = error in using the regression line to predict y, regression SS = how much less error there is in predicting y using the regression line compared to using $\bar{y}$ **(b)** Sum of squares around mean divided by n-1 is 192,787/56 = 3442.625, and its square root is 58.7. This estimates the overall standard deviation of y values whereas the residual s estimates the standard deviation of y values at a fixed value of x. **(c)** The F test statistic is 92.87; its square root is the t statistic of 9.64. **11.47 (a)** MS values: 200,000, 22,580.6; F: 8.86 **(b)** $H_0: \beta = 0$ against $H_a: \beta \neq 0$

11.49

Source	DF	SS	MS	F	P
Regression	1	1.9577	1.9577	19.52	0.000
Residual Error	57	5.7158	0.1003		
Total	58	7.6736			

(a) Total = 7.6736 = 1.9577 + 5.7158, where residual SS = 5.7158 = error in using the regression line to predict y, regression SS = 1.9577 = how much less error there is in predicting y using the regression line compared to using $\bar{y}$ **(b)** 0.32; estimates standard deviation of y at fixed value of x, and describes typical size of the residuals about the regression line. **(c)** 0.36; this describes variability about the overall mean of $\bar{y}$ for those at all values of x, not just those at a specific value of x. **11.51 (a)** $(1.072)^{10} = 2.0$; **(b)** $(1.10)^{10} = 2.59$; the effect here is multiplicative, not additive. **11.53 (a)** $1.15^5 = 2.0$; **(b)** $1.15^{10} = 4.0$; **(c)** $1.15^{20} = 16.4$ **11.55 (a)** Scatterplot shows the relation between variables is curvilinear. **(b)** -16.8; no, a weight cannot be negative. **(c)** Yes **(d)** (i) 80.6; (ii) 1.3 **(e)** indicates that predicted weight multiplies by 0.813 each week **11.57 (a)** At a fixed value of x there is variability in the values of y so we can't specify individual y values using x but we can try to specify the mean of those values and how that mean changes as x changes. **(b)** Since y values vary at a fixed value of x, the model has a σ parameter to describe the spread of the conditional distribution of those y values at each fixed x. **11.59** Response: height of children; explanatory: height of parents. Children tend to be short but not as short as parents. Because of regression toward the mean, the value of y tends to be not so far from its mean as the x value is from its mean. **11.61 (a)** $\hat{y} = 6.66 - 0.0648x$; (i) 6.66; (ii) 0.18; the difference is 100 times the slope. **(b)** 3.36; -2.06 **(c)** The observation for Belgium was 2.97 standard errors below the predicted value. **11.63 (a)** 210,991. This house sold for $210,991 more than would have been predicted. **(b)** This observation is 4.02 standard errors higher than predicted. **11.65 (a)** residual standard deviation of y: spread of the y values at a particular x value; standard deviation: spread of all of the y values **(b)** Spread of y values at a given x is about the same as spread of all y observations. $r^2 = 0.13$. **11.67 (a)** It tells us that the sample slope falls nearly 18 standard errors above the null hypothesis value of 0 for the population slope. **(b)** If H_0 were true that the population slope $\beta = 0$, it would be extremely unusual to get a sample slope as far from 0 as the observed one of $b = 1.23$. **11.69 (a)** The plausible values range from 337.65 to 364.84 for the *mean* of y values for *all* female high school athletes having $x = 80$. **(b)** For *all* female high school athletes with a maximum bench press of 80, we predict that 95% of them have maximum leg press between about 247.70 and 454.80 pounds. The 95% PI is for a single observation y, whereas the confidence interval is for the *mean* of y. **11.71 (a)** $1000 \times 2^5 = 32,000$; **(b)** $1000 \times 2^{10} = 1,024,000$; **(c)** 1000×2^x **11.73 (a)** 1900: $\hat{y} = 1.4193 \times 1.014^0 = 1.42$ billion; 2000: $\hat{y} = 1.4193 \times 1.014^{100} = 5.70$ billion **(b)** Because multiplying by 1.014 adds an additional 1.4% each year **(c)** (i) Because $1.014^{50} = 2.0$, the number by which we'd multiply the original population size (ii) $1.014^{100} = 4.0$ **(d)** Exponential regression model; the log of the population size and the year number are more highly correlated than are the population size and the year number. **11.75 (a)** The 3 outlying points represent outliers — values more than $1.5 \times$ IQR beyond either Q1 or Q3. **(b)** From software: Diff $= -9.125 + 1.178$ Runs; Difference is positive when $-9.125 + 1.178$ (runs) > 0, which is equivalent to runs $> 9.125/1.178 = 7.7$. **(c)** Runs, hits, and difference are positively associated with one another. Errors are negatively associated with those three variables.

	Run	Hits	Errors
Hits	0.819		
Errors	-0.259	-0.154	
Difference	0.818	0.657	-0.501

(d) From software:

Predictor	Coef	SE Coef	T	P
Constant	-9.1254	0.5934	-15.38	0.000
Run	1.17844	0.04989	23.62	0.000

11.77 Report would interpret results from:
high_sch_GPA $= 3.44 - 0.0183$ TV
S $= 0.446707$ R-Sq $= 7.2\%$

Source	DF	SS	MS	F	P
Regression	1	0.8921	0.8921	4.47	0.039
Residual Error	58	11.5737	0.1995		
Total	59	12.4658			

11.79 Explain how a player or team that had a particularly good or bad year had results in the following year that were not so extreme (i.e., regression toward the mean). **11.81** As the range of values reflected by each sample is restricted, the correlation tends to decrease when we consider just students of a restricted range of ages. **11.83 (a)** The slope would be 2 times the original slope. **(b)** The correlation would not change because it is independent of units. **(c)** The t statistic would not change because although the slope doubles, so does its standard error. (The result of a test should not depend on the units we use.) **11.85 (a)** There are two parameters α and β, and so $df = n - 2$. **(b)** There is only one parameter, and therefore, $df = n - 1$. **11.87 (a)** The percentage would likely fluctuate over time, and this would not be a linear relationship. **(b)** Annual medical expenses would likely be quite high at low ages, then lower in the middle, then high again, forming a parabolic, rather than linear, relationship. **(c)** The relation between these variables is likely curvilinear. Life expectancy increases for awhile as per capita income increases, then gradually levels off. **11.89 (a)** The statement is referring to additive growth, but this is multiplicative growth. **(b)** 150,896; about 50.9% growth **11.91** (b) **11.93** (d) **11.95** (a) **11.97 (a)** At 0 impact velocity, there would be 0 putting distance. The line would pass through the point having coordinates $(0, 0)$. **(b)** If x doubles, then x^2 (and hence the mean of y) quadruples; e.g., if x goes from 2 to 4, then x^2 goes from 4 to 16. **11.99** Because

$$r^2 = \frac{\sum(y - \bar{y})^2 - \sum(y - \hat{y})^2}{\sum(y - \bar{y})^2}, \text{ it represents the relative}$$

difference between the quantity used to summarize the overall variability of the y values and the quantity used to summarize the residual variability. These go in the numerator of the respective variance estimates. Therefore, the estimated variance of the conditional distribution of y for a given x is approximately 30% smaller than that of the marginal distribution of y. **11.101 (a)** Error is calculated by subtracting the mean from the actual score, y. If this difference is positive, then the observation must fall above the mean. **(b)** $\varepsilon = 0$ when the observation falls exactly at the mean. **(c)** Since the residual $e = y - \hat{y}$, we have $y = \hat{y} + e = a + bx + e$. As $\hat{y}$ is an estimate of the population mean, e is an estimate of ε. **(d)** It does not make sense to use the simpler model, $y = \alpha + \beta x$ that does not have an error term because it is improbable that every observation will fall exactly on the regression line.

Chapter 12

12.1 (a) $\hat{y} = 134.3$ **(b)** -19.3. The actual total body weight is 19.3 pounds lower than predicted. **12.3 (a)** (i) 3.80; (ii) 1.60 **(b)** $\hat{y} = 0.20 + 0.50x_1 + 0.002(500) = 0.20 + 0.50x_1 + 1 = 1.20 + 0.50x_1$ **(c)** $\hat{y} = 0.20 + 0.50x_1 + 0.002(600) = 0.20 + 0.50x_1 + 1.2 = 1.40 + 0.50x_1$ **12.5 (a)** (i) 18.3; (ii) 12.4 **(b)** When education goes up $10 = 80 - 70$, predicted crime rate changes by ten multiplied by the slope, $10(-0.58) = -5.8$ **(c)** (i) $\hat{y} = 59.12 - 0.5834x_1 + 0.6825(0) = 59.12 - 0.5834x_1$ (ii) $\hat{y} = 59.12 - 0.5834x_1 + 0.6825(50) = 93.2 - 0.5834x_1$ (iii) $\hat{y} = 59.12 - 0.5834x_1 + 0.6825(100) = 127.4 - 0.5834x_1$ **(d)** The line passing through the points having urbanization $= 50$ has a negative slope. The line passing through all the data points has a positive slope. Simpson's paradox occurs because the association between crime rate and education is positive overall but is negative at each fixed value of urbanization. It happens because urbanization is positively associated with crime rate and with education. As urbanization increases, so do crime rate and education tend to increase, giving an overall positive association between crime rate and education. **12.7 (a)** 289.8 **(b)** The predicted total score will increase by 0.85 for each increase of 1 in the number of putts taken, when controlling for the other variables in the model. **(c)** Increasing GIR by 10 led to the largest predicted decrease (-13.1) **12.9** We don't need to control x_2 if it's not associated with x_1. Changes in x_2 will not have an impact on the effect of x_1 on y. **12.11 (a)**

$$R^2 = \frac{\Sigma(y - \bar{y})^2 - \Sigma(y - \hat{y})^2}{\Sigma(y - \bar{y})^2} = (1886 - 287)/1886 = 0.848$$

(b) Using these variables together to predict total score reduces the prediction error by 84.8%, relative to using $\bar{y}$ alone to predict total score. Yes, the prediction is much better. **(c)** 0.921; strong association between the observed total score and the predicted total score **12.13** Height is not strongly correlated with either body fat or age. **12.15** Because x_1 and x_2 are themselves highly correlated, once one of them is in the model, the remaining one does not help much in adding to the predictive power. **12.17 (a)** The correlation between predicted house selling price and actual house selling price is 0.84. **(b)** If selling price is measured in thousands of dollars, each y value would be divided by 1,000. For example, 145,000 dollars would become 145 thousands of dollars. Each slope also would be divided by 1,000 (e.g., a slope of 53.8 for house size on selling price in dollars corresponds to 0.0538 in thousands of dollars for $\hat{y}$. **(c)** The multiple correlation would not change because it is not dependent on units. **12.19 (a)** It means that, in the population of all students, high school GPA doesn't predict college GPA for students having any given value for study time. **(b) 1.** *Assumptions*: We assume a random sample and that the model holds (each explanatory variable has a straight-line relation with μ_y, with same slope for all combinations of values of other predictors in model, and there is a normal distribution for

y with the same standard deviation at each combination of values of the predictors). Here, the 59 students were a convenience sample, not a random sample, so inferences are tentative. **2.** *Hypotheses*: H_0: $\beta_1 = 0$; H_a: $\beta_1 \neq 0$ **3.** *Test statistic*: $t = (b_1 - 0)/ se = 0.6434/0.1458 = 4.41$ **4.** *P-value*: 0.000 **5.** *Conclusion*: The P-value of 0.000 gives very strong evidence against the null hypothesis that $\beta_1 = 0$. At common significance levels, such as 0.05, we can reject H_0. **12.21 (a)** 0.32 describes the typical size of the residuals and also estimates the standard deviation of y at fixed values of the predictors. **(b)** between 2.97 and 4.25 **12.23** The first test analyzes the effect of LP_200 at any given fixed value of BP_60, whereas the second test describes the overall effect of LP_200 ignoring other variables. These are different effects, so one can exist when the other does not. In this case, it is likely that LP_200 and BP_60 are strongly associated with one another, and the effect of LP_200 is weaker once we control for BP_60. **12.25 (a)** H_0: $\beta_1 = \beta_2 = 0$; the null hypothesis states that neither of the two explanatory variables have an effect on the response variable y. **(b)** 3.17 **(c)** 51.39 with a P-value of 0.000. Extremely strong evidence against the null, and we can reject H_0. At least one of the two explanatory variables has an effect on BP. **12.27 (a)** $(0.04, 0.17)$ **(b)** The confidence interval gives plausible values for the slope, the amount that predicted mental impairment will increase when life events score increases by one, when controlling for SES. For an increase of 100 units in life events, we multiply each endpoint of the confidence interval by 100, giving $(4, 17)$. **12.29 (a)** H_0: $\beta_1 = \beta_2 = \beta_3 = 0$ means that house selling price is independent of size of home, lot size, and real estate tax. **(b)** The large F value and small P-value provide strong evidence that at least one of the three explanatory variables has an effect on selling price. **(c)** The results of the t tests tell us that each of the three explanatory variables contributes to the prediction of selling price. **12.31 (a)** Conditional distribution **(b)** That the distribution may be skewed to the right rather than normal **12.33** One might think this suggests less variability at low levels and even less at high levels of BP_60, but this may merely reflect fewer points in those regions. Overall, it seems OK. **12.37 (a)** Upper right and middle right plots; strong, positive associations **(b)** $\hat{y} = 55.012 + 0.16676$LBM $+ 1.6575$REP_BP; 1.6575 is the amount that maximum bench press changes for a one unit increase in number of repetitions, controlling for lean body mass. **(c)** 0.832; using these variables together to predict BP reduces the prediction error by 83.2%, relative to using $\bar{y}$ alone to predict BP. **(d)** 0.912; there is a strong association between the observed BPs and the predicted BPs. **(e)** $F = (7641.5/50.7) = 150.75$; P-value is 0.000; sufficient evidence to conclude that BP is not independent of these two predictors **(f) 1.** *Assumptions*: The 64 athletes were a convenience sample, not a random sample (inferences are tentative). We must assume that each explanatory variable has a straight-line relation with μ_y, with same slope for all combinations of values of other predictors in model, and that there is a normal distribution for y with the same standard deviation at each combination of values of the other predictors in the model. **2.** *Hypotheses*: H_0: $\beta_1 = 0$; H_a: $\beta_1 \neq 0$ **3.** *Test statistic*: $t = 2.22$ **4.** *P-value*: 0.030 **5.** *Conclusion*:

The P-value of 0.030 gives sufficient evidence against the null hypothesis that $\beta_1 = 0$. At common significance levels, such as 0.05, we can reject H_0. **(g)** The histogram suggests that the residuals are roughly bell-shaped about 0. They fall between about -3 and $+3$. The shape suggests that the conditional distribution of the response variable is roughly normal. **(h)** The plot of residuals against values of REP_BP describes the degree to which the response variable is linearly related to this particular explanatory variable. It suggests that the residuals are less variable at smaller values of REP_BP than at larger values of REP_BP. **(i)** The individual with REP_BP around 32 and standardized residual around -3 had a BP value considerably lower than expected. **12.39** Answer is example of possible plot **12.41 (a)** For each increase in one bid, predicted selling price decreases by 0.723 for the same buying option. **(b)** When the buy-it-now option was used, predicted selling price was 2.77 lower. **12.43 (a)** (i) Minimum: $61.3 + 0.35(12) = 65.5$; (ii) Maximum: $61.2 + 0.35(54) = 80.2$ **(a)** When controlling for region, an increase of one hour leads to a decrease in predicted defects of 0.78 per 100 cars. Japanese facilities had 36 fewer predicted defects, on average, than did other facilities. **(b)** Because the direction of the association between time and defects changed with the variable of whether facility is Japanese was added **(c)** Because, overall, Japanese facilities have fewer defects and take less time; other facilities have more defects and take more time. When the data are looked at together, this leads to an overall positive association between defects and time. (answer includes scatterplot) **12.45 (a)** From software: price = 43014 + 45.3 Taxes + 10814 NW. NW: price = 53,828 + 45.3 Taxes; not in NW: price = 43,014 + 45.3 Taxes **(b)** Indicates that the predicted selling price for houses in the NW is 10,814 higher than the predicted selling price elsewhere **12.47 (a)** Interaction **(b)** Males: $\hat{y} = 1.98 + 0.00570(312) = 3.758$; Females: $\hat{y} = 1.58 + 0.00698(312) = 3.758$ **(c)** No; it is dangerous to extrapolate beyond our existing data. We do not know whether associations among variables will change in the future.

12.49 $\hat{p} = \dfrac{e^{-3.52+0.105x}}{1 + e^{-3.52+0.105x}} = \dfrac{e^{-3.52+0.105(25)}}{1 + e^{-3.52+0.105(25)}} =$ $(0.4086/1.4086) = 0.290$ **12.51 (a)** Q1: 0.484; Q3: 0.814 **(b)** $0.814 - 0.484 = 0.33$ **12.53 (a)** 0.310; **(b)** 0.731; probability of voting Republican increases as income increases **12.55 (a)** As family income increases, people are more likely to vote Republican. As number of years of education increases, people are more likely to vote Republican. Men are more likely to vote Republican than are women. **(b)** (a) $\hat{p} = 0.470$; (b) $\hat{p} = 0.421$ **12.57 (a)** White **(b)** Black defendants who had white victims; because the coefficient for defendant race is negative [0 (black) would lead to a higher predicted proportion than would 1 (white)] and the coefficient for victim's race is positive [1 (white) would lead to a higher predicted proportion than would 0 (black)] **12.59 (a)** (Answer includes scatterplot matrix); The plots that pertain to selling price as a response variable are those across the top row. The highly discrete nature of x_2 and x_3 limits the number of values these variables can take on. **(b)** From software: price = 13596 + 74.1 size − 11164 Bedrooms + 17651 Baths. When number of bedrooms and number of bathrooms are fixed, an increase of one in size of home leads to an increase of 74.1 in price. **(c)** $R^2 = [(3.14 \times 10^{11}) -$

$(1.25 \times 10^{11})]/(3.14 \times 10^{11}) = 0.602$; indicates that predictions are 60.2% better when using the prediction equation instead of using the sample mean $\bar{y}$ to predict y. **(d)** 0.776 is the correlation between the observed y values and the predicted $\hat{y}$ values. **(e) 1.** *Assumptions*: multiple regression equation holds, data gathered randomly, normal distribution for y with same standard deviation at each combination of predictors. **2.** *Hypotheses*: H_0: $\beta_1 = \beta_2 = \beta_3 = 0$; H_0: At least one β parameter differs from 0. **3.** *Test statistic*: $F = 63{,}148{,}663{,}440/(1{,}301{,}943{,}013 = 48.50$ **4.** *P-value*: for $df(3,96)$; 0.000 **5.** *Conclusion*: The small P-value indicates that we have very strong evidence to conclude that at least one explanatory variable has an effect on y. **(f)** The t statistic is -1.64 with a one-sided P-value of $0.103/2 = 0.0515$. We do not have sufficient evidence to conclude that the number of bedrooms has an effect on selling price. Number of bedrooms might be associated with selling price on its own, but might not provide additional predictive information over and above the other explanatory variables. **(g)** (answer includes histogram of standardized residuals) The histogram of the standardized residuals shows the sizes of the residuals, $y - \hat{y}$, for each observed variable in the sample. This histogram suggests that the residuals are negatively skewed. There is an outlier that is lower than -3. **(h)** (answer includes residuals plotted against size of home) This plot depicts the size of the residuals for the different house sizes observed in this sample. It indicates greater residual variability as house size increases. **12.61 (a)** Difference = $-5.00 + 0.934$Hits $- 1.61$ Errors; for each increase of one hit, the predicted difference increases by 0.934, and for each increase of one error, the predicted difference decreases by 1.61. **(b)** six **12.63** Slope with x_3 in the model: effect of poverty when controlling for percentage of single-parent families, as well as percent living in urban areas; slope without x_3 in the model: effect of poverty when controlling only for percent living in urban areas **12.65 (a)** When two explanatory variables are included, the correlation between the observed y values and the predicted $\hat{y}$ values is 0.787. **(b)** No; the multiple correlation increased only by 0.001. **(c)** No; life expectancy might be highly associated with GDP, and therefore, does not add much predictive power to the model. **(d)** A large positive standardized residual indicates that the observed CO2 is a good deal higher than the predicted CO2. Australia's CO2 is 2.84 standard errors above what would be predicted from this model. **12.67 (a)** $F = 20.742/0.484 = 42.855$; P-value: 0.000; sufficient evidence to conclude that at least one of these explanatory variables predicts y better than the sample mean does. **(b)** No; if we were not interested in nations beyond those in the study, the group of nations that we studied would be a population and not a sample. **12.69** Because the probability that a child obtains a high school degree is categorical (binary). **12.71 (a)** Negative; if an individual used AZT, the probability of developing symptoms was lower. **(b)** Black/yes: $\hat{p} = 0.143$; black/no: $\hat{p} = 0.255$ **(c) 1.** *Assumptions*: The data were generated randomly. The response variable is binary. **2.** *Hypotheses*: H_0: $\beta_1 = 0$; H_a: $\beta_1 \neq 0$ **3.** *Test statistic*: -2.58 **4.** P-*value:* 0.0099 **5.** *Conclusion*: The P-value of 0.0099 gives very strong evidence against the null hypothesis that $\beta_1 = 0$. At common significance levels, such as 0.05, we can reject H_0. **12.73** The reports will be different for each student, but should include the following findings, along with associated graphs.

regression equation: college_GPA = 2.83 + 0.203
high_sch_GPA − 0.0092 sports

Predictor	Coef	SE Coef	T	P
Constant	2.8293	0.3385	8.36	0.000
high_sch_GPA	0.20309	0.09753	2.08	0.042
sports	−0.00922	0.01161	−0.79	0.430

S = 0.341590   R-Sq = 8.8%   R-Sq(adj) = 5.6%

Analysis of Variance

Source	DF	SS	MS	F	P
Regression	2	0.6384	0.3192	2.74	0.073
Residual Error	57	6.6510	0.1167		
Total	59	7.2893			

Unusual Observations

Obs	high_sch_GPA	college_GPA	Fit	SE Fit	Residual	St Resid
11	2.30	2.6000	3.1580	0.1476	−0.5580	−1.81 x
42	2.00	3.0000	3.1616	0.1351	−0.1616	−0.52 x
50	3.00	4.0000	3.3002	0.1224	0.6998	2.19R
60	3.40	3.0000	3.3722	0.1345	−0.3722	−1.19 x

12.75 One-page reports will be different for each student, but should indicate that the three non-categorical explanatory variables—years of experience, degree quality, and research productivity—are all positively associated with salaries, given the other explanatory variables in the model. For the categorical variables, salary is higher among those with a Ph.D. as a terminal degree, as well as those who are married, white, or male. **12.77** (d) **12.79 (a)** False; the multiple correlation is at least as large as the ordinary correlations. **(b)** False; it falls between 0 and 1. **(c)** False; $0 \le R^2 \le 1$ and it is not the sum of the individual r^2 s. **(d)** True **12.81 (a)** True; the slope for this variable is positive in both equations. **(b)** False; such a change occurs only when we ignore x_2. **(c)** True; the slope for x_2 is 0.003 when controlling for x_1. 0.003 multiplied by 100 is 0.30. **12.83** Examples will be different for each student, but should be examples of Simpson's paradox. **12.85** When there are many explanatory variables, doing the F test first provides protection from doing lots of t tests and having one of them be significant merely by random variation when, in fact, there truly are no effects in the population. **12.87** When $x = 0$, you likely can't afford a home no matter what. Similarly, when $x = 50,000$, you likely can afford a home no matter what. Only in the middle is the extra income likely to "push" you over the income level at which you can afford a home. In such a case, a linear regression model would not be appropriate, although a logistic regression model would. **12.89** If two variables are correlated with each other at a level of 0.77, for example, and we add another variable, that variable can only increase the multiple correlation. If it does not add anything to the model, its coefficient will be 0, and the multiple correlation will remain 0.77. **12.91** answer is scatterplot **12.93 (a)** When $p = 0.5$, $p(1 − p) = 0.5(1 − 0.5) = 0.25$; 0.25 multiplied by β is the same as $\beta/4$. **(b)** $0.1(1 − 0.1) = 0.09$; $0.3(1 − 0.3) = 0.21$; $0.7(1 − 0.7) = 0.21$; $0.9(1 − 0.9) = 0.09$; as p gets closer and closer to 1, the slope approaches 0. (answer includes sketched curve) **12.95** The responses will be different for each class.

Chapter 13

13.1 (a) Response variable: performance gap; factor: which hotel the guest stayed in; categories: five hotels **(b)** H₀: $\mu_1 = \mu_2 = \mu_3 = \mu_4 = \mu_5$; Hₐ: at least two of the population means are unequal. **(c)** $df_1 = g − 1 = 5 − 1 = 4$; $df_2 = N − g = 125 − 5 = 120$ **(d)** From a table or software, $F = 2.45$ **13.3 (a) 1.** *Assumptions*: Independent random samples, normal population distributions with equal standard deviations **2.** *Hypotheses*: H₀: $\mu_1 = \mu_2 = \mu_3$; Hₐ: at least two of the population means are unequal **3.** *Test statistic*: $F = 2.50$ ($df_1 = 2$, $df_2 = 5$) **4.** P-*value* = 0.18 **5.** *Conclusion*: It is plausible that H₀ is true. **(b)** Because the numerator of the between-groups variance estimate involves multiplying by the sample size n for each group, a smaller n leads to a smaller overall between-groups estimate of variance and a smaller test statistic value. **(c)** Observational; a lurking variable might be school GPA. Perhaps higher GPA students are more likely to have previously studied a language and higher GPA students also tend to do better on quizzes than other students. **13.5 (a)** H₀: $\mu_1 = \mu_2 = \mu_3$; μ_1 represents the population mean satisfaction rating for San Jose, μ_2 for Toronto, and μ_3 for Bangalore. **(b)** $F = 27.6 = 13.00/0.47$; $df_1 = 2$ and $df_2 = 297$. **(c)** Very small P-value of 0.000 gives very strong evidence against the null; with a 0.05 significance level, we would reject the null hypothesis and conclude at least two population means differ. **13.7 (a)** μ_1 represents the population mean ideal number of kids for Protestants; μ_2 represents the population mean for Catholics; μ_3 for Jewish people; μ_4 for those of another religion; and μ_5 for those of no religion; H₀: $\mu_1 = \mu_2 = \mu_3 = \mu_4 = \mu_5$ **(b)** Independent random samples; normal population distributions with equal standard deviations **(c)** $F = 5.48$; P-value = 0.001. Very strong evidence that at least two of the population means differ. **(d)** No; ANOVA tests only whether at least two population means are different. **13.9 (a)**

Variable	gender	N	Mean	StDev
sports	f	31	4.129	3.640
	m	29	6.310	3.828

(b) Hypotheses: H₀: $\mu_1 = \mu_2$; Hₐ: $\mu_1 \ne \mu_2$. $F = 5.12$; P-value = 0.027. Strong evidence that the population means differ. **(c)** The t statistic would be the square root of the F statistic, and the P-values would be identical. **13.11 (a)**

Level	N	Mean	StDev
famchange	17	7.265	7.157
cogchange	29	3.007	7.309
conchange	26	−0.450	7.989

The means of these groups are somewhat different. The standard deviations are similar. **(b)** $F = 5.42$; P-value = 0.006; strong evidence that at least two population means are different. **(c)** The assumptions are that there are independent random samples and normal population distributions with equal standard deviations. There is evidence of skew, but the test is robust with respect to this assumption. The subjects were randomly assigned to treatments but are not a random sample of subjects suffering from anorexia, so results are tentative. **13.13** (3.55, 8.25); because 0 does not fall in this confidence interval, we can infer that the population means are different (higher for the high religiosity group). **13.15 (a)** Classical music and Muzak **(b)** Because the Tukey method uses an overall confidence level of 95% for the entire set of intervals **13.17 (a)** $x_1 = 1$ for observations from the first group and $=0$ otherwise; $x_2 = 1$ for observations from the second group and $=0$ otherwise.

(b) $H_0: \mu_1 = \mu_2 = \mu_3$; $H_0: \beta_1 = \beta_2 = 0$ **(c)** The intercept estimate is the mean for the third group, which is 12. The first estimated regression coefficient is the difference between the means for the first and third groups, or $18 - 12 = 6$, and the second is the difference between the means for the second and third groups, or $15 - 12 = 3$. **13.19 (a)** $x_1 = 1$ for observations from San Jose and $=0$ otherwise; $x_2 = 1$ for observations from Toronto and $=0$ otherwise. **(b)** 7.1 is the sample mean for Bangalore, 0.5 is the difference between the sample means for San Jose and Bangalore, and 0.7 is the difference between the sample means for Toronto and Bangalore. **13.21 (a)** Group 2—Group 1: $(-8.7, 2.7)$ Group 3—Group 1: $(-3.1, 7.1)$ Group 3—Group 2: $(-0.7, 10.7)$ Because 0 falls in all three confidence intervals, we cannot infer that any of the pairs of population means are different. **(b)** Note: Statistical software such as Minitab is needed to complete this solution. Group 2—Group 1: $(-10.3, 4.3)$ Group 3—Group 1: $(-4.5, 8.5)$ Group 3—Group 2: $(-2.3, 12.3)$ Again, we cannot infer that any of the pairs of population means are different. The intervals are wider because we are now using a 95% confidence level for the overall set of intervals. **13.25 (a)** $F = 9128/2542 = 3.59$ **(b)** The small P-value of 0.017 provides strong evidence that the population mean house selling price depends on the region of the city. **13.27 (a)** H_0: Population mean holding time is equal for the three types of messages, for each fixed level of repeat time. **(b)** $F = 74.60/10.52 = 7.09$; the small P-value of 0.011 provides strong evidence that the population mean holding time depends on the type of message. **(c)** Population distribution for each group is normal, the population standard deviations are identical, and the data result from a random sample or randomized experiment **13.29 (a)** H_0: no interaction; $F = 0.67$; P-value $= 0.535$ **(b)** Large P-value; it is plausible that there is no interaction. This lends validity to the previous analyses that assumed a lack of interaction. **13.31 (a)** The means for women of each race are almost the same, whereas the mean for white males is about 0.5 higher than the mean for black males **(b)** The gender effect was such that women had the higher mean for blacks but men had the higher mean for whites. Overall, ignoring race, the means may be quite similar for men and for women, and the one-way ANOVA for testing the gender effect may not be significant. **(c)** From a two-way ANOVA, we learn that the effect of gender differs based on race (as described above), but we do not learn this from the one-way ANOVA. **13.33 (a)** $F = 13.90$; P-value $= 0.000$; very strong evidence that the mean weight gain depends on the protein level. **(b)** H_0: no interaction; $F = 2.75$; P-value: 0.073; the P-value is not smaller than 0.05, so we cannot reject the null hypothesis. No interaction is plausible. **(c)** $(7.7, 33.9)$ **13.35 (a)** Population means, μ_1 for married, μ_2 for widowed, μ_3 for divorced, μ_4 for separated, and μ_5 for never married. $H_0: \mu_1 = \mu_2 = \mu_3 = \mu_4 = \mu_5$. H_a: at least two of the population means are different. **(b)** No; large values of F contradict the null, and when the null is true the expected value of the F statistic is approximately 1. **(c)** The null hypothesis is plausible. **13.37 (a)** 1.27 **(b)** The sample standard deviations are similar.

(c) $(4.83 - 3.65) \pm 1.962(1.27)\sqrt{\dfrac{1}{465} + \dfrac{1}{596}} = 1.18 \pm 0.15$, which is $(1.03, 1.33)$. Because 0 does not fall in this confidence interval, we can infer that there is a difference in the political ideology scores of Democrats and Republicans, with Republicans tending to be more conservative. **13.39 (a)** $F = 249.93/6.41 = 39.0$; this test statistic has a P-value far smaller than 0.05; we can reject the null hypothesis. **(b)** The mean differences between blacks and whites and between blacks and other; it appears that blacks watch more hours of TV per day than either whites or those in the category "other."

(c) $(4.14 - 2.75) \pm 1.961(2.53)\sqrt{\dfrac{1}{305} + \dfrac{1}{1435}} = 1.39 \pm 0.31$, which is $(1.1, 1.7)$. **(d)** wider; it uses a 95% confidence level for the overall set of intervals.

13.41 (a) $3.182(2.00)\sqrt{\dfrac{1}{2} + \dfrac{1}{2}} = 6.4$ **(b)** The confidence interval formed using the Tukey 95% multiple comparison uses a 95% confidence level for the overall set of intervals. **(c)** Let $x_1 = 1$ for Bumper A and 0 otherwise, $x_2 = 1$ for Bumper B and 0 otherwise, and $x_1 = x_2 = 0$ for Bumper C. **(d)** 13 is the predicted mean damage cost for Bumper C, -11 is the difference between the estimated mean damage costs between Bumpers A and C, and -10 is the difference between the estimated mean damage costs between Bumpers B and C. **13.43 (a)** From software: 15.1 **(b)** All intervals contain 0; therefore, no pair is significantly different. **13.45 (a)** Software gives the following 95% confidence intervals: control and cog: $(-7.5, 0.6)$ family and cog: $(-8.8, 0.3)$ family and control: $(-12.4, -3.0)$ Only the interval for family and control does not include 0. We can infer that the weight change is greater among those who receive family therapy than among those in the control group. **(b)** Software gives the following Tukey 95% multiple comparison confidence intervals: Difference between control and cognitive: $(-8.3, 1.4)$ Difference between family and cognitive: $(-9.8, 1.3)$ Difference between family and control: $(-13.3, -2.1)$ The interpretations are the same as for the intervals in (a). The intervals are wider because the 95% confidence level is for the entire set of intervals rather than for each interval separately. **13.47 (a)** Let $x_1 = 1$ for NW and 0 otherwise. From software: price $= 106836 + 26483$ NW. The intercept, 106,836, is the estimated mean selling price when a house in not in the NW region, and 26,483 is the difference in estimated mean selling price between a house in a NW region and a house not in the NW. **(b)** From software:

Predictor	Coef	SE Coef	T	P
NW	26483	12805	2.07	0.041

Considerable evidence of a region effect. **(c)** From software:

Source	DF	SS	MS	F	P
NW	1	13149968133	13149968133	4.28	0.041
Error	98	3.01283E+11	3074311750		
Total	99	3.14433E+11			

Same result **(d)** The value of t in (b) is the square root of the value of F in (c). **13.49 (a)** The coefficient of f (0.04) is the difference between the estimated means for men and women for each level of race. The fact that it is close to 0 indicates that there is little difference between the means for men and women, given race. **(b)** Black female: 2.83, White female: 2.46; Black male: 2.79; White male: 2.42 **(c)** From the P-value of 0.000 for race, we can conclude that on the average blacks report a higher ideal number of children than whites do (by 0.37).

13.51 (a) The mean estimate for men is 0.2 higher than the mean estimate for women at each fixed level of religion. **(b)** The mean estimate for Protestants is 0.5 higher than the mean estimate for other religions at each fixed level of gender. **(c)** $\mu_y = \alpha + \beta_1 g + \beta_2 r1 + \beta_3 r2 + \beta_4 r3$; for the response variable to be independent of religion, for each gender, β_2, β_3, and β_4 would need to equal 0. **13.53** Smoking status has a different impact at different fixed levels of age. There is a bigger mean difference between smokers and non-smokers among older women than among younger women. **13.55 (a)** At fixed levels of race, men have higher estimated mean salaries than women by 0.8 (thousands of dollars). **(b)** (i) 31.0; (ii) 30.8 **13.57** The report can include and interpret the F test statistic of 0.81 and the P-value of 0.52. **13.59** An example is means 10, 17, 24 for A, B, C. **13.61 (a)**

	Female	Male
Humanities	56,000	55,000
Science	70,000	69,000

Women: 58,333.33; men: 63,400

(b) A one-way comparison would reveal that men have a higher mean. A two-way comparison would show that at fixed levels of university divisions, women have the higher mean. **13.63 (c) 13.65 (c) 13.67 (a)** Identical sample means **(b)** No variability (standard deviation of 0) within each sample. **13.69 (b)** $0.05/35 = 0.0014$ **13.71 (a)** (i) Number of bathrooms would be treated as a categorical variable (ii) Number of bathrooms would be a treated as a quantitative variable and we'd be assuming a linear trend **(b)** The straight-line regression approach would allow us to know whether increasing numbers of bathrooms led to an increasing house selling price. The ANOVA would only let us know that the house price was different at each of the three categories. Moreover, we could not use the ANOVA as a prediction tool for other numbers of bathrooms. **(c)** Mean selling price $150,000 for 1 bathroom, $100,000 for 2 bathrooms, $200,000 for 3 bathrooms. (There is not an increasing or decreasing overall trend.)

Chapter 14

14.1 (a)

Lotion	(1,2)	(1,3)	(1,4)	(2,3)	(2,4)	(3,4)
Studio	(3,4)	(2,4)	(2,3)	(1,4)	(1,3)	(1,2)

(b)

Lotion mean rank	1.5	2.0	2.5	2.5	3.0	3.5
Studio mean rank	3.5	3.0	2.5	2.5	2.0	1.5
Difference of mean ranks	−2.0	−1.0	0.0	0.0	1.0	2.0

(c) $P(-2.0) = 1/6$, $P(-1.0) = 1/6$, $P(0.0) = 2/6$, $P(1.0) = 1/6$, $P(2.0) = 1/6$

14.3 (a) and **(b)** together: example given for seven combinations only; there would be 20.

Treatment	Ranks
Therapy 1	(1,2,3) (1,2,4) (1,2,5) (1,2,6) (1,3,4) (1,3,5) (1,3,6)
Therapy 2	(4,5,6) (3,5,6) (3,4,6) (3,4,5) (2,5,6) (2,4,6) (2,4,5)

Therapy 1 mean rank	2.0	2.33	2.67	3.0	2.67	3.0	3.33
Therapy 2 mean rank	5.0	4.67	4.33	4.0	4.33	4.0	3.67

Difference of mean ranks −3.0 −2.33 −1.67 −1.0 −1.67 −1.0 −0.33 **(c)** Differences between mean ranks followed by probabilities: $(-3.00, 1/20)(-2.33, 1/20)(-1.67, 2/20)(-1.00, 3/20)$ $(-0.33, 3/20)$ $(0.33, 3/20)(1.00, 3/20)(1.67, 2/20)(2.33, 1/20)$ $(3.00, 1/20)$ **(d)** The P-value = 0.20; if the treatments had identical effects, the probability would be 0.10 of getting a sample like we observed, or even more extreme, in either direction. **14.5 (a)** −0.515 is an estimate of the difference between the population medians for the two groups. **(b)** $(-1.01, 0.02)$ estimates that the population median for the control group is between 1.01 below and 0.02 above the population median for the treatment group. **14.7 (a)** The difference between the population median weight change for the cognitive behavioral treatment group and the population median weight change for the control group is 3.05. **(b)** The confidence interval estimates that the population median weight change for the cognitive-behavioral group is between −0.60 below and 8.10 above the population median weight change for the treatment group. **(c)** P-value = 0.11; the alternative hypothesis is: The population medians are not equal. There is not much evidence against the null hypothesis. It is plausible that the distributions are identical. **14.9 (a)** Group 1 ranks: 2, 5, 6; Group 2 ranks: 1, 3.5; Group 3 ranks: 3.5, 7, 8 The mean rank for Group $1 = (2 + 5 + 6)/3 = 4.33$ **(b)** P-value = 0.209; it is plausible that the population mean quiz score is the same for each group. **14.11 (a)** dependent; all students receive both treatments. **(b)** 0.81 **(c)** $z = 3.55$; P-value = 0.0002. strong evidence that population mean reaction time is faster when not using the cell phone. **(d)** matched-pairs t test; the sign test uses merely the information about *which* response is higher, not the quantitative information about *how much* higher. **14.13** H_0: $p = 0.50$. H_a: $p > 0.50$; $P(3) = (0.50)^3 = 0.125$; The evidence is not strong that walking lowers blood pressure (but can't get a small P-value with such a small n for this test). **14.15 (a)** H_0: population median of difference scores is 0; H_a: population median of difference scores is >0. **(c)** P-value = 1/16 = 0.06; this P-value is not less than a typical significance level such as 0.05, so not enough evidence to reject the null. **14.17 (a)** Bumper A: ranks are 4, 6, 5; mean is 5; Bumper B: ranks are 1, 2, 3; mean is 2. **(b)** The following shows each set of possible ranks for Bumper A, and below it the corresponding ranks for Bumper B.

Treatment	Ranks
Bumper A	(1,2,3) (1,2,4) (1,2,5) (1,2,6) (1,3,4) (1,3,5) (1,3,6) (1,4,5) (1,4,6) (1,5,6)
Bumper B	(4,5,6) (3,5,6) (3,4,6) (3,4,5) (2,5,6) (2,4,6) (2,4,5) (2,3,6) (2,3,5) (2,3,4)

Treatment	Ranks
Bumper A	(2,3,4) (2,3,5) (2,3,6) (2,4,5) (2,4,6) (2,5,6) (3,4,5) (3,4,6) (3,5,6) (4,5,6)
Bumper B	(1,5,6) (1,4,6) (1,4,5) (1,3,6) (1,3,5) (1,3,4) (1,2,6) (1,2,5) (1,2,4) (1,2,3)

(c) There are only two ways in which the ranks are as extreme as in this sample: Bumper A with 1,2,3 and B with 4,5,6, or Bumper A with 4,5,6 and A with 1,2,3. **(d)** The P-value is 0.10 because out of 20 possibilities, only two are this extreme. 2/20 = 0.10.

14.19 (a)

Group	Ranks	Mean Rank
Muzak	1,2,4,5,3	3.0
Classical	9,8,7,10,6	8.0

(b) There are only two cases this extreme (i.e., Muzak has ranks 1–5 or Muzak has ranks 6–10). Thus, the P-value is the probability that one of these two cases would occur out of the 252 possible allocations of rankings. The probability would be 0.008 of getting a sample like we observed or even more extreme. **14.21 (a)** Kruskal-Wallis test **(b)** All examples would have one group with ranks 1–3, one with ranks 4–6, and one with ranks 7–9. **14.23 (a)** (i) H_0: $p = 0.50$; H_a: $p \neq 0.50$ (ii) $z = 4.96$ (iii) P-value $< .001$; extremely strong evidence that majority of countries have more cell-phone use than Internet use. **(b)** No; you would know the population parameters so inference would not be relevant. **14.25 (a)** We could use the sign-test for matched pairs or the Wilcoxon signed-ranks test. **(b)** One reason for using a nonparametric method is if we suspected that the population distributions were not normal, for example, possibly highly skewed, because we have a one-sided alternative and parametric methods are not then robust. **(c)** From software (used CGPA-HSGPA):

Test of median = 0.000000 versus median < 0.000000

N	N for Test	Wilcoxon Statistic	P	Estimated Median
59	55	268.5	0.000	−0.1850

The P-value of 0.000 gives very strong evidence that population mean high school GPA is higher than population mean college GPA. **14.27 (a)** H_0: population median of difference scores is 0; H_a: population median of difference scores >0. P-value = 0.25. It is plausible that the null hypothesis is true. **(b)** The results are identical to those in Example 8. Outliers do not have an effect on this nonparametric statistical method. **14.29** One example is a situation in which the population distribution is likely to be highly skewed and the researcher wants to use a one-sided test. **14.31** Kruskal-Wallis **14.33** False **14.35 (a)** The proportions are calculated by pairing up the subjects in every possible way, and then counting the number of pairs for which the tanning studio gave a better tan. **(b)** The P-value would be 2/10. The probability of an observed sample proportion of 5/6 or more extreme (i.e., 6/6) is 2/10 = 0.20. **14.37** No; we are taking the median of all slopes, and the median is not susceptible to outliers.

Index

Photo Credits

Data Files and Applet Usage

DATA FILES	EXAMPLES	EXERCISES Required	Referenced
Al_team_statistics	3-9		
Anorexia	8-7	7.28, 8.35, 8.36, 13.11, 13.45, 14.22	8.28, 8.30, 8.41, 8.42, 9.71, 9.72, 9.75, 14.7
asthma		2.114, 8.31	
Baseball's_hr_hitters		2.112	
Buchanan_ and_the_ butterfly_ballot	3-6, 3-10	3.21	
Car_weight_and_mileage			3.67, 11.1, 11.2, 11.23
Cell_phones	9-9, 9-12, 9-13, 14-3		3.10, 14.11, 14.15, 14.16
Central_park_yearly_temps	2-9, 3-12	2.103	2.25, 2.65
cereal	2-4, 2-5, 2-7, 2-10, 2-16, 2-17	2.20, 3.45, 3.74	2.14, 2.17, 2.18, 2.19, 2.24, 2.33, 2.101, 2.115, 3.29, 3.30
Cigarette_tax_file		2.100	2.50, 2.62
College_atheletes	12-4, 12-5, 12-6, 12-7		2.49, 2.60, 11.67, 12.1, 12.12, 12.13, 12.31, 12.37
Ebay_auctions	7-7	7.29, 7.80, 9.16, 9.17, 14.6	2.15, 7.29, 7.33, 7.80, 9.16, 9.17, 9.91, 12.41, 12.68
enchiladas		3.20	2.87, 2.88
energy_eu	2-18		2.68, 2.70
European_union_unemployment		2.55	2.58, 2.69
Fl_crime	1-3, 3-14, 9-16		3.50, 3.61, 11.60, 12.5, 12.6
Fl_student_survey		1.19, 1.34, 1.35, 2.73, 8.88, 9.24, 9.90, 10.57, 11.77, 12.73, 13.9, 14.28	3.12, 7.34, 9.31, 11.15, 13.50, 14.10
Georgia_student_survey	11-13, 11-14, 14.-6	2.113, 3.86, 11.7, 11.8, 11.36, 11.48, 11.49, 11.68, 12.18, 13.9, 14.24, 14.25	11.18, 11.38, 12.2, 12.19, 13.44
heights	2-15, 6-6		2.46, 2.71, 2.99
High_jump	3-11, 12-11	3.37	3.80, 12.40, 12.47
High_school_female_athletes	11-1, 11-2, 11-3, 11-6, 11-9, 11-11, 11-12, 11-16	11.27, 11.78, 12.85, 1286	2.48, 2.59, 3.23, 11.14, 11.31, 11.37, 11.43-11.45, 11.69, 12.22, 12.32, 12.33, 12.60
House_selling_prices	12-1, 12-2, 12-3, 12-8, 12-9, 12-10	12.10, 12.30, 12.38, 12.45, 12.59, 13.47	2.44, 3.24, 11.29, 11.30, 11.40, 11.63-65, 12.29, 12.39, 12.42, 13.12, 13.25
Hs_graduation_rates		3.79	2.107
Human_development	2-11, 3-4, 3-5, 3-7, 11-5, 11-18	3.70	2.29, 3.10, 3.11, 3.13, 3.25, 3.26, 3.27, 3.43, 3.68, 3.71, 11.11, 11.61, 12.14, 12.64-67, 14.23
Income_in_public_housing		7.81	
Long_jump			3.39
Mountain_bike		3.35, 3.36, 7.79	3.19
Newnan_ga_temps		2.27, 3.87	
nl_team_statistics		3.85	3.34, 12.50
protein_and_weight_gain		13.33, 13.34	
quality_and_productivity		12.43	
softball		11.75, 11.76, 12.16, 12.61	12.16
tv_europe		3.46	3.82
us_statewide_crime	3-13	3.44, 3.76, 3.77, 3.78, 11.26	3.38, 3.41, 3.42, 3.62, 3.75, 6.24, 12.62, 12.63
us_temperatures		3.40	
whooping_cough		3.58	2.26

APPLETS:	ACTIVITY	EXAMPLE	EXERCISES
Sample from a population	1-2		1.20, 1.21, 1.33, 4.104
Sampling distributions	6-1		6.52, 6.104, 6.105, 7.56
Random numbers	6-2	4-5, 5-16	4.16, 4.17, 5.57, 5.107, 5.108, 5.109, 6.123
Long run probability demonstrations	5-1	5-2, 5-4	5.10, 5.11, 5.12, 5.85
Mean versus median	2-1		2.36, 2.130
Standard deviation			2.56
Confidence intervals for a proportion	7-2		7.23, 7.24
Confidence intervals for a mean (the impact of confidence level)			7.41
Hypothesis tests for a proportion	8-2		8.59
Hypothesis tests for a mean			
Correlation by eye			3.104, 3.105
Regression by eye			3.105